10th International Conference on Compressors and Their Systems 2017

IOP Conference Series: Materials Science and Engineering
Volume 232

London, United Kingdom
11 – 13 September 2017

ISBN: 978-1-5108-4775-0
ISSN: 1757-8981

Printed from e-media with permission by:

Curran Associates, Inc.
57 Morehouse Lane
Red Hook, NY 12571

Some format issues inherent in the e-media version may also appear in this print version.

Printed by Curran Associates, Inc. (2017)

For permission requests, please contact the Institute of Physics at the address below.

Institute of Physics
Dirac House, Temple Back
Bristol BS1 6BE UK

Phone: 44 1 17 929 7481
Fax: 44 1 17 920 0979

techtracking@iop.org

Additional copies of this publication are available from:

Curran Associates, Inc.
57 Morehouse Lane
Red Hook, NY 12571 USA
Phone: 845-758-0400
Fax: 845-758-2633
Email: curran@proceedings.com
Web: www.proceedings.com

Table of contents

Volume 232

10th International Conference on Compressors and their Systems
11–13 September 2017, City, University of London, United Kingdom

Accepted papers received: 17 August 2017
Published online: 6 September 2017

Preface

Keynote

Reciprocating Compressors

Screw Compressors

Valves

Turbocompressors

Dynamic Machines

Scroll Compressors

Vane Compressors

Novel Compressors

Compressor systems

Optimization

Expanders

Modelling

Energy efficiency

Measurements and Control

Lubricants

IOP Conf. Series: Materials Science and Engineering 232 (2017) 011001 doi:10.1088/1757-899X/232/1/011001

Foreword

The 10th International Conference on Compressors and their Systems will be held at City, University of London from 9th to 13th September 2017. As Conference Chairman, I wish you a warm welcome to this event in our newly refurbished campus in Northampton Square.

This series of conferences started in 1999, then organised by the Fluid Machinery Group of the Institution of Mechanical Engineers (IMechE). Since 2009 it has been managed by City, University of London. This year, the conference has been organised, not only with the IMechE, but also with the IOR (Institute of Refrigeration) and the IIR (International Institute of Refrigeration). On behalf of the Organising Committee I would like to thank these organisations for their commitment and continuing support in promoting this event.

After a rigorous reviewing process, we have accepted 95 papers for presentation at the technical sessions of the conference, the highest number in the conference history. In addition, in this book, you will find abstracts of the 3 keynote presentations, 7 Industry presentations and 13 presentations for the 3rd Short course/Forum on CFD in positive displacement machines. I would like to extend my special thanks to our 138 reviewers listed on the pages below for their hard work and for vital input in the difficult process of selecting interesting and high standard papers for the conference.

The technical papers will be presented in three parallel sessions over the first two days of the conference. These are published in the open access conference proceedings by the IOP (Institute of Physics) publications. The authors of the best papers from the conference will be invited to resubmit them in enhanced form for publication in the highly reputable Journal of Process Mechanical Engineering.

Following a very positive reaction from the attendees to our last Conference, the Industry Day is now established as the 3rd day of the conference. Industry will have an opportunity to consider the business and market influences on compressor design and manufacture, including the impact of new technologies and changes in economic, environmental and regulatory conditions. We have selected 7 exciting presentations for that event, which we hope will promote fruitful discussion.

Let me also draw your attention to the 3rd Short course/forum on Computational Fluid Dynamics in rotating positive displacement machines. This event follows two very successful short courses in 2013 and 2015. Our focus this year is on speed and efficiency in calculating multiphase flows.

I would like to extend my special thanks to our main sponsors, Holroyd PTG, Howden and Kapp Niels for their continuing support for the conference and to welcome our new sponsors Hoerbiger and Samputensili. As a result of their generous contributions we have managed to keep the conference fees low and thereby to attract a wider range of attendees.

On this occasion, I must thank all those who have helped to make this event happen. Special thanks go to Lucy Bennett and Louise Gordon of the City events team, Dr Matthew Read who brilliantly chaired the Reviewing Committee and Dr Sham Rane who organised the Short Course. I would also like to thank all members of the organising committee and the international liaison committee.

Please do not forget to complete your assessment of the conference before leaving, and indicate how you think it can be improved next time. All the members of the Conference Committee hope that you find the event to be both worthwhile and enjoyable and we look forward to welcoming you to City, University of London.

Professor Ahmed Kovacevic

Conference Chairman

Centre for Compressor Technology
School of Mathematics, Computer Science and Engineering
City, University of London, United Kingdom

IOP Conf. Series: Materials Science and Engineering **232** (2017) 011002 doi:10.1088/1757-899X/232/1/011002

Sponsors

City, University of London would like to thank the following sponsors and exhibitors

HOLROYD MACHINE TOOLS & PRECISION COMPONENTS

Platinum Sponsor

Howden

Gold Sponsor

KAPP NILES

Silver Sponsor

SAMPUTENSILI

Silver Sponsor

HOERBIGER

Welcome Reception and Exhibitor

Ansys Ltd.
Exhibitor

Convergent Science
Exhibitor

Gamma Technologies
Exhibitor

PDM Analysis
Exhibitor

IOP Conf. Series: Materials Science and Engineering **232** (2017) 011003 doi:10.1088/1757-899X/232/1/011003

Conference Committees

Organising Committee 2017

Lucy Bennett	City, University of London, UK
Graeme Cook	Howden Compressors, UK
Didier Coulomb	International Institute of Refrigeration, France
Louise Gordon	City, University of London, UK
Chris Holmes	PTG Holroyd Advanced Developments Ltd, UK
Ahmed Kovacevic (Chair)	City, University of London, UK
Deonie Lambert	International Institute of Refrigeration, France
Sham Rane	City, University of London, UK
Matthew Read	City, University of London, UK
Ivor Rhodes	ACE Cranfield, UK
Miriam Rodway	Institute of Refrigeration, UK
Ian Smith	City, University of London, UK
Christopher Smout	IMechE, UK
Nikola Stosic	City, University of London, UK

International Liaison Committee 2017

Matthew Cambio,	Ingersoll Rand/Trane, USA
Cesar J. Deschamps	University of Santa Catarina, Brazil
Mitsushiro Fukuta,	Shizuoka University, Japan
Pega Hrnjak,	University of Illinois, USA
Guy Hundy	Aleph Zero, United Kingdom
Matthew Irons,	Emerson Climate Technologies GmBH, Belgium
Liansheng Li,	Hefei General Machinery Research Insitute, China
Gunther Machu	HOEBIGER Ventilwerke GmbH & Co. KG, Austria
Kim Tiow Ooi,	Nanyang Technology University, Singapore
Joaquim Rigola	Polytechnic University of Catalonia, Spain
Kristina V. Soldatova,	St Petersburg University, Russia
Igor Vaisman	Rolls-Royce Corporation, USA
Xiaolin Wang,	University of Tasmania, Australia

IOP Conf. Series: Materials Science and Engineering **232** (2017) 011004 doi:10.1088/1757-899X/232/1/011004

Scientific Review Committee 2017

Dr. Momin Abdalla	University of Khartoum, South Sudan
Dr. Adrian Alford	TPG D&T, United Kingdom
Dr. Jafar Alzaili	City, university of London, United Kingdom
Dr. Araz Ardehali Barani	Zapp Precision Metals GmbH, Germany
Dr. Bachir Bachir	Emerson Climate Technologies, Germany
Dr. Yurii Baidak	Odessa National Academy of Food Technologies, Ukraine
Prof.Wang Baolong	Tsinghua University, China
Dr. Oleg Baturin	Samara State Aerospace University, Russia
Dr. Giuseppe Bianchi	Brunel University London, United Kingdom
Dr. Nasir Bilal	Purdue University, United States
Prof. Andreas Bruemmer	TU Dortmund, Germany
Mr. David Buckney	Howden Compressors, United Kingdom
Dr. Xiaodan Cai	United Technologies Research Center, USA
Mr. Matthew Cambio	Ingersoll Rand/Trane, United States
Dr. Xuefei Chen	Xi'an Jiaotong University, China
Dr. Mallikarjuna Sarma	Chitti v4Econscious LLC, United States
Dr. Ekaterina Chukanova	MashTEK Ltd, Russia
Prof. Roberto Cipollone	University of L'Aquila, Italy
Mr. Graeme Cook	Howden Compressors Ltd, United Kingdom
Dr. Diogo Londero Da Silva	Federal University of Santa Catarina, Brazil
Prof. Cesar J Deschamps	Federal University of Santa Catarina, Brazil
Dr. Philippe Dewitte	United States
Dr. Hui Ding	Simerics Inc., United States
Mr. Marco Diniz	POLO – UFSC, Brazil
Mr. Peter Duineveld	Hoerbiger, Netherlands
Mr. Thiago Dutra	POLO - UFSC, Brazil
Prof. Ahmed Mohamed Reda	El Baz The British University in Egypt, Egypt
Dr. Ferran Escanes	Huayi Compressor Barcelona, S.L., Spain
Prof. Daniel Favrat	EPFL, Switzerland
Prof. Jianmei Feng	Xi'an Jiaotong University, China
Mr. Kenneth Ian Fitzsimmonds	Retired, United Kingdom
Dr. Mitsuhiro Fukuta	Shizuoka university, Japan
Dr. Andrea Fusi	Compression Service Technology, Italy
Prof. Yuri Galerkin	Peter the Great St. Petersburg University, Russia
Dr. Haiyang Gao	Simerics, United States
Dr. Wenkai Gao	China National Petroleum Corporation, China
Prof. Jos Luiz Gasche	Unesp-Ilha Solteira, Brazil
Dr. Alessandro Gattoronchieri	TMP Termomeccanica Pompe, Italy
Prof. Adel Ghenaiet	University of Science and Technology, Algeria
Mr. Manuel Grieb	TU Dortmund University, Germany
Prof. Pengcheng Guo	Xi'an University of Technology, China
Dr. Martin Heinrich	TU Bergakademie Freiberg, Germany
Prof. Christopher S. Holmes	PTG Holroyd, United Kingdom
Prof. W. Travis Horton	Purdue University, United States
Dr. Bukac Hubert	Lttle Dynamics, Inc.,United States
Dr. Guy Hundy	Aleph Zero Ltd, United Kingdom
Dr. Ermin Husak	University of Bihac, Bosnia and Herzegovina

IOP Conf. Series: Materials Science and Engineering **232** (2017) 011004 doi:10.1088/1757-899X/232/1/011004

Mr. Harshad Inamdar	Rheem Manufacturing Company, United States
Dr. Xiaohan Jia	Xi'an Jiaotong Univercity, China
Prof. Miroslav Jícha	Brno University of Technology, Czech Republic
Dr. Georgi Kalitzin	United Technologies Research Center, USA
Prof. Hirotaka Kameya	Ishinomaki Senshu University, Japan
Mr. Joseph Anthony Karnaz	Shrieve Chemical Products, United States
Prof. Michael Khmelniuk	Odessa National Academy of Food, Ukraine
Prof. Ahmed Kovacevic	City, University of London, United Kingdom
Dr. Orkan Kurtulus	Purdue University, United States
Prof. Geonho Lee	Doowon Technical College, South Korea
Dr. Gilbong Lee	Korea Institute of Energy Research, South Korea
Prof. Vincent Lemort	University of Liege, Belgium
Mr. Yujun Leng	Purdue Univeristy, United States
Dr. Daqing Li	Carrier Corporation, United States
Dr. Hongqi Li	Beijing University of Technology, China
Prof. Liansheng Li	Hefei General Machinery Research Institute, China
Dr. Jiangyan Liu Huazhong	University of Science and Technology,China
Dr. Koen Lodefier	Atlas Copco Airpower, Belgium
Dr. Qiao Lu	B/E Aerospace Inc., United States
Mr. Yang Lu	Xi'an Jiaotong University, China
Mr. Cerdoun Mahfoudh	Military Polytechnic School, Algeria
Dr. Duane McCormick	United Technologies Research Center, USA
Dr. William James Milligan	Howden Compressors, United Kingdom
Dr. Elvedin Mujic	BITZER Kühlmaschinenbau GmbH, Germany
Dr. Stefano Murgia	Ing. Enea Mattei SpA, Italy
Dr. Timur Mustafin	Kazan National Research Tehnological University, Russia
Mr. Kai Nadler	Leybold GmbH,Germany
Dr. Kashif Nawaz	Oak Ridge National Lab, United States
Mr. Alexander Nikolov	TU Dortmund University, Germany
Dr. Tsutomu Nozaki	Hitachi Appliances, Japan
Dr. Ali Bahadir	Olcay Yeditepe University, Turkey
Dr. Abdelaziz Omar	Oak Ridge National Laboratory, United States
Dr. Kim Tiow Ooi	Nanyang Technological University, Singapore
Dr. Ali Osama	Omdurman Islamic University, Sudan
Mr. Ahmet Ozdemir	ARCELIK A.S., Turkey
Dr. Michael Perevozchikov	Emerson Climate Technologies, United States
Prof. Milovan Peric	CD-adapco, Germany
Mr. Stefan Posch	Graz University of Technology, Austria
Prof. Keith Pullen	City, University of London, United Kingdom
Dr. Suxin Qian	Xi'an Jiaotong University, China
Prof. Feng Quanke,	Xian Jiaotong University,China, China
Dr. Sham Rane	City, University of London, United Kingdom
Dr. Matthew Read	City, University of London, United Kingdom
Mr. Ivor Rhodes	ACE Cranfield Ltd, United Kingdom
Dr. Guilherme Ribeiro	Institute for Advanced Studies,Brazil
Dr. Melanie Rigby,	SMT, United Kingdom
Prof. Joaquim Rigola	CTTC – UPC, Spain
Dr. David Rowinski	Convergent Science, Inc., United States
Dr. Hajime Sato	Mitsubishi Heavy Industries, Ltd., Japan
Mr. Jack Sauls	IngersollRand / Trane, United States

IOP Conf. Series: Materials Science and Engineering **232** (2017) 011004 doi:10.1088/1757-899X/232/1/011004

Prof. Abdulnaser Sayma	City, University of London, United Kingdom
Dr. Valentin Silivestru	National R&D Institute for Gas Turbines, Romania
Mr. Ernane Silva	POLO/UFSC, Brazil
Prof. Ian K Smith	City, University of London, United Kingdom
Dr. Kristina Soldatova	Peter the Great St.Petersburg University, Russia
Dr. Andreas Spille-Kohoff	CFX Berlin Software GmbH, Germany
Prof. Nikola Stosic	City, University of London,United Kingdom
Dr. Alison Subiantoro	TUM CREATE, Singapore
Dr. Swaminathan Subramanian	Eaton, United States
Dr. Shuaihui Sun	Xi'an University of Technology, China
Dr. Kok Ming Tan	SANDEN INTERNATIONAL PTE LTD, Singapore
Dr. Chengzhi Tang	Helmer Scientific, United States
Dr. Hong Tao	Shanghai Hitachi Electrical Appliances Co., China
Dr. Christiane Thomas	TU Dresden, Germany
Mr. Kenji Tojo	TOJO R&D Design Office/Waseda University, Japan
Mr. Matthias Utri	TU Dortmund, Germany
Prof. Gianluca Valenti	Politecnico di Milano, Italy
Mr. Hanushan Vasuthevan	TU Dortmund, Germany
Mr. Diego Vittorini	University of L'Aquila, Italy
Dr. Sergey Vizgalov	Kazan National Research Technological University, Russia
Dr. Xianfa Wang	Holroyd Machine Tool, United Kingdom
Dr. Tao Wang	Xi'an Jiaotong University, China
Dr. James Willie	Gardner Denver Schopfheim GmbH, Germany
Dr. Maria Wilson	Howden Compressors, United Kingdom
Prof. Weifeng Wu	Xi'an Jiaotong University, China
Prof. Ziwen Xing	Xi'an Jiaotong University, China
Mr. Jiu Xu	University of Illinois, United States
Mr. Di Yan	Chongqing University, China
Dr. Binyan Yu	Chang'an University, China
Dr. Xiaoling YU	Xi'an Jiaotong University, China
Dr. Dmyto Zaytsev	GEA Refrigeration Germany GmbH,Germany
Prof. Zaoxiao Zhang	Xi'an Jiaotong University, China
Dr. Jinjie Zhang	Beijing University of Chemical Technology, China
Mr. Xinye Zhang	Purdue University, United States
Dr. Yuanyang Zhao	Hefei General Machinery Research Institute, China
Dr. Wan Zhong	Tuthill Vaccum & Blower Systems, United States,
Dr. Yury Zhuk	Hardide Coatings Ltd, United Kingdom
Dr. Nataliia Zhykhareva	Odessa National Academy of Food Technolog., Ukraine
Mr. Davide Ziviani	Ghent University, Belgium
Dr. Yang Zou	Creative Thermal Solutions, United States
Mrs. Huiming Zou	Technical institute of physics and chemistry, China
Dr. Yang Zou	Creative Thermal Solutions, USA
Mrs. Huiming Zou	Technicial institute of physics and chemistry, CAS, China

IOP Conf. Series: Materials Science and Engineering **232** (2017) 011005 doi:10.1088/1757-899X/232/1/011005

Peer review statement

All papers published in this volume of *IOP Conference Series: Materials Science and Engineering* have been peer reviewed through processes administered by the proceedings Editors. Reviews were conducted by expert referees to the professional and scientific standards expected of a proceedings journal published by IOP Publishing.

IOP Conf. Series: Materials Science and Engineering **232** (2017) 012001 doi:10.1088/1757-899X/232/1/012001

New concept single screw compressors and their manufacture technology

Q Feng[1*], F Liu[1,2], L Chang[1], C Feng[1], C Peng[1], J Xie[1] and M van den Broek[2]

[1] School of Energy and Power Engineering, Xi'an Jiaotong University, Xi'an, China
[2] Department of Flow, Heat, and Combustion Mechanics, Ghent University, Gent, Belgium

E-mail*: qkfeng@mail.xjtu.edu.cn

Abstract. Single screw compressors were generally acknowledged as one of the nearly perfect machines by compressor researchers and manufacturers. However the rapid wear of the star-wheel in a single screw compressor during operation is a key reason why it hasn't previously joined the main current compressors' market. After more than ten years of effective work, the authors of this paper have proposed a new concept single screw compressor whose mesh-couple profile is enveloped with multi-column. Also a new design method and manufacture equipment for this kind of compressor have been developed and are described in this paper. A lot of prototype tests and a long period of industrial operations under full loading conditions have shown that the mesh-couple profiles of the new concept single compressors have excellent anti-wearness.

1. Introduction

A single screw compressor (SSC) is consist of a screw rotor, two star-wheels and a housing. SSC belongs to a kind of rotary compressors. Considering its structure and working property, it is mainly used in compression, transport of air, petrol gas and refrigeration media. Its application is the same as or overlaps that of reciprocating compressors, screw compressors, scroll compressors etc. shown in figure 1.

Especially, in some new and high-tech engineering fields SSC can't be substituted. For example, gas transportation engineering with large flow-rate and high compression ratio, air compression engineering with high compression ratio and oil free, some lower noise facilities in naval ships.

Figure 1. The structure of a single screw compressor.

In 1960, Zimmern opened his patent to public of SSC, and in 1962 the first SSC prototype was produced in Europe [1, 2]. In past over 50 years, a lot of this traditional SSC, in which single straight line and single column are used to envelope engaging profiles of rotors and star-wheels, were produced in many companies from England, Germany, America, Austria, Japan and so on. This made SSC played a role in the world compressor market.

China began to develop SSC in 1970s. Beijing First General Machinery Factory, Xi'an Jiaotong University and Shanghai 704 Institute then developed prototypes of this traditional SSC [3]. But star-wheel's rapid wear increased leakage seriously within a short time, which of course, hindered SSC's coming into a wilder market.

Since 1990's, along with the improvements of Chinese machinery manufacture level and the broad applications of PEEK material, star-wheel wear-resistant has been improved obviously. This made traditional SSC met the needs of common engineering in the whole. In this way, many kinds of SSC products from Guangdong GANEY, Shanghai FHOG, Shanghai GANRS etc. came into the market. In 2012, a report from Chinese Compressor Association says that SSC's output is over 8000~10000 every year in China. But in the same time screw compressor's output is up to about 100~130 thousand.

Outstanding advantages of a single screw compressor are as follows:

1) Axial and radial compound forces acted on a screw rotor by gas in compression chamber is balanced with zero. There is only a gravity force of a rotor acting on the bearing.
2) The forces acting on a star-wheel is very small, as it is only about one fourth of that acting on the screw rotor while operating in same conditions.
3) There is no leakage triangle area in an engaging pair of SSC, which always exists in a screw compressor.
4) Engaging forces between a rotor tooth and a star-wheel tooth is very small.
5) Noise level of SSC is lower as 3~5dB (A) than that in screw compressor.

All these advantages encourage a lot of researchers and customers to study and develop SSC.

2. Research of new conception engage profile

Authors of this paper examined several SSC's star-wheels which are produced by a famous international compressor company. Its star-wheel is only put into operation for 3~5 hours. The star-wheel's back and front surfaces are shown respectively in figure 2a and figure 2b. It is obviously that there is a straight edge in each surface. Figure 3 shows engaging situation of a star-wheel tooth with a rotor groove in different rotation positions. At any position on the star-wheel' engaging profile, there is only one edge point which is always contacting with the rotor tooth groove flank. In other words, the edge can be easily wear out and increase the leakage.

a) Back surface b) Front surface

Figure 2. Star-wheel enveloped with single straight line (3~5 hours operation).

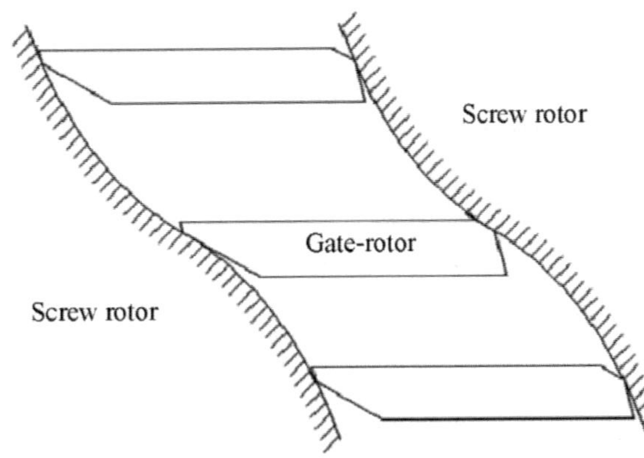

Figure 3. Engaging profile enveloped with single straight line.

Figure 4 expresses the enveloping principle of single column profile (bar milling) for a rotor groove. After processing of the rotor groove with a bar milling machine, the surface on a bar milling cutter contacting with the rotor groove flank is then characterized as the star-wheel tooth's engaging profile. Because engaging profile of a star-wheel tooth for single column is very small compared to the whole flank surface, friction and wear will still take place seriously. This result can't satisfy the requirements of most engineering situations.

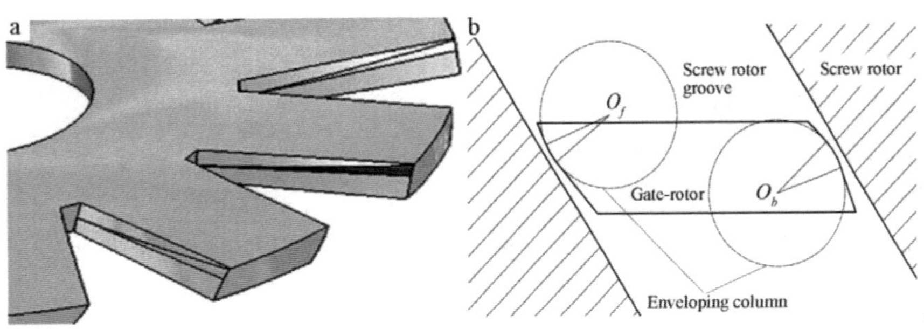

a) Profile b) Formation principle

Figure 4. Profile enveloped with single column and its formation principle.

Engagement pair enveloped with single straight line or single column both belongs to traditional profile. Its main disadvantages are: star-wheel tooth is easy to wear out and cause more leakage. It takes a long time to machine rotor groove with a cutter. The material of star-wheel is quite expensive, furthermore, it is easy to lose its shape and sometimes to be scraped and broken. In past 50 years many researchers made every effort to find a machinable enveloping profile for the engagement pair, with which different areas of a star-wheel tooth surface can touch the rotor groove flank in turn[4,5].

The authors had proposed the conception and theory of multi-straight line and multi-column compound enveloping engagement pair of SSC 10 years ago [6, 7]. Compound envelopment means the cutter or milling cutter is used repeatedly to process the tooth of a star-wheel in different directions and positions (figure 5). At the same time, according to the space arrangement of these milling cutters above, the part surface of a bar milling in every space arrangement, which contacts screw rotor groove flank, can be taken in their original positions and be connected with each other by several tangent planes. Thus, an engaging profile on star-wheel tooth is constructed. Figure 6 shows a star-wheel tooth flank profile enveloped by multi-column. Figure 6a shows the flank profile at the back side and figure 6b is the front flank profile. It should be pointed out that the multi-columns used for the enveloping engagement pair does not have to be the same in length. Figure 7 describes screw rotor groove flank profile which is enveloped with non-equal length's multi-columns. Figure7a shows screw rotor groove profile which is enveloped by the star-wheel tooth profile in figure 6a. Figure7b shows screw rotor groove profile which is enveloped by the profile in figure 6b. It is obvious that compound-enveloped profile with multi-columns is not a regular shape as the common traditional one [8].

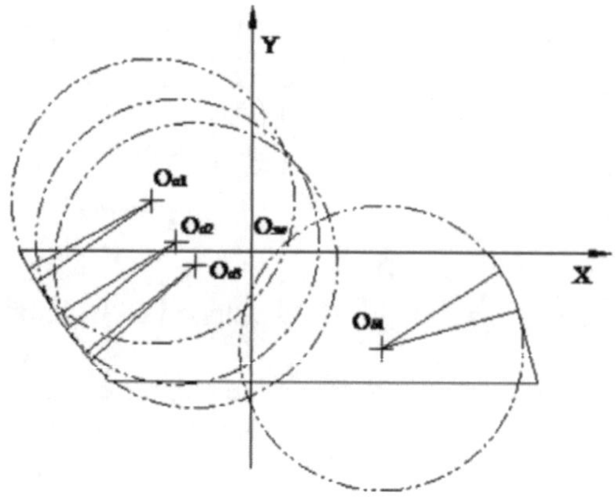

Figure 5. Locations of multi-column milling cutters.

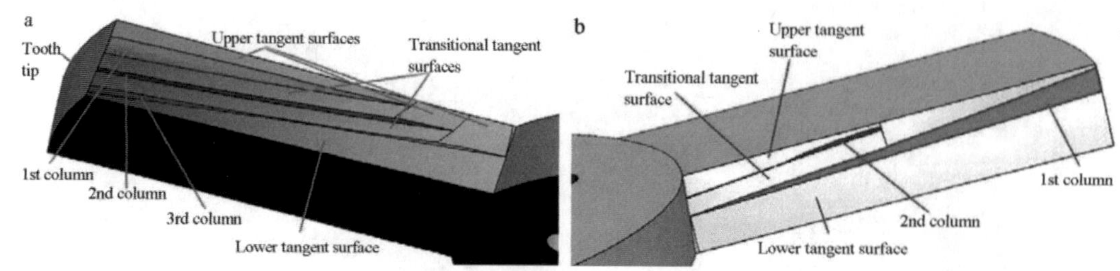

a) b)

Figure 6. Numerical simulated envelopment profile constructed by multi-column on star-wheel tooth surface.

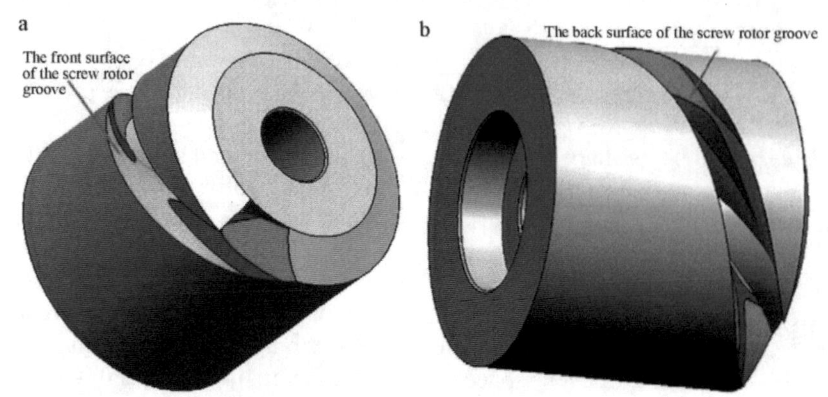

a) b)

Figure 7. Numerical simulated profile enveloped by multi-column on the screw rotor groove surface.

IOP Conf. Series: Materials Science and Engineering **232** (2017) 012001 doi:10.1088/1757-899X/232/1/012001

3. Research of new conception engage profile

In order to precisely and efficiently process the screw rotor and star-wheel of new concept SSC, the authors make numerical simulation of their processing repeatedly. Finally a scheme of screw rotor processing was achieved and a special milling machine was developed as shown in figure 8, in which milling cutter can move in the X and Y coordinates direction. Milling cutter axis can also rotate in this plane. Actually this milling machine is built with six coordinates. This machine can process screw rotor with diameter around 100~ 400mm. According to the profile, the NC programs for this milling machine was also studied and developed. After 3 years' processing tests and improvements, the range of processing error for this machine is brought down to 0.015mm. Simultaneously, the authors researched and developed a special grinding machine, as well as the NC programs related to the profile, for processing star-wheel tooth profile. In this machine, shown in figure 9, large diameter grinding wheel composed of diamonds are used to eliminate the error of grinding wheel wear. Processing error is reduced within 0.01mm.

Figure 8. Screw-rotor processing machine.

Figure 9. Star-wheel processing machine.

4. Development of new concept SSC prototypes and their performance tests

Since 2013, there are 6 companies have been cooperated with the author's group, and many kinds of new concept SSCs have been developed, such as two injection types with oil lubricated and water lubricated, several power models as 22kW,37kW, 75kW, 110kW, 132kW, 250kW and etc.. In order to show the advantages of new concept SSC, nodular cast iron is used as material for star-wheel with oil lubricated, see figure 10.

There were many researchers took nodular cast iron as the material of star-wheels in past. However, just after several hours' operation those star-wheels in traditional SSC were worn out, and the compressor's flow-rate went down over 10%. In development of water-lubricated compressors recent years, PEEK is generally used as the material of star-wheels. These kinds of star-wheels can be put into operation for thousands hours at most. Still, flow-rates of the compressors' decrease down apparently after that.

a) Left star-wheel b) Right star-wheel

Figure 10. Nodular cast iron star-wheel after 2000h's operation.

Authors separately designed and manufactured 3 kinds of the new concept SSCs (two with oil lubricated, one with water lubricated), as well as their testing systems. 2000 hours' performance tests were then conducted. Discharge pressure of those were all kept at 0.8 MPa (G) in the testing operation. The flow-rates of them were measured every 24 hours. Table 1 shows the flow-rate results measured at some time nodes for the two oil lubricated prototypes.

Table 1. Tested flow-rate results of oil lubricated compressors.

operation time (h)	5	400	800	1200	1600	2000
	flow-rate (m^3/min)					
Oil lub1#	2.85	2.88	2.87	2.86	2.87	2.86
Oil lub2#	7.98	7.99	8.00	8.00	7.98	7.98

For the water lubricated one, copper is used as material for the screw rotor and PEEK for the star-wheel, see figure 11.The designed flow-rate is 6 m^3/min, with pure water used as the lubricating fluid.

a) Copper screw rotor b) PEEK star-wheel

Figure 11. Water lubricated screw rotor and star-wheel.

The testing set-up is shown in figure 12. The air was first purified by the air inlet filter and then transported to suction chamber for compressing. The compressed air together with the lubricated water were discharged to the separator. The compressed air was then available through the air discharge valve. The water converging at the separator bottom was transported to the cooler as the lubrication fluid by the pressure difference between the suction chamber (inlet pressure) and the separator (separator pressure). The injection water was cooled down by the cooling water and went through two water filters to the injection end.

The testing results is shown in figure 13. Obviously, there are some changes at the beginning of the operation, which is because the whole system needs some time to run-in. A big drop of flow rate happened around 150 hours, the machine was stopped and checked. There were some changes on the position of both the star-wheels and the screw rotor through their axial direction. Besides, the environment temperature change can directly influence the suction temperature, which will lead to undulations of the flow-rate. Both the star-wheels' and screw rotor' positions were reset. After about 500 hours' operation, the flow-rate keeps steady around 5.37 m3/min till the end of the testing (2350hours). The average volume efficiency is 89.5%. The new concept profile avoids the problem of wearing out and reaches a really high sealing property.

All results show that there are no big changes on the flow-rates for these 3 kinds of new concept compressors in 2000 hours' testing operation, which proves the new concept profiles all work well with a long time operation.

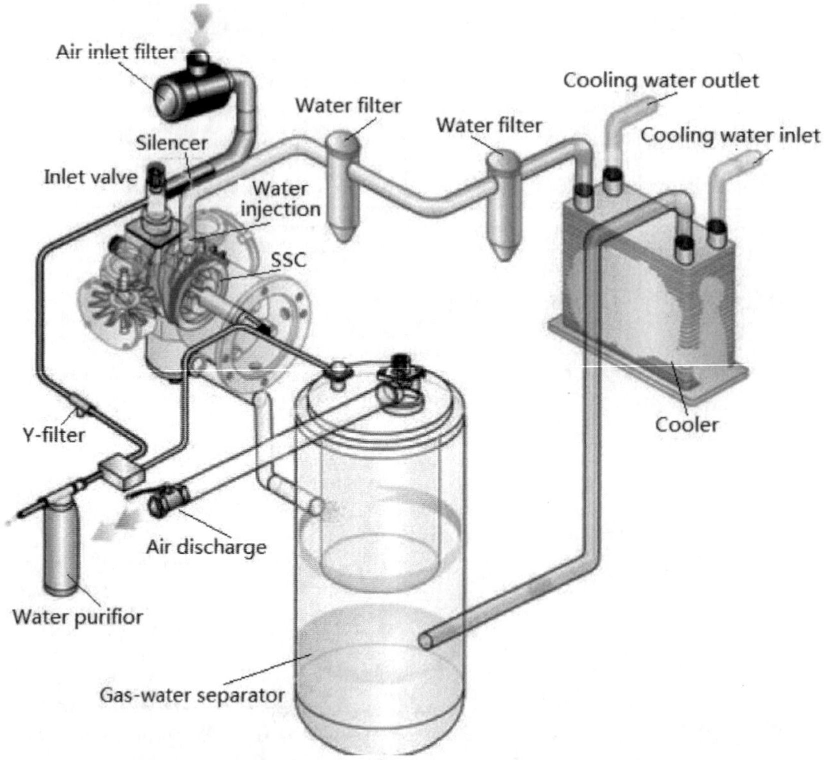

Figure 12. Water lubricated prototype compressor testing set-up.

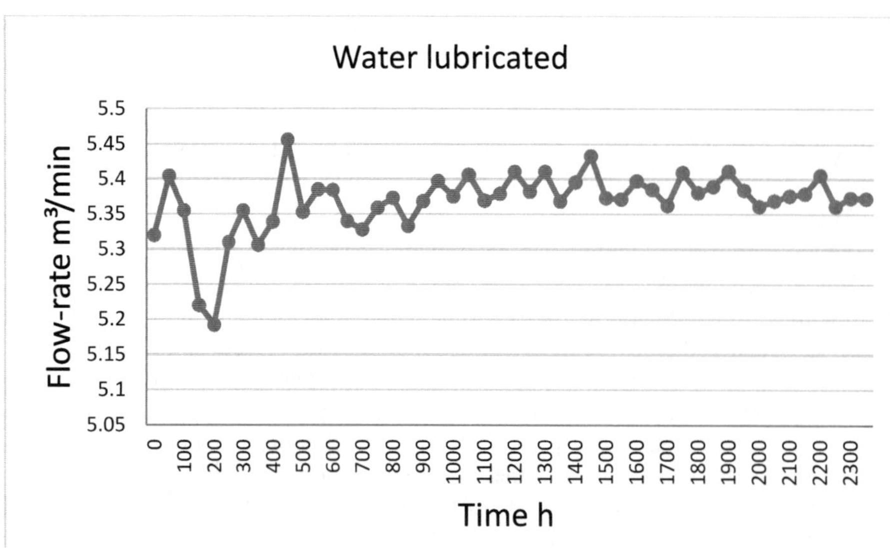

Figure 13. Testing results for the water lubricated prototype compressor.

5. Conclusions

1) The paper reveals wear mechanism of a star-wheel in a traditional SSC that star-wheel tooth flank profile enveloped with single straight line can't offer an essential lubricated condition for compressor operation.
2) Multi-columns enveloping profiles for engaging pairs of SSCs are proposed, which is designed to achieve a liquid floated engagement in screw rotor and star-wheel system.
3) Precise and effective milling machines are developed for processing screw rotor and star-wheel engaging pairs with multi-columns enveloping profiles respectively.
4) Many kinds of new concept SSC and their prototypes are developed and tested. Especially the water-lubricated SSC, which use water instead of oil working as the injection fluid. It is more pro-environmental and can also meets the requirements of oil-free working conditions. Experiments proved these new concept SSCs can work for a long time period without any obvious decrease of the flow-rates. The wear out problems of the star-wheels are solved.

References

[1] Zimmern B 1965 Worm rotary compressors with liquid joints (USA) 3180565
[2] Zimmern B 1984 From water to refrigerant: twenty years to develop the oil injection-free single screw compressor *The 1984 Purdue Compressor Technology Conference: Purdue University, West Lafayette, IN, USA* pp 513-18
[3] Jin G X and Tang Y 1986 *A study of the single screw compressor profile* (Purdue, West Lafayette, IN, USA, 1986)
[4] Zhao B 2010 Fault diagnosis for the gate rotor of the single screw compressor based on wavelet finite element method *Journal of Computational Information Systems* **6**(11) pp 3849–58
[5] Guohui L and Xiaonan P 2004 *Research on three performances of a single screw compressor engaging pair machinary* **05** pp 29–31
[6] Quanke F, Bei G and Cun Z 2005 Profile design method of multi-straight line enveloped engage pair in a single screw compressor *Compressor technology* **03** pp 1–5
[7] Wu W F and Feng Q K 2009 A multicolumn envelope meshing pair for single screw compressors *Journal of Mechanical Design, Transactions of the ASME* **131**(7) 0745051-0745053
[8] Cun Z, Weifeng W and Quanke F 2010 *Constructing method of engage pair profile enveloped with non equal length multi-column in a single screw compressor* (China, ZL) 201010293319.1. 2010
[9] Li J, Feng Q, Liu F *et al* 2013 Experimental studies of the tooth wear resistance with different profiles in single screw compressor *Tribology International* **57** pp 210–15

IOP Conf. Series: Materials Science and Engineering **232** (2017) 012002 doi:10.1088/1757-899X/232/1/012002

Thermal and Fluid Mechanical Investigation of an Internally Cooled Piston Rod

K Klotsche[1], C Thomas[1] and U Hesse[1]

[1] Technische Universität Dresden, Institute of Power Engineering, Bitzer Chair of Refrigeration, Cryogenics and Compressor Technology, 01062 Dresden, Germany

konrad.klotsche@tu-dresden.de

Abstract. The Internal Cooling of Reciprocating Compressor Parts (ICRC) is a promising technology to reduce the temperature of the thermally stressed piston and piston rod of process gas compressors. The underlying heat transport is based on the flow of a two-phase cooling medium that is contained in the hollow reciprocating assembly. The reciprocating motion forces the phases to mix, enabling an enhanced heat transfer.

In order to investigate this heat transfer, experimental results from a vertically reciprocating hollow rod are presented that show the influence of different liquid charges for different working temperatures. In addition, pressure sensors are used for a crank angle dependent analysis of the fluid mechanical processes inside the rod. The results serve to investigate the two-phase flow in terms of the velocity and distribution of the liquid and vapour phase for different liquid fractions.

1. Motivation

Thermodynamics and heat transfer play a key role during the compression process, particularly with regard to the energy and volumetric efficiency but also to the durability of sealing elements. The latter applies particularly to non-lubricated materials. Several studies[1,2,3] have addressed the effect of cooling on the abovementioned aspects and the actual potential was rated differently. However, the impact of compressor cooling depends on the specific compressor setup and especially on the realisation of the cooling concept. Thus, it is needed to be reassessed from case to case.

Heat is typically considered to be rejected from the outer cylinder parts either by a cooling system within the cylinder walls or by a cooled packing at the piston rod of crosshead-type piston compressors. In contrast, a novel cooling technique – the Internal Cooling of Reciprocating Compressor Parts (ICRC) – constitutes a promising and convenient possibility to reject heat through the reciprocating assembly. For this, a two-phase fluid contained inside the reciprocating assembly is intermixed by reciprocating motion, thus, transferring heat to the crosshead and finally rejecting it to the crank case oil.

Several experimental as well as theoretical investigations had been performed over the last years in order to gain more information about the potential heat transfer of the ICRC. It appeared that the heat transfer through the piston rod can be increased by a couple of orders of magnitude compared to solid designs. Those studies were mainly focused on the thermodynamic behaviour. For a deeper understanding of the occurring phenomena and for the pre-calculation of the heat transfer of the ICRC for a specific compressor configuration, improved knowledge about the two-phase flow of the contained fluid, *i. e.* the velocity and the distribution of the liquid and vapour phase is required. Therefore, based on the thermodynamic results and pressure measurements, a first insight into the fluid mechanical processes of the two-phase flow inside a vertical reciprocating hollow rod is provided. Furthermore, the suitability of the cooling fluid used in previous studies[4] is reviewed for different filling amounts and particularly for varying temperatures.

2. Internal Cooling of Reciprocating Compressor Parts (ICRC)

The idea of an internal cooling of the reciprocating assembly of piston compressors dates back to 2007 as Quack and Nickl[5] had patented this concept. The following section provides an overview of the principle of operation and previous research.

2.1. Principle of operation

The ICRC aims at cooling of compressor components that are thermally stressed due to their direct contact with the working fluid and due to friction at the sealing gaps. For this purpose, a two-phase coolant is contained within the hollow reciprocating assembly, *e. g.* the piston and the piston rod, see Fig. 1. Owing to the reciprocating motion, both phases are generally intermixed and, in doing so, coolant flows back and forth between both ends of the cavity. Thereby, heat is absorbed from warmer parts, *e. g.* from the piston or from the packing area, transported through the reciprocating assembly and released at the colder oil-lubricated crosshead.

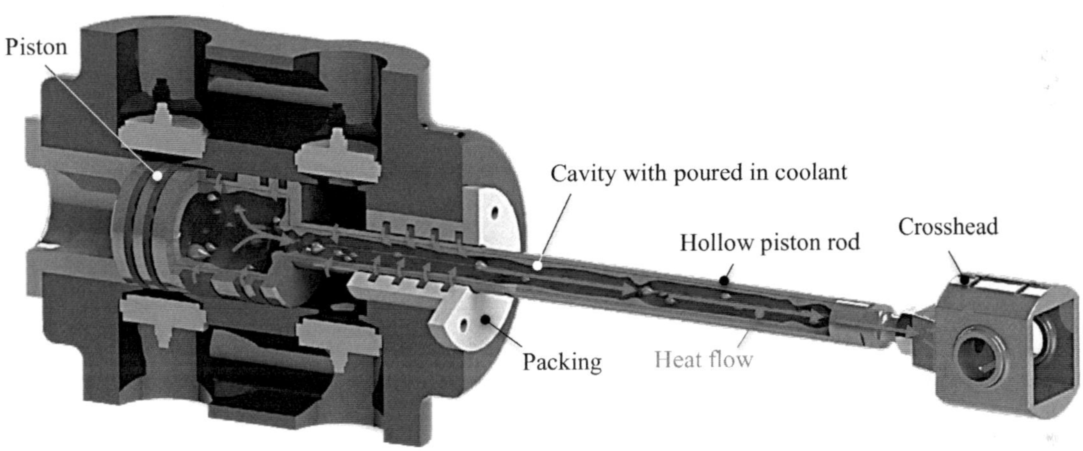

Figure 1. Design and principle of operation of the ICRC

Obviously, the temperature difference between the cylinder area and the crosshead as thermal drive is also present within a conventional solid design of the reciprocating components. But, the fundamental idea and main benefit of this cooling technology is to make use of the intensified fluid flow inside the cavity caused by the high-frequency reciprocating motion. This engenders a much lower thermal resistance of the reciprocating assembly, which results in an increased cooling effect for the cylinder area.

2.2. Previous research

Since the first reference of the ICRC idea, several experimental and theoretical studies were conducted, focusing at first on its feasibility and general applicability. Thomas[4] presented extensive results from both experimental and theoretical investigations which show the effectiveness of the internal cooling at three different test rigs, including a single-stage double-acting balanced-opposed compressor. In comparison to a solid piston rod, the implementation of an internally cooled piston rod had led to a reduction of the gas temperature inside the packing of up to 80 K.

In 2016 Klotsche *et al*[6] have shown the influence of the internal diameter on the heat transfer of a vertical configuration of the ICRC by means of experimental investigations. The axial heat flux could be quantified by a subsequent numerical analysis in the range of 13 W/cm² to 30 W/cm². For the same (vertical) configuration and a similar manner, the effect of different potential coolants on the heat transfer was examined by Klotsche *et al*[7]. Additionally, for the most promising coolant the optimal filling amount could be found.

3. Experimental investigation

3.1. Test rig setup

For the experimental investigation of the heat transfer inside a reciprocating hollow rod a test rig is available at the Technische Universität Dresden (see Fig. 2). It comprises a vertically reciprocating hollow rod actuated by a crank drive that is driven by an electric motor. The stroke s is 100 mm and the length of the connecting rod l is 250 mm. The rod is mounted onto a temperature-controlled crosshead. This crosshead comprises a cooling duct connected to an external cooling circuit providing the entire heat to be rejected at the rod's lower end, *i. e.* the temperature at this location can be held constant for all operation points. At the upper end of the rod a cylindrically shaped electrical heater is installed to provide a precisely adjustable heat input. At the top the rod is closed by an end cap. All relevant components except the crosshead are insulated against the ambient environment to guide the majority of the heat input through the fluid. In order to connect any electrical signals to the measurement system and to guide the crosshead coolant from the test rig to the cooling unit of the laboratory, an energy chain is attached to the rod's cap.

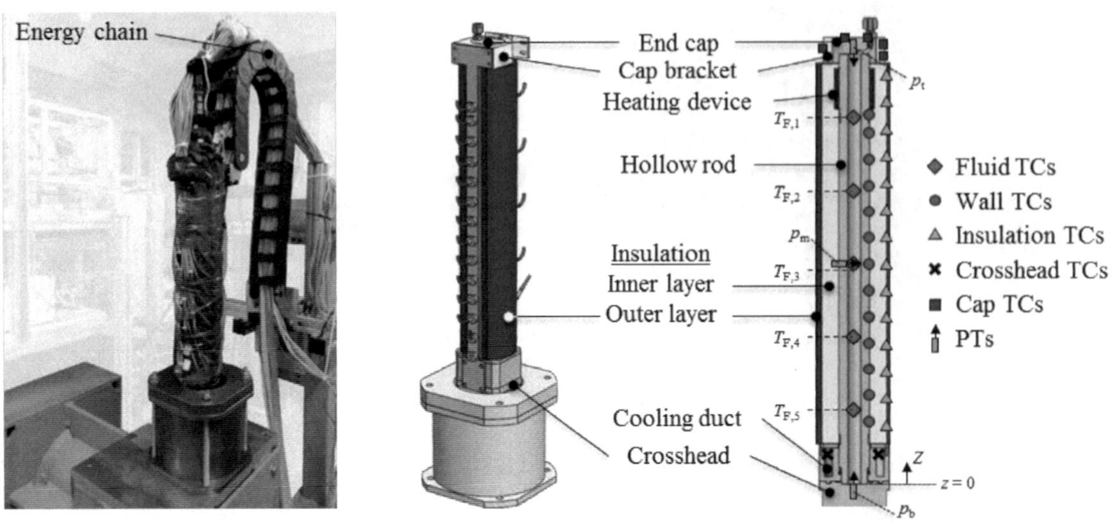

Figure 2. Setup of the test rig (a), CAD representation (b), and sectional view with installed thermocouples (TCs) and pressure transducers (PTs) with measuring direction (c)

To record the temperature field of all relevant components, 39 type K thermocouples (TCs) are installed whose signals are sampled at 1 Hz. Their locations can be seen in Fig. 2 (c). Five TCs ($T_{F,1}$...$T_{F,5}$) positioned inside the cavity are distributed along the rod's axis z measuring the fluid temperature. Thirteen temperature sensors are positioned 2 mm below the rod's external radial surface capturing the wall temperature. They are distributed along the z-axis similar to the fluid TCs. Additionally, two TCs are located at the rod's cap and 3 TCs at the lateral surface of the cap bracket as well as 2 TCs inside the cooling duct of the crosshead. The rod is equipped with two insulation layers. Fourteen TCs within the insulation layers measure the temperature along the rod's axis. As described in a previous work[6] the cap, insulation and crosshead TCs can be used as boundary conditions for a simulation of the temperature field by the FEM program Ansys.

For the investigation of the dynamic fluid behaviour, 3 pressure transducer (PTs, Kulite XTEL-190) with a possible sampling frequency of up to 100 kHz are positioned at the upper and lower end face of the cavity as well as at medium height. The measuring direction of the PTs, which is perpendicular to the mounting wall, is depicted in Fig. 2. The high-frequency pressure signals are referenced to the position of the rod, *i. e.* bottom dead centre (BDC) and top dead centre (TDC), by a rotary angle sensor.

3.2. Measurement procedure
At first, the cavity, which is opened to ambient air, is filled by the coolant. Then the hollow rod is closed. This results in a coolant-air mixture at ambient conditions. Then the rotational speed is adjusted and the flow through the crosshead's cooling circuit as well as the power supply of the heater is switched on. After a transient period in which the temperatures are converging towards a steady-state, the subsequent interval is used to record the final temperatures which are averaged over 5 minutes. In this way, long term transient phenomena are excluded. In the quasi-stationary state, also the dynamic pressure measurements were conducted for 10 crank revolutions at a sampling rate of 50 kHz.

3.3. Measurement series
Previous studies[7] with a natural coolant showed that a liquid volume fraction ε of 30% exhibits the highest heat transfer rates in comparison to 10%, 20%, 40%, and 50%. It cannot be assumed that there is another optimum outside this range of liquid fractions. Therefore, this study, as a first step, is intended to investigate the heat transfer and the occurring temperatures for a finer division (5% steps) in the range of 20% to 40%.

Furthermore, it is known from the heat pipe literature[8] that the heat transfer capability for a certain cooling fluid – typically represented as Merit number – changes with temperature due to the change of fluid properties. In order to analyse whether the temperature affects the found optimum of the filling amount, different liquid fractions were tested with varying heat loads (0 W to 250 W in steps of 50 W).

For all shown measurements, the rotational speed n was set to 500 min^{-1} and the inlet temperature of the crosshead coolant T_C to 10 °C. All measurements were carried out under laboratory conditions ensuring the ambient temperature to be sufficiently constant over time. Hence, all (thermal) boundary conditions are consistent, except the heat input that was varied for the different measurement series.

For all experiments the same rod (internal diameter $Ød_i$ = 25 mm, outer diameter $Ød_o$ = 35 mm, internal length l = 500 mm) was used.

3.4. Temperature measurements
Among the 39 thermocouples only the averaged fluid temperatures are presented in Fig. 3 (a-e), since they are the most meaningful with regard to the fluid's heat transfer capability. Unfortunately, no results for a liquid fraction of 40% at 250 W can be presented because the temperatures of the rest rig exceeded the allowable maximum.

Fig. 3 (a-e) shows the lowest temperatures in the upper region of the rod for a liquid volume fraction of 25% for any heat input. This means, that the optimum heat transfer occurs for a liquid-to-gas ratio of 1:3 independent of the temperature in the range of 25 °C to 100 °C. The temperatures in the lower part of the rod rise with increasing charge, indicating a worse heat rejection from the fluid to the cooled wall.

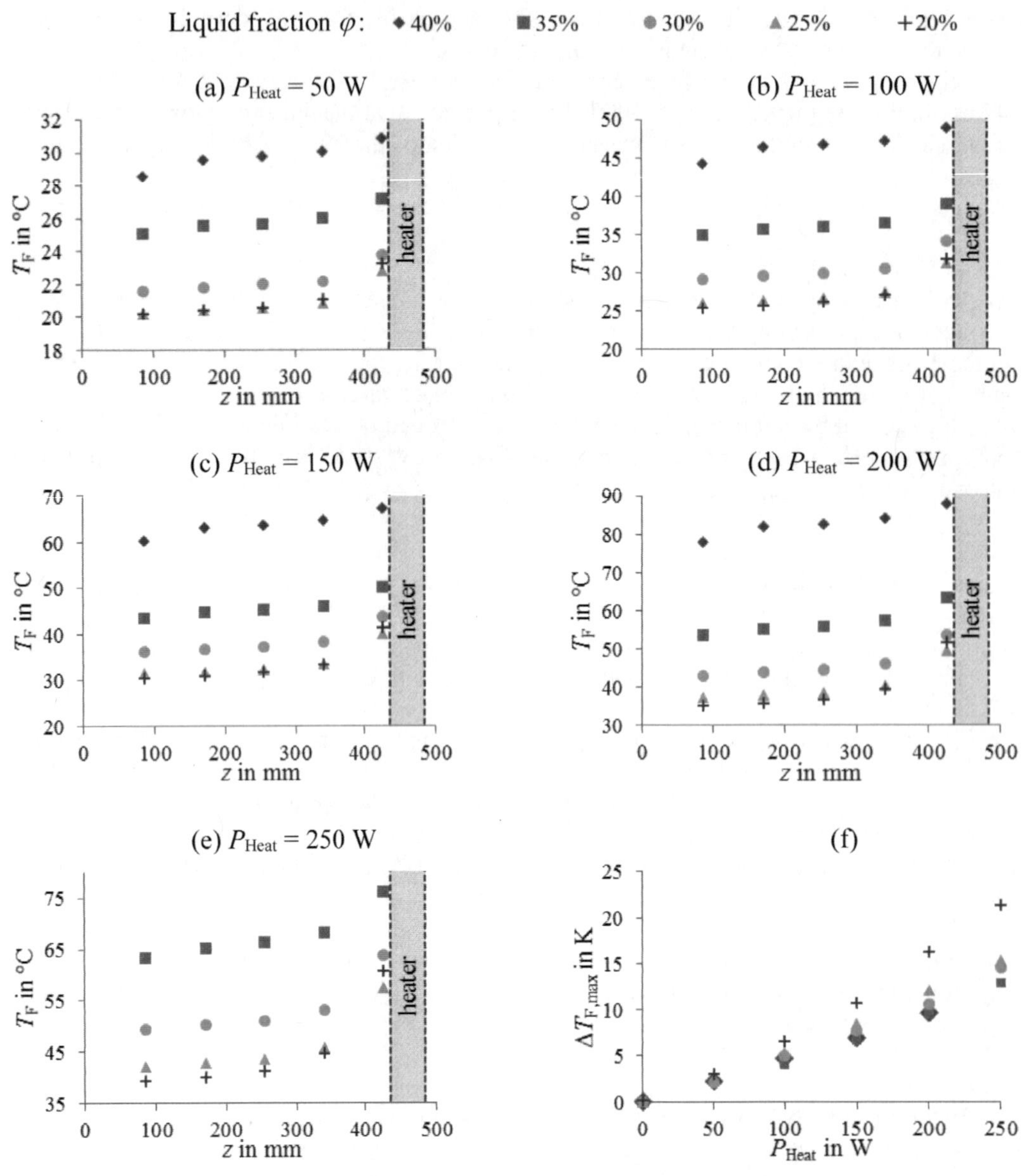

Figure 3. Averaged fluid temperatures of the steady-state period (a-e) and maximum fluid temperature differences (f) for varying liquid fractions and heat inputs

In Fig. 3 (f) the maximum temperature difference ($\Delta T_{F,max}$ = $T_{F,1}$-$T_{F,5}$) are presented for every measurement. This gives access to the thermal resistance within the fluid. It can be seen that the highest temperature differences occur for the smallest liquid fraction of 20% which is mostly caused by large gradients in the upper section of the rod. This can be explained based on the too small liquid charge which only insufficiently reaches the upper parts of the rod. The smallest difference and by that the lowest thermal resistance between the upper- and undermost fluid TC is given for the liquid fractions of 35% and 40% which do not differ significantly. In addition, these values deviate also only slightly from the measurement run with the optimal overall heat transfer (ε= 25%). This leads to the assumption that for the examined configurations, the heat transfer within the fluid plays only a minor role compared to the heat transfer from the fluid to the wall at the cold end.

IOP Conf. Series: Materials Science and Engineering **232** (2017) 012002 doi:10.1088/1757-899X/232/1/012002

3.5. Thermal simulation

In accordance with the simulation procedure presented in a previous work[6], the temperature field for all measurements was calculated. For this, the measured temperatures from the rod's end cap, from within the insulation layers and from the crosshead are used as boundary conditions and the thermal behaviour of the filling is varied in an iterative procedure to match the fluid and the wall temperatures of the simulation with the experimental results. The simulation based on the finite element method is used to derive the axial heat flux of the fluid cross section at medium rod height as a quantitative heat transfer measure (see Tab. 1).

Table 1. Axial heat flux $|\hat{q}_z|$ in $\frac{W}{cm^2}$ and axial heat flow $|\dot{Q}_z|$ in W (in parentheses) from the simulation for the fluid cross section at medium rod height ($z = 0.25$ m)

		Liquid volume fraction ε in %					Maximum axial heat flux $|\hat{q}_{z,max}|$ in $\frac{W}{cm^2}$
		20	25	30	35	40	
Heat input P_{Heat} in W	50	8.5 (41.7)	8.6 (42.2)	8.5 (41.7)	7.4 (36.3)	6.8 (33.4)	10.2
	100	16.3 (80.0)	16.5 (81.0)	16.2 (79.5)	15.3 (75.1)	13.4 (65.8)	20.4
	150	24.2 (118.8)	24.4 (119.8)	23.9 (117.3)	22.6 (110.9)	20.4 (100.1)	30.6
	200	31.9 (156.6)	32.2 (158.1)	31.4 (154.1)	30.0 (147.3)	24.7 (121.2)	40.7
	250	40.7 (199.8)	40.8 (200.3)	40.1 (196.8)	37.8 (185.6)	-	50.9

The values for the axial heat flux in Tab. 1 show that ca. 50 to 85% of the applied head load could be transferred through the rod. For the highest heat input (250 W) and the optimal liquid fraction (25%) about 220 W have been transferred through the fluid's cross section in axial direction, which corresponds to a temperature gradient $\Delta T/\Delta z$ of 45 K/m between the upper- and undermost FT. Though this is even a small temperature gradient in comparison to real applications, these results show that the ICRC can significantly contribute to reject heat from the packing and from the cylinder area.

3.6. Pressure measurements

For the investigation of fluid mechanical processes the pressure signals are shown in Fig. 4 for a heat input of 100 W and for $n = 500$ min^{-1}. On the left hand side the original signals of all 3 pressure sensors are shown for 10 crank turns. In a first step, these data was reduced to one representative period by averaging all pressure values related to the same crank angle value. That means, that *e. g.* exactly all values that correspond to the TDC position (0°, 360°, etc.) are averaged to one final value. In a second step, this period is processed by a rolling window (rolling average) analysis with a window width of 800 data points and window type "triang". This is done in order to remove small fluctuations for obtaining *e. g.* comparable peak values. More information on the rolling window processing, which has been executed via the Python package "pandas" can be found in [9]. The results of both data reducing steps can be seen in the diagrams at the right hand site of Fig. 4 where the index end "per" represents the period averaged time series and "per win" represents the time series after both processing steps (only shown for the p_b sensor).

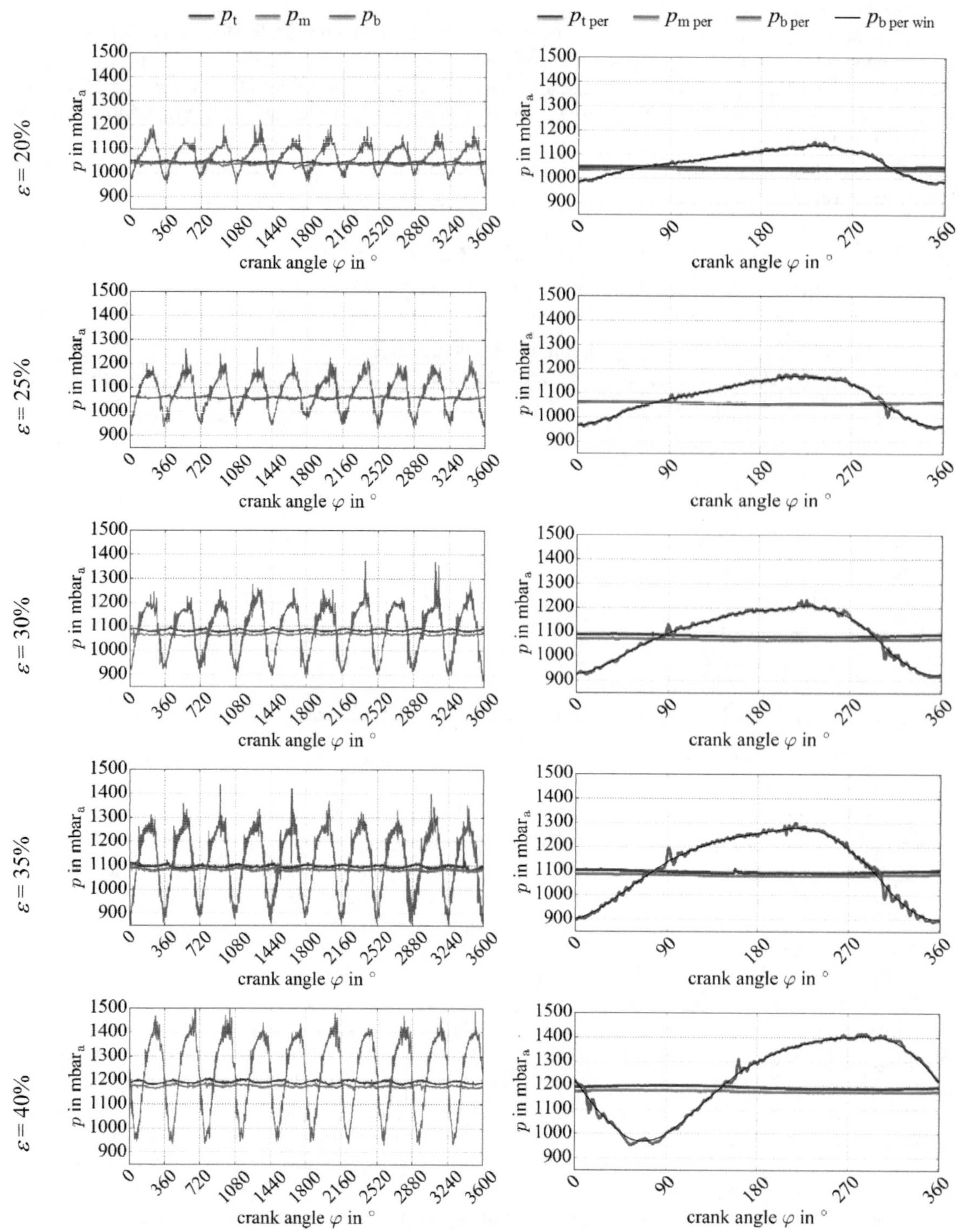

Figure 4. Original pressure signals (left) and processed pressure signals (right) (P_{Heat} = 100 W, n = 500 min^{-1}).

Fig. 4 shows that the pressure sensors at middle height (p_m) and at the top end of the rod (p_t) show only small fluctuations over time. Particularly, p_t happens to experience almost no liquid contact since no pressure peak compared to the values of the pressure sensor at the bottom of the cavity (p_b) is visible. Hence, p_t presumably show just the change of volume for the gaseous phase in the upper part of the rod which is quiet small. This is not surprising, if one assumes both phases to be a continuum of which the liquid one is not compressible, thus, almost no compression of the other (gaseous) phase can be excited. Equivalent and in-phase time series can be seen for p_m which also see little influence of the liquid phase. Since the temperature gradients of Fig. 3 indicate a wetting at the location of p_m this could be explained by minor flow components in radial direction. The slight difference of the mean values of both signals originates from temperature differences between both measuring positions.

Most interesting, the p_b signal provides the best insight into the flow characteristics of the liquid phase. Its time course is mainly determined by the periodical movement of the rod since the highest fluctuations correspond to the frequency of the reciprocating motion. So the flow of the liquid phase mainly follows the rod's motion. In contrast, the p_b curves differ from the sinusoidal motion of the rod with regard to the peaks. The minima matches with the TDC position (except for $\varepsilon = 40\%$), but the maxima are delayed and do not correspond with the BDC position ($\varphi = 180°$). However, all 10 periods of each measurement run have quiet similar characteristics in terms of frequency and extrema. They differ only with regard to minor temporal phenomena within each period, which supports the abovementioned period-averaging approach.

3.7. *Interpretation of pressure signals*

The obtained data, particularly for the lower pressure sensor, can be used to draw conclusions about the periodical liquid behaviour over one crank revolution. One approach is to approximate the liquid phase to exist in two forms: On the one hand a sump at the bottom end of the rod that more or less follows the rod's motion. And on the other hand a liquid film, which originates from liquid that leaves the sump and which mainly sticks to the rod's wall due to surface tension forces. The volumes of the sump and the film change over one crank revolution.

For the analysis of the sump behaviour, the p_b signal is divided into two intervals. The first one is the interval where the pressure values increase. In this period, the fluid sump has the tendency to move downwards with respect to the rod's motion, thus, creating a higher load for the pressure sensor. The second one is the rest of the period where the pressure is decreasing. This strain relief of the pressure sensor occurs when a part of the liquid sump shows the tendency to move upwards with respect to the rod's motion. Then, a liquid portion leaves the sump mainly as liquid film at the wall quiet similar to what is known as annular flow. Tab. 2 summarises the relevant data deduced from the processed pressure signals ($p_{b\ per\ win}$).

Table 2. Deduced data from the $p_{b\ per\ win}$ time series (P_{Heat} = 100 W, n = 500 min^{-1}).

	Liquid volume fraction ε in %				
	20	25	30	35	40
p_{max} in mbara	1136	1169	1208	1277	1404
p_{min} in mbara	987	966	925	902	969
$p_{peak\text{-}to\text{-}peak}$ in mbar	150	203	282	375	435
$\Delta\varphi_{incr}$ in °	242	220	231	222	219
$\Delta\varphi_{decr}$ in °	118	140	129	138	141
Δt_{incr} in ms	81	73	77	74	73
Δt_{decr} in ms	39	47	43	46	47

The peak-to-peak values for the $p_{b\ per\ win}$ time series from Tab. 2 increase with increasing liquid fraction almost linearly. These peak-to-peak values show that the sump height over one crank rotation is increasing with larger amounts of liquid, since a larger liquid sump is acting on the pressure transducer. This illustrates that not the entire liquid charge is forced to leave the sump end. With increasing overall liquid charge, those liquid portion, which does not transfer heat from the rod's upper to its lower part, is getting larger. It just blocks the lower end of the cavity and degrades the heat rejection, there.

Tab. 2 also shows that the length of intervals for increasing pressure values Δt_{incr} is about 1.5 to 2.0 times longer than the intervals for decreasing pressure values. If one assumes the increasing pressure interval to be the time period in which the liquid film returns to the sump, this time interval represents the period in which heat is actually transported from the upper end to the lower end. If the main heat flow is provided by the liquid motion, which means that the heat transfer through the gas phase is negligible, this axial heat transfer can be stated as following:

$$|\dot{Q}_z| = \dot{m}_l * c_{p,l} * \Delta T \quad (1)$$

Where $\dot{m}_l$ is the liquid mass flow, $c_{p,l}$ is the specific heat capacity of the liquid (for isobaric change of state) and ΔT is the corresponding temperature difference. When applying this formula on the middle section of the rod, where an almost linear temperature gradient can be found between $T_{F,2}$ and $T_{F,4}$ (see Fig. 3), the necessary liquid mass flow can be calculated based on the simulated heat flow values taken from Tab. 1. For a first approximation, this mass flow is assumed to be constant over time. The corresponding mean velocity v_l to this mass flow is calculated according to:

$$v_l = \Delta z_{T2\text{-}T4}/\Delta t_{incr} \quad (2)$$

where $\Delta z_{T2\text{-}T4} = 170$ mm is the vertical distance between the fluid TCs 2 and 4. Making use of the liquid mass flow $\dot{m}_l$, the liquid density ρ_l, and the mean velocity v_l the liquid flow area can be expressed as:

$$A_l = \dot{m}_l/(\rho_l * v_l). \quad (3)$$

If the liquid mass flow is furthermore assumed to appear as rotationally symmetric annular flow (circular ring) the film thickness of the liquid film is given as:

$$\delta_l = 0.5 * \{\text{Ø}d_i - [\text{Ø}d_i^2 - (4A_l/\pi)]^{1/2}\}. \quad (4)$$

The resultant values for this calculation are summarized in Tab. 3 for a heat input of 100 W. All necessary fluid properties are provided by REFPROP[10].

Table 3. Deduced data from the $p_{b\ per\ win}$ time series (P_{Heat} = 100 W, n = 500 min^{-1}).

	Liquid volume fraction ε in %				
	20	25	30	35	40
$\dot{m}_l$ in g/s	53.5	73.9	79.9	94.5	73.9
v_l in m/s	2.1	2.3	2.2	2.3	2.3
A_l in mm^2	25.4	31.9	36.4	41.4	32.0
δ_l in mm	0.3	0.4	0.5	0.5	0.4

4. Summary

To access the heat transfer capability of the ICRC, experimental results from a vertically reciprocating internally cooled rod, which is heated at its top end, were presented and the heat transfer was investigated. It could be shown that for a natural coolant used in previous studies[5] a liquid/gas volume ratio of 1:3 shows the best cooling effect for the temperature range of 25 °C to 100 °C. Temperature measurements in conjunction with a finite element simulation have been used to quantify the axial heat flux of the fluid's cross section. The results prove that the ICRC can significantly contribute to reject heat from the packing and cylinder area of process gas compressors through the reciprocating assembly. Pressure measurements conducted simultaneously serve to interpret the fluid mechanical processes of the fluid mixture and conclusions could be drawn about the distribution and the flow of the liquid phase.

5. Outlook

Since the pressure measurements provide only local insight, an accurate examination of the flow inside the hollow rod is difficult to achieve. Therefore, optical measurements would be a promising option to investigate the occurring fluid mechanical phenomena in a satisfying way. Moreover and based on such visual results, a model to predict the flow is needed for precise calculations of the heat transfer inside a reciprocating hollow rod.

References

[1] Adair R P *et al* 1972 *Instantaneous heat transfer to the cylinder wall in reciprocating compressors* International Compressor Engineering Conference at Purdue, West Lafayette, USA

[2] Brok S W *et al* 1984 *Modelling of cylinder heat transfer – large effort, little effect?* International Compressor Engineering Conference at Purdue, West Lafayette, USA

[3] Shiva Prasad B G 1998 *Heat Transfer in Reciprocating Compressors – A Review* International Compressor Engineering Conference at Purdue, West Lafayette, USA

[4] Thomas (née Hammer) C 2013 *Innenkühlung der Kolbenstange von trockenlaufenden Kolbenverdichtern* (Doctoral dissertation). Technische Universität Dresden, Germany

[5] Patent DE 10 2007 000 652 A1

[6] Klotsche K *et al* 2016 *Experimental and numerical investigation of the heat transfer inside a hollow piston rod* International Compressor Engineering Conference at Purdue, West Lafayette, USA

[7] Klotsche K *et al* 2016 *Experimental and numerical investigation of the heat transfer inside a hollow piston rod* 10th EFRC Conference, Düsseldorf, Germany

[8] Reay D, McGlen R, Kew P 2013 *Heat pipes: theory, design and applications* Butterworth-Heinemann

[9] McKinney W 2010 *Data Structures for Statistical Computing in Python* Python for Scientific Computing Conference, Austin, USA

[10] Lemmon E W, Huber M L, McLinden M O 2013 *NIST Standard Reference Database 23: Reference Fluid Thermodynamic and Transport Properties-REFPROP*, Version 9.1, National Institute of Standards and Technology, Standard Reference Data Program, Gaithersburg, USA

IOP Conf. Series: Materials Science and Engineering **232** (2017) 012003 doi:10.1088/1757-899X/232/1/012003

Characterization and Performance Testing of Natural Gas Compressors for Residential and Commercial Applications

Xinye Zhang[1,a], **Eckhard A. Groll**[1,b], **Dylan Bethel**[2,c]

[1] Purdue University, School of Mechanical Engineering, Ray W. Herrick Laboratories, West Lafayette, IN, USA
[2] BlackPak Inc. San Leandro, CA, USA

E-mail: [a]xinye@purdue.edu, [b]groll@purdue.edu, [c]dbethel@blackpaktech.com

Abstract. Relatively little information is available in the literature with respect to the performance of compressors used during the dynamic charging process of a tank. Therefore, work presented in this paper shows the measurement results of performance testing of a natural gas compressor and analyses the compressor characterization based on the experimental data. Initial tests were conducted using air and carbon dioxide given the thermodynamic similarities between these fluids and natural gas. Finally, a new test stand was specifically designed and built for compressor dynamic testing using pipeline natural gas (NG) and the compressor reliability has been evaluated inside an explosion-proof engine test cell. Reliability tests at standard operating conditions monitored the performance consistency of the compressors over the testing period and the testing consisted of a series of tank charges aimed at evaluating the maximum operating temperature as well as the mass flow rate in the system.

1. Introduction

Limited studies have been conducted with respect to experiments and simulation models to monitor and predict the dynamic process of reciprocating compressor. The dynamic testing of a two-stage reciprocating air compressor with inter- and after-cooling during steady and transient state operation has been reported by Nakagawa et al. [1]. The application of micro-thermocouples for measuring the instant temperatures and pressures in cylinders was mainly discussed. A specific dynamic model for a small refrigerating cycle was presented by Janssen et al. [2]. Experiments and simulations for the start-up and pull-down phase of a freezer were reported as well. However, in these papers, the working fluids were conventional refrigerants or ambient air, and the experimental considerations were not comparable to the dynamic testing with natural gas, which has more safety regulations. Kim and Lee [3] conducted several dynamic tests to verify the reliability of high cycle fatigue of a gas compressor blade, and analyzed the fatigue testing results. Norman Shade [4] presented several cases of existing reciprocating compression assets and especially showed optimization of the testing rigs using pipeline natural gas. Moreover, many simulation models have been developed and analyzed. Chi and Didion [5] as well as Rajendran and Pate [6] both came up with transient simulation models for a heat pump and vapor compression cycle, respectively, but the experimental validations were not discussed in these two papers. Considering the combustible property of natural gas and safety aspects necessary, little work has been found in the open literature relative to the experimental testing of the dynamic charging process with natural gas. This paper focuses on reporting experimental data to evaluate the dynamic performance of a natural gas compressor during the charging process of a tank. In addition, reliability tests of the compressor system are also discussed.

2. Description of systems

2.1 Overview of all testing systems

To evaluate the efficiency, performance, and safety characteristics of natural gas compressors used in residential and commercial applications, a three-step sequence of testing was conducted and three test stands were built or modified for experimental performance testing of a two-stage reciprocating compressor system. The compressor performance tests were conducted using a combination of open loop and closed loop test setups with different working fluids that have thermodynamic similarities with natural gas. Initial open loop tests were conducted using air as a working fluid, because air is readily available, nontoxic, and nonflammable and carbon dioxide was chosen as the working fluid for the closed loop system, because the ratio of specific heat is very similar to that of natural gas. The open loop system was used to investigate the dynamic compressor performance when the discharge pressure increases due to pressurizing ambient air into a tank. The closed loop system was used to evaluate the compressor performance at specific discharge pressures that the compressor encountered during steady-state operation using carbon dioxide as the working fluid. Fig. 1 shows two test stands for the first two stages of testing: the left side shows the open loop test stand and the right side shows the closed loop hot-gas load stand. Detailed experimental results obtained with both test stands have been discussed in Xinye et al. [7] and [8].

Figure 1. Open loop and closed loop system test stand setups

In the next step, a new test stand was specifically designed based on the experiences gained during the first two stages of testing and was built for compressor dynamic testing using pipeline natural gas (NG). In addition, the compressor reliability has been evaluated in this stage of testing. Furthermore, some improvements were made to the final design of the NG compressor box for safety purposes and better performance based on the test results from the first two test stands.

During all tests, the following compressor measurements were recorded: compressor mass flow rate, suction and discharge temperatures, suction and discharge pressures, intermediate pressure and temperature, and compressor power consumption. Using these measurements, the compressor performance was evaluated. An overview of the completed tests is shown in Table 1.

2.2 Natural gas testing system

The schematic of the natural gas test stand is shown in Fig. 2. All tests with natural gas were conducted inside an explosion-proof engine test cell. The test stand was designed to withstand pressures of up to 14 MPa by using stainless steel piping. A strainer, filled with desiccant, is used at the inlet of the test stand and installed before the compressor to collect moisture and dust contamination inside the pipeline natural gas. Three solenoid valves, located in the NG supply line (SV100), the suction line (SV01) and the discharge line (SV02), respectively, are controlled from the control room remotely for safety purpose. One manual ball valve is located after the strainer to shut off the natural gas flowing into the compressor. A check valve is also used to prevent any natural gas backflow from the pressurized tank. Furthermore, a mass flow meter is installed to measure the mass flow rate of natural gas after the compressor. At the end of the test stand, two pressure vessels, equipped with one pressure gauge, one pressure relief valve, one manual valve and one flexible hose, are connected to the test stand by two three-way valves.

Table 1. Overview of the full completed tests

Test Type	System	Working Fluid	Number of Tests	Comments
Dynamic charging	Open loop	Air	10	New test stand was built
Steady-state			5	
Steady-state	Closed loop	CO_2	5	Hot-gas load stand was modified
Dynamic charging	NG test system	Natural gas	300	New test stand was built

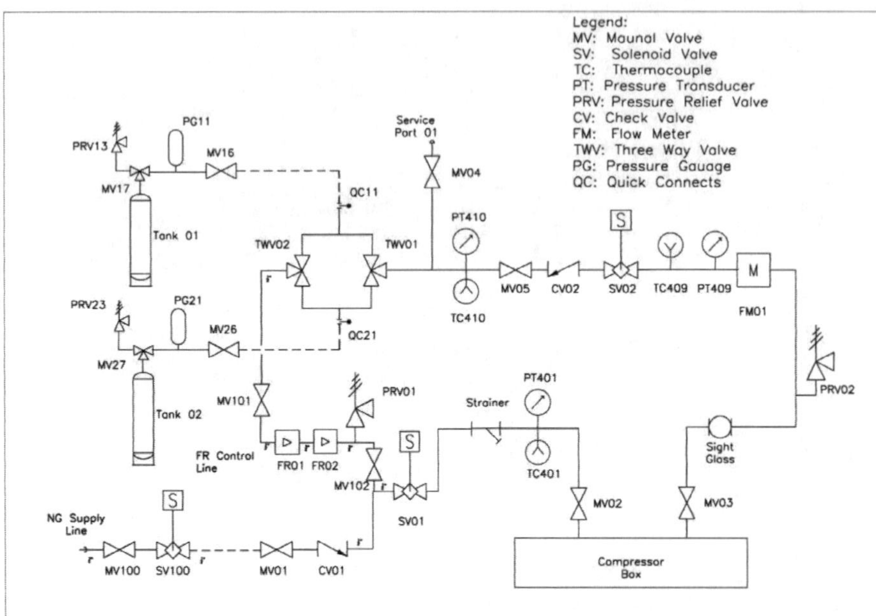

Figure 2. Schematic of the natural gas system

With this configuration of three-way valves, two methods were used during the natural gas testing: pipeline charging and tank-to-tank charging. Using the pipeline charging method, natural gas passes through TWV01 to either of two tanks when TWV02 is closed. Pipeline natural gas is used to charge the empty tank until the target pressure was reached. In this case, MV101 is closed and TWV01is opened to fill the target tank (Tank01 or Tank02). The tank-to-tank charging method was utilized to charge an empty Tank02 from a full Tank01 or vice versa. In this case, the pressurized gas passes through two flow regulators, FR01 and FR02, to throttle the compressed natural gas to the compressor suction pressure. The benefit of this approach is to significantly expedite the testing time, as there is no need to empty the natural gas after two tests as is the case with pipeline charging.

A photo of the natural gas test stand located in the explosion-proof engine test cell is shown in Fig. 3. A flammable gas detector is installed close to the ceiling and the power would be shut off once a gas leak is detected. All tests were conducted by the operator in the control room, shown in Fig. 5, from where the solenoid valves and compressor box were controlled by a remote-control panel. All measured data was obtained and displayed using an appropriate data acquisition system. An emergency stop was used to shut down the entire system in case of any unexpected issues.

To dispose of the compressed natural gas, a natural gas heater setup, consisting of two flow regulators and a patio-comfort heater, was used to burn the compressed natural gas. The movable tank as well as the heater setup were moved outside of the engine cell and connected to the heater setup to complete the natural gas burning process. A photo of the heater setup can be seen in Fig. 4.

Figure 3. Natural gas system test stand setup (engine cell)

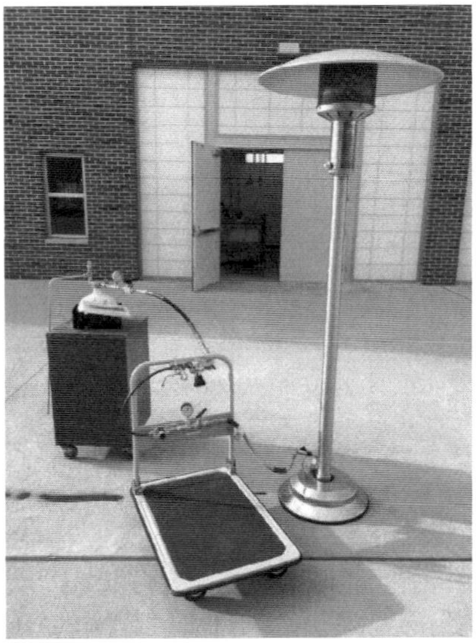

Figure 4. Natural gas system heater setup

An NG compressor box, as shown in Fig. 6, consisting of a two-stage reciprocating compressor to achieve the system target pressure in the storage tank, was connected to the test stand. An electronic box is installed inside the compressor box, which contains an ON/OFF switch, status indicator lights, and a remote-control button. Another flammable gas detector with a lower gas concentration setting, then the test cell, was installed inside the compressor box to shut-off the compressor in case of gas leaks. Built-in thermal protection has been included in the compressor control program to set the limits of compressor shell temperature as well as gas discharge temperature.

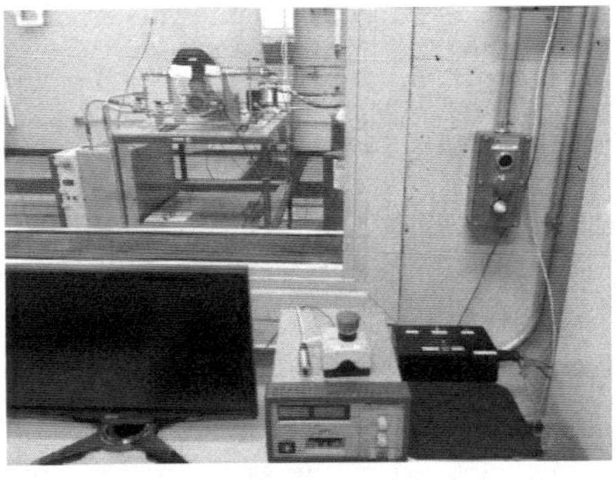

Figure 5. Natural gas system test stand setup (control room)

Figure 6. Natural gas compact compressor box

A schematic of the compressor box is shown in Fig. 7. The compressor is fitted with an oil management system, cooling fans, and an electronic control system. In addition, the compressor box includes both pressure switches and a pressure relief valve to ensure that a safe operating pressure is maintained. The target tank charging pressure of the compressor is 4500 kPa. When the compressor discharge pressure reaches 4500 kPa, the compressor electronics automatically shut down the compressor. The compressor is fitted with numerus thermocouples and pressure transducers to collect operational and reliability data.

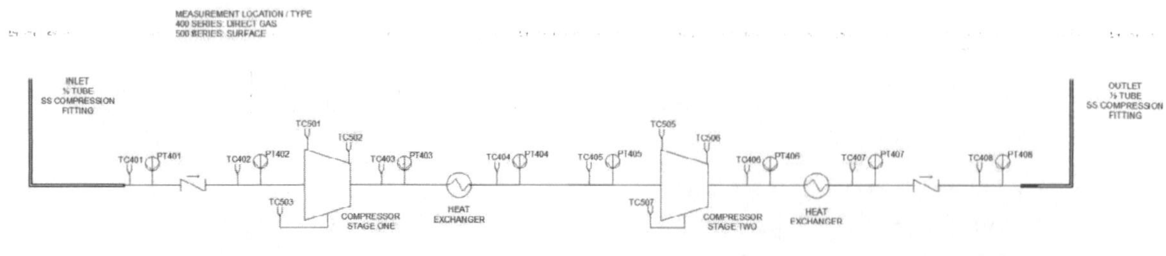

Figure 7. Schematic of the compressor box

2.3 Test stand instrumentation

All measuring instrumentation is indicated in the schematics shown in Fig. 2 and Fig. 7. All temperatures are measured with T-type thermocouples with ±0.25°C accuracy. The inlet pressure is measured with an absolute pressure sensor having an accuracy of ±0.25% of the reading. Other pressures are measured with gauge pressure sensors having an accuracy of ±0.13% of full scale. A micro-motion mass flow meter with accuracy of ±0.1% of the reading is installed after the compressor box to measure the gas mass flow rate. The electrical power consumption of the compressor box is measured with power meters that have an accuracy of ±0.04%. A data acquisition system is used to convert the incoming voltages from the measuring instrumentation to digital signals and then to transfer the signals to a personal computer. The computer uses a proper data reduction program for further data analysis.

2.4 Test matrix

A test matrix outlining the operating conditions for the natural gas system testing is shown in Table 2. The natural gas compressor testing consisted of a series of tank charges aimed at evaluating compressor performance and reliability. The following compressor measurements were recorded for each single dynamic test: compressor mass flow rate, temperature and pressure of all significant points and power consumption.

Table 2. Test matrix of natural gas system tests

Suction Pressure [*kPa*]	Target Pressure [*kPa*]	Method	Number of Tests [-]	Recorded Data
105	4500	Pipeline charging	20	Temperature, pressure, power, mass flow rate
105	4500	Tank to tank	280	

3. Experimental results

3.1 Experimental results

A dynamic test with 4500 kPa target pressure was conducted using the tank-to-tank charging method. Fig. 8 shows the variation of natural gas pressures as a function of time for test No.1, which means the first test in the sequence of test during the day.

Due to the nature of the charging process, it can be noticed that the intermediate pressure, PT_{405}, shows gradual rise during the start-up phase. This rate of pressure increase becomes smaller after the system has been running for 3 minutes. The discharge pressure of the compressor box, PT_{407}, increases significantly until it reaches the target pressure. Also, it is found that the intermediate pressure has a marked increase to an equilibrium pressure at compressor shutdown, which is due to the backflow through the second compressor stage.

By contrast, Fig. 9 shows the variation of natural gas pressures as a function of time for test No.10 which is the 10th test on the same day. It is observed that there have been distinct differences of pressure variation between test No.1 and No.10 during the start-up phase. Unlike the first test, it only takes approximately 1 minute in test No.10 for the rate of intermedate pressure (PT_{405}) rise to stablize. The shape of the discharge pressure (PT_{408}) curve remains markedly similar from test No.1 to No. 10.

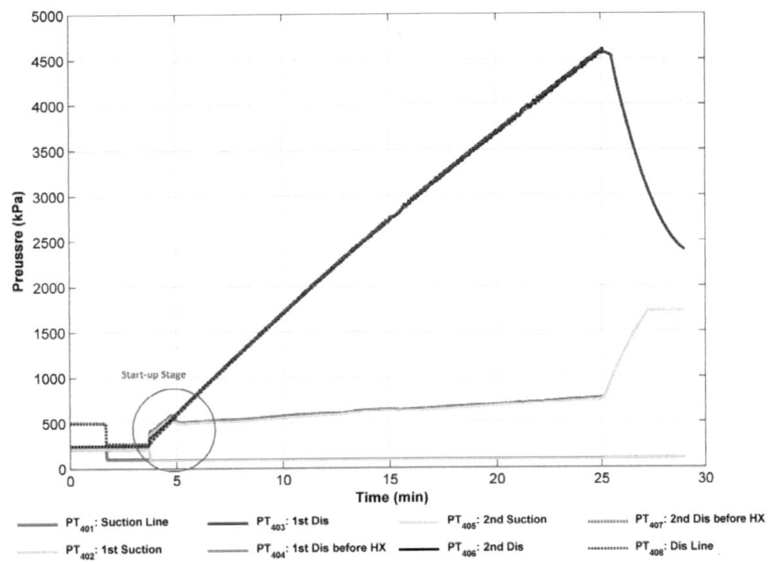

Figure 8. Pressure inside the compressor box (test No.1)

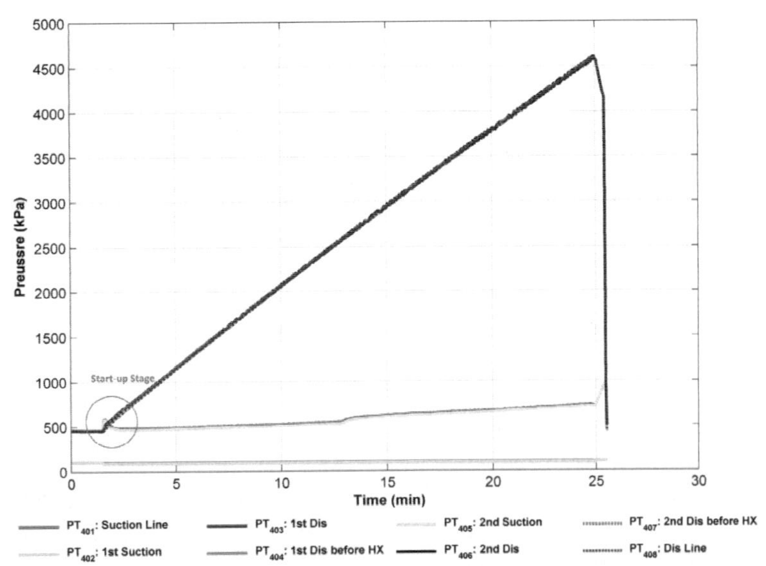

Figure 9. Pressure inside the compressor box (test No.10)

Fig. 10 shows the variation of natural gas temperatures as a function of time for test No.1. After the compressor system starts, it is found that the first-stage discharge temperature shows notable rise at the beginning. The second-stage discharge temperature (TC_{406}) has a delay, through the compressor start-up phase, before the temperature increases significantly until it reaches to a final value of 90°C.

Another obvious conclusion that can be drawn from Fig. 10 is that the each stage discharge temperatures show large jumps when the stage contribute to the system pressure rise. However, the rate at which the temperature of the second-stage gas temperature rises is larger than the rate at which the first-stage temperature rises. In addition, the second-stage temperature exceeds the first-stage temperature after the system is running for approximately 7 minutes. During this test, the maximum attainable temperature for the second stage is up to 90°C and for the first stage is 70°C. Also, it can be noticed that the temperature variations after the heat exchanger in both stages are relatively small, which are close to the ambient temperature during the entire testing. In addition, a small jump of the second-stage discharge temperature and oil return temperature can be seen in the right corner, which is due to the cooling fans being shut off with heat remaining in the compressor system after shut-down.

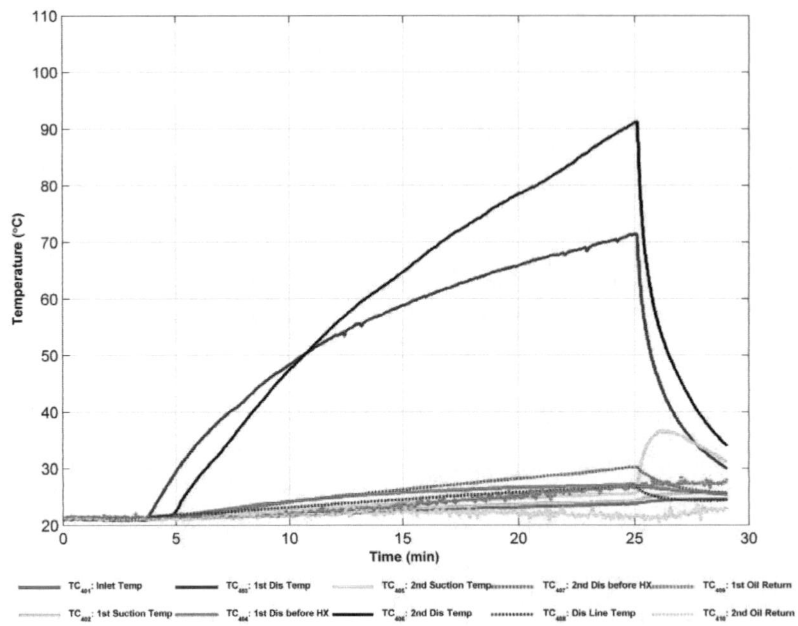

Figure 10. Temperature inside the compressor box (test No.1)

Similarly, a comparison of the system temperature is made between two tests on the same testing date. Fig. 11 shows the variation of the natural gas temperatures as a function of time for test No.10. A major difference between the No.1 and No.10 tests is the hike of the second-stage discharge temperature during the start-up phase, which jumps to 80°C immediately. There is only a very short time lag for the second-stage temperature rise, unlike 7 minutes in test No.1. Also, the rate at which the discharge temperature of the first stage rises is smaller than the previous test after 5 minutes of testing, the temperature increasing only from 88°C to 91°C. In addition, as shown in Fig. 11, the maximum attainable temperature of the second-stage is approximately up to 105°C, only 15°C higher compared with the temperature in test No.1. Moreover, except for the two discharge temperatures, all other system temperatures only show minor changes during the entire testing, in contrast to the gradual increase in test No.1. Furthermore, the variance is within 3°C, regardless of the ambient temperature change.

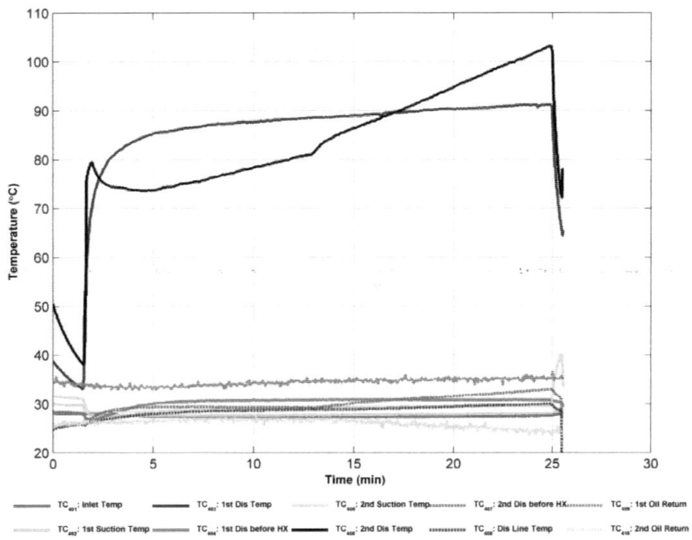

Figure 11. Temperature inside the compressor box (test No.10)

3.2 Reliability results analysis

Fig. 12 and Fig. 13 show that the variation of the measured compressed natural gas pressure and temperature is small in multiple dynamic charging tests with identical target pressure. This is significant given that there is only a two to five minutes pause between tests. The identical target pressure was set to 4500 kPa and suction pressure was kept stable at 105 kPa. The running time for each test is approximately 25 minutes. It can be seen that the variation of the measured compressed natural gas temperature in these continuous tests increases along with the testing time but the cooling fans worked efficiently to cool down the system. The maximum attainable temperature of the system was found to be 105°C, which is reached within six tests in series. Also, it is noted that the maximum temperature no longer increase after the third test in the continuous tests series when the system is under a periodic cycling.

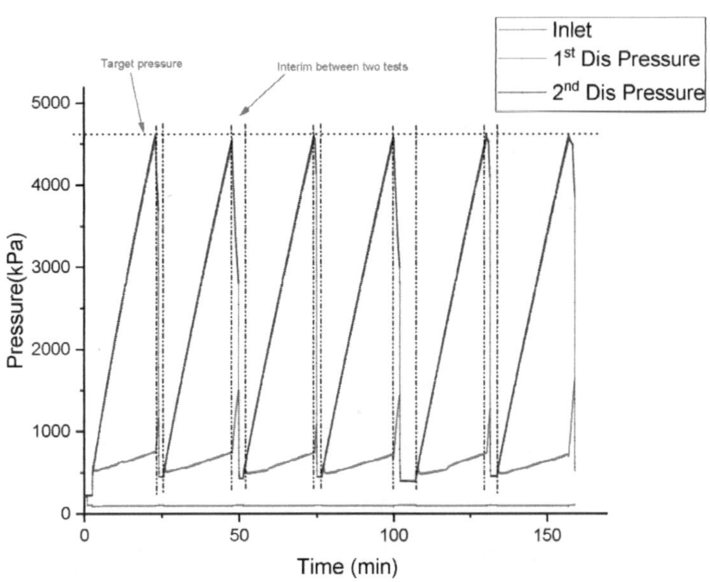

Figure 12. Variation of pressure in continuous dynamic charging test

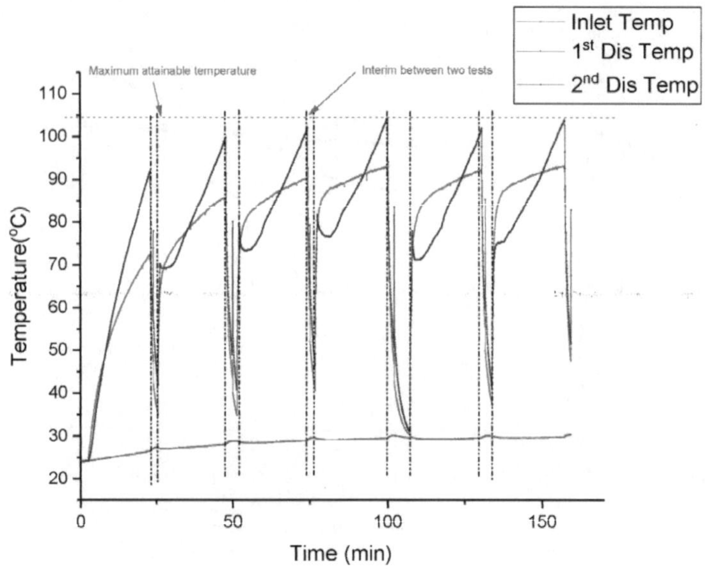

Figure 13. Variation of temperature in continuous dynamic charging test

In summary, 300 natural gas dynamic charging tests have been completed to evaluate the performance and safety characteristic of the compressor system. The differences between different continuous tests have been noticed and analyzed, which is instructive for the performance improvement and system design. The reliability analysis shows that the compressor system is robust and meets the discharge pressure and temperature design goals. Furthermore, the maximum operating temperatures are within allowable ranges and no thermal protection issues showed up. The operation of the NG compressor system was proven to be reliable and predictable.

4. Conclusions and future work

The paper demonstrates the measurement results of performance testing of a natural gas compressor and the compressor characterization are analyzed based on the experimental data. Following the previous two stages of testing, a new test stand was specially designed and built for compressor dynamic testing using pipeline natural gas (NG) and the compressor reliability has been evaluated. The results obtained from different tests appear to be plausible and significant to show the different performance of the compressor during a dynamic charging process. The development of the simulation model has recently been completed and the compressor dynamic performance will next be predicted and evaluated. Future development for this dynamic model will provide an accurate prediction of the intermediate pressure and the total charging time. Also, the mass flow rate and compressor characteristic will be calculated and validated by the experimental data in future work.

5. Acknowledgement

The authors would like to thank BlackPak Inc. for sponsoring and providing compressor for our testing.

References

[1] Nakagawa K, Tanaka S and Kaneko J, 1992. An aerodynamic investigation of a centrifugal compressor for HCFC123, *International Refrigeration and Air Conditioning Conference at Purdue*. Paper 15, West Lafayette, IN.

[2] Janssen M, Kuijpers L and de Witt J, 1988. Theoretical and experimental investigation of a dynamic model for small refrigerating systems, *International Refrigeration and Air Conditioning Conference at Purdue.* Paper 77, West Lafayette, IN.

[3] Kim K and Lee Y S, 2014. Dynamic test and fatigue life evaluation of compressor blades, *Journal of Mechanical Science and Technology*, 28(10), 4049-4056.

[4] W. Norman Shade, 2009. Optimization & Revitalization of Existing Reciprocating Compression Assets, *Optimization and Troubleshooting of Compressors and Turbines Conference,* Aberdeen, U.K

[5] Chi J and Didion D, 1982 A simulation model of the transient performance of a heat pump, *International Journal of Refrigeration.* 5, 176-184

[6] Rajendran H and Pate M, 1986. A computer model of the start-up transients in a vapor compression refrigeration system, *International Refrigeration and Air Conditioning Conference at Purdue.* Paper 17, West Lafayette, IN.

[7] Zhang X, Yang B, Osorio A, Bethel D, Kurtulus O and Groll E A, 2016. Characterization and Performance Testing of Two-Stage Reciprocating Compressors during the Dynamic Charging of a Tank with Air, *International Compressor Engineering Conference at Purdue.* Paper 1531, West Lafayette, IN.

[8] Zhang X, Yang B, Osorio A, Bethel D, Kurtulus O and Groll E A, 2016. Characterization and Performance Testing of Two-Stage Reciprocating Compressors using a Hot-Gas Load Stand with Carbon Dioxide, *International Compressor Engineering Conference at Purdue.* Paper 1533, West Lafayette, IN.

IOP Conf. Series: Materials Science and Engineering **232** (2017) 012004 doi:10.1088/1757-899X/232/1/012004

Novel Long Stroke Reciprocating Compressor for Energy Efficient Jaggery Making

M V Rane*, D B Uphade

Heat Pump Laboratory, Mechanical Engineering Department, IIT Bombay, Powai, Mumbai, INDIA 400 076

*Email: ranemv@iitb.ac.in

Abstract: Novel Long Stroke Reciprocating Compressor is analysed for jaggery making while avoiding burning of bagasse for concentrating juice. Heat of evaporated water vapour along with small compressor work is recycled to enable boiling of juice. Condensate formed during heating of juice is pure water, as oil-less compressor is used. Superheat of compressor is suppressed by flow of superheated vapours through condensate. It limits heating surface temperature and avoids caramelization of sugar. Thereby improves quality of jaggery and eliminates need to use chemicals for colour improvement. Stroke to bore ratio is 0.6 to 1.2 in conventional reciprocating drives. Long stroke in reciprocating compressors enhances heat dissipation to surrounding by providing large surface area and increases isentropic efficiency by reducing compressor outlet temperature. Longer stroke increases inlet and exit valve operation timings, which reduces inertial effects substantially. Thereby allowing use of sturdier valves. This enables handling liquid along with vapour in compressors. Thereby supressing the superheat and reducing compressor power input. Longer stroke increases stroke to clearance ratios which increases volumetric efficiency and ability of compressor to compress through higher pressure ratios efficiently. Stress-strain simulation is performed in SolidWorks for gear drive. Long Stroke Reciprocating Compressor is developed at Heat Pump Laboratory, stroke/bore 292 mm/32 mm. It is operated and tested successfully at different speeds for operational stability of components. Theoretical volumetric efficiency is 93.9% at pressure ratio 2.0. Specific energy consumption is 108.3 kWh_e/m^3 separated water, considering free run power.

1. Introduction

Jaggery is the traditional sweetener in India. It is preferred for consumption due to its medicinal values. It is conventionally prepared by evaporation of water for concentration of sugarcane juice in open pan boiling. Bagasse, which is residue of juice extraction, is used as fuel. Pan surface temperature is high due to hot flue gases, which has temperature about 950 to 1050°C at chimney end (Sardeshpande *et al.*, 2010). Hot spots are formed due to high temperature pool boiling of sugarcane juice in mild steel or stainless steel pan. It causes caramelization of sugar and gives dark brown colour to jaggery. Dark brown jaggery has less market value than golden yellow jaggery (Guddadamath *et al.*, 2014). Chemical clarificants like lime, sodium hydroxide, phosphoric acid are used to clarify juice and improve colour of jaggery (Shiralkar *et al.,* 2014). Traces of these chemicals are harmful to human health. Hence, chemical free or organic jaggery should be produced. Chemical free or organic jaggery can be produced if pan surface temperature maintained to or below 100°C and clarify juice using vegetative clarificants only. These issues are addressed by energy efficient vapour recompression for jaggery making using Long Stroke Reciprocating Compressor.

This paper mainly reports development of small capacity Long Stroke Reciprocating Steam Compressor and its use in Vapour Recompression System for jaggery making. This paper also reports the performance of Vapour Recompression System for jaggery making, which addresses the issues of conventional jaggery making.

1.1. Vapour recompression based jaggery making

Vapour Recompression System, VRCS, consists of Latent Heat Exchanger, LHE, and compressor. LHE comprises horizontal flat oval tube, which is kept inside a horizontal circular tube, alternatively a rectangular tube to accommodate more number of flat oval tubes. Inside space of flat oval tube serves as evaporator. It is filled with raw juice, generally it occupies 50% of internal volume of tube, as shown in figure 1. Space between flat oval tube and circular tube is condenser. Compressed vapours introduced in condenser where they are condensed by delivering its latent heat of condensation to evaporate water from juice in the evaporator.

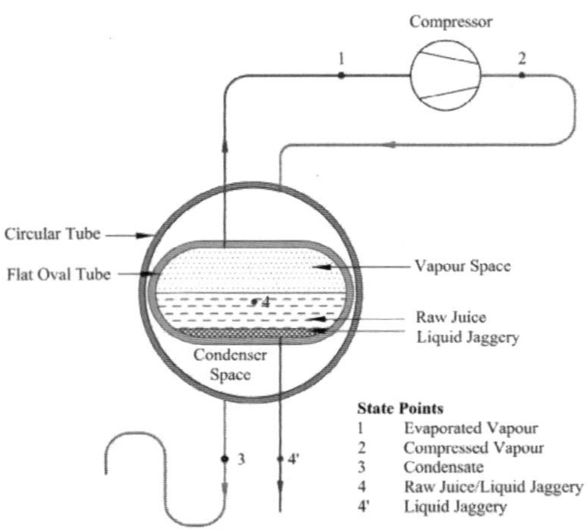

Figure 1: VRCS for Sugarcane Juice Concentration

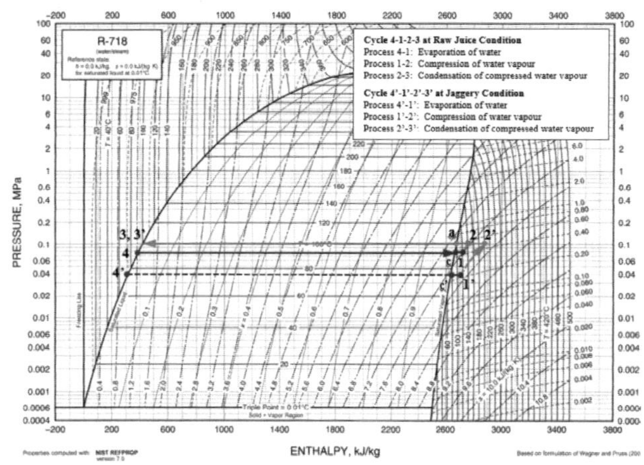

Figure 2: Heat Pump Cycle on p-h diagram R718

1.1.1. Working

Initially, saturated air is evacuated from evaporator and then compressed using Long Stroke Reciprocating Compressor. Due to low pressure in evaporator, moisture carrying capacity of air increases, thus, more moisture is removed. It increases temperature and pressure of compressed air. At high pressure, moisture carrying capacity of air reduces. Simultaneously, heat transfer from condenser to evaporator increases temperature of juice and reduces condenser temperature. This helps to condense out moisture, and increase the evaporator temperature, increases moisture carrying capacity of air in evaporator. Air evacuation from evaporator occurs for very short duration. After this, compressor compresses wet or saturated water vapour and it is condensed in the condenser in later phase. Condensate continuously drained through steam trap valve. Working cycle is represented on the R718 p-h chart, figure 2. Remaining solution gets enrich in jaggery constituents. Discharge pressure is atmospheric pressure, process 2-a-3 or 2'-a-3'. Evaporator pressure reduces from atmospheric pressure, process 4-c-1 for raw juice to certain lower pressure, process 4'-c'-1'. It depends on boiling point elevation, heat exchange approach.

1.2. Necessity of Long Stroke Reciprocating Compressor

Concentration of juice increases as the water gets evaporated from raw juice. It increases the boiling point elevation of juice, from 0.5 to 18°C for 20 to 90°Brix change. Water removal rate is higher during initial phase of water removal due to low boiling point elevation and high heat transfer coefficient, 1.58 kW/(m^2.K). The worst case is at jaggery condition. Here, temperature lift is enlarged due to cumulative effect of 18°C boiling point elevation and higher temperature approach required for heat exchange, due to very low heat transfer coefficient, 0.06 kW/(m^2.K). Suction pressure reduces much below that of raw juice condition. Hence, compressor power required during final stage of jaggery making is higher than raw juice condition. Small capacity steam compressors are not available in market. Judicious design of

small capacity steam compressor saves power. Pressure ratio across the compressor is 1.05 to 2.8. Compressor with long stroke improves isentropic and volumetric efficiencies. Isentropic efficiency is the ratio of ideal work required during isentropic compression to the actual work required. Isentropic efficiency of compressor depends upon pressure ratio, working fluid, number of compression stages and use of intercooler in case of multistage compressor. Isentropic efficiency can be improved by heat transfer from cylinder wall during compression process. It is possible to construct compressor with isentropic efficiency greater than 100% (Hanlon, 2001). Heat transfer from cylinder wall can be improved either by providing fins or providing long stroke with or without fins on cylinder wall. Crank and connecting rod mechanism limits stroke to bore ratio in conventional reciprocating compressors. Similarly, volumetric efficiency of reciprocating compressor is limited due to inertial forces.
In this paper, development of long stroke reciprocating compressor is presented with bore 32 mm and stroke 292 mm. Stroke to bore ratio is 9.125. It increases surface area of cylinder by 9.1 times, similarly heat transfer rate also increases by 9.1 times compare to conventional compressors. Valve operation time gets increased due to longer stroke. It enables to use of sturdier valves and improves volumetric efficiency.

1.2.1. Stroke to bore ratio of reciprocating drives
Stroke to bore ratios in some engines are listed in table 1. Stroke to bore ratio is 0.6 to 1.2 in conventional

Table 1: Stroke to Bore Ratio and Piston Speed of Some Engines (Greenman, 1996)

Make/Model	Bore mm	Stroke/Bore Ratio	Mean Piston Speed m/s	Power kW	Application
Tee-Dee 0.1	6.0	0.95	6.10	0.08	R/C aircraft
Arden	12.6	1.04	4.98	0.40	R/C aircraft
Dooling 61	25.8	0.74	9.19	4.47	R/C aircraft
Olsson & Rice	31.8	0.87	11.68	2.57	Handtool
Fichtel & Sachs	38.1	**1.15**	15.14	23.72	Motorcycle
Evinrude	39.6	0.88	4.67	4.44	Outboard motor
Power Products	53.1	**0.60**	7.41	5.22	Home use
Chrysler 610	55.6	0.78	10.97	23.91	Outboard motor
MZ ES 150	61.0	0.98	14.99	**53.77**	Motorcycle
Chrysler 610	64.3	0.64	12.50	30.03	Outboard motor

1 bhp = 0.746 kW , R/C: Radio Controlled

reciprocating drives. It is also applicable for double acting opposed piston reciprocating compressors, as shown in figure 3.

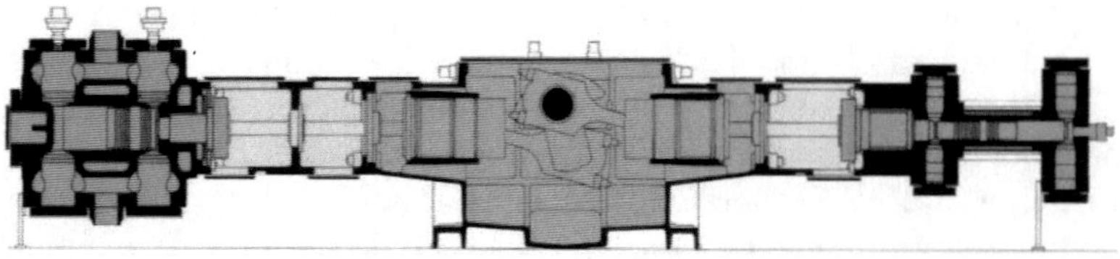

Figure 3: Conventional Opposed Piston Reciprocating Compressor (Bloach, 2006)

Reciprocating compressors can be operated more efficiently by increasing the stroke to bore ratio. In conventional machines, stroke to bore is limited as connecting rod strikes the cylinder wall at high stroke to bore ratio, while stroke accommodation is the issue at lower stroke to bore ratios. Long stroke in reciprocating compressors allows high outer surface area of cylinder. It enhances the heat dissipation to surrounding and increases isentropic efficiency by reducing compressor outlet temperature. Superheat of compressed water vapour is suppressed by condensate formed inside the cylinder. Longer strokes increases the inlet and exit valve operation timings, which reduces inertial effects substantially. Thereby allowing use of sturdier valves. This enables handling liquid along with vapour in the compressors. Thereby supressing the superheat and reducing the compressor power input. Longer stroke increases the stroke to clearance ratios which increases the volumetric efficiency and ability of compressor to compress through higher pressure ratios efficiently.

2. Research Methodology

Initially, long stroke reciprocating drive is developed and then, it is tested for operational stability of components. Stress and strain analysis of pinion and rack assembly is performed in SolidWORKS. Theoretical performance of vapour recompression system using LSRC is simulated in MathCAD for eight equal stages of water removal for jaggery making. This paper presents the development of long stroke reciprocating steam compressor and its use for jaggery making. Development of compressor has been started with design and development of long stroke reciprocating drive for suitably selected long stroke cylinders which will assembled with the cylinders.

3. Novel Long Stroke Reciprocating Compressor

Long Stroke Reciprocating Compressor involves double acting opposed piston-cylinders, Long Stroke Gear Drive, LSGD, base frame, motor frame, flexible shaft, BLDC motor, check valves and connecting tubes. LSGD consists of rack assembly and pinion. Rack assembly is designed for known stroke of reciprocating piston-cylinder. Maximum force applied on tooth is estimated using known speed, power of motor and pitch circle diameter of pinion, which is initially taken close to piston diameter. Stresses are more critical at fillet of shaft and pinion and/or at the tooth base of pinion, rack or semi-circular internal gear. Stresses and deflection in pinion and rack assembly are simulated in SolidWorks.

3.1. Reciprocating cylinders

Reciprocating piston-cylinders suitable for compression of steam were selected. These are oil-free compression cylinders. Isometric view is shown in figure 4 and cut section is shown in figure 5.

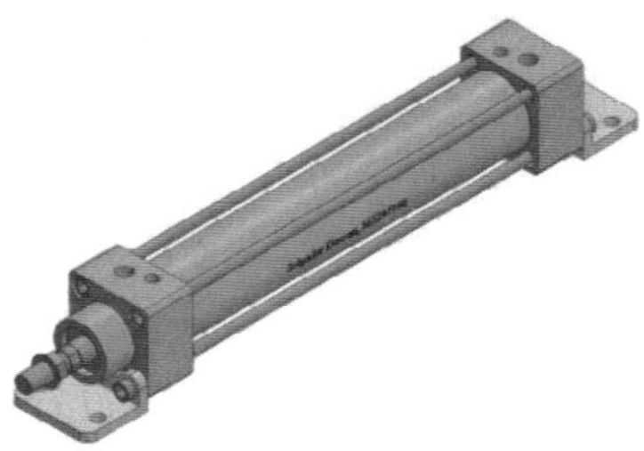

Figure 4: Cylinder

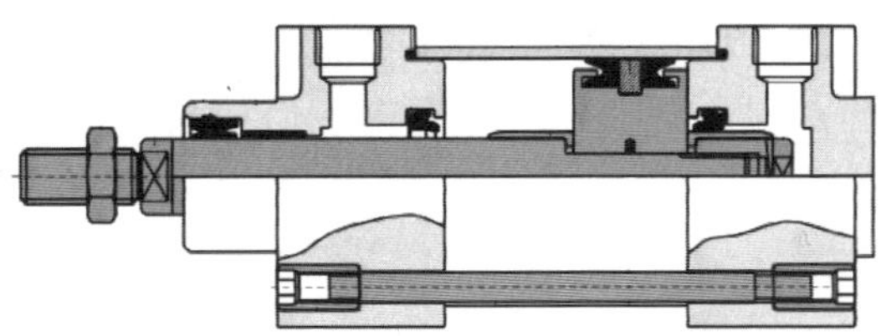

Figure 5: Cut Section of Cylinder

Available strokes of cylinder are 40, 50, 80, 100, 160, 250 and 300 mm. These cylinders can handle vapour in the range of -10 to 180°C. High temperature viton seals are used. Stroke of selected cylinder is 300 mm and bore is 32 mm. Stroke of rack assembly is 292 mm for mild steel gear drive. It is limited to 292 mm, to avoid locking of rack assembly and pinion due to interference of strokes of gear drive and cylinders. If number of tooth on rack assembly increased by one, stroke of gear drive becomes higher than stroke of cylinder. It increases interference and movement of pinion gets restricted. Hence, gear drive stroke must be slightly lesser than cylinders.

3.2. Long stroke reciprocating drive

Patented Long Stroke Reciprocating Drive involves rack assembly, pinion, two pinion guides and three bearings (Rane and Metange, 2014, Rane and Uphade, 2016). Rack assembly is combination of two racks and two semi-circular internal gears, figure 6. One pinion guide, PG, is integral part of gear drive. It guides the pinion by rollover on racks and internal gears. Second pinion guide is integral part in the motor frame. It guides the pinion in transverse direction, along the offset length. Bottom bearing supports the bottom end of pinion and its outer race rolls on outer guide of pinion guide in rack assembly. Middle bearing supports the pinion at middle and its outer race rolls on internal top guide of pinion guide in rack assembly. Top bearing supports the flexible shaft end at top end of pinion shaft and its outer race rolls on pinion guide in motor frame.

When pinion rotates about shaft, it moves the rack and its movement is transferred to the connecting

IOP Conf. Series: Materials Science and Engineering **232** (2017) 012004 doi:10.1088/1757-899X/232/1/012004

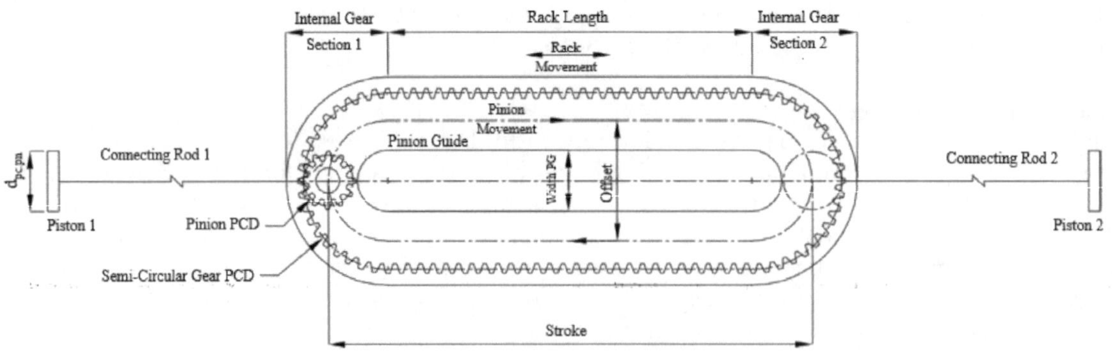

Figure 6: Long Stroke Reciprocating Drive

rods, which in turns moves the pistons. At the end of first rack, pinion smoothly rolls over the semi-circular internal gear from one rack to another rack and moves the piston by driving the second rack. Same cycle repeats for next rollover of pinion from rack to internal gear and vice-versa. Pinion is rotated about shaft using BLDC motor and pinion also oscillates along linear offset during rollover on semi-circular internal gears, while the rack assembly only reciprocates. In this way, rotary motion of motor is converted into long stroke reciprocating motion. Flexible shaft is an important component. It is rotated by motor at motor end. It rotates and oscillates at pinion end. Vapour recompression based jaggery maker is shown in figure 7. Details of developed LSRC compressor are listed in table 2.

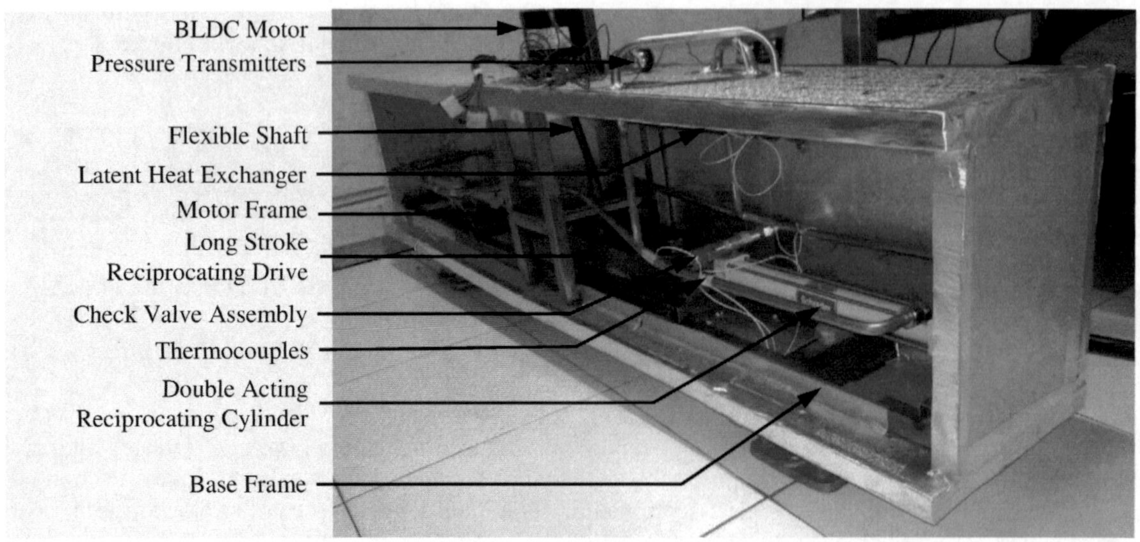

Figure 7: Vapour Recompression based Jaggery Maker

Table 2: Details of Long Stroke Reciprocating Compressor

Parameters	Abbreviation	Value	Unit
Speed of pinion	N_{pn}	720	rpm
Velocity of pinion and gear drive, $v_{pc.pn} = \pi d_{pn} N_{pn} / 60$	$v_{pc.pn}$	1.13	m/s
Bore	d_p	32	mm
Stroke of cylinder	l_{stk}	292	mm
Swept volume	vol_{stk}	234.8	cc
Clearance volume	vol_{cle}	7.36	cc
Volume of flow rate	$volf_{stk.ideal}$	1.409	L/s
Thrust required for pull and push single piston	$F_{pull/push}$	78.7	N
Vacuum developed	p_{suc}	720	mm_Hg vac

4. Heat Transfer Coefficients

Water is evaporated from sugarcane juice in stainless steel flat oval tube. Heating source is extracted vapours from juice, compressed to atmospheric pressure. Overall heat transfer is 1.58 kW/(m^2.K) at raw juice condition and 0.06 kW/(m^2.K) at jaggery condition (Hugot, 1986) during boiling of sugarcane juice in vertical flooded tube evaporator using low pressure steam. Pool boiling heat transfer coefficient is 4.089 to 12.59 kW/(m^2.K) for 0.7 to 1.3 kW$_h$ (Aldana *et al.*, 2015), less than 100 kW/(m^2.K) (Madrid *et al.*, 2016). Overall heat transfer coefficients reported by Hugot (1986) are used in simulation as these are more conservative.

5. Performance for Jaggery Making

Performance of jaggery making is estimated for developed Long Stroke Reciprocating Compressor capacity of 1.409 L/s at 720 rpm. Latent Heat Exchanger, LHE, is judiciously designed to maintain low temperature approach during water evaporation at jaggery condition.

5.1. Capacity of Latent Heat Exchanger

Capacity of prototype for jaggery making unit is 0.6 kg/h. Weighted average water evaporation rate is about 2.6 kg/h. Cane crushing rate for same flow rate of juice is 4.6 kg/h. Considering 20 h/d operation of prototype requires 92 kg/d cane crushing rate. Weighted average water evaporation rate is 52 kg/d and jaggery making rate is 12 kg/d.

5.2. Calculations

Total quantity of water to be evaporated is calculated using mass balance for concentration change from raw juice and jaggery concentration. Concentration of raw juice, c_{rj} is 20°Brix and of jaggery, c_{jg} 90°Brix. Mass of water evaporated from 20 to 90°Brix is calculated using mass balance equation,

$$m_{tot.w.evp} = m_{rj}\left(1 - \frac{c_{rj}}{c_{jg}}\right) \quad \ldots (1)$$

Performance of VRCS is determined by assuming, water is evaporated in equal quantity, in eight stages, $n_{stg} = 8$. Quantity of jaggery produced by evaporation of water, m_{jg},

$$m_{jg} = m_{rj} - m_{tot.evp.w} \quad \ldots (2)$$

Concentration at outlet of nth stage by evaporation of water is determined using concentration balance,

$$c_{jc.stg.n.o} = \frac{c_{jc.stg.n.i}}{\left(1 - \frac{m_{w.evp.stg.n}}{m_{rj}}\right)} \quad \ldots (3)$$

Boiling point elevation, $dt_{bpe.jc.stg.n.o}$, calculated using Hugot (1986) formula,

$$dt_{bpe.jc.stg.n.av} = \frac{2c_{jc.stg.n.o}}{100 - 2c_{jc.stg.n.o}} \quad \ldots (4)$$

Temperature of juice at stage n, $t_{jc.stg.n}$,

$$t_{jc.stg.n} = t_{cnd} - dt_{app.he.stg.n} \quad \ldots (5)$$

Saturation temperature of water at stage n, $t_{sat.w.stg.n}$,

$t_{sat.w.stg.n} = t_{w.stg.n} - dt_{bpe.jc.stg.n}$ using this corresponding saturation pressure was found

Volumetric efficiency is estimated using total clearance volume, η_{vol},

$$vol_{4.exp} = vol_{3.cl.cyl} \cdot PR_{cmp}^{\frac{\gamma-1}{\gamma}} \quad \ldots (6)$$

$$\eta_{vol} = \frac{vol_{eff.swp}}{vol_{tot.cyl}} \quad \ldots (7)$$

Temperature approach for heat exchange, $dt_{1a.stg.n.cal}$, is calculated as follow,

$$dt_{1a.stg.n.cal} = \frac{Q_{w.v.cmp.1}}{U_{o.he.stg.n} \cdot A_{he.stg.n}} \quad \ldots (8)$$

$dt_{he.stg.n.assume} = dt_{he.stg.n.cal}$, if this condition satisfied, following parameters were calculated.

Time required for first stage evaporation

$time_{req.evp.stg.n} = m_{w.evp.p.s} / mf_{w.v.cmp.n}$ ... (9)

Specific Moisture Extraction Rate,

$SMER_{stg.n} = mf_{w.v.cmp.n} / W_{cmp}$ ... (10)

Specific work of compressor,

$W_{sp.cmp.w.stg.n} = W_{cmp} / mf_{w.v.cmp.n}$ ... (11)

Each parameter is weighted with mass flow of water to estimate average parameters for different stages.

5.3. Flow chart of simulation

Performance of jaggery making is predicted using the MathCAD program as shown in figure 8.

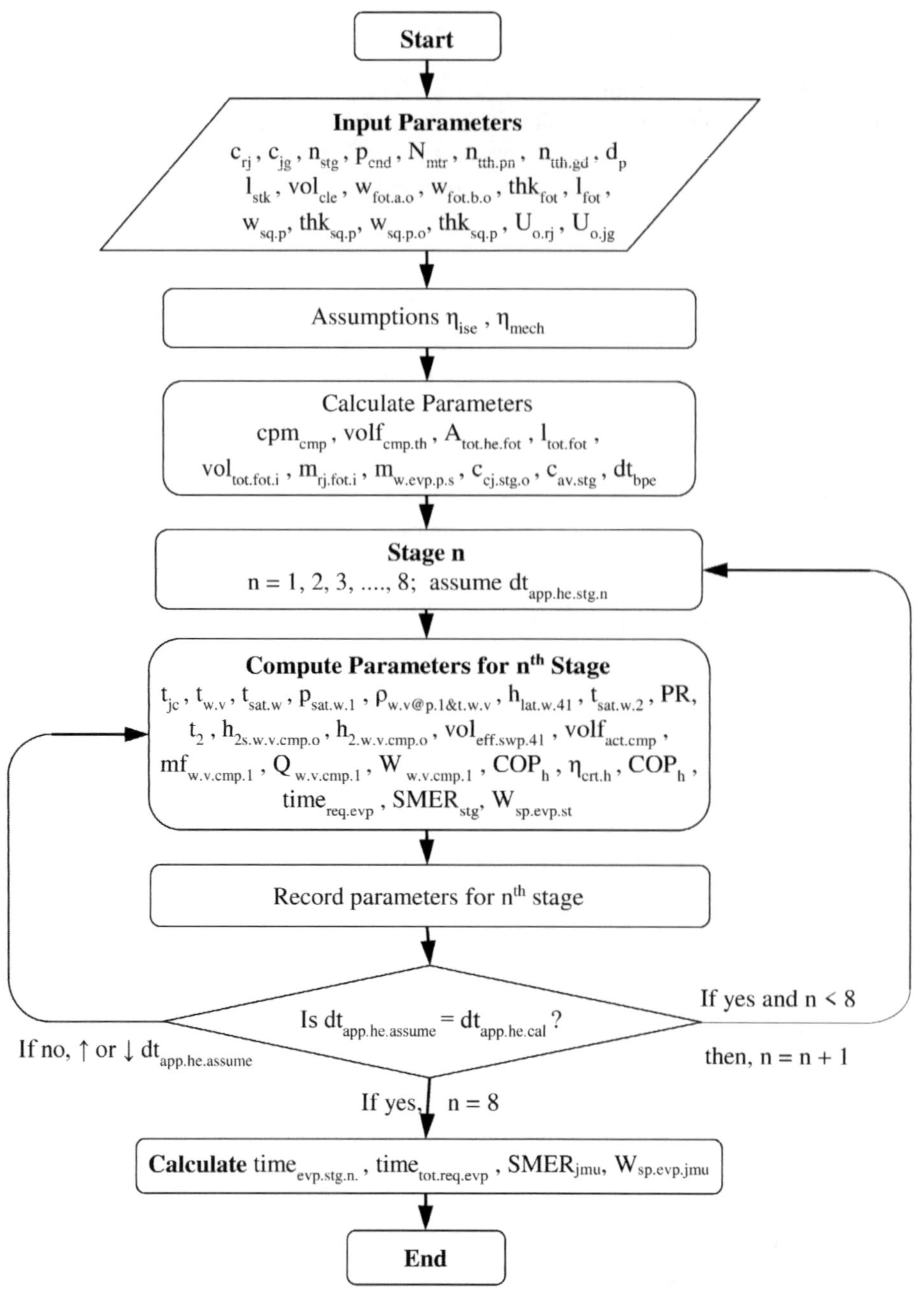

Figure 8: Flow Diagram for Performance of VRCS

6. Results and Discussions

Isothermal compression enhances the volumetric efficiency in conventional compressors. In case of Long Stroke Reciprocating Compressor for jaggery making, discharge pressure is limited to atmospheric

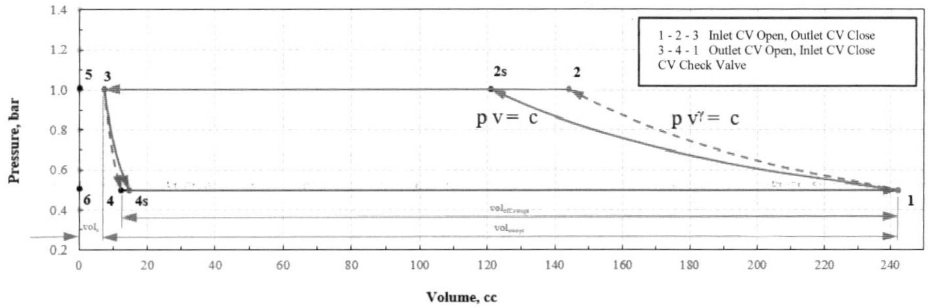

Figure 9: Theoretical p-V Diagram for Reciprocating Cylinder

pressure and suction pressure is determined by boiling point elevation and approach for heat exchange. Once the water vapour pressure in evaporator reaches the discharge pressure, outlet valve get opened and discharge pressure is maintained throughout the piston travel. Volumetric efficiency is 93.9% for suction pressure of 0.5 bar. Clearance volume is 7.36 cc. Total cylinder volume 242.2 cc reduces to effective swept volume 227.5 cc, as shown in figure 9.

6.1. Stress and strain analysis

Stress and strain were simulated in SolidWorks. Maximum force is 200 N. Maximum stress in pinion is 10.7 MPa at fillet between spur gear and shaft, as shown in figure 10. Maximum allowable stress is 144 MPa for mild steel. Design safety factor is 13.5. Maximum deflection is 1 micron at addendum circle, as shown in figure 11. Maximum stress in rack assembly is 1.9 MPa at rolling contact of bearing, as shown in figure 12. Factor of safety is 74.8. Maximum deflection in rack assembly is 1 micron in rack. It is tolerable in gears.

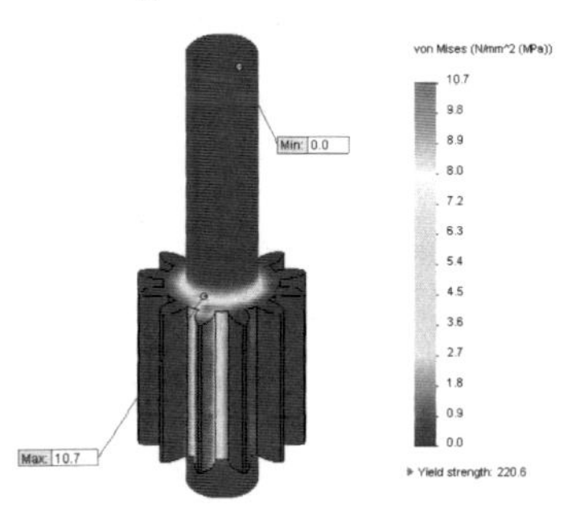

Figure 10: Max Stress 10.7 MPa in Pinion

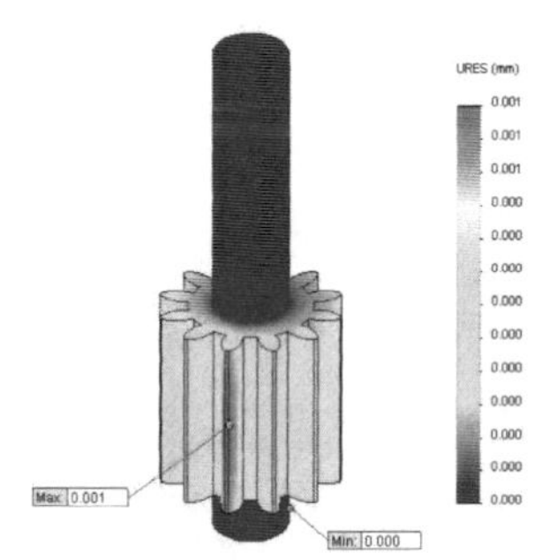

Figure 11: Max Deflection 1 micron in Pinion

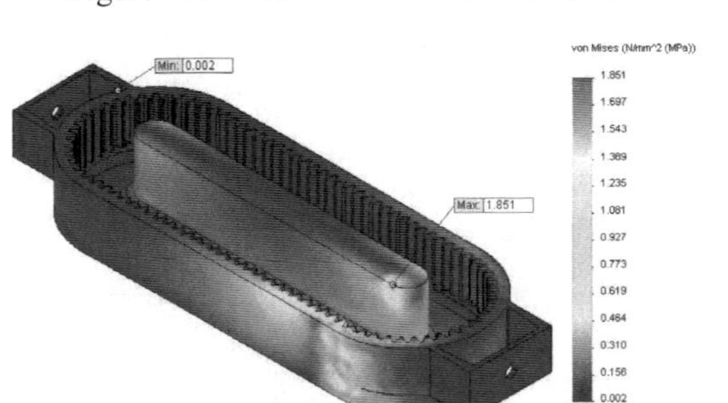

Figure 12: Max Stress 1.9 MPa at Rolling Contact in Pinion Guide

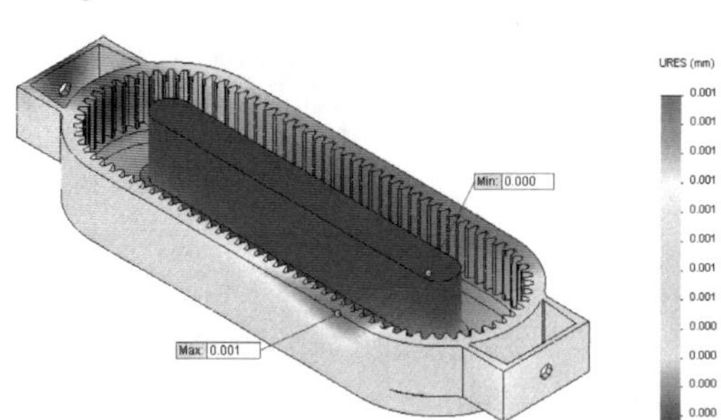

Figure 13: Max Deflection 1 micron at Rolling Contact on Pinion Guide

Factor of safety for pinion is 13.5 and for pinion guide is 74.8 for mild steel as material of construction. Deflections are tolerable. Hence, design of gear drive is very strong and safe.

IOP Conf. Series: Materials Science and Engineering **232** (2017) 012004 doi:10.1088/1757-899X/232/1/012004

6.2. Free Run Test of LSRC

Free run test of Long Stroke Reciprocating Compressor has been conducted to measure the power required to overcome the friction between piston and cylinders-two sets, pinion and semi-circular gears/racks as well as velocity head loss during back and forth motion of pistons. Inlet and outlet ports of both cylinders were kept open. This gives the minimum power required to run the compressor at desired speed. Starting power requirement is 0.067 kW at 160 rpm. Power required to run the LSRC with open inlet and outlet ports is 0.270 kW at 752.1 rpm of motor. Figure 14 shows that frictional power increases linearly as speed increases. Oscillations of LSRC were observed after 447.3 rpm. Free run power requirement at 720 rpm is 0.255 kW. Compressor power varies during jaggery making by evaporating water. It is due to variation in boiling point variation and heat transfer coefficient.

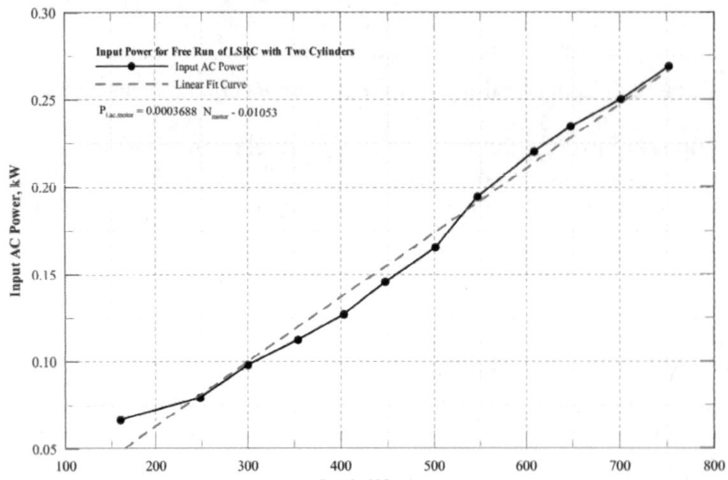

Figure 14: Free Run Input AC Power at Different Speed of Motor

Evaporator pressure reduces from 1.013 bar to 0.36 bar as shown in figure 15. Total water removal is equally divided in eight stages. Stages 6, 7 and 8 are further divided into 2, 2 and 4 equal parts respectively, as the concentration change is high. Theoretical performance of VRCS is estimated for

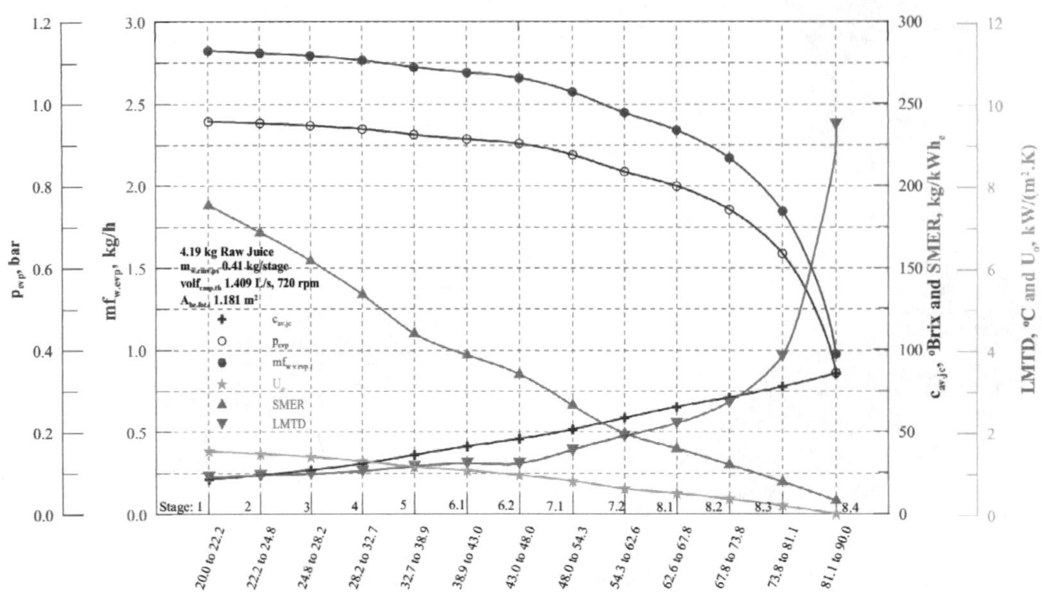

Figure 15: Performance Parameters of VRCS

each stage and then mass flow weighted average of each parameter is calculated. Compressor capacity reduces from 1.406 to 1.322 L/s. Specific energy consumption is 8.2 kWh_e/m^3 water removal. SMER is 122 kg/kWh_e. COP_h is 78 for 20 to 90°Brix concentration change. Total time required for jaggery making is 1 hour 18 min. Power requirement is 28.8 W for 2.62 kg/h water removal rate. Jaggery making rate is 0.6 kg/h. Experimental performance of VRCS will be tested for evaporation of water and then for sugarcane juice concentration. Considering, dry mode power consumption, actual power consumption is 0.284 kW, specific energy consumption is 108.3 kWh_e/m^3 water removal and SMER is 9.3 kg/kWh_e.

7. Conclusions

Novel Long stroke reciprocating drive has been developed with 292 mm stroke and bore 32 mm. Stroke to bore ratio is 9.125. It is successfully tested for operational stability of components. Flexible shaft is the crucial component in power transmission from BLDC motor to the piston through pinion of gear drive. Novel Long Stroke Reciprocating Compressor has been analysed for jaggery making while avoiding burning of bagasse for concentrating juice. Surface temperature of condenser tube is limited to 100°C by maintaining discharge pressure to atmospheric pressure. Hence, juice temperature is always less than 100°C. It reduces the caramelization of sugar and improves quality of jaggery. Theoretical specific energy consumption is 8.2 kWh_e/m^3 of water. Actual specific energy consumption of compressor is 108.3 kWh_e/m^3 of water removed for jaggery making, considering free run power. It eliminates bagasse firing and relives operators from strenuous environmental conditions. Saved bagasse can be composted on farm. It increases fertility of soil by increasing carbon content and sugarcane production gets significantly improved. Hence, LSRC is promising development in modernization of jaggery making for energy efficient vapour recompression. In future, next version of Long Stroke Gear Drive will be designed with lower tolerances to reduce frictional power and it will improve SMER.

Acknowledgements

This work was funded by Tata Centre for Technology and Design, IIT Bombay, Mumbai, India through grant # TCTD/DON/2015/125835. Support of Tata Centre for Technology and Design, IIT Bombay is gratefully acknowledged.

Nomenclature

Abbreviations		cpm	cycle per minute	PR	pressure ratio	tth	teeth, #	t	temperature, °C	v	velocity, m/s
c	concentration, °Brix	CV	check valve	mf	mass flow rate, kg/s	Q	heat duty, kW	thk	thickness, mm	volf	volumetric flow, m^3/s
COP	Coefficient of Performance	d	diameter, mm	p	pressure, bar	stk	stroke, mm	U	overall heat transfer coefficient, $W/(m^2.K)$	W	power, kW
Subscripts		cj	concentrated juice	evp	evaporator	jc	juice	o	outlet	sq	square
act	actual	cle	clearance	gd	gear drive	jg	jaggery	pn	pinion	stg	stage
amb	ambient	cmp	compressor	h	heating	jmu	jaggery making unit	req	required	swp	swept
app	approach	cnd	condenser	he	heat exchange	lat	latent	rj	raw juice	tot	total
c	cooling	cw	cold water	i	inlet	mech	mechanical	sat	saturated	v	vapour
cal	calculated	eff	effective	ise	isentropic	mtr	motor	sp	specific	w	water
Greek Letters		η	efficiency, %	ρ	density, kg/m^3	γ	adiabatic index				

References

[1] Aldana D M, Paul V Y, Madrid R L, 2015, Experimental Analysis of the Pool Boiling Phenomenon of Sugarcane Juice, *Applied Mechanics and Materials,* vol 789-790, pp 489-495.

[2] Bloach H P (2006), Compressors and Modern Process Applications, *John-Wiley and Sons Inc Publ*, p 4.

[3] Greenman D M, 1996, Design and construction of a Miniature IC Engines, *MS Dissertation*, MIT.

[4] Guddadamath S G, Sanjay B. Patil S B, Khadi B M, Chandrashekar C P (2014), Genetic Enhancement of Sugarcane for the Production of Organic Jaggery, *Springer Sugar Tech*, vol 16(1), pp 86-91.

[5] Hanlon P C, 2001, Handbook of Compressors, *McGraw Hill Publication*, SBN 0-07-026005-2, p 1.7.

[6] Hugot, E (1986), Handbook of Cane Sugar Engg, published by Elsevier Science, pp 495, 555-561.

[7] Rane M V and Metange N S (2014) An Apparatus for Converting Rotary Motion to Reciprocating Motion and Reciprocating Motion to Rotary Motion, Indian patent 1847/MUM/2014.

[8] Rane M V and Uphade D B (2016) A Single Stage or a Multi Stage Vapour Recompression System for Recovering Solvent and Process Thereof, Filing of Indian patent 201721008251 is in process.

[9] Sardeshpande V R, Shendage D J, Pillai I R (2010), "Thermal Performance Evaluation of a Four Pan Jaggery Processing Furnace for Improvement in Energy Utilization", *J of Energy Engg*, Vol 35, pp 4740-4747.

[10] Shiralkar K Y, Kancharla S K, Shah N G, Mahajani S M (2014), Energy Improvements in Jaggery Making Process, *Journal of Energy for Sustainable* Development, vol 18, pp 36-48.

Investigation on the variation of pressure in the cylinder of the refrigerator compressor based on FSI model

Zhilong He, Zhifang Jian, Tao Wang, Dantong Li, Xueyuan Peng

School of Energy and Power Engineering, Xi'an Jiaotong University, Xi'an 710049, China

The corresponding author's e-mail address: taowang@stu.xjtu.edu.cn

Abstract. The performance of refrigerator compressor has a significant influence on the coefficient of performance of the refrigerator. In order to find out the main cause for the variation of the performance, this paper presents a three-dimensional fluid structure interaction (FSI) model of the refrigerator compressor and an experimental investigation to verify the model. To record the p-V indicator diagram, a refrigerator compressor was modified and the pressure sensor was installed. Based on the FSI model, the variation of pressure in the cylinder that influenced the performance of compressor was identified by varying working condition, rotating speed and refrigerant. The FSI model and result could provide useful information for both performance testing and optimization of compressor performance.

1 Introduction

In recent years, domestic refrigerators with low power consumption and eco-friendly are in high demand because the amount of electricity consumed by refrigerators is among the highest for all consumer household products. The reduction of the power consumption of refrigerators is greatly dependent on a highly efficient compressor.

To develop a highly efficient compressor, it is important to have an in-depth research on the variation of pressure in the cylinder. A number of factors affect the variation of pressure in the cylinder of a reciprocating compressor. To fully understand the inner workings of a reciprocating compressor and to improve the compressor efficiency, the numerical simulation and experimental investigation are the main research methods.

The numerical tools give much help to understand the dynamic characteristics of compressor. Winandy et al. used a numerical model for the estimation of compressor performance [1]. Elhaj et al. applied the same compressor modeling to a two-stage reciprocating compressor for condition monitoring [2]. Yang et al. presented a comprehensive simulation model for a semi-hermetic CO_2 reciprocating compressor in which the leakages and the frictions are considered [3]. Farzaneh-Gord et al. developed a mathematical model based on real gas model to simulate natural gas reciprocating compressors [4].

With the development of Computational fluid dynamics (CFD) technology, CFD has been widely used as a design tool in many engineering applications. Yu wang et al presented a numerical simulation of the thermodynamic process in the cylinder and dynamics of the self-acting valves for an air reciprocating compressor [5]. Akira and Kenji showed the CFD applications in suction muffler and compressor valve design [6].

The latest research hotspots are the interaction of different physical fields, especially fluid structure interaction, as that is a typical fluid structure interaction problem. Hyeong-Sik et al showed the FSI in impact analysis of compressor discharge valves [7]. Aditya and Josep simulate numerically the FSI for flow through valves of a hermetic compressor using immersed boundary method [8]. Qin Tan et al proposed a three-dimensional fluid–structure interaction model for the discharge valve movement in a rotary compressor [9].

All the above mentioned numerical researches need to be verified with experimental data. Yuan Ma et al conducted an experimental investigation of the discharge valve dynamics in a reciprocating compressor for trans-critical CO2 refrigeration cycle [10]. José Luiz GASCHE et al measured the instantaneous position of the reed by using a very accurate optical sensor [11]. Dr. Kemal Sarıoğlu et al performed detailed temperature and pressure measurements as the boundary conditions for numerical study [12].

In this work, a hermetic reciprocating compressor was modified to record the p-V diagram in the cylinder. A three-dimensional fluid structure interaction (FSI) model was built based on the hermetic reciprocating compressor. Comparison of simulation results with experimental data implies that present model can well describe the working process of compressor. This FSI model could be used for a new compressor development.

2 The FSI model of refrigerator compressor

It is well known that the flow of refrigerant gas and the valve movement are coupled with each other in the working process of the valve in refrigerator compressor. Valve will produce bending deformation under refrigerant fluid load, and the deformation of valve in turn affects the flow of refrigerant. This kind of problems is much more complicated, because not only the equations of structure and CFD must be solved but also the interaction between them should be considered. The fluid and solid equations are solved individually using the latest data provided from each other. In this analysis, compressor model is divided into fluid domain and corresponding structural domain.

2.1 The fluid domain

Fluid domain of compressor model is divided into three primary parts, the cylinder part, the discharge volume and the suction pipe. Between those parts, gap boundary condition is imposed to segregate discharge volume and cylinder, also to segregate cylinder and suction pipe. FSI boundary is imposed on corresponding area where discharge valve and suction valve are placed in the coordinate system.

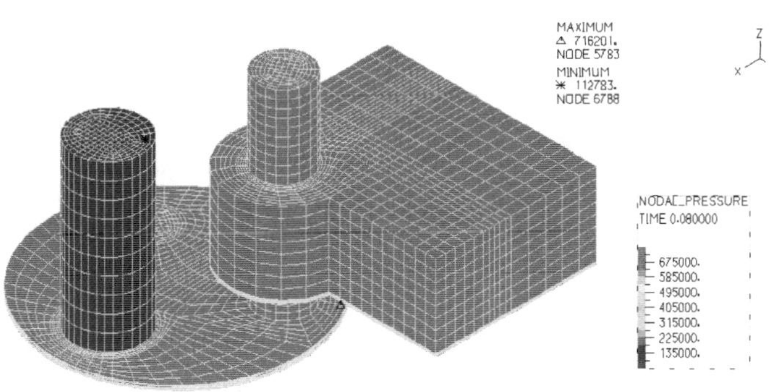

Fig.1. model of compressor fluid domain

The working fluid is R600A, and material properties of fluid are shown in Table 1. The fluid is assumed as low speed compressible flow model. In the fluid domain, the initial pressure and temperature in the cylinder and discharge chamber are given as 0.68MPa and 308.15 K, respectively. The pressure at the suction pipe (suction pressure) is given as 0.0672MPa.

The RNG model is employed to simulate the turbulent flow in the valve and the cylinder. The fluid domain and the structural domain are all discredited using 8 node 3D element. Solve time step is 1E-5 second.

2.2 *The structural domain*

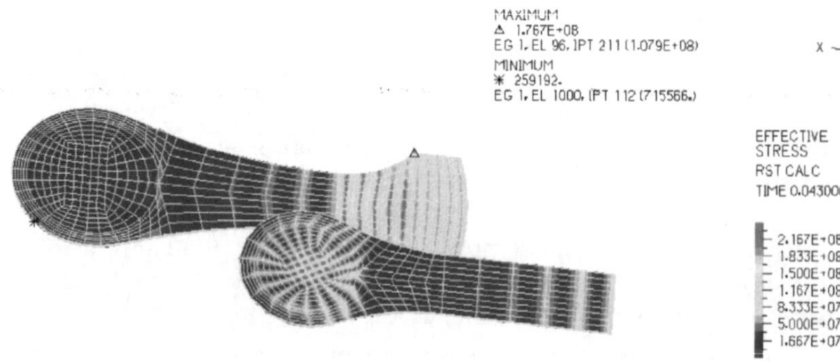

Fig.2. Finite element model of compressor structural domain

The structural domain of compressor model includes the suction valve, the discharge valve, the valve seat and the retainer with the discharge valve. The contact condition is introduced between the reed valve and the valve seat. FSI boundary is imposed on the surface of the discharge valve and suction valve. Material properties used in this model is constant properties with E = 210GPa, v = 0.25, and p = 7800 kg/m^3.

2.3 *The moving wall*

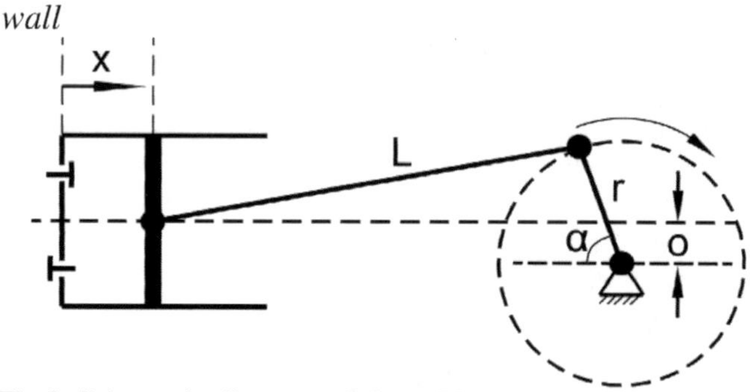

Fig.3. Schematic diagram of the refrigerator compressor

The piston movement is determined by the compressor crank as Figure 3. The piston position in dependence of crank drive parameters as equation (1-1) is considered as moving wall in the FSI model. In this paper, the offset of the piston which makes the lateral force applied on the piston well-distributed has been toke into account, and the piston move from the TDC (top dead center) to the BDC (bottom dead center) at start time.

$$x = r\left[(1-\cos\alpha) + \frac{\lambda}{4}(1-\cos 2\alpha) - k\lambda\sin\alpha\right] \qquad (1\text{-}1)$$

Where x is the piston displacement from the TDC, α is the crank angle of crank, r is crank radius, λ is ratio of crank radius to length of connecting rod, k is eccentric distance.

3 Experimental method

To verify the FSI model of the refrigerator compressor, experimental verification was carried out. The p-V indicator diagram, the pressure pulsation in suction pipe and discharge pipe and the suction valve displacement have been obtained by the experimental facilities. The valve dynamics is a strong fluid-structure coupling because the valve motion is a response to the pressure force it experiences, and the pressure evolves based on the mass influx/efflux through the valve at a particular lift.

3.1 Refrigeration system

The refrigeration system has been built up to adjust the operation condition of the compressor, as Fig.4 showed. The main refrigeration loop consisted of a compressor, condenser, manual throttling valve, evaporator and other additional components. There was a heating belt to heating the evaporator. The balance valve was used to connect the evaporator and the condenser when charging refrigerant. The operation condition can be coordinative adjusted by the heating belt, the throttle valve and the fan.

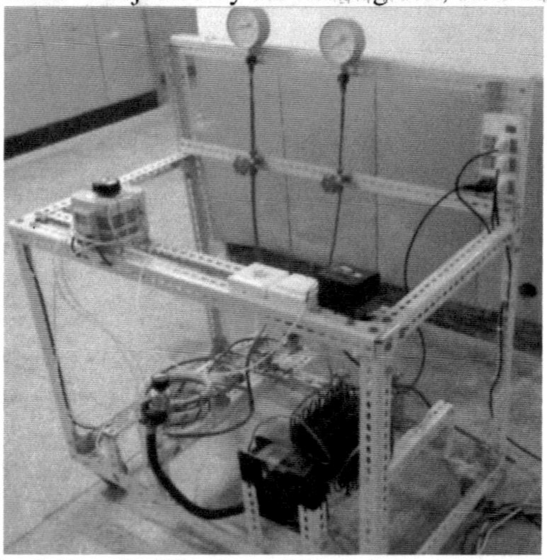

Fig.4. Test facility for refrigeration system

3.2 Modification of compressor to record the p-V indicator diagram

Conventionally, a hermetically sealed compressor shell comprises of two semi-shells welded together. Since a frequent modification of the system takes place, a modified flange shell is recommended for such experimentations. Due to the small size of the refrigerator compressor, the experimental study on recording the p-V diagram was complex. In order to measure the transient pressure together with the rotation angle, a high precision pressure sensor was adopted. Considering the practical compressor structure and sensor dimensions, the sensor was installed as shown in Fig. 5(a). The pressure sensor was embedded in the valve seat Fig. 5(b). And the signal wire was led out.

Fig.5. The modification of Compressor

IOP Conf. Series: Materials Science and Engineering **232** (2017) 012005 doi:10.1088/1757-899X/232/1/012005

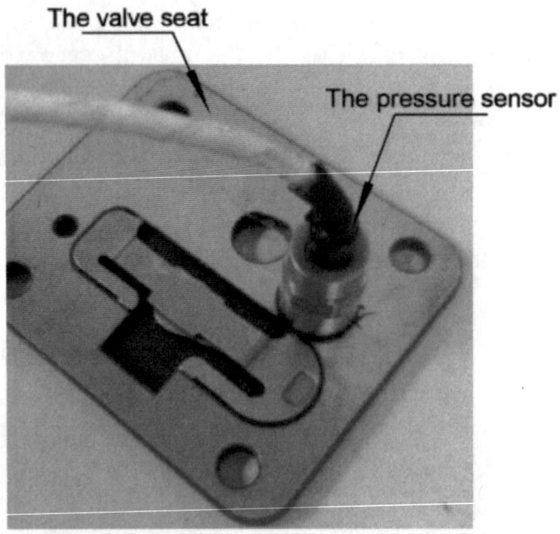

Fig.6. Schematic of sensor installation

It was difficult to obtain the positions of the piston directly, but the position of the piston is determined by the compressor crank as the equation (1-1). Thus, an approach switch was mounted on the frame directly facing to the big end of connecting rod, as shown in Fig.5 and Fig.6. When the piston was at the BDC (bottom dead center), the big end met the approach switch during the compressor was running, and a pulse signal was generated.
The dynamic pressure sensor uncertainty was less than ±0.1% and its response frequency was as high as 1400 kHz. The maximum working pressure was 1.9MPa, and the range of working temperatures was between -55°C and 175°C

4 Results and discussion

4.1 Comparison of experiment and simulation

Firstly, for validation the FSI model in this study, the results of numerical method have been compared with available experimental results; generally, there are a good agreement between the measured and numerical values. One way to verify the valve efficiency is through the p-V diagram.

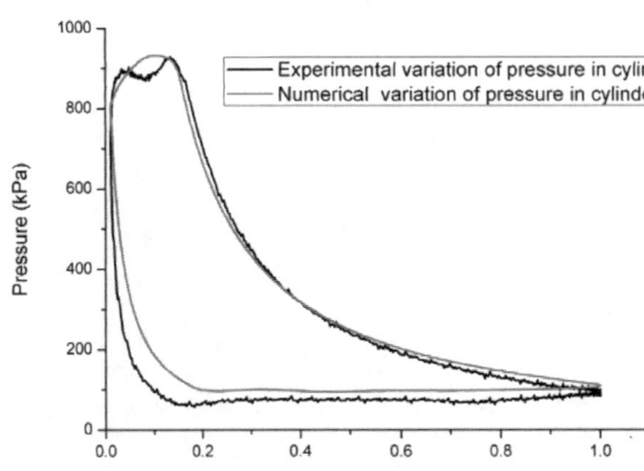

Fig.7. p-V indicator diagram

The understanding the power consumption on the refrigerating cycle and its losses can contribute to obtain better compressors. By the p-V indicator diagram it is possible to obtain the work exerted on the gas during the compression cycle and also the losses on the suction and discharge valves.

Comparison of the simulation and experimental p-V, the result of the FSI model was essentially in agreement with that obtained by experiment, in particular, the compression process completely overlap. According to the calculation mothed of power compression, numerical integral experiment and simulation of the compression work is 209.67W and 184.36W, respectively. It is proved that the calculation result of the FSI model is reliable.

4.2 Effect of variable working condition on compressor performance

The output of the simulation model includes the characteristic parameters' trend during the compression cycle. Fig.8 shows the compression cycle on the p-V diagram under the different operative conditions studied: it is possible to observe the compression phase, the discharge phase with the characteristic pressure fluctuation, the re-expansion of the refrigerant in the clearance volume and, finally, the suction phase.

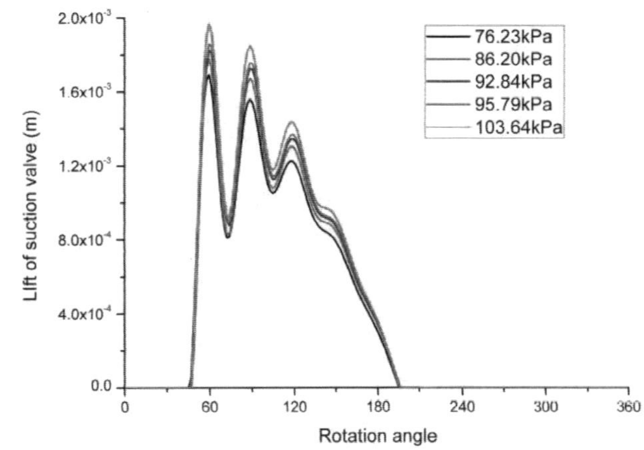

Fig.8. The lift of suction valve under different suction pressure

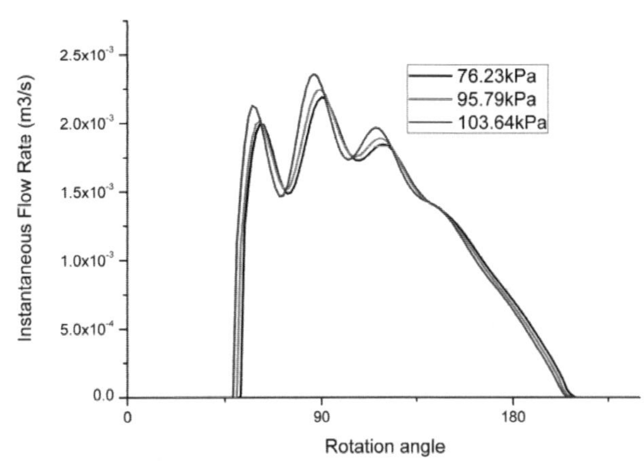

Fig.9. Instantaneous flow rate in suction phase under different suction pressure

IOP Conf. Series: Materials Science and Engineering **232** (2017) 012005 doi:10.1088/1757-899X/232/1/012005

The suction pressure significantly influences the motion of the suction reed valve. Under the same pressure ratio, the movement of the suction valve with different suction conditions were shown in Fig.8. As the thickness and stiffness of the valve plate remains the same, the time of valve opening and closing is consistent. With the increase of suction pressure, the maximum displacement of suction valve increased. And the refrigerant flow gets a larger flow area. When the suction pressure decreases, the gas force is smaller and the valve requires a relatively larger pressure difference to open, which leads to a larger flow loss and reduces the pressure of the gas refrigerant in the cylinder at the suction end, resulting in lower volumetric efficiency.

Variation of instantaneous flow rate with suction pressure was shown in the Fig.9, The instantaneous flow rate increases with the increase of suction pressure, since the suction valve get a larger flow area under the larger suction pressure.

4.3 Effect of rotating speed on compressor

Based on the FSI model, the effect of rotating speed on compressor has been studied at the same operating condition. As shown in the Fig.9, the speed has little effect on the suction phase. And the higher speed leads to higher pressure in compression phase, since the discharge valve flow area is relatively insufficient at the higher speed.

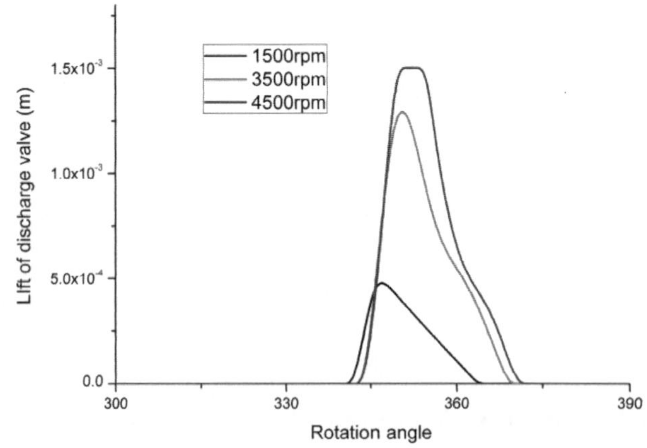

Fig.10. The lift of discharge valve under different rotational speed

The lifts of the discharge valve under different rotational speed were shown in Fig.10. Due to the higher rotating speed, the pressure in the cylinder increases rapidly while the flow capacity of the discharge valve is relative deficiency, which lead to higher pressure difference on the discharge valve in compression phase.

4.4 Effect of refrigerant on compressor

The variation of pressure in the cylinder with different refrigerants is studied, and the p-V diagram is shown in Fig.11. Since the polytropic exponent of R600a is close to that of R134a which is 1.0944 as shown in Fig.12, the compression and expansion processes of R600a coincide with that of R134a. The polytropic exponent of CO2, namely 1.2602, is bigger than that of R600a and R134a , therefore, the indicated work of CO2 is more than that of R600a and R134a.

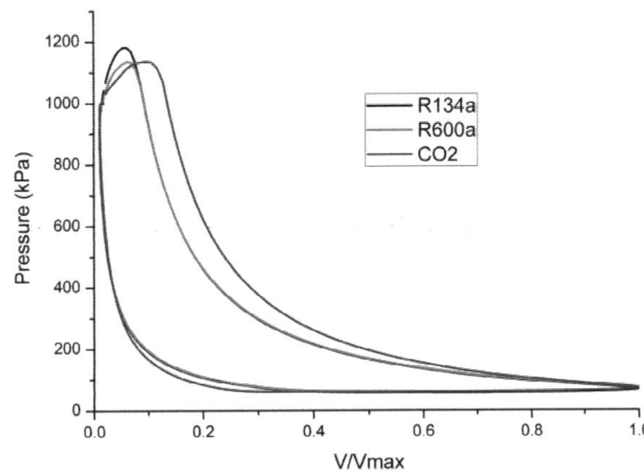

Fig.11. The p-V diagram with different refrigerants

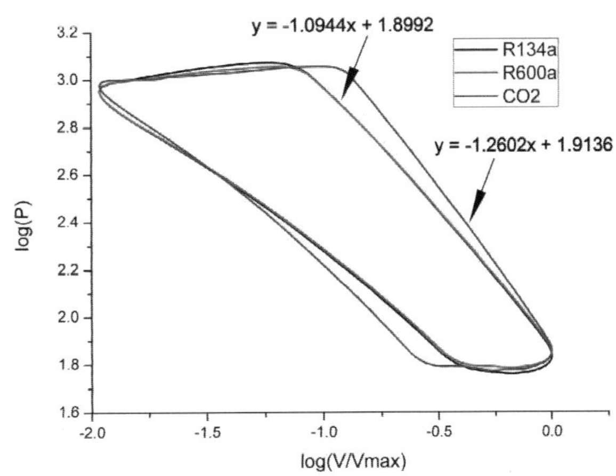

Fig.12. The log(p)-log(V) diagram with different refrigerants

Due to the density of R134a is larger than R600a, the pressure in the compression phase of R134a is higher than that of R600a. The property of refrigerant is shown in the Table 1.

Tab.1. The property of refrigerant (at P=101.23kPa, T=328.2K)

refrigerant	Density kg/m3	Viscosity (Pa·s)	Heat conductivity mW/(m·K)	Coefficient of volume expansion (1/K)	Cp kJ/(kg·K)	Cv kJ/(kg·K)
R600a	2.1985	8.2308E-6	20.152	0.0032426	1.8247	1.6689
R134a	3.8402	12.987E-6	15.793	0.0032102	0.89758	0.80972
CO_2	1.6386	16.374 E-6	19.103	0.0030846	0.87922	0.68701

5 Conclusions

This study provides an understanding of the Fluid Structure Interaction model of compressor. The simulation and experimental tools presented in this study offer promising opportunities for the development of high performance compressor.

Based on the FSI model, by varying the parameters such as the working condition, rotating speed and refrigerant, the main factors that influence the variation of pressure in the cylinder were identified and several conclusions have been made.

1. The three-dimensional fluid structure interaction (FSI) model which taking account of the offset crank mechanism has been established and experimental verified. And the FSI model is a useful tool for compressor performance analysis.
2. The instantaneous flow rate increases with the increase of suction pressure, since the suction valve get a larger flow area under the larger suction pressure.
3. The higher speed leads to higher pressure in compression phase, since the discharge valve flow area is relatively insufficient at the higher speed
4. The effect of polytropic exponent and density of refrigerant on the working process of compressor is studied.

References

[1] Winandy E, Saavedra C, Lebrun J. Simplified modelling of an open-type reciprocating compressor[J]. International journal of thermal sciences, 2002, 41(2): 183-192.
[2] Elhaj M, Gu F, Ball A D, et al. Numerical simulation and experimental study of a two-stage reciprocating compressor for condition monitoring[J]. Mechanical Systems and Signal Processing, 2008, 22(2): 374-389.
[3] Yang B, Bradshaw C R, Groll E A. Modeling of a semi-hermetic CO2 reciprocating compressor including lubrication submodels for piston rings and bearings[J]. International Journal of Refrigeration, 2013, 36(7): 1925-1937.
[4] Farzaneh-Gord M, Niazmand A, Deymi-Dashtebayaz M, et al. Thermodynamic analysis of natural gas reciprocating compressors based on real and ideal gas models[J]. International Journal of Refrigeration, 2015, 56: 186-197.
[5] Wang Y, Feng J, Zhang B, et al. Modeling the valve dynamics in a reciprocating compressor based on two-dimensional computational fluid dynamic numerical simulation[J]. Proceedings of the Institution of Mechanical Engineers, Part E: Journal of Process Mechanical Engineering, 2013, 227(4): 295-308.
[6] Nakano A, Kinjo K. CFD applications for development of reciprocating compressor[J]. 2008.
[7] Kim H, Ahn J, Kim D. Fluid structure interaction and impact analyses of reciprocating compressor discharge valve[J]. 2008.
[8] Lacerda J F, Gasche J L. Analysis of the flow in hermetic compressor valves using the immersed boundary method[J]. 2010.
[9] Tan Q, Pan S, Feng Q, et al. Fluid–structure interaction model of dynamic behavior of the discharge valve in a rotary compressor[J]. Proceedings of the Institution of Mechanical Engineers, Part E: Journal of Process Mechanical Engineering, 2015, 229(4): 280-289.
[10] Ma Y, He Z, Peng X, et al. Experimental investigation of the discharge valve dynamics in a reciprocating compressor for trans-critical CO 2 refrigeration cycle[J]. Applied Thermal Engineering, 2012, 32: 13-21.
[11] Gasche J L, Arantes D M, Andreotti T. Experimental analysis of the fluid structure interaction in a suction valve model[J]. 2014.
[12] Sarioglu D, Ozdemir A R, Oguz E, et al. An experimental and numerical analysis of refrigerant flow inside the suction muffler of hermetic reciprocating compressor[J]. 2012.

IOP Conf. Series: Materials Science and Engineering **232** (2017) 012006 doi:10.1088/1757-899X/232/1/012006

Numerical analysis of gas leakage in the piston-cylinder clearance of reciprocating compressors considering compressibility effects

V M Braga and C J Deschamps
POLO Research Labs for Emerging Technologies in Cooling and Thermophysics
Federal University of Santa Catarina, 88040-900, Florianopolis, SC, Brazil

deschamps@polo.ufsc.br

Abstract. Leakage is a major source of inefficiency in low-capacity reciprocating compressors. Not only does it lower the mass flow rate provided by the compressor, reducing its volumetric efficiency, but also gives rise to outflux of energy that decreases the isentropic efficiency. Leakage in the piston-cylinder clearance of reciprocating compressors is driven by the piston motion and pressure difference between the compression chamber and the shell internal environment. In compressors adopted for domestic refrigeration, such a clearance is usually filled by a mixture of refrigerant and lubricating oil. Besides its lubricating function, the oil also acts as sealing element for the piston-cylinder clearance, and hence leakage is expected to be more detrimental to oil-free compressors. This paper presents a model based on the Reynolds equation for compressible fluid flow to predict leakage in oil-free reciprocating compressors. The model is solved throughout the compression cycle so as to assess the effect of the clearance geometry and piston velocity on leakage and compressor efficiency. The results show that compressible fluid flow formulation must be considered for predictions of gas leakage in the cylinder-piston clearance.

1. Introduction

Gas leakage is a major source of inefficiencies in refrigeration compressors. Not only does it lower the mass flow rate provided by the compressor, reducing its volumetric efficiency, but also gives rise to outflux of energy due to gas leakage, which decreases the isentropic efficiency. According to Silva and Deschamps [1], a gap of 1 μm between the valve and its seat can reduce the volumetric and isentropic efficiencies of small reciprocating compressors by 2.7% and 4.4%, respectively. In fact, with respect to reciprocating compressors, the two main sources of gas leakage are: (i) the gap between the valve and the valve seat and (ii) the piston-cylinder clearance.

Two aspects bring about the flow in the piston-cylinder clearance: (i) the difference of pressure between the compression chamber and the internal environment of the compressor shell, (ii) the motion of the piston. In most reciprocating compressors, such clearance is filled by a mixture of refrigerant gas and lubricating oil. Besides its lubricating function, the oil also acts as a sealing element for the piston-cylinder clearance. Therefore, the leakage in the piston-cylinder clearance is more detrimental to oil-free compressors. Moreover, leakage is more critical in low-capacity compressors, since it may be comparable to the mass flow rate of such compressors. The reduction of the clearance results in higher manufacturing costs and may considerably increase the energy loss due to friction in the clearance.

Leakage in the clearance between the cylinder and the piston has been the focus of several studies. Zuk and Smith [2] proposed an analytical model to estimate leakage in small gaps between parallel plates by applying the Reynolds equation for compressible flow with static boundaries. Ferreira and Lilie [3] presented an analytical model to assess the leakage in the piston-cylinder gap with a moving piston, by assuming an incompressible flow formulation and disregarding inertia forces. Another analytical model was proposed by Yuan et. al [4], which considers both inertial and viscous forces as well as gas compressibility effects, but neglects the piston velocity effects. The authors observed that inertia effects are negligible for clearances smaller than 6 μm.

Rigola *et al.* [5] applied a numerical model based on the two-dimensional incompressible Reynolds equation including the piston velocity. They reported an exponential relationship between the size of the gap and the leakage rate, and analysed different compressors and refrigerant fluids. Lohn and Pereira [6] developed a three-dimensional numerical model for predictions of leakage in the piston-cylinder clearance, which were compared with the results of two analytical models: (i) Zuk and Smith [2] and (ii) Ferreira and Lilie [3]. Lohn and Pereira [6] verified that the mass flow rates obtained via a compressible flow formulation were considerably different from those calculated with the incompressible flow formulation, pointing out that the compressible formulation is more suitable for this type of flow. They also showed that the piston velocity significantly affects the gas leakage rate when the pressure difference is kept constant.

Although recent studies have shed some light on the effects of piston velocity on the gas leakage rate, the models available in the literature are either accurate and computationally too expensive, such as two and three-dimensional numerical models, or too simple analytical models that do not take into account compressibility effects or the piston velocity. The present paper reports a numerical model based on the one-dimensional compressible Reynolds equation, including the piston motion. The model considers only the refrigerant fluid in the clearance and can predict the gas leakage at a low computational cost, allowing its use in a coupled manner with a simulation model for the compression cycle. Thus, the leakage in the piston-cylinder clearance is calculated for different piston positions and pressure differences during the compression cycle. The role of the piston motion on the gas leakage rate is investigated throughout the entire compression cycle.

2. Numerical model and solution procedure

The flow geometry addressed herein is illustrated in figure 1. The region between the piston and the cylinder of length L and clearance δ represents the solution domain. The pressure in the compression chamber is higher than the pressure in the compressor internal environment for most of the compression cycle. This pressure difference and the piston motion bring about the flow in the piston-cylinder clearance. When the piston moves towards the cylinder head, gas is dragged into the compression chamber and vice-versa when the piston moves in the other direction.

The following hypotheses are assumed for the flow in the flow: i) inertial effects are negligible; ii) the pressure varies only in the x direction; iii) the piston-cylinder clearance δ is much smaller than the piston and cylinder diameters, so the annular gap can be treated as a flat channel; iv) the gas behaves as an ideal gas; v) dynamic viscosity does not vary with x (although it varies from one crank angle to the next); vi) isothermal flow and vii) piston and cylinder are concentric, so δ is constant; viii) subsonic flow condition.

As shown by Hamrock *et al.* [7], the governing equation for compressible flow in such a geometry can be modelled by equation (1), which is a simplified form of the Reynolds equation, where p is the static pressure, μ is the fluid dynamic viscosity, ρ is the fluid density, and V_{p} the piston velocity.

$$\frac{\mathrm{d}}{\mathrm{d}x}\left(\frac{\rho\delta^3}{12\mu}\frac{\mathrm{d}p}{\mathrm{d}x}\right) = \frac{\mathrm{d}}{\mathrm{d}x}\frac{\rho\delta V_{\mathrm{p}}}{2} \tag{1}$$

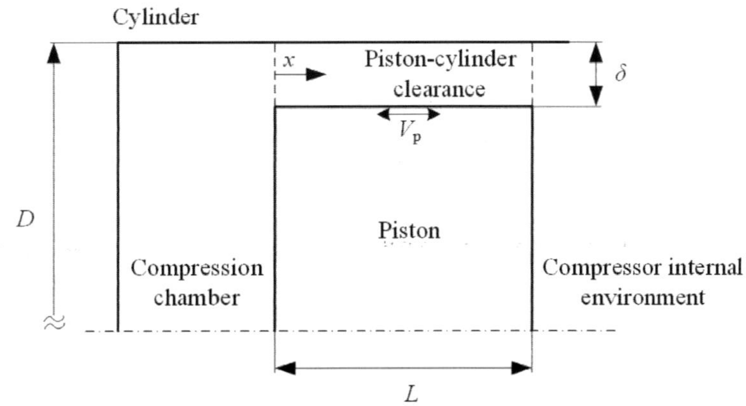

Figure 1. Flow geometry.

With the hypotheses i)-viii), and by replacing the density by pressure via the isothermal flow condition of an ideal gas (i.e., p/ρ = constant), equation (1) becomes

$$\frac{\mathrm{d}}{\mathrm{d}x}\left(p\frac{\mathrm{d}p}{\mathrm{d}x}\right) = \frac{6\mu V_\mathrm{p}}{\delta^2}\frac{\mathrm{d}p}{\mathrm{d}x} \tag{2}$$

If the Couette term on the right-hand side of equation (2) is eliminated, an analytical solution can be obtained, as indicated by Zuk and Smith [2].

Fluid refrigerants do not behave as an ideal gas in typical operating conditions found in household refrigeration. However, this hypothesis was adopted herein for simplicity since the main objective of our study is to analyze the effect of gas compressibility on leakage. Naturally, a formulation of real gas must be adopted if the interest is to accurately predict the impact of leakage on compressor performance.

Equation (2) is a differential equation that cannot be solved analytically and, for this reason, we applied the finite volume method (FVM) for the numerical solution. The solution domain is divided into equally spaced control volumes to form a computational mesh. The differential equation is discretized in each control volume, resulting a system of algebraic equations.

The pressures at the boundaries of the volumes are estimated via arithmetic mean of the pressures at the neighbouring points, allowing the pressure field to be calculated by solving the system of algebraic equations via the solver provided by the Eigen library [8]. Since equation (2) is non-linear, an initial pressure field is required to determine the coefficients of the algebraic equations. An iterative procedure is then applied, with the coefficients being updated and the calculations repeated until the convergence criterion in achieved.

Once the pressure field is known, the leakage mass flow rate

$$\dot{m}_\mathrm{l} = \pi D\left(-\frac{\rho\delta^3}{12\mu}\frac{\mathrm{d}p}{\mathrm{d}x} + \frac{\rho V_\mathrm{p}\delta}{2}\right) , \tag{3}$$

associated with this Couette-Poiseuille flow can be obtained at any position of the solution domain, but we chose the cross section adjacent to the compression chamber, i.e., $x = 0$. The first and second terms inside the parentheses are the contributions to leakage due to the pressure difference and piston motion. The pressure derivative in equation (3) was obtained via a second order approximation.

Two pressure boundary conditions are needed in the numerical model. The first one is the pressure in the compressor internal environment, which is constant throughout the compression cycle. The second is the pressure at $x = 0$, which is evaluated in the compression chamber (p_{cc}) at each crank angle (wt) with a simulation model described by Link and Deschamps [9]. This simulation model also evaluates the piston velocity, as well as the dynamic viscosity and density that are required in equation (3) for $x =$

0 at each crank angle. In turn, the model for the flow in the piston-cylinder clearance gives the leakage required to evaluate the compression cycle. Figure 2 illustrates the coupling between these two models.

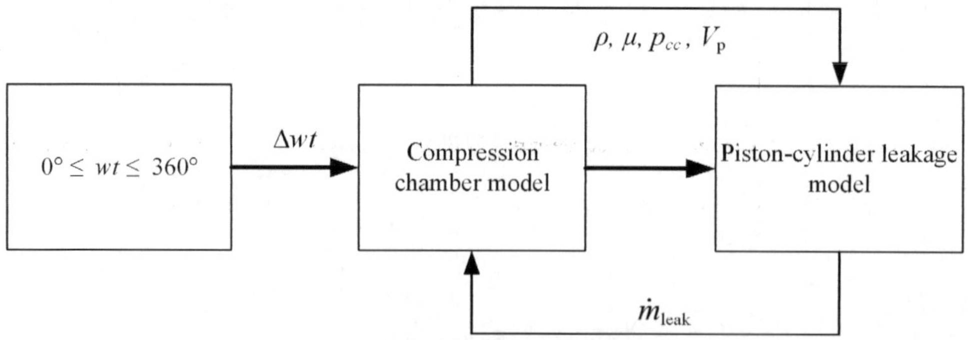

Figure 2. Schematic of the solution procedure.

3. Results and discussion

The developed model was used to investigate the effects of several parameters on gas leakage, such as compressor operating conditions, compressor speed (represented by the drive frequency) and clearance dimensions. Additionally, compressibility effects were also assessed by comparing the results obtained with formulations of compressible and incompressible flows.

The results presented hereafter were obtained for a low-capacity reciprocating compressor with displacement of 3 cm^3 operating with isobutane (R600a). The compression cycle was solved for small increments of crank angle ($\Delta wt = 0.001°$). A total of 250 control volumes were used to discretize the solution domain, with a difference less than 0.01% between predictions and estimates via the Richardson extrapolation. It should be mentioned that the model employed to simulate the compression cycle does not consider valve dynamics and heat transfer at the walls of the compression chamber. This simplification is adopted for convenience, since the study is an assessment of compressibility effects on leakage rather than simulation of compressor performance.

The model was initially validated through comparisons between its predictions and the results of a three-dimensional model developed with a CFD commercial code (Ansys FLUENT v.14), considering a single piston position with fixed pressure difference $\Delta p = 7$ bar, diametric clearance $c = 8$ μm and different instantaneous piston velocities. The 3-D model employed the same assumptions used in the simplified model, i.e., isothermal, compressible, laminar flow of an ideal gas, subjected to the same boundary conditions. The computational mesh had approximately 8 and 250,000 volumes in the transversal and longitudinal directions. Figure 3 shows the results for the leakage mass flow rate ($\dot{m}_l$); it reveals that both models are in good agreement with differences within 1%. Positive velocities represent that the piston is moving towards the bottom dead center, and the opposite for negative values.

The compressor operating conditions are defined by the evaporating and condensing temperatures of the refrigeration cycle, which defines the compressor suction and discharge pressures. Reference operating conditions used to test refrigeration compressors are the low-back pressure (LBP), medium-back pressure (MBP) and high-back pressure (HBP), as shown in table 1. The effect of such operating conditions on gas leakage is also investigated.

The results for $c = 9$ μm and $f = 60$ Hz shown in figure 4 reveal that the leakage mass flow rate in the compressor increases with the evaporating pressure (p_{evap}). This may seem unexpected since the pressure difference decreases as the evaporating pressure increases, keeping the same condensing pressure (7.62 bar). Nevertheless, as indicated by equation (3), the leakage mass flow rate also depends on the gas density, which is highest for the HBP condition, as shown in figure 5, because the gas temperature in the discharge process reaches its lowest value. This aspect compensates for the smaller pressure difference of the HBP condition. The relative leakage mass flow rate ($\dot{m}_l^*$) is defined as the leakage mass flow divided by the compressor mass flow rate disregarding the leakage through the piston-cylinder clearance. As opposed to the leakage mass rate ($\dot{m}_l$), the relative leakage ($\dot{m}_l^*$) decreases with evaporating pressure, which shows that the relative influence of the gas leakage is more significant for the LBP case.

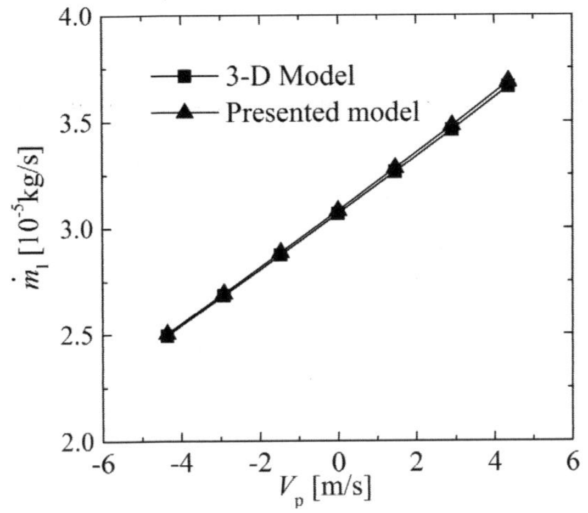

Figure 3. Comparison between simplified and 3-D models.

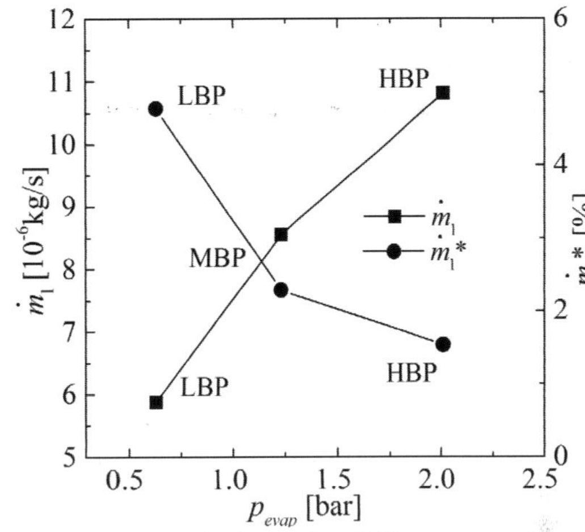

Figure 4. Gas leakage for different compressor operating conditions.

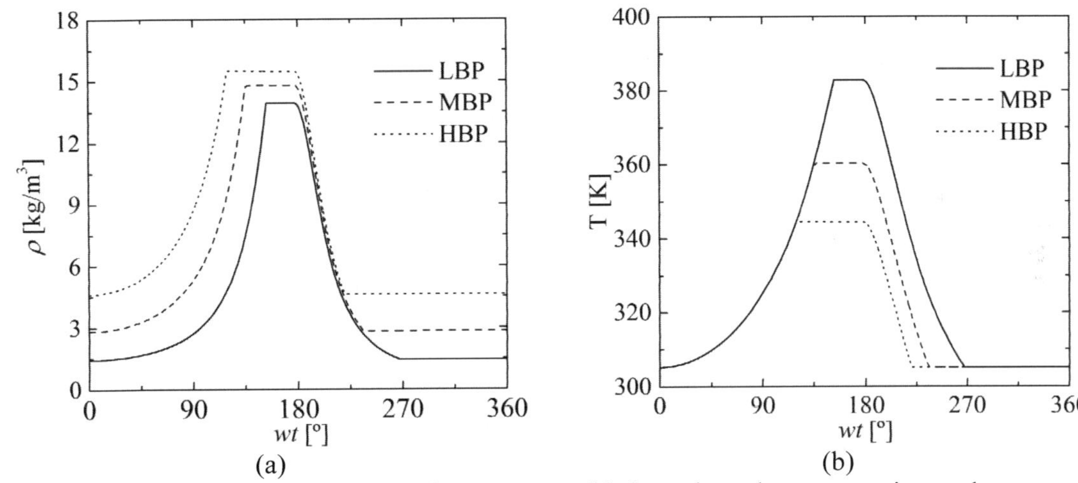

Figure 5. Gas density (a) and temperature (b) throughout the compression cycle.

Table 1. Compressor operating conditions.

Denomination	Evaporating temperature	Evaporating pressure	Condensing temperature	Condensing pressure
LBP	-23.3°C	0.629 bar		
MBP	-6.7°C	1.229 bar	54.4°C	7.62 bar
HBP	7.2°C	2.011 bar		

The piston velocity is a direct consequence of the drive frequency. The model was applied with different values of frequency in order to investigate its effects on gas leakage throughout the compression cycle. Figure 6 shows the results of simulations for $c = 5$ μm and LBP condition. The 'No velocity' curve represents the results of equation (3) without the Couette term. Figure 6 shows that the instantaneous leakage rate is greatly affected by piston velocity, being reduced in the compression and discharge processes and decreased in the expansion and suction processes as the piston velocity is increased. Positive values of $\dot{m}_l$ mean that the gas is leaving the compression chamber, whereas negative values imply that gas is actually being admitted to the compression chamber. In the compression process, the piston is moving towards the top dead center and therefore it carries gas to the compression chamber through viscous effects, reducing the instantaneous gas leakage rate. The opposite effect occurs when the piston moves towards the bottom dead center.

The most significant difference between the results of leakage for the three velocity conditions occurs at the crank angle around 135°, when the Couette term reaches its maximum value, acting towards leakage reduction. The total mass of leakage in the cycle, obtained by integrating the curves in figure 6 over time, is reduced by roughly 5% when f = 90 Hz.

The instantaneous gas leakage rate for two diametric clearances (c) is shown in figure 7. The simulations were carried out with a frequency of 60 Hz and LBP condition. As can be seen, the influence of the piston velocity represented by the Couette term is more significant in the smaller clearance. The results also suggest that the pressure term in equation (3) becomes increasingly important as the clearance is increased.

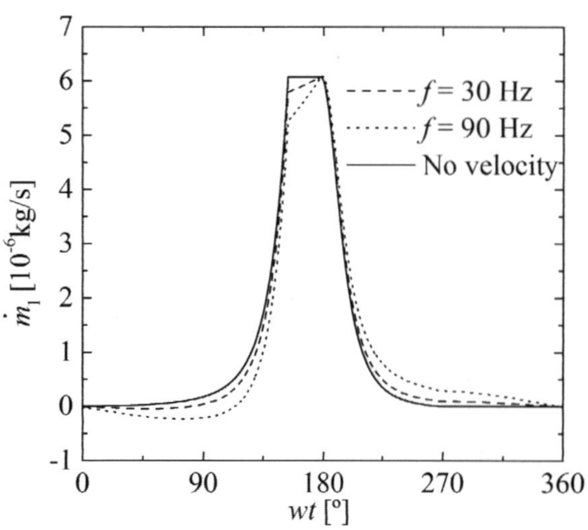

Figure 6. Instantaneous gas leakage rate during the compression cycle for different compressor speeds; c = 5 μ m and LBP condition.

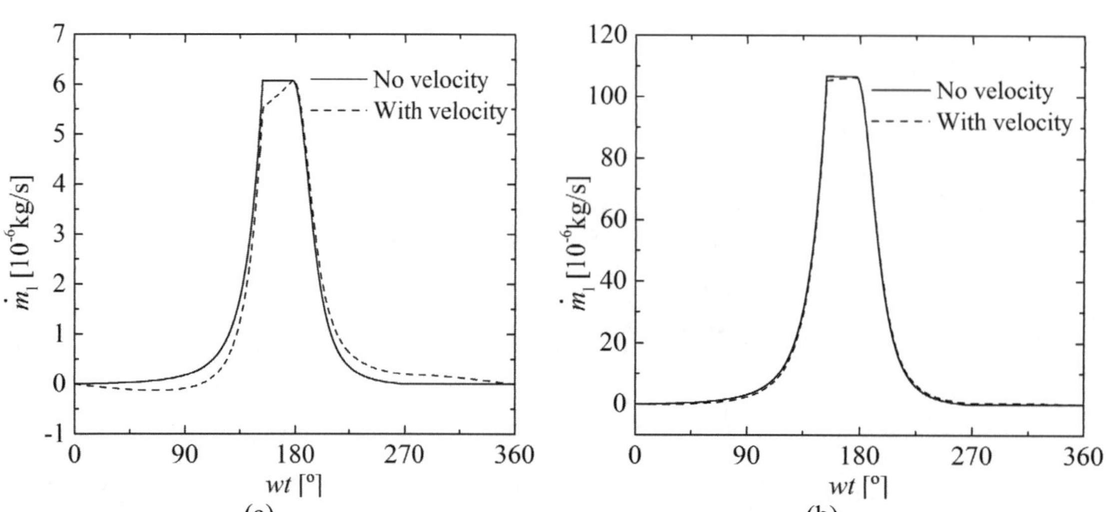

Figure 7. Instantaneous gas leakage rate for different clearances; f = 60 Hz, LBP condition: (a) c = 5 μm; (b) c = 13 μm.

In order to investigate the effects of compressibility on gas leakage, the model was compared to the incompressible fluid flow model presented by Ferreira and Lilie [3]. The calculations considered different drive frequencies, two clearances and LBP condition. Figure 8 shows the relative leakage mass flow rate in each case. The results indicate that, although both curves present the same shape, the incompressible fluid model curve overestimates leakage in every case. For low frequencies the leakage predicted by the incompressible fluid model is twice as big as the value predicted by the compressible fluid model, reaching more than 70% of the compressor mass flow rate when c = 13 μm.

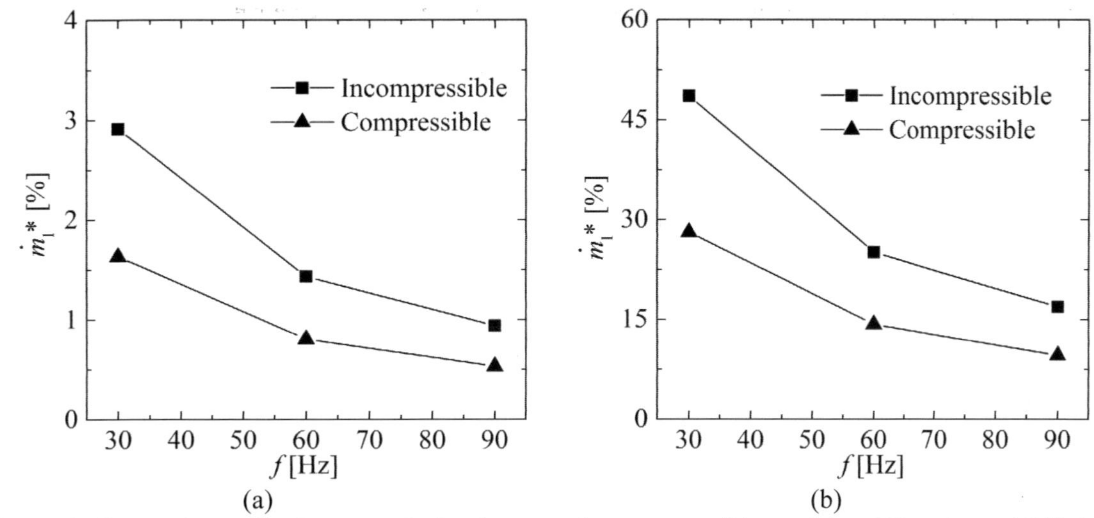

Figure 8. Comparison of leakage prediction between the compressible model and Ferreira and Lilie's incompressible model [3] for different clearances: (a) c = 5 μm; (b) c = 13 μm.

The volumetric efficiency and isentropic efficiency are two parameters commonly used to evaluate the compressor performance. The first is defined as the ratio between the actual mass flow and the ideal mass flow rate, as indicated in equation (4). The ideal mass flow rate is obtained for the hypothetical case with no gas leakage, cylinder clearance volume, flow restriction and backflow in valves and suction gas superheating.

$$\eta_v = \frac{\dot{m}}{\dot{m}_{th}} \tag{4}$$

The isentropic efficiency is defined as the ratio between the compression power required by an isentropic process and the actual indicated power, i.e.:

$$\eta_s = \frac{\dot{W}_{th}}{\dot{W}} \tag{5}$$

Figure 9 shows the results of efficiency reduction due to leakage, considering different drive frequencies and clearances for the LBP condition. The efficiency reduction is evaluated as the difference between the compressor efficiencies with and without gas leakage (subscript 'nl'), as indicated by equation (6). Figure 9 reveals that gas leakage is more detrimental to both efficiencies for smaller drive frequencies and larger clearances. In the worst case, the efficiency reduction can reach more than 8% for the volumetric efficiency and more than 22% for the isentropic efficiency.

$$\Delta\eta_v = |\eta_v - \eta_{v,nl}|, \quad \Delta\eta_s = |\eta_s - \eta_{s,nl}| \tag{6}$$

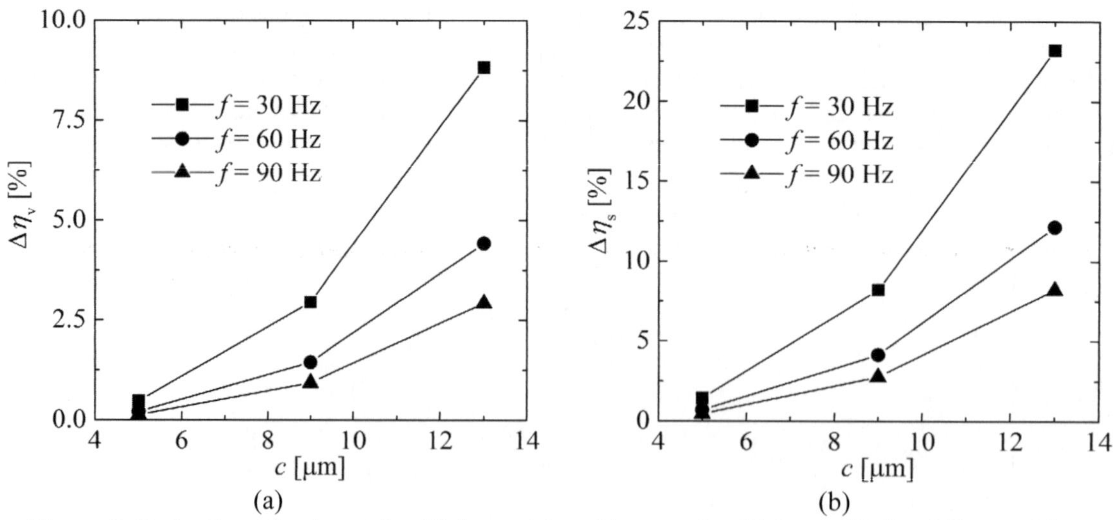

Figure 9. Reduction of volumetric efficiency (a) and isentropic efficiency (b) for different drive frequencies.

4. Conclusions

A simulation model based on the compressible fluid flow formulation of the Reynolds equation was developed to predict gas leakage in the piston-cylinder clearance of small reciprocating compressors adopted for domestic refrigeration. The model was validated by comparing its predictions with results of a three-dimensional model. The model was then coupled with a simulation model of the compression cycle to allow predictions of leakage throughout the compression cycle for different parameters, such as operating conditions, compressor speed and piston-cylinder clearance. The results show that leakage increases with the evaporating pressure when the condensing pressure is kept constant. Nonetheless, leakage becomes more detrimental as the evaporating pressure is decreased because the compressor mass flow rate decreases to greater extent when the evaporating pressure is lowered. We found that the piston motion plays a minor role on leakage, becoming negligible in comparison with the effect of pressure difference as the clearance is increased. The present study also showed that compressible fluid flow formulation must be considered to accurately predict gas leakage in the piston-cylinder clearance.

Acknowledgements

The present study was developed as part of a technical-scientific cooperation program between the Federal University of Santa Catarina and EMBRACO. The authors are also thankful for the support of CAPES and EMBRAPII Unit Polo/UFSC.

References

[1] Silva L R and Deschamps C J 2015 Modeling of gas leakage through compressor valves *Int. J. Refrigeration* **53** 195–205

[2] Zuk J and Smith P J 1969 Computer program for viscous isothermal compressible flow across a sealing dam with tilt angle *NASA*

[3] Ferreira R T S and Lilie D E B 1984 Evaluation of the leakage through the clearance between piston and cylinder in hermetic compressors *Proc. Int. Compressor Eng. Conf. at Purdue (West Lafayette, IN, USA)* p 424

[4] Yuan X, Chen Z and Fan Z 1992 Calculating model and experimental investigation of gas leakage *Proc. Int. Compressor Eng. Conf. at Purdue (West Lafayette, IN, USA)* p 919

[5] Rigola J, Pérez-Segarra C D and Oliva A 2009 Numerical simulation of piston leakage over hermetic reciprocating compressors behavior *Proc. Int. Conf. on Compressors and Their Systems (London, UK)* pp 375-84

[6] Lohn S K and Pereira E L L 2014 Numerical investigation of the gas leakage through the piston-cylinder clearance of reciprocating compressors *Proc. Int. Compressor Eng. Conf. at Purdue (West Lafayette, IN, USA)* p 2295

[7] Hamrock B J, Schmid S R and Jacobson B O 2004 *Fundamentals of Fluid Film Lubrication* (Marcel Dekker, New York) p 347

[8] Guennebaud G and Jacob B 2010 Eigen v3: a C++ template library for linear algebra. *http://eigen.tuxfamily.org/ (retrieved May 5, 2016)*

[9] Link R and Deschamps C J 2011 Numerical modeling of startup and shutdown transients in reciprocating compressors *Int. J. Refrigeration* 34 1398-1414

IOP Conf. Series: Materials Science and Engineering **232** (2017) 012007 doi:10.1088/1757-899X/232/1/012007

Comprehensive model of a hermetic reciprocating compressor

B Yang[1]*, D Ziviani[2] and E A Groll[3]
Purdue University, School of Mechanical Engineering,
Ray W. Herrick Laboratories, West Lafayette, IN, USA

Email: [1]yang62@purdue.edu, [2]dziviani@purdue.edu, [3]groll@purdue.edu

Abstract: A comprehensive simulation model is presented to predict the performance of a hermetic reciprocating compressor and to reveal the underlying mechanisms when the compressor is running. The presented model is composed of sub-models simulating the in-cylinder compression process, piston ring/journal bearing frictional power loss, single phase induction motor and the overall compressor energy balance among different compressor components. The valve model, leakage through piston ring model and in-cylinder heat transfer model are also incorporated into the in-cylinder compression process model. A numerical algorithm solving the model is introduced. The predicted results of the compressor mass flow rate and input power consumption are compared to the published compressor map values. Future work will focus on detailed experimental validation of the model and parametric studies investigating the effects of structural parameters, including the stroke-to-bore ratio, on the compressor performance.

1. Introduction

Many simulation models for reciprocating compressors are presented in the literature. Navarro et al. [1] proposed a phenomenological model for a hermetic reciprocating compressor to predict the overall isentropic efficiency and volumetric efficiency. The efficiencies were expressed as a function of ten parameters representing various kinds of the main losses in the compressor, including the refrigerant superheating by motor cooling, mechanical loss dissipation, and pressure losses at inlet and outlet valves. Perez-Segarra et al. [2] presented a detailed model for the thermodynamic efficiencies to characterize hermetic reciprocating compressors. The authors split the volumetric efficiency, isentropic efficiency and combined mechanical-electrical efficiency into several partial efficiencies so as to denote effects of different thermodynamic phenomena. Winandy et al. [3] presented a simplified steady-state model for a reciprocating compressor to identify the processes which have an influence on refrigerant mass flow rate, compressor power and discharge temperature. Estupinan and Santos [4] presented a dynamic model of a hermetic reciprocating compressor that considered the lubrication of the journal bearings, but the model

IOP Conf. Series: Materials Science and Engineering **232** (2017) 012007 doi:10.1088/1757-899X/232/1/012007

did not include the frictional loss between the piston rings and cylinder wall. Todescat et al. [5] developed a model for the thermodynamic energy analysis in a hermetic reciprocating compressor. However, their model did not include the motor model.

In this paper, a comprehensive model for a hermetic R410A reciprocating compressor is presented. This model includes sub-models for the kinematics, compression process, frictional power loss and overall energy balance. In addition, the in-cylinder heat transfer sub-model, leakage sub-model and valve sub-model are included in the compression process model. Both the frictional power losses at the piston ring and bearings are considered in the current model. Moreover, a single-phase induction motor model similar to that of Dutra and Deschamps [6] is included. The algorithm of the solution to the whole compressor model is also provided. This comprehensive model can be very helpful for the compressor designer by investigating the effect of varying different parameters, including the reed valve shape and stiffness, on compressor performance.

2. Geometry and Kinematics Model

A schematic of the geometry and kinematics of the reciprocating compressor is shown in Fig. 1. The bottom dead center (B.D.C.) is the starting point of the crank angle θ. The piston displacement x, velocity U, and acceleration a, are expressed as a function of the crank angle as follows,

$$x = \frac{S}{2}[1 - \cos(\theta + \pi) + \frac{\lambda}{4}(1 - \cos 2(\theta + \pi))] \tag{1}$$

$$U = \frac{S}{2}\omega(\sin\theta + \frac{\lambda}{2}\sin 2\theta) \tag{2}$$

$$a = \frac{S}{2}\omega^2(\cos\theta + \lambda\cos 2\theta) \tag{3}$$

Where, λ is the ratio of the crank radius to the connecting rod length; ω is the compressor rotational speed in rad/s; S is the piston stroke.

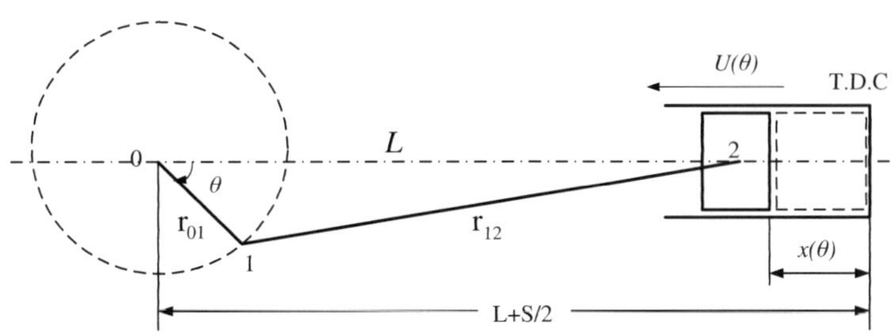

Figure 1: Schematic of geometry and kinematics of a reciprocating compressor

3. Compression process model

3.1 Governing equations

The refrigerant inside the cylinder is assumed to be at uniform state, which means the thermal properties of the refrigerant are uniform anywhere in the cylinder. The refrigerant thermal properties are evaluated using REFPROP [7]. According to the law of continuity, the mass change of refrigerant in the cylinder can be expressed by,

$$\frac{dm_c}{dt} = \frac{dm_{suc}}{dt} + \frac{dm_{li}}{dt} - \frac{dm_{dis}}{dt} - \frac{dm_{lo}}{dt} \tag{4}$$

Where, m_c refers to the refrigerant mass in the cylinder. The conservation equation of energy can be derived by employing the first law of thermodynamics,

$$\frac{dQ}{dt} + \frac{dW}{dt} = \frac{dm_{dis}}{dt}h_{dis} + \frac{dm_{lo}}{dt}h_{lo} + \frac{d(m_c u_c)}{dt} - \frac{dm_{suc}}{dt}h_{suc} - \frac{dm_{li}}{dt}h_{li} \tag{5}$$

Eq. (5) can be further transformed into the following equation which describes the refrigerant temperature in the cylinder as a function of crank angle,

$$\frac{dT_c}{dt} = \frac{1}{m_c C_v}\left[\frac{dQ}{dt} + \frac{dm_{suc}}{dt}(h_{suc} - h_c) + \frac{dm_{li}}{dt}(h_{li} - h_c) - T_c\left(\frac{\partial P_c}{\partial T_c}\right)_v\left(A_p U - v_c\frac{dm_c}{dt}\right)\right] \tag{6}$$

The instantaneous heat transfer rate between the refrigerant and the cylinder wall is determined using the heat transfer correlation proposed by Adair et al. [8],

$$Nu = 0.053\mathrm{Re}^{0.8}\,\mathrm{Pr}^{0.6} \tag{7}$$

Then, the heat transfer rate between refrigerant and cylinder wall is calculated,

$$\dot{Q}_{heat} = \frac{dQ}{dt} = h_{heat}A_{heat}(T_c - T_{wall}) \tag{8}$$

The heat transfer coefficient h_{heat} is determined based on the Nusselt number calculated in Eq. (7). Additionally, Adair et al. [8] reported that the cylinder wall temperature changed little in a reciprocating compresssor compared with its time averaged value in a cycle. Thus, the cylinder wall temperature is assumed to be constant, which is calculated by incorporating the overall energy balance model with the compression process model.

3.2 Valve model

The refrigerant mass flow rates through the suction and discharge valves are determined by assuming an isentropic, compressible flow,

$$\dot{m}_{valve} = \frac{dm_{valve}}{dt} = C_{valve}A_{valve}(2\rho_{high}P_{high})^{\frac{1}{2}}\left\{\frac{\kappa}{\kappa-1}\left[\left(\frac{P_{low}}{P_{high}}\right)^{\frac{2}{\kappa}} - \left(\frac{P_{low}}{P_{high}}\right)^{\frac{\kappa+1}{\kappa}}\right]\right\}^{\frac{1}{2}} \tag{9}$$

The discharge coefficient C_{valve} of 0.58 is used for the suction valve and a coefficient of 0.6 is used for the discharge valve [9]. A_{valve} denotes the valve opening area for the suction and discharge valves. For a valve, there are two dominant regions, pressure dominant region and mass flux dominant region. In the pressure dominant region, the valve motion is dominated by the pressure difference across the valve. However, beyond the transition point the assumption that there is a significant pressure difference across the valve is no longer applicable and the valve motion is driven by the gas mass-flux. The transitional valve plate lift, below which the pressure force dominates, can be calculated by [10],

$$h_{tr} = \frac{1}{4}\frac{D_{port}^2}{D_{valve}} \tag{10}$$

Where, D_{port} and D_{valve} are the diameter of valve port and valve plate, respectively. A_{valve} is the valve flow area. Both the suction valves and discharge valves are reed-type. The motion of the reed valve under the force due to the gas pressure difference across the valve can be approximated as a motion of a cantilever beam exerted by the same force. According to Soedel [11], the transverse displacement of a cantilever beam can be described by,

$$EI\frac{\partial^4 u}{\partial x^4} + \rho A\ddot{u} = q \tag{11}$$

Where, E is the Young's modulus; I is the moment of inertia of the beam cross section. ρ is the density of valve material; A is the cross-sectional area; q is the load per unit length; u is the instantaneous transverse displacement at position, x.

The solution of Eq. (11) can be expressed as the combination of infinite modal series,

$$u(x,t) = \sum_{k=1}^{\infty}\eta_k(t)U_k(x) \tag{12}$$

Where, $U_k(x)$ is the k-th natural vibration mode; $\eta_k(t)$ is the time varying weight factor corresponding to kth natural vibration mode. The k-th natural vibration mode is expressed as,

$$U_k(x) = \cosh(\frac{\lambda_k x}{L}) - \cos(\frac{\lambda_k x}{L}) - \frac{\cosh\lambda_k + \cos\lambda_k}{\sinh\lambda_k + \sin\lambda_k}[\sinh(\frac{\lambda_k x}{L}) - \sin(\frac{\lambda_k x}{L})] \tag{13}$$

Where, L is the length of the beam. For the reed valve presented in this paper, only the first mode shape is considered, i.e., k=1, λ_1=1.875. Substituting Eq. (12) into Eq. (11) and utilizing orthogonality condition,

$$\ddot{\eta}_k + 2\zeta_k\omega_k\dot{\eta}_k + \omega_k^2\eta_k = F_k \tag{14}$$

Where,

$$\omega_k = \frac{\lambda_k^2}{L^2}\sqrt{\frac{EI}{\rho A}},\ \zeta_k = \frac{\lambda_k}{2\rho h\omega_k},\ N_k = \int_x U_k^2(x)dx,\ F_k = \frac{F_{valve}}{\rho AN_k}, F_{valve}\ \text{is the flow force} \tag{15}$$

The solution to the Eq. (14) can be obtained by

$$\eta_k = e^{-\zeta_k\omega_k t}[A\cos(\gamma_k t) + B\sin(\gamma_k t)] + \frac{F_k}{\omega_k^2} \tag{16}$$

Where,

$$\gamma_k = \omega_k\sqrt{1-\zeta_k^2} \tag{17}$$

Differentiate the equation with respect to time once and twice respectively,

$$\dot{\eta}_k = -\zeta_k\omega_k e^{-\zeta_k\omega_k t}[A\cos(\gamma_k t) + B\sin(\gamma_k t)] + e^{-\zeta_k\omega_k t}[-A\gamma_k\sin(\gamma_k t) + B\gamma_k\cos(\gamma_k t)] \tag{18}$$

$$\ddot{\eta}_k = e^{-\zeta_k \omega_k t}[(-\zeta_k \omega_k M - N\gamma_k)\cos(\gamma_k t) + (\zeta_k \omega_k N - M\gamma_k)\sin(\gamma_k t)] \tag{19}$$

Where,

$$A = \eta_k(0) - \frac{F_k}{\omega_k^2},\ B = \frac{\dot{\eta}_k(0)\omega_k + \eta_k(0)\zeta_k \omega_k^2 - F_k \zeta_k}{\gamma_k \omega_k},\ M = B\gamma_k - A\zeta_k \omega_k,\ N = A\gamma_k + B\zeta_k \omega_k \tag{20}$$

Using Eq. (15), Eq. (18), Eq. (20) and Eq. (21), the instantaneous transverse displacement, velocity and acceleration of a cantilever beam at position x can be determined.

3.3 Leakage sub-model

The gas leaking through the piston ring gap is modeled as an isentropic, compressible fluid flowing through an orifice. The mass flow rate of the gas leakage is calculated by,

$$\begin{cases} \dot{m}_{gap} = \dfrac{dm_{gap}}{dt} = C_{gap} A_{gap} P_u \left(\dfrac{2}{ZRT_u}\right)^{\frac{1}{2}} \left\{\dfrac{\kappa}{\kappa - 1}\left[\left(\dfrac{P_d}{P_u}\right)^{\frac{2}{\kappa}} - \left(\dfrac{P_d}{P_u}\right)^{\frac{\kappa+1}{\kappa}}\right]\right\}^{\frac{1}{2}}, \dfrac{P_d}{P_u} > 0.54 \\ \dot{m}_{gap} = \dfrac{dm_{gap}}{dt} = C_{gap} A_{gap} P_u \left(\dfrac{\kappa}{ZRT_u}\right)^{\frac{1}{2}} \left[\left(\dfrac{2}{\kappa + 1}\right)^{\frac{\kappa+1}{\kappa-1}}\right]^{\frac{1}{2}}, \dfrac{P_d}{P_u} \le 0.54, choked \end{cases} \tag{21}$$

Where the discharge coefficient C_{gap} is assumed to be 0.86 [12] and the compressible factor Z is determined by using REFPROP.

4. Dynamics and Frictional Power Loss Model

The force analysis on the crank-connecting rod driving mechanism during the up-stroke and down-stroke are given in Fig. 2 and Fig. 3 respectively. Employing the force balance on the system, all the force loads are determined which can be used to calculate the frictional power loss on the main and crank pin journal bearings.

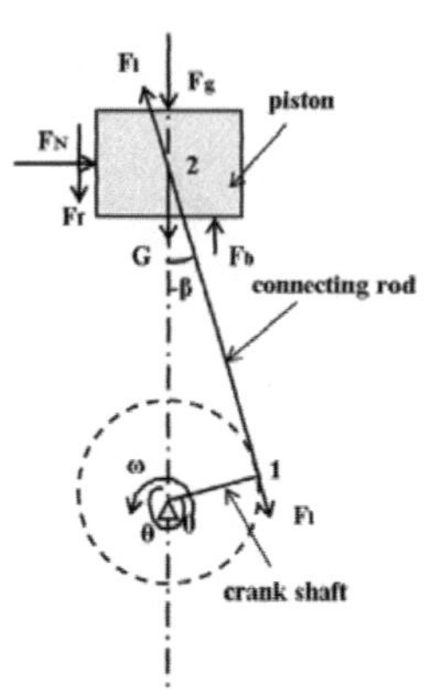

Figure 2: Schematic of force and moment analysis when the piston is in up-stroke

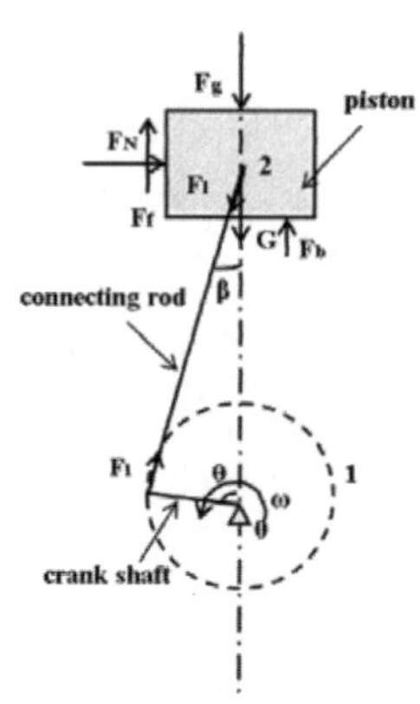

Figure 3: Schematic of force and moment analysis when the piston is in down-stroke

4.1 Piston Ring Lubrication Model

The main frictional power loss in the reciprocating compressor can be located at the piston ring frictional power loss and journal bearing frictional power loss. There is a piston ring located between the piston and cylinder wall to seal the in-cylinder refrigerant. Considering the effect of the surface roughness on the lubrication performance, the model proposed by Yang and Zhao[13] is used to get the frictional power loss at the piston ring-cylinder wall interface.

4.2 Bearing Frictional Power Loss Model

As mentioned above, with the determined force loads from the force balance analysis, the frictional power losses at the main and crank pin journal bearings can be predicted using the regression method proposed by Stachowiak and Batchelor [14],

$$H_{friciton} = 3.9307e3\nu_{1,oil}^{-0,706}\nu_{2,oil}^{1.577}L_{bearing}^{0.477}D_{journal}^{2.240}N_j^{1.287}c^{0.249}T_s^{-0.204}\left(1+\ln W^*\right)^{1.324} \tag{22}$$

Where, $\nu_{1,oil}$, $\nu_{2,oil}$ are the kinematic viscosities of the oil at 37.8°C and 93.3°C. T_s is the oil temperature in °F. The dimensionless load capacity W^* and clearance c are calculated by,

$$W^* = \frac{\sqrt{F_x^2 + F_y^2 + F_z^2}\,\overline{c}^2}{\mu_{oil} U L_{bearing} R_{journal}^2} \tag{23}$$

$$c = R_{bush} - R_{journal} \tag{24}$$

5. Single-Phase Motor Model

A single-phase induction electric motor is used in this conventional reciprocating compressor. As pointed out by Dutra and Deschamps [6], the single-phase induction electric motor can be analyzed using an equivalent circuit. Fig. 4 shows a typical equivalent circuit for a single-phase induction electric motor.

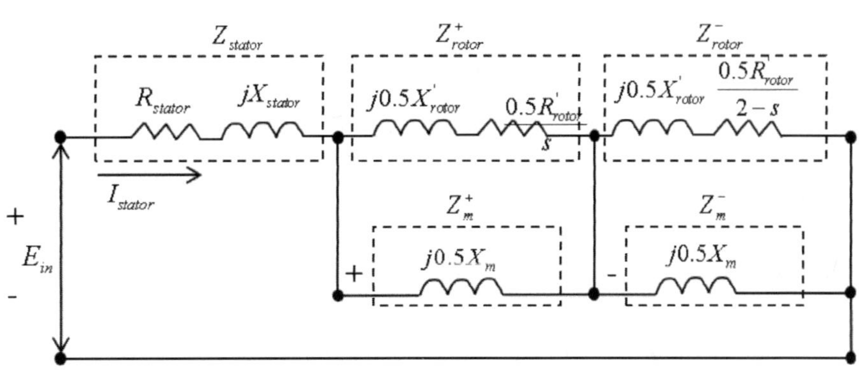

Figure 4: Equivalent circuit of the single phase induction motor [6]

Z_{stator}, Z_{rotor}, Z_m are stator impedance, rotor impedance and magnetized impedance. R and X are resistance and reactance respectively. The positive and negative sign refer to the forward loop and backward loop. Given the input voltage, resistance, reactance and motor slip ratio, the motor shaft angular speed, motor efficiency and power consumption can be determined.

6. Overall Energy Balance Model

The energy flow in the conventional reciprocating refrigeration compressor is shown in Fig. 5. It is assumed that there is no temperature gradient in the individual compressor components, i.e., the lumped temperature analysis method is used.

Fig. 6 shows the thermal resistance network for the overall energy flow in the compressor. The heat transfer rate between two compressor components is calculated as the ratio of the temperature difference to the thermal resistance. In order to find the thermal resistance, empirically correlated averaged Nusselt numbers are determined [10].

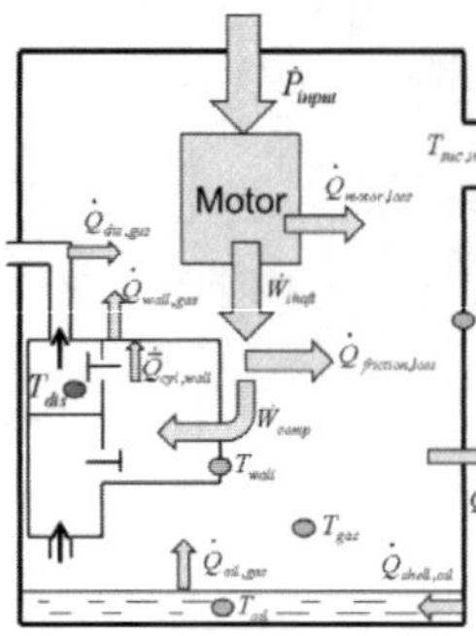

Figure 5: Energy balance of the conventional reciprocating piston compressor

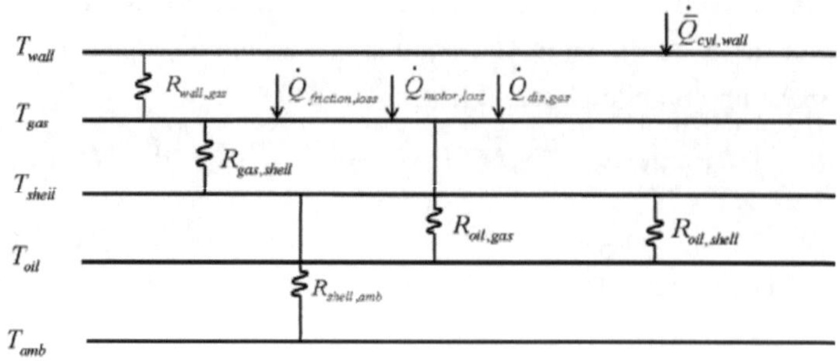

Figure 6: Thermal resistance network of the overall energy balance

7. Numerical Algorithm

The flow chart of the algorithm used to solve the presented comprehensive compressor model is given in Fig. 7. Given the initial guess values for the temperatures of different compressor components, in-cylinder refrigerant mass/temperature and the suction valve state, the determination of the instantaneous in-cylinder refrigerant pressure and temperature becomes an initial value problem, which can be solved by a variable step size method RKF45. However, the end states of both the in-cylinder refrigerant and suction valve after one cycle need to match the initial state at the very beginning of the cycle. Moreover, the temperatures of the compressor components are also required to meet the overall energy balance. Therefore, Broyden's method is employed to solve the nonlinear system of equations.

8. Results and Discussion

The hermetic reciprocating compressor investigated in this paper uses R410A as its working fluid. There are two cylinders located in the shell. Since the compression process of the refrigerant in each cylinder is the same, except for the compression process phase difference, the in-cylinder process in one of the cylinders is investigated. The bottom dead center (B.D.C.) is chosen as the starting point of the piston. Fig. 8 shows the P-V indicated diagram at different pressure ratios for one of the cylinders. The suction pressure stays constant and there is a 11°C superheat at the compressor inlet. As it is shown, the duration of the re-expansion process for a higher discharge pressure case is longer than that for a lower discharge pressure case. This is one of the reasons accounting for the decrease of the volumetric efficiency with an increment

IOP Conf. Series: Materials Science and Engineering **232** (2017) 012007 doi:10.1088/1757-899X/232/1/012007

in pressure ratio, since less volume is available for the suction gas through the suction valve if a higher discharge pressure is applied.

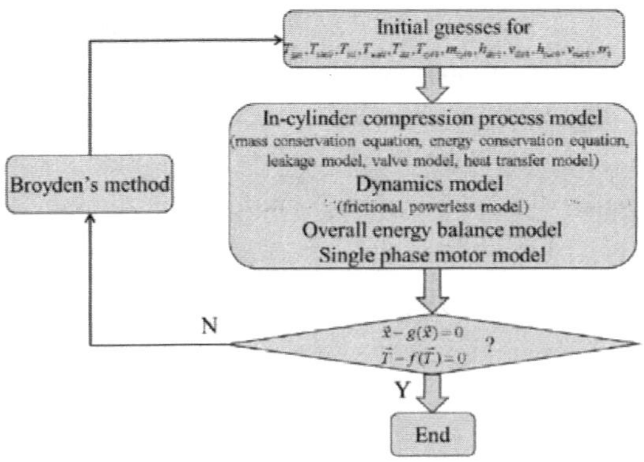

Figure 7: The flow chart of algorithm for the conventional reciprocating piston compressor model

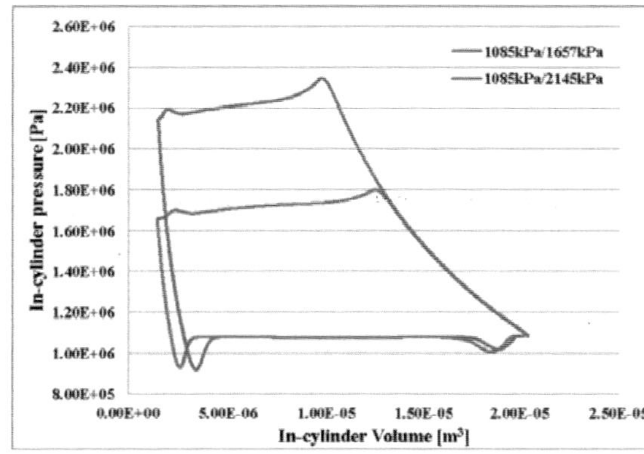

Figure 8: P-V indicated diagram for one cylinder

Fig. 9 shows the variation of the discharge valve displacement with respect to the compressor crank shaft rotation angle for different discharge pressures. It is seen that the discharge valve is still open when the re-expansion process starts. Back flow occurs at the discharge valve due to the late closing of the valve plate, which can be seen in Fig. 10.

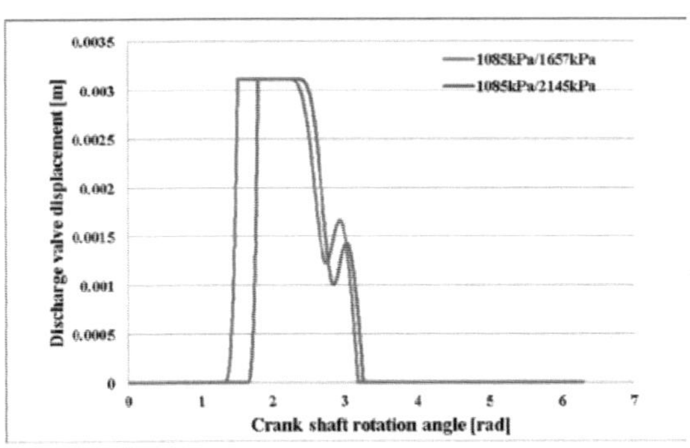

Figure 9: Discharge valve plate displacement v.s. crank shaft rotation angle

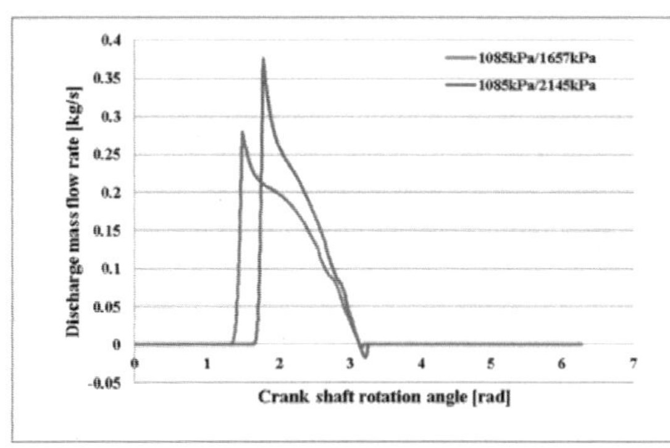

Figure 10: Discharge valve mass flow rate v.s. crank shaft rotation angle

The comparison between the predicted mass flow rate/input power consumption and the online performance table values are given in Table 1. The relative error of the mass flow rate and input power are around 5% and 10%, respectively.

Table 1: Comparison of predicted mass flow rate and input power to compressor map values

T_{evap}	T_{cond}	T_{sp}	mass flow rate [kg/s]			input power [W]		
[°C]	[°C]	[°C]	map value	predicted	error (%)	map value	predicted	error (%)
10	25	11	0.078	0.081	3.3	1442	1415	-1.9
10	35	11	0.074	0.077	5.1	2025	2269	12.0
5	25	11	0.065	0.068	5.2	1593	1739	9.2
5	30	11	0.063	0.067	6.1	1830	2035	11.2

IOP Conf. Series: Materials Science and Engineering **232** (2017) 012007 doi:10.1088/1757-899X/232/1/012007

9. Conclusions

A comprehensive model for a hermetic R410A reciprocating compressor is presented. This model consists of kinematics model, in-cylinder compression process model, frictional power loss model, single phase induction motor model and overall energy balance model. The valve, leakage and heat transfer model are included in the compression process model. A numerical algorithm used to solve the whole compressor model is introduced.

The P-V diagram is given for one cylinder. The discharge valve is found to be still open even when the re-expansion process starts which brings the back flow of refrigerant from the discharge plenum to the in-cylinder control volume. The predicted mass flow rate and input compressor power consumption have a relative error of 5% and 10% respectively compared to published table values.

10. Nomenclature

A	area
C	coefficient, specific heat capacity
D	diameter
E	Young's modulus
H	Power
I	moment of inertia of the beam cross section area
L	Length
P	Pressure
Q	heat transfer rate
R	gas constant, radius
S	piston stroke
T	temperature
U	velocity
V	volume
W	Load
Z	compact factor, impedance
H	unit enthalpy, or valve lift
M	mass
Q	load per unit length of beam
U	transverse displacement of the beam
N	specific volume
X	piston displacement

Greek letters

A	Relative clearance volume
B	incline angle
θ	crank angle
μ	dynamic viscosity
v	kinematic viscosity
η_k	weight factor of k-th vibration mode
κ	ratio of specific heat capacity
λ	ratio of crank radius to connecting rod length
ρ	density
ω	compressor angular speed

Subscripts

c	refrigerant in control volume
cond	condensing
cyl	cylinder
d	downstream
dis	discharge
evap	evaporating
high	high pressure side
li, lo	leak in and out
low	low pressure side
mean	mean velocity of piston
oil	lubricating oil
p	piston
sp	superheat
suc	suction
u	upstream
valve	valve

11. Acknowledgement
The authors would like to thank the Center of High Performance Building (CHPB) for sponsoring this project.

12. References
[1] Navarro, E., Granryd, E., Urchueguia, J.F., Corberan, J.M., 2007. A phenomenological model for analyzing reciprocating compressors. Int. J. Refrigeration 30, 1254-1265.

[2] Perez-Segarra, C.D., Rigola, J., Soria, M., Oliva, A., 2005. Detailed thermodynamic characterization of hermetic reciprocating compressors. Int. J. Refrigeration 28, 579-593.

[3] Winandy, E., Saavedra, O.C., Lebrun, J., 2002. Simplified modeling of an open-type reciprocating compressor. Int. J. Therm. Sci. 41, 183-192.

[4] Estupinan, E.A., Santos, I.F., 2007. Dynamic modeling of hermetic reciprocating compressors, combining multibody dynamics, finite elements method and fluid film lubrication. Int. J. Mechanics 1 (4), 36-43.

[5] Todescat, M. L., Fagotti, F., Prata, A. T., Ferreira, R. T. S. 1992. Thermal energy analysis in reciprocating hermetic compressors. *Proc. of the Int. Comp. Eng. Conf.* Purdue University, West Lafayette, IN USA, pp. 1419-1428.

[6] Dutra, T., Deschamps, C. J., 2015. A simulation approach for hermetic reciprocating compressors including electrical motor modeling. Int. J. Refrigeration 59, 168-181.

[7] NIST. 2010. Reference fluid thermodynamic and transport properties (REFPROP) Version 9.0.

[8] Adair, R. P., Qvale, E. R., Pearson, J. T., 1972. Instantaneous heat transfer to the cylinder wall in reciprocating compressors. In: Proceedings of the International Compressor Engineering Conference. Purdue University, West Lafayette, IN USA, pp. 521-526.

[9] Lin, M., Sun, S.Y., 2006. Theory of the piston compressor. Xi'an Jiaotong University Press. Xi'an, China.

[10] Kim, J-H, Groll, E.A., 2007. Feasibility study of a bowtie compressor with novel capacity modulation. Int. J. Refrigeration 30, 1427-1438.

[11] Soedel, W., Sound and vibrations of positive displacement compressors. 2007. CRC Press, 162-168.

[12] Furuhama, S., Tada, T., 1961. On the flow of the gas through the piston-rings (1st report, the discharge coefficient and temperature of leakage gas). Bulletin of JSME 4 (16), 684-690.

[13] Yang, B., Zhao, Y., 2011. Piston ring-cylinder liner lubrication analysis in a CO_2 refrigeration reciprocating compressor. Inst. of Mech. Eng., Part C: J. Mech. Eng. Sci., 225(11), pp. 2638-2648.

[14] Stachowiak, G.W., Batchelor, A.W., 2001. Engineering Tribology, 2nd ed. Butterworth Heinemann, Boston.

Comprehensive 3D-elastohydrodynamic simulation of hermetic compressor crank drive

S Posch[1], J Hopfgartner[1], E Berger[1], B Zuber[1], R Almbauer[1] and P Schöllauf[2]

[1] Graz University of Technology, 8010 Graz, Austria
[2] Secop Austria GmbH, 8280 Fürstenfeld, Austria

posch@ivt.tugraz.at

Abstract: Mechanical, electrical and thermodynamic losses form the major loss mechanisms of hermetic compressors for refrigeration application. The present work deals with the investigation of the mechanical losses of a hermetic compressor crank drive. Focus is on 3d-elastohydrodynamic (EHD) modelling of the journal bearings, piston-liner contact and piston secondary motion in combination with multi-body and structural dynamics of the crank drive elements. A detailed description of the model development within the commercial software AVL EXCITE Power Unit is given in the work. The model is used to create a comprehensive analysis of the mechanical losses of a hermetic compressor. Further on, a parametric study concerning oil viscosity and compressor speed is carried out which shows the possibilities of the usage of the model in the development process of hermetic compressors for refrigeration application. Additionally, the usage of the results in an overall thermal network for the determination of the thermal compressor behaviour is discussed.

Nomenclature

Roman symbols		Greek symbols	
h	fluid film thickness	μ	dynamic viscosity
p	pressure	ϕ_x, ϕ_y, ϕ_s	flow factors
t	time	ρ	density
u	velocity	σ_δ	standard deviation of the combined roughness
x, y	coordinate		

1. Introduction

The compressor is the core element of the cooling circuit of a refrigeration appliance. In addition to the improvement of the entire refrigeration cycle in terms of energy efficiency to meet the energy standards, the further development of the single components like the compressor is driven forward. The determination of the loss mechanisms namely electrical, mechanical and thermodynamic is indispensable in the development of innovative efficient hermetic reciprocating compressors. Due to the hermetic design of the compressors, the different loss mechanisms interact with each other and thus a holistic consideration of the overall thermal compressor behaviour is necessary. Several studies presented thermal compressor models trying to use as much physical information as possible. In [1] and [2], the authors combined 3d heat conduction formulation for solid parts and lumped formulation of the gas path with experimental data or correlations from the literature to model the heat transfer coefficients. A more detailed method is used in [3]. Additionally to the 3d heat conduction formulation of the solid parts the refrigerant gas path was simulated 3d using commercial CFD software. Mechanical and electrical losses were considered via heat sources in the particular part or in the lubrication oil. The determination of the mechanical losses in the compressor, mainly the journal bearing losses and the piston losses, has been published in several studies. A majority of the studies uses the similar general simulation strategy embedded in self coded simulation codes. The simulation strategy can be summarized in the following steps: (i) determination of the dynamic loads by applying multi body simulation, (ii) calculation of the hydrodynamic forces solving the Reynolds equation with numerical methods, (iii) calculation of displacements using a Newton-Raphson algorithm and (iv) calculation of the friction losses based on the displacements. A short overview of some studies and their statements concerning the modelling of the Reynolds equation and the model accuracy should be given at this point: [4] applied three different approaches to simulate the motion of the crank shaft and furthermore the journal bearing forces, whereby one approach takes into account lateral and tilting oscillations of the crank shaft. The modelling of the journal bearing forces was carried out applying the so-called short bearing approximation. The authors stated that no significant difference in the estimation of the journal bearing forces and the fluid film pressure between the different cases was found. A comparison between short and finite bearing approach can be found in [5]. As the authors concluded, the short bearing approach can be used as a first approximation to predict the bearing performance. In order to predict good results for piston cylinder contact, the finite bearing approach should be used. A more detailed calculation of the journal bearings of a hermetic compressor by using the averaged Reynolds equation introduced in [6] and [7] was presented in [8]. By applying the finite bearing approximation, the authors determined the friction power losses for the two journal bearings and investigated the influence of the oil viscosity and the gap width in the bearing. The method was verified by simple analytical friction loss calculations based on Couette shear stress.

The present study deals with the detailed calculation of the friction losses in a hermetic reciprocating compressor using the commercial software package AVL EXCITE Power Unit. An example of an application of the software can be found in [9]. The study analyses the piston secondary movement and its impact on the structure-borne noise behaviour of a direct injection Diesel engine. Although the software is usually used in engine development, the present study should show the ability of the software for the usage in reciprocating compressor investigations. The software is able to simulate the friction forces taking into account the elastic deformation of the crank drive parts. A general workflow for the use of AVL EXCITE Power Unit in the simulation of the mechanical losses of a hermetic reciprocating compressor is given and the model is utilised to investigate the influence of compressor speed and oil viscosity.

The compressor used in the present study works with R600a (isobutane) and contains a displacement of 5.5 cm^3. Its COP at ASHRAE test conditions (-23.3 °C/55 °C) is approximately 1.8.

2. Simulation

The software AVL EXCITE Power Unit is an advanced tool for the simulation of rigid and flexible multi-body dynamics for the application in combustion engine and drivetrain development. Due to the similar structure of reciprocating compressors and combustion engines, the usage of the software for the calculation of mechanical losses in refrigeration compressors is useful. An overview of the workflow for setting up the simulation of the compressor crank drive with AVL EXCITE is given in this chapter. Furthermore, the pre-processing steps concerning condensation of the compressor parts and the theoretical background of the hydrodynamic force modelling are given.

2.1 Layout

An illustration of the simulation layout within the used software can be seen in Figure 1. The model is built up with subsystems representing several compressor parts connected with relevant types of joints. In the present study, a mechanical loss calculation of the overall compressor crank drive is carried out. Therefore, all crank drive parts and the crankcase are considered. Additionally, detailed modelling of the friction behaviour in the journal bearings of the crank shaft, connecting rod big end and the piston-liner with elastohydrodynamic (EHD) contacts is carried out. Joints with little influence on the overall mechanical losses like the piston-piston pin and the connecting rod-piston pin connection are modelled with simple revolution joints by setting linear or non-linear spring/damper functions [10].

Geometric input data of the crank drive enable the calculation of the compressor dynamics via multi body simulation.

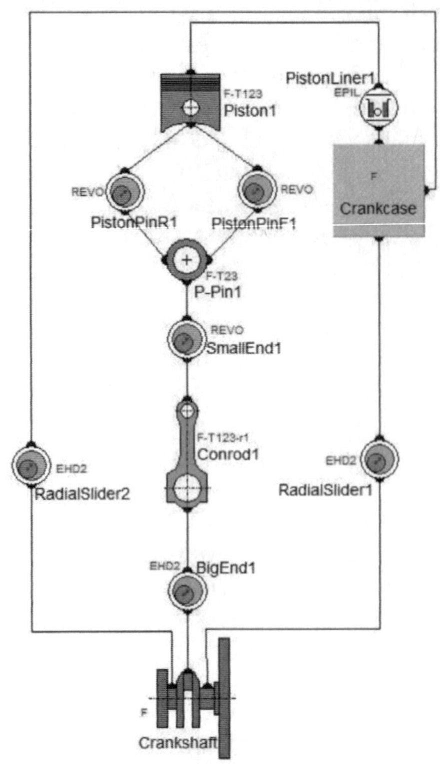

Figure 1. Layout of the simulation model.

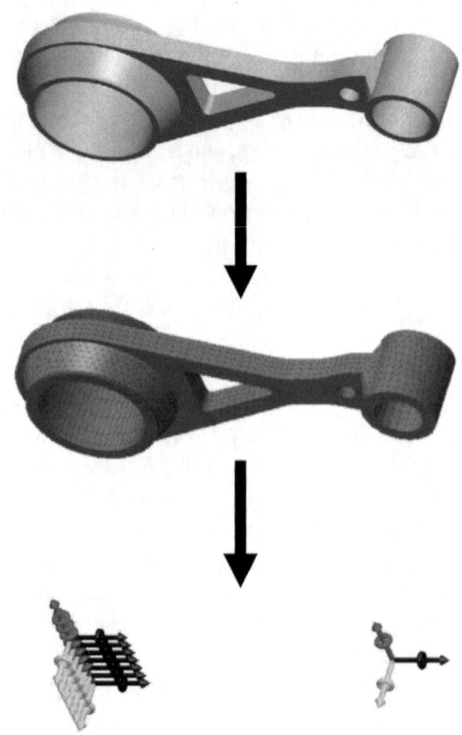

Figure 2. Condensation process.

2.2 Pre-processing

EHD simulation requires detailed information about the elastic behaviour of the considered parts. In the pre-processing phase of the simulation, the body input data for the use in AVL EXCITE Power Unit is generated. Starting from the 3d-CAD model, a finite element model with defined degrees of freedom is generated in the commercial software SIMULIA Abaqus. A predefined interface between the two software packages ensures the implementation of the input data in correct syntax. Figure 2 shows exemplarily the condensation process of the connecting rod. At the big end of the connecting rod seven node points with defined degrees of freedom are created which have to be aligned with the appropriate nodes on the crank shaft. Due to the simple modelling of the connecting rod small end, the part is condensed to only one node with defined degrees of freedom.

2.3 Simulation handling

After the several compressor parts are loaded into the software and are linked by appropriate contacts, the compressor load must be defined. In the present study, a pressure profile gained by CFD simulation is used. The CFD simulation was carried out in the commercial software ANSYS Fluent at the previous mentioned ASHRAE test conditions modelling the valve motions by the use of ideal moving

IOP Conf. Series: Materials Science and Engineering **232** (2017) 012008 doi:10.1088/1757-899X/232/1/012008

flat plates. Assuming equal pressure profile for every crankshaft revolution, the pressure is set to be periodic. Depending on the surface contact model, the roughness values of the several bearing partners must be specified and the material properties are necessary. Since the software AVL EXCITE Power Unit is mainly used for the simulation of combustion engines, the oil database does not contain the used compressor oil, therefore, the oil properties must be specified by the user. Gravitational forces are neglected to avoid the implementation of axial bearings.

2.4 Lubrication theory

The calculation of the hydrodynamic forces is carried out by solving the average Reynolds equation introduced in [6] and [7] which links the hydrodynamic pressure to the fluid film thickness.

$$\frac{\partial}{\partial x}\left(\phi_x \frac{\rho\, h^3}{12\mu}\frac{\partial \overline{p_h}}{\partial x}\right)+\frac{\partial}{\partial y}\left(\phi_y \frac{\rho\, h^3}{12\mu}\frac{\partial \overline{p_h}}{\partial y}\right)=\frac{(u_I+u_{II})}{2}\left(\frac{\partial\left(\rho \overline{h_T}\right)}{\partial x}+\sigma_\delta\frac{\partial \phi_S}{\partial x}\right)+\frac{\partial\left(\rho \overline{h_T}\right)}{\partial t} \tag{1}$$

The terms in the average Reynolds equation can be interpreted as follows: the term on the left-hand side represents the Poiseuille flow, the terms on the right-hand side represent the Couette flow and the time dependent displacement flow, respectively. Compared to the original Reynolds equation, several flow factors which depend on the roughness properties of the surfaces are introduced. Additionally to the hydrodynamic contact, an elastic-plastic asperity contact can occur which is considered with the model presented in [11]. The oil supply of the regarded contacts is not modelled specially, therefore, the contacts are assumed to be filled with oil. To fulfil the continuity balance, a cavitation approach similar to the model in [12] is used in the software.

3. Results

Before presenting the results of the comparison in terms of friction power loss, results of the predefined standard case are shown. The standard case is defined as the previous described compressor working in steady-state mode for R600a at a suction pressure of 0.629 bar and a discharge pressure of 6.04 bar corresponding to -23.3 °C evaporating temperature and 45 °C condensing temperature, respectively. Compressor speed is set to 3000 rpm and the oil viscosity is assumed to be constant at 8 cSt. The usage of the commercial software AVL EXCITE Power Unit enables easy post-processing. A visualisation of the hydraulic pressure in the regarded EHD contacts is shown in Figure 3. It can be seen that the region in the piston-liner contact shows two peaks in the hydrodynamic pressure field, indicating increased wear in this region.

IOP Conf. Series: Materials Science and Engineering **232** (2017) 012008 doi:10.1088/1757-899X/232/1/012008

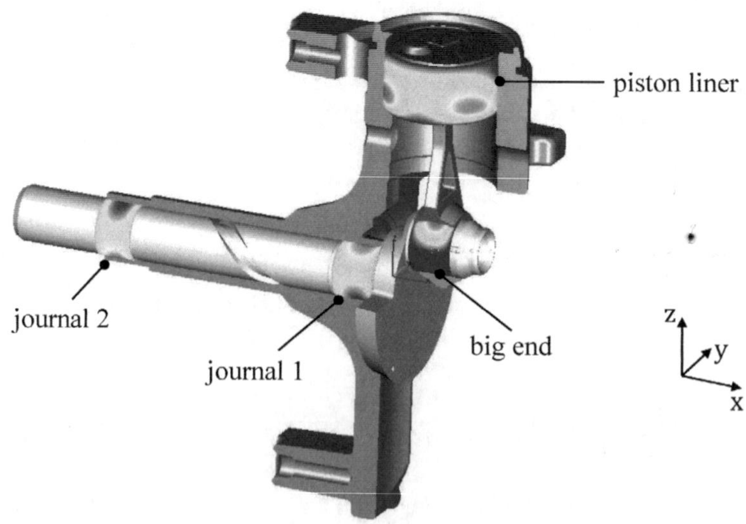

Figure 3. Visualisation of the hydraulic pressure in the EHD contacts.

The simulation of the standard case is carried out until steady-state conditions between two consecutive time steps are reached. For the standard case five crankshaft revolutions are needed to reach steady-state conditions yielding in a computational time of 20 hours on a quad core 3.50 MHz processor.

Figure 4 and Figure 5 illustrate results in terms of the reaction forces in the particular EHD contacts for one crankshaft revolution at steady-state conditions. The force curves reflect the crank mechanism behaviour and show the high forces at the connecting rod big end joint. Furthermore, the reaction forces at the piston liner contact and at the connecting rod big end joint have approximately sinusoidal character. In contrast, the amplitude reaction forces at the journal bearings of the crankshaft are smaller compared to the connecting rod big end and the curves do not show sinusoidal character.

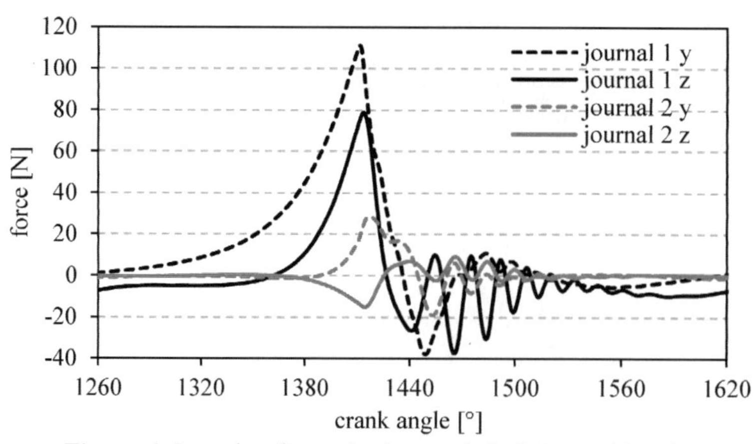

Figure 4. Reaction forces in the crankshaft journal bearings.

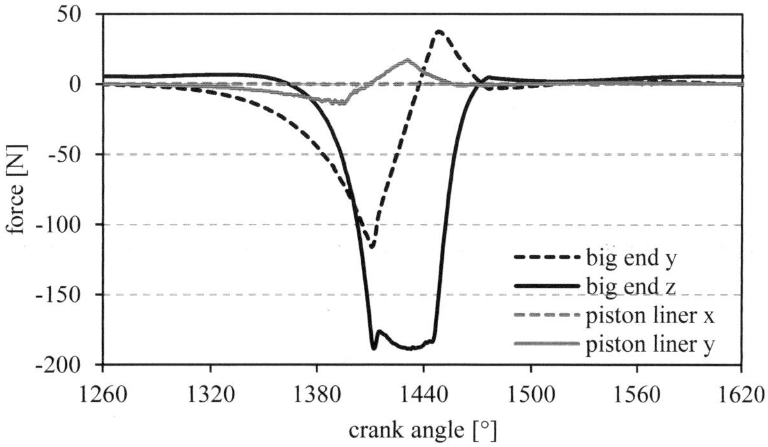

Figure 5. Reaction forces in the connecting rod big end and the piston liner contact.

Simulation results of the body kinematics can be seen in Figure 6 and Figure 7. The combination of piston pin offset and partly one-sided piston side load yields in tilting angle curve according Figure 6. As the orbital paths of the crankshaft journal bearings show, the displacement of the crankshaft at journal bearing 2 is smaller compared to bearing 1 due to the lower force amplitude acting on it. Furthermore, the higher loads according to Figure 4 at bearing 1 lead to a not perfectly cyclical orbital path. The lower force amplitude also yields in an almost circular displacement curve of journal bearing 2 and furthermore, to a faster achievement of the final shape of the orbital path compared to bearing 1.

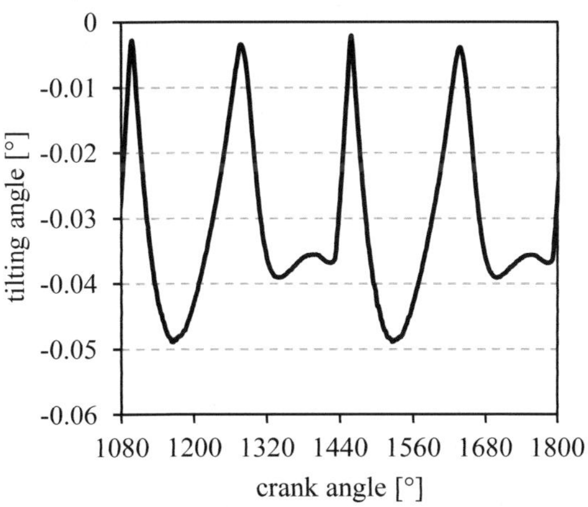

Figure 6. Piston tilting angle.

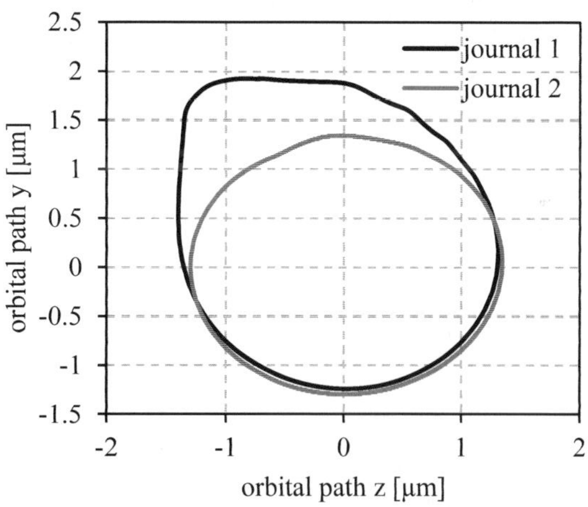

Figure 7. Orbital path of the crankshaft journal bearings.

The investigation of the friction loss behaviour of the considered joints is carried out by varying the oil viscosity and the rotational speed of the crankshaft. Figure 8 and Figure 9 illustrate the calculated friction power loss according to their occurrence. It is assumed, that the cylinder pressure gained by CFD results can be obtained for every variant. The viscosity variation (at constant speed of 3000 rpm) shows almost linear dependency of the friction power loss in the several joints and thus the overall friction loss.

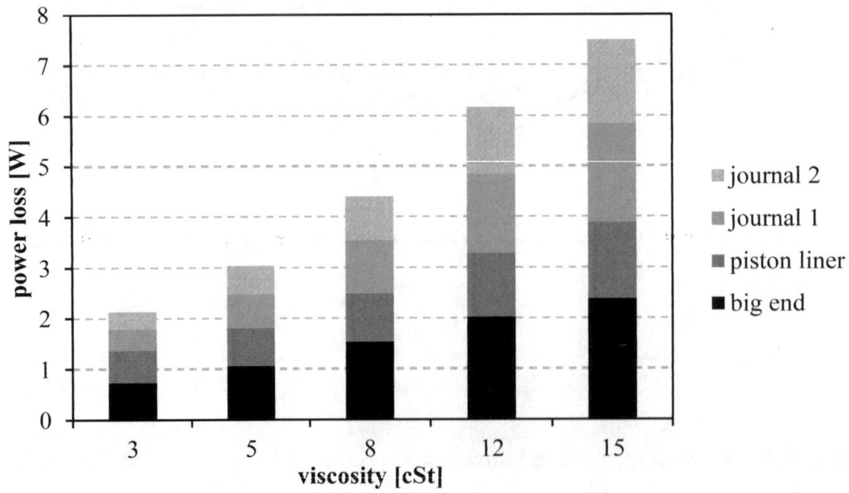

Figure 8. Viscosity influence on the friction power loss.

Similar results are obtained for the variation of the compressor speed (at constant viscosity of 8 cSt). The comparison in Figure 9 shows an almost linear dependence between compressor speed and friction power loss. Another aspect which can be derived from the simulation results is the big influence of the reaction forces on the friction loss compared to the relative speed in the regarded bearing. By comparing the results of the connecting rod big end and the crankshaft journal bearings, the significant influence of the reaction forces can be seen. The share of the power loss in the connecting rod big end joint on the overall power loss is higher compared to the individual crankshaft journal bearings although the relative speed between the bearing gap walls in the connecting rod big end is smaller compared to the relative speed in the crankshaft journal bearings.

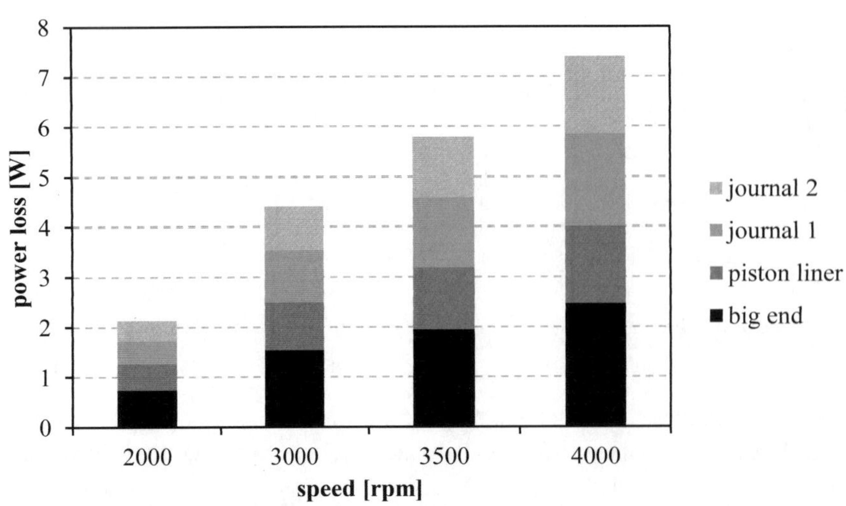

Figure 9. Rotational speed influence on the friction power loss.

Additionally to the investigation of the compressor crank mechanism dynamics and kinematics, the obtained results can be used in thermal modelling of the compressor. The results can serve as input data for modelling approaches like presented in [12]. By taking into account the produced heat due to friction in the individual joint and considering the heating of the oil, the overall accuracy of such models can be increased.

4. Conclusion

The presented paper deals with the investigation of the friction behaviour of a reciprocating hermetic compressor for refrigeration application. For this purpose a 3d-elastohydrodynamic simulation using the commercial software AVL EXCITE Power Unit was carried out. Focus was laid on the simulation of the connections where the highest friction losses occur which are the two crankshaft journal bearings, the connecting rod big end and the piston liner contact. The simulation model was used to determine the dynamics and kinematic behaviour of the compressor crank mechanism. Furthermore, the influence of the oil viscosity and the compressor speed on the friction power loss was investigated. Based on a predefined standard configuration (3000 rpm, 5 cSt) the friction power loss according to their occurrence and the overall friction power loss of the several configurations were compared. The results show an approximately linear dependency of rotational speed and oil viscosity on the friction loss. It can be pointed out that the connecting rod big end joint produces higher friction losses compared to the other connections due to the high reaction forces although the relative speed between the journal walls is quite small.

The gained results can be used to estimate the usage possibility of the regarded crank mechanism design in a variable speed compressor. According to the presented results, the friction power loss at lower compressor speeds will decrease but this effect can be cancelled out due to the lower temperature in the compressor at lower speeds yielding in higher oil viscosities and therefore in increasing friction losses. This circumstance shows a trade-off in the design of variable speed compressors. The choice of the lubrication oil usually depends on the compressor modes in which the overall compressor temperature is high like in the refrigerator pull-down mode. By linking simulation models of the friction behaviour and thermal compressor models, a holistic approach of modelling the oil influence on the compressor performance can be achieved.

Acknowledgements

This work has been carried out within the framework of ECO-COOL, a research project initiated and funded by the FFG (Austrian Research Promotion Agency). Furthermore the authors particularly acknowledge the technical support by Secop Austria GmbH, formerly ACC Austria GmbH and Liebherr-Hausgeräte Lienz GmbH.

References

[1] Sanvezzo J Jr and Deschamps C J 2012 A Heat Transfer Model Combining Differential and Integral Formulations for Thermal Analysis of Reciprocating Compressors *Proc. Int. Compr. Eng. Conf. at Purdue* **2104**

[2] Lohn S K, Diniz M C and Deschamps C J 2015 A Thermal Model for Analysis of Hermetic Reciprocating Compressors und the On-Off Cycling Operation Condition *Proc. Int. Conf. Compr. and their Sys.(London)* **012068**

[3] Posch S, Hopfgartner J, Heimel M, Berger E, Almbauer R and Stangl S 2016 Thermal Analysis of a Hermetic Reciprocating Compressor using Numerical Methods *Proc. Int. Compr. Eng. Conf. at Purdue* **1215**

[4] Estupinan E A, Santo I F 2009 Modelling Hermetic Compressors Using Different Constraint Equations to Accommodate Multibody Dynamics and Hydrodynamic Lubrication *Journal of the Braz. Soc. of Mech. Science and Eng.* **31** 35-46

[5] Chieh H, Couto P R C and Prat A T 2007 An Integral Analysis and Simulation of the Complete Bearing System in Reciprocating Hermetic Compressors *Proc. Int. Conf. Compr. and their Sys.(London)* **658054**

[6] Patir N and Cheng H S 1978 An average model for determining effects of three-dimensional roughness on partial hydrodynamic lubrication *Journal of Lubr. Tec.* **100** 12-17

[7] Patir N and Cheng H S 1979 Application of average flow model to lubrication between rough sliding surfaces *Journal of Lubr. Tec.* **101** 220-230

[8] Posch S, Hopfgartner J, Heimel M, Berger E, Almbauer R and Schöllauf P 2016 A Numerical Friction Loss Analysis of the Journal Bearings in a Hermetic Reciprocating Compressor *Proc. Int. Compr. Eng. Conf. at Purdue* **1214**

[9] Seifriedsberger J, Wichtl R and Eichlseder H 2016 3d-Elastohydrodynamic Simulation Model for Structure-Borne Noise Analysis of a DI Diesel Engine *SAE Technical Paper* **2016-01-1854**

[10] AVL List GmbH 2011 User Guide – AVL EXCITE Power Unit *AVL List GmbH* **2011.1**

[11] Greenwood J A and Tripp J H 1970 The contact of two nominally flat rough surfaces *Proc. of the Inst. of Mech. Sc. and Eng.* **31** 35-46

[11] Elrod H G 1981 Cavitation Algorithm *Journal of Lubr. Techn.* **103** 350-354

[12] Posch S, Hopfgartner J, Heimel M, Berger E, Almbauer R and Schöllauf S 2016 A Numerical Investigation of the Oil Pump Suction Behaviour in a Hermetic Reciprocating Compressor *Proc. Int. Compr. Eng. Conf. at Purdue* **1215**

IOP Conf. Series: Materials Science and Engineering **232** (2017) 012009 doi:10.1088/1757-899X/232/1/012009

Determination of the oil distribution in a hermetic compressor using numerical simulation

S Posch[1], J Hopfgartner[1], E Berger[1], B Zuber[1], R Almbauer[1] and P Schöllauf[2]

[1] Graz University of Technology, 8010 Graz, Austria
[2] Secop Austria GmbH, 8280 Fürstenfeld, Austria

posch@ivt.tugraz.at

Abstract: In addition to the reduction of friction the oil in a hermetic compressor is very important for the transfer of heat from hot parts to the compressor shell. The simulation of the oil distribution in a hermetic reciprocating compressor for refrigeration application is shown in the present work. Using the commercial Computational Fluid Dynamics (CFD) software ANSYS Fluent, the oil flow inside the compressor shell from the oil pump outlet to the oil sump is calculated. A comprehensive overview of the used models and the boundary conditions is given. After reaching steady-state conditions the oil covered surfaces are analysed concerning heat transfer coefficients. The gained heat transfer coefficients are used as input parameters for a thermal model of a hermetic compressor. An increase in accuracy of the thermal model with the simulated heat transfer coefficients compared to values from literature is shown by model validation with experimental data.

Nomenclature

Roman symbols

c_p	specific heat capacity
$\boldsymbol{f}_b$	body force vector
p	pressure
$\dot{q}_V$	rate of heat source/sink per volume
t	time
T	temperature
$\boldsymbol{v}$	velocity vector

Greek symbols

λ	thermal conductivity
μ	dynamic viscosity
ρ	density
Φ	viscous dissipation term

1. Introduction

The use of simulation in different fields of engineering has been increased in the last two decades due to the new possibilities of applying numerical methods in combination with raising computational performance. Additionally to the early pioneers of using simulation in the development process like the automotive or aeronautic industry the decreasing costs allow the use of simulation tools in other fields of mechanical engineering. As an example, the traditionally experimentally driven development of hermetic reciprocating compressors for refrigeration appliances can be mentioned. To meet future regulations concerning the energy consumption of refrigeration devices comprehensive research on the level of the overall cooling cycle and on the component level is needed. Especially for the hermetic compressor which serves as the core element of a refrigeration cycle, considerable efforts have to be made to find energy saving potentials. The development of innovative compressors requires the knowledge of its thermal behaviour and occurring loss mechanisms. Due to the hermetic shell an individual consideration of the electrical, mechanical and thermodynamic losses is not expedient. Electrical and mechanical losses produce heat which influences the thermodynamic behaviour of the entire compressor. Several studies have been published and can be found in the open literature which deal with the thermal modelling of hermetic reciprocating compressors. The following survey of thermal compressor models should give an overview without any claim of completeness.

Generally, thermal compressor models can be characterized due to their complexity and geometrical resolution. Studies presented in [1] or [2] split up the compressor in several lumped volumes representing the appropriate compressor part. Using the first law of thermodynamics the temperature field is calculated. Heat transfer coefficients are determined either by simple correlations or experimental data. These studies represent the early stage of thermal compressor modelling and show the limitations due to low computational performance in the late 80s and early 90s. The models are only able to give a rough estimation of the temperature field.
The Thermal Network approach (TNW) can be considered as more flexible compared to the previous described models. Examples of TNW for thermal compressor modelling can be found in [3] and [4]. Using the Lumped Conductance Method the heat transfer between mass points representing the several compressor parts is modelled. The authors calculate the convective heat transfer with correlations for forced or natural convection based on the Nusselt number which can cause uncertainties especially in regions with transient flow like in the suction or discharge muffler. Low geometric resolution is also critical for the validation of the models using experimental data due to significant temperature differences in parts with a low thermal conductivity.

Modern thermal models integrate Computational Fluid Dynamics (CFD) resulting in so called hybrid models. Hybrid models combine 3d numerical simulation (e.g. solid part conduction) and simple correlation (e.g. convective heat transfer) in one model to gain high geometrical resolution by keeping the calculation time at acceptable limits. Examples for hybrid models using 3d heat conduction of the solid parts in combination with lumped volume formulation for the compressor gas path can be found in [5], [6] and [7]. The modelling of the heat transfer in these models was carried out either by using experimental data or correlations from the literature. In [8] special care was laid on the cylinder of the compressor. The authors combined 1d flow simulation of the compressor gas line, 3d formulation of the compressor cylinder and lumped formulation of the remaining compressor parts.

A more detailed hybrid model using 3d formulation of the full gas line and 3d formulation of the solid components can be found in [9]. The authors introduced a simulation algorithm to combine transient flow calculation with steady-state heat conduction calculation. Heat transfer coefficients calculation between the oil and the oil covered walls is carried out with correlations from literature. The authors remark that the results of thermal compressor models should not be trusted without considering the modelling of the heat transfer coefficients especially between oil and oil covered walls. Generally, considerable influence on the thermodynamic compressor behaviour is caused by the lubrication oil. Additionally to the lubrication of the moving parts and reduction of wear losses, the oil acts as heat transfer media between the hot compressor parts, the shell and hence the ambience. Based on the thermal compressor model presented in [9] the present study increases the accuracy of the model by calculating the heat transfer coefficients between oil and compressor shell applying 3d CFD. Therefore, a simulation of the oil distribution inside the compressor shell is performed in the commercial software ANSYS Fluent. The two phase flow (oil/refrigerant) is modelled with the Volume of Fluid (VOF) approach. A comparison between the results of the thermal model with heat transfer gained with correlations from the literature and by CFD, respectively, is carried out and the results are validated with experimental data of a calorimeter test bench.

The compressor used in the present study works with R600a (isobutane) and has a displacement of 5.5 ccm at a rotational speed of 2950 rpm. Its COP at ASHRAE test conditions (-23.3 °C/55 °C) is approximately 1.8.

IOP Conf. Series: Materials Science and Engineering **232** (2017) 012009 doi:10.1088/1757-899X/232/1/012009

2. Modelling

An overview of the used CFD methods and solution strategy is given in the following chapter. The fluid flow of the distributed oil is assumed to be a continuous jet which can also be seen in experimental tests using a transparent compressor shell and a stroboscope light, therefore the volume of fluid approach (VOF) is used for fluid flow simulation.

2.1. Theoretical background

The basic idea of CFD is the transformation of governing equations from the partial differential equation form into an algebraic form using finite volume discretization. For the present study the governing equations for mass, momentum and energy are essential [10]:

$$\frac{\partial \rho}{\partial t} + \nabla \cdot [\rho \boldsymbol{v}] = 0 \tag{1}$$

$$\frac{\partial}{\partial t}[\rho \boldsymbol{v}] + \nabla \cdot \{\rho \boldsymbol{v}\boldsymbol{v}\} = -\nabla p + \mu \nabla^2 \boldsymbol{v} + \boldsymbol{f}_b \tag{2}$$

$$\frac{\partial}{\partial t}[\rho c_p T] + \nabla \cdot \{\rho c_p \boldsymbol{v} T\} = \nabla \cdot \{\lambda \nabla T\} + \frac{\mathrm{D}p}{\mathrm{D}t} + \mu \Phi + \dot{q}_V \tag{3}$$

Commercial software packages like ANSYS Fluent discretize the governing equations automatically on the regarded flow domain and solves the system of equations by means of appropriate numerical methods. The modelling of the oil flow in the gaseous environment inside the compressor shell, the previous mentioned VOF approach presented in [11] is used thus an additional value, the volume fraction χ is introduced. It describes the volumetric amount of oil in a finite volume of the solution domain. The use of χ requires the solution of an additional governing equation and the volume fraction is used to determine density and viscosity of a partial filled finite volume by applying the following equations:

$$\frac{\partial \chi}{\partial t} + \nabla \cdot [\chi \boldsymbol{v}] = \boldsymbol{0} \tag{4}$$

$$\rho = \rho_g \chi + \rho_l (1 - \chi) \tag{5}$$

$$\mu = \mu_g \chi + \mu_l (1 - \chi) \tag{6}$$

For more details about the VOF approach the interested reader is referred to [11] and [12].

2.2. Solution domain and computational settings

The gas volume inside the compressor shell is meshed for the present simulation with approximately 700.000 mainly tetrahedral cells with hexahedral cells adjacent to the shell wall. To keep the number of cells in reasonable boarders regions which are covered by compressor parts and thus no oil is transported are left out. Moving mesh techniques like layering are taken to model the piston and crankpin movement to consider the induced secondary movement of the gas in the shell which influences the oil jet.

Oil mass flow rate at the oil pump outlet is determined by simulating the lower pump geometry according to [13] assuming constant oil mass flow rate in the rest of the pump and thus at the pump outlet. The properties of the oil are assumed to be constant over the entire simulation. Values for density and viscosity are set to 832 kg m^{-3} and 6.7 x 10^{-3} Pa s, respectively. Surface tension is also set constant at 0.028 N m^{-1}. Additionally, the following assumptions are made for the simulation of the oil distribution:

- Wall temperatures are constant.
- Flow is isothermal.

- Laminar flow.
- Dissolving of gaseous phase in liquid phase is neglected.

The oil distribution simulation is carried out with the following solver settings:

- free surface model: implicit VOF
- time discretization: bounded second order implicit, time step 10^{-4} s
- spatial discretization: second order upwind for momentum [10], compressive scheme for volume fraction [12]
- pressure-velocity coupling: PISO [14]
- double precision mode

3. Results

The presentation of the results is divided into the individual work steps until the oil distribution results can be used in the thermal compressor model.

3.1. Oil flow distribution

The simulation of the oil flow distribution is carried out until steady-state conditions are reached. In the present work steady-state means that the difference in the absolute oil mass flow rate into the compressor oil sump between two time steps becomes approximately constant which is obtained after 32 crankshaft revolutions. Illustrations of the simulation results with the starting oil flow can be seen in Figure 1.

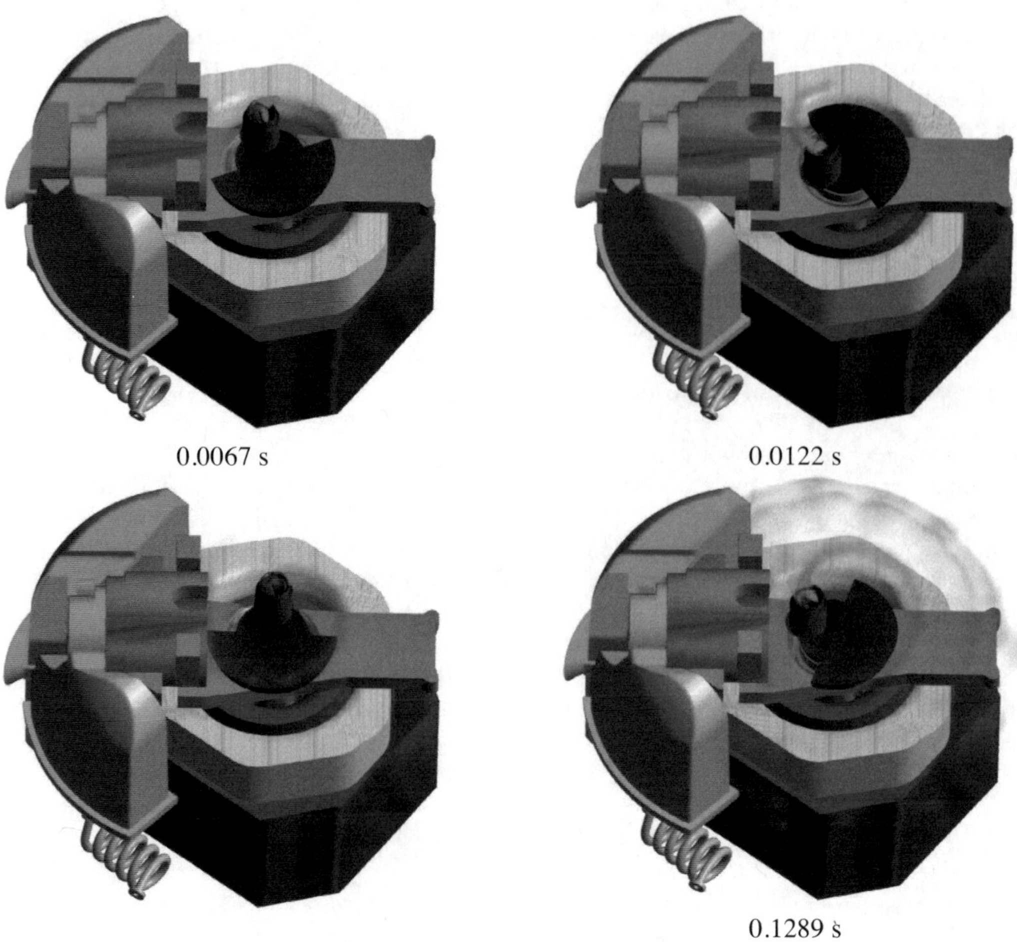

0.0244 s

Figure 1. Oil distribution inside the compressor shell at several time steps.

The oil film thickness as a result of the oil distribution simulation can be seen in Figure 2 (only a characteristic area is shown to enable a proper overview). Figure 2a shows the iso-surface of the oil volume fraction and Figure 2b the oil film thickness value. The oil film thickness is not constant along the fall film flow direction which can be explained by the periodic impingement of the oil. By the use of an user defined function (UDF) the mean oil film thickness of 1.4 mm was calculated.

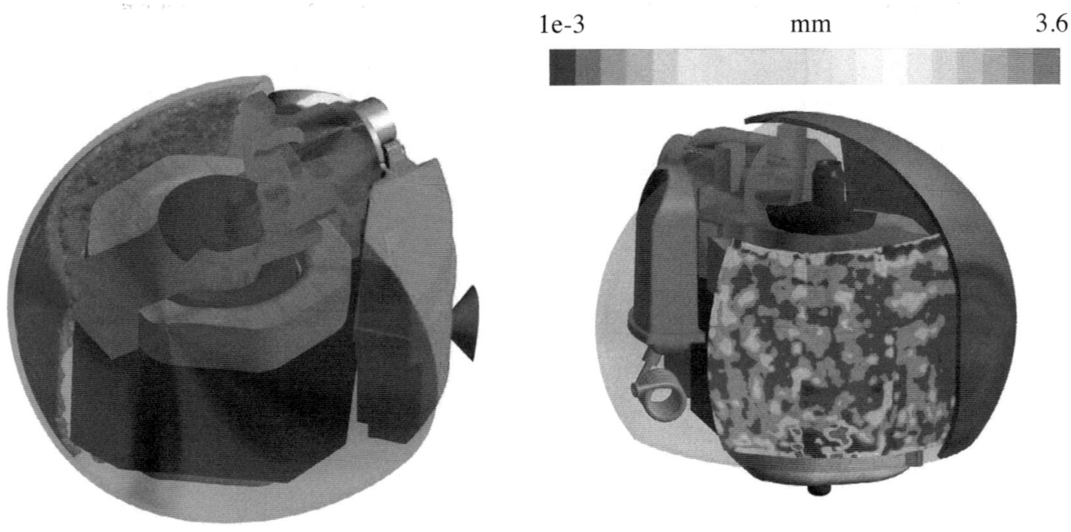

(a) (b)

Figure 2. Oil film (a) and oil film thickness (b).

3.2. Heat transfer coefficient

The simulation of the oil distribution inside the compressor and furthermore on the compressor shell enables the determination of heat transfer coefficients on the oil covered shell walls. Additionally to the main task of lubricating the moving compressor parts the oil acts as the main heat transfer media between the hot compressor parts and the compressor shell and thus to the ambience. High heat transfer coefficients result in increased heat transfer and improved compressor performance by decreasing the overall compressor temperature.

Results of the determination of the heat transfer coefficient on the oil covered shell walls can be seen in Figure 3a. Areas on which the oil jet hits the walls show higher heat transfer coefficients compared to areas where the heat transfer is affected by the falling oil film.

To enable the use of the oil distribution simulation results in the thermal model, the oil covered wall of the compressor shell is divided into 24 elements. The division of the area in circumferential direction is carried out with eight equidistant fields; in vertical direction the area is divided in three fields whereby the middle field represents the shell area where the oil jet mainly hits the shell wall. Figure 3b shows the distribution of the heat transfer coefficient on the shell after the interpolation process.

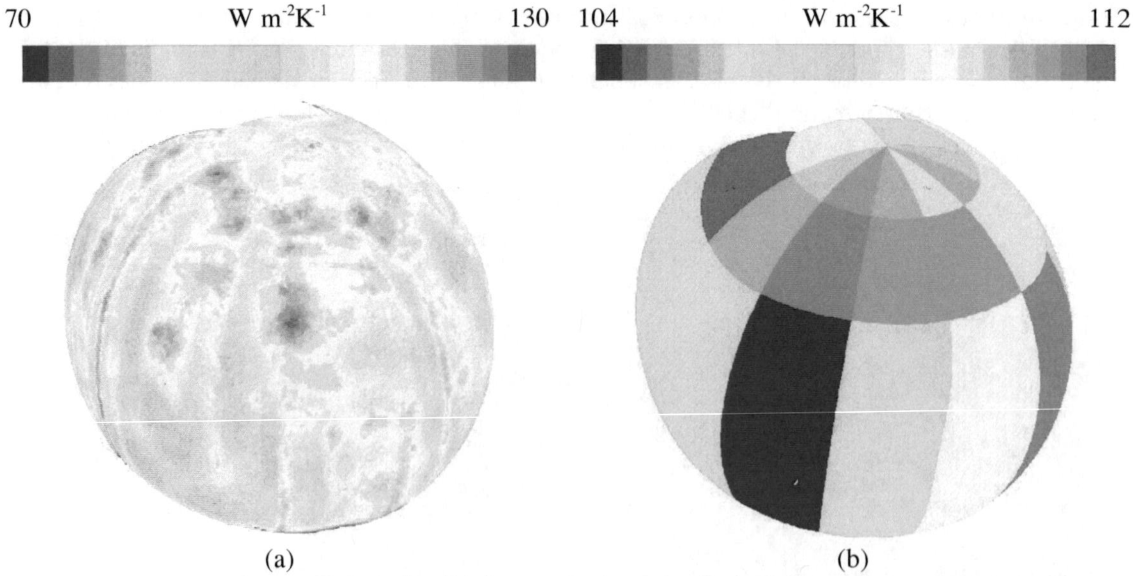

Figure 3. Heat transfer coefficient distribution on the inner surface of the compressor shell before (a) and after (b) the interpolation.

3.3. Thermal model

The determination of the temperature field in the hermetic reciprocating compressor using the thermal modelling approach published in [9] is carried out with the heat transfer values gained either by oil distribution CFD results and correlations from the literature. To value the prediction accuracy of the two methods the simulation results are compared with temperature measurements on twelve significant measuring points. The positon of the measuring points in the compressor can be seen in [9]. For the validation of the simulation the experimental data are gained in a calorimeter test bench with -23.3 °C evaporating temperature, 45 °C condensing temperature and 32 °C ambient temperature, respectively.

The results of the comparison between the simulation results and the experiment can be seen in the diagram in Figure 4 (the temperatures are referred to the temperature of the shell gas in the experiment). As the diagram shows, the temperatures of solid parts inside the compressor, shell gas and oil sump can be predicted accurately. Due to the generally lower heat transfer coefficients gained by literature correlations compared to those determined by CFD, the heat transfer between the hot compressor parts to the ambience is smaller. So the literature correlation case results in higher temperatures of the inner solid parts compared to the case with CFD heat transfer values. The highest deviation between experiment and simulation with CFD oil distribution regarding the inner compressor parts occurs at the discharge muffler surface with a value of 4.2 K. Using similar heat transfer coefficients between compressor shell and ambient air, the temperature deviation of the shell between the two simulation cases is very small which shows the significant influence of the heat transfer coefficient on the shell outer surface. Generally, the deviation between shell temperatures determined by simulation and experiment is higher compared to the deviation of the inner compressor parts. The maximum temperature deviation occurs at the top of the shell with 4.6 K. As the diagram in Figure 3 illustrates, the prediction accuracy is improved using heat transfer coefficients of the compressor oil gained by CFD compared to correlations of the literature.

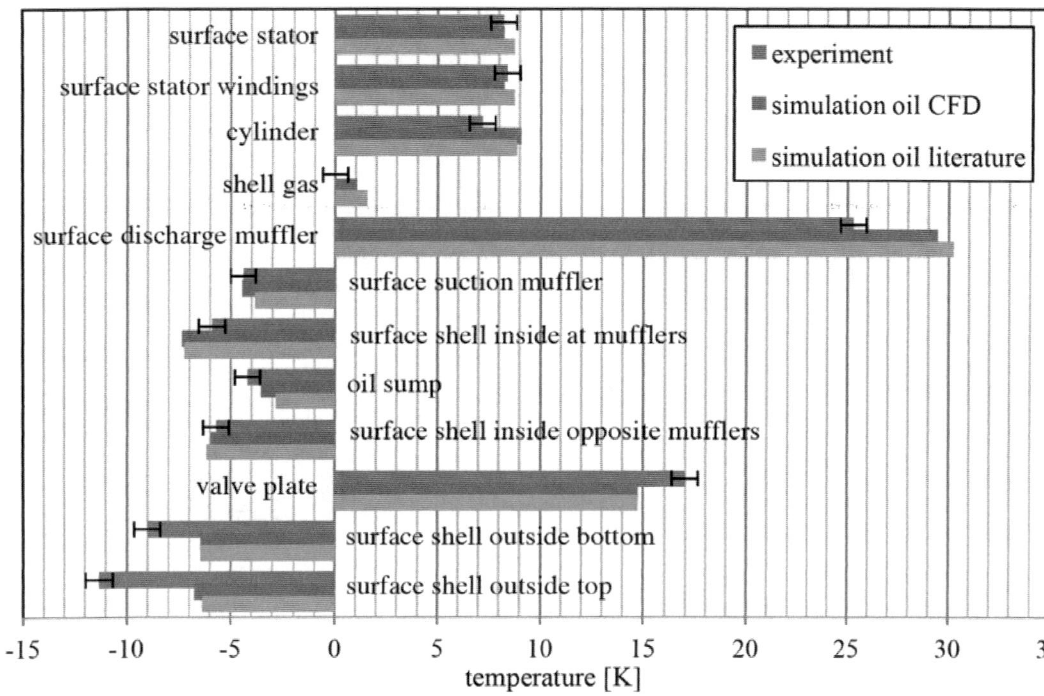

Figure 4. Comparison between simulation results and experiment (related to shell gas experiment temperature).

Figure 5 shows the results of the simulation with CFD heat transfer correlations for the compressor oil in terms of solid parts temperature field. The highest temperature values occur at the discharge muffler and not as often expected at the valve plate which can be explained which can be explained by the relatively long duration of the hot refrigerant gas in the discharge muffler and the cooling of the valve plate by the suction gas. As the shell temperature field shows, the temperature values at the oil covered walls are higher than in the vicinity of the mufflers where the walls are not covered with oil.

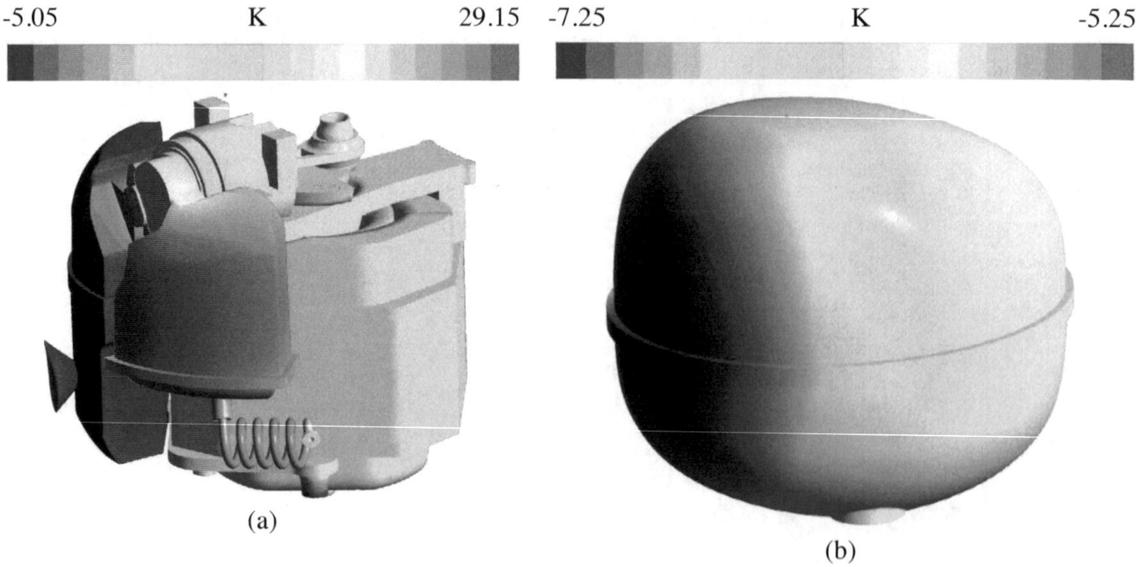

Figure 5. Solid part temperature field (related to shell gas experiment temperature) inside the compressor (a) and on the compressor shell (b).

The accuracy of the temperature determination of the inner compressor parts could be increased by introducing a higher number of zones with different heat flux values at parts with significant temperature gradients which is the case for the suction and discharge muffler. Furthermore, the use of temperature dependent heat transfer coefficients in combination with several zones at the shell outer surface could increase the accuracy of the thermal model.

An important aspect which should be mentioned at this point, is the application of the present model to different operating conditions. If the compressor load is increased and thus the temperature field inside the compressor rises also the oil and shell temperature increase. Assuming similar temperature difference between oil and shell, the determined heat transfer coefficients could be used for other operating points. More accurate results at other operating points can be gained by carrying out the simulation procedure again.

Due to the investigation of a constant speed compressor the oil distribution at several speeds is not part of the present study. Since the oil transport is highly influenced by the rotational speed of the compressor, the oil distribution would be different and thus the heat transfer.

4. Conclusion

In the presented study the simulation of the oil distribution in a hermetic reciprocating compressor for refrigeration application using commercial CFD software ANSYS Fluent is shown. The simulation was carried out by applying the VOF method to calculate the oil jet flow in the refrigerant environment inside the compressor shell. Furthermore, the CFD results were analysed in terms of heat transfer coefficient between the compressor oil and the oil covered shell walls. The calculated heat transfer coefficients were then used in a thermal model of the entire compressor according to [9]. A comparison of the determined temperature fields with the results of a simulation with heat transfer values gained by literature correlations was carried out and furthermore, the prediction accuracy of the simulations was analysed by a validation with experimental results of a calorimeter test bench. The maximum temperature deviation between simulation results with CFD heat transfer coefficients and experiment was 4.6 K which shows generally good agreement. Due to the lower heat transfer coefficients of literature correlations, the temperature level inside the compressor is predicted too high compared to the CFD method. Although the temperature prediction of the case using CFD heat transfer coefficients for oil covered shell walls is better compared to the case using literature correlations, the application of this approach has to be reviewed closely in terms of benefit and time. Due to the high computational effort for the CFD simulation this method can only be applied in research and not in the development phase of a compressor. Additionally, the fact that the heat transfer coefficients between solid parts and shell gas were still determined with literature correlations gives scope for further improvements.

Acknowledgements
This work has been carried out within the framework of ECO-COOL, a research project initiated and funded by the FFG (Austrian Research Promotion Agency). Furthermore the authors particularly acknowledge the technical support by Secop Austria GmbH, formerly ACC Austria GmbH and Liebherr-Hausgeräte Lienz GmbH.

References

[1] Todescat M L, Fagotti F, Prata A T and Ferreira R T S 1992 Thermal Energy Analysis in Reciprocating Hermetic Compressors *Proc. Int. Compr. Eng. Conf. at Purdue* **936**

[2] Meyer W A and Thompson E 1998 An Analytical Model of Heat Transfer to the Suction Gas in a Low-Side Hermetic Refrigeration Compressor *Proc. Int. Compr. Eng. Conf. at Purdue* **662**

[3] Ooi K T 2003 Heat Transfer Study of a Hermetic Refrigeration Compressor *Appl. Thermal Eng.* **23** 1931-1945

[4] Sim Y H, Youn Y and Min M K 2000 A Study on Heat Transfer and Performance Analysis of Hermetic Reciprocating Compressor for Refrigeration *Proc. Int. Compr. Eng. Conf. at Purdue* **1390**

[5] Sanvezzo J Jr and Deschamps C J 2012 A Heat Transfer Model Combining Differential and Integral Formulations for Thermal Analysis of Reciprocating Compressors *Proc. Int. Compr. Eng. Conf. at Purdue* **2104**

[6] Lohn S K, Diniz M C and Deschamps C J 2015 A Thermal Model for Analysis of Hermetic Reciprocating Compressors und the On-Off Cycling Operation Condition *Proc. Int. Conf. Compr. and their Sys.(London)* **012068**

[7] Ribas Jr F A 2007 Thermal Analysis of Reciprocating Compressors *Proc. Int. Conf. Compr. and their Sys.(London)* **C65804207**

[8] Almbauer R, Burgstaller A, Abisid Z and Nagy D 2006 3-Dimensional Simulation for Obtaining the Heat Transfer Correlations pf a Thermal Network Calculation for a Hermetic Reciprocating Compressor *Proc. Int. Compr. Eng. Conf. at Purdue* **1794**

[9] Posch S, Hopfgartner J, Heimel M, Berger E, Almbauer R and Stangl S 2016 Thermal Analysis of a Hermetic Reciprocating Compressor using Numerical Methods *Proc. Int. Compr. Eng. Conf. at Purdue* **1215**

[10] Moukalled F, Mangani L and Darwish M 2015 The Finite Volume Method in Computational Fluid Dynamics (Springer)

[11] Scardovelli R and Zaleski S 1999 Direct numerical simulation of free-surface and interfacial flow *Annual Review Fluid Dynamics* **31** 567-603

[12] Fluent 14 2011 User Guide (Fluent Inc.)

[13] Posch S, Hopfgartner J, Heimel M, Berger E, Almbauer R and Schöllauf S 2016 A Numerical Investigation of the Oil Pump Suction Behaviour in a Hermetic Reciprocating Compressor *Proc. Int. Compr. Eng. Conf. at Purdue* **1215**

[14] Isaa R I 1986 Solution of the Implicitly Discretised Fluid Flow Equations by Operator-Splitting *J. of Comp. Physics* **62** 40-65

IOP Conf. Series: Materials Science and Engineering **232** (2017) 012010 doi:10.1088/1757-899X/232/1/012010

A conjugate heat transfer analysis of a hermetic reciprocating compressor

M DİNCER, K SARIOĞLU, H GÜNEŞ

Arcelik A.Ş. Compressor Plant, Organize Sanayi, 26110, Eskisehir, TURKEY
Arcelik A.Ş. Research and Development Center, Tuzla, 34950, Istanbul, TURKEY
Department of Mechanical Engineering, Istanbul Technical University, Gumussuyu 34437, Istanbul, TURKEY

onurdincer@arcelik.com

Abstract. Thermodynamic efficiency of a household refrigeration compressor is considerably affected by superheating that occurs inside the compressor. This phenomenon can be defined as the temperature increment of the refrigerant before entering the compression volume. On its flow path; refrigerant gains heat from suction pipe, suction muffler, suction chamber and related compressor components, which are at higher temperatures. In purpose of investigating conjugated heat transfer mechanism inside a hermetic reciprocating compressor a detailed numerical model is presented. The numerical conjugate heat and flow model is formed both with fluid domain (refrigerant) and solid domain (compressor components). Effects of using different materials on temperature distribution of some key components such as the crankcase, cylinder head and the valve plate are investigated. In addition to steady state analysis, transient CFD analysis is performed in order to understand fluid flow characteristics and its influence on superheating of the refrigerant.

1. Introduction

Detailed temperature measurement is a significant study to understand heat transfer network and temperature distribution of various components in compressors. This knowledge is necessary for designing compressor components and choosing their materials. However instrumentation of experimental test setup, which requires hard efforts, causes deviations from the original situation. Experiments are also to be repeated in order to see the effect of every prototype component. A valid numerical model, serving for thermally mapping the compressor, would be an appreciable alternative against aforementioned drawbacks.

CFD applications are widely used for improvement of compressor technologies. Raja et al. [1] simulated heat transfer network of a compressor numerically and presented the temperature distribution. Padhy et al. [8] developed a heat transfer model for rotary compressors. Nakano et al. [3] performed CFD analysis specifically on suction muffler. Sarıoğlu et al. [6] studied transient pressure variations in suction muffler of a hermetic compressor both experimentally and numerically. Kara et al. [4] achieved a numerical model for crankcase of a compressor with the support of experimental temperature measurements and convective heat transfer correlations. Almbauer et al. [9] performed 3D simulations to obtain heat transfer correlations. Dutra et al. [10] investigated heat transfer in compressor components experimentally. Deschamps et al. [11] developed heat transfer model of a reciprocating compressor by combining differential and integral formulations. Pereira et al. [2] studied the fluid flow in cylinder region numerically.

Kawai et al. [7] studied the relation between suction gas temperature and compressor efficiency. Morriesen et al. [5] express superheating as the dominant factor in thermodynamic losses with 49%, while viscous effects during suction and discharge processes and cylinder-piston leakage are responsible for 47% and 4% respectively. It is seen that most of the studies concern heat transfer mechanisms between compressor components or fluid flow characteristics in any refrigeration path separately. This eliminates the necessity of including all compressor components in numerical model in a detailed way. On the other hand, developing a detailed compressor model, which would be capable of investigating conjugate heat transfer, might provide more realistic view in basis of superheating phenomenon.

This study presents detailed non-isothermal CFD analyses of a hermetic reciprocating compressor for several cases including the base situation as well as modifications for lowering suction gas temperature with the aim of increasing volumetric efficiency.

2. Experimental studies

Experimental studies are discussed in two parts which are temperature and pV measurements. These studies are carried out to obtain boundary condition data for numerical analyses and to provide experimental data to validate the numerical model.

2.1. Temperature Measurements

Temperature distribution inside the compressor is investigated primarily. The compressor being investigated is designed for R600a and has a cooling capacity of 170W at ASHRAE conditions. The compressor is instrumented with several T-type thin thermocouples. Sensitivity of the thermocouples are ±5°C. Temperature values are recorded while the compressor is operating on a fully automated calorimeter system with conditions of -23.3 °C evaporation temperature and 54.4 °C condensation temperatures. Sub-cooling, superheating and ambient temperatures were 32.2 °C. Sample images of temperature measurement setup are given in Figure 1.

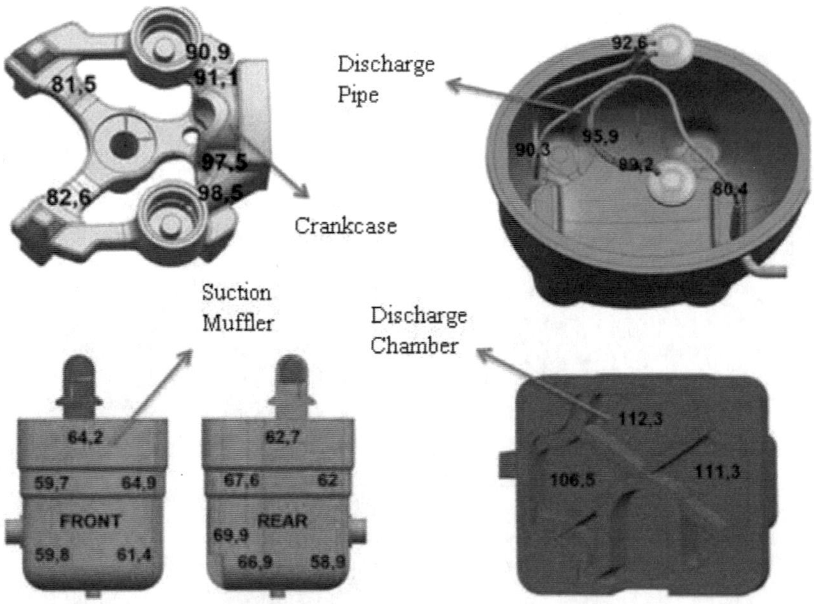

Figure 1. Temperature values on crankcase, discharge pipe, suction muffler and discharge chamber surfaces.

It is seen that the cylinder head has the highest temperature compared to other components of the compressor. Besides the cylinder head, the crankcase and discharge pipe are other heat sources for inner ambient of reciprocating compressors. With respect to rising temperature of the inner gas, heat transfer to the suction muffler by means of convection increases. The suction muffler also gains heat from the valve plate and cylinder head by conduction and the stator of the motor by radiation. The measured temperature values on the refrigerant path are indicated on a CAD model in Figure 2. The measurements show that the temperature of the refrigerant increases approximately 46 °C from the suction muffler inlet to the suction chamber. Thus, the density of the refrigerant decreases due to rising temperature which results in decreased mass flow rate and volumetric efficiency. It is clear that in addition to the acoustic damping, the suction muffler and its interaction with surroundings play an important role on the performance of the compressor.

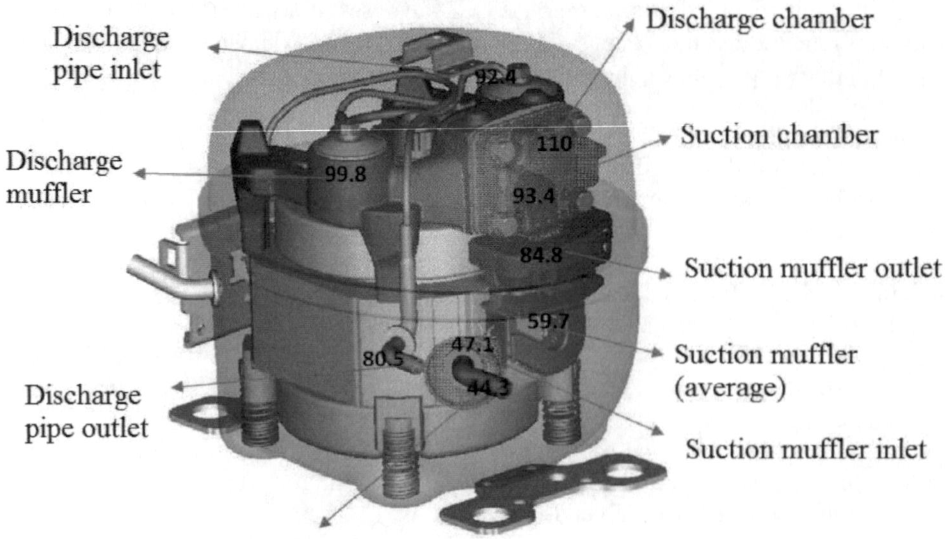

Figure 2. The measured temperature values on the refrigerant path of the compressor.

2.2. PV Measurements

Time-dependent pressure values in the suction and discharge chambers and inside the cylinder are obtained with pressure transducers. An optical encoder was instrumented on top of the crankshaft. The optical encoder generates a signal per shaft angle and another signal per cycle. With this test setup, crankshaft angle and pressure values can be measured in a coupled manner during the operation of the compressor. Locations of the pressure transducers and the optical encoder are given in figures 3 and 4, respectively.

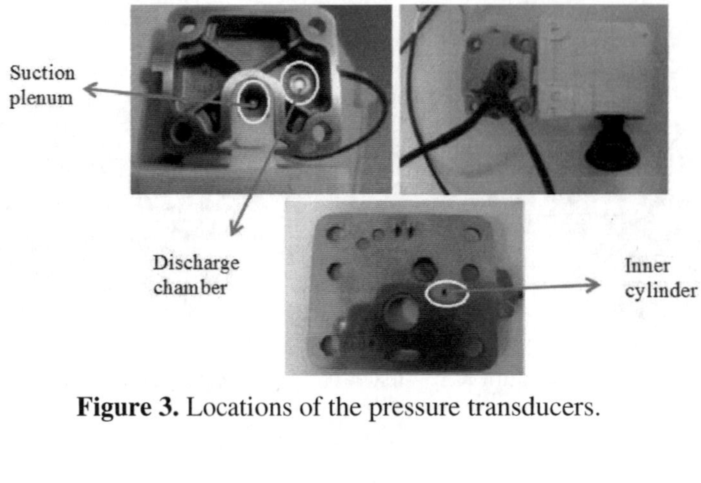

Figure 3. Locations of the pressure transducers.

Figure 4. Location of the optical encoder.

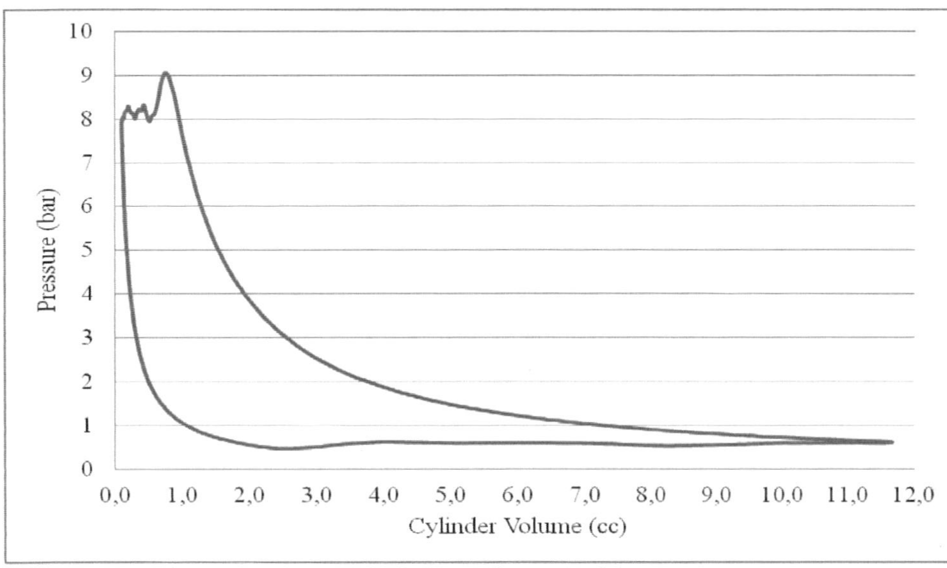

Figure 5. The PV (indicator) diagram of the compressor.

The PV diagram of the compressor is given in Figure 5. Compression work is calculated from this diagram and it will be used in deciding thermal boundary conditions of the cylinder bore. Experimentally obtained pressure profiles are used as boundary conditions in order to simulate transient fluid flow on suction and discharge lines in numerical studies as appropriate.

3. Numerical studies

Numerical studies will be discussed in three sub-sections which are the pure heat conduction model (consisted of the compressor solid parts only), the steady conjugate heat transfer model (includes both solid-fluid domains) and finally the transient conjugate heat transfer model. The CAD model of the hermetic reciprocating compressor is obtained with the commercially available software (NX 7.5). The mesh consisting of approximately 5 million elements is obtained with ANSYS meshing software. As it can be seen in Figure 6, all components of the compressor are included in the numerical model. Fluid domain and mesh structure are also given in Figure 6. Note that the refrigerant gas inside the compressor shell and the compressor oil do not exist in the solution domain. In addition, the expansion and the compression processes of the refrigerant inside the cylinder are not taken into account. The experimentally obtained transient pressure data of the suction chamber and the cylinder is used as boundary conditions. A commercially available CFD code (ANSYS Fluent v14.5) is used for the analyses. Momentum and energy equations are discretized with second order upwind approximation. Realizable k-epsilon is used as turbulence model.

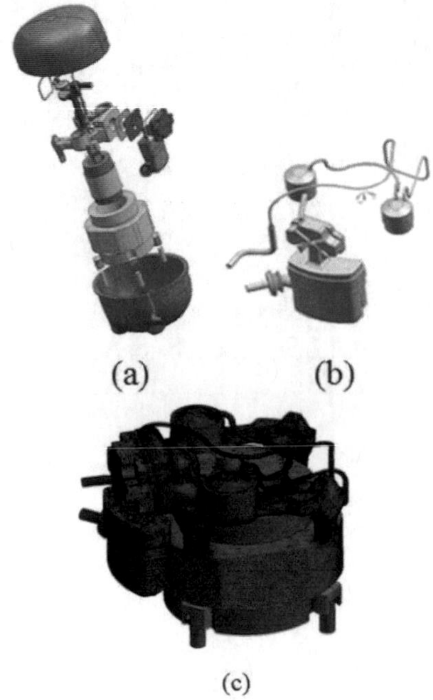

Figure 6. (a) Solid parts, (b) fluid domain (refrigerant), (c) mesh structure

3.1. Pure Heat Conduction Model

In this model, the refrigerant domain was deactivated, so the pure heat conduction analysis was performed for the solid components of the compressor. The measured temperature values from the refrigerant flow path are used as boundary conditions and the experimental temperature distribution on the solid parts are compared with numerical results obtained with the pure heat conduction model while optimizing the local heat transfer coefficients on the outer surfaces of the components. Previous to this study mechanical and electrical losses of the compressor were measured experimentally. In the numerical model, the mechanical losses and electric motor losses are assumed to be converted into heat and integrated into the model as a heat generation (source) term for the corresponding component volume. The difference between real and isentropic compression work is assumed to be converted into heat. The heat generation due to compression process inside the cylinder is simulated by the heat flux boundary condition on the cylinder surface. Prescribed heat sources in the numerical model are given in Table 1.

Table 1. Heat sources

Geometry	Heat Flux [W/m^2]	Heat Generation [W/m^3]
Cylinder bore	5546	-
Electric motor	-	23249
Crank bearing clearance	-	1,44x10^8

Figure 7 shows that experimentally measured surface temperatures and the numerical results are in good agreement. The solution of the pure heat conduction model pointed out that the heat transfer coefficients varies between 8 and 40 W/m^{2}K depending on the location of components inside the compressor.

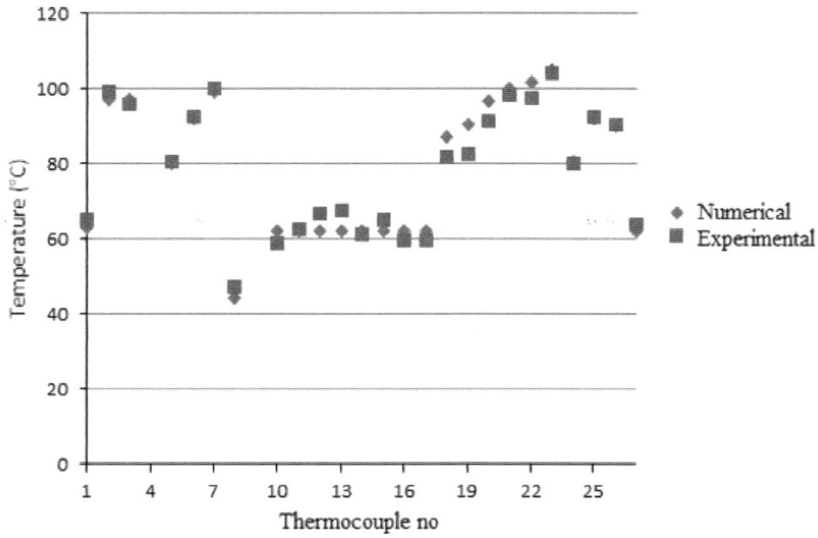

Figure 7. Comparison of experimental and numerical results

3.2. Steady Conjugate Heat Transfer Model

The fluid (refrigerant) domain was activated for this model in order to solve the steady state conjugated heat transfer problem employing the heat transfer coefficients obtained in the pure heat conduction model as discussed previously. The boundary conditions at inlet and outlet of the suction and discharge lines are given in Table 2.

Table 2. Boundary conditions

	Pressure[bar]	Temperature[°C]	Mass Flow Rate[g/s]
Suction inlet	-	44	0,54
Suction outlet	0,624	-	-
Discharge inlet	-	113	0,54
Discharge outlet	7,6	-	-

Temperature contours of the compressor components at steady state operation are given in Figure 8. The heat generation during the compression process inside the cylinder causes the heat transfer between the crankcase, the valve plate and cylinder head. This region attains the highest temperatures inside the compressor. Nearly all hermetic reciprocating compressor designs require suction muffler to be located in this high temperature region. As a result, the suction muffler gains heat from these hot components. This fact will also be discussed when the results of the transient conjugate heat transfer model are shown with figures taken from different phases of a compression cycle.

IOP Conf. Series: Materials Science and Engineering **232** (2017) 012010 doi:10.1088/1757-899X/232/1/012010

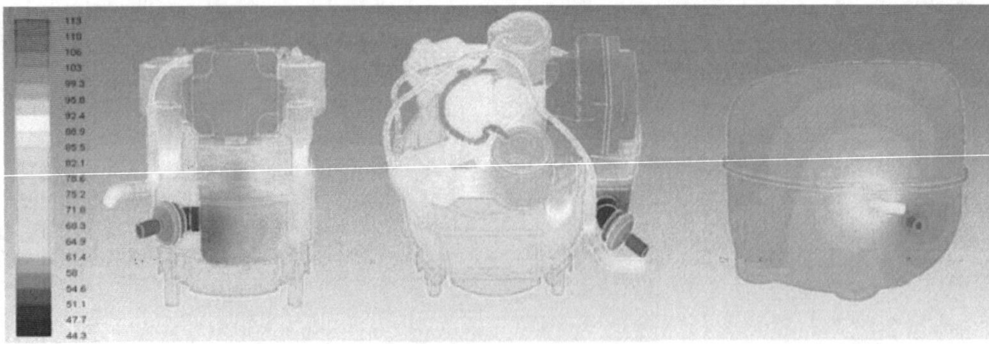

Figure 8. Temperature contours of the compressor components and shell.

A section view of the temperature contours is given in Figure 9. It is seen that the suction muffler is heated by the valve plate and cylinder head especially from the contact surfaces. This rises the temperature of the refrigerant at the start of the compression process. In order to keep the refrigerant at lower temperatures, the materials of selected components are replaced and CFD analyses are repeated for each case. Materials of the crankcase, valve plate and cylinder head are replaced with aluminium, POM (Polyoxymethylene) and PBT (Polybutylene terephthalate) respectively. The valve plate, cylinder head and the suction muffler temperatures are compared for each case with the original case.

The material of the crankcase was changed from cast iron to aluminium at first case. Temperature contours of original situation (cast iron) and aluminium crankcase is given in Figure 10. This case leads to a more homogenous temperature distribution on the crankcase and lower temperature values were obtained around the cylinder region. As the average temperature of the crankcase decreases, the inner ambient of the compressor, which is a common heat source for suction muffler, remains at lower temperatures.

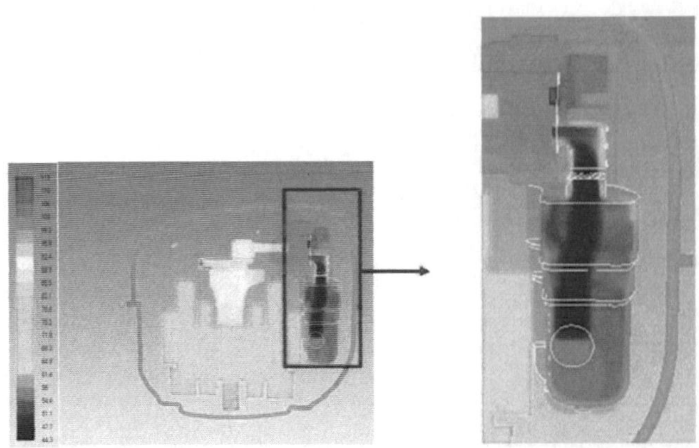

Figure 9. A section view of temperature contours.

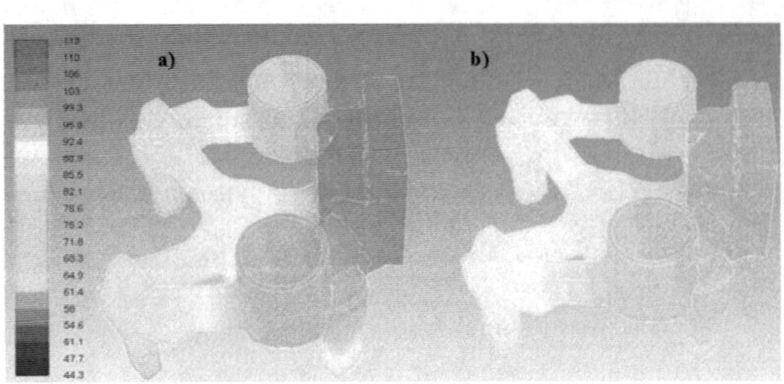

Figure 10. Temperature distribution of crankcase a) Original (cast iron) b) Aluminium

Comparison of the valve plate temperature contours for different circumstances are given in Figure 11. Using an aluminium crankcase lowers the average temperature of valve plate. Replacing the material of the valve plate to an insulative (POM) reduced heat conduction from cylinder region to the suction muffler, thus the refrigerant temperature inside the suction muffler is reduced. The average temperature of the valve plate increases, in the case of changing the material of the cylinder head from aluminium to PBT.

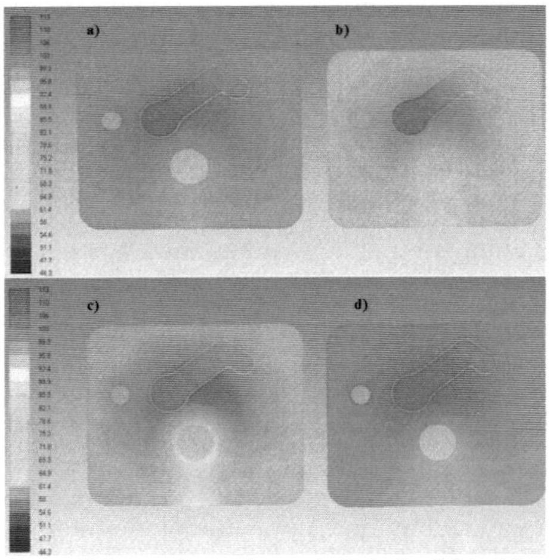

Figure 11. Temperature distribution of the valve-plate a) Original b) Aluminium crankcase c) POM valve-plate d) PBT cylinder head

Temperature distributions on cylinder heads for all cases are given in Figure 12. The effect of aluminium crankcase on the cylinder head is similar to the valve plate. POM valve plate avoids heat conduction from crankcase to the cylinder head and the suction muffler. Plastic based cylinder head shows a significant change on the suction chamber wall temperature. In virtue of the low thermal conductivity of the material, the suction chamber region remains at lower temperatures.

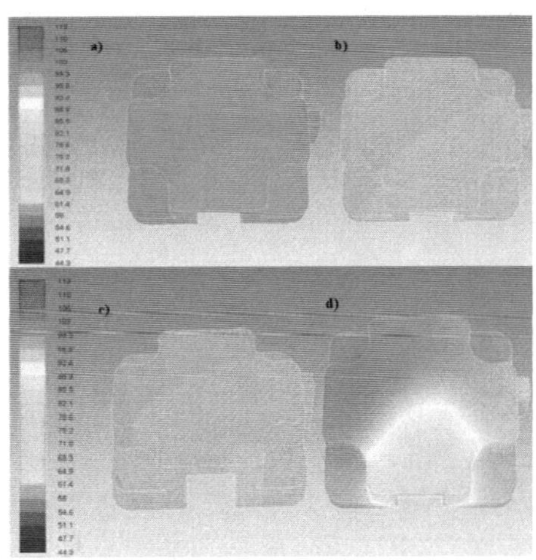

Figure 12. Temperature distribution of the cylinder a) Original b) Aluminium crankcase c) POM valve-plate d) PBT cylinder head

Temperature contours of the suction muffler are given in Figure 13. Variations between temperature distributions of the suction muffler on different cases are specified especially on the suction chamber region. Aluminium crankcase and POM valve plate provide a better result for the purpose of reducing wall temperatures of the suction chamber. On the other hand, PBT originated cylinder head shows the highest impact on the suction chamber wall temperature with 10°C decrease.

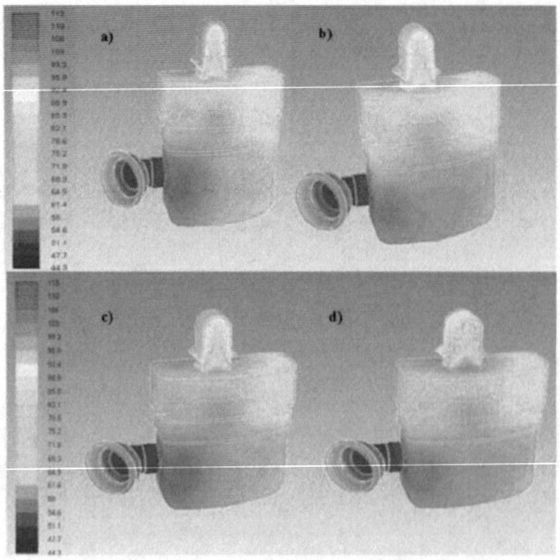

Figure 13. Temperature distribution of suction muffler a) Original b) Aluminium crankcase c) POM valve-plate d) PBT cylinder head

3.3. Transient Conjugate Heat Transfer Model

Transient non-isothermal CFD analysis is performed in order to reveal the fluid flow and its influence on superheating of the refrigerant. Time-dependent pressure profiles, which were obtained experimentally, are used as boundary conditions at the inlets of suction and discharge lines. Constant pressure outlet boundary conditions are used at the outlets of suction and discharge lines. Time step is 0,00005 seconds and calculations are performed for 400 time steps which corresponds to 0,02 seconds total cycle time of a compressor running at 50Hz. Figure 14 shows instantaneous views of streamlines corresponding to the discharge phase, the beginning of suction phase, the last phase of suction and the compression phase respectively. It is seen that refrigerant is heated by convection inside the suction muffler during the periods in which the suction valve is closed. Refrigerant which exists inside the suction muffler at higher temperature enters the cylinder at the beginning of the suction period. Fresh refrigerant replaces the heated refrigerant and fills the cylinder volume as suction process continues. Refrigerant starts to gain heat beginning from the suction chamber during the compression process.

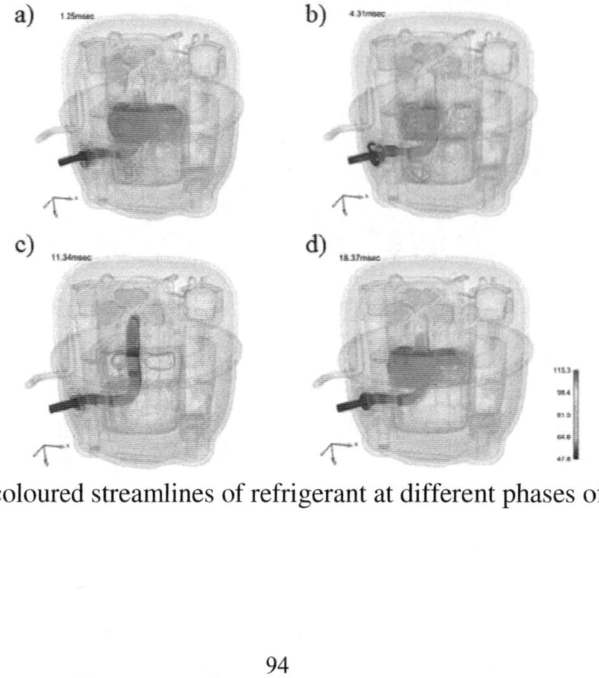

Figure 14. Temperature coloured streamlines of refrigerant at different phases of a compression cycle.

IOP Conf. Series: Materials Science and Engineering **232** (2017) 012010 doi:10.1088/1757-899X/232/1/012010

4. Conclusion

Effects of using different materials on the temperature distribution of selected key components such as the crankcase, cylinder head and the valve plate are investigated. The material of the crankcase was changed from cast iron to aluminium at first case. This case lead to a nearly homogenous temperature distribution on the crankcase and lower temperature values around the cylinder region. The other modification was changing the material of the valve plate to an insulative. This change reduced heat conduction from cylinder region to the suction muffler. In the last case, a plastic based material for the cylinder head was used which resulted in a significantly decreased wall temperature of suction chamber. In virtue of the low thermal conductivity of the material, the suction muffler was remained at lower temperatures. The transient conjugate heat transfer analysis showed that refrigerant is heated by convection inside the suction muffler especially during the periods in which the suction valve is closed. Temperature of the refrigerant at the start of the compression process is critical for the thermodynamic efficiency of the compressor. The temperature distribution on the components of the compressor determines the heat gain of the refrigerant until the beginning of the compression. The numerical model presented in this study is used for analysing the effects of component materials of the compressor on its temperature distribution and the results are in good agreement with the experimental studies. The model also provides information, before prototyping, for structural changes on compressor components from the thermodynamic efficiency perspective.

References

[1] Raja B., Joseph Sekhar D., et al. (2003). Numerical Model for Thermal Mapping in a Hermetically Sealed Reciprocating Refrigerant, Compressor International Journal of Refrigeration.

[2] Pereira E.L.L., Santos C.J., (2012). A Simplified CFD Model for Suction Process of Reciprocating Compressor, Compressor Engineering Conference, Purdue.

[3] Nakano A., Kinjo K., (2008). CFD Applications for Development of Reciprocating Compressor, Compressor Engineering Conference, Purdue.

[4] Kara S., Oğuz E., (2010). Thermal Analysis of a Small Hermetic Reciprocating Compressor, Compressor Engineering Conference, Purdue.

[5] Morriesen A., Deschamps C.J. (2012). Experimental Investigation of Transient Fluid Flow and Superheating in the Suction Chamber of a Refrigeration Reciprocating Compressor, Applied Thermal Engineering,41 (Sf.61-70)

[6] Sarıoğlu K., Özdemir A.R., (2012) An experimental and numerical analysis of refrigerant flow inside the suction muffler of hermetic reciprocating compressor, Compressor Engineering Conference, Purdue.

[7] Kawai H., (1982). The Development of High Efficiency Compressor by Reducing Suction Gas Temperature, Compressor Engineering Conference, Purdue.

[8] Padhy S.K., (1992). Heat Transfer Model of a Rotary Compressor, Compressor Engineering Conference, Purdue.

[9] Almbauer, R.A., (2006). 3-Dimensional Simulation for Obtaining the Heat Transfer Correlations of a Thermal Network Calculation for a Hermetic Reciprocating Compressor, Compressor Engineering Conference, Purdue.

[10] Dutra, T., Deschamps, C.J., (2010). Experimental Investigation of Heat Transfer in Components of a Hermetic Reciprocating Compressor, Compressor Engineering Conference, Purdue.

[11] Deschamps, C.J., (2012). A Heat Transfer Model Combining Differential and Integral Formulations for Thermal Analysis of Reciprocating Compressors, Compressor Engineering Conference, Purdue.

IOP Conf. Series: Materials Science and Engineering **232** (2017) 012011 doi:10.1088/1757-899X/232/1/012011

Heat Transfer Simulation for Reciprocating Compressor with Experimental Validation

Ruixin Zhou[1], *Bei Guo[1], Xiaole Chen[1], Jinliang Tuo[1], Rui Wu[1], Fabian Fagotti[2], Yali Zhao[2], Song Yang[2], Bo Xu[2]
[1]School of Energy and Power Engineering, Xi'an Jiaotong University;
[2]Beijing Embraco Snowflake Compressor Company Ltd

*Corresponding Author Email: guobei@mail.xjtu.edu.cn

Abstract: The efficiency of reciprocating compressor can be influenced by heat transfer and the reliability can be also affected by the temperature distribution in compressor. In consideration of the complex relationship of heat transfer, the compressor is divided into six control volumes including the suction muffler, the cylinder, the discharge chamber, the discharge muffler, the discharge line and the compressor shell. The steady state energy balance equations of the open system for each control volume are built up after the crankshaft rotates one cycle. The heat flux of the cylinder is calculated by the existing correlation. The heat transfer coefficient correlations in energy equations are chosen in references and revised by experimental results. Three same type reciprocating compressors used in R290 system installed with themocouples are tested under some planed conditions in order to ensure accuracy. The simulation results are compared with the experimental results. It shows that the simplified method presented in this paper is effective.

1.Introduction

Compressor is the heart of refrigeration system. The performance of the refrigeration system is greatly dependent on the performance of the compressor and the working conditions. As to the 1-D simulation of refrigeration compressor, Wu Shoufei[1][2] simulated a certain refrigeration compressor based on the 4 basic equations of the 1-D flow. Ma Pengcheng[3] got the performance of a household refrigeration compressor by building the mathematic model of valve and considering the leakage in the compressor. Bei Guo[4] studied the valve movement and simulated the thermal performance of reciprocating compressor with the refrigerant R600a experimental validation, but the heat transfer, which can influenced the simulation results, was not considered in that program.

A lot of researchers had been carried out on the heat transfer simulation of refrigeration compressor. Todescat and Fabian Fagotti[5] studied the heat transfer simulation based on lumped-parameter models

and finite volume method to get the temperature distribution in compressor, and experiments with the refrigerant R134a was also carried out. The heat transfer coefficients of each control volume are determined by experiments. Willianm A. Meyer[6] Kim, Tiow Ooi[7] and Cesar[8] have adopted classical heat transfer correlations in literatures based on lumped-parameter models and finite volume method. This method that heat transfer coefficients are from experiments is more accurate than the method that coefficients are determined by classical heat transfer correlations in literatures. In order to insure the practicability and precision, This method that heat transfer coefficients are from experiments is adopted. Moreover, R290 is widely used in commercial compressor, such as the drinks refrigerator and wind screen refrigerator which are used in supermarkets, but the refrigeration compressor with R290 was not studied and its experiment was not carried out. In order to optimize the R290 compressor and enlarge the practical scope of simulation, the experiment is carried out on R290 commercial compressors.
In this paper, the heat transfer model is established which calculates the heat flux between different control volumes in the compressor. In consideration of the complex relationship of heat transfer, the compressor is divided into 6 control volumes including the suction muffler, the cylinder, the discharge chamber, the discharge muffler, the discharge line and the compressor shell. The heat transfer coefficient of each control volume is measured by experiment and the heat flux of the cylinder is calculated by Lawton model[9][10]. The experiments are also carried out and the thermalcouples are installed at its exact position shown in following considering the insulation and leakage problems. The experiments results indicate that the heat transfer model of the refrigerator compressor can accurately calculate the thermodynamic performances and the comparable errors are reasonable.

2.Heat transfer simulation

Based on the simulation of performance parameters of reciprocating compressor [7], numerical simulation method to simulate heat transfer in reciprocating compressor is applied. In order to achieve the heat transfer subroutine, assumptions are shown below.

1) The process of heat transfer is a steady state.

2) The pressure of the suction side is the saturation pressure corresponding to the evaporating temperature, and the pressure of the exhaust side is the saturation pressure corresponding to the condensing temperature.

3) The internal temperature of a control volume is the average temperature of the inlet's and outlet's.

4) For the adjacent control volumes, outlet temperature of one control volume is equal to inlet's of the next control volume.

5) The compressor shell temperature, internal environment of shell and each control volume is average.

2.1 Energy balance equation of control volumes

As shown in Fig 1, I is the suction muffler, II is the cylinder III is the discharge chamber. IV is the discharge muffler. V is the discharge line. VI is the motor. T_{smw}, T_{dcw}, T_{dmw}, T_{cw}, T_{core} and T_{shell} are all the wall temperature of each compressor parts. m_0 is the compressor mass rate and m_1 is the compressor outlet mass rate and the m_1 is the mass rate of gas flow from inner environment to the suction muffler. According to the law of mass conservation, equation (1) and (2) are established.

$$\dot{m}\_0 = \dot{m}\_dis \tag{1}$$

$$\dot{m}\_suc = \dot{m}\_0 + \dot{m}\_1 \tag{2}$$

IOP Conf. Series: Materials Science and Engineering **232** (2017) 012011 doi:10.1088/1757-899X/232/1/012011

The compressor was divided into 6 control volumes, which are the suction muffler, the cylinder, the discharge chamber, the discharge muffler, the discharge line and the compressor shell. The heat transfer simulation was carried out on these 6 control volumes. Each control volume equation was solved simultaneously, and then the wall temperatures and the internal environment temperature were worked out. What must be pointed out is that the equation set should be calculated when the crankshaft rotates 360°angle which is different from the calculation of $\frac{dQ}{d\theta}$ The $\frac{dQ}{d\theta}$ term need be calculated when the crankshaft rotate $d\theta$ angle and integrate the heat transfer value. As the same, the compress power W_p ,the motor power W_e and all the mass rate are all integrate among 360° angle.

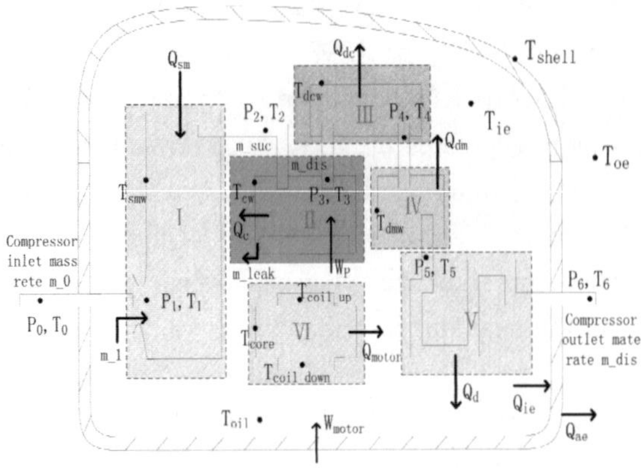

Fig 1. The control volume sketch of semi direct intake reciprocating compressor

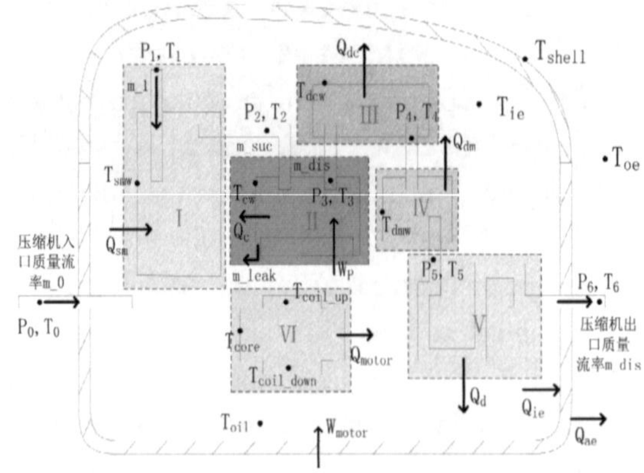

Fig 2. The control volume sketch of indirect intake reciprocating compressor

The gas intake method of the suction muffler can be divided into 2 types in current refrigeration compressors. One is the semi direct intake method and it is shown in Fig 1. Another is indirect intake method, shown in Fig 2, whose inlet tube of compressor is not aimed to the suction muffler inlet, so the inlet gas would be warmed up by the motor and discharge parts. Both of the gas intake method can be expressed by an intake mixing coefficient ξ and m_1 is can calculate by equation (3).

$$\dot{m}\_1 = \xi \cdot \dot{m}\_suc \tag{3}$$

The semi direct intake method of the suction muffler is shown in Fig1. Its coefficient ξ is shown in equation (4).

$$\xi = \dot{m}\_leak \,/\, \dot{m}\_suc \tag{4}$$

But as to the indirect intake method of the suction muffler, its coefficient ξ is equal to 1.

Above all the equations on mass rate, the energy balance equation of the suction muffler is shown in equation (5) and equation (6).

$$\dot{m}\_suc * h_1 = \xi * \dot{m}\_suc * h_{ie} + (1-\xi) * \dot{m}\_suc * h_0 \tag{5}$$

$$\dot{m}\_suc * (h_2 - h_1) = \alpha_{sm} F_{sm} \left(\frac{T_1 + T_2}{2} - T_{ie} \right) \tag{6}$$

According to the research of Fabian[5], Lawton model was adopted in this thesis to calculate the heat transfer value between the gas in cylinder and the cylinder wall. The energy equation of the cylinder is shown in equation (7).

IOP Conf. Series: Materials Science and Engineering **232** (2017) 012011 doi:10.1088/1757-899X/232/1/012011

$$\dot{m}\_dis * h_3 - \dot{m}\_suc * h_2 + m\_leak * hie = -\sum\left[\frac{k(T_g(t)-T_{cw})}{D}(0.28\,\mathrm{Re}(t)^{0.65} - 0.25L(t)\frac{T_{cw}}{T_g(t)-T_{cw}})\cdot S_c \cdot \frac{d\theta}{dt}\right] + W_p \quad (7)$$

As shown in Fig 1, the discharge chamber, the discharge muffler and the discharge line are all treated as control volumes separately and their energy equations are shown in equation (8), equation (9) and equation (10).

$$\dot{m}\_dis * (h_3 - h_4) = \alpha_{dc} F_{dc}(\frac{T_3 + T_4}{2} - T_{ie}) \quad (8)$$

$$\dot{m}\_dis * (h_4 - h_5) = \alpha_{dm} F_{dm}(\frac{T_4 + T_5}{2} - T_{ie}) \quad (9)$$

$$\dot{m}\_dis * (h_5 - h_6) = \alpha_{dl} F_{dl}(\frac{T_6 + T_5}{2} - T_{ie}) \quad (10)$$

The shell temperature is easy to measure and it is also an important value to judge the characteristic of condition case. So the shell temperature is listed in the energy equation of compressor shell. The heat transfer capacity of the inner shell surface Q_{si} and outer shell surface Q_{so} are equal. The energy equation between the internal environment and inner shell surface is shown in equation (11) and the energy equation between the internal environment and outer shell surface is shown in equation (12).

$$\dot{m}\_dis * h_6 - m\_0 * h_0 = \alpha_{si} F_{si}(T_{ie} - T_{shell}) + w_e \quad (11)$$

$$\alpha_{si} F_{si}(T_{ie} - T_{shell}) = \alpha_{so} F_{so}(T_{shell} - T_{oe}) \quad (12)$$

The simultaneous equation set composed by equation (1) to (12) is calculated to get the variables with red color. The αF term is the product value of heat transfer coefficient and heat transfer area. The mass rate term m_0, m_dis, m_suc, the discharge gas temperature and the compress power W_p are all calculated by the former simulation program[4] which used to calculate the compressor performance parameters. Motor efficiency curve is tested on the R290 commercial reciprocating compressor, and it is used to calculate the motor power W_e. h terms in the equation set shows that the equation set is nonlinear and the calculation method is to call NIST REFPROP function library and use the Newton's iterative method to resolve.

2.2 Heat transfer coefficients correlations

Ooi[7] and Joel[8] choose the empirical correlation in references and use them directly without revising and this must cause big errors in simulation. In this thesis, the heat transfer coefficient of each control volume is obtained by combining the theoretical heat transfer correlation with the experimental data. The coefficient correlations of simple control volumes are chosen in references and revised by experimetal results. The 3-D flow field simulation of the discharge chamber and discharge muffler, where the gas flow is complecated, is carried out. The corresponding heat transfer correlation type is selected according to the 3-D simulation results. Because the structure of the suction muffler is extremely complex, its heat transfer coefficients correlation is fitted by experimental results. Table 1. shows the revised heat transfer coefficient correlations used in the heat transfer simulation.

Table 1. The revised heat transfer coefficient correlations

Control volume	Surface	Correlation

Compressor shell	Inner	$\overline{Nu} = 0.019\,\mathrm{Re}^{0.8}\,\mathrm{Pr}^{1/3} \quad (\mathrm{Pr} \geq 0.5)$
	Outer	$\overline{\mathrm{Nu}} = \left(2 + \left(0.8 \times \mathrm{Re}^{\frac{1}{2}} + 0.23 \times \mathrm{Re}^{\frac{2}{3}}\right)\mathrm{Pr}^{0.4}\right) \times \left(\frac{\mu_\infty}{\mu_w}\right)^{\frac{1}{4}} \quad (3.5 < \mathrm{Re} < 7.6 \times 10^4, 0.71 < \mathrm{Pr} < 380)$
Discharge line	Inner	$\overline{\mathrm{Nu}} = 0.023\,\mathrm{Re}^{0.8}\,\mathrm{Pr}^{0.3}$
	Outer	$\overline{Nu} = 0.18 + \frac{0.372\,\mathrm{Re}^{1/2}\,\mathrm{Pr}^{1/3}}{\left[1 + (0.4/\mathrm{Pr})^{2/3}\right]^{1/4}}\left[1 + \left(\frac{\mathrm{Re}}{282000}\right)^{5/8}\right]^{4/5} \quad \mathrm{Re} \times \mathrm{Pr} > 0.2$
Discharge muffler	Inner	$\overline{\mathrm{Nu}} = 0.023\,\mathrm{Re}^{0.8}\,\mathrm{Pr}^{0.3}$
	Outer	$\overline{Nu} = 0.3 + \frac{0.62\,\mathrm{Re}^{1/2}\,\mathrm{Pr}^{1/3}}{\left[1 + (0.4/\mathrm{Pr})^{2/3}\right]^{1/4}}\left[1 + \left(\frac{\mathrm{Re}}{282000}\right)^{5/8}\right]^{4/5} \quad \mathrm{Re} \times \mathrm{Pr} > 0.2$
Discharge chamber	Inner	$\overline{Nu} = \frac{(f/8)(\mathrm{Re}_D - 1000)\mathrm{Pr}}{1 + 12.7(f/8)^{1/2}(\mathrm{Pr}^{2/3} - 1)} \quad (\mathrm{Pr} \geq 0.5; 3 \times 10^3 \leq \mathrm{Re} \leq 5 \times 10^6)$ $f = (0.79 \ln \mathrm{Re} - 1.64)^{-2}$
	Outer	$\overline{Nu} = 0.037\,\mathrm{Re}^{0.8}\,\mathrm{Pr}^{1/3} \quad (\mathrm{Pr} \geq 0.5)$
Suction muffler	Total	$\alpha_{sm} F_{sm} = -1.19516 - 5492.97 * \dot{m}$
Cylinder	Inner	$Nu = \frac{q_w D}{k(T_g(t) - T_w)} = 0.28\,\mathrm{Re}(t)^{0.65} + 0.25 L(t)\frac{T_w}{T_g - T_w}$

3.Experiments

3.1 Arrangements

Fig 3. is the experiment system sketch. According to the requirement of the standard testing condition, the 3m/s wind blow from a fan is adopted.

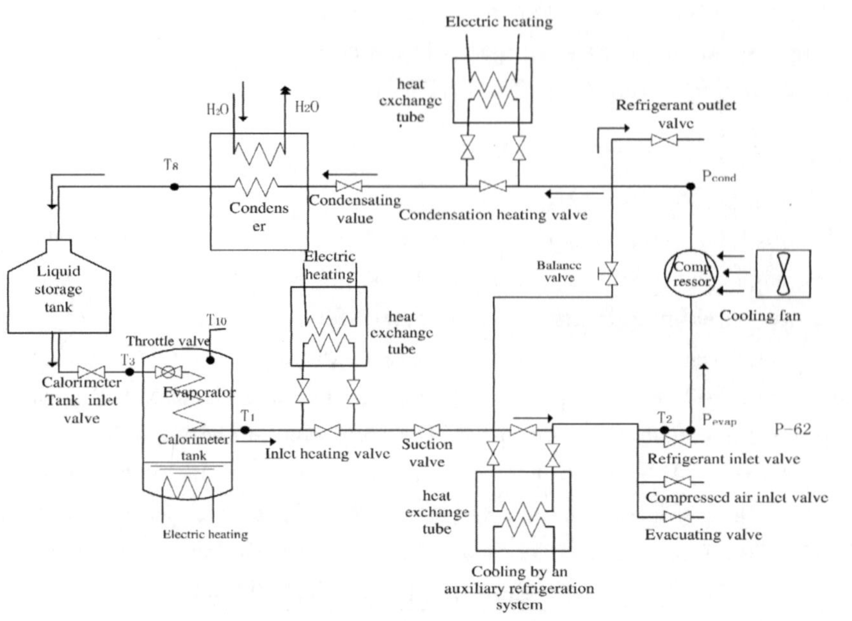

Fig 3. The experiment system sketch

In order to reach the unusual condensing temperature, the radiation heating installation and water-cooled tube are installed on condenser side. Also in order to reach limiting evaporating temperature, the water-cooled device and small refrigeration device with an extra refrigeration cycle are installed on the evaporator side. The automatic regulating valve is adopted as the throttling valve.

Firstly, connect all the solder joints of themocouples installed in compressor to the temperature acquisition equipment. Then, connect the temperature acquisition equipment to computer. Lastly, open the software to test the themocouples' accuracy. Fig 4 shows the compressor installed all thermocouples and shows the solder joints on the shell.

Fig 4. The compressor sample installed thermocouples

The detailed information of the themocouples is shown in Table 2. The data acquisition frequency is once per 10 seconds. The temperature of each test point under some condition is the average value of the experimental data which acquired after the system is stable.

Table 2. The detail of the themocouples

No.	Name	Short name	Position
1	T_{up}	Temperature of the upper shell	Middle of the upper shell
2	T_{bottom}	Temperature of the bottom shell	Middle of the bottom shell
3	T_{left}	Temperature of the left shell	When facing to socket, middle of the left shell
4	T_{right}	Temperature of the right shell	When facing to socket, middle of the left shell
5	T_{front}	Temperature of the front shell	When facing to socket, middle of the front shell
6	T_{behind}	Temperature of the behind shell	When facing to socket, middle of the behind shell
7	T_{ie1}	The environment temperature NO.1 in compressor	When facing to socket,The right side in compressor and between the shell and motor
8	T_{ie2}	The environment temperature NO.2 in compressor	When facing to socket,The left side in compressor and between the shell and motor
9	T_0	Gas temperature at suction pipe	In the middle of the suction pipe
10	T_1	Gas temperature at suction muffler inlet	In the middle of the suction muffler inlet
11	T_2	Gas temperature at suction muffler outlet	In the middle of the suction muffler outlet
12	T_3	Discharge gas temperature of cylinder	In the discharge chamber and facing to the discharge valve
13	T4	Gas temperature at discharge muffler inlet	In the discharge muffler inlet
14	T_5	Gas temperature at discharge muffler outlet	In the discharge muffler outlet
15	T_6	Gas temperature at the outlet pipe	In the middle of the discharge pipe

16	T_{oil}	Temperature of oil	In the middle of the oil pool
17	T_{coil}	Temperature of coil	The coil temperature
18	T_{shaft}	Temperature of shaft	Drilling a hole at a distance of 3mm to the upper shaft, and stick the themocouple in the hole
19	T_{cw}	Temperature of cylinder wall	Drilling a hole at a distance of 3mm to piston, and stick the themocouple in the hole.
20	T_{oe}	Temperature of ambient	One themocouple in ambient

3.2 Experiment results

R290 is used in this experiment. Because the R290 commercial refrigerator is high efficiency when the refrigeration case is in HBP（high back pressure） conditions, the ASHRAREHBP46 case is selected to analyze the heat transfer value of each compressor part.

The detail of the condition: the evaporating temperature =7.2℃ ; The condensing temperature=54.4℃ ; The suction temperature=35℃ ; The ambient temperature=35℃ ; The subcooled temperature=35℃.

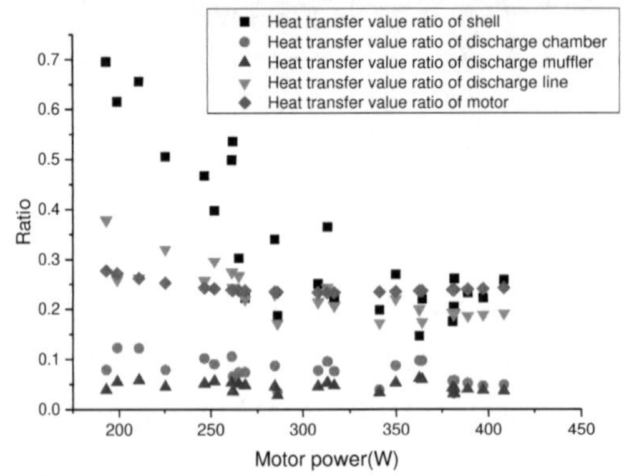

Fig 5 Heat transfer value ratio of compressor parts

The heat transfer valve ratio is the ratio between the heat transfer value of each compressor part and the motor power W_e. The heat transfer value of discharge line is 3-5 times as large as the value of discharge chamber or discharge muffler. The heat transfer value ratio of motor is about 23%. The inner environment of compressor is heat up by the discharge parts and the motor, but heat transfer of the shell can cool down it. The heat transfer valve ratio of the shell is larger than 50% when the motor power is lower than 220W, because the mass rate is smaller when the refrigeration system is in LBP (low back pressure) cases and the pressure ratio is higher than those in HBP.

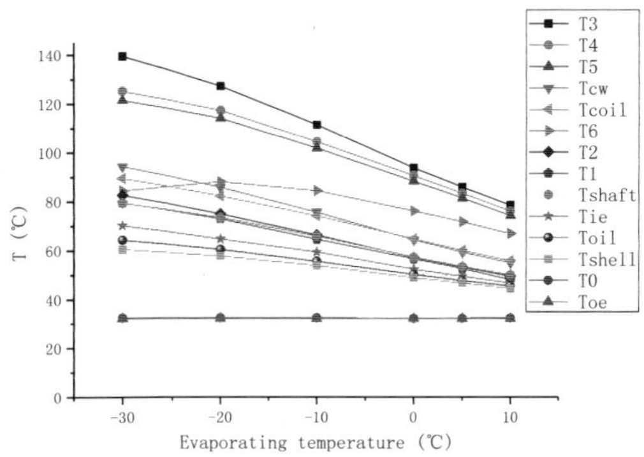

Fig 6. The tested temperatures vary with the evaporating temperature

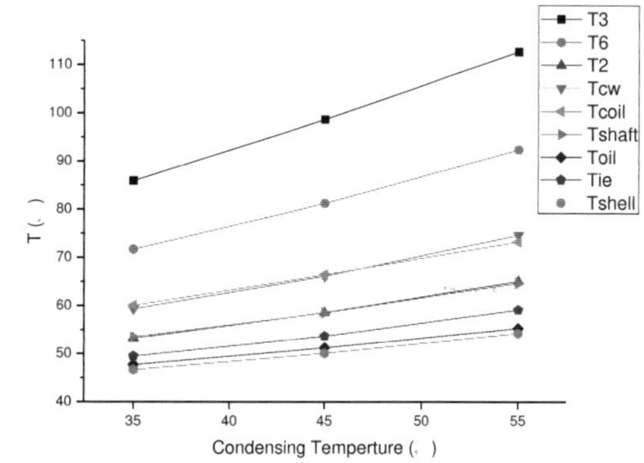

Fig 7. The tested temperatures vary with the condensing temperature

Fig 6 shows the tested temperatures vary with evaporating temperatures when the condensing temperature is 35℃,the suction temperature is 32.2℃ and the ambient temperature is 32.2℃. Fig 7 shows the tested temperatures vary with condensing temperatures when the evaporating temperature is 5℃,the suction temperature is 32.2℃ and the ambient temperature is 32.2℃. The temperature T_{ie} and T_{shell} are all the average value of their testing points.

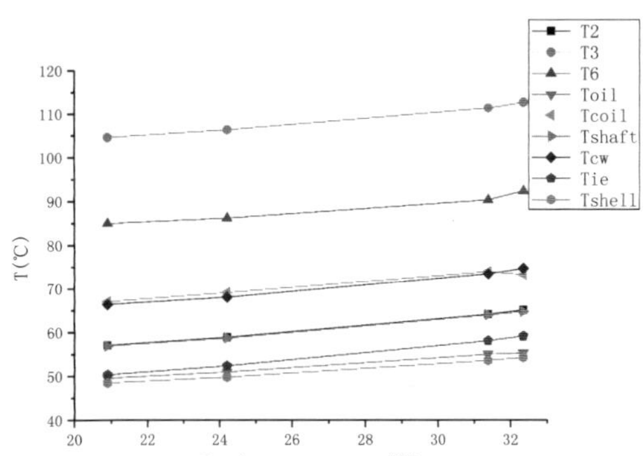

Fig 8. The tested temperatures vary with the suction gas temperature

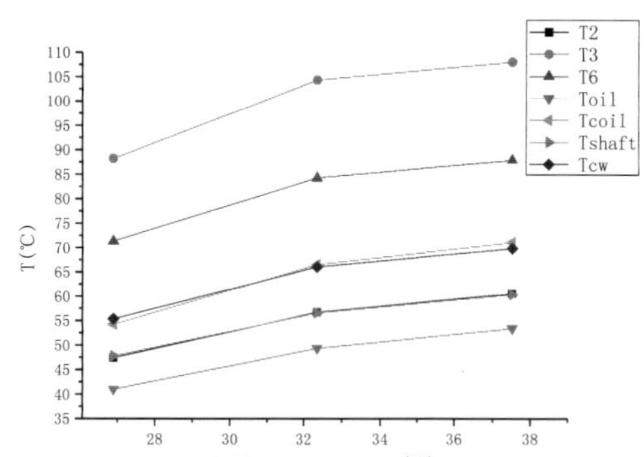

Fig 9. The tested temperatures vary with the ambient temperature

Fig 8 shows the tested temperatures vary with suction temperatures when the evaporating temperature is 5℃,the condensing temperature is 54.4℃ and the ambient temperature is 32.2℃. Fig 9 shows the tested temperatures vary with ambient temperatures when the evaporating temperature is 5℃,the condensing temperature is 54.4℃ and the ambient temperature is 20℃.

4. Comparison

As shown in Table 3, three refrigeration cases are chosen to compare the simulation results and the experimental results. Case No 1.(ASHRARELBP46) can express the LBP conditions. Case No 2.(ASHRAREHBP46) and No 3. can express the HBP conditions.

Table 3. Refrigeration cases chosen to compare the simulation and experiment

Case No.	Evap Temp(°C)	Cond Temp(°C)	Ambi Temp(°C)	Suct Temp(°C)
1	-23.3	54.4	32.2	32.2
2	7.2	54.4	35	35
3	5	60	32.2	32.2

Table 4. shows the results comparison between simulation and experiment, and the error is the calculate using equation (13).

$$\text{Relative error}=\frac{\text{Experiment value- Simulation value}}{\text{Experiment value}}*100\% \tag{13}$$

Table 4. Results comparison between simulation and experiment

Parameters	Case No.1			Case No.2			Case No.3		
	Simu	Test	Error (%)	Simu	Test	Error (%)	Simu	Test	Error (%)
T0(°C)	32.2	32.7	-1.53	35.0	35.1	-0.28	32.2	32.6	-1.23
T1(°C)	73.8	90.0	-18.00	60.1	65.5	-8.24	62.3	68.5	-9.05
T2(°C)	78.8	91.6	-13.97	64.3	66.6	-3.45	62.3	69.5	-10.36
T3(°C)	163.2	156.3	4.41	116.8	111.2	5.04	124.7	119.5	4.35
T4(°C)	157.9	148.6	6.26	111.9	108.5	3.13	119.9	116.6	2.83
T5(°C)	152.1	144.4	5.33	109.4	106.4	2.82	116.1	114.4	1.49
T6(°C)	122.8	115.7	6.14	98.7	96.7	2.07	104.5	103.0	1.46
Tcw(°C)	100.7	111.2	-9.44	70.85	75.7	-6.41	75.3	80.4	-6.34
Tie(°C)	73.8	78.8	-6.35	60.1	60.6	-0.83	62.3	62.6	-0.48
Tshell(°C)	55.3	60.6	-8.75	52.2	53.2	-1.88	51.8	52.9	-2.08
Tdcw(°C)	131.1	126.4	3.72	89.2	88.1	1.25	96.2	94.4	1.91
Tdmw(°C)	114.5	108.8	5.24	76.3	74.4	2.55	81.0	78.9	2.66
Wp (W)	176.66	——	——	317.69	——	——	328.20	——	——
Motor efficiency	75.5%	——	——	74.9%	——	——	74.6%	——	——
We (W)	248.84	247.36	0.60	424.28	388.69	9.16	439.96	408.36	7.74
m_suc(kg/s)	1.11* 10-3	——	——	3.54* 10-3	——	——	3.13* 10-3	——	——
m_dis (kg/s)	-9.844 *10-4	-9.688 *10-4	1.61	-3.316 *10-3	-3.365 *10-3	-1.45	-2.872 *10-3	-2.921 *10-3	-1.68
Capacity(W)	348.93	344.41	1.31	1022.37	1007.5	1.48	974.27	889.52	-1.71
COP	1.402	1.392	0.71	2.410	2.592	-7.02	1.987	2.178	-8.77

Results shows that the temperature error between simulation and experiment is below ±20%. The compressor performance parameters error (power, capacity and mass rate) between simulation and experiment is below±10%.

5. Conclusion

The experiments of R290 refrigeration compressor are carried out. The heat transfer value of discharge line is 3-5 times as large as the value of discharge chamber or discharge muffler. The heat transfer value ratio of motor is about 23%. There are linear relationships between the temperatures in compressor and the conditions temperature. The heat transfer simulation of reciprocating compressor is studied and the heat transfer coefficients are chosen from reference and revised by experiment data. The relative error of temperatures between simulation and experiment is below ±20% and the relative error of compressor performance parameters is below ± 10%. Therefore, the numerical simulation of reciprocating compressor can simulate its working process well.

Reference:

[1] Wu S, Wang Z, Xiong X, et al. Research on System Simulation of Refrigerator Compressor[J]. China Appliance, 455~458, 2012.

[2] Wu Shoufei, Wang Zonghuai, Wang Zhen. Fluid Structure Interaction Based Refrigerator Compressor Exhaust System Simulation [J]. China Appliance, 307~312, 2011.

[3] Ma Pengcheng. The working process simulation and the performance analysis of the R600a refrigeration compressor [D]. Xi an, Xi'an Jiao Tong university, 2011.

[4] Zhou Ruixin, Guo Bei, Lu yang, Chang Yunfeng, Zhao, Yali, Wang Wei, Fabian Fagotti. Numerical simulation for household refrigeration compressor with experimental validation. The 7th International Conference on Compressor and Refrigeration(ICCR2015), 2015.

[5] Todescat, M.L., Fagotti, F., Prata, A.T., Ferreira, R.T.. Thermalenergy analysis in reciprocating hermetic compressors. Proceedings International Compressor EngineeringConference at Purdue. West Lafayette, USA, pp. 1419–1428, 1992

[6] Meyer, W.A., Thompson, H.D.,. An analytical model of heattransfer to the suction gas in a low-side hermeticrefrigeration compressor. InternationalCompressor Engineering Conference at Purdue. WestLafayette, USA, pp. 898–907,1990.

[7] Ooi, K.T.. Heat transfer study of a hermetic refrigerationcompressor. Appl. Therm. Eng. 23, 1931–1945, 2003.

[8] Joel SANVEZZO Jr.,Cesar J. DESCHAMPS. A heat transfer model combining differential and integral formulations for thermal analysis of reciprocating compressors. International Compressor Engineering Conference at Purdue. West Lafayette, USA, 2012.

[9]B Lawton, Bsc(Eng), PhD, CEng, MIMechE. Effect of compression and expansion on instantaneous heat transfer in reciprocating internal combustion engines. Pro Instn Mech ENgrs Vol 201 No A3(IMechE 1987), 175-186, 1987.

[10] F. Fogotti, M.L. Todescat, R.T.D.S. Ferreira, A.T. Prata, Heat transfer modelling in a reciprocating compressor,in: Proceedings of the 1994 Purdue University Compressor Engineering Conference, pp. 605–610, 1994.

Acoustical contribution calculation and analysis of compressor shell based on acoustic transfer vector method

Xiaol Chen[1], Bei Guo[1], Jinliang Tuo[1], Ruixin Zhou[1], Yang Lu[1]

[1]Xi'AnJiaoTong University, People's Republic of China

E-mail： guobei@mail.xjtu.edu.cn

Abstract. Nowadays, people are paying more and more attention to the noise reduction of household refrigerator compressor. This paper established a sound field bounded by compressor shell and ISO3744 standard field points. The Acoustic Transfer Vector (ATV) in the sound field radiated by a refrigerator compressor shell were calculated which fits the test result preferably. Then the compressor shell surface is divided into several parts. Based on Acoustic Transfer Vector approach, the sound pressure contribution to the field points and the sound power contribution to the sound field of each part were calculated. To obtain the noise radiation in the sound field, the sound pressure cloud charts were analyzed, and the contribution curves in different frequency of each part were acquired. Meanwhile, the sound power contribution of each part in different frequency was analyzed, to ensure those parts where contributes larger sound power. Through the analysis of acoustic contribution, those parts where radiate larger noise on the compressor shell were determined. This paper provides a credible and effective approach on the structure optimal design of refrigerator compressor shell, which is meaningful in the noise and vibration reduction.

1. Introduction

Refrigerator is one of the widely-used household appliances. The hermetic reciprocating compressor is the main noise source so that vibration and noise reduction of compressor is the key of refrigerator noise reduction.

As so far, many researchers have studied many different methods to reduce the refrigerator noise. Zhou and Jiang[1] used fast multipole BEM to find the weak part of compressor shell where some optimal methods were used to reduce the radiation noise of compressor shell; Zhong and Jiang[2] found two biggest noise source in compressor through the block test method and determined that the piston mechanism is main cause of low-frequency noise radiation; Li and Huang[3] used theoretical method to establish the vibration velocity formulas on compressor shell which can be used to describe the relationship between vibration velocity on shell and the pressure on field point; Philippe Dugast[4] introduced an interior source finite element method to simulate the compressor noise which has been used as an optimal method to reduce compressor noise about 2dB.

Compared with these methods used in compressor noise analysis mentioned above, the Acoustic Transfer Vector approach used in this paper has some advantages: firstly, it merely depends on the inherent attribute of compressor structural, the position of field points and the medium in the sound field while doesn't depend on the load fluctuation on compressor and the structural response of compressor; then, it just need be calculated once when obtaining the response of one certain compressor in different operating condition; therefore, it saves large time on calculation and increases the research efficiency.

In this paper, the compressor shell is divided into several parts, and the acoustic contribution of each part is obtained basing on Acoustic Transfer Vector approach. After analysing the acoustic contribution of each part, some parts on compressor shell where radiating biggest noise were determined. With the help of this method, we can design the compressor shell reasonably to reduce noise.

2. Acoustic theory basis

2.1 The Acoustic Transfer Vector approach

Acoustic Transfer Vector is the inherent attribute of acoustic system. If the surface shape of vibrating body, the acoustic medium and the field points are fixed, the ATV in this sound field is fixed. So for a sound field, the ATV is calculated one time that it can be used to predict sound pressure causing by vibrating surface on different loadings.

The surface of radiating body is discretized into several units which can be regarded as a vibration source. The vibration on these units cause the pressure fluctuation in medium then the sound pressure on one field point is linear superposed by sound pressure caused by each unit on this point. This relationship in one frequency can be described as formula (2-1) [5]:

$$p(\omega) = ATV(\omega) \cdot v_n(\omega) \tag{2-1}$$

In this formula: $p(\omega)$ is the sound pressure; $ATV(\omega)$ is the Acoustic Transfer Vector; $v_n(\omega)$ is the vibration velocity in normal direction of structure surface; ω is the angular frequency。
When using the boundary element to obtain discretization, the value of grid node can be obtained as the matrix following:

$$[A]\{P_b\} = [B]\{v_b\} \tag{2-2}$$

Corner mark b stands for the boundary.
The sound pressure on field point can be calculated by formula (2-3):

$$p = \{d\}^T \{P_b\} + \{m\}^T \{v_b\} \tag{2-3}$$

Combining formula（2-2）and formula（2-3）obtains this formula:

$$p = \{ATV\}^T \{v_b\} \tag{2-4}$$

Above, $\{ATV\}^T$ can be expressed as this formula:

$$\{ATV\}^T = \{d\}^T [A]^{-1} [B] + \{m\}^T \tag{2-5}$$

It is obvious that the ATV is the transfer function between the vibration velocity of vibrating surface in normal direction and the sound pressure on filed point. When the sound pressure is calculated in several positions, formula (2-4) can be replaced by formula (2-6)

$$p = \{ATM\}^T \{v_b\} \tag{2-6}$$

Above, $\{ATM\}^T$ is consist of different Acoustic Transfer Vectors。

2.2 Acoustic contribution

2.2.1 Sound pressure contribution

Basing on ATV method, the sound pressure contribution of each grid unit can be expressed as formula (2-9):

$$p = \left[ATV(\omega)\right]^T \left[v_e(\omega)\right] = \sum_{i=1}^{N} ATV_i(\omega) \cdot v_{e,i}(\omega) = \sum_{i=1}^{N} p_{e,i}(\omega) \tag{2-7}$$

In this formula: N is the total number of units; $ATV_i(\omega)$ is the element of matrix $\left[ATV_i(\omega)\right]$; $v_{e,i}(\omega)$ is the velocity in normal direction of unit i; $p_{e,i}(\omega)$ is the sound pressure contribution, which can be described as this formula:

$$p_{e,i}(\omega) = ATV_i(\omega) \cdot v_{e,i}(\omega) \text{。} \tag{2-8}$$

In order to analyse the sound pressure contribution of each unit, the value of $v_{e,i}(\omega)$ and $ATV_i(\omega)$ must be known advanced.

Panel is consist of a set of units, so that the sound pressure contribution of one panel is the summation of the sound pressure contribution of each unit on this panel, which can be described as this formula [6]:

$$p = \sum ATV_i(\omega) \cdot v_{e,i}(\omega) = \sum p_{e,i}(\omega) \tag{2-9}$$

The sound pressure on field point is the vector addition of the sound pressure caused by each panel, shown as this formula:

$$\vec{p} = \sum_{i=1}^{n} \vec{p}_i \tag{2-10}$$

Above: $\vec{p}_i$ is the sound pressure contribution to the field point of panel i; n is the total number of panel。

2.2.2 Acoustical power contribution

Panel sound pressure contribution can only describe the contribution to one field point, so in order to describe the noise radiation of one panel contribute to whole sound field, the acoustical power contribution is introduced as following:

$$W_c(\omega) = \frac{S}{\rho_0 c}\left(\sum_{i=1}^{n_p} \{ATV_i(\omega)\}^T \{v_i^n(\omega)\}\right)^2 \tag{2-11}$$

$$W = W_{c1} + W_{c2} + \cdots + W_{cn} \quad , \quad n = 1,2,3\cdots \tag{2-12}$$

Above: W_{ci} is the acoustical power contribution of one panel; $W_c(\omega)$ is the panel acoustical power contribution in frequency ω; W is the total acoustical power in the sound field; n is the total number of panel。

With the help of panel acoustical power contribution, the larger panel acoustical power contribution will be found quickly and some optimal methods will be used in these panels.

3. Sound field establishment and compressor shell division

In noise experiment, the test points are arranged in the place where is 300mm distance to each surface of compressor. In order to verify the accuracy of calculation, five field points on the same position to text experiment are established in LMS Virtual.Lab software (the distance to ground is 150mm and the distance to compressor centre is 300mm of S1、S2、S3、S4; S5 is right above compressor and its distance to ground is 300mm), which shown as figure 3-1.

For explaining ATV calculation of sound field further more, ISO3744 standard field points was estabilshed in this paper. This paper used LMS Virtual.Lab software to establish ISO3744 standard field points on Fine Sphere Mesh type, which refines the standard field points. Under the situation of symmetry panel, there is a hemisphere suface field points grid consist of 19 nodes and 18 units around compressor shell when esabilishing the standard field points. The standard field points model is shown as figure 3-2.

IOP Conf. Series: Materials Science and Engineering **232** (2017) 012012 doi:10.1088/1757-899X/232/1/012012

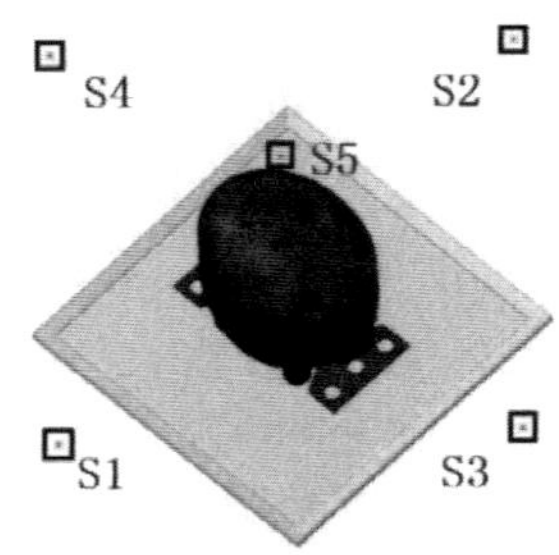

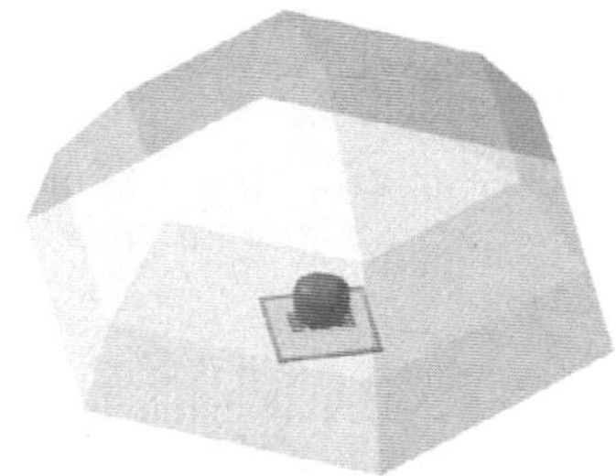

Figure 3-1. Field points arrangement **Figure 3-2.** ISO 3744 standard field points

After establishing the sound field, the compressor shell is divided into several parts which can be regarded as vibrating panel according to different structure in compressor. So the compressor shell is divided into upper shell, below shell and engine base. The upper shell is divided into four symmetrical parts; the below shell is divided into five parts where four parts above engine base are symmetrical parts same to upper shell and below the engine base is one part; the engine base is divided into two parts: left engine base and right engine base. Above all, there are eleven parts on the compressor shell surface shown as figure 3-3.

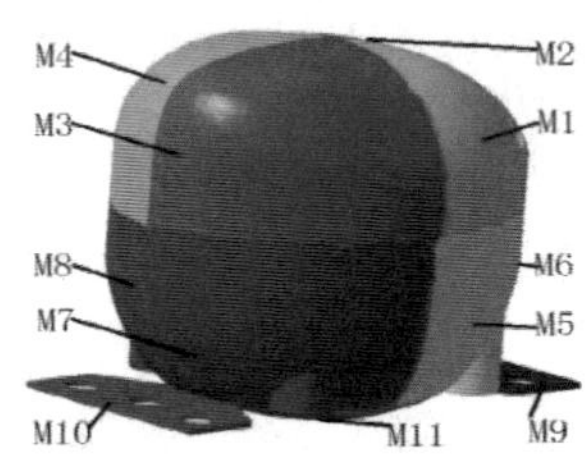

Figure 3-3. The division of compressor shell surface

4. The correctness verification of ATV calculation

After ATV calculation by LMS Virtual.Lab software, these points: ISO:1、ISO:6、ISO:14 and ISO:19 were selected to analyze. The following figures show the ATV cloud picture of these four points in 630Hz:

Figure 4-1. ATV cloud picture of point ISO:1 in 630Hz

Figure 4-2. ATV cloud picture of point ISO:6 in 630Hz

Figure 4-3. ATV cloud picture of point ISO:14 in 630Hz

Figure 4-4. ATV cloud picture of point ISO:19 in 630Hz

Figure 4-1 to Figure 4-4 show the sound transfer of compressor shell to sound filed points as well as the sound contribution of compressor shell to sound filed points. Point ISO:1 is head against nameplate and close to the bottom where the nameplate on compressor shell contributes larger sound transfer shown as Figure 4-1; point ISO:6 is close to the opposite of junction box where the compressor engine base and spring pedestal make lager sound contribution shown as Figure 4-2; point ISO:14 is close to compressor junction box where the compressor engine base and junction box make lager sound contribution shown as Figure 4-3; point ISO:19 is right above compressor where the top surface of compressor shell and the ground reflection make larger sound contribution to shown as Figure 4-4.

In order to verify the correctness of ATV calculation on compressor shell, 5 field points same to those points in noise test experiment were defined in LMS Virtual.Lab software when the standard sound field was established. Figure 4-5 to Figure 4-9 show the 1/3 octave sound pressure level comparison of simulation value to experiment value of these 5 field points.

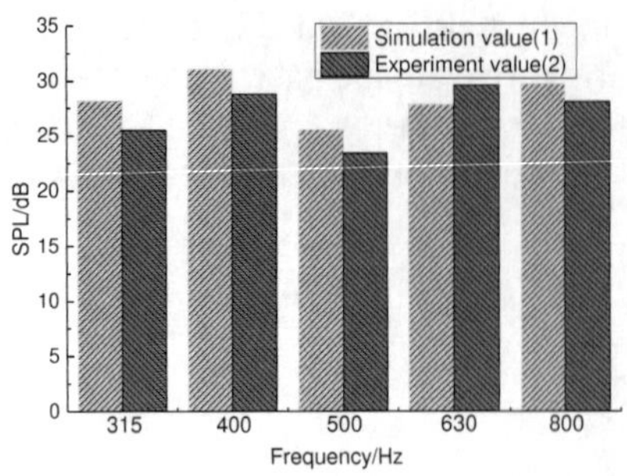

Figure 4-5. The 1/3 octave sound pressure level comparison of point S1

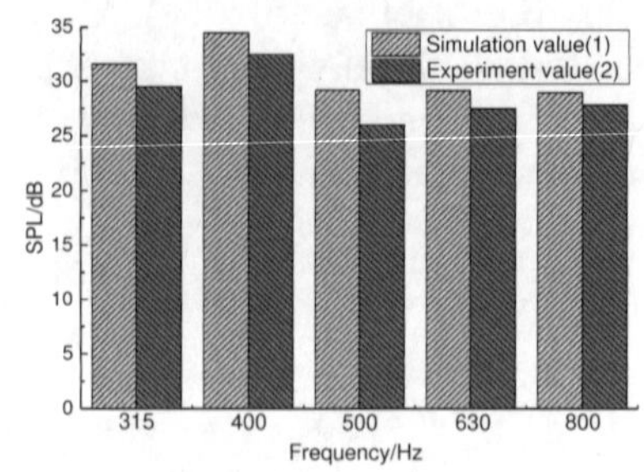

Figure 4-6. The 1/3 octave sound pressure level comparison of point S2

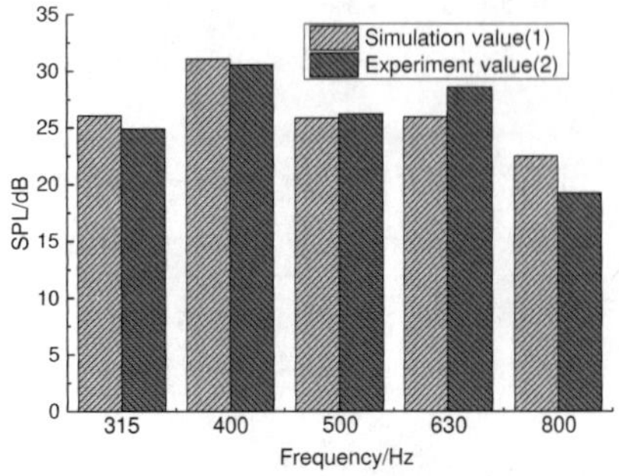

Figure 4-7. The 1/3 octave sound pressure level comparison of point S3

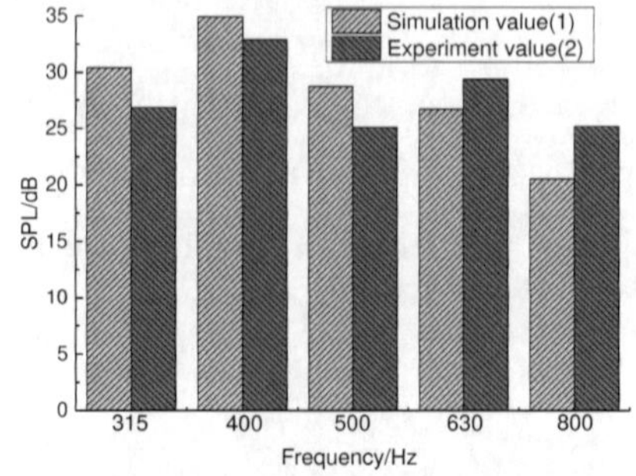

Figure 4-8. The 1/3 octave sound pressure level comparison of point S4

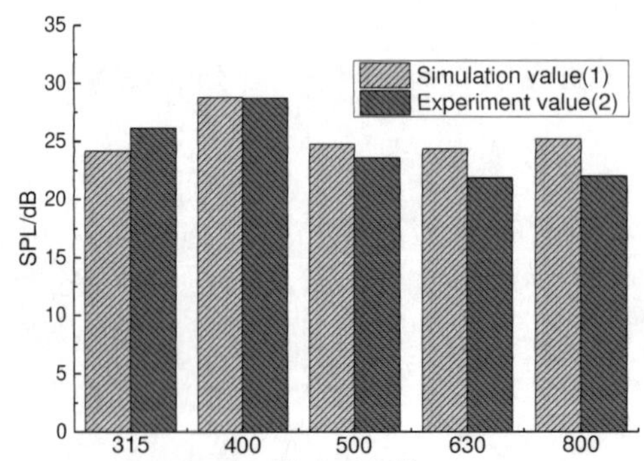

Figure 4-9. The 1/3 octave sound pressure level comparison of point S5

Shown as figures above, points S1、S2、S3 and S4 have larger 1/3 octave sound pressure level in both 400 Hz and 630 Hz, and point S5 is right above compressor which has larger 1/3 octave sound pressure level in 400 Hz. Comparing the simulation value to the experiment value of these points, their errors are almost within 12%, but there are still a few errors are beyond 12%. These errors are produced by the measurement errors in experiment, calculation errors in data processing and the error caused by different situation set in simulation and experiment while the reflection of ground is total reflection in simulation but the reflection of ground is not total reflection in experiment which makes the values in simulation are larger than experiment. In spite of the errors exist in simulation, we can also easily obtain the sound pressure variation tendency in different frequency and find the larger sound pressure in some frequencies. So the ATV method can be used in prediction of compressor shell radiation noise.

5. Acoustic contribution analysis

5.1 Panel sound pressure contribution analysis in ISO 3744 standard sound field

The distance of all points in ISO 3744 standard sound field is 1m to the shell which is beneficial to determine the larger noise resource on the shell. Because of the symmetry of compressor shell, the sound pressure cloud pictures of panel 1、4、6、7、9、11 in 400Hz were selected to analyse, shown as following figures:

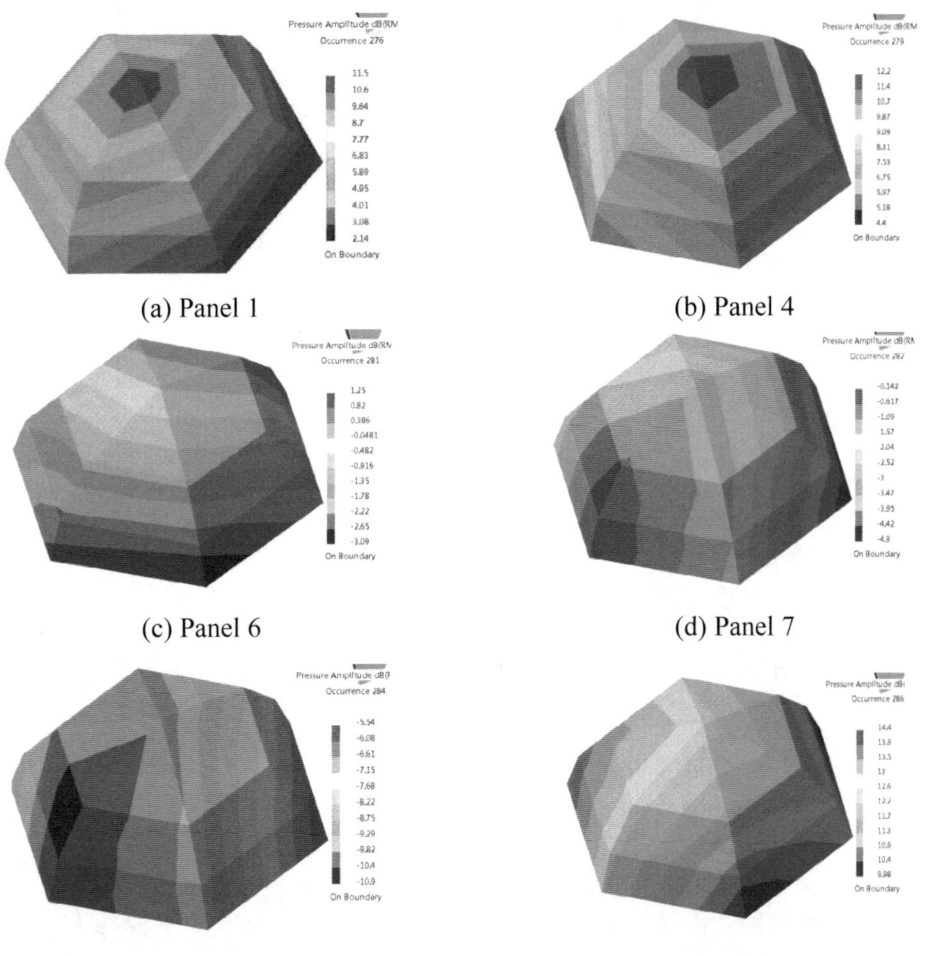

(a) Panel 1 (b) Panel 4

(c) Panel 6 (d) Panel 7

(e) Panel 9 (f) Panel 11

Figure 5-1. Panel sound pressure contribution cloud pictures in 400Hz

The sound pressure contribution made by different panel areas can been straight seen in the sound pressure contribution cloud pictures: the radiated noise in 400Hz of Panel 1 and Panel 4 make larger influence to nearby areas shown as Figure (a) and Figure (b); the contribution of Panel 6, Panel 7 and Panel 9 made to the sound field is negative shown as Figure (c), Figure (d) and Figure (e) which means that these panels weaken the sound level in the sound field; Panel 11 is the bottom of compressor shell and it radiates larger noise level to the sound field shown as Figure (f) which is relative to the ground reflection. From sound pressure contribution cloud pictures, we can tentatively determine that the suction panel, discharge panel and the bottom panel on the shell make larger sound pressure contribution to sound field in 400Hz.

Researchers may easily acquire the sound radiation of each panel and find the larger contribution among these panels from sound pressure contribution cloud pictures. However, the frequency in which the radiated noise is larger may be focused on in compressor noise control. Shown as Figure 5-2, the contribution curves in different frequency of each panel are listed where the red imaginary line stands for the contribution of whole panels made to the field point and the green full line stands for the contribution of one panel made to the field point. In Figure 5-2, the marks in frequency axis are the value of target frequency we want to study which are 250 Hz, 315 Hz, 400 Hz, 500 Hz, 630 Hz and 800 Hz. The frequency around the target frequency hasn't been studied in this paper hence we think the sound pressure value in these frequencies is equal to the target frequency. Then the frequency axis is divided into equal parts where those target frequency are the midpoint in these parts. So the Picture (a) to Picture (f) in Figure 5-2 is shown as step plots that researchers can find the larger contribution made by each panel in the target frequency.

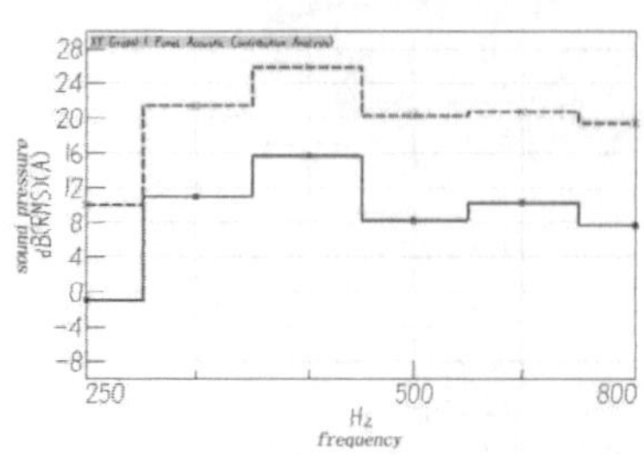

(a) The sound pressure contribution curve of Panel 1 made to point ISO:2

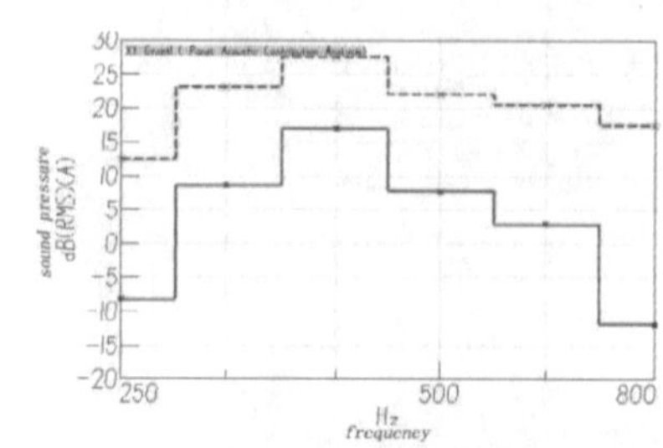

(b) The sound pressure contribution curve of Panel 4 made to point ISO:5

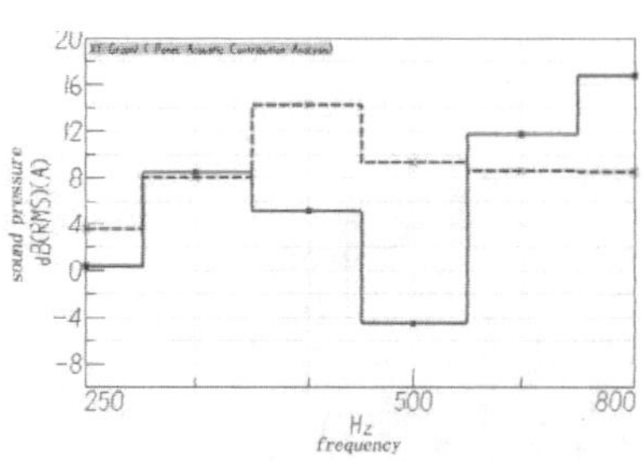

(c) The sound pressure contribution curve of Panel 6 made to point ISO:3

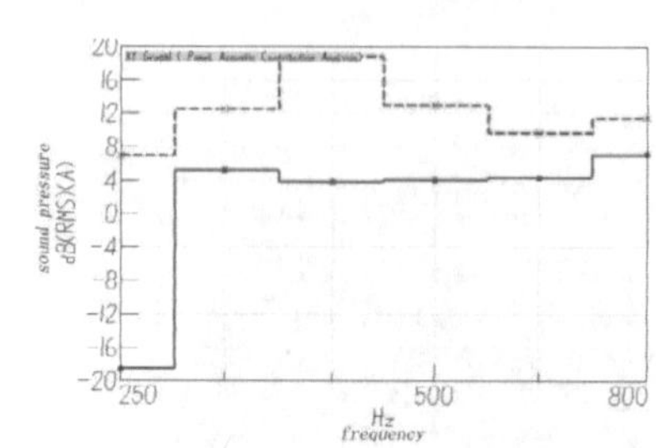

(d) The sound pressure contribution curve of Panel 7 made to point ISO:6

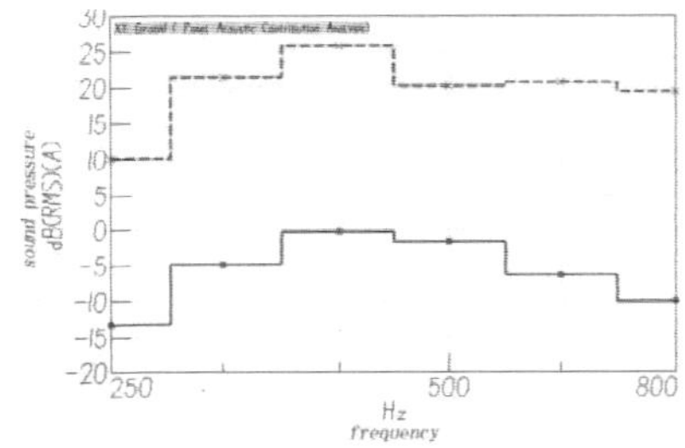

(e) The sound pressure contribution curve of Panel 9 made to point ISO:2

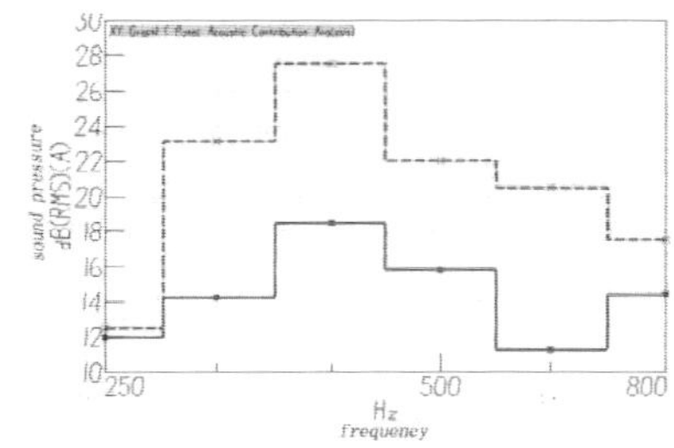

(f) The sound pressure contribution curve of Panel 11 made to point ISO:5

Figure 5-2. The sound pressure contribution curves of Panels made to standard points

Compared to cloud pictures, the sound pressure contribution curves will help researchers understand the frequency characteristic of sound pressure more clearly. The 6 figures in Figure 5-2 show that all these 6 panels radiate larger noise in 400 Hz which is related to the structure of compressor shell. The main radiated frequency of Panel 1 is in 400 Hz and 630 Hz shown as Figure (a); the main radiated frequency of Panel 4 is in 400Hz shown as Figure (b); the main radiated frequency of Panel 6 is in 630 Hz and 800 Hz shown as Figure (b); Panel 7 and Panel 9 make little influence to the field point and even weaken the sound pressure on the field point shown as Figure (d) and Figure (e); the main radiated frequency of Panel 11 is in 400 Hz shown as Figure (f).

5.2 Panel acoustical power contribution analysis in ISO 3744 standard sound field

Mentioned above, panel sound pressure contribution can only account for the contribution made to one field point. But in compressor noise control, the contribution made by compressor shell to whole sound field need to be studied, so that the panel acoustical power contribution was selected to be analysed which can reflect the radiation characteristic of compressor shell as a whole.

In LMS Virtual.Lab software, there are several modes to analyse the acoustical power contribution, including: the acoustical power contribution curves, the contribution histogram of each panel in one frequency and so on. There are different advantages of different modes: the peak frequency may be seen clearly in curve graph which help researchers find the larger contribution in some frequency of different panels; the contribution level of different panels in one frequency may be seen clearly in the contribution histogram. The acoustical power contribution curves made by each panel to the sound field are listed in Figure 5-3.

In Figure 5-3, the curve SUM stands for the acoustical power contribution made by whole compressor shell to the sound field and it has largest contribution in 400Hz which is fit with the peak noise tested in noise experiment preferably. As a whole, each panel has larger contribution in 400Hz. However, the acoustical power contribution of Panel 4, Panel 8and Panel 11 are larger than other panels obviously in 400Hz because of suction panel, discharge panel and bottom panel on compressor. The larger acoustical power contribution of Panel 5, Panel 8 and Panel 11 are in 630Hz which is due to the motor junction box, compressor suction and discharge line and bottom panel of compressor. Aiming at these two frequencies which have larger acoustical power contribution, the contribution histograms in these two frequencies are acquired in LMS Virtual.Lab which describe the contribution level of different panels in these two frequencies shown as Figure 5-4 and Figure 5-5.

IOP Conf. Series: Materials Science and Engineering **232** (2017) 012012 doi:10.1088/1757-899X/232/1/012012

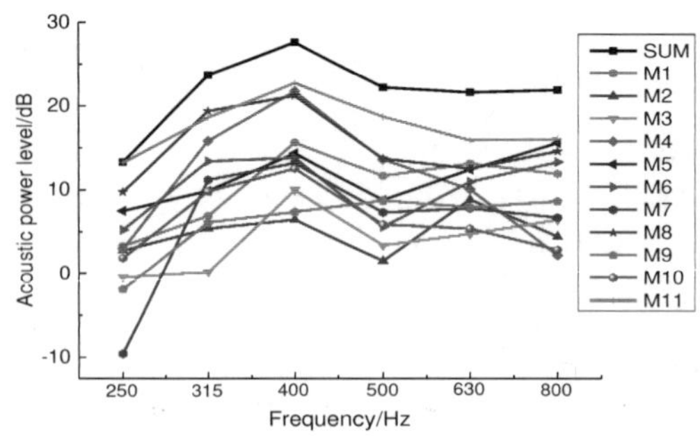

Figure 5-3. The 1/3 octave acoustical power curves of each panel

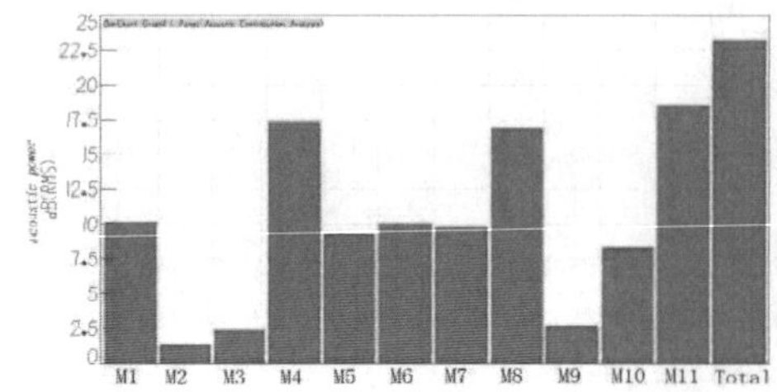

Figure 5-4. Panel acoustical power contribution in 400Hz

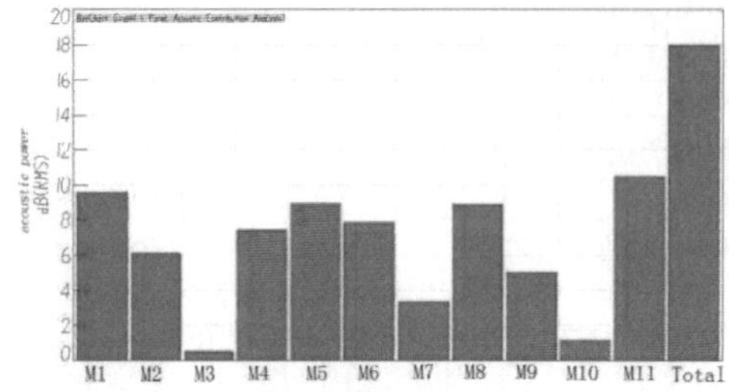

Figure 5-5. Panel acoustical power contribution in 630Hz

The result of contribution histograms analysis is fit with the analysis conclusion that Panel 4, Panel 8 and Panel 11 make larger acoustical power contribution in 400Hz and Panel 1, Panel 4, Panel 5, Panel 8 and Panel 11 make larger acoustical power contribution in 630Hz. Above all, these panel areas may be emphatic focused on in freezer compressor noise control, meanwhile, the analysis of panel acoustic contribution provide a direction to determine the noise source inside the compressor.

6. Conclusion

The ISO 3744 standard sound field surrounding compressor shell surface was established in this paper and the ATV on the sound field was calculated. The ATV method was used to predict the sound pressure on field points which was compared with the sound pressure on the same field points tested in noise experiment. After ATV calculation, the area on compressor shell was divided into several panels on which the panel sound pressure contribution and acoustical power contribution was analysed. Some conclusions were acquired following:

(1) The acoustical contribution of compressor shell was predicted through ATV calculation. The sound pressure predicted was compared with the sound pressure tested in experiment and the result fit preferably. The result of comparison indicates that ATV method can be used in prediction of compressor shell radiation noise.

(2) The panel acoustical contribution of different parts on compressor shell was analyzed which indicates that some panels can enhance the sound pressure on field points and others weaken the sound pressure. The acoustical contribution spectrogram of each panel was acquired from which some conclusions were obtained: the discharge and suction line of compressor and the

IOP Conf. Series: Materials Science and Engineering **232** (2017) 012012 doi:10.1088/1757-899X/232/1/012012

bottom panel of compressor shell made larger noise contribution to the sound field in 400Hz; the discharge and suction line of compressor, the panel closed to motor junction box and the bottom panel of compressor shell made larger noise contribution to the sound field in 630Hz; the engine base make negative contribution to the sound field both in 400Hz and 630Hz. The contribution characteristic of compressor shell made to sound field was acquired through panel acoustical power contribution analysis which confirms the conclusions above.

(3) The ATV method can predict the radiation characteristic of compressor shell accurately. The analysis of panel sound pressure contribution and panel acoustical contribution based on ATV method may help researchers acquire the acoustical contribution made by compressor shell to whole sound field and the areas where made the larger contribution. A new method was presented in this paper to confirm the noise source inside the compressor which can enhance the efficiency of noise research on freezer compressor as well.

References

[1] Zhou Xinfu. The Procedure of the Optimal Design of Compressor Shell to Reduce Acoustic Noise [D]. ShangHai: ShangHai Jiao Tong University, 2012.

[2] ZhongChongming. Research on Vibration and Noise of Compressor[D]. ShangHai: ShangHai Jiao Tong University, 2010.

[3] Li Hongliang, Wu Chengjun, Huang Xieqing. The Acoustic Radiation Characteristic Parameter Research of Cylindrical Compressor Shell [J]. Refrigeration Journal, 2002, 04: 5-9.

[4] Dugast P. Global acoustical finite-element simulation of scroll compressor noise from internal sources[C]. Nineteenth International Compressor Engineering Conference at Purdue University, 2008.

[5] Zhan Fuliang, XuJunwei. The Learning of Acoustic Simulating Calculation Soft Virtual.Lab Acoustics from Introduction to Master[M]. Xi'an: NorthwesternPolytechnical University Press, 2013.

[6] Lu Dong, The Research on Modal Analysis of Mechanical Structure Test System [D]. Chongqing: Chongqing University , 2007.

Pulsation damping of the reciprocating compressor with Helmholtz resonator

W Wang, Y Zhang, Q Zhou, X Peng, J Feng and X Jia

Xi'an Jiaotong University, Xi'an, Shannxi, China

E-mail: xypeng@mail.xjtu.edu.cn

Abstract. Research presented in this paper investigated the mounting of a Helmholtz resonator near the valve chamber of a reciprocating compressor to attenuate the gas pulsation in the valve chamber as well as the pipeline downstream. Its attenuation characteristics were simulated with the plane wave theory together with the transfer matrix method, and the damping effect was checked by comparing the pressure pulsation levels before and after mounting the resonator. The results show that the Helmholtz resonator was effective in attenuating the gas pulsation in the valve chamber and piping downstream, and the pulsation level was decreased by 40% in the valve chamber and 30% at maximum in the piping downstream. The damping effect of the resonator was sensitive to its resonant frequency, and various resonators working simultaneously didn't interfere with each other. When two resonators were mounted in parallel, with resonant frequencies equal to the second and fourth harmonic frequencies, the pressure pulsation components corresponding to the resonant frequencies were remarkably decreased at the same time, while the pulsation levels at other harmonic frequencies kept almost unchanged. After a series of simulations and experiments a design criterion of chock tube and volume parameter has been proposed for the targeted frequencies to be damped. Furthermore, the frequency-adjustable Helmholtz resonator which was applied to the variable speed compressor was investigated.

Keywords: Helmholtz resonator; gas pulsation; frequency-adjustable; parameters investigation

1. Introduction

The intermittent suction and discharge of reciprocating compressors would lead the gas pulsation and piping vibration in the system, which degrades the compressor performance, impede the working conditions of the valves and excites the piping vibration [1]. Therefore, it is very important to prediction and control the gas pulsation and piping vibration in a compressor piping system [2]. Currently, installing an orifice plate at the suction and discharge cylinder flange connections [3] is widely used to attenuate the pulsation in the cylinder nozzle as well as the piping system thanks to its simple structure and convenient installation. However, the orifice plate would generate an obvious pressure drop and consume a large amount of horsepower, which might significantly reduce compressor capacity and efficiency. Thus, controlling gas pulsations while maintaining acceptable compressor performance has been a key issue in reciprocating compressors for the last decades.

Helmholtz resonator had been proven to be effective for solving the problem of high amplitude compressor cylinder pulsations which is the source of the piping pulsation downstream. The SwRI (Southwest Research Institute) [4-6] studied many advanced pulsation control technologies including Helmholtz resonator, volume-choke-volume filter, infinite nozzle, and etc. Their research showed that the SBA (side branch absorber) could be installed on both low-speed and high-speed compressors to

IOP Conf. Series: Materials Science and Engineering **232** (2017) 012013 doi:10.1088/1757-899X/232/1/012013

replace any cylinder nozzle orifices that are consuming large amounts of horsepower. The location of installation is either the suction or discharge side of the cylinder or both side of the cylinder. Anderson [7] investigated a single side branch Helmholtz resonator in a circular duct and obtained the resonant frequency from the transfer function of the resonator. Broerman et al. [8] studied the VO (virtual orifice) to replace the compressor nozzle orifice. Their experiments indicated that the pulsation amplitudes in the cylinder nozzle were approximately 50% lower with the virtual orifice than they were with the normal orifice. For the double-acting compressor, Liu Boxiang [9] clarified the installation positions of Helmholtz resonator had a significantly influence on the attenuation effectiveness, and the best obtained when two resonators were installed at both the HE and the CE. Chen Zhicai [10] proposed an extended neck or a spiral neck to take the place of the traditional straight neck of a Helmholtz resonator when space was limited. Jean-Michel Coulon [11] investigated the role of distance between multiple Helmholtz resonator side branch openings on the whole array attenuation which could form a wide band silencer.

The analysis of the gas pulsation was based on the plan wave theory which has been proved accurate in the low-frequency pulsation analysis [12-13]. In this article, a model, based on the plan wave theory, was built to simulate the gas pulsation which contained reciprocating compressor, nozzle, chamber bottle, scrubber and heat exchanger. To verify the simulation, series of experiments were conducted with a reciprocating compressor. Based on the experimental results, the non-interference characteristics of multiple Helmholtz resonators was analysed, as well as the transmission properties of Helmholtz resonator with/without the piping system. For a wider absorption bandwidth, a frequency-adjustable Helmholtz resonator was proposed and investigated. Meanwhile, for the Helmholtz resonator parameters diversity of the same characteristic frequency, the further studies focused on the choke tube and volume's parameters analysis had been carried out, and a design criterion in practice application was recommended.

2. Theoretical studies on the frequency characteristic

As shown in figure 1, the Helmholtz resonator is composed of a choke tube and a volume, where V is the volume of the container, S is the cross-section area of the choke tube, L is the length of choke tube, d is the diameter of the choke tube.

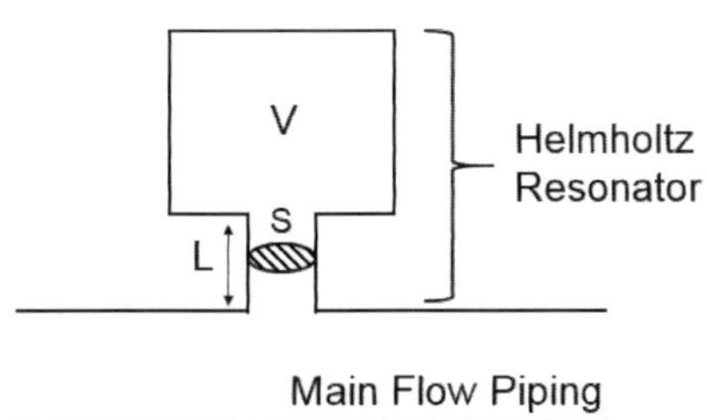

Figure 1. Illustration of Helmholtz resonator.

The characteristic frequency of the Helmholtz resonator could be calculated by following Equation,

$$f_0 = \frac{a}{2\pi}\sqrt{\frac{S_b}{L_b V}} \tag{1}$$

where c is the local speed of sound, Lb is the acoustic length of the choke tube, which could be obtained from Eq. (2),

$$L_b = l + 0.7d \tag{2}$$

where l is the physical length of the choke tube.

To get the frequency characteristic of the Helmholtz resonator in the piping system, the study of acoustic propagation property in branch piping is necessary. Figure 2 shows the branch pipe system, where pi is the main piping incoming wave pressure, pr is the reflected wave pressure, pt is the transmitted wave pressure and p2 is the side branch wave pressure.

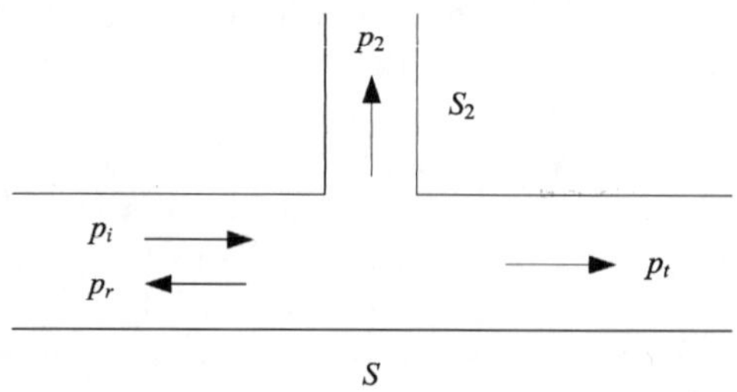

Figure 2. Acoustic propagation in branch piping.

Selected the intersection of the branch pipe and the main pipe as the origin of coordinates, based on the plane wave theory, the pressures and velocities could be obtained from Eq. (3):

$$\begin{cases} p_i = p_{ai}e^{j\omega t},\ v_i = \dfrac{p_i}{\rho_0 c_0} \\ p_r = p_{ar}e^{j\omega t},\ v_r = -\dfrac{p_r}{\rho_0 c_0} \\ p_t = p_{at}e^{j\omega t},\ v_t = \dfrac{p_t}{\rho_0 c_0} \\ p_2 = p_{a2}e^{j\omega t},\ v_2 = \dfrac{p_2}{S_2 Z_{a2}} \end{cases} \tag{3}$$

Considering the continuity of the pressure and velocity at the intersection point, Eq. (3) could be deduced to Eq. (4) and (5):

$$p_i + p_r = p_t = p_2 \tag{4}$$

$$Sv_i + Sv_r = Sv_t + S_2 v_2 \tag{5}$$

And the reflection coefficient of pressure $|r_p|$ could be derived from Eq. (3) and (5):

$$|r_p| = \left|\frac{p_{ar}}{p_{ai}}\right| = \left|\frac{-\rho_0 c_0 / 2S}{\rho_0 c_0 / 2S + Z_{a2}}\right| \tag{6}$$

The transmission coefficient of pressure could be obtained from Eq. (4) and (6):

$$\alpha_t = \frac{R_{a2}^2 + X_{a2}^2}{(\dfrac{\rho_0 c_0}{2S} + R_{a2})^2 + X_{a2}^2} \tag{7}$$

The impedance of Helmholtz resonator could be calculated by following equation:

$$Z_{a1} = j(\omega M_a - \frac{1}{\omega C_a}) = jX_a \tag{8}$$

where M_a is the sound mass, C_a is the acoustic capacitance.

Generally, the transmission coefficient is used to denote the frequency characteristic of the Helmholtz resonator, and the transmission coefficient of the branch pipe with the Helmholtz resonator could be derived from Eq. (7) and (8):

$$\alpha_t = \frac{1}{1+\dfrac{\rho_0^2 c_0^2}{4S^2(\omega M_a - \dfrac{1}{\omega C_a})^2}} = \frac{1}{1+\dfrac{c_0^2}{4S^2(\dfrac{\omega L_b'}{S_b} - \dfrac{c_0^2}{\omega V})^2}} \tag{9}$$

In order to study the frequency characteristic of a Helmholtz resonator, a series of simulations and experiments were conducted with an air reciprocating compressor and piping system. The compressor rated speed was 750 rpm, and a frequency transformer was applied to adjust rotational speed. Considering the fundamental frequency of the compressor was 12.5 Hz, two Helmholtz resonators with characteristic frequency of 25 Hz and 50 Hz were designed and processed, which were named 2x resonator and 4x resonator. The parameters of resonators were shown in Table 1.

Table 1. Parameter of the resonators.

	2x resonator	4x resonator
Diameter of choke tube/m	0.051	0.051
Length of choke tube /m	0.95	0.4
Diameter of volume/m	0.152	0.152
Length of volume/m	0.6	0.35
Volume/m^3	0.011	0.007

The transmission coefficient of the 2x resonator can be obtained by Eq. (9), as shown in figure 3. The transmission coefficient decreased sharply to 0 around 25 Hz. Beyond this narrow bandwidth, the transmission coefficient remained at 1.0. It indicated that the 2x resonator merely worked efficiently at its characteristic frequency. When the frequency to be suppressed coincides with the characteristic frequency, the Helmholtz resonator could attenuate gas pulsation effectively. Otherwise, the resonator had no effect.

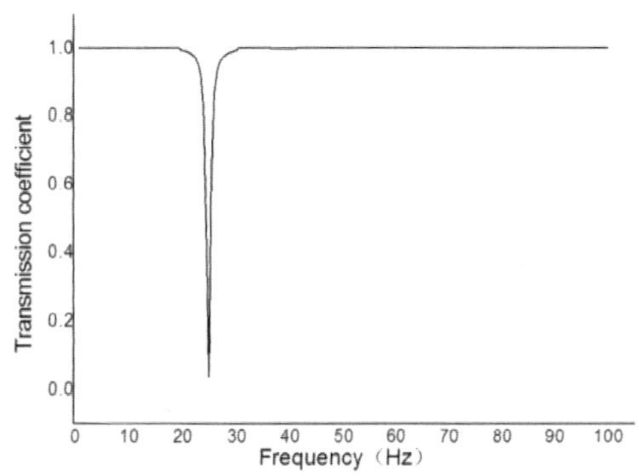

Figure 3. Theoretical transmission coefficient of 2x Helmholtz resonator.

The 2x Helmholtz resonator connected with the main piping was also simulated, as shown in figure 4. A constant velocity was given as a motivation at the left end of the main piping. And the pressure at the right end, which was an anechoic end, could be obtained by the frequency-swept method. The simulation result of the 2x resonator transmission coefficient was shown in figure 5. It was noted that there was an obvious difference between the transmission coefficients in figures 3 and 5. The analysis would be discussed after completed the experimental investigation.

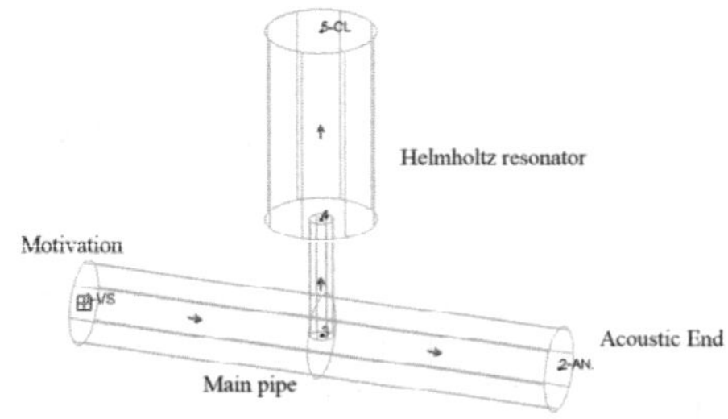

Figure 4. Simulation model of Helmholtz resonator.

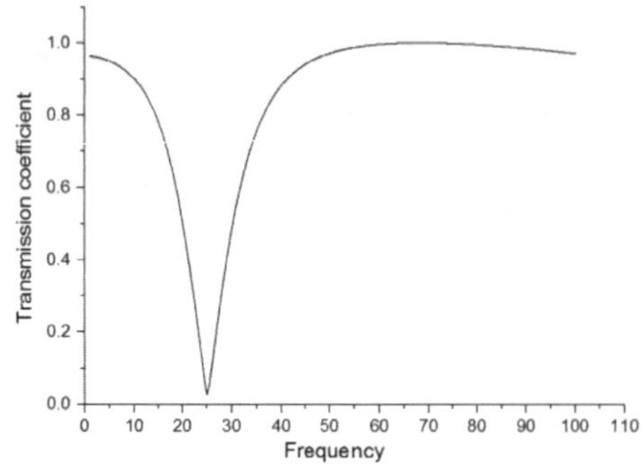

Figure 5. Simulated transmission coefficient of 2x Helmholtz resonator.

In order to study the pulsation attenuation of a Helmholtz resonator, an entire interstage piping system of the double acting air reciprocating compressor adopted for the experiments was built. As shown in figure 6, the model contained the 1st discharge nozzle, 1st discharge chamber, heat exchanger, scrubber, 2nd suction tank and 2nd suction nozzle. The 2x Helmholtz resonator was mounted near the 1st discharge valve chamber. The boundary conditions contained the reciprocating compressor motivation and the closed end. A series of simulations were carried out at various compressor rotational speed. The rotational speed was changed from 650 rpm to 800 rpm, and the interval was set as 50 rpm. Nine nodes were monitored to record the pulsation pressure. Node 1 to 3 were near the 1st discharge cylinder, including the Head End and Crank End. Node 4 and 5 were on the 1st discharge nozzle. Node 6 and 7 were on the 1st discharge tank. Node 8 and 9 were on the pipe after the buffer tank.

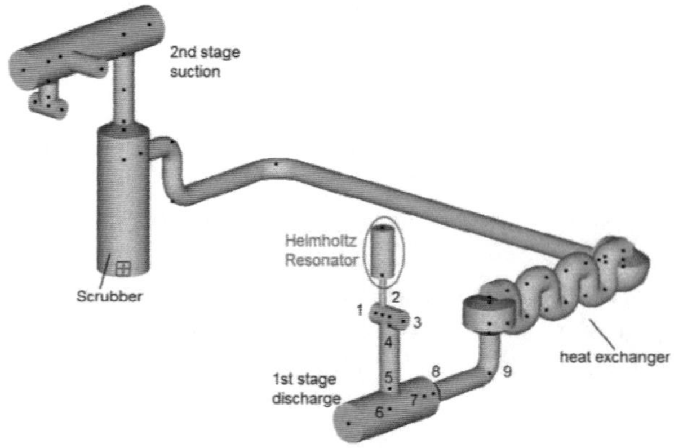

Figure 6. Simulation model of the interstage piping.

IOP Conf. Series: Materials Science and Engineering **232** (2017) 012013 doi:10.1088/1757-899X/232/1/012013

Figure7 illustrated the pressure pulsation amplitude of the nine nodes at 2' order under different compressor rotational speed. As shown in figure 7, thanks to the attenuation effect of the 1st discharge tank, the pressure pulsation amplitude of nodes 5 to 9 were much less than that of nodes 1 to 4. Figure 8 showed the pressure pulsation attenuation rate of the nine nodes at 2' order under different rotational speed. It was obvious that the 2x Helmholtz resonator had no influence on the pulsation under 650 rpm and 700 rpm, but the maximal attenuation rate was reached 40% while the compressor rotational speed was at 750 rpm. The 2' order frequency of 750 rpm was equal to the characteristic frequency of 2x Helmholtz resonator, so the transmission coefficient of the resonator reached the minimum value, which led to a maximal pulsation attenuation. Meanwhile, the pressure pulsation attenuation rate under 750 rpm declined from 40% to 20% with the nodes changed from 1 to 9. It indicated that a Helmholtz resonator could attenuate pressure pulsation at not only the mounting position but also the pipeline downstream, but the farther away from the mounting position, the less attenuation effect of the resonator.

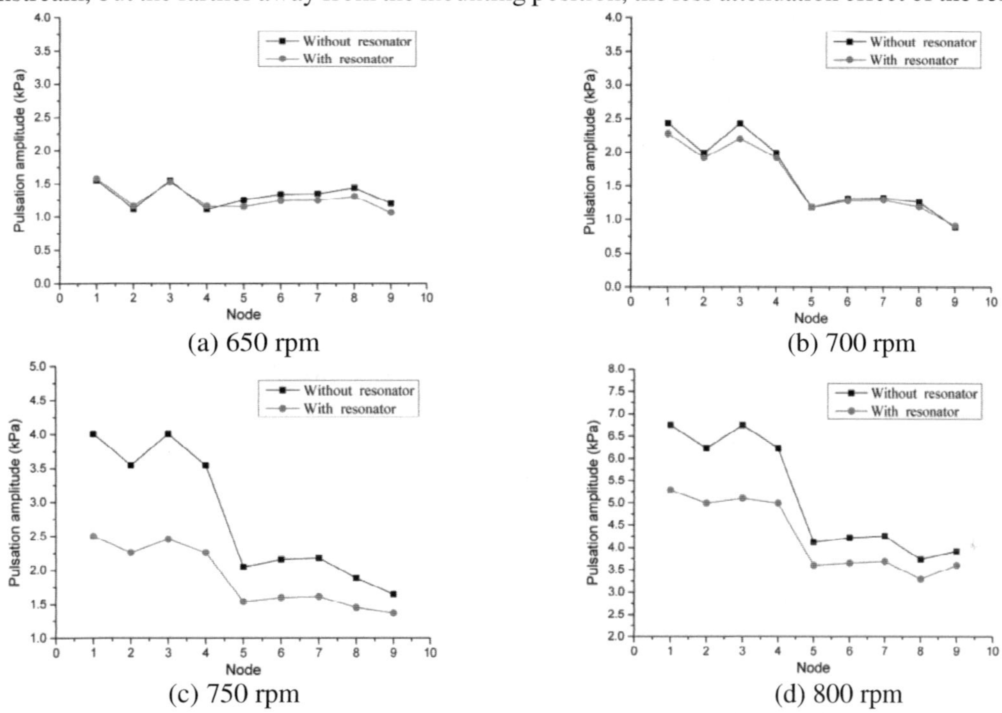

Figure 7. Pressure pulsation amplitude at different rotational speed.

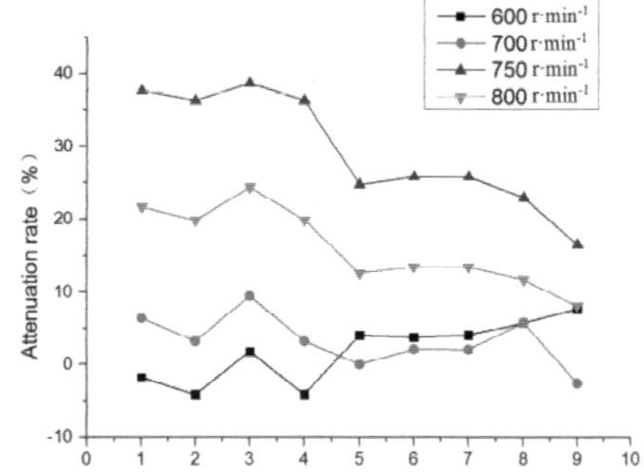

Figure 8. Pressure pulsation attenuation rate at different rotational speed.

3. Experimental study

To verify the attenuation effect and the frequency sensibility of Helmholtz resonator, the 2x resonator and 4x resonator whose characteristic frequency were 25 Hz and 50 Hz were mounted on the cylinder valve gap, as shown in figure 9. A gate valve was employed as a switch to control the resonator quickly and easily. Two piezoresistive pressure sensors were used to collect the pressure signals at different position, and the signals were transferred into the data acquisition system for further processing and recording. The sensor 1 was installed in the valve chamber and the sensor 2 was installed in the pipeline after the 1st discharge tank.

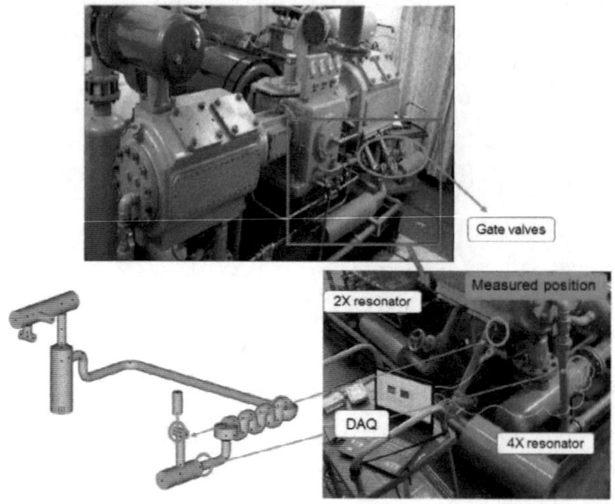

Figure 9. Schematic diagram of the test rig.

Figures 10 and 11 illustrated the first six order's pressure pulsation amplitude of the sensor 1 and 2 while the compressor rotational speed kept at 750 rpm. It can be seen that as the 2x resonator connected to the valve chamber, the pressure pulsation of 2' order declined from 3.75 kPa to 2.25 kPa at the sensor 1 position, and at sensor 2 position, the amplitude declined from 3.40 kPa to 2.40 kPa. However, the pressure pulsations of other orders were nearly unchanged at all sensor positions no matter the 2x resonator was connected or not. When the 4x resonator was connected to the valve chamber, the pressure pulsations of 4' order were decreased from 1.65 and 1.20 kPa to 0.95 and 0.80 kPa at two sensor positions, respectively. Similarly, connecting the 4x resonator also had no influence on the pressure pulsation to the other orders. It was worth noted that while the two resonators were connected to the valve chamber in parallel, the pressure pulsation amplitudes of 2' and 4' orders were both decreased simultaneously, and the attenuation rates were as same as the resonators worked separately. Meanwhile, the pressure pulsations of the other orders were still almost unchanged.

Based on the experimental results, it can be draw that employing a Helmholtz resonator could attenuate the pressure pulsation effectively in the valve chamber and pipeline downstream. However, the pulsation attenuation rates of 2' and 4' order were 40.0% and 42.4%at valve chamber, and 29.4% and 33.3% at pipeline downstream. Furthermore, employing multiple resonators could attenuate the multi-frequency pulsation synchronously, which proposed an effective way to attenuate the pressure pulsation which was resulted from more than one high-amplitude orders.

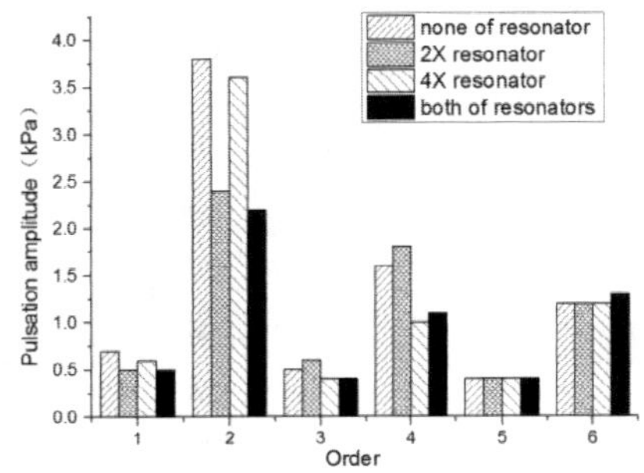

Figure 10. Pressure pulsation amplitude of sensor 1.

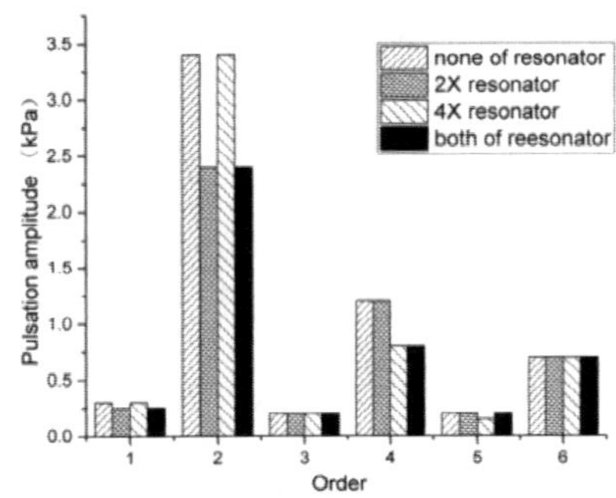

Figure 11. Pressure pulsation amplitude of sensor 2.

Adjusting the compressor rotational speed from 600 rpm to 810 rpm and connecting the 2x resonator to the valve chamber. The pressure pulsation attenuation rates of the 2' order at sensor 1 position were shown in figure 12. It can be seen that attenuation rate increased obviously as the rotational speed changed from 600 rpm to 750 rpm, then decreased from 40.0% to 33.8% when the speed was increased continuously. The maximum attenuation effect occurred at the speed of 750 rpm, whose corresponding frequency was equal to the characteristic frequency of the 2x resonator. It was indicated that employing a Helmholtz resonator could attenuate the pressure pulsation in a bandwidth, but the best effect occurred at its characteristic frequency.

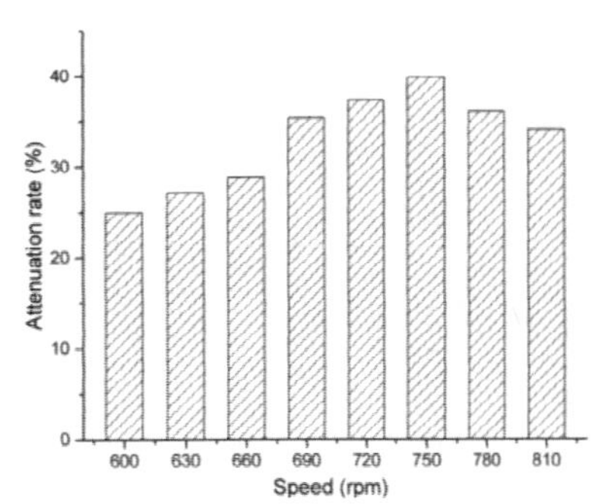

Figure 12. Pressure pulsation attenuation rate at different rotational speed.

4. Frequency-adjustable Helmholtz resonator

As discussed above, the pressure pulsation attenuation effect of the Helmholtz resonator was not satisfied when its characteristic frequency was not equal to the targeted attenuation frequency, which was a common phenomenon for the variable speed compressor. Therefore, a kind of frequency-adjustable Helmholtz resonator was designed and manufactured as shown in figure 13(a). The choke tube was composed of the fixed choke and adjustable choke. By rotating the adjustable lever, the length of the choke tube was changed, so as the characteristic frequency. Thus, the adjustable Helmholtz resonator would have a maximal attenuation effect within a range in the frequency.

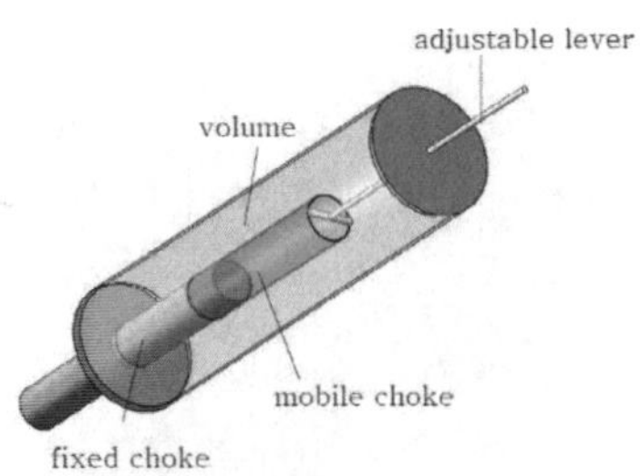

(a) Physical model

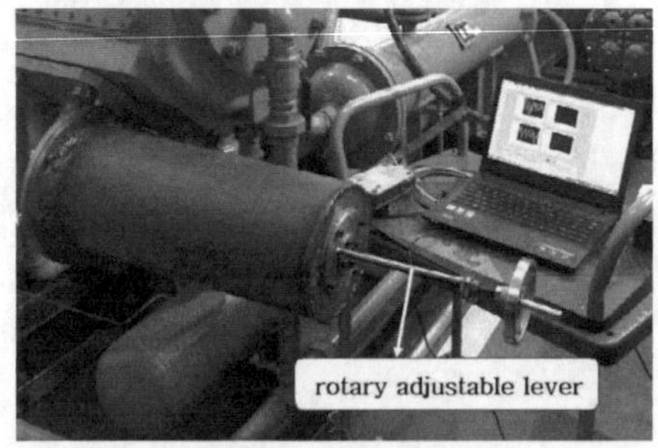

(b) Experiment test

Figure 13. Frequency-adjustable Helmholtz resonator.

By rotating the adjustable lever, the characteristic frequency of this adjustable Helmholtz resonator could be changed from 20 Hz to 26.5 Hz. To verify the attenuation effect of the adjustable resonator, the compressor rotational speed was changed from 600 rpm to 795 rpm, and the characteristic frequency of the adjustable resonator was adapted to motivation frequency at each rotational speed. Figure 14 illustrated the comparison of the transmission coefficient between the 2x resonator and the adjustable Helmholtz resonator. It can be seen that the transmission coefficient of the 2x resonator decreased obviously with the motivation frequency increased from 20 Hz to 25 Hz (compressor rotational speed increased from 600 rpm to 750 rpm), and then increased with continuing to increase the motivation frequency. However, the transmission coefficient of the adjustable Helmholtz resonator was around 0.63 as the motivation frequency was changed for 20 Hz to 26.5 Hz. It indicated that a satisfied attenuation effect could be achieved in a range of the compressor rotational speed by using the adjustable Helmholtz resonator.

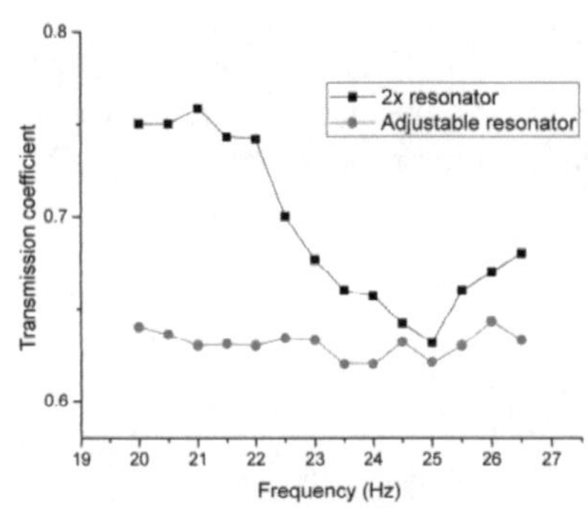

Figure 14. Comparison of the transmission coefficient between 2x resonator and adjustable resonator.

5. Parameter Analysis

According to Eq. (1), there were several combinations of the Helmholtz resonator parameters for one specific characteristic frequency, including the volume, the length and the diameter of the choke tube. However, there was few studies of the Helmholtz resonator parameters' influence on the attenuation effect. Thus, the further simulation and experiment investigations were carried out to analyse the influence of the resonator parameters. In the simulation investigations, the interstage piping model was the same model as shown in figure 6, except the Helmholtz resonator. The compressor rotational speed was set at 750 rpm. The characteristic frequency of the Helmholtz resonators was kept at 25 Hz, which was equal to the characteristic frequency of the 2x resonator. The d (the diameter of the choke tube) was kept at 50 mm, the l (the length of the choke tube) was serialization from 0.2 mm to1.4 mm. The other key non-dimensional parameter was d/D (D was the diameter of the volume), which indicated the diameter rate of the resonator. It was serialization from 0.15 to 0.45, and then the L (the length of the volume) could be calculated from the Eq. (1).

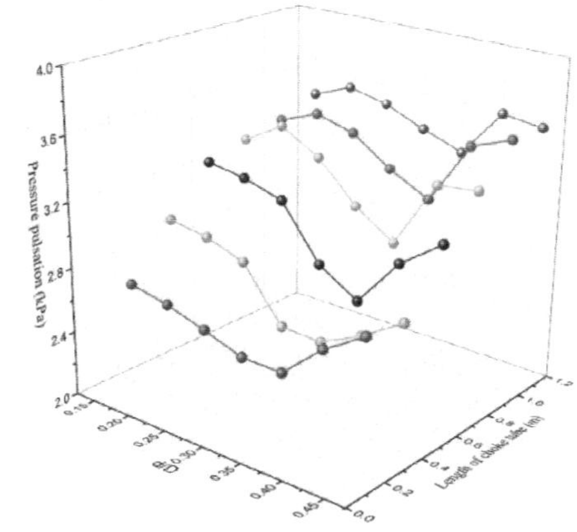

Figure 15. 2' order pressure pulsations in the valve chamber.

Figure 15 illustrated the 2' order pressure pulsations in the valve chamber where the Helmholtz resonator was mounted on. The 2' order pressure pulsation in same position without the resonator was 3.8 kPa. It was worth noting that although the characteristic frequencies of the various resonators were same, the pressure pulsations had a remarkable difference, the maximum pressure pulsation was probably twice that of the minimum. It can be seen that for the same length of the choke tube, the pressure pulsation decreased at first and then raised up with increasing of the d/D. And no matter how much the length of the choke tube was, the minimum pressure pulsations were always obtained when the d/D was 0.35. It indicated that compared to the diameter of the volume, the diameter of the choke tube could be neither too big nor too small when design a Helmholtz resonator. Meanwhile, when the diameters of the volume and the choke tube were confirmed, the pressure pulsation fell obviously with decreasing of the length of the choke tube. It meant that the bigger volume had a better pressure pulsation attenuation effect.

Figure 16. Three resonators with the different parameters.

Table 2. Parameter of three resonators.

	HR1	HR2	HR3
Diameter of choke /m	0.051	0.034	0.034
Diameter of volume/m	0.152	0.2	0.103
Length of Choke tube/m	0.95	0.4	0.95
d/D	0.33	0.17	0.33
Characteristic frequency/Hz	25	25	25

In order to verify the simulation results, three different Helmholtz resonators were designed and processed. These resonators had the same characteristic frequency and the other different parameters, as listed in Table 2. The HR1 and HR3 had the same d/D (the diameter rate of the Helmholtz resonator) and the length of choke tube, but the volume of HR1 was larger than that of HR3. The volume of HR2 was equal to that of HR1, but the diameter rate of HR2 was only 0.17, which was much smaller than HR1. The compressor rotational speed was kept at 750 rpm, so the 2' order of the compressor motivation frequency was 25 Hz, which coincided with the resonators characteristic frequencies. Figure 17 illustrated the 2' order pressure pulsation amplitude in the valve chamber with different resonators. It can be seen that the pressure pulsation decreased from 3.8 kPa to 2.4 kPa when the HR1 was mounted, the attenuation rate of the HR1 achieved 36.8%. However, when the HR3 was mounted, the pressure pulsation was 3.1 kPa, and the attenuation rate was 18.4%. It indicated that a Helmholtz resonator with a larger volume had a better attenuation effect. Meanwhile, although the volume of the HR2 was equal to that of the HR1, the attenuation rate of the HR2 was only 5.3%. It suggested that the diameter rate (d/D) of the Helmholtz resonator had a significant influence on the attenuation effect. According to the experiment and simulation results, a design criterion of the Helmholtz resonator in practice application could be recommended, the diameter rate is the most important resonator parameter, which should be set as 0.3 to 0.35, meanwhile the volume of the resonator should be designed large enough to ensure a better attenuation effect.

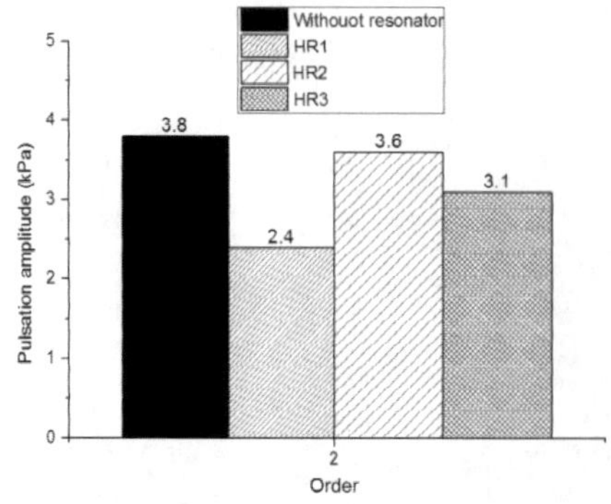

Figure 17. 2' order pressure pulsation amplitude with different resonators.

6. Conclusions

To investigate the attenuation effect of a Helmholtz resonator, the transmission coefficient in side branch piping system was studied on the basis of the plane wave theory with transfer matrix method. A series of simulations and experiments were carried out. The following conclusions can be drawn:

(1) Mounting a Helmholtz resonator near the discharge valve chamber could significantly attenuate the pressure pulsation in the valve chamber and the pipeline downstream.

(2) Multiple Helmholtz resonators can be mounted in parallel to attenuate the multi-frequency pulsation synchronously and effectively, which proposes a solution to attenuate the pressure pulsation which is resulted from more than one high-amplitude orders.

(3) The frequency-adjustable Helmholtz resonator whose choke tube was composed of fixed choke and mobile choke proved valid in the variable speed compressor. It eliminated the inconvenience of the resonator on account of its frequency sensibility.

(4) For a specific characteristic frequency, the Helmholtz resonators with various parameters have different attenuation rates. To obtain the best attenuation effect, the diameter rate of the Helmholtz resonator should be set as 0.3 to 0.35, and the volume of the resonator should be designed as large as possible.

Acknowledgements
Supported by Natural Science Foundation of Shaanxi Province of China (No.2016JQ5114).

References

[1] American Petroleum Institute 2007 *Reciprocating Compressors for Petroleum, Chemical, and Gas Industry Services, API Standard No. 618* 5th ed (Washington, DC)
[2] Jia X H, Liu B X, Feng J M and Peng X Y 2013 Influence of an orifice plate on gas pulsation in a reciprocating compressor piping system *Proc IMechE Part E: J Process Mechanical Engineering* **229**(1) pp 64–77
[3] Iocco D E 1974 Air piping system design for reciprocating compressor *Proceedings of the 1974 Purdue Compressor Technology Conference (West Lafayette, Indiana)* pp 140–43
[4] Gehri C M and Harris R E 1999 High speed reciprocating compressors - the importance of interactive modeling *Proceedings of the 1999 GMRC Gas Machinery Conference (Houston, Texas)*
[5] Deffenbaugh D M, et al. 2005 Advanced reciprocating compressor technology, *Department of Energy, National Energy Technology Laboratory, Pittsburgh, PA, Final Report* SwRI Project No. 18.11052
[6] Nored M, Broerman E, Deffenbaugh D and McKee R 2007 Advancement in pulsation control technology, *Proceedings of the 2007 GMRC Gas Machinery Conference, Dallas, Texas*
[7] Anderson J S 1977 The effect of an air flow on a single side branch Helmholtz resonator in a circular duct *Journal of Sound and Vibration* **52**(3) pp 423–31
[8] Broerman E, McKee R, Scrivner C and Bourn G 2008 Advancement in pulsation control for reciprocating compressors *Proceedings of the 6th Conference of the EFRC (Dusseldorf, Germany)*
[9] Liu B X 2012 *Attenuation of Gas Pulsation in the Valve Chamber of a Reciprocating Compressor Using the Helmholtz Resonator, PhD thesis* (Xi'an Jiaotong University, Xi'an, China)
[10] Cai C, Mak C M and Shi X, 2017 An extended neck versus a spiral neck of the Helmholtz resonator *Applied Acoustics* **115** pp 74–80
[11] Coulon J M and Atalla N 2016 Optimization of concentric array resonators for wide band noise reduction *Applied Acoustics* **113** pp 109–15
[12] Grover S 1966 Analysis of pressure pulsations in reciprocating compressor piping systems *Transactions of the ASME, Journal of Engineering for Industry* **88**(2) pp 164–71
[13] Abe T, Fujikawa T and Ito S 1970 A calculating method of pulsation in a piping system *Bulletin of the JSME (The Japan Society of Mechanical Engineers)* **13**(59) pp 678–87

IOP Conf. Series: Materials Science and Engineering **232** (2017) 012014 doi:10.1088/1757-899X/232/1/012014

Research on Noise Reduction of Reed Valves of a Hermetic Refrigerator Compressor

Zhilong He[1], **Qian Chen**[1], **Dantong Li**[1], **Ju Wang**[2], **Pu Xia**[3], **Tao Wang**[1*]
[1] Xi'an Jiaotong University, Xi'an, China
[2] Nanjing Engineering Institute of Aircraft Systems, Nanjing, China
[3] Carrier SRDC, Shanghai, China
[*]Corresponding Author
E-mail: taowang@stu.xjtu.edu.cn
Abstract. The noise level of the refrigerator compressor has received more and more attention in recent years. As the key component of a compressor, reed valve is its main noise source. In this paper, a new noise reduction technology of coating on reed valve surface is proposed and verified by experiments. Firstly, the reed valves were coated, and their surface characteristics were checked. Then, the refrigerator compressor p-V diagram test was carried out to investigate the influence of doped diamond-like carbon (DLC) coating on power consumption. Finally, the noise test rig for the refrigerator compressor was set up. Based on the standard test method, noise spectrum was measured in a semi-anechoic room under standard working condition. Research results showed that the compressor noise was significantly reduced by 1.8dB (A) after coating. Moreover, the effect of aerodynamic noise reduction at suction side is better than that at discharge side. However, the influence of the film thickness on noise reduction value is little. The COP was reduced by 0.6% as compared to the compressor with uncoated reed valves.

Key words: Refrigerator compressor; Reed valve; Noise; Coating; p-V diagram

1. Introduction

Valve noise often acts as the main source of compressor noise in a hermetic refrigerator compressor. It consists of mechanical and aerodynamic noise generated by the intermittent movement of the reed valve. The suction and discharge valves repeatedly hit the valve seat or lift guard producing mechanical noise. Additionally, the periodic motion of reed valves results in the pulsation of refrigerant pressure and velocity, leading to aerodynamic noise. It is visible that the intermittent movement of the reed valve has a significant influence on the mechanical and aerodynamic noise of a compressor. The major findings and contributions of the compressor noise reduction are summarized as follows:

Soedel classified valve noise sources as gas pulsations, valve-to-stop noise, and impact noise. He also proposed some noise reduction measures: using mufflers at suction and discharge sides, optimizing the valve port, changing the surface characteristics of the valve seat [1]. Kral PJ optimized reed thickness, stiffness and other parameters of the compressor through numerical simulation and experimental verification [2]. The noise can be reduced by decreasing the rigidity, thickness and density of the reed valve. In China, Zou Qiang introduced a reed valve made of polyether-ether-ketone plastic which not only reduces noise but also improves the life span of a reed valve[3]. Guan Hua et al. stated the reed valve noise was mainly caused by its vibration under the high-speed flow of working medium and suggested it can be reduced via optimizing the valve structure and thickness [4].

From above research work, it can be concluded that the noise reduction technology of the reed valve is well achieved by changing its basic structure, material and surface characteristics. Nevertheless, the optimization of reed valve and lift guard change its shape or weight and also affect suction and discharge efficiency or other performances of the compressor. Little attention has been paid to this aspect.

Therefore, this study introduces a new noise reduction technology: doped diamond-like carbon coating on the surface of suction and discharge valves of a compressor.

2. Experiment of film coating on reed valve surface

In this study, the doped diamond-like carbon film was prepared on the substrate of reed valves by unbalanced magnetron sputter equipment. Figure 1 and 2 present the conventional uncoated reed valves and the improved coated reed valves, the left side for the improved samples, the right side for the conventional samples.

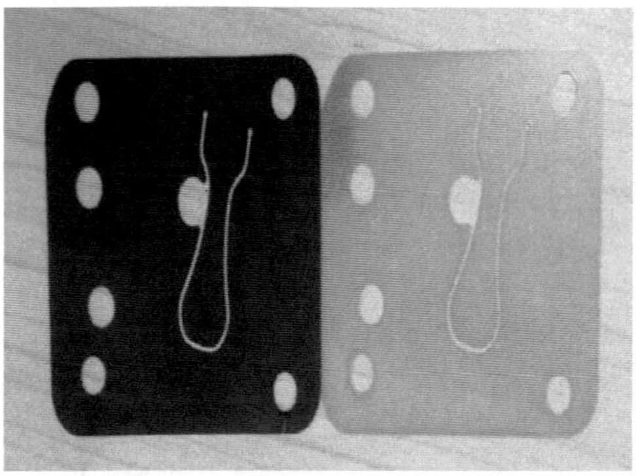

Figure 1. Comparison of conventional and improved suction valves.

Figure 2. Comparison of conventional and improved discharge valves.

2.1. Friction and wear test

Figure 3 presents the friction coefficient vibration of conventional and improved samples. Pin-on-disc experiment showed that DLC films possessed excellent wear resistance and presented low values of friction coefficient (fc=0.2) when it was sliding against SiC grinding disc.

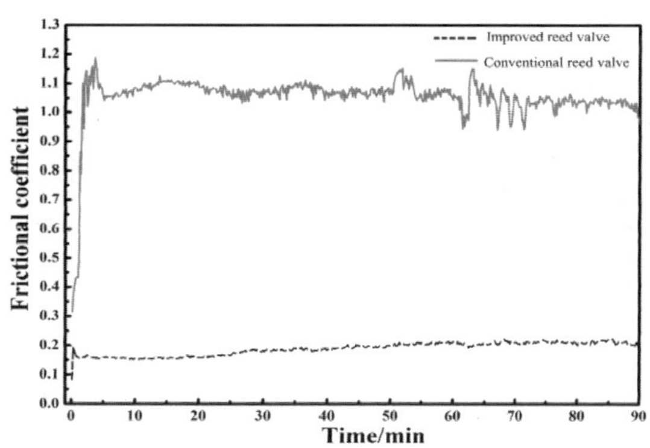

Figure 3. Friction coefficient vibration of conventional and improved samples.

In Figure 3, it also has been manifested that the DLC films provided a self-lubricating mechanism, which is critical for the tribological performance of the nano-composite DLC coatings [5, 6]. Compared with the improved sample, the conventional one exhibited high values of friction coefficient, accompanying a severe fluctuation, which means that DLC coated reed valves have advantages of both wear resistance and antifriction properties.

IOP Conf. Series: Materials Science and Engineering **232** (2017) 012014 doi:10.1088/1757-899X/232/1/012014

If the suction valve of a refrigerator compressor is regarded as a throttle element, it can be derived the following equation from the empirical formula [7]. The relationship between the SPL and pressure loss is defined by the formula:

$$L_p = 80 + 20\lg\frac{\Delta p^2}{0.5(p_s^2 - \Delta p^2)} + 20\lg D \tag{1}$$

Where Lp is the SPL (dB), p_1 is the suction pressure (Pa), Δp is the valve resistance pressure (Pa), D is the suction port diameter (m).

When the reed valve friction coefficient is reduced, the valve pressure drop is declined. So the reed valve noise can be indirectly decreased.

2.2. Scanning electron microscopy (SEM)

SEM images are listed as follows (Figure 4 and 5) to analyze the surface microstructure of samples. It can be obtained from these pictures that the surface roughness of the conventional reed valve is poor, also existing many cracks. These defects can be considered the failure source of the suction and discharge valves of the refrigerator compressor. By contrast, the surface of the coated reed valve is smooth, and the grain of diamond-like carbon film is evenly and densely distributed on its surface. Furthermore, no obvious pores are detected in the films, indicating the films are dense, promising for a better adhesion [8].

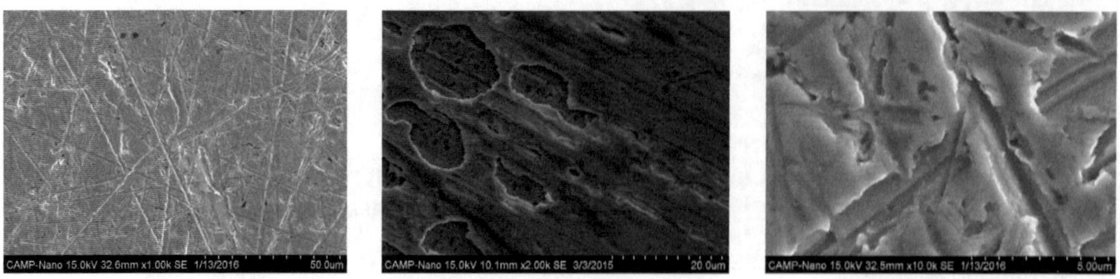

(a) 1000 times (b) 5000 times (c) 10000 times

Figure 4. Surface morphology of conventional samples.

(a) 1000 times (b) 5000 times (c) 10000 times

Figure 5. Surface morphology of improved samples.

3. Noise test

To analyze the valve noise characteristics of a refrigeration compressor, the noise test of the refrigerator compressor was carried out in the noise test rig.

3.1. Experiment setup

Figure 6 shows the schematic diagram of the experimental setup and main test rig components. The compressor used in this experiment is a hermetic refrigerator compressor of Qianjiang Wansheng, and its model is 153YV.

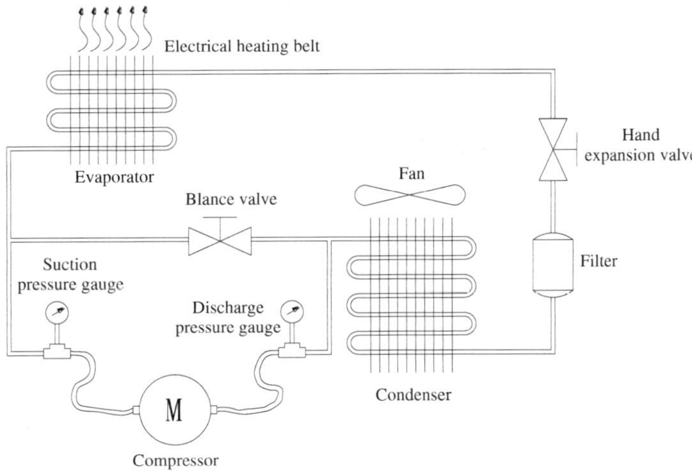

Figure 6. Noise test rig.

Based on the standard test method, noise spectrum on ten specified points around the compressor was measured in a semi-anechoic room under a standard condition for the pressure of 96kPa ± 10kPa, the temperature at 20°C ± 5°C. According to Chinese national standard "GB/T 9098-2008 Hermetic motor-compressors for refrigerators", the noise test condition is presented in Table 1.

Table 1. Noise test condition.

Refrigerant	Discharge pressure /MPa	Suction pressure/MPa	Reflux temperature/℃
R600a	0.68±0.05	0.072±0.01	32.2±3.0

According to the obtained spectral characteristics of 1/3 octave sound power level, the noise reduction target frequency was identified. Figure 7 shows the location of noise test points and Figure 8 shows a semi-anechoic room of noise test.

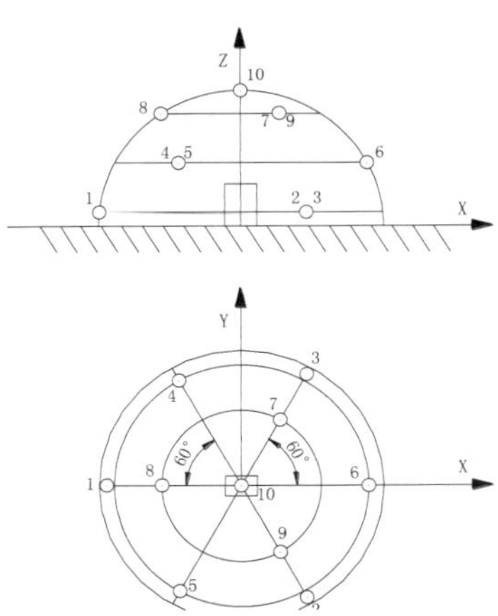

Figure 7. Location of noise test points.

Figure 8. A semi-anechoic room of noise test.

3.2. Noise reduction target frequency

Figure 9 illustrates the 1/3 octave band frequency spectra of A-weighted sound power level. It can be seen that the noise of a domestic refrigerator compressor is most prominent in the 500Hz-800Hz and 3150Hz-6300Hz frequency ranges. Generally, it is believed that the mechanical noise was caused by crankshaft unbalanced force and the electromagnetic noise was caused by the motor， which are both distributed below 400Hz. The aerodynamic noise and mechanical noise are generated by the cyclical movement of suction and discharge valves, which are mainly distributed at the central frequency of 630Hz, 800Hz, 3150Hz and 4000Hz [9]. Therefore, the noise reduction target of the paper is in these frequency band areas.

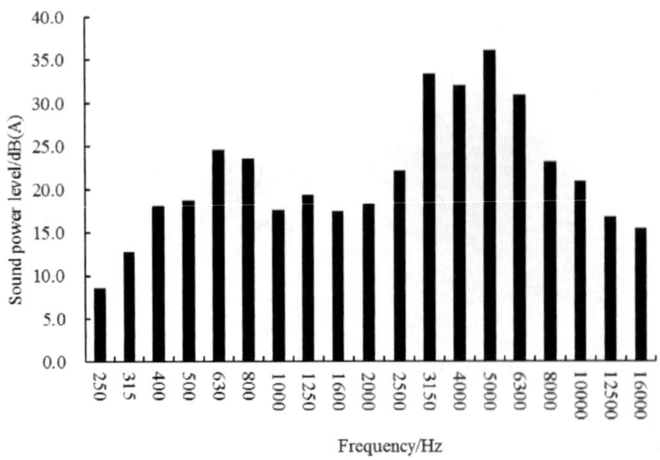

Figure 9. 1/3 octave band frequency spectra of A-weighted sound power level.

4. p-V diagram analysis

An experimental study on recording the p-V diagram was conducted, aimed to investigate the effect of film coating on power consumption. This experiment was performed under the standard condition when the film thickness is 0nm, 425nm and 850nm, respectively. The evolution of the pressure with the volume is obtained from the p-V test system of a refrigerator compressor, as shown in Figure 10.

From Figure 10, it can be seen that there are some slight differences among these p-V diagrams, which may be the influence of the hardness and stiffness of improved reed valves. These differences were mainly reflected in suction and discharge pressure.

As shown in Figure 11, the p-V diagram is divided into three parts: suction, compression and discharge. Where p_s is the evaporation pressure, p_d is the condenser pressure, V_ideal is the theoretical displacement of the compressor, V_exp is the pre-expansion gas volume, V_loss is the displacement loss, V_real is the actual displacement of the compressor.

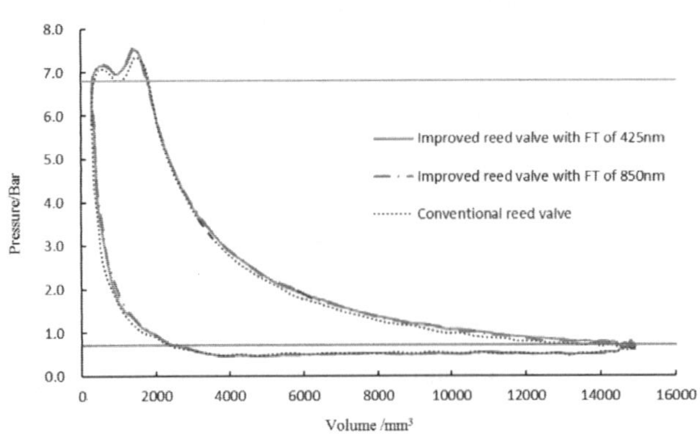

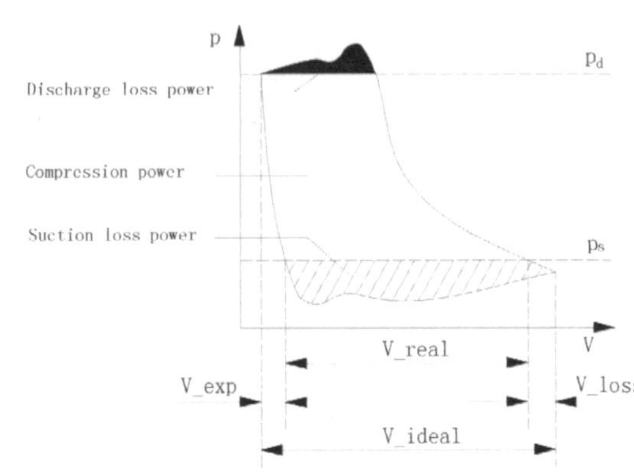

Figure 10. p-V diagram of compressor under different conditions.

Figure 11. Analysis of the compressor p-V diagram.

The analysis of pressure loss at suction and discharge sides is presented in Table 2. It can be concluded that pressure loss at suction side changes a little, while pressure loss at discharge side is increased by 0.19Bar after reed valves were coated.

Table 2. Analysis of pressure loss at suction and discharge sides.

Valve type	Pressure loss at suction side /Bar	Relative pressure loss at suction side	Pressure loss at discharge side/Bar	Relative pressure loss at discharge side
Conventional	0.24	33.33%	0.54	7.94%
Improved (film thickness of 425nm)	0.25	34.72%	0.73	10.73%
Improved (film thickness of 850nm)	0.25	34.72%	0.80	11.76%

According to the previous analysis, the noise at suction and discharge side is related to the pressure loss of suction and discharge reed valve. However, the increase of pressure loss at discharge side does not result in larger noise. To some extent, it is considered that the aerodynamic noise of the refrigerator compressor is focused on the suction side, instead of discharge side. Thus, the effect of coating on the discharge valve surface on noise reduction and compressor performance is not ideal. It provides a theoretical basis for the future study of film coating optimization of the refrigerator compressor reed valve.

Table 3 shows that the suction, discharge and compressed gas power consumption of the compressor. In addition, the compressed gas power consumption represents the indicated power without the suction and discharge power consumption. It can be observed that the valve power consumption at discharge side is lower than that at suction side. Because the coating increases the reed valve hardness and stiffness, so the suction, discharge, compressed gas power consumption and valve power loss also increase, especially the discharge power consumption. The discharge power consumption of an improved compressor with the film thickness of 425nm is increased by 52.30% compared with the conventional compressor. In the end, the indicated power consumption of the whole machine is increased by 5.31W than that of a conventional compressor. The three power consumption is also growing slightly with the increase of the film thickness.

Table 3. The power consumption of conventional and improved compressor.

Valve type	Suction power consumption/W	Discharge power consumption/W	Compressed gas power consumption/W	Indicated power /W	valve power loss
Conventional	11.75	1.74	102.83	116.32	13.1%
Improved(film thickness of 425nm)	12.44	2.65	106.54	121.63	14.2%
Improved(film thickness of 850nm)	13.04	2.71	106.80	122.55	14.7%

Table 4 shows the performance analysis of the conventional refrigeration system with uncoated reed valves and the improved refrigeration system with coated reed valves. It can be obtained that the actual displacement of the compressor is improved although its each part of power consumption becomes larger after the reed valves were coated. Consequently, the COP of the coated one is decreased by 0.6% as compared with the conventional system. Hence, it can be considered that the COP of the compressor nearly remains unchanged.

Table 4. Performance analysis of the refrigeration system.

Valve type	V_ideal/cc	V_exp/cc	V_loss/cc	V_real/cc	Cooling capacity/W	COP
Conventional	14.62	2.05	1.04	11.53	Q	ε
Improved(film thickness of 425nm)	14.62	2.18	0.45	11.99	1.04Q	0.994ε
Improved(film thickness of 850nm)	14.62	2.23	0.57	11.82	1.03Q	0.972ε

5. Results and discussion

The experiment was performed to investigate the influence of film coating, film thickness and discharge pressure on compressor noise. It was also carried out in the noise test rig according to the above test method.

5.1. Noise performance analysis of a compressor

In this study, the suction and discharge valves of the refrigeration compressor were coated, and other components were not coated. Three experiments were carried out, aimed to obtain the compressor noise level under different conditions.

The first group experiment is to test the noise level of a conventional compressor and an improved compressor with the film thickness of 850nm under standard condition.

The second group experiment is to test the noise level of an improved compressor with the film thickness of 425nm and 850nm under standard condition.

The third group experiment is to test the noise level of an improved compressor with the film thickness of 425nm under discharge pressure for 0.68MPa, 0.6MPa and 0.5MPa, respectively.

5.2. The first group test results.

Table 5 shows the A-weighted sound pressure level of each test point under the first condition. It can be observed that the A-weighted sound power level of the conventional compressor is 32.86dB (A), and that of the improved compressor is 31.06dB (A). Thus, the total noise level is significantly reduced by 1.8dB (A) when the reed valves were coated.

Table 5. The A-weighted sound pressure level of each test point under the first condition.

	A-weighted sound pressure level of each test point /dB(A)									
Location	1	2	3	4	5	6	7	8	9	10
SPL(conventional)	23.7	23.9	25.0	23.2	23.4	23.7	27.1	24.1	23	28.1
SPL(improved)	23.3	23.8	23.6	22.1	22.4	23.4	21.9	22.1	21.9	25.1

The 1/3 octave band frequency spectra of A-weighted sound power level under the second condition, as shown in Figure 12. It can be seen that the sound power level of the compressor noise is remarkably reduced at 630 Hz, 3150 Hz and 4000 Hz, especially at 3150 Hz and 4000 Hz.

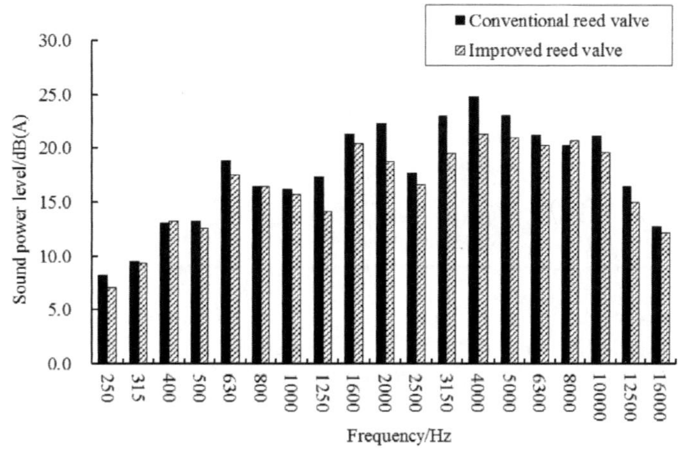

Figure 12. 1/3 octave band frequency spectra of A-weighted sound power level under the first condition.

5.3. The second group test results.

Table 6 presents the A-weighted sound pressure level of each test point under different film thicknesses. The A-weighted sound power level of the improved compressor characterized by the film thickness of 850nm is 31.06dB (A), and that of the other improved compressor is 31.03dB (A). It can be seen that the total noise level is only decreased by 0.03dB (A), indicating that the film thickness has little effect on noise reduction value.

The 1/3 octave band frequency spectra of the A-weighted sound power level of the improved reed valves is depicted in Figure 13. It can be obtained that the amplitude of the sound power level in the four target bands changes a little. The film thickness has little effect on mechanical noise of the reed valve, although the hardness of the reed valves was improved. However, it has a greater impact on the aerodynamic noise.

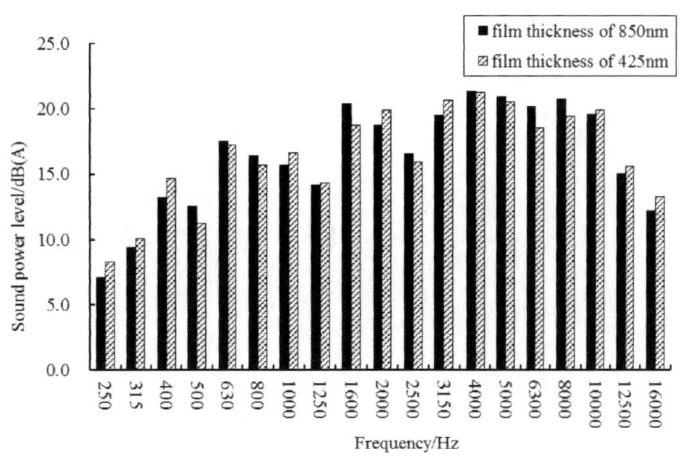

Figure 13. 1/3 octave band frequency spectra of A-weighted sound power level under the second condition.

Table 6. The A-weighted sound pressure level of each test point under the second condition.

	A-weighted sound pressure level of each test point /dB(A)									
Location	1	2	3	4	5	6	7	8	9	10
SPL(film thickness of 425nm)	23.7	23.7	24.8	22.7	22.1	23.1	21.6	20.9	21.8	24.4
SPL(film thickness of 850nm)	23.3	23.8	23.6	22.1	22.4	23.4	21.9	22.1	21.9	25.1

5.4. The third group test results.
Table 7 shows the measured SPL under different discharge pressure. The A-weighted sound power level was 31.03dB (A), 30.87dB (A) and 30.83d (A) when the compressor was tested under the pressure of 0.68MPa, 0.6MPa and 0.5MPa, respectively. Besides, the smaller the discharge pressure is, the lower the compressor noise is. But the noise reduction effect is not obvious with the lower discharge pressure. It can be considered that the discharge side noise plays an unimportant role in the compressor noise source.

Figure 14 illustrates that the amplitude of the sound power level at the four target frequency bands is decreased when the discharge pressure is reduced, especially at the central frequency of 3150Hz and 4000Hz.

Table 7. The A-weighted sound pressure level of each test point under the third condition.

	A-weighted sound pressure level of each test point /dB(A)									
Location	1	2	3	4	5	6	7	8	9	10
SPL(P=0.68MPa)	23.7	23.7	24.8	22.7	22.1	23.1	21.6	20.9	21.8	24.4
SPL(P=0.60MPa)	23.8	23.4	25.3	21.7	22.5	22.0	22.6	20.9	22.8	22.3
SPL(P=050MPa)	23.2	23.1	25.4	20.6	23.5	22.9	21.6	21.1	22.7	22.4

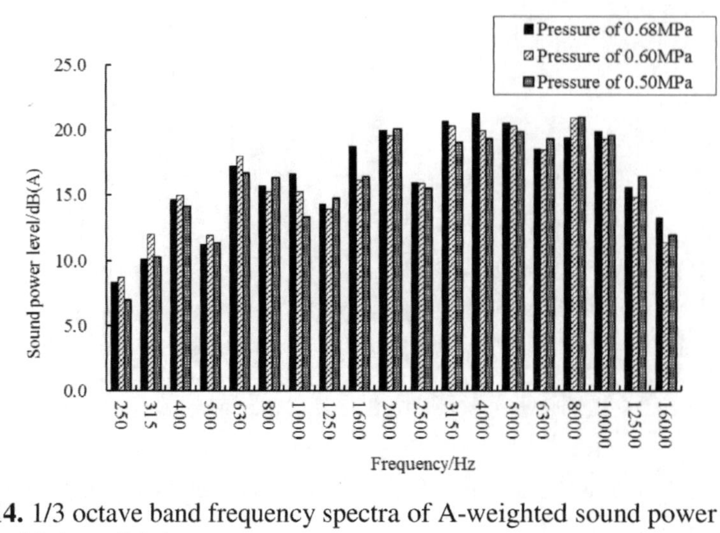

Figure 14. 1/3 octave band frequency spectra of A-weighted sound power level under the third condition.

6. Conclusions

In this study, a new noise reduction technology was proposed based on the theoretical and experimental analysis. The noise is reduced by doped diamond-like carbon coating on the surface of suction and discharge valves of a refrigeration compressor. On the other way, it also improves the reliability and life of the valve. The conclusions are as follows:

- The noise of the compressor was significantly reduced by 1.8dB (A) when the reed valves were coated. The effect of the aerodynamic noise reduction at the suction side was more obvious than that at discharge side. However, the influence of the film thickness of the noise reduction can be ignored.
- The actual displacement and cooling capacity were raised although the compressor indicated power was increased by 5.31W after coating. Consequently, the COP nearly remains unchanged.
- The friction of reed valve was reduced and wear performance of reed valve was dramatically enhanced after coating. The surface defects of the reed valve were also improved, which can effectively prevent the generation of crack on reed valve surface.

References

[1] W Soedel 1984 "Designs and Mechanics of Compressor Valves" (Short Course Text, Purdue University).

[2] Kral PJ 2008 Influence of constructional parameters of small reciprocating compressors on sound power emissions. *Journal of the Acoustical Society of America*, 123(5):3316.

[3] Qiang Z 2002 Polyether ether ketone(PEEK) and its application in compressor valves. *Compressor Technology*, (5): 26-29.

[4] Hua G and Jiahua Y 1996 Experimental analysis for lowering the noise of refrigerator compressors. *Journal of Beijing University of Technology*, 22(2): 92-98.

[5] Voevodin A.A, O'Neill J P and Zabinski J S 1999 Tribological performance and tribochemistry of nanocrystalline WC/amorphous diamond-like carbon composites. *Thin Solid Films,* 342(1–2):194-200.

[6] Liu Y, Erdemir A and Meletis E I 1997 Influence of environmental parameters on the frictional behavior of DLC coatings. *Surface & Coatings Technology,* s 94–95(545): 463-468.

[7] Dayou M, Peizi L, Genhau D and Hongyu W 1978 Pressure dependence of turbulent Tet Noise. *Chinese Journal of Physics*, 27(2): 121-125.

[8] Sun F H, Zhang ZM, Chen M and Shen H S 2003 Improvement of adhesive strength and surface roughness of diamond films on co-cemented tungsten carbide tools. *Diamond & Related Materials*, 12(3–7): 711-718.

[9] Chongming Z 2010 Research on vibration and noise of compressor (Doctoral dissertation, Shanghai: Shanghai Jiao Tong University).

Dynamic analysis of the reciprocating compressor package on the offshore platform

Y Zhao, W Wang, Q Zhou, J Feng, X Jia and X Peng

Xi'an Jiaotong University, Xi'an, Shannxi, China

E-mail: xypeng@mail.xjtu.edu.cn

Abstract. Reciprocating compressor packages are key equipment for the offshore platforms which are used for the pressurization and transportation of natural gas. Dynamic analyses of high-speed and flexible supported offshore compressor packages are more complex than that of compressors on land. The dynamic analysis techniques of the offshore reciprocating compressor package were studied including numerical modelling, excitation forces calculation, support boundary determination of the compressor package and load transmission between the compressor package and the offshore platform. The finite element model of the compressor package with multiple types of elements such as the pipe, beam and shell elements were established. Excitation forces mainly including gas forces inside cylinders, crosshead forces, acoustic shaking forces and unbalanced inertia forces were calculated. In order to investigate the influence of the flexibility of the platform on dynamic characteristics of the compressor package and set reasonable boundaries for the support structure, three kinds of support boundaries were compared. The support points of columns and vertical beams on the layer of deck where the compressor package is installed were determined to be the most suitable ones. Super-element method was applied to implement the load transmission from the compressor package to the platform. The analysis techniques herein were successfully applied to the dynamic analysis of an offshore reciprocating compressor package in an engineering project.

1. Introduction

Reciprocating compressor packages are key equipment for offshore platforms. Dynamic loads such as unbalanced inertial forces and gas forces inside cylinders can induce vibrations on the compressor and the offshore platform as well [1]. Severe vibration will reduce the reliability of the offshore platform [2]. Dynamic analyses of high-speed and flexible supported offshore compressor packages are more complex than that of compressor packages on land. Therefore, it is of great importance to study the vibration characteristics of offshore compressor packages.

Dynamic analysis of the compressor package has drawn much attention in recent years. Eberle *et al.* [3] conducted numerical analysis for a reciprocating compressor, and found that the deflection of the compressor frame due to gas forces may have significant effects on dynamic responses of the compressor package. Giacomelli *et al.* [4] developed a finite element model of a reciprocating compressor's cylinder manifold, and calculated the vibration velocity responses under the acting of acoustic shaking forces and cylinder gas loads. Passeri *et al.* [5] generated a super-element of the compressor frame based on the sub-structuring technique to achieve the balance between calculation accuracy and computation time. However, little attention has been focused on the high-speed and flexible supported reciprocating compressors packages on the offshore platform.

In this paper, the dynamic characteristics of the offshore reciprocating compressor package were studied from four aspects. Firstly, the numerical models of the compressor package and offshore platform were established and coupled. Secondly, excitation forces in the compressor package were calculated. Thirdly, three kinds of support boundaries were compared to determine reasonable support boundary for the compressor package. Finally, the super-element method was applied to implement the load transmission from the compressor package to the platform.

2. Numerical modelling

Appropriate element types were chosen for different components of the structure to improve calculation efficiency. Numerical models of the compressor package and offshore platform with multiple element types were established and coupled.

2.1. Numerical model of the compressor package

The geometric model of the compressor package as shown in figure 1(a) was simplified before numerical modelling due to its complexity. Instrumentations, oil tubes, cooling water pipes and non-load-bearing components such as the fence and handrails were removed. The simplified geometry model mainly consists of the skid, pedestal, compressor, motor, scrubbers, buffer tanks, coolers and main pipelines. The numerical model of the compressor package with multiple element types were established, as shown in figure 1(b). The skid, pedestal, scrubbers, buffer tanks and coolers were simplified as shell elements. Included pipes were modelled with pipe elements as appropriate. The compressor, motor and pipe supports were modelled with beam elements. Masses of the removed components were considered by distributing mass elements over the skid area. Based on the assumption that all welds are 100% strength, different components were connected with rigid elements. A total of 46281 elements and 40742 nodes were generated in this model.

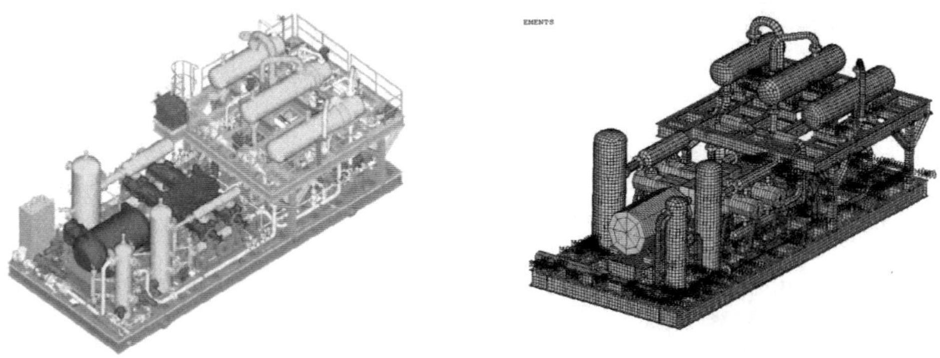

(a) Geometrical model. (b) Numerical model.

Figure 1. The compressor package.

2.2. Numerical model of the offshore platform

The jacket offshore platform consists of the upper structures and foundation structures (jackets). The foundation structures were ignored to reduce computational complexity. The upper structures were modelled and the bottom of the eight columns were fixed constrained, as shown in figure 2. The beams and decks were modelled as beam and shell elements, respectively. Equipment on the offshore platform except the compressor package was simplified as mass elements. A total of 32540 elements and 14482 nodes were generated in this model.

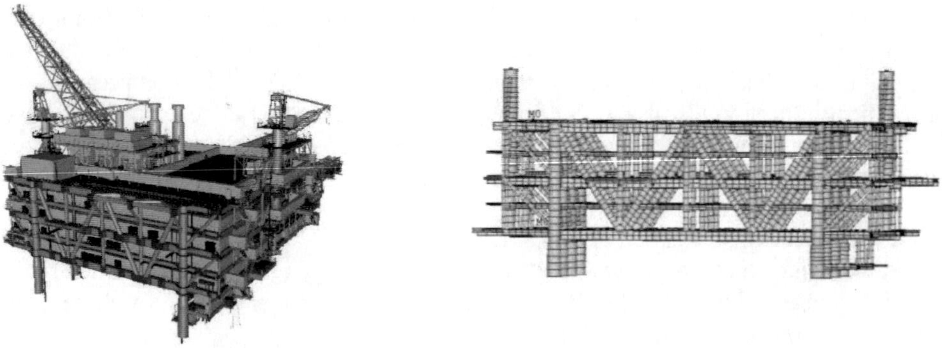

(a) Geometrical model. (b) Numerical model.

Figure 2. Upper structures of the offshore platform.

2.3. The coupled model

The numerical model of the compressor package and offshore platform were coupled by creating rigid regions between the skid of the compressor package and the beams of the platform, as shown in figure 3. A total of 78821 elements and 55224 nodes were included in the coupled model.

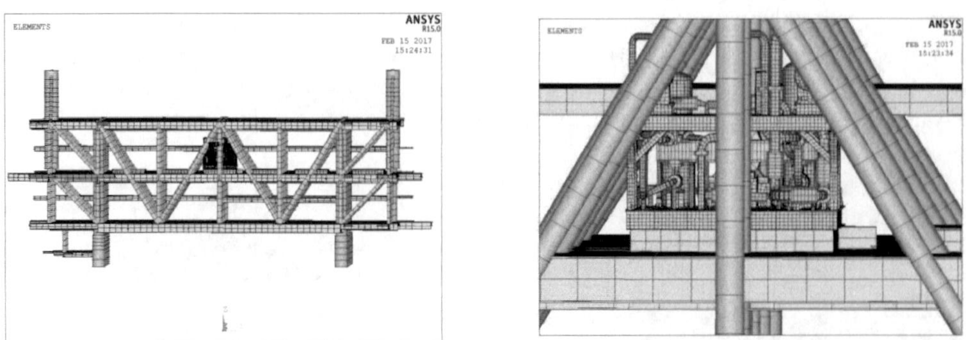

Figure 3. The coupled model of the compressor package and offshore platform.

3. Excitation forces calculation

The vibration excitation forces of reciprocating compressor package mainly include unbalance inertia forces and moments, gas forces, crosshead forces and acoustic shaking forces. All of these forces consist of multiple harmonics which can be obtained by the Fourier decomposition. In this paper, the speed of the compressor is 994 rpm, thus the first order excitation frequency is 16.6 Hz and its multiples are also the excitation frequencies. The first order harmonic of excitation forces were used for subsequent dynamic response analyses.

3.1. Unbalanced inertia forces and moments

For each throw of cylinder, the reciprocating inertia force F_{Is} on reciprocating components and rotating inertia force F_r on rotating components are given by

$$F_{Is} = m_s a = m_s r\omega^2(\cos\theta + \lambda\cos 2\theta) \tag{1}$$

$$F_r = m_r a_r = m_r r\omega^2 \tag{2}$$

where m_s is the equivalent reciprocating mass, m_r is the equivalent rotating mass, λ is the ratio of the crank radius and connecting rod length, θ is the crank rotation angle and ω is the angular velocity. The balanced-opposed reciprocating compressor discussed here has four throws and two stages, and inertia forces and moments of all throws didn't completely balance. Unbalanced inertial forces and moments were supplied by the compressor manufacturer.

3.2. Gas forces and cross-head guide forces

The gas force inside cylinders will make the cylinder suffer cyclical tensile or compressive loads and induce vibration of the compressor. The gas force, reciprocating inertia force and reciprocating frictional force are all along the axis of the cylinder, and their resultant force is called the piston force F_p. The piston force transmitted to the cross-head through the piston rod was divided into two forces—the connecting rod force and cross-head guide force. The direction of the crosshead guide force alternated as the crankshaft rotating, resulting in continuous percussion between the crosshead and the crosshead guide. The crosshead guide force F_N can be represented as

$$F_N = F_p \tan\beta = \frac{F_p \lambda \sin\theta}{\sqrt{1-\lambda^2 \sin^2\theta}} \tag{3}$$

3.3. Acoustic shaking forces

The periodic suction and discharge of the reciprocating compressor will cause gas pulsation in the piping system, and induce acoustic shaking forces on the elbow, valve, reducer etc. The transfer matrix method was used to calculate acoustic natural frequencies and shaking forces in the piping system. The piping system was divided into some basic elements such as pipe, valve, tee and volume. Each element has a transfer matrix M. The transfer matrix of the overall piping system is formulated by multiplying the transfer matrixes of all the piping elements, as shown below.

$$\begin{bmatrix} p_n \\ u_n \end{bmatrix} = M_{n,n-1} \cdot M_{n-1,n-2} \cdots M_{2,3} \cdot M_{1,2} \cdot \begin{bmatrix} p_1 \\ u_1 \end{bmatrix} \tag{4}$$

where p_1 and u_1 are the amplitude of perturbation pressure and velocity of the first node in the piping system, p_n and u_n are the amplitude of perturbation pressure and velocity of the last node in the piping system, respectively. The acoustic natural frequency, pressure unevenness and acoustic shaking forces can be obtained with relevant boundary conditions of the piping system.

4. Support boundary determination of the compressor package

Support boundary will affect dynamics of the structure. As for the compressor package on land, the foundation is generally ignored and the bottom of the skid is assumed as fixed constrained. However, the flexibility of support structures should be considered while investigating dynamics of the offshore compressor package. The whole coupled model of the offshore platform and compressor package as shown in figure 3 approaches reality well, but has the advantage of low computational efficiency. Ignoring the platform and applying fixed constraint directly on the compressor skid can significantly reduce calculation time, but will cause large errors. Therefore, determining a reasonable support boundary of the compressor package to shorten calculation time on the premise of ensuring the accuracy is of great importance in the vibration analysis of the offshore compressor package.

4.1. Three kinds of support boundaries of the compressor package

Three kinds of support boundaries were compared to check the effects of support structures to the dynamic characteristics of the offshore compressor package and determine a reasonable support boundary for the offshore compressor package. The harmonic response analysis was conducted and the displacement responses of the compressor package at 16.6 Hz were extracted. The results of the boundary A were considered as the baseline.

- Boundary A: The whole coupled model of the compressor and the offshore platform was used and the bottom of eight columns were fixed constrained, as shown in figure 3.

IOP Conf. Series: Materials Science and Engineering **232** (2017) 012015 doi:10.1088/1757-899X/232/1/012015

- Boundary B: A part of the offshore platform was included to represent the flexibility of the support structure, and the support points of columns and vertical beams on the layer of deck were fixed constrained, as shown in figure 4.
- Boundary C: The offshore platform was removed, and the welded points on the compressor skid were fixed constrained, as shown in figure 5.

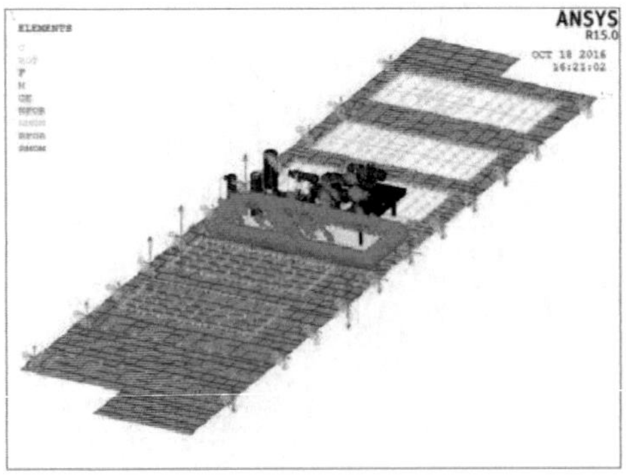

Figure 4. Support boundary B for the offshore compressor package.

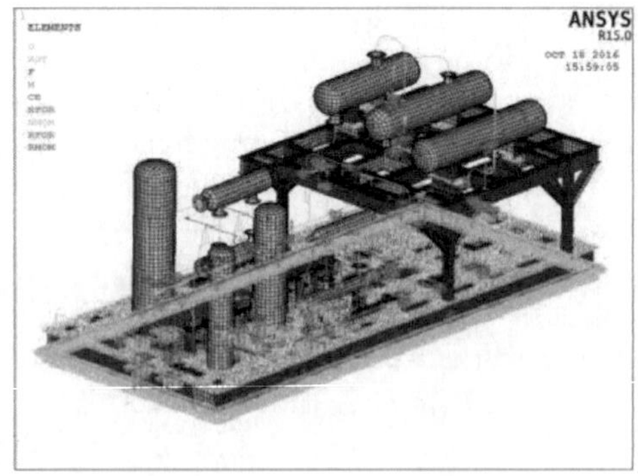

Figure 5. Support boundary C for the offshore compressor package.

4.2. Comparison of displacement responses

The displacement distributions of the compressor package with three support boundaries were shown in figure 6, figure 7 and figure 8, respectively. It can be seen that the maximum displacements of the compressor with boundary A, B and C were 1.437 mm, 1.246 mm and 0.103 mm, respectively. The displacements of all nodes on the compressor package were extracted and compared. The results of boundary A were considered as the baseline, the relative displacement error at each node of boundary B and C was calculated. The relative displacement errors were divided into six intervals: 0-10%, 10%-20%, 20%-30%, 30%-40%, 40%-50%, and >50%. The percentage of node number in each interval was shown in figure 9. It can be seen that the relative displacement errors at 67.9% nodes of the compressor package with boundary B were less than 20%, and the average relative error was 17.1%. The relative displacement errors at all nodes of the compressor package with boundary C were larger than 50%, and the average relative error was 93.2%. It was indicated that results of boundary B agreed well with the baseline, while results of boundary C had large discrepancy with the baseline.

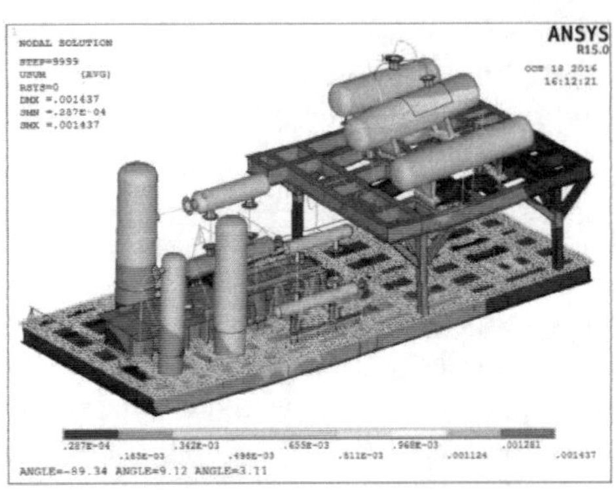

Figure 6. Displacement of the compressor package with support boundary A.

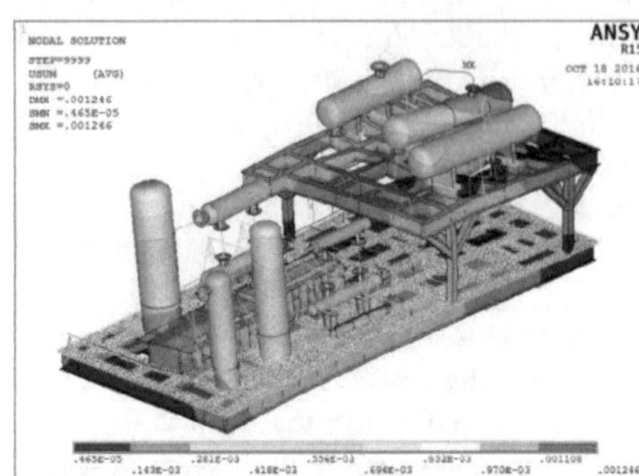

Figure 7. Displacement of the compressor package with support boundary B.

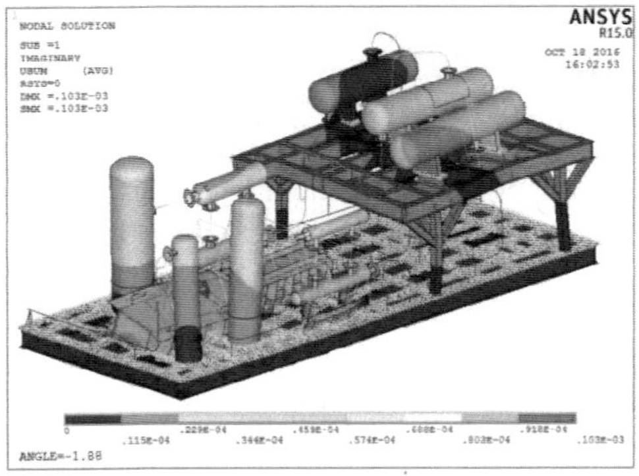

Figure 8. Displacement of the compressor package with support boundary C.

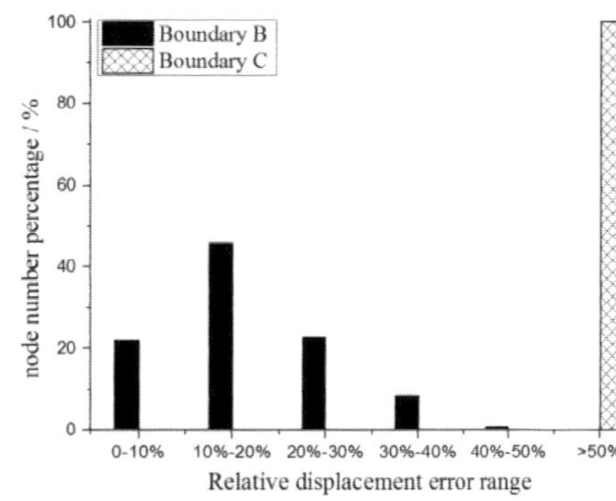

Figure 9. Relative displacement errors of support boundary B and C.

Table 1 shows the comparison of calculation time and storage space of three support boundaries. It can be seen that compared with boundary A, boundary B can save 83.7% of calculation time and 67.8% of storage space, and boundary C can save 98.6% of calculation time and 89.7% of storage space.

Table 1. The comparison of calculation time and memory space of three support boundaries.

	Boundary A (baseline)	Boundary B	Boundary C
Calculation time (min)	215	35	3
Storage space (GB)	8.70	2.8	0.9

From above analyses, we can see that the support structure had great impact on the dynamics of the offshore compressor package and shouldn't be ignored while conducting dynamic analysis of the offshore compressor package. Containing parts of the platform structure and applying fixed constraints on the support points of columns and vertical beams on the layer of deck can greatly improve computational efficiency on the premise of good accuracy, thus was recommended as the support boundary of the offshore compressor package.

5. Load transmission between the compressor package and the offshore platform

Since the excitation loads on the compressor package will affect the dynamics of the offshore platform, the load transmission method between the compressor package and the offshore platform is of great importance while investigating dynamics of the offshore platform. Using the whole coupled model of the offshore platform and compressor package is closest to reality, but will lead to the problem of large calculation time and storage space.

5.1. Three kinds of load transmission methods

Three kinds of load transmission methods between the compressor package and the platform were compared to check the effects of compressor package to the dynamic characteristics of the offshore platform and determine reasonable load transmission method.

- Method A: The whole coupled model of the offshore platform and compressor package as shown in figure 3 were used and set as the baseline. The excitation loads were exerted on the compressor package. The results of the boundary A were considered as the baseline.
- Method B: The coupled model of the offshore platform and mass elements which located on the installation of the compressor package was used to study the dynamic response of the

compressor package, as shown in figure 10. Since the compressor package wasn't modelled in detail, the excitation loads couldn't applied directly to the model. A harmonic response analysis was conducted to the compressor package model in which the fixed constrained were applied on the skid welding nodes and excitation loads were applied on the compressor and pipeline, and the reaction forces and moments at the skid welded nodes were extracted. The reaction forces and moments at the skid welded nodes obtained from the harmonic response analysis of the compressor package were applied to the coupled model of offshore platform and mass elements.

- Method C: The coupled model of the offshore platform and a super-element were used to obtain the vibration responses of the compressor package, as shown in figure 11. This simplified method is based on the substructure technique [6-7]. Firstly, the complete model of the compressor package was condensed to a super-element, the skid welding nodes and the loads applying nodes were selected as master nodes of the super-element, and a substructure analysis was conducted to generate the condensed mass and stiffness matrixes of the super-element. Then, the model of the offshore platform were coupled with the super-element which only contained the DOFs of the master nodes, and the dynamic loads of the compressor package were applied to the master nodes of the super-element.

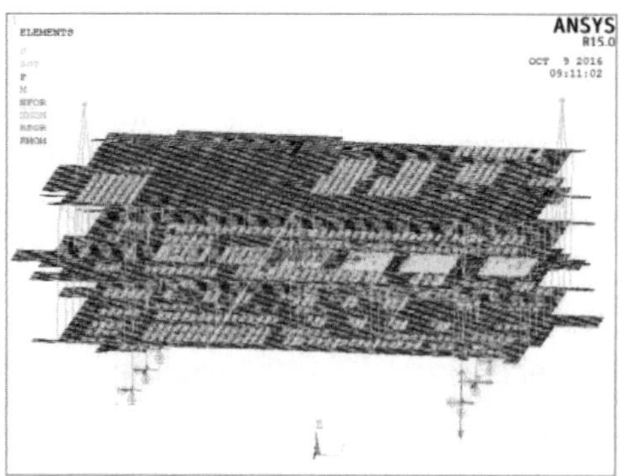

Figure 10. Load transmission method B between compressor package and offshore platform.

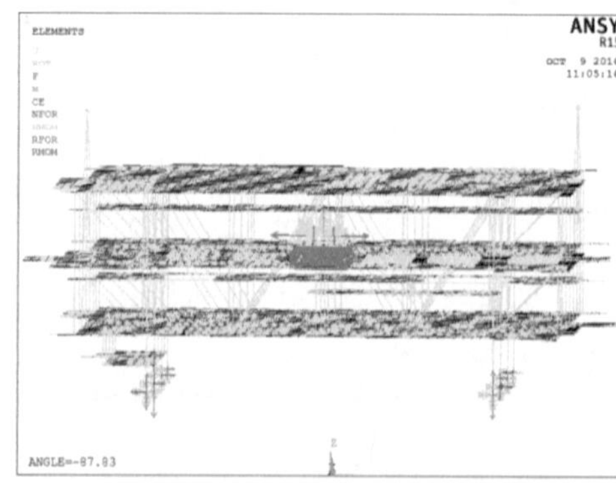

Figure 11. Load transmission method C between compressor package and offshore platform.

5.2. Comparison of displacement responses

The displacement distributions of the offshore platform at 16.6 Hz were shown in figure 12, figure 13 and figure 14, respectively. The maximum displacement the offshore platform with load transmission method A, B and C were 0.661 mm, 0.196 mm and 0.582 mm, respectively. The displacements on all nodes (14482 nodes) of the offshore platform at 16.6 Hz were extracted and compared. The results of load transmission method A were considered as the baseline, the relative displacement errors on each nodes of the offshore platform with load transmission method B and C were calculated. The relative displacement errors were divided into six intervals: 0-10%, 10%-20%, 20%-30%, 30%-40%, 40%-50%, and >50%. The percentage of node number in each interval was shown in figure 15. It can be seen that the relative displacement errors at 42.6% nodes of the offshore platform with load transmission method B were larger than 50%, and the average relative error of method B was 68.2%. The relative displacement error at 99.3% nodes of the offshore platform with load transmission method C were lower than 10%, and the average relative error of method C was 3.2%. It was indicated that the results of load transmission method C agreed well with the baseline, while results of method B had large discrepancy with the baseline.

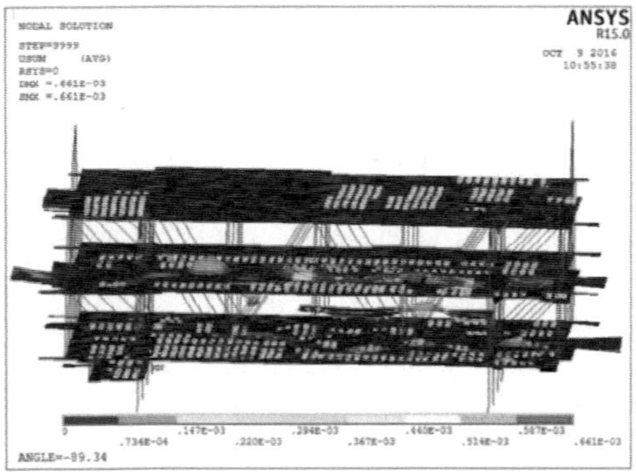

Figure 12. Displacement of the offshore platform with load transmission method A.

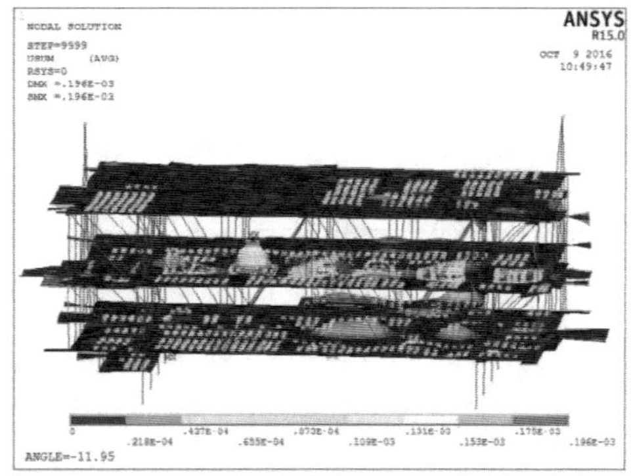

Figure 13. Displacement of the offshore platform with load transmission method B.

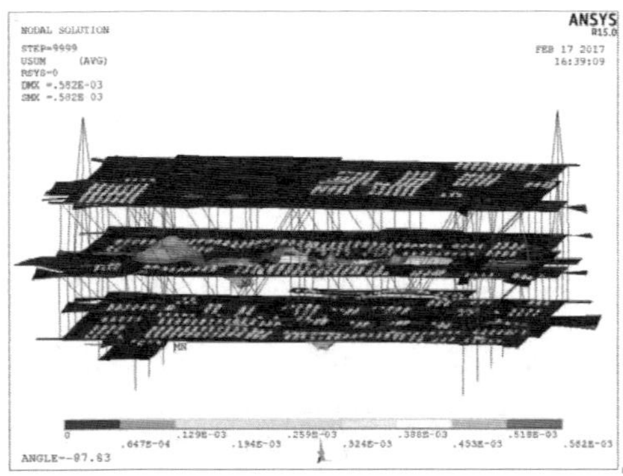

Figure 14. Displacement of the offshore platform with load transmission method C.

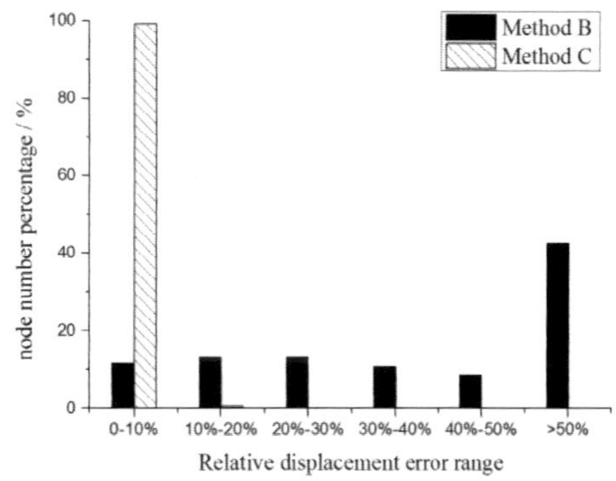

Figure 15. Relative displacement errors of load transmission method B and C.

Table 2 shows the comparison of calculation time and storage space of three kinds of load transmission methods. It can be seen that compared with load transmission method A, load transmission method B can save 42.3% of calculation time and 56.3% of storage space, and method C can save 31.6% of calculation time and 35.6% of storage space.

Table 2. The comparison of calculation time and memory space of three load transmission methods.

	Method A (baseline)	Method B	Method C
Calculation time (min)	215	124	147
Storage space (GB)	8.70	3.8	5.6

From above analysis, we can see that the load transmission method C in which the compressor package was simplified as a super-element was more accurate than method B in which the compressor package was simplified as mass elements. It maybe because both of the mass and stiffness of the compressor package were taken into account in the super-element of method C, but the stiffness of the compressor package was not considered in method B. Besides, simplifying the compressor package as a super-element can greatly save the computational time compared to the whole coupled model of the offshore platform and compressor package. Therefore, it is advisable to simplify the compressor package to a super-element when performing vibration analysis for the offshore platform.

6. Conclusions

The most important conclusions of the present paper are as follows.

- Choosing appropriate element type for each component and establishing the numerical model with multiple types of element such as the beam, pipe, shell and mass elements is an efficient way to improve the efficiency of dynamic analysis of the compressor package and offshore platform.
- Unbalanced inertia forces and moments of the compressor were supplied by the compressor manufacture. The gas forces inside cylinders and crosshead guide forces were calculated through dynamic calculation of the compressor. The acoustic shaking forces in the piping system were calculated by an acoustic analysis.
- The flexibility of the offshore platform will greatly affect the dynamics of the offshore compressor package. Containing parts of the platform structure near the compressor package and applying fixed constraints on the support points of columns and vertical beams on the layer of deck can greatly improve computational efficiency on the premise of good accuracy, thus was recommended as the support boundary of the offshore compressor package.
- The super-element method is an effective method to implement the load transmission from the compressor package to the platform while performing vibration analysis of the offshore platform.

References

[1] American Petroleum Institute 2007 *API standard 618 5th ed.: Reciprocating compressors for petroleum, chemical, and gas industry services* (Washington, D.C: API Publishing Services)

[2] Gas Machinery Research Council 2013 *GMRC guideline for high-speed reciprocating compressor packages for natural gas transmission and storage applications* (Ohio: ACI Services Inc.)

[3] Eberle K and Harper C 2007 Dynamic analysis of reciprocating compressors on FPSO topside modules *Compressor Tech Two* pp 10–16

[4] Giacomelli E *et al* 2006 Forced response of cylinder manifold for reciprocating compressor applications *ASME 8th Biennial Conference on Engineering Systems Design and Analysis* pp 939–48

[5] Passeri M, Bagagli R and Maggi C 2015 Compressor general arrangement dynamic design guided by preliminary cylinder manifold forced response *ASME 2015 Pressure Vessels and Piping Conference* V005T05A002

[6] Kim J S *et al* 2012 Development and application of the added fluid mass and substructure techniques for integrated pressurized water reactor assembly *Nuclear Engineering and Design* **250** pp 68–80

[7] Kong X *et al* 2012 New strategy of substructure method to model long-span hybrid cable-stayed bridges under vehicle-induced vibration *Engineering Structures* **34** pp 421–35

IOP Conf. Series: Materials Science and Engineering **232** (2017) 012016 doi:10.1088/1757-899X/232/1/012016

Vibration analysis in reciprocating compressors

V Kacani

Leobersdorfer Maschinenfabrik GmbH, Vienna, Austria

Vasillaq.Kacani@LMF.at

Abstract. This paper presents the influence of modelling on the mechanical natural frequencies, the effect of inertia loads on the structure vibration, the impact of the crank gear damping on speed fluctuation to ensure a safe operation and increasing the reliability of reciprocating compressors. In this paper it is shown, that conventional way of modelling is not sufficient. For best results it is required to include the whole system (bare block, frame, coupling, main driver, vessels, pipe work, etc.) in the model (see results in Table 1).

1. Introduction

The main dynamic loads on reciprocating compressors are: the shaking forces due to pressure pulsation, inertia forces and gas forces, and additional forces due to torsion vibration of the drive train. The shaking forces occur typically when there is a change in pipe direction or area, at cylinder passages, on the ends of the pulsation dampers, elbows, reducers, orifices, tee pieces and on other pipe components. The inertia forces are caused by oscillating of reciprocating masses and inertia forces act through the cross head on the structures and on the main bearing of the crankshaft. The cylinder gas forces are generated within the cylinder during the working cycle: suction, compression, discharge and expansion. The gas forces are generally the highest one and act on cylinder covers and through the crank mechanism. They also act on crosshead guide and on crankshaft main bearing. These loads lead to vibration of the compressor components. Modeling has a direct impact on the modal analysis and the calculation of the forced vibration. The modeling criteria as well as the pre-stressed effects of the structures due to internal pressure, temperature and the misalignments during the assembly of the components, are of importance for the calculation of natural mechanical frequencies and the system's response under dynamic loads. Moreover, modelling of the structure, modelling of the crank gear, the damping factors, and the damping ratio of the mechanical structure are decisive for the dynamic behavior of the drive train and also for the design of other mechanical components.

The influence of inertia and cylinder gas forces, in addition to the pulsation shaking forces are included in the forced-mechanical-response analysis of the compressor mechanical model. This requires an adequate finite element model of reciprocating compressors, specifically of the crankshaft, of the crankcase, of the distance pieces and of the pulsation suppression devices. For these components 2D and 3D-modeling is used. This model also enables an undertaking of the lateral dynamic analysis of the crankshaft and the stress evaluation under torsion and bending [3], [11]. The calculation of the torsional/lateral vibration mode shapes and the mechanical natural frequencies of crankshaft with flywheel will be performed with a separate 3D-model.

2. Vibration analysis of the complete compressor unit

The compressor unit consists of all in-skid components such as compressor manifold, piping system, e-motor / engine, pipe, base frame, pipe/vessel supports, anti-vibration mounts etc. The skid is connected on its termination points to the off-skid piping system. In different regulations and standards [2], [3], [11] are included recommendations and proposal regarding the scope of supply, limits of the system as well as the exciting loads for the vibration analysis of the compressor mechanical model. For many reciprocating compressor units, such as off-shore applications or compressors mounted on a skid, the recommended limits of the system should be extended, at least, up to base frame - foundation interface. If available, the anti-vibration mounts should be included in the analysis. Not only the cylinder gas forces but also the inertia loads must be taken into consideration while performing the mechanical response analysis of the reciprocating compressor systems. The goal of vibration analysis is the determination of the dynamic load on all components of compressor unit and a comparison with the maximum allowable values [2],[11]. This requires an accurate modeling of all parts and adequate definition of the connecting zones (interfaces) between all components involved.

2.1. Modeling

The modeling of reciprocating compressors requires much experience and represents a balancing act. On the one hand, the model should be as simple as possible to reduce calculation time and on the other hand as fine as necessary to get the correct technical solutions. The preprocessing is the most time consuming step while undertaking vibration analysis. In this paper an automatic parametric modeling of the components of the reciprocating compressor is shown. The script language APDL [5] (ANSYS Parametric Design Language) is used for modelling of all components. Special interfaces are defined to get the dynamic force members on bolts or contact areas/zones for all connecting parts. Generally, the crankcase is rarely modelled. The modelling effort for the crankcase may be expensive but it is absolutely necessary for the vibration analysis. Using the parametric design language, the modeling effort can be considerably reduced. The correct influence of cylinder gas forces and crank gear mass forces on vibrations of the machine can only be achieved by an adequate modelling of the crankcase, cross head, distance piece, and pulsation suppression devices. The use of 1D pipe or beam elements is only for some components such as pipes, tee-pieces support beams [6] admissible. Crankcase, distance piece, base frame and pulsation suppression devices should be modeled at least with 2D shell-elements. The recommended system limits, specified in the API 618 [11] needs to be expanded with all components up to the base frame, including if available the anti-vibration mounts. The following figures shows the finite element models for crankcase, crankshaft, flywheel, cylinder, cylinder supports, piping systems, pulsation suppression device, e-motor, base frame, piping and vessel supports. The bolts are modeled with pretension elements and are connected to the structure via spider with MPC-elements [5]. Figure 1 shows the finite element model of the base frame, including the pedestal for the reciprocating compressor and e-motor as well as the base plates for the vertical separators: a) on-shore application, the skid is connected with bolts to the concrete fundament b) off-shore application, the skid is isolated, supported on anti-vibration mounts (AVM). The bolts and AVM are part of base frame model. Figure 2 shows the crankcase manifold including the crankshaft, flywheel and the distance piece. The presented model is also suitable for the dynamic lateral analysis of the crankshaft. For the crankshaft main bearings, spring elements with appropriate stiffness and damping were used.

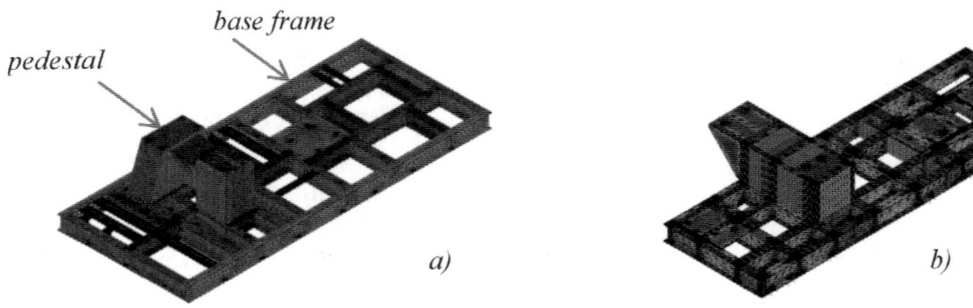

Figure 1. *FE-Model of base frame and pedestal: a) onshore-application b) offshore- application.*

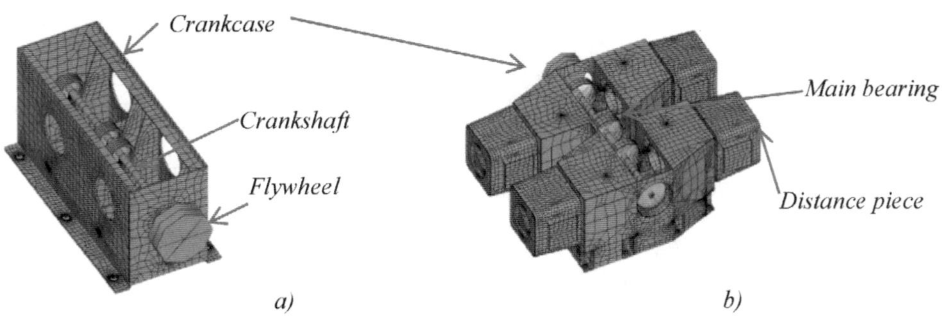

Figure 2. *Crankcase, flywheel, crankshaft, main bearing, distance piece*

Based on the parametric technique described above, Figure 3 shows the finite element model of the two-stage offshore reciprocating compressor, horizontal arrangement, type B154, with a power of 1MW, 720 rpm, 45ton weight. The finite element model includes all components of the compressor unit: crankcase, crankshaft, flywheel, double compartment distance pieces, cylinders, cylinder supports, e-motor, pulsation suppression devices, coolers, separators, piping system, base frame including pedestal, anti-vibrating mounts and all corresponding supports.

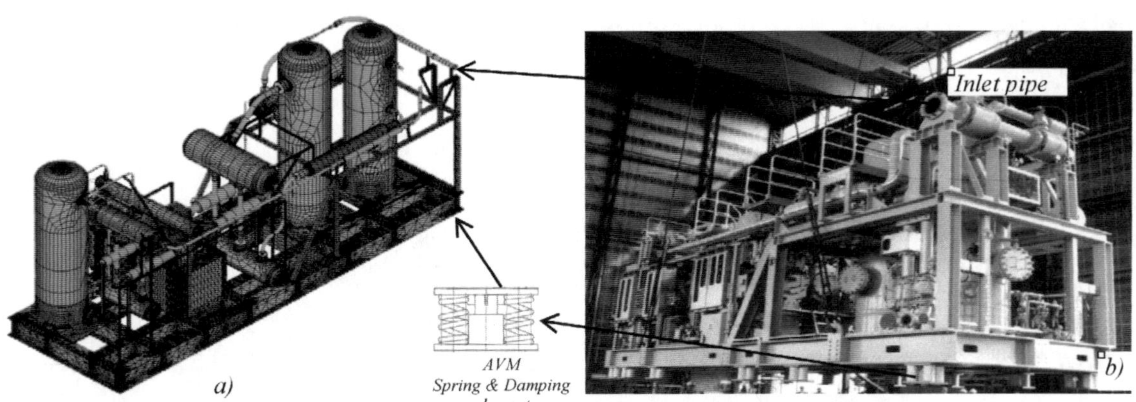

Figure 3 *a) FEM of compressor unit, b) photograph of the offshore application with anti-vibrating mounts.*

IOP Conf. Series: Materials Science and Engineering **232** (2017) 012016 doi:10.1088/1757-899X/232/1/012016

2.2. Modal analysis.

After modeling the compressor, modal analysis was performed. The effect of pre-load conditions – temperature [7], [10]., internal pressure, misalignment - on the Natural Mechanical Frequencies (NMF's) must be considered. Under pre-stressed conditions the mechanical system is stiffer - hence the NMF's are higher. In this paper the influence of the internal pressure on the NMF's is investigated. Figure 4b) shows the circumferential mode of vibration and the corresponding natural frequency of a vertical vessel – length L=3000mm, Diameter OD=609mm, shell thickness t=15mm – depending on the internal pressure. The curve in figure 4 c) represents the ration of the frequency f(p) under internal pressure p to the frequency f0 at pressure zero. The difference between the two natural frequencies f(p) and f0 is up to 25%, and depends on internal pressure.

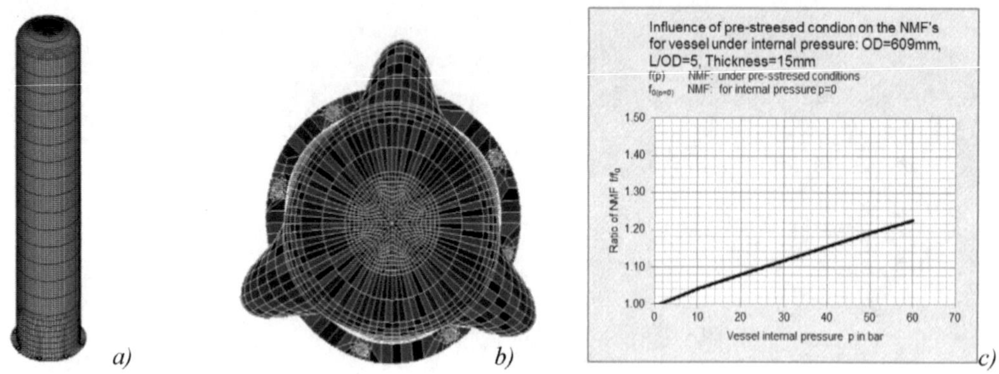

***Figure 4**. a) Vertical separator D=609mm,t=15mm, b) vibration mode, c) frequency ratio $f_{(p)}/f_0$.*

Further the system modelling limits have a major influence on the Natural Mechanical Frequencies (NMF's) of compressor units. In Figure 5 three different models of a compressor unit are shown: *a)* pulsation suppression devices and piping system, *b)* crankcase manifold, piping system, cylinder and cylinder support, and *c)* complete compressor unit including e-motor, base-frame with anchor bolts on concrete foundation. Figures 5a) and 5b) represent a simplified model that is commonly used to carry out pulsation studies in reciprocating compressors. The third model is closer to the real system. This model must be used for simulation calculation.

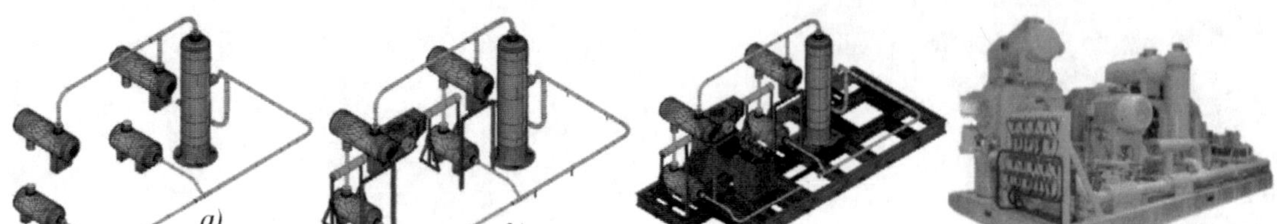

Figure 5.Three different models for the same compressor unit: a), b,) c) and photograph d), onshore application.

IOP Conf. Series: Materials Science and Engineering **232** (2017) 012016 doi:10.1088/1757-899X/232/1/012016

The following table 1 contains the first 10 MNF's, calculated for the three above models: a), b) and c). The differences between the natural frequencies are more than 100%. The first model is practically unsuitable for conducting vibration analysis. The model is very simple and is far away from the real system, photograph d). Only the third model can provide useful results. With this model both the dynamic response analysis and the thermal study can be performed immediately one after another. So, quickly and easy an optimal solution can be found, as the pipe thermal stress analysis is in conflict with the goals of the mechanical dynamic design.

Table 1

NMF's: Natural mechanical frequencies			
Number	Model *a)*	Model *b)*	Model *c)*
[-]	[Hz]	[Hz]	[Hz]
1	28.5	27.7	16.2
2	29.6	28.2	17.9
3	40.4	29.4	22.1
4	59.6	40.4	25.5
5	63.1	41.9	26.3
6	67.0	43.1	31.0
7	74.8	60.8	34.5
8	84.7	61.1	35.2
9	86.7	63.0	38.0
10	92.7	63.9	39.0

The difference between the NMF's ~75%

2.3. Forced mechanical response analysis

Next, the forced response analysis shall be performed. The exciter forces on the structure are the inertia (mass) forces, gas (stretching) forces in cylinders as well as the shaking forces on valve passages, cylinder passage, heads of the pressure pulsation devices, separators, coolers, pipe work components such as elbows, tee's, reducers, safety valves, etc. Figure 6 shows harmonic amplitudes of the cylinder gas force $F_{Gas,axial}$ acting in cylinder in axial direction (cylinder axis), gas and mass force $F_{crosshead\_vertical}$ acting on the crosshead guide in vertical direction (vertical to cylinder axis), gas and mass force $F_{Crank\_pin\_vertical}$, $F_{Crank\_pin\_axial}$ acting on the crank pin in normal and axial direction. Thus there are four different dynamic exciter forces for each axis of the reciprocating compressor. Further, additional forces, resulting from the torsion should also be considered. The exciter forces must be determined for all operating conditions of the reciprocating compressor. For unloaded start-up and shut-down, only the mass forces of oscillated and rotated parts are acting on the compressor components.

IOP Conf. Series: Materials Science and Engineering **232** (2017) 012016 doi:10.1088/1757-899X/232/1/012016

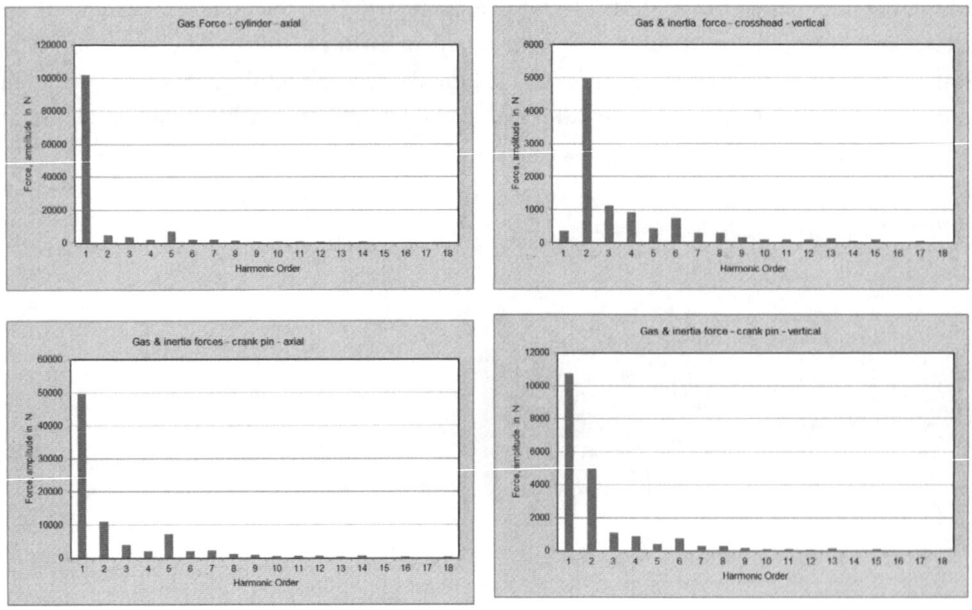

Figure 6. Exciting forces of reciprocating compressor: $F_{Gas,axial}$, $F_{crosshead\ vertical}$, $F_{Crank\ pin\ vertical}$, $F_{Crank\ pin\ axial}$

Figure 7 shows the results of the simulations during the unloaded start-up only under inertia forces. The results are for the offshore application shown in Figure 3. Figure 7 *a)* shows the dynamic loads on the Anti-Vibrating Mount (AVM) in three directions: axial *Fx*, vertical *Fy* and horizontal *Fz*. Figure 7 *b)* contains the vibration velocity at inlet pipe, at cylinder 1, and at cylinder 2. Figure 8 shows the results for the onshore application: *a)* crankcase bolt forces and *b)* vibration velocities. The investigation can be done for all specified operating condition of reciprocating compressor. The last step is the thermal stress analysis which can be performed with the same model. The simulation steps - modeling, modal analysis, dynamic analysis, thermal analysis - can be repeated until the required specified criteria are met. During the simulation process also different standards, regulations and specifications [1], [2], [3], [4], [11], [12] must be considered.

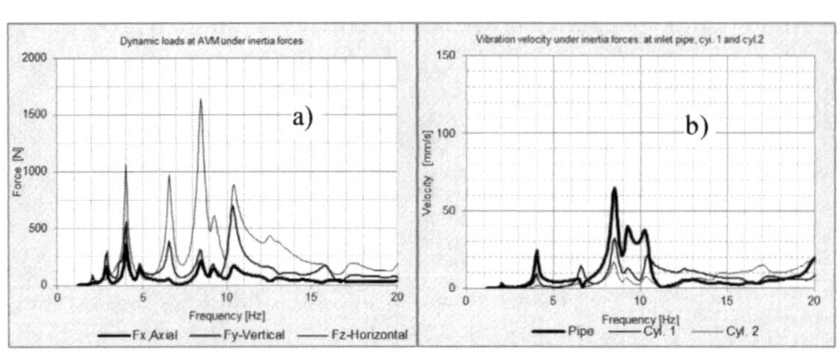

Figure 7. a) Dynamic loads on AVM b) vibration velocities at inlet pipe, cylinder 1, and cylinder 2 for the offshore application in Figure 3.

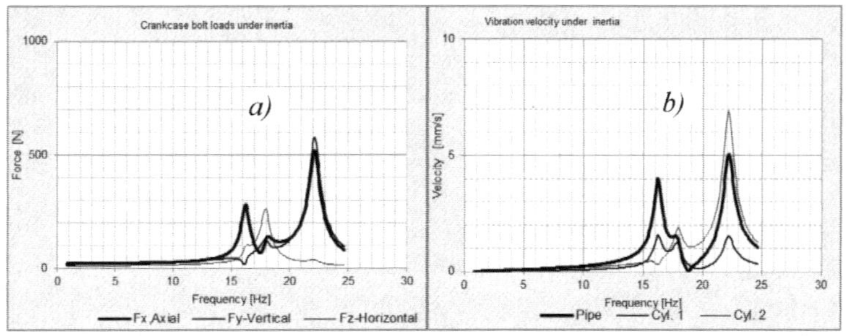

Figure 8. a) Dynamic loads on crankcase bolt b) vibration velocities at inlet pipe, cylinder 1, and cylinder2 for the onshore application in Figure 5.

3. Torsional vibration

The reciprocating compressors have an extensive drive train. There are very complex components such as, crankshaft, flywheel, electric machine, turbine, engine, fan, screw compressor, blower, viscus damper, coupling, gear unit, V-belt etc. The correct determination of all necessary data and parameters for the Torsional Vibration Analysis (TVA) is very important. This has very high impact on the quality of the calculation, on the reliability and the availability of reciprocating machine train. The main parameters of the components are the torsional stiffness, the damping factors and the excitations. The calculation of stiffness of the crankshaft of reciprocating compressors is described in detail in [8], [9]. Figure 9 shows the Finite Element (FE) model of the crankshaft and flywheel of a four cylinder 3MW horizontal compressor. The calculated 1st and 2nd Natural Torsional Frequencies (NTF) are f_1=114 and f_2=323Hz. The mass moment of inertia J of the crankshaft is ~ 25kgm². The stiffness C_1 and C_2 for one or two mass spring torsional system for the crank shaft – Figure 9 - can be determined from previously calculated NTF's (f_1=114 and f_2=323Hz) and J's based on the corresponding frequency equations (Eq. 1), (Eq. 2) and (Eq. 3) below [13]. For two mass spring model - with $J_1=J_2=\sim J/2=12.5$kgm² -the calculated torsional stiffness are (from Eq. 2 and Eq. 3) C_1=21138976Nm/rad, C_2=15619613Nm/rad.

Single masss – spring system :

$$4\pi^2 \cdot f_1^2 = \frac{C_1}{J_1} \qquad (Eq.1)$$

Two mass – spring System :

$$4\pi^2 \cdot f_1^2 = \frac{1}{2}\cdot\left[\left(\frac{C_1+C_2}{J_2}+\frac{C_1}{J_1}\right)-\sqrt{\left(\frac{C_1+C_2}{J_2}+\frac{C_1}{J_1}\right)^2-\frac{4\cdot C_1\cdot C_2}{J_1\cdot J_2}}\right] \qquad (Eq.2)$$

$$4\pi^2 \cdot f_2^2 = \frac{1}{2}\cdot\left[\left(\frac{C_1+C_2}{J_2}+\frac{C_1}{J_1}\right)-\sqrt{\left(\frac{C_1+C_2}{J_2}+\frac{C_1}{J_1}\right)^2-\frac{4\cdot C_1\cdot C_2}{J_1\cdot J_2}}\right] \qquad (Eq.3)$$

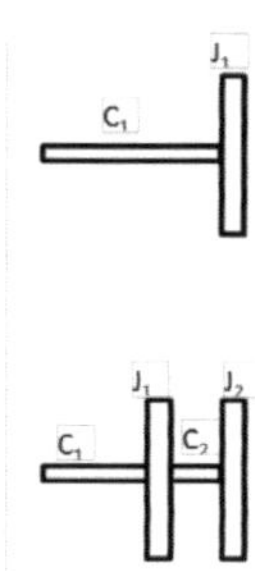

Figure 9 Crankshaft torsional model

The relative shaft damping coefficient k_i (various couplings or shaft damping) depends on the Lehr's damping factor and the corresponding natural torsional frequency [14]. For the absolute damping coefficient k_a, there are data available only for diesel engines [13], [14]. The absolute damping factors depends on piston area A, crank radius r, and the type of engine represented through coefficient μ which are included in the following equation (4):

$k_a = A \cdot r \cdot \mu \quad (Eq.4)$

A Piston Area :
r Crank radius
μ Diesel engine type

The mathematical model of each throw [9] of the crankshaft of reciprocating compressor is (Eq. 5):

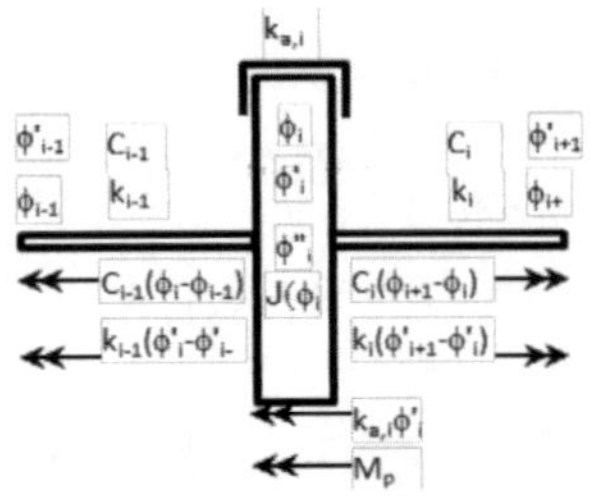

$$J(\phi_i)\cdot\phi_i'' + \frac{1}{2}\cdot\frac{dJ(\phi_i)}{d\phi_i}\cdot\phi_i'^2 + k_{a,i}\cdot\phi_i' + k_{i-1}\cdot(\phi_i' - \phi_{i-1}') + C_{i-1}\cdot(\phi_i - \phi_{i-1})$$
$$- k_i\cdot(\phi_{i+1}' - \phi_i') + C_i\cdot(\phi_{i+1} - \phi_i) + M_p(\phi_i, \phi_i') = 0 \quad (Eq.5)$$

Figure 10. *Crank gear equation of motion for the throw **i**: throw angle ϕ_i, angular velocity ϕ_i'.*

For the simulation under resonance conditions the equation of motions in [8] and [9] should be extended with the equation for connecting rod and oscillating parts (crosshead, piston rod and piston). Therefore the mass, the mass moment of inertia and the center of gravity (CG) location of the connecting rod are required. The equations of motion can be derived from the kinetic energy of the system. The required torques on the shafts of driven train and the angular velocities of each mass can be obtained from the numerical solution of differential equations [15, Page 360]. The torsional vibration analysis should be performed for all operation cases of the compressor unit including: continuous operation, part-loads, start-up, shut-down under loaded and unloaded condition, emergency shut-down etc. Generally the emergency shut-down represents a critical case. Furthermore, the operation of the system under resonance condition must be taken into consideration. In this paper the calculation of speed fluctuation of the crankshaft throw, and dynamic torques under the resonance conditions – high frequency resonance with up to ~15th harmonic - of the reciprocating compressor, is described. Near the resonance, high dynamic shaft torques and a high fluctuation of rated speed is expected. This will result in higher rod load and cross head normal force. Furthermore this can affect the cylinder PV-Curve and it can influence the power consumption, capacity, and the pressure pulsation. Since the absolute damping of crank gear has a strong influence on the speed fluctuation, their precise determination is of major importance. The following figures show the simulation results of torsional vibration analysis.

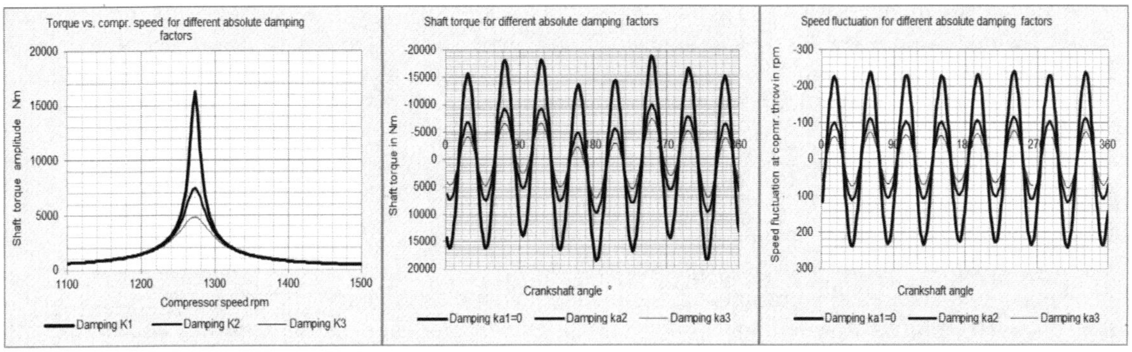

a) *b)* *c)*

Figure 11. a), b) Torque amplitude, c)Speed fluctuation of crankshaft throw.

Figure 11 *a)* shows the torque amplitude over compressor speed for three different absolute damping factors k_a of crank gear. Near the resonance the dynamic toque at the crankshaft, as expected, is very high. To show the influence of the damping the calculation was performed for three different damping coefficients: k_{a1}=0, k_{a2}=5Nms/rad and k_{a3}=10Nms/rad. Damping coefficient k_{a2} corresponds approximately to the above Equation (4) for diesel engine. Depending on the calculated torque loads, generally a separation margin from the critical speed is required.

Figure 11 *b)* and 11 *c)* shows the dynamic torque amplitude and the speed fluctuation versus crankshaft angle, at resonance condition for different damping coefficients.

4. Summary

- The finite element model of compressor units should include at least the following components: crankcase, crankshaft, distance piece, e-motor, pipe work, pressure pulsation devices and base-frame. A combination of 1D, 2D and 3D elements is required to realize a proper model. The crankcase, the distance piece, the pressure pulsation devices and the base-frame must be modeled with 2D elements. For some local zones 3D elements are required. The modeling methods and the scope of the system should be carefully selected. Depending on the scope of the model, the natural frequencies differ from each other up to 100% and sometimes more.
- The modal analysis as well as the forced mechanical response analysis, must be carried out at pre-stressed condition of the compressor unit. In the pre-stressed state, the system is stiffer, and consequently the natural frequencies are higher. For vessel under internal pressure the NMF's are up to 25% higher compared to the natural mechanical frequencies for the unloaded conditions!
- The forced vibration analysis must be performed, not only under the shaking-forces and cylinder- gas-forces as specified in API 618, but also under inertia-forces.
- With the presented FE model - Figure 5c) - can be performed successively, the forced response mechanical analysis and the thermal analysis (flexibility analysis) of the reciprocating compressor systems. So quickly and easy an optimal design of the system can be found ([11], Page 61, 62).
- The presented model allows a lateral dynamic analysis of the crankshaft, under inertia forces, gas forces as well as forces due to torsion.

- The torsional vibration analysis should be performed for all operation cases of the compressor unit including: continuous operation, part-loads, start-up, shut-down under loaded and unloaded condition, emergency shut-down. The calculation must be carried out also for the expected torsional critical speeds (resonance condition). At the critical speed or near the critical speed the torque amplitude and the speed fluctuation are very high. This causes higher rod load and cross head normal force. This can also affect the cylinder- PV-Curve and so the power consumption, capacity, and the pressure pulsations of the machine.

References

[1] VDI 2230 Systematic calculation of high duty bolted joints.
[2] VDI 3842 Vibration in piping systems
[3] VDI 2049 Torsional vibration of driveline - Calculation, measurement, reduction
[4] ISO 10816-8:2014, Mechanical vibration - Evaluation of machine vibration by measurements on non-rotating parts - Part 8: Reciprocating compressor systems
[5] ANSYS Computer-aided engineering software.
[6] Kacani V., Huttar E., Stibi H. Torsionsschwingungen in Kolbenkompressoranlagen. Der Kolbenkompressor- eine zeitgemäße Arbeitsmaschine 4.-5. November, 1999, Dresden
[7] Kacani V., Huttar E., Economical pipe stress analysis using parametric modelling and design standards. 4th Conference of the EFRC, 2005, Antwerpen
[8] Kacani V. Huttar E., Heumesser T. Simulation of Reciprocating Compressor Start-Up. 6th Conference of the EFRC 2008, Düsseldorf
[9] Kacani V. Simulation of Reciprocating Compressor Start-Up and Shut down under Loaded and Unloaded Conditions. International Compressor Engineering Conference at Purdue, July 16-19, 2012
[10] Kacani V., Huttar H., Ognar, G. State of the Art Design and calculation of Vessels and Piping Systems in Reciprocating Compressors. ICCR, Xi'an, China.
[11] API Standard 618, Fifth Edition, 2007, Reciprocating Compressors for Petroleum, Chemical, and Gas Industry Services.
[12] FKM Rechnerischer Festigkeitsnachweis für Maschinenbauteile. VDMA Verlag
[13] Holzweißig F., Dresig H. Lehrbuch der Maschinendynamik. Springer Verlag 1979
[14] Andreas Laschet. Simulation von Antriebssystemen. Springer Verlag. Springer Verlag 1988
[15] Julien C. Sprott. Numerical Recipes. Cambrige University Press 1991

IOP Conf. Series: Materials Science and Engineering **232** (2017) 012017 doi:10.1088/1757-899X/232/1/012017

Application of numerical grid generation for improved CFD analysis of multiphase screw machines

S Rane and A Kovačević

City, University of London, Centre for Compressor Technology, London, UK.
Email: sham.rane@city.ac.uk

Abstract. Algebraic grid generation is widely used for discretization of the working domain of twin screw machines. Algebraic grid generation is fast and has good control over the placement of grid nodes. However, the desired qualities of grid which should be able to handle multiphase flows such as oil injection, may be difficult to achieve at times. In order to obtain fast solution of multiphase screw machines, it is important to further improve the quality and robustness of the computational grid. In this paper, a deforming grid of a twin screw machine is generated using algebraic transfinite interpolation to produce initial mesh upon which an elliptic partial differential equations (PDE) of the Poisson's form is solved numerically to produce smooth final computational mesh. The quality of numerical cells and their distribution obtained by the differential method is greatly improved. In addition, a similar procedure was introduced to fully smoothen the transition of the partitioning rack curve between the rotors thus improving continuous movement of grid nodes and in turn improve robustness and speed of the Computational Fluid Dynamic (CFD) solver. Analysis of an oil injected twin screw compressor is presented to compare the improvements in grid quality factors in the regions of importance such as interlobe space, radial tip and the core of the rotor. The proposed method that combines algebraic and differential grid generation offer significant improvement in grid quality and robustness of numerical solution.

1 Introduction

The CFD analysis of twin screw machines is important in order to fully understand the mechanics of compression, gas-oil interaction, study the impact of oil injection design and modifications in particular and to accurately predict the performance of the machines due to variation in operating condition. Classically used lumped parameter models face a limitation in providing high resolution details of flow in such machines due to geometric and kinematic simplifications. But in the application of CFD models for calculation of flow in screw machines a major challenge is the treatment of highly deforming computational meshes. Kovačević et al [4], Kovačević [5] have successfully used an algebraic grid generation method with boundary adaptation and transfinite interpolation which has been implemented in the grid generation tool SCORG – Screw Compressor Rotor Grid Generator. In [4], Kovačević et al have reported CFD simulations of twin screw machines to predict flow, heat transfer, fluid-structure interaction, etc. A test case study of an oil injected compressor using source terms in transport equations with a segregated pressure based solver has also been presented. Pascu et al [7] have reported optimization of the discharge port of a Twin screw compressor using SCORG grid generator. Optimization was based on the selection of the port geometry by relative comparison of flow field predicted by the CFD models. In the same field of analysis, Vande Voorde et al [13] used an algorithm for generating block structured mesh from the solution of the Laplace equation for twin screw

compressors and pumps using differential methods. Reports on analysis of dry air screw compressors and twin screw expanders with real gas models are available in literature using these techniques [8]. Although several attempts have been made in the recent past to extend the CFD technology to oil injected compressors, it has proven to be difficult to achieve the desired grid structure and the modelling conditions that can provide stability to the numerical solvers. Kovačević and Rane have reported in [6] results from the analysis of a dry air expander and it was found that CFD models deviated in prediction of leakage flow much more under the operating condition of low rotor speed and higher pressure difference then at higher rotor speeds with low pressure difference. On the other hand, Rane et al in [9] analyzed a twin screw compressor and found that refinement of the rotor grid in circumferential direction has a direct influence on the prediction of the mass flow rate. Rane [10] proposed a new analytical approach for grid generation that can independently refine the interlobe region of the screw rotors. It was demonstrated that such grid refinement improves the prediction of mass flow rates. The same algebraic algorithm has been extended to produce a rotor grid that eliminates the interface between two rotors and thereby providing a desirable grid for oil injected or multiphase modelling.
In this paper, deforming grid of the twin screw machine is generated using these algebraic techniques and treated as an initial mesh upon which Elliptic PDE of the Poisson's form is solved. The resulting differential grid has highly improved cell quality and distribution. In addition, a special procedure has been introduced that completely smooths the transition of the partitioning rack curve between the two rotors thus improving grid node movement and robustness of the CFD solver. A sample analysis of an oil injected twin screw compressor using the new Elliptic PDE grids has been presented to compare the improvements in grid quality factors in the regions of importance such as interlobe space, radial tip and core of the rotors. A significant improvement in the grid quality and robustness of numerical solver with higher order schemes has been obtained with this differential implementation of the deforming mesh.

2 Differential grid generation principle

In structured grid generation, techniques based on solutions of Partial Differential Equations to define coordinate transformation are widely used. The idea of using elliptic PDE like Laplace equation or Poisson equation is based on the work of *Crowley* [1] and *Winslow* [14]. Elliptic PDE's have certain beneficial properties in their solution that make them preferable for body fitted curvilinear grids:
Extremum principle – It is possible to formulate PDE's such that the extrema of their analytical solutions cannot be within the domain. Hence there are fewer tendencies for folding of mesh lines.
Inherent smoothness – The resulting coordinate curves in the interior are smooth and some discontinuities over the boundaries do not get propagated into the interior of the domains.
Boundary – The coordinate points on physical boundaries can be exactly specified as boundary conditions in the computational space.
But the numerical grid generation is computationally expensive and preferably an initial grid based on algebraic methods needs to be prescribed in order for the required derivatives to be used efficiently for solution of PDE's.

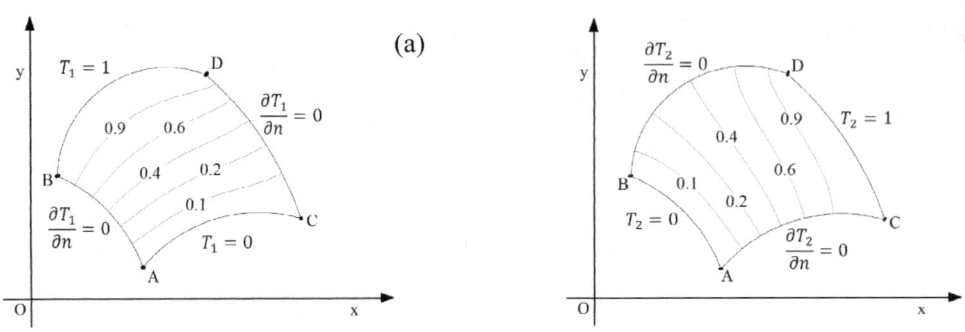

Figure 1. Differential construction of a 2D curvilinear grid.

Consider a 2D domain as shown in Figure 1. The region is bounded by four curves AB, BD, CD and AC. To demonstrate the principle, a heat conduction problem will be used with boundary conditions T1 and T2 as shown in Figure 1 a and b.
For domain (a)

$$\frac{\partial^2 T_1}{\partial x^2} + \frac{\partial^2 T_1}{\partial y^2} = \frac{q''_1}{k},$$

$$T_1 = 0, at\ AC\ and\ T_1 = 1, at\ BD \tag{1}$$

$$\frac{\partial T_1}{\partial n} = 0, \qquad AB\ and\ CD$$

where n is the Normal to the boundary

Similarly for domain (b)

$$\frac{\partial^2 T_2}{\partial x^2} + \frac{\partial^2 T_2}{\partial y^2} = \frac{q''_2}{k},$$

$$T_2 = 0, at\ AB\ and\ T_2 = 1, at\ CD \tag{2}$$

$$\frac{\partial T_2}{\partial n} = 0, \qquad AC\ and\ BD$$

The solution to systems (a) and (b) provides isotherms which can be written as

$$T_1 = T_1(x,y), \qquad T_2 = T_2(x,y) \tag{3}$$

Each isotherm T_1 and T_2 thus represents a set of (x,y) and the two solutions get superimposed as a body-fitted curvilinear grid in the domain. Equation (1) and (2) can be re-written by interchanging roles of T and (x,y), with correspondence of $T_1: \xi$ and $T_2: \eta$.

$$\nabla^2 \xi = \frac{\partial^2 \xi}{\partial x^2} + \frac{\partial^2 \xi}{\partial y^2} = P(\xi,\eta),$$

$$\nabla^2 \eta = \frac{\partial^2 \eta}{\partial x^2} + \frac{\partial^2 \eta}{\partial y^2} = Q(\xi,\eta), \tag{4}$$

Where (ξ,η) are curvilinear coordinates and $P(\xi,\eta)$ and $Q(\xi,\eta)$ are stretching functions and the system is a basic Poisson system of PDE. In order to determine Cartesian coordinates from equation (4) they must be inverted. The inverted equation in 2D reads as [2]:

$$\nabla^2 \emptyset = \frac{1}{J^2}\left[\alpha \frac{\partial^2 \emptyset}{\partial \xi^2} - 2\beta \frac{\partial^2 \emptyset}{\partial \xi\, \partial \eta} + \gamma \frac{\partial^2 \emptyset}{\partial \eta^2}\right] + P\frac{\partial\, \emptyset}{\partial \xi} + Q\frac{\partial\, \emptyset}{\partial \xi} \tag{5}$$

Replacing $\emptyset$ by x and y and $\nabla^2 x = \nabla^2 y = 0$,

$$\alpha \frac{\partial^2 x}{\partial \xi^2} - 2\beta \frac{\partial^2 x}{\partial \xi\, \partial \eta} + \gamma \frac{\partial^2 x}{\partial \eta^2} = -J^2\left[P\frac{\partial\, x}{\partial \xi} + Q\frac{\partial\, x}{\partial \xi}\right] \tag{6}$$

$$\alpha \frac{\partial^2 y}{\partial \xi^2} - 2\beta \frac{\partial^2 y}{\partial \xi\, \partial \eta} + \gamma \frac{\partial^2 y}{\partial \eta^2} = -J^2\left[P\frac{\partial\, y}{\partial \xi} + Q\frac{\partial\, y}{\partial \xi}\right] \tag{7}$$

$$J = \frac{\partial x}{\partial \xi}\frac{\partial y}{\partial \eta} - \frac{\partial y}{\partial \xi}\frac{\partial x}{\partial \eta}, \quad \alpha = \left[\left(\frac{\partial x}{\partial \eta}\right)^2 + \left(\frac{\partial y}{\partial \eta}\right)^2\right]$$

$$\beta = \left[\frac{\partial x}{\partial \xi}\frac{\partial x}{\partial \eta} + \frac{\partial y}{\partial \xi}\frac{\partial y}{\partial \eta}\right], \quad \gamma = \left[\left(\frac{\partial x}{\partial \xi}\right)^2 + \left(\frac{\partial y}{\partial \xi}\right)^2\right]$$

Equation (6) and (7) are coupled and nonlinear because of α, β *and* γ. They are solved simultaneously using schemes like Tri-diagonal Matrix Algorithm. The boundary conditions as grid coordinates are specified at $\xi = 0, \xi = \xi_{max}$ *and* $\eta = 0, \eta = \eta_{max}$. One of the difficulties with elliptic PDE's is that at the boundary either Dirichlet or Neumann boundary conditions are required.
Thompson, Thames and Mastin [12] have proposed the following stretching function in order to control grid clustering.

$$P(\xi,\eta) = -\sum_{n=1}^{N} a_n \frac{(\xi-\xi_n)}{|\xi-\xi_n|} e^{-c_n|\xi-\xi_n|} - \sum_{i=1}^{I} b_i \frac{(\xi-\xi_i)}{|\xi-\xi_i|} e^{-d_i\left[(\xi-\xi_i)^2+(\eta-\eta_i)^2\right]^{1/2}} \quad (8)$$

$$Q(\xi,\eta) = -\sum_{n=1}^{N} a_n \frac{(\eta-\eta_n)}{|\eta-\eta_n|} e^{-c_n|\eta-\eta_n|} - \sum_{i=1}^{I} b_i \frac{(\eta-\eta_i)}{|\eta-\eta_i|} e^{-d_i\left[(\xi-\xi_i)^2+(\eta-\eta_i)^2\right]^{1/2}} \quad (9)$$

Here N is the number of clustering grid lines (coordinate lines $\xi = \xi_n$ *and* $\eta = \eta_n$) and I is the number of clustering points (with $\xi = \xi_i,\ \eta = \eta_i,\ 0 \le \xi_i,\ \eta_i \le 1$) to which the grid is to be attracted. a_n, c_n, b_i *and* d_i are positive control parameters. The first term in $P(\xi,\eta)$ has the effect with amplitude a_n of attracting $\xi - lines$ towards curves $\xi = \xi_n$ in the physical domain, while the second term attracts $\xi - lines$ with amplitude b_i towards points. Similarly the two terms in $Q(\xi,\eta)$ control the grid line spacing. In each case the attractive effect decays with the distance in the computational space from the lines or points according to the decay parameters c_n *and* d_i.
The functions $\frac{(\xi-\xi_n)}{|\xi-\xi_n|}$ *and* $\frac{(\eta-\eta_n)}{|\eta-\eta_n|}$ are function which can take only the value ± 1 and ensure the clustering to take place on both sides of $\xi_n - lines$ *and* $\eta_n - lines$. The amplitude can be set negative to turn attractive effects into repulsive ones.

3 Differential grid generation for twin screw rotor grids

Algebraic methods are convenient for quick adjustment and recalculation of the deforming grid. Kovačević et al [4], Kovačević [5], Kovačević and Rane [6] and Rane et al [9] have successfully used an algebraic grid generation method together with boundary adaptation and transfinite interpolation. Vande Voorde and Vierendeels [13] have implemented a grid conversion algorithm for unstructured to block structured mesh from solution of Laplace equation for twin screw compressors and pumps using the differential methods. Based on this grid generation, flow in a double tooth compressor and a twin screw compressor has been analyzed and the results have been compared with experimental data over a range of discharge pressure and rotor speeds. A detailed description of the algebraic grid generation for twin screw rotors can be found in Kovačević [5]. By use of a background blocking technique [10], the algebraic grid generation is able to produce a single domain structured grid for the twin rotor domains. With this algebraically generated single domain grid as an initial framework it is possible to solve the PDE system presented in Section 2 in order to produce a better quality differential mesh.

3.1 Initial Algebraic grid generation

The procedure of analytical grid generation of screw machine working domain is explained in Kovačević [5]. In order to achieve a conformal single domain mesh, a new approach of background blocking has been presented in Rane [10, 15 and 16]. In this procedure, outer boundary in each background block is defined as a combination of the rack segment and the casing circle segment. The rack segment stretches between the bottom and top cusp points and is closed by the casing. The distribution obtained on the outer boundaries of the two blocks is used as a reference for rotor profile distribution as shown in Figure 2a.

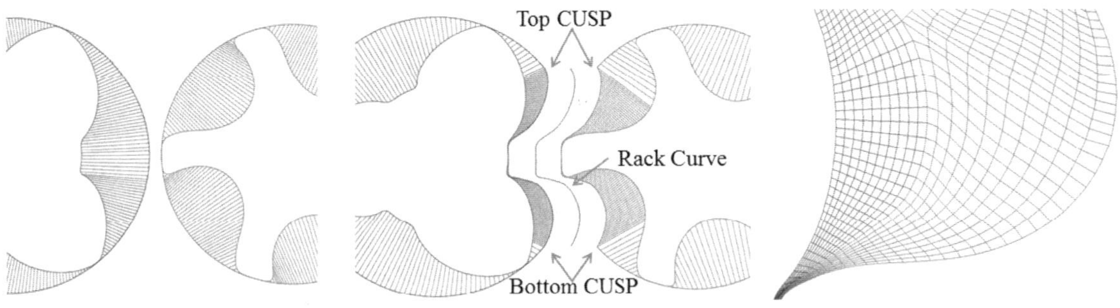

Figure 2. Background blocking for regular boundary discretization and conformal algebraic mesh.

Figure 2b shows how the rack curve is used to partition the two rotor domains and the boundary distribution so obtained is used to generate the 2D mesh using transfinite interpolation. Using the blocking approach allows both conformal and non-conformal boundary map to produce fully hexahedral 3D grid. Figure 2c shows the nodes on the rack segment between the main and the gate rotor grids with conformal boundary map. The 3D mesh generated from such 2D cross sections allows the rotor domains of the male and female rotors to be combined into a single rotor mesh. This avoids inaccuracies and instabilities that may arise due to the interface mismatch in non-conformal boundary map. The resultant grids are recommended for oil injected multiphase flow modelling. But during operation when the rotors turn further, the rack curve has to transition and the current algebraic methods have this transition as a step change at certain positions. One of the objectives of elliptic PDE mesh generation presented in this paper was to improve the time transition of the partitioning rack curve between the two rotor domains.

3.2 Applying the Elliptic PDE solver to improve algebraic grid

The treatment described in Section 2 for differential mesh generation requires the solution of the coupled PDE equations (6) and (7). Schemes like Tri-diagonal Matrix Algorithm can be used for the solution. The boundary conditions are specified as grid coordinates at $\xi = 0, \xi = \xi_{max}$ and $\eta = 0, \eta = \eta_{max}$.

$$P(\xi, \eta) = 0 \tag{10}$$

$$Q(\xi, \eta) = -Q_0 \, sgn(r) e^{-|r|} \ , \qquad r = (\eta - \eta_s) \ and \ sgn(r) = \frac{r}{|r|} \tag{11}$$

This means that for the generation of grid in twin screw rotor domain the coordinates (x, y) of the boundary nodes need to be prescribed as boundary condition to the equations (6) and (7). At this stage the grid generated by algebraic methods acts like the initial condition and also provides the boundary condition. In the presented work, an O grid topology has been implemented for the differential solver using successive-over-relaxation procedures as described in Knupp and Steinberg [3]. Also the stretching functions as described in equations (8) and (9) have been simplified to use only two input parameters based on the test function given by [3] and are listed in equations (10) and (11). η_s is called as the Radial Bias Factor and Q_0 is called the Radial Bias Intensity. Both these factors contribute to the inflation layer control in the rotor meshes. System of equations (6) and (7) are solved over an initial grid generated by TFI. The control parameters of relaxation factor and tolerance factor are used to control the convergence of the SOR solver [3]. Finally the two O grids generated by the solution of the differential solver are merged to produce a single domain mesh for the two rotors as seen in Figure 3 – Final PDE Mesh.

3.3 Smoothing the rack curve time transition to improve boundaries

The PDE Mesher described in Section 3.2 is used in a selective interlobe area bounded by the top and bottom cusp radial nodes of the mesh to convert the rack curve into a smooth transitioning curve across the specified number of angular positions of the rotor. This results into a gradually changing partition between the two O grids of the male and female rotors as shown in Figure 3.

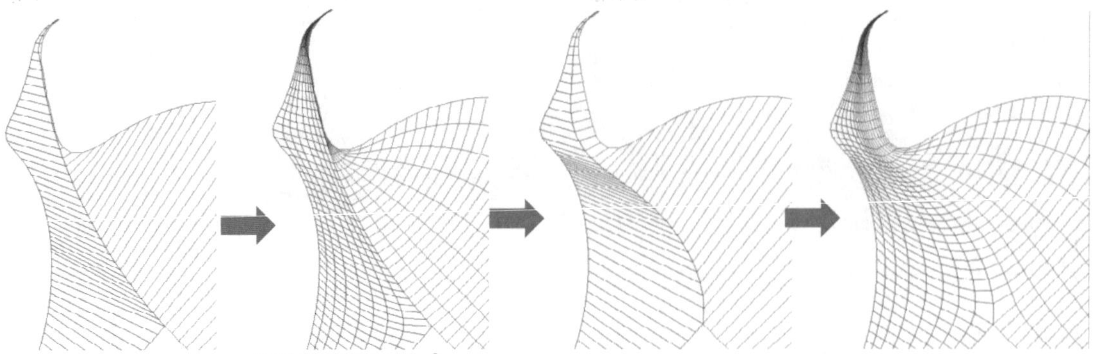

Figure 3. Four-step procedure for improving time transition of the rack curve.

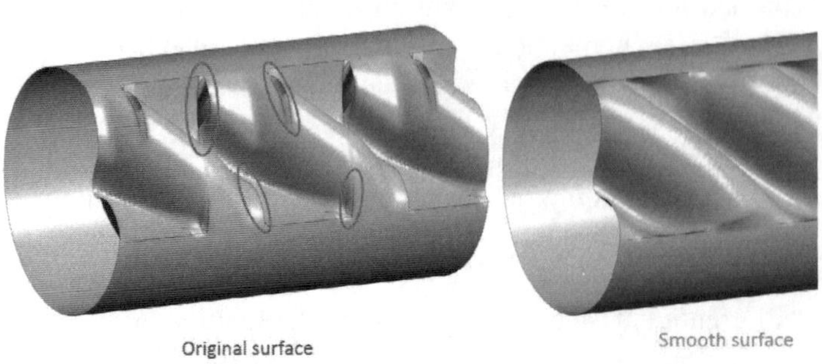

Figure 4. Improvement in the 3D rack surface.

This smooth rack is then supplied back to a second stage of boundary distribution calculation resulting in a new conformal distribution. New conformal distribution is further used as boundary condition for final differential mesh generation. As a result a significant improvement in the mesh quality is achieved.

This four-step procedure is very quick and the time required for calculation of a representative 2D mesh with 5000 node count, across 50 rotor positions is less than 15 seconds. Figure 4 shows the smooth rack surface generated on the 3D mesh. The sharp sections observed with original surface (Initial algebraic mesh) are eliminated and a completely smooth surface deformation with time was achieved.

3.4 Meshing parameters for PDE Mesh generation

Meshing parameters listed in Table 1 are used to control the stretching functions of the Poisson system of the PDE mesh and SOR solver convergence. Smoothness of the conformal grid described in Section 3.3 also needs a control over the number of cells to smooth adjacent to the rack curve in the final 2D mesh. Using this control continuity can be achieved on the mesh lines and cell expansion factor can be improved by specifying a higher interlobe smoothing factor. Some examples of the outputs with different control factors are shown in Figure 5 (a-d). Figure 5a shows a grid generated by specifying a relaxation factor = 0. This produces a pure TFI based algebraic mesh with no smoothing and orthogonalisation.

IOP Conf. Series: Materials Science and Engineering **232** (2017) 012017 doi:10.1088/1757-899X/232/1/012017

Table 1. Meshing parameters – Differential Mesher

Mesh Orthogonality and Smoothing Factor	
Relaxation Factor	Relaxation factor controls mesh orthogonalisation and smoothing by relaxing the increment in each consecutive iteration.
Tolerance Factor	Convergence criteria for PDE solver.
Inflation Layer Control	
Radial Bias Factor - η_s	Factor is used to select boundary for bias. 1 – Rotor, 0 – Casing.
Radial Bias Intensity - $\boldsymbol{Q_0}$	Intensity of stretching the nodes towards the boundaries.
Conformal Mesh Control	
Interlobe Smoothing Factor	Factor is used to smooth layers of mesh adjacent to the rack.

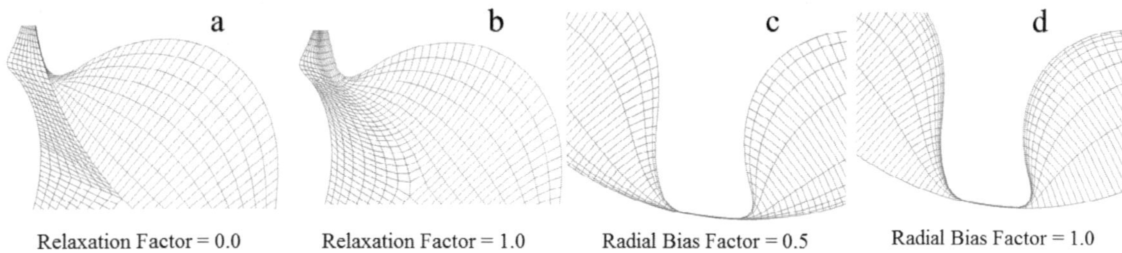

Figure 5. Response of the PDE meshing parameters in a representative 2D section.

Figure 5b is a grid generated by a high relaxation factor = 1. In combination with a higher tolerance factor makes the grid lines smooth and the rack transition becomes gradual. Radial Bias Factor is used to control the stretching of nodes towards the boundaries. In Figure 5c with a Radial Bias Intensity = 2 and a Radial Bias Factor = 0.5, the radial nodes are pulled towards the rotor and the casing. The value of $\boldsymbol{\eta_s}$ can range from 0 to 1. 1 pulls nodes towards the rotor while 0 pulls the nodes towards casing. An intermediate value can thus be used to control the inflation layers in the rotor domain.

3.5 Improvement in grid quality

A few examples of improvement in grid quality by the differential grid generation are presented here.

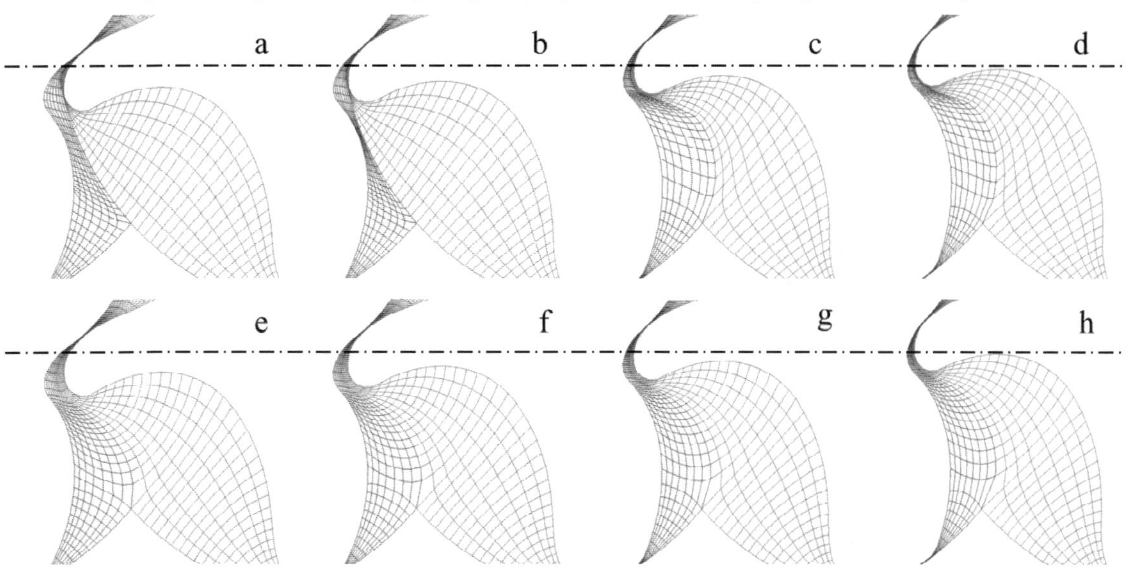

Figure 6. Comparison of grid and rack transition between Algebraic (a-d) and Elliptic (e-f) meshing.

Figure 6 shows a comparison of the grid and rack transition between algebraic and elliptic meshes. Figure 6 a-d are algebraic meshes for four successive rotor positions and Figure 6 e-h are the elliptic meshes for the respective rotor positions. Between position Figure 6b and c a sudden transition in the rack curve is noticeable. On the other hand, in the respective Figure 6f and g the rack node transition is gradually. Also Figure 6e and h highlight the characteristics of the elliptic mesh that the rack nodes in the earlier and next rotor positions were in smooth transition. Figure 6 e-h also show that the interior cell structures have improved and radial mesh lines between the two rotor domains have continuity.

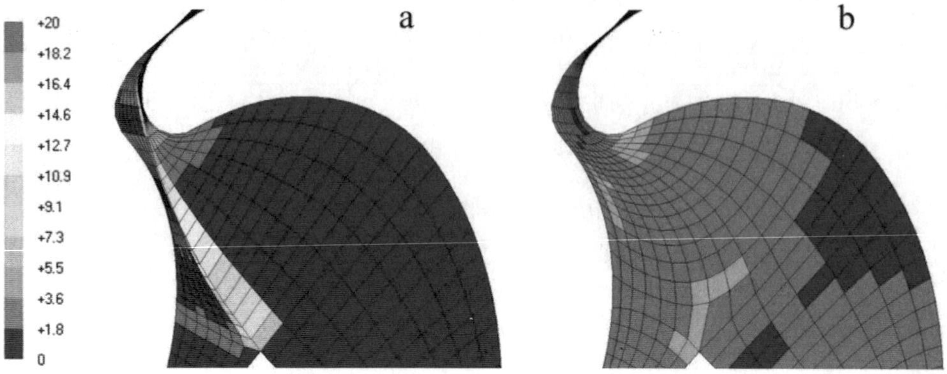

Figure 7. Comparison of cell expansion factor between Algebraic (a) and Elliptic (b) meshing.

Figure 7 a and b present a comparison of the cell expansion factor. Figure 7a corresponds to Figure 6b and Figure 7b corresponds to Figure 6f which are the positions just before rack transitions in the algebraic mesh. A high cell expansion factor of about 20 is seen in the algebraic mesh in this rotor position. In case of elliptic mesh, due to rack smoothing, the cell expansion factor is less than 3.6

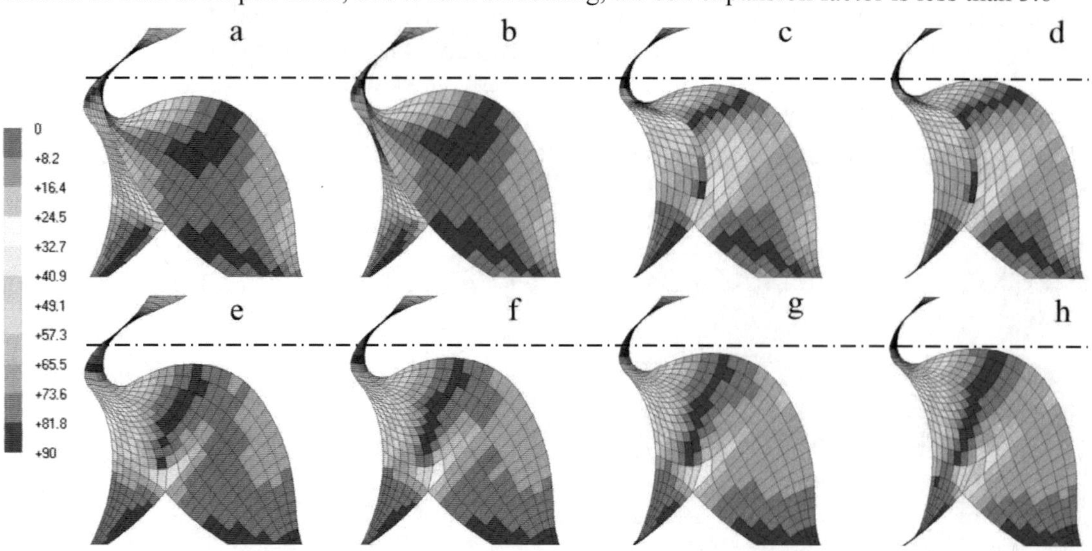

Figure 8. Comparison of cell orthogonal quality between Algebraic (a-d) and Elliptic (e-f) meshing.

Figure 8 presents a comparison of the cell orthogonal quality between the algebraic meshes and the elliptic meshes. Figure 8(a-d) are algebraic meshes and Figure 8(e-h) are elliptic meshes in the respective rotor positions. The rotor positions are identical to the representations in Figure 6. It is seen that the minimum orthogonal angle in case of the algebraic meshes goes to about 8 degree after the rack curve transitions from position b to c. Many cells are in the range of 40 – 60 degree. Low orthogonality values are also seen in position d. In case of elliptic meshes, there is improvement in the overall orthogonality quality and also the minimum value is not below 25 degree. Most of the cells are in the range of 75 – 90 degree. One particular cell at the bottom cusp shows low orthogonality of about 15 degree in both the algebraic and the elliptic mesh. This is mainly due to the abrupt topology change that happens at the cusp point.

4 Application of Elliptic grid for oil injected twin screw compressor analysis

Flow in an oil flooded twin screw compressor with a combined axial and radial suction and an axial discharge port was calculated with the new elliptic PDE grid. The male rotor has four lobes and the female rotor has five lobes with 'N' profiles. The nominal interlobe, radial and axial leakage gaps are 60 μm. The operating speed of the machine is between 3000 rpm and 6000 rpm, discharge pressure can vary between 4.0 bar and 12.0 bar. A complete description of the model numerical setup can be found in Rane et al [11].

4.1 Oil distribution inside the compression chamber

The fluctuating nature of the oil injection was well captured by the CFD model. Figure 9 shows the distribution of oil and its interaction in the discharge port. Iso-surface of 10% oil volume fraction shows an accumulation of oil in the rotor tips and also high concentration on the discharge port walls. Similarly oil accumulates in the interlobe gaps and is transported forward by the rotors. The oil distribution pattern in the discharge port is highly unsteady but cyclically repeating.

Figure 9. Oil flow distribution near the discharge port.

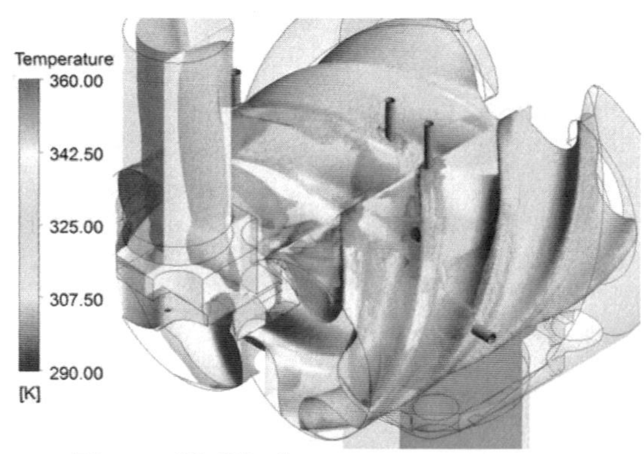

Figure 10. Discharge gas temperature distribution at 6000rpm, 8.0bar pressure.

4.2 Temperature distribution

Figure 10 shows the distribution of air temperature in the domain at 8.0bar discharge pressure, 6000rpm with an iso-surface of 10% oil volume fraction. With built-in V_i of 3.6 for the machine, the adiabatic discharge temperature without oil injection would be about 500 K but due to leakages and recompression effects, the temperature rise would be much higher. Due the injected oil at 323 K, the gas temperature is substantially lowered and does not exceed 340 K. It can be seen that the local maximum temperature reaches to about 360 K. Also the gas temperature is not uniformly distributed. This non-uniformity is due to the non-homogeneous distribution of oil in the compression chamber.

5. CONCLUSIONS

Algebraic grid generation is widely used for discretization of the working chamber of twin screw machines. In order to improve the speed and robustness of the numerical solution of multiphase flows, a new technique which combines differential grid generation procedure with the existing algebraic grid generation is presented in this paper. The deforming grid of a twin screw machine is initially generated using algebraic grid generation. This mesh is treated as the initial mesh upon which Elliptic PDE of the Poisson's form is solved. The resulting grid has highly improved quality and distribution of computational cells. In addition, a special four-step procedure has been introduced that completely smoothens the transition of the algebraically generated partitioning rack curve between the two rotors thus smoothing the grid node movement in time and space. These grid features have improved the robustness of the solution in CFD solvers.

Analysis of an oil injected twin screw compressor is presented in order to compare the improvements in the grid quality factors in the regions of importance such as interlobe space, radial tip and core of the rotors. Significant improvement in the grid quality and robustness of numerical solver with higher order advection schemes has been obtained with this new deforming mesh. The results from the simulation provided an exceptional visualization of the interaction of gas and oil inside the compression chamber. The interaction of the phases, distribution of oil, heat transfer between gas and oil and also effects on sealing due to high oil concentration in leakage gaps were well captured.

Nomenclature

PDE	- Partial Differential Equation	q''	- Heat flux
ξ, η	- computational coordinate	x, y, z	- Cartesian coordinates
r, θ	- polar coordinates of a point	$\boldsymbol{r}$	- radius vector
$P(\xi, \eta)$	- Control function for x stretching	$Q(\xi, \eta)$	- Control function for y stretching
α, β, γ	- Coefficients of elliptic PDE	J	- Transformation Jacobian Matrix
η_s	- Radial bias factor	Q_0	- Radial bias intensity
TFI	- Transfinite Interpolation	SOR	- Successive over-relaxation

References

[1] Crowley W. P., 1962. An equipotential zoner on a quadrilateral mesh. Memo, Lawrence Livermore Lab.

[2] Date W. A., 2005. Introduction to computational fluid dynamics. ISBN-10 0-521-85326-5. Cambridge Uni Press.

[3] Knupp P. and Steinberg S., 2002. The Fundamentals of Grid Generation. ISBN 9780849389870. CRC Press.

[4] Kovačević A., Stošić N. and Smith I. K., 2007. *Screw compressors - Three dimensional computational fluid dynamics and solid fluid interaction*, ISBN 3-540-36302-5, Springer-Verlag New York.

[5] Kovačević A., 2002, Three-Dimensional Numerical Analysis for Flow Prediction in Positive Displacement Screw Machines, Thesis, City, University of London.

[6] Kovačević A. and Rane S., 2013. 3D CFD analysis of a twin screw expander, *8th International conference on compressors and their systems*, London, p. 417.

[7] Pascu M., Kovačević A. and Udo N., 2012. Performance Optimization of Screw Compressors Based on Numerical Investigation of the Flow Behaviour in the Discharge Chamber. Proc. Int. Compressor Conf. at Purdue. Paper 1145.

[8] Papes, I., Degroote, J. and Vierendeels, J., 2013, 3D CFD analysis of an oil injected twin screw expander. Pro of the ASME 2013 International Mechanical Engineering Congress and Exposition, San Diego, USA: ASME IMECE 2013.

[9] Rane S., Kovačević A., Stošić N. and Kethidi M., 2014. Deforming grid generation and CFD analysis of variable geometry screw compressors, Computers and Fluids, 99, p. 124–141.

[10] Rane, S., 2015. Grid Generation and CFD analysis of Variable Geometry Screw Machines, Thesis, City, University of London.

[11] Rane S., Kovačević A. and Stošić N., 2016. CFD Analysis of Oil Flooded Twin Screw Compressors. Proc. Int. Compressor Conf. at Purdue. Paper 2392. http://docs.lib.purdue.edu/icec/2392

[12] Thompson J. F., Thames F. C. and Mastin C. W., 1974. Automatic Numerical Generation of Body-Fitted curvilinear coordinate system for field containing any number of arbitrary two-dimensional bodies. J. Comput. Phys., vol. 15, pp. 299-319.

[13] Voorde Vande J., Vierendeels J., 2005. A grid manipulation algorithm for ALE calculations in screw compressors. *17th AIAA Computational Fluid Dynamics Conference*, Canada, AIAA 2005-4701.

[14] Winslow A., 1966. Numerical solution of the quasilinear poisson equation in a nonuniform triangle mesh. J. Comput. Physics, 1:149-172.

[15] Rane S. and Kovačević A., 2017, Algebraic generation of single domain computational grid for twin screw machines. Part I. Implementation, Advances in Engineering software, Vol 107, Pg 38–50, d.o.i: 10.1016/j.advengsoft.2017.02.003

[16] Kovačević A. and Rane S., 2017, Algebraic generation of single domain computational grid for twin screw machines Part II – Validation, Advances in Engineering software, Vol 109, Pg 31–43, d.o.i: 10.1016/j.advengsoft.2017.03.001

IOP Conf. Series: Materials Science and Engineering **232** (2017) 012018 doi:10.1088/1757-899X/232/1/012018

Energy potential of dual lead rotors for twin screw compressors

Matthias Utri, Andreas Brümmer

TU Dortmund University, Chair of Fluidics
Leonhard-Euler-Str. 5, 44227 Dortmund, Germany

matthias.utri@tu-dortmund.de, andreas.bruemmer@tu-dortmund.de

Abstract. The paper deals with a theoretical investigation into the ways in which fluid properties influence the optimal geometrical parameters of dry running twin screw compressors with uniform and dual lead rotors. Dual lead rotors possess two segments with different rotor leads in order to optimise the progression of chamber volume, clearance size and outlet area. Optimal geometry parameters are determined by means of a multi-chamber model simulation. Fluids are treated as an ideal gas and fluid properties are varied systematically in order to provide guidance on how the machine should be designed for a particular application. The optimal geometry parameters mainly depend on the speed of sound and isentropic exponent of the fluid, because they influence outlet throttling during discharge, and pressure progression due to volumetric compression. In particular, the values for the rotor wrap angle and the internal volume ratio need to be adjusted individually for each compressor application. The results reveal that dual lead compressors possess most potential for fluids with low speed of sound and isentropic exponent.

1. Introduction

Screw machines used for gas compression can be found in numerous engineering applications due to their reliable and low-maintenance operation. Compression is achieved by means of two intermeshing rotors which rotate in a close-fitting casing. The shape of the rotors influences the performance of the compressor and the optimal configuration depends to a large extent on the intended use. In particular, the rotor wrapping characteristics have a decisive impact on the operation of the machine because they influence the size of the clearances and the performance during the discharge phase. Weckes [1] optimised the geometry of a screw air supercharger and estimated the best energy efficiency for a wrap angle of 100 degrees. You et al [2] investigated a screw refrigeration compressor with R22 and identified an optimal wrap angle of 330 degrees for this application. These opposing values are not only a result of the different pressure ratios and rotor profiles under examination, but of the different thermodynamic properties of the fluids.

Newer attempts try to combine the advantages of low and high rotor wrap angles by using variable lead rotors, Fig. 1. So far, an increase in efficiency for variable lead rotors is predicted for high pressure applications [3] and in the field of refrigeration compressors [4]. The improvement is based on the optimised progression of volume curve, clearances and discharge area and is a result of the reduction in the overall wrapping of the rotors. This reduction has positive effects on screw machine performance:

IOP Conf. Series: Materials Science and Engineering **232** (2017) 012018 doi:10.1088/1757-899X/232/1/012018

- Decrease in working cycle duration (less time for clearance mass flow results in higher volumetric efficiency)
- Increase in maximum chamber volume leading to higher mass flow for a machine of a given size
- Reduction in throttling during discharge

So far no detailed investigation of possible fields of application where benefits may be expected by using variable rotor leads has been carried out.

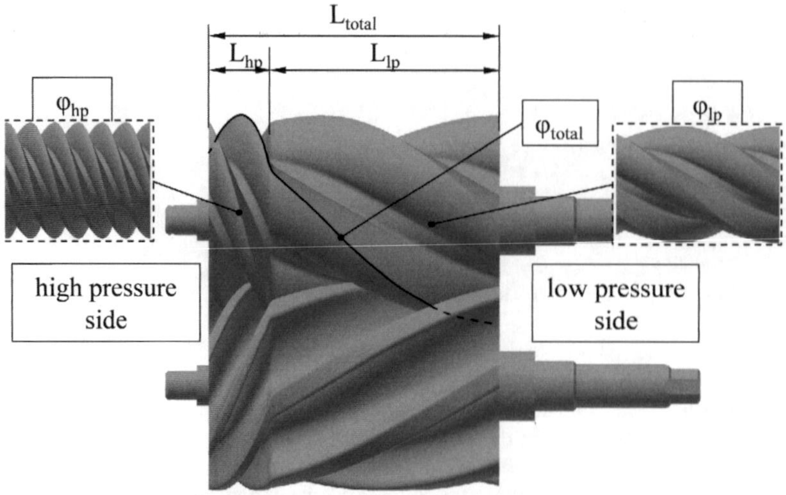

Figure 1. Representative screw machine with dual lead rotors [4]

2. Multi-chamber model simulation

A suitable simulation principle to predict the performance of positive displacement machines is a multi-chamber model simulation which is nowadays used in several applications [5]. The simulation is based on the conservation of mass and energy and calculates the fluid state inside all working chambers simultaneously on the basis of a zero-dimensional chamber model. This chamber model includes the time-dependent progressions of the relevant properties of the machine under examination, e.g. chamber volumes, clearance and port areas. The clearances and port areas maintain an exchange of mass and energy between chambers and ports, thus influencing the fluid pressure inside the working chambers. This interaction necessitates an iterative calculation of the fluid states, which are assumed to be homogeneous in each working chamber. Basically, the change in fluid state inside a working chamber is sectioned in two steps that are calculated consecutively for a very small time step: a change in chamber volume and an exchange of mass and energy via ports and clearances. For the assumption of an ideal gas these two phases will be explained in the following.

A new pressure p_2, that is the result of a change in chamber volume from V_1 to V_2, is calculated assuming an adiabatic change of fluid state with the isentropic exponent κ:

$$p_2 = p_1 \cdot \left(\frac{V_1}{V_2}\right)^{\kappa} \tag{1}$$

The exchange of mass flow between two volumes is calculated by means of the formula introduced by St. Venant and Wantzel [6], which assumes an adiabatic and frictionless flow of an ideal gas from a volume with higher pressure (index hp) to a volume with lower pressure (index lp) through a flow cross-section A (narrowest cross section with index *):

$$\dot{m} = \rho_* \cdot A_* \cdot c_* = p_{hp} \cdot A_* \cdot \sqrt{\frac{2 \cdot \kappa}{(\kappa - 1) \cdot R \cdot T_{hp}} \cdot \left(\Pi^{\frac{2}{\kappa}} - \Pi^{\frac{\kappa+1}{\kappa}}\right)} \tag{2}$$

The equation assumes the low pressure p_{lp} to be present in the flow cross section A_* and is valid for subcritical pressure ratios $\Pi = p_{lp} / p_{hp}$. Mass flow also depends on the isentropic exponent κ and the specific gas constant $R = \mathcal{R} / M$, with the universal gas constant $\mathcal{R}$ and the molar mass M of the fluid. Assuming a pressure ratio below the critical pressure ratio $\Pi \leq \Pi_{crit}$ the fluid reaches the speed of sound in the flow cross-section and the mass flow does not depend on the pressure ratio any more. The choked mass flow can be calculated by:

$$\dot{m}_{crit} = \rho_* \cdot A_* \cdot a_* = \rho_* \cdot A_* \cdot \sqrt{\kappa \cdot R \cdot T_*} = \frac{p_{hp}}{\sqrt{T_{hp}}} \cdot A_* \cdot \frac{1}{\sqrt{R}} \cdot \sqrt{\kappa \cdot \left(\frac{2}{\kappa + 1}\right)^{\frac{\kappa+1}{\kappa-1}}} \tag{3}$$

The exchanged mass flow is charged to the fluid state inside the working chamber, thereby changing the pressure and energy of the fluid. The estimated mass flow in equation (2) and (3) is usually reduced by a friction coefficient to take friction into account. In this paper friction is ignored and the friction coefficient is set to one.

3. Energy efficiency and delivery rate

As a result of the multi-chamber simulation the required internal compressor power (not including any mechanical power loss) and the displaced mass flow are determined. They are used to calculate the internal isentropic efficiency, which is defined as the ratio between the power in the case of an isentropic change in fluid state, and the internal compressor power

$$\eta_{is} = \frac{P_s}{P_i} = \frac{\dot{m}_{real} \cdot \Delta h_s}{-\oint p \, dV \cdot n_{MR} \cdot z_{MR}}, \tag{4}$$

and the delivery rate

$$\lambda_L = \frac{\dot{m}_{real}}{\dot{m}_{th}} = \frac{\dot{m}_{real}}{V_{max} \cdot \rho_{in} \cdot n_{MR} \cdot z_{MR}}, \tag{5}$$

which is used to evaluate the amount of clearance flow inside the compressor.

4. Similarity law for screw machines

Geometrical and physical similarity need to be achieved in order to scale a dry-running screw compressor or change the working fluid without a change in internal isentropic efficiency and delivery rate. For example, geometrical similarity implies a constant ratio of clearance heights to rotor diameter. Physical similarity mainly requires the following coefficients to remain constant:

- Isentropic exponent $\kappa = c_p / c_v$
- Pressure ratio $\Pi = p_{in} / p_{out}$
- Circumferential Mach number $Ma_u = u_{MR} / a_{in}$

Here u_{MR} is the male rotor tip speed and the speed of sound at the suction side is

$$a_{in} = \sqrt{\kappa \cdot R \cdot T_{in}} = \sqrt{\frac{\kappa \cdot \mathcal{R} \cdot T_{in}}{M}}. \tag{6}$$

The minor important characteristic numbers (e.g. the circumferential Reynolds number) are not considered in the following investigation.

5. Relevant fluid properties

It can be seen from equation (1) to (3) that the pressure inside the working chamber depends on the geometry of the machine (volumes and available flow cross-sections) and the fluid properties, isentropic exponent and molar mass as a part of the specific gas constant. Fig. 2 shows an overview of typical values of isentropic exponent and molar mass and how they affect the speed of sound at a given temperature. Although fluids with higher molar mass, such as most refrigerants, are not suitable for exact simulations employing the ideal gas law, the treatment of all fluids as ideal gas simplifies the examination of the effects of the fluid properties on optimal screw compressor parameters and the analysis of general dependencies.

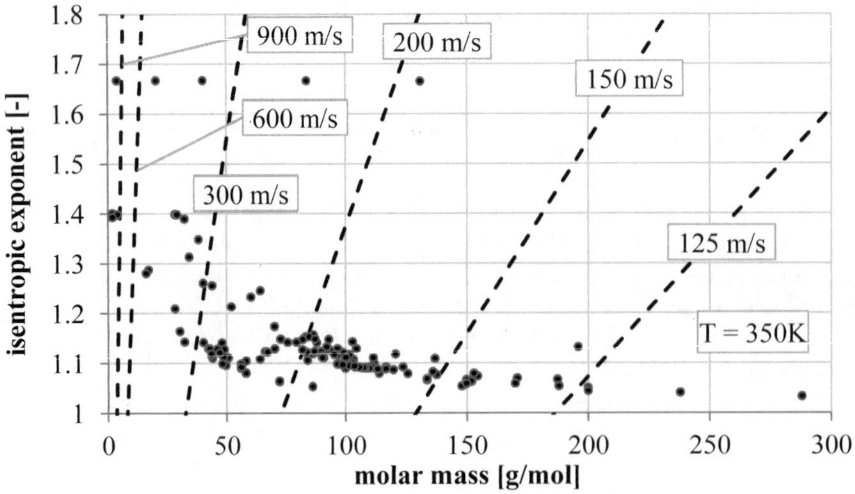

Figure 2. Isentropic exponent and molar mass of vaporous fluids and mixtures from the REFPROP [7] data base with lines of constant speed of sound for T = 350K

6. Variation of fluid properties and boundary conditions

Fluid properties are varied systematically in order to show their individual influence on the optimal geometry of screw machines and demonstrate ways in which efficiency can be improved by using dual lead rotors as presented in [4]. Fig. 3 shows the fluids under investigation. The starting point is fluid #1 with a molar mass of 150 g/mol, an isentropic exponent of 1.05 and a temperature of 300 K. At this point the fluid has a speed of sound of 132 m/s. Fluid #2 possesses a higher speed of sound and a lower density due to an increase in inlet temperature. Fluid #3 has a higher molar mass, but the inlet temperature was adjusted so that the speed of sound is equal to the initial fluid #1. Fluid #4 has a higher isentropic exponent, while temperature was also adjusted to guarantee constant speed of sound. The fluid properties under examination are applied to the suction side of the compressors. Although the fluid properties are not based on real operating conditions, the variations presented show the individual influence of the various fluid properties and provide relevant general information for screw compressor design.

Table 1. Constant boundary conditions

Parameter [dimension]	**Value**
Cycle parameters:	
- Compressor inlet pressure [bar]	1
- Compressor outlet pressure [bar]	5
Machine parameters:	
- Lobe count combination [-]	4+6
- Male rotor crown circle [mm]	72
- Rotor length [mm]	130
- Clearance heights [mm]	0.07
Operating parameters:	
- Circumferential speed [m/s]	80

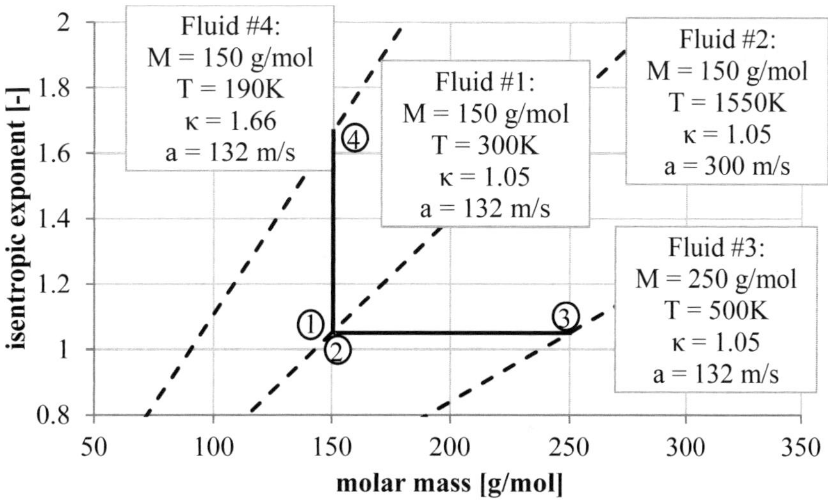

Figure 3. Fluid properties investigated

Other boundary conditions remaining constant throughout all simulations are summarised in Table 1. The rotor profile under investigation is the asymmetric SRM-profile with four male and six female rotor lobes with equal diameters for both rotors. The length-to-diameter ratio is set to 1.8. Analytical functions for the profile can be found in [8]. Dry-running screw machines are simulated by means of the multi-chamber simulation tool KaSim [5], assuming adiabatic flows inside the compressor.

7. Influence of fluid properties on the pressure-volume-diagram

Idealised and real screw compressor cycles are shown in Fig. 4. The cycle can be sectioned in three characteristic phases: During suction the fluid flows from the low pressure port into the emerging working chamber, until maximum chamber volume V_{max} is reached. At this point the chamber becomes disconnected from the low pressure port and the compression phase begins. Once the fluid state inside the chamber arrives at discharge conditions the chamber opens to the high pressure port and the fluid is expelled through the outlet area. Maximum chamber volume and the chamber volume when discharge begins are used to calculate the internal volume ratio of the compressor, equation (7). The change in chamber volume which is necessary for the idealised working cycle to achieve high pressure conditions inside the working chamber as the high pressure port opens, can be estimated by:

$$v_i = \frac{V_{max}}{V_{compr,end}} = \left(\frac{p_{out}}{p_{in}}\right)^{1/\kappa} \quad (7)$$

The influence of molar mass, the speed of sound and the isentropic exponent on the screw compressor cycles is discussed in the following for a representative screw compressor geometry.

7.1. Influence of fluid properties on the pressure-volume-diagram

For the ideal screw compressor working cycle, clearance flow and throttling effects are ignored, so that the load change is isobaric, and the compression is isentropic. Fluids #1, #2 and #3 possess the same isentropic exponent and have the same idealised working cycle so that no isolated influence of molar mass and speed of sound can be observed. Compared to fluid #4 they need a greater change in chamber volume in order to achieve high pressure conditions, equation (7). The lower the isentropic exponent, the higher the internal volume ratio which is necessary to compress the fluid to high pressure conditions. In Fig. 4 the internal volume ratio was adjusted to match the high pressure conditions inside the working chamber when the idealised compression phase has been completed. The internal volume ratio influences the size of the outlet area which is available for discharge, so that a smaller outlet area is usually available for fluids with a lower isentropic exponent as shown for the axial outlet area in Fig. 4. In addition, the discharge temperature and therefore the speed of sound are reduced for these fluids. This in turn influences the discharge flow which is crucial for the real compressor cycle which will be investigated next.

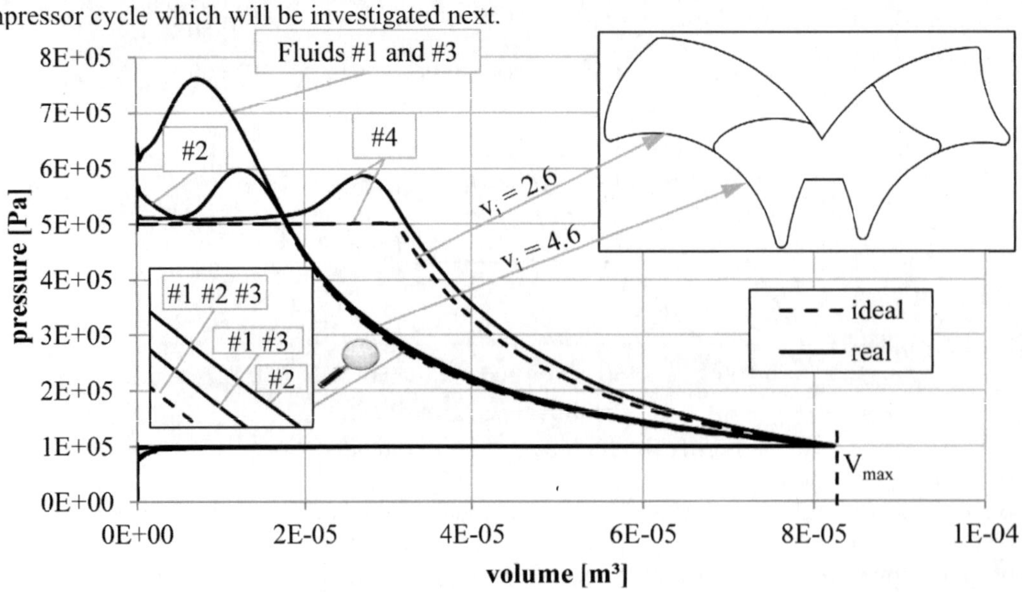

Figure 4. Screw compressor cycles for fluids #1 - #4 and corresponding contours of axial discharge area (wrap angle φ = 300°)

7.2. Influence of fluid properties on the real pressure-volume-diagram

In contrast to the ideal cycle the real screw compressor cycle includes throttling and clearance flows. Analogous to equation (3) the maximum exchanged volume flow can be calculated by:

$$\dot{V}_{crit} = A_* \cdot a_* = A_* \cdot \sqrt{\kappa \cdot R \cdot T_*} = A_* \cdot \sqrt{T_{hp}} \cdot \sqrt{R} \cdot \sqrt{\frac{2\kappa}{\kappa + 1}} \tag{8}$$

Critical volume flow for fluid #4 (κ = 1.66) is 10% higher than for fluid #1 (κ = 1.05) although the speed of sound of the non-flowing fluid at the inlet side is equal for both fluids. Furthermore, greater volumes can be displaced with a larger outlet area and a higher temperature at the discharge port. For the low isentropic exponent of fluids #1 - #3 the temperature when attaining the high pressure conditions is comparatively low. In practice this is an advantage because less cooling of the compressor is necessary, but the lower temperature results in a lower speed of sound during discharge. Although the speed of sound for fluids #1 and #4 is the same at the inlet port, the speed of sound rises to a value of 137 m/s for fluid #1 and to 182 m/s for fluid #4 on the high pressure side, as calculated via the isentropic equations. In turn, the speed of sound increases clearance flow, pre-heating the fluid even more during compression, thus amplifying the effect referred to. In accordance with the faster compression and the resulting larger outlet area discussed in the last section, fluids with a higher isentropic exponent can discharge greater volumes and are less at risk of throttling at the outlet if the internal volume ratio is well adjusted (Fig. 4). On the other hand, fluid #4 results in lower volumetric efficiency because clearance mass flow is also comparatively high, which can be seen from the difference between the idealised and the real working cycle during the compression phase.

A higher speed of sound (fluid #2) decreases throttling at the outlet, because more volume can be displaced in the same time (equation (8), Fig. 4). Clearance flow is also increased, so that the pressure progress during the compression phase is slightly increased and volumetric efficiency is reduced.

The pressure progression for fluids #1 and #3 are equal. Compared with fluid #1, fluid #3 possesses a greater molar mass while isentropic exponent and speed of sound remain constant by adjusting the temperature of the fluid. Due to the equality of speed of sound and isentropic exponent the similarity laws are complied with, so that no individual influence of molar mass, inlet temperature or inlet density on the screw compressor cycle can be observed.

8. Influence of fluid properties on optimal design parameters

Results for an optimisation of the rotor wrap angles for the different fluids are discussed in the following, for constant and dual lead compressors. For constant lead the wrap angle is varied between 150° and 500° to determine the optimal geometry. For dual lead the rotors consist of two segments, each with a constant rotor lead as shown in Fig. 1. For constant overall wrap angles φ_{total} the wrappings of the segments φ_{hp} and φ_{lp} can be varied according to the following equations [4]:

$$L_{total} \cdot \varphi_{total} = L_{hp} \cdot \varphi_{hp} + L_{lp} \cdot \varphi_{lp} \tag{9}$$

$$L_{total} = L_{hp} + L_{lp} \tag{10}$$

Each overall wrapping creates hundreds of possible combinations of the two rotor segments which need to be simulated in order to optimise the geometry. The overall wrap angle is varied between 175° and 350° whereas results are only shown for optimised geometries for each overall wrapping. For each compressor (constant or dual lead) the internal volume ratio is optimised individually for steps of 0.25, so that only maximum efficiency values are shown for each rotor wrapping.

8.1. Influence of the speed of sound

The influence of the fluid properties on performance is investigated by comparing their effect on the optimal design parameters of the screw compressors. Simulation results for fluid #1 are shown in Fig. 5 and reveal that maximum efficiency for constant rotor lead is attained at comparatively high wrap angles. Due to the low speed of sound, discharge speed is limited so that a high degree of wrapping is necessary to avoid throttling during discharge. Higher wrap angles reduce the time-dependent change in chamber volume so that less volume flow needs to be displaced during discharge. Furthermore, at the same internal volume ratio, the discharge area increases with higher wrap angles, and outlet throttling is also reduced [9], so that optimal values for the internal volume ratio are increased with higher wrap angles. For low wrap angles the chamber connects to the high pressure port even before the fluid reaches high pressure conditions, but the quasi-isochoric compression under high pressure conditions is less detrimental than throttling at the discharge port caused by a smaller outlet area. On the other hand, high wrap angles extend the duration of the working cycle and clearance areas are larger, so that volumetric efficiency is reduced. Because of the low speed of sound the effect of clearance mass flow is insignificant, so that the maximum efficiency which can be achieved is very high and only slightly reduced if wrap angles are enlarged. With dual lead screw compressors maximal isentropic efficiency can be increased by about 1.5 percentage points and even more for lower rotor wrappings. Delivery rate is somewhat decreased, depending mainly on the blowhole area, which depends entirely on the rotor lead. Optimal energy efficiency is attained with higher internal volume ratios due to the optimised progressions of volume curve and outlet area.

In contrast, the higher speed of sound of fluid #2 reduces the optimal value of the wrap angle from 400° to 300°, Fig. 6. The higher speed of sound increases maximum possible volume flow, so that outlet throttling is reduced, resulting in higher values for the optimal internal volume ratio. On the other hand, clearance mass flow is also higher, resulting in a lower delivery rate, so that maximum overall efficiency is reduced by almost seven percentage points. Delivery rate drops sharply for higher wrap angles. With dual lead rotors no significant benefit in maximal isentropic efficiency is possible. An enhancement in maximal isentropic efficiency in case of dual lead rotors is therefore more likely for fluids with a low speed of sound.

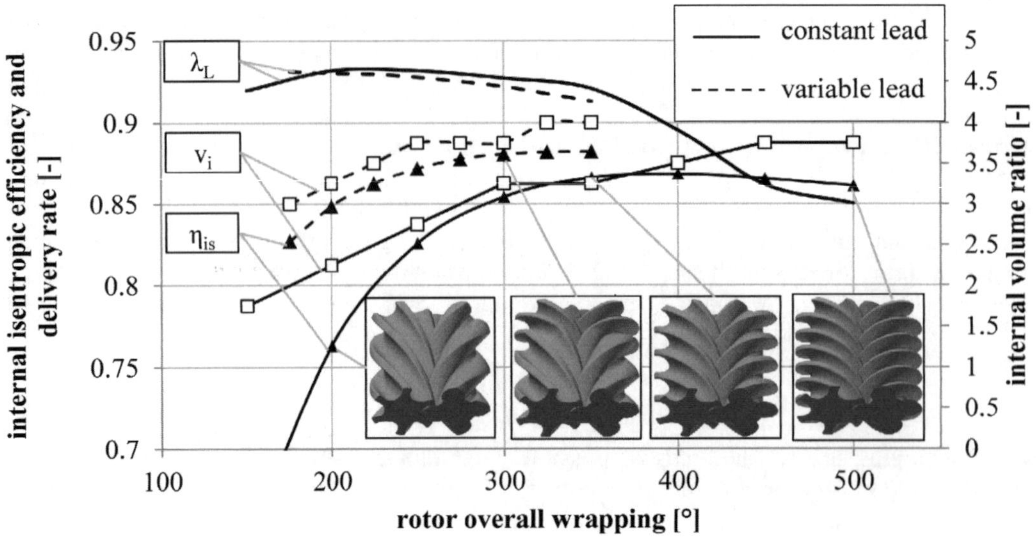

Figure 5. Internal isentropic efficiency, corresponding delivery rate and optimised internal volume ratio for different rotor overall wrappings for fluid #1

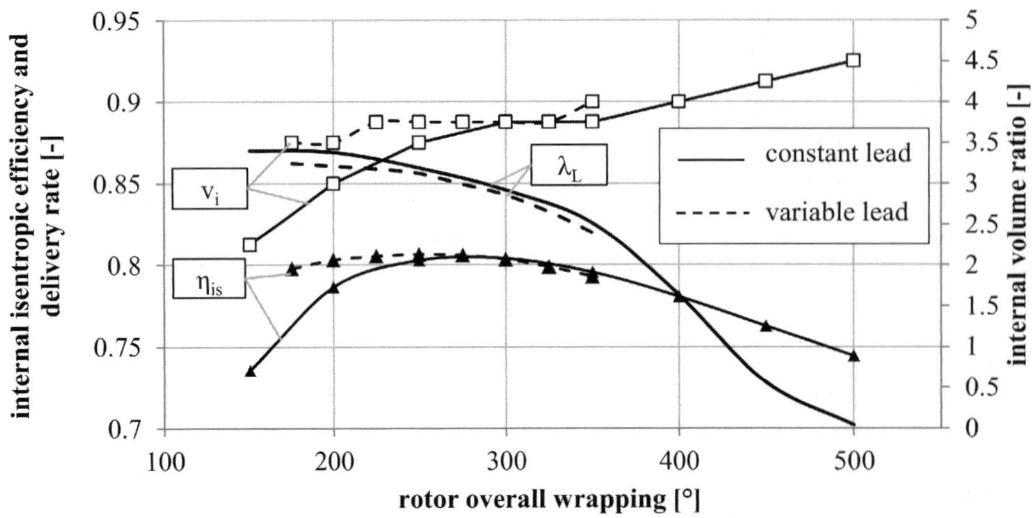

Figure 6. Internal isentropic efficiency, corresponding delivery rate and optimised internal volume ratio for different rotor overall wrappings for fluid #2

8.2. Influence of inlet density and temperature

Comparison of fluids #1 and #2 reveals that the density at the compressors inlet varies from 6 to 1.2 kg/m³. Density could be held constant without varying the speed of sound, molar mass and isentropic exponent by simply changing the inlet pressure while the pressure ratio of the compressor remains constant. However, the exchanged volume (equation (8)) does not depend on the absolute pressure of the fluid, and in the case of subcritical flow it depends entirely on the pressure ratio. Furthermore, the change in temperature due to the volumetric compression would remain the same because it depends entirely on the volume ratio. Thus, a change in density at the inlet - e.g. due to the unrealistically high inlet temperature of fluid #2 - would change absolute values such as mass flow and internal power of the compressor, but delivery rate and isentropic efficiency would remain exactly the same. A comparison of fluids with different densities, e.g. by using different inlet temperatures, is therefore reasonable, because density is not a crucial parameter for the design parameters of screw compressors.

8.3. Influence of the isentropic exponent

Fig. 7 shows the results for fluid #4, which obtains a higher isentropic exponent of 1.66 while the speed of sound is the same as for fluid #1, due to an adjustment in the temperature. Results for the optimal internal volume ratio differ considerably from the other fluids. Less change in volume is necessary to compress the fluid so that high pressure conditions are attained at lower internal volume ratios.

As discussed in section 7, the results are comparable with an increase in the speed of sound (Fig. 6) and maximum possible efficiency, delivery rate and optimal wrap angle are reduced. As for fluid #2, no significant benefit concerning maximum isentropic efficiency or delivery rate is achievable with dual lead rotors. Besides speed of sound, the isentropic exponent is the second major parameter that influences the optimal geometry parameters.

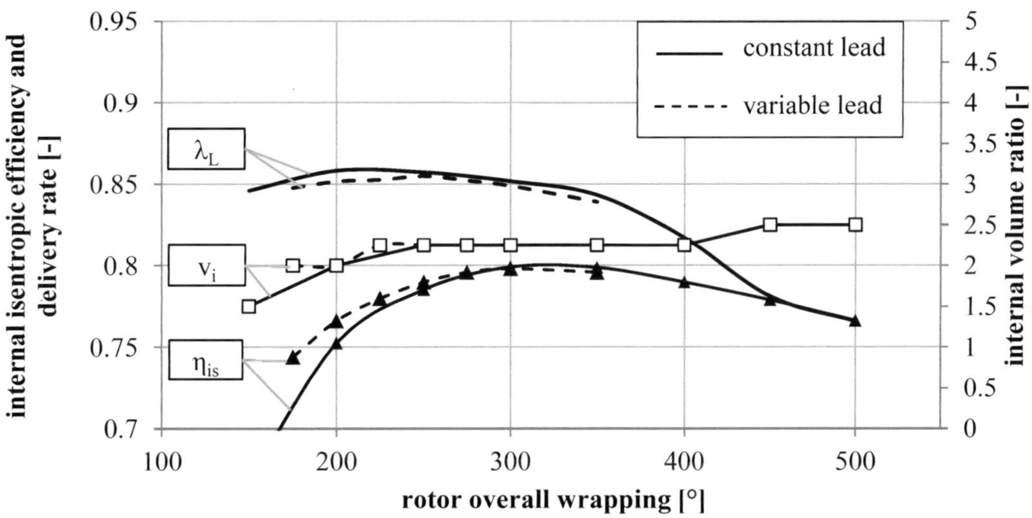

Figure 7. Internal isentropic efficiency, corresponding delivery rate and optimised internal volume ratio for different rotor overall wrappings for fluid #4

9. Conclusion

The present study indicates the ways in which the relevant fluid properties influence the screw compressor working cycle and the optimal geometry parameters of twin screw machines for uniform and dual lead rotors. Speed of sound and the isentropic exponent are the major parameters influencing the design of screw machines, while no isolated influence of molar mass, density and inlet temperature was observed. Potential for dual lead compressors exists for fluids with a low speed of sound and a low isentropic exponent. In this case the optimal wrap angle for constant lead needs to be very large in order to avoid throttling during discharge. An optimised progression of volume curve and outlet area enables the screw machine to operate with a lower overall wrap angle and results in higher efficiency due to decreased throttling at the outlet. Besides, the reduced overall wrapping of dual lead rotors may increase maximum chamber volume and therefore significantly increase the absolute mass flow for a given compressor size.

For wet-running screw machines, which are not investigated in this paper, the injected fluid reduces the temperature of the working fluid as well as the effective size of the available flow cross section and is therefore comparable to a reduction in speed of sound, so that an enhancement in efficiency for this application can be expected too.

References

[1] Weckes N 1994 *Ein Beitrag zur Optimierung geometrischer und thermodynamischer Kenngrößen von Schraubenladern* Dissertation University of Dortmund
[2] You C X, Tang Y and Fleming J S 1996 *International compressor engineering conference at*

Purdue paper 1074
[3] Rane S, Kovacevic A, Stosic N and Kethidi M 2014 *Computers & Fluids* **99** 124-141
[4] Utri M and Brümmer A 2016 *International compressor engineering conference at Purdue* paper 1428
[5] Kauder K, Janicki M, Rohe A, Kliem B and Temming J 2002 Thermodynamic Simulation of Rotary Displacement Machines *VDI-Berichte 1715* (Düsseldorf: VDI Verlag) 1-16
[6] Sigloch H 2009 *Technische Fluidmechanik* (Heidelberg, London, New York: Springer-Verlag) 7th ed
[7] Lemmon E W et al 2013 NIST Standard Reference Database 23: Reference Fluid Thermodynamic and Transport Properties-REFPROP, Version 9.1, National Institute of Standards and Technology
[8] Rinder L 1979 *Schraubenverdichter* (Wien, New York: Springer-Verlag)
[9] Utri M and Brümmer A 2014 *International Conference on Screw Machines 2014* (Düsseldorf: VDI Verlag) 249-266.

Nomenclature

A	Area	(m^2)
a	Speed of sound	(m/s)
c	Flow velocity	(m/s)
c_p	Isobaric heat capacity	(J/kg/K)
c_v	Isochoric heat capacity	(J/kg/K)
h	Specific enthalpy	(J/kg)
L	Length	(m)
M	Molar mass	(kg/mol)
Ma_u	Circumferential Mach number	(-)
$\dot{m}$	Mass flow	(kg/s)
n	Rotational speed	(1/s)
P	Power	(W)
p	Pressure	(Pa)
R	Specific gas constant	(J/kg/K)
$\mathcal{R}$	Universal gas constant	(J/mol/K)
T	Temperature	(K)
u	Circumferential speed	(m/s)
V	Volume	(m^3)
$\dot{V}$	Volume flow	(m^3/s)
v_i	Internal volume ratio	(-)
z	Number of lobes	(-)
Δ	Difference	(-)

η_{is}	Internal isentropic efficiency	(-)
κ	Isentropic exponent	(-)
φ	Wrap angle	(°)
λ_L	Delivery rate	(-)
Π	Pressure ratio	(-)
ρ	Density	(kg/m^3)

Subscript

1, 2	Fluid states
compr,end	End of compression
crit	Critical
hp	High pressure
i	Internal
in	Suction / inlet side
lp	Low pressure
MR	Male rotor
max	Maximum
out	Discharge / outlet side
real	Real flow
th	Theoretical flow
total	Total wrapping
s	Constant entropy
*	Critical cross section

Acknowledgement
The work leading to these results has received funding from the European Community's Horizon 2020 Programme (2014-2020) under grant agreement no 678727. The opinions expressed in the document are of the authors only and no way reflect the European Commission's opinions. The European Union is not liable for any use that may be made of the information.

IOP Conf. Series: Materials Science and Engineering **232** (2017) 012019 doi:10.1088/1757-899X/232/1/012019

Research on the injected effect by the narrow-slit injection orifice in screw compressors

J Xie, J Feng, F Liu, C Peng and Q Feng*

School of Energy and Power, Xi'an Jiaotong University,

Xi'an 710049, China

E-mail*: qkfeng@mail.xjtu.edu.cn

Abstract. A narrow-slit injection orifice in screw compressors (mainly in single screw compressors) is described in this paper. In contrast to the traditional injection orifice, the section is slit shaped but not circular. Liquid which is injected by the narrow-slit injection orifice appears as a curtain shape. The heat transfer area is increased so that the cooling effect is enhanced and the efficiency of the compressor is improved. The injected effect was simulated through the CFD software, and then the experiments were carried out. The study verified the accuracy of the simulation results and will provide a theoretical foundation for the design of a narrow-slit injection orifice in screw compressors.

List of symbols

W_e	Weber number
ρ	Density of the lubricant liquid
v	Liquid injection rate
l	Characteristic length of the injection orifice
σ	Surface tension coefficient of the lubricant liquid
SMD	Sauter mean diameter
σ_0	Surface tension coefficient of the lubricant liquid
μ_0	Velocity of the liquid jet
$\dot{m}_0$	Weight of the liquid per diameter range
ρ_0	Density of the medium around
γ	Half cone angle
d_{or}	Diameter of the injection orifice,
Δp	Pressure difference of nozzle
σ_K, σ_ε, $C_{1\varepsilon}$, $C_{2\varepsilon}$	Experimental constant

1. Introduction

Liquid injected screw compressors have been widely used in industry because of its high efficiency, stable performance and so on. While gas is compressed in the screw compressor, a certain amount of liquid with a certain pressure is injected into the compression cavity generally. The liquid contacts with compressed gas directly and it mainly to play a role of cooling and sealing. Usually it needs to spray 8-15L liquid into per cubic meter gas, the volume ratio is about 1:100 [1]. In general, the lubricant fluid flows in the loop depend on the pressure difference between the compressor's exhaust pressure and the spray pressure. For large screw compressor, the oil pump is added the circuit. The injected liquid droplets are mixed with the compressed gas. The extremely heat transfer surface absorbs the heat of gas rapidly, so the compression process is similar to the isothermal process. A reduction on the energy consumptions is up to 7% with a positive spray cooling [2]. The accuracy requirements for machining of screw rotors are reduced through a sprayed liquid injection technology, also the performance of the compressor is improved.

The main injection parameters have been investigated, such as injection pressure and temperature, liquid flow rate, orifice diameter, position of the injector, etc. Many scholars have studied the injection technology of screw compressor in the world. Stosic *et al.* (1988) [3] presented some results of mathematical modeling of the oil injection into the screw compressor and its influence upon the thermodynamic process of the engine and analyzed the effect of various parameters upon the compressor physical cycle. Pawan J. and Bowman, J. L. (1986) [4] described a mathematical model to calculate the oil-gas heat transfer, assuming that the oil is injected in the form of non-interacting spherical droplets. The model was used to calculate the effect of the heat transfer on compressor performance. N. Seshaiah *et al.*(2007) [5] carried out the mathematical analysis of oil injected twin-screw compressor and effect of some of the compressor operating and design parameters on power and volumetric efficiencies had been analyzed and presented. Li *et al.* (1992) [2] analyzed the effect of oil atomization and its effect on the exhaust gas temperature of single screw compressor. The influence of the main structural parameters of atomizer on lubricant oil atomization effect was analyzed, and the exhaust temperature of single screw compressor without atomization was compared. Finally, it was concluded that the performance of the compressor is improved effectively by the atomization of lubricant.

The section crosses of the injection orifice are all circle, and the liquid is sprayed into compression cavity in a column sharp. Because the injection path is so short, its effect on liquid injection optimization is limited.

2. Injection orifice and its position

In the compressor without volume flow rate regulation, the position of injection orifice is on the side of compression gas and in the section of the compressor that volume has been out of the suction port and compression process starts. The lubricant liquid could be injected through a larger spray injection orifice or through a number of smaller orifices, usually it can obtain better performance of compressor by using a number of small injection orifice. In the large and medium sized injection screw refrigeration machine and process compressor, inner-placed volume flow refrigeration valve device is used usually. When the compressor working under different conditions, the working length of the rotor is changed and the inspiratory end position moves to the exhaust end with the slide valve. Generally, the injection orifice is opened in the slide valve in this type of compressor, the injection orifice will move with the slide valve together. So it is ensure that the lubricant liquid is always sprayed after the end of the suction. In order to improve the volumetric efficiency, sometimes the injection orifices are opened in the middle of the compression process for small and high pressure ratio compressor.
In twin-screw compressor, the injection orifice is opened in or near the intersection of two rotor's tooth addendum circle [6].

Vertical liquid spray method is currently used in the single screw compressor, that is, the nozzle axis is perpendicular to the surface of the star wheel. The lubricant liquid is sprayed where the star-wheel teeth cutting into the depths of the screw, the liquid injection rate is equal to the circumferential speed of the screw. Another injection method is that lubricant liquid is sprayed into basic capability along the rotation direction of the screw tooth [7]. The orifices are opened in front of projection of the outer edge of the spiral screw that is in front of the star wheel on cylinder when the basic capability is formed.

The injection orifices connect to the channel for injection liquid. Liquid channel is arranged on the casing by drilling hole along the axial direction. In part of the compressor, the orifice is opened along the radial of the casing directly.

3. Narrow-slit shaped injection orifice

Liquid is sprayed into the cogging space in column sharp by the above injection orifice, and the injection path in the cogging is so short. Column sharped liquid is too late to spread for atomization when it attaches to the metal wall of the casing. Therefore, the heat transfer area of the lubricant liquid and the compressed gas is very small. Due to the cooling effect is not ideal, the inner wall of the compression cavity is scaled. Scaling phenomenon is shown in figure 1. It is a photo for the casing of a brand compressor when the compressor running some time, the cross section of the injection orifice is circular.

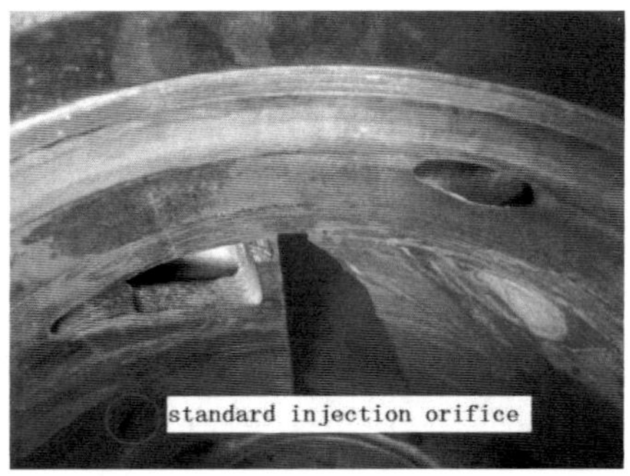

Figure 1. Scaling of casing and the standard orifice.

The heat exchange between the compressed gas and the liquid is less when the size of the injection orifice is larger. That is because the large size of the injection orifice will lead to a lower injection rate and thus format large liquid droplets, heat transfer area decreases. In addition, the increase of liquid droplets will lead to a shorter contact time between compressed gas and lubricant liquid, and the heat transfer effect is weakened more. In order to enhance the heat transfer and promote that the compression process is more similar to isothermal compression of the most power saving theoretically, the lubricant liquid should have a larger surface area in the same amount of liquid. The narrow-slit injection orifice is proposed in this paper. It is opened by arranging appropriate width and length of the narrow slit on the air chamber, and injecting at an appropriate location. The lubricant liquid is sprayed into the cogging space in a curtain shaped, then the span of the lubricant liquid and gas mixed along the axial is larger. And the lubricant liquid in the casing is uniformly distributed in the compression cavity. As a result of the increase of the contact area between the gas and the liquid injection, cooling of the compression process is enhanced.

In the process of injection, the viscous force of the lubricating liquid and the air resistance and the surface tension of the lubricating liquid play an important role in the crushing of lubricant liquid. Larger liquid droplets that are injected in compression cavity twist under air resistance, and generate friction under the relative velocity of liquid and air. Due to the inner wall of the liquid droplet, a part of the surface protrude out of the liquid droplet is devised and then a smaller liquid drop is formed under the action of the liquid surface tension. At the same time, the liquid droplets will distorted and deformed when the resultant force of surface tension and viscous force is overcome by the air resistance. It continues to that the air resistance and the internal force of the liquid drop regain the equilibrium state. Based on oil atomizing theory, the parameter of Weber number is put forward for evaluating the atomizing level of the oil in compression cavity. The flows of liquid atomization that resulted by different Weber number are divided into five forms [8]: liquid drop ($We<0.05$), smooth liquid flow ($0.5<We<5$), ripple current ($5<We<20$), spray flow ($We>20$).

Weber number is estimated according to Equation (1).

$$W_e = \frac{\rho v^2 l}{\sigma} \tag{1}$$

Where ρ the density of the lubricant liquid, v is the liquid injection rate, l is the characteristic length of the injection orifice, σ is the surface tension coefficient of the lubricant liquid.

The better effect of atomization, the more compressor power consumption is saved. So it's the better that the atomization of the compressor working cavity is the form of spray flow, that is, liquid droplets are fine particles. The liquid droplets of the spray flow are smaller size droplets when Weber number is greater than 20. The compressor which speed is 3000r/min is the main object in reference [9], the experiment is carried out when the diameter is ϕ3mm、ϕ2mm and ϕ1.6mm. Weber number and the diameter of lubricant liquid drop is 43.4, 60.5, 118.3 and ϕ 3.8mm, ϕ 1.46mm, ϕ 0.75mm according the analysis of experimental data. The atomization of lubricant liquid injected by these three kinds of injection orifice is the spray form through Weber number. However, the diameter of liquid drop is too large when diameter of the injection orifice is ϕ 3.8mm through the drop diameter, quality of atomization is poor. Therefore the Sauter mean diameter (SMD) is introduced, also it is known as surface area mean diameter or Sauter mean diameter. SMD can be estimated by Equation (2).

$$SMD = 4.52\left(\frac{\sigma_0 \mu_0^2}{\rho_a \Delta p}\right)^{0.25}\left[2.7\left(\frac{d_{or} m_0 \mu_0}{\rho_a \Delta p}\right)\cos\gamma\right]^{0.25} + 0.39\left(\frac{\rho_0 \mu_0}{\rho_a \Delta p}\right)^{0.25}\left[2.7\left(\frac{d_{or} m_0 \mu_0}{\rho_a \Delta p}\right)\cos\gamma\right]^{0.75} \tag{2}$$

Where σ_0 is the surface tension coefficient of the lubricant liquid, u_0 is the velocity of the liquid jet, m_0 is the weight of the liquid per diameter range, ρ_0 is the density of the medium around, γ is half cone angle, d_{or} is the diameter of the injection orifice, Δp is pressure difference of nozzle.

The smaller the SMD value, the better the atomization quality is. Five different types of nozzles under the premise of the same mass flow rate were tested, including three diameter of holes with diameter of ϕ3mm、ϕ2.5mm and ϕ2mm. The Sauter mean diameters of the drops are ϕ3.53mm、ϕ1.97mm、and ϕ0.96mm.

It is concluded that the Sauter mean diameter of droplet will decrease with the decreasing of orifice diameter. Narrow-slit sharped injection orifice can be regarded as a lot of smaller circular injection orifice, and its role is equivalent to an atomizer.

4. Injection velocity simulation

The working process of screw compressor is complex, so the model must be assumed and simplified appropriately to receive results quickly in the premise of sufficient accuracy. Simplifications are as follows: 1. Property of lubricating oil for working medium has not changed; 2. On-way resistance of the lubricant liquid flow process is negligible; 3. The screw cogging is a regular shape.

Steady flow field calculation is achieved by software Ansys 14.0. The grid division is achieved by CFD14.0 ICEM, the results shown in figure 2.

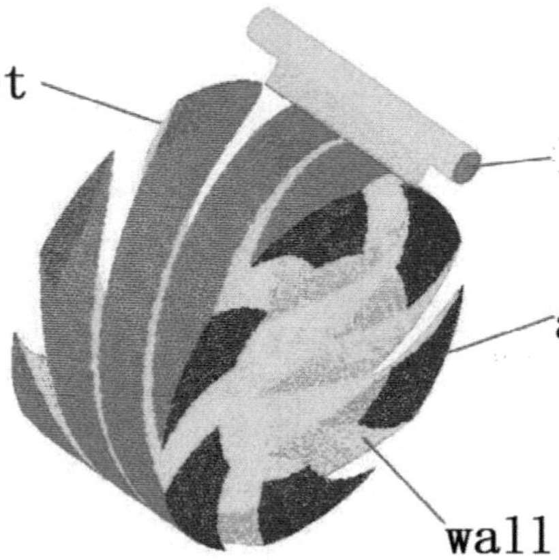

Figure 2. Grid division.

The flow control equations are composed of incompressible flow Reynolds equation (3), turbulent kinetic energy equation (κ equation)(4) and the turbulent dissipation rate equation (ε equation)(5).

$$\frac{\partial}{\partial x}\left(\frac{ph^2}{\eta}\frac{\partial p}{\partial x}\right)+\frac{\partial}{\partial y}\left(\frac{ph^2}{\eta}\frac{\partial p}{\partial y}\right)=6\left(U\frac{\partial ph}{\partial x}+V\frac{\partial ph}{\partial y}+2\frac{\partial ph}{\partial t}\right) \quad (3)$$

$$\rho U_i\frac{\partial k}{\partial x_i}=\mu_t\left(\frac{\partial U_j}{\partial x_i}+\frac{\partial U_i}{\partial x_j}\right)\frac{\partial U_j}{\partial x_i}+\frac{\partial}{\partial x_i}\left\{(\mu_t/\sigma_k)\frac{\partial k}{\partial x_i}\right\}-\rho\varepsilon \quad (4)$$

$$\rho U_i\frac{\partial \varepsilon}{\partial x_i}=C_{1\varepsilon}\left(\frac{\varepsilon}{k}\right)\mu_t\left(\frac{\partial U_j}{\partial x_i}+\frac{\partial U_i}{\partial x_j}\right)\frac{\partial U_j}{\partial x_i}+\frac{\partial}{\partial x_i}\left\{(\mu_t/\sigma_k)\frac{\partial k}{\partial x_i}\right\}-C_{2\varepsilon}\rho\left(\frac{\varepsilon^2}{k}\right) \quad (5)$$

Where σ_K, σ_ε, $C_{1\varepsilon}$, $C_{2\varepsilon}$ are experimental constant.

Firstly, the boundary conditions and the correlation of the model must be set up when the flow distribution of lubricating liquid is simulated by fluent software. In the model, the setting of boundary at the inlet and outlet of the lubricant liquid is the key to influence whether the flow agree with actual condition of the lubricant liquid. Boundary conditions are set in table 1.

Table 1. Boundary conditions set.

Boundary definition	**Liquid inlet**	**Liquid outlet**	**Air inlet**	**Outlet**	**Other wall**
Boundary type	Pressure inlet	Interior	Velocity inlet	Pressure outlet	Wall

The simulation result is showed in figure 3. It is a section of the simulation that liquid oil is injected into the screw cogging.

IOP Conf. Series: Materials Science and Engineering **232** (2017) 012019 doi:10.1088/1757-899X/232/1/012019

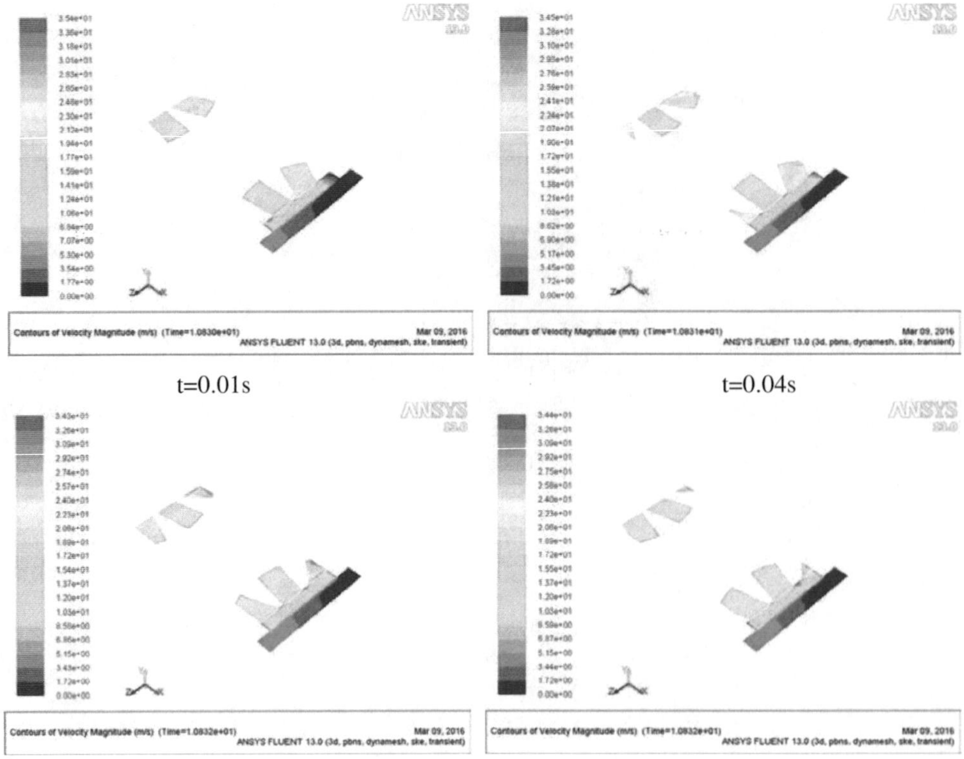

t=0.01s t=0.04s

t=0.07s t=0.10s

Figure 3. Simulation result.

It is velocity gradient section with speed 3000r/min and spray pressure 500kPa. It is obviously that the velocity will rise or fall as the tooth moves. The reason for this phenomenon is the periodicity compression of oil by the cogging.

Simulation is carried out when the crack length is 40mm and width is 0.4mm. The injected velocity is shown in table 2.

Table 2. Simulation results.

Pressure(MPa)	0.05	0.1	0.15	0.2	0.25	0.3	0.35	0.4
Velocity(L/(mm² · min))	0.088	0.176	0.247	0.317	0.382	0.441	0.503	0.556

Maximum pressure condition is 0.4MPa because of the limit of the experimental equipment. It is indicated that the injected velocity increases with the increase of pressure in the range of pressure. The simulation results will be verified by experiment.

5. Experiment equipment

Experiment equipment is shown in figure 4. The experimental oil pump is a gear pump which type is KCB-33.3 with design flow 33.3L/min, pressure 1.45MPa and speed 1420r/min. The oil pump is mainly used for all kinds of machinery and equipment of lubricant system to convey lubricant oil. It is suitable for conveying the lubricating oil which viscosity for $5\times10^{-6}\sim1.5\times10^{-3}m^2/s$ (5-1500cSt) and the temperature under 300° C. The nozzle shape is obtained by machining and low tolerances need to be guaranteed. The accuracy is particularly important because of the small size of the slit. In the experiment, the narrow slit size of the

IOP Conf. Series: Materials Science and Engineering **232** (2017) 012019 doi:10.1088/1757-899X/232/1/012019

nozzle is 40mmx0.4mmx3mm. The experimental medium is ANDEROL3032M, an ISO32 compressor lubricant formulated from synthetic base fluid utilizing synthetic hydrocarbon (PAO) with specially developed additive systems.

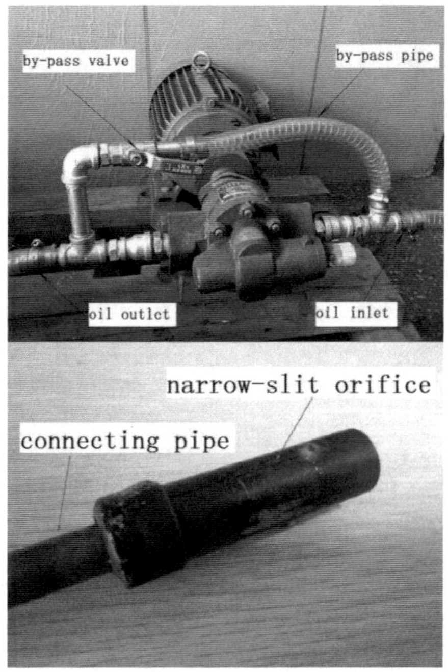

Figure 4. Experiment equipment.

The injected velocity obtained by weighing method. Experiment results and errors are shown in table 3.

Table 3. Experiment results and errors.

Pressure(MPa)	0.05	0.1	0.15	0.2	0.25	0.3	0.35	0.4
Velocity(L/(mm^2 · min))	0.05	0.177	0.244	0.307	0.375	0.431	0.484	0.517
Error (%)	7.7	0.4	1.2	3.6	2.0	2.4	4.0	7.5

Error is the ratio of the difference between simulated results and experiment results to experiment results. It is shown that the simulation results is accordance with the real situation in the experiment because of all errors are less than 10%. The simulation lays a theoretical foundation for the design.

In order to observe the experimental results, the experimental media is water when the simulation of injection flow in the compressor is carried out. When the velocity of air is 35m/s (approximate value of average velocity in the compression chamber), the injection diagram is shown in figure 5.

Figure 5. Injection diagram.

It is shown that the injection is the spray flow. The atomization effect is better. The jet is blown into many small droplets; the heat exchange area increases a lot. The atomizing form is beneficial to the gas-liquid mixture uniformly, and the heat exchange effect is enhanced. It is verified the feasibility of narrow-slit injection orifice.

6. Conclusions

For screw compressor, the curtain jet that is formed by the narrow-slit orifice proposed in this paper promotes increase of the heat transfer greatly. In this paper, the model of narrow slot injection is established. The injection is simulated by Ansys. Corresponding experiments are carried out to get the relation between the flow rate and parameters of the injection orifice. The accuracy of the simulation and the correctness of the model are verified by the comparison between simulation results and experimental results. The theoretical analysis lays a theoretical foundation for the design of the narrow-slit injection orifice.

Next the proper size of the narrow-slit injection orifice aiming at a volume model will be designed and the casing will be processed. The real machine experiments will be carried out to test and verify whether the narrow-slit orifice has its superiority.

References

[1] Deng D and Shu P 1982 *Rotary compressor* (Beijing, MA: China Machine Press, in China)

[2] Li H, Feng J and Jin G 1992 Study on the effect of oil atomization and its effect on the exhaust gas temperature of single screw compressor *Int. J. Fluid Engineering* **20**(8) pp 14–16

[3] Stosic N, Kovaevi Ac A, Hanjali K and Milutinovi L 1988 Mathematical modeling of the oil influence upon the working cycle of screw compressors *Proceedings of the International Compressor Engineering Conference, Indiana* (Purdue University)

[4] Singh P J and Bowman J L 1986 Heat transfer in oil-flooded screw compressors *International Compressor Engineering Conference, Indiana* (Purdue University) 521

[5] Seshaiah N, Ghosh S Kr, Sahoo R K and Sarangi S Kr 2007 Mathematical modeling of the working cycle of oil injected rotary twin screw compressor *Int. J. Applied Thermal Engineering* **27** pp 145–55

[6] Xing Z 2000 *Theory, design and application of screw compressor* (Beijing, MA: China Machine Press, in China)

[7] Yu Y, Jiang P and Sun S 2012 *Compressor engineering manual* (Beijing, MA: China Petrochemical Press, in China)

[8] Bein T W and Hamilton J F 1982 Computer modeling of an oil flooded single screw air compressor *International Compressor Engineering Conference, Indiana* (Purdue University) 383

[9] Lin Q, Fan L and Jin G 2000 Theoretical analysis and experimental study of oil atomizing in a single screw compressor *Int. J. Journal of X'ian Petroleum Institute* (Natural Science Edition) **15**(2) pp 41–44

IOP Conf. Series: Materials Science and Engineering **232** (2017) 012020 doi:10.1088/1757-899X/232/1/012020

CFD simulation of a screw compressor with oil injection

Hui Ding, and Yu Jiang

Simerics Inc.,
1750 112th Ave NE Ste C250,
Bellevue, WA 98004, USA

E-mail: hd@simerics.com

Abstract. In this paper, a full 3D transient CFD model of a twin screw compressor with oil injection will be described in detail. The Volume Of Fraction (VOF) approach was used for two phase flow of gas and liquid. The numeric method and simulation conditions will be explained. Simulation results will be presented with discussion. The cooling and sealing effects of oil injection will be evaluated by comparing the simulation results for the cases with and without oil. The imbalance of mass and energy of the compression system in simulation results will be rigorously checked to prove the methodologies used in the simulation are fully conservative and consistent. The efficiency, speed, and robustness of the proposed approach will also be demonstrated through the test case.

1. Introduction

Besides lubricating the gears, oil injection has been commonly used in different types of positive displacement compressors to cool the compressed gases and to seal the leakage gaps to improve compressor gas flow rate and efficiency [1]. CFD simulation provides valuable insights to help compressor designers to verify, to analyze, and to improve the performance of compressors. It is very beneficial if CFD can be applied to the analysis of compressors with oil injection.

In recent years, CFD has been widely used to analyze performance of varies Positive Displacement (PD) compressors. Those models cover different types of PD compressors with/without intake and/or discharge valves but mostly solving single gas phase flow.

Solving two phase flow using CFD with complex moving parts and small leakage gaps is a very challenging task. Even for relatively simple two-phase flow problems, high density ratio between liquid and gas, sophisticated interaction among the phases, and the interface tracking with complex shape make the problem difficult to solve. It is even more difficult to solve a multiphase flow in a PD compressor. In such case, gas phase has to be treated as compressible, heat transfer effects has to be included, and interface tracking has to be done inside moving, deforming volumes. The major issues users have experienced with many CFD solvers in multiphase simulations are poor convergence, very long simulation time, and unsatisfactory mass/energy conservation. Generally speaking, CFD simulation of compressors with liquid injection is a very difficult task.

As a test case for Simerics improved VOF model, Ding and Jiang (2016)[2] has used Simerics-MP+ (also known as PumpLinx) solved an oil flooded scroll compressor. In the study, the oil was assumed to be uniformly mixed with the gas at the compressor inlet with up to 18% oil mass fraction. The simulation results demonstrated good convergence, fast calculation speed, and excellent conservation of mass and energy. However, the simulated conditions are relatively ideal.

In this paper, a screw compressor with oil injection is investigated using VOF model. Compared with previous work by Ding and Jiang [2], this study uses a more realistic model with much higher oil mass fraction of up to 57%. In the following sections, the proposed CFD model of the screw compressor with oil injection will be described in detail. Simulation results will be presented and discussed. The conservation of mass and energy as well as the flow and thermodynamic cycles inside the compression chamber will be subtracted from simulation results. The overall oil injection effects on the compressor performance will be evaluated by comparing the simulation results with/without oil injection.

2. CFD solver and governing equations

2.1. *Conservation equations for gas liquid mixture*

The CFD package, Simerics-MP+, used in this study solves conservation equations of mass, momentum, and energy of a compressible fluid using a finite volume approach. Those conservation laws can be written in integral representation as

$$\frac{\partial}{\partial t}\int_{\Omega(t)} \rho d\Omega + \int_{\sigma} \rho(v - v_{\sigma}) \cdot n d\sigma = 0 \tag{1}$$

$$\frac{\partial}{\partial t}\int_{\Omega(t)} \rho \boldsymbol{v} d\Omega + \int_{\sigma} \rho\big((\boldsymbol{v} - \boldsymbol{v}_{\sigma}) \cdot \boldsymbol{n}\big)\boldsymbol{v} d\sigma = \int_{\sigma} \tilde{\boldsymbol{\tau}} \cdot \boldsymbol{n} d\sigma - \int_{\sigma} p\boldsymbol{n} d\sigma + \int_{\Omega} f d\Omega \tag{2}$$

$$\frac{\partial}{\partial t}\int_{\Omega(t)} \rho E d\Omega + \int_{\sigma} \rho\big((\boldsymbol{v} - \boldsymbol{v}_{\sigma}) \cdot \boldsymbol{n}\big)E d\sigma = \int_{\sigma} k\nabla T \cdot \boldsymbol{n} d\sigma - \int_{\sigma} p\boldsymbol{v} \cdot \boldsymbol{n} d\sigma + \int_{\sigma} (\boldsymbol{v} \cdot \tilde{\boldsymbol{\tau}}) \cdot \boldsymbol{n} d\sigma + \int_{\Omega} \boldsymbol{f} \cdot \boldsymbol{v} d\Omega \tag{3}$$

The standard $k - \varepsilon$ two-equation model [3] with wall function is used to account for turbulence,

$$\frac{\partial}{\partial t}\int_{\Omega(t)} \rho k d\Omega + \int_{\sigma} \rho\big((\boldsymbol{v} - \boldsymbol{v}_{\sigma}) \cdot \boldsymbol{n}\big)k d\sigma = \int_{\sigma} (\mu + \frac{\mu_t}{\sigma_k})(\nabla k \cdot \boldsymbol{n}) d\sigma + \int_{\Omega} (G_t - \rho\epsilon) d\Omega \tag{4}$$

$$\frac{\partial}{\partial t}\int_{\Omega(t)} \rho \varepsilon d\Omega + \int_{\sigma} \rho\big((\boldsymbol{v} - \boldsymbol{v}_{\sigma}) \cdot \boldsymbol{n}\big)\varepsilon d\sigma = \int_{\sigma} (\mu + \frac{\mu_t}{\sigma_\varepsilon})(\nabla \epsilon \cdot \boldsymbol{n}) d\sigma + \int_{\Omega} (c_1 G_t \frac{\varepsilon}{k} - c_2 \rho \frac{\varepsilon^2}{k}) d\Omega \tag{5}$$

Together with equation of state, where properties are functions of temperature and pressure, to form a closed system:

$$\rho = f(p, T) \tag{6}$$

In the solver, each of the fluid properties will be a function of local pressure and temperature, and can be prescribed as an analytical formula or in a table format.

2.2. *VOF model for multiphase*

VOF models are widely used in simulation of two phase flow [4] [5]. VOF solves a set of scalar transport equations representing the volume fraction of each fluid component occupies in every computational cell. The transport equation of the volume fraction for each fluid component can be written as:

$$\frac{\partial}{\partial t}\int_{\Omega(t)} \rho_i F_i d\Omega + \int_{\sigma} \rho_i (v - v_{\sigma}) \cdot n F_i d\sigma = 0 \tag{7}$$

Where F_i is the volume fraction of the *ith* fluid component, and ρ_i is the local density of *ith* fluid component. The weighted mixture density of the fluid in equation (1) to (5) are then calculated as:

$$\rho = \sum \rho_i F_i \tag{8}$$

Both implicit and explicit methods can be used to solve the equation. High Resolution Interface Capturing (HRIC) scheme can be used for the convective term in the transport equation.

This software package has been validated against many different types of compressors including: centrifugal compressor, lobe compressor, twin screw compressor [6], scroll compressor [7] [8], rolling piston [9], and reciprocating compressor [10] for single phase compression of air, refrigerants, and other type of gases. The VOF model has also been validated against many industrial applications for multiphase flow without solving heat transfer directly [11] [12], and for multiphase flow solving together with heat transfer [2].

3. Twin screw compressor test case

The compressor model used in this study was originally designed as an oil-free twin screw compressor with a 3/5 lobe arrangement and 'N'rotor profile rotors [6]. The operating speed on the male rotor varies from 6000 to 14000 rpm. The male rotor diameter is 127.45 mm; the female rotor diameter is 120.02 mm while the center distance between the two rotors is 93.00 mm. The length to diameter ratio of the rotors is 1.6 and the male rotor has a wrap angle 285.0 deg. In this study, two oil injection pipes were added to both male and female rotor. The discharge port was also modified in order to generate higher compression ratio, such that the oil cooling becomes important.

The rotor part of the twin screw was meshed as a single domain using a grid generation software SCORG [6]. SCORG creates a series of mesh files for the rotor at different rotation angles. The rotor mesh files were read into the solver through Simerics - SCORG mesh interface. The inlet and outlet ports of fluid volumes are meshed using Simerics binary tree unstructured mesher. All the fluid volumes are connected together using Mis-matched Grid Interface (MGI). The total number of cells is around 0.86 million. Figure 1 and 2 show the complete simulation fluid domain and the mesh in 3D view.

The gas inlet is set to a fixed total pressure, and a fixed total temperature boundary condition while the outlet is set to a fixed static pressure boundary condition. The gas phase of the fluid is air, modelled using ideal gas law. The male rotor rotation speed is 8000 RPM. The oil is assumed to be incompressible with a density of 950 Kg/m3, and a heat capacity of 1670 J/kgK. In order to demonstrate the effects of oil injection, a similar case with the same parameters but without oil was also simulated for comparison. Simulation time is about 4 hours per revolution for the simulation with oil injection and about 2 hours per revolution without oil on a Dell Precision Mobile Workstation with Intel i7-4800MQ CPU @ 2.70GHz.

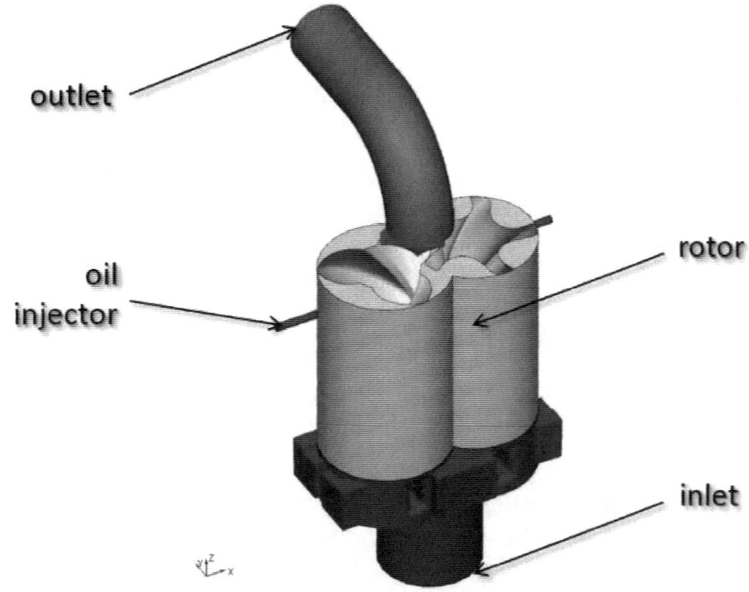

Figure 1: All fluid volumes

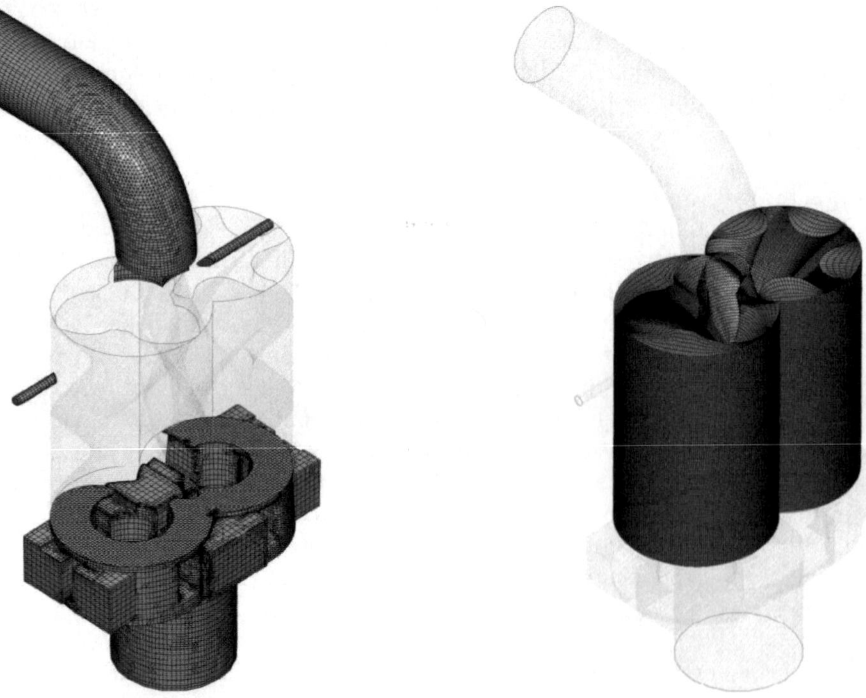

(a) (b)

Figure 2: Mesh: (a) ports mesh (b) rotor mesh

4. Results and discussion

In the simulation, the inlet total pressure and the outlet static pressure are set to 1 atm and 5 bar respectively. The inlet total temperature is set to 300K. The oil is injected from two pipes with 2 gpm flow rate from each pipe and with inlet temperature of 320K. Both rotors are assigned with correct rotational speed. Table 1 summarizes simulation parameters used for the oil injection simulation.

Table 1: Simulation parameters

Parameters	Values
Gas	Air (using ideal gas law)
Gas inlet total pressure	1 atm
Gas inlet total temperature	300 K
Outlet static pressure	5 bar absolute
Oil density	950 kg/m^3
Oil viscosity	0.008 PaS
Oil heat capacity	1670 J/KgK
Oil injection rate	2 x 2 gpm
Oil injection temperature	320 K
Compressor speed	8000 rpm (male rotor)

During the simulation, results become periodic after around 6 revolutions. Figure 3 shows typical pressure contour at 5 different male rotor crankshaft angles for one complete tooth rotation. The colour map ranges from 1 bar to 6.5 bar with magenta represents high pressure and blue represents low pressure. During the operation, isolated fluid "pockets" move up from inlet port to the discharge port. The pressure in each pocket keeps increasing with crankshaft angle due to the continuous volume reduction of the pocket until it reaches outlet.

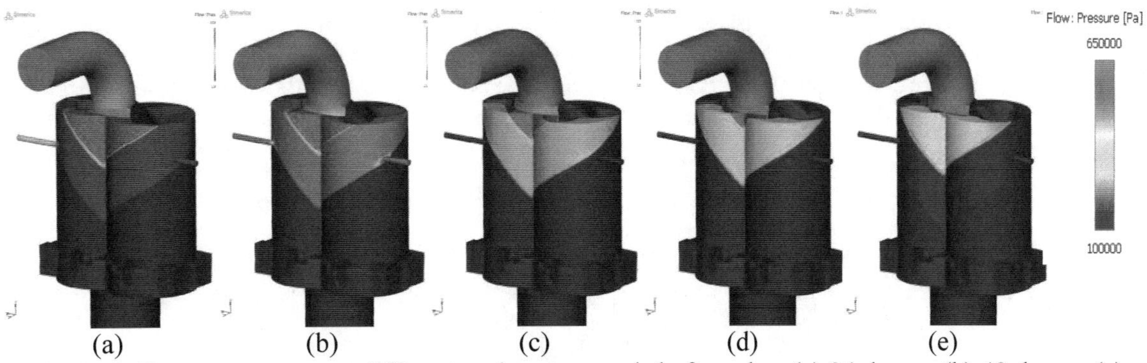

Figure 3: Pressure contour at different male rotor crankshaft angles: (a) 24 degree (b) 48 degree (c) 72 degree (d) 96 degree (e) 120 degree

Figure 4 shows temperature contour at 5 crankshaft angles. The colour map ranges from 300 K to 550 K with magenta represents high temperature and blue represents low temperature. Temperature in the pocket follows a similar trend. However, unlike the pressure, the temperature inside each pocket is highly non-uniform due to the cooling effects of injected oil. The region close to the oil injection has significantly lower temperature.

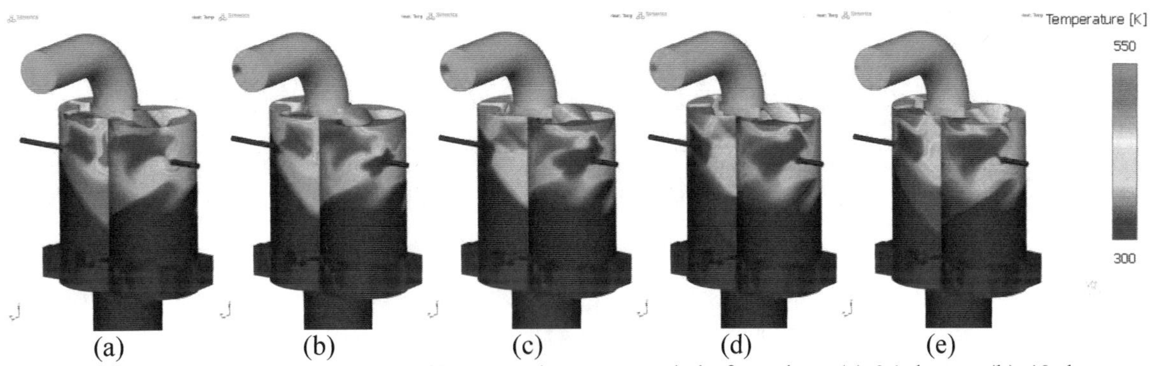

Figure 4: Temperature contour at different male rotor crankshaft angles: (a) 24 degree (b) 48 degree (c) 72 degree (d) 96 degree (e) 120 degree

Figure 5 shows oil concentration contour at 5 crankshaft angles. The colour map ranges from 0% to 10% volume fraction of oil with magenta represents high concentration and blue represents low concentration. Oil has higher concentration close to the injector and then been carried away by the pocket.

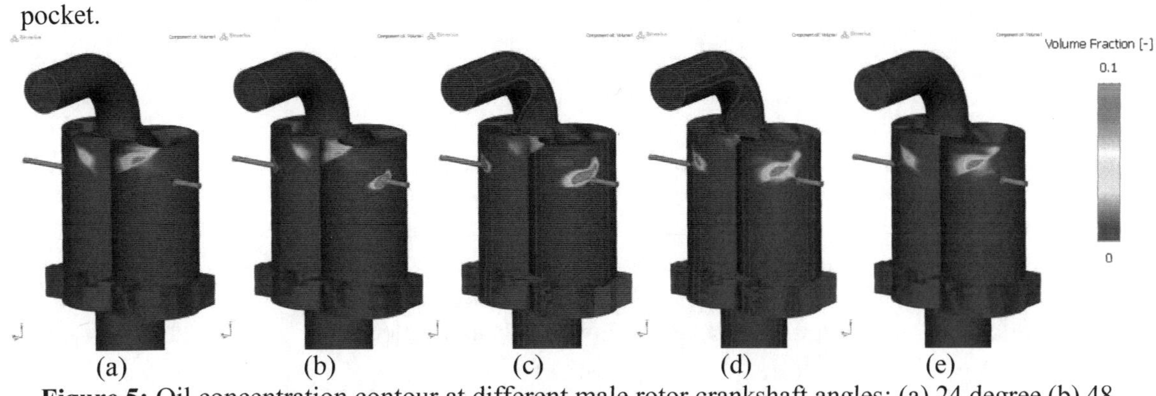

Figure 5: Oil concentration contour at different male rotor crankshaft angles: (a) 24 degree (b) 48 degree (c) 72 degree (d) 96 degree (e) 120 degree

Figure 6 shows oil concentration iso-surface at 5 crankshaft angles with rotor surface coloured by temperature. After injected into the rotor, the oil is not well mixed with the gas. Instead, the oil is still hold together to form very non-uniform concentration in the rotor chamber. Also from the picture, the oil cooling effects are demonstrated clearly as the areas around the oil have significantly lower temperature.

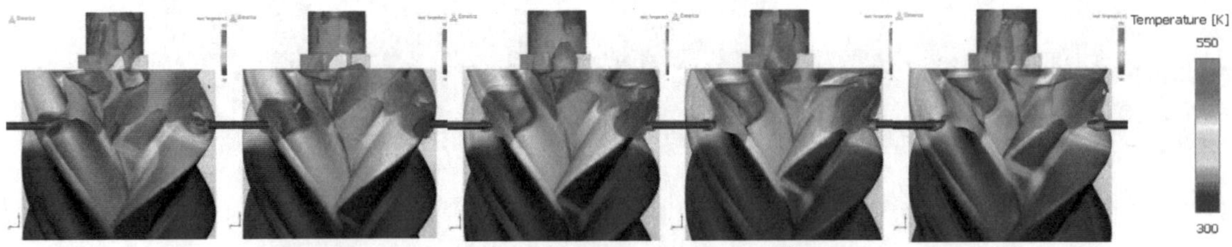

Figure 6: Oil iso-surface with rotor colored by temperature at different male rotor crankshaft angles: (a) 24 degree (b) 48 degree (c) 72 degree (d) 96 degree (e) 120 degree

For the case with oil injection, the average gas mass flux of the compressor under simulation condition is 0.1794 kg/s. The oil mass fraction, defined as average oil mass flow rate divide by total mass flow rate, is about 57%. The average outlet temperature is about 391 K. Table 2 summarizes mass and energy balances over one male rotor revolution after solution becomes periodic.

Table 2: Mass and energy balance

	Gas inlet	Oil injector	Outlet	Rotor power	Imbalance
Gas mass flow rate (kg/s)	*0.1794*	*0*	*0.1808*	*N/A*	*0.8%*
Oil mass flow rate (kg/s)	*0*	*0.2398*	*0.2385*	*N/A*	*0.5%*
Heat flux/Power (kw)	*53.0*	*128.2*	*226.5*	*46.4*	*0.5%*

Figure 7 plots instantaneous gas flow rate at inlet and outlet vs. male rotor crankshaft angle for two male rotor revolutions after solution becomes periodic. The results show perfect periodic shape repeated for every rotor tooth rotation. Outlet flow rate shows more oscillation than inlet flow rate, and has a short period of reverse flow.

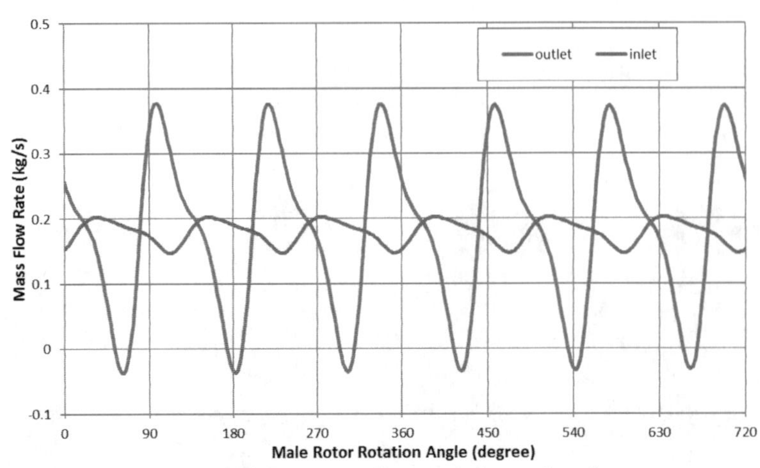

Figure 7. Gas mass flow at inlet and outlet

In order to evaluate oil injection effects on compressor performance, a similar case with the same condition but without oil injection has also been simulated. For the gas only case, the simulation imbalance of gas mass flowrate is about 0.4%, and the energy imbalance is about 0.2%. Figure 8 compares pressure distributions for the case with and without oil at the same crankshaft angle. Although the discharge pressures are the same, the internal pressure for the case without oil is significantly higher than the case with oil. Figure 9 compares temperature distributions for the case with and without oil at the same crankshaft angle. The temperature for the case without oil is much higher than the case with oil.

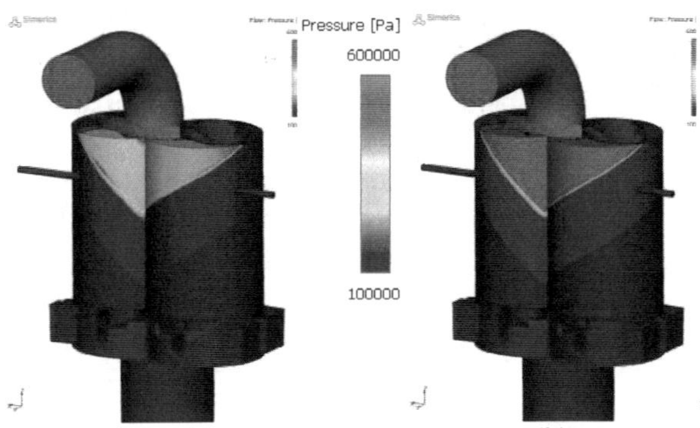

(a) (b)

Figure 8: Comparison of pressure: (a) with oil (b) without oil

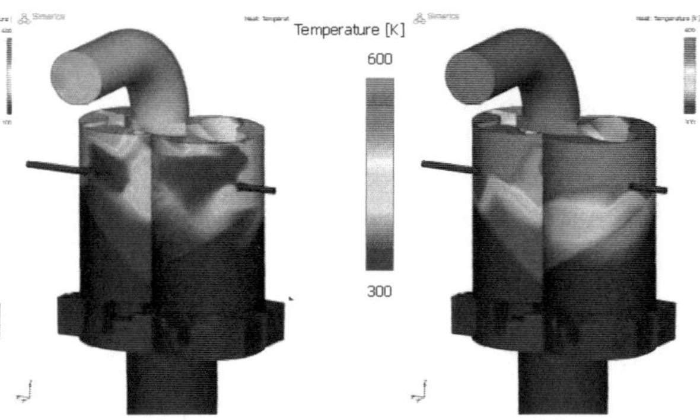

(a) (b)

Figure 9: Comparison of temperature: (a) with oil (b) without oil

Table 3 summarizes differences in compressor performance with/without oil injection. With oil injection, the compressor running under the same condition has a 23% increasing in gas mass flow rate, a 74% reduction of gas temperature rise at outlet, and a 12% reduction in rotor power compared with the "dry" (without oil injection) operation. The performance improvement of compressor with oil injection could be contributed from the combined effects of oil cooling and oil sealing.

Table 3: Effects of oil injection on compressor performance

	Gas mass flow rate (kg/s)	Gas temperature rise[a] (K)	Rotor power (kw)
With oil injection	0.1808	91	46.4
Gas only	0.1465	353	52.8
Differences	+23%	-74%	-12%

[a] With reference to inlet gas temperature of 300 K.

Figure 10 plots the pressure history of one compression pocket for both cases with and without oil injection for two revolutions of male rotor. The plotted pocket data was obtained from a monitor point moving together with the pocket.

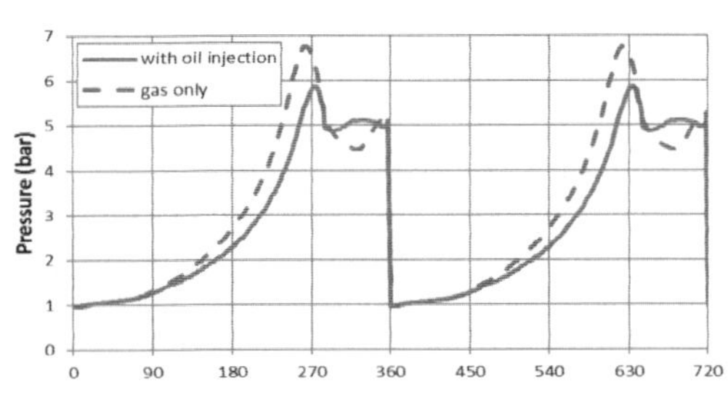

Figure 10: Pressure history of a pocket

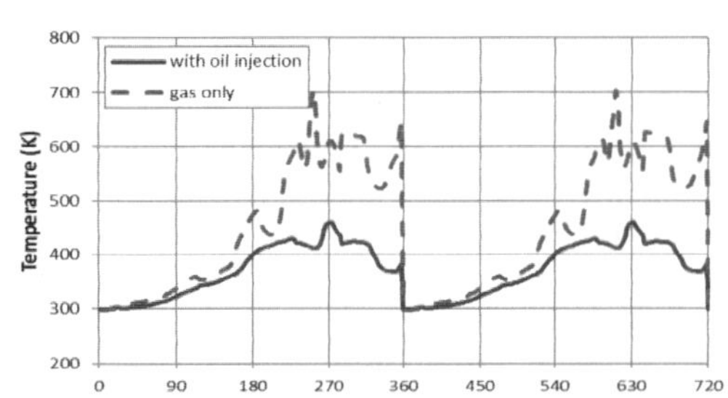

Figure 11: Temperature history of a pocket

The history curves show clear periodic patterns. The case without oil injection has a higher over compression than the one with oil. Figure 11 plots the temperature history of the same pocket. Due to the non-uniformity of the temperature field, the temperature results obtained from the monitor point have much more oscillation than the pressure results. The case without oil has a much higher temperature.

5. Summary

A newly improved VOF multiphase model has been successfully applied to the modelling of a twin screw compressor with oil injection. Simulation results predicted correct trends of performance improvement due to oil injection by comparing with a similar model operating without oil. Simulation results also show excellent mass and energy conservations. Detailed flow patterns and thermodynamic history of compression process with oil injection have been revealed from simulation results, which will help compressor designer and analyst to better understand and to improve the design. Overall, this new model has demonstrated great potential to be a very useful tool for the analysis of two phase flow problems in PD compressors.

Acknowledgement
The authors would like to express our gratitude to Dr. Kovacevic and Dr. Rane of City University London for letting us use their twin screw rotor geometry and mesh in this study.

References

[1] Bell, I., 2011. Theoretical and Experimental Analysis of Liquid Flooded Compression in Scroll Compressors. Ph.D. thesis, Purdue University. Full-text: http://docs.lib.purdue.edu/herrick/2/
[2] Ding, H. and Jiang, Y., "CFD Simulation of An Oil Flooded Scroll Compressor Using VOF Approach" (2016). International Compressor Engineering Conference. Paper 2508
[3] Launder, B.E., and Spalding, D.B., 1974, The numerical computation of turbulent flows, Comput. Methods Appl. Mech. Eng., 3, pp. 269-289.
[4] Ubbink, O., 1977. "Numerical prediction of two fluid systems with sharp interfaces". Ph.D. Thesis, Department of Mechanical Engineering, Imperial College, University of London.
[5] Hirt, C.W. and Nichols, B.D., 1981, Volume of fluid (VOF) method for the dynamics of free boundaries. J. Comput. Phys., Vol. 39, p. 201-225.
[6] Kovacevic, A., Rane, S., Stosic, N., Jiang, Y., and Lowry, S., 2014, Influence of approaches in CFD Solvers on Performance Prediction in Screw Compressors, *Int. Compressor Engrg. Conf. at Purdue*.
[7] Gao, H., and Jiang, Y., 2014, Numerical Simulation of Unsteady Flow in a Scroll Compressor, *Int. Compressor Engrg. Conf. at Purdue*.
[8] Gao H., Ding, H., and Jiang, Y., 2015 "3D Transient CFD Simulation of Scroll Compressors with the Tip Seal,"IOP Conference Series: Materials Science and Engineering. Vol. 90. No. 1. IOP Publishing, 2015.
[9] Ding, H. and Gao, H., 2014,"3-D Transient CFD Model For A Rolling Piston Compressor With A Dynamic Reed Valve", 2014, *22nd International Compressor Engineering Conference*, Paper 1548.
[10] Dhar, S., Ding, H., and Lacerda, J., 2016, "A 3-D Transient CFD Model of a Piston Reciprocating Compressor with Dynamic Port Flip Valves," *23rd Int. Compressor Engrg. Conf. at Purdue*. Paper 1392.
[11] Ding, H., Jiang, Y., Wu, H., and Wang, J., 2015, "Two Phase Flow Simulation of Water Ring Vacuum Pump Using VOF Model," ASME/JSME/KSME 2015 Joint Fluids Engineering Conference, Seoul, South Korea, July 26–31, 2015
[12] Kucinschi, B. and Shieh, T., "Estimation of Oil Supply Time during Engine Start-Up at Very Low Temperatures," SAE Int. J. Fuels Lubr. 9(2):2016, doi:10.4271/2016-01-0893.

IOP Conf. Series: Materials Science and Engineering **232** (2017) 012021 doi:10.1088/1757-899X/232/1/012021

Use of CFD to predict trapped gas excitation as source of vibration and noise in screw compressors

James Willie

Gardner Denver Schopfheim GmbH, Johann-Sutter-Straße 6+8, 79650 Schopfheim, Germany

E-mail: james.willie@gardnerdenver.com

Abstract. This paper investigates the source of noise in oil free screw compressors mounted on highway trucks and driven by a power take-off (PTO) transmission system. Trapped gas at the discharge side is suggested as possible source of the excitation of low frequency torsional resonance in these compressors that can lead to noise and vibration. Measurements and lumped mass torsional models have shown low frequency torsional resonance in the drive train of these compressors when they are mounted on trucks. This results in high torque peak at the compressor input shaft and in part to pulsating noise inside the machine. The severity of the torque peak depends on the amplitude of the input torque fluctuation from the drive (electric motor or truck engine). This in turn depends on the prop-shaft angle. However, the source of the excitation of this low torsional resonance inside the machine is unknown. Using CFD with mesh motion at every 1° rotation of the rotors, it is shown that the absence of a pressure equalizing chamber at the discharge can lead to trapped gas creation, which can lead to over-compression, over-heating of the rotors, and to high pressure pulsations at the discharge. Over-compression can lead to shock wave generation at the discharge plenum and the pulsation in pressure can lead to noise generation. In addition, if the frequency of the pressure pulsation in the low frequency range coincides with the first torsional frequency of the drive train the first torsional resonance mode can be excited.

1. Introduction

Simulation technique is playing an increasing role in the development process of screw compressors at Gardner Denver (GD). This reduces development time by reducing expensive tests in the laboratory. The availability of computing power at a relatively lower cost also makes modeling larger cases using finer meshes possible in industry. This paper presents one such example of how CFD can be used in industry to help in the development process of screw compressors. Using this technique in the modeling of turbo-compressors is commonplace but in the case of screw compressors, which are positive displacement machines, this is not the case. Some of the few and most recent cases that report the modeling of screw compressors can be found in [1, 2, 3, 4]. The reason is due to the complex nature of the screw profile that must be simulated and the very small gaps/clearances in these machines that must be modelled. The screw mesh in this paper was generated using TwinMesh from CFX-Berlin together with the CFX-solver. The Junction Box routine is used to enable mesh motion of the screw. Because of the complicated nature of the screw, the rotation angle for generating each mesh is 1°. The fine mesh generated makes it possible to capture the complicated physics taking place inside the compressor. For example, it was possible to capture the over-compression at the discharge

in the absence of the axial gap/equalizing chamber. It is shown that in this case, a trapped gas region is created that leads to over-compression at the discharge. To mitigate the effect of the pulsation created by this trapped gas a pressure equalizing chamber or groove in the discharge side cover is used. It is seen that using this groove or chamber reduces the pressure pulsation significantly. This is the case because the axial discharge port has a delay in closing and this can create a residual volume, V_r, when the port closes. As a result, V_r is still significant when the discharge port closes, especially when the number of male rotor lobes are few and as the rotor rotation continues it leads to a very high pressure of the fluid contained in it. When left unchecked, it can lead to increase in compressor power and a loss of efficiency. The prediction by CFD reported here is validated with measurement data. The CFD model is further validated by comparing the compressor torque/power and the pressure and mass flow rates to those determined in measurements.
One of the main findings reported in this paper is the possibility that the low frequency torsional modes observed in screw compressors that are mounted on trucks and driven via the truck PTO could be excited by the pulsation of the trapped gas reported in this work. The prediction and measurement of torsional resonance in screw compressors has been reported in [5, 6, 7]. In them, the authors showed that based on sensitivity analysis, the compressor behaves as a rigid body and that the most sensitive element responsible for the first torsional mode excitation is the prop-shaft in the case of the electric motor drive and the thin PTO shaft in the case of the truck PTO drive. One manifestation of this torsional resonance is the presence of side bands at all the gear meshing frequencies and their harmonics and the presence of pulsating noise that was found to be due to amplitude modulation and not frequency modulation [6, 7]. The element providing the excitation within the machine was not reported. Based on the current work, it is shown that when trapped gas is present, over-compression is seen in the pressure data and Fast Fourier Transform of the pressure time series data showed clear side-bands at the lobe passing frequencies and harmonics of the compressor. When the pressure equalizing chamber is used, this is not the case. Furthermore, the amplitude at the low side-band frequency is much higher in the pressure time series data without the pressure equalizing chamber than with the pressure equalizing chamber.
This paper begins with the description of the case that is simulated, including the measurement point that is modeled and the machine layout. After this, the geometry and the mesh of the test case simulated are presented and this is followed by the presentation of the case setup in CFX-pre and the computer hardware on which the case is run. This is proceeded by the presentation of the results and finally, the conclusions are given.

2. Measurement and layout of the machine

The schematic showing the setup of the measurement and a picture of the measurement setup is depicted in Figure 1.

In addition to performance measurement, waterfall plots are generated using the torque run-up test described in [7] to check if any torsional mode is excited. The Brüel & Kjaer (B&K) pulse noise & vibration analysis system for torque measurement is used. The torque is measured by mounting the torque flange on the compressor input shaft. In Figure 1, the axial, profile and housing gaps are sketched. One key aspect in designing screw compressors for transport is noise. To limit noise, the suction and discharge ports are sized in such a way that their velocities are low. This is critical because based on wind tunnel test, the sound pressure level, $SPL \sim (Ma)^n$, where Ma is the Mach number and n is a real number. After finding the inlet duct diameter based on the volume flow rate to limit the velocity to a low value, conservation of mass and continuity equation are used to determine the size of the discharge port. The inlet and discharge ports are also aerodynamically optimized in order to limit the pressure drop in the machine. One

IOP Conf. Series: Materials Science and Engineering **232** (2017) 012021 doi:10.1088/1757-899X/232/1/012021

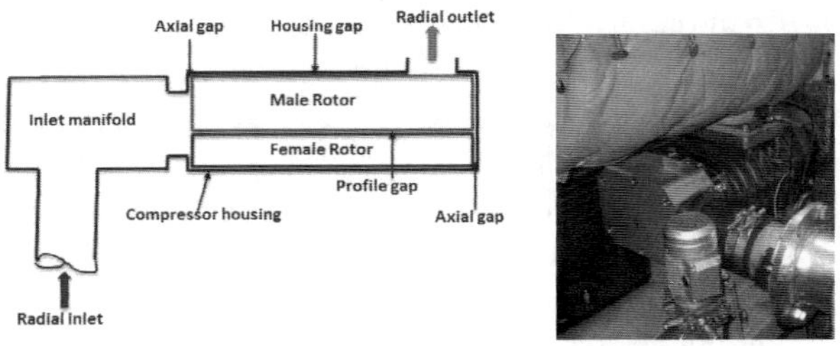

(a) Schematic layout (b) Picture of measurement setup

Figure 1. Schematic of compressor layout and picture of the setup used in the measurements

of the key findings is that during the operation of screw compressors, most of the compression work is done by the male rotor while the female rotor transports most the fluid. Hence, the direction of rotation of the female rotor dictates the flow direction of the air exiting at the discharge. The fluid leaves the rotor with a given angular momentum that is given by mvr, where m is the fluid mass, v is its velocity and r is the tip radius of the female rotor and this dictates the flow direction. This information is helpful in deciding on where to place the discharge and how to optimize the flow at the discharge.

The L/D of the rotors profile used is 2.06 and the Center Distance (CD) is 100mm. To limit noise, the maximum male rotor speed is low while the input speed range is 1000-1800 rpm. The female rotor has more lobes than the male rotor as in most screw machines. The rotors used are large enough to limit any large torsional response (angular deflection) and the profile backlash are designed to ensure no torque fluctuation (torque reversal) at the rotors.

2.1. Measurement point used in the simulation

The case corresponding to the maximum male rotor speed at input speed of 1800 rpm is selected and used in the CFD simulation. The measurement point simulated is shown in Table 1. In this table, the volume flow rate is normalized with the maximum design value and the female and male rotor speeds are normalized with the male rotor maximum design speed.

Table 1. Screw compressor test parameters used in the CFD simulation.

Parameter	Measured value
Inlet pressure (p_1)	1.002 bar
Inlet temperature (T_1)	21° C
Discharge pressure (p_2)	2.2bar g
Discharge temperature (T_2)	181°C
Normalized volume flow rate ($\dot{V}$)	1.041
Normalized female rotor speed (n_f)	0.6
Normalized male rotor speed (n_m)	1.0
Compressor power (P)	69.26kW

3. Geometry and mesh generation using TwinMesh

The screw compressor simulated is comprised of the following key parts used in the simulation of the air from the suction to the discharge: inlet manifold, rotors (male and female), discharge cover and the discharge plenum and duct as shown in Figure 1. The mesh is divided into two parts: The rotor and the stator. The rotor mesh is made of the screws. The stator mesh is made of the suction fluid and the discharge fluid and it is generated using Ansys Meshing. The two regions are imported into CFX-pre together with the defined interfaces that makes transfer of information between the static and moving regions of the geometry possible. Two cases are simulated. In the first case, denoted as Case 1, the axial gap and equalizing chamber or groove at the discharge are omitted in the geometry and in the second case labelled as Case 2, the axial gap and equalizing chamber are included at the discharge. In both cases, the stator mesh at the suction side and rotor mesh remained the same.

3.1. Mesh generation

The stator mesh is an unstructured tetrahedral mesh as shown in Figures 2 and 3. In Figure 2, a sectional view of the mesh at the suction side is shown. The mesh has a total of 1 615 484 cells.

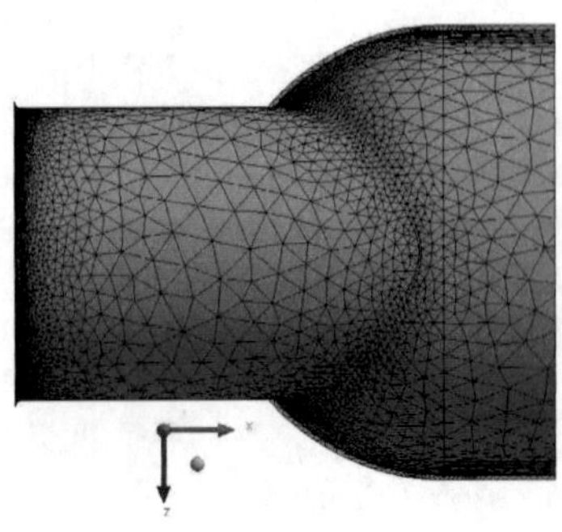

Figure 2. Sectional view of the suction stator mesh, 1 615 484 cells, inlet located left

In Figure 3(a), a sectional view of the stator mesh at the discharge in Case 1 is shown and in Figure 3(b), a sectional view of the stator mesh at the discharge in Case 2 is shown.

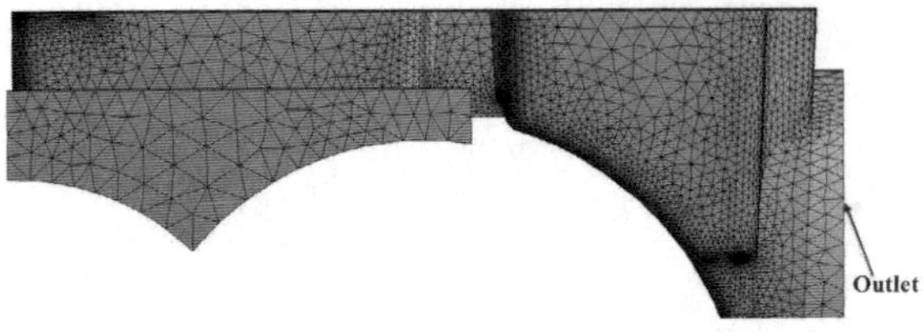

(a)) Case 1, 507470 tetra cells

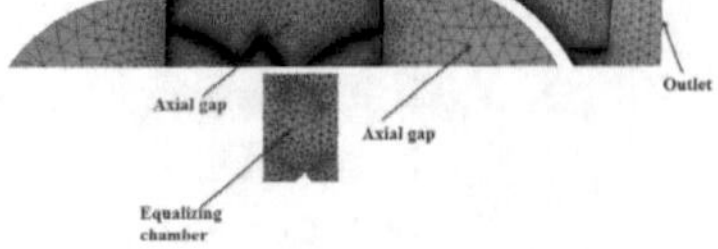

(b) Case 2, 667222 tetra cells

Figure 3. Sectional view of the discharge stator mesh used in the CFD simulation

To generate the rotor mesh, TwinMesh is used. It is a template based approach used for generating structured meshes for screw rotors. It is developed by CFX Berlin GmbH [3, 4]. In the case presented here, the 2D structured mesh is generated for every 1° rotation of the rotors. The rotor curves are imported into TwinMesh as shown in Figure 4 using a cut sectional view.

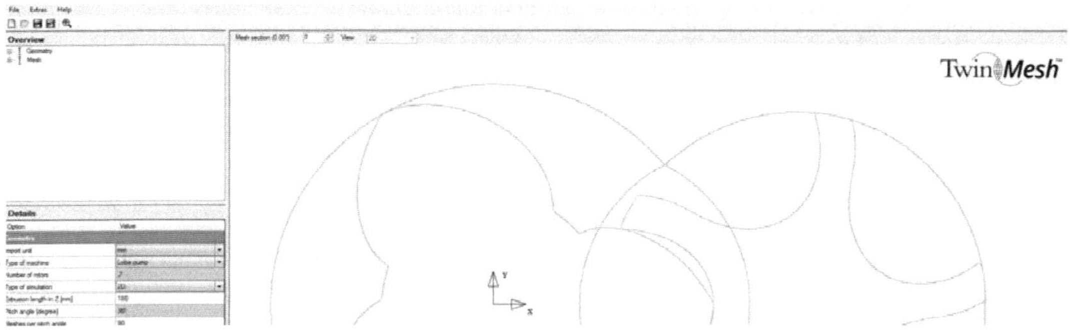

Figure 4. Cut sectional view showing a cut through the rotor curves imported into TwinMesh

The cold gaps values for the housing, the profile and the axial gaps at the suction and discharge are included. After defining the mesh parameters, the 2D mesh is generated as shown in Figure 5 for a rotation angle of 0°. This is repeated for every angle up to 360° after which the mesh is generated in 3D by projecting the 2D mesh in the axial direction along the rotor axis for the length of the rotor. The mesh information is as follows: The male rotor has 6340 quad elements in 2D and 634000 hexahedral elements in 3D. The female rotor has 5540 quad elements in 2D and 554000 hexahedral elements in 3D. For mesh quality, it is recommended a minimum angle of 24 is maintained. Rotor 1 is the male rotor and Rotor 2 is the female rotor. When viewed from the discharge size, the male rotor rotates in the counterclockwise direction while the female rotor rotates in the clockwise direction.

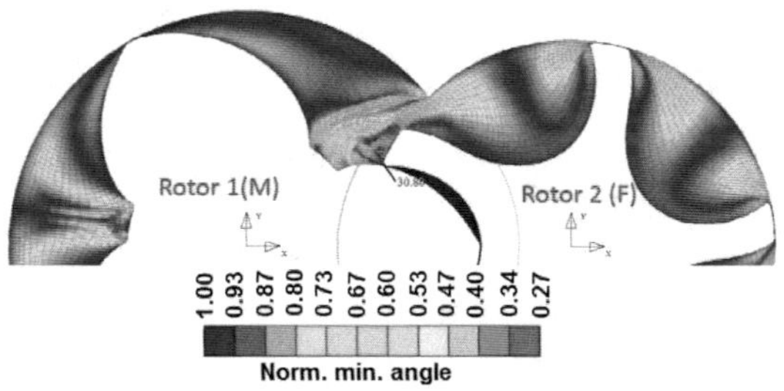

Figure 5. Cut sectional view showing TwinMesh 2D grid of the male and female rotors at 0°

3.2. Case setup in CFX-pre

The setup is done by importing the rotor and stator meshes into CFX-pre. The boundary conditions used are: suction inlet is taken as an Opening and under the Mass and Momentum Option, Opening Pressure and Direction is used and at the discharge an Opening is also used with Entrainment under the Mass and Momentum Option and the Turbulence Option is set to Zero Gradient in order to specify the flow direction as required. This way, the solver computes the mass flow rate based on the pressure and temperature defined from measurements. Interfaces are defined between the static and dynamic mesh regions of the suction and discharge sides. The Shear Stress Transport (SST) turbulence model is used [8]. It combines the k-ω and the k-ε turbulence models such that the former is used in the inner region of the boundary layer and then switches to the the latter in the free shear flow. The Total Energy option is used for heat transfer and air ideal gas is used for the fluid material since the flow is compressible and for the mesh motion, the Junction Box Routine is used. The male and female rotors are given the appropriate rotational speeds and directions. Heat transfer within the solid rotors are not accounted for in the simulation.

Cases 1 and 2 are run on a HP Z820 16 Core Computer Workstation with a RAM of 64GB and an Intel Xeon processor. During the run, 8 of the 16 cores are used and in both cases, a minimum of three revolutions of the rotors are simulated.

4. Results and discussions

The results section begins with the discussion of the flow optimization at the discharge in order to reduce the pressure drop inside the machine. This is followed by the discussion of the two Cases 1 and 2 considered. Three design variations with regards to the optimization at the discharge are considered, but only the third is presented here. One aim in the development of turbo and screw compressors is the reduction of the pressure drop in the machine, especially at the discharge. It has implication for noise, compressor power and in some cases for pressure pulsation as it will be shown in this paper. In the analysis performed here, attempts are made to eliminate flow recirculation at the discharge when the air exits at the discharge port. Figure 6 shows the velocity vectors in a plane at the discharge with and without the flow recirculation. In screw compressors, heat generation is mainly due to gas compression and expansion at the discharge port. Creating a flow re-circulation at the discharge can have other localized effects like increased thermal gradients.

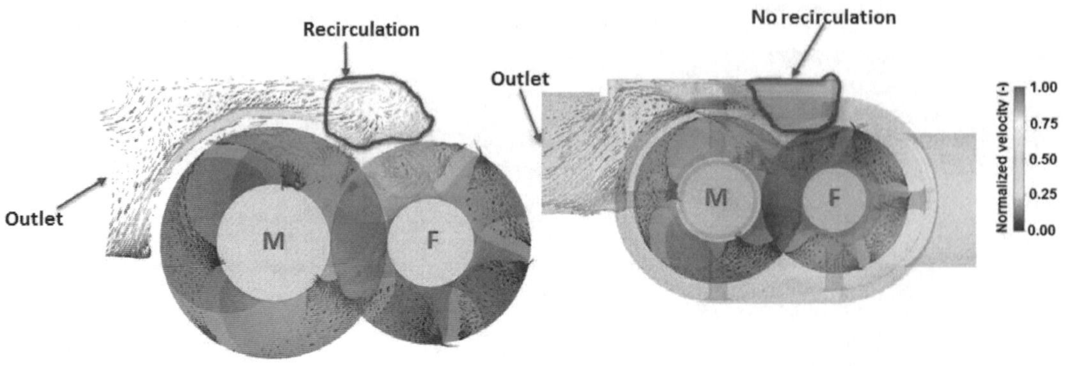

Figure 6. Velocity vectors in a plane at the discharge (Left is without flow deflector & right is with the flow deflector)

4.1. Performance comparison

The performance of Cases 1 and 2 are compared and validated with measurement data. The following performance parameters are compared: average maximum pressure at the discharge, the male and female rotor torques, the compressor power and the mass flow rate as shown in Table 2. The mass flow rate is normalized using the measured value.

The pressure data shows that in Case 1, the pressure is much higher and we have over-compression as a result of the trapped gas (See Figure 7). In Case 2, there is no over-compression and the pressure in the trapped gas region or residual volume region is the same as the pressure at the discharge.

IOP Conf. Series: Materials Science and Engineering **232** (2017) 012021 doi:10.1088/1757-899X/232/1/012021

Table 2. Performance comparison of cases simulated and validation with measurement.

Parameter	Case 1	Case 2	Measurement
Max. avg. press. at discharge (bar abs.)	4.6	3.6	3.2
Male rotor avg. torque (Nm)	84.11	79.5	-
Female rotor avg. torque (Nm)	1.11	2.997	-
Compressor power (kW)	73.88	70.88	69.26
Norm. average mass flow rate (-)	0.72	0.77	1.00

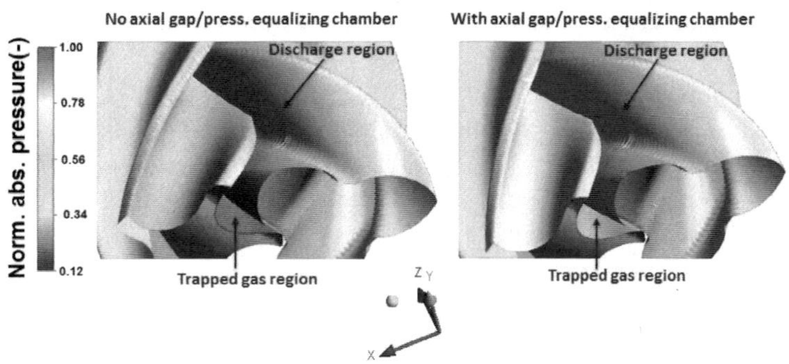

Figure 7. Normalized pressure contours of regions at the discharge showing pressure on rotors surfaces (Left=Case 1 and Right=Case 2)

As already mentioned, the compression work is done by the male rotor as can be seen from the male rotor torque value when compared to the female rotor torque. In the two cases considered, it is seen that the male rotor torque and power are larger in Case 1. Due to the over-compression in Case 1 the rotor must do more work to overcome the over-compression. The compressor power in Case 2 is closer to the measured value than in Case 1. The mass flow rate deviates a lot more from the measured value due to leakages in the gaps. The gaps values used in the mesh generation are values for the cold flow. In reality, the hot gaps values are smaller than the cold gaps values due to the thermal expansion of the rotors. The hot gap values are however more difficult to measure when the machine is in operation. Due to the large pressure at the discharge, leakages are also possible which leads to losses in the volumetric efficiency of the machine. In both cases, mass conservation is achieved and the periodic mass flow showed statistical convergence. The flow in the gaps are however treated as stationary instead of rotating. Due to over-compression, the corresponding rotor temperature in Case 1 is higher as shown in the contour plot in Figure 8. The measured air temperature at the discharge is 454K and the average temperature at the discharge in Case 1 is 657K while that in Case 2 638K. The rotor temperatures in the CFD are larger because the rotor solid and heat transfer through the solid rotors are not accounted for in the CFD; only the surface of the rotors are accounted for. Also, the cooling of the rotors through convection are not accounted for in the CFD. During measurement,the measured temperature of the cooling oil is 10°C lower in Case 2 when compared to Case 1, with both machines always having an axial gap.

Compared to other methods like the Immersed Solid Method [3], TwinMesh is able to accurately predict the physics of the flow taking place inside screw machines. With Immersed Solid, a momentum source term is used to force the fluid to move with the rotor velocity and it does not account for mesh motion. However, it is able to reproduce the velocity trends and

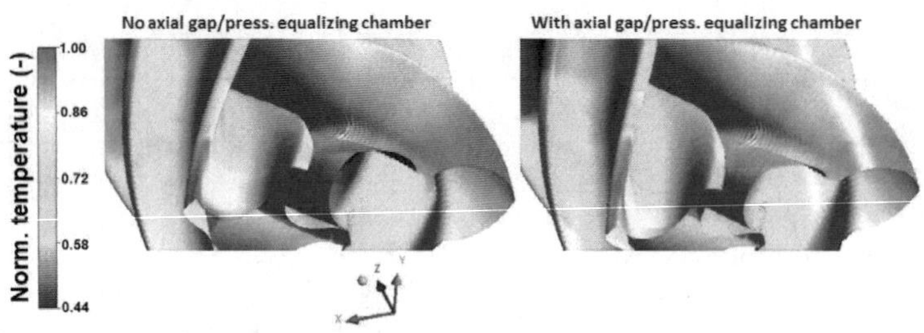

Figure 8. : Normalized temperature contours of regions around discharge showing temperature on the rotors surfaces (Left=Case 1 and Right=Case 2)

pressure drop trend and so it can be useful in doing Design of Experiments (DoE) during the initial stages of design to compare different design cases. It is also less memory intensive when compared to TwinMesh that has to store mesh data for every 1 degree rotation of the rotors. The TwinMesh approach is also computationally more intensive.

4.2. Pressure pulsations

To study the effect of pulsation on the machine in Cases 1 and 2, pressure time series data is taken at the discharge and Fast Fourier Transform (FFT) is performed and plotted as shown in Figure 9. The ordinate axes have the power spectra density (PSD) that is given by $PSD = 20log10(p)$, since in this case the measured pressure is not referenced to $p_{ref} = 2 \times 10^{-5} Pa$. If referencing is done, then $PSD = 20log10(\frac{p}{p_{ref}})$. The sampling rate of the pressure data is 10 kHz and 10 000 samples points are taken. In both cases presented, the machine was run at an input speed of 1800rpm, 2bar(g) using a 90kW 4 pole asynchronous motor. The data shows the fundamental lobe passing frequency at around 400Hz and the first harmonic as expected. But in the data of Case 1 we have sidebands of 30Hz at the lobe passing frequency. When the data is zoomed to show the lower frequency range it is seen that at 30Hz we have a peak suggesting that 30Hz could be providing the excitation to drive torsional resonance originating from the prop-shaft as shown in Figure 10. This frequency is 1 $\times$ input speed at 1800rpm and therefore synchronous as previously reported by the authors [5, 6, 7]. In the paper reported in [5], it was shown that the torsional frequency occurring in the screw compressor driven by an electric motor was occurring at 1 $\times$ input speed. In Figure 10, the peak amplitude of in Case 1 of 30Hz is 38dB while that in Case 2 is 28dB.

The peak in the FFT data of Case 1 is clearer in Figure 10, where the frequency range has been narrowed to 50Hz.

Pressure time series is also taken in the CFD and plotted as shown in Figure 11 for the two cases. The pressure is normalized using the maximum pressure Case 1. Clearly, the pressure amplitude in Case 1 is larger than in Case 2.

The small difference in the plots shown in Figure 9 and 10 is because the pressure transducer is placed at the discharge pipe and not in the trapped gas region. When the air exits the compressor, there is a sudden expansion, which leads to a decrease in pressure in the case with over-compression. The difference between them and the CFD plot shown in Figure 11 is mainly because in the CFD, the maximum pressure at the discharge is monitored, which is in the trapped gas region. Also, in Case 1 no axial gap is included in the CFD, whereas in the measurement, there is an axial gap but no equalizing chamber. The CFD plot is also done using a linear scale whereas in the measurement a non-linear scale is used. The trends between the CFD and the measurement data are however similar, as expected.

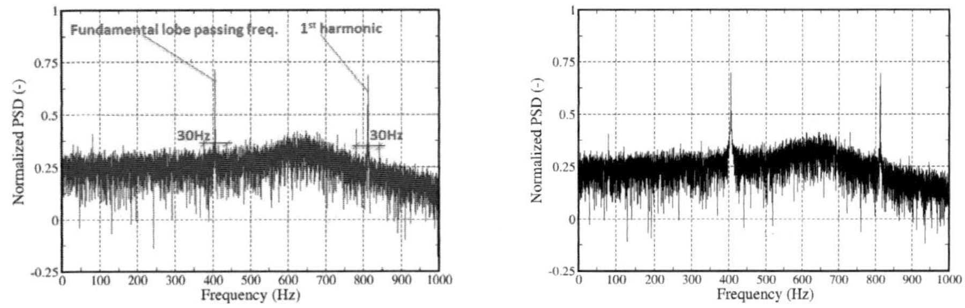

(a) No axial gap/equalizing chamber (b) With axial gap/equalizing chamber

Figure 9. Plot of measured normalized PSD of pressure taken at discharge of the compressor

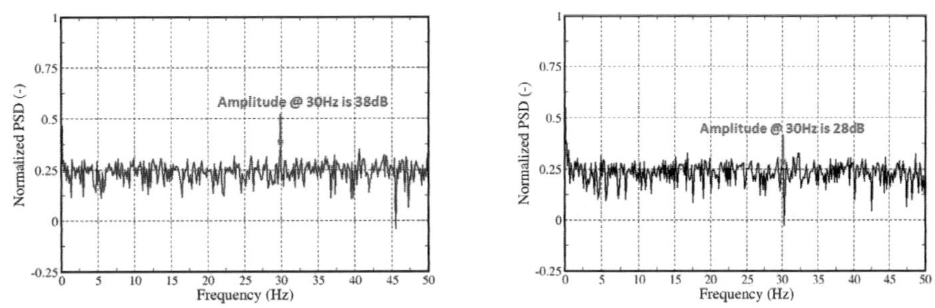

(a) No axial gap/equalizing chamber (b) With axial gap/equalizing chamber

Figure 10. Plot of measured normalized PSD of pressure taken at discharge of the compressor

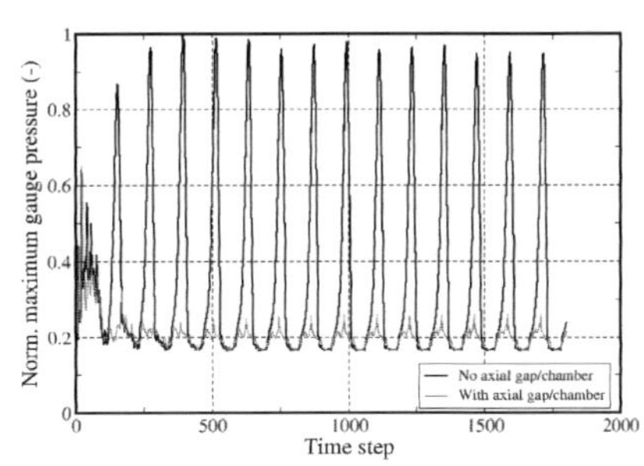

Figure 11. : Normalized CFD maximum pressure time series data of the compressor

The difference in pressure pulsation in both cases has large implication for design. For example, Case 1 will demand more motor power to be able to run at 1800rpm input speed. Noise and vibration will also be more in this case and the generation of flow instability like shock waves is possible in addition to having larger thermal gradients.

5. Conclusions

This paper reports the successful simulation of the flow inside a screw compressor built for use in bulk transport on trucks. The compressor is powered using PTO on trucks and it is susceptible to noise and vibration and torsional resonance. In particular, the first torsional mode is easily excited in the PTO shaft. The response to this excitation inside the machine can lead to sidebands at all gear meshing frequencies and their harmonics and pulsating noise is possible [5, 6, 7]. Sidebands in the pressure data at the rotors lobe passing frequency and harmonics is also visible. The following conclusions can be drawn:

- Using CFD it is possible to reproduce the flow and physics taking place inside a screw compressor. This is possible due to the successful integration of TwinMesh for the structured mesh generation of the rotors and the CFX solver using the Junction Box Technique for the mesh motion.
- Flow optimization can lead to pressure drop reduction and to reduced noise and reduced thermal gradients at the discharge of such compressors. The power consumption is also improved.
- The inclusion of a groove/pressure equalizing chamber at the discharge can lead to reduced pressure pulsations and noise and thermal gradient inside the machine.
- The pulsation of the gas inside the trapped pocket may provide the necessary excitation for the first torsional mode outside the compressor, if a frequency of the pulsation coincides with the first torsional frequency of the compressor. This can lead to pulsating noise at the gears inside the compressor if care is not taken to optimize the gears at the micro-geometry level in order to ensure the right backlash and the mid-flank contact that is needed.

Acknowledgment

The authors would like to thank Francesco Maraini and Steve Munt, who carried out the measurements reported at GD Robuschi, Parma, Italy, and at GD DRUM, Bradford, UK, respectively. Their valuable contribution through discussions is also acknowledged.

References

[1] Hauser J and Schulze-Beckinghausen P 2016 *Performance Prediction of Oil-Free Screw Compressor using Computational Fluid Dynamics* (VDMA Compressors Users Int. Forum 2016, Düsseldorf, Germany), September 14-15, pp. 303-312

[2] Rane S, Kovacevic A, and Stosic, N 2015 *Analytical Grid Generation for Accurate Representation of Clearances in CFD for Screw Machines* (International Conference on Compressors and their Systems, City University of London, London, UK) September 7-9, http://iopscience.iop.org/article/10.1088/1757-899X/90/1/012008

[3] Spille-Kohoff A, Hesse J and El Shorbagy A 2015 *CFD Simulation of a Screw Compressor including Leakage Flows and Rotor Heating* (Int. Conference on Compressors and their Systems, City University of London, London, UK) September 7-9, http://iopscience.iop.org/article/10.1088/1757-899X/90/1/012009

[4] Hesse J, Spille-Kohoff A, Hauser J and Schulze-Beckinghausen P 2014 *Structured Meshes and Reliable CFD Simulations: TwinMesh for Positive Displacement Machines* (Int. Conf. on Screw Machines 2014, TU Dortmund University, Dortmund, Germany) September 23-24 2014 (VDI-Berichte 2228) pp. 297-105

[5] Willie J, Asal W and Sachs R 2014 *Analysis of the Noise and Vibration of a Dry Screw Compressor* (Int. Conf. on Screw Machines 2014, TU Dortmund University, Dortmund, Germany) September 23-24 2014 (VDI-Berichte 2228) pp 91-308

[6] Willie J and Sachs R 2015 *Structural and Torsional Vibration Analysis of a Dry Screw Compressor* (9th Int. Conference on Compressors and their Systems, City University of London, London, UK) September 7-9 2015, http://iopscience.iop.org/article/10.1088/1757-899X/90/1/012005

[7] Willie J and Sachs R 2016 *Structural and Torsional Vibration and Noise Analysis of a Dry Screw Compressor* (Proc IMechE Part E. J Process Mechanical Engineering 0) vol 0, DOI: 10.1177/0954408916648989

[8] Pope S B 2000 *Turbulence Flows*, ed 1 (Cambridge University Press)

IOP Conf. Series: Materials Science and Engineering **232** (2017) 012022 doi:10.1088/1757-899X/232/1/012022

Manufacturing method and performance assessment for variable lead vacuum rotors

Holmes C. S.[1], Rane S[2], Wang X.[1]

[1]PTG Holroyd, United Kingdom; [2]City University of London, United Kingdom

Chris.holmes@outlook.com, Sham.Rane@city.ac.uk, xianfa.wang@holroyd.com

Abstract. In recent years variable lead rotors have been produced, mainly for vacuum applications, involving a multiple pass manufacturing process which is necessarily time-consuming. A faster method of manufacturing such rotors uses a full profiled disc-type milling or grinding tool but involves clearance variations along the length. These effects have been assessed by computer modelling to quantify any disadvantages. The results indicate that the effects on performance are negligible and the profiled disk tool process is suitable for such components.

1. Introduction

For certain types of twin-screw compressor, notably those used in vacuum applications, there has been a requirement for a reducing inter-lobe spacing along the length of the rotors. This is referred to as 'variable lead', where lead is defined as the axial advance per 360 degrees progression around a helical flank.

One method which has been used in production has been a rotating 'end mill' machine principle, where the flank is generated by multiple passes along the length of the rotating part. The successive engagement positions between the tool and the part are calculated such that the envelope of the individual facets so produced will approximate to the desired profile. The 'end' tools used have a relatively simple shape such as a sphere, since an attempt at a dedicated profile would be adding cost without any benefit.

This method of manufacture can require cycle times of several hours for a medium-sized rotor, and if larger production quantities are required, the cost per part may be high compared with disk-type tool which produces the entire profile in a single pass.

It has been assumed that the variable lead is an obstacle to using full-profiled wheel or disk type tool, because although the tool path can be progressively changed during a cycle, the profile which results may be correct for one end of the helix, but will be distorted at the other end, causing a mismatch between the two engaged profiles, with a consequent increase in leakage and reduction in efficiency.

It was decided to optimise the tool path to minimize the mismatch and leakage effects (Section 2), and then to investigate, by CFD modelling, how much this mismatch impacts the final performance of the compressor (Section 3).

Figure 1. Variable lead screws made using a full-profile disc tool

2. Manufacturing process and simulation for a full-profile disk tool

The method of machining a rotor with a variable-lead screw was developed many years ago and is described in Davey [1]. An example is shown in Figure 1. The method uses a full-profile disk tool with adjustments of some or all of the following: head angle, approach angle, work rotation and infeed. The rotors engaged and rotated smoothly.

Assuming that the clearances at the discharge end are more important when determining leakage, it is reasonable to keep these as initially designed, and in this case the deviations will increase towards the inlet/suction end, which will therefore have slightly greater clearances and might be expected to give more leakage. In order to estimate the commercial viability of using a fast manufacturing process using disk-type milling, it is necessary to estimate the 'leakage cost' of the compromise profile clearances which result.

Figure 2. Combined constant and variable lead model, discharge end to the right

Figure 2 shows the rotor design used in this study of the effect of small clearance mismatch on performance. It has two segments, with constant lead at the discharge end, and variable lead at the suction end. The main body dimensions were:

Outside diameter	156 mm
Short lead	21 mm
Longest lead	72 mm
Total length	190 mm
Inter-lobe clearance	0.2 mm

The transverse profile of the rotor, and the tool used to produce the constant lead segment are shown in Figure 3. This tool produces correct clearances over the constant lead segment only.

The same tool was used to make the variable lead segment, i.e. there was no modification of the tool profile. There is no doubt that a more favourable profile would be achieved if movements of head angle, approach angle, infeed and work rotation are applied. In this case, to reduce the complexity imposed on the process, the profile of the variable lead segment was simulated without infeed or approach angle movement. Also, it will be obvious that where variable lead is required, opposite flanks must be manufactured separately.

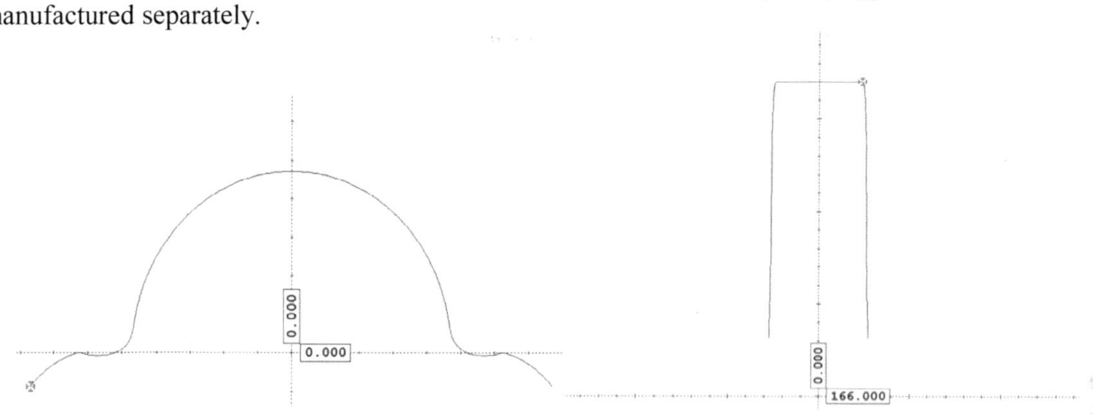

Figure 3. Transverse section of rotor (excluding O.D.) and tool

The variable lead segment was divided into five sub-segments, and the profile of each sub-segment was calculated for input to a CAD model. The intention was to keep the minimum inter-lobe clearance close to its nominal value. An iterative process was used to find the best tool paths. The program calculated the rotor profile with estimated tool path adjustments, next the clearance was calculated using a pair of rotors with this actual profile, then feedback was used to give further adjustments to the profile calculation to optimise the clearances. The clearances at the suction end were optimised by adjusting the tool head angle and work rotation, using an iterative procedure. The effect of changing head angle is shown in Figure 4.

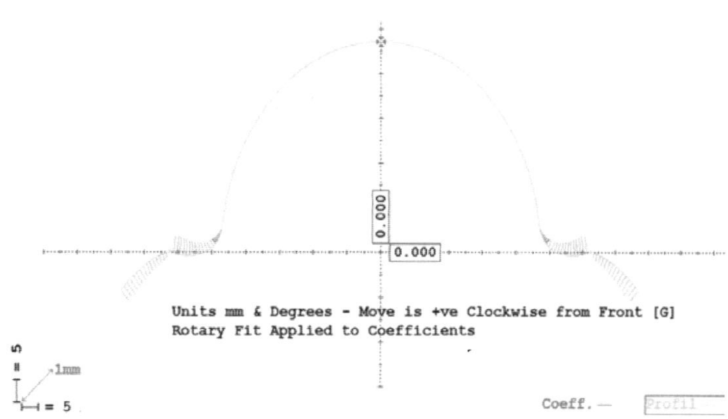

Figure 4. The effect of changing tool head angle

Figure 5 shows the target clearances as designed for the full length of the rotors. Figure 6 shows the calculated clearance distribution at the suction end using the tool profile which is correct for the discharge end. It can be seen the clearance is no longer uniform. The inter-lobe clearance has increased significantly at the intake end. Figure 7 shows the two suction end profiles superimposed, green being the target, and red being actual.

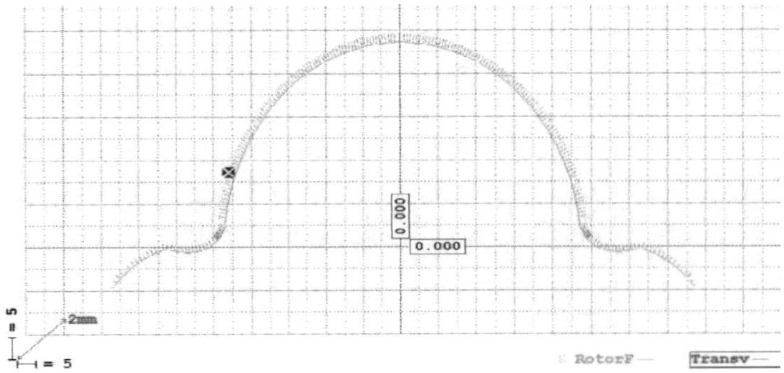

Figure 5. Designed clearance distribution over full length

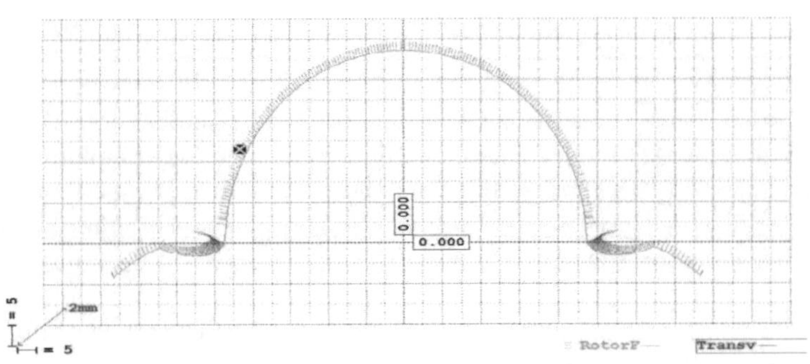

Figure 6. Calculated suction end clearance distribution using same tool profile as for discharge end

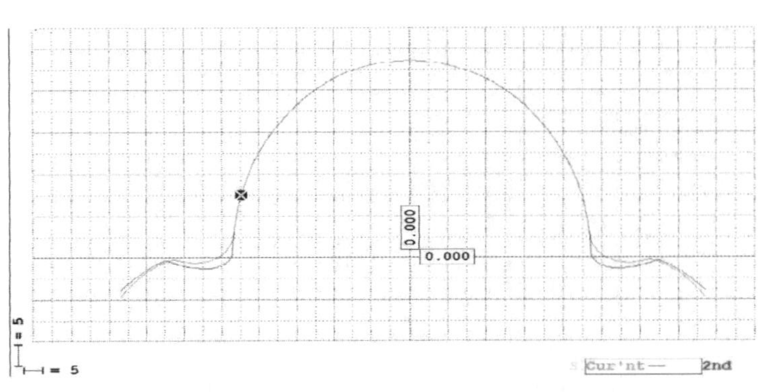

Figure 7. Suction end profiles superimposed, green= target, red=actual

Two CAD models were created, one with uniform clearances over the whole length based on the reference profiles, and the other using the simulated profiles with non-uniform clearance. In both cases the reference profile was used over the uniform lead section

3. CFD analysis for comparison of rotor performance

Various modelling approaches are available for the design and analysis of thermodynamic and flow processes inside the working chamber of twin screw machines such as compressors and expanders. For a long time these models have been formulated based on lumped parameters and solution of equations of conservation of mass and internal energy of the working fluid. Although very fast, such quasi-one dimensional models need fine tuning of important geometrical parameters such as leakage gaps, port areas and the chamber volume variation with time. In recent times, a more accurate representation of the real 3D rotor geometry has been made possible by the use of computational fluid dynamics. Although currently employed CFD models are enormously slower in comparison to lumped parameter models, they have high resolution and capture the flow process in its entirety, so were felt to be more reliable. (The model is based on conservation of mass, momentum and energy equations applied within continuous fields with negligible molecular interactions, and absolute vacuums cannot be solved by these methods, so this is a limitation. It will be valuable to compare the two methods in a future study.)

The major challenge in CFD modelling for twin-screw rotors is the treatment of highly deforming domains that need a mesh to represent the geometry. Studies such as Kovačević et al [2], Vande Voorde et al [5], and Rane et al [3] have used customised grid generation tools in order to address this challenge of deforming domains for CFD analysis of twin screw compressors, hook and claw type pumps, oil injected twin screw compressors etc. In the application of these models to the analysis of twin screw vacuum pumps with variable lead rotors, it is possible to create such a customised model for grid generation and solve the flow. Rane [4] has presented a full development procedure for variable twin-screw geometry rotor grid generation and an application to dry air, low-pressure compressor performance has been evaluated.

The challenge required for the presented study, i.e. the analysis of variable lead rotors from a manufacturing and performance perspective, was of a different nature. The variable lead rotors to be manufactured using the proposed method had small deviations in rotor to rotor clearances and the impact on pump performance was required to be estimated at the desired operating conditions. Analysis of such a variable lead screw rotor with variation of clearance has not been reported in available literature. Existing CFD modelling techniques have used uniform clearances in the representative mesh for calculations. There are reports which demonstrate the change in clearance and performance impact but these clearance changes are uniform along the rotor length [4].

Figure 8 shows the construction of the rotor geometry with variable lead and variable clearance. Five rotor length segments were identified and cross section profiles were varied in each of these lengths to achieve the lead and clearance variation. For the CFD model shown in Figure 9 the fluid volume with leakage gaps is required. This was extracted from the CAD model by subtracting the rotor volume from the bore volume. This operation captured the rotor-to-rotor and rotor-to-bore leakage gaps. The entire fluid volume was then meshed using ANSYS meshing with a refined tetrahedral grid. The total number of nodes in the mesh was 1,324,250. Another geometry was constructed with the same lead variation but with ideal clearances i.e. without the manufacturing deviation. The objective of the CFD analysis was now to compare the performance of the two rotor geometries under same operating conditions. The performance has been compared using the parameters of leakage flow, pressure distribution, torque and thrust forces.

Important assumptions of the model were: the rotors are non-rotating to simplify the model, the casing and rotor surfaces are smooth walls i.e. no roughness, the gas is air with ideal gas properties for density variation, the high-pressure side gas temperature was set at 300K which is the leakage gas temperature, the flow equations used in the CFD solver for continuum are assumed to be valid even under very low density conditions, the rotors do not contact (friction between the rotors which is a big source of heat and mechanical torque has been ignored), and no thermal deformation occurs. The leakage flow solution was calculated for both the rotor geometries at a set of decreasing low pressure levels which went from 75000 Pa down to 50Pa.

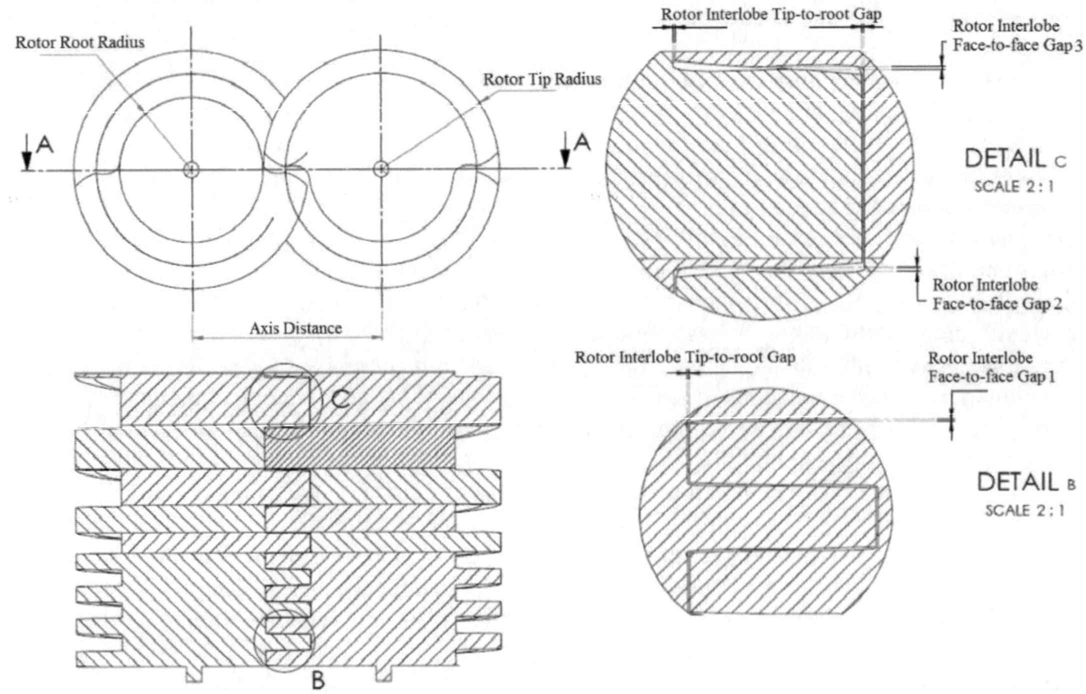

Figure 8. CAD model of variable lead variable clearance rotor geometry

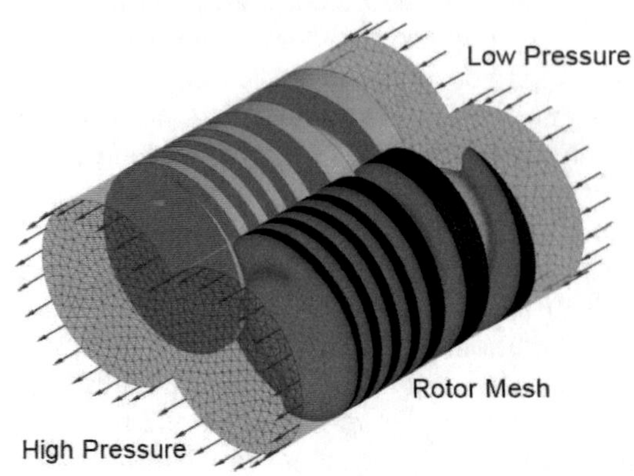

Figure 9. CFD model grid and boundaries of the fluid geometry

3.1. Pressure Distribution

Figure 10 shows the pressure variation in rotor pair 1, having uniform clearances for the full set of suction pressure conditions. It is seen that the distribution of pressure follows a stepped variation corresponding to the rotor lobes. As the suction pressure is lowered from 75000Ps to about 10000Pa, a noticeable pressure drop is observed. But beyond 10,000Pa and lower pressure there is very insignificant drop in pressure.

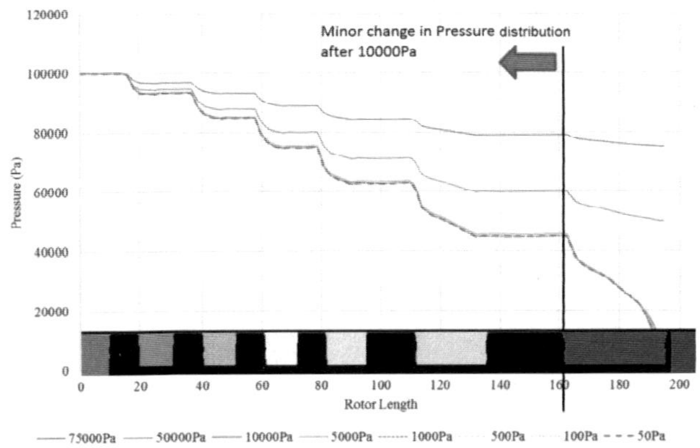

Figure 10. Comparison of pressure distribution along the length of rotor pair 1 for a set of suction pressure conditions

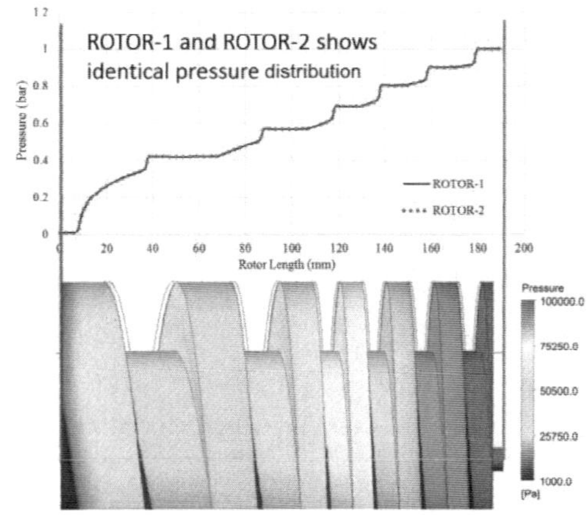

Figure 11. Comparison of pressure distribution on rotor pairs 1 and 2 at 1000Pa suction pressure condition

Figure 11 shows a comparison of rotor pair 1 with uniform clearance, and rotor pair 2 with variable clearance caused by the manufacturing deviation. It can be seen that both rotors produced identical pressure distribution.

3.2. Leakage flow

Figure 12 shows the comparison of leakage flow through the rotor pairs 1 and 2 under a set of reducing suction pressure conditions. In both rotor sets it is seen that initially when the suction pressure drops, the leakage flow increases. But after about 10,000 Pa at the suction end the increase in leakage flow ceases. This trend can be attributed to the effect of choking. It happens in the leakage gaps between the first and second rotor chambers which have large pressure distribution. This flow field trend also corresponds to the observation in pressure distribution results. With non-uniform clearance in rotor pair 2, leakage flow increases by 4.5 – 5.6 % over the entire operating suction pressure range.

Although there is an increase in leakage flow observed with the variable clearance rotor pair 2, volumetric efficiency analysis presented in Table 1 helps in evaluating the magnitude of the impact. Here the efficiency of ideal rotor pair 1 is given a set of probable values from the best case of 95% to a worst case of 70%. As seen the decrease in volumetric efficiency of rotor pair 2 produced by the proposed manufacturing method is very small and even in the worst case only about 1.5%.

(The angular position of the rotor affects the port positions, and this will affect the axial pressure and hence the torque, but due to the large wrap angle, the leakage is unchanged.)

Table 1. Volumetric efficiency analysis for the two rotors

Rotor pair 1 Efficiency	**Inefficiency**	**Effect of 5% higher leakage**	**Rotor pair 2 Efficiency**	**Decrease in Efficiency of Rotor pair 2**
x%	(1-x)%	y=1.05*(1-x)%	z=(1-y)%	(x-z)%
95	5	5.25	94.75	0.25
90	10	10.5	89.5	0.5
85	15	15.75	84.25	0.75
70	30	31.5	68.5	1.5

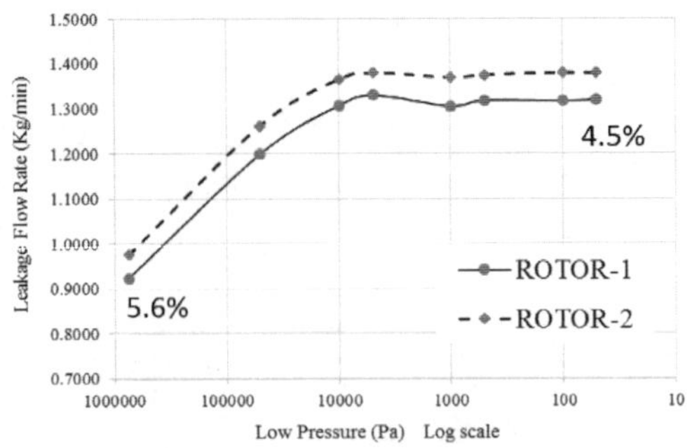

Figure 12. Comparison of leakage flows

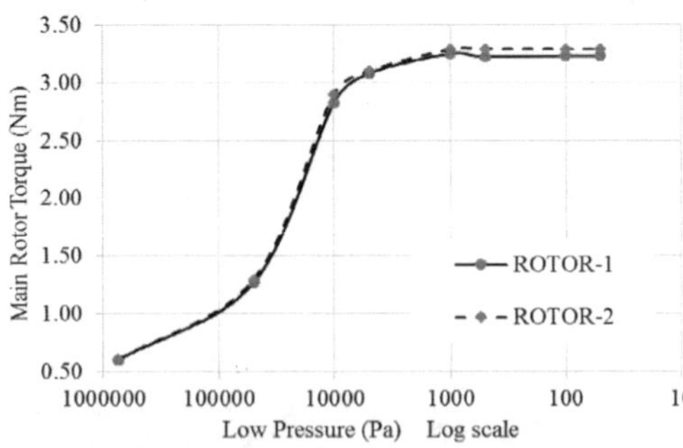

Figure 13. Comparison of torques

3.3. Rotor Torque

Figure 13 presents the comparison of torque on the two rotor types for different reducing suction pressure conditions. The thrust on the rotors is calculated from the integration of the pressure solution. Hence the trend in thrust variation also closely follows the trend in leakage flow. In both the rotors as the suction pressure starts dropping the rotor thrust starts to increase. After about 10,000 Pa suction pressure and further drop, the thrust magnitude ceases to increase. Unlike the observation in leakage flow, the thrust and also axial force variation (not presented here) remained identical between rotor pairs 1 and 2, signifying that indicated power will be of similar order in both the rotors.

In summary, the CFD model was able to predict the leakage flow, pressure distribution, the axial thrust and pressure torque acting on the rotors under different suction pressure conditions. With rotating effects in real operation, flow conditions will have differences and the presented results should be used only for a relative comparison. A reduced volumetric efficiency due to increase in leakage caused by manufacturing deviation is likely, but as presented in Table 1 the relative impact is very low.

4. Conclusions

It is shown that with careful calculation and iterative design of tool paths it is possible to use a profiled tool designed for a uniform lead rotor to manufacture a rotor (or rotor segment) having non-uniform lead. The clearances between such rotor pairs deviate slightly from the design clearances, and the effect on the performance of the two rotor types has been compared using CFD. The predicted differences are very small, suggesting that the disadvantages of using a full-profile disc-type tool process for variable lead rotor production are negligible, whilst the economic advantages may be significant, especially where volume production is required. The CFD modelling inevitably used some assumptions and simplifications, and experimental verification would now be valuable.

The above work relates to multi-wrap vacuum screw machines with timing gears, and no conclusions may be drawn for oil-injected screw compressors with traditional profiles and variable lead.

References

[1] Davey D, Method of machining a rotor with variable-lead screw, WO 2014091232 A1, 2014
[2] Kovačević A., Stošić N. and Smith I. K., 2007. Screw compressors - Three dimensional computational fluid dynamics and solid fluid interaction, ISBN 3-540-36302-5, Springer-Verlag Berlin Heidelberg New York.
[3] Rane S, Kovačević A., and Stošić N., 2016. CFD Analysis of Oil Flooded Twin Screw Compressors, International Compressor Engineering Conference – Purdue. Paper 2392.
[4] Rane S., Grid Generation and CFD analysis of Variable Geometry Screw Machines, 2015, PhD Thesis, City, University of London, URL: openaccess.city.ac.uk/id/eprint/15129.
[5] Vande Voorde, J, Vierendeels J., and Dick E., 2004. A Grid Generator for Flow Calculations in Rotary Volumetric Compressors. In *Proceedings of the Fourth European Congress on Computational Methods in Applied Sciences and Engineering*.

Determining of actual profile clearances and screw compressor rotor positions depending on working conditions

T N Mustafin[1], R R Yakupov[1], M S Khamidullin[1], I G Khisameev[1], L H Uybekova[1], O Y Paranina[1,2]

[1]Kazan National Research Tehnological University, Russian Federation;

[2]HMS Group NIIturbokompressor, Russian Federation

Email: mustimur@rambler.ru

Abstract. In general, screw compressor characteristics depend on their working chamber clearances. One of the main of clearances in the working chamber of screw compressors are the clearances between rotors' profiles, which are produced due to a reducing of theoretical rotor profiles. This reducing is made to compensate thermal deformation, manufacturing errors and other negative factors, which occur during the screw compressor working cycles and may cause a rotors' teeth interference. This also may result in a transmission error, which increases the rotor profile clearance on one side of rotor tooth and reduces on other. A screw compressor rotor gearing mathematical model which can predict the actual rotor position and value of the transmission error was developed. An analysis of compressor working conditions influence on the transmission error value is done. Results of the presented work are used for working process mathematical models for improving their accuracy. They also may be used as basis for development of reducing theoretical profiles methods with the aim to reducing of the rotor profiles' clearances and improve screw compressor characteristics.

1. Introduction

The profiled rotor clearances are one of determining values that influence the energy characteristics of screw-type compressors. Many works are devoted to their determination and research [1–4]. Generally, they also determine the smoothness of transmission and affect to a certain extent the values of vibration. However, it should be noted that the majority of works determine these clearances in the rotors nominal position without taking into account the possible transmission error. Thus, the neglect of this factor results in overstating the clearances, on the one hand, and in understating the error introduced in the calculation, on the other hand. The calculation procedure itself of profiled clearances remains unchanged, while the calculation procedure of the relative position of the rotors requires clarification. Just this is the subject of this article.

2. Characteristic of rotors in operating mode of the transmission process

The positions of a free rotor are determined by the extreme positions upon contact of rotors along one of the sides or by some intermediate positions in the transition to one or another extreme position. For their determination, it is necessary to determine the frames, wherein a free rotor has the liberty to rotate until the contact with the female rotor is established (transmission error). This problem was considered in [3, 4]. In order to determine the shift angle of the female rotor until the contact with the free rotor is established, we consider the position of the rotors when the male rotor is rotated at an angle θ_1 (figure 1), whereby the female rotor, respectively, should be rotated nominally at an angle:

$$\theta_2 = \theta_1 \frac{z_1}{z_2}, \tag{1}$$

where z_1, z_2 – the number of teeth of the male and female rotors.

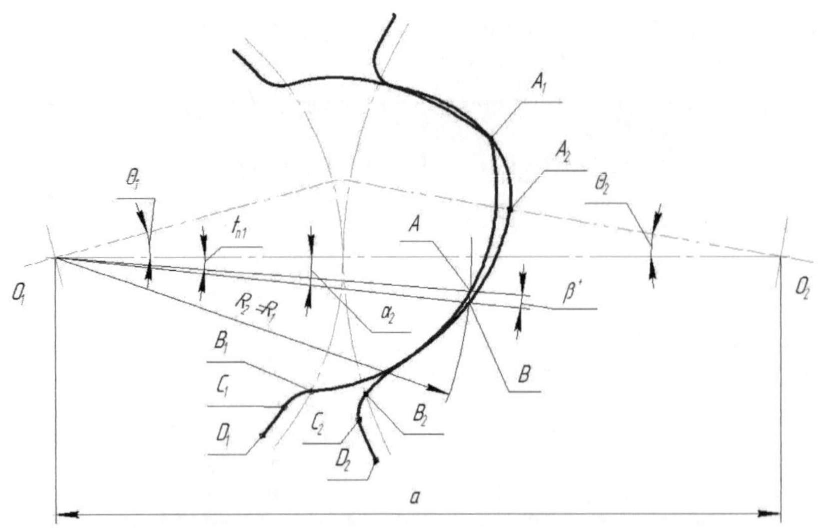

Figure 1. Diagram of shift angle determination of the male rotor until contacting the female rotor

Select in the A_1D_1 segment the arbitrary point A with the angular coordinate α_1. Find on the front part of the profile of the female rotor (A_2D_2 segment), rotated by the angle θ_2, the point B, whose radial coordinate R_2 is equal to R_1. The difference β of the angular coordinates of points A and B will give the value of the additional-turn angle of the male rotor until the point A of the female rotor profile is contacted

$$\beta = \alpha_2 - t_{n1}. \tag{2}$$

After finding the minimum of the function (2) by the variable α_2, we will obtain the value of the additional turn of the male rotor until its contact with the female rotor profile for the considered position:

$$\beta^{+}_{\min \alpha} = f(\theta_1). \tag{3}$$

From the relation (3), determine the minimum shift angle of the male rotor $\beta^{+}_{\min}$ for the entire rotation range θ_1 within the existence of the contact line along the front side of the rotor tooth profile.

Similarly, the additional-turn angle $\beta^{-}_{\min}$ is calculated.

The rotor profile should be accepted in the calculation by taking into account the thermal deformations, whose method of calculation is illustrated in details in [3, 4]. The thermal load on the compressor rotors is cyclical. However, the period of a single cycle is short enough, so the temperature pattern of the rotor remains practically unchanged over time. Then, the thermal state of the rotors is determined by the stationary heat-transfer equation:

$$\nabla^2 T = 0 , \tag{4}$$

where T – temperature, ∇ – nabla operator. The thermal load on compressor rotors is cyclical.

The boundary conditions for the equation (4) are determined by the calculation model (figure 2) on the basis of equation $\lambda_M \frac{\partial T}{\partial n} = q$, where λ_M – thermal conductivity of the rotor material, $\frac{\partial T}{\partial n}$ – derivative on the normal to the rotor's surface affected by the heat flow q.

Thereby, the heat gains resulting from the heat transfer of the gas-oil medium with the rotors' surface in the working space should be separated from the heat transfer of the end surfaces of the rotors with gas-oil medium in the increasing spaces through the end surfaces of the suction and discharge ports; the heat gain resulting from the friction of the rotors in the gas-oil medium in the end and radial clearances, as well as the heat gain along the shaft areas from the bearings. The procedure of their determination is presented in [5, 6].

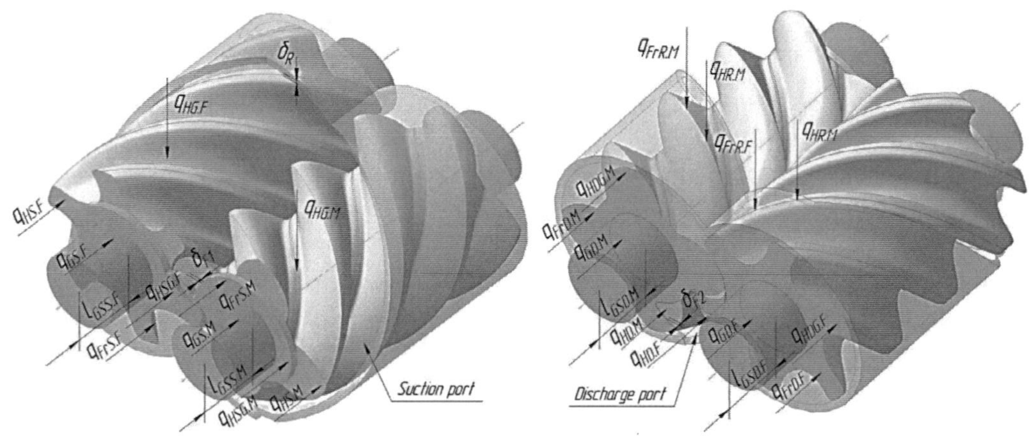

Figure 2. Calculation model for determining the boundary conditions

Equations describing the ratio of motions and strains were used or calculation of thermal deformations:

$$\left(\varepsilon_X - \beta_R \cdot (T - T_{St})\right) = \frac{\sigma_X - \nu \cdot (\sigma_Y + \sigma_Z)}{E}, \tag{5}$$

$$\left(\varepsilon_Y - \beta_R \cdot (T - T_{St})\right) = \frac{\sigma_Y - \nu \cdot (\sigma_X + \sigma_Z)}{E}, \tag{6}$$

$$\left(\varepsilon_Z - \beta_R \cdot (T - T_{St})\right) = \frac{\sigma_Z - \nu \cdot (\sigma_X + \sigma_Y)}{E}, \tag{7}$$

$$\gamma_{XY} = \frac{\tau_{XY}}{G}, \tag{8}$$

$$\gamma_{XZ} = \frac{\tau_{XZ}}{G}, \tag{9}$$

$$\gamma_{YZ} = \frac{\tau_{YZ}}{G}, \tag{10}$$

where x, y, z – direction of the coordinate axes; β_R – coefficient of volumetric expansion of the rotor material, G – elasticity modulus of the rotor material in shear; E – elasticity modulus of the rotor material under tension; ν – Poisson ratio; σ, τ – normal and shearing stresses, accordingly; ε, γ – relative elongation and relative shear deformation, T_{St} – standard rotor temperature; it was accepted to be equal to 293 °K. Thus, the absence of axial movement of the rotor mid-section and the absence of rotors deformation by external forces (they were neglected because they were small [5, 6]) were set as boundary conditions.

3. Dynamics of the free rotor and the possibility of its recovery from direct contact with the female rotor

The change in the additional-turn angle of the free rotor leads to it being forced to rotate with certain acceleration. This can be caused both by a direct contact of the rotors and by gas forces rotation torque. The screw-type compressor running at nominal rating typically encounters the last one: when driving by the male rotor - additional turn of the female rotor until the contact with the main rotor in the direction of rotors rotation. The value of this additional-turn angle can be calculated by the foregoing procedure. Therefore, for the rotors to continue being in transmission, the possible acceleration of rotor due to gas forces must exceed the required acceleration related to the variation of the additional-turn angle. This is performed upon the following condition:

$$\frac{d^2\theta_2}{d\tau^2} \geq \frac{d^2\beta_2}{d\theta_1^2} \cdot \omega_1^2 . \tag{11}$$

The approach to determining $\frac{d^2\theta_2}{d\tau^2}$ is considered in [5]; the following development of this approach is presented further on. From the dynamics equation, it can be written:

$$\frac{d^2\theta_2}{d\tau^2} = \frac{M_{GF} + M_{AF} - M_R - M_{F1} - M_{F2} - M_{MEC}}{J_2}, \tag{12}$$

where J_2 – moment of inertia of a free rotor; M_{GF} – the moment produced by gas forces is determined on the basis of the indicator (P-V) diagram by using the "subtend chords" method [3, 7]; M_{MEC} – moment of resistance of the mechanical components of the compressor such as seals, balance pistons and bearings (its value is mainly determined by the design of the mechanical components of the rotor); M_R, $M_{F1,2}$ – friction torques in the compressed medium in the radial and axial clearances, respectively; M_{AF} – moment created by adhesion forces. It is not difficult to show that in this case the term "adhesion" may be replaced by the term "cohesion" that is close to it. Then, this moment will be proportional to the surface tension coefficient, the contact area and the wetting angle; at the same time it should be noted that this moment will be much less than the other moments of forces, which in turn permits to ignore them in the further calculations: $M_{AF} \approx 0$.

According to the calculation diagram, the moments of friction in the compressed medium in the end clearances depend on the operating mode:

$$M_{F1,2} = \begin{cases} \dfrac{\pi \cdot \omega_2 \cdot \eta_{MIX} \cdot R_{eq}^4}{2 \cdot \delta_{F1,2}} \cdot \left(1 - \dfrac{r_R^4}{R_{eq}^4}\right), & if \quad Re_{F1,2} < 10^4 \\ \dfrac{\rho_{MIX} \cdot \omega_2^3}{4} \cdot \left(c'_{M1,2} \cdot R_{eq}^5 - c''_{M1,2} \cdot r_R^5\right), & if \quad 10^6 \geq Re_{F1,2} \geq 10^4 \end{cases}, \tag{13}$$

where $c_{\mathrm{M}1,2} = 0{,}0277 \cdot \mathrm{Re}_{\mathrm{F}}^{-0,2} \cdot \left(\frac{\delta_{\mathrm{F}1,2}}{R}\right)^{-0,2}$ – friction moment coefficient; for determining c''_{M} as the characteristic value, r_{R} $(R = r_{\mathrm{R}})$ is used, in all other cases – R_{eq} $(R = R_{\mathrm{eq}})$; $\mathrm{Re}_{\mathrm{F}} = \frac{R^2 \cdot \omega_2 \cdot \rho_{\mathrm{MIX}}}{\eta_{\mathrm{MIX}}}$ – Reynolds number; $\omega_2 = \omega_1 \cdot \frac{z_1}{z_2}$ – nominal angular speed of a free rotor; ρ_{MIX}, η_{MIX} – respectively, the density and kinematic viscosity of the gas-oil mixture in the working space of the compressor; $\delta_{\mathrm{F}1}$, $\delta_{\mathrm{F}2}$ – end clearances between the rotor and the housing on the suction and discharge sides, respectively. The equivalent radius is determined as $R_{\mathrm{eq}} = \left(\frac{1}{2 \cdot \pi} \int_0^{2\cdot\pi} \left(r(t;0)\right)^4 \cdot dt\right)^{\frac{1}{4}}$, where $r(t;0)$ is determined for a certain fixed rotation angle θ, for which it is expedient to take the initial position: $\theta = 0$.

Similarly, the friction torques in the compressed medium in the radial clearance can be written as:

$$M_{\mathrm{R}} = \begin{cases} \dfrac{2 \cdot \pi \cdot \xi \cdot \omega_2 \cdot \eta_{\mathrm{MIX}} \cdot R_{2*}^3 \cdot z_2 \cdot b_{\mathrm{R}}}{(1+\xi) \cdot \delta_{\mathrm{R}}}, & if \quad \mathrm{Re}_{\mathrm{R}} < 2500 \\ \pi \cdot c_{\mathrm{R}} \cdot \rho_{\mathrm{MIX}} \cdot \omega_2^2 \cdot z_2 \cdot b_{\mathrm{R}} \cdot R_{2*}^4, & if \quad 10^5 \geq \mathrm{Re}_{\mathrm{R}} \geq 2500 \end{cases}, \tag{14}$$

where $c_{\mathrm{R}} \approx 0{,}0076 \cdot \mathrm{Re}_{\mathrm{R}}^{-0,25}$ – friction coefficient, $\mathrm{Re}_{\mathrm{R}} = \frac{\omega_2 \cdot R_{2*} \cdot \delta_{\mathrm{R}} \cdot \rho_{\mathrm{MIX}}}{\eta_{\mathrm{MIX}}}$ – Reynolds number, b_{R} – rotor tooth thickness in the direction of Z axis, R_{2*} – outer diameter of the rotor, δ_{R} – radial clearance between the rotor and the housing of the compressor, $\xi = \frac{m_{\mathrm{OIL}}}{m_{\mathrm{GAS}}}$ – gas-oil ratio.

The thermodynamic parameters of the mixture are determined by the following relations:

$$\eta_{\mathrm{MIX}} = \left(\frac{1}{1+\xi}\right) \cdot \eta_{\mathrm{GAS}} + \left(\frac{\xi}{1+\xi}\right) \cdot \eta_{\mathrm{OIL}}, \tag{15}$$

$$\rho_{\mathrm{MIX}} = (1+\xi) \cdot \rho_{\mathrm{GAS}}. \tag{16}$$

Thereby the suction temperature for the clearance on the suction side and the discharge temperature for the clearance on the discharge side can be taken as a determining temperature for the equation (19) in the preliminary calculations. An arithmetic average temperature between suction and discharge can be taken as a determining temperature for the equation (14). In order to simplify the calculations, the value ξ can be assumed constant for all parts of the rotor.

In compliance with the inequality (11), the design angle of rotation of the female rotor is determined as:

$$\theta_2 = \theta_1 \frac{z_1}{z_2} + \beta_{\min a}^{+} \tag{17}$$

At the moment, when the compliance with the inequality (11) ends, the position of the female rotor should be determined by integrating the differential equation (12) with the following boundary conditions, taking the time of separation as a reference point:

$$\begin{cases} \theta_2(0) = \theta_1 \dfrac{z_1}{z_2} + \beta^{+}_{\min a} \\ \dfrac{d\theta_2}{d\tau}(0) = \omega_1 \dfrac{z_1}{z_2} \end{cases}. \quad (18)$$

The relation between the integration time τ and the angle of rotation of the male rotor is determined by the expression

$$\omega_1 = \frac{d\theta_1}{d\tau}. \quad (19)$$

The integration is carried out until the rotors contact again. Thereby, at the moment of contact due to the elasticity of impact a rebound will occur of the rotors relatively to each other. However, its value is determined by virtue of the low relative velocity of the contact points and the high value of the damping capacity of the greasy layer. Therefore, the conditions for continuation of the integration will have the form:

$$\theta_1 \frac{z_1}{z_2} - \beta^{-}_{\min a} < \theta_2 < \theta_1 \frac{z_1}{z_2} + \beta^{+}_{\min a}. \quad (20)$$

When coming into contact on the back side of tooth, the recess condition (11) will change its sign to the opposite:

$$\frac{d^2\theta_2}{d\tau^2} \le \frac{d^2\beta_2}{d\theta_1^2} \cdot \omega_1^2 \quad (21)$$

It should be noted also the fact that, under normal operating conditions, the rotors loss of contact occurs only at the moment of teeth disengagement. In this case, the dynamics of the female rotor in a free state can be neglected and its position is determined by the maximum angle of the possible deviation, which is the additional-turn angle.

4. Example of the use of the presented procedure and analysis of the obtained results

A standard screw compressor was selected as the object of the research, designed in V.B. Shnepp NIIturbokompressor (HMS GROUP) CJSC (the successor of SKBK (Special Boiler Design Bureau)) with 250 mm rotor diameter, 340 mm length of the profiled section of the rotors, 440 mm and 660 mm screw pitch for the male and female rotors, respectively, and transmission ratio of 4/6, which is analogous to the profiles described in [8]. The following values were taken for end clearances between the rotor and the housing: on the suction side – 0.5 mm, on the discharge side – 0.06 mm. The value of 0.2 mm was taken as the radial clearance. As a mode being studied, the following parameters were selected: the mode of compressor operation in the air with compression rate from 1 to 9 at (pressure ratio P = 9), suction temperature 25 °C, discharge temperature 90 °C, temperature of the injected oil 40 °C and volume ratio equal to 4.5. The gas-oil ratio was taken equal to 3.8.

The calculation results of the temperature patterns of the female rotor, as well as the dependency diagrams characterizing the transmission in the compressor operation, are shown in figure 3 and figure 4. The calculations are presented for the case where the compressor design does not provide a suction port on the radial housing boring

IOP Conf. Series: Materials Science and Engineering **232** (2017) 012023 doi:10.1088/1757-899X/232/1/012023

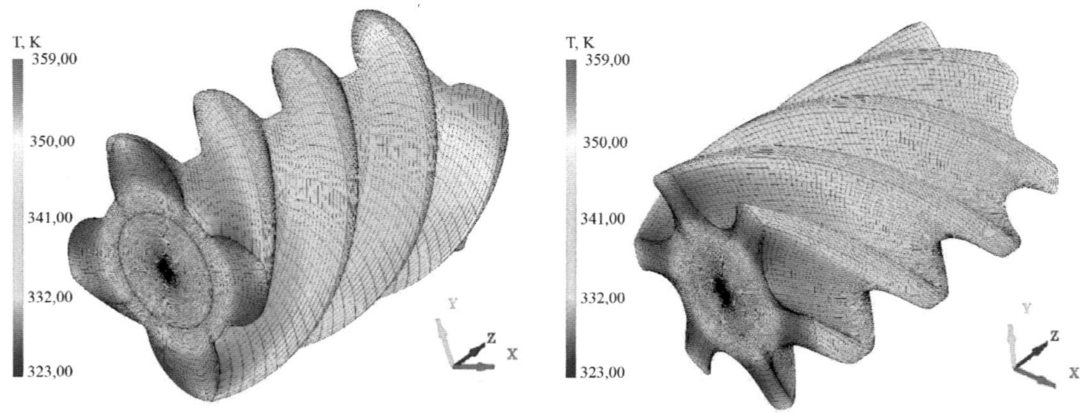

Figure 3. Design temperature patterns of the rotors

The dependences characterizing the transmission are shown in figure 5. This figure shows that the second derivative of the shift angle has two gaps. The first gap is at the moment of teeth relocation in the transmission, the second one – at the moment when the contact point starts running along the rotor axis (at the initial moment the contact occurs in the initial cross section of the rotor). Analyzing this diagram, it can be concluded that the process of contact loss in transmission for the mode being studied occurs only at the moment of teeth relocation. For the rest of the time, the female rotor is pressed to the male rotor. It provides, in turn, that the required adjustment of the female rotor rotation angle, characterizing the transmission error, is determined by the $\beta^{+}_{\min\alpha}$ values. The dependence in its final form is shown in figure 6.

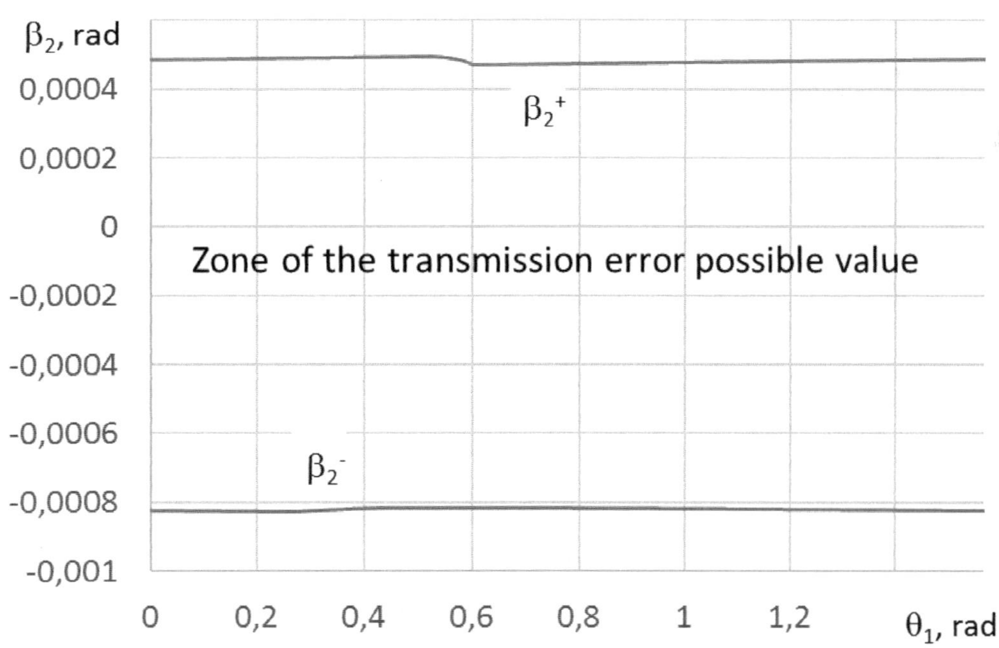

Figure 4. Possible deviations of the rotation angle of the female rotor

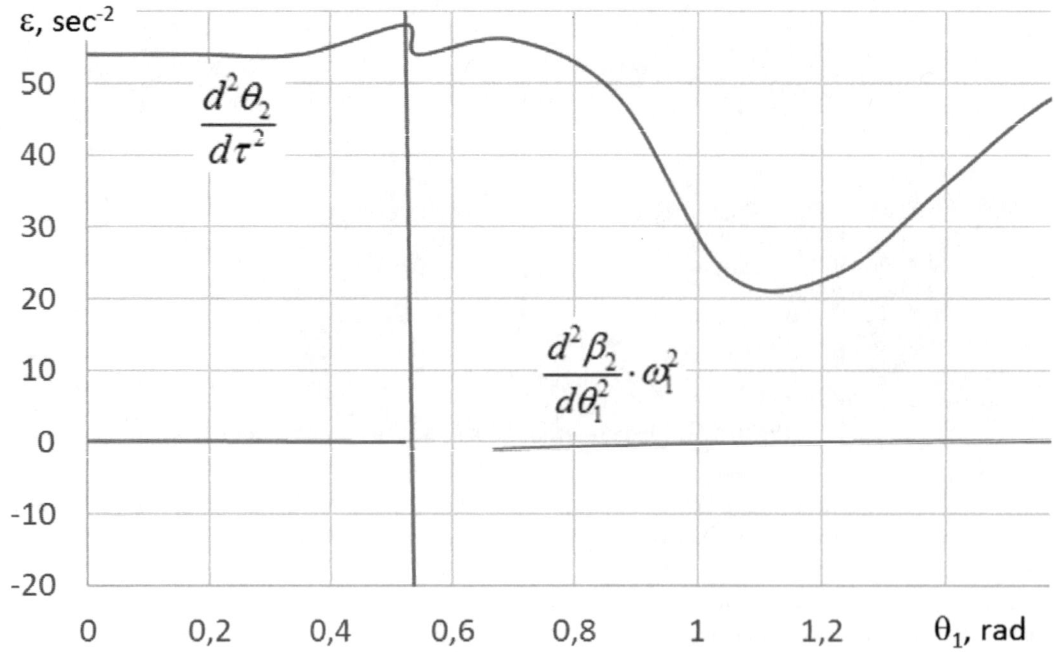

Figure 5. Transmission characteristic curve

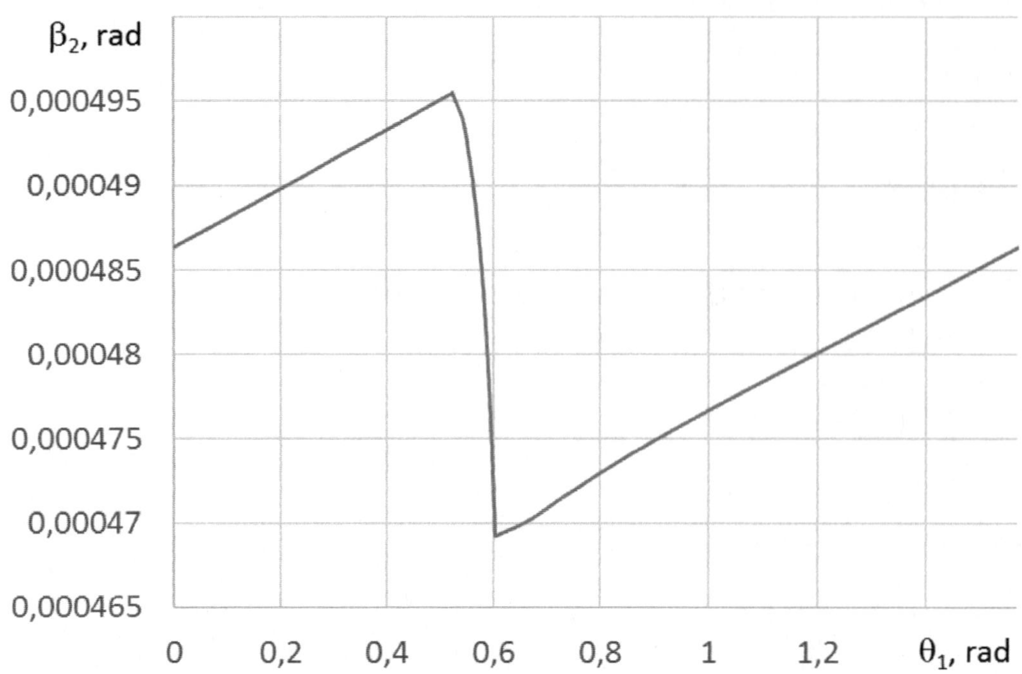

Figure 6. Real deviation value of the rotation angle of the female rotor

IOP Conf. Series: Materials Science and Engineering **232** (2017) 012023 doi:10.1088/1757-899X/232/1/012023

5. Conclusion

The presented methodology permits to do forecasting by calculating the real position of the rotors in the transmission. Its development may clarify the calculation of the operating process of the screw-type compressor and form the basis for calculation of the contact stresses and for vibration forecasting. At the same time, it can be used already at the design study stage.

Acknowledgments

This study is a part of scientific investigations supported and funded by V.B. Shnepp NIIturbokompressor (HMS GROUP) CJSC.

References

[1] Buckney D, Kovacevic A, Stosic N 2015 Rotor clearance design and evaluation for an oil injected twin screw compressor *IOP Conf. Ser.: Mater. Sci. Eng.*, 90

[2] Buckney D, Kovacevic A, Stosic N 2014 Accounting for Local Thermal Distortions in a Chamber Model for Twin Screw Compressors *International Compressor Engineering Conference*. Paper 2306.

[3] Yakupov R R, Mustafin T N, Nalimov V N, Khamidullin M S and Khisameev I G 2013 Discussion of actual profile clearances' calculation method in rotary compressors in the absence of rotor timing units *8th Int. Conf. on Compressors and their Systems (London: City University)* 209

[4] Yakupov R R, Mustafin T N, Nalimov V N, Khamidullin M S and Khisameev I G 2014 Analysis of transmission error depending on compressor working conditions *J of Proc. Mechanical Engineering (Proc. of the Institution of Mechanical Engineers) Part E (London: SAGE Publications)*

[5] Mustafin T N, Yakupov R R, BurmistrovA V, Khamidullin M S, Khisameev I G 2015 Analysis of the screw compressor rotors' non-uniform thermal field effect on transmission error *IOP Conf. Ser.: Mater. Sci. Eng.* 90

[6] Mustafin T N and etc. 2016 Analysis of influence of screw compressor construction parameters and working condition on rotor temperature fields *Procedia Engineering* **152**,

[7] Adams G P and Soedel W 1994 Dynamic simulation of rotor Contact Forces in Twin Screw Compressors *Proc. Int. Compressor Engineering Conference*, (West Lafayette: Purdue University)

[8] Khisameev I G and Maksimov V A 2000 *Dvukhrotornye vintovye I pryamozubye kompressory. Teoriya, raschetiproyektirovanie (Twin rotors screw and spur compressors. Theory, calculation and design (in Russian))* (Kazan: Fen)

IOP Conf. Series: Materials Science and Engineering **232** (2017) 012024 doi:10.1088/1757-899X/232/1/012024

Statistical Capability Study of a Helical Grinding Machine Producing Screw Rotors

C. S. Holmes, M. Headley, P. W. Hart

PTG Advanced Developments, Holroyd, Harbour Lane North, OL16 3LQ, UK

chris.holmes@holroyd.com

Abstract. Screw compressors depend for their efficiency and reliability on the accuracy of the rotors, and therefore on the machinery used in their production. The machinery has evolved over more than half a century in response to customer demands for production accuracy, efficiency, and flexibility, and is now at a high level on all three criteria. Production equipment and processes must be capable of maintaining accuracy over a production run, and this must be assessed statistically under strictly controlled conditions. This paper gives numerical data from such a study of an innovative machine tool and shows that it is possible to meet the demanding statistical capability requirements.

1. Introduction

The performance of a screw compressor depends heavily on the machinery used in the production of the screw rotor pair. The machinery has evolved over more than half a century in response to customer demands for production accuracy, efficiency, and flexibility. To show fitness for purpose, production equipment and processes must be capable of maintaining accuracy over a production run, and this is best assessed statistically under controlled conditions. Some data for the accuracy of such machines has been published [1], [2], but this has not extended to statistical capability.

The performance behaviour depends primarily on the accuracy of the rotor pairs, as described by Sauls [3]. Methods of measuring rotor pairs have been developed which reveal their important characteristics. Specifically, the inter-lobe contact patterns, clearances and transmission errors are the key factors in determining efficiency, reliability, and noise. A full description of these measurement methods is given in [4]. Characteristics of interest include profile shape, depth, outside diameter, lobe divide (pitch), and lobe helix form and lead. Such measurements may also be used as inputs to a pair simulation program in order to give an approximate indication of the pair quality.

For process capability assessment, the measurement of individual rotors, e.g. by CMM, is preferable [5], [6]. The characteristics to be used for the statistical processing will depend on the individual user's preferred methods of specifying tolerances, profile fitting, etc.

2. Machine and part used

The machine used for this study was a Holroyd Zenith 400, a general description of which has been given previously by the author [7]. The model used was the Mk 2 version, in dressable grinding mode. Measurements were made with a Leitz PMM C portal-frame CMM using a rotary table. The compressor rotor dimensions, and the features measured, are as shown in Figs. 1 & 2.

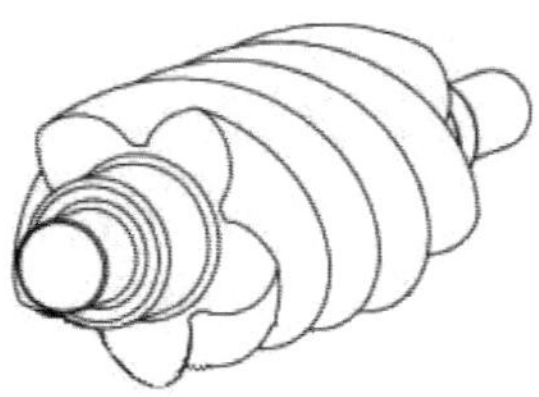

Outside diameter:	220 mm
Pitch circle diameter:	150 mm
Root diameter:	140 mm
Body length:	320 mm
Helix lead, left hand:	340 mm
No. of flutes:	5

Fig. 1. Rotor and main dimensions

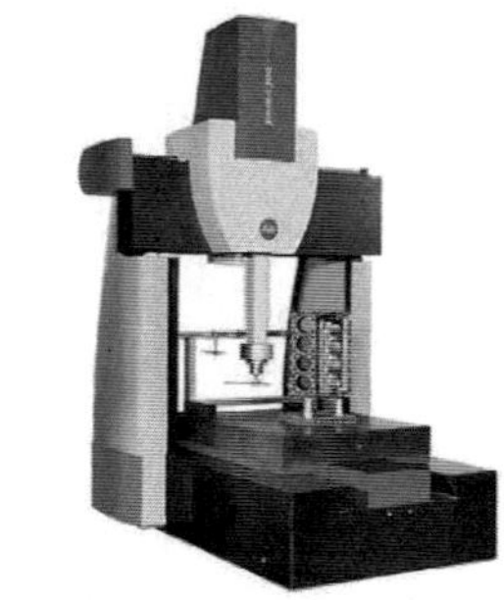

- Mean Profile (5 points)
- Mean Lead (left and right flank independently)
- Divide

Fig. 2. CMM and features measured

3. Statistical Methods

The statistical methods owe much to the mass production developments in the automobile industry, and only came to be applied to the rotor industry during the mid-1990s. Clear definitions and procedures are given in [8].

The process variability (Cp), and the Process Capability (Cpk) are useful in distinguishing between basic repeatability and the centralisation adjustment. Both are dimensionless ratios relating the available tolerance on a feature to the variability achieved. Assuming the underlying data is normally distributed, the definitions are as follows (see also Fig. 3):

Cp = Total tolerance band / (6 x standard deviation)

Cpk adapts this to the condition where the distribution of achieved values is not centered in the tolerance band, and is defined as:

Cpk = Margin from mean achieved value to closest tolerance limit / (3 x standard deviation)

In both cases, a higher number represents a more closely-controlled process. A value of 1 corresponds to a 0.26% probability of a given part being out-of-tolerance. A value of 1.33 is often used as a pass/fail criterion, representing a probability of error of 6 parts per 100,000.

For most dimensional features of a rotor, the feature can deviate above or below the target value and normal distribution is common. There are well-defined and documented tests for normality of data, and methods for converting non-normal data into alternative forms which can be analysed as normal [8],[9]. However, there are parameters such as surface finish, or divide as described below, which have a lower bound of zero. These cannot be analysed in the same simple manner.

Capability trials follow procedures which are designed to reveal the behaviour of the process only and avoid extraneous influences. Infinite batch sizes may be desirable but are not feasible, and the specialist statistical software (Minitab) applies a statistical compensation for the actual batch size. Unnecessary interventions such as modifications to machine input settings are to be avoided, unless these are an essential part of the designated process. Also, abnormal events which have a clearly identifiable cause, initial warm-up parts, or a faulty part being used in error, should be excluded from the analysis. A careful log of the run is essential for this to be done legitimately. The log should include part number, time started, blank condition, and any observed events before, during, or after the cycle.

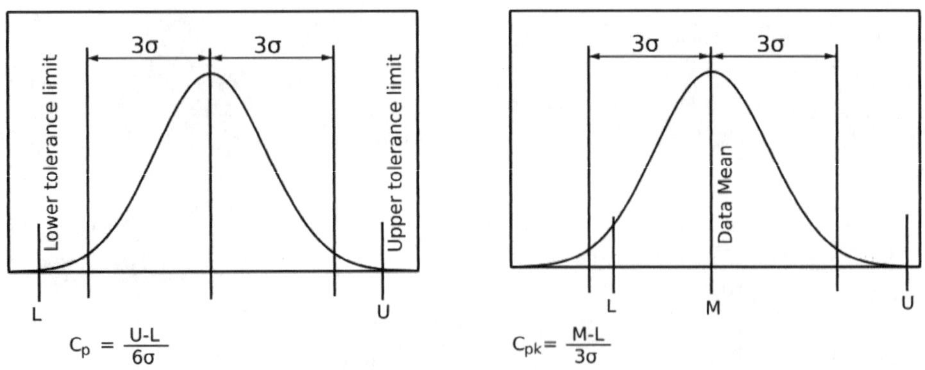

Fig. 3. Graphical illustrations of the basis for capability formulae

4. Conditions for the production run

In this instance the batch of parts was a production batch, and the additional controls were pared to a minimum to facilitate the process. In particular:

1. The machine was not in a temperature-controlled environment.
2. The order of flutes during machining and measurement was kept consistent.
3. Machined workpieces were stabilised in a temperature-controlled environment of 20°C ±1°C for 24 hours before measurement.
4. Profile errors were measured normal to the surface. Profile deviations were then minimised by a single bodily rotation, or 'best fit'.
5. Blank run-out was measured before and after the grinding.

As permitted under the procedure, two initial warm-up parts were dropped from the analysis regardless of whether they were within tolerance. Also, the run was unavoidably interrupted, and a second warm up was required, with those parts also being omitted.

Unfortunately, time did not permit the ordering of the ideal wheel type for the cast iron material, and the actual wheel used was prone to 'clogging' or 'loading'. Nevertheless, the results will be presented here for their learning value.

5. Results

5.1. Profile

A typical CMM scan of a profile is shown in Fig. 4, with points used in Fig. 5. Errors are measured normal to the metal surface. Profile data for the run is shown in Fig. 6, the error distribution in Fig. 7, and profile capabilities in Table 1. With one exception, the capability results are well above the 1.33 target, showing excellent machine suitability for consistent rotor production quality. The Cpk value for the root of the profile was slightly below the target, and this is revealing. On each rotor there was a progressive change in the profile depth from flute to flute, which followed the same order as the grinding sequence, and was quite consistent throughout the batch. This was diagnosed as being the result of 'push-off' of the rotor body as the wheel loaded with debris. This also affected the divide values.

Fig. 4. Transverse profile

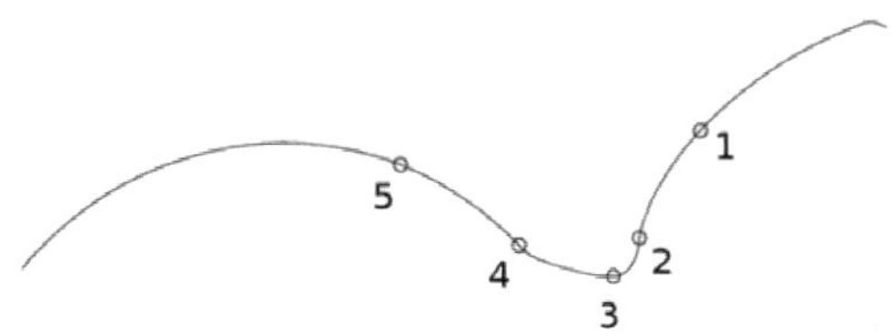

Fig. 5. Profile points

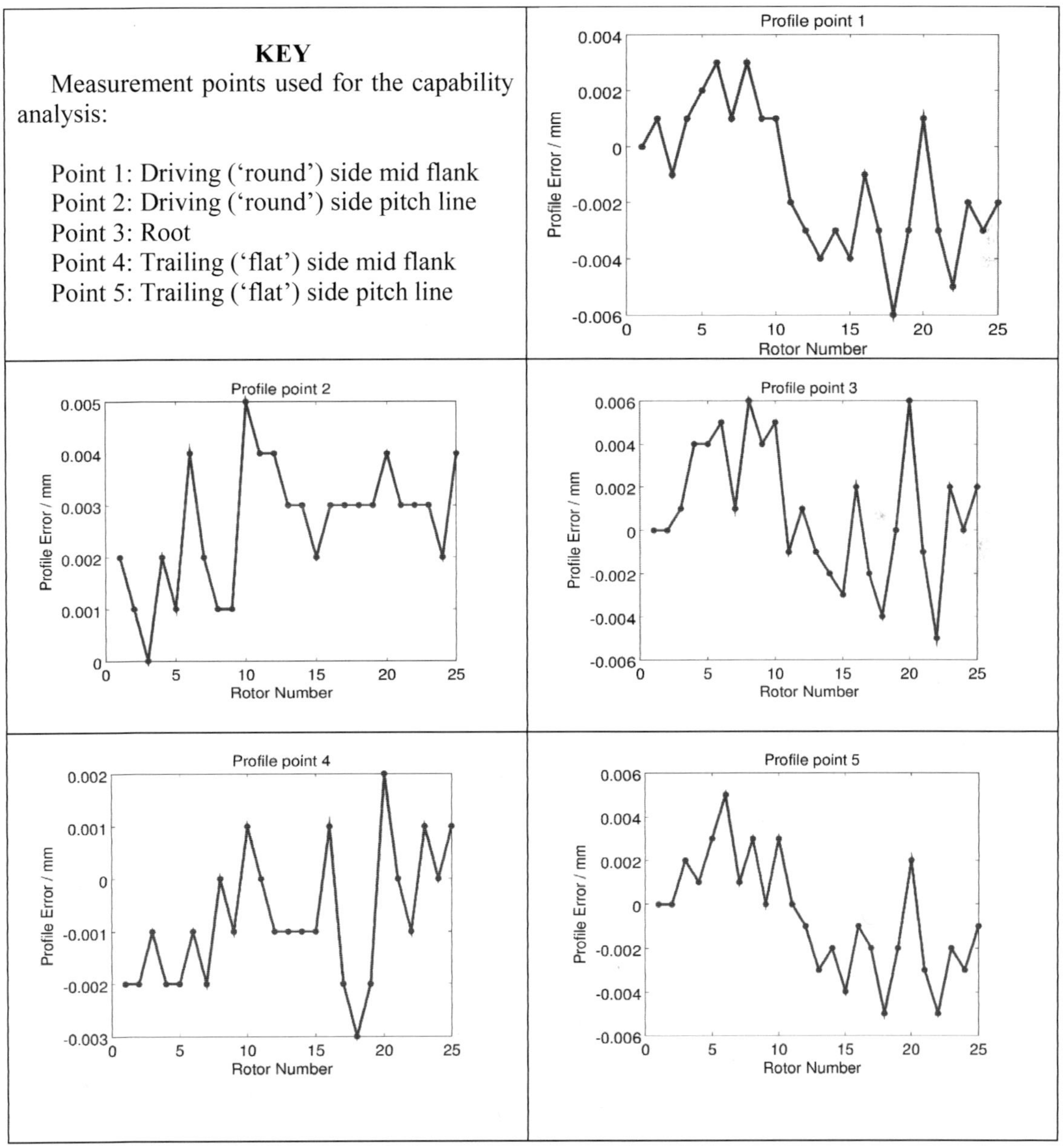

Fig. 6. Profile data

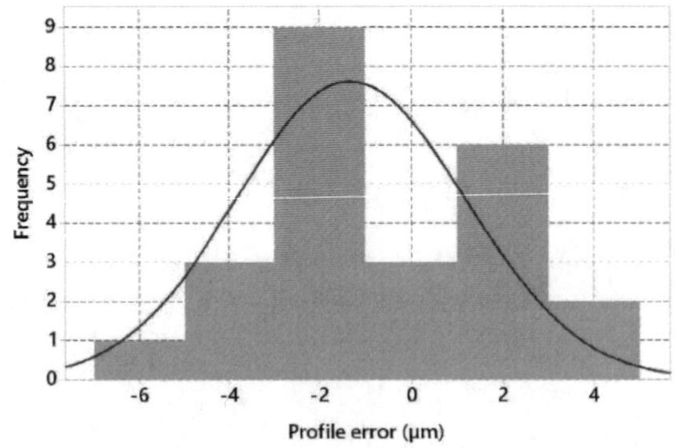

Fig. 7. Error frequency distribution for profile point 1 data set

Table 1. Profile capabilities

Profile point & Tolerance	*Cp*	*Cpk*
1. Profile Mean, Round side, Mid flank, ±12.5µm	2.40	2.15
2. Profile Mean, Round side, Pitch radius ±12.5µm	4.91	3.85
3. Profile Mean, Root, ±12.5µm	1.33	1.23
4. Profile Mean, Flat side, Pitch radius ±12.5µm	3.60	3.39
5. Profile Mean, Flat side, Mid flank ±12.5µm	1.97	1.87

5.2. Lead

Lead error is derived from surface normal deviations from the theoretical helix over the length of the flute at the pitch radius. The lead was scanned on each flank independently and the results for each flank were averaged over all flutes. Lead error Ha was the parameter chosen for analysis, being the slope error of the best-fit straight-line through the data. It must be remembered that the normal errors are magnified twice: firstly by projection of the surface normal error into the axial direction, and secondly by further axial projection from the body length to a 'full wrap' of the helix over 360 degrees. Thus surface errors might be 3 times larger when expressed as lead.

CMM scans of all flutes on one rotor are shown in Fig. 8. The raw data and frequency distribution are shown in Fig. 9 and Fig. 10. The capability values as shown in Table 2 are high, and this will have beneficial results for the noise behaviour of the rotors, as described in [10].

IOP Conf. Series: Materials Science and Engineering **232** (2017) 012024 doi:10.1088/1757-899X/232/1/012024

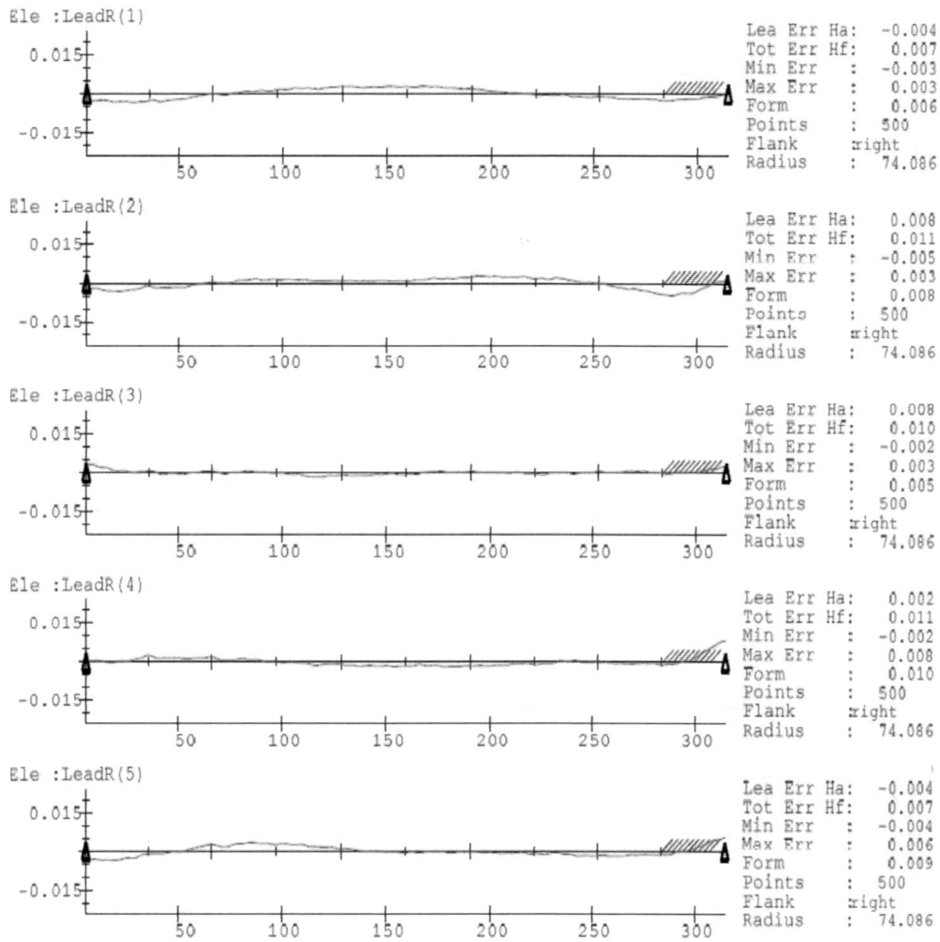

Fig. 8 Lead scans for all lobes on a typical rotor

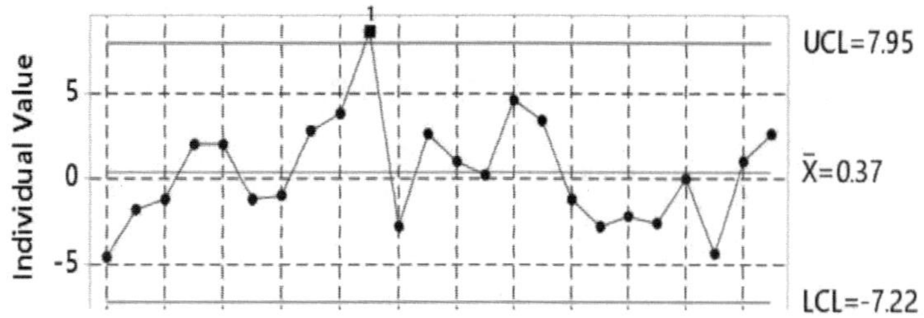

Fig. 9. Round side lead data

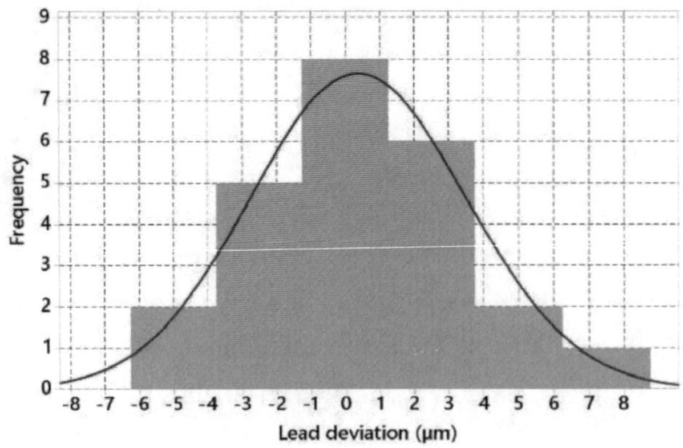

Fig. 10. Round side lead error frequency analysis

Table 2. Lead capability

Lead measurement position	*Cp*	*Cpk*
Lead Mean, Round side. Tolerance ±15μm	1.98	1.93
Lead Mean, Flat side. Tolerance ±15μm	1.95	1.53

5.3. Divide (Angular Pitch)

Divide is determined from one selected point in the smoothed-profile of each flute. The parameter analysed is fu, the total range of individual pitch deviations measured tangential to the pitch circle in the transverse plane. This is an absolute value and not normally-distributed, so the standard calculation of Cpk can therefore not be used. As explained above, the divide was adversely affected by the wheel being unsuitable for the cast iron material. The results are shown in Fig. 11. This may be compared with data collected from a different trial using the same machine, but with a grinding wheel suited to the rotor material, as shown in Fig. 12.

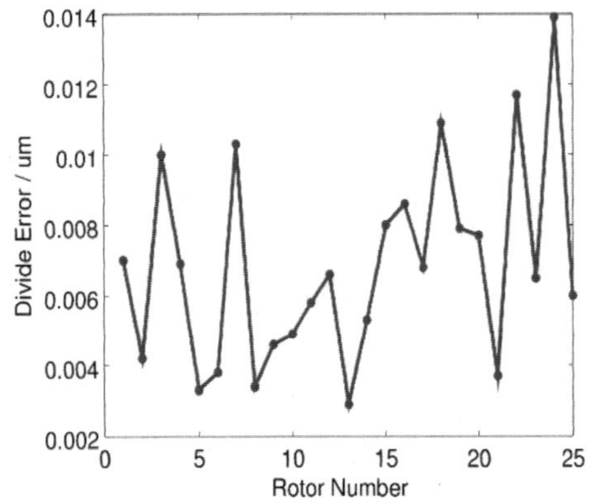

Fig. 11. Divide data

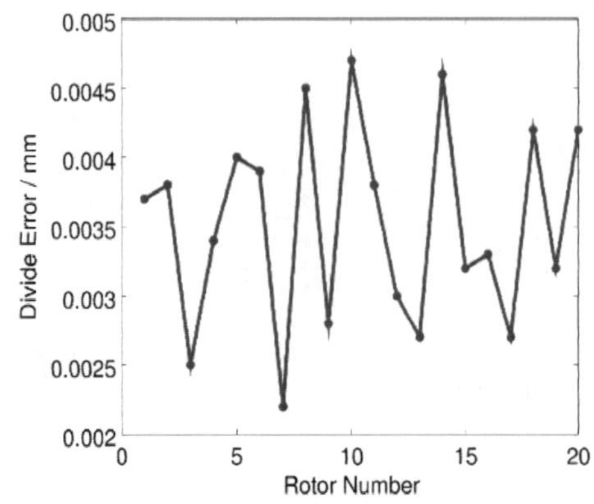

Fig. 12. Divide data from a different trial, without wheel 'loading'

6. Conclusions

Although such testing as described above is time-consuming and costly, the results, in addition to providing the machine user with confidence in its fitness for purpose, can also be very revealing about the many factors which affect the process. Besides wheel selection, the condition of blanks, centre holes, gripping journals, coolant application, thermal compensation, and other factors may be revealed as significant and their effects quantified by comparing different trials.

In this case, the lead and profile capabilities of the machine itself proved more than ample for the compressor application, but the wheel used for the trial was found to be unsuitable because its pores were prone to clogging by the cast iron material. Tests with a more porous wheel using the same machine but a different rotor size gave good results, the lesson being that the performance of a capable machine tool can be degraded significantly by an unsuitable tool.

References

[1] Holmes, C. S., September 2007. Form Grinding of Precision Helical Rotors and Gears Using Closed-loop Profile Control.
Proceedings of the 10th ASME International Power Transmission and Gearing Conference.

[2] Lorenz, M., 1994. Fertigung von Rotoren Schraubenmaschinen.
Schrauben-maschinen Proceedings, VDI Berichte 1135, VDI Verlag.

[3] Sauls, J. 1998. An analytical study of the effects of manufacturing variations on screw rotor profiles and rotor pair clearances.
Schrauben-maschinen proceedings, VDI Verlag.

[4] Holmes, C. S. 2004. Inspection of Screw Compressor Rotors for the Prediction of Compressor Performance, Reliability, and Noise.
7th International Compressor Engineering Conference at Purdue. Purdue University, USA.

[5] Holmes, C. S. 1994. Towards a core program for the measurement of screw rotor bodies by co-ordinate measuring machine.
Schrauben-maschinen '94 proceedings, VDI Berichte 1135 p. 331, VDI Verlag.

[6] Muller, H., Wahl, K. Messen, Prüfen und Analysieren von Schraubenkompressoren unter Einsatz eines 3D-Koordinatenmeßgerätes.
*Schraubenmaschien '94.*VDI Berichte 1135

[7] Holmes, C. S., 2011. Developments in the productive grinding of screw rotors and other helical parts in response to user feedback.
7th International Conference on Compressors and their Systems. City University, London. IMechE 2011 ISBN 978-0-85709-208-3

[8] *DataMyte Handbook.*
Allen-Bradley Co, 1992.

[9] Box, G. E. P., Cox, D. R. 1964. An Analysis of Transformations.
Journal of the Royal Statistical Society. Series B (Methodological), Vol. 26, No. 2 (1964).

[10] Matsuo, K, and Holmes, C. S. 2013 Experimental Study of Noise and Vibration Reduction in a Medium-size Oil-flooded Twin-screw Compressor by the Application of Helix Relief
8th International Conference on Compressors and their Systems. City University, London. IMechE 2013 ISBN 978-1-78242-169-6

IOP Conf. Series: Materials Science and Engineering **232** (2017) 012025 doi:10.1088/1757-899X/232/1/012025

Comparative study of bearing loads for different twin screw compressor rotor configurations

D Buckney and C Anderson

Howden Compressors, Glasgow, UK

david.buckney@howden.com

Abstract. Designing rotor geometry is a critical stage in the design of a twin screw compressor which has a significant impact on: capacity; leakage characteristics; thermodynamics; rotor stiffness; dynamics; and loading on the bearings. The focus of this paper is on bearing loads. In order to design screw compressors that can operate at higher pressures the bearings quickly become a limiting factor. With the need to house the bearings adjacent to one another on each of the parallel rotor shafts at a given centre distance there is an inherent limit to the bearing geometry envelope. In this investigation the 'rotor configuration' refers to the rotor lobe combination, length to diameter ratio (L/D), and wrap angle. The geometry of the transverse rotor profiles is kept constant, as far as possible, allowing conclusions to be drawn based on a manageable number of variables. A procedure to calculate bearing specific loads based on results from a thermodynamic chamber model is presented and results for a range of rotor configurations are discussed.

1. Introduction

Efficient procedures for calculation of twin screw compressor rotor forces due to pressure loading are well documented [1, 2]. With the availability of FEA / CFD software packages, numerical approaches to force calculation are also commonplace [3, 4]. The current work is based on a numerical procedure that builds on previous work by the authors [5] that allows the pressure curve from a chamber model to be mapped onto the 3D rotor surfaces. Additional calculation steps to calculate bearing forces are presented here.

This investigation is part of wider ranging design work that compared rotor geometry envelopes, weights, lateral stiffness, torsional stiffness, leakage areas, port areas, thermodynamic performance and forces for various rotor configurations. In this paper selected results are presented in order to demonstrate important considerations for high gas load applications.

In order to allow clear presentation of results a nominal reference rotor configuration will be used with a 4/6 lobe combination, 1.375 L/D, and 300° wrap angle. Each of these three variables will be modified one at a time.

2. Rotor design

2.1. Profile generation

Different profile co-ordinates were generated for each lobe combination. Whilst it is easy to linearly scale any existing profile, changing the number of lobes on each profile can potentially cause unwanted changes to the nature of the profile. Some of the profile parameters that were carefully controlled in this investigation are detailed in Figure 1.

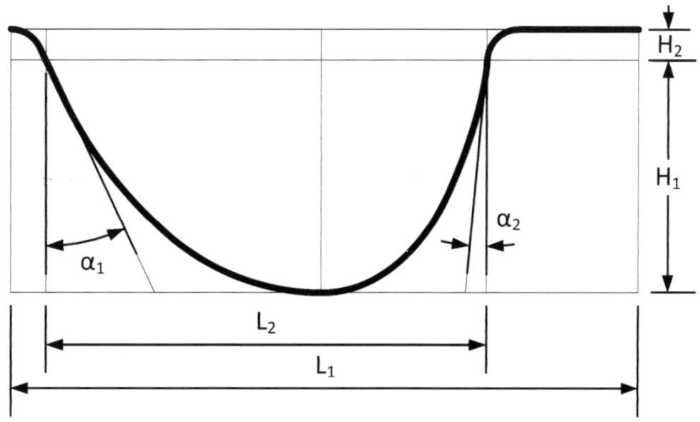

Figure 1. Fixed rotor profile parameters

The parameters α_1 and α_2, which describe the flank angles at the pitch circles, have been kept constant. In order to preserve the relative width of the male and female rotors the ratio L_2/L_1 has been kept constant. This ratio is achieved by adjusting the parameter H_1. When changing the lobe combination, the appearance of the rotors changes a fair amount and preserving the relative width of the lobes at the pitch circles provides a tangible measure that the profile is as similar as possible.

From an optimisation point of view, minimising the parameter H_2 can essentially eliminate the blow hole leakage path. In addition to increasing flow, this parameter is generally non-zero for practical purposes: to maintain a good 'contact' band either side of the pitch radius; and for manufacturability. Therefore it seemed appropriate to keep the ratio H_2/r_{2o} constant, where r_{2o} is the outer radius on the female rotor.

2.2. Rotor normalisation

For each rotor lobe combination the generated profile was used to create rotor bodies for the various combinations of wrap angle and L/D ratio. Rotor models were created using 'SCORG' software [6] in order to calculate rotor characteristics including the volume displacement per revolution of the male rotor, V_c. Each rotor pair was then scaled, or normalised, in order to target V_c equal to 10 litres/rev.

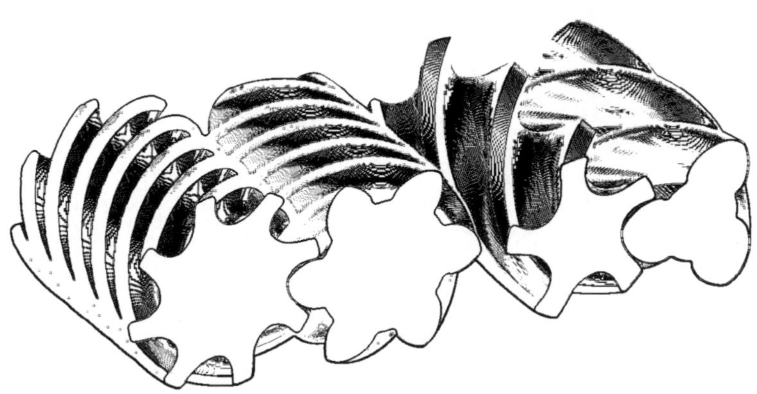

Figure 2. Normalised rotors shown to scale

Examples of two rotor pairs with different lobe combination, wrap angle, and L/D are shown in Figure 2. The 6/8 combination on the left has small compression chambers but more of them resulting in the

IOP Conf. Series: Materials Science and Engineering **232** (2017) 012025 doi:10.1088/1757-899X/232/1/012025

same displacement per revolution of the male rotor. The normalised rotor tip diameters (OD) and rotor root diameters (ID) for a selected range of rotor configurations are presented in the results section.

3. Simulation

For each rotor configuration model generated in SCORG, the thermodynamic performance was simulated using a chamber model included in the software package. This model was developed by Stosic et al. [6]. The chosen duty used in this investigation is for oil free compression of methane gas which is representative of common hydrocarbon compression applications. The operating pressures investigated at this stage in the design are actually fairly moderate but are nonetheless adequate to allow relative comparisons between different rotor configurations in terms of thermodynamic performance and pressure loading in this investigation. In the wider study it is planned to repeat comparisons at more specific duties to confirm the current findings and explore the operating limits in more detail. At this stage the simulated power is based only on indicated power of compression and no attempt has been made to model bearing or other parasitic power losses. The adiabatic efficiencies presented in this study are consequently high. The thermodynamic model input parameters are summarised in Table 1.

Table 1. Thermodynamic model parameters

Parameter	Value
Interlobe clearance (mm)	0.15
Radial clearance (mm)	0.15
Axial clearance (mm)	0.1
Volume index, Vi	2.6
Speed, N (rpm)	6000
Suction pressure, p_1 (Bara)	2
Discharge pressure, p_2 (Bara)	7
Suction temperature, T_1 (°C)	40
Ratio of specific heats, γ	1.3
Gas constant, R (J/(kgK))	518.3
Compressibility factor, Z	1

4. Pressure forces

4.1. Pressure distribution

The pressure curve of the instantaneous internal pressure against the male rotor angle is calculated by the chamber model. This pressure distribution can be mapped onto the 3D surface of the rotors using a procedure detailed in previous publications [5]. This local pressure can be described as a function of: the spatial co-ordinates X, Y and Z; and a reference cycle angle, θ, which describes the male rotor position. The resulting pressure distribution on the rotor surfaces, in Pascals, is shown in Figure 3.

4.2. Pressure area and force calculation

The numerical procedure used to calculate the resulting loads utilised discrete force vectors at nodal positions. The location of the node is known along with the pressure at the node therefore the effective pressure area is required in order to calculate the point force.

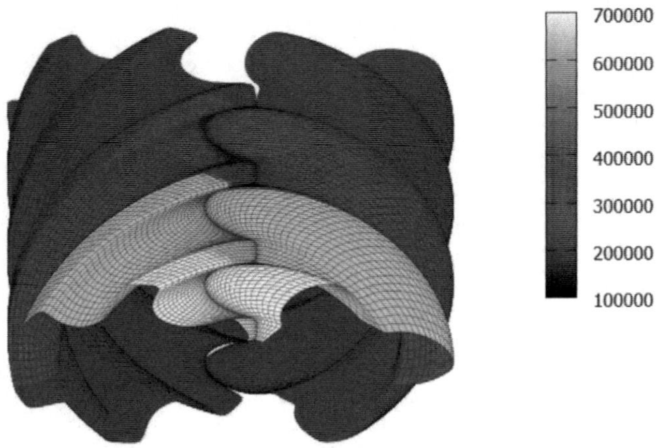

Figure 3. Pressure distribution on rotor surfaces

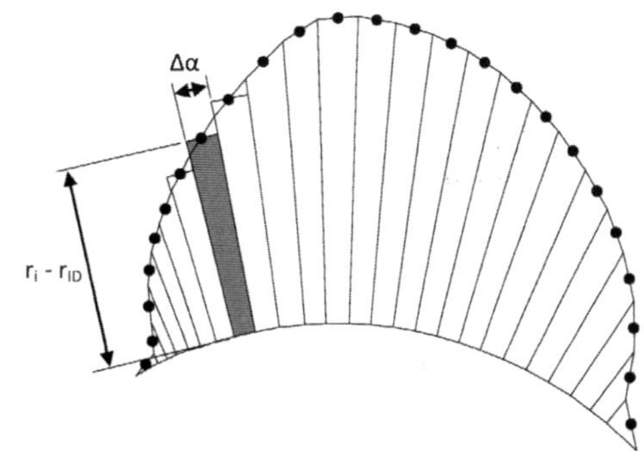

Figure 4. End face pressures

The nodes are equidistant around the transverse rotor profile so length of the profile curve and the number of lobes gives dimension Δs. The nodes are also equidistant along the length so similarly the dimension ΔL is easily calculated. The effective area is calculated in equation (1) where the length must be corrected using the angle, φ_S, which is measured between the surface normal and the transverse plane; this describes the slope when travelling axially along the rotor surface. This angle is affected by both local helix angle and local flank angle on the rotor surface. The force normal to the rotor surface, F_N, is simply calculated using the local pressure and the area as in equation (2).

$$A = \Delta s \frac{\Delta L}{\cos(\varphi_S)} \tag{1}$$

$$F_N = pA \tag{2}$$

The end face pressure forces were approximated using the pressure results and nodes on the end planes of Figure 3. The pressure area over which it was assumed the pressure at each node acts on the end face is shown in Figure 4. The effective rotor body pressure area has been truncated at the root of the rotor for the purposes of this investigation. This assumes that pressure faces on the end faces of the protruding shafts is similar at each end of the rotors and can be neglected. These geometry assumptions can later be readily adapted with preliminary or known rotor shaft details. The total pressure force calculated for each end face is taken as the sum of the individual forces calculated from each local pressure and area:

$$F_{END\ FACE} = \sum_{i=1}^{n} \frac{\Delta\alpha_i}{2}(r_i^2 - r_{ID}^2)p_i \tag{3}$$

4.3. Calculation of force components

A given point on the rotor surface, at a given rotor position, for which the polar co-ordinate angle, β, and the surface normal angle, φ_{PA}, can be readily calculated is shown in Figure 5. Using these angles, the local transverse flank angle, φ_F, can be found using equation (4).

The force component F_C describes the transverse force tangent to the 'circle' of rotation and F_R is the transverse, radial force. The local helix angle, φ_H, is defined on the plane viewed on A-A which is the plane tangent to the 'cylinder' of rotation. This is calculated using the local radius and the lead length, h, using equation (5).

Applying equations common for gear theory, the transverse force tangent to the circle of rotation can be found using equation (6). Knowing F_C, basic trigonometric relationships can be applied to find F_x, F_y, and F_A.

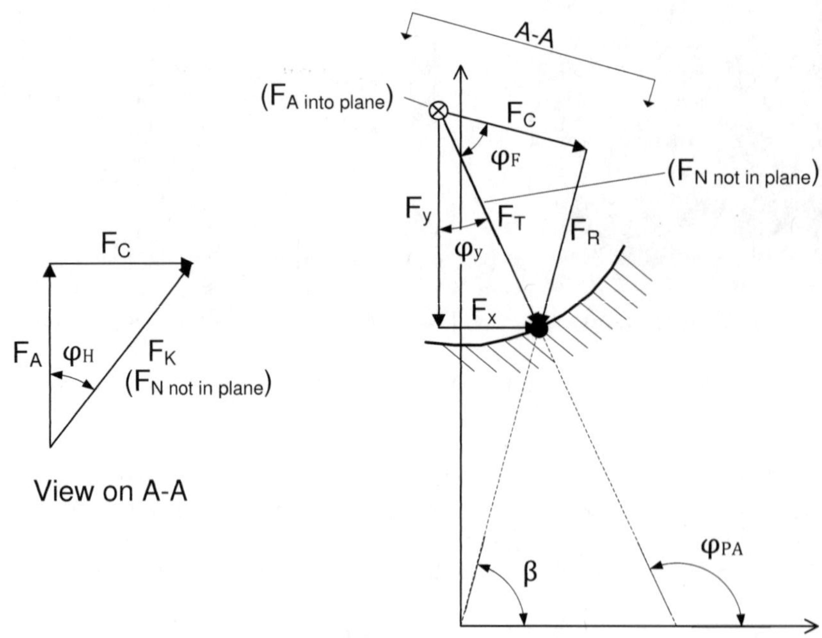

Figure 5. Force vectors on transverse plane

$$\varphi_F = 90 - (\varphi_{PA} - \beta) \tag{4}$$

$$\varphi_H = tan^{-1}\left(\frac{2\pi r}{h}\right) \tag{5}$$

$$F_C = F_N \cos(tan^{-1}(\tan(\varphi_F)\cos(\varphi_H)))\cos(\varphi_H) \tag{6}$$

4.4. Resolving reaction forces

Net forces are readily resolved using static analysis of the rotor bodies once the components F_x, F_y, and F_A, which are aligned to the global Cartesian co-ordinate system, are known. While the rotor torque is only dependent on the geometry of the rotor bodies, the axial and radial forces are somewhat dependent on additional geometry features of the rotor shafts. The assumptions for the end face pressure areas already presented will have a significant effect on the net axial forces. The radial forces are dependent on the axial position of the bearings. In order to evaluate the rotor configurations without assuming shaft dimensions the radial pressure forces were resolved at planes co-incident with the inlet and outlet rotor end faces. This will result in the largest outlet radial forces since moving the bearings away from the rotor body allows the radial load to be shared more evenly between the inlet and outlet end bearings.

IOP Conf. Series: Materials Science and Engineering **232** (2017) 012025 doi:10.1088/1757-899X/232/1/012025

5. Results

5.1. Rotor diameters

Normalisation of each rotor pair, whereby the geometry was scaled until the displacement constant was 10 l/rev, resulted in the different rotor sizes as summarized in Figure 6. Generally, increasing the overall number of rotors lobes results in rotors with bigger diameters and unsurprisingly, increasing the relative length (or L/D) results in smaller diameters. Wrap angle has little effect on rotor size. 4/5 and 5/6 lobes combinations result in a considerably smaller female rotor for the same displacement.

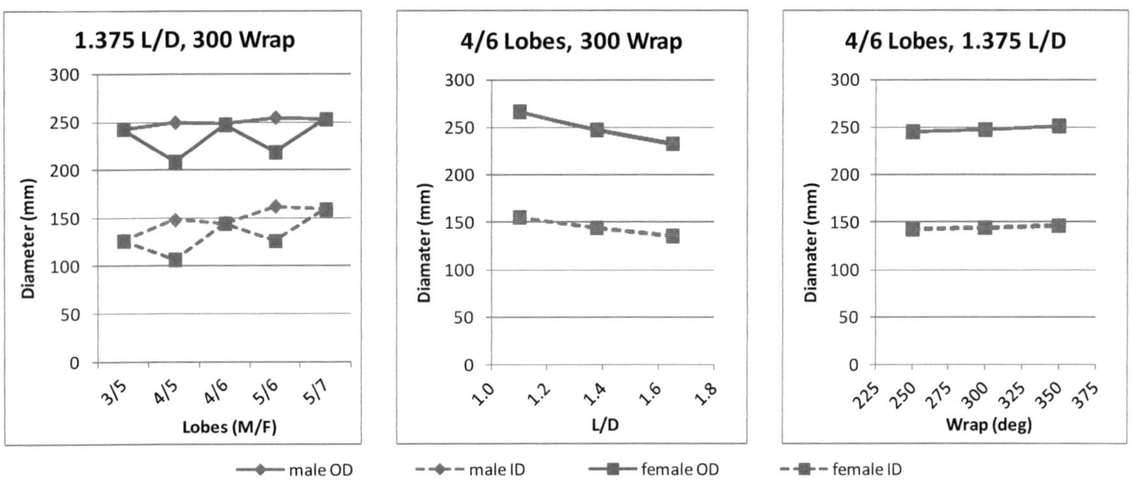

Figure 6. Rotor diameters

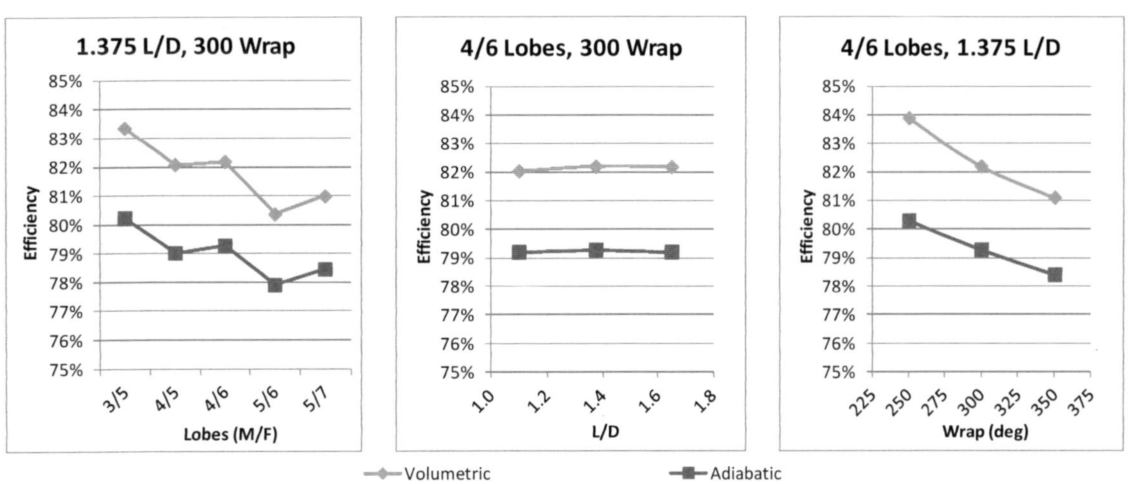

Figure 7. Thermodynamic performance

5.2. Thermodynamic performance

Results from the chamber model are shown in Figure 7. L/D has little effect on thermodynamic performance for the simulated duty. Smaller lobe combinations generally improve efficiency but interestingly 4/6 was slightly better that 4/5 and 5/7 was slightly better than 5/6. Smaller rotor wrap angles shorten the overall cycle time therefore there is less time during which leakage can occur; this is thought to be the reason for the better performance with smaller wrap angles.

IOP Conf. Series: Materials Science and Engineering **232** (2017) 012025 doi:10.1088/1757-899X/232/1/012025

5.3. Pressure forces

A summary of the resultant radial forces at the high pressure end is shown in Figure 8. Due to the way the pressure distribution changes as the rotors turn, the resulting forces are not constant therefore the minimum and maximum forces are shown for each rotor. This shows that a smaller number of male rotor lobes or smaller wrap angle results in larger variation in the radial force. In all cases the largest radial force occurs on the female rotor. Between 3/5, 4/6, and 5/7 lobe combinations there is little difference in the maximum radial force whilst in 4/5 and 5/6 this force is smaller. As expected, smaller L/D ratios result in smaller radial forces on both rotors.

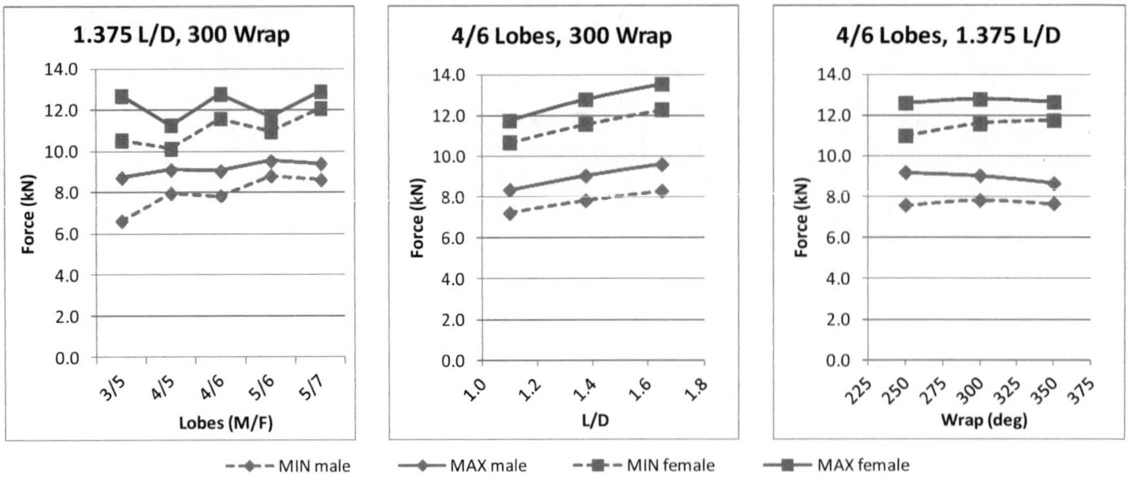

Figure 8. Radial forces at outlet rotor end face

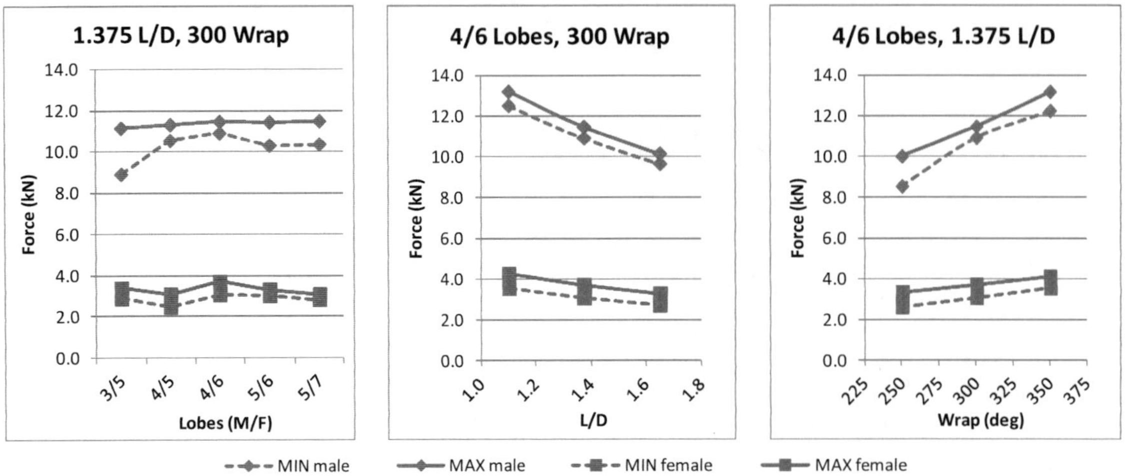

Figure 9. Axial forces

Resultant axial forces are summarised in Figure 9; these are considerably higher for the male rotor than for the female rotor. The lobe combination has little effect on the maximum axial force however the 4/6 lobe combination results in the smallest variation in the axial forces which would provide improved dynamic stability. The reduction in radial forces with reduced L/D (Figure 8) comes at the expense of having larger axial forces (Figure 9). Larger wrap angles increase the axial force; this stands to reason as rotors with zero wrap angle would have zero net axial load.

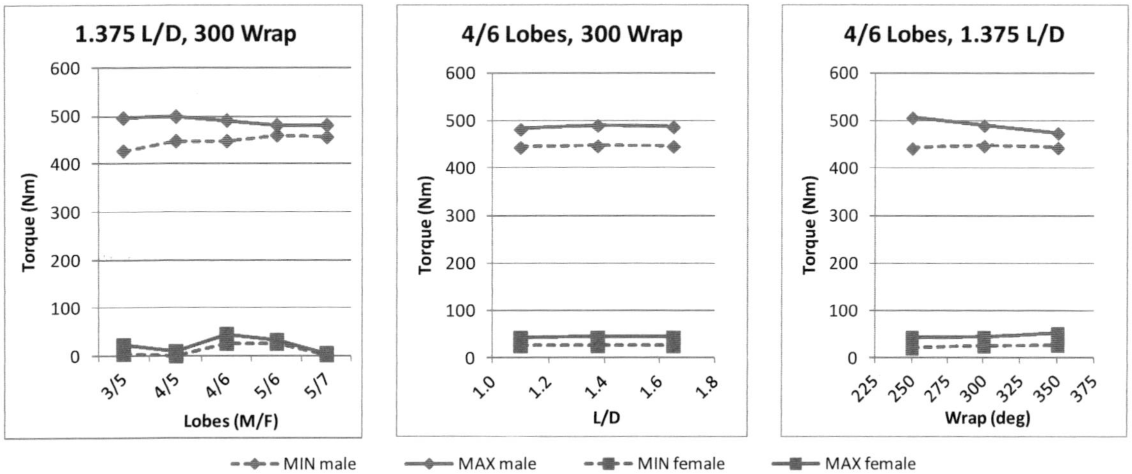

Figure 10. Torque

The rotor torque for each rotor configuration is shown in Figure 10. There isn't a great deal of change in the maximum torque however, as with the radial forces, having more lobes or increasing the wrap angle reduces the variation in the torque. This may be an important factor on high pressure, and consequently high torque, operating duties.

5.4. Bearing specific load

Each rotor configuration introduces inherent limits on the bearing geometry hence the size of bearings that can be accommodated should be considered. Applying some assumptions about the bearing type and geometry will allow some relative comparisons of the bearings for each rotor configuration.

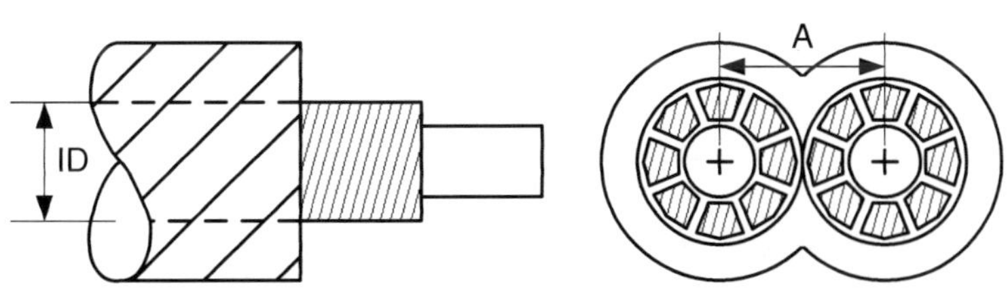

Figure 11. Driving dimensions for bearing area calculation

Hydrodynamic journal bearings are assumed at the outlet radial bearings. The diameter of the journal will be taken as the root diameter, ID, of each respective rotor as shown in Figure 11. In practice, the journal diameter is likely to be a bit smaller for practical reasons. The length of the journal will be taken to be the same as the diameter resulting in an aspect ratio of 1; it is not usually recommended to increase the length of the journal beyond this limit. The male and female root diameters in Figure 6 were simply squared to give the projected area.

The main limitation on the size of the thrust bearings is the outer diameter of the bearing housing, or carrier ring, which, assuming the male and female axial bearings are positioned side by side, will be limited by the rotor centre distance, A. By assuming a typical tilting pad thrust bearing design, the effective thrust surface can be calculated using a factor of 0.42 of the area of the carrier ring, which in this case will have a maximum diameter equal to the centre distance.

These areas for the radial and axial bearings were used, along with the forces in Figure 8 and Figure 9 respectively, to calculate the specific load on the bearings for each rotor configuration. Note that since the radial loads were calculated at the end face of the rotor, and not at the centre of pressure of the assumed bearing, the load is slightly higher but for the purposes of comparing different rotor configurations this is not significant.

Presenting specific load rather than radial forces in Figure 12 has little change on the trends against L/D and wrap angle however there is a significant change in the results against different lobe combinations: The lower forces in the 4/5 and 5/6 rotors do not offset the reduction in pressure area which is significantly smaller on these female rotors and as a result the specific load is much higher.

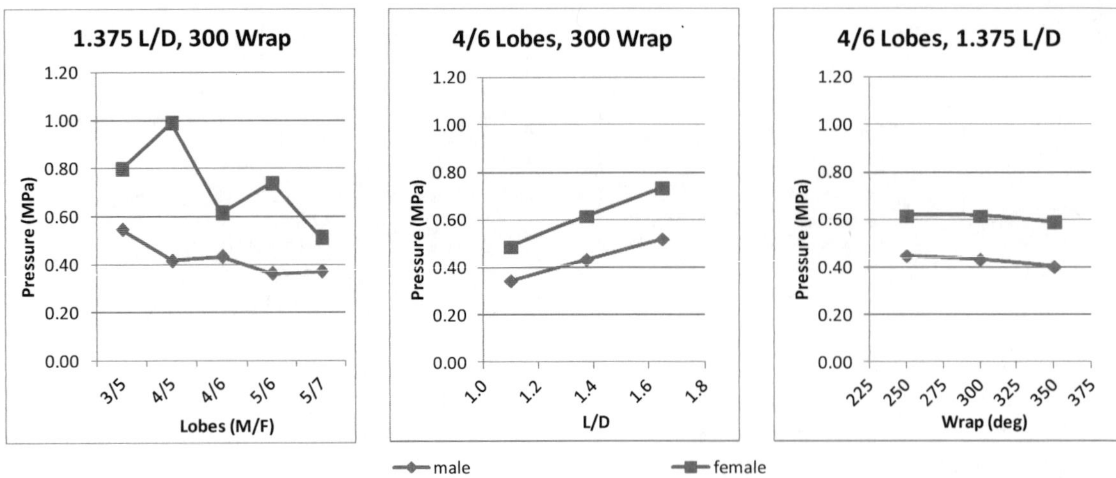

Figure 12. Radial bearing specific loads

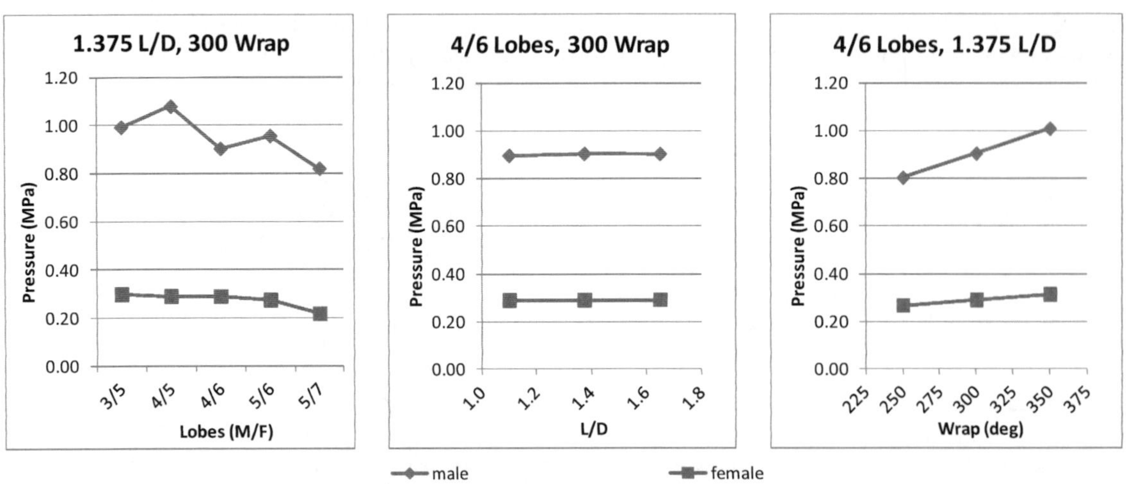

Figure 13. Axial bearing specific loads

Whereas the maximum axial forces were not sensitive to the lobe combination, having more rotor lobes generally reduces the specific load and 4/5 and 5/6 combinations stand out as poor performers as shown in Figure 13. These results suggest that the L/D can safely be reduced, in order to limit the radial bearing pressure, without any increase in the axial bearing pressures since the larger centre distance and bearing size compensate for the higher axial forces predicted in Figure 9. Larger wrap angles increase the specific load on the axial thrust bearings.

6. Conclusions

The female radial bearing at compressor outlet and the male axial thrust bearing are the most critical for design. With the duty investigated in this study, the bearing pressures are all well within realistic operating values. Rather than exploring the actual pressure limitations for a specific rotor configuration this study has provided a general evaluation of the relative performance of different rotor configurations, this is suited for use at an early conceptual design phase.

A key finding is that rotor pairs with only one more female rotor lobe number than male rotor lobe number show both disadvantages in terms of predicted thermodynamic performance and bearing loading which allows these to be ruled out for high pressure design.

Reducing rotor L/D can reduce the radial bearing pressure load without any apparent detriment to the thermodynamic performance and does not result in an increase in the specific load on the thrust bearings. The downside of excessively reducing L/D would be oversized rotor diameters and axial bearings diameters which could result in excessive costs and larger bearing losses as well as higher rotor tip speeds which would ultimately limit the operating speed and capacity.

The results generally seem to suggest smaller rotor wrap angles are better however it may be the case that the current model doesn't adequately capture kinetic losses that would occur due to the higher flow velocities that are a feature of low wrap angle designs. Some refinement of the chamber model setup/ use of CFD would be interesting to shed more light on this.

In general the results predict that having a smaller number of rotor lobes is good in terms of thermodynamic performance but having more lobes is better to reduce bearing loads and load variations. The best lobe combination for a given design will be a trade off based on the specific operating envelope of the designed compressor.

Verification of the bearing forces calculated in this work has so far been limited to comparison to other available software. Work is planned to acquire measured bearing forces from a test compressor.

Further work planned but not available at the time of writing is to include the effect of radial and axial loads introduced by timing gears and also to model the bearing power losses. Appropriately designed gears can be used to mitigate, to some extent, the imbalance between the axial loads on the male and female rotors shown in Figure 13. This imbalance can be further mitigated by decreasing the size of the female thrust bearing in order to accommodate a larger male thrust bearing. The inclusion of power losses will change the adiabatic efficiency results presented in Figure 7 and would be useful to allow quantification of performance penalties.

References

[1] Stosic N, Smith IK and Kovacevic A 2005 *Screw Compressors: Mathematical Modelling and Performance Calculation* London

[2] You CX, Tang Y, Fleming JS, and Tramschek AB 1994 A generalized computer program for calculating the bearing loads in helical twin screw compressors *International Compressor Engineering Conference at Purdue* Paper 1051

[3] Hsu RH, Wu YR, Wu WT, and Tran VT 2015 Rotor load analysis method for twin screw compressors with considering gaseous pressure and working temperature *14th International Federation for the Promotion of Mechanism and Machine Science* Taipei

[4] Wu HG, Ma Y, and Xing ZW 2004 Theoretical and experimental investigation of compressor loads in twin screw compressor *International Compressor Engineering Conference at Purdue* Paper 1701

[5] Buckney D, Kovacevic A, and Stosic N 2016 Design and evaluation of rotor clearances for oil-injected screw compressors *Proc IMechE Part E: J Process and Mechanical Engineering* **231** pp 26-37

[6] Hanjalic K and Stosic N 1997 Development and optimization of screw machines with a simulation model - Part II: thermodynamic performance simulation and design optimization *J Fluids Eng Trans ASME* **119** pp 664-670

[7] SCORG online documentation *http://pdmanalysis.co.uk/online-documentation* 24/05/2017

IOP Conf. Series: Materials Science and Engineering **232** (2017) 012026 doi:10.1088/1757-899X/232/1/012026

Numerical investigation of flow characteristics in twin-screw pump under cavitating conditions

D Yan[1, 2], **A Kovacevic**[2], **Q Tang**[1], **S Rane**[2]
[1] State Key Laboratory of Mechanical Transmission, Chongqing University, 400044, China;
[2]Centre for Compressor Technology, City University of London, EC1V0HB, UK

Abstract. In order to investigate the flow characteristics and the formation process of cavitation in twin-screw pumps, three-dimensional CFD (Computational Fluid Dynamics) numerical analysis has been carried out. A conformal structured moving mesh generated by an in-house code SCORG was applied for the rotor domain. The VOF (Volume of Fluid) Method has been adopted for dealing with the liquid-gas two-phase flow, while the bubble dynamics was handled by a homogenous cavitation model. By changing the rotation speed and discharge pressure, the intensity, distribution area and variation of cavitation at different rotor angle were obtained. The effects of rotation speed and discharge pressure on cavitation characteristics have been analysed. Calculation results with cavitation model are compared with the results without cavitation and the experimentally obtained values. The influence of cavitation on the performance of a screw pump in terms of the mass flow rate, pressure distribution, rotor torque and the shaft power have been analysed and discussed.

1. Introduction

Twin screw pumps are positive displacement machines widely used in petrochemical, shipping, energy and food industries due to their reliability, self-priming capability and excellent performance in single phase or multiphase operation. Typical arrangement of a twin screw liquid pump is given in Figure 1, showing rotors synchronised by timing gears and enclosed in the casing.

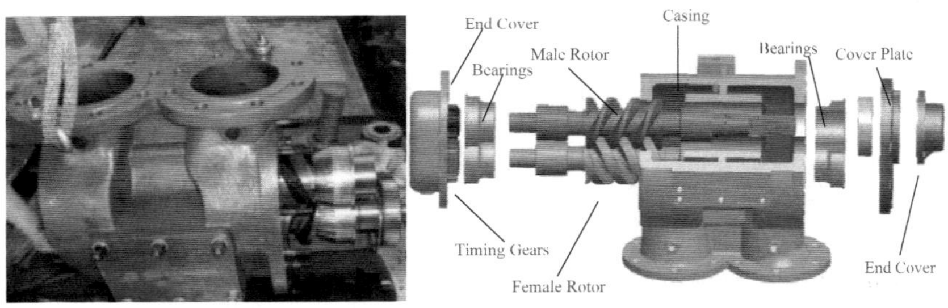

Figure 1. Structure and components of twin-screw pump

With the trend of increasing rotational speed, delivery pressure and power in screw pumps, the cavitation becomes a phenomenon unavoidable in such kinds of machines. Nowadays, the research of cavitation is mainly focused on centrifugal pumps, hydrofoils, propellers and hydraulic valves, etc.[1]. The lobes of a positive displacement screw pump are much thicker than in turbo machines, while the clearances of screw pumps are much smaller than in dynamic pumps. Positive displacement machines also rotate slower than equivalent size dynamic machines. Therefore such machines are less sensitive to cavitation, and usually no obvious cavitation erosion is observed on screw rotor surfaces. For that reason less attention is paid in the literature to the cavitation in screw pumps [2]. However, during the rotation of screw rotors, cavitation develops through initiation, growth and collapse of bubbles, which will directly influence the performance of a screw pump, such as increasing vibration and noise, eroding surfaces of rotors and casing wall, lowering the volumetric efficiency, and so on [3]. Therefore, in order to improve the performance and stability of screw pump and expand its application fields, cavitation cannot be ignored.

In this paper, the unique grid generation software SCORG is used for grid generation of a structured moving mesh around screw pump rotors [4]. The mesh for stationary domains such as ports and pipes is generated by use of a commercial grid generator built into the CCM (Computational Continuum Mechanics) solver STAR-CCM+ [5]. Handling of the mesh generated by SCORG in STAR-CCM+ solver is managed by use of the UDF made specially for handling conformal rotor mesh and it will be described later in the paper. This method has been validated by experiment in single-phase CFD modelling of screw pump [6][7].

The aim of this paper is to study the flow characteristics of twin-screw pumps under cavitating conditions. Based on the structured moving mesh, full CFD simulations with and without cavitation have been carried out and will be explained in detail in this paper. The real-time mass flow rate, rotor torque, pressure distribution, velocity field, power consumption were obtained. In cavitation modelling, Volume of Fluid (VOF) model [8] was applied for dealing with the liquid-gas two-phase flow. The Sauer-Schnerr model was used for modelling bubble dynamics. By changing rotation speed and discharge pressure separately, the accordingly variation of cavitation was acquired. The influence of cavitation on performance of a screw pump was analysed and discussed.

2. Mathematical Model

Positive displacement screw pumps operate on the basis of changing the size and position of a working domain which consequently causes change in the pressure of the working domain thereby transporting the fluid. To calculate performance of a screw pump, quantities such as mass, momentum, energy etc. need to be modelled. The governing equations required for the solution a closely coupled, time dependent set of partial differential equations (PDE's) and often employ a Finite Volume Method (FVM) to be solved [9][10].

2.1. Governing Equations

The cavitating flow can be treated as a homogeneous mixture of vapour, air and liquid, in which vapour and air are gas. So it is liquid-gas two-phase flow. VOF model considers a single effective fluid whose properties vary according to volume fraction of individual fluids:

$$\alpha_i = \frac{V_i}{V} \tag{1}$$

The mass conservation equation for fluid *i* reads:

$$\frac{\partial(\alpha_i \rho_i)}{\partial t} + \nabla \cdot (\alpha_i \rho_i \boldsymbol{v}) = \rho_i s_{\alpha_i} \tag{2}$$

It can be rearranged into an equation in integral form:

$$\frac{\partial}{\partial t}\int_V \alpha_i dV + \int_S \alpha_i \boldsymbol{v} \cdot \boldsymbol{n} dS = \int_V \left(S_{\alpha_i} - \frac{\alpha_i}{\rho_i}\frac{D_{\rho_i}}{Dt}\right) dV \tag{3}$$

where S_{α_i} is the source of the phase i

This equation is used to compute the transport of volume fraction α_i

The mass conservation equation for the effective fluid is obtained by summing up all component equations and using the condition:

$$\sum_i \alpha_i = 1 \tag{4}$$

The integral form of mass conservation equation (used to compute pressure correction) reads:

$$\int_S \boldsymbol{v} \cdot \boldsymbol{n} dS = \sum_i \int_V \left(S_{\alpha_i} - \frac{\alpha_i}{\rho_i}\frac{D_{\rho_i}}{Dt}\right) dV \tag{5}$$

The properties of effective fluid are computed according to volume fractions:

$$\rho = \sum_i \alpha_i \rho_i \tag{6}$$

The VOF model description assumes that all immiscible fluid phases present in a control volume share velocity, pressure, and temperature fields. Therefore, the same set of basic governing equations describing momentum, mass, and energy transport as in a single-phase flow is solved.

The equations are solved for an equivalent fluid whose physical properties are calculated as functions of the physical properties of its constituent phases and their volume fractions.

2.2. Cavitation Model

The cavitation model describes the growth of bubbles which includes the phase change between liquid and its vapour. In this calculation, a homogenous distribution of bubble seeds in the liquid was applied. Based on the Rayleigh-Plesset equation [11], the vapour volume fraction can be calculated as follows:

$$\alpha_v = \frac{V_v}{V_l + V_v} = \frac{n_0 V_l \frac{4}{3}\pi R^3}{V_l + n_0 V_l \frac{4}{3}\pi R^3} = \frac{n_0 \frac{4}{3}\pi R^3}{1 + n_0 \frac{4}{3}\pi R^3} \tag{7}$$

The bubble growth rate reads:

$$\dot{R} = \frac{dR}{dt} = \sqrt{\frac{2}{3}\frac{P_B - P_\infty}{\rho_l}} \tag{8}$$

then the vapour production rate reads:

$$\frac{d\alpha}{dt} = (1-\alpha)\frac{4\pi n_0 R^2}{1+\frac{4}{3}n_0\pi R^3}\dot{R} \tag{9}$$

The continuity equation reads:

$$\nabla \cdot \boldsymbol{v} = -\frac{1}{\rho}\left(\frac{\partial \rho}{\partial t} + \boldsymbol{v}\cdot\nabla\rho\right) = -\frac{1}{\rho}\frac{d\rho}{dt} = \frac{\rho_l - \rho_v}{\rho}\frac{d\alpha}{dt} \tag{10}$$

Then the transport equation reads:

$$\frac{d\alpha}{dt} + \nabla\cdot(\alpha\boldsymbol{v}) = \frac{d\alpha}{dt} + \alpha\nabla\cdot\boldsymbol{v} = \frac{(1-\alpha)\rho_l}{(1-\alpha)\rho_l + \alpha\rho_v}\frac{n_0}{1+n_0\frac{4}{3}\pi R^3}\frac{d}{dt}\left(\frac{4}{3}\pi R^3\right) \tag{11}$$

3. Setup for Calculation of the Screw Pump

The screw pump used for this study shown in Figure 1 is a twin-screw pump with a 2/3 lobe arrangement and A-type profile rotors. The operating speed of the male rotor changes from 630 to 2100 rpm while keeping the discharge pressure as 0.85 MPa. By controlling the discharge valve, the discharge pressure varies from 0.35 to 0.85 MPa while keeping rotation speed of the male rotor at 2100 rpm. The male and the female rotors have 140.00 mm diameter and the centre distance of 105.00 mm. The length of the rotors is 200.00 mm and the male rotor has a wrap angle of 590.0°.

3.1. Grid Generation

Grid generation is a process of discretising a working domain of the screw pump in control volumes for which a solution of local fluid properties is to be found. It may be numerical, analytical or variational [12]. The results obtained in this research work used grids generated by analytical grid generation. Applying the principles of analytical grid generation through transfinite interpolation with adaptive meshing, the authors have derived a general, fast and reliable algorithm for automatic numerical mapping of arbitrary twin screw machine geometry built into an in-house grid generation code SCORG[13][14].

The rotor profile of the screw pump is shown in Figure 2. Table 1 shows the geometry parameters of rotors. The numerical grid for the working fluid domain between two rotors is shown as Figure 3.

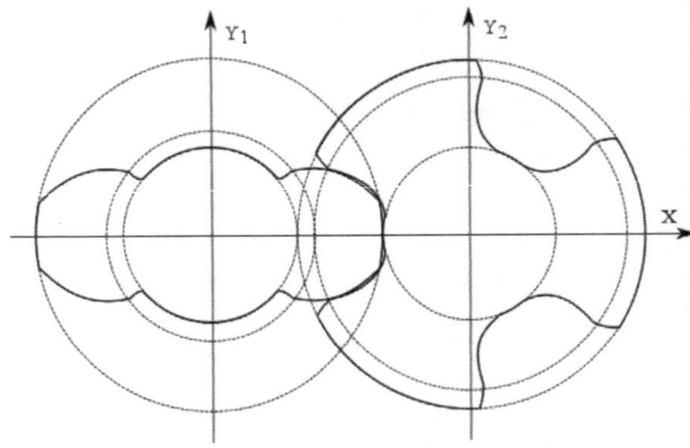

Figure 2. A-type tooth profile used in this paper

Table 1. Geometry parameters of screw rotors used in study

	Number of Lobes	Pitch Radius (*mm*)	Root Radius (*mm*)	Tip Radius (*mm*)
Male Rotor	2	42	35	70
Female Rotor	3	63	35	70
Centre distance		105*mm*		
Thread pitch		61*mm*		
Radial clearance		0.24*mm*		
Inter-lobe clearance		0.12*mm*		

Figure 3. Grids in the fluid domainleft) in cross section of rotors; middle) on the rotor surface; right) polyhedral mesh of suction and discharge ports

Polyhedral mesh of the inlet port and outlet port is generated using STAR-CCM+ grid generator as shown in Figure 3(right). The grid independency study for this calculation model has already been performed in another publication [6]. A numerical mesh used in this study compromises 1068864 cells of which 775180 cells represent the fluid domain between the rotors while 293684 cells represent the two ports.

3.2. Numerical Methods

The Star-CCM+ pressure-based solver is used for the calculation of screw pump. In order to solve the pressure-velocity coupling, the implicit unsteady segregated flow scheme has been applied. The second-

IOP Conf. Series: Materials Science and Engineering **232** (2017) 012026 doi:10.1088/1757-899X/232/1/012026

order upwind discretization scheme is applied. Gauss-Seidel node is used for the relaxation scheme which provides better convergence by iteratively correcting (relaxing) the linear equation during multigrid cycling.

The main rotor rotates 2.4° per step. The mesh is updated before commencing solution for each time step. The time-step is defined as follows:

$$\Delta t = \frac{DPTS}{6 \cdot RPM} \tag{12}$$

whereby, $DPTS$ is the degree per time step, RPM is the rotation speed of male rotor.

When the $DPTS$ is not small enough, it will make the simulation run with a relatively large time-step which may cause the divergence of calculation. Here, Δt is inversely proportional to RPM, which means that the mesh has to be changed for different RPM in order to keep the ratio of time and spatial step constant. It's not always essential to keep the ratio constant but it is limited by Courant stability condition.

The k-ε turbulence model is adopted in the calculation. Stagnation inlet and pressure outlet are used respectively for the inlet and outlet boundaries. The pressure of the inlet port is 0 Pa. The discharge pressure ranges from 0.35 to 0.85MPa while the rotation speed of the male rotor ranges from 630 to 2100 rpm. The initial pressure and initial velocity are 0 Pa and 0 m/s respectively. The turbulence intensity is 1% and the turbulence viscosity ratio is 10.

4. Numerical Results

The calculations were carried out in a computer powered by 4 Intel 3.00 GHz processors and 8 GB memory. Screw pump rotation was simulated by means of 75 time steps for one interlobe rotation, which was equivalent to 150 time steps for one full rotation of the male rotor. The time step length was synchronised with a rotation speed of 630 to 2100 rpm. An error reduction of 4 orders of magnitude was required, and achieved in 30 inner iterations at each time step. The overall performance parameters such as chamber pressure, velocity distribution, rotor torque, mass flow rate and shaft power were then calculated.

4.1. Cavitation under Different Boundary Conditions

The start-up process of a screw pump is highly unsteady leading to the large area of cavitation in the pump chamber in a short period of time. As time progresses, the flow field gradually reaches a steady state condition when the domain in which fluid cavitates becomes reduced and repeats periodically. Figure 4 shows the volume fraction of vapour observed in the working domain as the function of the rotation angle under different rotation speed and different discharge pressure. It can be observed from Figure 4(left) it can be observed that, when cavitation stabilized, the volume fraction of cavitation and its fluctuation amplitude repeat and are higher for higher rotation speeds.

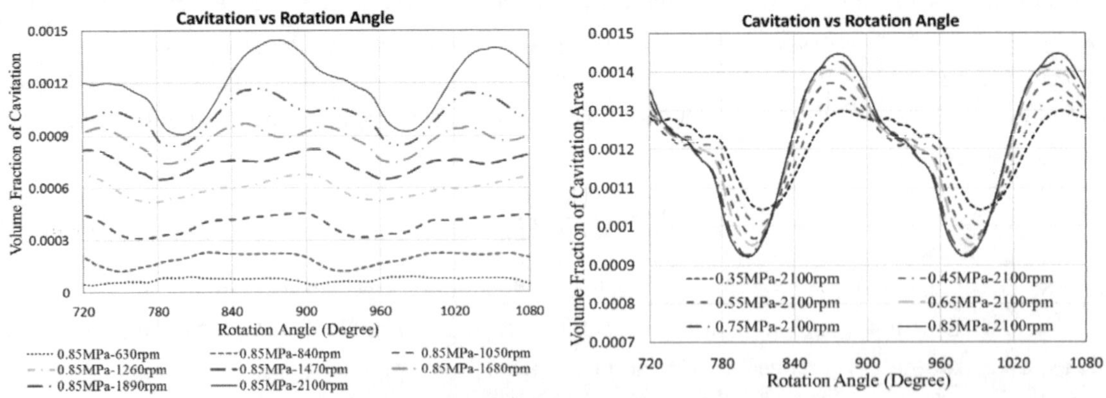

Figure 4. Cavitation changes at different rotation angles: (left) under different rotation speeds; (right) under different discharge pressure

Figure 4 (right) shows the changes of the maximum volume fraction of vapour with the constant rotation speed at different discharge pressures. When the pump reached steady state conditions, the average volume fraction of cavitation nearly remains the same under different discharge pressure, while the amplitude of volume fraction remains higher for higher pressure difference between the suction and discharge pressures. Compared with Figure 4(left), it can be observed that the effect of the discharge pressure on cavitation is not as significant as in the case of increased rotation speed.

Figure 5 shows the location of the cavitation area at different rotor angle when the screw pump reaches a relatively stable state at 2100rpm and 0.85MPa. The increased volume fraction of vapour is mainly located in the radial and inter-lobe clearances near the suction port, and its value changes periodically with the rotation of the rotors.

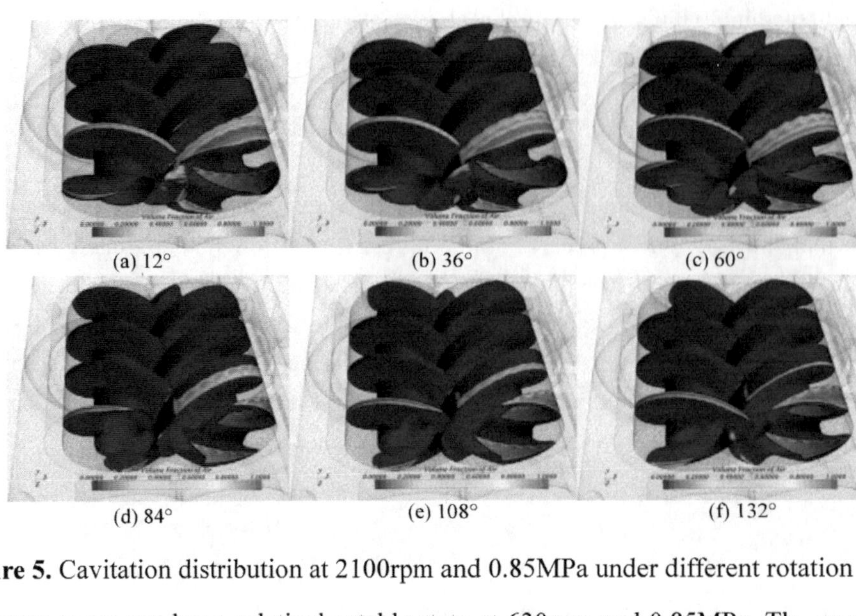

Figure 5. Cavitation distribution at 2100rpm and 0.85MPa under different rotation angle

When the screw pump reaches a relatively stable state at 630rpm and 0.85MPa. The area where fluid cavitates is the same as for the higher speed but its amplitude is much smaller especially on the female rotor. When comparing with values in Figure 5 for speed of 2100 rpm, it can be concluded that the cavitation weakens with the decrease of rotation speed.

To investigate what will happen at much higher speeds, a calculation is performed for 4 times the maximum speed of the liquid pump, in this case 8400 rpm. In such case, cavitation is not limited just to the clearance area, it extends to the entire chamber and also will spread to other chambers in the axial direction from the suction. It would take much longer for cavitation to reduce. The mean value of the volume fraction of vapour and its amplitude would both increase compared to lower speeds shown in previous figures.

From the analysis above, it can be concluded that by controlling the rotation speed to a reasonable range, cavitation and its influence in screw pumps can be limited without the need to change the structure of the pumps. However, when designing high-speed screw pump, the inside structure of the pump needs to be designed carefully to reduce cavitation.

4.2. Mass Flow Rate and Rotor Torque

In order to investigate effects of cavitation on the flow through the pump, two separate calculations were performed at 2100rpm and 0.85MPa discharge pressure, one without cavitation and another which includes modelling of the cavitation.

Figure 6(left) shows the calculated instantaneous mass flow rate of the screw pump with and without cavitation at 2100rpm and 0.85MPa. It can be seen from the comparison that the mass flow pulsations with cavitation included in prediction model are larger than that without cavitation, the fluctuation amplitude without cavitation is 0.62kg/s while under cavitation it is 1.03kg/s. The averaged mass flow rate under cavitation is similar to that without cavitation.

Figure 6(right) shows the rotor torque of screw pump with and without cavitation at 2100rpm and 0.85MPa. The periods of torque fluctuations with and without cavitation are the same. Cavitation brings a slightly increased amplitude of torque value. For the male rotor, the peak-to-peak amplitude of torque without cavitation is 11.83N-m, while which increases to 18.43N-m when with cavitation. The averaged torque of male rotor with and without cavitation are the same, while cavitation causes an increase of 3.17N-m which is 26.04% in averaged female rotor torque.

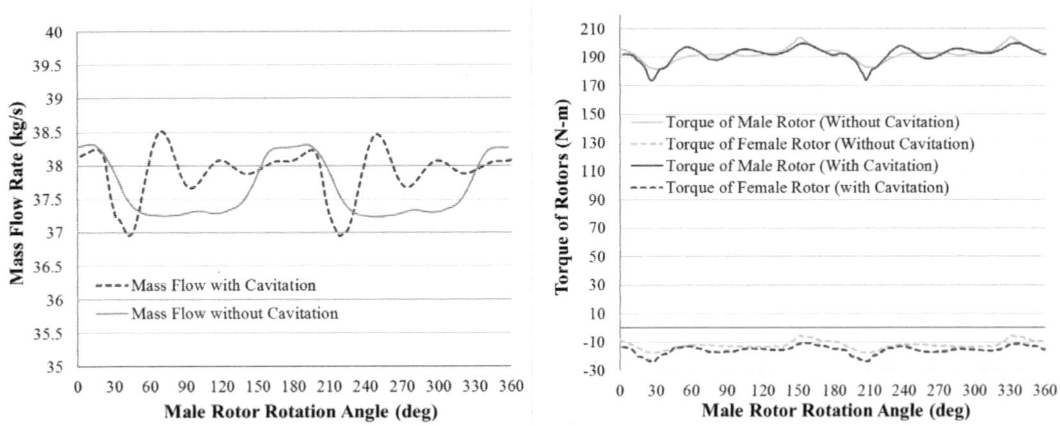

Figure 6. (left) Mass flow rate; (right) Rotor torque with and without cavitation under different rotation angle

5. Comparison with experiment and discussion

An experiment has been carried out to validate the calculation[6].The working medium used in the experiment is CD40 lubricating oil. The density is 889kg/m^3 and the dynamic viscosity is 5.25×10^{-2} Pa·S. The temperature of oil is 50 to 70℃. The nominal interlobe clearance (between two rotors) is 0.12 mm, the nominal radial clearance (between rotor and casing) is 0.24 mm.

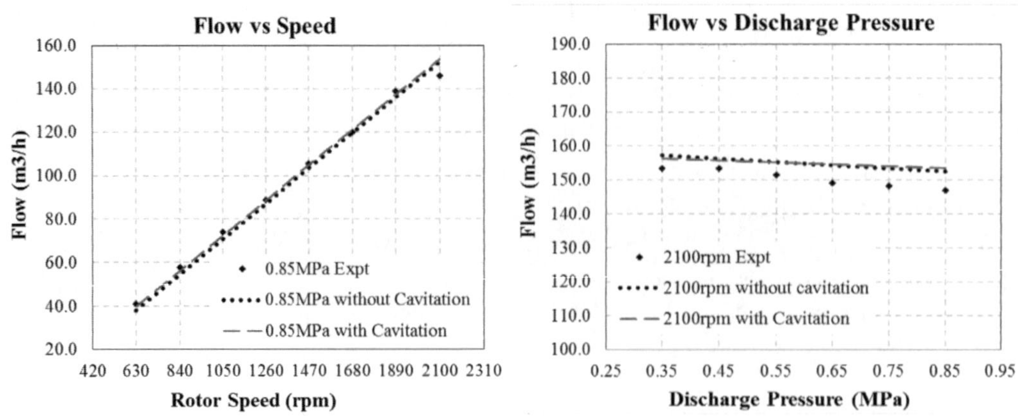

Figure 7. Flow rate: (left) variable rotational speed; (right) variable discharge pressure

Figure 7(left) shows the flow rate of cavitation and non-cavitation condition with the variation of rotation speed and (under constant discharge pressure 0.85MPa). By comparison, it can be found that with cavitation the flow rate is 4.90% higher than that without cavitation.

From the distribution of cavitation in Figure 5, it can be seen that under a relatively low rotation speed cavitation mainly appeared in clearance area. Clearance cavitation leads to a decrease of flow rate through clearances, which means smaller leakage rate. It is observed by analysing modelling results that forming of the vapour in clearances effects sealing and it brings a slightly increased volumetric efficiency of a screw pump.

Figure 7(right) shows the flow rate of cavitation and non-cavitation condition with the variation of discharge pressure and (under constant rotation speed 2100rpm). It can be found that the flow rates of both are very close. With the increase of discharge pressure, the flow rate under cavitation gradually exceeds that without cavitation in a small degree.

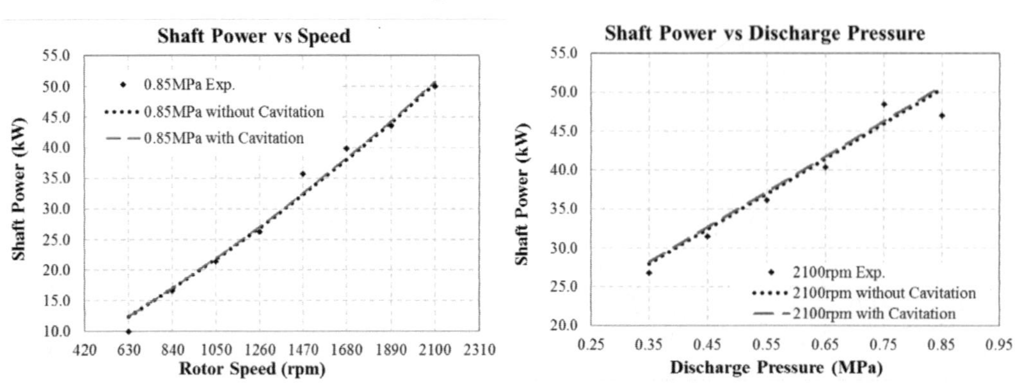

Figure 8. Shaft power: (left) variable rotational speed; (right) variable discharge pressure

Figure 8(left, right) shows the shaft power of screw pump under cavitation and non-cavitation with the variation of rotation speed and discharge pressure. It can be found that cavitation and non-cavitation have a very close shaft power while cavitation brings a slightly increase of 1.21% in shaft power. The main reason caused higher power consumption is that cavitation influenced the pressure distribution in chamber and brought fluctuation and increase in rotor torque.

6. Conclusions

A full 3-D CFD simulation of twin-screw pump has been carried out using structured moving numerical mesh with the conformal interface between the rotor domains. The cavitation intensity and distribution area were predicted based on the combination of VOF method and the Sauer-Scherr model. Through analysis and discussion, it can be concluded as follows:

(1) If it appears, the cavitation will mainly happen in the area close to the suction port and will be distributed in the outward surfaces of rotors and clearances when the rotation speed is not too high. With the increase of rotation speeds beyond the current limit of pumps, the cavitation area will expand to the whole chamber along the rotor from suction to discharge.

(2) The increase of rotation speed will enlarge the intensity of cavitation observably in screw pump. The increase of the discharge pressure at constant speed of rotation will increase the amplitude of volume fraction, while the average of volume fraction nearly remains the same.

(3) The cavitation in screw pumps has many influence on the pump performance, which is manifested in larger pressure impact in screw rotors, increase of rotor torque and mass flowrate pulsation and larger power consumption.

The study of use of CFD for screw pumps provides better understanding of the internal flow field and cavitation characteristics, which provides a good basis for the next-step research on multiphase screw pump under different gas volume fraction (GVF).

Acknowledgments: This research was supported by the National Natural Science Foundation of China [grant number 51575069], State Key Program of National Natural Science of China [grant number 51635003] and Program of International S&T Corporation [grant number 2014DFA73030]. The project was performed at City University of London.

References

[1] Luca d'Agostino, Maria Salvetti, CISM International Centre for Mechanical Sciences: Fluid Dynamics of Cavitation and Cavitating Turbopumps, Udine: Springer-Verlag Wien, 2007. ISBN 978-3-211-76668-2.

[2] Ruschel Andrej, Fluiddynamische Effekte in Schraubenspindelpumpen bei der Multiphasenförderung (Fluiddynamic effects in screw pumps at multiphase operation), Shaker Verlag GmbH, PhD thesis, Erlangen-Nürnberg University, 2014.

[3] Brennen Christopher, Cavitation and Bubble Dynamics, New York: Oxford University Press, 1995. ISBN 0-19-509409-3.

[4] Sham Rane, Ahmed Kovacevic, Nikola Stosic. Analytical Grid Generation for accurate representation of clearances in CFD for Screw Machines, in 9th International Conference on Compressors and their Systems, London, 2015.

[5] STAR-CCM+ Version 9.06.009 Tutorial Guide.

[6] Di Yan, Ahmed Kovacevic, Qian Tang, Sham Rane, Wenhua Zhang, Numerical Modelling of Twin-screw Pumps Based on Computational Fluid Dynamics, Proceedings of IMechE, Part C: Journal of Mechanical Engineering Science, vol. 0, no. 0, pp. 1-18, 2016.

[7] Ahmed Kovacevic, Di Yan, Qian Tang, Sham Rane. Numerical Investigation of Twin-Screw Pump by Use of CFD, Proceeding of 13th Fluid Machinery Congress, Hague, 3-4 October 2016. ISBN 978-0-9956263-0-0

[8] C.W Hirt, B.D Nichols, Volume of fluid (VOF) method for the dynamics of free boundaries, Journal of Computational Physics, vol. 39, no. 1, pp. 201-225, 1981.

[9] I. Demirdžić, S. Muzaferija, Numerical Method for Coupled Fluid Flow, Heat Transfer and Stress Analysis Using Unstructured Moving Mesh with Cells of Arbitrary Topology, Computer Methods in Applied Mechanics and Engineering , vol. 125, no. 1-4, pp. 235-255, 1995.

[10] Joel H. Ferziger, Milovan Peric, Computational Methods for Fluid Dynamics, 3rd Edition ed., Berlin Heidelberg: Springer-Verlag, 2002. ISBN 3-540-42074-6.

[11] P. S. Epstein, M. S. Plesset. On the Stability of Gas Bubbles in Liquid-Gas Solutions, vol. 18, no. 11, pp. 1505-1509, 1950.

[12] Sham Rane, Grid Generation and CFD analysis of variable Geometry Screw Machines, PhD thesis, City University London, London, 2015.

[13] Rane S, Kovacevic A. Algebraic generation of single domain computational grid for twin screw machines. Part I. Implementation[J]. Advances in Engineering Software, vol. 107, pp.38-50, 2017.

[14] Kovacevic A, Rane S. Algebraic generation of single domain computational grid for twin screw machines Part II – Validation[J]. Advances in Engineering Software, 2017.

IOP Conf. Series: Materials Science and Engineering **232** (2017) 012027 doi:10.1088/1757-899X/232/1/012027

Numerical Simulation of the Working Process in the Twin Screw Vacuum Pump

Yang Lu [1], Yu Fu [2], Bei Guo[1], Lijuan Fu[1], Qingqing Zhang[1], Xiaole Chen[1]

[1]Xi'an Jiaotong University, Xi'an,China
[2] Xi'an Aerospace Propulsion Institure, Xi'an,China

E-mail：guobei@mail.xjtu.edu.cn

Abstract. Twin screw vacuum pumps inherit the advantages of screw machinery, such as high reliability, stable medium conveying, small vibration, simple and compact structures, convenient operation, etc, which have been widely used in petrochemical and air industry. On the basis of previous studies, this study analyzed the geometric features of variable pitch of the twin screw vacuum pump such as the sealing line, the meshing line and the volume between teeth. The mathematical model of numerical simulation of the twin screw vacuum pump was established. The leakage paths of the working volume including the sealing line and the addendum arc were comprehensively considered. The corresponding simplified geometric model of leakage flow was built up for different leak paths and the flow coefficients were calculated. The flow coefficient value range of different leak paths was given. The results showed that the flow coefficient of different leak paths can be taken as constant value for the studied geometry. The analysis of recorded indicator diagrams showed that the increasing rotational speed can dramatically decrease the exhaust pressure and the lower rotational speed can lead to over-compression. The pressure of the isentropic process which was affected by leakage was higher than the theoretical process.

1. Introduction

Compared with the experimental results, the numerical simulation has the advantages of a short cycle, low cost, wide application range and reliable performance prediction. In order to obtain better thermal performance of the working process of the compressor, the optimal design of structural parameters is needed. The optimal design is based on the numerical simulation of the internal working process. At the same time, the numerical simulation of thermal process is an important way to predict the performance of the prototype and to shorten the product cycle.

The key to establish the vacuum pump model is to confirm the leakage and heat transfer in the process after obtaining the structural parameters of the rotor. The study of the screw vacuum pump can be used for reference. At present, the study on the leakage of the screw compressor is mainly based on the relevant leakage model[1]. The leakage of the screw compressor can be calculated according to the experience to select the appropriate flow correction coefficient. At home and abroad, a lot of experimental study and numerical simulation on the thermal process of the twin screw compressor are carried out in order to optimize the design of the screw compressor rotor profile, the suction port and the discharge ports. Zhanhua Feng[2]and Feichuan Huang[3] analyzed the deformation of vacuum pump rotors and the performance of the screw vacuum pump. Compared with the experimental results, the calculated results obtained with the performance prediction method can agree well. Northeastern

University Zhao[4]and Zhejiang University Wang[5] derived a theoretical equation to calculate the leakage of the twin screw vacuum pump, but there was no experimental verification. Ohbayashi[5] proposed a method to calculate the pumping speed of the vacuum pump, which is based on the theory of pumping speed, actual pumping speed and leakage. Dirk Stratmann[6]in Germany measured the pumping speed curve of the vacuum pump, and a simplified thermodynamic model of the vacuum pump was established.

In this study, the working process of the twin screw vacuum pump is analyzed. According to the principle of variable mass thermodynamics and energy conservation equation, considering the effect of leakage on the thermal performance mass conservation equation and gas state equation, the mathematical model was established to describe the thermodynamic model of the twin screw vacuum pump working process.

2. Geometrical parameters

2.1. Calculation of geometrical parameters

Many kinds of rotor profiles used for twin screw vacuum pumps were designed and patented by our team, and in this paper, we used the most universal profile in the current industry consisting of arc, pseudo Archimedes, pseudoArchimedes line envelope and cycloid. The equations of the profile can be found in the paper by Yang[7] published in 2015.

2.1.1. Rotor Parameters. The male and female structure parameters are shown in Table 1 commonly because their parameters are completely the same. The rotor pitch is divided into two sections, one section of the pitch is 88mm, the other section of the pitch is 60 mm.

Table 1. Geometry parameters of screw rotor

Tip Radius/ *mm*	130	Root Radius/ *mm*	66
Rotor Length/ *mm*	340	Centre Distance/ *mm*	98
First Rotor Pitch/*mm*	88	Second Rotor Pitch/ *mm*	60
First Rotor Length/ *mm*	200	Second Rotor Length/ *mm*	140

Vacuum pump rotor structure shown in figure 1. As can be seen from the figure, the pitch of screw rotor is variable, and divided into two parts. This design can reduce power consumption.

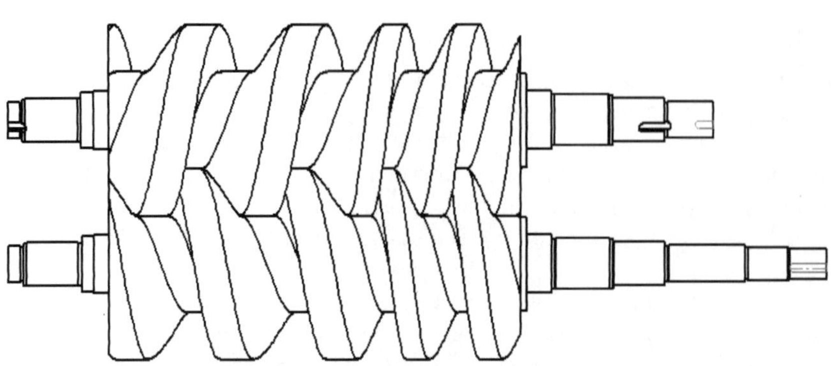

Figure 1. Structure of screw rotors.

*2.1.2. Calculation of tooth area and chamber volume.*The tooth area of the rotor varies with the rotational angle and can be calculated using the analytical method. The maximum tooth area was 8700mm2 as shown in figure 2.

IOP Conf. Series: Materials Science and Engineering **232** (2017) 012027 doi:10.1088/1757-899X/232/1/012027

With the twin screw vacuum pump being representative of the screw machinery, there are generally more than two rotor leads. During the operation of the pump, the chamber volume is completely enclosed by the rotor and machine frame. The change of the chamber volume with the rotational angle can be described as the change of the chamber area with the rotational angle. In this study, the basic chamber volume needs to complete six processes including the suction, transportation, the first compression, transportation, the second compression and exhaust as shown in figure 3. The maximum inter tooth volume was $0.767L$.

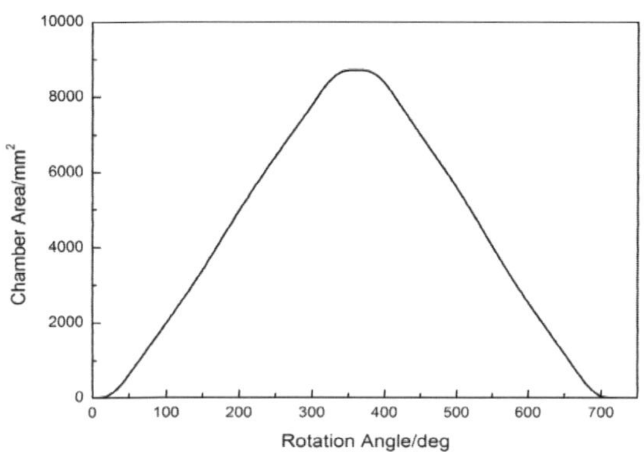

Figure 2. Tooth area varies with the rotational angle.

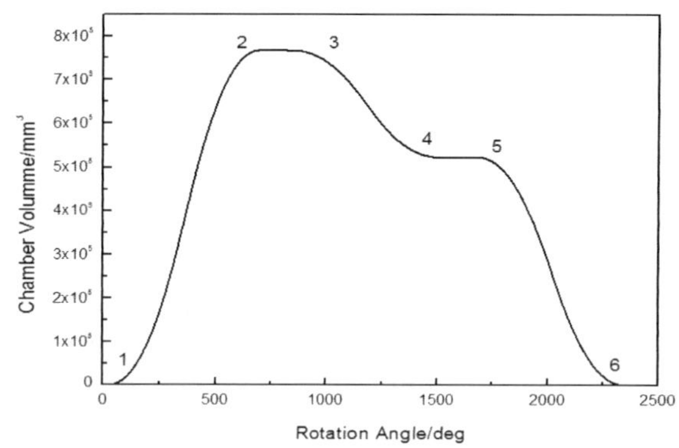

Figure 3. Chamber volume varies with the rotational angle.

*2.1.3. Calculation of the sealing line.*The length has a significant influence on the thermal performance of the twin screw vacuum pump. The shorter the length of the sealing line, the lower magnitude of leakage would be. The sealing line equation can be obtained by the simultaneous of meshing condition and the rotor tooth surface equation. Figure 4 shows the complete sealing line of the rotor profile, where Z is the axial length of the rotor and Y is the upright length of the rotor. The section ab and ef are generated by the meshing of point and the cycloid curve; the section bc and de are generated by the meshing of the tip and root circle; the section cd is generated by the meshing of the archimedes spiral and the Archimedes spiral envelope conjugate.

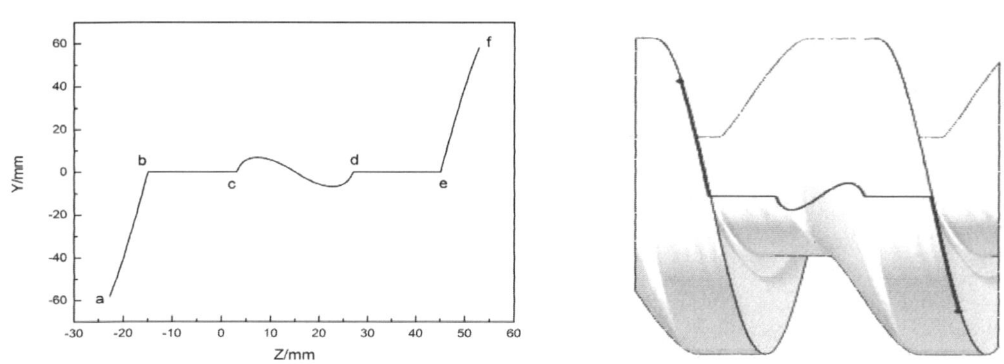

Figure 4. Sealing line.

The rotor profile of the vacuum pump is composed of a plurality of sections, and the length of the contact is calculated by combining every component. The rotor profile consists of a point, the cycloid curve, Archimedes spiral and Archimedes spiral conjugate envelope . The calculation of length should be calculated according to the position of the sealing line segment. The length of the sealing line varies with the rotational angle as shown in figure 5. The rotor pitch is divided into two sections, one section of the pitch is 88 *mm*, the corresponding complete sealing line length is 198 *mm*, the other section of the pitch is 60 *mm*, and the corresponding contact length is 173 *mm*.

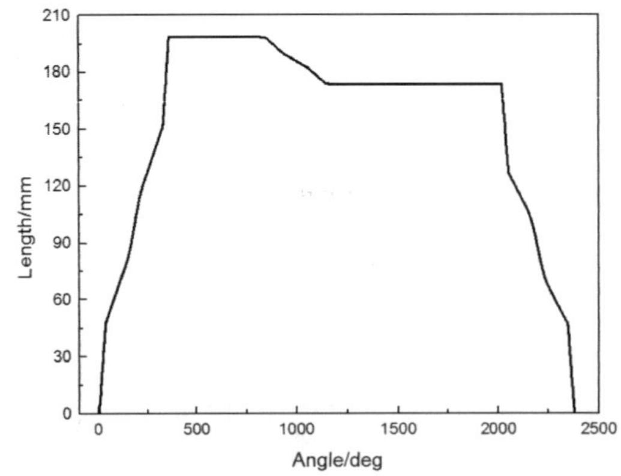

Figure 5. Length of sealing line varies with the rotational angle.

3. Mathematical model of vacuum pump performance

In this study, the working process of the vacuum pump can be divided into six processes, such as suction, transportation, internal compression, transportation, compression and exhaust as shown in figure 6.

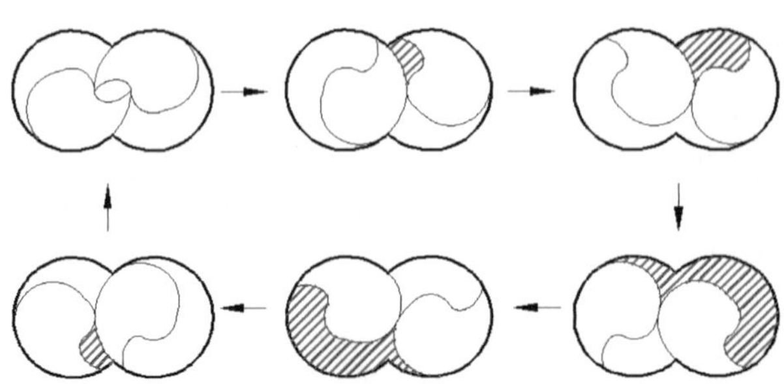

Figure 6. Vacuum pump working process.
（Shaded area is tooth area）

3.1. Certain assumptions

Certain assumptions had to be introduced to ensure efficient computation.These do not impose any limitations on the model nor cause significant departures from the real processes and are as follows[8]:

(1) The state of the control volume is uniform, that is, each point of the control volume has the same stated parameters;

(2) Gas inflow to and outflow from the pump ports are assumed to be isentropic;

(3)The volume of suction and the exhaust chamber are infinite. It is considered that there is no flow resistance loss and no air flow pulsation, and the suction and exhaust pressure in the working chamber is corresponding to be nominal.

3.2. Fundamental equations

According to the principle of variable mass thermodynamics and the first law of thermodynamics, the following forms of the conservation equations have been employed in the model:

$$dU = dQ + dE_i - dE_o - dW \tag{1}$$

Which can be the following forms:

$$d(mu) = \sum dm_{\mathrm{i}}(h_{\mathrm{i}} + \frac{v_{\mathrm{i}}^2}{2} + z_{\mathrm{i}}g) - \sum dm_{\mathrm{o}}(h_{\mathrm{o}} + \frac{v_{\mathrm{o}}^2}{2} + z_{\mathrm{o}}g) + \sum dQ - dW \tag{2}$$

The kinetic energy and the potential energy of the fluid are ignored in the equation, and the equation (2) is written as the change rate of the rotational angle:

$$\frac{d(mu)}{d\varphi} = \sum \frac{dm_{\mathrm{i}}}{d\varphi} h_{\mathrm{i}} - \sum \frac{dm_{\mathrm{o}}}{d\varphi} h_{\mathrm{o}} + \frac{dW}{d\varphi} - \frac{dQ}{d\varphi} \tag{3}$$

Taking into account $u = h - pv$ and $v = V_c / m$:

$$\frac{d(mu)}{d\varphi} = m\frac{dh}{d\varphi} + h\frac{dm}{d\varphi} - p\frac{dV_c}{d\varphi} - V_c\frac{dp}{d\varphi} \tag{4}$$

According to the relationship between the thermal state parameters of gas:

$$\frac{dh}{d\varphi} = \left(\frac{\partial h}{\partial \varphi}\right)_T \frac{dv}{d\varphi} + \left(\frac{\partial h}{\partial T}\right)_v \frac{dT}{d\varphi} \tag{5}$$

Taking equation(5) into the equation (3) and (4), and considering:

$$\frac{dp}{d\varphi} = \frac{1}{v}\left[\left(\frac{\partial h}{\partial v}\right)_T \frac{dv}{d\varphi} + \left(\frac{\partial h}{\partial T}\right)_v \frac{dT}{d\varphi}\right] - \frac{1}{V_c}\left[\sum \frac{dm_i}{d\varphi}(h_i - h) - \frac{dQ}{d\varphi}\right] \tag{6}$$

According to the relationship between the thermal state parameters of gas:

$$\frac{dp}{d\varphi} = \left(\frac{\partial p}{\partial v}\right)_T \frac{dv}{d\varphi} + \left(\frac{\partial p}{\partial T}\right)_v \frac{dT}{d\varphi} \tag{7}$$

$\frac{\mathrm{d}T}{d\varphi}$ can be calculated in the equation (7) and brought into (6), the pressure varies within the rotational angle is given by the differential equation:

$$\frac{dp}{dt} = \frac{\frac{1}{v}\left[\left(\frac{\partial h}{\partial v}\right)_T - \frac{(\partial h / \partial T)_v (\partial p / \partial v)_T}{(\partial p / \partial T)_v}\right]\frac{dv}{d\varphi} - \frac{1}{V_c}\left[\sum \frac{dm_i}{d\varphi}(h_i - h)\right]}{1 - \frac{1}{v}\frac{(\partial h / \partial T)_v}{(\partial p / \partial T)_v}} \tag{8}$$

Equation (8) is the basic equation of the numerical simulation of the working process of the screw vacuum pump. This equation takes into account the effect of the leakage on the pressure. The leakage model is introduced in detail.

3.3. *Leakage model*

3.3.1. Gas leakages. The gas leakages in the screw vacuum pump can be divided into internal and external leakage. The internal leakages refers to that gas leaks into the low-pressure chamber from the high-pressure chamber through the leakage clearances, which increase the power consumption, and reduce the adiabatic efficiency of the vacuum pump. The external leakages refer to that the gas leaks into the low pressure chamber connected with the suction port from the high-pressure chamber through the rotor sealing line, which reduces the volumetric efficiency of the vacuum pump. The leakage clearances of the vacuum pump can be divided into circumferential gap, sealing line gap, end face gap and leakage triangle gap[9], as shown in figure 7.

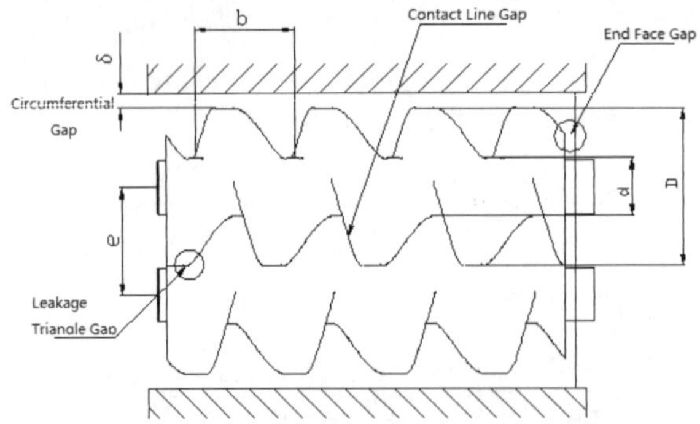

Figure 7. Leakage clearances.

For the oil free twin screw vacuum pump, the leakage is mainly calculated by the isentropic flow nozzle model, which not only simplifies the model, but also has high accuracy:

when $\left(\frac{2}{k+1}\right)^{\frac{k}{k-1}} \le \frac{p_2}{p_1} \le 1$:

$$\frac{dm}{d\varphi} = \frac{CAp_1}{\omega}\sqrt{\frac{2k}{(k-1)RT_1}\left(\left(\frac{p_2}{p_1}\right)^{\frac{2}{k}} + \left(\frac{p_2}{p_1}\right)^{\frac{k+1}{k}}\right)} \tag{9}$$

when $0 \le \frac{p_2}{p_1} \le \left(\frac{2}{k+1}\right)^{\frac{k}{k-1}}$:

$$\frac{dm}{d\varphi} = \frac{CAp_1}{\omega}\sqrt{\frac{k}{(k-1)R}\left(\frac{2}{k+1}\right)^{\frac{k+1}{k-1}}} \tag{10}$$

where p_1 is high side pressure, p_2 is low side pressure, T_1 is temperature of high pressure, A is leakage area, C is flow coefficient, k is adiabatic exponent, R is universal gas constant, ω is angular velocity.

3.3.2. Calculation of flow coefficient. The sealing line of the rotor profile in this study can reach the intersection of male and female of the rotor cylinder, so the leakage triangle can not be considered in the calculation of leakage. The leakage from the sealing line gap, the circumferential gap and the end face gap are considered in the calculation. Because of the adaptability the leakage, it is calculated by the isentropic flow nozzle model[10]. In the leakage model, the choice of leakage coefficient is the key to the correct calculation. The leakage from the sealing line gap and the circumferential gap are the main leakage clearances of the twin screw vacuum pump. This study focuses on the calculation of the flow coefficient of the leakage pass through the two gaps. The circumferential gap is deemed to the flat leakage model, called flat leakage. As mentioned above, the sealing line can be divided into ab(generated by the meshing of point and cycloid curve), bc(generated by the meshing of the tip and the root circle) and cd(generated by the meshing of the Archimedes spiral and the Archimedes spiral envelope conjugate) According to the geometric characteristics of contact, the first leakage clearance can be designated as circular leakage and the later two leakage clearance can be designated as cuspate leakage.

In this study, the geometric models of the leakage from sealing line gap and circumferential gap were set up, and the corresponding flow coefficient of the leakage clearance was calculated. The geometric models were divided into structure grids by using ICEM®. The leakage coefficient was calculated using Fluent®. The same boundary conditions were set up for different models. The inlet and outlet are pressure boundary conditions. The inlet pressure was divided into 100*kPa*, 50*kPa*, 25*kPa* and 10*kPa* respectively, and the outlet pressures were set according to the pressure ratio of 1.11, 1.25, 1.43, 1.67 and 1.82 respectively.

（1）Leakage through sealing line gap

The geometric of circular leakage model and the cuspate leakage model are shown in figure 8.

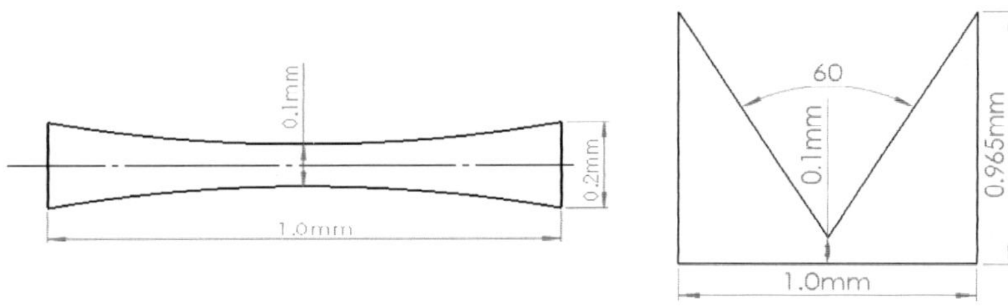

（a）Circular leakage model （b）Cuspate leakage model

Figure 8. Leakage model.

The flow coefficients in the circular leakage model varied with pressure ratio as can be seen from the figure 9(a). Flow coefficient changes little with pressure ratio which can be taken as 0.35-0.4 in the model.

The flow coefficients in the cuspate leakage model varied with pressure ratio can be seen from the figure 9(b). Flow coefficient changes little with pressure ratio which can be taken as 0.7-0.8 in the model.

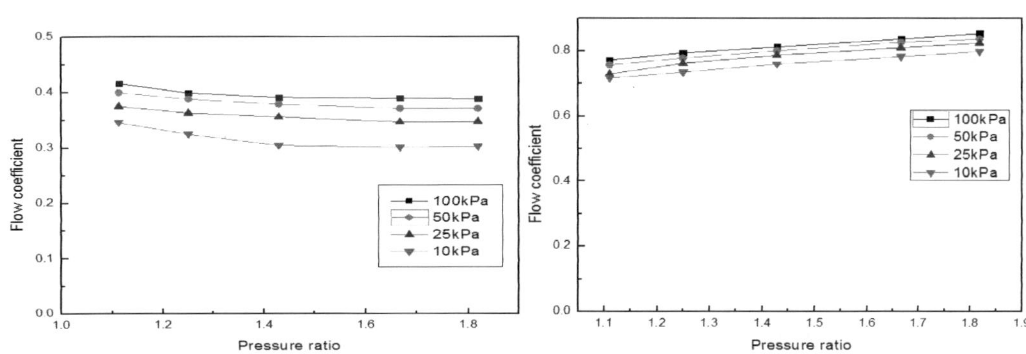

（a）Circular leakage model （b）Cuspate leakage model

Figure 9. Flow coefficient varies with the pressure ratio.

（2）Leakage through circumferential gap

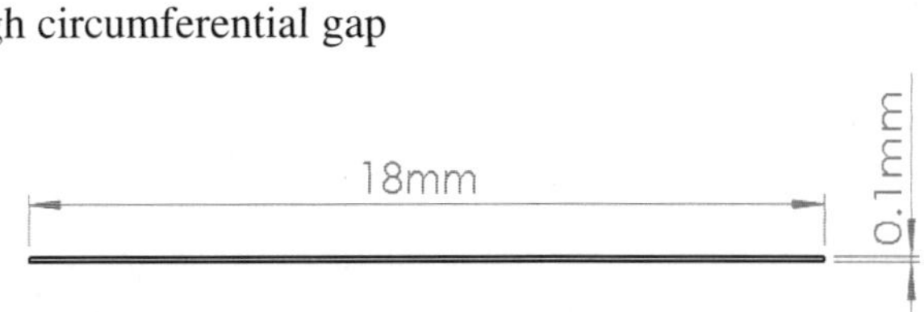

Figure 10. Flat leakage model.

The geometrics of the flat leakage model is shown in figure 10.

The flow coefficients in the flat leakage model varied with the pressure ratio can be seen from figure11. Flow coefficient changes little with pressure ratio which can be taken as 0.08-0.12 in the model.

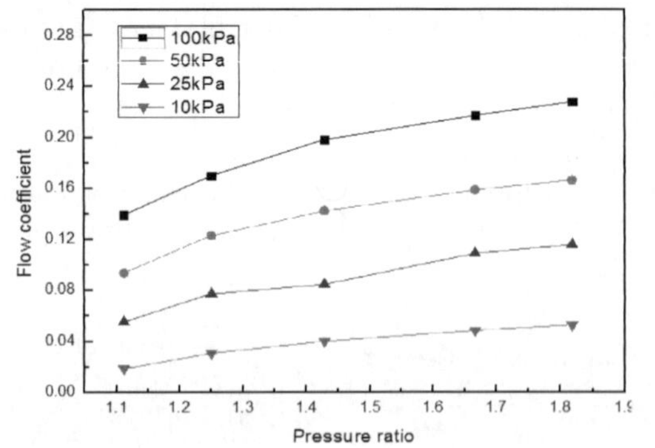

Figure 11. Flow coefficient varies with the pressure ratio.

The calculation results show that the three flow coefficients change little with the pressure ratio. Under the given conditions, the flow coefficient of the circular leakage model can be taken as 0.35-0.4, the flow coefficient of cuspate leakage model can be taken as 0.7-0.8, the flow coefficient of the flat leakage model can be taken as 0.08-0.12. For the model of this paper, the flow coefficient of the uspate leakage model is 0.4, the flow coefficient of cuspate leakage model is 0.75, the flow coefficient of the flat leakage model is 0.08.

4. The equation solution and the results analysis

4.1. The equation solution

The mathematical equations describing the working process of the twin-screw vacuum pump have been established, and then were performed numerically by means of the Runge-Kutta 4th order method, with appropriate initial and boundary conditions. First, the no-leakage process was calculated on the assumption that the model was thermally insulated, and each thermodynamic parameter changing with the rotational angle in the whole working process was determined, including initial pressure and initial temperature. Second, considering the leakage influence, the procedure ran with the initial pressure and temperature as given values. Then new pressure and temperature were obtained by the iterative computations. If the relative error between the new values and the last values of pressure and temperature was greater than ε, then the procedure continued to run with the new values of pressure and temperature as given values until the relative error was less than ε.

4.2. The result of the analysis

Based on the mathematical model of the working process in the screw vacuum pumps, the p-θ diagrams which describe the pressure changing with the rotary angle were calculated at three different rotational speeds of $2400r{\cdot}min^{-1}$, $2700r{\cdot}min^{-1}$ and $3000r{\cdot}min^{-1}$ when the suction pressure was $3kPa$, as shown in Figure 12. Figure 12 indicates that the pressure varies greatly between the ideal condition and the actual condition. The main reason is that the mesh clearance of the oil-free twin screw vacuum pump is relatively large, plus the clearance gap seal, so that the leakage is more serious. The pressure curve is roughly divided into two sections at the point of 1500 degrees. Before 1500 degrees, the pressure rises slowly, and there is no significant difference in pressure at different speeds; after 1500 degrees, the control volume reaches the exhaust end and the exhaust pressure is 1.0 *bar*, the adjacent control volumes have a larger pressure difference and there is a large leakage.

The exhaust pressures under these three speeds were 103.5*kPa*, 87.0*kPa* and 71.5*kPa* respectively, which shows that the leakage can be reduced greatly by increasing the speed. There is the phenomenon of over compression at the speed of 2400$r\cdot min^{-1}$ which means the pressure is already greater than the back pressure of the system before the control volume connected with the exhaust port. The simulation results have guiding significance for the design of the screw pitch and the exhaust port.

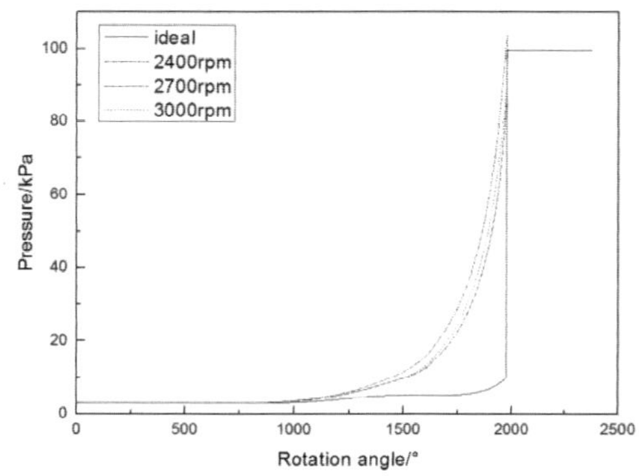

Figure 12 *P-θ* diagrams at different rotational speeds.

The figure 13 shows *p-V* diagrams turned by the *p-θ* diagrams, and the working process inside the vacuum pump can be seen clearly. There is a constant volume stage in the working process, the reason for the increasing pressure is the leakage.

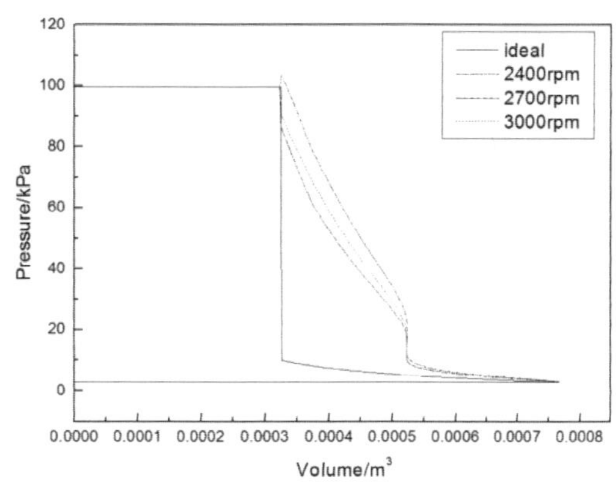

Figure 13 *P-V* diagrams at different rotational speeds.

According to the *p-V* diagram, the pressures of three speeds changed in the same trend, the leakage reduced as the speed increased, and the pressure rose slower and slower. Because of the constant volume stage there is a segment in which pressure rises perpendicularly. Overall, the variable pitch rotor is beneficial to seal, and reduce the leakage from the exhaust port to the suction port. The power consumption under different conditions can be seen directly from the *p-V* diagram, and the power consumption at high speed is relatively low.

5. Conclusion

The working process characteristics of the oil-free twin screw vacuum pump were analyzed, including suction, compression, transmission and exhaust. The basic control equations of the working process were established according to the energy conservation equation and the basic principle of variable mass thermodynamics, also considering the effect of leakage on the process. The main conclusions can be summarized as following:

（1）The thermodynamic model describing the working process of twin-screw vacuum pump was established according to the basic principle of variable mass thermodynamics, the energy conservation equation, the mass conservation equation and the gas state equation.

（2）The leakage through the sealing line and the meshing line were analyzed, and the corresponding mathematical model was established. ICEM®was used to build the structure grid of the geometric model, then the flow coefficients of each path were obtained by Fluent®. The flow coefficient can be seen as a constant for the leakage models with a given geometry. For the model of this paper, the flow coefficient of the cuspate leakage model is 0.4, the flow coefficient of the cuspate leakage model is 0.75, the flow coefficient of the flat leakage model is 0.08.

（3）The Runge-Kutta method was applied to calculate the control equations of the working process in the vacuum pump. The p-θ diagrams at three different rotational speeds of $2400 r{\cdot}min^{-1}$, $2700 r{\cdot}min^{-1}$, $3000 \mathrm{r{\cdot}min^{-1}}$ were calculated when the suction pressure was $3kPa$. The results show that the pressure rise is small in the first half of the screw rotor and the pressure rise is larger in the second half. There is the phenomenon of over-compression at the speed of $2400 r{\cdot}min^{-1}$. It is found that The calculated results are instructive to the design of the screw rotor.

Acknowledgements
The authors would like to express appreciation to the Department of Compressor Engineering of the Xi'an Jiaotong University for supporting and encouraging our efforts.

References

[1] Lee B, Yanagisawa T, Fukuta M and Choi S 2002. A study on the leakage characteristics of tip seal mechanism in the scroll compressor. *16th Int. Compressor Engineering Conf. at Purdue* pp1586-94.
[2] Z H Fong, F C Huang 2006. Evaluating the interlobe clearance and determining the sizes and shapes of all the leakage paths for twin-screw vacuum pump. *Mechanical engineering science*, **220** Part C pp499-506.
[3] FeiChuan H 2007. Study on the leakage evaluation of the non-conjugate rotor profiles for a twin-screw vacuum pump, *National Chuang Cheng University*.
[4] Yu Z 2008. The structure and properties of the rotors of dry screw vacuum pump. *Northeastern University*.
[5] Ohbayashi T, Sawada T, Hamaguchi M and Miyamura H 2001. Study on the performance prediction of screw vacuum pump. Applied Surface Science,**169** pp768-771.
[6] Stramann D, Kauder K 2008. Operating performance of screw vacuum pumps. *Vacuum in Forschung and Praxis*, **20** pp19-25.
[7] Yang L, Bei G, Maofei G 2015. Study on design of totor profile for the twin screw vacuum pump with single thread tooth. IOP Conf. Ser.: Mater. Sci. Eng. **90** 012007
[8] Yang L, Bei G, Ruixin Z 2015. Numerical simulation for working process in twin screw vacuum pump. *Journal of Xi'an Jiaotong University*, **49**(7) pp67-71.
[9] Egashir K, Shoda S, TochikAWA T and Furukawa A 1998. Backflow in twin-screw-type multiphase pump. *SPE production & facilities* **13**(01) pp64-69.
[10] Feng C, Ziwen X, Pengcheng S 2001. Study on the working process of twin screw multiphase pump. *Chinese Journal of Mechanical Engineering* ,**37**(3) pp73-77.

A CFD study of Screw Compressor Motor Cooling Analysis

S Branch
Ingersoll Rand – Trane
Compressor Development, La Crosse, WI, USA

E-mail: sbranch@trane.com

Abstract. Screw compressors use electric motors to drive the male screw rotor. They are cooled by the suction refrigerant vapor that flows around the motor. The thermal conditions of the motor can dramatically influence the performance and reliability of the compressor. The more optimized this flow path is, the better the motor performance. For that reason it is important to understand the flow characteristics around the motor and the motor temperatures. Computational fluid dynamics (CFD) can be used to provide a detailed analysis of the refrigerant's flow behavior and motor temperatures to identify the undesirable hot spots in the motor. CFD analysis can be used further to optimize the flow path and determine the reduction of hot spots and cooling effect. This study compares the CFD solutions of a motor cooling model to a motor installed with thermocouples measured in the lab. The compressor considered for this study is an R134a screw compressor. The CFD simulation of the motor consists of a detailed breakdown of the stator and rotor components. Orthotropic thermal conductivity material properties are used to represent the simplified motor geometry. In addition, the analysis includes the motor casings of the compressor to draw heat away from the motor by conduction. The study will look at different operating conditions and motor speeds. Finally, the CFD study will investigate the predicted motor temperature change by varying the vapor mass flow rates and motor speed. Recommendations for CFD modeling of such intricate heat transfer phenomenon have thus been proposed.

1. Introduction

There are two methods to consider for hermetic motor cooling. The first is liquid using either gravity, spray, or pool distribution. The second is vapor. A third method using an external air cooled motor isn't practical or cost effective compared to hermetic motor cooling methods. There are several benefits to vapor cooling in refrigeration screw compressors. The vapor temperature is controlled by the evaporator in a temperature range that is easily applicable to cooling the motor. The vapor is readily available as it travels into the compression process. Using the relatively cool suction gas allows for a straight forward cooling flow path design, a cool running motor, and increases the motor power density for a given motor size.

Temperature distribution of the motor can influence the performance of a screw compressor. If the insulation gets too hot, it breaks down and reduces the life of the motor. To understand the temperature distribution of the screw compressor motor, thermocouples are added to the stator and then evaluated in the lab on a test stand. In Figure 1, the large dots illustrate the locations where these thermocouples are placed. Even though many thermocouples are added to the stator portion of the motor, only a small percentage of the motor is being monitored.

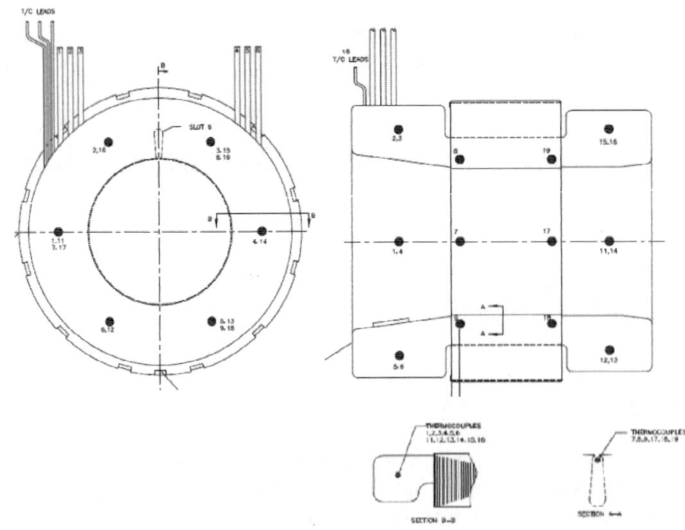

Figure 1. Thermocouple Locations for the Stator

By using CFD to evaluate the thermal profile of the motor, it can be done without thermocoupling. CFD can provide a more detailed distribution of the temperature profile than a thermocoupled motor could provide and more easily discover hot spots in the motor that can cause the motor to fatigue and shorten its life. CFD can also look at the flow path of the fluid around and through the motor which can be used to reduce the thermal hot spots in the motor, increase its life, and improve compressor efficiency. Literature review found two papers that looked at modelling a compressor motor; one from Chin (1) in 2012, and one from Tang (2) in 2016. Chin (1) used the lump parameter method by describing the heat transfer characteristics on a screw compressor by modelling a thermal equivalent circuit. Tang (2) performed a CFD analysis on a hermetic motor for a scroll compressor.

Motors consist of many parts, some of which consist of composite materials. Orthotropic thermal conductivity material properties were used during this study to simplify the composite materials into one volume. A more accurate representation of the material properties could then be used without having to create multiple volumes to represent the composite materials. A journal from Pietrak (3) in 2015 wrote about a Rayleigh model that addresses a process to calculate composite materials. A process described by Williams (4) in 2003 describes how to use orthotropic thermal conductivity in ANSYS CFX.

2. Motor Power Distribution

In the paper by Tang (2), motor efficiency FEA simulations were done to help determine the motor power. For this study, the motor power values were provided by the motor vendor. A simple motor power formula is **Watt = Volts x Amps x** $\sqrt{3}$ **x Motor Power Factor**. Motor power factor, is a ratio of the active power to apparent power. Active power is the real power doing the work. Apparent power is the vector sum of active and reactive powers. Reactive power is the power required for magnetization of the motor; it has no action and is influenced by the inductance of the motor.

These simulations used a uniform power distribution that did not change with temperature, and were applied to the following motor volumes:

- Rotor Aluminium
- Stator Laminate
- Inside and End Windings.

3. Review of Modelling Approach

CFD simulations were conducted using ANSYS CFX version 16.2 and were built with several domains. The laminates and windings used orthotropic thermal conductivity material properties. Most domains were stationary except for the rotor and the shaft. Figure 2 shows a cross section of the simulation setup. The motor rotor consisted of the aluminium rotor and the rotor laminate. The stator was created with many parts, including the liner, windings, wedges, and the laminate. The liner goes in between the outside of the copper windings and wedges on the inside of the stator laminate slots. The liner was simulated using a 2-D feature which was modelled as a thin material. The stator laminate, air gap, and inside copper windings are shown in Figure 3. Figure 4 highlights the inside copper windings and copper end windings.

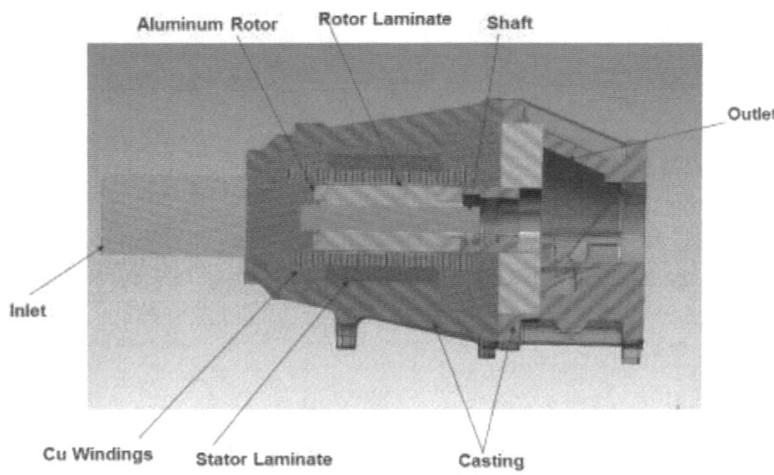

Figure 2. Simulation Setup

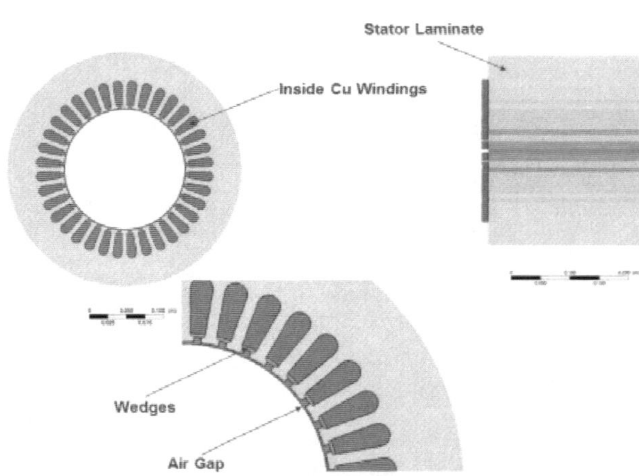

Figure 3. Simulation Setup

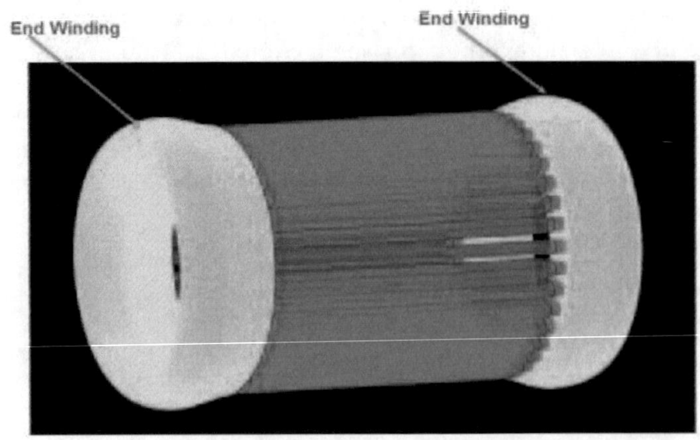

Figure 4. Simulation Setup

The flow path for the simulation went from the inlet to the radial and axial entrances to the rotors. This is shown in Figure 5, which includes the air gap between the stator and the rotor.

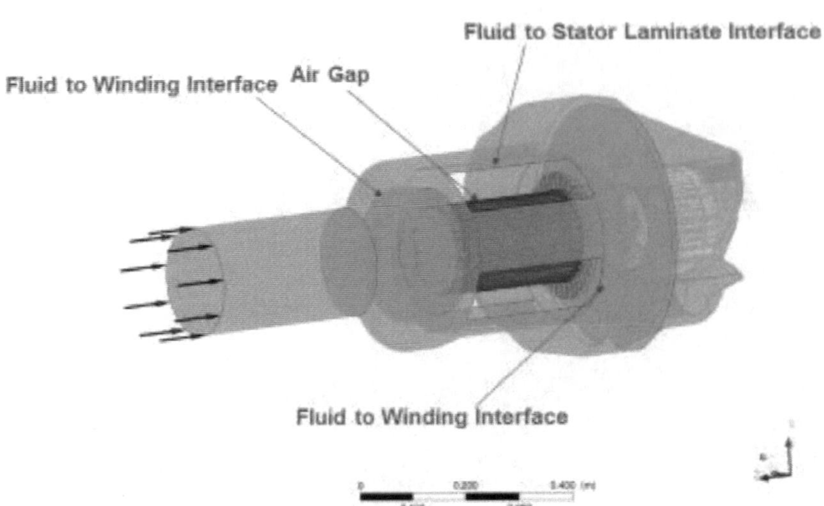

Figure 5. Simulation Setup

IOP Conf. Series: Materials Science and Engineering **232** (2017) 012028 doi:10.1088/1757-899X/232/1/012028

A rough heat transfer coefficient was applied to the outside of the casting using a flat plate heat transfer coefficient equation. The CFD setup assumptions are shown below.

- ANSYS CFX Version 16.2 was used for the CFD Analysis
- Steady State Simulation
- Heat Transfer = Total Energy w/ Vis Work Term.
- Turbulence Model = SST
- The outside walls had a heat transfer coefficient applied.
- RPM varied (3532 RPM, 864 RPM, 1679 RPM)
- Fluid = R134a Gas
- No oil
- Radiation is not considered.
- Orthotropic thermal conductivity applied to
  - rotor laminate
  - stator laminate
  - windings
- Mesh size = 77.9 million nodes
- Interfaces = 449

The simulations were run on a cluster using 162 cores with several E5-2697 v3 CPU's. Simulation times ranged from one to six days depending on the differences in the operating conditions from the initial conditions.

The orthotropic material properties were calculated using the Rayleigh models which came from the Journal "A review of models for effective thermal conductivity of composite materials", on page 16 from Pietrak (3). The equations used are shown below where k_l is treated as a cylinder of the metal and k_m = the lamination or varnish in the windings, and $\emptyset$ = volume fraction of the materials.

Assuming z is the axis of the cylinder, the formulas are

- $\frac{k_{eff,ZZ}}{k_m} = 1 + \left(\frac{k_1 - k_m}{k_m}\right)\emptyset$
- $\frac{k_{eff,XX}}{k_m} = \frac{k_{eff,YY}}{k_m} = 1 + \frac{2\emptyset}{C_1 - \emptyset + C_2(0.30584\emptyset^4 + 0.013363\emptyset^8 + \ldots)}$
- $C_1 = \frac{k_1 + k_m}{k_1 - k_m}, C_2 = \frac{k_1 - k_m}{k_1 + k_m}$

Williams (4) describes the setup and how to incorporate an orthotropic material property in cylindrical or Cartesian coordinates for ANSYS CFX. He also performs a test case that shows how well it works. In CFX, this is a beta feature. To be able to use beta features, that option needs to be turned on. The CEL expression in cylindrical coordinates for how to set this up in CFX are shown in Figure 6.

```
MATERIAL: Material
  Material Group = User
  Option = Pure Substance
  Thermodynamic State = Solid
  PROPERTIES:
    Option = General Material
    EQUATION OF STATE:
      Density = XX.X [kg m^-3]
      Molar Mass = X [kg kmol^-1]
      Option = Value
    END
    REFERENCE STATE:
      Option = Specified Point
      Reference Specific Enthalpy = X [J kg^-1]
      Reference Temperature = XX [C]
    END
    SPECIFIC HEAT CAPACITY:
      Option = Value
      Specific Heat Capacity = X.XX [J kg^-1 K^-1]
    END
    THERMAL CONDUCTIVITY:
      Option = Orthotropic Cylindrical Components
      Thermal Conductivity Axial Component = XX.X [W m^-1 K^-1]
      Thermal Conductivity Theta Component = XX.X [W m^-1 K^-1]
      Thermal Conductivity r Component = XX.X [W m^-1 K^-1]
      AXIS DEFINITION:
        Option = Coordinate Axis
        Rotation Axis = Coord X.X
      END
    END
  END
END
```

Figure 6. Simulation Setup

4. Test and Analysis

Before the CFD test cases were conducted, a thermal boundary layer study was performed to determine the correct inflation layer for the simulations. It is important to note that a viscous boundary layer is not the same as a thermal boundary layer, but the relationship is the Prandtl number. After archiving mesh independence, the simulations were started. All simulations were run until the RMS residuals converged to at least 1e-4 and all of the imbalances were converged to at least one percent.

The results from the first three cases had a motor installed with several thermocouples on the stator as shown in Figure 1, which were monitored and measured in the lab. Results from these cases have been normalized by setting the lowest temperature value in each experiment to zero and setting the largest temperature value in each experiment to one. In all cases, the inlet was a pressure and temperature boundary condition, and the outlet was a mass flow rate boundary condition. For cases 1 through 3 the volumetric average, minimum and maximum temperature values came from the CFD simulations. The thermocouple values came from the average values of the thermocouples on the motor from lab testing.

For case 4, the motor speed and mass flow rates were doubled from case 3, and the motor temperature drop was predicted. Case 4 was not tested in the lab, so no thermocouple data was available. The operating conditions are provided for all cases in the format of saturated suction temperature, saturated discharge temperature, and super heat.

For all cases, the flow path of R134a fluid goes towards the negative Z direction. The rotor side end windings are downstream from the suction side end windings. The temperature profile of one of the simulations is seen in Figure 7.

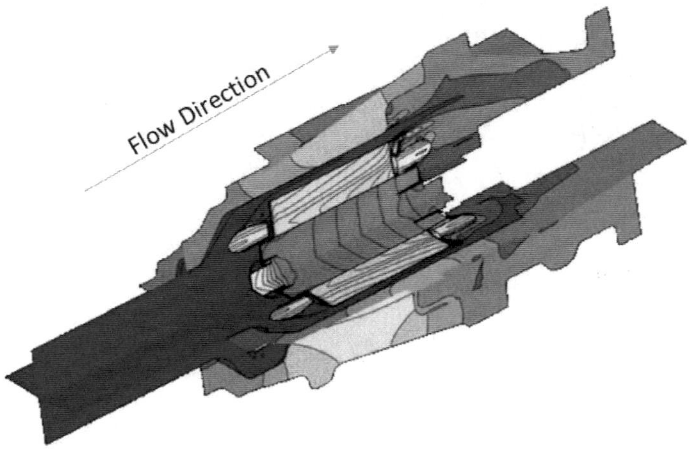

Figure 7. Temperature Profile of the Simulation

Even though the stator geometry is symmetrical, the stator contour temperature profiles is not symmetrical because the flow path around the motor is not symmetrical. For Case 1, the compressor was operated at a (40/100/0) °F operating condition. The motor was rotating at 3532 rev/min. The inlet density of the fluid is 0.8247 lbm/ft^3. The CFD contour plot of the motor is shown in Figure 8. The normalized numerical results are in Table 1. From the data in Table 1 the lowest temperature is the suction side winding and the largest temperature is the rotor.

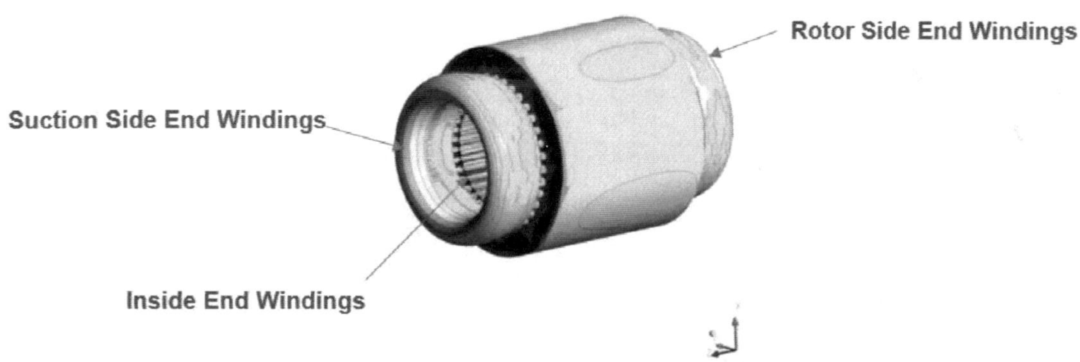

Figure 8. Case 1 Simulation Results

Table 1. Normalized Case 1 Temperature Results

	Volumetric Average	Min.	Max.	Ave. Thermocouple
Suction Side End Windings	0.57	0.00	0.89	0.37
Inside End Windings	0.60	0.05	0.91	0.44
Rotor Side End Windings	0.66	0.03	0.84	0.58
Rotor Aluminum	0.73	0.51	1.00	

For Case 2, the compressor was operated at a lower pressure ratio than Case 1, and a much lower speed. The operating condition was (40/75/0) °F, and the motor speed was 864 rev/min. The inlet density of this simulation is the same as in Case 1. The lower speed reduced the mass flow rates of the R134a by 4.5x. Along with lower mass flow rates, the power distribution in the motor is 3.5x lower than in Case 1. The CFD contour plot of the motor is shown in Figure 9. The normalized numerical results are in Table 2.

Figure 9. Case 2 Simulation Results

Table 2. Normalized Case 2 Results

	Volumetric Average	Min.	Max.	Ave. Thermocouple
Suction Side End Windings	0.30	0.00	0.42	0.29
Inside End Windings	0.33	0.01	0.50	0.24
Rotor Side End Windings	0.40	0.02	0.53	0.43
Rotor Aluminum	0.85	0.70	1.00	

For Case 3, the compressor was operated at the largest pressure ratio of the three cases and at a medium speed. An operating condition of (17.3/116/0) °F was tested. The motor speed was 1679 rev/min and the inlet density of the fluid is 0.3586 lbm/ft^3. The mass flow rates of this case is 4.53x slower than Case 1. The main differences in Case 3 over the other two is that the mass flow rates are lower like Case 2 but the power is increased by 1.4x over Case 1. The increase in power is due to the higher pressure ratio, which caused the increase in torque. The CFD contour plot of the motor is shown in Figure 10. Normalized numerical results are listed in Table 3.

IOP Conf. Series: Materials Science and Engineering **232** (2017) 012028 doi:10.1088/1757-899X/232/1/012028

Figure 10. Case 3 Simulation Results

Table 3. Normalized Case 3 Temperature Results

	Volumetric Average	Min.	Max.	Ave. Thermocouple
Suction Side End Windings	0.13	0.00	0.23	0.04
Inside End Windings	0.14	0.01	0.29	0.08
Rotor Side End Windings	0.22	0.03	0.68	0.25
Rotor Aluminum	0.62	0.39	1.00	

Table 4 was created to compare Cases 1-3. Only the volumetric average values were used in this comparison. The numbers were normalized with the smallest values being set to 0 and the largest values being set to one.

Table 4. Normalized Cases 1,2,3 Volumetric Averaged Temperature Results

	Volumetric Average		
	Case 1	Case 2	Case 3
Suction Side End Windings	0.06	0.00	0.41
Inside End Windings	0.06	0.00	0.42
Rotor Side End Windings	0.07	0.01	0.52
Rotor Aluminum	0.09	0.08	1.00

Case 4 was not tested in the lab, but was simulated to predict the differences in motor temperature drop by doubling the motor speed and mass flow rates from Case 3. The motor speed for this simulation was 3358 rev/min. The operating condition was the same as Case 3 at (17.3/116/0) °F. Table 5 shows the percent reduction in motor temperature.

Table 5. Average Volumetric Percent Reduction

	Average Volume % Reduction
Suction Side End Winding	15.7
Inside End Winding	19.8
Rotor Side End Winding	20.4
Rotor Aluminum	38.8

5. Conclusion

- The CFD study from the first three test cases looked at various operating conditions at different speeds. These CFD solutions were compared with the lab results where the motor was installed with thermocouples and measured in the lab. The CFD results were consistent with the thermocoupled measured data.
  - The coolest parts of the motor were the end windings and the inside windings. The hottest part of the motor was the rotor.
- When all three cases were compared to each other, Cases 1 and 2 were very similar. With the decreased mass flow rates along with the increased power in Case 3, the temperature of the motor was increased considerably.
- Case 4 looked at motor response when the mass flow rates were doubled along with the rotational speed from Case 3. The largest reduction was the rotor at 38.8%. The motor temperature drop improved as the gas moved through the motor from the suction side to the rotor side.
- In time, additional improvement could be gained by incorporating the motor efficiency FEA simulations methods done from Tang (2).

Acknowledgement

The author acknowledges Rick James for informing me about the orthotropic material property method in CFX. The author also acknowledges Lucas Fabian, Chris Espelien, and the Motor Vendor for helping me with the motor power section of this paper.

References

[1] Chen, W, Shen, J, He, Z, Wu, H, Xing, Z, 2012, "Thermal Modelling for the Motor in Semi-hermetic Screw Refrigeration Compressor under Part-load Conditions", Proceedings of the International Compressor Engineering Conference at Purdue, West Lafayette, IN, 1246

[2] Tang, W, Wang, L, Yao, L, Lin, J, 2016, "The Improvement of Motor Cooling Through Stator Profile Optimization using CFD Analysis in Hermetic Scroll Compressors", Proceedings of the International Compressor Engineering Conference at Purdue, West Lafayette, IN, 1370

[3] Pietrak, K, Wisniewski, T, 2015, "A review of models for effective thermal conductivity of composite materials", Institute of Heat Engineering, Warsaw University of Technology, 21/25 Nowowiejska Street, 00-665 Warsaw, Poland

[4] Williams, Dan, 2003, "Orthotropic Conductivity in CFX-5", CFX International 8.19 Harwell, Didcot Oxford OX11 ORA, United Kingdom

Optimization analysis of the motor cooling method in semi-closed single screw refrigeration compressor

Z L Wang*, Y F Shen, Z B Wang and J Wang

College of Chemical Engineering, China University of Petroleum, Qingdao 266580, China
Email address: wangzengli25@ upc.edu.cn
* Corresponding Author

Abstract. Semi-closed single screw refrigeration compressors (SSRC) are widely used in refrigeration and air conditioning systems owing to the advantages of simple structure, balanced forces on the rotor, high volumetric efficiency and so on. In semi-closed SSRCs, motor is often cooled by suction gas or injected refrigerant liquid. Motor cooling method will changes the suction gas temperature, this to a certain extent, is an important factor influencing the thermal dynamic performance of a compressor. Thus the effects of motor cooling method on the performance of the compressor must be studied. In this paper mathematical models of motor cooling process by using these two methods were established. Influences of motor cooling parameters such as suction gas temperature, suction gas quantity, temperature of the injected refrigerant liquid and quantity of the injected refrigerant liquid on the thermal dynamic performance of the compressor were analyzed. The performances of the compressor using these two kinds of motor cooling methods were compared. The motor cooling capacity of the injected refrigerant liquid is proved to be better than the suction gas. All analysis results obtained can be useful for optimum design of the motor cooling process to improve the efficiency and the energy efficiency of the compressor.

1. Introduction

As a rotary type compressor, the single screw compressor is firstly invented by Zimmen B in 1960s [1]. Due to the advantages of simple structure, small vibration, good dynamic balance, low noise and high reliability, the single screw compressor has always been regarded as a good theoretical performance compressor. It was first used in air compression field, and then was introduced to the field of refrigeration soon in the mid of 1970s [2,3]. The first opened single screw refrigeration compressor was manufactured in 1975 [4], and the semi-closed single screw refrigeration compressor (SSRC) was researched and developed in 1982 [5]. Now semi-closed SSRCs are widely used in refrigeration and air conditioning systems. The semi-closed SSRC mainly consists of one screw rotor and two star-wheels which symmetrically located with the screw rotor as shown in figure 1. The motor is connected with the screw rotor directly. During the compressor operating process, the copper loss, iron loss, excitation loss, ventilation, friction loss and so on will be generated by the motor and be converted into heat energy which will cause the motor heating. In order to cool the motor, motor cooling methods using suction gas, injected refrigerant gas/liquid or exhaust gas were put forward [6]. In semi-closed SSRCs, motor cooling method will changes the suction gas temperature, this to a certain extent, is an important factor influencing the thermal dynamic performance of the compressor.

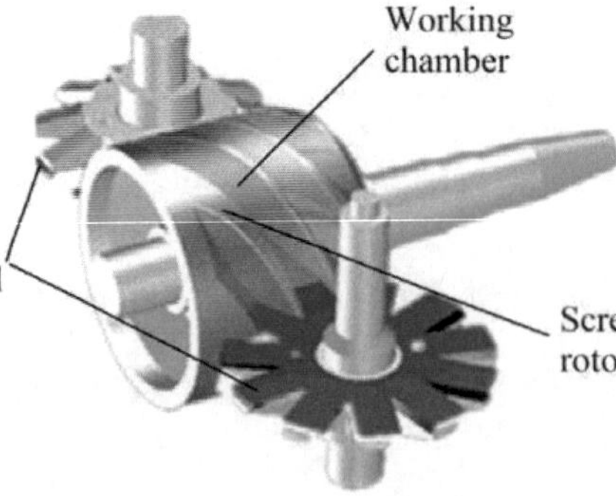

Figure 1. The core working parts of a single screw compressor

Over the years, a series of studies have been carried out for the motor heat dissipation, the motor cooling method and the influence of the motor cooling process on the compressor. Chen et al. built a mathematical model describing flow and heat transfer characteristics of the motor in semi-hermetic twin screw refrigeration compressor under part-load conditions based on fluid network and thermal equivalent circuit theories [7,8]. In their study, the temperature and pressure distribution inside the motor and their influences on performance of the compressor were analyzed. Wen et al. investigated the coil temperature field of electromotor in screw refrigeration compressor. On basis of this, the effects of working condition, velocity of refrigerant flow and electromotor power upon coil temperature were analyzed [9]. He et al. developed a comprehensive model to study the flow and thermal behavior of the motor cooling process through suction refrigerant in semi-hermetic twin screw refrigeration compressors. By using this model, the temperature distribution inside the motor was accurately predicted [10]. He et al. analyzed the temperature distribution of motor in the semi-hermetic screw refrigerant compressor which was applied in high temperature heat pump system [11]. Liang et al researched the influence of the motor cooling process on the enclosed type compressor and the semi-closed compressor with motor cooling methods using suction gas and injected refrigerant gas/liquid [12]. Dutra et al. presented a simulation approach for hermetic reciprocating compressors including electrical motor modeling, and studied the effect of the motor on the performance of the compressor [13]. For the semi-closed SSRC, Wang et al. established the mathematical model to analysis the thermodynamic performance of the semi-closed SSRC with the consideration of motor cooling process using suction gas [14]. Although many scholars have done a lot of work, so far no work has been done about the contrastive analysis of different motor cooling method in semi-closed SSRCs. In semi-closed SSRCs, motor cooling method is an important factor influencing the thermal dynamic performance of the compressor. Thus the effects of motor cooling method on the dynamic performance of semi-closed SSRCs must be studied.

In this paper, mathematical models of the motor cooling process by using suction gas or injected refrigerant liquid were established. On the basis of these models, the influences of the motor cooling parameters such as suction gas temperature, suction gas quantity, temperature of the injected refrigerant liquid and quantity of the injected refrigerant liquid on the thermal dynamic performance of the compressor were analyzed. The performances of the compressor using these two kinds of the motor cooling methods were compared. All analysis results obtained in this paper can be useful for optimum design of the motor cooling process in the semi-closed SSRC to improve the efficiency and the energy efficiency of the compressor.

2. Motor cooling principle in the SEMI-CLOSED SSRC

At present, the common methods used in semi-closed SSRC for cooling the built-in motor are the suction gas cooling method, the injection gas cooling method and the exhaust gas cooling method. In this study, the motor cooling method using the suction gas and the injected refrigerant liquid were investigated. The motor cooling principles of these two methods are shown as follows.

2.1. Motor cooling method using the suction gas

Figure 2 shows the structure of the semi-closed SSRC with the motor cooling method using suction gas (case 1). As shown in figure 2, the semi-closed SSRC is composed by three parts: motor, working component and the oil-gas separator. The screw rotor of the work component is driven by motor directly. When the semi-closed SSRC begins to run, the refrigerant gas with lower temperature and pressure will flows into the suction chamber through the motor after filtering. After this process, the motor windings will be cooled, and the refrigerant gas will be overheated. Then the overheating refrigerant gas will enters into the compression chamber of the working component and be compressed to a high pressure.

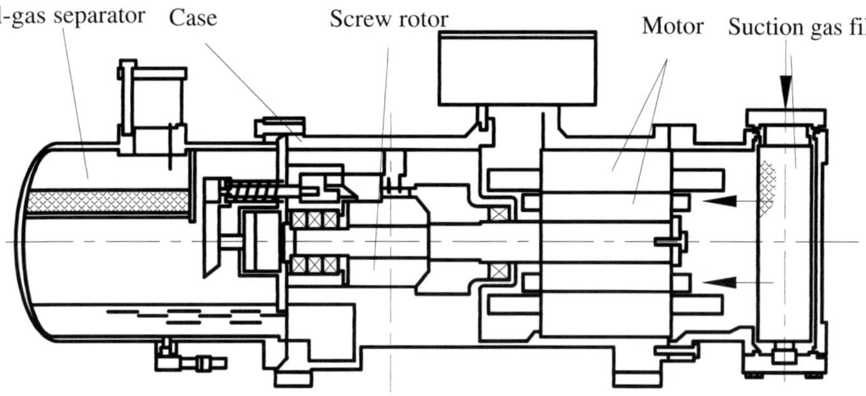

Figure 2. Semi-closed SSRC with the motor cooling method using suction gas

As shown in figure 3, the refrigerant gas with lower temperature and pressure at the state point 1' will be overheated to the state point 1 after cooling the motor. The larger suction superheat will increases, the specific volume of the suction refrigerant gas which will cause the decrease of the compressor volume flow per unit time, and will eventually reduce the compressor working performance. So in this study, the influence of the refrigerant gas overheating caused by the motor cooling process has been analyzed.

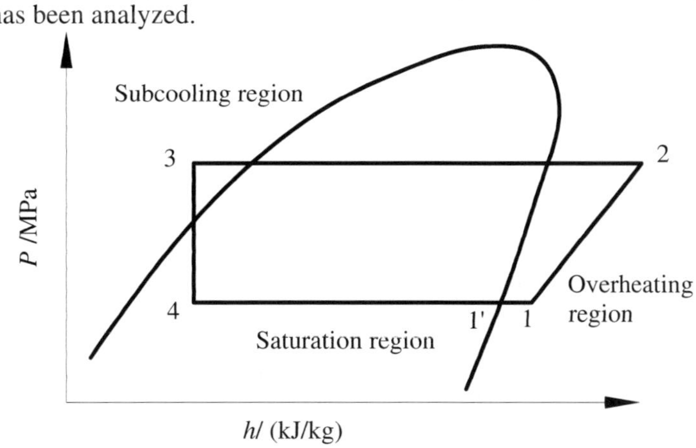

Figure 3. Semi-closed SSRC work cycle with the motor cooling method using suction gas

2.2. Motor cooling method using injected refrigerant liquid

Figure 4 shows the structure of the semi-closed SSRC with the motor cooling method using injected refrigerant liquid (case 2). In this method, the liquid refrigerant from the condenser is divided into two parts. One part of the liquid refrigerant with high pressure and low temperature enters into the evaporator after flowing through the expansion valve to participate in the refrigeration cycle process, the other part of the liquid refrigerant with high pressure and low temperature is directly injected into the motor chamber after flowing through the throttling capillary to realize the cooling of the motor windings. Then this part of the refrigerant used for cooling the motor will be mixed with the rest of the refrigerant and entered into the working chamber to participate in the refrigeration cycle process. By using this method, the temperature of the suction refrigerant gas will be decreased than that of the case 1. The influence of cooling process using injected refrigerant liquid also has been analyzed as follows.

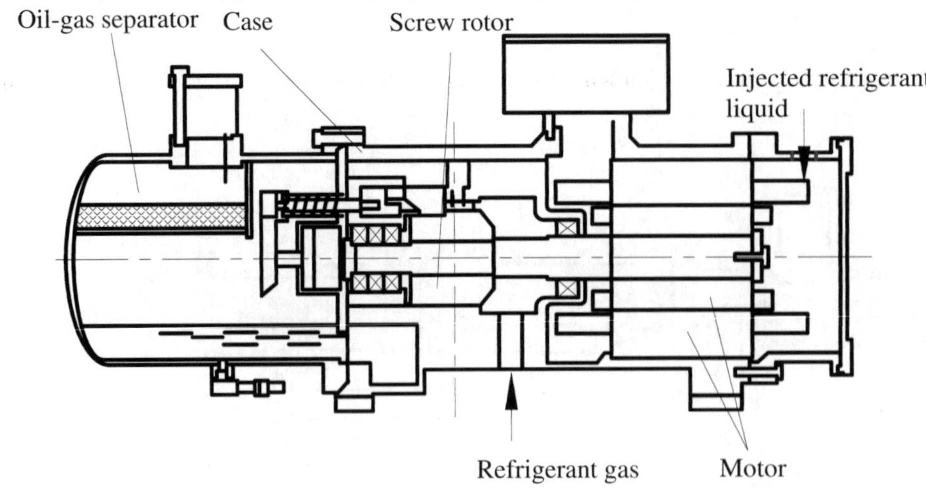

Figure 4. Semi-closed SSRC with the motor cooling method using injected refrigerant liquid

3. Mathematical model of the motor cooling process

3.1. Energy loss calculation model for the motor

According to the motor cooling principle in the semi-closed SSRC, the dissipation heat generated by the motor must be carried off by the refrigerant. So in order to analyze the motor cooling process, the dissipation heat generated by the motor under the steady state must be gained. By the motor design theory, dissipation heat generated by the motor due to the energy loss in the process of electrical energy converted into mechanical energy mainly include the basic iron loss, the electric loss and the stray loss [15]. The energy loss calculation model for the fixed frequency motor is shown as follows.

Basic iron loss of the fixed frequency motor contains the hysteresis loss and eddy current loss caused by the main magnetic field changes in the core, the calculation formula of these basic iron losses is:

$$P_{Fe} = K_a P_{he} G \sum_i P_{hei} G_i \tag{1}$$

Where K_a is the experience factor which is used to represent the basic iron losses increase caused by the processing and of the steel, the magnetic flux density distribution and change, P_{hei} is the loss coefficient, G_i is the quantity which is suffered by the magnetizing effect.

The electric loss of the fixed frequency motor includes the copper (aluminum) consumption loss of each winding and the contact consumption loss between the brush, collector ring and the commutator, the calculation formula of these basic iron losses is:

$$P_{\mathrm{Cu(Al)1}} = m I_x^2 R_x \tag{2}$$

$$P_{\mathrm{cb}} = \Delta U_b I_b \tag{3}$$

Where m is the phase number, I_x is the electric current in the winding, R_x is the resistance of the winding, ΔU_b is the contact voltage drop of the brush, I_b is the electric current flow through the brush.

The leakage magnetic field (including harmonic magnetic field) which is generated by the working current in the stator and rotor will cause the loss in the winding and iron core and structures. These losses compose the stray loss and make the stray loss very complex. In view of this, the stray loss of the fixed frequency motor is difficult to calculate accurately. At present, motor design scholars at home and abroad tested the stray loss of different types and specifications induction motor under load condition. Through summarizes the result of the test, the relationship between the load stray loss and the motor rated power was obtained. For the cast aluminum rotor, the load stray loss in proportion to the rated power of motor is shown in table 1. So the stray loss can be calculated by the rated power of motor.

Table 1. The load stray loss in proportion to the rated power

Motor series	2	4	6	8/10
P_{s1} / P_N	0.025	0.02	0.015	0.01

In addition, the friction loss and ventilation loss also vary with the change of the variable frequency motor rotation speed. Generally speaking, the friction loss of rolling bearing is proportional to the speed, and the ventilation loss is proportional to the three times of the speed. By solving the losses mentioned above, the heat generated by the motor used in the semi-closed SSRC can be obtained.

3.2. The heat exchange process calculation model

In the case 1, the suction refrigerant gas with lower temperature and pressure will flows into the suction chamber through the motor. Then the heat exchange between the suction refrigerant gas and the motor will occurs which will cause the heat of the refrigerant gas. This inspiratory preheating phenomenon will changes the thermodynamic properties of the suction refrigerant gas, and eventually result in the decrease of capacity and efficiency of the semi-closed SSRC. As a consequence, in semi-hermetic SSRCs, the dissipation heat generated by the motor has bigger influence on the suction refrigerant gas. According to the principle of conservation of energy, the dissipation heat generated by the motor can be divided into two parts. One part of it makes the motor temperature rise, and the other part of it will be distributed to the refrigerant gas flows through the motor. So the distribution of the heat generated by the motor can be obtained:

$$\dot{Q}\mathrm{d}t = C\mathrm{d}T + \alpha_{\mathrm{m}} ST\mathrm{d}t \tag{4}$$

Where $\dot{Q}$ is heat generated by the fixed frequency motor per unit time, C is the heat capacity of the motor, α_{m} is the heat transfer coefficient of the motor, S is the heat transfer area of the motor, T is the temperature rise of the motor.

When the motor running to steady state, the dissipation heat generated by the motor will all is passed to the suction refrigerant gas which will cause the overheating of the gas. And then a steady state temperature rise will be achieved. The temperature rise of the suction refrigerant gas caused by the motor is:

$$\dot{Q}\mathrm{d}t = c_{\mathrm{g}} m_{\mathrm{g}} \left(T_{\mathrm{sd}} - T_{\mathrm{s}}\right) \tag{5}$$

Where c_{g} is the specific heat of refrigerant, m_{g} is the mass of refrigerant, T_{s} is the suction temperature of refrigerant.

In the case 2, the part of the liquid refrigerant with high pressure and low temperature is directly injected into the motor to take away the dissipation loss of the motor. After the heat transfer process, the liquid refrigerant absorbs heat and becomes superheated steam. Then the superheated refrigerant gas will enters into the suction chamber and be mixed with the rest of the refrigerant. The mixed refrigerant gas finally enters into the working chamber to participate in the refrigeration cycle process. In this method, there are two times of heat exchange process: the first process is the heat exchange between the liquid refrigerant and the motor, the second process is the heat exchange between the superheated refrigerant and the refrigerant comes from the evaporator.

According to the energy conservation equation, the heat exchange process between the liquid refrigerant and the motor can be expressed as follows:

$$\dot{Q}\mathrm{d}t + m_1 h_1^{'} = m_1 h_1 \tag{6}$$

The heat exchange process between the superheated refrigerant and the refrigerant comes from the evaporator can be expressed as:

$$m_1 h_1 + m_2 h_2 = (m_1 + m_2) h_s \tag{7}$$

Where m_1 is the mass of the injected liquid refrigerant, $h_1^{'}$ is the enthalpy of the injected liquid refrigerant, h_1 is the enthalpy of superheated steam, m_2 is the mass of the refrigerant comes from the evaporator, h_2 is the enthalpy of the refrigerant comes from the evaporator, h_s is the enthalpy of the suction refrigerant gas.

These calculation models were integrated into the mathematical model describing the working process of the SSRC as a subroutine in the form of the preheating of the suction refrigerant gas caused by the motor cooling process [14]. a step-by-step calculation procedure as shown in ref [14]with the fourth order Runge-Kutta method was compiled to solve the presented theoretical calculation model. Then influences of motor cooling parameters on the thermal dynamic performance of the compressor can be analyzed. In order to compare accurately, lubricating oil heat exchange effect was ignored in this study.

4. Results and discussion

In this paper, a typical semi-closed SSRC used in refrigeration and air conditioning field with the structure and operation parameters as shown in table 2 was chosen for the calculate example. By using this semi-closed SSRC, the influence of the motor cooling parameters such as suction gas temperature, suction gas quantity, temperature of the injected refrigerant liquid and quantity of the injected refrigerant liquid on the thermal dynamic performance of the compressor were analyzed. The performances of the compressor using these two kinds of the motor cooling methods were compared. The analysis results are shown as follows.

Table 2. The main parameters

Parameters(and unit)	Symbols	Value	Parameters(and unit)	Symbols	Value
Gear ratio	ε	11/6	Length of the screw (mm)	L	155.7
Radius of the screw (mm)	R_1	181	Star wheel thickness (mm)	t	7.25
Star wheel radius (mm)	R_2	181	Refrigerant/Lubricating oil	Re/ oil	R22/S uniso 4GS
Center distance (mm)	A	144.8	Evaporation temperature (K)	T_e	278.15
Breadth of the star wheel (mm)	b	26.5	Condensation temperature (K)	T_c	313.15
Rotate speed (r.min^{-1})	n	2880	Superheat temperature (K)	ΔT	5

For the case 1, the influence of the suction gas quantity on the thermal dynamic performance of the compressor is analyzed by changing the rotate speed of the compressor and is shown in figure 5. The variation tendency of the curves in figure 5 shows that the suction gas temperature of the compressor will rises due to the dissipation heat generated by the motor. That will cause the increase of the specific volume and the decrease of the displacement of the compressor. The temperature rise of the suction gas and the specific volume increment will decrease with the increase of the compressor rotate speed. But the lower degree of displacement will increases with the increase of the rotate speed.

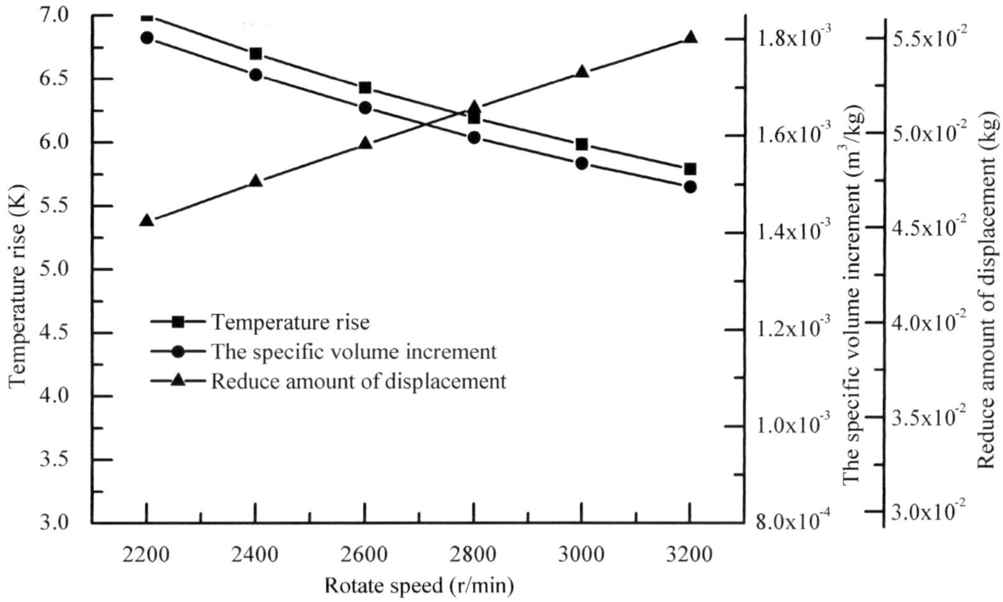

Figure 5. The influence of the suction gas quantity in the suction gas cooling method

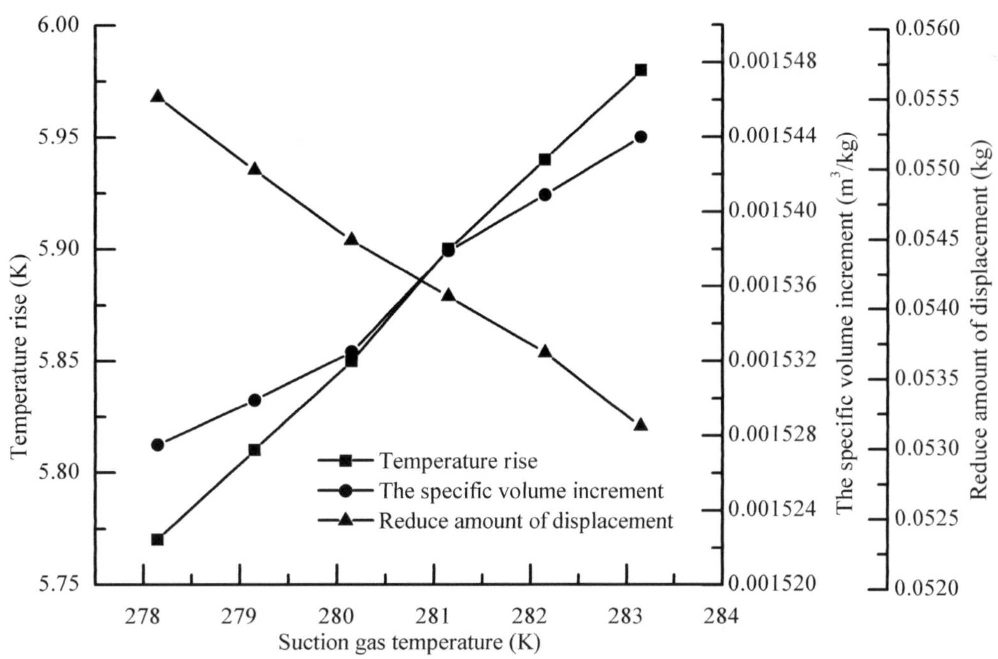

Figure 6. The influence of the suction gas temperature in the suction gas cooling method

IOP Conf. Series: Materials Science and Engineering **232** (2017) 012029 doi:10.1088/1757-899X/232/1/012029

Figure 6 shows the influence of the suction gas temperature on the thermal dynamic performance of the case 1. As the variation tendency of the curves shown in figure 6, the temperature rise of the suction gas, the specific volume and the displacement of the compressor are all affected by the suction gas temperature. The temperature rise of the suction gas and the specific volume increment will increases with the increase of the suction gas temperature. But the lower degree of displacement will decreases with the increase of the suction gas temperature.

After researching the influence of the motor cooling parameters on the thermal dynamic performance of the case 1, the influence of the motor cooling parameters for the case 2 were also analyzed. The analysis results are shown as follows.

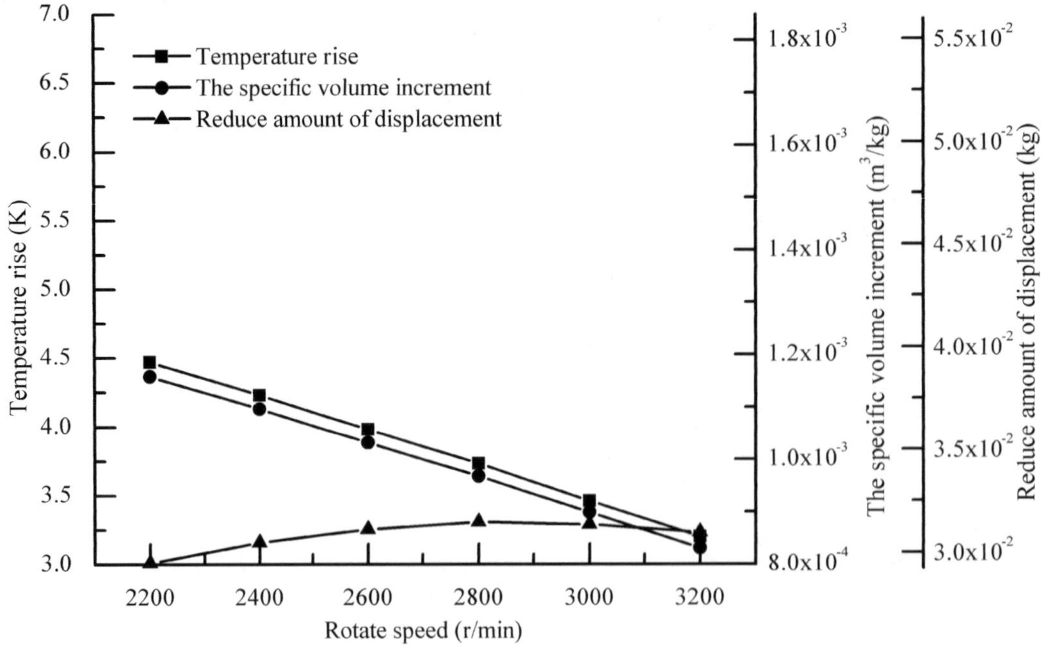

Figure 7. The influence of the suction gas quantity in the refrigerant liquid cooling method

The influence of the rotate speed on the thermal dynamic performance of the compressor is analyzed and shown in figure 7. The contrast of the curves in figure 5 and figure 7 shows that the temperature rise of the suction gas caused by the motor cooling process in case 2 is lower than that in the case 1. The lower temperature of the suction gas will cause the lower specific volume increment and the lower reduce amount of displacement. That means the motor cooling capacity of the injected refrigerant liquid is better than the suction gas. The variation tendency of the curves in figure 7 shows that the temperature rise of the suction gas will decreases with the increase of the compressor rotate speed. The specific volume increment will increases with the increase of the rotate speed. But the lower degree of displacement will increases firstly and then decreases with the increase of the rotate speed.

The influences of the liquid refrigerant temperature and the liquid refrigerant quantity on the thermal dynamic performance of the case 2 are shown in figure 8 and figure 9 respectively. Through the variation tendency of the curves in figure 8, the temperature rise of the suction gas, the specific volume increment and the lower degree of displacement will all increase with the increase of the liquid refrigerant temperature. But the curves in figure 9 show that the temperature rise of the suction gas, the specific volume increment and the lower degree of displacement will all increase firstly and then decrease with the increase of the liquid refrigerant quantity.

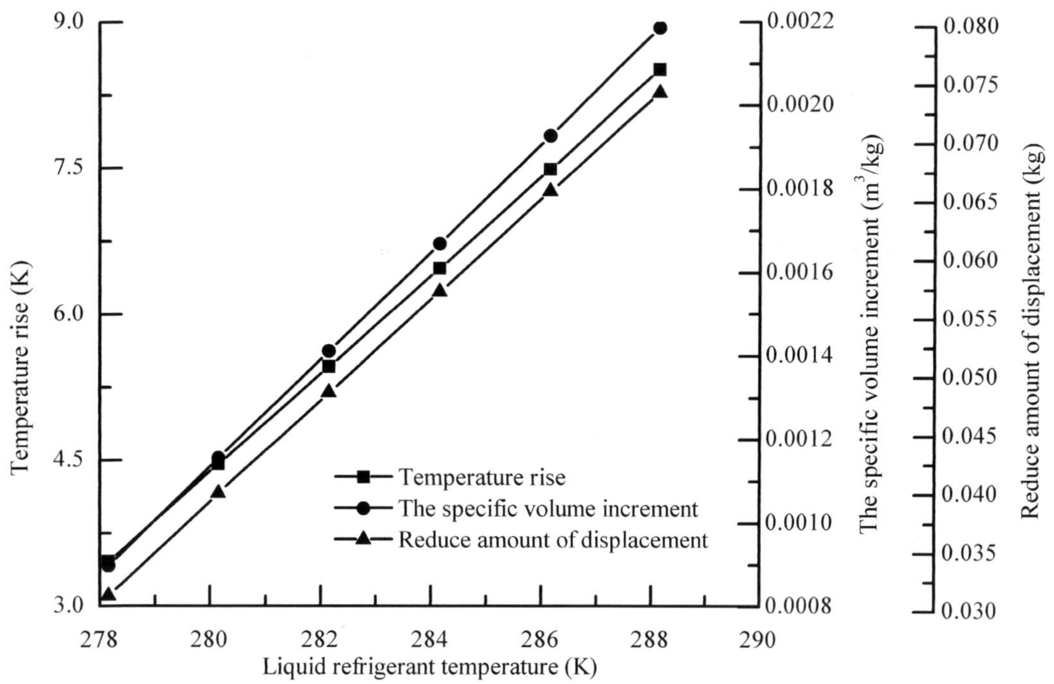

Figure 8. The influence of the liquid refrigerant temperature

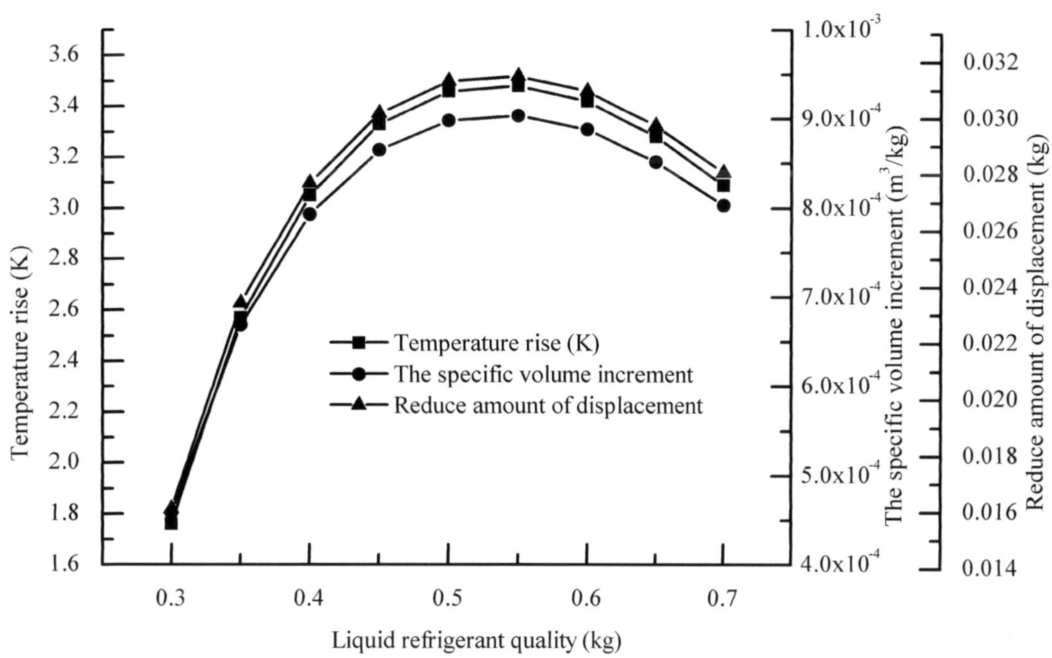

Figure 9. The influence of the liquid refrigerant quantity

5. Conclusion

1) All of motor cooling methods by using suction gas and injected refrigerant liquid will affect the thermal dynamic performance of the semi-closed SSRC. The motor cooling capacity of the injected refrigerant liquid is better than the suction gas.
2) For the semi-closed SSRC with the motor cooling method using suction gas, the increase of the compressor rotate speed will makes the decrease of the suction gas temperature rise and the specific volume increment but the increase of the displacement lower degree. But the influence of suction gas temperature is just the opposite.
3) For the semi-closed SSRC with the motor cooling method using injected refrigerant liquid, the increase of the compressor rotate speed will results in the decrease of the suction gas temperature rise and the increase of the specific volume increment. The lower degree of displacement will increases firstly and then decrease with the increase of the rotate speed.
4) In the semi-closed SSRC with the motor cooling method using injected refrigerant liquid, the temperature rise of the suction gas, the specific volume increment and the lower degree of displacement will all increase with the increase of the liquid refrigerant temperature. But the temperature rise of the suction gas, the specific volume increment and the lower degree of displacement will all increase firstly and then decrease with the increase of the liquid refrigerant quantity.

References

[1] Zimmern B. Worm rotary compressors with liquid joints: USA, 3,180,565[P], 1965.

[2] Constant L C, Hundy G F, Chan C Y, et al. Single-screw compressor for refrigeration and other duties, London, UK: Mech Eng Publ Ltd, 1981: 117-126.

[3] Zimmern B. From water to refrigerant: twenty years to develop the oil injection-free single screw compressor, in: Proceedings of the 1984 International Compressor Engineering Conference, West Lafayette, USA, p513-518.

[4] Wang S H. Single screw refrigeration compressor [J]. Chemical Industry and General Machinery, 1978, (07):50-52.

[5] Otsuki L. Semi-closed single screw refrigeration [J]. Fluid Engineering, 1989, (06):49-55.

[6] Li J. Comparison of motor cooling modes for hermetic screw compressor [J]. Refrigeration and Air-conditioning, 2006, 6(02):79-81.

[7] Chen W Q, Shen J B, Wu H G, et al. Flow and Heat Transfer analysis of the motor in screw compressor [J]. Refrigeration Technology, 2012, 32(01):4-8, 23.

[8] Chen W Q, Xing Z W, Shen J B, et al. Thermal modeling for the motor in semi-hermetic screw refrigeration compressors under part-load conditions [J]. International Journal of Refrigeration, 2013, 36(7):1874-1882.

[9] Wen M Q, Wang H D, Chen Q D, et al. Research on coil temperature field of electromotor in screw refrigeration compressors [J]. Refrigeration Technology, 2013, (03):61-64.

[10] He Z L, Xing Z W, Chen W Q, et al. Thermal and hydraulic analysis on the flow around the motor in semi-hermetic twin screw refrigeration compressors [J]. Applied Thermal Engineering, 2013, 58(1-2), 114-124.

[11] He Y N, Jin L, Deng W C, et al. Numerical study of refrigerant flow in screw compressor motor for high temperature heat pump [J]. Fluid Machinery, 2015, 43(6):29-33.

[12] Liang H J, Liang X Y, Guan H Y, et al. built-in motor Cooling problems of the central air conditioning compressor [J]. China High Tech Enterprises, 2015, (02):82-84.

[13] Dutra T, Deschamps C J. A simulation approach for hermetic reciprocating compressors including electrical motor modeling [J]. International Journal of Refrigeration,59(2015), 168-181.

[14] Wang Z L, Wang Z B, Wang J, et al. Theoretical and experimental study on thermodynamic performance of single screw refrigeration compressor with Multicolumn Envelope Meshing Pair. Applied Thermal Engineering, 103(2016): 139-149.

[15] Yang L, Dai W J, et al., Electric machine design theory and practice. Tsinghua University Press, Beijing, 2013, pp. 53-77.

IOP Conf. Series: Materials Science and Engineering **232** (2017) 012030 doi:10.1088/1757-899X/232/1/012030

Fully coupled fluid-structure interaction model of reed valves in a multi-cylinder reciprocating piston compressor

F Xie[1], J Nieter[2], A Lifson[3], R Reba[2] and V Sishtla[3]
[1] United Technologies Research Center, Shanghai, China
[2] United Technologies Research Center, East Hartford, CT
[3] United Technologies Climate Controls & Security, Carrier Corp, Syracuse, NY

E-mail: xief@utrc.utc.com

Abstract. For years compressor researchers have tried to account for the fluid interaction effect of the working fluid on valve motion in displacement compressors. In recent years, the computing capacities and available CFD and FEA programs have allowed fully coupled interaction of fluids and moving structures to be modelled more comprehensively. This paper describes our experience and results from developing a model of a multi-cylinder reciprocating piston compressor with suction and discharge valve systems that are fully coupled with the pressure pulsation in the adjacent plenum. Valve dynamics are captured by the model that affects compressor performance. The results show that higher running speed causes more discharge valve delay on closing due to higher pressure pulsation in discharge plenum. The acoustic property of the discharge plenum as it relates to valve motion is studied by the developed cost-effective standalone model.

1. Introduction

The main components of a reciprocating compressor that affect thermodynamics and flow are the cylinder, piston, connecting rods, valve plate, valves, and cylinder head. Valve behaviour controls the suction and discharge process and may have a substantial impact on coefficient of performance (COP) for refrigerant applications, which is defined by the ratio of refrigerating capacity to power consumption. Consequently, the numerical study of valve behaviour has become an important area in reciprocating compressor design, and has been widely researched in recent years. In this study, we model coupling effect of the reed valve dynamics and gas pulsations in the adjacent plenum for R410a 3 cylinder compressor. Another issue affecting compressor performance is the reed valve leakage, and Rezende [1] and Silva [2] developed numerical tools to predict the leakage gap of the compressor valve. The deformation of the lubricating oil film between the valve and the valve seat causes the stiction effect, which generates the valve delay. Rodrigo [3] developed a theoretical model for solving the dynamic behaviour of a ring-shaped lubricant oil film between a discharge valve and the seat.

Valve dynamics is very important in affecting mass flow rate through the ports. Discharge valve closing delay causes the backflow from discharge plenum to the cylinder, and the suction valve closing delay pushes the compressed gas to flow back to the suction plenum. Delay in discharge valve closing for example effectively increases the compressor clearance volume, as the start of re-expansion process is delayed. Both suction valve delay and discharge valve delay decrease the mass flow rate (capacity), and consequently result in lower COP. The study of valve dynamics is critical to design of a reciprocating compressor, and it involves the complex interaction between fluid and structural dynamics. Unfortunately, the small size and harsh environment inside a compressor, coupled with the material compatibility of sensors, makes the experimental study [4,5,6] of valve dynamics very difficult, and the numerical study becomes a more viable option. Further, modelling results often provide more descriptive details and understanding of the physical valve and fluid behaviour that is difficult or impossible to obtain from experimental testing. By coupling one-dimensional (1D) lumped structural model with 1D unsteady fluid model, Boswirth [7] developed an efficient tool, KV-DYN, to calculate the valve lift, exit velocity and pressure difference through valves. The late closing of suction valve and pressure pulsation in suction plenum were observed, meanwhile the mechanism of late closing due to pulsations in the plenum was also illustrated by Boswirth [20].

With the advancement of high performance computing and computer clusters, the computational cost of complex models becomes acceptable, and the complicated coupling models needed to study the valve dynamics can be developed. The coupling of the three dimensional CFD solver and 1D lumped structural model [8,9,10] was developed for better prediction of flow field. Mayer [11], Ajay [12] and Gasche [13] constructed a fully coupled three dimensional (3D) CFD/FEA models and compared them with 1D/lumped models. The results showed that the 1D models needed to be calibrated by the fully coupled 3D CFD/FEA models. Barbi [14] utilized an IBM (Immersed Boundary method) to handle the moving valve boundary in his work. David [15] adopted the Cartesian Cut-Cell Method to generate the deforming mesh of the valves. More complicated CFD models by Estruch [16] and Gonzales [17] were applied to the coupling with structural models. In order to shift the resonance frequency of the plenum away from running operation, recently more researchers [18,19] started to evaluate the effect of pressure pulsation in the plenum on compressor vibration and noise, where typically the acoustic property of the three dimensional plenum was simplified as 1D impedance transfer matrix. However to date the authors are not aware of any CFD studies that coupled the pressure pulsation in the plenum with the valve dynamics. Furthermore, the prior studies of valve dynamics were mainly focused on the reciprocating compressor with a single cylinder [9, 16, 17], whereas this paper is focused on 3D simulation of a three-cylinder reciprocating compressor system where pulsation effects in the plenum are considered in conjunction with the fluid and valve dynamics.

In this study, the computational time of considering the fluid-structure interaction of coupling the fluid in all three cylinders with the movement of all reed valves was approximated to be at 15 days which is time prohibitive. Thus a three step sequential modelling approach was developed. First, the model was constructed having only one physical cylinder and one *portion* of the discharge plenum to analyse the compressor PV diagram. As the next step, to capture the effect of the plenum geometry on pressure pulsation in discharge plenum, the model was extended to include the *full* discharge plenum with the one physical cylinder. In reality three cylinders work together with a specified crank angle difference. Consequently, a third model was developed with the *full* discharge plenum connected to one physical cylinder and two virtual cylinders (the virtual cylinders are used to contribute mass flow from those cylinders to the pulsations in the full plenum and not for specifics of valve movement, whereas the one physical cylinder included the fully coupled valve movement to the pulsations in the plenum). The results and analysis in this paper are based on this third model where the fully coupled method of CFD/FEA is employed to study the reed valve dynamics in relation to the pressure pulsations in the plenum. This study concentrated on the behaviour of the discharge valve, as calculated results show a significant effect of pulsations in the compressor discharge plenum on the delay of the discharge valve closure and subsequent reduction in the mass flow rate through the compressor.

2. Methodology

The commercial software CFX and ANSYS Workbench are selected to solve the fluid and solid equations respectively, and the coupling between fluid and solid setting is executed in CFX-Pre by importing the mechanical output file. The flowchart is shown in Figure 1.

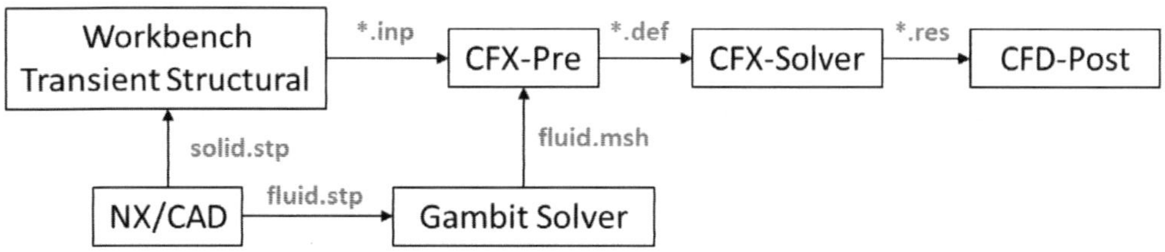

Figure 1. The flowchart of the simulation

2.1. Modeling of the fluid

The fluid in the reciprocating compressor system is compressible where the governing equations of compressible flow include mass, momentum and energy balances, and the equation of state closes the system of equations. All equations are discretized and solved at the cell nodes under the finite volume method. High resolution is adopted for the advection scheme, and the K-Omega model is selected to simulate turbulence. A deforming mesh is set up in CFX Pre-processor to account for the changes in the volume of the fluid zones in the compression process. The current study contains 36 fluid zones, and the information communication between adjacent zones is necessary where the different interface conditions are carefully set. To investigate the influence of pressure pulsation in the discharge plenum on the valve behaviour, we assumed non-reflective boundary at the discharge plenum outlet (this is a common and appropriate boundary condition that avoids unwanted downstream system effects).

2.2. Modeling of the solid

Virtual work and virtual displacement theorem generates the structural dynamics equation based on nodal displacement. The assembled equation based on the finite element method is shown as follows:

$$\mathbf{m}\ddot{\mathbf{x}}(\mathrm{t}) + \mathbf{c}\dot{\mathbf{x}}(\mathrm{t}) + \mathbf{k}\mathbf{x}(\mathrm{t}) = \mathbf{f} \tag{1}$$

$\mathbf{x}(\mathrm{t})$: nodal déplacement matrix, **m**: mass matrix, **c**: damping matrix, **k**: stiffness matrix, **f**: external force matrix. With the specified boundary conditions, the nodal displacement can be solved, and then the stress and strain calculated. In the current study, the non-linear contact is considered when valves impact the seat or backer.

2.3. The coupling method

The mechanical analysis in ANSYS Workbench generates the output file which is imported into CFX-Pre, and all FSI interfaces are shown in the CFX-Pre boundary condition setting cards. The communication property of interfaces between fluid and solid can be set up in CFX-Pre.

3. Case Definition

The main components of the reciprocating compressor are the housing, cylinders, connecting rods, pistons, valve plate, valves, and cylinder head, along with bolts and gaskets. The current compressor model is composed of three cylinders. The common cylinder head contains three connected chambers within the suction plenum and three connected chambers within the discharge plenum. The suction valves and discharge valves are reed type valves with one end fixed on the valve plate. The retainer (backer) is designed to limit the maximum displacement of the discharge valves. The properties of the working fluid (refrigerant) are dependent on pressure and temperature, and can be interpolated from the property table. The motion of the piston is pre-defined by the driveline geometry and running speed. Figure 2 shows the structure of the compressor in this study.

IOP Conf. Series: Materials Science and Engineering **232** (2017) 012030 doi:10.1088/1757-899X/232/1/012030

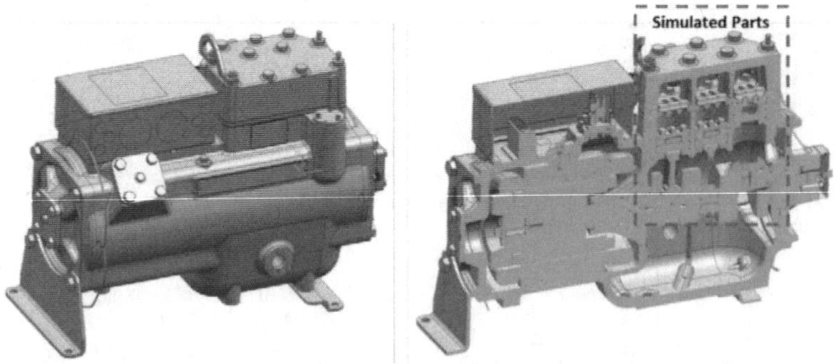

Figure 2. The structure of the compressor

3.1. The fully coupled model

The fluid zone starts from the entrance to the suction plenum and ends at the exit of the discharge plenum. The emphasis of this study is on the valve behaviour and pressure pulsation in discharge plenum. The compressor contains three cylinders with identical sets of discharge valves and suction valves. Each cylinder has three individual sets of suction and discharge ports covered by reed valves. The middle cylinder and associated valves are selected for fully coupled physical modelling, while the behaviour of the cylinders and valved ports on either side of the middle cylinder are simplified using the enforced mass flow rate which was calculated by a separate single cylinder model having the same plenum. The fluid zone and solid zone for the fully coupled simulation are shown in Figure 3. One section is sliced from the model in Figure 3, and displayed in Figure 4 where more details of the model are shown.

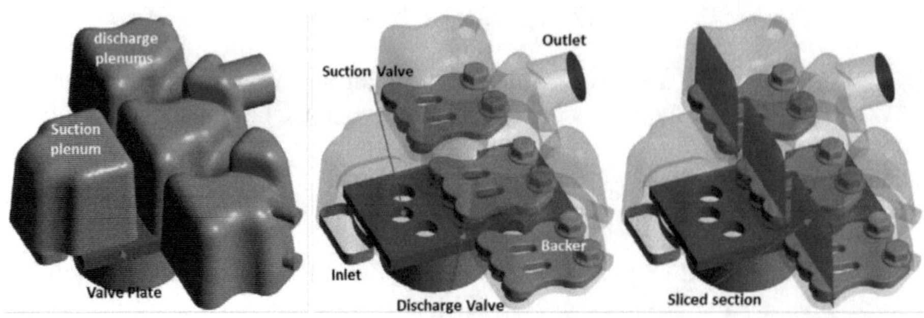

Figure 3. 3D fully coupled model

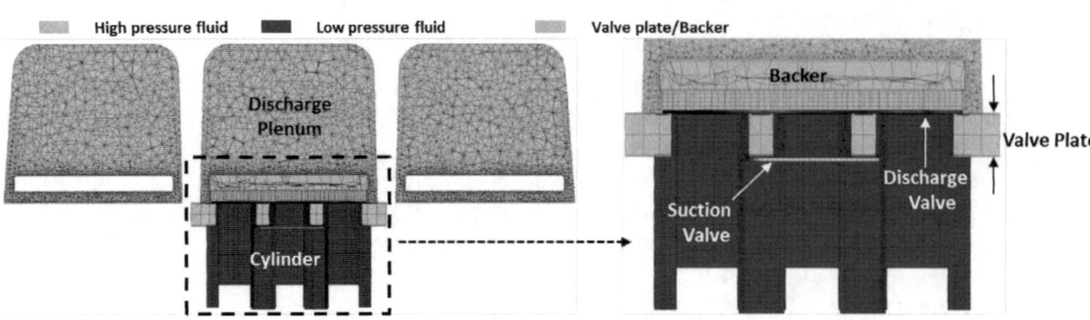

Figure 4. Sliced section from the 3D fully coupled model in Fig. 3

3.2. The standalone model

In the fully coupled model, both fluid zone and structural zone are discretized and solved. This method has a relatively high computational cost. Since our main interest here is to evaluate the discharge valve behaviour coupled to pressure pulsation response in the discharge plenum, we also developed a standalone model consisting of only the discharge plenum as shown in Figure 5. The mass flow through the discharge ports used by this standalone model is calculated by the fully coupled model, and the discharge plenum is treated as the only fluid zone. Consequently there is no moving solid zone in the standalone discharge plenum model, and investigating plenum geometry changes can be more easily accomplished. The fluid zone of standalone model is shown on the right in Figure 5, and the mass flow rates of all nine ports (3 per cylinder for 3 cylinders) are shown on the left.

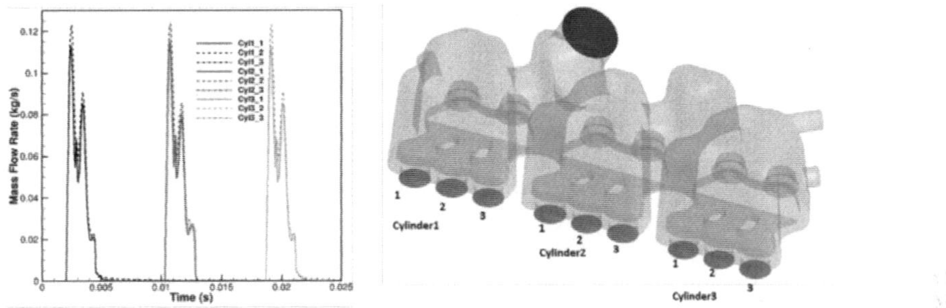

Figure 5. Standalone model of discharge plenum.

4. Results and Discussion

4.1. Results calculated by fully coupled model

We consider several "monitoring" points, where we extract pressure or valve displacement data from the model. One pressure monitoring point is located in the cylinder, and one pressure monitoring point is located in the discharge plenum. Two displacement monitoring points are located on the discharge and suction valves respectively to determine valve movement. All monitoring points are shown in Figure 6 as indicated by probe locations on the left and P1 on the right.

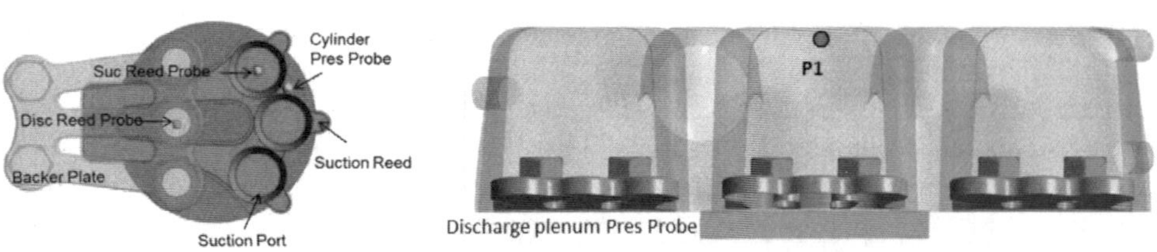

Figure 6. The location of monitoring points

The cases of fully coupled model were run on one computing node with 28 cores (CPU type: Intel E5-2680v4, 14 cores for fluid dynamics and 14 cores for structure dynamics) of Linux cluster. The high resolution scheme is employed for spatial accuracy. The time step size can capture fluctuations up to 8000 Hz, and the spatial size guarantees there are at least ten cells within the shortest wave length. One operational cycle took 3.5 days. We considered two cases, with the running speed of the compressor at rotating speeds of 29 Hz and 40 Hz, which represents a typical lower and higher end of the operating range for this compressor. For the first case of 29 Hz, Figure 7 shows the cylinder pressure versus cylinder volume (PV diagram).

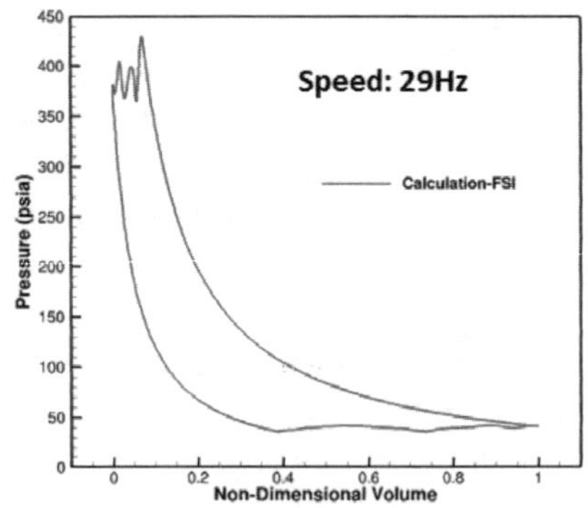

Figure 7. Cylinder pressure vs volume

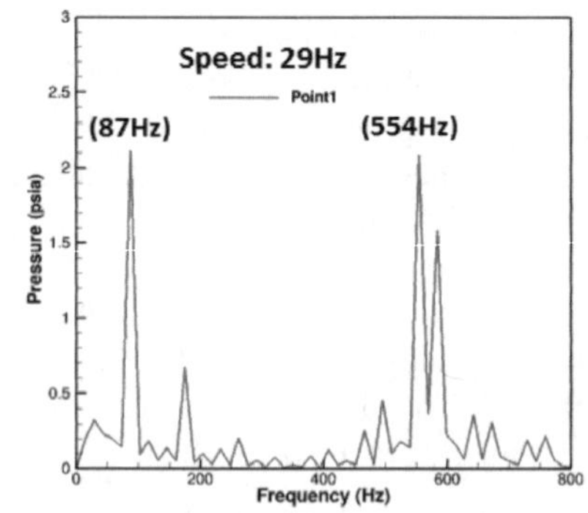

Figure 8. Plenum pressure vs frequency

The Fourier transform of discharge plenum pressure pulsations for this first case is shown in Figure 8 where two significant peaks at 87 Hz and 554 Hz are observed. These two peaks are located at three times and nineteen times of the basic running speed. The first peak is a result of the superposition of the basic excitation frequency from three cylinders (3x29), and the second frequency is the result of fluid acoustic resonance in the discharge plenum.

Figure 9 shows the pressure in cylinder and discharge plenum versus CA (crank angle), along with the discharge valve displacement. Pressure labelled as Pres_Disc, represents pressure pulsations in the plenum. Pressure in the cylinder, labelled as Pres_Cyl, represents pressure during compression, discharge, expansion, and suction processes. For the ideal case there should be no valve closing delay with respect to the piston at the top dead center (TDC) position. However, the current simulation shows phase difference between the TDC and the discharge valve closing. From the data shown in Figure 9 and the details in Figure 10, point A is at the TDC location where the discharge valve displacement is not zero yet but is still nearly 0.2mm from the valve seat. Point B is the location where the discharge valve finally closes and 5 degrees of phase difference (delay) can be calculated. In addition to substantial delay in valve closing due to pulsations in the plenum (discussed in details later in the paper), we can also observe additional delay due to the valve bounce between point A and B, which is initiated when the valve first contacts the seat. At crank angles beyond point B we can also observe a discharge valve deflection into the port caused by increasing pressure differential across the valve during the re-expansion process. It should be noted that a small delay in the discharge valve closing causes much more substantial delay in the suction valve opening (as the re-expansion process would also be delayed). This delay in the suction valve opening results in reduced compressor mass flow as the start of the suction process is delayed. Reduction in the mass flow rate results in the loss of system capacity and COP. As indicated earlier, to save on computational time, we only analysed the discharge valve movement for the middle cylinder (physical cylinder) and the other two cylinders are only considered as sources of mass flow (virtual cylinders) to the plenum. The mass flow rate through the middle discharge port labelled by Cyl2_2 in Figure 5 for the simulated middle physical cylinder is shown in Figure 11 and Figure 12 (showing local details of Figure 11 near TDC). Because of the delayed discharge valve closing we can observe the negative (backward) flow into the cylinder from the plenum.

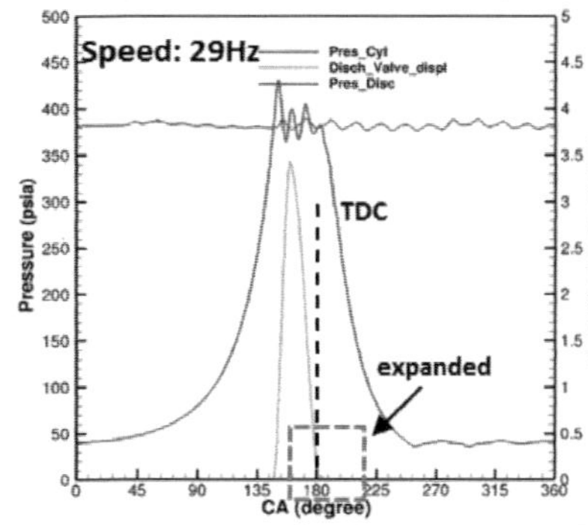

Figure 9. Pressure and displacement vs. CA

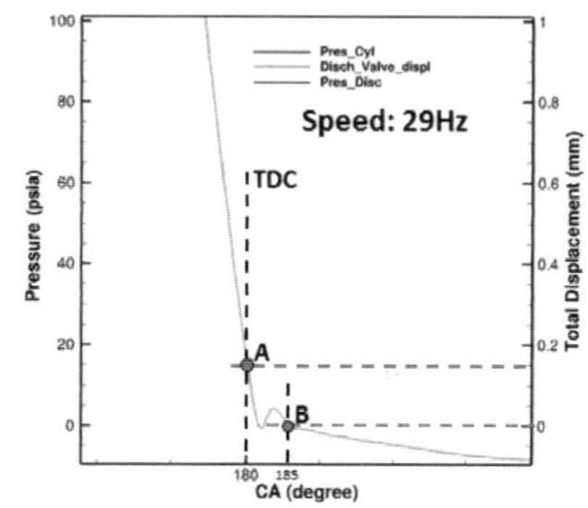

Figure 10. Displacement vs. CA (expanded)

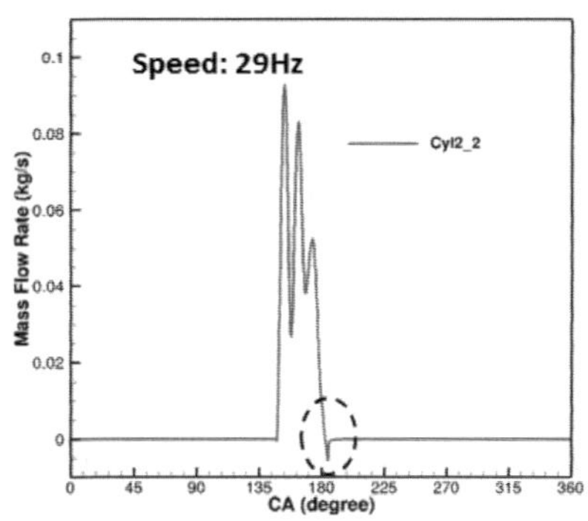

Figure 11. Flow rate through valve port

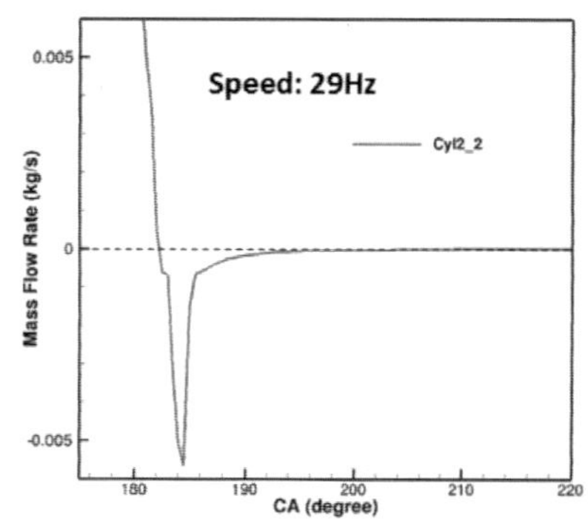

Figure 12. Flow rate through valve port (expanded)

In order to study the effect of running speed on the valve delay, we also performed a simulation for a second case of running the compressor at higher speed of 40 Hz. Figure 13 and Figure 14 (expanded with local details) compare cylinder pressure versus cylinder volume between running speeds at 29 Hz and 40 Hz.

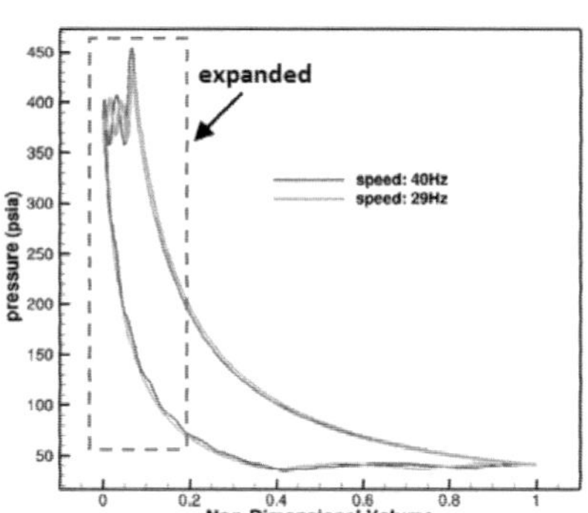

Figure 13. Cylinder pressure vs. volume

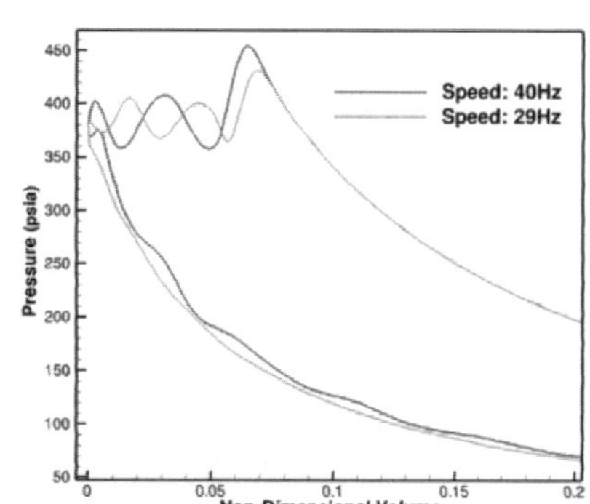

Figure 14. Cylinder pressure vs. volume (expanded)

Figure 15 shows the discharge plenum pressure amplitude versus frequency for the 40 Hz case, where two peaks can be observed at the frequencies of 120 Hz and 560 Hz. Similar to the case of 29 Hz, the first frequency is three times the basic running speed, generated by superimposing the motion of the three cylinders (3x40 Hz). The second frequency is fourteen times the basic running speed, and is the result of fluid resonance in the plenum. It is noted that the second frequency for the 40 Hz running speed case is quite close to the second frequency observed for the 29 Hz running speed case. Both the frequency at 554 Hz for 29 Hz running speed and 560 Hz for 40 Hz running speed are due to the resonant frequency of the discharge plenum (small variations in resonant frequency can be expected due to changes in fluid properties and changes in the frequency of the forced pulsations). Figure 16 shows the pressure and valve displacement versus CA (crank angle) for 40 Hz operation. The closing delay of the discharge valve at 40 Hz is now 9.5 degree as compared to only 5 degrees for 29 Hz operation, and details for discharge valve closing delay can be observed in Figure 17 for the 40 Hz speed. The increased valve closing delay between the case of 40 Hz vs. 29 Hz operation can be mainly attributed to larger pressure amplitude at the plenum resonant frequency in the discharge plenum (compare amplitude of 2.06 psi in Figure 8 vs. 3.46 psi in Figure 15). And to a smaller extend the delay can be due to a bounce of the valve on its seat due to impact (small peak near point B). If the "bounce" effect is subtracted from the delay, then the effect of pulsations alone on the discharge valve closing delay becomes even more pronounced (roughly 2 degrees for 29 Hz vs. 7 degrees for 40 Hz).

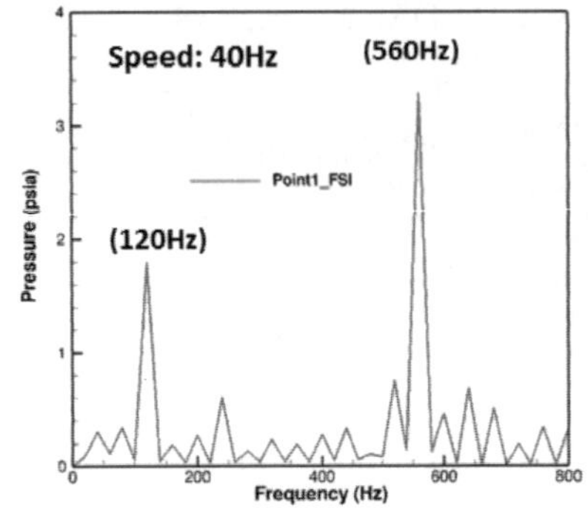

Figure 15. Plenum pressure vs frequency

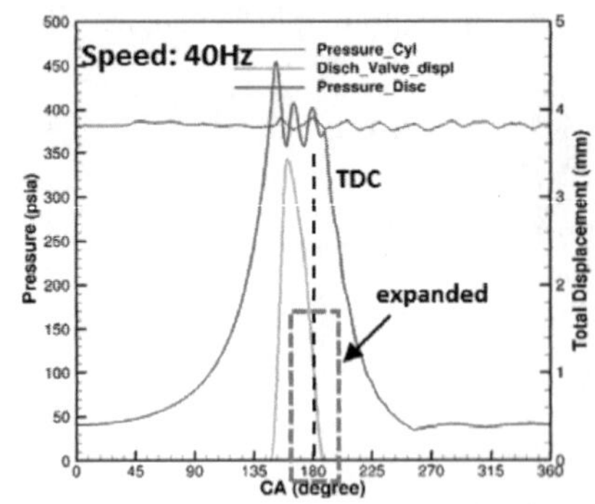

Figure 16. Pressure and displacement vs CA

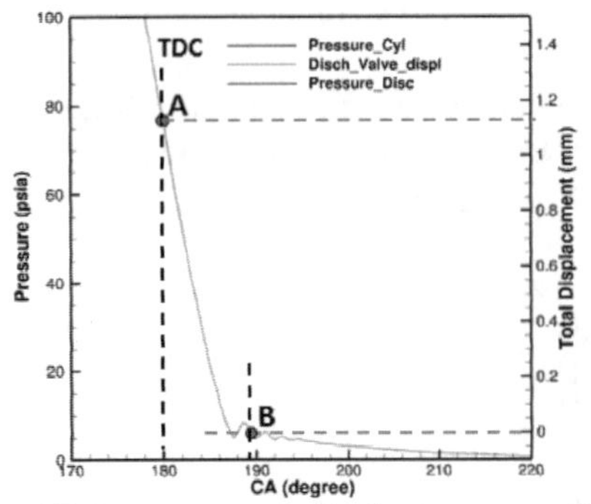

Figure 17. Pressure and displacement vs CA expanded

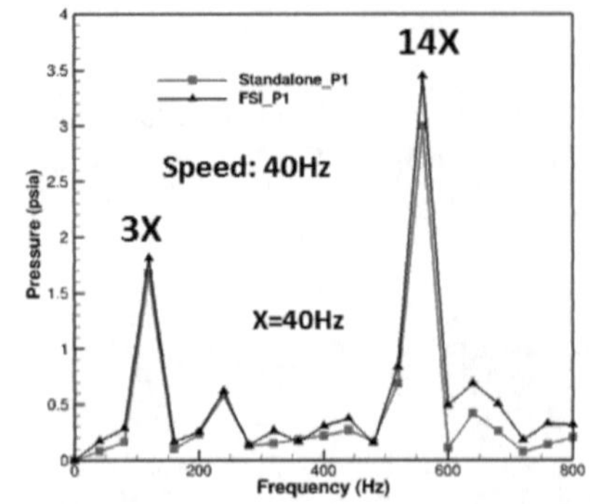

Figure 18. Amplitude versus frequency

Results from 29 Hz and 40 Hz running speed simulation are summarized in Table 1. As stated earlier, in both cases there is a delay in the discharge valve closing. The case with higher speed generates more discharge valve closing delay resulting in more substantial decrease in mass flow rate as compared to no delay case. Reduction in mass flow rate causes subsequent drop in the cooling or heating capacity. One of the reasons for more pronounced discharge valve closing delay at 40 Hz vs 29 Hz operation is the higher pressure pulsation amplitude at plenum acoustic resonance frequency (near 560 Hz).

Table 1 comparison between different running speeds

running speed	first pressure pulsation peak (psi)	second pressure pulsation peak (psi)	discharge valve delay (degree)	re-expansion in cylinder	frequency peak (Hz)
29Hz	2.01	2.06	5	low	87/554
40Hz	1.81	3.46	9.5	high	120/560

4.2. Results calculated by standalone model

To address the effects of pressure pulsations in the discharge plenum and its effect on discharge valve closing delay, a plenum standalone model is developed with the mass flow rate applied to the nine discharge ports. Since the purpose of the plenum standalone model is to only calculate pressure pulsation amplitude in the plenum, which is mainly affected by refrigerant mass flow rate from each port and not by intricacies of the valve movement, we do not need to employ the computationally intensive fully coupled FSI model for this task. The cases with the standalone model were run on the same computing resources as for the fully coupled model, as described above in Section 4.1, and one operation cycle takes only 10 minutes here as compared to the 3.5 days needed with the fully coupled model. In Figure 18 we see good agreement between the two models, which gives us confidence that the standalone model can be used for predicting pressure pulsations while providing a significant reduction in computation time. Figure 19 shows the pressure pulsation amplitude distribution in the discharge plenum at the 560 Hz frequency, which is fourteen times the running speed (14x40 Hz), and the amplitude in the middle chamber of the plenum has the largest amplitude. Figure 20 shows that the side chambers are 180 degrees out of phase with the middle chamber. This indicates that the pulsations in the discharge plenum at this frequency are a result of an acoustic plenum resonance.

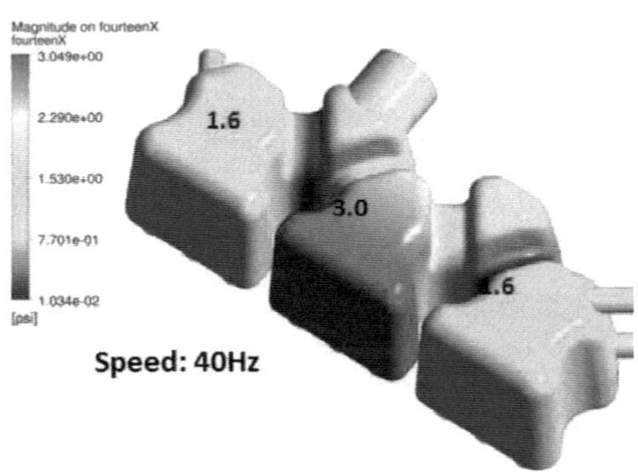

Figure 19. Amplitude distribution at 14X frequency

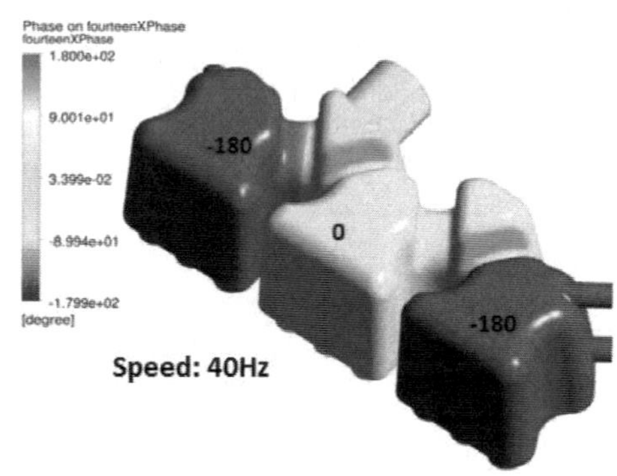

Figure 20. Phase distribution at 14X frequency

5. Conclusion

This paper described a fully coupled CFD/FEA method to simulate the valve dynamics and pressure pulsations in a multi-cylinder reciprocating compressor. The results show that pressure pulsations in the discharge plenum significantly contribute to the delay in discharge valve closing after the cylinder reaches top dead center. Higher running speed generates higher pressure pulsations in the discharge plenum, causing more pronounced delay in discharge valve closing. This delay can significantly decrease compressor mass flow rate and reduce system cooling or heating capacity. It would also decrease compressor operating efficiency. This occurs because a delay in the discharge valve closing results in the delay in the suction valve opening, which reduces volumetric efficiency of the compressor. Acoustic resonance in the discharge plenum is the main contributor to pressure pulsations in the discharge plenum. To analyze the acoustic property of the discharge plenum, a computationally effective plenum standalone model was developed. The standalone model compares well with the fully coupled FSI model and can be used for efficient mitigation of pressure pulsations in the plenum.

6. Reference

[1] Gustavo C R, Ernane S, Cesar J D 2016 A combined experimental-numerical procedure to estimate leakage gap of compressor valves *International Compressor Engineering Conf. (Purdue Univ.)* p 1313

[2] Silva L R and Deschamps C J 2012 Modeling of gas leakage through compressor valves *International Compressor Engineering Conf. (Purdue Univ.)* p 1344

[3] Pizarro-Recabarren R A, Barbosa J Jr., Deschamps C J 2012 Modeling the stiction effect in automatic compressor valves *Internat. Compressor Engineering Conf. (Purdue Univ.)* p 1126
[4] Gasche J L, Arantes D M, Andreotti D 2014 Experimental analysis of the fluid structure interaction in a suction valve model *International Compressor Engineering Conf. (Purdue Univ.)* p 1217
[5] Lohn S K, Pereira E L, Camara H F, Deschamps C J 2016 Experimental investigation of damping coefficient for compressor reed valves, *International Compressor Engineering Conf. (Purdue Univ.)* p 1228
[6] Lemke N C, König M, Hennig J, Försterling S, Köhler J 2016 Transient experimental and 3D-FSI investigation of flapper valve dynamics for refrigerant compressors, *International Compressor Engineering Conf. (Purdue Univ.)* p 1143
[7] Boswirth L 1996 A new valve dynamics simulation program and its use for the design of valves, *International Compressor Engineering Conf. (*Purdue Univ.) p 1133
[8] Dhar S, Ding H, Lacerda J 2016 A 3-D transient CFD model of a reciprocating piston compressor with dynamic port flip valves, *International Compressor Engineering Conf. (Purdue Univ.)* p 1392
[9] Pereira E L, Santos C J, Deschamps C J, Kremer R 2012 A simplified CFD model for simulation of the suction process of reciprocating compressors, *International Compressor Engineering Conf. (Purdue Univ.)* p 1276
[10] Bhakta A, Dhar S, Bahadur V, Angadi S, Dey S 2012 A valve design methodology for improved reciprocating compressor performance, *International Compressor Engineering Conf. (Purdue Univ.)* p 1252
[11] Mayer J, Bjerre P, Brune F 2014 A comparative study of different numerical models for flapper valve motion, *International Compressor Engineering Conf. (Purdue Univ.)* p 1315
[12] Parihar A, Myszka D, Robinet B, Hodapp T 2016 Integrating numerical models for efficient simulation of compressor valves, *International Compressor Engineering Conf. (Purdue Univ.)* p 1238
[13] Gasche J L, Lima Dias A D S, Bueno D D, Lacerda J F 2016 Numerical simulation of a suction valve: comparison between a 3D complete model and a 1D model, *International Compressor Engineering Conf. (Purdue Univ.)* p 1358
[14] Gasche J L, Barbi F, Villar M M 2012 An efficient immersed boundary method for solving the unsteady flow through actual geometries of reed valves, *International Compressor Engineering Conf. (Purdue Univ.)* p 1283
[15] Rowinski D H, Davis K E 2016 Modeling reciprocating compressors using a Cartesian cut-cell method with automatic mesh generation, *International Compressor Engineering Conf. (Purdue Univ.)* p 1658
[16] Estruch O, Lehmkuhl O, Rigola J, Pérez-Segarra C D 2014 Fluid-structure interaction of a reed type valve subjected to piston displacement, *International Compressor Engineering Conf. (Purdue Univ.)* p 1659
[17] Gonzalez I, Lehmkuhl O, Naseri A, Rigola J, Oliva A 2016 Fluid-structure interaction of a reed type valve, *International Compressor Engineering Conf. (Purdue Univ.)* p 1498
[18] Novak K, Sauls J 2012 Comparing FEM transfer matrix simulated compressor plenum pressure pulsations to measured pressure pulsations and to CFD results, *International Compress Engineering Conf. (Purdue Univ.)* p 1120
[19] Sauls J, Novak K 2012 Incorporating 3D suction or discharge plenum geometry into a 1D compressor simulation program to calculate compressor pulsations, *International Compressor Engineering Conf. (Purdue Univ.)* p 1121
[20] Boswirth L 1994 Flow and Valve Plate Dynamics in Compressor Valves (in German, published by author 1994, reprint 1996) p 400

IOP Conf. Series: Materials Science and Engineering **232** (2017) 012031 doi:10.1088/1757-899X/232/1/012031

Investigations on pneumatically forced-actuated compressor valves

Christian Stöckel, Christiane Thomas, Jörg Nickl and Ullrich Hesse

Technische Universität Dresden, Institute of Power Engineering, Bitzer Chair of Refrigeration, Cryogenics and Compressor Technology, 01062 Dresden, Germany

christian.stoeckel@tu-dresden.de

Abstract. In the present paper the performance of a novel designed valve for reciprocating piston machines is investigated, which makes existing compressors utilizable for operating as expander. Three design parameters were identified as critical for the valves performance particularly in forced actuated mode. Within a numerical simulation a study on the crucial geometrical parameters, the influence could be observed. Afterwards the experimental setup for the integral test of the valve design is presented and also additional tests for single valve components.

1. Introduction

Intermediate storage of natural gas is necessary in regional and national provision, because of the distinction between an almost constant supply from the gas fields and the seasonal fluctuation of the gas demand for both industrial and domestic heat as well as electricity supply. The pressure levels in the most common used underground storage sites (UGS) are varying from less than 100 bar up to more than 300 bar [1]. A significant amount of energy has to be spent to compress the natural gas from the pipeline pressure to the storage's pressure level. Contrary the nowadays usual way of unloading the gas storage is the expansion by the use of a simple throttling device, which means a waste of exergy. Previous investigations [1] have shown a potential power output of about 1 MW up to 4 MW when using an expansion machine instead of the simple throttling device for the storage unloading process. An even more complex plant is described in [2]: the necessary energy for preheating the expansion gas is provided by a cogeneration plant (CHP). The fuel for the CHP is the natural gas itself; only 0.5 % of the expansion gas is needed. For that configuration the produced electrical energy of CHP and expansion machine is 6.9 MW. However, the extension of an existing UGS to a power generating plant at the unloading process requires an additional gas expander. Alternatively in [3] a novel valve system was presented, which could retrofit an existing reciprocating piston compressor to a reversible usable expansion machine. Since the UGS anyhow needs the compression stages for loading the storage, the use of a retrofitable valve seems promising with respect to investment costs and space consumption.

To understand the working principle of the pneumatic actuated system a numerical simulation is a reasonable method. However, for the evaluation of the simulation a small scale compressor test with low pressure using air is appropriate.

2. Pneumatically actuated poppet valve

2.1. Valve design

In [3] the most promising design concept for a combined valve for both compression and expansion was found: the poppet type. The shape of the sealing element allows the control of the valve lift by the use of the pneumatic pressure difference provided by the process itself. That control pressure is applied to the backside of the sealing element in the poppet control chamber. By changing the position of the additionally installed electromagnetically actuated slider valve, the pressure level in the control chamber could be altered between suction and discharge pressure. In case of the valve on the high pressure side of the reciprocating machine (HP valve) following both modes can be performed: when the poppet control chamber is connected to the discharge chamber the valve works self-actuated. Connecting the pressure control chamber to the suction side a lifting force is acting on the sealing element caused by the pressure difference. The proposed design for the discharge valve of the reciprocating machine is shown in Figure 1.

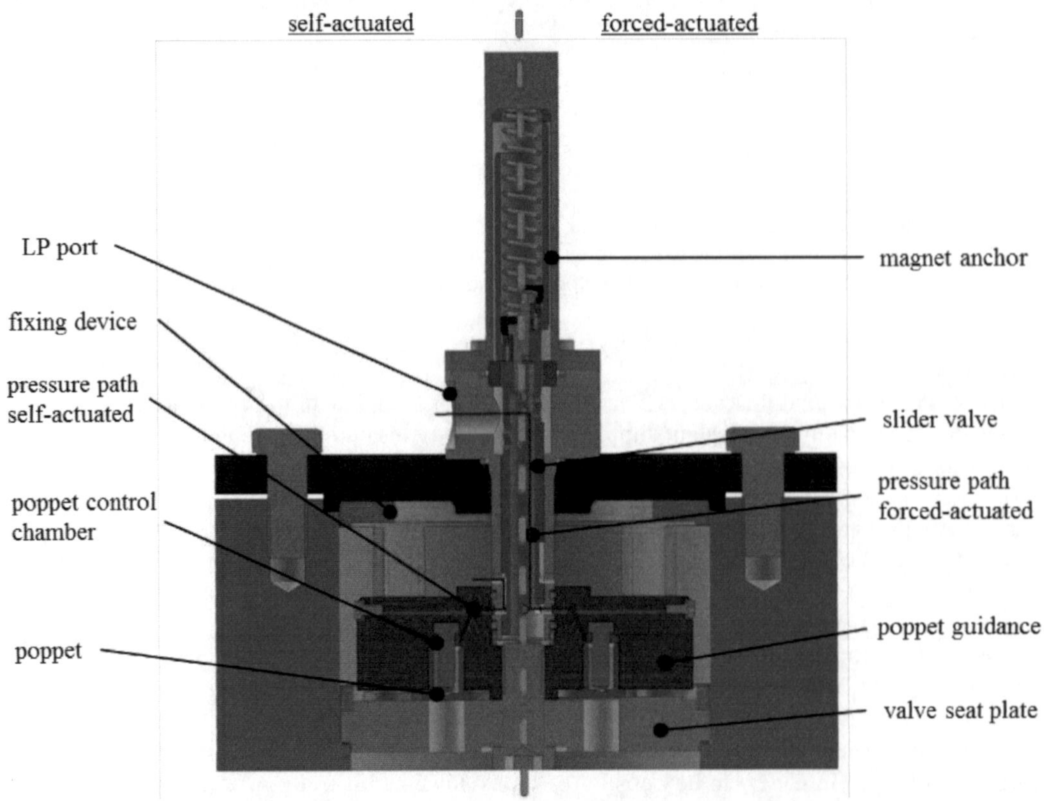

Figure 1: CAD-model of the pneumatically actuated discharge valve (valve springs hidden) [3].

2.2. Important design aspects

Using a valve system like described above leads to additional aspects which have to be considered during the design process. The following control signal chain is looked at:

- Crank angle position delivered by sensor
- Signal processing by programmable control unit
- Activation of the electromagnetically actuated slider
- Build-up of the magnetic field until positional change of the slider valve
- Change of the pressure level inside the poppet control chamber
- Poppet lift response caused by the pressure difference

IOP Conf. Series: Materials Science and Engineering **232** (2017) 012031 doi:10.1088/1757-899X/232/1/012031

The time delay of the electronical signal between the measurement of the crank angle position until the application of the voltage to the coil of the magnet anchor is insignificantly small in this case. The first relevant lag is generated by the interaction between the magnet field and the movement of the magnet anchor. This switching time depends on the electrical current of the magnet coil. Furthermore the inertia of the magnetic anchor and the spring rate of the return spring have an influence on the slider movement. However, multiple valve system manufacturers have already introduced electromagnetic actuator systems which can be operated within the required circuit times and also with the switching frequencies.

The most influencing time delay is expected by the pneumatic control itself. The poppet control chamber is a finite volume filled with the compressible process gas. Changing the pressure level on the chamber's inlet leads to a dynamic response, depending on the size of the volume and also on the pressure loss between the inlet and the control chamber. This behavior is superimposed by the change of the size of the control chamber by the poppet movement itself. Another important detail for that consideration is the leakage over the sealing elements between poppet control chamber and the suction respectively the discharge chamber.

Another notable point to be observed is the pressure loss over the valves. When retrofitting an existing machine with new valves, the overall efficiency should not decrease. Also the backward directed flow in the expansion mode has to be considered, since the efficiency during the expansion process has a significant influence on the amount of energy which could be recovered and is associated to the economical profitability.

Summarizing the above mentioned aspects it can be concluded that for the proper functionality of the pneumatically controlled valve, two important points have to be investigated: The electrodynamic behavior of the magnet which operates the slider valve and the extended valve dynamics caused by the poppets control chamber. For those observations numerical simulation and experimental testing are required.

3. Numerical observation

3.1. Numerical model

For preliminary observations of the system's functionality, the valve dynamics and the overall reciprocating machine's operational behavior a numerical model was set up. A simple thermodynamic multi chamber model was defined, as already described in [3]. The geometrical data for the numerical model as well as the thermodynamic boundary conditions were taken from a real reciprocating compressor, which will be equipped for the validation of the model. This compressor is described in chapter 4.1. The specific valve parameters like geometry, valve spring rate and mass of the sealing elements are known from the initial valve design fitted to the existing compressor.

The pressure loss caused by the flow through the valve has a strong influence on the dynamic valve behavior. That pressure loss is defined by the flow speed inside the valve and the pressure loss coefficient ζ. A larger flow cross section leads to lower velocities; the ζ-value depends on the valve lift and the overall valve geometry. Pressure loss coefficients for poppet valves used in internal combustion engines are well described in the literature. Nutting and Lewis [4] for example observed already in 1918 the pressure loss for different geometries for stationary flow experimentally. Waldron [5] found that the values derived from stationary testing are also applicable to intermittent flow. However, the flow pattern over the valve into the working chamber in engines is quite different to the one in compressors. Idelchik [6] gives resistance coefficients for various types of valves like single plates with and without guide or conically shaped sealing elements, which are geometrically very alike poppets. Similar observations were done by Vaughan [7] by the use of CFD-simulation; he also found specific points in the valve lift, where flow separation occurred and thus the function of the valve loss coefficient became discontinuous. Important geometrical parameters are the valve lift, the sealing element diameter, but also the shape of the valve seat and the sealing element. From Pohlenz [8] we can derive coefficients particularly for compressor ring and plate valves. In more recent publications, e.g. [9], [10] or [11], ζ-values for compressor poppet valves are given. However, the coefficients of those publications are mostly afflicted by one unknown geometrical parameter, hence the conversion to our specific problem is not possible. An overview of pressure loss coefficients depending on the valve opening ratio is given in Figure 2, were x denotes the valve lift and d the poppet diameter.

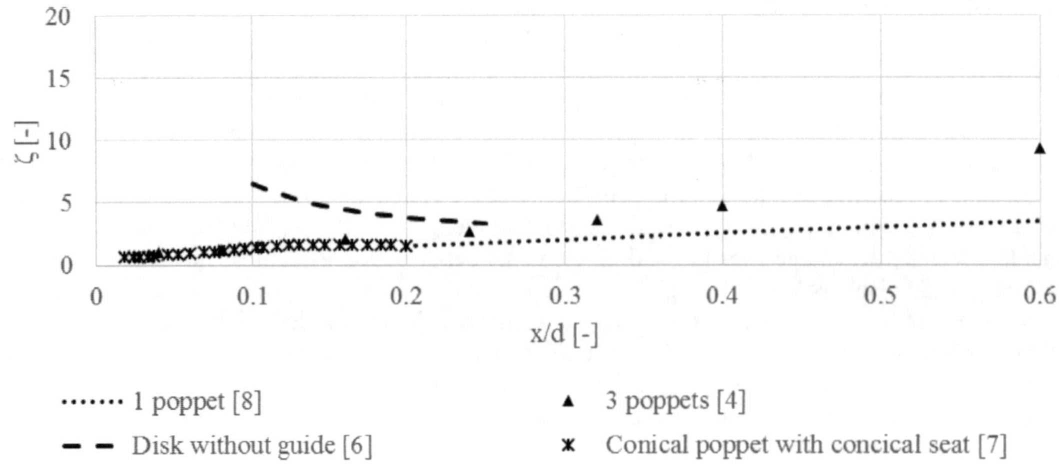

Figure 2: pressure loss coefficient ζ of poppet and similar valves over the dimensionless valve lift x/d.

It can be seen, that apart from the data of Idelchik [6] the ζ-values follow similar trends and are actually almost equal for small valve lifts. However, apart from the data of Nutting the values are only valid for single poppets. The ζ-value for the flow through multiple poppets might be larger since the mutual influence of the flow through each single poppet. Also Schulz [12] showed that each single sealing element will open differently from the others, caused by the valve pockets shape and internal pressure waves inside the cylinder. For the initial calculations an approximation of the curve presented by Pohlenz [8] is used:

$$\zeta = 0.5 + 4.5 \cdot x/d \tag{1}$$

A resolution of the crank angel revolution of 0.1° was defined, which is a temporal resolution of about 28.5 µs at the chosen drive speed of 585 rpm. The calculation was conducted with 10 iterations and the mass flow balance for in and outlet and also the converging of the mass flow was checked, to ensure a stationary point of operation. Using the preliminary valve design the geometrical parameters of the valve assembly were varied to observe the influence on the sealing elements motion.

3.2. Results from simulation

For the first study the fit between the sealing element and its guide was changed. A clearance fit is necessary to facilitate the poppet's movement along its axis. However, a leakage path between poppet control chamber and suction respectively discharge chamber occurs caused by this clearance. The simulation was performed with different common tolerance fields, which are in accordance with DIN ISO 286. The gap was assumed evenly distributed over the circumference and the maximum width resulting from the clearance fit was set.

The relative leakage is defined as the mean value of the absolute leakage of the poppet normalized to the mass flow of the machine. Figure 3 shows this value during expansion mode. On the secondary y-axis the area of the gap of a single sealing element is presented. The leakage increases during the forced actuated mode as expected with a looser fit. The increase follows a similar trend like the growth of the gap area. Since the actuation time for the low pressure side is longer, the relative leakage is higher than for the high pressure valve.

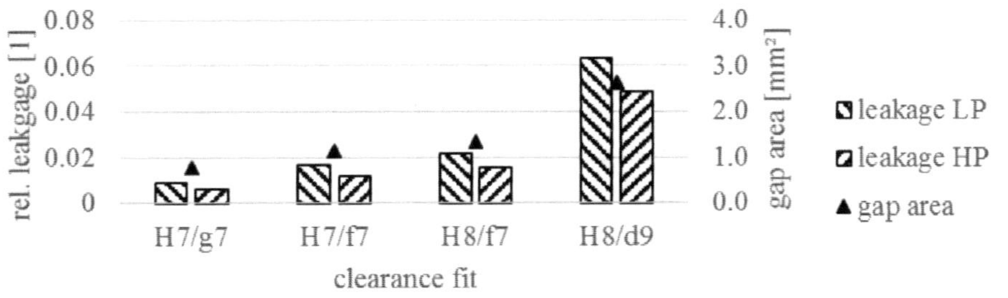

Figure 3: Mean value of the relative leakage (flow through the gap between poppet and guide normalized to the compressor mass flow) and gap area for a single poppet.

The influence of the leakage on the poppet movement of the high pressure valve is shown in Figure *4*. The increasing leakage at larger clearances leads to a slower valve response when forcing the poppet to move. The forcing starts at a crank angle of 325° and is completed about 17.5° later for the smallest calculated tolerance of "H7/g7". For the largest observed fit the calculation leads to an additional delay of 5° of the crank angle. Contrary when stopping the forcing the sealing element with the looser fit moves faster to the closed position. This takes about 12.5° for the "H8/d9" and about 15° for the "H7/g7" respectively. Simulations with even larger clearances were performed, but the resulting calculated pressure levels in the control chamber did not converge. The poppet sealing element's movement of the low pressure valve follows a similar trend and is hence not presented within that paper. Since the influence of the gap is higher during forced operation mode it can be concluded, that the loose fit should be chosen as small as possible.

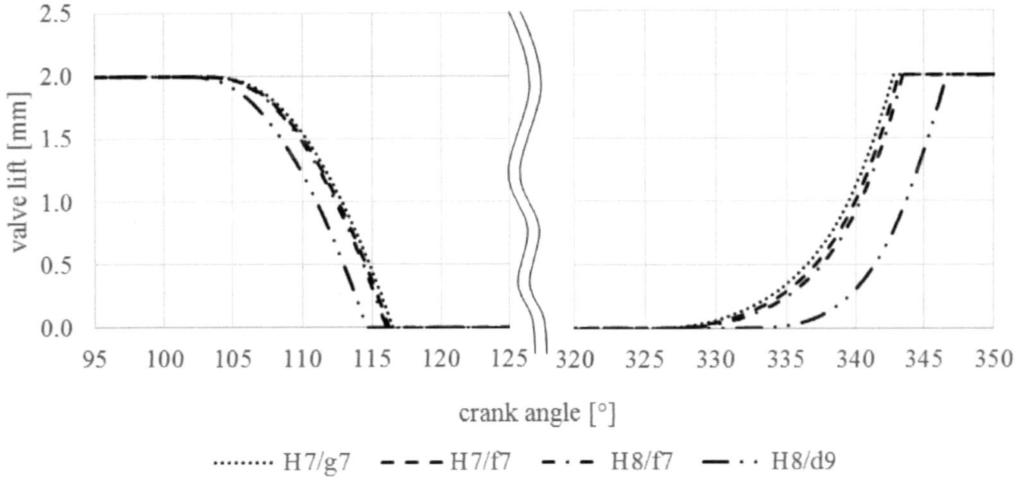

Figure 4: Influence of the clearance fit between poppet and its guide on the high pressure valve movement during expansion mode.

In a second observation the size of the poppet control chamber was varied. Based on the volume of the preliminary design the controlled volume was simply multiplied by a factor. Two limits for that factor were found for the domain of convergence of the calculations. The lower limit of the mentioned factor of 0.35 is defined by the clearance for the poppet movement itself. Smaller volumes lead to a collision between the sealing element and its guide, hence the calculation aborted. Above the upper limit of 3 the gas inside the control volume reacts too slowly, so the poppets do not react on the applied pressure change.

The results for the forced actuated high pressure valve for factors between the above discussed limits are shown in Figure 5. For the initial design (factor = 1) a delay of the valve response between the signal and the movement to the end position of about 15° and 20° can be seen. Minimizing the control volume to factor 0.35, leads to a decrease of the valve delay of about 7°. Increasing the volume leads to a significant increase of the delay. Also the starting point of the valve movement relocates to higher crank angles substantially. The calculated time delays have to be considered during the forced actuation to achieve the desired charge. From these observations it can be concluded that the volume of the control chamber has to be chosen as small as possible. This can be achieved by positioning the surface of the control volume, were the pressure change is applied, as close as possible to the sealing elements. Hence the ideal design would mean a pressure switch above each sealing element. However, in respect to the economic effort the usage of a single slider valve which is connected to a manifold, as already presented in Figure 1, is more reasonable.

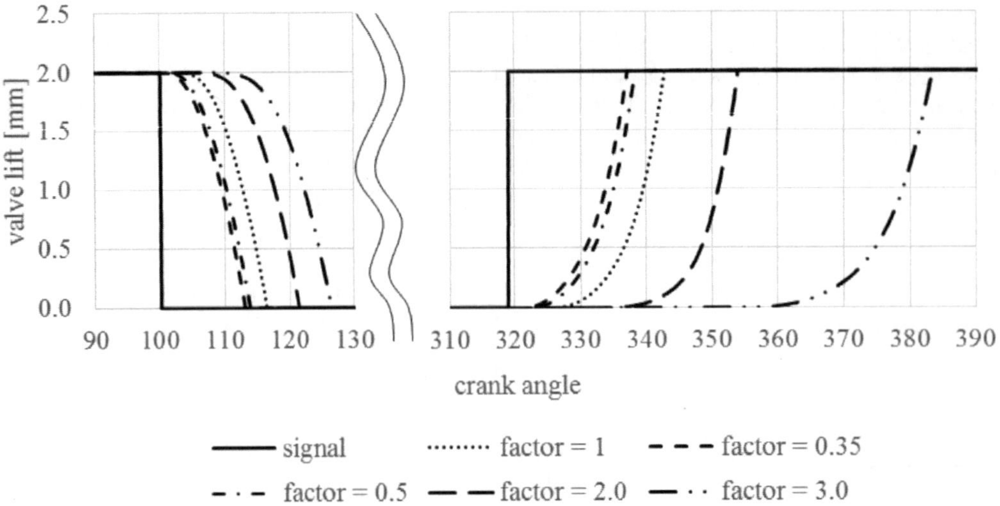

Figure 5: Influence of the volume of the poppet control chamber on the high pressure valve movement during expansion mode.

4. Experimental setup

4.1. Test compressor

For the functionality proof of the proposed design and validation of the data derived from the numerical model a test compressor is needed. A two-stage L-shaped Atlas Copco AR1 compressor (year of manufacture about 1955) is used. Each cylinder is double acting with each two suction and discharge valves per working chamber. The machine data is given in Table 1.

Table 1: Machine data of test compressor Atlas Copco AR1.

	1st stage	2nd stage	unit
compressor speed	585		min^{-1}
free air delivery in normal operation	8.6		m^3/min
max. shaft input	46		kW
cylinder bore	285	170	mm
stroke	150	150	mm
normal working pressure	1.77	6.87	bar
piston rod length	295	295	mm

IOP Conf. Series: Materials Science and Engineering **232** (2017) 012031 doi:10.1088/1757-899X/232/1/012031

4.2. Instrumentation

To investigate the valve behaviour and the overall machine performance, an extended pressure indicator setup was intended. Piezo-resistive pressure transducers are installed in the working, suction and discharge chamber and additionally in the poppet control chambers. In Figure 6 the positioning of the sensors for the high pressure valve can be seen. The crank angle position will be determined by an inductive dead point sensor. The test equipment for the high dynamic measurements will be completed by a positional sensor for one single sealing element using a method based on eddy current. Thus it is possible to evaluate the ζ-value of the valve with the help of pressure drop and the valve lift.

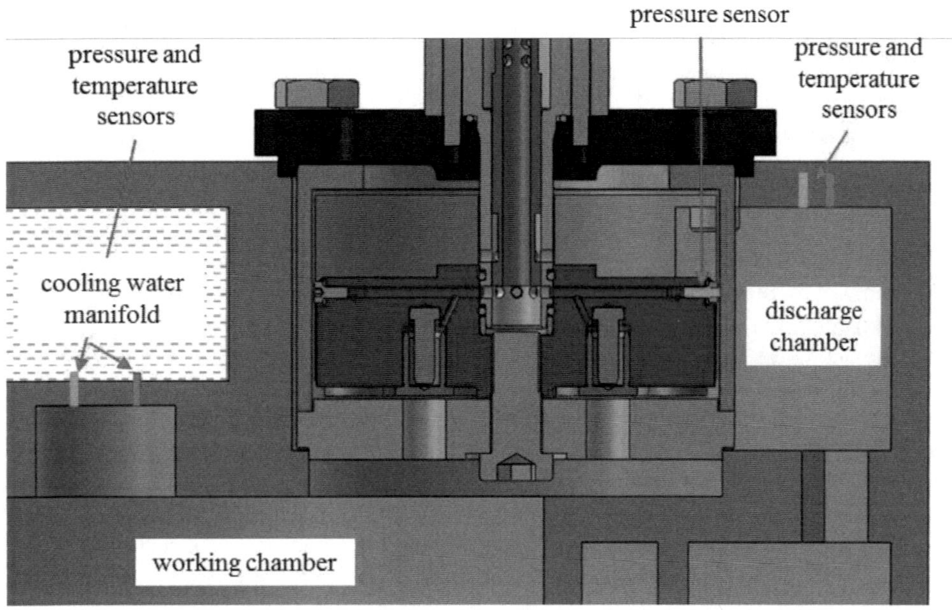

Figure 6: Positions of temperature and pressure probes for dynamic valve performance testing.

The volume flow will be measured by means of a calorimetric principle. Additionally installed pressure and temperature sensors are used for the acquisition of the different operation points. Overall the instrumentation allows the measurement of the compressor performance in a stationary operating point, but also the evaluation of the highly dynamical, transient processes.

4.3. Testing procedure

The first testing series will be performed using the modified high pressure valves for the first compression stage only. The reciprocating machine will be operated as a compressor, but forced opening of the discharge valve will be applied to check the valve performance. Simultaneously the possibility of a compressor capacity control by the rather uncommon forced opening of the discharge valve is checked. With the achieved data the numerical model will be adjusted and optimizations can be adapted to the low pressure valve of the first stage and also for the valves of the second stage.

Afterwards the compressor will be fully equipped with the controllable valves and the first stage will be operated as a compressor and the second stage as an expansion machine. Using the pressure indicator diagrams, the efficiency of both compression and expansion stage can be illustrated. Also the readout of the electric power of the frequency converter controlling the machine's motor will be used for the assessment of the compressor and expander stage. By changing the machines drive speed and varying the valve timing different operation points can be observed and the limits of the valve control can be determined.

4.4. Complementary testing

Like already mentioned in section 2.1. , the time delay caused by the build-up of the electromagnetic field in the slider is very important for the functionality of the valves. Thus a simple test setup is planned to determine the time between the two end positions of the slider. The electromagnet is controlled by a pulse generator. Simultaneously the position of the slider is measured by usage of an optical distance sensor. By synchronizing the pulse generator and distance sensor the slider valve delay can be determined. The pulse generator allows different switching frequencies, so a change of the reciprocating machines drive speed can be simulated. Another focus for this testing lies on the applied electric current on the magnetic coil. By increasing the current a faster slider movement should be possible.

The knowledge of the pressure loss coefficient of the valve is another important parameter to know. For this special case the pressure loss for both flow directions is of interest. In cooperation with Mario Cozzani Srl stationary flow tests will be performed to evaluate the necessary flow coefficients.

5. Conclusions and Outlook

In the present paper a critical view on design aspects for a force-actuated compressor valve was formulated. The proposed design allows the reciprocating machine valves to operate in the typical self-actuated compressor mode but also in a forced actuated mode, to operate the compressor reversible as expansion machine. By the use of a simulation model it could be shown, that the influence of the fit between the sealing element and its guide is small, as long as common tolerances for the clearance fit are applied. As expected the leakage of the pneumatic control chamber increases with larger tolerances and leads to a slower valve response. The size of the dead volume for the control chamber, which is unavoidable to ensure the flow paths for the control, has a much higher influence on the valve performance. An upper limit of the volume's size where valve control is no longer possible was shown, since the gas volume becomes too inert.

The next step is the completion of the experimental test setups, which have been also presented in this paper. The expected results will give important data to improve the numerical model. Additionally the influence of the flow path to the poppets control chamber has to be observed more in detail. Especially strong pressure drops caused by flow deflection and sudden changes in the cross section might lead to pressure pulsations inside the control chamber and worse valve performance.

Acknowledgment

This work was supported by the European Forum for Reciprocation Compressors (EFRC) in the scope of the project "Valves systems for Piston Reversible Compression-Expansion Machines". We would also like to thank Borsig ZM Compression for the donation of the Atlas Copco Compressor for the further valve testing and Dott. Ing. Mario Cozzani for the manufacturing of the valves.

References

[1] U. Hesse and P. Goldmann, "Survey on valve systems for piston expansion machines - Phase 1," Dresden, 2013.
[2] S. Grill and S. Zacherl, "Energieeffiziente Massnahmen zur Abwarmenutzung und Stromerzeugung bei Erdgasspeicher," *3R - Fachzeitschrift für sichere und effiziente Rohrleitungssysteme*.
[3] C. Stöckel and C. Thomas, "A novel valve design for combined reciprocating piston expansion and compression machines," in *Proceedings of the 10th EFRC Conference*, 2016, pp. 211–223.
[4] E. M. Nutting and G. W. Lewis, "Air flow through poppet valves," 1918.
[5] C. D. Waldron, "Intermittent-flow coefficients of a poppet valve," 1939.
[6] I. E. Idelchik, *Handbook of hydraulic resistance (3rd edition)*. 1966.
[7] N. Vaughan, N. Johnston, and K. Edge, "Numerical simulation of fluid flow in poppet valves,"

J. Eng. Mech. Sci., vol. 210, no. 8, pp. 259–266, 1992.
[8] W. Pohlenz (Hrsg.), *Pumpen für Flüssigkeiten und Gase: Bauteile für Pumpen*. Berlin: VEB Verlag Technik, 1983.
[9] E. Giacomelli, F. Falciani, G. Volterrani, R. Fani, and L. Galli, "Simulation of Cylinder Valves for Reciprocating Compressors," in *Proceedings of ESDA2006*, 2006.
[10] E. Giacomelli, M. Schiavone, F. Manfrone, and A. Raggi, "Flow Coefficient Evaluation of Poppet Valves Used in Reciprocating," in *Proceedings of PVP2008*, 2008.
[11] W. Eifler, E. Schlücker, U. Spicher, and G. Will, *Küttner Kolbenmaschinen*. 2009.
[12] R. Schulz, "Interactions between valve flow characteristics, valve pocket geometry and compressor performance," in *Proceedings of the 10th EFRC Conference*, 2016, pp. 467–480.

IOP Conf. Series: Materials Science and Engineering **232** (2017) 012032 doi:10.1088/1757-899X/232/1/012032

A fluid-structure interaction solver for the fluid flow through reed type valves

I González, A Naseri, J Rigola, C D Pérez-Segarra and A Oliva
Heat and Mass Transfer Technological Center (CTTC), Universitat Politècnica de Catalunya - BarcelonaTech (UPC), ESEIAAT, Colom 11, 08222, Terrassa (Barcelona), Spain
E-mail: nacho@cttc.upc.edu

Abstract. Suction and discharge processes with self actuated valves have a major influence in efficiency and reliability of hermetic reciprocating compressors. Understanding the operation completely in order to enhance compressor's design needs precise prediction of the fluid-structure interaction complexities involved in these processes. This paper presents a comprehensive description of a numerical methodology to account for the coupled behaviour of a reed valve and a turbulent flow. The method is based on a partitioned semi-implicit scheme that only strongly couples the fluid pressure term to the structural solver. A three-dimensional CFD analysis with LES turbulence modelling is used for the flow while a combination of plate theory and mode summation method is used for the solid. The dynamically changing domains are tackled by means of lagrangian and arbitrary lagrangian-eulerian approaches for the solid and the fluid, respectively. The whole model is compared with experimental data at Reynolds number 10,000, showing good agreement in lift amplitude and deformation fluctuations. Finally, as an illustrative case, results regarding lift, pressures, force and effective areas are compared with those of a valve with wider gland.

1. Introduction

Design of reed valves is a challenging task and is of great relevance for the overall performance of hermetic reciprocating compressors [1, 2]. They are involved in suction and discharge processes, which are responsible for nearly half of the total thermodynamic losses of the compressor [3]. But besides efficiency, reliability is also influenced, since the repetitive loading of the valves is the most common cause of failure of the machine [4, 5].

The problem presents different complexities such as high pressure and velocity gradients, flow detachment, turbulence, collisions and fluid-structure multi-physics coupling. Traditionally the process has been tackled by means of one-dimensional lumped models [6], which have been proved to work reasonably well for broad design studies. However, a more detailed approach needs to be considered in order to fully understand the flow topology and optimise valve design. In the last years, following an enhance in computational resources, a growing research on 3D computational fluid dynamics (CFD) simulations for valve flow has been carried out. A verified fluid-structure interaction (FSI) model, which strongly couples both 3D CFD and finite element structural solvers, was thoroughly explained and used for reed valves in [7]. Besides, different CFD analysis considering FSI have been performed which include comparison with experimental data. Some examples can be found in [8, 9, 10, 11]. They mostly focused on the valve response rather than on the description of the numerical methodology or the results regarding the flow behaviour.

The objective of this work is to provide an insight into a numerical methodology which is able to perform a complete three-dimensional analysis of the turbulent valve flow. The method is based on a new strong fluid-structure coupling algorithm which ensures numerical stability with a moderate increase in computational cost. First, an experimental test is numerically replicated. Then the case is simulated with a variation in the original design of the valve to evaluate the sensitivity of the system and to prove the potential of the numerical procedure to optimise valve designs.

2. Numerical methodology

A partitioned approach is followed to solve the FSI problem, which uses independent solvers for fluid and structural sub-problems and adopts a coupling scheme to account for their interaction. The elastic reed motion is modelled using the normal mode summation method, within a Lagrangian mesh, thus taking the local loads and bending deformations into account with negligible computational effort. On the other hand, the fluid flow is solved with a parallel CFD code on a 3D unstructured grid, which will deform according to the dynamic domain defined by the solid position. Throughout valve opening, a gap will appear between the valve and the seat, making necessary a new set of fluid elements in that zone. In order to avoid a remeshing procedure, there will be moving elements which first perform as solid, i.e. the valve seat, and later as fluid, i.e. the gap in between. This change in the nature of the elements is achieved using an immersed boundary method (IBM) [12].

Next, the governing equations of both sub-domains as well as the moving mesh technique and the coupling algorithm are described.

2.1. Governing equations

The flow is governed by the incompressible Navier-Stokes and continuity equations, formulated within an arbitrary lagrangian-eulerian moving domain framework,

$$\frac{\partial \vec{u}}{\partial t} + (\vec{u} - \vec{u}_w) \cdot \nabla \vec{u} = -\frac{1}{\rho_f}\nabla p + \frac{\mu}{\rho_f}\Delta \vec{u} \quad (1)$$

$$\nabla \cdot \vec{u} = 0 \quad (2)$$

where p and $\vec{u}$ are the fluid pressure and velocity, respectively. A fractional step projection method is adopted to decouple the unknowns and solve equations 1 and 2 [13]. Considering an explicit time advancement, the projection method yields to a three step solution procedure which can be written as:

$$\vec{u}^p = \vec{u}^n - \Delta t\left[(\vec{u}^n - \vec{u}_w) \cdot \nabla \vec{u}^n - \frac{\mu}{\rho_f}\Delta \vec{u}^n\right] \quad (3)$$

$$\Delta p^{n+1} = \frac{\rho_f}{\Delta t}\nabla \cdot \vec{u}^p \quad (4)$$

$$\vec{u}^{n+1} = \vec{u}^p - \frac{\Delta t}{\rho_f}\nabla p^{n+1} \quad (5)$$

with $\vec{u}^p$ denoting the predicted velocity. Fully conservative symmetry-preserving schemes [14] are used for the spatial discretization, whereas the wall-adapting local-eddy viscosity model (WALE) is used to describe the turbulent flow.

In regards to the reed structure, the momentum equation for an isotropic plate on (x, y) under transverse loading, having considered the assumptions of Kirchhoff-Love theory, is:

$$D\nabla^4 d_z(x,y,t) + 2\rho_s h \ddot{d}_z(x,y,t) = \sigma_{\Gamma,z}(x,y,t) \quad (6)$$

where D is the bending stiffness, d_z the normal displacement, h the uniform thickness and $\sigma_{\Gamma,z}$ the normal component of the traction exerted on the plate surface Γ. According to the mode superposition method [15, 16], the motion of a vibrating system can be approximated by the sum of a limited number of normal modes m, weighted by generalised coordinates q_m:

$$d_z\left(x,y,t\right) \approx \sum_{m=1}^{N} q_m(t)\phi_m(x,y) \tag{7}$$

where $\phi_m(x,y)$ are the deformation patterns of the vibration modes, in z direction.

The combination of equations 6 and 7 leads to one generalised momentum equation for each mode [2],

$$\ddot{q}_m(t) + \omega_m^2 q_m(t) = \frac{\int_\Gamma \phi_m(x,y)\sigma_{\Gamma,z}(x,y,t)d\Gamma}{\rho_s h \int_\Gamma \phi_m^2(x,y)d\Gamma} \tag{8}$$

where ω_m are the natural radial frequencies, which can be found, along with the mode shapes, previously to the simulation with a normal mode analysis. Therefore, once the loads on the structure are known, the generalised coordinates q_m can be calculated with a numerical time integration scheme, in this case with the Newmark one.

Finally, the impact between valve and seat is estimated as a change in its velocity regulated by a coefficient of restitution C_r. As soon as contact is detected, the generalised velocities are corrected as

$$\dot{q}_m(t^{n+1}) = -C_r\dot{q}_m(t^n) \tag{9}$$

When C_r is set to one, all the kinetic energy is preserved after impact. Whilst this simple approach avoids body overlapping and controls the energy dissipation, it can be inaccurate from a continuum medium perspective, since it does not take into account the effect of local consecutive impact events.

2.2. Moving mesh technique

The fluid mesh adaptation is governed by the equations of linear elasticity, similarly to [17] but with a finite-volume discretization. By introducing a volume-based stiffening directly in the Young modulus of the elements

$$E_i = V_i^{-\chi}, \quad \chi \geq 0 \tag{10}$$

the major distortion will be absorbed by the larger elements, hence maintaining a good quality in the smaller elements which are located near the solid boundaries.

The velocity of the frame of reference $\vec{u}_w$ is here evaluated in the control volume faces. To calculate these velocities, the volume swept by a face between two time steps can be exactly computed while respecting the conditions imposed by the geometric conservation law [18], which yields:

$$\frac{d}{dt}\int_\Omega dV + \oint_{\partial\Omega} \vec{u}_w \cdot \vec{n}dS = 0 \tag{11}$$

2.3. Fluid-structure coupling scheme

On the fluid-structure interface Γ, there is a Dirichlet boundary condition for the fluid and a traction (Neumann condition) for the structure. The former as a function of the structure displacement $\vec{u}_\Gamma = \partial\vec{d}/\partial t$, and the latter as a function of fluid velocity and pressure $\vec{\sigma}_\Gamma\left(\vec{u},p\right)$.

Having updated the boundary velocity $\vec{u}_\Gamma$ with $\vec{d}$, the fluid solver is able to find p and $\vec{u}$ in the fluid domain and, in this manner, evaluate the traction distribution on the solid surface: $\vec{\sigma}_\Gamma = \vec{F}(\vec{d})$. Likewise, the structural solver will find the appropriate $\vec{d}$ derived from the fluid load: $\vec{d} = \vec{S}(\vec{\sigma}_\Gamma)$. Therefore, the coupled fluid-structure system of equations reduces into an interface problem of the form:

$$\vec{S} \circ \vec{F}(\vec{d}) - \vec{d} = 0 \tag{12}$$

In most of the previous works, the coupling is treated explicitly for simulating suction/discharge processes, which means solving each domain by employing the other-domain data from the previous time step. Nevertheless, it has been shown that explicit schemes may cause numerical instability when both physics are strongly coupled [19]. In order to overcome this issue, a novel semi-implicit coupling scheme has been developed [20]. The new approach leverages the projection method to implicitly couple only the fluid pressure term to the structure, instead of the whole fluid set of equations. By doing so, the algorithm ensure an accurate and stable resolution of the coupled problem avoiding excessive computational cost.

Therefore, at the beginning of the time step the fluid mesh is adapted in accordance with the last deformation found for the solid, and the velocities of the faces $\vec{u}_w$ are evaluated. With these updates, the first step of the projection method (Equations 3) can be used to find the fluid predicted velocities $\vec{u}^p$. Then, the strong FSI coupling starts by iteratively solving Equations 4 and 8. As soon as the convergence criterion is met, the fluid velocity correction (Equation 5) is evaluated to finally finish the time step simulation.

To solve the nonlinear FSI iterative system, a fixed-point solver with a block Gauss-Seidel method and Aitken's dynamic relaxation has been chosen for this study. Aitken's dynamic relaxation mitigates possible convergence problems of the Gauss-Seidel method and reduces the number of required iterations.

3. Definition of the case

Aiming at validating the numerical methodology, here a comparative study of the experimental case carried out in [21] is presented. It was focused on the valve motion under the load of a developed and stationary air flux. The set-up consists of a feeding tube with $d = 34.9$ mm in diameter, a plate where the reed valve lies and an empty space downstream. In Figure 1 there are the dimensions of the valve as well as those of the cylindrical domain chosen for the simulations. The valve diameter D is 1.3 times the orifice diameter d. Apart from the original valve geometry, a modified version with a wider gland has been considered so as to highlight the differences regarding the coupled system performance. The former will be called valve A whereas the latter valve B.

The temperature of the air is 25 °C and ambient pressure is defined at the outlet. The physical properties of the flexible valve are set to be: $\rho_s = 7850$ kg/m^3, $\nu = 0.3$ and $E = 2 \cdot 10^{11}$ Pa. An average coefficient of restitution of $C_r = 0.3$ has been fixed for the seat impact dissipation according to [22]. The inlet velocity of the fluid is determined so that the Reynolds number is 10,000:

$$Re = \frac{\rho_f \bar{u}_z d}{\mu} \tag{13}$$

where $\bar{u}_z$ is the axial bulk velocity. The inlet boundary condition is implemented as a developed turbulent pipe flow. The averaged velocity profile has been extracted from [23] and is depicted in Figure 2. In addition, synthetic inflow conditions are prescribed at the boundary like in [24] to reproduce the transient turbulent nature of the flow (Figure 3). To do so, the turbulent statistics for wall flow at $Re = 11,700$ found in [25] have been used.

The results presented in this section have been obtained with a computational mesh of around $3 \cdot 10^5$ control volumes. This discretization has been proved to provide a suitable converged solution for the present flow configuration.

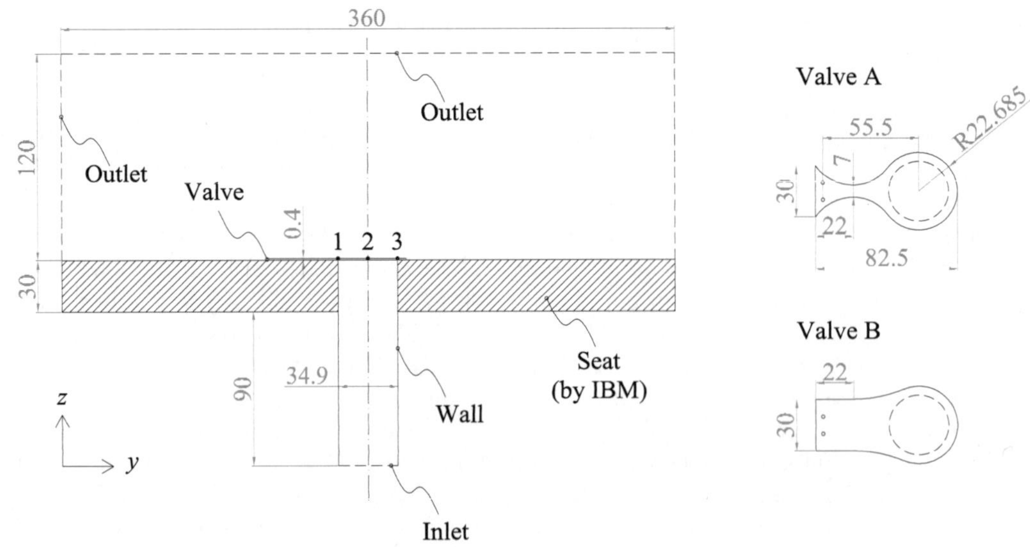

Figure 1: Section of the computational domain and geometry of the valves (dimensions in mm). Points 1, 2 and 3 for lift measurements.

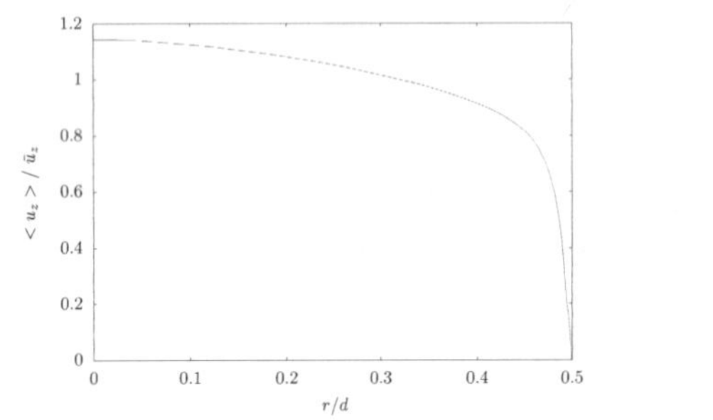

Figure 2: Averaged inlet velocity profile.

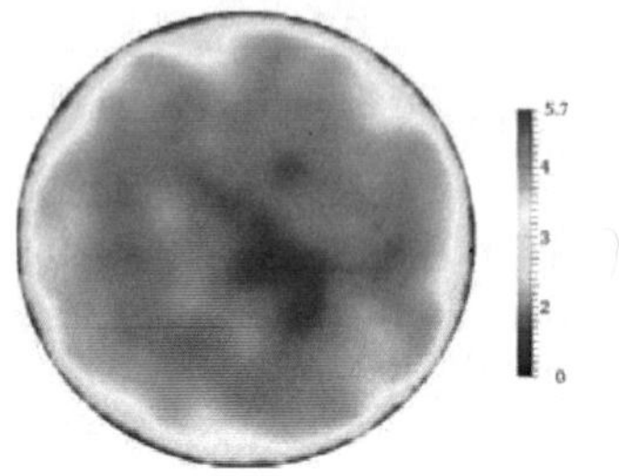

Figure 3: Instantaneous inlet velocity map in [m/s].

4. Numerical results

First, a linear modal analysis has been carried out for the reeds with a clamped boundary condition in their root. The results, obtained with a commercial structural software, are shown in Figure 4. Only the first four modes have been taken into account for this study. As can be seen, the first natural frequency of the design A is very close to the value determined in [21], 31.4 Hz. The geometry B evidently has larger mass and stiffness as a result of the gland widening. Observing the frequencies, the major rise is in the stiffness, since all but the fourth mode vibrate at higher rates than the original valve.

Under the fluid load, the valve bends and starts fluttering with impacts on the seat. The system rapidly reaches a periodic steady state of operation with constant amplitude in lift (h) and pressures (p). This cyclic response is shown in Figure 5, where the displacement of point 1 (see Figure 1) of valve A is plotted along with the experimental measurements. Although the resultant prominent frequency is pretty similar (f_A is 58.6 Hz numerically while it was 57.9 Hz experimentally) there is a considerable deviation in the oscillation peak. The model over-predicts the maximum aperture by approximately 16 %. However, the amplitude error falls down to 6 % and the perturbations within a smooth sinusoidal motion agree with the experiment.

IOP Conf. Series: Materials Science and Engineering **232** (2017) 012032 doi:10.1088/1757-899X/232/1/012032

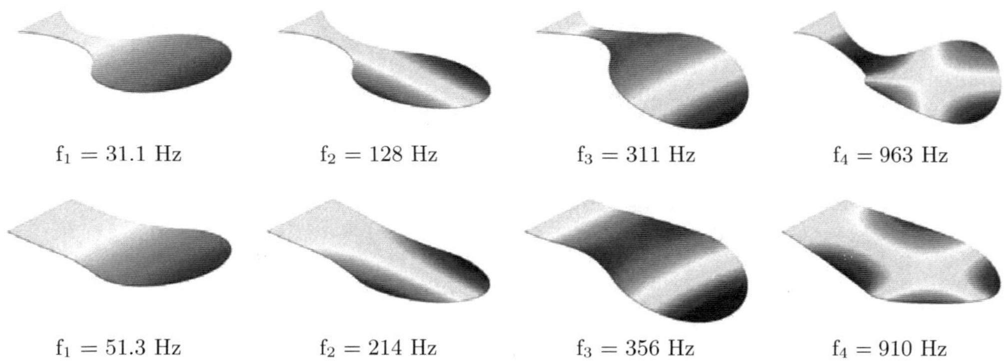

Figure 4: Normal modes of vibration of valves A (top) and B (bottom).

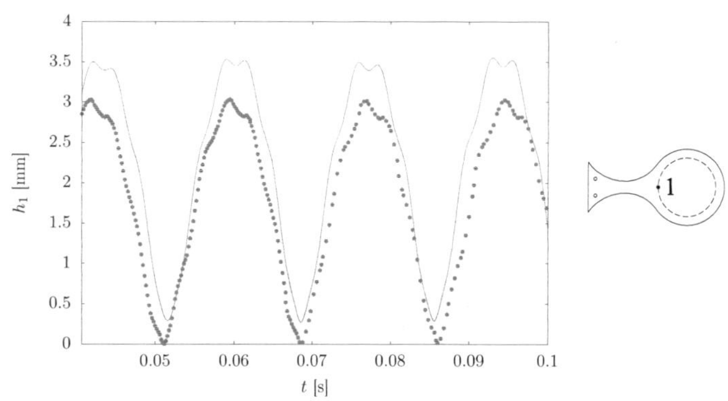

Figure 5: Valve lift with design A in point 1. Numerical results (——) and measures from Dr. José Luis Gasche (●).

Observing the displacements of the three valve control points (Figure 6 (a)), one may point out the impact evaluation as the most plausible cause of the mentioned discrepancies. The graph shows there is not a flat contact with the seat, because each of the three points would reach the seat at a different time. This is due to the effect of higher vibration modes on the valve transient deformation. With the approximation used for the impact (Equation 9), not all valve points are colliding with the seat and the valve eventually rises earlier than expected. Thus, it would be worth implementing a more accurate model for this critical event. On the other hand, the configuration B does not exhibit such an issue regarding the collision. Even though there is also an influence of other modes of vibration apart from the first –noticeable in the fluctuation of points 1 and 3 of Figure 6 (b)– all three points hit the seat at the same time, giving rise to an almost flat contact.

The greater stiffness of option B is responsible for the decrease in lift amplitude (~ 1.4 times lower) and the increase in flutter frequency ($f_B = 90.7$ Hz). The momentum carried by the flow is unable to open valve B as much as valve A, and a greater force is required in order to bend the former by the same angle as the latter. This fact can be distinctly observed by comparing the flow pressure exerted on the valve for a common valve height. In Figure 7 this pressure is depicted at instant t_1, when both valves share the same displacement measured in the port axis (see Figure 6). As expected, the pressure is considerably higher in case B. Besides, the pressure

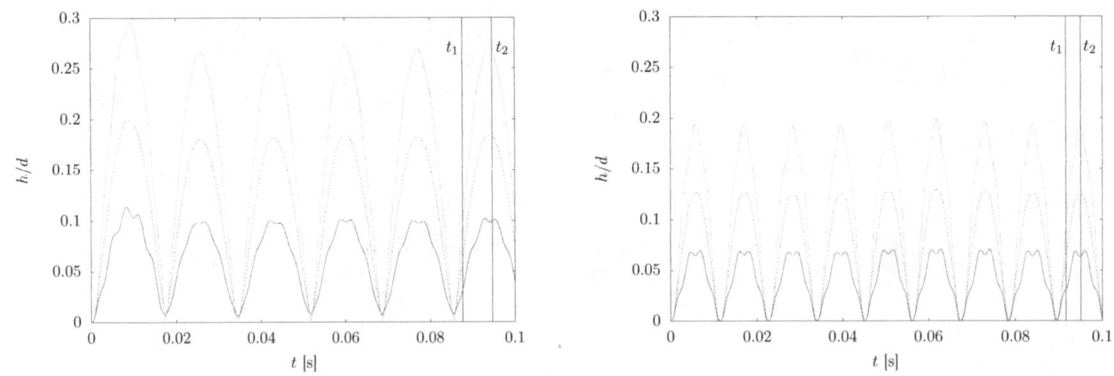

(a) Valve A (b) Valve B

Figure 6: Valve lift in points 1 (——), 2 (- - - -) and 3 (· · · · · ·).

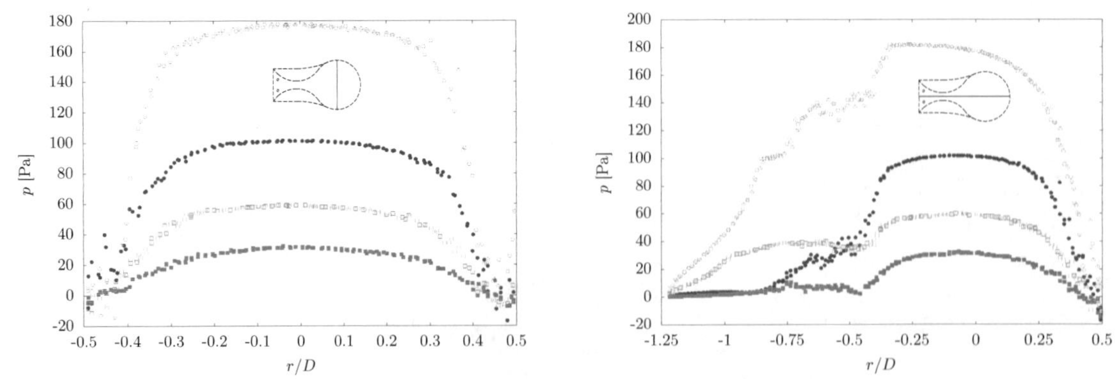

Figure 7: Pressure over the lower face of the valve at time t_1 (● A and ○ B) and t_2 (■ A and □ B).

along the longitudinal line shows that this variable is not symmetric with respect to the channel axis, as a result of the valve geometry and its deformed shape.

The same profiles at instant t_2 are presented in Figure 7 as well, which corresponds to the time when the maximum valve lift is achieved by each geometry. They show the clear effect of height and flow area on the fluid pressure. Having a higher valve lift, valve A allows the flow to change its direction sooner and reduce the axial momentum carried when it impinges into the valve. In addition, the fluid velocity crossing the larger gap between valve and valve seat should not accelerate as much as with valve B, as can be seen in the velocity field of Figure 9. It can also be identified the separation bubble which is generated at the outside edge of the valve seat.

In order to check the influence of the wider gland on the outflow area, the effective flow area defined for incompressible flow as:

$$(KA)_e = \frac{\dot{m}}{\sqrt{2\rho_f(p_{inlet} - p_{outlet})}} \tag{14}$$

is depicted in Figure 10 (a) as a function of valve height. The value has been averaged taking into account all the periodic oscillations simulated. Due to the incompressibility assumption, the effective flow area is only geometry dependent and a proportional relation is found with

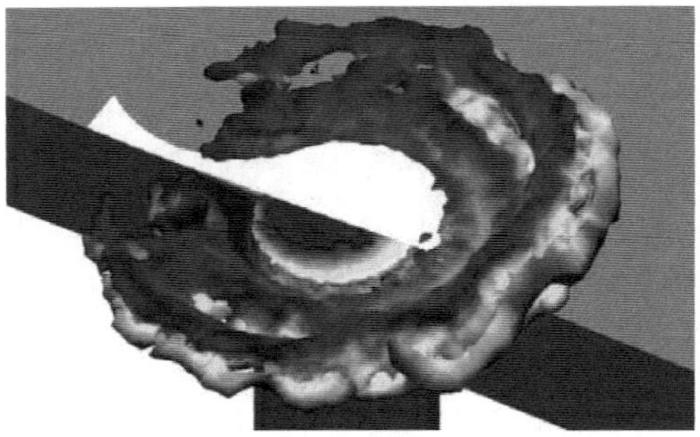

(a) Valve A

(b) Valve B

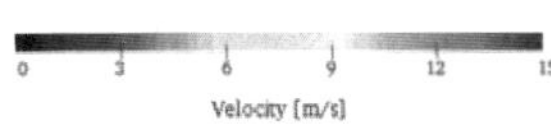

Figure 8: Contour surface for 1000 rad/s of vorticity at t_1.

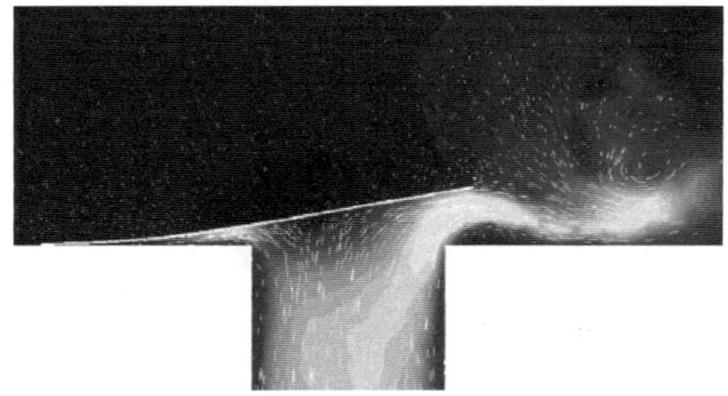

(a) Valve A

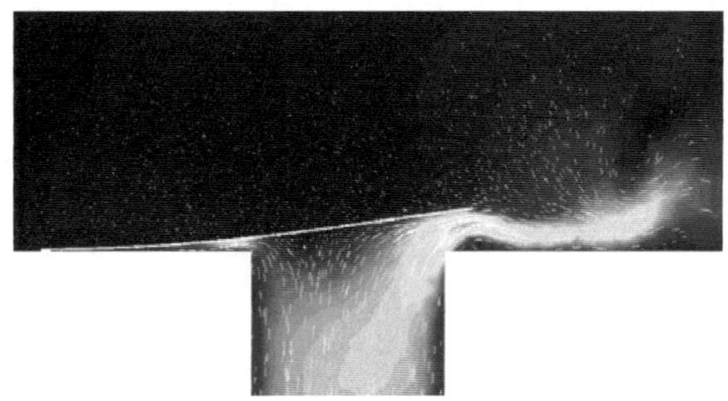

(b) Valve B

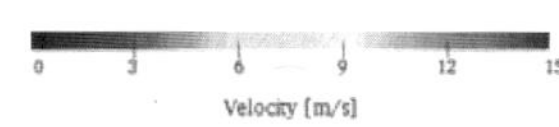

Figure 9: Velocity magnitude map and velocity vectors at t_2.

the valve lift. As can be observed, the wider body of valve B does not generate a significant difference in the effective flow area compared with A. Since the mass flow $\dot{m}$ does not vary, this implies that both valves produce almost the same pressure drop for a given displacement. In fact, the air flow pattern is pretty similar in A and B at the highlighted time instant t_1, when both are equally opened, as can be seen in the vorticity maps of Figure 8.

In the same manner, in Figure 10 (b) the averaged effective force area is presented. It has been calculated as the fraction between thrust force F and pressure drop:

$$A_F = \frac{F}{p_{inlet} - p_{outlet}} \tag{15}$$

The graph shows how the pressure rises to a different extent as the fluid reaches each valve, because of their different stiffness. In both cases the effective force area decreases with increasing valve height. This is explained by the flow detachment and the separation bubble developed between valve and seat, which create a negative pressure that counteracts the positive pressure over the port and diminishes the net force. A similar drop in the effective force area has been previously detected for small displacements of the valve and strong Bernoulli effect in [26, 27, 28] for example.

It has been stated above that valves A and B cause the same instantaneous pressure drop with a common valve height. However, this does not imply that the process efficiency is equivalent for both valves, since they have different oscillation rates and lift peaks. In this case, one can analyse the total pressure drop per time unit in order to compare the difficulty the air must

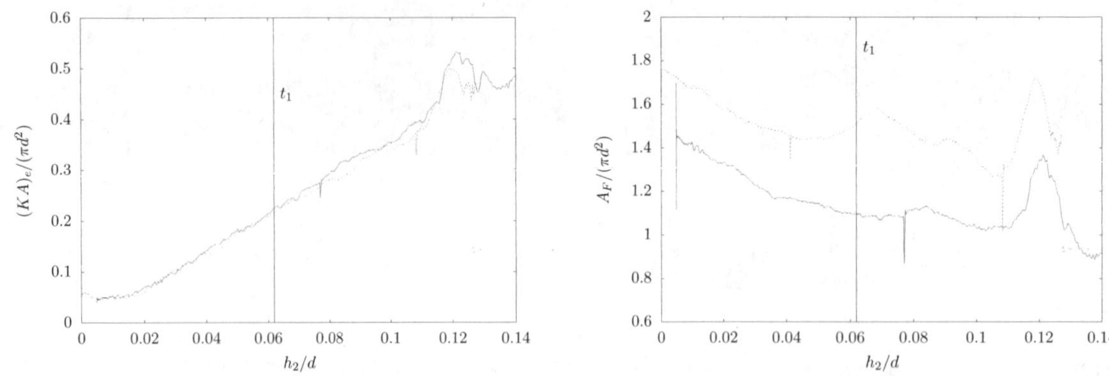

(a) Effective flow area.

(b) Effective force area.

Figure 10: Cycle-averaged parameters of case A (——) and B (- - - -) as a function of displacement of point 2 during the opening process.

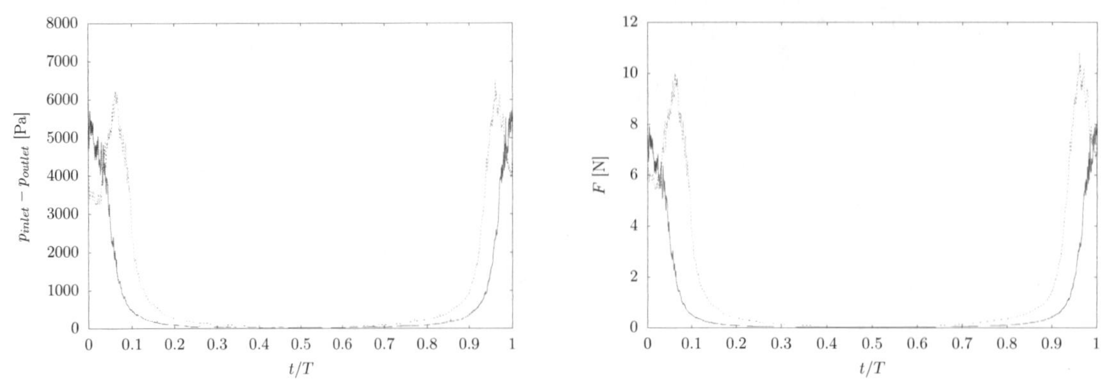

(a) Pressure drop.

(b) Net force on the valve.

Figure 11: Cycle-averaged variables of case A (——) and B (- - - -) as a function of dimensionless time.

withstand to flow out the channel. Pressure drop per oscillation period is presented in Figure 11 (a). The integral of this graph indicates that 573 and 1020 Pa per cycle are needed in cases A and B, respectively. If this value is multiplied by the cycle frequency, then the total pressure drop can be obtained: 33,600 Pa/s with valve A whereas 92,000 Pa/s with valve B. Therefore, the outflow efficiency of this steady flow is far better with valve A, basically due to its higher flexibility.

Finally, in Figure 11 (b) the thrust force on the valve plate is presented. It shows the rise in the fluid force as soon as the valve closes, i.e. as soon as the outflow area diminishes and the valve surface and jet velocity become more perpendicular.

5. Conclusions

An in-house numerical methodology to solve the fluid-structure interaction of the turbulent flow through self actuated reed valves has been presented. The platform strongly couples a three-dimensional CFD solver with turbulence modeling to a structural model for vibrating plates. A moving mesh technique along with the arbitrary lagrangian-eulerian formulation deforms the fluid domain according to the valve deformation. In order to avoid remeshing, the immersed boundary method defines the position of the valve seat. Validation of the model is carried out by comparing the fluttering of a suction valve under a stationary flow at Reynolds number 10,000. The results show that the method is able to yield a representative description of the flow field surrounding the valve. However, a more detailed analysis of the impact between reed and other bodies is necessary to resolve the discrepancies detected in the valve lift. The same set-up is used to test a similar valve with a wider gland body in order to showcase the potentiality of the proposed methodology to optimise valve designs. Although lower stresses are expected with this modification, its higher stiffness reduces the maximum effective flow area and increases the pressure drop, hence worsens the process efficiency.

Acknowledgments

We would like to thank Dr. José Luis Gasche and his team at Universidade Estadual Paulista for providing the experimental measures and for their valuable comments. This work has been financially supported by the Ministerio de Educación, Cultura y Deporte, Spain (FPU13 fellowships).

References

[1] Soedel W 1984 *Design and Mechanics of Compressor Valves* (Purdue University)
[2] Soedel W 1992 *Mechanics, Simulation and Design of Compressor Valves, Gas Passages and Pulsation Mufflers* (Purdue University)
[3] Ribas F A, Deschamps C J, Fagotti F, Morriesen A and Dutra T 2008 *Int. Compressor Engineering Conf.* (Indiana)
[4] Leonard S M 1996 *Hydrocarb. Process.* **75** 67 – 74
[5] Woo S, O'Neal D L and Pecht M 2010 *Eng. Fail. Anal.* **17** 979 – 991
[6] Boeswirth L 1996 *Int. Compressor Engineering Conf.* (Indiana)
[7] Schildhauer M and Spille-Kohoff A 2014 *Prog. in Comp. Fluid Dy.* **14** 38–48
[8] Mayer J, Bjerre P and Brune F 2014 *Int. Compressor Engineering Conf.* (Indiana)
[9] Lemke N C, König M, Hennig J, Försterling S and Köhler J 2016 *Int. Compressor Engineering Conf.* (Indiana)
[10] Parihar A, Myszka D, Robinet B and Hodapp T 2016 *Int. Compressor Engineering Conf.* (Indiana)
[11] Gasche J L, de Lima Dias A D S, Bueno D D and Lacerda J F 2016 *Int. Compressor Engineering Conf.* (Indiana)
[12] Favre F, Antepara O, Lehmkuhl O, Borrell R and Oliva A 2014 *Joint WCCM - ECCM - ECFD 2014 Congress* (Barcelona)
[13] Jofre L, Lehmkuhl O, Ventosa J, Trias F X and Oliva A 2014 *Numer. Heat Transf. Part B Fundam.* **65** 53–79
[14] Trias F, Lehmkuhl O, Oliva A, Pérez-Segarra C and Verstappen R 2014 *J. Comput. Phys.* **258** 246 – 267
[15] Thomson W 1998 *Theory of Vibration with Applications (5th Edition)* (Taylor & Francis)
[16] Bathe K J 1996 *Finite Element Procedures* (Prentice Hall)
[17] Smith R 2011 A pde-based mesh update method for moving and deforming high reynolds number meshes *49th AIAA Aerospace Sciences Meeting* (Orlando)
[18] Lesoinne M and Farhat C 1996 *Comput. Methods in Appl. Mech. Eng.* **134** 71 – 90
[19] Causin P, Gerbeau J F and Nobile F 2005 *Comput. Methods in Appl. Mech. Eng.* **194** 4506 – 4527
[20] Naseri A, Lehmkuhl O, Gonzalez I and Oliva A 2016 *J. Phys.: Conf. Series* **745** 032020
[21] Gasche J L, Arantes D M and Andreotti T 2014 *Int. Compressor Engineering Conf.* (Indiana)
[22] Habing R A and Peters M C A M 2006 *Journal of Fluids and Structures* **22** 683 – 697
[23] Kim J, Moin P and Moser R 1987 *J. Fluid Mech.* **177** 133 – 166
[24] Rigola J, Aljure D, Lehmkuhl O, Pérez-Segarra C D and Oliva A 2015 *IOP Conf. Ser.: Mater. Sci. Eng.* **90** 012026
[25] El Khoury G K, Schlatter P, Noorani A, Fischer P F, Brethouwer G and Johansson A V 2013 *Flow Turbul. Combust.* **91** 475 – 495
[26] Schwerzler D D and Hamilton J F 1972 *Int. Compressor Engineering Conf.* (Indiana)
[27] Ferreira T S and Lainor J 1986 *Int. Compressor Engineering Conf.* (Indiana)
[28] Deschamps C J, Ferreira R T S and Prata A T 1988 *Int. Compressor Engineering Conf.* (Indiana)

IOP Conf. Series: Materials Science and Engineering **232** (2017) 012033 doi:10.1088/1757-899X/232/1/012033

Stainless steel valves with enhanced performance through microstructure optimization

A A Barani[1], M Boukhattam[2], M Haggeney[1], S Güler[2]

[1] Zapp Precision Metals GmbH | Hochstraße 32 | 59425 Unna | Germany
[2] University of Duisburg and Essen | Chair of Materials Science and Engineering | Lotharstr. 1 | 47057 Duisburg, Germany

Abstract. Compressor valves are made of hardened and tempered martensitic steels. The main design criterion for the material selection is the fatigue performance of the material under bending loads. In some cases impact loads and corrosive atmospheres additionally act on the part. For the first time, the microstructure of the most commonly used stainless steel and its influence on the properties relevant for flapper valves is presented and described in this paper. It is demonstrated how the tensile properties of a martensitic stainless steel can be enhanced by tailoring the microstructure. Electron back scatter diffraction method is carried out to explain the changes in monotonic mechanical properties. Through a modified heat treatment the martensite microstructure is refined resulting in an increase of yield and ultimate tensile strength and at the same time a significant increase of elongation.

1. Introduction

Stainless martensitic steels were discovered in the year 1912 by Harry Brearly in Sheffield England and by Elwood Haynes in the USA [1, 2]. These alloys combine high hardness and wear resistance with reasonable toughness and are resistant to corrosion under ambient atmospheric conditions. Since then various alloys have been developed and mainly used for blades in household, medical, industrial and consumer applications. Because of the high strength, corrosion resistance, good toughness, and very good fatigue properties they are used as well in compressors as valves [3-5]. Numerous researchers have reported about the performance and properties of the martensitic stainless steels used in valves [3-8]. Manifold commercial names are available for the stainless chromium steel X38CrMo14 used for the production of valves. It contains approximately 13.5 weight percent of chromium, 0.38 weight percent of carbon, and around 1 weight percent of molybdenum.

However, few reports are found on the correlation between the microstructure of the stainless martensitic steels itself and the mechanical properties. Toro et. al. described the influence of chromium nitrides on the corrosion-erosion resistance of AISI 410S [9]. Yuan has thoroughly investigated a chromium steel with 0.44 weight percent of carbon and demonstrated how through control of the heat treatment parameters the tensile properties of X44Cr13 (1.4034, AISI420) can be tailored (Yuan 2009). The TRIP effect described in the paper is related to reverted austenite. The influence of martensite packet size, lath or needle geometry on mechanical properties remains open for further research. Especially, for flapper valves where fatigue behavior is of utmost importance the morphology of the martensite could play a crucial role for the fatigue behavior. It has been demonstrated for carbon steels that the

IOP Conf. Series: Materials Science and Engineering **232** (2017) 012033 doi:10.1088/1757-899X/232/1/012033

toughness and fatigue properties can be controlled by the martensite block size [11, 12]. In the present paper for the first time the influence of the martensite microstructure on tensile properties is investigated for the alloy X38CrMo14 (1.4419, UNS420026, Zapp 1.4028MO) used in flapper valves since decades.

Table 1 - Chemical composition of precision strip used for the investigation (Zapp 1.4028MO, X38CrMo14, 1.4419, UNS420026)

	C	Si	Mn	Cr	Mo
Content in weight percent	0.38	0.30	0.46	13.46	0.93

2. Experimental

Precision strip with a thickness of 0.41mm made out of the alloy X38CrMo14 with the composition given in table 1 was austenitized, quenched, and tempered in a continuous hardening line. The continuous hardening line consists of an austenitization furnace, quenching section, and tempering furnace. The heat treatment is schematically shown in figure 1.

In order to obtain various microstructures the heat treatment was modified. The strip selected for the test was divided into two strips. Two set of heat treatment parameters were defined to investigate the influence of the microstructure on the tensile properties. The first set of parameters consists of the standard heat treatment parameters used at Zapp Precision Metals for the production of quenched and tempered Zapp 1.4028MO. The second set of parameters was defined to reduce the size of the martensite units. The austenitization and tempering temperature and time in each process step were kept constant. The sample taken from the first quenched and tempered strip heat treated with the standard set of parameters are designated as modification A. The strip heat treated to obtain a refined martensite microstructure and the samples thereof are called modification B.

The microstructure was observed by light microscopy using various etchants to reveal the different phases and microstructural features. Electron backscatter diffraction technique was used to characterize the microstructure of the two modifications. EBSD samples in rolling direction were prepared using standard grinding and polishing procedures. An area of 80 micrometer by 80 micrometer was mapped with a step size of 50nm.

The tensile properties were determined in the longitudinal direction according to standard practice on 10 samples taken from start and end of the heat treated strips.

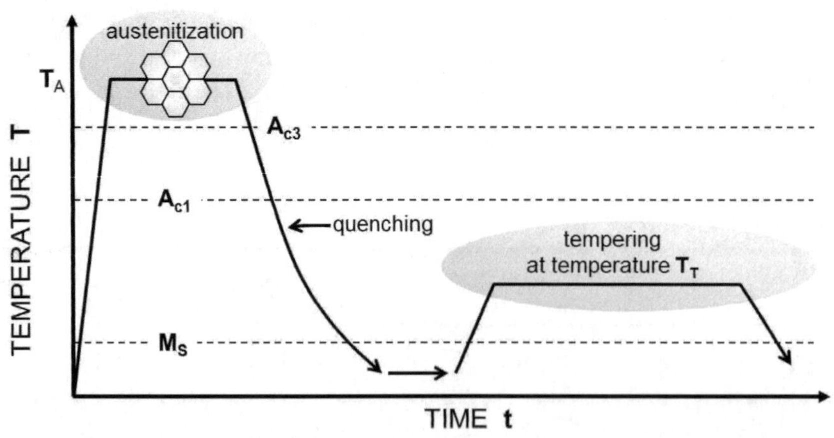

Figure 1 - Schematic presentation of the heat treatment

3. Results

The samples were examined by light microscopy. Significant differences with respect to carbide dispersion and carbide size distribution could not be identified using various etchants with a magnification up to 1000 times.

The fraction of retained austenite was determined using the EBSD results. Both modifications exhibit approximately one percent of retained austenite.

Figure 2 presents grain boundary maps of the two modifications. Low angle grain boundaries with a rotation angle between 3 and 15 degrees are shown as blue lines, high angle grain boundaries with rotation above 15 degrees are shown as black lines. The coherent twin boundaries (Sigma 3) are presented as red lines. The comparison of maps shows that the microstructure of sample B is much finer. Because areas enclosed by high angle grain boundaries and their size influences the mechanical properties, both the strength and ductility, an effective grain size was defined to be an area enclosed by a boundary with at least 15 degrees misorientation to the adjacent area and having at least 10 measuring points. The effective grain size area was determined for both modifications and an average diameter was derived assuming the area is circle. The average diameter defined in such a way was reduced from 0.74 (sample A) to 0.54 (sample B) micrometer. Acknowledging the fact the martensite block and lath are elongated and more elliptical the width of the blocks constrained by a high angle grain boundary was determined. The result is presented in figure 3. It is seen that 75 percent of the blocks of modification B have a smaller width than 0.5 micrometer and hence the median of modification A.

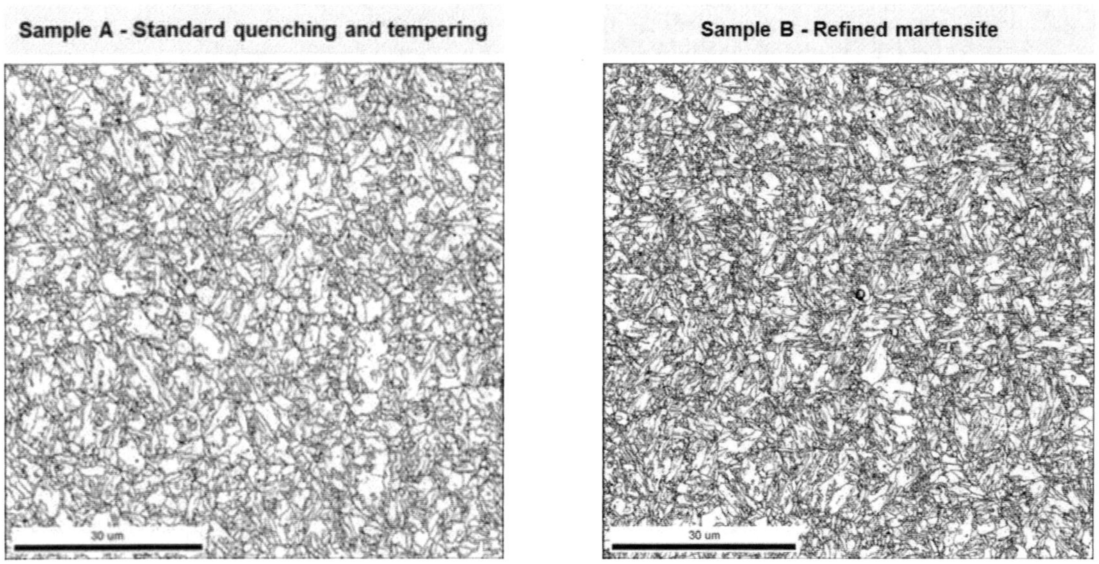

Figure 2 - Grain boundary maps of the two modifications A and B. Black line: high angle grain boundary (15 to 62.8°); blue line: low angle grain boundaries (3 to 15°); red line: sigma 3 grain boundary with a tolerance of 8.7°.

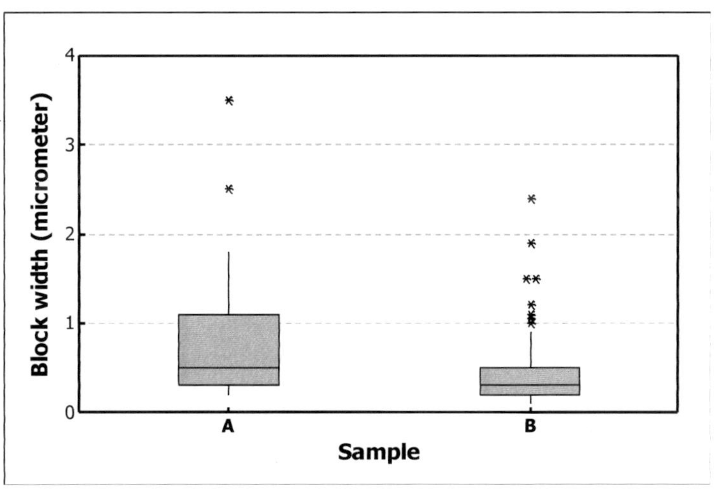

Figure 3 - Width distribution of martensite blocks limited by high angle grain boundaries for modifications A and B.

Modification B exhibits higher yield and ultimate tensile strength than modification A (figure 4). While the yield strength is approximately 50 MPa higher, the ultimate tensile strength is almost 150 MPa higher. At the same time the elongation is significantly increased (figure 4).

4. Discussion

The present work demonstrates that the microstructure of stainless martensitic steel X38CrMo14 can be modified to be finer resulting in higher tensile yield and ultimate strength and at the same time higher elongation. The structural unit responsible for the improved mechanical properties is the area enclosed by high angle grain boundaries. Due to the shape of martensitic blocks the width of these units has been extracted as proposed by Naylor [11]. A finer microstructure including high angle grain boundaries increases the yield strength and as well the elongation [11, 12]. It is as well beneficial for dynamic properties like fracture toughness or fatigue limit.

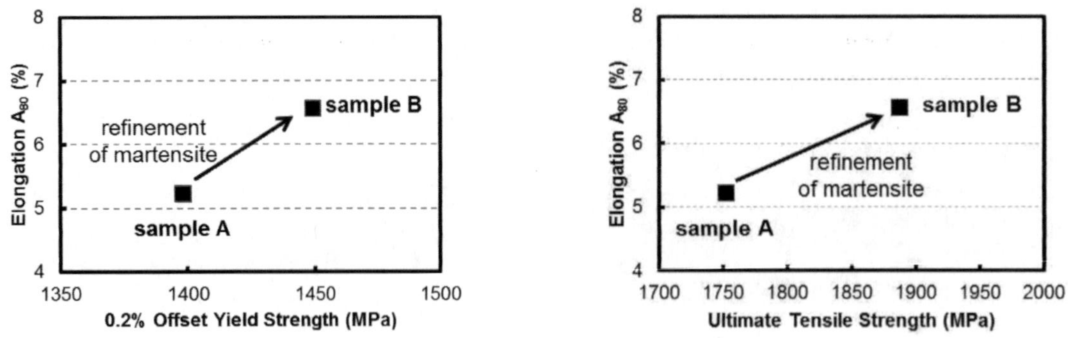

Figure 4 – Average tensile properties in rolling direction of the two treatment, i.e. standard treatment (modification A) and treatment for finer martensite (modification B)

However, the effect of the improved tensile properties can not only be correlated to the structural refinement only. The finer martensite morphology is corresponding at the same time to an increased grain boundary area. As reported earlier for carbon steels an increase of grain boundary area can lead as well to finer carbides [12].

5. Conclusion

The present study shows that the performance of flapper valve could be improved by making the martensite finer. It is known that an increase of the tensile strength values and the elongation at the same time is connected to a higher fatigue life. The microstructure of the stainless martensitic steel X38CrMo14 (Zapp 1.4028MO) is refined by modification of the heat treatment resulting in finer martensite microstructure with smaller blocks. The refinement leads to higher 0.2 percent offset yield strength and ultimate tensile strength while increasing the elongation. The concept of the effective grain size and the effective width of martensitic units are validated for the first time for martensitic stainless steels. The improvement of the reported mechanical values enables the compressor designers to consider a higher fatigue limit or higher loads for the valves without comprising on other attributes.

6. Acknowledgement

The EBSD investigations are carried out at the Salzgitter Mannesmann Research Center in Duisburg. The authors are grateful to Mr. Matthias Frommert and his colleagues for carrying out the investigations.

References

[1] Lippold J C and Kotecki J D 2005 *Welding Metallurgy and Weldability of Stainless Steels* (John Wiley and Sons)

[2] Cobb H B et al 2010 *The History of Stainless Steel* (ASM International).
[3] Chai G C et al 2004 *International Compressor Engineering conference at Purdue* Paper 1702
[4] Hareland M 2014 *International Compressor Engineering conference at Purdue* Paper 2330
[5] Nadar V.S 2014 *Fatigue testing of scratched flapper valve steel*
[6] Lof A 2016 *International Compressor Engineering conference at Purdue* Paper 2439
[7] Johansson R 1975 *International Compressor Engineering conference at Purdue* Paper 173
[8] Mueller M 2006 *International Compressor Engineering conference at Purdue* Paper 1800
[9] Toro A 2003 *Acta Materialia* **51** 3363-3374
[10] Yuan L 2012 *Acta Materialia* **60** 2790-2804
[11] Naylor J P 1979 *Met. Trans.* A **10A** 861-873
[12] Ardehali Barani A 2008 *Optimization of the Critical Content of Tramp Elements in Ultra-High Strength Silicon Chromium Spring Steels through Thermomechanical Treatment*

IOP Conf. Series: Materials Science and Engineering **232** (2017) 012034 doi:10.1088/1757-899X/232/1/012034

Reduction of the suction losses through reed valves in hermetic reciprocating compressors using a magnet coil

J Hopfgartner[1], S Posch[1], B Zuber[1], R Almbauer[1], K Krischan[1] and S Stangl[2]

[1] Graz University of Technology, 8010 Graz, Austria
[2] Secop Austria GmbH, 8280 Fürstenfeld, Austria

hopfgartner@ivt.tugraz.at

Abstract. Reed valves are widely used in hermetic reciprocating compressors and are responsible for a large part of the thermodynamic losses. Especially, the suction valve, which is opened nearly during the whole suction stroke, has a big potential for improvement. Usually, suction valves are opened only by vacuum created by the moving piston and should be closed before the compression stroke starts to avoid a reversed mass-flow through the valve. Therefore, the valves are prestressed, which results on the other hand in a higher flow resistance.
In this work, a suction valve is investigated, which is not closed by the preload of the valve but by an electromagnetic coil located in the suction muffler neck. Shortly before the piston reaches its bottom dead centre, voltage is applied to the coil and a magnetic force is generated which pulls the valve shut. Thereby, the flow resistance through the valve can be reduced by changing the preload on the reed valve because it is no longer needed to close the valve.
The investigation of this adapted valve and the electromagnetic coil is firstly done by numerical simulations including fluid structure interactions of the reed valves of a reciprocating compressor and secondly by experiments made on a calorimeter test bench.

1. Introduction

Reed valves play a key role in hermetic reciprocating compressors used for domestic refrigerating appliances. Beneath simplicity, their high reliability and low costs are the main reasons why reed valves are so extensively used in hermetic reciprocating compressors. Reed valves open and close depending on the pressure difference between cylinder and suction/discharge muffler thus reed valves are self-actuating. The piston, moving from the top dead centre to the bottom dead centre, decreases the pressure in the cylinder and creates a vacuum, which forces the suction valve to open. When the forces due to the vacuum become bigger than the valve preload and the oil "stiction" forces, the suction valve starts to open. In reality, the suction valve in a hermetic compressor opens and closes several times during one suction stroke depending on the operating conditions of the compressor and the parameters of the reed valve. It may happen that the suction valve has not closed yet, when the piston starts to move upward again and thus gas can flow back from the cylinder into the suction muffler. To avoid this reversed mass-flow, the preload must force the valve to close exactly at the end of the gas movement into the cylinder.

Many different studies, which investigate the suction and discharge losses inside reciprocating compressors, were done in the past using either experiments, simulations or both. [1] measured the cylinder pressure and valve lift curves of a reciprocating compressor simultaneously to investigate the losses. Special attention was paid to the opening and closing delay of the valves as well as back-flow effects. [2] analysed the characteristic of valve dynamics and introduced a 1D model, used for the prediction of the valve performance of different designs of the reed valve. The authors showed a big potential for improvement of the suction valve which can mainly be attributed to a timely closing leading to a reduction of the gas back-flow through the suction port. Similar results can be observed in the work of [3]. The authors changed the rotational speed of the compressor while measuring the lift of the suction valve. Since the movement of the valve is mainly a function of its resonant frequency, the closure time varies related to the bottom dead centre of the compressor at different rotational speeds. This results in different gas back-flow rates and in a fluctuation of the volumetric efficiency. In the work of [4] the influence of different suction valve parameters on the efficiency of the reciprocating compressor was investigated. It was shown, that a reduction of the valve spring preload results in a significant reduction of the indicated power consumption during the suction phase. Nevertheless, a slightly higher back-flow rate of the gas could be observed.

Summarizing, it can be pointed out, that a perfectly designed valve should open as fast as possible, stay open at a maximum lift and close at the right time with low velocities to enlarge the durability. In order to use the big advantage of a lowered preload of the suction valve, this work investigates a compressor with a suction valve, which is negatively prestressed. This means, the valve is slightly open at equalized pressure. To close the suction valve in time, an electromagnetic coil which is located in the suction muffler neck is used instead. The coil starts to pull the suction valve in order to shut shortly before the piston reaches its bottom dead centre. The basic idea using electromagnetic forces to close reed valves was already published by [5].

The investigation of this closing mechanism is firstly done by numerical simulations including fluid structure interactions of the reed valves and secondly by experiments made on a calorimeter test bench. For the measurements inside the calorimeter, stabilized operating conditions according to ASHRAE have been applied. Therefore, the R600a-compressor is operated at a fixed rotational speed of 3000 rpm. Simulations as well as measurements show a general potential for improvement of the COP between 0.4 to 1.7 %.

2. Experimental Setup

A conventional suction valve is a prestressed thin piece of steel, which is closed at equalized pressure. In this work, a reed valve with a negative preload, which is closed by a magnetic coil, is investigated. Therefore, a commercially available reciprocating compressor with a displacement of approximately 6 ccm is equipped with extensive measurement instrumentation and a magnetic coil.

Figure 1 shows schematically the magnetic arrangement and a CAD-drawing where the assembly situation of the magnetic coil and the suction valve can be seen. The coil consists of an iron core and a copper winding with approximately 200 turns. It is located above the suction muffler neck and its iron core feeds through the muffler neck to the reed valve. To close the magnetic circuit, an iron cramp is mounted which connects the iron core with the ferromagnetic parts fixing the valve.

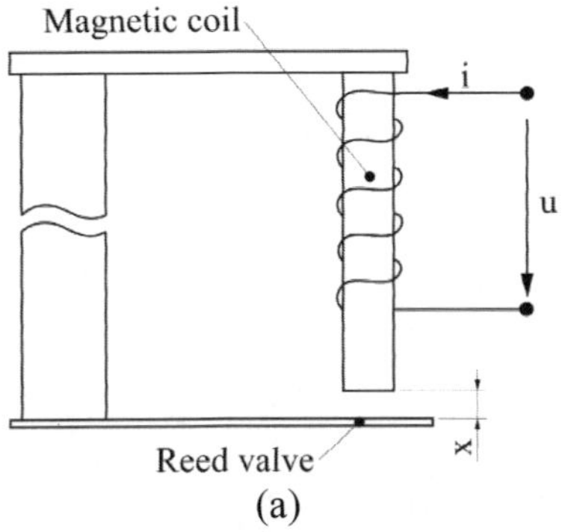

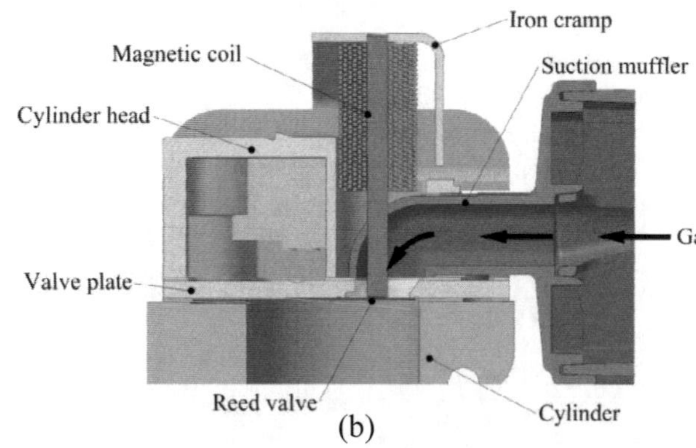

Figure 1. Magnetic arrangement (a) and assembling situation of the electromagnetic coil (b).

2.1. Electromagnetic coil

To close the suction valve when the piston reaches its bottom dead centre, an electric current is applied to a magnetic coil. The resulting magnetic field closes the suction valve. After the valve has shut and the pressure in the cylinder has risen, the electromagnetic coil is demagnetized by applying a negative voltage to avoid a delay of the suction valve opening and thus an increasing of the suction losses.

To estimate the magnetic force acting between the iron core and the reed valve, the approach described in [6] is used. With the magnetomotive force $\Theta = N \cdot i$, where N is the number of turns and i is the current, the magnetic flux through the coil Φ_c is given with:

$$\Phi_c = \frac{\Theta}{\mathcal{R}} \tag{1}$$

Therein, $\mathcal{R}$ is the equivalent reluctance of the whole system consisting of the effective reluctance of the iron core and the iron cramp $\mathcal{R}_c$, the effective reluctance of the valve $\mathcal{R}_v$, the effective reluctance of the gas gap $\mathcal{R}_g$ (between valve and iron core) and the reluctance $\mathcal{R}_l$ which takes the leakage flux into consideration. The equivalent reluctance can finally be written as:

$$\mathcal{R} = \mathcal{R}_c + \mathcal{R}_v + \frac{\mathcal{R}_l \cdot \mathcal{R}_g}{\mathcal{R}_l + \mathcal{R}_g} = \frac{l_c}{\mu_0 \cdot \mu_{rc} \cdot A_c} + \frac{l_v}{\mu_0 \cdot \mu_{rv} \cdot A_v} + \frac{\left(\frac{l_l}{\mu_0 \cdot A_l}\right) \cdot \left(\frac{\alpha \cdot x}{\mu_0 \cdot A_g}\right)}{\left(\frac{l_l}{\mu_0 \cdot A_l}\right) + \left(\frac{\alpha \cdot x}{\mu_0 \cdot A_g}\right)} \tag{2}$$

To simplify this approach, the relative permeabilities μ_r are assumed to be constant. l and A are the effective lengths and effective areas of the particular reluctances. With the flux linkage of the coil $\psi = N \cdot \Phi_c$ and Faraday's law of induction, the voltage u_L can be written as:

$$u_L = \frac{d\psi}{dt} \tag{3}$$

Using the coil current $i_{(x,\psi)}$ expressed in terms of the flux linkage and the gas gap, the magnetic energy is:

$$W_m = \int_0^{\psi} i_{(x,\bar{\psi})} d\bar{\psi} = \int_0^{\psi} \frac{\mathcal{R}(x)}{N^2} \cdot \bar{\psi}\, d\bar{\psi} = \frac{\mathcal{R}(x)}{2 \cdot N^2} \cdot \psi^2 = \frac{N^2 \cdot i^2 \cdot \left(\frac{\alpha \cdot x}{\mu_0 \cdot A_g} + \mathcal{R}_l\right)}{2 \cdot \left[(\mathcal{R}_c + \mathcal{R}_v + \mathcal{R}_l)\frac{\alpha \cdot x}{\mu_0 \cdot A_g} + \mathcal{R}_l \cdot (\mathcal{R}_c + \mathcal{R}_v)\right]} \tag{4}$$

With the derivative of the magnetic energy with respect to the gas gap, the magnetic force F_m is given by:

$$F_m = \frac{\partial W_m}{\partial x} = -\frac{N^2 \cdot i^2 \cdot \frac{\alpha}{\mu_0 \cdot A_g} \cdot \mathcal{R}_l^2}{2 \cdot \left[(\mathcal{R}_c + \mathcal{R}_v + \mathcal{R}_l) \cdot \frac{\alpha \cdot x}{\mu_0 \cdot A_g} + \mathcal{R}_l \cdot (\mathcal{R}_c + \mathcal{R}_v)\right]^2} \tag{5}$$

The design of the magnetic coil is based on the requirement, that the "negative" preload and the magnetic force acting on the valve should be equal to the preload of the standard reed valve. With a valve deflection of 0.4 mm in its neutral position, a magnetic force of 0.212 N is needed. With the values stated in Table 1, the required number of turns can be estimated based on equation (5):

$$N_{min} = \sqrt{\frac{2 \cdot F_m \cdot \mu_0 \cdot A_g}{\alpha}} \cdot \frac{\mathcal{R}_g \cdot (\mathcal{R}_c + \mathcal{R}_v + \mathcal{R}_l) + \mathcal{R}_l \cdot (\mathcal{R}_c + \mathcal{R}_v)}{i_{max} \cdot \mathcal{R}_l} \approx 190 \tag{6}$$

Table 1. Basic assumptions for the design of the electromagnetic coil.

	value	unit		value	unit
x	0.4	mm	A_g	3.14	mm^2
l_c	50	mm	A_c	3.14	mm^2
l_v	20	mm	A_v	0.5	mm^2
l_l	0.4	mm	A_l	7.1	mm^2
α	2	-	μ_0	1.26 e-6	Vs/Am
i_{max}	2.3	A	μ_{rc}	1000	-
F_{req}	0.212	N	μ_{rv}	850	-

It should be noted at this point, that equation (6) is influenced by many different estimated values with a high level of uncertainty, since the magnetic flux in the real application cannot be predicted very well. With this rough estimate and taking several constraints due to the lack of space in the region of the cylinder head into consideration, the parameters for the investigated electromagnetic coil can be determined (Table 2).

Table 2. Parameters of the investigated electromagnetic coil.

	unit	value
number of turns	-	190 up to 205
length of the core	mm	31.8
iron core diameter	mm	2.5
outer diameter of the coil	mm	9.3
diameter of the wire	mm	0.5

2.2. Settings for the electric circuit

In order to reduce the suction valve losses, the magnetic coil should close the suction valve nearby the bottom dead centre of the piston. Therefore, the position of the piston on the one hand, and the delay between applying voltage to the coil and the completely closing of the valve on the other, must be known. To control the magnetic coil, a full bridge circuit is used which has two control inputs to change the direction of the supply voltage.

2.2.1. Determination of the bottom dead centre. For the determination of the bottom dead centre, a sensor based on the Hall-effect is used. It detects a permanent magnet integrated in the crank shaft and generates a periodic signal with a significant peak when the magnet passes the sensor. To determine the delay between the peak of the signal and the exact position of the bottom dead centre, the movement of the piston at different pressure ratios is measured with a Laser-Doppler vibrometer in a pretest.

2.2.2. Response behaviour of the closing mechanism. The closing behaviour of the adapted suction valve (0.41 mm opened in neutral position) was investigated on a dummy compressor without a piston. The closing time of the valve was measured by a Laser-Doppler vibrometer which was directed at the backside of the valve (middle of the valve plate) while it was shut by the magnetic coil. Figure 2 shows a diagram of the measured valve lifts over time for different coil supply voltages from 2 up to 8 V. The delay between applying voltage to the coil and closing of the reed valve varies between 0.95 and 1.55 ms.

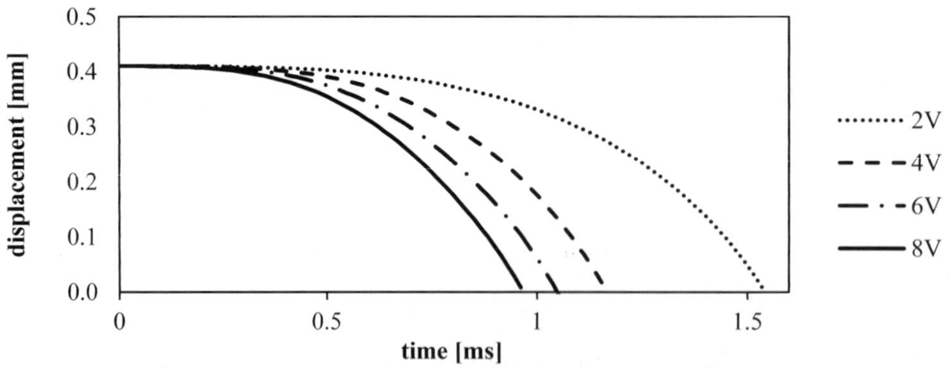

Figure 2. Delay between applying voltage and closing the valve.

2.3. Further measurement instrumentation (temperature, pressure)

Beside the measurement instrumentation necessary for controlling the electromagnetic coil, thermocouples and pressure transducers are installed in the compressor. Several thermocouples distributed on the compressor shell and inside the compressor are used to investigate the influence of the waste heat from the coil on the thermal behaviour of the compressor. The pressure sensors are installed in the suction muffler neck and cylinder head. Figure 3 shows pictures of the installed measurement instrumentation and the electromagnetic coil.

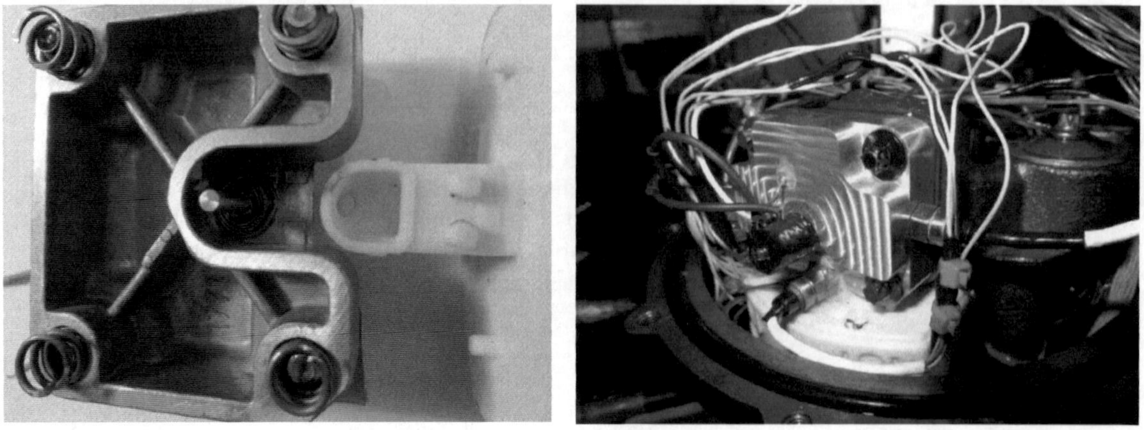

Figure 3. Electromagnetic coil assembled in the cylinder head (left). Assembling situation of pressure transducers, thermocouples and electromagnetic coil (right).

3. Simulation Model

Numerical simulation in this work is conducted using the commercial CFD software package Fluent from ANSYS. The simulation domain, which can be seen in Figure 4, includes suction muffler, suction and discharge valve, cylinder, cylinder head and discharge line. The mesh size of the domain varies between 1.9 and 5.2 million cells depending on the position of the piston and the reed valves. The movement of the piston and the reed valves requires a dynamic mesh which is enabled by using the remeshing, layering and smoothing techniques provided by the software. While the forced piston movement can be modelled easily, the movement of the reed valves must be modelled with a more complex fluid structure interaction which is described in a section below.

For the coupling of pressure and velocity, the pressure based coupled solver is applied. Turbulence is considered with the realizable k-ε model using the standard wall function for the near-wall treatment.

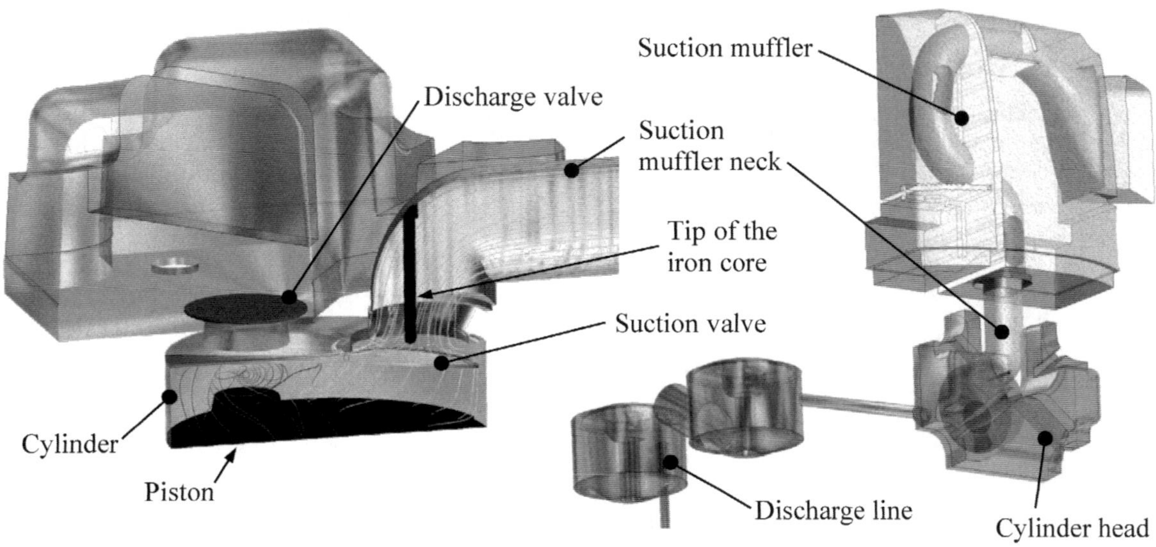

Figure 4. Simulation domain of the investigated hermetic compressor.

3.1. Boundary conditions

The numerical simulations in this work are based on the operating conditions according to ASHRAE (-23 °C / 55 °C). At the inlet of the suction muffler, a pressure inlet boundary condition has been set with a pressure of 62770 Pa and a temperature of 323 K. At the outlet of the discharge line, a pressure outlet boundary of 786000 Pa has been set. The temperature boundary conditions used in the simulations are based on the measurements of [7] and are shown in Table 3.

Table 3. Temperature boundary conditions.

	temperature [K]
suction muffler	332.1
valve plate	360.0
cylinder	357.2
cylinder head	362.7
discharge line	352.0

3.2. Modelling of the reed valve movement

Fluid structure interactions (FSI) must be taken into consideration, when reed valves are modelled. Therefore, subroutines are hooked to Fluent via so called User Defined Functions (UDF) which includes the oil stiction effect based on [8].

3.2.1. Discharge valve. In order to reduce the simulation effort, the discharge valve is assumed to be a parallel moving flat plate (see Figure 4). The equation of motion of this spring-mass system, including the preload and the oil stiction effect during the opening, is solved by using an explicit Runge-Kutta method of fourth order:

$$\ddot{x} = \frac{1}{m^*}\left(F_{pre} + F_{gas} - c \cdot x - d \cdot \dot{x} - \frac{c_t}{x} - \dot{x} \cdot \frac{c_v}{x^3}\right) \tag{7}$$

3.2.2. Suction valve. A more accurate model must be used for the suction valve, since the suction losses are strongly influenced by the deflection and motion of the reed valve. Therefore, a Finite Element Approach is used to solve the nonlinear dynamic structure analysis problem. Forces due to the oil stiction effect, valve preload, electromagnetic forces as well as forces due to contact with surrounding parts are considered. The electromagnetic force is assumed to act in the middle of the valve plate. The contact of the reed valve with surrounding parts is modelled based on a penalty-method neglecting friction. The time integration of the equation of motion shown in equation (8) is based on a Newmark algorithm coupled applying a Newton-Raphson iteration and its resulting linear equation system is solved with Cholesky decomposition using sparse matrices. Damping is considered with the so called Rayleigh-damping. Further details concerning the structure analysis can be found in [9].

$$\left(a_0 \cdot \boldsymbol{M} + a_1 \cdot \boldsymbol{D} + {}^{t+\Delta t}\boldsymbol{K}^{(i)}\right) \cdot \boldsymbol{\Delta u}^{(i)} = RHS$$

$$RHS = {}^{t+\Delta t}\boldsymbol{r} + \boldsymbol{M} \cdot \left[a_0 \cdot \left({}^{t}\boldsymbol{u} - {}^{t+\Delta t}\boldsymbol{u}^{(i)}\right) + a_2 \cdot {}^{t}\dot{\boldsymbol{u}} + a_3 \cdot {}^{t}\ddot{\boldsymbol{u}}\right] \tag{8}$$

$$+\boldsymbol{D} \cdot \left[a_1 \cdot \left({}^{t}\boldsymbol{u} - {}^{t+\Delta t}\boldsymbol{u}^{(i)}\right) + a_4 \cdot {}^{t}\dot{\boldsymbol{u}} + a_5 \cdot {}^{t}\ddot{\boldsymbol{u}}\right] - {}^{t+\Delta t}\boldsymbol{f}^{(i)} + {}^{t+\Delta t}\boldsymbol{f}_{pen}^{(i)}$$

The parameters a_0 - a_5 from the upper equation can be written as:

$$a_0 = \frac{1}{\alpha \cdot \Delta t^2}; \quad a_1 = \frac{\delta}{\alpha \cdot \Delta t}; \quad a_2 = \frac{1}{\alpha \cdot \Delta t};$$
$$a_3 = \frac{1}{2 \cdot \alpha} - 1; \quad a_4 = \frac{\delta}{\alpha} - 1; \quad a_5 = \frac{\Delta t}{2} \cdot \left(\frac{\delta}{\alpha} - 2\right) \tag{9}$$

3.3. Electromagnetic force

The electrical circuit of the magnetic coil can be compared with a resistor-inductor circuit shown in Figure 5. When a voltage u is applied to the circuit, the inductor stores a certain amount of energy which takes time. Therefore, the voltage and the current of the inductor cannot change instantaneously. For such circuits, the temporal profile of the current is well known and can be given with:

$$i_{(t)} = \frac{u}{R} \cdot \left(1 - \mathrm{e}^{-\frac{t}{\tau}}\right) \tag{10}$$

In equation (10), u is the voltage applied to the coil and $\tau = L_{(x)}/R$ is the characteristic time constant with the electrical resistance R and the inductance $L_{(x)}$. With this expression of the current, equation (5) can finally be written as:

$$F_{\mathrm{m}} = -\frac{N^2 \cdot \frac{\alpha}{\mu_0 \cdot A_{\mathrm{g}}} \cdot \mathcal{R}_{\mathrm{l}}^2}{2 \cdot \left[(\mathcal{R}_{\mathrm{c}} + \mathcal{R}_{\mathrm{v}} + \mathcal{R}_{\mathrm{l}}) \cdot \mathcal{R}_{\mathrm{g}} + \mathcal{R}_{\mathrm{l}} \cdot (\mathcal{R}_{\mathrm{c}} + \mathcal{R}_{\mathrm{v}})\right]^2} \cdot \frac{u^2}{R^2} \cdot \left(1 - \mathrm{e}^{-\frac{t \cdot R}{L_{(x)}}}\right)^2 \tag{11}$$

Since equation (11) contains several parameters which cannot be measured or predicted with high accuracy, the reluctance of the leakage flux $\mathcal{R}_{\mathrm{l}}$ as well as the parameter α are fitted so that the simulated closing of the suction valve meets the measured closing behaviour shown in Figure 2.

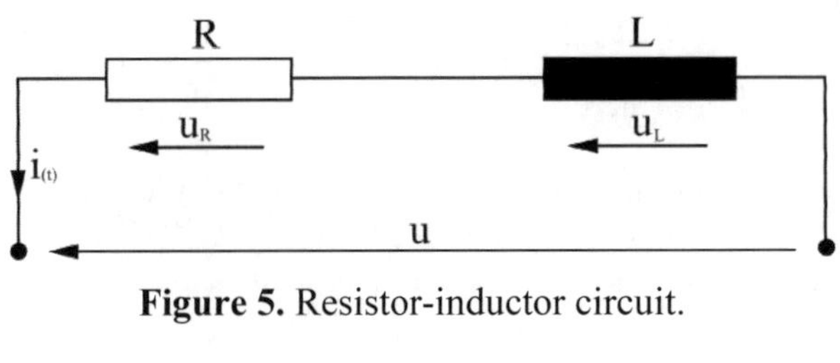

Figure 5. Resistor-inductor circuit.

IOP Conf. Series: Materials Science and Engineering **232** (2017) 012034 doi:10.1088/1757-899X/232/1/012034

4. Results

To validate the calculated reed valve motion, measurements from [10] are used. Therein, the valve lift measurements for a standard compressor of the same type as it is investigated in the present paper were carried out by using the Laser Doppler vibrometer method. A comparison between the simulated and the measured valve lift can be seen in Figure 6. The agreement between simulations and measurements using the standard suction valve is satisfying and therefore, the models of the reed valves are expected to be well suited for simulations with the adapted suction valve too.

Figure 7 shows the simulated pressure-volume diagram over 360 degree crank angle using the standard suction valve. On the right hand side the low pressure range is enlarged.

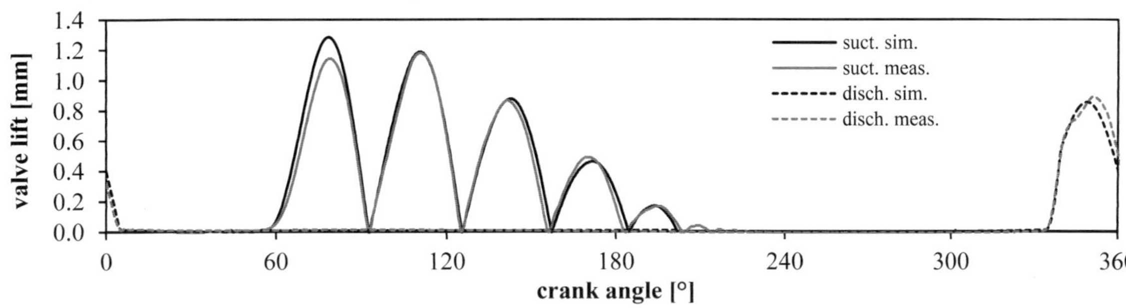

Figure 6. Comparison between simulated and measured valve lift of the suction and discharge valve.

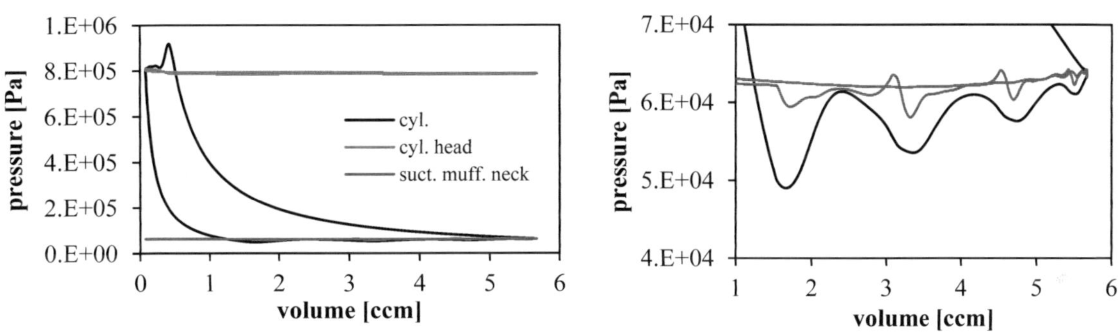

Figure 7. Simulated pressure-volume diagram using the standard suction valve.

Results of the simulation using the adapted suction valve and the electromagnetic coil are shown in Figure 8. On the left side the pressure-volume diagram enlarged at low pressures and on the right side a diagram of the valve lift can be seen. For better comparison between the two investigated cases, the simulations using the standard suction valve are drawn in additionally. The reduced preload of the adapted suction valve results besides a slightly earlier opening in a significantly increased maximum valve lift. No significant differences could be seen in the movement of the discharge valve.

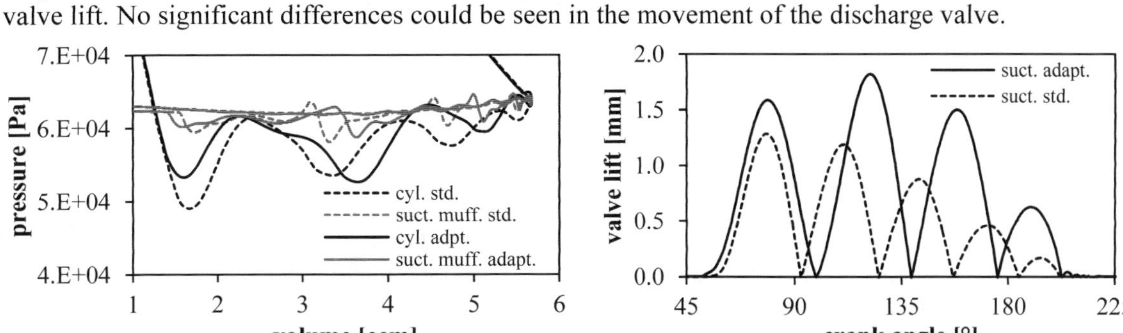

Figure 8. Simulated pressure-volume diagram and suction valve lift using the adapted valve with the electromagnetic coil.

Table 4 shows the calculated values of the indicated power consumption, the suction loss, the mass-flow rate and the reversed mass-flow of the suction gas at the end of the suction stroke. The reversed mass flow is given in percent related to the mass-flow rate.

Table 4. Comparison of calculated values between standard and adapted suction valve.

valve	indicated power [W]	suction loss [W]	mass-flow rate [g/s]	reversed mass-flow [%]	reversed mass-flow @ CA [°]
standard	40.76	0.84	0.283	-0.26	188.1 – 201.4
adapted	40.49	0.60	0.282	-1.60	192.8 – 202.5

To verify the results of the numerical simulations, several measurements are carried out at steady-state conditions in a standardized calorimeter. The calorimeter was running at operating conditions according to the ASHRAE cycle with an ambient temperature of 32 °C, an evaporation temperature of -23.3 °C and a condensation temperature of 55 °C. Additionally to this operating condition, further measurements changing the condensation temperature to 45 °C were carried out. To get measurements which can be compared as good as possible, the same compressor was used for all measurements (only with different suction valves).

The electromagnetic coil is controlled with a full bridge circuit using two control inputs to demagnetize the coil. Voltage is applied to the coil approximately 1 ms before the piston reaches its bottom dead centre. After 3 ms the circuit is opened again. Results of previous investigations show, that the electric power consumption of the electromagnetic coil, operated at 8 V and a duty cycle ratio of 15 %, is about 2.1 W. This leads to increased temperatures inside the hermetic shell of approximately 2.5 K. Based on measurements investigating cooled compressors, a decreasing of about 0.25 % of the COP must be expected due to the waste heat of the electromagnetic coil.

Table 5 shows the differences between the reference measurements and the measurements with the adapted valve. The values stated in this table represent the percentage changes which are based on the reference measurements with the standard suction valve. For both ASHRAE cycles, a decreasing of the power consumption of approximately 0.7 % can be seen which results from the reduction of the suction losses. As it can be seen in the results of the numerical simulation, the adapted suction valve leads to a completely different dynamic behaviour during the suction stroke. The decreased cooling capacity of the measurement with a condensation temperature of 55 °C, can be explained due to the reversed mass-flow through the suction port because the suction valve could not be closed in time. When changing the condensation temperature to 45 °C, it seems that the reversed mass-flow has decreased.

Table 5. Effect of the adapted suction valve in percent related to the reference measurements.

ASHRAE cycle	COP	cooling capacity	power consumption
-23 °C / 45 °C	+1.72	+0.98	-0.73
-23 °C / 55 °C	+0.39	-0.32	-0.71

5. Conclusion

The reduction of the suction losses using a negative prestressed suction valve in conjunction with an electromagnetic coil was investigated. Numerical simulations at operating conditions according to the ASHRAE cycle (-23 °C/55 °C) showed that the indicated power consumption decreases by 0.68 %. The reversed mass-flow rate through the suction port increases from 0.26 to 1.6 % and thus the overall mass-flow rate decreases by 0.31 %. Measurements confirmed the results from the simulation (decrease of the power consumption by 0.71 %; decrease of the cooling capacity by 0.32 %) and showed an increased COP of 0.39 %. Additionally, a second operating condition was investigated by measurements which showed an increased COP of 1.72 %.

At this point it should be mentioned, that the power consumption of the electromagnetic coil should normally also be taken into consideration when comparing the COP, but here the effects of the adapted suction valve were in the centre of attention.

Summarizing it can be stated, that the adapted suction valve with the electromagnetic coil shows a general improvement of the performance of the compressor. Nevertheless, the effect depends on the operating conditions and therefore, the suction valve should be completely redesigned.

Nomenclature

Roman symbols

A	area	R	resistance
c	stiffness	$\mathcal{R}$	reluctance
c_t	capillary coefficient oil stiction	t	time
c_v	viscous coefficient oil stiction	u	voltage
d	damping coefficient	$\boldsymbol{u}$	displacement vector
$\boldsymbol{D}$	damping matrix	$\dot{\boldsymbol{u}}$	velocity vector
$\boldsymbol{f}$	vector of nodal point forces	$\ddot{\boldsymbol{u}}$	acceleration vector
F_{gas}	resulting gas force	x	gas gap, displacement
F_m	electromagnetic force	$\dot{x}$	velocity
$\boldsymbol{f}_{pen}$	force vector due to contact	$\ddot{x}$	acceleration
F_{pre}	preload	Greek symbols	
i	current	α	correction factor, Newmark param.
$\boldsymbol{K}$	stiffness matrix	δ	Newmark parameter
l	length	Θ	magnetomotive force
L	inductance	μ_0	vacuum permeability
m^*	equivalent mass	μ_r	relative permeability
$\boldsymbol{M}$	mass matrix	τ	characteristic time constant
N	number of turns	Φ	magnetic flux
$\boldsymbol{r}$	vector of externally applied loads	ψ	flux linkage

Acknowledgements

This work has been carried out within the framework of ECO-COOL, a research project initiated and funded by the FFG (Austrian Research Promotion Agency). Furthermore the authors particularly acknowledge the technical support by Secop Austria GmbH, formerly ACC Austria GmbH and Liebherr-Hausgeräte Lienz GmbH.

References

[1] Real M A and Pereira E A G 2010 *10th Int. Compressor Engineering Conf. at Purdue* **1225** 1-8
[2] Bhakta A, Dhar S, Bahadur V, Angadi S and Dey S 2012 *11th Int. Compressor Engineering Conf. at Purdue* **1252** 1-10
[3] Nagata S, Nozaki T and Akizawa T 2010 *10th Int. Compressor Engineering Conf. at Purdue* **1273** 1-8
[4] Burgstaller A, Nagy D, Almbauer R and Lang W 2008 *9th Int. Compressor Engineering Conf. at Purdue* **1315** 1-8
[5] Yoshida T, Sato K 1978 *United States Patent* Patent Number 4089348
[6] Glück T, Kemmetmüller W and Kugi A 2011 *Proc. of the 18th World Congress The International Federation of Automatic Control Milano* 11532-37
[7] Hopfgartner J, Heimel M, Berger E, Posch S, Almbauer R and Stangl S 2016 *23rd Int. Compressor Engineering Conf. at Purdue* **1216** 1-8
[8] Khalifa H E and Liu X 1998 *Proc. Int. Compressor Eng. Conf. Purdue* vol.1 87-92
[9] Bathe K J 1996 *Finite Element Procedures* (Prentice Hall, Upper Saddle River, New Jersey)
[10] Nagy D, Almbauer R, Lang W and Burgstaller A 2008 *19th Int. Compressor Engineering Conf. at Purdue* **1274** 1-8

IOP Conf. Series: Materials Science and Engineering **232** (2017) 012035 doi:10.1088/1757-899X/232/1/012035

Determination of Pressure Fluctuations in Rotor Bundle of Centrifugal Compressor at Critical Conditions of Operation

V A Levashov[1] and K Y Lyubchenko[2]

[1]Leading Researcher, Sumy NPO PJSC, Sumy, Ukraine
[2]Researcher, Sumy NPO PJSC, Sumy, Ukraine

E-mail: ljubchenko.konstantin@gmail.com

Abstract. This article describes the physical processes that occur in the stage flow part of the compressor while it is operating and can create conditions for the occurrence of forced vibrations, which in turn can lead to the destruction of the impellers. Critical conditions of compressor operation are determined. To understand that critical condition of operation is cause of the destruction of the impellers, transient CFD analysis was carried for test stage of compressor. The obtained pressure fluctuation amplitudes allow to evaluate the critical conditions of compressor operation.

1. Introduction

One of the existing problems is unsteady flow phenomena in centrifugal compressors. These phenomena result in significant dynamic loads, which in turn can lead to equipment failures.

Due to fatigue failure of the impellers in the presence of variable aerodynamic loads, there is a necessity of research unsteady processes during the operation the centrifugal compressor. Among the known unsteady phenomena in centrifugal compressors an uneven distribution of flow parameters on the circumferential coordinate near impeller has greatest negative influence on the fatigue strength of the impeller. This is due to the complex flow structure at the outlet of the impeller, influence of the vaned diffuser (VD), inlet guide vanes (IGV) and the return guide vanes (RGV). At rotor rotation circumferential unevenness of flow parameters, stationary relative to the stator is transformed into the unsteady pressure field and velocity relative to the rotating impeller. Importantly, this kind of non-stationarity exists at all operating conditions of the centrifugal compressor, which are allowed during operation. Furthermore, there is a possibility of resonance when the frequency of the exciting aerodynamic load coincides with one of the natural frequencies of impeller, which can lead to a dangerous dynamic stress and fatigue fractures. The main source of forced vibrations in centrifugal compressors are so-called Tyler-Sofrin modes (TS) from the interaction of the rotor and stator elements of the compressor [1]. Also as a result of instability of the gas flow acoustic eigenmodes of the flowing part stage elements can be excited. Critical conditions of centrifugal compressors operation occur if frequencies of Tyler-Sofrin modes coincide with acoustic eigenfrequencies and natural frequencies of structural elements. In addition, Tyler-Sofrin-modes, structural and acoustic modes should be similar in shape.

There is a small number of works devoted to the study of non-stationary processes in the flowing part centrifugal compressors, aimed at identifying the causes of the impeller destruction. However, among the companies that design centrifugal compressors, there are those who have made progress in this direction. [2, 3, 4]. Special attention in these works is paid to acoustic processes taking place in the side cavities of impeller. It is shown that the resonant acoustic modes can cause intense excitation of forced vibrations in the impeller.

2. Determination of critical conditions of compressor operation

When analyzing the causes of the destruction of the impellers critical conditions of operation the centrifugal compressor were determined. We used the so-called resonance approach [5] based on a comparison of the TS modes frequencies, the natural frequencies of the impeller and acoustic eigenfrequencies.

Research results of the impeller destruction in the third stage of centrifugal compressor are presented below.

Figure 1 shows the object of study - a five-stage centrifugal compressor; design scheme of the third stage, and also fragments of the impeller destruction.

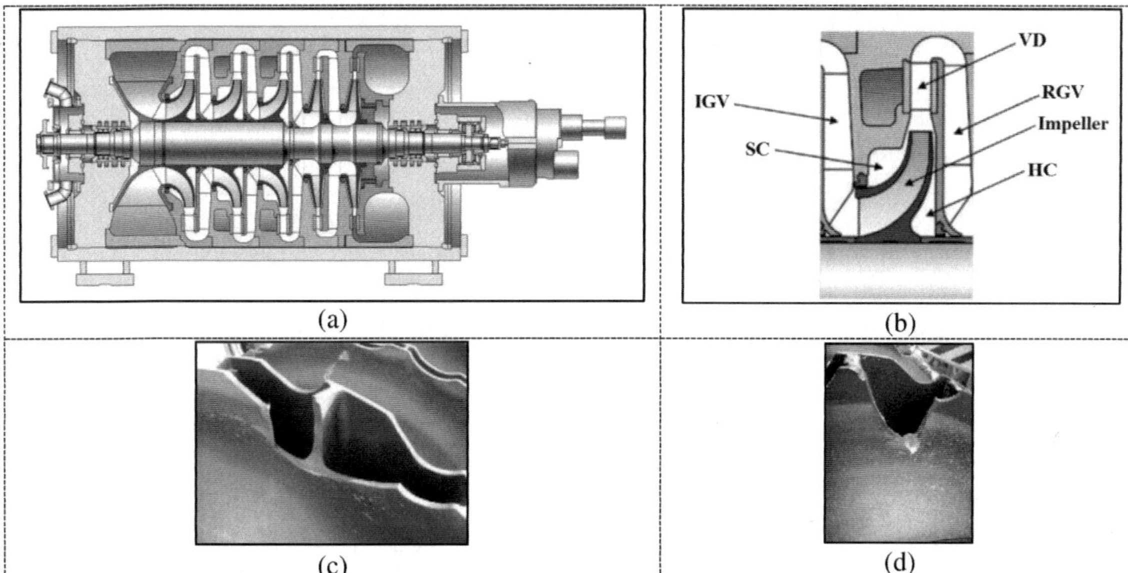

Figure 1. The object of study: (a) - centrifugal compressor; (b) - stage design scheme; impeller; VD – vaned diffuser; IGV - inlet guide vanes; RGV - return guide vanes; SC - side cavity between shroud and housing; HC - side cavity between hub and RGV; (c) and (d) - fracture fragments of impeller.

The main stages of determining the critical conditions of operation the centrifugal compressor are as follows:

- Static stress analysis of the impeller;
- Modal analysis of the impeller;
- Determination of the order of circumferential modes resulting from the interaction between the rotor and the stator, predicting the resonance frequency of rotation;
- Calculation of acoustic eigenmodes in side cavities;
- Plot of Analysis diagrams of harmonic excitatory impacts.

All numerical studies have been performed using ANSYS software.

2.1 Static stress analysis of the impeller

Calculation of stress-strain state of the impeller performed for rated speed n=5200 rpm with interference fit of impeller on the shaft. Figure 2 shows a 3D model of the impeller and calculated circumferential stress in the impeller.

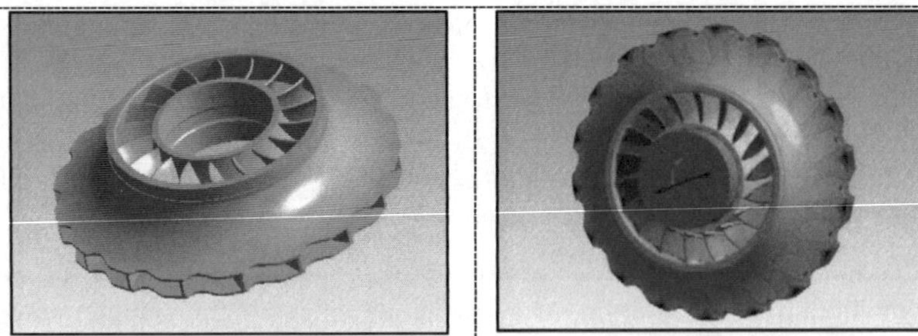

(a) (b)

Figure 2. Compressor impeller: (a) – 3D model; (b) – static circumferential stress.

Analysis of the results of calculations shows that the shroud on the outer diameter has static stress at which the strength of the impeller is provided in accordance with regulatory requirements. This means that intensive forced vibrations should provide the main influence on the fatigue failure of the impeller on this section.

2.2 Modal analysis of the impeller

Calculations were performed to determine the natural frequencies and mode shapes of the impeller.

Table 1 shows two natural frequencies for zero "umbrella" mode shape and mode shapes with 2, 4, 6 and 8 nodal diameters (ND = m_{struct}).

Table 1. The natural frequencies of the impeller

Nodal diameters (m_{struct})	Mode	Natural frequencies, Hz
0	1	1103
	2	2945
2	1	1010
	2	2489
4	1	1635
	2	2369
6	1	2002
	2	2215
8	1	1973
	2	2284

Figure 3 shows the mode shapes of the impeller with 0, 2, 4, 6 and 8 nodal diameters.

Figure 3. Mode shapes of the impeller with 0 (second frequency), 2, 4, 6 and 8 nodal diameters.

2.3 Determination of the order of circumferential modes resulting from the interaction between rotor and stator, predicting the resonance frequency of rotation

Number of nodal diameters forced TS modes is characterized by the order of circumferential modes m_{TS}, which can be determined from the following equation [2]:

$$m_{TS}= h_B Z_B+ h_{VD} Z_{VD}+ h_{IGV} Z_{IGV} +h_{RGV} Z_{RGV} \quad (1)$$

where z_B – number of impeller blades; z_{VD} – number of VD vanes; z_{RGV} – number of RGV vanes; z_{IGV} – number of IGV vanes; h_B – index, which covers the positive integers (values are taken for the calculation - 1,2,3), h_{VD}, h_{RGV}, h_{IGV} – indices covering the positive and negative integers (values are taken for calculation - -3, -2, -1, 0, 1, 2, 3).

TS modes are characterized by a frequency in the rotating reference frame, which are determined by the formula:

$$\omega^R_{TS}= \omega^S_{shaft}\cdot| h_{VD} Z_{VD}+ h_{IGV} Z_{IGV} +h_{RGV} Z_{RGV} |, \quad (2)$$

and order of harmonic impact frequency component, which is defined by the following formula:

$$m^R_{TS} = h_{VD} Z_{VD}+ h_{IGV} Z_{IGV} +h_{RGV} Z_{RGV}, \quad (3)$$

where the parameter ω^S_{shaft} refers to the angular velocity of the rotor shaft in a stationary reference frame. Basic data for the calculation of the order and frequency TS modes are as follows: the number of impeller blades - 18; the number of VD vanes - 36; the number of RGV vanes - 16; number of IGV vanes - 16. The calculation was performed for the compressor condition of operation on the rotation frequency is 4900 rpm. Table 2 shows some of the results of the calculation mode order m_{TS}, order harmonic frequency impact m^R_{TS} and exciting frequency ω^R_{TS} at rotation frequency of the rotor 4900 rpm, as well as the predicted resonance rotation frequencies $n_{shaft,rez}$.

Table 2. Main mode orders, harmonic impact frequency orders and harmonic impact rotation frequency at n=4900 rpm and predicted resonance rotation frequencies

m_{TS}	h_B	h_{VD}	h_{RGV}	h_{IGV}	m^R_{TS}	ω^R_{TS}, Hz	f_{st}, Hz	$n_{shaft,rez}$, rpm
-6	1	-2	0	3	24	**1960,00**	**2002**	**5005**
-6	1	-2	1	2	24	**1960,00**	**2002**	**5005**
2	1	0	-1	0	16	1306,67	1010	3787,5
-2	1	-1	0	1	20	1633,33	1010	3030
8	2	-3	3	2	28	2286,67	1973	4227,86
8	2	-3	3	2	28	**2286,67**	**2284**	**4894**
4	2	0	-2	0	32	2613,33	1652,3	3098,06
0	2	-1	0	0	36	**2940,00**	**2944,7**	**4907,83**

Predicted resonant frequencies are determined by the following formula:

$$n_{shaft,rez}= 60\cdot f_{st}/m^R_{TS} \quad (4)$$

where f_{st} - natural frequency of the impeller with the number of nodal diameters corresponding to the order of the TS modes m_{TS} in modulus.

When analyzing the predicted resonance rotation frequencies it can be noted that the resonance frequencies for TS modes orders m_{TS} = -6, m_{TS} = 8 and m_{TS} = 0 are close to the operating rotation speed n=4900 rpm, therefore they are the main source of excitation of the impeller. Table 2 shows that the resonance rotation frequencies for mode shapes m_{TS} = -6 and m_{TS} = 8 occur in the complex interaction

IOP Conf. Series: Materials Science and Engineering **232** (2017) 012035 doi:10.1088/1757-899X/232/1/012035

of the impeller (interaction of the impeller with multiple blade elements running of the compressor) with the diffuser vanes, reverse guide vanes and inlet guide vanes. At the same time for zero mode shape rotation frequency occurs in interaction of the impeller with the VD only (simple interaction).

Acoustic eigenmodes in the side cavities at the impeller can also be an additional excitation source of impeller natural frequencies.

2.4 Calculation of acoustic eigenmodes in side cavities

In ANSYS Workbench software calculation of acoustic eigenmodes in the side cavities at the impeller was performed. Example of acoustic eigenmodes is shown in Figure 4.

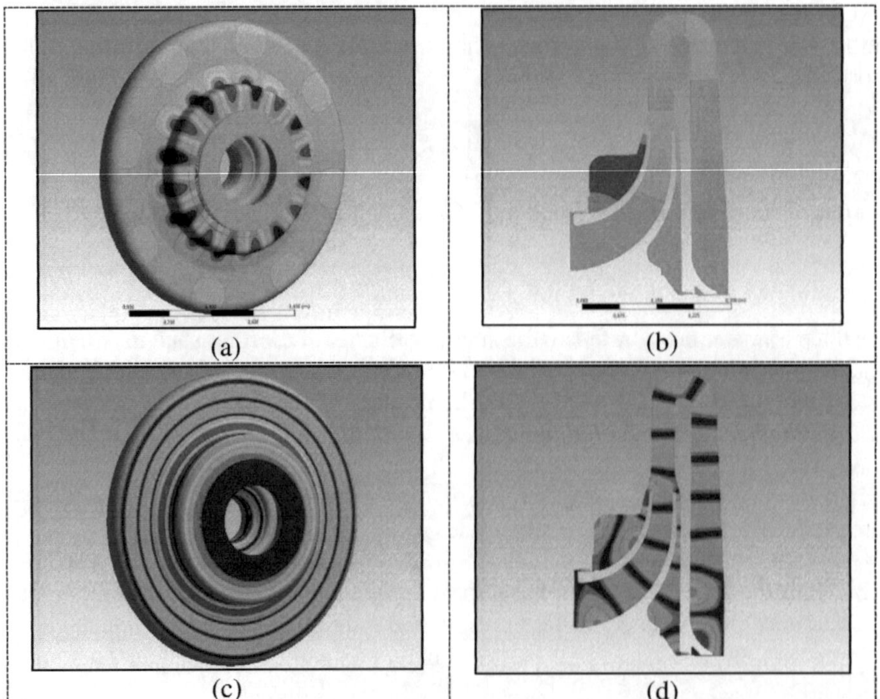

Figure 4. The mode shape with 8 nodal diameters $m_{ac} = 8$: a - axonometry, b - a plane section, and zero mode shape $m_{ac} = 0$: c - axonometry, d - flat section.

The rotation of the gas in the side cavities has a significant influence on the value of the acoustic eigen frequencies. Usually it is sufficient to assume that the core gas flow rotates at a circumferential speed being equal to equal to half the circumferential speed of the rotor ω^S_{shaft}, that is ω^S_{swirl}=0,5 ω^S_{shaft} [2].

The resonance condition must be satisfied in the same reference frame, namely in a rotating reference frame. In this case, the frequency of the acoustic eigenmodes must be determined in a rotating reference frame [2]:

$$\omega^R_{ac} = \omega^A_{ac} + m_{ac} \cdot (\omega^S_{swirl} - \omega^S_{shaft}) \quad (5)$$

where ω^A_{ac} - frequency of acoustic eigenmodes in the reference frame associated with the rotating flow core;

m_{ac} – order (number of node diameters) of acoustic eigenmodes.

The values of acoustic eigenfrequencies in side cavity for different rotational speeds are shown in Table 3.

Table 3. Acoustic eigenmodes.

		Acoustic eigenfrequencies, Hz			
		n=3640 rpm	n=4900 rpm	n=5200 rpm	n=5460 rpm
m_{ac}= -6	1	1863	1926	1941	1954
	2	1996	2059	2074	2087
	3	2298	2361	2376	2389
m_{ac}= 8	1	1484	1400	1380	1363
	2	2305	2221	2201	2184
	3	2377	2293	2273	2256
m_{ac}= 0	1	2565			
	2	2983			
	3	3117			

2.5 Plot Analysis diagrams of harmonic excitatory impacts

Analysis of the interaction between the mechanical natural frequencies, TS modes and acoustic modes is convenient to do using analysis diagrams of harmonic excitatory impacts.

Analysis diagrams of harmonic excitatory impacts are construct for mode orders m_{TS}= -6 and m_{TS}= 8 and shown in Figure 5.

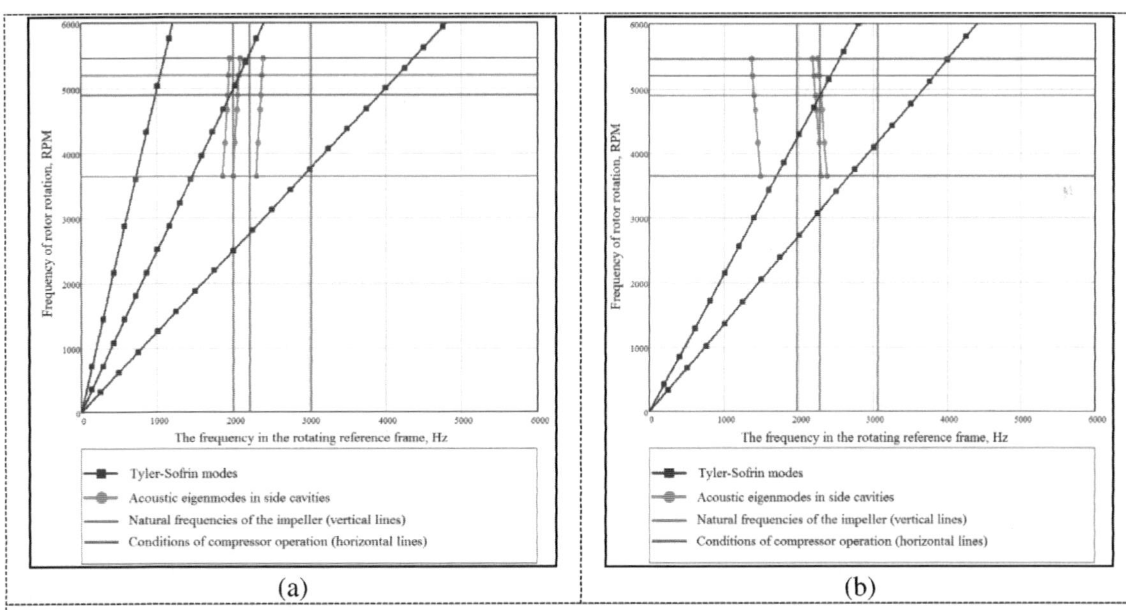

Figure 5. Analysis diagrams of harmonic excitatory impacts for mode orders m_{TS}= -6 and m_{TS}= 8.

The frequency in the rotation reference frame is displayed on the horizontal axis of abscissas. The rotor speed is shown on the vertical ordinate axis. The red vertical lines indicate the natural frequencies of the impeller, which are assumed to be constant for different rotor speeds. Blue lines indicate the frequency of exciting TS modes for the first, second and third harmonic component. Pink lines indicate the frequency of acoustic eigenmodes in the rotation reference frame in the side cavity SC. Green lines limit the range of compressor operation (n_{min}=3640 rpm and n_{max}=5460 rpm) and two conditions of operation (n=5200 rpm and n=4900 rpm).

Analysis diagram of harmonic excitatory impacts for zero mode shapes is shown in Figure 6.

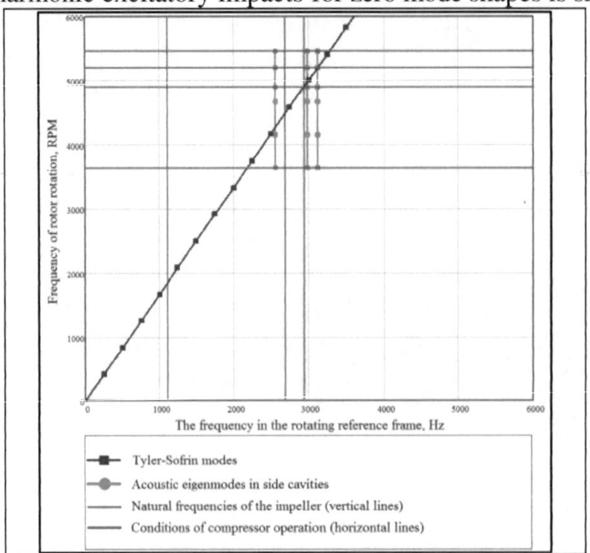

Figure 6. Analysis diagram of harmonic excitatory impacts for zero mode shapes.

Critical condition of compressor operation corresponds to point of intersection of the blue and red lines. The diagram indicated in figure 5a shows that for mode shape with order m_{struct}=6 condition of compressor operation is operation on the rotor speed of 5005 rpm, which is very close to the operating speed of 4900 rpm. In this case, the first natural frequency of the impeller for sixth mode shape is excited. For mode shape with 8th order m_{struct}=8 (see figure 5b) the critical condition of operation is operation on the rotor speed of 4894 rpm, but second natural frequency is excited. Analysis diagrams in figure 5a and 5b show that the additional excitatory impact on the natural frequencies of the impeller have acoustic eigenfrequency (pink line) in the side cavity, which lie near the critical conditions of operation compressor.

For zero mode shape (see figure 6) a critical condition of compressor operation is also operation on the rotor speed of 4900 rpm and second natural frequency is excited. It also shows that acoustic eigenmodes create additional impact in the side cavity SC. It is also necessary to note that critical conditions of compressor operation appeared on one of the rotor speed - 4900 rpm, and this significantly reduced the number of fluid dynamics calculations in transient formulation.

3. Transient CFD Analysis of compressor stage

Existing critical conditions of compressor operation suggest a possible cause of impeller destruction, but not assert that the cause of the destruction is found. Therefore, transient problem of viscous flow in the compressor stage has been solved that evaluate risk of critical operation conditions. The goal solving the problem is the definition of pressure fluctuations and finding the amplitude and frequency of the fluctuation.

Transient CFD problem was solved using ANSYS-CFX software. To solve the problem the third stage of the compressor was taken (see. figure 1b), in which radial vanes of RGB from second stage were used as an IGV. 3D model of the third stage jointly with second stage RGV vanes is shown in figure 7.

IOP Conf. Series: Materials Science and Engineering **232** (2017) 012035 doi:10.1088/1757-899X/232/1/012035

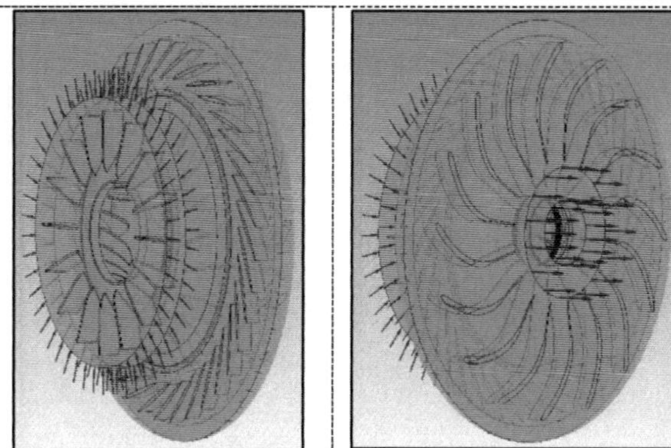

(a) (b)

Figure 7. 3D model of the third stage: a - view from the inlet of the gas flow; b - view from the outlet of the gas flow.

The total pressure of 1,114 MPa and the total temperature of 288 K are set as the boundary conditions at the inlet to the stage. The mass flow rate of 62.23 kg/s was set in the outlet stage. Methane gas similar in properties to a real gas was used for the working fluid.

Finite element characteristic of stage is shown in figure 8.

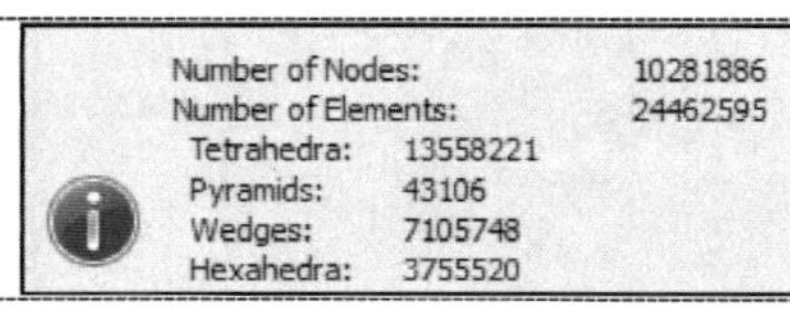

Figure 8. Finite element characteristic.

Transient CFD calculation was performed for 15 of the impeller revolutions. Time interval discreteness corresponds to one degree of rotation angle of the rotor to provide sufficient resolution when analyzing the pressure distribution within a single revolution of the rotor.

Several weeks of computer calculations were required to produce stable solution for the selected finite element mesh breakdown.

Figure 9 shows the instantaneous pressure distribution for the meridian section of the test stage of the compressor.

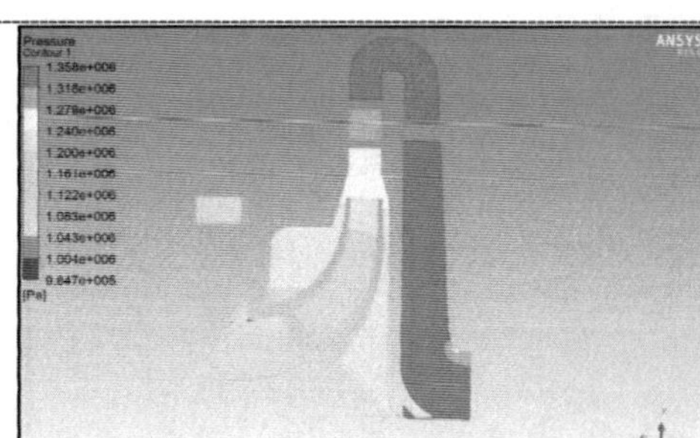

Figure 9. The instantaneous pressure distribution for the meridian section of the test stage of the compressor.

Figure 10 shows the instantaneous pressure distribution for outer surface of the impeller shroud and instantaneous pressure distribution for the middle section of the impeller and the VD.

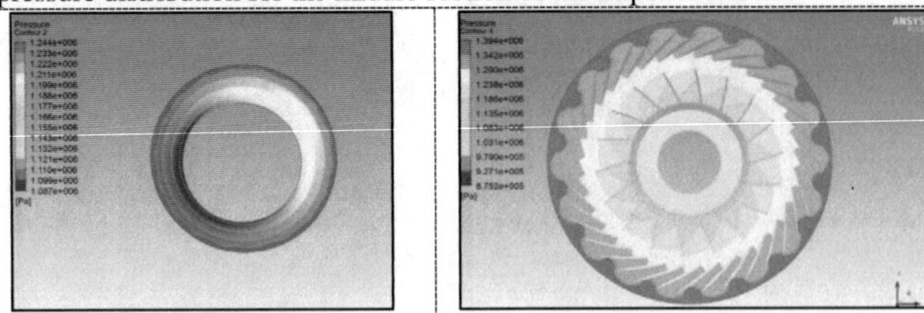

(a) (b)

Figure 10. Instantaneous pressure distribution: a - for the outer surface of impeller shroud; b - for the middle section of the impeller and VD.

Pressure fluctuations for two points in the stage cavity was analyzed as a result of the transition solution, as shown in figure 11. The first point is located at a distance 0,97D2 in the side cavity at the shroud, and the second point is located above the impeller at a distance 1,01D2.

Figure 11. Location points in the stage cavity for the analysis of pressure fluctuations.

Fast Fourier Transform procedure for the instantaneous pressure distribution at the same time interval discreteness for these points was performed for numerical analysis of pressure fluctuations.

The resulting amplitude for each circumferential mode order from 1 to 64 is shown in figure 12a and 12b for the two points, respectively.

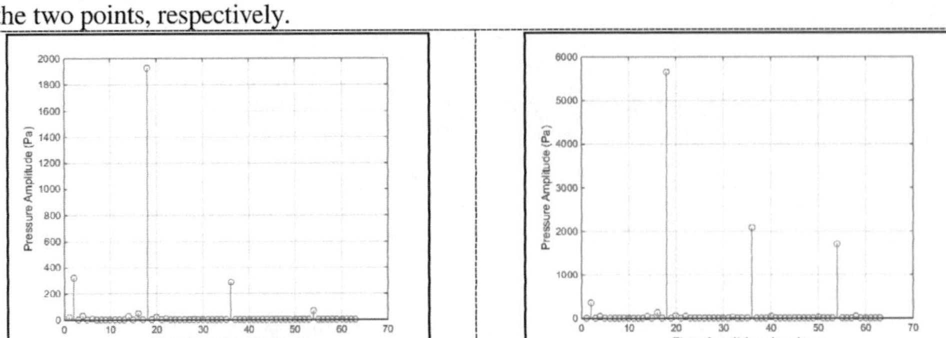

(a) (b)

Figure 12. Harmonic analysis results for the two points obtained by the calculation data in accordance with transient CFD.

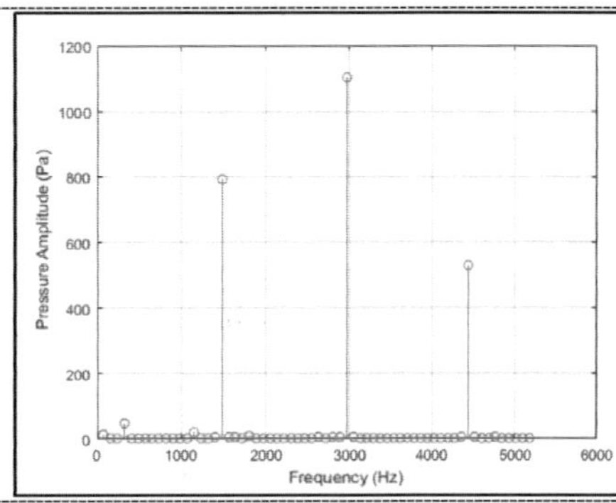

Figure 13. Harmonic analysis results for the one points obtained by the calculation data in accordance with transient CFD.

As the results of harmonic analysis (see figure 13), there is practically no pressure fluctuation amplitude for the mode orders m_{TS}=-6 and m_{TS}=8. Low levels of amplitudes can be explained by the fact that the forced TS modes of these orders occur from a complex combination of several elements of the stator and rotor components of the stage, as shown in the Table 2.

Analysis of the results showed that most dangerous critical conditions of compressor operation is critical conditions of operation for zero mode order m_{TS}=0. The frequency of forced vibrations is equal to a multiple of the first blade blade frequency of 2940 Hz (see figure 13) and practically coincides with the acoustic eigenfrequency of 2983 Hz and natural frequency of the impeller of 2944.7 Hz. It is also worth noting that the zero mode shape of forced TS modes occurs from the interaction of impeller (h_b = 2) and the VD (h_{VD} = -1) only.

4. Conclusions

Using the Fast Fourier Transform, the processing of the results of numerical calculation in accordance with the procedure CFD, determination of the amplitude of the pressure fluctuations was performed. This amplitude affects the impeller for the finding a critical condition of compressor operation, the value of which will be used to evaluate the fatigue strength of the impeller.

5. References

[1] Tyler J M, Sofrin T G 1961 Axial flow compressor noise studies *SAE Transactions* 70 pp 309-332

[2] König S, Petry N, Wagner N G 2009 Aeroacoustic Phenomena in High Pressure Centrifugal Compressors – A Possible Root Cause For Impeller Failures *Proceedings of 38th Turbomachinery Symposium* pp 103-121

[3] Petry N, Benra F K, König S, Woiczinski C 2009 Interaction Between Aerodynamic Phenomena and Impeller Structure of High Preeure Radial Compressors *Proceedings of 8th European Turbomachinery Conference* pp 757-771

[4] Petry N, Benra F K, König, S. 2010 Experimental Study of Acoustic Resonances in the Side Cavities of a High-Pressure Centrifugal Compressor Excited by Rotor/Stator Interaction. *Proceedings of ASME Turbo Expo* Nr. GT2010-22054

[5] Gadyaka V G, Levashov V A, Lyubchenko K Yu 2015 On the causes of the destruction of the impellers of centrifugal compressors *J. The vibrations in technique and technology* 1(77) pp 82-89.

IOP Conf. Series: Materials Science and Engineering **232** (2017) 012036 doi:10.1088/1757-899X/232/1/012036

The numerical study of the rake angle of impeller blade in centrifugal compressor

A Drozdov, Y Galerkin
R&D Laboratory "Gas dynamics of turbo machines" Peter the Great St.Petersburg Polytechnic University, Polytechnical st. 29, St.Petersburg, Russia

Email: buck02@list.ru.

Abstract: Investigated impellers have blade surfaces formed by straight generatrix. Blade profiles on shroud and disc surfaces are optimized by velocity diagram control (inviscid, quasi-three dimensional calculations). The blade profiles at hub and shroud blade-to-blade surfaces must be coordinated. A designer can choose the generatrix position at a trailing edge for it. The position is defined by the rake angle that is the angle between a trailing edge generatrix and a meridional plane. Two stages with 3D impellers, vaneless diffusers and return channels were investigated. Seven candidates of impellers of these stages with rake angles in range plus-minus 30 degrees were designed and investigated by quasi-three-dimensional inviscid calculation. CFD-calculations were made for the stages with these impellers. The optimal rake angle is minus 20 degrees for the high flow rate impeller due to lesser blade surface area and favorable meridian velocity field. Zero rake angle is optimal for the medium flow rate impeller where blade surface area is not so important. The combination of inviscid and viscid calculations is the informative instrument for further studies.

Nomenclature

b	blade height
c	absolute velocity
c_m	meridional component of absolute velocity
c'	velocity in view of a blade blockage factor
C_p	isobaric heat capacity
D	diameter
k	isentropic coefficient
$\bar{l}_m$	relative length in meridional plane
$\bar{L}_m$	relative axial length of impeller
$\bar{m}$	mass flow rate
M_u	blade Mach number
n	RPM
p	pressure
R	gas constant
R_h	hub radius of curvature
R_s	shroud radius of curvature

Re_u	impeller diameter Reynolds number
S	surface area
T	temperature
u_2	blade velocity
w	relative velocity
$\Delta\overline{w}$	blade load coefficient
z	number of stages
z_{imp}	number of blades
α_2	flow angle at an impeller exit
β_{bl}	blade angle to tangential direction
ε_1'	densities' ratio at inlets of impeller and of an impeller
$\overline{\delta}_{bl}$	relative blade thickness
τ	blade blockage factor
η	polytrophic efficiency
ζ	loss coefficient
ψ_T	loading factor
χ	rake angle
φ	flow coefficient
Φ	flow rate coefficient

Subscripts

1, 2, 4, 0, 0′ indices of control sections

av	average
bl	blade
des	design
fr	friction
h	hub
imp	impeller
inl	inlet
m	meridian
mix	mixing
min	minimum
opt	optimum
out	outlet
s	shroud
t	total parameter of gas
narr	narrowing

Abbreviation

VLD	vaneless diffuser.
RCH	return channel

1. Research aim

Impellers with three-dimensional blade cascades (3D impellers) have complicated and not completely studied configuration. Figure 1 illustrates geometry parameters of a 3D impeller meridional plane. The meridian configuration of the impeller in Figure 1 is a combination of arcs and straight lines. Figure 2 illustrates an impeller geometry as it is presented by the PC program for quasi-three inviscid calculations [1, 2]. A meridian configuration and quasi-orthogonal lines (red) to calculate axis-symmetric flow are shown in the Figure 2 on the left, a blade cascade view and blade angles at three blade-to-blade surfaces are presented amidst. Inviscid velocity diagrams on three blade-to-blade surfaces (hub, mean, shroud) are shown on the right.

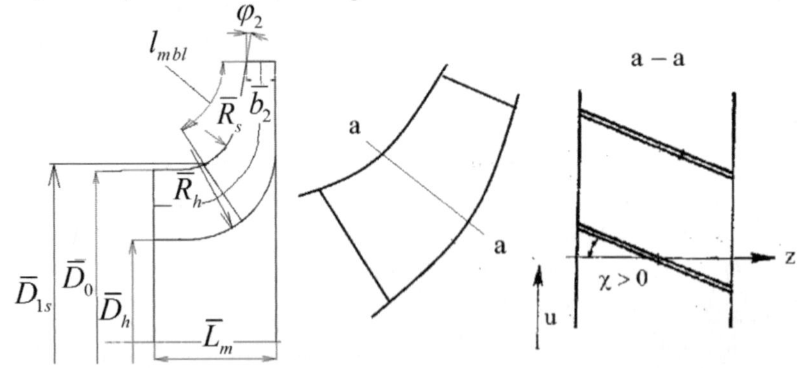

Figure 1. Meridional dimensions of 3D impeller (left), Blade meridional cross section (amidst) and blade generatrix position on a conical surface "a-a" (right).

3D impellers under investigation have blade surfaces formed by straight generatrix. Blade profile on shroud and disc surfaces are optimized by velocity diagram control. In accordance with [1, 2, 3] a maximum velocity on a suction side must be minimized. A flow deceleration and non-dimensional blade load must be chosen in accordance with an impeller parameters Φ_{des}, $\psi_{T\,des}$. The blade profiles at hub and shroud must be coordinated in 3D space. An angle χ_2 of the generatrix at a trailing edge can be chosen by a designer. Rake angle χ (figure 1) is the angle between a generatrix and a meridional surface.

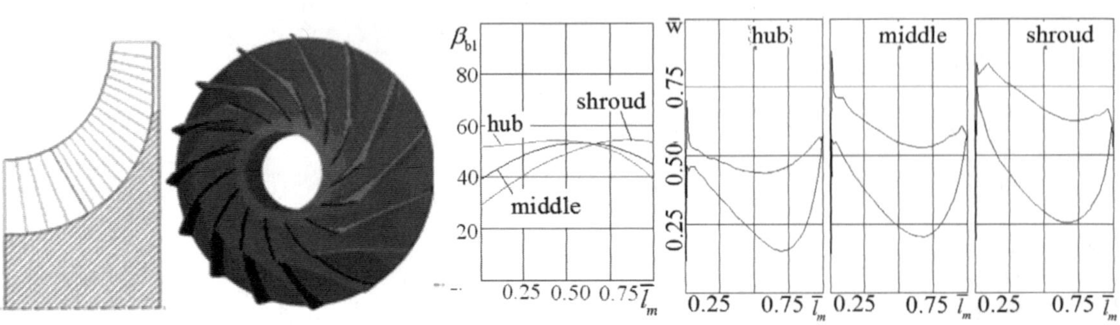

Figure 2. Graphic information on a 3D impeller #1 in course of design (PC program 3DM.023 [1])

Rake angle is positive - $\chi > 0$ - if a generatrix is tilted in the direction of an impeller rotation. The geometric equation presented in [4] connects a meridian configuration, blade angles and rake angle:

$$\frac{1}{\cos^2\chi}\frac{\partial\chi}{\partial l_m} = -\frac{\dfrac{\partial\beta_{bl}}{\partial b} + \sin\beta_{bl}\cos\beta_{bl}\left(\dfrac{1}{R_m} + \dfrac{\cos\gamma}{r}\right)}{\sin^2\beta_{bl}}. \quad (1)$$

Here R_m - curvature radius of a blade-to-blade surface, γ - angle between a radial direction and a normal to blade-to-blade surface.

The rake angle values along the meridional coordinate are function of the meridional shape and the blade shape that are chosen in a course of design. The chosen value of the rake angle χ_2 defines rake angles along the entire blades surface in accordance with the Eq. (1). The equilibrium equation for the axis-symmetric meridional flow is presented in [5]:

$$\frac{\partial c_m}{\partial b} = \frac{c_m}{r} + \frac{ctg\beta}{r}\frac{\partial (c_u r)}{\partial b} \quad \frac{P_u}{c_m} tg\chi \cdot \tag{2}$$

Here the force $P_u = \frac{z_{bl}}{2\pi r}\frac{p_s - p_p}{\rho}$ that represents influence of a blade load $p_s - p_p$ on meridian flow in a real three-dimensional flow.

The meridional non-uniformity of the flow is caused by change of its direction in a meridional plane. In accordance with Eq. (2) the negative rake angle reduces the velocity gradient. This can be considered as a positive factor. Uniform flow has less kinetic energy, which should be considered as an advantage. But in case $\chi \neq 0$ the blade surface is bigger:

$$S_{bl} = \int_1^2 \delta_{bl} \frac{b}{\cos\chi} dl_m \text{ (m)}. \tag{3}$$

It increases the friction losses. Blade blockade factor is smaller. It is a negative effect too:

$$\tau = 1 - \frac{z_{imp}\delta_{bl} tg\beta_{bl}}{\pi D \cos\chi}. \tag{4}$$

There are other, less obvious factors probably.

Design experience shows that if zero rake angle χ_2 is chosen by a designer, a rake angle of a leading edge is positive and can reach up to 40-50^0 and more. Blade height and velocity are the highest at the impeller entrance. There's also the greatest flow blockade by blades. It is possible that better flow uniformity if $\chi_1 > 0$ gives greater positive effect than listed above negative ones. Aim of the research is to study rake angle χ_2 choice as design recommendation.

In the paper [6] the research of a high-pressure aviation centrifugal compressor with a different rake angle χ_2 is carried out. The values of the angle χ_2 within the limits of +45 - -45 degrees were investigated. However, the outlet blade angle varied on its height, depending on the rake angle. For an angle χ_2 of 45 degrees, the angle β_{bl2s} at the shroud is less than the angle on the hub by about 10 degrees. For an angle χ_2 of -45 degrees, the angle β_{bl2s} is bigger than the angle on the hub by 5 degrees. Calculations were made by the ANSYS CFX program. The results had shown a high efficiency of the compressor with rake angle χ_2= +45 degrees and a larger stall margin for a compressor with a χ_2= -45 degrees.

2. Research objects

Two stages with different design parameters were chosen as research objects. Both stages have 3D impellers, vaneless diffusers and return channels. The stage #1 has a high flow rate coefficient Φ_{des} = 0.105. The stage #2 has a medium flow rate coefficient Φ_{des} = 0.067. The similarity criteria are M_u = 0.7, Re_u= 5600000, k = 1.4. The values are typical for industrial compressor model stage tests. Stages' parameters are presented in table 1.

Table 1. Gas dynamic and geometry parameters of impellers ## 1, 2

Impeller	#1	#2

Φ_{des}	0.105	0.065
$\psi_{T\,des}$	0.56	0.5
$\bar{D}_h$	0.28	0.325
$\bar{D}_0$	0.6	0.55
$\bar{b}_2$	0.0743	0.0525
$\bar{L}_m$	0.191	0.105
z_{imp}	12	15
$\bar{l}_m$	0.8	0.67
$\bar{R}_s$	0.215	0.075
$\bar{R}_h$	0.257	0.1

Despite the development of computational fluid dynamics, the world's turbomachinery manufacturers and scientists [8, 9] insist that mean line calculations are obligatory as a first step of design. The stages were designed by means of the so-called "Universal modelling method" [1-2, 7]. Mean line programs based on the mathematical model are applied to optimize stage main dimensions. Program 3DM.023 mentioned above (Figure 2) was applied for blade cascade design. The samples of this design method successful application are presented in [10].

A large number of CFD-calculations for candidate stages with 3D impellers were made by the authors of [3, 11]. The following are design recommendations:

- inlet diameter must be somewhat smaller than calculated by formula $\bar{D}_{0w\min} = \sqrt{\bar{D}_h^2 + 2^{\frac{1}{3}}\left(\frac{\Phi_{des}}{\varepsilon'_1 \tau_1}\right)^{\frac{2}{3}}}$.

  The relative diameter $\bar{D}_{0w\min}$ corresponds to minimal inlet velocity at a shroud blade-to-blade surface,
- leading edge of the blades is located at the distance 0.25% of meridian length from the section "0",
- exit blades angle must be increased from hub to shroud: $\beta_{bl2s} > \beta_{bl2h}$.

The information on the impeller #1 with Φ_{des} = 0.105 is shown in Figure 2 above. The candidate impellers ## 1 and 2 have different rake angles χ_2: 30, 20, 10, 0, -10, -20, -30 degrees. All candidates have identical meridional configuration and blade cascade shape. Only the rake angles χ_2 are different.

3. Q3D flow analysis

Results of flow parameters measurement in rotating impellers are presented in [1, 5]. Good agreement between calculated and measured velocity diagrams determines the successful application of Q3D calculations in design practice [10]. The analysis of inviscid flow structure provides useful information. The information on the candidates – inviscid and viscid calculations - is presented in Table 2.

Table 2. Some results of inviscid Q3D and viscid 3D calculations

1	2	3	4	5	6	7	8
Imp.	χ_2	χ_1	$\bar{c}_{ms} - \bar{c}_{mh}$	$S_{bl\Sigma} / S_o$	$\psi_{T\,des}$	$\zeta_{3D\,imp}$	η_{des}
1	-30	32,54	0,1863	1,23545	0,630	0,148	86,88
1	-20	35,03	0,1310	1,24616	0,633	0,148	87,03
1	-10	37,37	0,1230	1,26098	0,634	0,151	87,01
1	0	39,08	0,1150	1,27888	0,632	0,163	86,7

1	10	40,81	0,1050	1,30085	0,635	0,159	86,65
1	20	42,51	0,0950	1,32826	0,634	0,164	86,43
1	30	44,47	0,0738	1,36435	0,632	0,173	86
2	-30	46,63	0,0625	0,68144	0,565	0,361	85,85
2	-20	47,96	0,0575	0,67707	0,570	0,352	86,44
2	-10	49,07	0,0485	0,67625	0,574	0,363	86,68
2	0	50,05	0,0288	0,67789	0,574	0,358	86,82
2	10	50,99	0,0268	0,68199	0,573	0,362	86,6
2	20	51,95	0,0130	0,68910	0,571	0,382	86,5
2	30	52,97	0,0070	0,70032	0,563	0,378	85,71

The rake angles χ_1 for the candidate impeller #1 are 32.5-41 degrees and are 46.6-53 degrees for the candidate impeller #2. The larger value of rake angles χ_1 in case of the impeller #2 takes place due to the greater curvature in the meridional plane – the term $1/R_m$ in equation (1).

The design parameter χ_2 is varied in the range 60 degrees. The variation of the rake angle at the entrance χ_1 is 8.5 degree for the candidate impeller #1 – 7. The variation of the rake angle at the entrance χ_1 is 6.4 degree for the candidate impeller #2.

Blades' surface of the candidate # 1 with $\chi_2 = 0$ is by 28% more than the mean blade-to-blade surface. Blades' surface of the candidate # 2 with $\chi_2 = 0$ is by 32% less than the mean blade-to-blade surface. Blades' surface varies in the range 1.24 – 1.36 of disk surface for the candidate impeller #1. Blades' surface variation is 12%. Blades' surface varies in the range 0.68-0.70 of the average blade-to-blade surface for the candidate impeller #2. There is practically no influence of χ_2 on blade surface in case of the medium flow rate impeller #2.

Graphic information of inviscid calculations is presented in Figure 3. Positive rake angles suppress meridional velocities non-uniformity. The difference of non-dimensional meridian velocities along a blade height in a middle of a blade channel is shown in the column 5 Table 1. This parameter is reduced from 18 to 6% for the impeller #1 candidates in the range of χ_2 -30 ...+30 degrees. It is important that the velocity $\bar{c}_{ms}$ is visibly lower at negative χ_2. Non-uniformity is reduced from 6 to 1% for the candidate impeller #2 in the same range of χ_2.

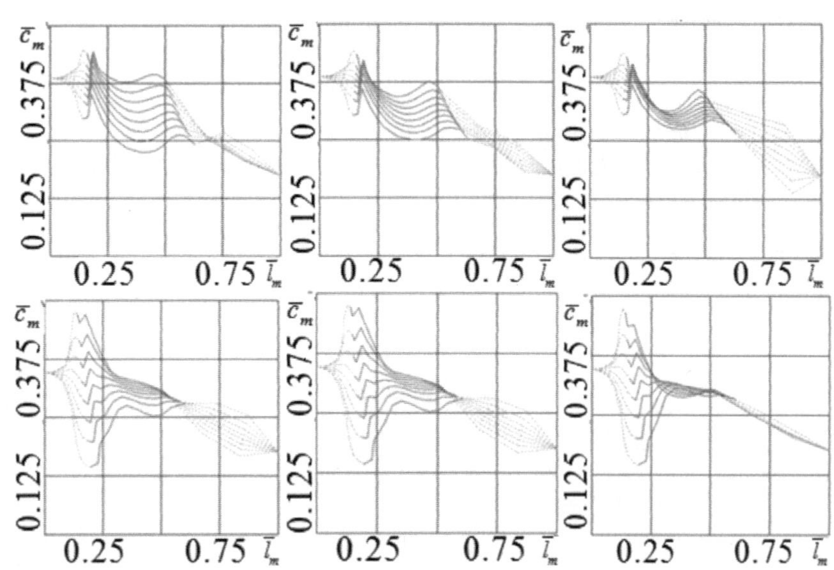

Figure 3. Impeller #1 (above) and impeller #2 (below). The meridional velocity of eight quasi orthogonals. χ_2 of the candidates are: -30^0 - left, 0^0 - amidst, 30^0 - right

Velocity diagrams on three blade-to-blade surfaces of impellers ## 1 and 2 with rake angles χ_2= -30, 0, +30 degrees are shown in Figure 4. The rake angle variation does not lead to significant changes of velocity diagrams. There is slight increase of a load on the leading edge of impeller #1 with χ_2= 30^0. It can lead to slight shift of performances to higher flow rates.

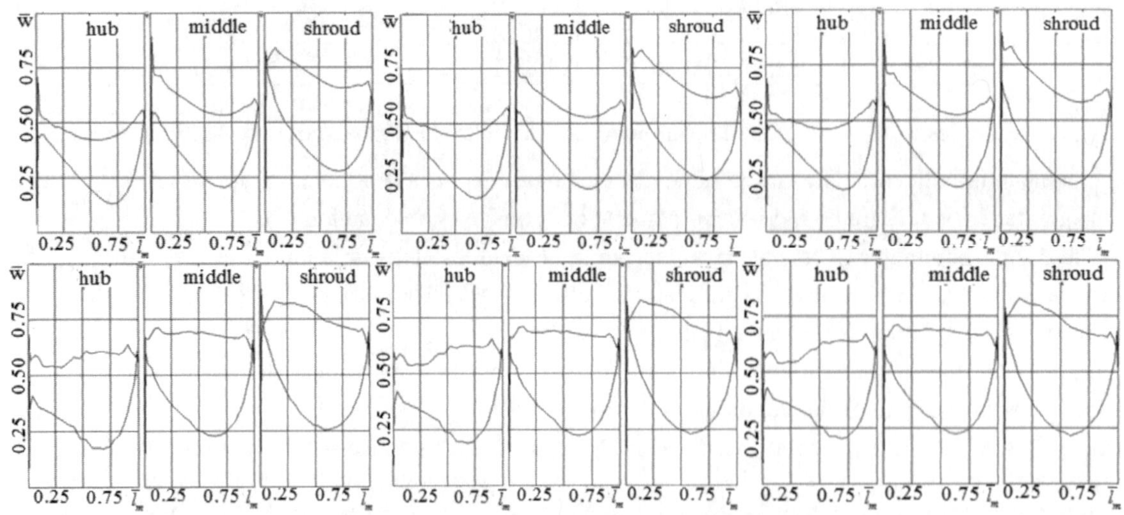

Figure 4. Impeller #1 (above) and impeller #2 (below). Velocity diagram on the hub on the mean surface and the shroud. χ_2= -30^0 left, 0^0 amidst, 30^0 right

4. Numerical research method

Authors' experience and known sources [3, 12-14] show that CFD-calculation satisfactorily predicts stage maximum efficiency. Loading factor is usually overestimated by 6-12% at a design flow rate. Performances at flow rates below design flow rate are not predicted satisfactorily.

Performances of two stages with vaneless diffusers and return channels were calculated. Each stage included seven candidates of the impellers analyzed above. Computational fluid dynamics program package NUMECA Fine/Turbo was applied:

- computational grid consists of 1.8 million cells;
- the turbulence model Spalart-Allmaras was applied;
- one blade channel was calculated;
- boundary conditions at a stage inlet: the total pressure and total temperature;
- boundary conditions at a stage exit: mass flow rate.

Disc friction losses and leakage through the labyrinth seals were not modeled. The boundary condition for the rotating and stationary coordinate systems – "mixing plane".

Non-dimensional gas dynamic performances are presented as functions the flow rate coefficient:

$$\Phi = \frac{4\bar{m}}{\rho_{t0}\,\pi\,D_2^2\,u_2}\,. \tag{5}$$

Loading factor is:

$$\psi_T = \frac{C_p\left(T_{t2} - T_{t0}\right)}{u_2^2}. \tag{6}$$

Total polytrophic efficiency is calculated as:

$$\eta_t = \frac{ln\left(\frac{p_{t\,out}}{p_{t\,inl}}\right)}{ln\left(\frac{T_{t\,out}}{T_{t\,inl}}\right)}\frac{k-1}{k}. \tag{7}$$

An impeller loss coefficient is:

$$\zeta_{imp} = \frac{2 \cdot \psi_T \cdot \left(1 - \eta_{t\,imp}\right)}{\left(w_1 / u_2\right)^2}. \tag{8}$$

5. Results of CFD-calculations

Efficiency performances of stages are presented in Figure 5. Efficiency of the candidate stage #1 increases monotonously within 1.2% in the range of χ_2 from plus 30 to minus 10 degrees. The maximum efficiency is the same for the candidates with rake angles minus 10 and minus 20 degrees. Evidently, blades surface area influences mainly efficiency level of the candidates.

Efficiency of the candidate stage #2 changes within 1.1% in the range of χ_2 from plus 30 to minus 30 degrees. The maximum efficiency has the candidate with $\chi_2 = 0$ degree. The candidates with $\chi_2 > 0$ are less effective. Obviously, the above recommendation $\chi_{2\text{opt}} < 0$ is not valid for impellers with medium specific speed. Loss coefficient performances of the candidate impeller are presented in Figure 6.

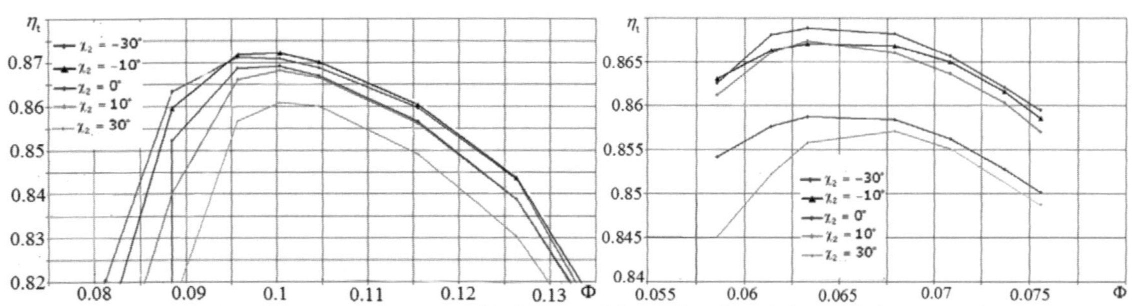

Figure 5. Candidates of the stage #1 (left) and impeller #2 (right) efficiency performances

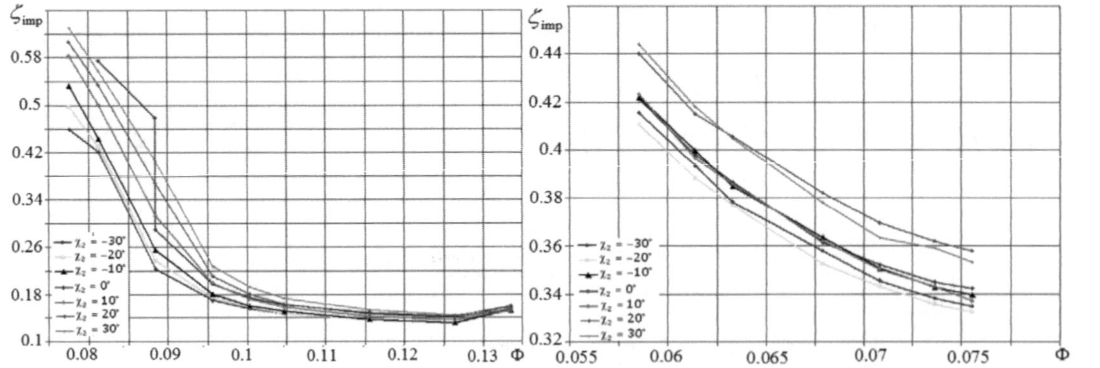

Figure 6. Candidates of the impeller #1 (left) and impeller #2 (right) loss coefficient performances

Loss coefficients (Eq. (8)) of the candidate impeller #1 vary within 0.145-0.175 at the design flow rate Φ_{des} = 0.105. Minimal loss coefficients have the candidate impellers of the most effective stages. It means that a rake angle χ_2 does not influence stator part efficiency but only an efficiency of an impeller. Loss coefficients of the candidate impeller #2 are significantly larger in absolute value $\zeta_{imp\,des}$ = 0.34 - 0.38. Velocity diagrams of all candidates indicate the favorable flow character. The blade profiling is OK. Perhaps large loss coefficients are the result of large friction losses on discs. The channels of the impeller #2 are narrower in comparison with the impeller #1. The lowest loss coefficient has the candidate with the rake angle minus 20 degrees. Rake angle has little effect on the surface area of the blades. The most effective stage has the impeller with zero rake angle. Its meridian non-uniformity is twice less (column 4 in Table 2). It may be important for stators.

The information on flow structure is presented in figures below. Exit flow angle along blade height for candidate impeller with three rake angles is shown in Figure 7. Flow angle is averaged on mass flow in the section "2".

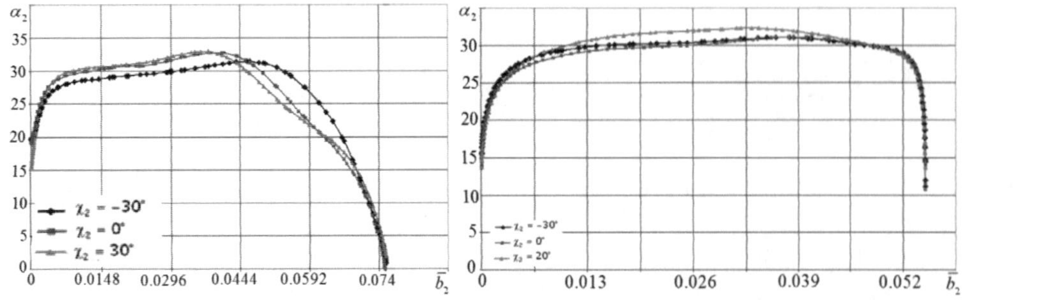

Figure 7. Flow exit angle along blade height (zero is at a hub). Impeller #1 (left), impeller #2 (right). Design flow rate

Uneven exit angle diagram with low angles near a shroud is typical for high flow rate impellers. Low angles lead to flow separation in a vaneless diffuser. The candidate impeller #1 with the negative rake angle 30 degree has the better diagram. This candidate's inviscid meridian velocities in accordance with Figure 3 are bigger near a shroud. It helps maybe to rise flow angles of a viscid flow. The candidate impeller #2 with lower Φ_{des} = 0.067 has practically the same angle diagrams that are rather uniform. The results confirm the conclusions drawn from the analysis of the Eq. (2).

Velocity fields in the blade cascades of the impellers ## 1 and 2 are presented in Figure 8. Flow seems to be well organized in all candidates. There are no low-energy zones near blades' suction side ("wake"). This is typical for well-designed impellers with low loading factor less than 0.55. Low-energy zone is on the blade pressure side in all candidates of the impeller #1. This can be due to

IOP Conf. Series: Materials Science and Engineering **232** (2017) 012036 doi:10.1088/1757-899X/232/1/012036

increased shear stress, as the large deviation of flow from its inertial direction. The intensity of this zone increases with algebraic increasing of the χ_2.

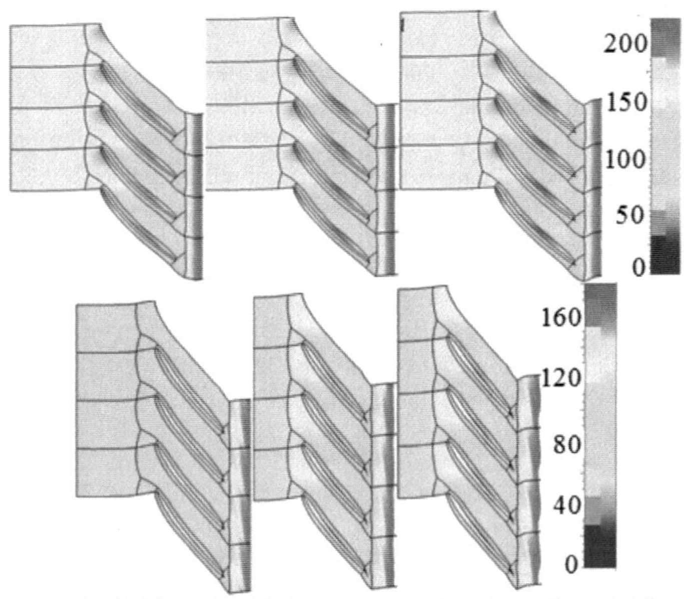

Figure 8. Relative velocity fields in the blade cascades of the impeller #1 (above) and impeller #2 (below). Blade-to-blade surface near the shroud. Design flow rate.
χ_2= -30^0 left, χ_2= 0^0 amidst, χ_2= 30^0 right

Loading factor performances calculated by Eq. (6) are shown in Figure 9. The design flow rate loading factor of the candidate impeller #1 is practically constant, with the exception of the candidate with χ_2= -30 degrees. The loading factor is bigger for the candidate impeller #2 with negative χ_2. The loading factor at the optimal zero rake angle is by 2% higher than at rake angle +30^0. This trend is not confirmed by the analysis of non-viscous flow velocity diagrams.

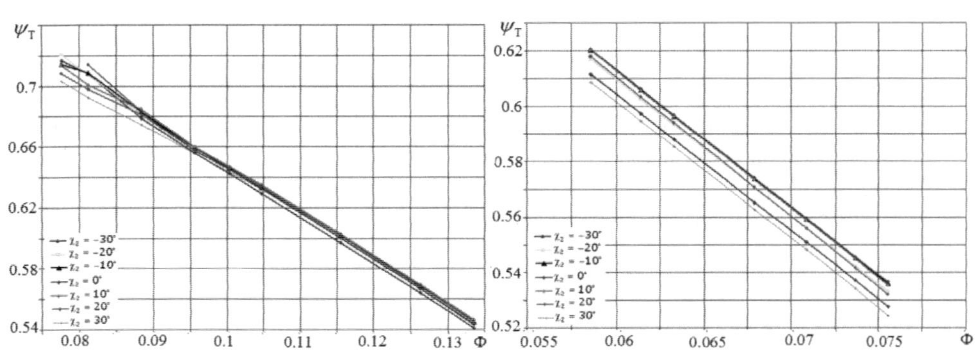

Figure 9. Loading factor performances of the candidate impeller #1 (left) and #2 (right)

Conclusion

In case of the stage with Φ_{des}= 0,105 the efficiency of candidates with χ_2= -10 … -20 degrees is by 0.7% higher against the candidate with χ_2= 0 degrees. The optimal candidates have smaller blade surface area and higher inviscid meridian velocities on a shroud surface. The last probably is a reason of more uniform viscid exit velocity angles near a shroud.

The optimal rake angle is zero for the medium flow rate stage with q_{des} = 0.067. The surface of blades is less than the surface of the discs and its role in the overall friction losses is less. Rake angle χ_2 has little effect on the surface of the blades in this medium flow rate impeller. The rake angle χ_2 influences little the loading factor performances of the candidate impeller #1. The loading factor of the optimal candidate impeller #2 is by 2% higher than that of the candidate with $\chi_2=30^0$. Inviscid velocity diagrams do not explain this fact. But the CFD-calculated efficiency of the candidate with $\chi_2=30^0$ is by 1% lower. An impeller's lower efficiency leads to lower loading factor. However, these considerations are not supported by the candidate impeller #1 performances.

As a whole the combination of inviscid and viscid calculations is a proper instrument for choice of better candidates of three-dimensional blade cascades. The authors recommend applying it in design optimization and will continue the study of the rake angles in impellers of different design parameters.

Acknowledgment
The article is written in the framework of the program "5-100-2020".

References
[1] Galerkin Y. Tubrocompressors. - SPbSTU. - SPb. - 2010. [Russian].
[2] Galerkin Y., Drozdov A. New generation of Universal modeling for centrifugal compressors calculation. Conference “Compressors and their Systems”. – London. – 2015.
[3] Galerkin Y., Drozdov A. Sample of CFD optimization of a centrigudal compressor stage. Conference “Compressors and their Systems”. – London. – 2015.
[4] Traupel W.Di theorie der stromung durch radial maschinen. Karlsrue. 1962.
[5] Seleznev K., Galerkin Y. Centrifugal compressors.- Leningrad. – 1982. - P. 271. [Russian].
[6] Arunachalam V.C., Nagpurwala Q.H., Deshpande M.D., Shancapal S.R. Numerical studies on the effect of impeller blade skew on centrifugal compressor performances. – sasTech. – 2088. Volume 7, No2. P. 33-40
[7] Galerkin Y., Soldatova K. Operational process modeling of industrial centrifugal compressors. Scientific bases, development stages, current state. Monograph. - SPbTU. – 2011. [Russian].
[8] http://www.pcaeng.co.uk/software/ [Электронный ресурс] / PCA Engineers Limited.
[9] Japikse, D. Centrifugal Compressor Design and Performance. - Concepts ETI, Inc. – 1996.
[10] Y. Galerkin, A. Rekstin, K. Soldatova, A. Drozdov Gas dynamic design of centrifugal compressors for gas industry. Proceedings of the ASME 2015 Gas Turbine India Conference.
[11] Galerkin Y., Soldatova K., Solovieva O. Numerical study of centrifugal compressor stage vaneless diffusers. Conference “Compressors and their Systems”. – London. – 2015.
[12] Melnicov V., Proscurov A. Using the software package FlowVision for flow calculation of the turbochargers elements in "SKBT. - CAD and graphics. – 2005. -№ 4. - P. 92-96. [Russian]
[13] Gamburger, D., Epifanov A ., Gaev A. Numerical investigation of three-dimensional viscous compressible gas in the centrifugal compressor impeller. Scientific and technical transactions of the TU SPb. – 2009. – № 2(78) – P. 76-82. [Russian]
[14] Lurie E., Van Slooten P., Medic G., Mulugeta J., Holley B., Feng J., Sharma O. Design of a high efficiency compact centrifugal compressor for rotorcraft applications. - American Helicopter Society 67th Annual Forum. – Virginia Beach. – May. – 2011.

IOP Conf. Series: Materials Science and Engineering **232** (2017) 012037 doi:10.1088/1757-899X/232/1/012037

Optimization of a centrifugal compressor impeller using CFD: the choice of simulation model parameters

V V Neverov, Y V Kozhukhov, A M Yablokov, A A Lebedev

Department of Compressor, Vacuum and Refrigeration Engineering, Peter the Great St. Petersburg Polytechnic University, 29 Politechnicheskaya St., St. Petersburg, 195251, Russia

neverovpu@gmail.com

Abstract. Nowadays the optimization using computational fluid dynamics (CFD) plays an important role in the design process of turbomachines. However, for the successful and productive optimization it is necessary to define a simulation model correctly and rationally. The article deals with the choice of a grid and computational domain parameters for optimization of centrifugal compressor impellers using computational fluid dynamics. Searching and applying optimal parameters of the grid model, the computational domain and solver settings allows engineers to carry out a high-accuracy modelling and to use computational capability effectively. The presented research was conducted using Numeca Fine/Turbo package with Spalart-Allmaras and Shear Stress Transport turbulence models. Two radial impellers was investigated: the high-pressure at ψ_T=0.71 and the low-pressure at ψ_T=0.43. The following parameters of the computational model were considered: the location of inlet and outlet boundaries, type of mesh topology, size of mesh and mesh parameter y^+. Results of the investigation demonstrate that the choice of optimal parameters leads to the significant reduction of the computational time. Optimal parameters in comparison with non-optimal but visually similar parameters can reduce the calculation time up to 4 times. Besides, it is established that some parameters have a major impact on the result of modelling.

Introduction

The design quality of the flowpath is one of the basic factors that determine the overall efficiency of centrifugal compressors. Automatic optimization techniques are used for the aerodynamic and mechanical design of turbomachine components. The optimization algorithm based on a sequence of 3D simulations for a large number of stage shape variations, which are generated to improve output parameters, as efficiency, pressure ratio, head etc. There are many approaches to optimization, which are good for optimal solution search [1]. Therefore, proper problem definition, the selection of the main parameters and criteria for optimization allows designers to significantly improve the performance of the final product [2].

Most computational fluid dynamic software packages include the full-cycle optimization algorithm with geometrical model parameterization, geometry and mesh regenerating tools. This allows CFD users to easily use geometric CFD optimization software to automatically generate design variants. These software packages often include a built-in optimization algorithm (in most cases it is genetic algorithm) or have the ability to use third-party code, as IOSO. Thus, technical aspects is well understood at the moment, and it is not problem to realize a sequence of calculations.

IOP Conf. Series: Materials Science and Engineering **232** (2017) 012037 doi:10.1088/1757-899X/232/1/012037

Despite the many advantages, their use is usually limited to simple applications in industrial practice, because of their high computational cost. During optimization computing systems work through many variants of a flowpath to achieve the desired result. If a powerful cluster with hundreds of cores carries out the calculations [3], the efficiency of optimization will not drop. This will increase only the workload of computing machine and the cost of electricity. In the case of low-power computing systems, performance of the works and the manufacture of products will be reduced. For this reason, one of the main problem that an iteration of optimization takes long time to compute. Therefore, one of the methods of increasing the optimization effectiveness is to decrease the time per one iteration while keeping the adequacy of calculation results. In this way, efficiency of the optimization mostly depends on the accuracy of the simulation model [4].

1. The aim of study and the object of research

The aim of the study is to investigate the influence of the computational domain and mesh model parameters on the integral parameters of the impeller and the rate of obtaining a converging solution. The best combination of these parameters will be used for future multiobjective optimization.

In this study, computation models of two radial impellers of the centrifugal compressor with cylindrical blades have been used:

The first one is a low-pressure impeller (IMP1) with the mean-camber curve of the blade formed by a circular arc. The flow rate coefficient F=0.06; the Euler work coefficient ψ_T=0.43; the impeller exit diameter is 0.82m; blade outlet angle β_{bl2}=32°; the number of blades z=13.

The second one is a high-pressure impeller (IMP2) impeller with the mean-camber curve of the blade formed by an s-shaped curve. The flow rate coefficient F=0.064; the Euler work coefficient ψ_T=0.71; the impeller exit diameter is 0.4m; blade outlet angle β_{bl2}=75°; the number of blades z=24.

Mesh model generation and subsequent computations were carried out using Numeca Fine/Turbo. The Fine/Turbo is a specialized CFD suite for rotating machinery with incompressible and compressible fluids that proved oneself as the fastest, accurate and convenient tool [5].

The computational domain and variations of the inlet and outlet boundary are shown in figure 1. The following boundary conditions were applied: steady conditions, air ideal gas as fluid, total pressure and temperature as inlet boundary condition, mass flow as outlet boundary condition. The following two turbulence models were compared: Spalart-Allmaras and Shear Stress Transport. Furthermore, the regions of labyrinth seals were not included in the computational model to speed up the calculations. The values of the polytropic efficiency presented below was corrected for estimated losses of the disk friction and leakage through labyrinth seals.

Figure 1. Control sections and computational domain variations

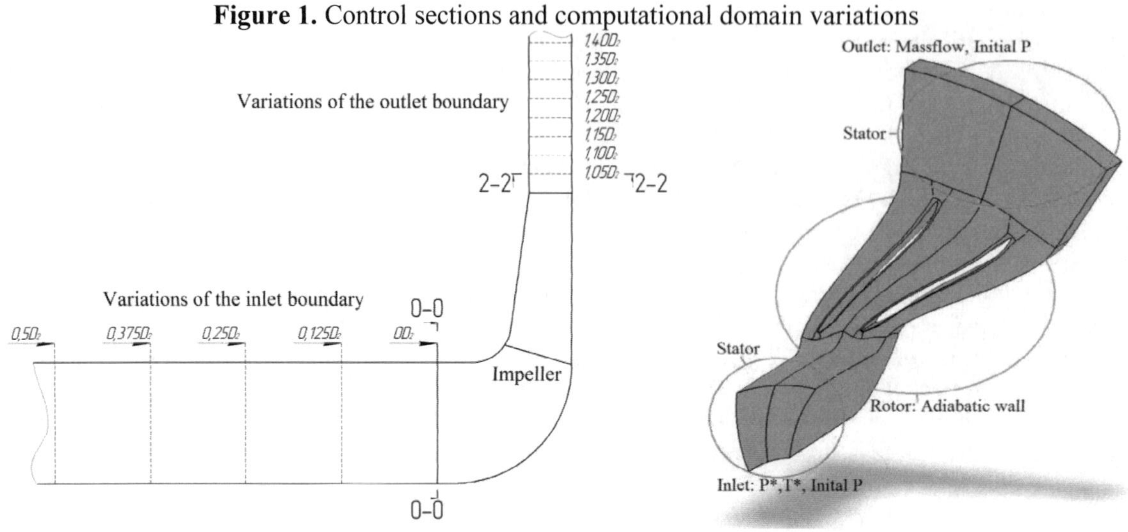

The following were taken as a result of calculation:

- The polytropic efficiency, which were calculated at cross-sections 0-0 (impeller inlet) and 2-2 (impeller exit) shown in figure 1.
- The number of iterations required to reach the convergence. The calculation were considered as converging when flow variables, which was monitoring at inlet and outlet boundaries, reach a steady-state value and the mean-squared residuals are dropped by four orders.
- The non-dimensional time per one iteration (normalized by a minimum time per one iteration in each series of calculations).
- The non-dimensional computing time required to reach the convergence (normalized by a minimum computing time in each series of calculations).

In general, the time per one iteration depends on the mesh size and the number of processor cores used in calculations. The computing time depends on both the number of iterations to reach convergence and the time per one iteration. In present paper, the overall computing time is considered as the main evaluation criteria.

2. The influence of the outlet boundary location

The flow separation and the so-called "trace" after impeller blades may have a significant impact on computational results if the outlet boundary is defined close to the impeller exit. The study was conducted with fixed inlet boundary by extending the part of the vaneless diffuser where width $b_3=b_2$. The overall number of mesh elements is increased in conjunction with the number of elements in vaneless diffuser. The computational domain and locations of the outlet boundary are shown in figure 1. To determine the efficiency, the flow parameters were derived from sections 0-0 and 2-2 ($=1.05D_2$). The calculation results for both impellers are shown in figures 2 and 3.

Figure 2. The influence of the outlet boundary location for the low-pressure impeller (IMP1)

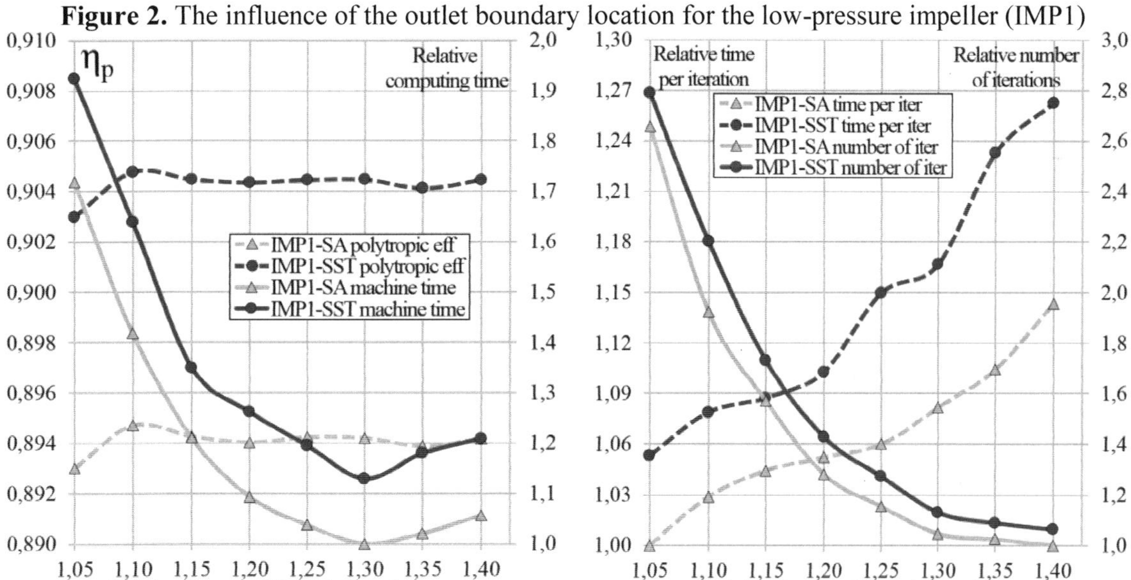

For the low-pressure impeller, the difference between obtained efficiencies for different location of the outlet boundary is small, and at $1.15D_2$ efficiency approaches a constant value. The number of iterations to reach convergence, which is described by the power-law function, is reduced while the outlet boundary is located at the range from $1.05D_2$ to $1.3D_2$. However, placing the outlet boundary further away from the section 2'-2', increases the total number of elements and the time per one iteration. Thus, from the point of view of computing time, the optimal location of the outlet boundary is at 1.3D_2. Further shifting of the outlet boundary from section 1.3D_2 is useless, because the number of iterations is not reduced and the time per 1 iteration is increased.

Figure 3. The influence of the outlet boundary location for the high-pressure impeller (IMP2)

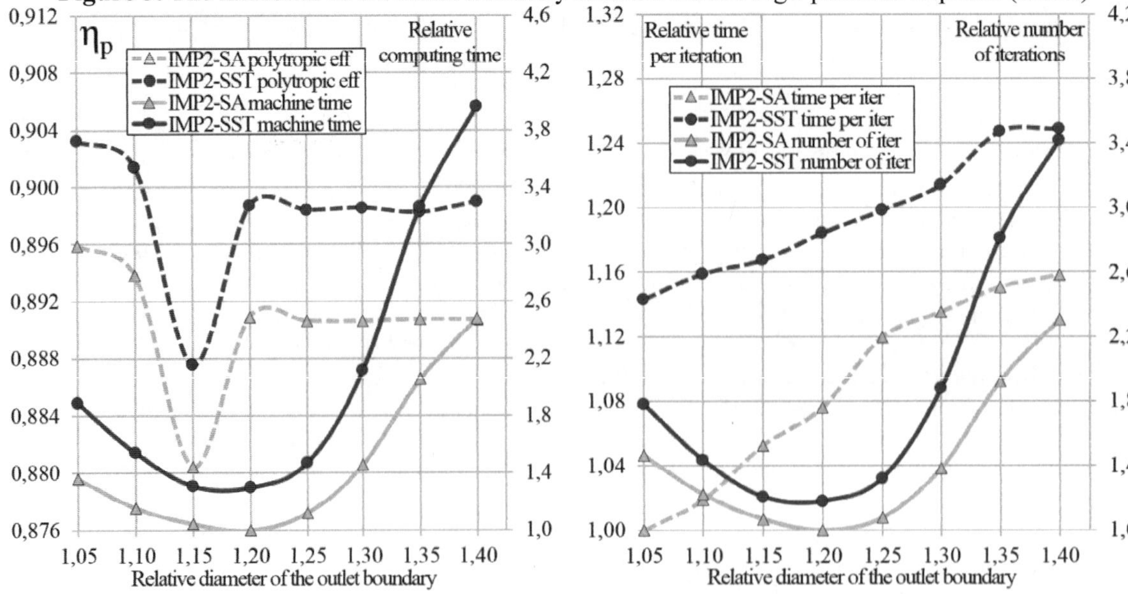

In case of the high-pressure impeller, the location of the outlet boundary has a strong impact on the integral parameters because heterogeneity of the flow behind blades is greater than that in the low-pressure impeller. For that reason alone, to calculate a high-pressure impeller it is necessary to place the outlet boundary at least at 1.25D_2. On the other hand, according to calculation results the convergence deteriorated after 1.2D_2. The number of required iterations for the SA turbulence model is increased by a factor of 2.3 when the outlet boundary is shifted from 1.2D_2 to 1.4D_2. This is due to the separated flow in the elongated sections of the diffuser, which leads to pulsations of the mass flow at the outlet boundary and slows down the convergence of the calculation. In figures 2 and 3, the width of the diffuser is equal to the width of the channel at the exit of the impeller, i.e., $b_3/b_2=1$. Increasing of the diffuser width contributes to the emergence of the separated flow in a diffuser. Therefore, in addition to the original case $b_3/b_2=1$, the case $b_3/b_2=0.8$ has been considered. Based on calculation results, narrowing of the diffuser had no effect. If the outlet boundary is placed at the distance greater than 1.2D_2, the convergence deteriorates in comparison with the original case. Therefore, the most appropriate option is to locate the outlet boundary at 1.25D_2, where the solution can be considered acceptable and the convergence rate is not critically increased.

To sum up, the influence of the outlet boundary location is as follows:

For the high-pressure impeller, the outlet boundary location has significant impact on result of the numerical experiment. In comparison with the low-pressure impeller, to reach the convergence more iterations and, therefore, greater computing time is needed because the flow is more heterogeneous. In the range from 1.05D_2 to 1.2D_2 number of required iterations is reduced by factor of 1.4. If the diameter of the outlet boundary is greater than 1.25D_2, the convergence deteriorates due to the separated flow and the unsteady phenomena in the vaneless diffuser.

For the low-pressure impeller, the effect of the outlet boundary location is relatively small; the convergence improves by elongation of the diffuser. As a result, in the presented case, computing time was reduced by more than 1.7 times. However, the effectiveness of the elongation is limited by the optimal ratio of the required number of iterations and the time per one iteration.

For all series of calculations, following should be noted: the calculation under SA turbulence model reaches convergence faster by 10-20% than under the SST model, which is consistent with the fact that the SST model is more complicated. Similarly, the time per one iteration is smaller by 5-15% than that for the SST model. The SST model exhibits the polytropic efficiency overestimation by 0.5-1% in absolute values in comparison with the SA model. This difference was also observed in other studies of the centrifugal compressors using Numeca Fine/Turbo [6]. In General, the performance curves for both turbulence models are equidistant.

3. The influence of the inlet boundary location

The location of the inlet boundary is often chosen randomly. It can also be based on the geometric shape of section before impeller, but at the same time, the inlet chamber or the return channel of the previous stage is not considered. However, the choice of the long entrance section could not be reasonable. The influence of the inlet chamber or the return channel of the previous stage could be preset by velocity, pressure and temperature profiles at the inlet boundary. But it brings its own problems in the implementation. Otherwise, setting the uniform field does not simulate the real flow in the inlet section, and therefore the rationality of increasing the grid model by extending the inlet section is not obvious. In order to save up computational power, the optimal location of the inlet boundary should be chosen.

The search of the optimum length of the entrance section in front of the impeller were conducted in the range from 0 to $0.75D_2$. The calculation results are shown in figures 4 and 5.

Figure 4. The influence of the inlet boundary location for the low-pressure impeller (IMP1)

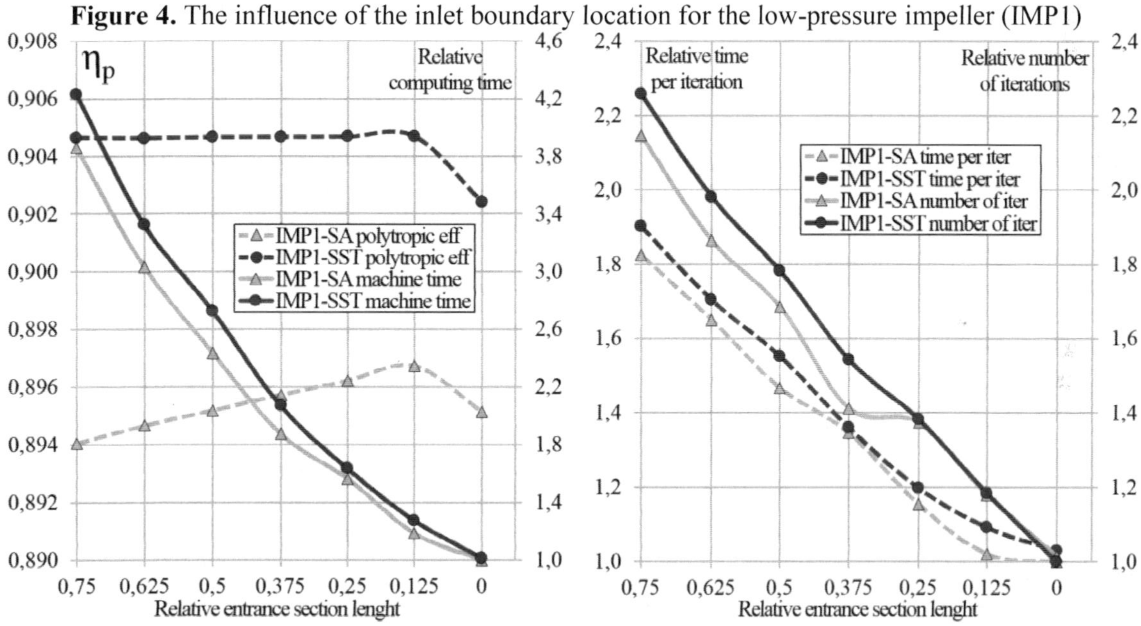

For the low-pressure impeller, decreasing of the entrance length led to the linear increase of efficiency under the SA turbulence model and to the constant efficiency under the SST model. The sharp drop of the efficiency by 0.2% occurs for both used turbulence models at the zero length. This drop is due to the fact that the velocity profile at this location already changed to enter impeller blades and it is different from the preset profile by the uniform field on this coordinate. It changes the flow around blades and leads to the difference in efficiency of impeller. Therefore, in spite of the continuing acceleration of the convergence, the choice of a short entrance section is precarious.

In General, the location, where the velocity profile is significantly changed, depends on many factors (for example: the width of the impeller, the length of the section under the shroud labyrinth seal, smoothness of the meridional turn and difference between areas at sections 0-0 and blade inlet 1-1).

Figure 5. The influence of the inlet boundary location for the high-pressure impeller (IMP2)

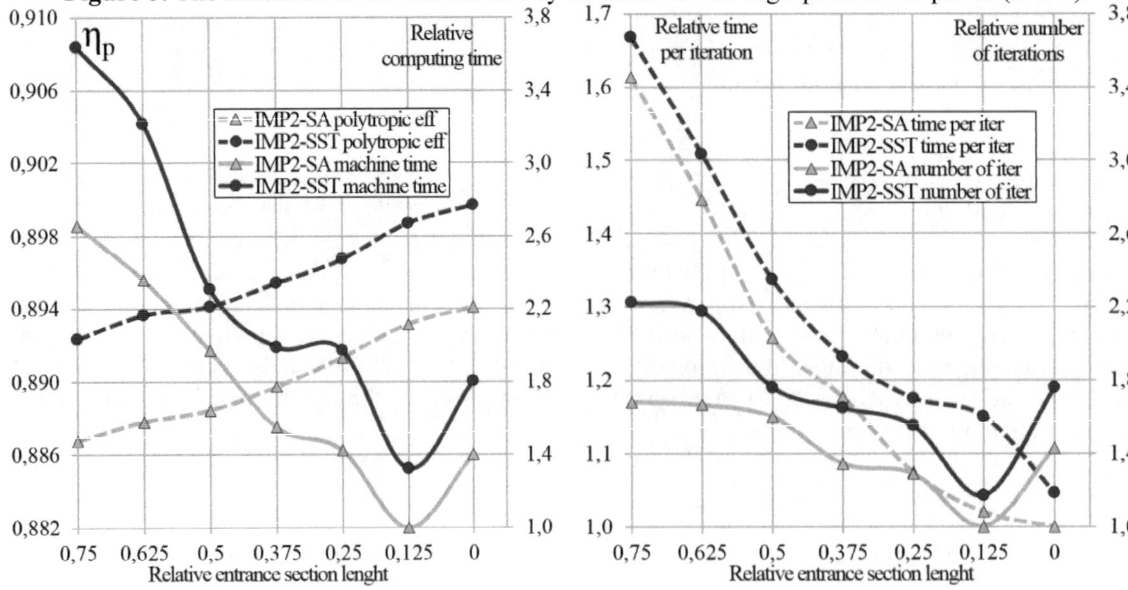

The high-pressure impeller has longer system of the labyrinth seals provided by design than the low-pressure impeller. Therefore, under the zero length of the entrance section, the velocity profile does not have an impact on flow around the leading edge, but it is resulted in a sharp deterioration of the convergence. The main factor that accelerate the convergence is reduction in the total number of mesh elements as described further. Thus, if the inlet chamber or the return channel of previous stage are not simulated, then it is pointless to choose long entrance section along with increasing the size of the grid. For both considered impeller the optimum inlet boundary is located at $0.125D_2$.

4. The influence of the mesh parameters

4.1. Mesh topology

The choice of the grid dimension for verification of the numerical experiments is solved by the standard examination of the grid independence. This study allows us to determine the number of mesh elements, starting from which increasing this number have no effect on calculation results. As for the optimization problem, it is reasonable to choose grid parameters more accurate, providing acceptable calculation results with minimal time spent on solving. Mesh generator Numeca AutoGrid 5 allows user to generate block-structured parameterized mesh. In the automatic mode for the radial impeller, it is usually possible to use two kinds of topology: O4H and H&I. O4H is a default topology composed by 5 blocks: an O block around the blade, 2 H-blocks upstream the leading and trailing edges of the blade and 2 H-blocks up and down to the blade section. The H&I topology is often used to obtain better mesh quality with multiblades configuration. The H&I topology is composed by: one block to mesh the blade passage, contrary to the default topology which creates a mesh around the blade; an optional skin block around the blade with two H-blocks before and after the skin block. The difference between grids with both topologies is shown in figure 6. In contrast to the O4H type, the grid-type H&I uses an intersectoral splitting with boundary near the blade surface. This splitting may lead to less intensive convergence due to averaging associated with a periodic boundary condition.

In general, the choice of the mesh topology is based on a grid model quality: the minimum skewness of the elements, the maximum expansion ratio of adjoining elements, etc. Therefore, both topologies cannot be right for all configurations of impellers. If it is possible to generate mesh for both topologies that equivalent in quality, it is reasonable to choose topology be guided by the speed of

IOP Conf. Series: Materials Science and Engineering **232** (2017) 012037 doi:10.1088/1757-899X/232/1/012037

solution obtaining. The mesh generated using both topologies was calculated and compared. The number of total grid elements for both impellers (IMP1 and IMP2) is 1.15 million, the minimum skewness for IMP1 - 45°, IMP2 - 36°. The time per one iteration for both topologies is the same, because the numbers of mesh elements are equal. The required number of iterations are presented in Table 1.

Figure 6. a) O4H mesh topology; b) H&I mesh topology (overall mesh size 1,15 million)

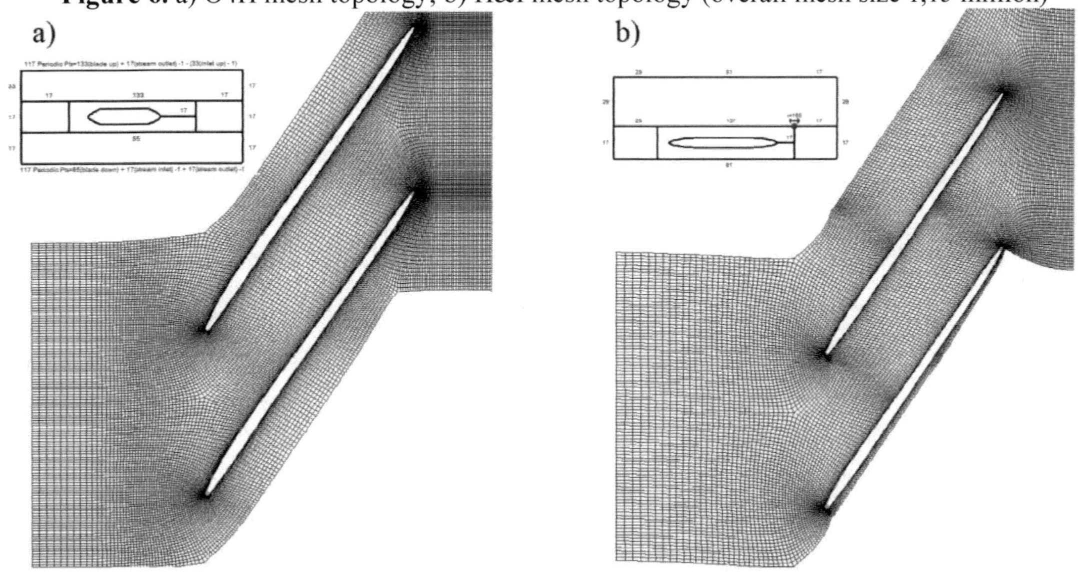

Table 1. Results of the calculation under mesh topologies

	IMP1				IMP2			
Turbulence model	SA		SST		SA		SST	
Mesh topology	O4H	H&I	O4H	H&I	O4H	H&I	O4H	H&I
Number of required iteration	696	835	830	1022	1070	1305	1390	1606

Thus, the optimal choice of the mesh topology allows to save up to 20% of computing time that is encountered when carrying out a large number of calculations.

4.2. Mesh refinement

Research on grid independence is a standard procedure for conducting a various numerical experiments. Findings of such researches are shown in many articles, for example [7].

The size of general mesh element is decreased with the total number of mesh elements increasing. The error that can be quantified by mesh refinement is known as the discretization error. Using an optimal mesh, the accuracy of the results are good enough to capture all the necessary flow features, their gradients and so forth. For both investigated impellers, the optimal number of elements were chosen for further optimization calculations. The result for the low-pressure impeller are shown in figure 7. The result for the high-pressure impeller correspond with result for the low-pressure impeller.

Starting with 0.8 million elements the efficiency of the low-pressure impeller almost ceases to change. The time per one iteration grows linearly and the number of required iteration is almost linear. As a result, the computing time required to reach convergence is described by non-linear function and it is considerably increases when the number of the mesh cells exceeds 1 million elements. Therefore, the optimal size of the mesh is the size that lies in the range of cell quantity from 0.8 to 1 million for one blade sector. This mesh size range for this impeller is good enough to achieve acceptable accuracy at the expense of minimum possible computational power. For the high pressure impeller the optimal number of elements is about 0.6-0.8 million, because this impeller has smaller size, larger number of blades and consequently smaller blade passage than the low-pressure impeller.

Figure 7. The influence of the number of mesh elements for the low-pressure impeller (IMP1)

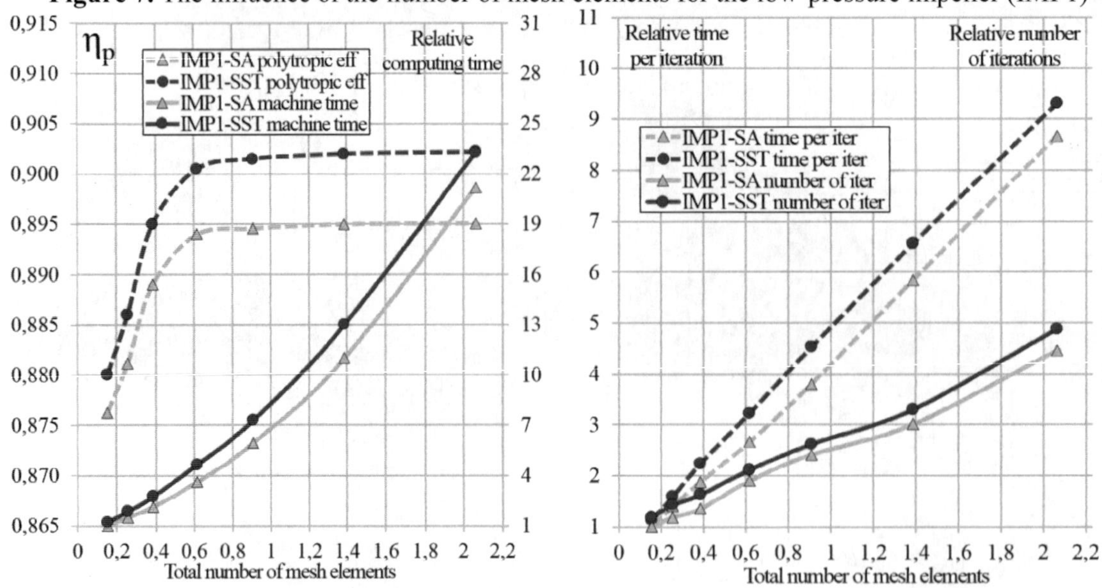

4.3. y^+ parameter

The mesh parameter y^+ is a non-dimensional distance between the first node and the wall. This parameter is important for the turbulence model selection (Low-Reynolds or High-Reynolds), because it determines the requirement to the grid. Under High-Reynolds models with standard wall-function, it is necessary to avoid values of y^+ less than about 20-30 and greater than 300 for the calculation of the gas flow. Whereas under Low-Reynolds models with scalable wall functions, the recommended range of y^+ is from 0.001 to 1, because using scalable wall functions requires to have a sufficiently thin layer of elements to decompose the boundary layer. A trial calculation is usually carried out to determine y^+ in the whole computational domain. Then the size of the first near-wall cell is corrected manually to meet the criterion. The influence of this parameter considered in article [8].

It is believed that Low-Reynolds models allow to predict points of local separations more accurate, however, such models make high demands on the grid dimension. In present paper Low-Reynolds models were used. The influence of the parameter y^+ is considered in the recommended range (from 0.001 to 1) and out of it (from 1 to 48) on integral parameters and the rate of convergence (see figure 8 and table 2).

Table 2. Variation of the cell height and values of the parameter y+

	Variations	1	2	3	4	5	6	7	8
IMP1	Height of cell, mm	0.001	0.0015	0.005	0.01	0.03	0.08	0.15	0.3
	Average y+	0.17	0.25	0.82	1.75	5.5	12.6	19.7	30.4
	Maximum y+	0.59	0.86	3.03	6.23	14.6	25.2	34.5	47.7
IMP2	Height of cell, mm	0.0005	0.0007	0.0012	0.005	0.01	0.03	0.08	0.15
	Average y+	0.12	0.165	0.276	1.17	2.5	7.3	16.0	24.4
	Maximum y+	0.38	0.53	0.92	4.08	7.8	16.1	28.1	39.9

As can be seen in figure 8, the y^+ in the recommended range does not have any noticeable effect on the convergence. That is, in this range the rate of convergence remains constant. Increasing in the average value of the parameter y^+ from 0.17 to 0.82, efficiency of the low-pressure impeller is increased by 0.5% and then at y^+=1.75 is dropped by 0.7%. For the high-pressure impeller results are similar.

Figure 8. The influence of the parameter y^+ for the low-pressure impeller (IMP1)

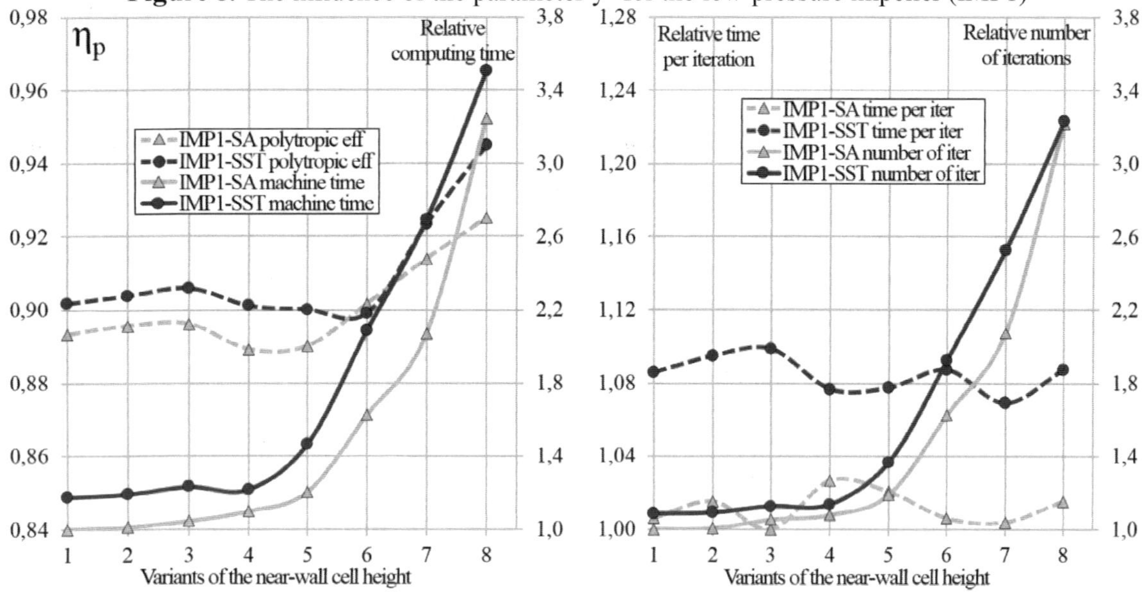

Increasing of the parameter y^+ out of the recommended range leads to an incorrect results, when the efficiency of the impeller increased. It is arise from the significant decrease of friction losses, because of the insufficient resolution of grid dimension in the boundary layer. Along with incorrect results, the number of iteration is significantly increased. With the equal time required per one iteration, the computing time is increased in proportion to the required number of iterations. In case of the high-pressure impeller, integral parameters and the convergence rate are correspond to those for the low-pressure impeller. However, the SST model makes exacting requirements for the y^+ parameter: at y^+>2.5 the calculation was interrupted. Thus, the interruption of the calculation under the SST model may be due not only to the overall quality of the mesh, but also due to an unacceptably large size of the first near-wall cell. In this case, under the SA model it is possible to reach the convergence, even at a significant deviation from the recommended range.

5. Potential saving of the computing power

To sum up, the optimization of investigated impellers under optimal and random parameters of the computational model gave the following results:

The random parameters: the inlet boundary located at 0.5D_2, the outlet boundary located at 1.1D_2; the selected grid topology is H&I; the height of near-wall cell is 0.01 mm (which is corresponding to the average value of y^+=1.75); the total number of cells – 1.15 million.

The optimal parameters: the inlet boundary located at 0.125D_2, the outlet boundary located at 1,3D_2; the selected grid topology is O4H; the height of near-wall cell is 0.001 mm (which is corresponding to the average value of y^+=0.17); the total number of cells – 0.92 million.

According to the result of the calculation, under non-optimal parameters, the time per one iteration on a single processor core was 3 seconds and the number of required iterations to reach the convergence was 775. Consequently, the total computing time of calculation for a single approximation of the optimization was 2322 seconds. As for calculation under optimal parameters, these values are respectively equal to 2.4 seconds per iteration and 227 iterations. Moreover, the total calculation time was 535 seconds, which is faster than the calculation with non-optimal parameters by 4.25 times. Thus, 100 optimization steps would be taken 64 hours under randomly chosen parameters and 15 hours under optimal parameters. This is a significant difference, considering the fact that computational domains and mesh models are visually similar for both variants.

Conclusion

As a result, presented research reveal the most important parameters of the computational model (the location of the outlet boundary and the overall size of the grid) during large series of calculation. These parameters with non-optimal values, regardless of the impeller type, increase the time of obtaining converging solutions by 2 times or more and have a significant influence on results of calculation (up to 1.7% of impeller efficiency for the outlet boundary and 1.5% for the grid model). This is unacceptable when optimizing the geometric shape of impellers in case of both adequacy result and the rational use of computing resources.

Besides, to obtain adequacy result, the grid parameter y^+ for selected turbulence models and the location of the inlet boundary should be observed. For the Low-Reynolds turbulence models, too large values of the grid criterion y^+ have the significant impact on the calculation result.

According to results of a preliminary assessment, optimal parameters of the computational model were determined. For the low-pressure impeller IMP1 – the outlet boundary is defined at $1.3D_2$; the inlet boundary located at $0.125D_2$; mesh topology is O4H; the size of the near-wall cell is 0.002 mm; the total number of grid elements is 0.9 million. For the high-pressure impeller IMP2 – the outlet boundary is defined at $1.25D_2$; the inlet boundary located at $0.125D_2$; mesh topology is O4H; the size of the near-wall cell is 0.001mm; the total number of grid elements is 0.7 million. These sets of computational model parameters provide the optimal computing time while keeping adequacy of results. Thus, determined values may be considered as a reference point to define a computational model of 2D-impellers. However, it is necessary to check overall quality of the computational model in each individual case.

References

[1] Frese F, Einzinger J, Will J 2012 Design optimization of an impeller with CFD and Meta-Model of optimal Prognosis (MoP) *10th International conference of turbocharges and turbocharging* 2012, London, pages 121-134

[2] Dickmann H P 2013 Shroud contour optimization for a turbocharger centrifugal compressor trim family *10th European Turbomachinery Conference* (Lappeenranta, Finland)

[3] Boldyrev Y, Rubtsov A, Kozhukhov Y, Lebedev A, Cheglakov I, Danilishin A 2015 Simulation of unsteady processes in turbomachines based on nonlinear harmonic NLH-method with the use of supercomputers *CEUR Workshop Proceedings* Volume 1482, Pages 273-279

[4] Galiando J, Hoyas S, Fajardo P, Navarro R 2013 Set-up analysis and optimization of CFD simulations for radial turbines *Engineering Applications of Computational Fluid Mechanics*, 7(4), 441-460. doi: 10.1080/19942060.2013.11015484

[5] Borm O, Kau H-P, Unsteady aerodynamics of a centrifugal compressor stage–validation of two different CFD solvers 2012 *Proceedings of ASME Turbo Expo 2012*, GT2012, Copenhagen, Denmark, June 11-15, GT2012-69636

[6] Borm O, Balassa B, Kau H-P Comparison of different numerical approaches at the centrifugal compressor radiver 2011 *ISABE Conference, International Society for Airbreathing Engines,* Gothenburg, Sweden, 12th-16th September 2011 ISABE-2011-1242

[7] Le Sausse P, Fabrie P, Arnou D, Clunet F 2013 CFD comparison with centrifugal compressor measurements on a wide operating range *EPJ Web of Conferences*, Vol. 45, p. 01059

[8] Aghaei Tog R, Tousi A M, Tourani A 2008 Comparison of turbulence methods in CFD analysis of compressible flows in radial turbomachines *Aircraft Engineering and Aerospace Technology* Vol. 80, Iss. 6, Pages 657-665 ISSN 1748-8842

IOP Conf. Series: Materials Science and Engineering **232** (2017) 012038 doi:10.1088/1757-899X/232/1/012038

Numerical analysis of flow in a centrifugal compressor with circumferential grooves: influence of groove location and number on flow instability

X Chen[1], G Qin[1], Z Ai[1,2], Y Ji[1]

[1] School of Energy and Power Engineering, Xi'an Jiaotong University, Xi'an 710049, China
[2] School of Mechanical and Electronic Engineering, Sanming University, Sanming 365004, China

E-mail: glqin@mail.xjtu.edu.cn

Abstract. As an effective and economic method for flow range enhancement, circumferential groove casing treatment (CGCT) is widely used to increase the stall margin of compressors. Different from traditional grooved casing treatments, in which the grooves are always located over the rotor in both axial and radial compressors, one or several circumferential grooves are located along the shroud side of the diffuser passage in this paper. Numerical investigations were conducted to predict the performance of a low flow rate centrifugal compressor with CGCT in diffuser. Computational fluid dynamics (CFD) analysis is performed under stage environment in order to find the optimum location of the circumferential casing groove in consideration of stall margin enhancement and efficiency gain at design point, and the impact of groove number to the effect of this grooved casing treatment configuration in enhancing the stall margin of the compressor stage is studied. The results indicate that the centrifugal compressor with circumferential groove in vaned diffuser can obtain obvious improvement in the stall margin with sacrificing design efficiency a little. Efforts were made to study blade level flow mechanisms to determine how the CGCT impacts the compressor's stall margin (SM) and performance. The flow structures in the passage, the tip gap, and the grooves as well as their mutual interactions were plotted and analysed.

1. Introduction

With the development of the modern engines, the compressor is required to have higher performance with fewer stages while satisfying ever stricter the safety and economical efficiency demands. From the viewpoint of compressor aerodynamics, deeper understanding of the stall process in rotor and system optimization is needed to provide better designs. A number of approaches have been proposed and adopted to enhance the compressor map width and overcome the problem of stall [1, 2].

Many efforts have been made in revealing the stall mechanism. Researchers have studied the flow structures near the endwall both experimentally and numerically to find possible stall triggers. Schlechtriem and Loetzerich [3], Hoffmann and Ballmann [4] and Yamada et al. [5] believe that the stall is triggered by the breakdown of the leading-edge blade tip leakage vortex (BTLV). However, Hah, Rabe and Wadia [6] find that for the forward swept rotor the vortex breakdown would not necessarily happen even if the compressor operates in a stalled condition, the shock oscillation and the BTLV oscillation induced by the shock-boundary layer interaction is a possible reason for the stall.

Those studies as well as some others' results indicate that the stall process in the compressor is quite complicated and may be affected by many factors.

Casing treatment as a passive flow control method to extend the stable operating range for tip critical rotor has been used in practice for years. While it can give rise to improved stall margin, there can be an associated efficiency penalty. This is due to the lack of good knowledge on the mechanism of casing treatment, and the design often largely depends on experience. CGCT is a type of casing treatment which is simple and effective. Many research works have been done in this area. Rabe and Hah [7] conducted experimental and numerical investigations to study the fundamental flow mechanisms of CGCT in a transonic axial compressor and the influence on the compressor SM. Experimental results showed that circumferential grooves increased the SM of the compressor at the tested operating condition. Detailed investigation of calculated flow fields indicated that the grooves increased the SM by reducing the flow incidence angle on the pressure side of the leading edge and losses were generated by interaction between the main passage flow and flow exiting the grooves. Wilke and Kau [8] conducted numerical study on CGCT on a HPC front stage. The circumferential grooves are reported to suppress the extension of the BTLV and improve the stall margin. Shabbir and Adamcyzk's work [9] studied the physical mechanism of the CGCT with an analysis on axial momentum budget on a low speed rotor. Müller et al [10] conducted experimental and numerical investigations of CGCT on an axial single stage transonic compressor. Four different CGCT configurations were considered. The configuration with deep grooves and large coverage showed the best results on the stall margin improvement in both design and off design speed.

Recently, researchers also use circumferential grooves to expand the SM of centrifugal compressors. Liu et al. [11] designed two different CGCTs to enhance the SM of a centrifugal compressor. The results indicated that the first treated casing could extend the calculated mass flow range of the compressor by 10.5% and experimental case by 17.9% at 50% design speed with no additional large negative loss. Hu et al. [12] numerically studied the performance of a centrifugal compressor with CGCT. Their work indicated that CGCT could reduce the reverse flow of the tip leakage to the impeller inlet and delay the occurrence of stall. Gao et al. [13] performed three-dimensional (3-D) numerical simulation for a centrifugal compressor with CGCT to study the mechanism of stall margin improvement (SMI). They concluded that circumferential grooves could restrain the development of tip leakage vortex and delay the occurrence of stall.

Traditionally, in the above-mentioned literature works, the circumferential grooves are always placed along the casing wall of the rotor part of the compressor. However, the mechanisms of the stall and casing treatment are still not well understood. More efforts are needed not only on the mechanisms but also on providing some guidelines for the turbomachinery design. With rotating stall, which has recently been observed in vaned diffusers of centrifugal compressors by Spakovszky et al [14, 15]. Thus, in this article, the CGCT configuration [16, 17] is proposed and investigated as a control method to increase the SM. In this configuration, CGCT is placed along the shroud side of the diffuser passage to enhance the SM of the compressor. CFD analysis is performed under stage environment in order to find the optimum location of the CGCT in consideration of SM enhancement and efficiency gain at design point, and the influence of groove number in enhancing the SM of the compressor stage is studied. Also, the effect of CGCT to the compressor performance and near-tip flow structure was evaluated based on stall margin improvement.

2. Analysis Model

As previously explained, all experimental and numerical work is conducted on the single-stage, high speed, low flow rate centrifugal compressor. The design specifications of the impeller and diffuser are summarized in Table 1. The past numerical simulations indicate that there is a vortex in the diffuser passage of the compressor with smooth wall casing in near stall condition [18]. The overall performance and flow passage data obtained by experiment were referred for the validation of numerical modeling, which will be presented below.

Table 1. The design parameters of the compressor.

Impeller		Diffuser	
Blade inlet diameter (mm)	140	The first vane inlet diameter (mm)	251.7
Blade outlet diameter (mm)	240	The first vane outlet diameter (mm)	296
Tip clearance (mm)	0.21	The second vane inlet diameter (mm)	300
Blade outlet width (mm)	12	The second vane outlet diameter (mm)	380
Number of rotor blades	9/9	Number of stator vanes	17
Rotation speed (rpm)	30215	Design mass flow rate (kg/s)	1.348

3. Numerical Method

3.1. Numerical Scheme

Three-dimensional flow calculations were performed in this article by using the software package NUMECA FINE/TurboTM to predict the performance of the compressor stage and to investigate the flow characteristics. The equations were discretized in space using a cell-centered finite volume formulation and in time using an explicit four-stage Runge-Kutta method. The one-equation Spalart-Allmaras turbulence model was adopted to evaluate the eddy viscosity. Local time stepping, implicit residual smoothing, and multi-grid techniques were used to reduce the computation cost.

The flow field inside the centrifugal compressor is inherently unsteady. Thus, in the state-of-the-art practice of CFD, unsteady simulations have become standard procedures in surge studies and can supply a precise prediction of the stall/surge point of the compressor. However, performing unsteady simulation is a really time consuming task, and unsteadiness had no significant effect on the performance of the low-loaded compressor. In this work, steady simulation is performed, and the flow communication between the rotor and stator uses a mixing plane method.

At the inlet, total pressure and total temperature with axial-flow direction is set as boundary condition. The ideal air is considered as working fluid. No-slip and no-heat transfer wall conditions are applied over all solid boundaries. Periodic conditions are applied for pitchwise boundaries other than the blade surfaces. Blade and wheel set as rotating parts, rotating speed as the design speed 30,215 RPM; wheel cover part set as stationary part. The mass flow rate is specified at the outlet of the compressor stage and is degressively reduced. When the stall point is impending, using the convergent solution with the lowest mass flow as the initial solution, one can move up the characteristics by decreasing the prescribed outlet mass flow in steps of 1g/s until the convergent solution no longer exists. The last converged or stable point is then referred to as the near-stall point.

3.2. Computational Grid

As the CGCTs are axially symmetrical, single passage simulations are enough to resolve all the

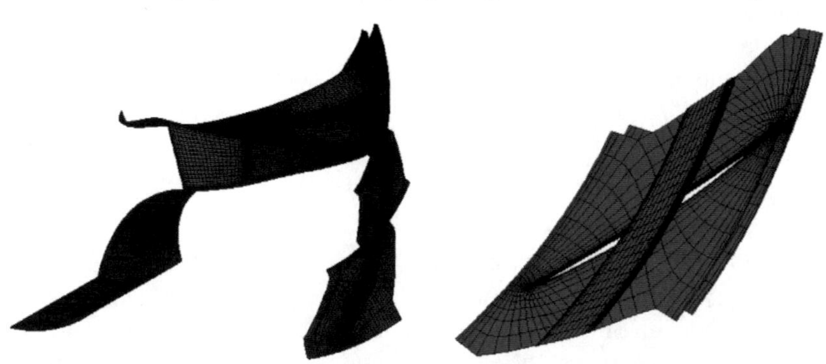

(a) Impeller/diffuser stage (b) Diffuser passage with one groove

Figure 1. Computational grid in detail

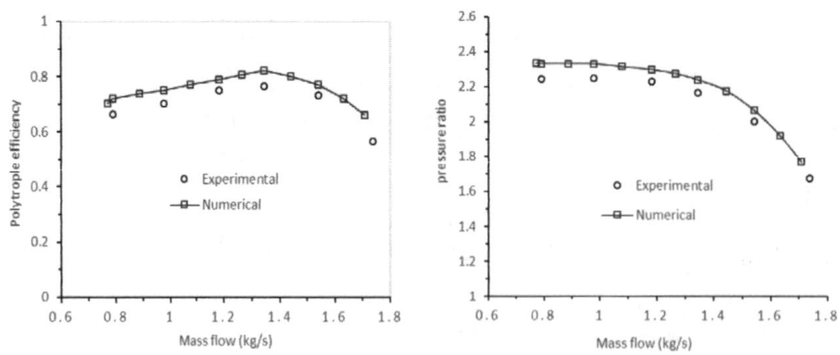

Figure 2. Characteristics plot of no-treated centrifugal compressor

important flow mechanisms associated with casing treatment configuration. The calculation was done for a single passage of the centrifugal compressor. The computational grid of the impeller/diffuser stage is generated with the auto-grid generation software IGG/AutoGrid of NUMECA. The H&I topology structure was used in grid generation of impeller, proper encryption in the blade leading edge and trailing edge, the number of nodes is 1,223,952. Two series of vane diffuser using the default topology of NUMECA, the grid number is 202,223 and 198,695 respectively. The grids of circumferential groove were generated by ZR Effect function in Autogrid5. The grids of circumferential groove and blade channel were connected by full no-matching boundary (FNMB). The grid node of circumferential groove casing treatment in three directions is 41, 37, 37, respectively, the total number of nodes is 56,129. The grid dependency analysis indicates that the computational grid is sufficient to obtain grid independent values for the figures of merit and can provide enough details for flow analysis. In order to properly evaluate the viscous fluxes at the walls, the minimum grid spacing away from the wall of the first node gives y+ to be less than 5 in the process of grid generating for all the computational cases and is sufficiently fine for the applied S-A turbulence model. Figure 1 displays the passage mesh of the impeller/diffuser stage, the diffuser passage with one circumferential groove.

3.3. Validation

The computational method was validated against the experimental data of the centrifugal compressor. The overall characteristics are shown with the corresponding experimental data in Figure 2. It can be seen from Figure 2 that simulation gives a qualitatively good agreement with experimental results. In numerical simulations, such as the insufficiency of the used turbulence model, the deviation of the real blade geometry, the simple hypothesis for ideal gas, without considering the leakage loss, wheel resistance loss, constancy hypothesis did not tally with the actual situation and other factors, the efficiency and pressure ratio of the experiment are lower than the calculated result, but the overall trend is apparent, the computational method is able to predict the flow with sufficient reliability.

4. Results and Discussion

4.1. Design of Circumferential Groove

In this paper, the circumferential groove slot positions were designed in the vane diffuser where stall inception occurred. Flow analysis of the solid casing stage showed that as the fluid diffuses in the impeller passage and interacts with the tip gap flow [16], it began to separate at the shroud side of the vaneless space and low momentum fluid accumulated near the shroud casing of the diffuser passage.

In order to determine the best slot position of the circumferential grooves, a series of radial location of circumferential grooves were designed along the wheel side of the diffuser and numerical simulations were carried out. The circumferential groove with radial width of 6mm and axial depth of 6mm. Figure 3 shows the slot location of circumferential grooves.

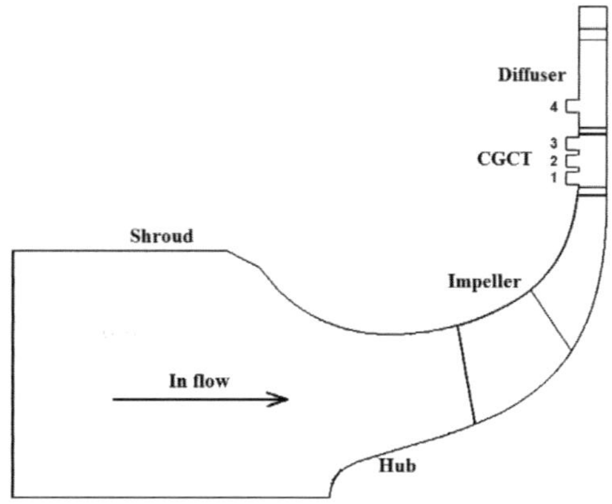

Figure 3. The slot location of CGCTs

In order to quantitative the improvement of CGCTs to stall margin and the influence of stage efficiency at design point, stable working range Φ, comprehensive improvement of stall margin ΔΦ, polytropic efficiency improvement at designed point $\Delta\eta_{pol}$ were evaluated in present work.

$$\Phi = \frac{Q_{m,des} - Q_{m,surge}}{Q_{m,des}} \times 100\%$$

$$\Delta\Phi = \frac{(Q_{m,surge})_{SW}}{(Q_{m,surge})_{CT}} \times \frac{(\varepsilon_{surge})_{CT}}{(\varepsilon_{surge})_{SW}} - 1 \qquad (1)$$

$$\Delta\eta_{pol} = \frac{(\eta_{pol,des})_{CT}}{(\eta_{pol,des})_{SW}} \quad 1$$

4.2. Impact of Groove Location on The Effect of Circumferential Groove

In Table 2, the term R_{CGCT} indicates the radius of the leading edge of the treated grooves, as shown in Figure 3. Table 2 lists the specific calculation results of different radial location of circumferential grooves as CGCT No.1/2/3/4. Figure 4 shows the influence of circumferential groove at different radial locations.

Compared with the results of the smooth wall, it can be found from Table 2 and Figure 4 that the CGCT No.1 reached the largest stable working range of 47.48% and the most comprehensive stall margin improvement of 9.10%, but also caused the efficiency of the compressor design points reduced by 0.28%. The CGCT No.2 had smaller efficiency to reduce by 0.23%, and stable working range is 47.11%, 8.37% stall margin improvement. The CGCT No.3 and No.4 had less stall margin improvement and efficiency decreased more. In addition, it can be found from Table 2 that the circumferential grooves located in the leading edge and middle of the first vane diffuser can get a better expanding stability than the other circumferential grooves.

Table 2. Results of different radial location circumferential grooves

Name	No. 1	No. 2	No.3	No. 4	SW
R_{CGCT}	128mm	136mm	144mm	156mm	-
$Q_{m,surge}$ (kg/s)	0.708	0.713	0.733	0.735	0.771
ε_{surge}	2.3686	2.3694	2.3397	2.3465	2.3642
$\eta_{pol,des}$	0.8181	0.8185	0.8179	0.8177	0.8204
Φ	47.48%	47.11%	45.62%	45.47%	42.80%
ΔΦ	9.10%	8.37%	4.09%	4.11%	-

$\Delta\eta_{pol}$	-0.28%	-0.23%	-0.30%	-0.33%	-

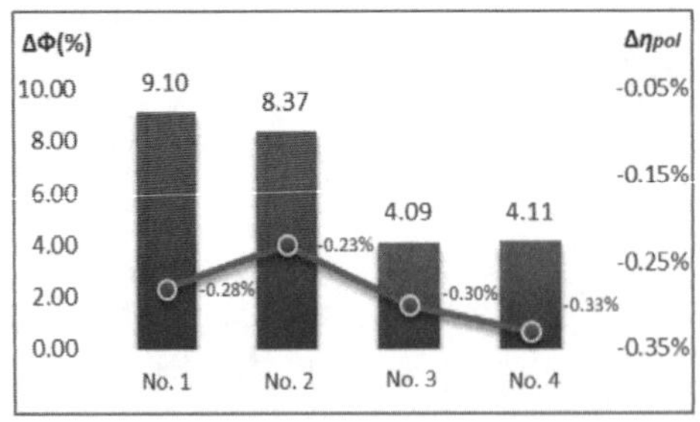

Figure 4. Effects of different radial location circumferential grooves

4.2.1. Flow field at design point. Figure 5 presents the entropy contours at 95%-span in the diffuser passage at design point. The increase in the area of the low momentum fluid region near the diffuser casing wall due to the existence of circumferential grooves which effects the steady flow at design point contributes to the increase in the losses in the diffuser passage. It can be seen from the figure that, for the treated casing stages, the entropy value at the downstream region of the treated groove increases dramatically. For the region covered by the treated groove, the entropy value of the fluid also increases dramatically. The entropy value of CGCT No.3 increases more than CGCT No.2, which makes the CGCT No.3 efficiency decreases more at design point.

4.2.2. Flow field at near-stall point. Figure 6 gives flow details and entropy distribution in the circumferential casing grooves. The stage with CGCT No.3 shows the higher entropy value and consequently the lower efficiency. As for the CGCT, there is one single regular vortex in the groove; the core of the vortex almost locates at the centre of the groove. For the treated casing diffusers, the existence of the CGCT supplies a recirculation passage for the low momentum fluid near the diffuser casing. The low momentum fluid is sucked into the groove and transports along the circumferential

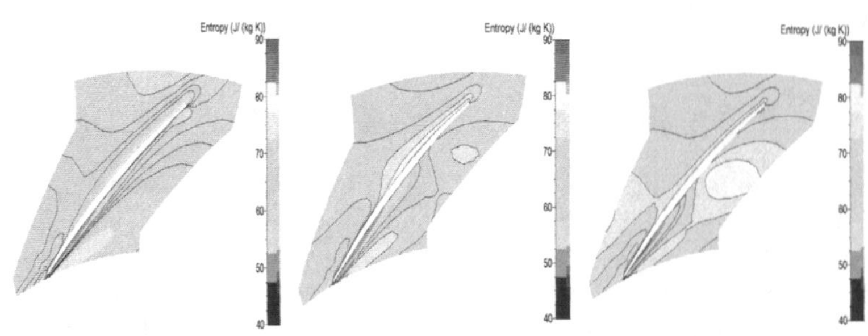

(a) SW (b) CGCT No. 2 (c) CGCT No. 3

Figure 5. Entropy distribution at 95%-span in the diffuser passage at design point

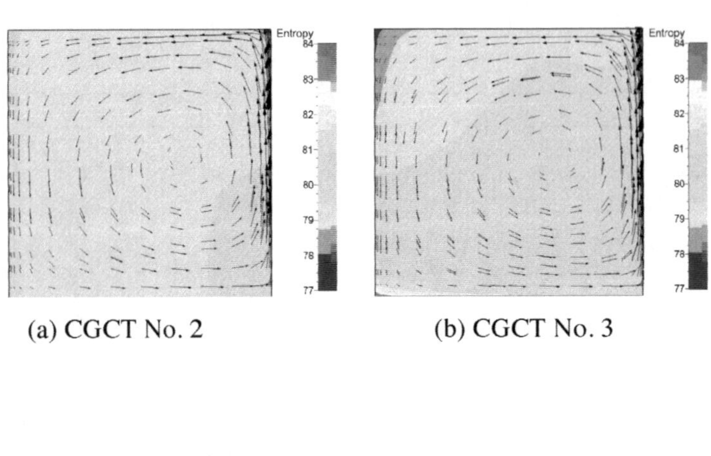

(a) CGCT No. 2 (b) CGCT No. 3

Figure 6. Relative velocity vector and entropy distribution in CGCTs at near-stall point

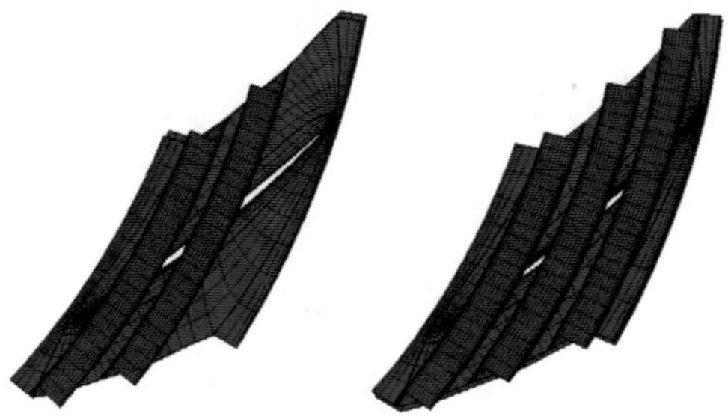

(a) CGCT-2 (b) CGCT-3

Figure 7. Computational grid of diffuser passage with different number grooves

direction in the groove. Thus, a single regular vortex locating near the centre of the grooves helps to improve the flow status in the stage, and contributes to the improvement of the stall limit of the stage.

4.3. Impact of Groove Number on The Effect of Circumferential Groove

Table 3 lists the specific calculation results of different numbers of circumferential grooves as CGCT-1 (No. 2), CGCT-2 (No.1 and No.2) and CGCT-3 (No.1, No.2 and No.3). Figure 7 presents the computational grid of diffuser passage with different number CGCTs. Figure 8 shows the influence of circumferential grooves with different number.

Compared with the results of the smooth wall, it can be found from Table 3 and Figure 8 that when the circumferential groove number increased from 1 to 2, stable working range and comprehensive stall margin improvement enhanced with static pressure ratio fell and the efficiency decreases obviously. When the circumferential groove number increase from 2 to 3, stable working range continue to increase, while the increasing amplitude is reduced, the comprehensive stall margin improvement decreased, and the static pressure continue to decline, efficiency decreases more.

Table 3. Results of different groove number of circumferential grooves

Name	CGCT-1	CGCT-2	CGCT-3	SW
	No. 2	No.1/No.2	No.1/No.2/No.3	
$Q_{m,surge}$ (kg/s)	0.713	0.701	0.699	0.771
ε_{surge}	2.3694	2.3468	2.3312	2.3642
$\eta_{pol,des}$	0.8185	0.8148	0.8063	0.8204
Φ	47.11%	48.00%	48.15%	42.80%
$\Delta\Phi$	8.37%	9.18%	8.76%	-
$\Delta\eta_{pol}$	-0.23%	-0.68%	-1.72%	-

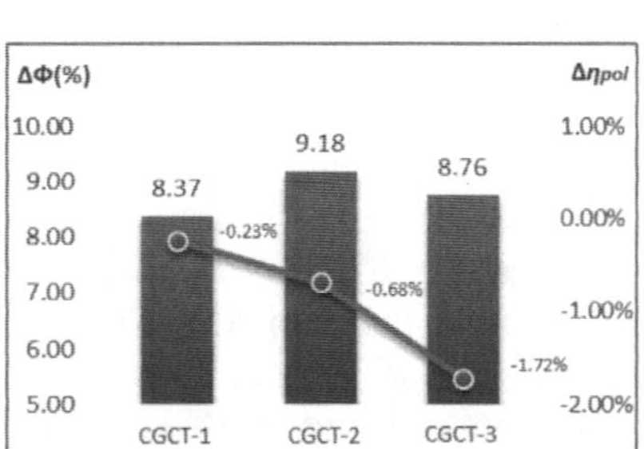

Figure 8. Effects of different number of circumferential grooves

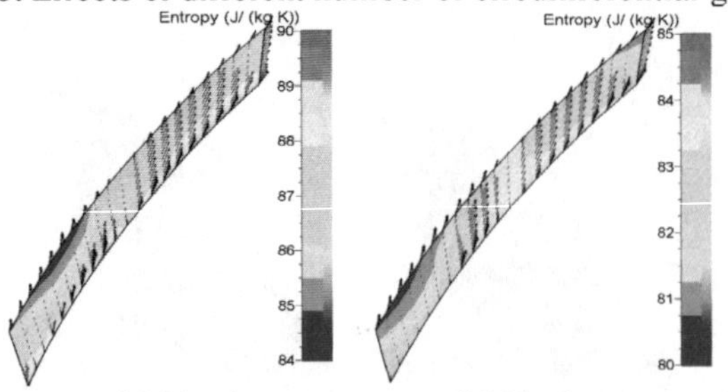

(a) No. 1 (b) No.2

Figure 9. Relative velocity vector and entropy distribution at inlet of CGCT-2 at design point

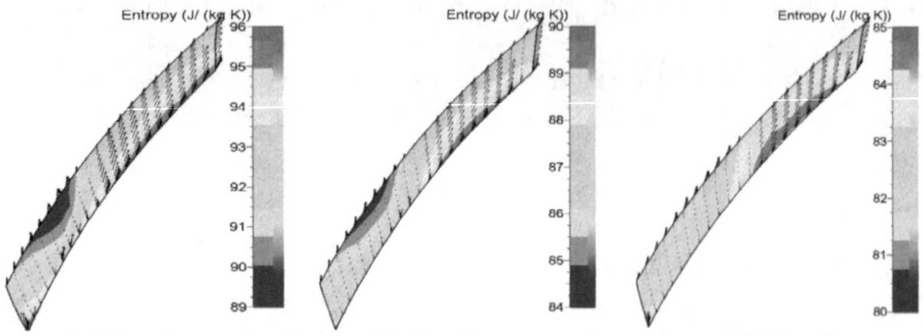

(a) No. 1 (b) No. 2 (c) No. 3

Figure 10. Relative velocity vector and entropy distribution at inlet of CGCT-3 at design point

4.3.1. Flow field at design point. Figure 9 and Figure 10 examine the relative velocity vector and entropy distribution at inlet of CGCTs at design point for the treated casing stages with 2 and 3 grooves. It can be seen that, as the number of grooves increases, the flow status at the inlet cross-section plane of the same groove has little difference in velocity vector. After the low momentum fluid is sucked into the groove, the fluid expresses as circumferential transportation from the pressure side of the diffuser vane to the suction side near the diffuser casing wall. As the number of grooves increases, the entropy at the inlet of CGCT-3 has obviously increase than CGCT-2, and the losses in the groove increase, which makes the efficiency of CGCT-3 decreased at design point.

4.3.2. Flow field at near-stall point. Figure 11 and Figure 12 present the axial velocity component distribution at inlet of CGCTs for the treated casing stages with 2 and 3 grooves at near-stall point. In the following figures, the positive axial velocity indicates that the fluid outflows from the groove, while the negative axial velocity indicates that the fluid inflows into the groove. It can be seen that, for the same groove No.1 and No.2, as the number of grooves increases, the extents of the positive and negative velocity core regions shrink obviously. The suction-reinjection capacity of the groove to the low momentum fluid near the diffuser casing wall decreases, and at the inlet cross-section plane of the groove, the flow status deteriorates. Comparing with the second groove, the axial velocity at the inlet cross-section plane of the third groove decreases obviously. It can be seen that, as the number of grooves increases, the suction-reinjection capacity of the single groove to the low momentum fluid decreases, but the stable working range increase with the whole capacity increase.

From the above analysis, we can conclude that the existence of the CGCTs supplies a recirculation passage for the low momentum fluid near the diffuser casing wall at near-stall operating conditions. The low momentum fluid is sucked into the treated casing groove and transports along the circumferential and streamwise directions in the groove. As only a part of the fluid reinjects into the main flow, the area of the low momentum fluid region shrinks and the velocity increases. Thus, the initiation of stall is delayed, and the overall contribution of each groove to SM gain depends on the compound effect of suction-reinjection.

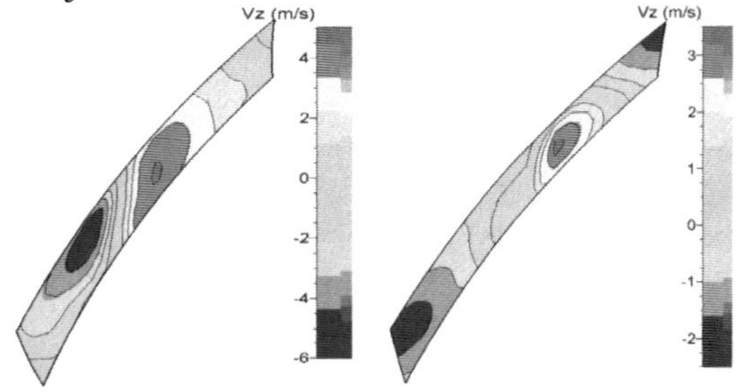

(a) No. 1 (b) No.2

Figure 11. Axial velocity component distribution at inlet of CGCT-2 at near-stall point

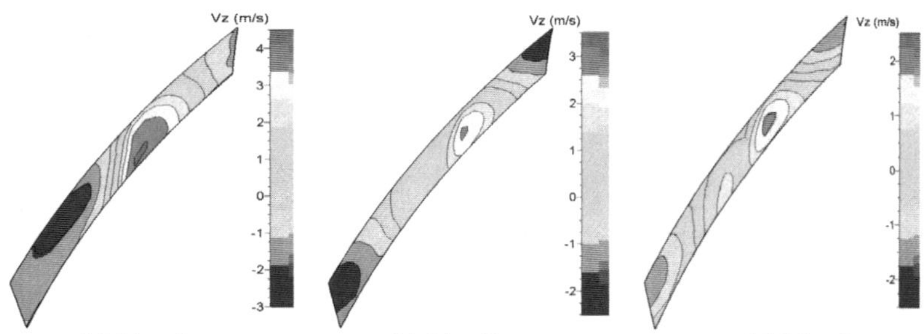

(a) No. 1 (b) No. 2 (c) No. 3

Figure 12. Axial velocity component distribution at inlet of CGCT-3 at near-stall point

5. Concluding Remarks

To evaluate the effects of radial locations and the number of CGCTs in vaned diffuser on the stall margin improvement (SMI) of a centrifugal compressor, steady three-dimensional Navier-Stokes flow simulations were conducted in this paper. The conclusions are summarized as follows:

(1) When the CGCTs is applied at the endwall of the diffuser, suitable position of circumferential groove can effectively expand the SM of the compressor, and the efficiency has fallen a little at the design point.

(2) The investigation based on CGCT-2, which consists of two circumferential grooves, indicates that CGCT-2 can generate maximum SMI, and CGCT-3 including three grooves provides more SMI than CGCT-1 including only the first groove of CGCT-2. The compressor with CGCT No.3 including only the third groove of CGCT-3 gains minimum SMI.

(3) The stage efficiency at the design point decreases as the groove number increases. The near-stall mass flow moves to a lower value and the SMI increases when increasing the groove number from 1 to 2. However, the SMI decreases when the groove number is increased to 3. The overall contribution of each groove to SMI depends on the compound effect of suction-reinjection.

(4) The evaluation based on stall margin improvement showed the optimal position for the groove to be located was indicated to exist near the leading edge of the diffuser, and a combination of geometry parameters and number of circumferential grooves that will maximize both SMI and efficiency.

Nomenclature

CGCT=circumferential groove casing treatment
SMI = stall margin improvement
Q = quantity of flow
ε = static pressure ratio

η = efficiency
Φ = stable working range
$\Delta\ \Phi$ = comprehensive improvement of stall margin
$\Delta\ \eta$ = efficiency improvement

Subscripts
m = mass
surge = surge point condition
pol = polytropic
des = design point
SW= smooth wall case
CT = casing treatment case

Acknowledgements
Financial support for the work presented is provided by the National Key Basic Research Program of China (Project No. 2012CB026000), this support is greatly appreciated.

References
[1] Ishida M, Surana T and Sakaguchi D 2005 Trans *ASME, J. Turbomach.* 127(1) 76-83.
[2] Choi K J, Kim J H and Kim K Y 2010 *ASME* Paper No. GT2010-22396.
[3] Schlechtriem S and Loetzerich M 1997 *ASME* Paper No. GT97-41.
[4] Hoffman W H and Ballman J 2003 *ISABE* Paper No. 2003-1223.
[5] Yamada K, Furukawa M, Inoue M and Funazaki K 2003 *IGTC* Paper No. 2003-095.
[6] Hah C, Rabe D C and Wadia A R 2004 *ASME* Paper No. GT2004-53867.
[7] Rabe D C and Hah C 2002 *ASME* Paper No. GT2002-30641.
[8] Wilke I, Kau H P 2003 *ASME* Paper No.GT2003-30642.
[9] Shabbir A and Adamczyk J J 2005 *ASME J. Turbomach.* 127 708-717.
[10] Müller M W, Schiffer H P and Hah C 2007 *ASME* Paper No. GT-2007-27365.
[11] Liu B, Wang Q W and Chen Y Y 2009 *ISABE* Paper No. ISABE2009-1268.
[12] Hu L J, Yang C, Qi M X and Ma C C 2009 *Chinese Journal of Mechanical Engineering* 45(7) 138-144.
[13] Gao P, Chu W L, Wu Y H and Deng W J 2007 *Journal of Propulsion Technology* 28(4) 383-387.
[14] Spakovszky Z S and Roduner C H 2009 *ASME J. Turbomach.* 131(3) 031012.
[15] Everitt J N and Spakovszky Z S 2013 *ASME J. Turbomach.* 135(1) 011025.
[16] Ma Y, Xi G and Wu G K 2011 *ASME* Paper No. GT2011-46015.
[17] Xi G, Ma Y, Wu G K, Zhang K Y, Xiao W and Mou Z J 2013 *Journal of Power and Energy* 227(4) 464-478.
[18] Chen X F, Qin G L and Ai Z J 2015 *IOP Conference Series: Materials Science and Engineering* 90(1) 012043.

IOP Conf. Series: Materials Science and Engineering **232** (2017) 012039 doi:10.1088/1757-899X/232/1/012039

An alternative method for centrifugal compressor loading factor modelling

Y Galerkin, A Drozdov, A Rekstin, K Soldatova
R&D Laboratory "Gas dynamics of turbo machines" Peter the Great St.Petersburg Polytechnic University, Polytechnical st. 29, St.Petersburg, Russia

Email: buck02@list.ru.

Abstract: The loading factor at design point is calculated by one or other empirical formula in classical design methods. Performance modelling as a whole is out of consideration. Test data of compressor stages demonstrates that loading factor versus flow coefficient at the impeller exit has a linear character independent of compressibility. Known Universal Modelling Method exploits this fact. Two points define the function – loading factor at design point and at zero flow rate. The proper formulae include empirical coefficients. A good modelling result is possible if the choice of coefficients is based on experience and close analogs. Earlier Y. Galerkin and K. Soldatova had proposed to define loading factor performance by the angle of its inclination to the ordinate axis and by the loading factor at zero flow rate. Simple and definite equations with four geometry parameters were proposed for loading factor performance calculated for inviscid flow. The authors of this publication have studied the test performance of thirteen stages of different types. The equations are proposed with universal empirical coefficients. The calculation error lies in the range of plus to minus 1,5%. The alternative model of a loading factor performance modelling is included in new versions of the Universal Modelling Method.

Nomenclature

b	width of channel
c_u	tangential component of absolute flow velocity
D	diameter
h_T	theoretical head, i.e. the head transferred to gas by impeller blades
K_{pd}	empirical coefficient of velocity diagram
K_μ	empirical coefficient of viscosity influence
l	length of blade
$\frac{l}{t}$	cascade solidity
$\bar{m}$	mass flow rate
M_u	blade Mach number
N_i	internal power, i.e. all power transferred to gas by an impeller
p	pressure
r	radius
u	tangential velocity

IOP Conf. Series: Materials Science and Engineering **232** (2017) 012039 doi:10.1088/1757-899X/232/1/012039

$X_{\psi t0}$ empirical coefficient
w relative velocity
Γ flow circulation
y empirical coefficient
z number of blades, empirical coefficient
β angle between the relative speed and the reverse district direction
β_{bl} blade angle
β_{df} non-dimensional coefficient of disk friction
β_{lk} non-dimensional leakage coefficient
β_T angle of a loading factor performance inclination
φ flow coefficient
η polytrophic efficiency
λ friction coefficient
ω rotation speed
Φ flow rate coefficient
ψ_T loading factor.

Subscripts
0 impeller inlet
1 impeller blade row inlet
2 impeller outlet
ni non incidence
inl inlet
inv inviscid flow
bl blade
boud boundary layer
imp impeller
des design
max maximum
pd center of pressure

1. Aim of the work

For flow path design, an instrument is necessary for calculation of centrifugal compressor gas dynamic performance. The authors use the primary design procedure described in [1]. The Universal Modelling Method presented in [2] is applied to calculate performance of a primary design and of possible better candidates. Calculated non-dimensional performance of a stage is presented in Figure 1.

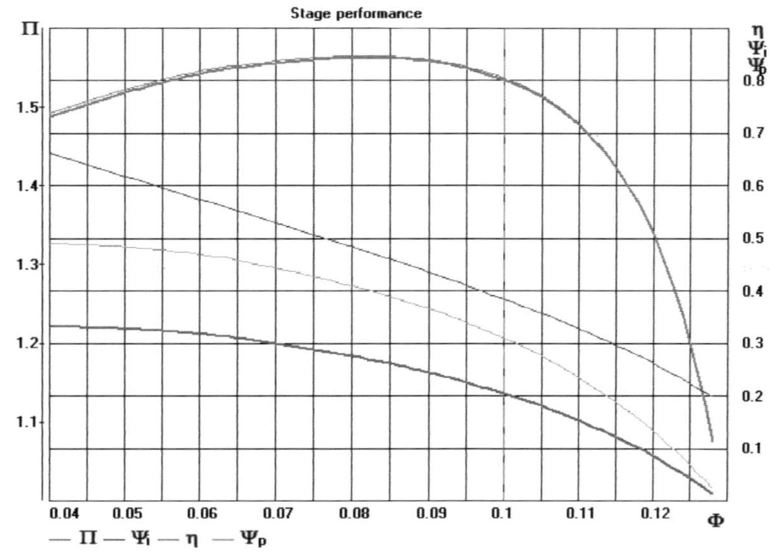

Figure 1. Calculated non-dimensional performance of a centrifugal compressor stage

Efficiency η and work coefficient ψ_i are necessary for calculation of polytrophic coefficient and pressure ratio:

$$\psi_p = \psi_i \times \eta \cdot \tag{1}$$

$$\pi = \left(1 + (k - 1)\psi_i M_u^2\right)^{\frac{k}{k-1}\eta} . \tag{2}$$

Efficiency prediction is possible with new versions of the Universal Modelling PC programs without special skill and experience [2-4]. Prediction of the work coefficient performance $\psi_i = f(\Phi)$ still requires experience and intuition. The authors' aim is to offer an alternative way of the work coefficient performance calculation that is simple and precise. Presented work is based on ideas and results of Y. Galerkin and K. Soldatova who had studied loading factor performances of impellers with inviscid flows.

2. Scheme of work coefficient modelling

In accordance with the scheme proposed previously in [5] the head transmitted to gas by an impeller h_i consists of three parts. The main part of engine's mechanical head h_T is transferred to gas by blades of the impeller. Two additional parts appear due to parasitic losses. The part of the head h_{lc} is lost in labyrinth seals. The friction on outer surfaces of hub and shroud transmits additional mechanical head h_{df} but this head does not increase gas pressure:

$$h_i = h_T + h_{df} + h_{lc} \cdot \tag{3}$$

Non-dimensional presentation of this equation:

$$\psi_i = \psi_T \left(1 + \beta_{df} + \beta_{lk}\right) \cdot \tag{4}$$

IOP Conf. Series: Materials Science and Engineering **232** (2017) 012039 doi:10.1088/1757-899X/232/1/012039

For an impeller with large flow coefficient $\Phi_{des} \approx 0{,}15$ the sum $\beta_{df} + \beta_{lk}$ is less than 0,01. For an impeller with small flow coefficient $\Phi_{des} \approx 0{,}015$ this sum is about 0,055-0,065. It is not large part of a head coefficient. Semi-empirical formulae in [1] and CFD-calculations [9] are good instruments for modelling of these coefficients. Therefore the main problem is modelling of a loading factor.

As a rule there is no velocity tangential component at an impeller inlet in industrial compressors. Then impeller blades transfer the head to a gas in accordance with the Euler equation:

$$h_T = c_{u2} u_2. \tag{5}$$

This head is presented by the non-dimensional loading factor:

$$\psi_T = c_{u2} / u_2. \tag{6}$$

The modelling of a loading factor performance is facilitated by the fact that it is a linear function – Figure 2.

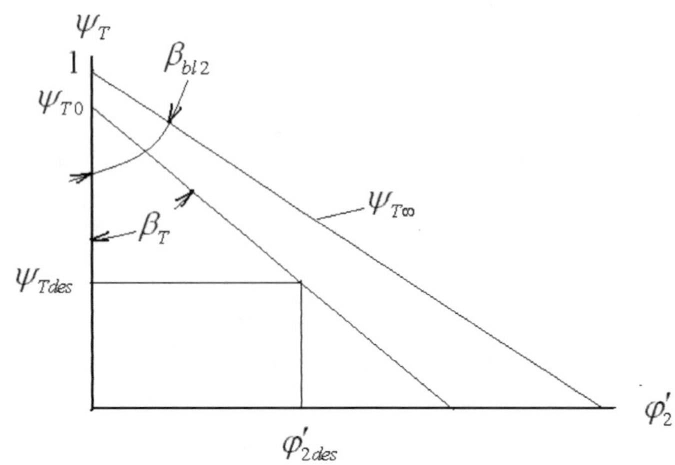

Figure 2. Linear function of a loading factor for ideal and real impellers

The linear nature of it is evident for an "ideal" impeller with infinite number of infinitely thin blades:

$$\psi_{T\infty} = 1 - \varphi_2' ctg \beta_{bl2} \tag{7}$$

The similar equation is valid for real impellers:

$$\psi_T = 1 - \varphi_2' ctg \beta_2. \tag{8}$$

The flow angle β_2 is not constant versus flow coefficient. Anyway test data for industrial compressors demonstrates practically a linear characteristic of work input versus flow rate [5]. Model stages' tests demonstrated linear characteristic of the function $\psi_T = f(\varphi_2')$ independent of blade Mach number [10].

The existing way of modelling [2-4] exploits this fact. Two values of a loading factor are necessary to determine a linear performance $\psi_T = f(\varphi_2')$.

In the method described in [1] these two values are a loading factor at a design flow rate $\psi_{T\,des}$ and at zero flow rate ψ_{T0}. The design flow rate corresponds to non-incidence inlet of the critical streamline. This condition is $\beta_{1cr} = \beta_{bl1}$. Exit and inlet velocity triangles demonstrates influence of blades' load and blockade on a critical streamline direction – Figure 3.

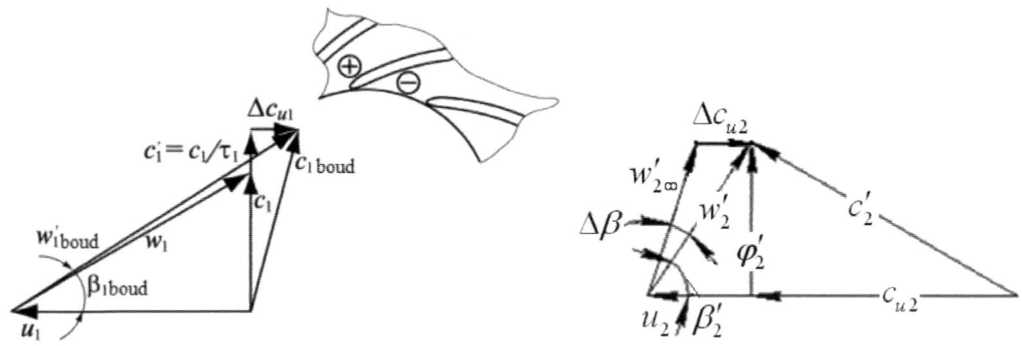

Figure 3. Inlet and exit velocity triangles at design flow rate

The absolute velocity c_1 is accelerating near a blade cascade inlet due to a blade blockade: $c_1' = c_1/\tau_1$. A critical streamline turns its direction to a suction side of a blade where pressure is less than on a pressure side. Symbols "plus/minus" in the Figure 3 demonstrate difference of pressures on blade surfaces. A critical streamline obtains a tangential velocity Δc_{u1}. The scheme of the Δc_{u1} calculation is based on the idea of representing a blade by a swirl with the circulation $\Gamma_{bl} = 2\pi r_2 c_{u2}/z$. The swirl induces velocities depending of distance from its center. The equation to calculate tangential velocity induced on the radius r_1 where the blades start is presented in non-dimensional way:

$$\Delta\bar{c}_{u1} = \frac{\psi_{\mathrm{T\,des}}}{z\left(\bar{r}_{pc} - \bar{r}_1\right)}. \tag{9}$$

Here $\bar{r}_{pc} = r_{pc}/r_2$ is a relative distance from an impeller axis to a center of a velocity diagram, Figure 4.

A loading factor at design flow rate is calculated on the same principle. The velocity triangle is shown above in the Figure 2. In accordance with Kutta-Zoukovsky postulate a critical streamline leaves blades in direction of an angle β_{bl2}. Due to a blade load a pitch-averaged velocity got a tangential velocity component:

$$\Delta\bar{c}_{u2des} = K_{\mu}\frac{\psi_{\mathrm{T}des}}{z\left(1 - \bar{r}_{pc}\right)}. \tag{10}$$

The equation for a loading factor is:

$$\psi_{T\,des} = \frac{1-\varphi'_{2des} ctg\beta_{bl2}}{1+K_{\mu}\dfrac{1}{z\left(1-\bar{r}_{pc}\right)}}. \tag{11}$$

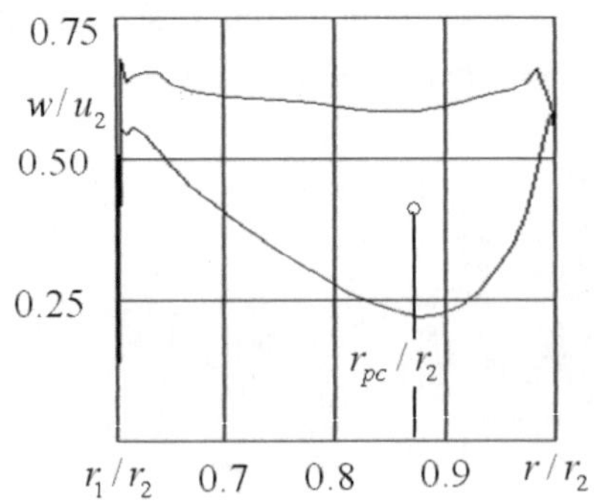

Figure 4. Velocity diagram at design flow rate (inviscid flow)

The empirical coefficient $K_{\mu}>1$ represents influence or real viscid character of flow. Its value for different types of impellers may be between 1,5-2,3. There is no satisfactory correlation with an impeller configuration. The close analog is necessary. The alternative is to model a loading factor performance by values of ψ_{T0} and angle β_T that are shown in Figure 2:

$$\psi_T = \psi_{T0} - \varphi'_2 ctg\beta_T. \tag{12}$$

The aim of this work is to model loading factor performance of impellers in real viscid flow on the basis of several impeller geometry parameters.

Test data for model stages and factory test data of several pipeline compressors were reduced. The information on the objects of modelling is presented in Table 1. All stages and compressors were tested at M_u=0,60 or 0,80. Model stage names mean the following. For example, 0,0604-0,527-0,290 means that the design flow rate coefficient of the stage is Φ_{des} = 0,0604, design loading factor is $\psi_{T\,des}$ = 0527, hub ratio is D_h/D_2 = 0,29. Names of compressors are given by their manufacturers on their own principles. Data on compressors are taken from [4]. Symbol * means that 2D impeller has an arc blade mean line. Mean lines are designed by the Method presented in [1] in other cases. In columns 3-9 geometry parameters of impellers are presented, they are included in the presented below equations for loading factor performances modelling.

Table 1. Parameters of stages and compressors used for modelling

1	2	3	4	5	6	7	8	9	10	11
	Parameter/ /stage	b_2/b_1	z	D_1/D_2	1/t	β^0_{bl1}	β^0_{bl2}	$\bar{\delta}_{bl}$	K_{μ}	$X_{\psi T0}$
1	0.0373-0.482-0.373*	0.754	15	0.514	3.423	25	30.3	0.0133	2.0	2.1
2	0.0455-0.539-0.350*	0.589	13	0.565	2.40	25	34	0.02	1.05	2.3
3	0.0480-0.49-0.290	0.573	11	0.534	2.461	23	30	0.014	1.65	1.3

4	0.0604-0.527-0.290	0.618	13	0.592	1.916	22	47	0.0115	1.45	0.9
5	0.0653-0.409-0.290	0.657	11	0.570	1.967	28	32	0.014	1.7	1.9
6	0.0692-0.476-0.290	0.707	11	0.570	2.062	25	32	0.0138	1.2	1.7
7	0.0685-0.68-0.345	0.539	17	0.577	1.864	27	79	0.0128	1.34	0.4
8	16-76-1.6*	0.693	11	0.5765	1.987	26	32	0.0121	0.95	2.17
9	108-5-1*	0.587	13	0.5651	2.399	21	38	0.02	1.5	1.02
10	GPA-76-1.7	0.614	13	0.583	2.227	24.8	35.8	0.0143	2.62	1.7
11	61-1.64	0.549	19	0.548	2.518	25.5	67	0.0138	1.62	0.95
12	650-1.37-76	0.711	11	0.570	1.967	28	32	0.0138	1.6	1.4
13	16-76-1.7	0.360	25	0.5	2.996	30	104	0.0089	1.34	0.4

The sample of data on model stages from the "IDENT" program is presented in Figure 5. The dimensions that are necessary for calculation of efficiency and a loading factor performances are presented in the database. The measured values of these parameters are black figures and lines. The efficiency calculated according to the 6th version of model with a universal set of coefficients are red figures and lines. The values $X_{\psi T0}$ (the coefficient for an empirical equation for ψ_{T0} calculation) and K_μ are picked up individually for each stage for the best compliance to the measured performances. These values are presented in columns 10, 11 in the table 1. The program calculates also values β_T and ψ_{T0} corresponding to preset values $X_{\psi T0}$ and K_μ.

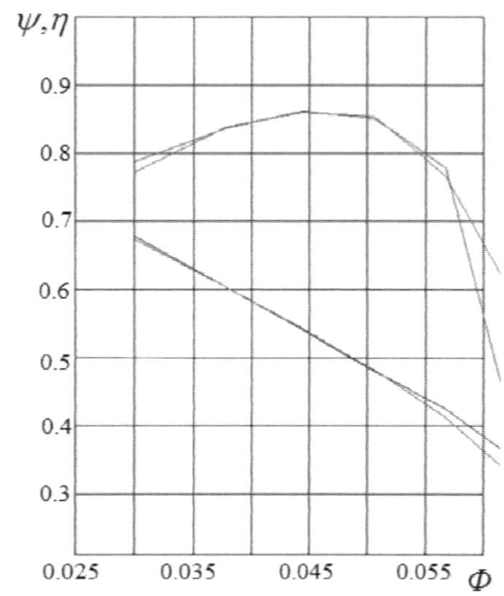

Figure 5. An example of information from the database of the IDENT program. (stage No. 3, from the table 1)

The authors have studied all impeller geometry parameters as possible arguments in functions $\beta_T, \psi_{T0} = f\left(\varphi_2', \bar{F}\right)$, where $\bar{F}$ - symbol of geometry parameters that define an impeller flow path. Main dimensions of a typical industrial impeller in meridian plane are shown in Figure 6.

Among dimensions shown in Figure 6 only the ratio b_2 / b_1 influences functions $\beta_T, \psi_{T0} = f\left(\varphi_2', \bar{F}\right)$. The most important parameter is a blade exit angle β_{bl2}. Ratio of a blade length l to average pitch t presents in functions too. This ratio is calculated by the formula from [5]:

$$1/t = z\frac{\lg\frac{D_2}{D_1}}{2.73\sin\left(\frac{\beta_{bl2}+\beta_{bl1}}{2}\right)}. \tag{13}$$

Most of 2D impellers designed by the authors have either an arc mean line of blades, or a mean line ~~is~~ optimized by analysis of velocity diagrams. Two mean lines and corresponding velocity diagrams are presented in Figure 7.

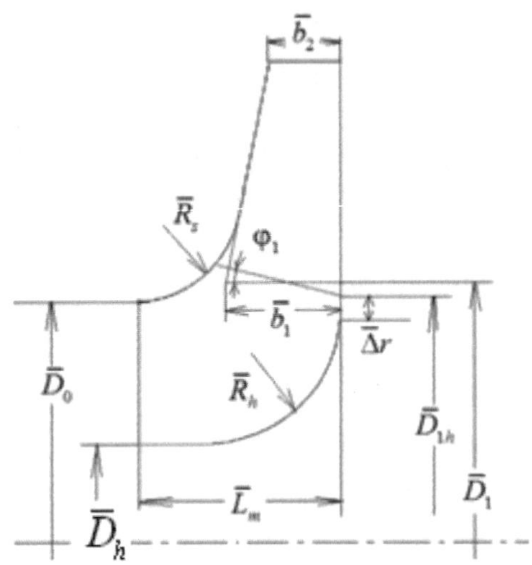

Figure 6. Main dimensions of a typical industrial impeller in meridian plane

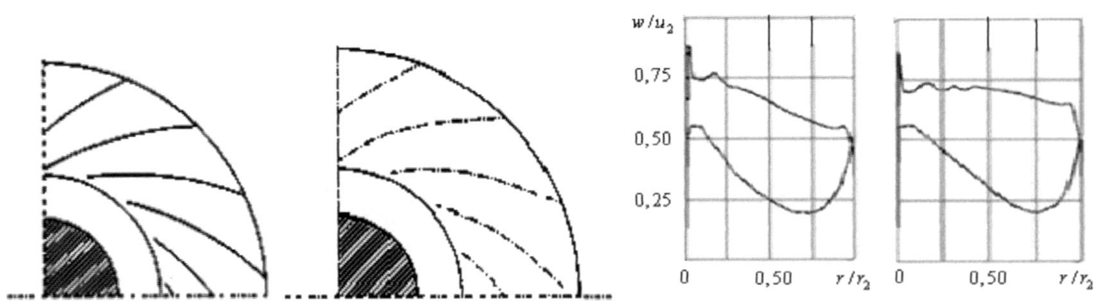

Figure 7. Mean lines and velocity diagrams of 2D impellers. Left – arc mean line, right – optimized mean line

Two groups of approximating formulae are developed for two types of blade mean line.

The loading factor at zero flow rate:
- for impellers with optimized mean line:

$$\psi_{T0} = 0.964 + 0{,}002\left(\frac{\beta_{bl2}}{40}\right)^4 - 0.068\left[\ln\left(\frac{l}{t}\right)\right]^3 - 0.00011\left[\ln\left(\frac{b_2}{b_1}\right)\right]^{-4}. \tag{14}$$

IOP Conf. Series: Materials Science and Engineering **232** (2017) 012039 doi:10.1088/1757-899X/232/1/012039

- for impellers with arc mean line:

$$\psi_{T0} = 0.747 + 0,097\left(\frac{\beta_{bl2}}{40}\right)^4 - 0.008\left[\ln\left(\frac{l}{t}\right)\right]^7 - 20.247\left[\ln\left(\frac{b_2}{b_1}\right)\right]^9. \tag{15}$$

Angle β_T^0:

- for impellers with optimized mean line:

$$\beta_T^0 = 0.534\frac{\beta_{bl2}}{40} + 0.028\left(\frac{l}{t}\right)^6 - 3.577\left[\ln\left(\frac{b_2}{b_1}\right)\right]^{-1}. \tag{16}$$

- for impellers with arc mean line:

$$\beta_T^0 = 20.311\left(\frac{\beta_{bl2}}{40}\right)^{-2} + 3.595\left(\frac{l}{t}\right) \; 0.106\left[\ln\left(\frac{b_2}{b_1}\right)\right]^{-4}. \tag{17}$$

Figures 8, 9 demonstrate accuracy of approximation.

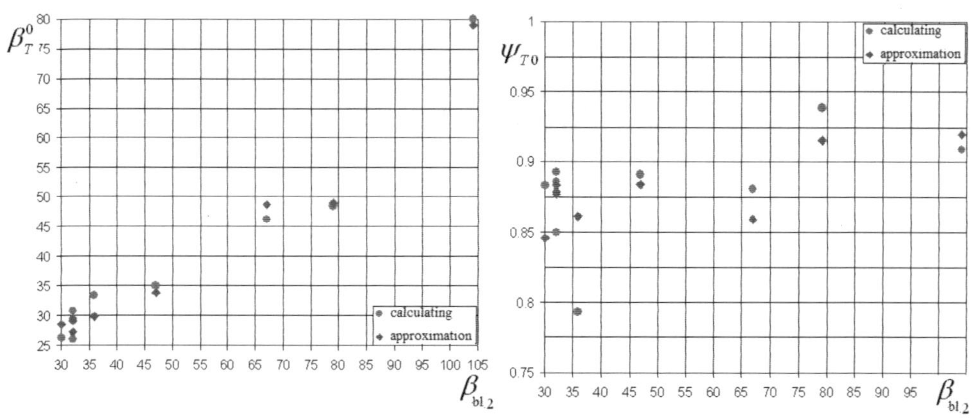

Figure 8. Comparison of the measured and calculated β_T and ψ_{T0} for impellers with different β_{bl2} (optimized mean line)

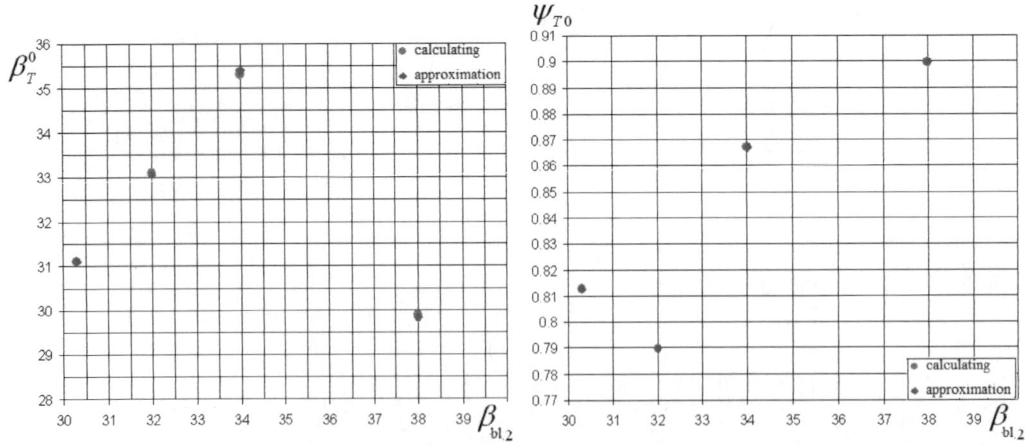

Figure 9. Comparison of the measured and calculated β_T and ψ_{T0} for impellers with different β_{bl2} (arc mean line)

Modelling accuracy is satisfactory for an exception of the 1st stage of 6-stage pipeline compressor GPA-76-1.7 (line 10 Table 1). For impellers of this compressor the empirical coefficient is K_μ= 2.62 (eq. 10). It is an unusually large value. The authors have no explanation for this fact. The modelling error in all other cases is within admissible limits. Figure 10 presents the comparison of performance of one of model stages.

Graphics with individual adjustment of the loading factor performance are presented on the left. Right – loading factor is calculated by eq. (14, 16). Efficiency performance is calculated by 6th version of the Math model in both cases. Despite visually noticeable difference of load performances, an error of calculation $\psi_{T\,des}$ is 1,2%. It means that an impeller designed by this method will have +1,2% of head input. It is acceptable for design practice.

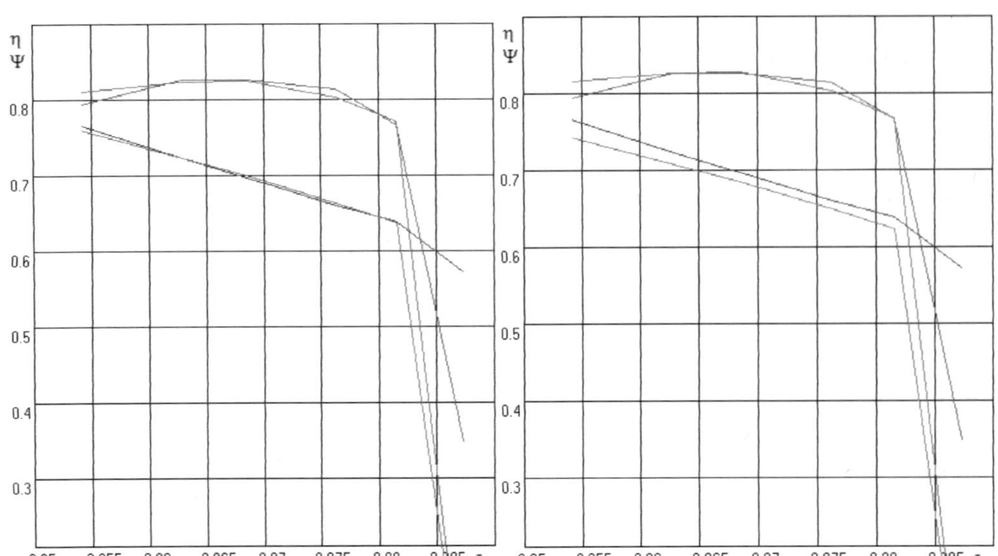

Figure 10. Model stage performance comparison (#3, β_{bl2}=79^0). Green – measured, red – calculated. Right – loading factor is calculated by eq. (15, 17)

Conclusion

The authors plan to apply the presented method of a loading factor performance calculation in parallel with the existing method in their future projects. In case of a positive result the new method will be incorporated into the Math models. This method is simple and does not require users' high experience and corrections with analogs.

Acknowledgment

The reported study was funded by RFBR according to the research project № 16-08-00624 A.

References

[1] Galerkin. Y.B. Turbocompressors. // LTD information and publishing center. - Moscow. – 2010. – P.650. [Russian].

[2] Galerkin Y., Soldatova K. The application of the Universal Modelling Method to development of centrifugal compressor model stages. Conference "Compressors and their Systems". – 477-487. – London. – 2013.

[3] Galerkin Y., Drozdov A. New generation of Universal modelling for centrifugal compressors calculation. Conference "Compressors and their Systems". – London. – 2015.

[4] Galerkin Y., Rekstin A., Soldatova K., Drozdov A. Universal modelling method – the instrument for centrifugal compressor gas dynamic design. ASME Gas Turbine India Conference. – 2015. - GTINDIA2015-1202.

[5] Ris V. Centrifugal compressors. // Leningrad. – 1964. [Russian].

IOP Conf. Series: Materials Science and Engineering **232** (2017) 012040 doi:10.1088/1757-899X/232/1/012040

2D and 3D impellers of centrifugal compressors – advantages, shortcomings and fields of application

Y Galerkin, A Reksrin, A Drozdov
R&D Laboratory "Gas dynamics of turbo machines" Peter the Great St. Petersburg Polytechnic University, Polytechnical st. 29, St.Petersburg, Russia

Email: buck02@list.ru.

Abstract: The simplified equations are presented for calculation of inlet dimensions and velocity values for impellers with three-dimensional blades located in axial and radial part of an impeller (3D impeller) and with two-dimensional blades in radial part (2D). Considerations concerning loss coefficients of 3D and 2D impellers at different design flow rate coefficients are given. The tendency of reduction of potential advantages of 3D impellers at medium and small design flow rate coefficients is shown. The data on high-efficiency compressors and stages with 2D impellers coefficients designed by the authors are presented. The reached efficiency level of 88 - 90% makes further increase of efficiency by the application of 3D impellers doubtful. CFD–analysis of stage candidates with medium flow rate coefficient with 3D and 2D impellers revealed specific problems. In some cases the constructive advantage of a 2D impeller is smaller hub ratio. It makes possible the reaching of higher efficiency. From other side, there is a positive tendency of gas turbine drive RPM increase. 3D impellers have no alternative for stages with high flow rate coefficients matching high-speed drive.

Nomenclature

b	width of channel
c	absolute velocity at an impeller exit
D	diameter
K_D	geometry parameter of 2D impeller
K_F	geometry parameter of 2D impeller
l	length
L_m	impeller meridian length
M	Mach number
$\overline{m}$	mass flow rate
u	blade velocity
w	relative velocity
α	flow angle related to tangent
β_{bl}	blade angle related to tangent
χ	angle between blade and meridian surfaces
ρ	gas density
ζ	loss coefficient
τ	blade blockage factor

IOP Conf. Series: Materials Science and Engineering **232** (2017) 012040 doi:10.1088/1757-899X/232/1/012040

ψ_T loading factor
φ flow coefficient
η efficiency
π pressure ratio
Φ flow rate coefficient

Subscripts
1 impeller inlet
2 impeller exit
h hub
des design flow rate
m meridian
min minimum
max maximum
s shroud
inl inlet

1. Main features of 2D and 3D impellers

Radial impellers with blades of a cylindrical form are used in industrial centrifugal compressors since the end of 19th century. Fig. 1 demonstrates 2D impeller with stamped blades riveted to impeller discs. These were widely applied in the past. Late decision is a 2D impeller with milled blades on the surface of the main disk. Process of blade milling is shown in Fig. 1.

Figure 1. Classical impeller with stamped and riveted blades and element of a stamped blade – left. Milling of blades with cylindrical surface – photo on the right [1]

Manufacturing technology of 2D impellers is convenient for individual and small-scale production. In general, in pipeline and industrial compressors 2D impellers are widely applied. However, from the aerodynamic point of view, the cylindrical shape of blades contradicts the three-dimensional character of a flow. Especially this contradiction seems important at an impeller inlet. Inlet flow angle varies in a wide range along blade leading edge. The design technique applied by authors [1], allows to solve this problem for 2D impellers with limited design flow rate coefficients.

The impellers with 3D blades and leading edges in an axial part were applied firstly in aviation gas-turbine engines. Their obvious advantage is the highest durability at high blade velocity. 3D impellers are more effective at high Mach number and big loading factor.

Five-coordinate milling machines with numerical control made possible application of 3D impellers in industrial compressors. Fig. 2 shows 3D impeller for an industrial compressor. A meridional configuration of 3D and 2D impellers is presented in Fig. 3.

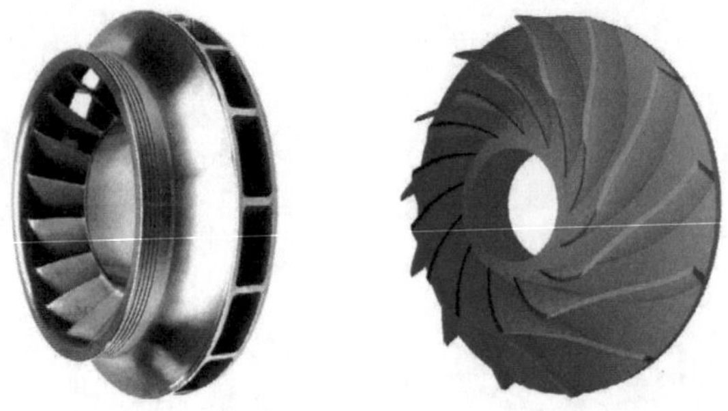

Figure 2. 3D impeller, closed type – left, blade cascade - right

Pipeline compressors operate at low blade velocity and low Mach numbers. They have low loading factors in most cases. Advantages of 3D impellers for pipeline compressors are not evident if flow rate coefficients are not too high. The efficiency of 3D and 2D impeller is analyzed below.

The loss of efficiency in the impeller depends on dimensionless relative velocity at an inlet taking into account blade blockage factor [1]:

$$\Delta\eta_{imp} = \zeta_{imp} \frac{\bar{w}_1'^2}{2\psi_{T des}} \tag{1}$$

Here $\bar{w}_1'$ is an average velocity along a blade leading edge.

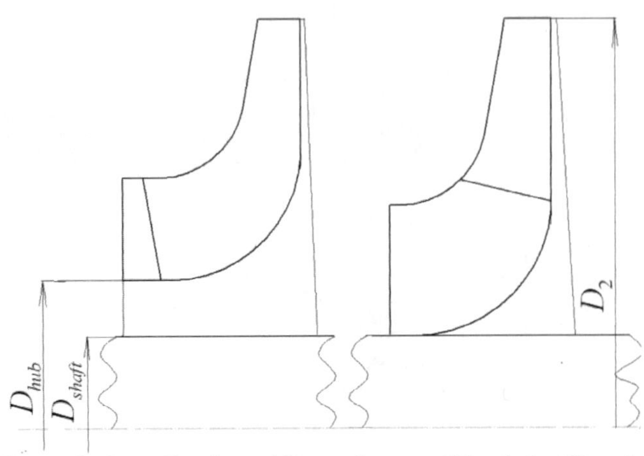

Figure 3. Impellers' meridian schemes. 3D – left, 2D – right

To minimize efficiency loss it is necessary to minimize $\bar{w}_1'$. The loss coefficient minimization depends on a design skill and impeller relative dimensions.

2. Inlet and outlet dimensions and flow velocities

If a leading edge of a 3D impeller is placed as shown in Fig. 3:

$$\bar{w}_0' = \sqrt{\bar{D}_0^2 + \varphi_0'^2}\ , \tag{2}$$

For 2D impeller:

$$\bar{w}_1' = \sqrt{\bar{D}_1^2 + \varphi_1'^2} \tag{2a}$$

Relative inlet diameter and flow rate coefficient are connected by the continuity equation.
For 3D impeller:

$\bar{m} = \rho_0^* \cdot \Phi_{des} u_2 \frac{\pi}{4} D_2^2 = \rho_0' \cdot \varphi_0' u_2 \cdot \tau_0 \frac{\pi}{4}\left(D_0^2 - D_{hb}^2\right) = \rho_0' \cdot \varphi_0' u_2 \cdot \tau_0 \frac{\pi}{4} 2b_0 \left(D_0 + D_{hb}\right)$, and:

$$\varphi_0' = \frac{\rho_0^*}{\rho_0'} \frac{\Phi_{des}}{\tau_0\left(\bar{D}_0^2 - \bar{D}_{hb}^2\right)} = \frac{\rho_0^*}{\rho_0'} \frac{\Phi_{des}}{\tau_0 2\bar{b}_1\left(\bar{D}_0 + \bar{D}_{hb}\right)} \tag{3}$$

For 2D impeller $\bar{m} = \rho_0^* \cdot \Phi_{des} \frac{\pi}{4} D_2^2 u_2 = \rho_1' \cdot \varphi_1' \cdot \tau_1 \pi D_1 b_1 u_2$, and:

$$\varphi_1' = \frac{\rho_0^*}{\rho_1'} \frac{\Phi_{des}}{4\tau_1 \bar{D}_1 \bar{b}_1} \tag{3a}$$

Here Φ_{des} - flow rate coefficient, which connects mass flow rate with an impeller diameter and blade velocity [1]:

$$\Phi_{des} = \frac{\bar{m}_{des} / \rho_0^*}{\pi \frac{D_2^2}{4} u_2} \tag{4}$$

The designer can choose the necessary design flow rate coefficient if RPM or number of stages are not constrained [1]. It is not so for most pipeline compressors.

According to the equations (2, 2a) to a bigger inlet diameter corresponds a smaller flow rate coefficient but bigger blade velocity $\bar{u}_1 = \bar{D}_1$. Minimal inlet velocities can be calculated by the formulae [1]. For 3D: $\bar{D}_{0w\min} = \sqrt{\bar{D}_{hb}^2 + 2^{\frac{1}{3}}\left(\frac{\Phi_{des}}{\frac{\rho_0'}{\rho_0^*}\tau_0}\right)^{\frac{2}{3}}}$. For 2D: $\bar{D}_{1w\min} = K_D\sqrt{\bar{D}_{hb}^2 + 2^{\frac{1}{3}}\left(\frac{\Phi_{des} K_F}{\frac{\rho_1'}{\rho_0^*}\tau_1 K_D}\right)^{\frac{2}{3}}}$, for the approximate analysis we will accept typical values of parameters of a secondary importance:

- ratio of static and total parameter density $\frac{\rho_1'}{\rho_0^*}$ is 0,9,
- blade blockage factor τ_1 is 0,85,
- geometry coefficient K_D [1] is 1,0,
- geometry coefficient K_F [1] is 0,85.
- inlet blade blockade factor τ_1 [1] is 0,85.

Then:

$$\bar{D}_{0w\min} \approx \left(\bar{D}_{hb}^2 + 1{,}5\Phi_{des}^{2/3}\right)^{0,5}, \tag{5}$$

$$\bar{D}_{1w\min} \approx \left(\bar{D}_{hb}^2 + 1,35\Phi_{des}^{2/3}\right)^{0,5}. \tag{5a}$$

In Fig. 3 above is shown that a shaft diameter can be equal to diameter of the hub of a 2D impeller. A hub of a 3D impeller sometimes is bigger. A shaft diameter must provide the necessary rigidity of a rotor. Bigger diameter of the hub doesn't increase rigidity, but worsens 3D impeller gas dynamic properties.

A flow coefficient in eq. (2), (2a) depends on an inlet blade height. From a geometrical ratio $D_0 - D_{hb} = 2b_0$:

$$\bar{b}_0 = \frac{\left(\bar{D}_{hb}^2 + 1,5\Phi_{des}^{2/3}\right)^{0,5} - \bar{D}_{hb}}{2}. \tag{6}$$

From a geometrical ratio $\left((D_1/K_D)^2 - D_{hb}^2\right) = K_F 4D_1b_1$:

$$\bar{b}_1 = 0,4\frac{\Phi_{des}^{2/3}}{\left(\bar{D}_{hb}^2 + 1,35\Phi_{des}^{2/3}\right)^{0,5}} \tag{6a}$$

From equations (5) and (3):

$$\varphi_0' \approx 0,70\Phi_{des}^{1/3} \tag{7}$$

From equations (5a) and (3a):

$$\varphi_1' \approx 0,82\Phi_{des}^{1/3}. \tag{7a}$$

Non- dimensional inlet velocity on a diameter D_0 of a 3D impeller:

$$\bar{w}_0' \approx \sqrt{\bar{D}_{hb}^2 + 2\Phi_{des}^{2/3}}. \tag{8}$$

Non-dimension inlet velocity in 2D impeller:

$$\bar{w}_1' \approx \sqrt{\bar{D}_{hb}^2 + 2\Phi_{des}^{2/3}} \tag{8a}$$

Equal values of velocities $\bar{w}_0'$ and $\bar{w}_1'$ in equations (8) and (8a) is a coincidence at the accepted typical values K_D=1,0, K_F=0,85: $\frac{\rho_0^*}{\rho_0'}$=0,9 and $\tau_0 = \tau_1$=0,85. But at the same time it is a demonstration of the fact that velocities $\bar{w}_0'$ and $\bar{w}_1'$ are close in principle. The next considerations must be taken into account:

- a leading edge of 3D impeller blades is located between the hub and shroud discs. Dimensionless relative velocity is maximal at a shroud and minimal at a hub due to change of a leading edge velocity. An average value of $\bar{w}'_0$ is less than practically constant value of $\bar{w}'_1$. 3D impellers have an advantage;

- hub ratio of 2D impeller can be smaller. Then $\bar{w}'_1 < \bar{w}'_0$ in accordance with equations (8) and (8a). 2D impellers have advantage.

3. Loss coefficients of 3D and 2D impellers. General assessment of application

The obvious opportunity to minimize profile losses due to 3D profiling induces some producers to apply this technology in all range of design flow rate coefficients. The 3D impeller family presented in paper [2].

Calculation analysis shows that at Φ_{des} < 0,05-0,06 disc friction losses prevail on profile losses [3]. Loss coefficient of 3D impellers $\zeta_{3D} = 0,5\frac{\bar{w}_0^2}{\psi_T}$ are more than 0.19 while loss coefficient of 2D impellers are usually less than 0.14. Here are meant impellers with Φ_{des} = 0,05-0,07, designed by the technique used by authors [1, 3]).

4. Examples of high-efficiency compressors and stages with 2D impellers

In [4-6] there is an information on factory tests or model tests of compressors designed by the team of Prof. Y. Galerkin, including other authors of the paper. In total more than 400 of designed compressors are constructed with a total power that exceeds 5 million kW. The performances of the four-stage pipeline compressor and results of their modeling by the 5th version of Universal modeling method are shown in Fig. 4. The maximum efficiency of the compressor is 0.875. The maximum efficiency of its 1st stage is 0.888. The stage design flow rate coefficient is Φ_{des} = 0,0633, the design loading factor $\psi_{T\,des}$ = 0.42.

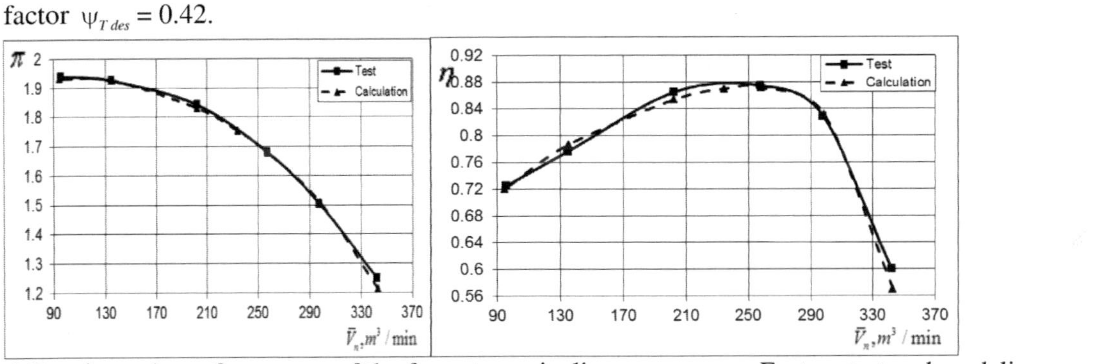

Figure 4 Gas dynamic performances of the four-stage pipeline compressor. Factory test and modeling calculation [4]

For 32 MW pipeline compressors the industrial partner offered high-speed drive with 5500 RPM and effective single-stage scheme. An axial inlet application, vaneless diffuser with optimal dimensions and design optimization procedure has led to high-efficiency solution that was proven by model tests, in scale 1:2. The model cross section, design and test performances are presented in Fig. 5.

Efficiency and polytrophic head coefficient performances are predicted with good accuracy by the Universal Modeling Method [7, 8]. The maximum efficiency equals 90%, is confirmed experimentally. 2D blade height on an inlet is equal to 0,11. The non-incidence flow at inlet along a leading edge is achieved anyway.

The examples support the authors' opinion that 3D impeller application hardly could rise pipeline and alike other industrial compressors efficiency if design flow rate coefficient of stages does not exceed 0,070 – 0,075.

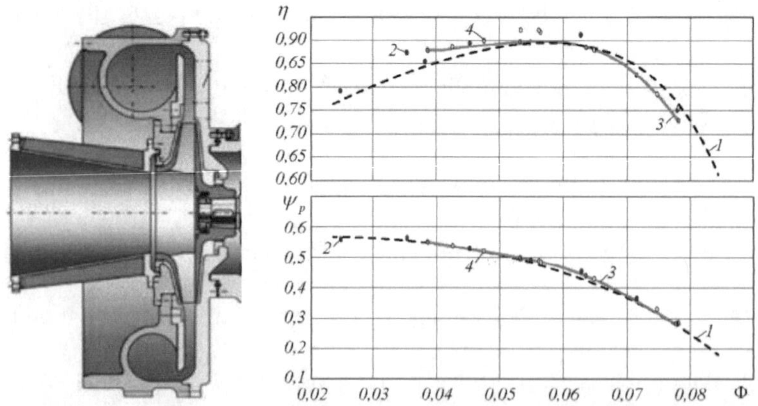

Figure 5. The model of the 32 MW pipeline compressor, in scale 1:2 (left). Test results and calculations (right) 1- design performance, 2- M_u = 0.705, 3- M_u = 0.710, 4- M_u = 0.700 [9]

5. Numerical analysis of candidate stages with 3D and 2D impellers

The stage with design parameters Φ_{des} = 0,068, $\psi_{T\,des}$ = 0,50 was chosen for efficiency comparison. Stator elements are a vaneless diffuser and a return channel.

The initial 3D candidate impeller was designed in accordance with recommendations proven for high flow rate 3D impellers [3].

Non-viscid Q3D approach is used for velocity calculation. The principles of Q3D non-viscid calculation and design procedure of an impeller on the basis of this approach are described in [1]. Some of design recommendations based on general principles and on high flow rate impeller numerical study [3]:

- inlet diameter D_0 is about 0.91 of a diameter calculated by the formula (5),
- a leading edge is located at a distance 0.25 of a meridian length from an inlet section "0",
- blade 3D configuration is based on velocity diagram analysis and optimization,
- an exit blade angle β_{bl2} is increased from hub to shroud,
- a blade surface generatrix at an impeller exit χ_2 is inclined to a direction of rotation to diminish blade surface area.

These principles were applied to the candidate #1 design. Information on this and other candidates is presented in the Table 1.

Symbol $\beta_{bl2}(b_2) = \text{var}$ means that its value at a shroud is bigger than at a hub. If a blade surface generatrix at an impeller exit χ_2 is inclined to a direction of rotation then $\chi_2 < 0$.

The NUMECA Fine Turbo program was applied for calculations. The grid consists of 1,8 million cells. The Spallart-Almaras model of turbulence was used. Boundary conditions:

- on an inlet: total pressure and total temperature;
- at the outlet: mass flow rate.

Friction of external surfaces of impeller discs and labyrinth seal leakage weren't modelled.

Table 1. Candidate impellers' main parameters 2D and 3D

Candidates	Type	$\bar{D}_{hub}$	$\bar{L}_m$	$\bar{c}_m(b)$	$\beta_{bl2}(b_2)$	χ_2	η_{max} %
1	3D	0,363	0.22	no unif.	var	-10	84,5
2	3D	0,363	0.28	no unif.	var	-10	84,5
3	3D	0,363	0.25	unif.	const	30	86.1
4	2D	0.363	-	unif.	const	0	86.2
5	2D	0,325	-	unif.	const	0	88.5
2D analog	2D	0.24	-	unif.	const	0	88.5 calc/87,9 test

Calculations overestimate a design loading factor by 6–12%. Calculated performances are shifted to bigger flow rate. But measured and calculated maximum efficiency coincides. The candidate stages are compared on the maximum efficiency therefore. Their efficiency performances are presented in Fig. 6. Candidates' numbers are the same as in table 1.

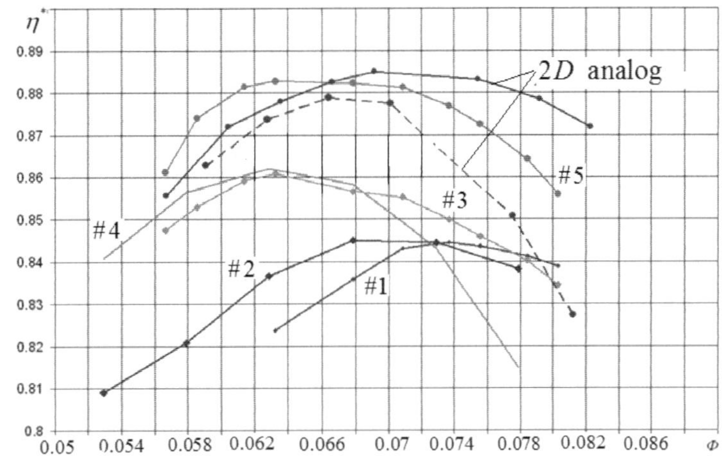

Figure 6. Efficiency performance curves of candidate stages and of 2D analog (stroke line – measurement)

There is information on the stage "2D analog" in the table 1. It is one of stages with 2D impeller from the database of the Laboratory of gas dynamics of turbomachines. Measured and calculated efficiency performances are presented in Fig. 6. The calculated performance is overestimate efficiency at high flow rates as in other cases. Parasitic losses were not taken into account at CFD-calculation. Meaning it, the CFD-calculation predicts maximum efficiency of a stage well enough.

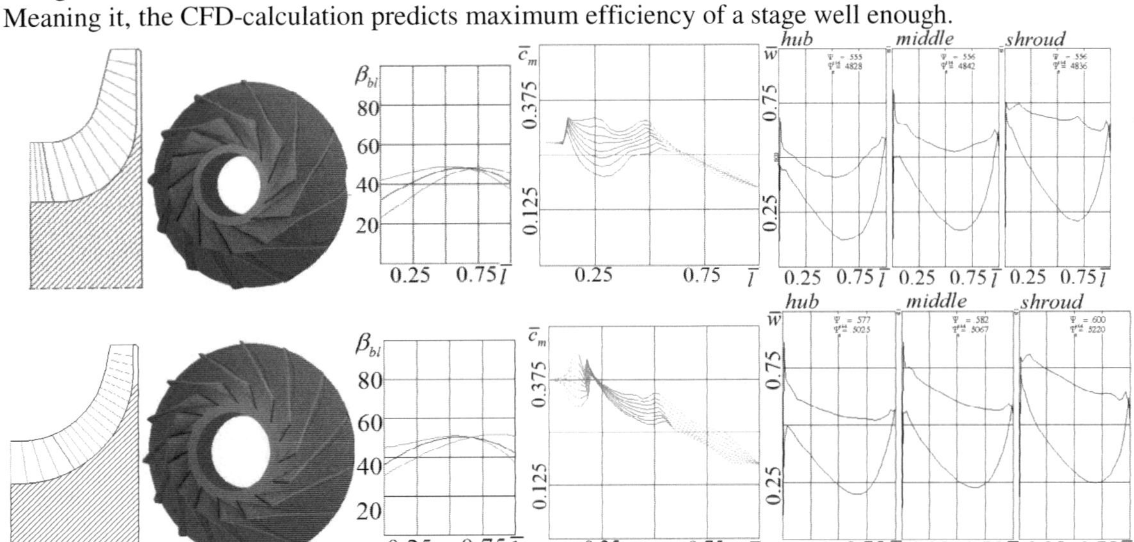

Figure 7. 3D Impeller candidate #1(above) and #2 (below). From left to right: meridian configuration, blade cascade view, blade angles on three blade to blade surfaces along meridian coordinate, meridian velocities on eight blade-to-blade surfaces, blade velocity diagrams on hub, mean, shroud blade to blade surfaces

The candidate #1 was designed by the principles that are mentioned above. The information on this candidate is presented in fig. 7.

The design flow rate coefficient of the candidate #1 and others is by 35% less than that of the analog in Fig. 6. Hub relative diameter is by 32% bigger. The meridional contour is made with a reduced axial dimension.

The maximum efficiency of the candidate #1 is 84.5%. Angles of absolute velocity at an impeller exit along a trailing edge $\alpha_2 = f(b/b_2)$ are shown in Fig. 8. One of the reasons of the stage low efficiency and poor performance left from a design point is negative flow angle near a shroud already at the design flow rate. It provokes flow separation in a vaneless diffuser.

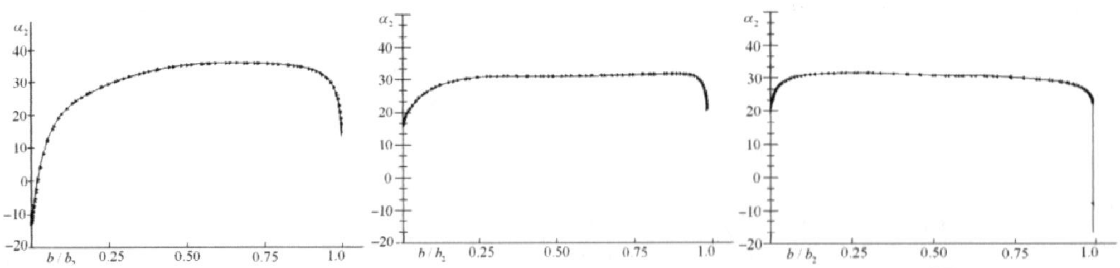

#1 #3 #5

Figure 8. Absolute velocity angles along trailing for candidates ## 1, 3, 5. Design flow rate. Shroud is at the left

The candidate #2 was designed by the same principles, as #1, but its meridional dimension is bigger and is closer to the analog with $\Phi_{des} = 0.105$. Fig. 7 provides information on the candidate #2. The exit flow angle diagram $\alpha_2 = f(b/b_2)$ in Fig. 8 is much better. The maximum efficiency is on the same level but the stage operates much better than the candidate #1 (Fig. 6).

The candidate #3 has a 2D impeller – Fig. 9. The maximum efficiency of the stage is by 1.5% higher than that of the candidates ##1 and 2 with 3D impellers. The diagram $\alpha_2 = f(b/b_2)$ in Fig. 10 demonstrates the best uniformity.

Velocity diagram comparison has pointed on different characters of meridian velocity diagrams in 3D and 2D impellers. In 2D impellers meridian flow non uniformity is high near a leading edge. The meridian velocities become uniform in the second part of a blade channel. Meridian velocities are sufficiently non uniform in all 3D blade channels. Diagrams $\bar{c}_m = f(l_{mi}/l_m)$ in Fig. 7 demonstrates it. Symbol "non uniform" points on the fact in the table 1.

The candidate #3 was designed meaning the goal to reach maximum possible for 3D impeller uniformity (symbol "unif." in table 1) of $\bar{c}_m = f(l_{mi}/l_m)$ diagram. The candidate has a positive angle $\chi_2 = +30^0$ and constant blade exit angle along a trailing edge. Information on the candidate #3 is presented in Fig. 9.

Meridian velocity diagram is not alike diagrams of 2D impellers. But it is the most uniform for 3D impeller of given parameters. The positive result is obvious. The maximum efficiency of the candidate #3 is the highest among 3D and is equal to efficiency of the candidate #4 with 2D impeller.

The candidate #5 realizes the constructive advantage of 2D impellers. Its non dimensional hub ratio is 0.325 against 0.363 of the candidates with 3D impellers. The principle of design is the same as for the candidate #4. The efficiency of this candidate is 88,2%. Advantage of the candidate #5 on efficiency is 2.1%. It is a result of not only lower kinetic energy of a flow at an inlet. The minimum loss coefficient of the candidate impeller #5 is 0,132 against 0,173 of the candidate #4. The authors can not explain this result of CFD calculation, as design principles and all parameters of the impellers #4 and #5 are similar.

CFD – calculations have provided extensive information on structure of a flow in the candidates. Samples of velocity fields are shown in Fig. 10 for the candidates ##1, 3 and 5. Fields of velocities are presented on two blade to blade surfaces– near the shroud and near the hub (design flow rate). Low-energy zones of various intensity take place in all three candidates: #1 – the lowest efficiency, #3 – the best of 3D, #5 – the best of all. It is remarkable that in the 2D impeller a favorable character of flow is identical on surfaces near a shroud and a hub. Visual impression is that the 2D candidate has identical blade cascades in near the shroud and the hub. On the contrary, blade cascades of 3D candidates seem too dense near a hub.

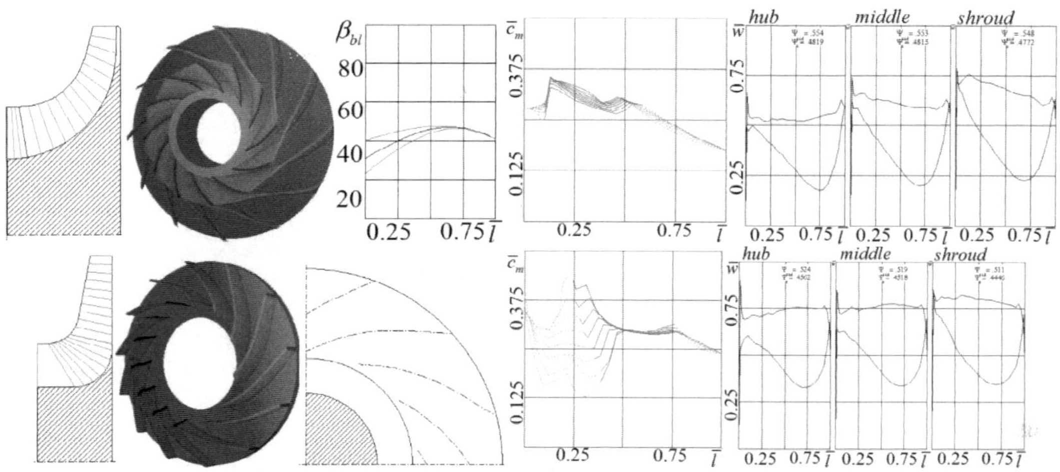

Figure 9. 2D impeller candidate #3 (above) and #4 (below). From left to right: meridian configuration, blade cascade view, blade angles on three blade to blade surfaces along meridian coordinate, meridian velocities, blade velocity diagrams on blade to blade surfaces

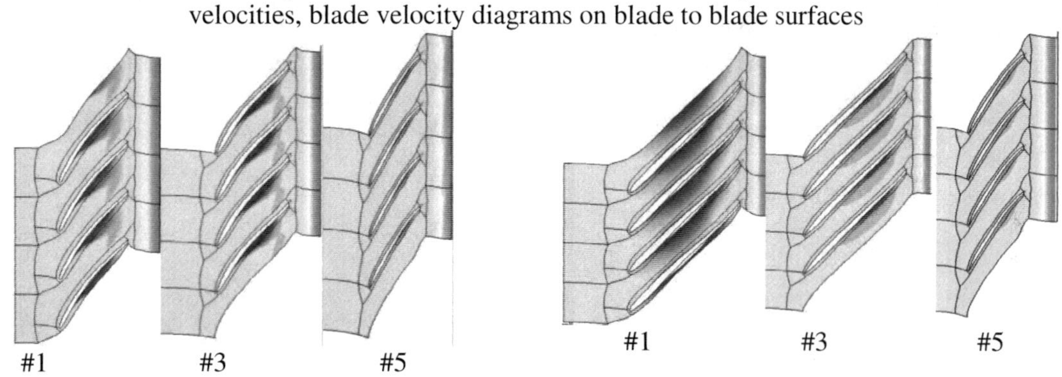

Figure 10. Flow fields on two blade to blade surfaces. Design flow rate. 10% from shroud – left, 10% from hub - right

Conclusion

The authors are sure that at low Mach number and medium and low loading factor 2D impellers are better if a design flow rate $\Phi_{des} < 0.07$-0.08. Compressor users insist on 3D impellers application in areas where they sometimes have no advantages. Properly designed compressor with 2D impellers can be cheaper and not less effective in many cases.

From the other side there are no doubts that the advantages of high flow rate stages with $\Phi_{des} \geq 0.07$-0.08 can be achieved only by application of 3D impellers. High flow rate stages effective design principles deserve proper attention.

Acknowledgment

The reported study was funded by RFBR according to the research project № 16-08-00624 A.

References

[1] Galerkin Y. Turbo compressors. [text] / Gale2Din, Y. // LTD information and publishing center. - Moscow. – KHT. - 2010. - P. 596. [Russian].

[2] Higauti, X. Development and use of gas injection compressor ultra-high pressure of 700 bar. / Tratsactions of the 19 international symposium «COMPRESSOR users-manufacturers – 2015». – SPb. – SPbTU. – 2015. – P. 12-21. [Russian].

[3] Galerkin Y., Drozdov A. Modeling of gas dynamic characteristics of centrifugal compressor stages with 3D impellers. Scientific and technical transactions of the TU SPb. - SPb. -2014. - № 3(202). - P. 45-53. [Russian].

[4] Galerkin Y., Soldatova K. Operational process modeling of industrial centrifugal compressors. Scientific bases, development stages, current state. Monograph. SPb – 2011. – P. 327. [Russian].

[5] Galerkin, Y.B., Soldatova, K.V. Working out of «virtual» modeling stages by means of programs of the 5th generation of the method of universal modeling. Scientific and technical transactions of the TU SPb. – 2011. – No. 4. – Page 241-248. [Russian].

[6] Soldatova K. Universal modeling method loss and wo2D input models verification on results of industrial compressors test. Scientific and technical transactions of the TU SPb. - 2010. – No. 4. – P. 300-308. [Russian].

[7] Galerkin, Y. B., Soldatova, K.V. Universal modeling method application for development of centrifugal compressor model stages. International Conference on Compressors and their systems. – London: City University. – UK. – 2013. – P. 477-487.

[8] Galerkin Y., Soldatova K., Drozdov A. Modern state of the universal modeling for centrifugal compressors. International Conference on Numerical Methods in Industrial Processes. World Academy of science, engineering and technology. Paris 2015 Conference. http://www.waset.org/Publications/?path=Publications. - Vol:9 No:01 2015. – № 242.

[9] A.V. Smirnov, V.G. Panenko, V.G. Gadiaka, V.P. Parafeynik, A.M. Borodenko (Sumy Frunze NPO PJSC). High Efficiency Centrifugal Compressor New Design for the GPA-C32/76-1.35 Compressor Package Applied at Gas Main Compressor Stations. // Compressors and pneumatics. - 2015. – № 3. – P. 12-16. [Russian].

IOP Conf. Series: Materials Science and Engineering **232** (2017) 012041 doi:10.1088/1757-899X/232/1/012041

Improving of the working process of axial compressors of gas turbine engines by using an optimization method

E Marchukov[1], I Egorov[1], G Popov[2], O Baturin[2], E Goriachkin[2], Y Novikova[2] and D Kolmakova[2]

[1] Lyulka Design Bureau, Moscow, Russia
[2] Samara National Research University, Samara, Russia

E-Mail: okb@okb.umpo.ru, egorov300657@yandex.ru, grishatty@gmail.com, evgeni0063@yandex.ru, y.d.novikova@gmail.com, kolmakova.daria@gmail.com

Abstract. The article presents one optimization method for improving of the working process of an axial compressor of gas turbine engine. Developed method allows to perform search for the best geometry of compressor blades automatically by using optimization software IOSO and CFD software NUMECA Fine/Turbo. Optimization was performed by changing the form of the middle line in the three sections of each blade and shifts of three sections of the guide vanes in the circumferential and axial directions. The calculation of the compressor parameters was performed for work and stall point of its performance map on each optimization step. Study was carried out for seven-stage high-pressure compressor and three-stage low-pressure compressors. As a result of optimization, improvement of efficiency was achieved for all investigated compressors.

1. Introduction

Compressor design process involves several stages from preliminary 1D calculation to the final 3D design with stacking blade sections along height. At the step of 3D design, software of numerical simulation of the workflow (CFD software systems) is usually used. They allow to predict reliably the characteristics of the compressor taking into account even the most minor features of the three-dimensional geometry of the flow part and to assess the impact of variation of any geometric or operational parameter on them.

Application of CFD software for the calculation of compressor operation does not cause any trouble when the designer has the geometry of the blades and the flow part. In case if the designer must find the geometry of the blades and flow part for specific operating parameters of the compressor, it is difficult to use CFD software. The main reason is that the flow part shape of axial compressor stages is described by large number of variables. At least two sections, each of which is determined by dozens of independent variables, must be determined to describe the shape of one blade. Furthermore, it is necessary to describe the relative position of the sections relative to each other, shape of meridional section of the flow part, etc. As a result, the number of independent variables for a complete description of one stage reaches several tens which affect contradictory compressor characteristics. Man is not capable physiologically to analyze the problem of such dimension. In addition, the search time for optimal combination of variables will be prohibitively high. The problem becomes proportionally larger if it is necessary to investigate the operation of multistage compressor.

In this case, the optimization program can help the engineer. Their use allows the automatization of the search for the optimal combination of independent variables by their automatic variation and analysis of the results based on optimization algorithms.
This article shows the technique for improving of the working process of axial compressor using optimization software IOSO and CFD software NUMECA Fine/Turbo and shows examples of the application of this technique on real seven-stage high-pressure compressor and three-stage low-pressure compressors of gas turbine engine (figure 1).

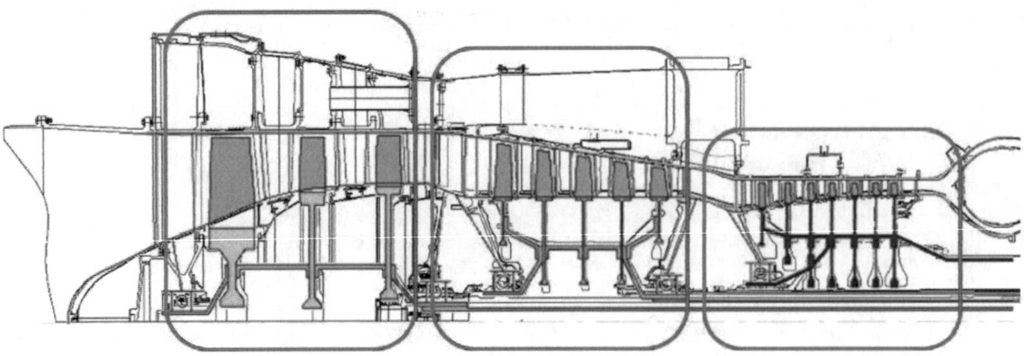

Figure 1. Investigated three-spool compressor

The solution of this problem was carried out in the study of upgrade options of gas turbine engine with power of 25 MW held in Samara University [5]. Compressor of the engine (figure 1) consists of three spools: a low-pressure compressor (LPC), an intermediate pressure compressor (IPC) and a high pressure compressor (HPC). Optimization of LPC and HPC are presented in the article. The aim of optimization was to investigate which maximum efficiency is possible to obtain in each compressor [3]. The criteria and constraints for the optimization problems of each compressor have been selected on the basis of thermodynamic calculations of the engine [1].

2. Developed method of optimization of axial compressors

Indirect Optimization on the base of Self-Organization (IOSO) [9, 10] is used in this research.
IOSO Technology is based on the response surface methodology approach. At each IOSO iteration the internally constructed response surface model for the objective is being optimized within the current search region. This step is followed by a direct call to the actual mathematical model of the system for the candidate optimal point obtained from optimizing internal response surface model. During IOSO operation, the information about the system behavior is stored for the points in the neighborhood of the extremum, so that the response surface model becomes more accurate for this search area. The following steps are internally taken while proceeding from one IOSO iteration to another:
- the modification of the experiment plan;
- the adaptive adjustment of the current search area;
- the function type choice (global or middle-range) for - the response surface model;
- the adjustment of the response surface model;
- the modification of both parameters and structure of the optimization algorithms;
- if necessary, the selection of the new promising points within the search area.

Specified actions are executed automatically by the program IOSO, and data exchange with a mathematical model of the process under investigation (in this case the numerical model of the working process in the compressor) is performed by text files.
The developed algorithm of the compressor optimization using IOSO is shown in figure 2.

At each step of the optimization, the optimizer IOSO creates a design of experiment and generates a Vector of variable parameters x_1, x_2, x_3, …, x_n. This vector is a set of variables that describe the geometry of the compressor blade in a parametric form. The Vector of variable parameters is transferred to the block of reprofiling. In this block, specialized «in-house» programs perform a conversion of compressor blades based on the Vector of variable parameters and save them in the form of geometry files (GF) of blades in a format **.geomturbo* [4] which is suitable for importing blade geometry in software systems of numerical simulation *Numeca FineTurbo* [6]. Used parametric models of the blades and reprofiling method will be described hereinafter.
At the next step, construction of computational mesh in the software *Numeca AutoGrid 5* is initially performed in the block of gas-dynamic parameters' calculation (figure 3) using the files with geometry of reprofiled compressor blades GF_1, GF_2 ... GF_n. Then, creation of computational model is carried out in the software *Numeca FineTurbo* and its calculation is performed. Processing of CFD-calculation results is performed in the software package *Numeca CFView*. As a result, several output files are created containing operation parameters of compressor (constraint vector and optimization criteria) in the text format. These parameters are passed to the optimizer *IOSO*. At the final stage, the vector of output parameters (optimization criteria and constraints) y_1, y_2 ... y_n, formed by the results of the running of the blocks of gas-dynamic and strength parameters' calculation is transferred to the optimizer *IOSO*. Analysis of obtained data is performed in the optimizer as it was indicated above in the description of one IOSO iteration. Then, a new Vector of variable parameters is generated. And the cycle (iteration) of optimization is repeated.

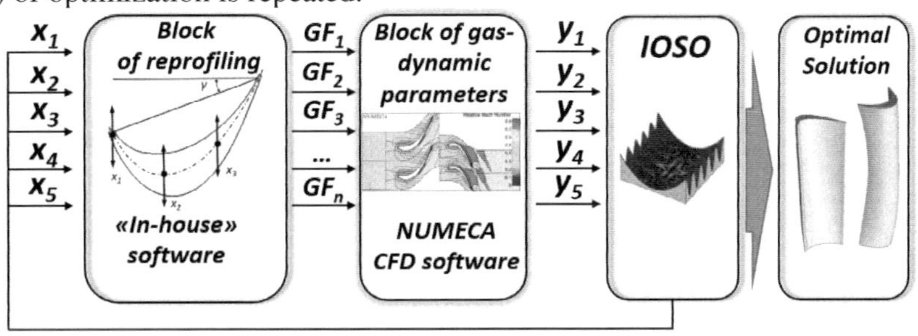

Figure 2. The algorithm for solving optimization problem of compressors.

3. Numerical model of compressors used for optimization

The numerical models of the compressors created by means of the mesh generator *NUMECA AutoGrid 5* were used to calculate operational parameters of HPC and LPC.
Geometrical model of computational domain of the flow in the investigated compressors was created based on design documentation of engine. Figure 3 shows created geometrical models of LPC and HPC.

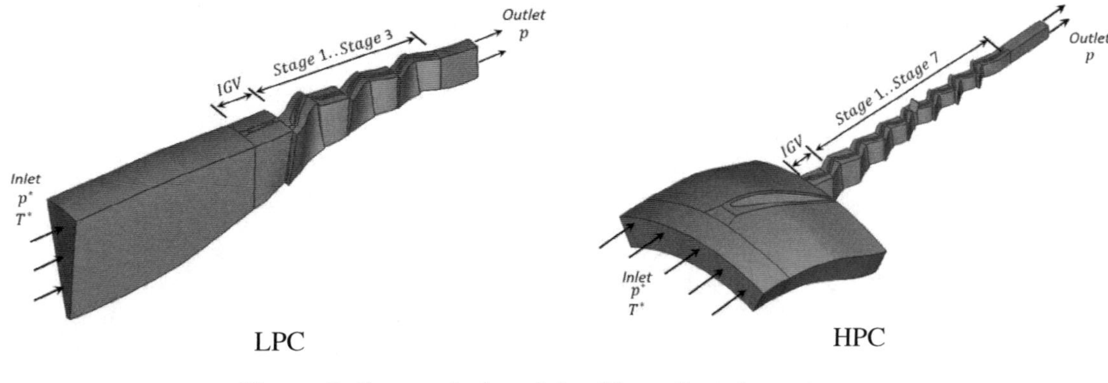

Figure 3. Geometrical models of investigated compressors

The created model was divided into finite volumes to block-structured grid by internal tools of the program *NUMECA*. Two computational models ("light" model and "heavy" model) were created for each compressor [11]. Light model contained 300 thousand elements per blade row on an average; the value of y + was 3-5. Heavy model contained 500 thousand elements per blade row on an average; the value of y + was 1-2.

For data transmission between the area of the GV and BW the built-in software package *NUMECA* interface *Full Non Matching Mixing Plane* is used. It averages the flow parameters in the circumferential direction in the upstream area and transmits as the boundary condition in the area located downstream.

As the boundary conditions at the inlet of the HPC the value of the total pressure was set equal to $p^*=101,325kPa$ and total temperature was equal to $T^* = 288,15K$. Parameters of turbulence at the inlet boundary are $k=5m^2/c_2$, $\varepsilon=30000\ m^2/c_3$.

In the calculation of characteristics the stall borderline was defined as the point with the minimum flow rate of the working fluid, in which it managed to get the coincided decision.

To prove the validity of the numerical models LPC (figure 5) and HPC (figure 6) characteristics were calculated at several operational modes. Then they were compared with the available experimental data for the compressors.

The calculated parameters of the compressor are presented in relative form:

$$\bar{n}_{cor} = n_{cor}/n_{corBASE} \cdot 100\%,$$

where $n_{corBASE}$ – the rotor speed at the primary operational mode of the base LPC;
n_{cor}- corrected rotor speed,

$$n_{cor} = n\sqrt{288,15/T_H^*},$$

where n - specific rotor speed;
n – physical rotor speed;
T_H^* - air temperature at engine inlet.
Normalized Efficiency – efficiency referred to efficiency at the primary operational mode of the base LPC.
Normalized Pressure Ratio – pressure ratio referred to efficiency at the primary operational mode of the base LPC.
Normalized Mass Flow – air flow rate referred to efficiency at the primary operational mode of the base LPC.

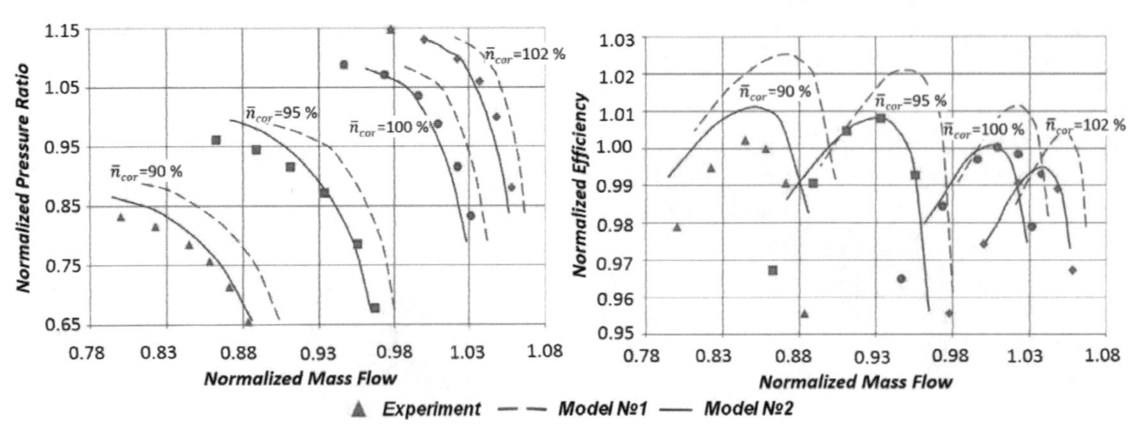

Figure 5. Comparison of calculated and experimental parameters of HPC.

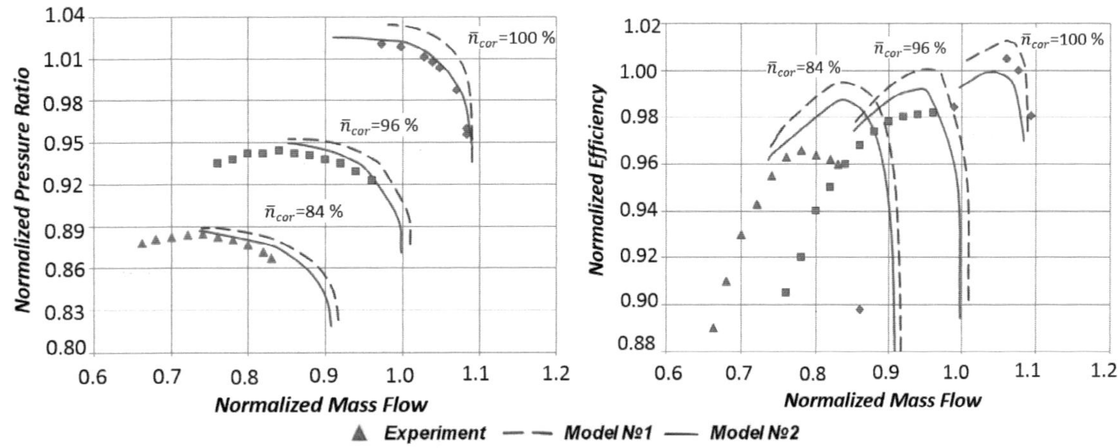

Figure 6. Comparison of calculated and experimental parameters of LPC.

As can be seen from figure 5, both created numerical models show good qualitative coincidence with the experimental results. However, the model №2 shows significantly better quantitative coincidence with experimental results. The difference of values for efficiency and the compression ratio is not more than 1%.

Apparently from figure 6, calculated pressure characteristics of the LPC qualitatively repeat the experimental ones. The maximum deviation of calculated values of the pressure ratio from the experimental value is 1% (abs) at the mode $\bar{n}_{cor} = 84$ %. Also, figure 6 shows that calculated curves of efficiency characteristic of the base LPC qualitatively repeat the experimental ones at all modes. The maximum difference between calculated and measured efficiency is 2.9% (abs.) at the mode $\bar{n}_{cor} = 84$ %. At the mode $\bar{n}_{cor} = 100$ %, the maximum deviation is 0.7% (abs.).

Based on the above, it was concluded that, despite some quantitative discrepancies with the available experimental data, created numerical model can adequately describe the operation of the LPC and can be used for search of configuration with maximum efficiency. For this reason, the model №2 was used for further studies.

4. Optimization of the HPC of engine

The optimization problem of the HPC was testing and was solved in order to prove the possibility of increasing the efficiency of the multistage compressor with the help of the developed optimization method. Objective functions were assigned to the values of efficiency at the modes corresponding to the rotational speed of 95% and 100% from the primary mode; and all stagger angles of the HPC blade rows were selected as varying variables.

The criteria for this optimization problem:

- increase in HPC efficiency at the mode corresponding to the rotational speed of 95%;
- increase in HPC efficiency at the mode corresponding to the rotational speed of 100%.

In order to prevent the shift of characteristics of the compressor, the following restrictions were set in the optimization:

flow rate of the working fluid through the HPC at relative frequency of rotation of 95% was not supposed to be different from the respective flow rate of the base compressor more than ±1,3%;

flow rate of the working fluid through the HPC at relative frequency of rotation of 100% was not supposed to be different from the respective flow rate of the base compressor more than ±0,6%;

value change of HPC pressure ratio compared with the basic compressor at points of the maximum efficiency at relative rotational speeds of 95% and 100% allowed within ±1,5%.

In formulating the optimization problem and the appointment of restrictions the changes in stall margin of HPC did not take into account, in order to reduce the time of receipt of the decision.

Evaluation of changes in stall margin of HPC is conducted at the analysis phase of optimization results.
Schematically, the optimization criteria and restrictions used in the formulation of the optimization problem, shown in figure 7.

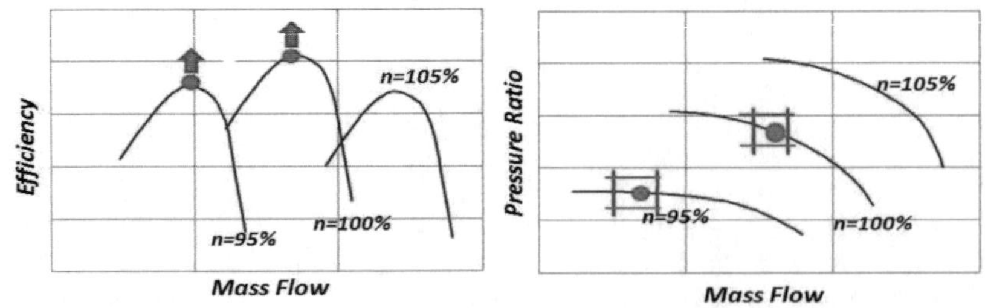

Figure 7. Formulation of the optimization problem by varying the stagger angles of compressor blades.

As varied variables the stagger angles of all rotor blades, guide vanes and IGV of the HPC were selected (figure 8). The range of change of stagger angles of the vanes of each blade row has been selected so that during the blades rotation their profiles fits into existing blade locks. The number of blades in the row has not changed. This solution allows to find the variant to increase the efficiency of the HPC, which would not require modification of the disk and the body parts of the compressor. The total number of changed variables was 15.

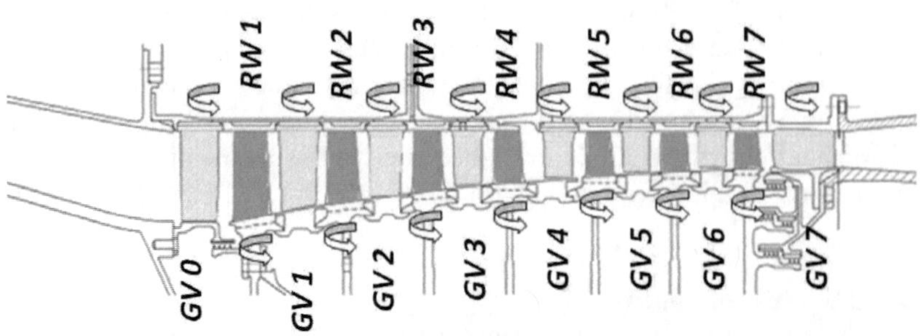

Figure 8. Blade rows changed during optimization

Optimization was performed according to the algorithm given in section 2. The whole optimization process was completely automated. To solve the formulated problem of optimization the software package *IOSO* had 446 calls to the numerical model of the HPC. Each call to the numerical model is calculation of two points on the characteristic of the HPC (points of maximum efficiency on the branches corresponding to the relative frequency of rotation of 95% and 100%) in the software package *NUMECA FineTurbo*.

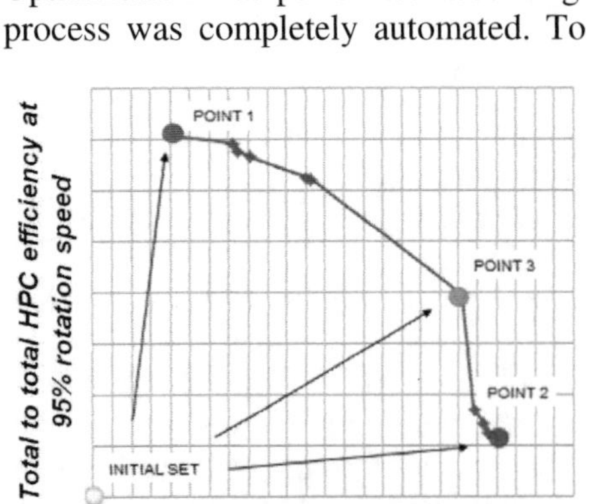

Figure 9. Pareto set of HPC optimization.

As a result, a lot of unimprovable solutions (Pareto set) were obtained, which is a compromise between increase of efficiency at the relative rotational speed of 95% and increase of efficiency on the relative rotational speed of 100% (figure 9). Each point of Pareto set corresponds to the unique geometry of the HPC represented as an array of stagger angles of all blade rows of HPC.
Analysis of the extreme points of Pareto set shown that at relative rotation frequency of 95% the highest increase of

maximum efficiency is 1.8% (abs.) at a substantially constant maximum efficiency at relative rotational speed of 100% (point 1 of Pareto set in figure 9). When the relative rotational speed of 100% the highest increase of maximum efficiency is 0.6% (abs.) at the increasing of maximum efficiency at relative rotational speed of 80% to 1% (point 2 Pareto set in figure 9). However, for further research one of midpoints of the set Pareto has been selected (point 3 in figure 9), providing increasing of the efficiency as at relative rotation frequency of 100% (0.5% (abs.)) and at the relative frequency rotation of 95 % (1.2% (abs.)).
To analyze the results of the optimization the numerical model of HPC variant was created, which corresponding to the selected point 3 of the Pareto set. By this numerical model the characteristics of optimized version of HPC at the relative rotational speeds of 95% and 100% were obtained as well as their comparison with the characteristics of HPC base version (figure 10) and with the searching results of the optimal combination of stagger angles at the first three stages, described above, performed.
The result of comparison revealed the following characteristics:
- stall margin of operation of optimized HPC compared to the base case at the investigated frequencies of rotation have changed slightly;
- changing the values of the air flow rate and compressor pressure ratio of the optimized HPC at points of maximum efficiency at the investigated frequencies of rotation is within the accepted limits;
- HPC efficiency at the relative rotation frequency of 95% has increased by 1.2% (abs.) and at relative rotation frequency of 100% increase of efficiency was 0.5% (abs.).

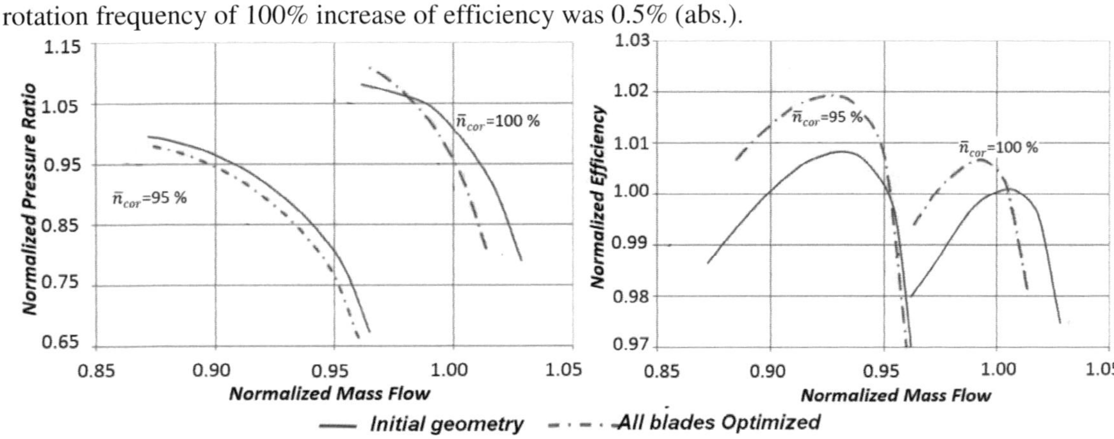

Figure 10. Comparison of the characteristics of optimized versions of the compressor and basic HPC

5. Optimization of the LPC of engine

The aim of LPC optimization was to reprofile the compressor for new operating conditions. Required parameters of the compressor operation have been obtained as a result of thermodynamic analysis of the engine. On conditions of the problem, it was necessary to ensure the growth of the LPC efficiency by 1.5% (absolute value) with 4% increase of pressure ratio, 2% of the rotor speed and decrease of the working fluid flow rate by 8% with respect to the parameters of the original engine. The goal is to be achieved with unchanged diametrical and axial dimensions of the compressor flow, while maintaining acceptable safety margins of all the rotor elements.
Optimization problem was solved in a two-criteria formulation. As optimization criteria were selected (figure 11):
- increase of LPC efficiency at rotor speed n = 102% (relative to rotor speed of the base compressor);
- decrease of the relative flow rate of air the through the compressor.

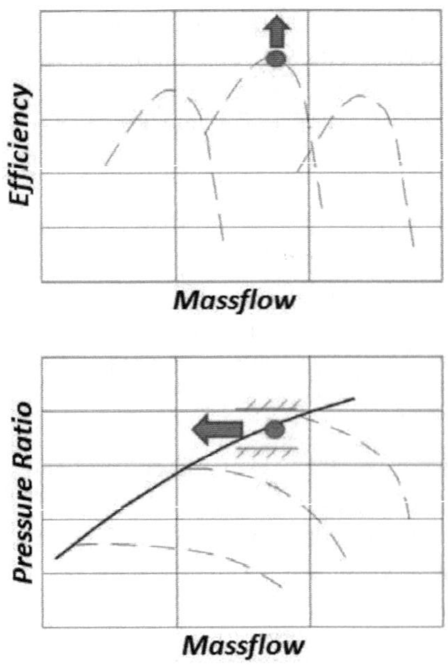

Figure 11. Schematic formulation of the optimization problem.

The constraints that determine the position of the working point on the LPC characteristic specified for the optimization problem:
- minimum value of the air flow rate in the design point is limited by the range: 0.91 ... 0.96·base;
- the total pressure ratio was maintained at the design point a predetermined range: 1.009 … 1.046·base;
- the flow angle changing at the LPC outlet was limited by the range: ±5 degrees.

Varying variables were selected as follow. Slightly modified approach from [7] was used to describe the shape of the camber line. Changing the camber line shape of the rotor blades and the relative position of the cross sections relative to each other was carried out by moving the spline middle control points in the circumferential direction in the global coordinate system, as well by varying the stagger angle (figure 12, a). This solution helped to keep the value of the blade chord in control sections, which is important in terms of maintaining the stress strain state. The guide vane profile was changed by moving the middle point of the spline in the circumferential direction and moving the point of the trailing edge along both coordinates (figure 12, b).

Reforming algorithm of the blade profile along the height with the approaches described above has been implemented in the «in-house» program, developed at the Department of Aircraft Engines Theory of SSAU [18, 19]. This program allows to convert a table of coordinates, which describes the blade shape in design drawing into text files of source data to build turbomachinery computational models in programs like Auto Grid and TurboGrid.

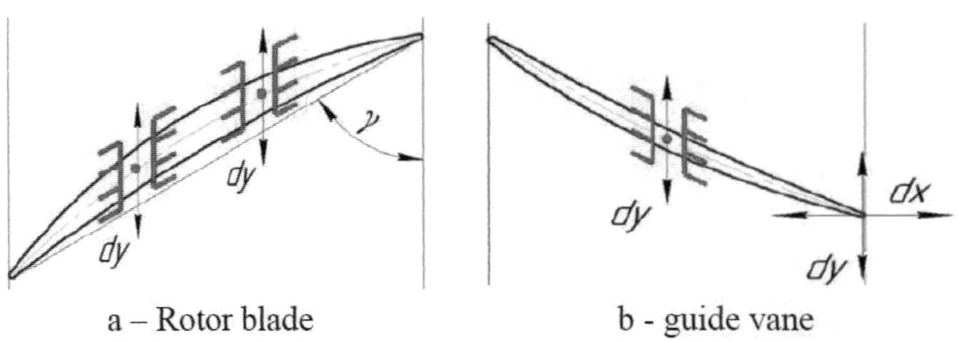

Figure 12. The changing scheme of the camber line shape of the blade profile

To solve the formulated problem of optimization the software package *IOSO* had 446 calls to the numerical model of the HPC. Each call to the numerical model is calculation of two points on the characteristic of the HPC (points of maximum efficiency on the branches corresponding to the relative frequency of rotation of 95% and 100%) in the software package *NUMECA FineTurbo*. As a result, the Pareto set of best possible criteria according to two criteria - the relative efficiency and the relative flow rate of the working fluid through the compressor were obtained (figure 13).
Three points were selected for the optimization results validation and analysis (figure 13):
The pressure characteristics were calculated for each point at the rotor speed of 102% of the base LPC speed and compared with the initial characteristic of the compressor (figure 14).

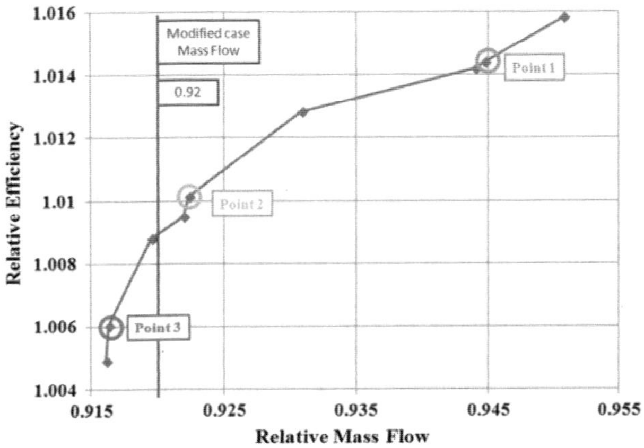

Figure 13. Pareto frontier of the problem.

Comparing the characteristics of the different variants, it can be seen that all the variants have approximately the same efficiency at the desired air flow rate through the compressor. Moreover, this value is greater than efficiency of the base compressor by 1.3% (absolute). Figure 17. Comparison of relative characteristics of the selected LPC variants with characteristic of original compressor Variant №2 was adopted as a final embodiment of the compressor, because it allows to obtain the desired pressure ratio (variant №3 has less value), has a higher efficiency than the variant №1 and stall margin for this variant does not differ from the original compressor stall margin.

Variant №2 was adopted as a final embodiment of the compressor, because it allows to obtain the desired pressure ratio (variant №3 has less value), has a higher efficiency than the variant №1 and stall margin for this variant does not differ from the original compressor stall margin.

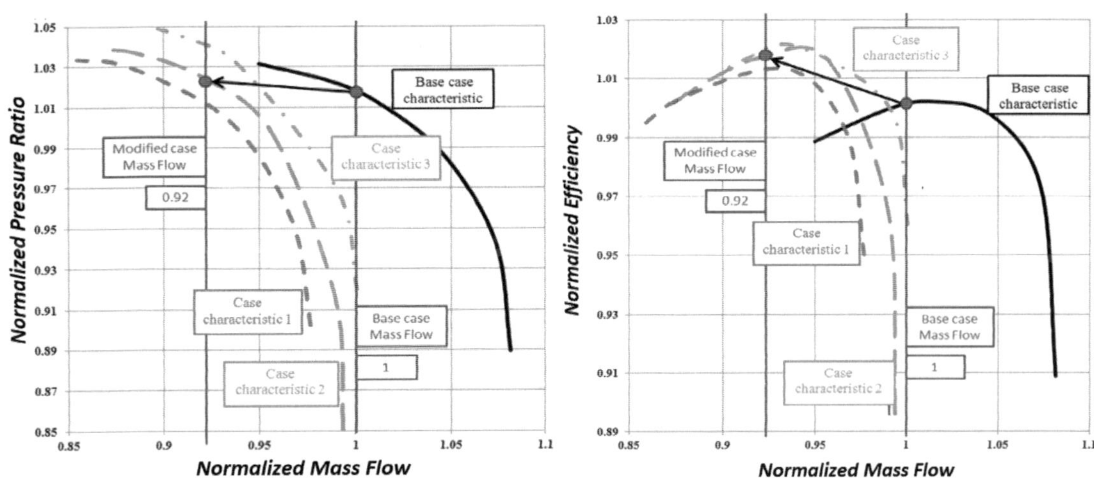

Figure 14. Comparison of characteristics of the selected LPC variants with characteristic of original compressor

5. Conclusion

Summing up the work done it can be concluded that the goal has been reached. The developed algorithm showed its effectiveness for the reprofiling multistage turbomachinery.

HPC variant has been found that provides an increase in the efficiency at the operation mode corresponding to 95% of the rotational speed by 1.2% (absolute) and at the mode corresponding to 100% of the rotational speed by 0.5% (absolute).

Also a variant of the shape of low pressure compressor blades has been found that provides an increase in efficiency by 1.3% (abs.), while increasing the pressure ratio by 4%, the rotational speed by 2% and decrease in the mass flow rate of working fluid flow by 8% relative to compressor of base engine.

IOP Conf. Series: Materials Science and Engineering **232** (2017) 012041 doi:10.1088/1757-899X/232/1/012041

In this case compressor parameters have been changed only by changing the shape of the blades, while maintaining unchanged the remaining design elements, including the blade number and shape of the grooves for blade attachment. During the search for improved shapes of blades, measures for the conservation of their stress strain state in the former borders were taken.
It would be difficult to obtain a similar result with traditional methods of designing compressor as the base variants of the compressor already had a high level of efficiency, and the requirements were contradictory.

References

[1] Kuzmichev V, Rybakov V, Tkacheno Y, Krupenich I 2014 Optimization of working process parameters of gas turbine engines line on the basis of unified engine core *ARPN Journal of Engineering and Applied Sciences* ed J Tarakzai Vol 9 pp 1873-1878

[2] Kulagin V 2002 *The theory, calculation and design of aircraft engines and power plants Fundamentals of the theory of the CCD. Workflow and thermodynamic analysis* p 616

[3] Gritsenki E, Danilchenko V, Lukachev S 2004 *Converting GTE in gas turbine plants for ground applications* p 266

[4] NUMECA 2008 *User Manual AutoGrid5 Release 8.4*

[5] *http://www.ssau.ru/english/*

[6] NUMECA 2008 *User Manual FINE/Turbo Release 8.4*

[7] Shablii L, Dmitrieva I 2014 Blade geometry transformation in optimization problems from the point cloud to the parametric form *Russian Aeronautics* Vol 57 ed v Firsov No 3 pp 276–282

[8] Matveev V, Popov G, Goryachkin E, Smirnova Y 2014 Effect of Accounting of Air Bleed from the Flow Passage of the Multi-Stage Axial Low Pressure Compressor on its Design Performances *Research Journal of Applied Sciences* Vol. 9 ed A Mtibaa (INSInet Publication) No11 pp 784-788

[9] *http://www.iosotech.com*

[10] Egorov I, Kretinin G, Leshchenko I, Kuptzov S 2002 Optimisation Toolkit - Novel Software to Create Better Design 9th *AIAA/ISSMO Symposium on Multidisciplinary Analysis and Optimisation*

[11] Komarov O, Sedunin V, Blinov V 2014 Application of Optimisation Techniques for New High-Turning Axial Compressor Profile Topology Design *ASME* J

IOP Conf. Series: Materials Science and Engineering **232** (2017) 012042 doi:10.1088/1757-899X/232/1/012042

Development of the virtual experimental bench on the basis of modernized research centrifugal compressor stage test unit with the 3D impeller.

A A Aksenov, A M Danilishin, A M Dubenko, Y V Kozhukov

Department of Compressor, Vacuum and Refrigeration Engineering, Peter the Great St. Petersburg Polytechnic University, 195251 St. Petersburg, Russia, Politechnicheskaya St., 29

E-mail: lexach@mail.ru, Danilishin_am@mail.ru, kozhukov_yv@mail.ru

Abstract. Design modernization of the centrifugal compressor stage test bench with three dimensional impeller blades was carried out for the possibility of holding a series of experimental studies of different 3D impeller models. The studies relates to the problem of joint work of the impeller and the stationary channels of the housing when carrying out works on modernization with the aim of improving the parameters of the volumetric capacity or pressure in the presence of design constraints. The object of study is the experimental single end centrifugal compressor stage with the 3D impeller. Compressor stage consists of the 3D impeller, vaneless diffuser (VLD), outlet collector - folded side scroll and downstream pipe. The drive is a DC motor 75 kW. The increase gear (multiplier) was set between the compressor and DC motor, gear ratio is i = 9.8. To obtain the characteristics of the compressor and the flow area the following values were measured: total pressure, static pressure, direction (angles) of the stream in different cross sections. Additional pneumometric probes on the front wall of the VLD of the test bench have been installed. Total pressure probes and foster holes for the measurement of total and static pressure by the new drainage scheme. This allowed carrying out full experimental studies for two elements of centrifugal compressor stage. After the experimental tests the comprehensive information about the performance of model stage were obtained. Was measured geometric parameters and the constructed virtual model of the experimental bench flow part with the help of Creo Parametric 3.0 and ANSYS v. 16.2. Conducted CFD calculations and verification with experimental data. Identifies the steps for further experimental and virtual works.

Nomenclature

b [-] dimensionless width
D [-] dimensionless diameter
h [J·kg^{-1}] head
$\dot{m}$ [kg·s^{-1}] mass flow
n [-] mode number
P [Pa] pressure
T [K] temperature
U [m·s^{-1}] peripheral velocity
y^+ [-] wall distance in normal direction
Z [-] number of blades
Π [-] total to total pressure ratio
β [-] loss coefficient
ρ [kg·m^{-3}] density
η [-] efficiency coefficient
ψ[-] head coefficient
Φ[-] conditional flow rate coefficient

Subscripts

0, 1, 2,... - specified section
p, d, i - polytropic, dynamic, internal
inl., out - inlet, outlet

Superscripts

*- total parameters

1. Introduction

One of the main industries is the power engineering industry, which includes energy machines such as centrifugal compressors. Pressure head and efficiency requirements increases on the modern centrifugal compressors [13]. Continues appears the need in the production of high-performance, high-pressure 3D impellers. The use of centrifugal compressors with 3D impellers is now very important to the linear compressor stations of the unified gas transportation system of JSC «Gazprom». It is known that over time the gas field is depleted , thus changing the pressure, flow rate and gas composition, in this connection, to reduce the capital costs are used replaceable flow parts, which do not require changes of housing parts of gas pumping unit. Also this is especially relevant for all enterprises nitric industry for ammonia production, as well as processes for gasoline reforming, diesel hydrotreater, gas topping, where multistage centrifugal compressors are used as is compressed machines. In some conditions specified by the customer, it is impossible to carry out a full replacement of replaceable flow parts with the stator part and limited to the replacement of impellers, cutting existing flow parts and the installation of additional elements, such as in [7.9]. Similarly these impellers are widely demand in units of the turbocharging of the combustion engines [11].

In the past decade, actively developing supercomputing cluster technology, which allowed the use the numerical simulation in the development of new machines and reduce the cost of subsequent experimental refinements, and in the ideal case, to abandon such at all. Have the development and of multi-parameter optimization methods, by which is automatically calculated dozens of choices of energy efficient flow parts of the centrifugal compressors [2]. It is also possible to use for the study of complex forms of inlet and outlet chambers and other elements, because the flow in them significantly effect on the entire flow in general [6]. However, today's methods of numerical calculation of the flow in a centrifugal compressor are not able to fully reflect the effects that repeatedly observed experimentally (for example, during the formation of "wake-flow" at the exit of the impeller). Often, to accelerate the procedure for calculating the elements of a compressor flowing part is carried out individually, that cannot account for their mutual influence on each other. To save computation time is also used averaging time formulation, therefore it is impossible to solve the time-dependent processes occurring in the flowing part. There is a need for a transient calculations on the performance ends, where is most pronounced flow separation and phenomena related to the loss of stability, as the local – prestall, rotating stall and global - surge. That requires higher computational resources, especially when solving problems with inlet and outlet chambers. However, there are developing methods, significantly accelerating the transient calculation, such as non-linear harmonic balance [1,4].

In studies [3,12] obtained the modeling of the "medium flow" centrifugal compressor stages performance, which of qualitatively and quantitatively the same as the full-scale experiment in the framework of permissible in engineering calculation error ± 2,5% in the zone of optimal performance mode. For other modes should further explore other turbulence models and methods of numerical solutions, allowing low energy losses associated with flow separation, as well as emerging non-stationary processes.

Summarizing the above, the aim of this work is to develop a virtual bench to perform comprehensive numerical calculations of the centrifugal compressor stage performance. The main development objectives of a virtual bench:

1. The ability to simulate the experiment.
2. The ability to simulate inconsistent modes of elements of a flowing part with a view to its correction.
3. The possibility of multicriterial and multiparameter optimization of the impeller.
4. The possibility of transient calculations.

The basis for the development of a virtual model of the bench is a full-scale training bench, upgraded to the possibility of carrying out research, which conducted an experiment for verification of numerical calculation.

IOP Conf. Series: Materials Science and Engineering **232** (2017) 012042 doi:10.1088/1757-899X/232/1/012042

2. The design procedure with constraints

If there are insuperable geometrical restrictions, it is offered to use a virtual bench concept for study of engineering decision variants. It allows conducting tests, researches of both model, and full-scale objects. There are similar experimental bench developments [8], but for the purpose of the transient phenomena studying in a flowing part. It is expected that with the help of modern methods of multi-criteria and multiparameter optimization to carry out optimum geometrical parameters selection. In cases of modernization, operating on ammonia compressors often seeks to establish a flowpath with a nominal consumption by 25-40% larger than the original, while not changing the compressor housing. It is necessary to not install new piping installation, sometimes you want the opposite effect with decreasing consumption and pressure in connection with the improvement of technological process of target units [5, 10].

In such instances, the inlet and outlet chambers executed in the housing are left invariable. By increase in compressor productivity in not upgraded inlet and outlet, devices there are additional losses. Occurs research problem of the uncoordinated work of the last stage and the existing outlet chamber. Collaboration modeling of last stage and outlet chamber during their uncoordinated work causes special difficulties in calculations. In this study, such option is investigated: the impeller has nominal productivity more than the output chamber for 40%.

The following actions should be made to this task:

1) Performing measurements of geometrical setting parameters (fig. 1 a, b).
2) Parametrical model development (fig. 1 with, d).
3) Numerical model solution of setting (fig. 1 e, f).
4) Carrying out verification of the initial machine. (fig. 1 g).
5) Carrying out optimization with the existing constrains.
6) Carrying out the confirming tests.

In the figure 1 the procedure of the similar analysis performed according to points 1-4 is submitted.

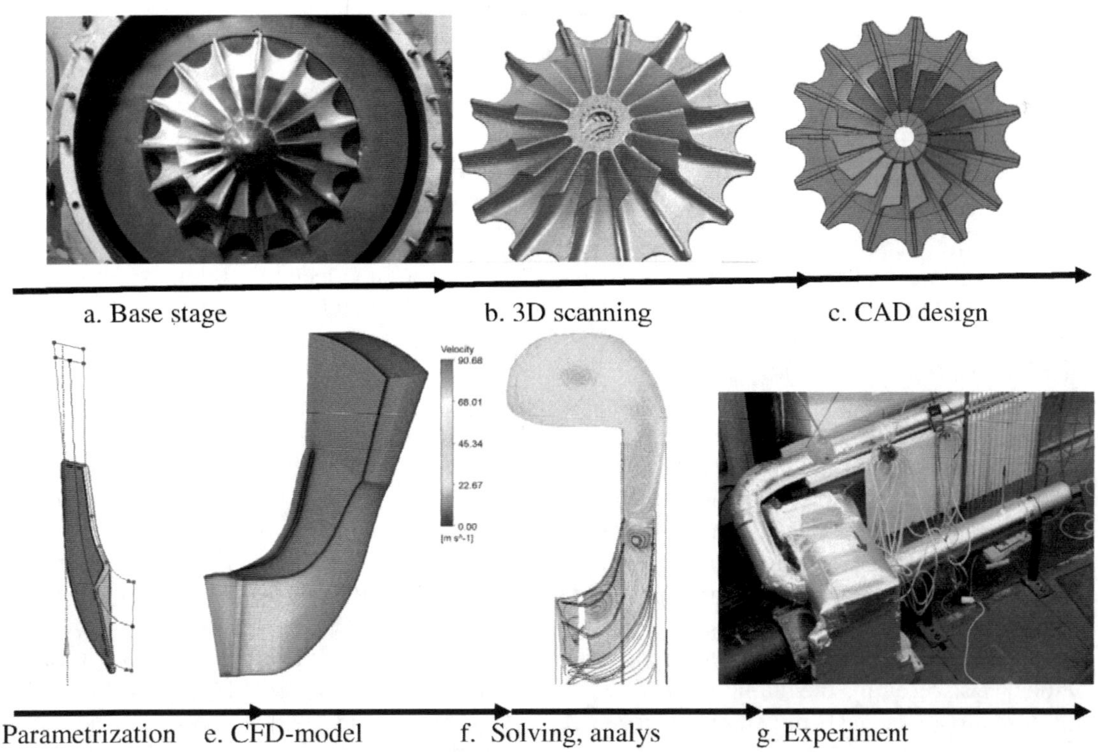

Fig. 1 The design procedure

IOP Conf. Series: Materials Science and Engineering **232** (2017) 012042 doi:10.1088/1757-899X/232/1/012042

According to the concept of the impeller was made using digitized laser 3d scanning. In the 3d scanning process the cloud of points which was transformed to the .stl format and then the solid-state CAD model was constructed. As the device of scanning was used the Faro scanner hand equipped to 7 joints and a mobile laser head of Metris Model Maker that allows to conduct 3d scanning of irregular shape products up to 2.5 meters in size with an accuracy from 0.04 mm. Measurement of flow parts and clearances was made by regular measuring tools. Then the parametrical model (CFD-model) about which it will be told below was constructed.

3. Subject of research

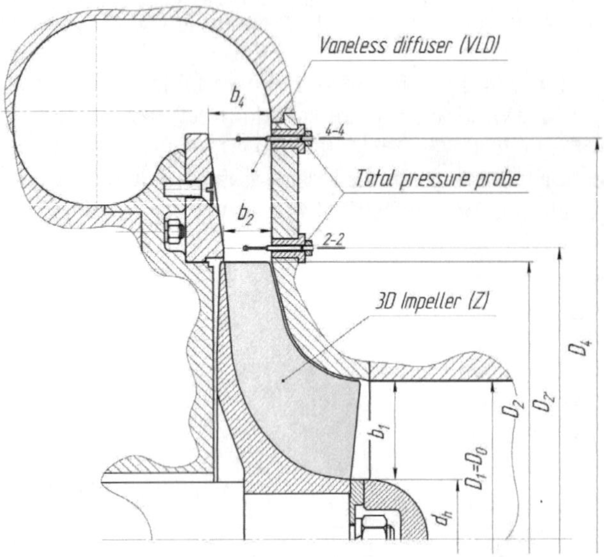

Fig. 2 Scheme of the object of study

Table 1. The basic dimensionless parameters of stage.

Parameter	Value
D_0	0.571
D_1	0.571
D_h	0.207
D_4	1.445
D_2'	1.056
b_1	0.182
$b_2 = b_3$	0.087
b_4	0.116
Z_2, pcs.	15

The subject of research is a high flow centrifugal compressor stage model (Fig.2), consisting of impeller with 3D blades and vaneless diffuser installed in the experimental bench. Integral parts of the bench are axial inlet and outlet chamber (the volute coiled to one side). The main dimensionless geometrical parameters of the flow part are provided in Table. 1.

4. Experimental bench

Training experimental bench is adapted to receive pressure data in 4 sections (Inlet, 2-2, 4-4, Outlet) and temperature parameters of the inlet and outlet of the compressor. Experimental Bench before the modernization had 1-point measurements of total and static pressure in sections 2-2 and 4-4, and had no insulation of the housing and pipes.

In the course of works on modernization, on the front wall of VLD were installed, additional receivers – total pressure probes and holes for measuring total and static pressure for a new drainage scheme Fig.3 b. This allows conducting full-scale experimental study of centrifugal compressor impeller and vaneless diffuser. Also was carried out the insulation of the compressor housing. Selection of the static pressure in the cross section 2-2 and 4-4 was carried out in six and five points, respectively. Total pressure measurements were made in four points along the width of the channel. The definition of the capacity was determined in the cross section 0-0 in accordance with the established measurement instruments in Fig. 3 a.

The drive unit for the bench is a synchronous DC motor with 75 kW. The electric motor through a multiplier lets you adjust the rotational speed of the bench in the range of 5800-11200 rev/min.

After commissioning test, measurement parameters were performed on five modes in the zone of optimal performance and appearance of the area of transient processes. The number of revolutions of the rotor in the tests was set at 8000 rev/min.

IOP Conf. Series: Materials Science and Engineering **232** (2017) 012042 doi:10.1088/1757-899X/232/1/012042

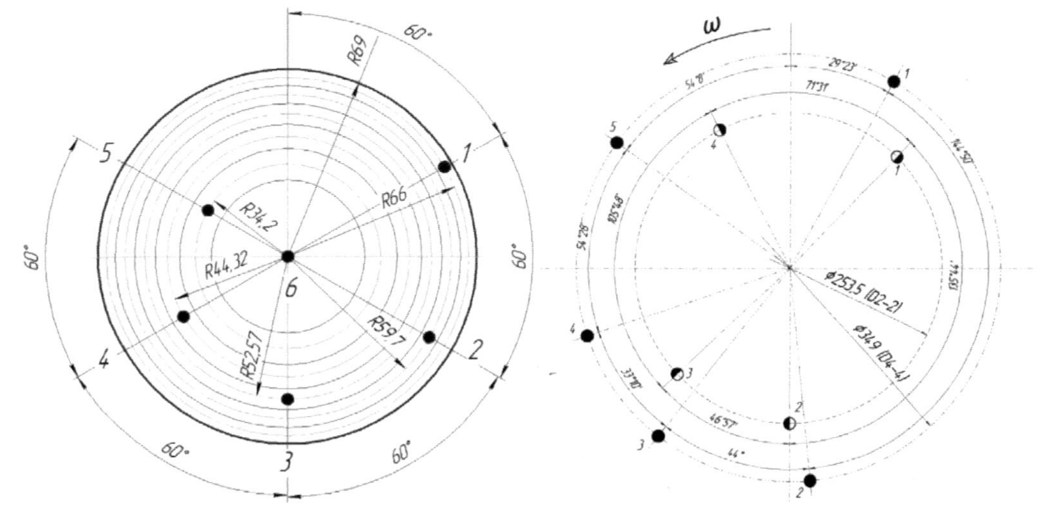

a) b)

Fig. 3 Total pressure probe locations in a) 0-0 section b) 2-2, 4-4 sections

Calculation of the characteristics was conducted in the following basic formulas:

Conditional flow rate coefficient is

$$\Phi = \frac{4 \cdot \dot{m}}{\rho_{inl}^{*} \pi D_2^2 U_2} \tag{1}$$

Polytropic efficiency at full range of parameters is

$$\eta_p^* = {(h_p + h_d)}/{h_i}, \ \eta_p = {h_p}/{h_i - h_d}. \tag{2}$$

Internal head coefficient is

$$\psi_i = h_i / u_2^2, \ h_i = c_p \left(T_{out}^* - T_{in}^*\right) \tag{3}$$

Theoretical head coefficient is

$$\psi_t = C_{u2} / u_2 \tag{4}$$

Total to total pressure ratio coefficient

$$\Pi^* = P_{out}^* / P_{inl}^* \quad \Pi = P_{out} / P_{inl} \tag{5}$$

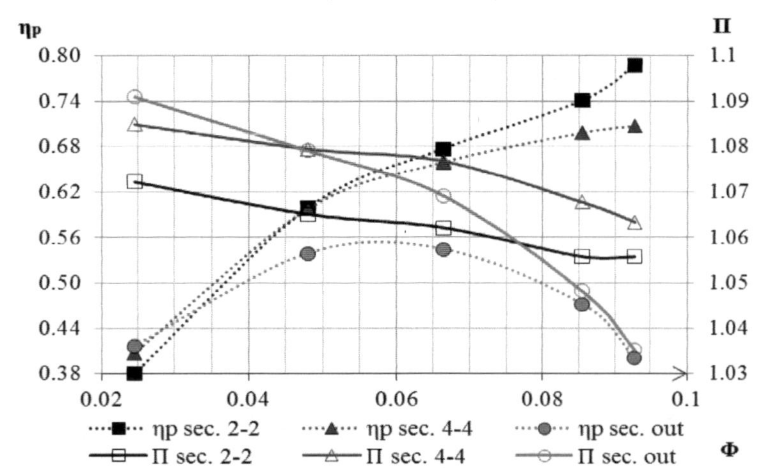

Fig. 4 Performance of centrifugal compressor

As expected the obtained performances (fig 4) of the mismatched elements, namely the optimal mode of operation of the impeller is in the range of Φ=0.095-0.11. Because of the stage "stiffness", right side of the impeller performance is cut.
Optimal work zone of the outlet chamber - volute in the range Φ = 0.05-0.065. This problem is typical for the last stages of the multistage compressors for modernization to increase their productivity at constant inlet and outlet chambers. To solve the problem, last stages should be designed so that the angle of entrance flow in the outlet chamber remained within (+-) 2-4 degrees from the nominal design. It is also recommended to unload on the head level, to reduce the flow rate at its output. These activities provide an acceptable mutual work of the modernized last stage and outlet chamber.
In our case, we use the concept of a virtual bench for studies on this problem with the help of multi-criteria methods and multi-parameter optimization, but first conduct verification of numerical calculation to determine the error for the object.

5. Virtual experimental bench

CFD model has full sweep angle of the elements with inlet and outlet pipes (fig. 5). Inter gap and dummy piston seals at the base disk have not been simulated. The calculation is based on RANS – method (Reynolds-averaged Navier-Stokes) with Low-Re turbulence model. Was built a detailed computational mesh in the boundary layer, where the 1st grid node should be in the viscous sublayer ($y^+ = 0 \ldots 2$). A CFD Solver of Ansys CFX 16.2 was applied. The work fluid is air (ideal gas). The outcome of the mesh independence studies provides the grid with app. 40 million cells.
Were selected the following boundary conditions (BC): the inflow direction is assumed as normal to the axial inlet of the computational domain. The simulations were run with P^*_{inl} and T^*_{inl} in inlet boundary and mass flow ṁout at outlet boundary. The walls were assumed as adiabatic. The inlet and outlet BL conditions are listed in Table. 2. The Table also provides the conditional coefficient of flow rate (Φ) and mass flow (ṁ). The BC were determined by the outcomes of the full-scale experiment.

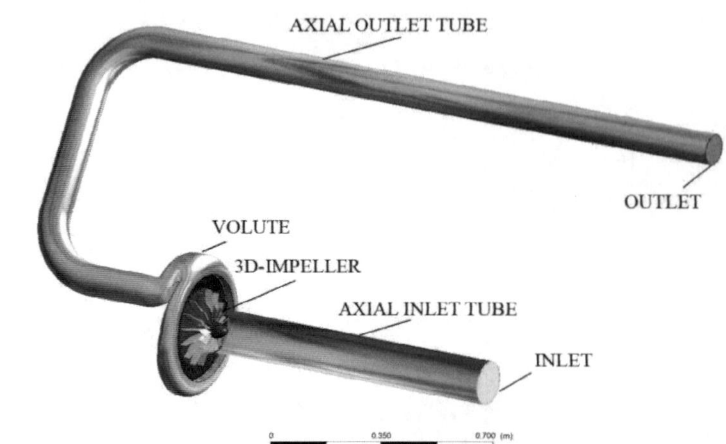

Fig. 5. CFD model and BC of virtual experimental bench

Table 2. The parameters set on the inlet and outlet BC of the computational domain

Mode, n	Φ	T^*_{inl}, K	P^*_{inl}, Pa	ṁ, $kg \cdot s^{-1}$
1	0.093	294.35	99051	0.504
2	0.085	294.60	99327	0.465
3	0.066	294.55	99830	0.363
4	0.048	294.05	100179	0.264
5	0.025	294.60	100650	0.136

IOP Conf. Series: Materials Science and Engineering **232** (2017) 012042 doi:10.1088/1757-899X/232/1/012042

The calculation was carried out at the supercomputer center "Polytechnic" with heterogeneous cluster HPC Tornado using 8 nodes. The parameters of one node of the cluster: RSC Tornado – 2 CPU with 14 cores (2xXeon E5-2697v3 2.6 GHz 64 GB RAM). In sum - 712 nodes (19936 cores).
Calculation of one mode is 21 hours.

6. Verification results

All modes developed the converging solutions (example of convergence history for mode 1 shown in Fig. 6), defined by the following parameters within the computational domain BC:

- the fall of the level of root-mean-square (RMS) residuals below 10^{-4},
- the relative error in mass and energy imbalance is $\sim 1 \cdot 10^{-2}$ %,

The controlled values of integral parameters in the control sections do not change from iteration to iteration.

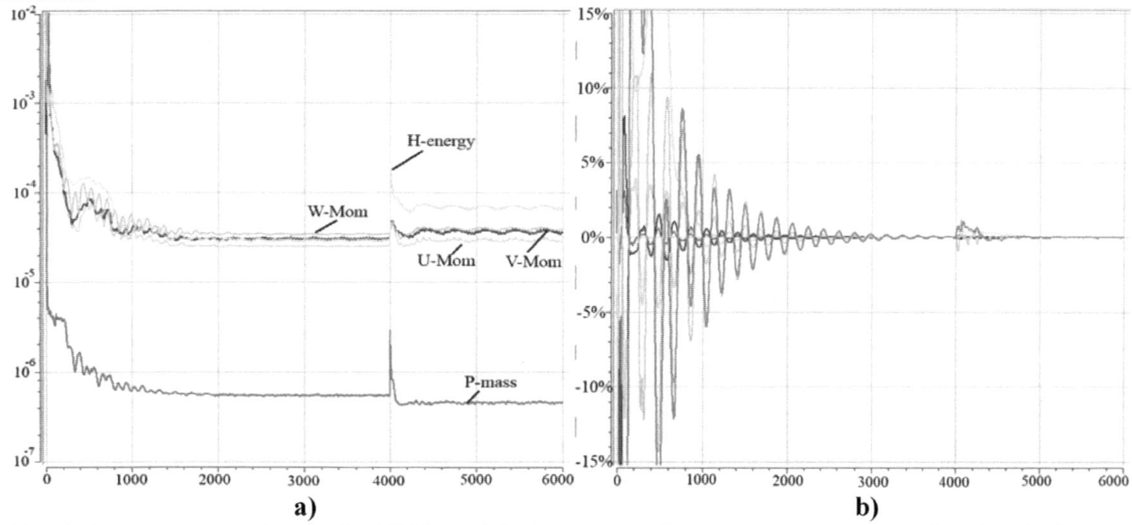

Fig. 6. Convergence history for RMS residuals mass and momentum, energy (a) and imbalances within computations domains (b)

In the figures 7,8,9 are shown experimental characteristics of object with comparison virtual bench numerical calculation of the characteristics.

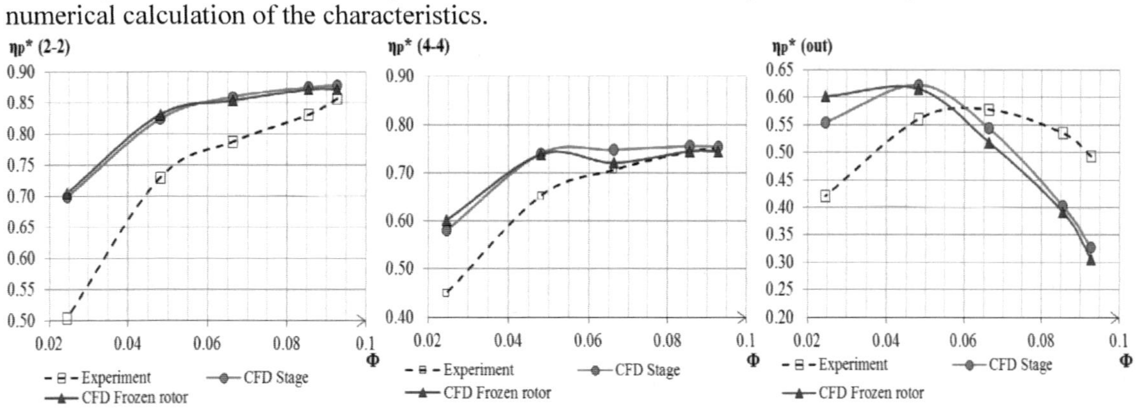

Fig. 7. Verification results for polytropic efficiency (total) at the outlet of the impeller, VLD, volute.

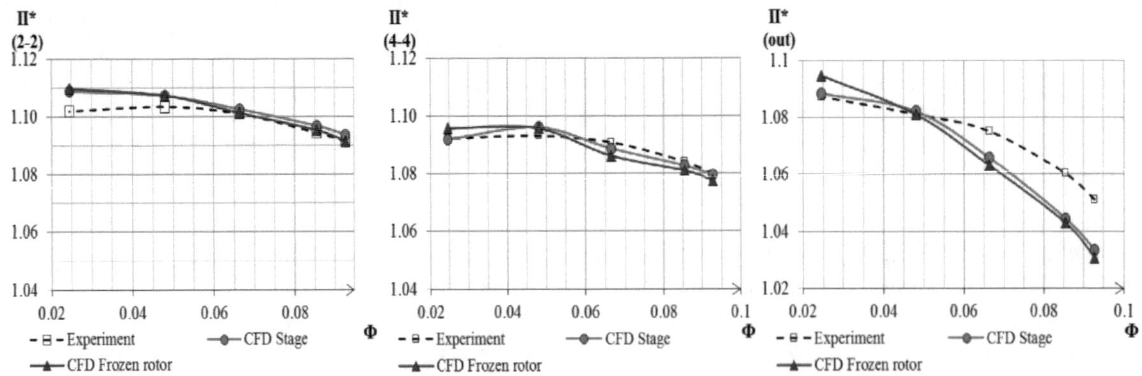

Fig. 8. Verification results for pressure ratio (total) at the outlet of the impeller, VLD, volute.

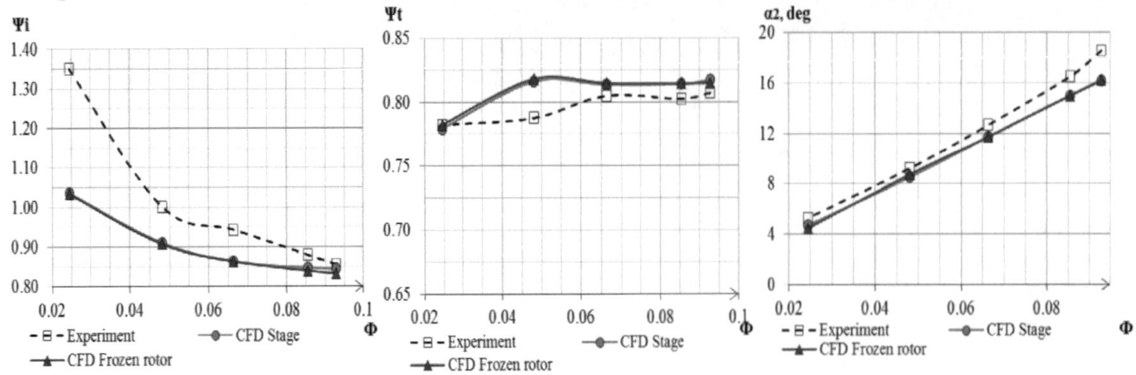

Fig. 9. Verification results for internal, theoretical head and flow angle at the outlet of the impeller, VLD, volute.

The calculation for the verification was conducted with interfaces "stage" and "frozen rotor" between the impeller and volute. The chart Fig. 7 shows good agreement between the results of calculation and experiment in the area of optimal mode Φ=0,093-of 0.085 for the impeller. However, when considering areas with low productivity errors increases significantly. This is due to the appearance of low-energy zones of separation flow associated with the deterioration of the flow conditions of the blade apparatus in these regimes. Also, there is an overestimation of the characteristics, because not taken into account the thermal effect associated with the occurrence of non-stationary processes (fig. 9). Determination of the optimal operation mode of the output device is complicated by the fact that starting with the 3 modes the impeller operates when $\alpha_2<16$ (fig.9), indicating that the early appearances of non-stationary phenomena so there needs a transient calculation. This thermal effect increases internal head in the impeller and give big error for polytrophic efficiency in stationary results, also it can be imperfect of the semi-empirical turbulence model.

In Fig. 8 shows plots of the relationship of the total pressure on the control sections, there is an exact match on the first three modes for Impeller and VLD, however, after the volute there is a big error, probably related to the increase of the error in the numerical calculation due to the volute.

7. Parametric model

To solve the problem a parameterized model of the impeller was developed (fig. 10). For the geometrical parameters of the model the variables names and were assigned and the ratios between them were adjusted. In total we used 16 variables defining the geometrical model of the the impeller flow part. The model allows to build semi-open and closed impellers. Ansys BladeModeler used for the parameterization of impeller flow part geometry parameters.

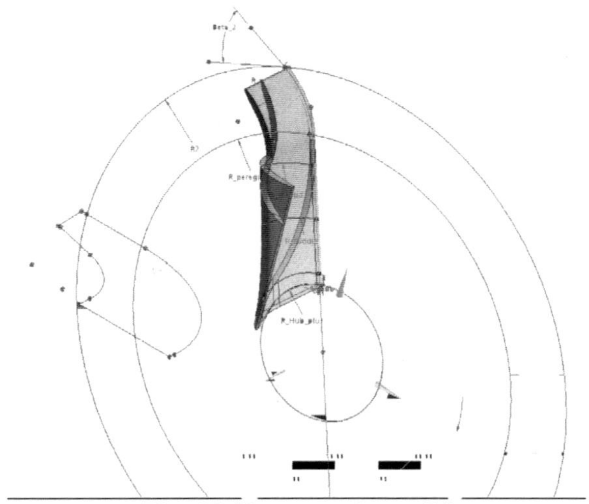

Fig. 10. Parametric model

The Optimization will perform on a heterogeneous cluster HPC tornado in SPbPU supercomputer center using the multi-objective optimization technology IOSO (Indirect Optimization based on Self-Organization). The following target parameters will select: impeller adiabatic efficiency, impeller pressure ratio, head coefficient, stage adiabatic efficiency, and stage pressure ratio. The ranges of limitations of the geometric parameters were determined in terms of the design constrains.

8. Conclusions

Because types of machines a lot, you should use original verification data, if there is no specialized experimental database. Error obtained during the verification can be used at optimum modes to evaluate the effectiveness of the every element. This is due to the fact that the existing methods of numerical analysis does not fully reflect the effects of flow separation and because of the mismatch of the optimal mode of elements. Thus one cannot say about the admissibility of the results of the inconsistent elements stationary calculation, therefore, it seems for a correct evaluation requires a transient calculation. However, the results are acceptable for verification of the characteristics of the impeller in the area of optimal mode at the stationary calculation, which requires much less computational resources compared to unsteady calculations.

Developed a parametric model will be used to conduct multi-criteria optimization to verify the concept design of the revamped compressors. The works will be conducting numerical and field experiment produced for the flowing part of the impeller and VLD, assessment of the effectiveness of the concept and development of recommendations on modernization of centrifugal compressors.

References

[1] Boldyrev, Y., Rubtsov, A., Kozhukhov, Y., Lebedev, A., Cheglakov, I., Danilishin, A. Simulation of unsteady processes in turbomachines based on nonlinear harmonic NLH-method with the use of supercomputers // CEUR Workshop Proceedings. Volume 1482, 2015, Pages 273-279.

[2] Danilishin A., Kozhukhov Y., Yun V, Multi-objective optimization for impeller shroud contour, the width of vane diffuser and the number of blades of the centrifugal compressor stage based on the CFD calculation. IOP Conference Series Materials Science and Engineering 08/2015; Volume 90(1):012047. DOI:10.1088/1757-899X/90/1/012046

[3] Danilishin A., Kozhukhov Y., Gileva L., Lebedev A. Verification of the CFD calculation on a supercomputer of medium flow model stages of centrifugal compressor. Russian

Supercomputing Days: Proceedings of the international conference (September 26-27, 2016, Moscow, Russia). Moscow State University, 2016.

[4] He L. and Ning W., "An Efficient Approach for Analysis of Unsteady Viscous Flows in Turbomachinery", AIAA Journal, Vol.36, No.11, 1998.

[5] Kovalevskiy N.L., Kuz'min V.Ye., Spirin N.Yu., Chernyavskiy L.K. 25% productivity increase of K1290-121-1 compressor for large-capacity ammonia production. Compressors & Pneumatics. 2009. №1 pp. 39-42.

[6] Kozhukhov Y.V., Yun V.K., Reshetnikova L.V., Prokopovich M.V. Numerical Investigation of Different Radial Inlet Forms for Centrifugal Compressor and Influence of the Deflectors Number by Means of Computational Fluid Dynamics Methods with Computational Model Validation. IOP Conference Series Materials Science and Engineering 90(1):012047 · August 2015. DOI: 10.1088/1757-899X/90/1/012047

[7] Kuzmin V.E., Primak A.N., Chernyavsky L.K. The most significant reconstructions and modernizations of centrifugal compressors, executed JSC «SIF «Nevintermash».Compressors & Pneumatics 2011 №1. pp.11-15.

[8] Lopulalan, H.D. Unsteady flow modeling using transient rotor – stator interface / R.A. Izmailov, H.D. Lopulalan, G.S. Norimarna / Proceedings of ASME Turbo Expo 2013 GT2013-95788 June 3-7, 2013, San Antonio, Texas, USA. pp. 1-8.

[9] Semakov A., Evdokimov V., Reprintcev A., Liubimov A. Modernization of centrifugal compressors on nitric production facilities. Compressors & Pneumatics. 2013. № 3. pp. 2-4.

[10] Semakov A., Evdokimov V., Reprintcev A., Liubimov A. Increase of Capacity by 35% and Efficiency Raise of the Technological Air Centrifugal Compressor 101J in Ammonia Production. Technical Preprints Nitrogen+Syngas 2015. 2015. pp. 415-424.

[11] Simonov A.M., Study of the efficacy and optimal design of high-pressure centrifugal compressor stages. pp. 164 – 188. / Works of scientific school of the compressor design SPbPU. Under the editorship of Professor Yu. b. Galerkin. – Ed. St. Petersburg State Polytechnical University, St. Petersburg, 2010 - 670 p.

[12] Yablokov A.M., Kozhukhov Y.V., A.A Lebedev An ansys CFX numerical study of a low-flow stages centrifugal compressor // Nauchno-tekhnicheskie Vedomosti SPbGPU. Vol. 4(231)/2015. Ed. SPbSPU. pp. 59 – 69.

[13] Yun V.K. The use of advanced technologies to improve the economic performance of the centrifugal compressor. Gas industry №9, 2014 pp. 68-71.

IOP Conf. Series: Materials Science and Engineering **232** (2017) 012043 doi:10.1088/1757-899X/232/1/012043

Impact of inlet coherent motions on compressor performance

Jacopo Forlese[1,2] and Giovanni Spoleti[1,3]

[1]GM Global Propulsion System Europe S.r.l., Corso Castelfidardo 36, 10129 Torino, Italy.

[2]jacopoforlese@gm.com, [3]giovannispoleti@gm.com.

Abstract. Automotive engine induction systems may be characterized by significant flow angularity and total pressure distortion at the compressor inlet. The impact of the swirl on compressor performance should be quantified to guide the design of the induction systems. In diesel engines, the presence of a valve for flow reduction and control of low pressure EGR recirculation could generate coherent motion and influence the performance of the compressor. Starting from experimental map, the compressor speed-lines have been simulated using a 3D CFD commercial code imposing different concept motion at the inlet. The swirl intensity, the direction and the number of vortices have been imposed in order to taking into account some combinations. Finally, a merit function has been defined to evaluate the performance of the compressor with the defined swirl concepts.
The aim of the current work is to obtain an indication on the effect of a swirling motion at the compressor inlet on the engine performance and provide a guideline to the induction system design.

1. Introduction

3D CFD simulation of the air induction system in internal combustion engines design is a process largely used at every production level. The addition of a centrifugal compressor leads to more long and complex analyses that have to deal with the needs of the time frame of design and production.

This work come from the necessity to find a correlation between coherent motions at the compressor inlet and the compressor behavior that affect the engine performance itself.

The maximization of boost and the turbo-compressor efficiency plays a very important role in the engine optimization in terms of performance and reduction of pollution. Bent and contorted under-hood pipes and EGR recirculation could lead to a very complex inlet system with negative effects on the compressor. These systems typically produce one or more changes in flow direction before the compressor inlet. Some systems produce strong swirl distortion able to influence the flow through the wheel and the stability. [1]

Compared to the other works available in literature, the peculiarity of this project is to study the compressor behavior modification due to theoretical shapes of inlet flow without referring to a specific induction system design. [2] [3]

The case of study is a mid-size diesel engine for automotive traction.

2. Discussion

2.1. Classifications of swirl motion

The flow angularity generated by inlet systems may have both radial and circumferential velocity components. In particular conditions is possible recognize coherent motions as bulk swirl or paired swirl.

Bulk swirl *Paired swirl*

Figure 1. Typical swirl pattern.

The bulk swirl consists of an inlet flow that is all rotating in the same direction, either with or opposed to the compressor rotation. In the *figure 1* is shown a typical bulk swirl pattern. In addition to the axial component the predominant velocity component is the tangential.

The second type of swirl pattern is called paired swirl and consists of two or more paired vortices rotating in opposite directions. In the figure is possible to see also two example of paired swirl. Nevertheless, the vortex pairs do not have to be of equal magnitude: one or more vortices can dominate a smaller set of vortices.

There are also other types of swirl beside bulk and paired, such as in a rectangular pipe, but typically, upstream the compressor wheel, the most common motions are the bulk swirl generated by bend and twin swirl generated by flap to reduce the flow.

2.2. Swirl description

In order to evaluate and compare the motion generated by the inlet system the swirl angle has been calculated as a function of the circumferential position at constant radius at the compressor inlet section. The swirl angle is defined as

$$\alpha = \text{atan}\left(\frac{v_{tangential}}{v_{axial}}\right) \quad (1)$$

The swirl description used in this document has been implemented in the commercial CFD 3D software in order to easily evaluate the value for each cell of the fluid domain and plot it for an external circumferential crown as in the figure 2 [4]. In the example, a T-junction with a paired swirl along with a plot showing how the swirl pattern look like a sinusoidal wave when the swirl angle is plotted out at a constant radius.

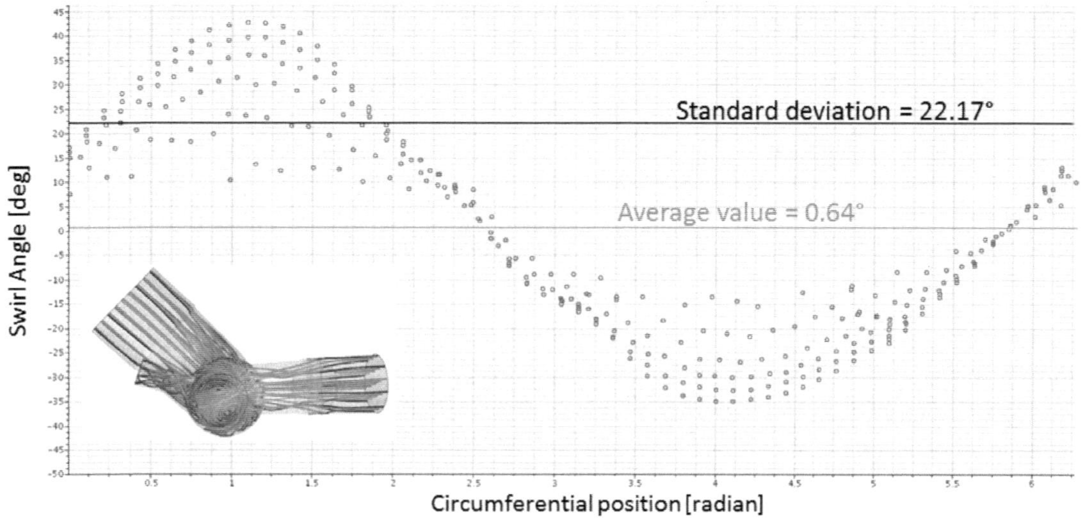

Figure 2. Example of plot of swirl angle distribution.

The coherent structure could be represented by the average and the standard deviation of the swirl angle value in the crown:

- A bulk swirl will have a high average value representing the intensity of the swirl;
- A paired swirl will have an almost zero average and the standard deviation will represent the intensity of the motions.

The flow has been considered positive if with the same rotation direction of the compressor wheel, otherwise negative.

2.3. Analysis setting and CFD model definition

The aim of the current work is to obtain an indication on the effect of a swirling motion on the engine performance and provide a guideline to the induction system design.

The work is based on the 3D Computational Fluid dynamic simulation of the operation of the compressor stand-alone. The idea is to compare the compressor maps of the compressor with and without a coherent motion imposed at the inlet.

The analysis has been led with the commercial software StarCCM+ by CD-Adapco - Siemens. A coupled implicit scheme has been choose for the steady-state simulations, the turbulent model was a SST-k-ω, the fluid is compressible and the properties were temperature dependent. The fluid domain was divided into almost 5 million of polyhedral cells, the mesh was generated using the StarCCM+ tools. A mesh refinement is present at the shroud, at the diffuser and at blades paying particular attention to catch the curvature with very good accuracy.

As the wall treatment was used the *All y+ Wall Treatment* that is a hybrid approach that seeks to recover the behaviors of the other two wall treatments in the limit of very fine or very coarse meshes. It is a design goal that this wall treatment should give results similar to the low-y+ treatment as y+ close to 0 and to the high-y+ treatment for y+ greater than 30. It gives reasonable results also for intermediate meshes where the cell centroid falls in the buffer layer [5].

In our model 12 prismatic cell layers has been used: with this settings the y+ is in all the flow conditions below 30 (mainly about 1) in order to take advantage of the more accurate low-Reynold approach. [6]

The impeller motions was modelled with a moving reference frame: it is a method commonly used in proprietary CFD codes, where it is referred to as *frozen rotor*. This approach resembles a snapshot of the flow, frozen in time, thus failing to predict the unsteadiness of the flow between the diffuser vanes. Thus different relative positions result in different solutions, which is a major disadvantage of this approach. It is a fast preliminary method, however, and gives satisfying results and represent a good compromise between accuracy and computational effort giving a solution that represents the time-averaged behavior of the flow. [7]

The choice of the MRF is mainly due to the high number of the simulation done in this project and the analysis points selections has included only stable points of the compressor map (not affected by evident surge and choke phenomena).

Inlet was set as a stagnation inlet and the outlet was set as a pressure outlet in the high mass flow conditions and as a mass flow outlet at the left part of the map.

This is due mainly to stability: setting the mass flow rate the simulations are more stable in low flow conditions, on the other hand setting the pressure ratio is more stable at the right part of the map. It is possible to notice that in this way reply an experimental point is more accurate because close to the choke the speed-lines are almost vertical (for example a small error setting the mass flow close to choke condition could lead to ambiguity in determine the correspondent pressure ratio).

The experimental compressor map has been used as starting point to set the boundary condition for each speed-line. To reduce the simulation time has been chosen to consider only 3 of the 8 speed-lines of the map as the most representative condition for the engine (100k, 150k and 200k rpm). Then, for each speed-line up to 7 point were simulated with more than 50 k iterations each one. Also the points characterized by incipient surge and choke phenomena are still in the stable portion of the map and they have been analyzed in order to consider the range of the map.

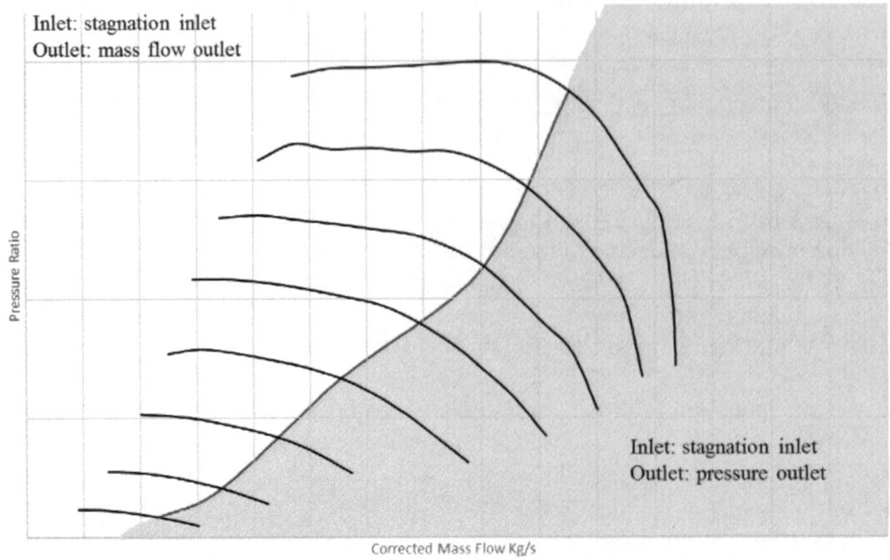

Figure 3. Setting of the boundary condition type depending from the map.

2.4. *Develop Concept*

In order to consider all the typical swirl, generalized coherent motions were set at the compressor inlet. The simple theoretical motions considered have been:

- The bulk swirl, where the entire flow field is rotating either in the same direction as wheel rotation (co-swirl, positive swirl angle) or in the opposite direction of wheel rotation (counter-swirl, negative swirl angle); both the direction has been considered.
- The twin swirl that is composed of two vortices rotating in opposite directions. One vortex is a co-swirl vortex while the other is a counter-swirl vortex; the two vortices are considered symmetric.

For each type of swirl and direction two level of intensity has been considered: low swirl angle value (6°) and high swirl angle value (10°).

Table 1. Definitions of coherent motions.

Number of swirl	Direction	Swirl angle value	Definition
1	Co-rotating	6°	1L +
1	Co-rotating	10°	1H +
1	Contra-rotating	6°	1L -
1	Contra –rotating	10°	1H -
2	-[a]	6°	2L
2	-[a]	10°	2H

[a] Changing the direction of the twin swirl does not provide any difference in the results.

The 3D CFD software used allow to set the flow velocity components at the inlet through field functions. Flow velocity components are set in a cylindrical coordinate system integral with the compressor inlet system. [4] In the bulk swirl, the tangential velocity has been as proportional to the pipe radius and the radial velocity is zero. In the twin swirl, the radial component is dependent from the angular coordinate θ:

$$v_{radial} = cos^3(\theta) \qquad (2)$$

The tangential component is dependent from the angular coordinate θ, the y-coordinate and the radial coordinate r. The tangential component has a maximum absolute value equal to 1 in order to be compatible with the radial component.

$$v_{tangential} = \begin{cases} 1 & if \quad w > 1 \\ -1 & if \quad w < -1 \\ & else \quad w \end{cases} \qquad (3)$$

where:

$$w = \begin{cases} [\sin\theta\,(0.4 - \cos(\pi y))] + r & if\ y > 0 \\ [\sin\theta\,(0.4 - \cos(\pi y))] + r & if\ y < 0 \end{cases} \qquad (4)$$

Setting the value of the axial component (greater than 1) is possible to define the swirl angle value.

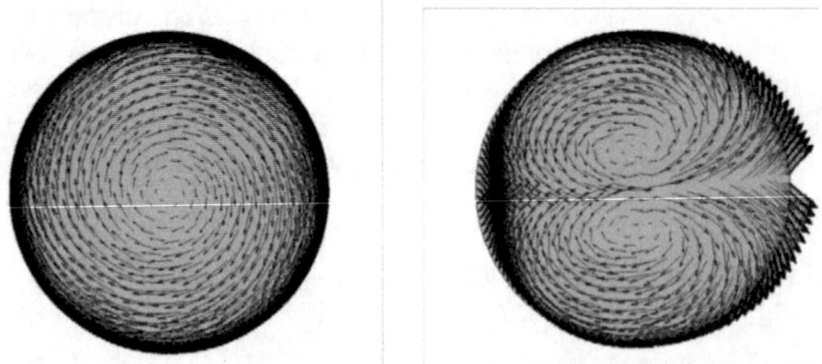

Figure 4. Tangential distribution of the vector at the fluid domain inlet.

2.5. Swirl Merit Function

In literature [2], [4] and [1] it's possible to find indication about the impact of swirl on compressor performance: a co-rotating swirl could reduce the flow and the pressure ratio but could increase the efficiency and the stall margin. Otherwise a counter-rotating swirl could have the opposite effects. Thus, in order to quantify the effects of swirl on compressor map, a merit function has been introduced called *Speed Line Efficiency Factor*.

The *SLEF* is calculate for each compressor speed-line and is defined as:

$$SLEF = R \cdot E_P \cdot E_S \cdot E_C \qquad \text{Eq. 5}$$

where:

- R is the map range (calculated as mass flow difference between the near-choke and the near surge point)
- Ep is the peak of efficiency,
- Es is the efficiency at the surge point,
- Ec is the efficiency at the choke point.

The calculated merit function can take into account the extension of the map and the increasing of efficiency at the same time.

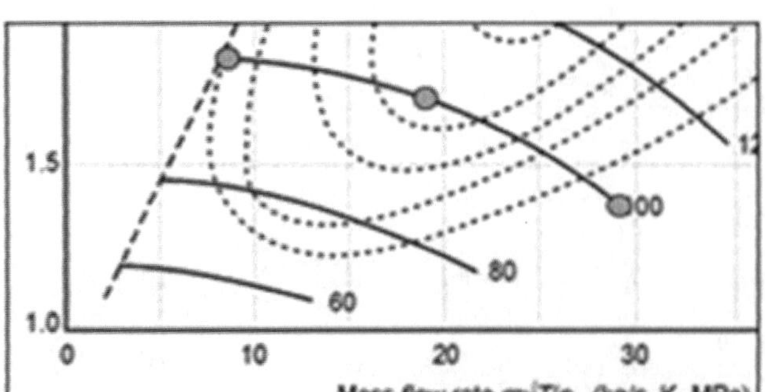

Figure 5. Merit Function calculation exhample.

3. Results

In the plots figures are plotted the 3D-CFD results in terms of compressor maps and *Speed Line Efficiency Factor*. The co-rotating swirl maximize the efficiency at all the swirl angle values simulated, but decreasing the mass flow rate: this lead to the highest value of *Speed Line Efficiency Factor*. The counter-rotating swirl can maximize the mass flow rate and the pressure-ratio values, but with a lack in terms of efficiency. The counter-rotating flow with low swirl angle (6°) value represent a good trade-

IOP Conf. Series: Materials Science and Engineering **232** (2017) 012043 doi:10.1088/1757-899X/232/1/012043

off between the efficiency and the extension of the compressor map (same *SLEF* of the single swirl co-rotating with low swirl angle). The twin swirl leads to a mass-flow-rate/pressure-ratio map very close to the baseline but with a loss of efficiency growing with the increasing of the swirl angle.

In the plot below it is possible to see the compressor map, the efficiency map for each imposed swirl type and the *Speed Line Efficiency Factor* calculated for each speed-line simulated.

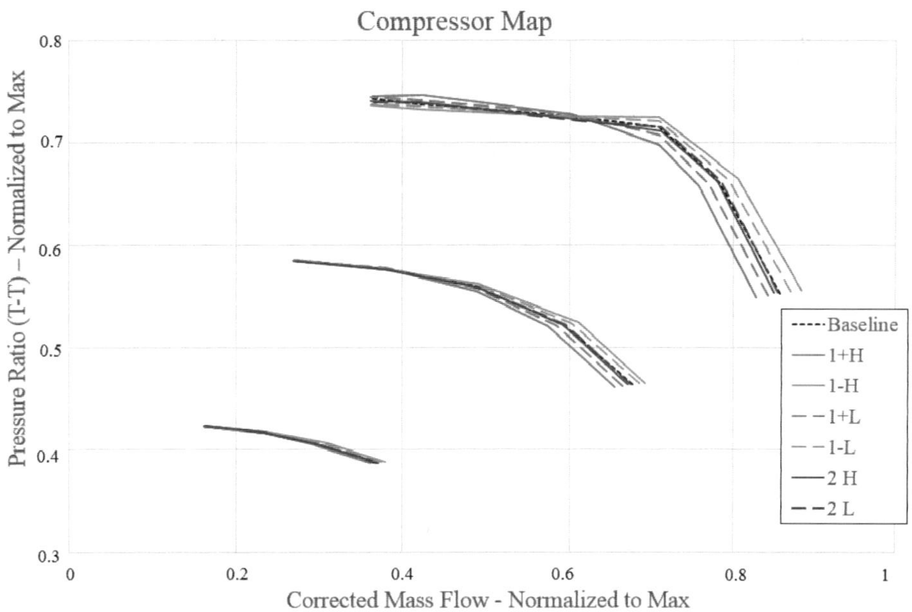

Figure 6. Compressor maps comparison.

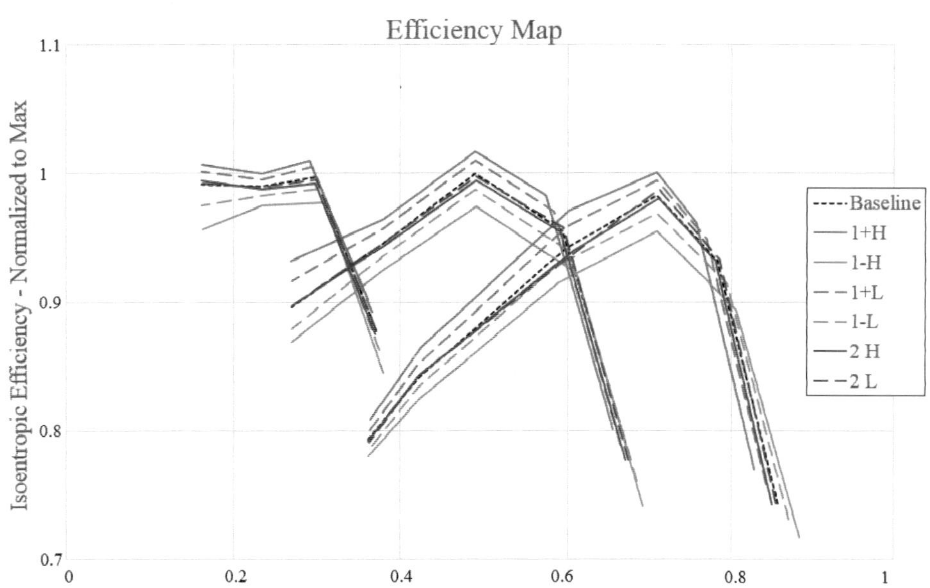

Figure 7. Efficiency maps comparison.

IOP Conf. Series: Materials Science and Engineering **232** (2017) 012043 doi:10.1088/1757-899X/232/1/012043

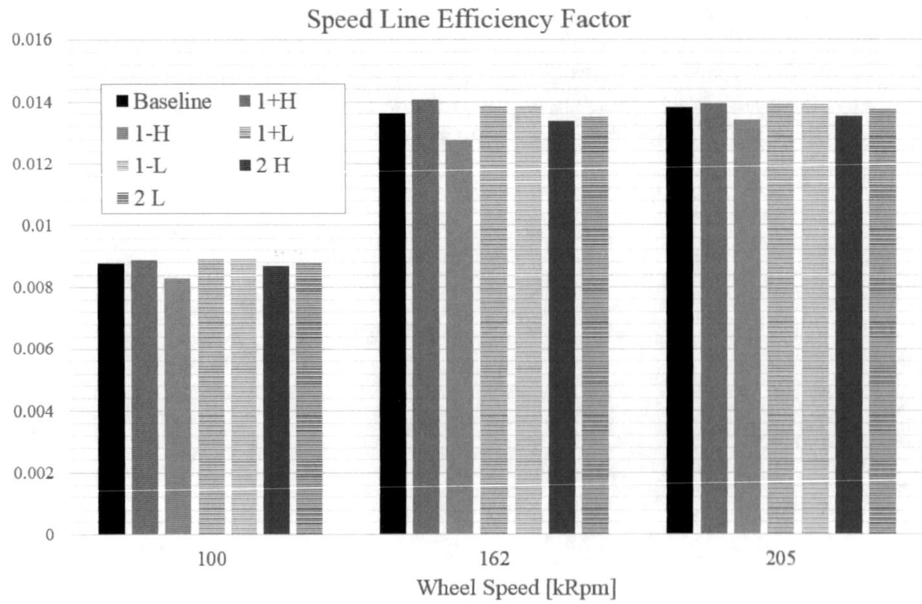

Figure 8. Speed Line Efficiency Factor comparison.

3.1. Engine impact evaluation

A complete compressor map obtained from the optimal swirl concept simulation has been put in a GT-Power 1D model in order to understand the impact on the engine performance as an additional verification.

Typically, 1D Diesel performance simulation are run with compressor boost and brake torque fixed with the aim of have the same combustion condition. Thus, the comparable output are the air-fuel-ratio and the brake specific fuel consumption (BSFC) at the full-load condition.

The result is that a co-rotating swirl at the compressor inlet with a high swirl angle value (10°) in our case could lead to a reduction of BSFC up to 1%.

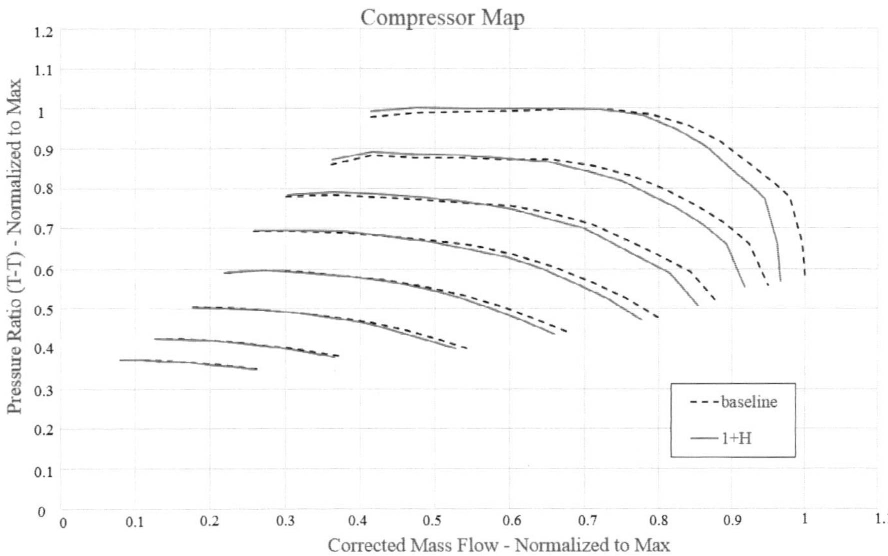

Figure 9. Compressor map comparison.

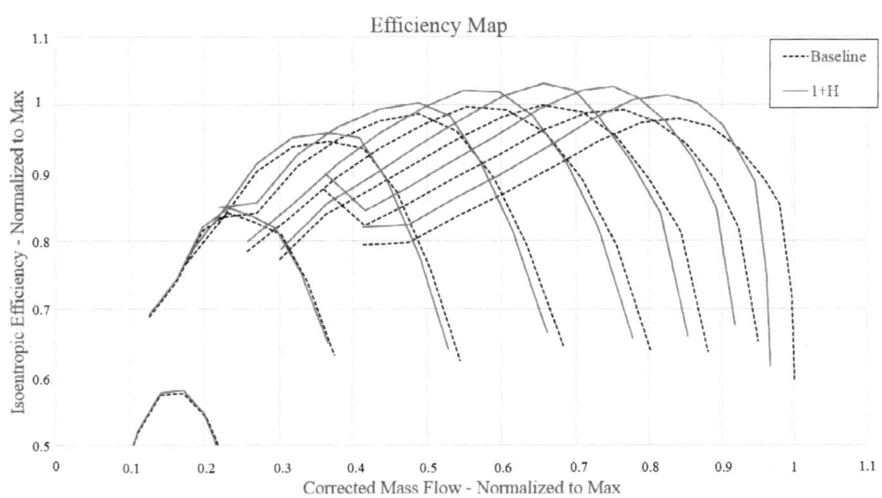

Figure 10. Efficiency map comparison.

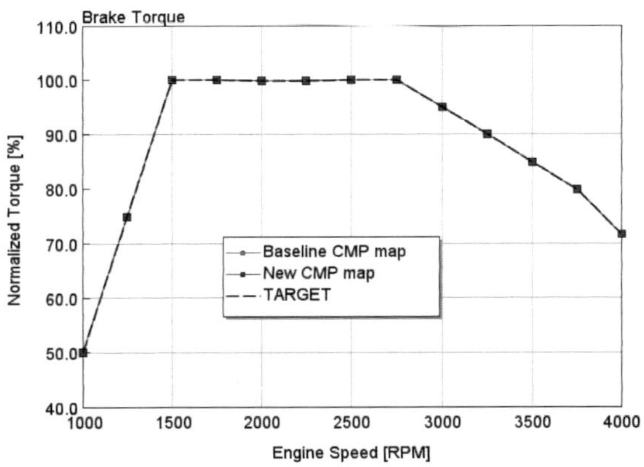

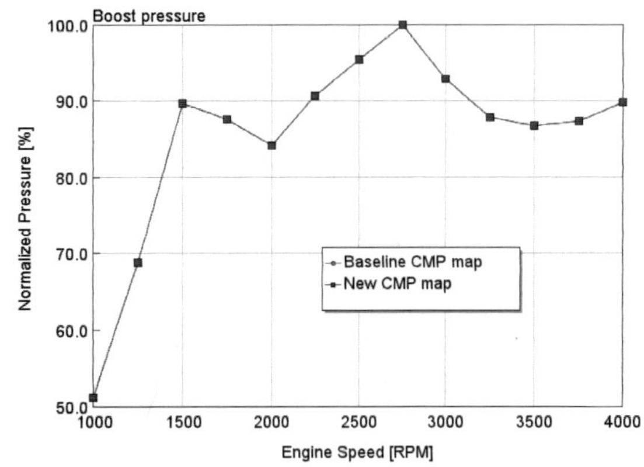

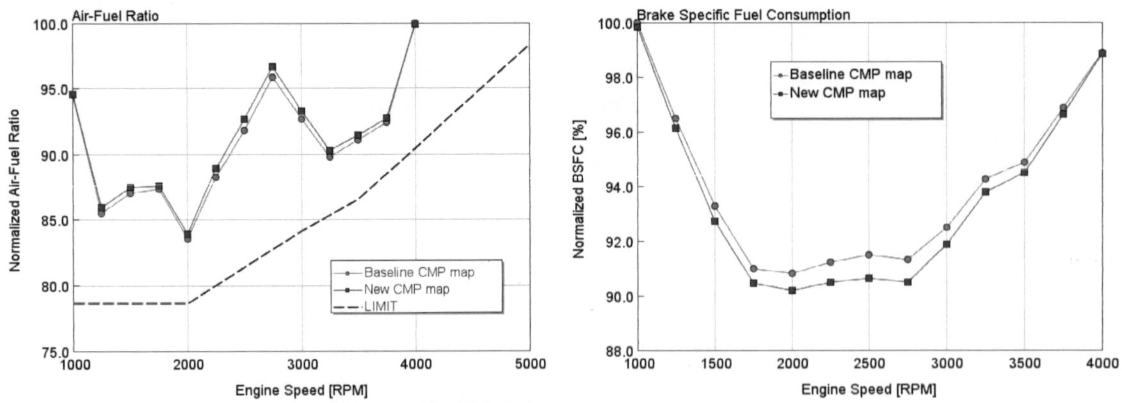

Figure 11. GT-Power outputs comparison.

4. Conclusion

In the compressor inlet design for automotive engine, there is the necessity to create a correlation of the possible coherent motions with the performance of the compressor end the engine itself. This paper presents a case of study of optimization of the induced swirl in order to maximize the overall performance. The merit function shows that the co-rotating swirl can increase the compressor performance with positive effect on the engine efficiency and the behavior rise increasing the swirl angle value.

The limitation of this study should be take into account evaluating the impact of swirl margin. Surge is typically an oscillating phenomenon and the steady state analysis cannot give an accurate forecast of the complex behavior. Nevertheless this work can be a starting point to transient analysis of the impact of coherent motions on compressor surge margin.

References

[1] Galindo J, Tiseira A, Navarro R, Tarí D and Meano C, 2017, Effect of the inlet geometry on performance, surge margin and noise emission of an automotive turbocharger compressor, *Applied Thermal Engineering*, Elsevier

[2] Kyrtatos N and Watson N, 1980. Application of aerodynamically induced prewhirl to small turbocharger compressor, *J. Eng. Power*, ASME

[3] Nikpour B, 2004, Turbocharger compressor flow range improvement for further heavy duty diesel engines. *Proceedings of Thiesel - 2004 Conference. Thermo and Fluid Dynamic Processes in Diesel Engines*

[4] Bouldin B and Sheoran Y, Inlet Flow Angularity Descriptors Proposed for Use With Gas Turbine Engines, 2002 *SAE Technical Paper* 2002-01-2919

[5] STAR-CCM+ version 10.06, *User Guide*, Cd-Adapco Siemens

[6] Versteeg H K and Malalasekera W, 2007, *An Introduction to Computational Fluid Dynamics*, Second edition, Pearson, 58-62

[7] Petit O and Nilsson H, Numerical investigations of unsteady flow in a centrifugal pump with a vaned diffuser, *International Journal of Rotating Machinery* 2013

[8] SAE S-16 Committee, ARP 1420, 1978 *Gas Turbine Inlet Flow Distortion Guidelines*, Society of Automotive Engineers

[9] Galindo J, Serrano J R ; Margot X, Tiseira A. Schorn N and Kindl H, 2007, Potential of flow pre-whirl at the compressor inlet of automotive engine turbochargers to enlarge surge margin and overcome packaging limitations, *International journal of heat and fluid flow*, Elsevier

IOP Conf. Series: Materials Science and Engineering **232** (2017) 012044 doi:10.1088/1757-899X/232/1/012044

Comparison of CFD-calculations of centrifugal compressor stages by NUMECA Fine Turbo and ANSYS CFX programs

Y B Galerkin[1], I B Voinov[2] and A A Drozdov[1*]

[1] R&D Laboratory "Gas dynamics of turbo machines" Peter the Great St. Petersburg Polytechnic University, Polytechnical st. 29, St. Petersburg, Russia

[2] Engineering Center "Center of Computer-Aided Engineering" Peter the Great St. Petersburg Polytechnic University, Polytechnical st. 29, St. Petersburg, Russia

E-mail*: A_drozdi@mail.ru

Abstract. Computational Fluid Dynamics (CFD) methods are widely used for centrifugal compressors design and flow analysis. The calculation results are dependent on the chosen software, turbulence models and solver settings. Two of the most widely applicable programs are NUMECA Fine Turbo and ANSYS CFX.

The objects of the study were two different stages. CFD-calculations were made for a single blade channel and for full 360-degree flow paths.

Stage 1 with 3D impeller and vaneless diffuser was tested experimentally. Its flow coefficient is 0.08 and loading factor is 0.74. For stage 1 calculations were performed with different grid quality, a different number of cells and different models of turbulence. The best results have demonstrated the Spalart-Allmaras model and mesh with 1.854 million cells.

Stage 2 with return channel, vaneless diffuser and 3D impeller with flow coefficient 0.15 and loading factor 0.5 was designed by the known Universal Modeling Method. Its performances were calculated by the well identified Math model. Stage 2 performances by CFD calculations shift to higher flow rate in comparison with design performances. The same result was obtained for stage 1 in comparison with measured performances. Calculated loading factor is higher in both cases for a single blade channel. Loading factor performance calculated for full flow path ("360 degrees") by ANSYS CFX is in satisfactory agreement with the stage 2 design performance. Maximum efficiency is predicted accurately by the ANSYS CFX "360 degrees" calculation. "Sector" calculation is less accurate. Further research is needed to solve the problem of performances mismatch.

1. Introduction

CFD-methods are applied to centrifugal compressors, both for research and design. Experience shows that results are highly dependent on turbulence model, computational grid quality and so on. To select the correct way of modeling is necessary to make a comparison of CFD- calculated and test data. R&D Laboratory "Gas dynamics of turbo machines" has extensive experience in CFD- calculations [1-6]. The stage 1 was tested in the Compressor problems R&D Laboratory [7]. Performances of 3D impeller and vaneless diffuser were compared with calculated performances by different turbulence models and grids. The stage 2 of a multistage industrial compressor consisting of 3D impeller, vaneless diffuser and return channel was designed and performances were calculated by the Universal modeling method [8]. Its performances were calculated by two different programs NUMECA Fine Turbo and ANSYS CFX. The fundamental equations that describe fluid flow behavior are the Navier Stokes equations.

This is a set of five partial differential equations that describe the conservation of mass, momentum and energy. Flow in gaps between an impeller and a stator were not modeled in some cases. Unstructured and structured computational meshs with different number of cells were constructed. Their number depends on the simulated elements of the flow part. The calculations were made in a stationary setting.

For the impeller and stator elements different computational meshes were constructed. At the interface between the rotating mesh of the impeller and the fixed mesh of the stator elements an interface defined to connect them. The stage averaging interface is used in both programs. The convergence criterion was the attainment of the value of quadratic discrepancies no higher than 1e-5. A summary of various aspects and properties of the NUMECA Fine Turbo and ANSYS CFX flow solvers can be found in their online documentation.

Authors' experience [1-6] and known sources [9-11] show that CFD-calculation satisfactorily predicts stage maximum efficiency. Loading factor is overestimated by 6-12% at a design flow rate usually. Performances at flow rates below design flow rate are not predicted satisfactorily. The results in paper [11] show that numerical simulation in stationary and nonstationary calculation does not lead to any significant difference in the modeling of the right-hand side of the performances. Unsteady calculation allows more accurate calculation of the surge.

In the paper [12] it was shown that calculation of a centrifugal compressor stage in a "360 degrees" setting and with considering leakage losses, gives good agreement with the experiment. It is not possible to estimate the accuracy of the coincidence of the calculation and the experiment results on the given graphs. The polytrophic head performances are in good agreement with the experimental ones on the inclination angle and quantity. In the paper [13] the stage of the industrial centrifugal compressor is calculated by the ANSYS CFX program. Comparison of the calculation and experiment results had shown that the calculated efficiency performances are shifted towards higher flow rate compared to the measured one. The curves look similarly to each other from the qualitative viewpoint. The quantity of the maximum efficiency is well modelled. The maximum calculated pressure ratio is less than the experimental order of 6%.

2. Performance parameters

Calculated and tested performances are presented as functions $\eta^*, \psi_i, \psi_T = f(\Phi)$. Flow rate coefficient Φ is:

$$\Phi = \frac{4\overline{m}}{\rho_0^* \pi D_2^2 u_2}, \tag{1}$$

there $\rho_0^* = p_0^* / RT_0^*$ - gas density by total parameters.

Total polytrophic efficiency η^* (adiabatic compression):

$$\eta^* = \frac{ln\left(\frac{p_{ex}^*}{p_{inl}^*}\right)}{ln\left(\frac{T_2^*}{T_{inl}^*}\right)} \frac{k-1}{k}. \tag{2}$$

Loading factor ψ_T calculated by CFD:

$$\psi_T = \frac{C_p\left(T_2^* - T_0^*\right)}{u_2^2}. \tag{3}$$

Work coefficient measured:

$$\psi_i = \frac{C_p\left(T_2^* - T_0^*\right)}{u_2^2}. \tag{4}$$

Impeller loss coefficient ζ_{imp} calculated by CFD:

$$\zeta_{imp} = \frac{2\cdot\psi_T\cdot\left(1-\eta_2^*\right)}{\left(w_1/u_2\right)^2}. \tag{5}$$

Stator elements loss coefficient ζ_{stat} is calculated by the formula:

$$\zeta_{(VLD,Rch)} = \frac{p_{\text{out}}^* - p_{\text{inl}}^*}{\left(\rho_{inl}c_{inl}^2\right)/2}. \tag{6}$$

Direct comparison of efficiency calculated by CFD and measured efficiency in case of the stage 1 is not quite correct. Parasitic losses, i.e. disc friction and labyrinth leakage losses were not taken into account by CFD-calculations. These losses lead to loss of efficiency not more than 1.5% in stage 1. I.e., if CFD -calculated efficiency is bigger than the measured efficiency inside 1.5% the prediction is quite correct.

3. Stage 1

The stage with 3D impeller, VLD and scroll was designed and tested in SPbPU [7]. The experimental 600-kW test rig was driven by an electric motor. Nominal motor speed was 18 000 rpm with the possibility of regulation. Test rig included a boosting mechanical transmission.

Its design parameters are: $\Phi_{des} = 0.08$, $\psi_{T des} = 0.74$, $M_u = 0.78$. The main part of blades is a radial plate. Its blade cascade is shown in figure 1. The VLD normalized radial length is $\bar{D}_4 = 1.6$. The computational domain was extended to $\bar{D}_4 = 2.4$. The gas flow in the gaps between the impeller and the stator elements is not modeled. Therefore, the loading factor was calculated – not a work coefficient.

Calculations were carried out by the software package NUMECA Fine Turbo. Unstructured computational grids with 1.083, 1.224 and 1.854 million elements were formatted by NUMECA Fine AutoGrid. The turbulence model Spalart-Allmaras was applied in accordance with NUMECA program guidelines. In order to evaluate the effect of turbulence models on the results the SST turbulence model and Spalart-Allmaras with extended wall function were used too. The surfaces of the hub, shroud and blades are modeled as a smooth wall with adhesion.

Experimental and calculated total polytrophic efficiency performances are presented in figures 2 and 3. Measured work coefficient performance and calculated loading factor performances are presented in figure 4.

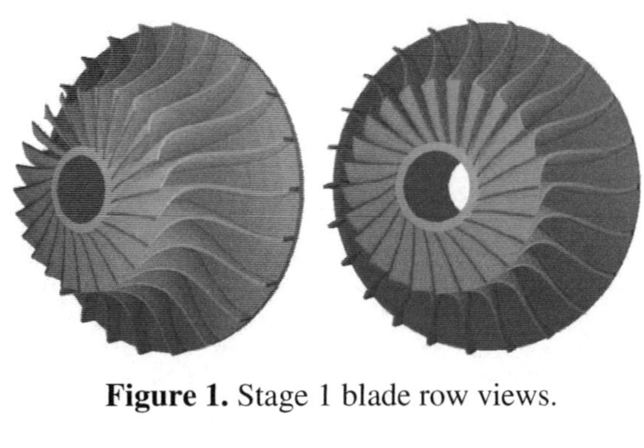

Figure 1. Stage 1 blade row views.

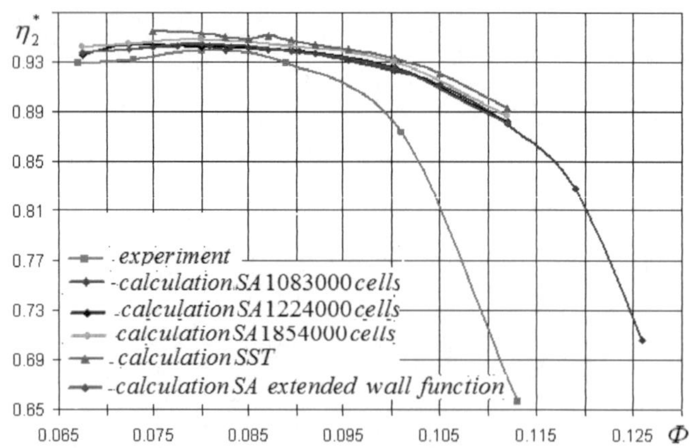

Figure 2. Measured and calculated total polytrophic efficiency of the stage 1 impeller.

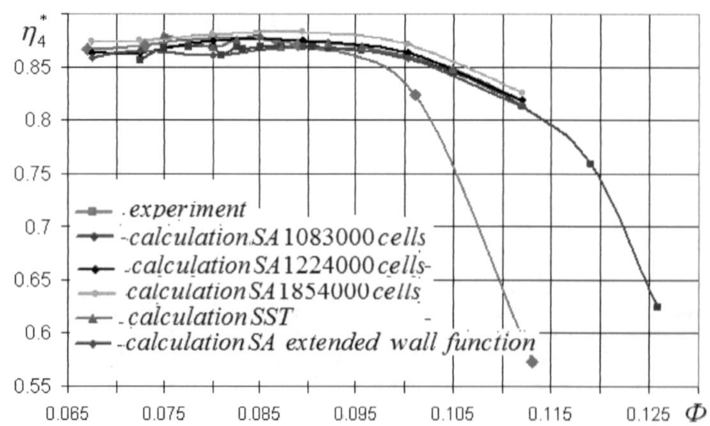

Figure 3. Measured and calculated total polytrophic efficiency of the stage 1 impeller and vaneless diffuser.

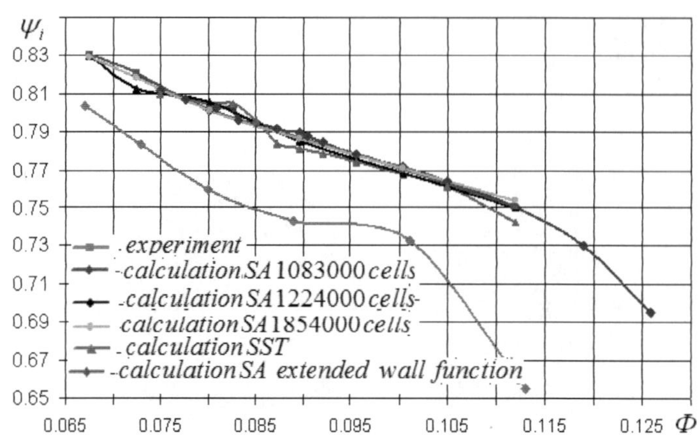

Figure 4. Measured performance of the work coefficient and calculated performances of the loading factor.

The visible result is that all calculations cannot predict efficiency of the impeller if $\Phi > 1.08\Phi_{des}$. The same is true for the stage "Impeller and VLD" if $\Phi > 1.19\Phi_{des}$ (that is not bad). The calculated efficiency of the impeller in the flow rate zone $\Phi < 1.08\Phi_{des}$ is bigger than the measured efficiency. The parasitic losses are about 1.5% for this impeller. These losses are neglected by the calculations. Meaning it the result of modeling is satisfactory in all cases. The performance calculated by the Spalart-Allmaras model with 1.224 million elements is more logical. The stage "impeller and VLD" performance is modeled the best of all by the Spalart-Allmaras model with 1.854 million elements. Calculation by different turbulence models with 1.084 million elements has not demonstrated significant difference. For further calculations the turbulence model Spalart-Allmaras was selected according to NUMECA recommendations. In all cases calculations with flow rate coefficients less than 0.070 have demonstrated problems with iterative processes.

Loading factor performances modeling is incorrect in all calculations. Calculated loading factor exceeds the measured work coefficient by 7.5% at Φ_{des} = 0.080. The calculated and measured efficiency at the design flow rate is quite satisfactory.

4. Design of the stage 2

The stage "impeller, VLD and return channel" has the following design parameters: $\Phi_{des} = 0.15$, $\psi_{T des} = 0.50$, $M_u = 0.55$. The stage was designed by the method presented in [14]. Inviscid velocity diagrams on three blade to blade surfaces and normalized meridional velocities on eight three blade to blade surfaces are presented in figure 5.

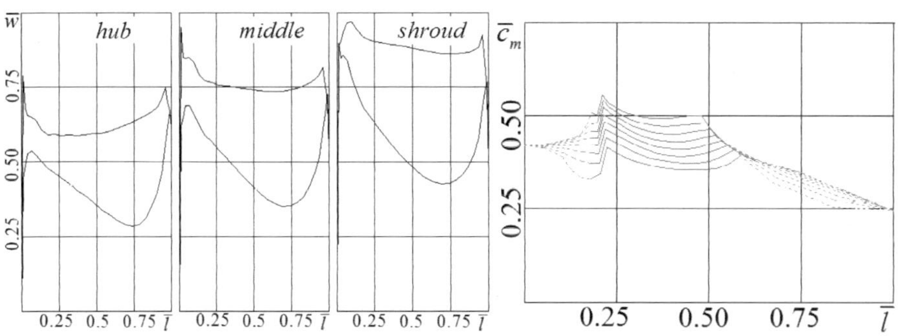

Figure 5. Stage 2. Inviscid velocity diagrams on three blade to blade surfaces and normalized meridional velocities on eight blade to blade surfaces.

Blade configuration is result of candidates' comparison with different inviscid velocity diagrams. The aim is to minimize velocity peaks and blade load near leading edges. Flow deceleration on a blade suction side must be minimized too. Its blade cascade is shown in figure 6.

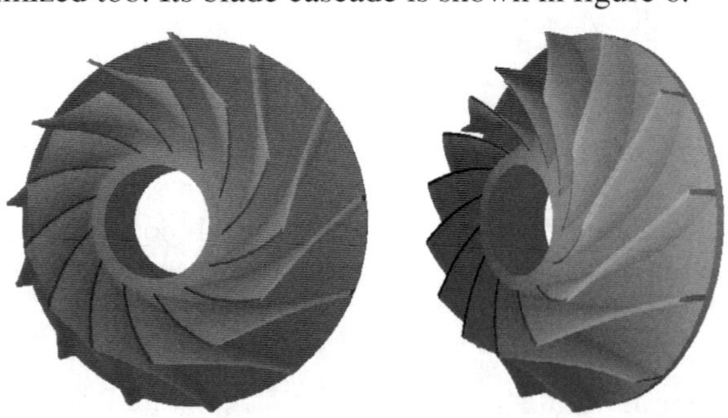

Figure 6. Stage 2 blade row views.

The calculations by the CFD-program NUMECA Fine Turbo were performed to optimize the stator part. The turbulence model Spallart-Almarras and a mesh with 2.5 million cells were used. Calculated flow visualization was applied to detect problematic zones in the stage's flow path. The calculation results have allowed correcting the shape of the stator elements. The meridional streamlines in original and corrected stages are shown in figure 7.

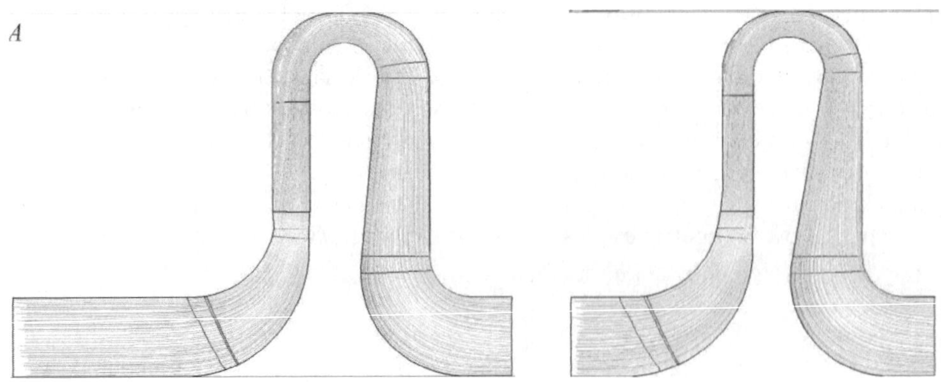

Figure 7. The meridional streamlines in two stage 2 candidates. Primary design - left, corrected by the results of CFD-calculations – right (NUMECA Fine Turbo).

VLD width was reduced to eliminate separation on the shroud wall of VLD. The RCH blades height at the entrance has been respectively reduced. It leads to the efficiency rise by 0.7%.

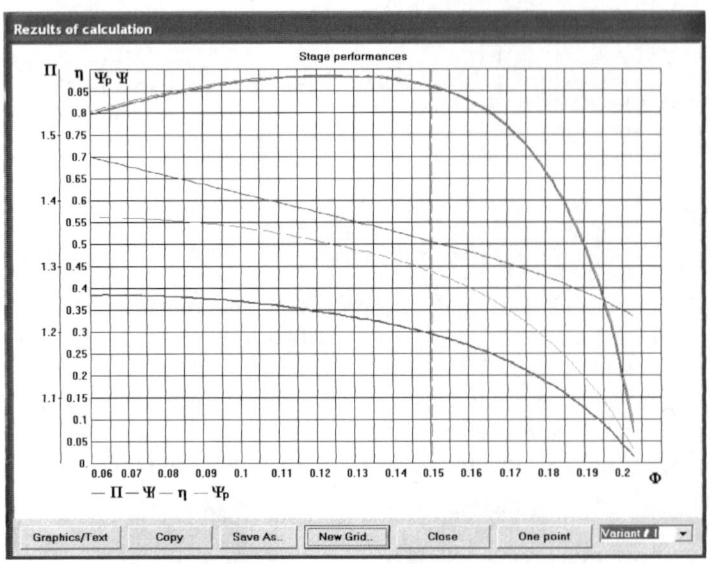

Figure 8. Stage 2 performances calculated by Universal modeling method.

Gas-dynamic performances were calculated by the Universal modeling method (figure 8). The high efficiency is expected because the design parameters $\Phi_{des} = 0.15$, $\psi_{T des} = 0.50$, $M_u = 0.55$. $\bar{D}_4 = 1.72$ are close to optimal.

5. Results of stage 2 performance modeling

The stage 2 performances were calculated by ANSYS CFX and by NUMECA Fine Turbo in four different variants. Results are presented in figures 9 and 10.

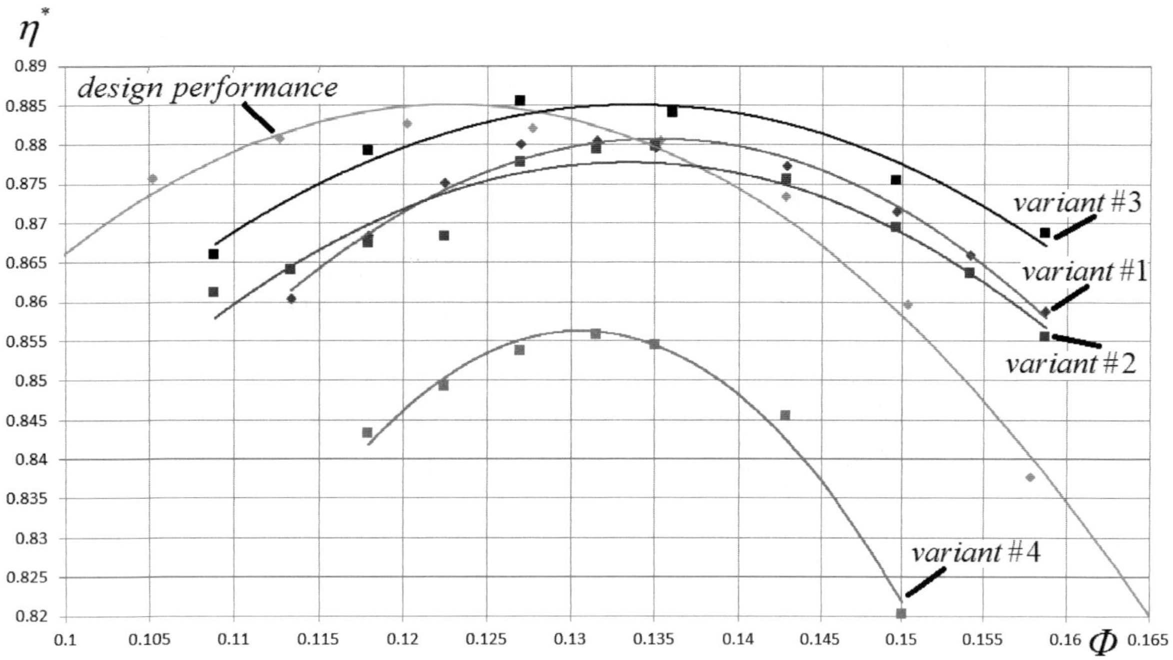

Figure 9. Stage 2 calculated efficiency performances.

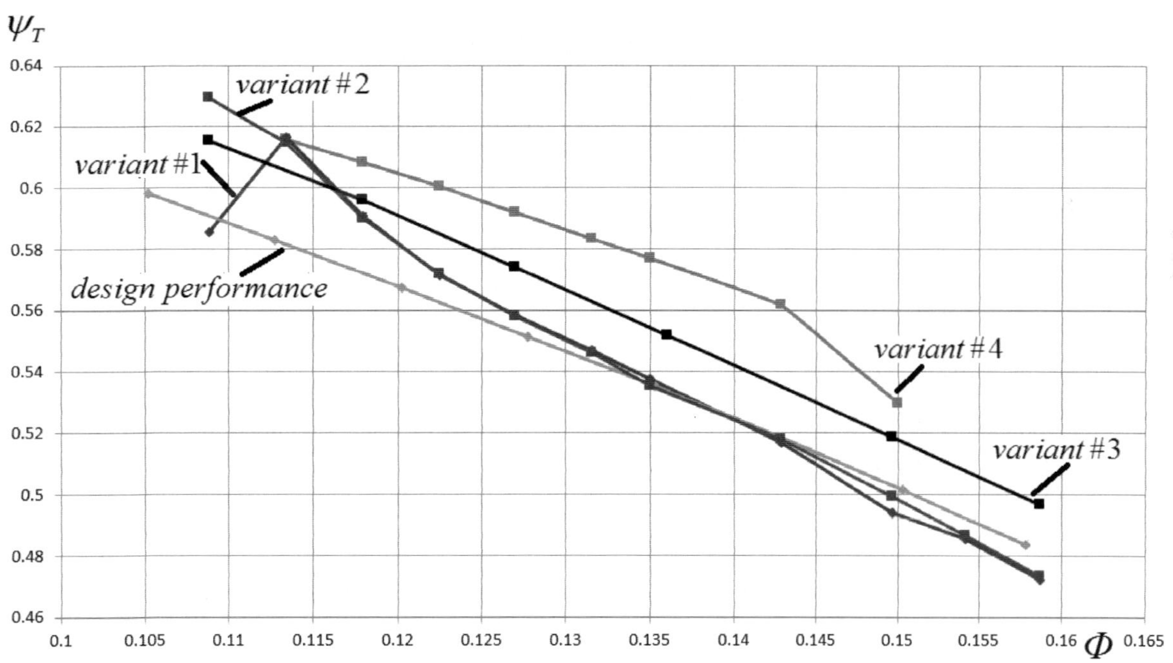

Figure 10. Stage 2 calculated loading factor performances.

The variants 1 includes radial inlet nozzle. The variant 2 has an axial inlet. Full flow path was calculated in both cases ("360 degrees"). Flow in “impeller – body” gaps was modeled and parasitic losses were calculated. The program ANSYS CFX uses the turbulence model SST. The design mesh for variants 1 contains 72 million cells, for variants 2 mesh contains 70 million cells.

Maximum efficiency is the same at $\Phi = 0.13$ that contradicts to practical experience. Tests demonstrate that the same stage with radial nozzle is less effective than with an axial nozzle by 2 - 4% [15]. The variant 3 was calculated without "impeller – body" gaps and one channel of the impellers and the RCH were modeled. The grids of different variants are presented in figure 11.

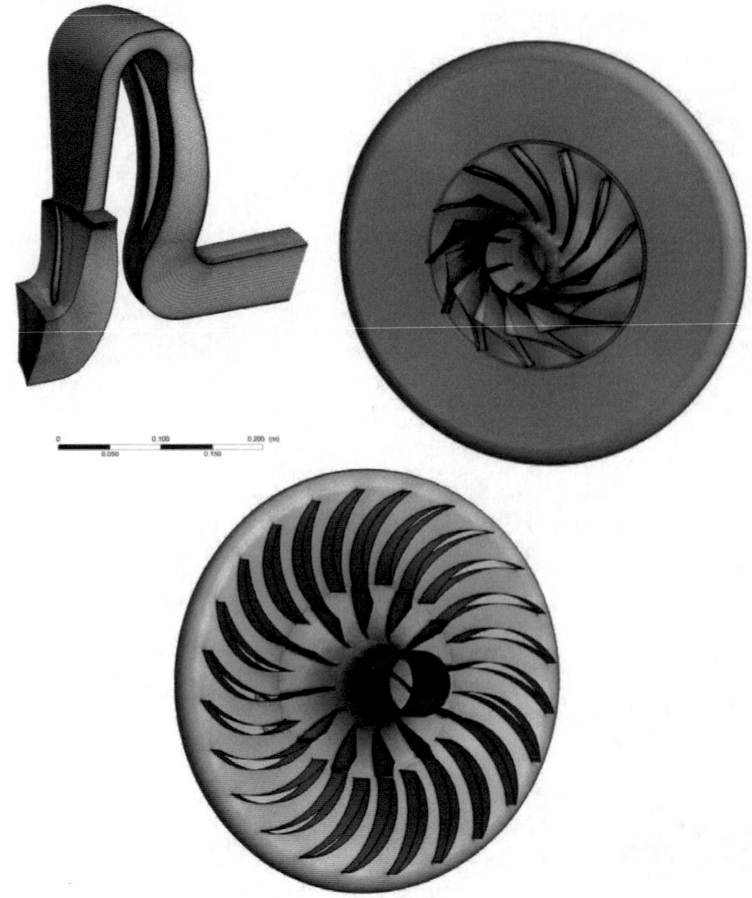

Figure 11. Computational grid for stage 2, variant 3 (left), variant 2 (amid and right).

To compare efficiency with the variant 2 parasitic losses were calculated by the Universal modeling method and calculated efficiency was diminished by about 1%. Maximum efficiency of the variant 3 is by 0.5% higher than that of the variant 2. It is equal to maximum efficiency calculated by the design program [14].

Variant 4 is the equivalent of the variant 3 but is calculated by NUMECA Fine Turbo. The efficiency is less by about 3% in comparison with ANSYS CFX calculation. The positive moment is that the loading factor performance calculated for full flow path ("360 degrees") is in satisfactory agreement with the design performance. For other variants, the difference between the calculated and design loading factor is achieved at 3.5% for variant 3 and at 8.7% for variant 4. The angle of inclination of loading factor performances is good match in all cases.

Several calculated parameters of the stage 2 variants are presented in table 1.

Table 1. Gas-dynamic parameters of stage on design flow rate and on the maximum efficiency.

	Universal modeling Method (design performance)	Variant 2 ANSYS «360 degrees»	Variant 3 ANSYS "sector"	Variant 4 NUMECA "sector"
η^*_{des}	0.860	0.869	0.875	0.820
$\psi_{T\,des}$	0.502	0.504	0.519	0.53
$\zeta_{imp\,des}$	0.0986	0.07	0.121	0.123
$\zeta_{VLD\,des}$	0.0600	0.127	0.03	0.09
$\zeta_{RCH\,des}$	0.366	0.275	0.29	0.322
$\eta^*_{\max}$	0.883	0.88	0.885	0.856

There is no visible regular tendency of the presented parameters variation which once more demonstrates problems of CFD-methods. Analysis of flow visualization leads also to contradictory results in some cases. As an instance figure 12 presents meridian velocity field in the variants 3 and 4 (design flow rate).

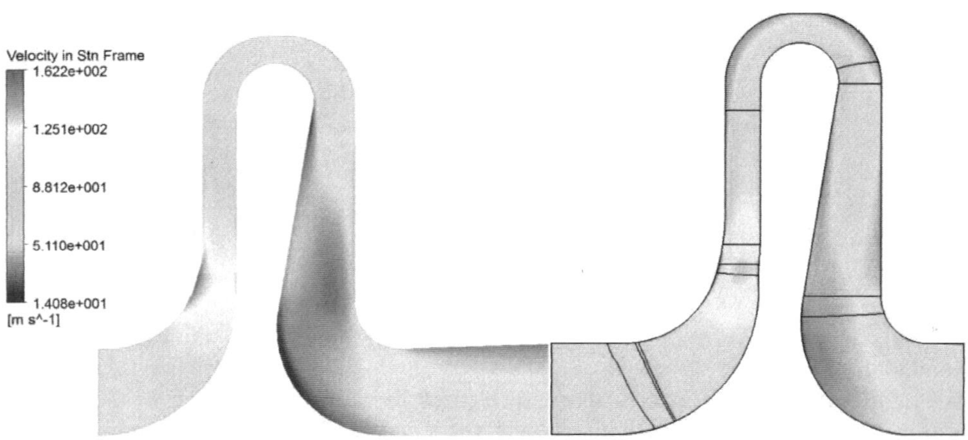

Figure 12. Meridian velocity field. Left – variant 3 (ANSYS CFX), right – variant 4 (NUMECA Fine Turbo).

Meridian flow seems more uniform in the variant 4, but impeller loss coefficients are practically equal in both cases. Flow in the VLD is more uniform in the variant 3 and its loss coefficient is much lower. On the contrary, flow is more uniform in the return channel of the variant 4 but its loss coefficient is higher.

6. Conclusions

1. The general tendency is that CFD-calculated performances shift to the right of real stage performances.
2. The connecting problem is that both programs cannot calculate flow rates far away of the design point. The modern stages with $\psi_{T\,des} \leq 0.50$ can operate with flow rate coefficients $\Phi = 0.5\Phi_{des}$ [16]. But CFD- calculation stops at $\Phi = 0.75\Phi_{des}$.
3. Maximum efficiency and design loading factor are predicted accurately by ANSYS CFX "360 degrees" calculation. "Sector" calculation is less accurate.

4. The program NUMECA Fine Turbo overestimates loading factor by 7–8% and underestimates efficiency by 3% in case of the stage with high flow rate coefficient $\Phi_{des} = 0.15$.

Nomenclature

C_p heat capacity, J/kg*K;
c absolute velocity, m/s;
D diameter, m;
$\bar{D} = \frac{D}{D_2}$ relative diameter;
$\bar{D}_h$ hub relative diameter;
k Isentropic coefficient;
$\bar{m}$ mass flow rate, kg/s;
M Mach number;
$M_u = \frac{u_2}{\sqrt{kRT_{inl\,tot}}}$;
p pressure, Pa
R gas constant, J/(kg*K);
T temperature, K
u_2 impeller periphery speed, m/s;
$\bar{w} = \frac{w}{u_2}$ relative velocity
β_{fr} disc friction coefficient;
β_{leac} shroud disc labyrinth seal leakage coefficient;
ρ gas density, kg/m^3;
ζ loss coefficient;
$\psi_T = c_{u2} / u_2$ - loading factor;
ψ_i work coefficient;
η^* total polytropic efficiency;
Φ flow rate coefficient;

Subscripts

0, 1, 2, 3, 4 indices of control sections
des design flow rate
inl inlet
ex exit
max maximum
m meridian
stat stator

Abbreviations

CFD Computational Fluid Dynamics
ST stator part of a stage
RCH return channel
VLD vaneless diffuser

Acknowledgments

The information about science work that is presented in the paper as result of the work of the R&D Laboratory "Gas dynamics of turbomachines", in science, research, technology, and development: "The creation of modern hi-tech production of design, production and installations tests, has compressing gaseous products for effective use in transport systems and technologies". (With financial support of the Government of the Russian Federation, Ministry of Education and Science of the Russian Federation. The contract 02.G25.31.0140 from 12/1/2015).

References

[1] Galerkin Y, Borovkov A I, Voinov I B, Gaev A V, Gamburder D M, Sofronova A A and Lozovaja N S 2007 Results of calculation of a viscous flow in motionless elements of centrifugal compressor stages by means of program ANSYS/CFX system *Compressors and pneumatics* **2** pp 10–16 [Russian]
[2] Galerkin Y B and Soldatova K V 2007 The analysis of gas flow in a gap "the covering disk – the case" a centrifugal compressor stage according to the FLUENT and CFX programs *Chemical and oil and gas mechanical engineering* **5** pp 27–29
[3] Galerkin Y B and Marenina L A 2014 Research and improvement of centrifugal stages fixed elements by methods of computational dynamics Part 1 *Compressors and pneumatics* **1** pp 30–36 [Russian]
[4] Galerkin Y B and Marenina L A 2014 Research and improvement of centrifugal stages fixed elements by methods of computational dynamics Part 2 *Compressors and pneumatics* **2** pp

10–15 [Russian]
[5] Galerkin Y, Soldatova K and Solovieva O 2015 Numerical study of centrifugal compressor stage vaneless diffusers *"Compressors and their Systems" Conference* (*London*)
[6] Galerkin Y and Drozdov A 2015 Sample of CFD optimization of a centrifugal compressor stage *"Compressors and their Systems" Conference (London)*
[7] Simonov A M 2000 Research and development of centrifugal compressor stages with increased loading factor *Proceedings of the scientific compressor school* (SPbPU) pp 189–212 [Russian]
[8] Galerkin Y 2010 *Transactions of the TU SPb compressor school* [Russian]
[9] Melnicov V and Proscurov A 2005 Using the software package FlowVision for flow calculation of the turbochargers elements in *SKBT CAD and graphics* **4** pp 92–96 [Russian]
[10] Gamburger D, Epifanov A and Gaev A 2009 Numerical investigation of three-dimensional viscous compressible gas in the centrifugal compressor impeller *Scientific and technical transactions of the TU SPb* **2**(78) pp 76–82 [Russian]
[11] Lurie E, Van Slooten P, Medic G, Mulugeta J, Holley B, Feng J and Sharma O 2011 Design of a high efficiency compact centrifugal compressor for rotorcraft applications *American Helicopter Society 67th Annual Forum* (*Virginia Beach May 2011*)
[12] Hazby H, Casey M, Robinson C and Spataro R 2017 The design of a family of process compressor stages *Proceedings of 12th European Conference on Turbomachinery Fluid dynamics & Thermodynamics* (*ETC12, April 3-7, 2017, Stockholm, Sweden*) Paper ID: ETC2017-134
[13] Kryłłowicz W, Kozanecki Z, and Kabalyk K 2017 Technical and aerodynamical aspects of a high pressure synthesis gas turbocompressor modernization *Proceedings of 12th European Conference on Turbomachinery Fluid dynamics & Thermodynamics* (*ETC12, April 3-7, 2017, Stockholm, Sweden*) Paper ID: ETC2017-171
[14] Galerkin Y 2010 *Tubrocompressors* (SPbSTU) 2010 [Russian]
[15] Seleznev K P and Galerkin Y B 1982 *Centrifugal compressors* (Leningrad) p 271 [Russian]
[16] Soldatova K 2016 Bank of model stages, with performances calculated based on the results of factory centrifugal compressors experiments *Compressors and pneumatics* **8** pp 20–24 [Russian]

IOP Conf. Series: Materials Science and Engineering **232** (2017) 012045 doi:10.1088/1757-899X/232/1/012045

CFD study of leakage flows in shroud cavities of a compressor impeller

K Soldatova
R&D Laboratory "Gas dynamics of turbo machines" Peter the Great St.Petersburg Polytechnic University, Polytechnical st. 29, St.Petersburg, Russia

Email: buck02@list.ru

Abstract: The flow character in a gap between shroud disc of an impeller and a stator surface (shroud cavity) influences disc friction loss, labyrinth seal loss (parasitic losses) and thrust force. Flow calculations inside the shroud cavity of a model of centrifugal compressor stage and its labyrinth seal in a range of flow rates and axial width and radial gap are presented. The results are presented in terms of non-dimensional coefficients of flow, disc friction and seal leakage losses coefficients and pressure coefficient. The distributions meridional and tangential flow velocities correspond to the continuity and equilibrium equations – flow radial circulation exists in wide cavity and is absent in narrow cavities. The radial pressure distributions as measured and calculated are not fully comparable. The possible reason is that CFD-calculated leakage coefficient is less than calculated by A.Stodola formula. The influence of a cavity width on the losses and the thrust force requires a balanced design.

Nomenclature

b	width of channel
$\bar{b}_z$	relative cavity width
c_m	meridional velocity
c_u	tangential component of absolute flow velocity
D	diameter
k	isentropic coefficient
$k-\varepsilon$	turbulence model
$\bar{m}$	mass flow rate
$\bar{m}_{leak}$	labyrinth seal leakage mass flow rate
M_u	blade Mach number
M_z	friction force moment
$P_{fr\,sc}$	disk friction power
P_T	power transmitted to gas by blades
p	pressure
r	radius
Re_u	impeller diameter Reynolds number
$\bar{R}_z$	axial force
R_{zx}	force in a shroud cavity
$\bar{R}_{zx}$	non-dimensional mode by axial force coefficient
T	temperature
u	peripheral velocity
w	relative velocity
Γ	flow circulation
Φ	flow rate coefficient
$\beta_{fr\,sc}$	non-dimensional coefficient of disk friction
β_{leak}	non-dimensional leakage coefficient
δ_r	rotor/stator gaps
$\bar{\delta}_r$	relative radial gaps
λ	friction coefficient
μ	empirical coefficient for the seal
ρ	density
τ	shear stress
ψ_T	loading factor
Δp	pressure differential

Subscripts

0	impeller inlet
1	impeller blade row inlet
2	impeller outlet
av	average

body case of compressor
imp impeller
des design
lb labyrinth
u peripheral velocity

Abbreviation
CFD Computational Fluid Dynamics

1. Object and goals

The object of this CFD analysis is a shroud cavity, i.e. an axial gap between a shroud disc and a body of a centrifugal stage. The model stage cross section and typical configuration of a shroud cavity are presented in Figure 1.

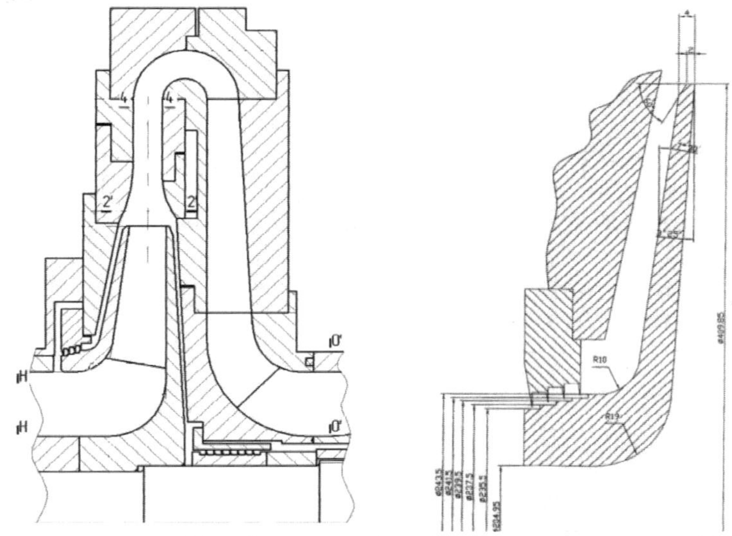

Figure 1. Compressor stage meridional cross-section with the shroud cavity and labyrinth seal [1-2]

Cavity flow results in so-called parasitic or winding losses. Static pressure difference in shroud and hub cavities influences ~~on~~ thrust force. Recent publications [3-4] point out the importance of cavity modeling for precise CFD calculation of centrifugal compressor performances. This work was done in 2004 – 2005 but has never been presented at international level. The Author is aware of recent study of the problem presented in several ASME and ATI publications and have decided to add that contribution to their results. It is very encouraging that recent numerical study and test verification presented in [4] are in good agreement with flow structure calculated by the Author.

The numerically studied shroud cavity belongs to one of tested model stages as shown in Fig. 1, above, left. The stage was tested in the TU SPb Federal Laboratory of Compressor problems in 1990th [5]. Boundary conditions at calculations correspond to measured flow parameters – pressures and temperatures at the impeller inlet and exit – 6 points with different flow coefficient inside a performance range. Similarity criteria are $M_u = 0.60$, $Re_u = 6E6$, $k = 1.4$. Three relative cavity widths were tested: $b_z / D_2 = 0.0145$ (as of the model stage), 0.0073 and 0.029. Relative radial gaps of the labyrinth seal were $\delta_r / D_2 = 6.1E\text{-}4$, 12.2E-4 (as of the model stage) and 24.4E-4. The minimal value $\delta_r / D_2 = 6.1E\text{-}4$ was increased twice to study seal wearing influence.

Pressure in five points along radius of the shroud cavity was measured altogether with overall performances of the model stage with $\Phi_{des} = 0.028$ – four tests. Goals of the numerical research were:

- To analyze the flow structures
- To compare the pressure distribution along the radius with the measured one,
- To estimate the influence of axial gap between the impeller disk and the stator wall,
- To estimate the influence of the radial gap in the labyrinth seal,

- To estimate the possible variation in the thrust force.

2. Cavity modeling

Initially calculations were performed with software package called FLUENT and later using package ANSYS CFX. The necessity to change software package was caused by access problems. Grids were created by the programs called TurboGrid. Turbulence model applied is $k-\varepsilon$. Flow is steady. The casing wall is stationary; the impeller wall is rotating with the velocity corresponding to $M_u = 0.60$.

Calculation results that are presented below, they are related to three aspects:

- Flow pattern in a cavity,
- Flow pattern in a labyrinth seal,
- Integral parameters as leakage coefficient, disk friction coefficient, thrust force coefficient.

3. Flow structure in the cavity

Streamlines on meridional plane are presented in Fig. 2, while meridional velocity vectors at inlet and outlet of the shroud cavity are shown in Fig. 3. Radial circulation appears due to different centrifugal forces of gas particles near rotor and stator surfaces: $dp_{cntr} = \rho \frac{c_u^2}{r}$.

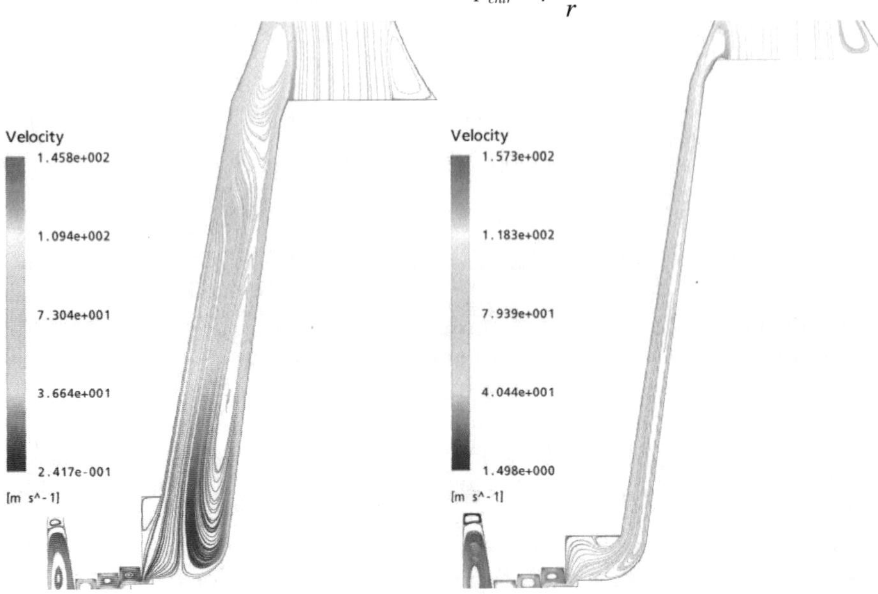

Figure 2. Meridional projections of streamlines. Design flow coefficient.
Left - $b_z / D_2 = 0.029$, right – 0.0073. FLUENT [2]

Peripheral velocity is close to zero near a stator surface. Peripheral velocity is close to $u = \omega \times r$ near an impeller disk surface.

Radial circulation is significantly active in the cavity with $b_z / D_2 = 0.029$ (it is two times higher than standard value $b_z / D_2 = 0.0145$). Streamlines are of the same character in the cavity with $b_z / D_2 = 0.0145$.

Flow radial circulation is suppressed to some extent in the cavity with $b_z / D_2 = 0.0073$. This width is too small to be in a real stage due to danger of rotor-stator contact.

As a result - meridional velocities are positive near an impeller and are negative near a stator wall. Fig. 4 demonstrates meridional velocities at different radii at design flow coefficient.

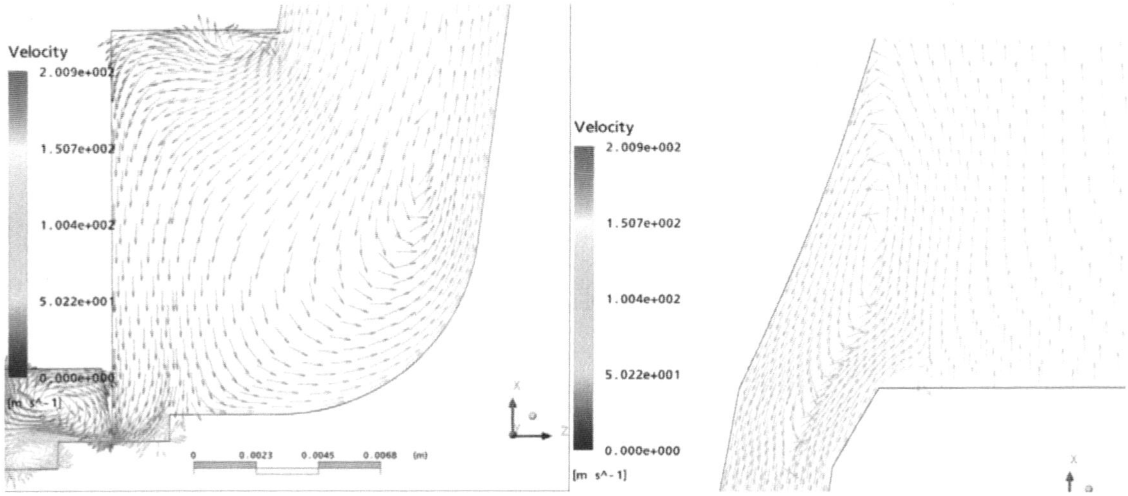

Figure 3. Meridional velocity vectors. Left – cavity inlet, right– cavity outlet. $b_z/D_2 = 0.0145$, $\Phi = \Phi_{des}$. FLUENT [2]

Meridional centripetal flow is also a result of a leakage in a labyrinth seal. Labyrinth seal leakage mass flow rate $\bar{m}_{leak}$ enters a shroud cavity with peripheral velocity $c_{u2} = \psi_T u_2$. Flow circulation $\Gamma = 2\pi r c_u = const$ would take place in a cavity of an ideal impeller without viscosity. "Inviscid" peripheral velocity increases in a cavity at smaller radii in accordance with $c_u = c_{u2}\frac{r_2}{r}$. In reality the friction forces are diminishing flow circulation and peripheral velocity: $d(c_u r)/dr = -dM_z$, where $dM_z = 2\pi r dr(\tau_{body} + \tau_{imp})$, $\tau_{body} = \lambda\rho\frac{\vec{c}_u^2}{2}$, $\tau_{imp} = \lambda\rho\frac{\vec{w}_u^2}{2}$.

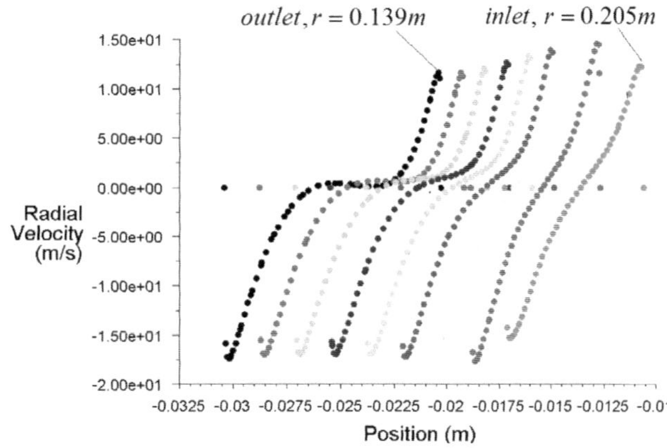

Figure 4. Meridional velocities along cavity width $c_m = f(b)$ at several radii. $\Phi = \Phi_{des}$, FLUENT [2]

Peripheral velocity distribution on several radii is shown in Fig. 5.

As loading factor ψ_T is larger at smaller flow coefficients, tangential velocities grow as well. Tangential velocities in rotational coordinates $\vec{w}_u$ are less than rotation velocity of an impeller disk. Disc surface transmits mechanical energy to a gas at all radii. Eventually tangential velocities sufficiently decrease from inlet to outlet of a cavity as result of friction forces on a stator wall.

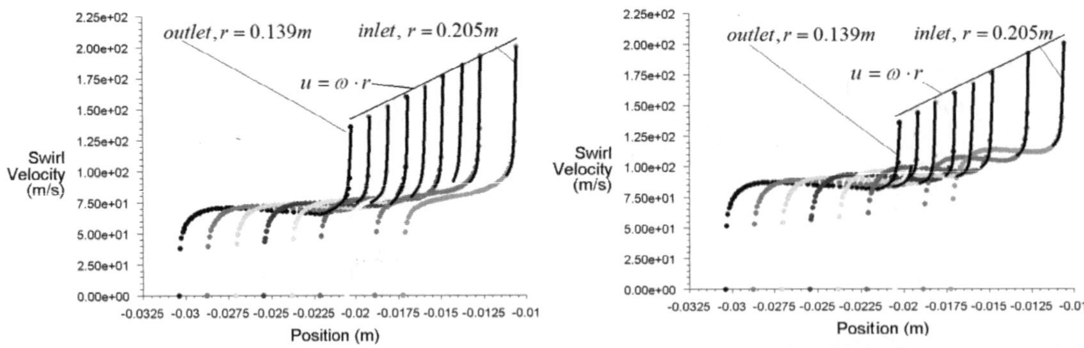

Figure 5. Peripheral velocity along cavity width $c_u = f(b)$ at several radii. Left - $\Phi = \Phi_{des}$, right - $\Phi = 0.6\Phi_{des}$. FLUENT [2]

4. Flow structure in the labyrinth seal

Velocity vectors in one element of the seal and meridional velocities in the seal are presented in Figure 6. The flow structures depict well the principle of a labyrinth seal operation [2].

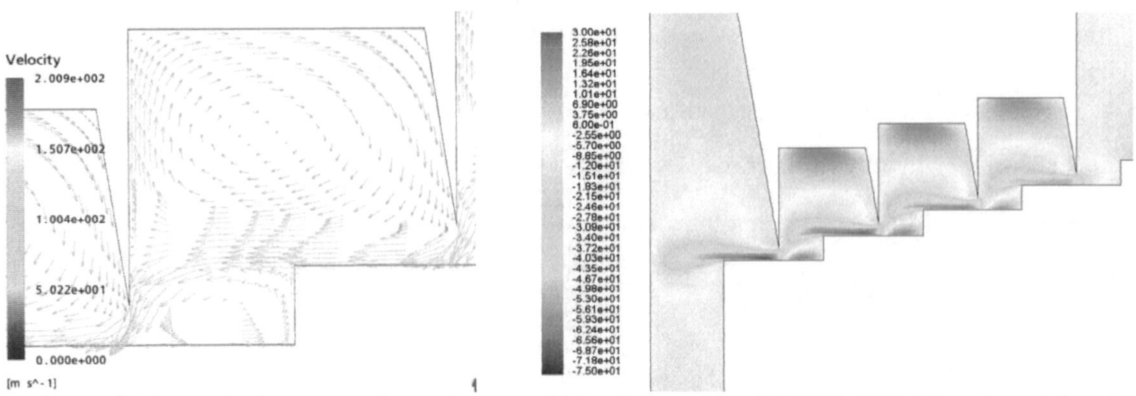

Figure 6. Flow velocity vectors in an element of labyrinth seal by ANSYS CFX [2] and meridional flow velocities in the labyrinth seal by FLUENT

5. Static pressure: measured and calculated

Calculation has demonstrated that static pressure in the cavity at the given radius is practically constant in axial direction. Its variation along the radius is due to flow rotation with peripheral velocity c_u: $p(r) = p_2 - \int_r^{r_2} \rho \frac{c_u^2}{r} dr$.

The measurements and calculations results are presented as non-dimensional pressure coefficients:

- $\Delta\overline{p}_{(r)}$ is a function of radius. It demonstrates pressure change in a cavity at given flow rate: $\Delta\overline{p}_{(r)} = \frac{p_2 - p_{(r)}}{\rho_{av} u_2^2}$.
- $\Delta\overline{p}_{lbr}$ demonstrates the pressure change at the inlet of labyrinth seal, which is a function of the flow rate coefficient or loading factor ψ_T (a loading factor ψ_T is a function of a flow rate coefficient Φ). It represents pressure drop in all cavity in non - dimensional mode: $\Delta\overline{p}_{lbr} = \frac{p_2 - p_{lbr}}{\rho_{av} u_2^2}$.

Results of measurements from four tests are presented in Fig. 7 as $\Delta\overline{p}_{(r)} = f(\Phi)$. Measurements were made during four tests. Results are presented as three groups of graphics with ψ_T = 0.33-0.44, 0.48-0.55 and 0.58-0.64.

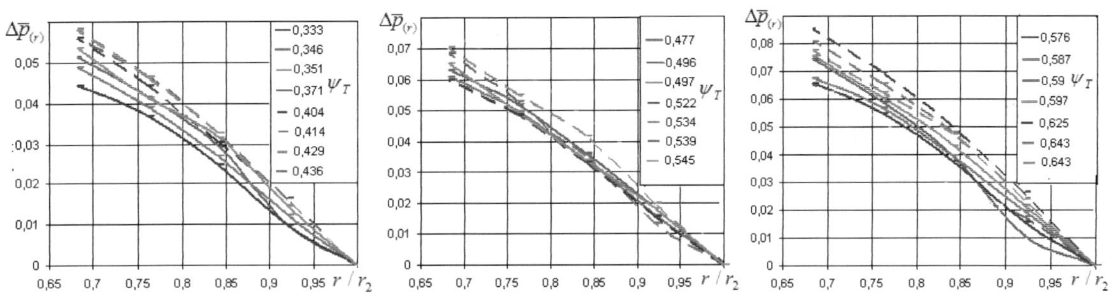

Figure 7. Pressure coefficient as function of radius at different loading factors. Dash – measurement, solid – calculation

The stage in question was tested several times with different purposes with proper assembling/disassembling. The seal radial gap was not controlled properly. It was approximately 0.5 ± 0.15 mm. Nevertheless, the general trend is evident: the higher the loading factor is – the more pressure drops in the cavity. Calculations also demonstrated this trend. Comparison of measurements and calculations is presented in Fig. 8.

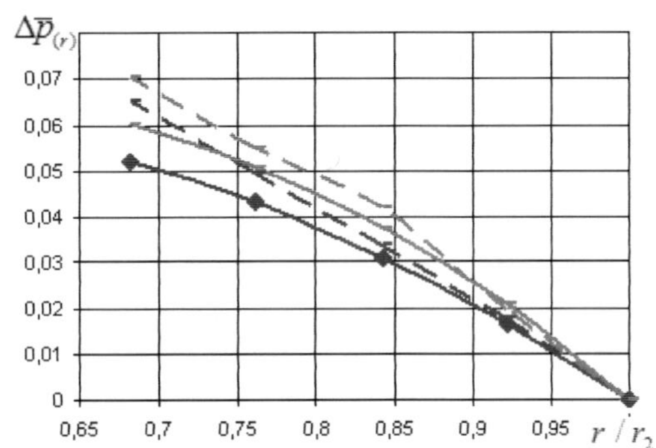

Figure 8. Measured and calculated pressure coefficient as function of radius. Dash – measurement, solid – calculation (ANSYS CFX). Red ψ_T = 0.545, blue – ψ_T = 0.495

The comparison is made at loading factors close to design regime. Matching is better at higher loading factors in this figure and in general.

Pressure coefficient $\Delta\overline{p}_{lbr} = \dfrac{p_2 - p_{lbr}}{\rho_{av} u_2^2}$ represents pressure drop in a cavity. Measured and calculated coefficients are compared in Fig. 9.

Calculated function $\Delta\overline{p}_{lbr} = f(\psi_T)$ is based on averaged values measured at four different tests. Calculated coefficients are visibly lower. It means that loss of flow circulation $c_u \times r$ is more than in reality. It is shown below that CFD calculated seal leakage is less than calculated by well proven A.Stodola formula. It also influences $\Delta\overline{p}_{lbr} = f(\psi_T)$ mismatch.

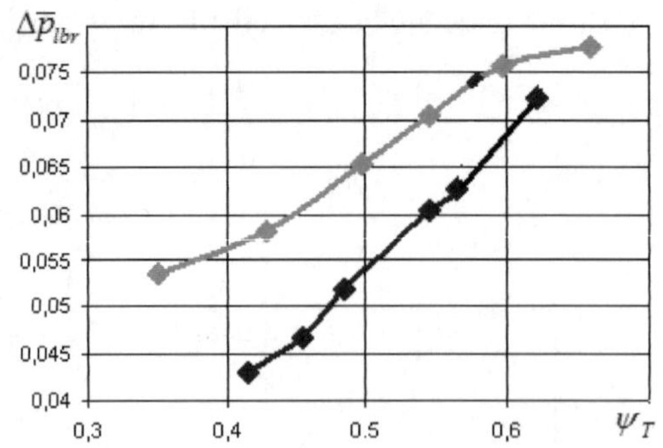

Figure 9. Measured and calculated pressure coefficient $\Delta\bar{p}_{lbr}$ as function of loading factor. Red – measurement, black – calculation (ANSYS CFX)

6. Disk friction losses

Non-dimensional coefficient of disk friction is a ratio of disk friction power and power transmitted to gas by blades. This coefficient for the shroud cavity is:

$$\beta_{fr\,sc} = \frac{P_{fr\,sc}}{P_T} = \frac{\omega \int_{r_{lbr}}^{r_2} \tau \cdot 2\pi r dr \cdot r}{\psi_T \cdot \rho_{0t} \cdot \Phi \frac{\pi}{4} D_2^2 u_2^3}.$$

Parameters in the denominator are parameters of the tested model stage. Shear stress τ is CFD calculated. Fig. 10 presents calculation results.

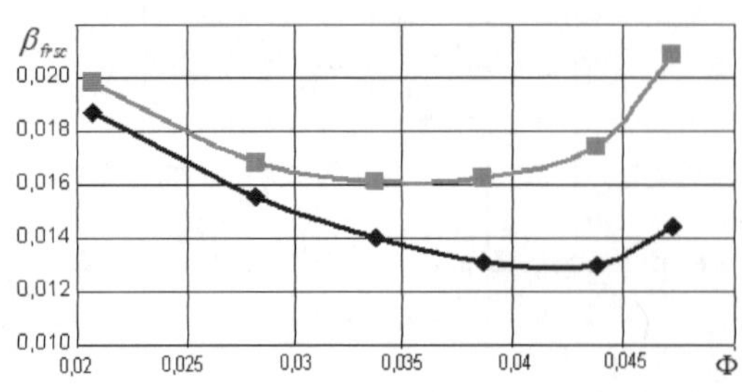

Figure 10. Friction coefficient of the shroud disk. Red – FLUENT, black –ANSYS CFX

All members of the equation for $\beta_{fr\,sc}$ calculation are the same for both program packages, except the shear stress coefficient τ calculated by both packages. The difference near design point Φ_{des} = 0.028 is approximately 8%. The author cannot explain the difference of results presented by FLUENT and ANSYS CFX packages.

The coefficient $\beta_{fr\,sc}$ is higher at high flow coefficients because of the loading factor diminishing.

Calculations in Fig. 11 are made for the shroud cavity with an axial width b_z / D_2 = 0.0145. For cavities with b_z / D_2 = 0.0073 and 0.029 the more the width the more is friction coefficient. In wide gaps radial circulation of flow is more intensive that leads to higher losses. The difference at the design flow rate is 9–10%. It is about 0.15% of a loading factor that is not insufficient. But minimization of friction losses by applying $b_z / D_2 <$ 0.0145 is impossible because of hazard of contacting rotor and stator.

Calculations have demonstrated negative correlation between radial gaps in a labyrinth seal and disk friction coefficient. But this results in stronger flow leakage and higher tangential velocities in the cavity. This is not the way to increase efficiency as higher leakage losses cannot be compensated.

7. Leakage losses

Non-dimensional leakage coefficient is: $\beta_{leak}=\frac{\overline{m}_{leak}}{\overline{m}}=\frac{\overline{m}_{leak}}{\Phi\cdot\rho_{0t}\cdot\frac{\pi}{4}D_2^2u_2}$.

The parameters in the denominator are obtained from the tested stage, while the flow rate of leakage is directly from CFD calculations. Leakage coefficient was calculated by ANSYS and FLUENT and compared with the result of Stodola. The formula for a shroud cavity is recommended in its simplified variant due to low compressibility effects in a cavity [7]:

$$\overline{m}_{leak}=\mu\cdot 2\pi D_{lb}\delta_{lb}\rho_{lb}\sqrt{\frac{2(p_{lb}-p_0)}{z_{lb}\rho_{lb}}}. \tag{1}$$

The empirical coefficient for the seal of the tested stage recommended value is μ = 0.7. Results of calculations are presented in Fig. 11.

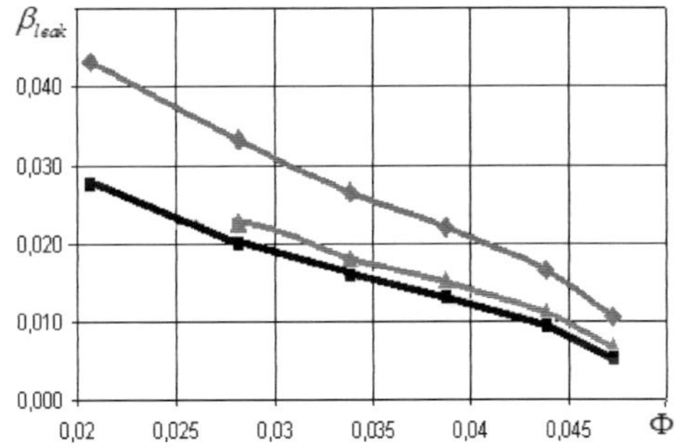

Figure 11. Calculated leakage loss coefficients. Red – Stodola, green – ANSYS CFX, black - FLUENT

CFD calculations predict character of $\beta_{leak}=f(\Phi)$ well enough. Calculated leakage flow is less than calculated by the A.Stodola formula. For a compressible flow it is recommended in [7-8]:

$$\overline{m}_{leak}=\mu\cdot 2\pi D_{lb}\delta_{lb}\sqrt{\frac{p_{lb}^2-p_0^2}{z_{lb}RT_{lb}}}. \tag{2}$$

Flow rate according to eq. (2) is 5% smaller than calculated by the eq. (1) for incompressible flow. Calculations by eq. (2) and by ANSYS CFX at different radial gaps are presented in Fig. 12.

Calculations eq. (2) were made with constant values of μ = 0.7. CFD calculations produced results not quite good. At δ_r = 0.25 the calculated coefficient is not far, but for δ_r = 0.5 mm it is less by about 30%, while for δ_r = 1.0 mm this coefficient is more by 15 – 20%.

The possible reason of Stodola and CFD calculation mismatch must be the following. CFD packages calculate flow with peripheral velocity up to 125 m/s in the labyrinth seal. Most of all that experiments for definition of coefficient μ in the Stodola formula were made for flow without peripheral velocity.

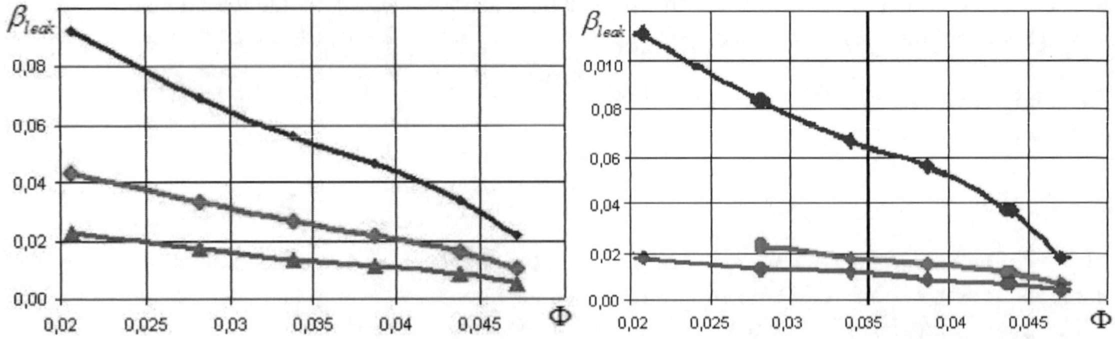

Figure 12. Leakage coefficient at different radial gaps. Above – eq. (2), below – ANSYS CXF. Green - $\delta_r = 0.25$ mm, red - $\delta_r = 0.5$ mm, blue - $\delta_r = 1.0$ mm ($D_2 = 410$ mm)

8. Thrust force coefficient

Gas pressure difference on impeller surfaces creates an axial force that is balanced by a thrust bearing. Precise calculation and minimization of the thrust force are important. The total force is directed to impeller inlet, while in the shroud cavity this force is opposite. A force in a shroud cavity R_{zsc} is directed against total force. Drop of pressure in a shroud cavity due to tangential velocities increases an axial force.

Results of calculation are presented in non-dimensional mode and are presented by axial force coefficient: $\overline{R}_{zsc} = \int_{r_{lb}/r_2}^{1} \frac{\Delta\overline{p}}{1-(D_{lb}/D_2)^2}\overline{r}d\overline{r}$. Axial force in a shroud cavity is: $R_{zsc} = \overline{R}_{zsc} \cdot \rho_2 u_2^2 \frac{\pi}{4}(D_2^2 - D_{lb}^2)$.

The more is R_{zsc} the less is total axial force. For axial force minimization bigger $\overline{R}_z$ are necessary. Calculation results by ANSYS CFX are presented in Fig. 13.

Low leakage flows lead to higher pressure in the shroud cavity and lower thrust force. Minimization of leakage flow is better for the compressor efficiency and for a thrust bearing.

Intense radial circulations in the wide shroud cavity ($b_z / D_2 = 0.029$) leads to a loss in flow momentum. Comparison of pressure coefficients for cavities of elevated $b_z / D_2 = 0.029$ and the recommended $b_z / D_2 = 0.0145$ is presented in Fig.14.

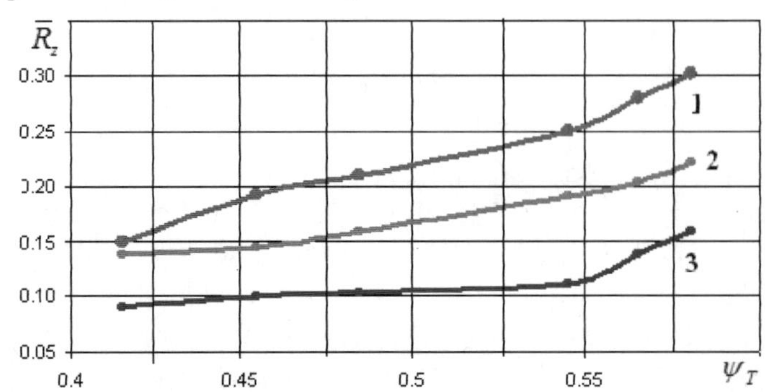

Figure 13. Thrust force coefficient as function of loading factor.
1 – $\delta_r = 0,25$ mm; 2 – $\delta_r = 0,50$ mm; 3 – $\delta_r = 1,0$ mm

The strong influence of the relative width on pressure distribution is evident. Low level of tangential velocities leads to higher disc friction, while larger pressure at the seal inlet leads to higher leakage. The disk friction and leakage coefficients are about 1% bigger in wide shroud cavity at design flow coefficient $\Phi_{des} = 0.028$ (ANSYS CFX), however, the loss in flow momentum is good for thrust force minimization. Thrust force coefficient versus loading factor for two shroud cavities are shown in Fig. 14. In case of necessity the designer can choose a wider shroud cavity for the thrust force minimization.

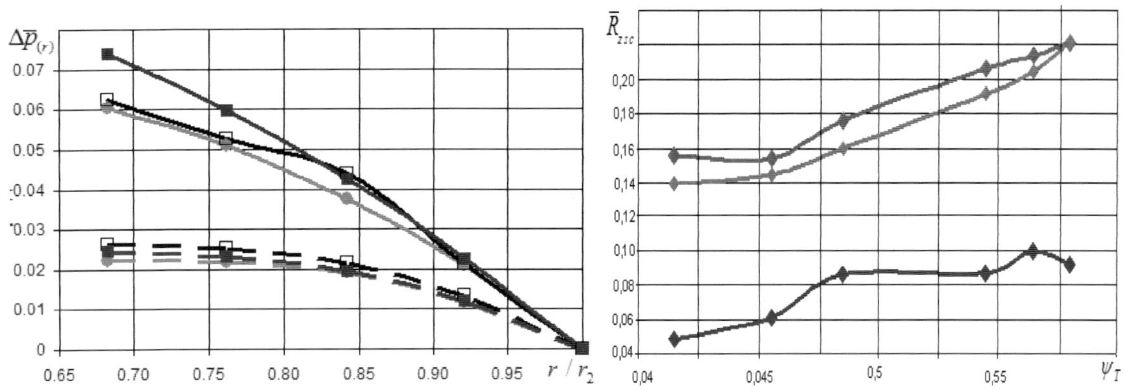

Figure 14. ANSYS CFX. Pressure coefficients in two shroud cavities. (Solid - $b_z / D_2 = 0.0145$, dashed - $b_z / D_2 = 0.029$. Data for three loading factors $\psi_T = 0.46, 0.53, 0.62$) and thrust force coefficient versus loading factor (green - $b_z / D_2 = 0.0076$, red - $b_z / D_2 = 0.0145$, blue - $b_z / D_2 = 0.029$)

Conclusion

CFD calculations present detailed information on flow character and important integral parameters as disc friction coefficient, seal leakage coefficient, thrust force coefficient. These parameters are essential for a compressor design and performance calculations. Calculated flow behavior at different seal gaps and cavity width corresponds to described already in [1] as result of 1-D analysis. Measured and calculated pressure distribution along the cavity is close by character but differ in values. CFD - calculated leakage coefficient is less than calculated by known Stodola formula. The problems require further study. The Author's recommendations to pay attention to axial gaps for loss minimization or the thrust force minimization are indirectly supported by results presented in [4]. Better understanding of flow behavior and following ideas influenced author's design practice positively. There is a hope that presented information and considerations will be useful for colleagues who are active in this area now.

References

[1] Seleznev, K.P., Galerkin, Y.B. Centrifugal compressors. Leningrad. – 1982. – P. 271. [Russian].
[2] Soldatova, K.V. Analysis of gas flow in a gap «shroud - casing» of a centrifugal compressor stage by numerical methods and design recommendations. Diss. Cand.Tech. sciences. SPbTU. – SPb. – 2007.
[3] Guidotti, E., Tapinassi, L., and Naldi, G., 2012. Infuence of seal leakages modeling on low flow coeffcient centrifugal compressor stages performance. In ATI Congress 2012.
[4] Guidotti, E. Towards centrifugal compressor stages virtual testing. Universita delgi studi di Bologna. – 2014.
[5] Transactions of the TU SPb compressor school. –TU SPb. – 2000. [Russian].
[6] Soldatova, K.V. Calculation of leakages in a shroud – casing gap and in a labyrinth seal of a centrifugal impeller by FLUENT package. Master. diss. - SPbTU. – SPb. – 2004.
[7] Ris V. Centrifugal compressors. Leningrad. – 1981. [Russian].
[8] Transactions of the TU SPb compressor school. Under Y.B. Galerkin's edition. - 2005 [Russian].

A network application for modeling a centrifugal compressor performance map

A Nikiforov[1], D Popova[1], K Soldatova[2]

[1] Department of Mechanization, Smolensk State Agricultural Academy, 214000, Smolensk, Bolshaja Sovetskaja, 10/2, Russia
[2] R&D Laboratory "Gas dynamics of turbo machines" Peter the Great St.Petersburg, Polytechnic University, Polytechnical st. 29, St.Petersburg, Russia

Email: buck02@list.ru.

Abstract: The approximation of aerodynamic performance of a centrifugal compressor stage and vaneless diffuser by neural networks is presented. Advantages, difficulties and specific features of the method are described. An example of a neural network and its structure is shown. The performances in terms of efficiency, pressure ratio and work coefficient of 39 model stages within the range of flow coefficient from 0.01 to 0.08 were modeled with mean squared error 1.5 %. In addition, the loss and friction coefficients of vaneless diffusers of relative widths 0.014-0.10 are modeled with mean squared error 2.45 %.

Nomenclature

$\bar{b}_1$	inlet relative blade height
$\bar{b}_2$	outlet relative blade height
$\bar{b}_3$	VLD relative width an inlet
$\bar{b}_4$	VLD relative width an outlet
$\bar{b}_5$	inlet relative vane height
$\bar{b}_6$	outlet relative vane height
$\bar{D}_0$	impeller inlet relative diameter
$\bar{D}_{0'}$	outlet of stage relative diameter
$\bar{D}_{1'}$	impeller blade cascade relative inlet diameter
$\bar{D}_2$	impeller diameter
$\bar{D}_3$	non-dimensional diameter in an inlet VLD
$\bar{D}_4$	VLD outlet relative diameter
$\bar{D}_5$	return channel vane cascade relative inlet diameter
$\bar{D}_6$	return channel vane cascade relative outlet diameter
$\bar{D}_h$	hub ratio
$\bar{D}_{seal}$	seal relative diameter

IOP Conf. Series: Materials Science and Engineering **232** (2017) 012046 doi:10.1088/1757-899X/232/1/012046

M	Mach number
$\bar{R}_3$	crossover relative external radius
$\bar{R}_4$	crossover relative internal radius
$\bar{R}_{s1}$	shroud relative curvature radius
$\bar{R}_{s2}$	non-dimensional edge radius cover impeller disk
Φ	flow coefficient
z_{imp}	number of impeller blades
z_{rch}	number of return channel blades
α_2	flow angle an inlet
β_{bl1}	inlet blade angle
β_{bl2}	outlet blade angle
β_{bl5}	return channel vane inlet angle
$\bar{\delta}_{bl}$	relative impeller blade thickness
$\bar{\delta}_{rch}$	relative vane thickness
η	polytropic efficiency
λ	friction coefficient
ψ	loading factor
ζ	loss coefficient

Abbreviation

CFD	Computational Fluid Dynamics
G	set of the parameters describing stage geometry
KPD	measured efficiency
KPD NS	efficiency calculated on model
PSI	measured pressure coefficient
PSI NS	pressure coefficient calculated on model
VLD	vaneless diffuser

1. Introduction

The complexity of the aerothermodynamic processes in centrifugal compressors does not allow defining their performances analytically. Until recently, any simple way to create high~~ly~~ effective designs of turbocompressors was very expensive and relied on experiments.

With the accumulation of experimental data and the development of math models, the mean line modeling tools started [1-3]. Nowadays these tools have already been developed and are provided by many scientific centers [4-7].

The mathematical modeling of centrifugal compressors is reduced to the identification of the models derived from experimental data. These models have rather complicated analytic representation defined by every researcher in view of the operating process and the selected physical model. For example, the authors Lunev A., Vyachkilev O., Drozdov Y. [4-5] use hydraulic analogues. The school of modeling [6] uses concepts of the boundary layer theory. The Universal modeling method [1-3] operates with loss coefficients of blades, vane cascades, and vaneless channels. Arguments of algebraic equations are tangential and normal velocity gradients.

The response surfaces of the objective functions are multi-extremal, thus the process of identification of models becomes complicated. The largest problems were related to the necessity of calculating the flow in the air-gas channel, the a priori assignment of the generalized analytical dependencies of the various losses on the velocity distribution characteristics, the assignment of the initial vector of the desired parameters and the search area. These problems are connected with need of calculation of a current for flow part, with an aprioristic task of the generalized analytical types of dependences of various losses on characteristics of distribution of velocities, a task of an initial vector of required parameters and areas of search of values.

Despite that, scientists managed to develop mathematical models for calculating gas dynamic performances. It opens a way to create computing methods of design and to offer producers high effective centrifugal compressors without long and expensive physical experiments. This gave confidence to researchers but the above difficulties of math modeling didn't manage to be overcome yet.

Software packages of neural networks application provide solutions to overcome difficulties of math model development and identification. Neural networks provide the universal mechanism of approximation adequate to multidimensional data files, are capable to be trained and arranged at changing conditions, can generalize the gained knowledge on the basis of what is considered as systems of artificial intelligence. The basis of functioning of neural networks is made by the algorithms of training allowing optimizing process of search of decisions [8-9].

2. Object and aim

Neural networks are used in various fields of science. They successfully solve tasks such as function approximation and optimization, management and forecasting processes. These capabilities are already being used in aerospace, military, medical, industrial, financial and banking fields, data transmission. Due to its efficiency and broad range of functions neural networks are becoming more common and useful for more complex problems.

Neural networks play a role of an universal approximator of function of several variables which realizes nonlinear function of a type $y = F(x)$, where x – an entrance vector (perhaps, an array), and y – the realized function of several variables. Many problems of modeling, identification and data processing can be solved by approximation with neural networks.

The process of approximation by means of neuronets consists of selecting the weight coefficients of defining the degree of importance of the communications between the neurons and this process is called "learning of neural networks". For an assessment of quality of training mean squared error is applied to minimize deviations from the entered experimental values.

Using of gradient methods became possible thanks to the introduction of the sigmoid activation function $f(x) = (1 + e^{-\beta x})^{-1}$, which is continuous, unipolar and differentiated. In practice it is most often used $\beta = 1$. But a user can independently select values β that influences a sigmoida form – at small β function is flat, at $\beta \rightarrow \infty$ it turns into step function. Such possibility of a variation increases flexibility of a neural network in general (Figure 1).

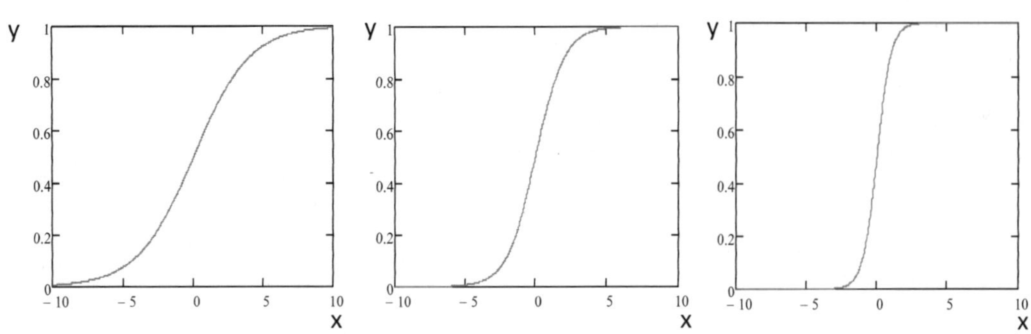

Figure 1. Sigmoid activation function of neurons: a) $\beta = 0.5$, b) $\beta = 1$, c) $\beta = 2$

Considered the simplest option of a multilayered neural network (Figure 2) with a direct signal transmission. The network consists of two layers, in the first layer - two neurons, in the second - one. Activation function of neurons is a logical sigmoid function.

Two arguments come to an input of the first layer come, each one is multiplied by weight w_{ij}, the weighed values are transmitted to the adder $\sum$ together with bias of b. Networks with bias allow to form more difficult communications between inputs and exits, than networks without bias, and to provide a nonzero exit of neurons. The resultant sum of n^1 serves as argument for function of activation of $f^1 : n_1^1 = w_{11} \times x_1 + w_{12} \times x_2 \ +b_1^1, n_{12}^1 = w_{21} \times x_1 + w_{2z} \times x_z \ +b_z^1$.

Exits of neurons of the first layer are passed to the second layer, which are described by the neuron equation with bias: $a_1^1 = \dfrac{1}{1+e^{-\beta(n_1^1)}} = \dfrac{1}{1+e^{-\beta(w_{11}\times x_1 + w_{12}\times x_2 \ b_1^1)}}$.

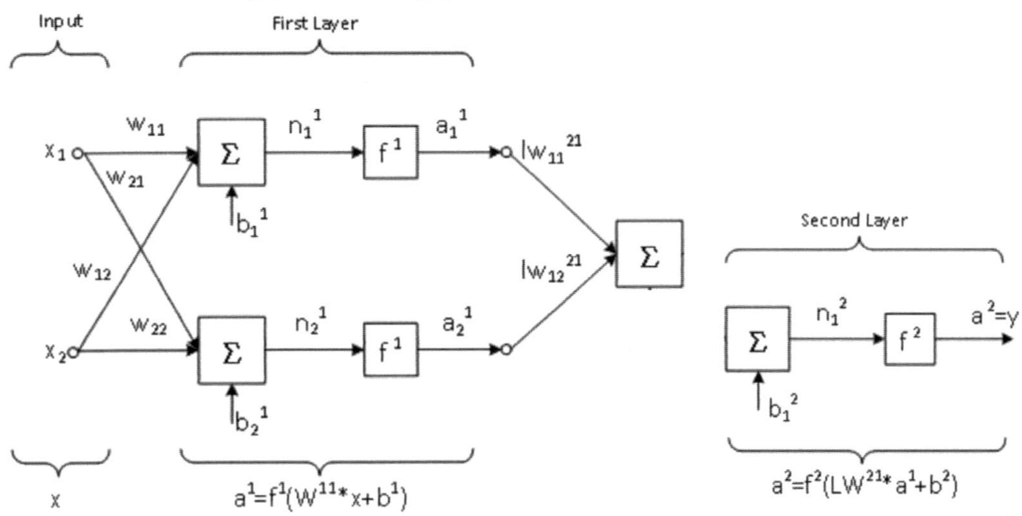

Figure 2. Scheme of simple option of a two-layer network

The scalar a_1^1 and a_2^1 as increased by weight coefficients of a layer of w_{ij} move on the adder of neuron of the second layer and finally are received at the exit of the second layer: $a^2 = \dfrac{1}{1+e^{-\beta(lw_{11}^{21}\times a_1^1 + lw_{12}^{21}\times a_2^1 \ b_1^2)+}}$ by knowing a_1^1 and a_2^1 we can writen as: $y = f(x_1, x_2) = a^2$,

where: $a^2 = 1/1+e^{-\beta\times\left(\dfrac{lw_{11}^{21}}{1+e^{-\beta(w_{11}\times x_1 + w_{12}\times x_2 + b_1^1)}} + \dfrac{lw_{12}^{21}}{1+e^{-\beta(w_{21}\times x_1 + w_{22}\times x_2 \ b_1^1)}} + b_1^2\right)}$. It is noticed that the exit of the second layer of a_2 can be designated as y, i.e. the exit of the last layer is a network exit. The type of total function is deprived of presentation, therefore neural networks are often called "a black box".

During learning procedure there is a control of the network and the weight-matrix W^{11}, which define extent of influence of this argument on a calculated value of function is formed: $W^{11} = \begin{bmatrix} w_{11} & w_{12} \\ w_{21} & w_{22} \end{bmatrix}$, where $w_{12} = W(1,2)-$ coefficient on which the second element of an entrance of x_2 is multiplied by transfer of value by the first neuron.

Also, during training process, the matrix of the weight coefficients of the output of the layer LW^{21} is formed, the indexes show interrelation between structural elements.

$LW^{21} = [lw_{11} \ \ lw_{12}]$ – coefficients are multiplied on which exits of the first layer before processing by neurons of the second layer.

The above example of equationsis shows inconvenience of use of analytical equations even for a simple case (two arguments, two layers). Addition of arguments, neurons or layers of a network repeatedly complicates analytical representation of mathematical model. If we add the third neuron in the first layer and one more argument of x_3, total number of weight coefficients will increase with 6 to 12. Therefore in practical use of neural network models people do not work with the equations, but with the files represented by computer programs.

The neural network can function in two modes:

[1] training, when weight adjustment occurs so that output signals most precisely correspond to the experimental;

[2] operating, when network takes input values and produces outputs.

The model is approached as much as possible to experimental data which are entered for training, but not to real process as the data, which are absolutely precisely reflecting real process, can't be obtained. Also at a stage of creation of mathematical model identification it is not possible to avoid data-entry errors except as increase the vigilance. Naturally "human" factor reduces the accuracy of mathematical model.

The main advantages of application of neural networks for modeling of characteristics of centrifugal compressors are:

[3] Creation of a model is simpler and quicker than methods, used earlier;

[4] Creation of structure of model is completely formalized and doesn't demand acceptance of preliminary hypotheses of the type of these or those dependences;

[5] Flexibility of model creation and possibility after learning in process of receipt of new data;

[6] The level of knowledge, necessary for successful application of neural networks, is significantly less than the level for traditional methods of modeling;

[7] Possibility of analysis of parameters influence on the studied characteristics.

Since 1970th, Y. Galerkin and his group created a number of math models of stages and compressors [10-11]. The Method of Universal modeling had been successfully applied in numerous designs of centrifugal compressors with power up to 32 MWt and delivering pressure up to 12.5 MPa (more than 400 compressors with total power 5 000 MWt). New models, versions 5th and 6th, demonstrate high precision of performance modeling [10], but the process of constant improvements points out the necessity of alternatives at the same time.

Measured gas dynamic of specially developed model stages of centrifugal compressors in databases are presented in the form of tables or in a graphic form. For use in the computerized gas dynamic calculations the performances have to be presented in analytical form. The methods based on artificial neural networks are an appropriate tool for this task. Their flexible structure and big functionality allow generalizing experimental data in a form, convenient for use [2-3, 8] (Figure 1).

Neural models of performances were developed for centrifugal compressor stages "impeller + vaneless diffuser + return channel" (Figure. 3). Information is provided by R&D Laboratory "Gas dynamics of turbomachines". Data on 39 stages was the object of neural model. Range of geometry parameters of stages is presented in Table 1.

At a preliminary stage some types of models were created with different structure of a neural network and parameters of training, from which the network with the smallest mean squared error is chosen.

For modeling of values of polytrophic efficiency and work coefficient was used the network consisting of two layers, 20 neurons in the first layer and 2 neurons in the second layer. The function that activates neuron layers in models was called «logsig» (a logical sigmoid).

Measured and calculated performances of efficiency and pressure coefficient are presented in Figure. 4-10 (KPD - measured efficiency, KPD NS – calculated efficiency, PSI – measured pressure coefficient, PSI NS – calculated pressure coefficient). Sample of Mu influence on the performances are depicted in Figure. 7.

The mean squared error for all 567 measured points is 2.5%, for pressure coefficient - 3%. The largest inaccuracy is for maximum flow coefficient (about 15%). Compressors never operate at these flow rates. For other points of performances the mean squared error for efficiency and for pressure coefficient is within 1.5%.

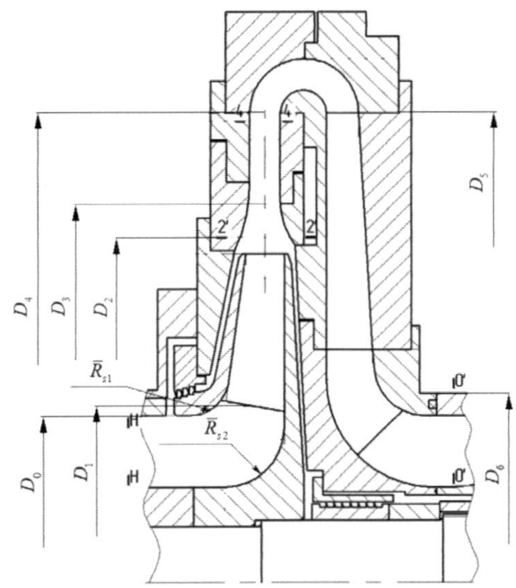

Figure 3. Meridional projection of a stage of the centrifugal compressor

Table 1. Range of geometry parameters of stages

Symbol	Min	Max	Symbol	Min	Max
$\bar{D}_h$	0,25	0,3916	$\bar{D}_4$	1,41	1,56
$\bar{D}_{seal}$	0,436	0,592	$\bar{b}_3$	0,0056	0,069
z_{imp}	13	21	$\bar{b}_4$	0,0056	0,069
$\bar{\delta}_{bl}$	0,007	0,017	$\bar{D}_5$	1,41	1,56
$\bar{D}_0$	0,425	0,573	$\bar{D}_6$	0,577	0,818
$\bar{D}_1$	0,502	0,668	$\bar{b}_5$	0,033	0,085
$\bar{b}_1$	0,0063	0,124	$\bar{b}_6$	0,033	0,085
$\bar{b}_2$	0,0063	0,069	$\bar{D}_{0'}$	0,475	0,625
β_{bl1}	24,39°	37,61°	$\bar{R}_3$	0,0716	0,159
β_{bl2}	22,5°	85,5°	$\bar{R}_4$	0,032	0,074
$\bar{R}_{s1}$	0	0,128	β_{bl5}	8,6	37

$\overline{R}_{s2}$	0	0,0653	$z_{r\,ch}$	16	32
$\overline{D}_3$	1,01	1,05	$\overline{\delta}_{r\,ch}$	0,028	0,057

One more sample of neural network application is related to vaneless diffuser performance. Numerical investigation by CFD (ANSYS CFX) of diffusers with relative width 0.014–0.10 at different Mach and Reynolds numbers is published in [12]. Diffusers were considered as isolated channels. Diffuser loss coefficient is presented as function of an inlet flow angle.

The object of modeling is a friction loss coefficient λ introduced by authors of [1] and used by authors of [12]. The loss coefficient ζ as calculated by ANSYS CFX is connected with friction loss coefficient [1] as follows:

$$\lambda = \left(\zeta \times \frac{b_2}{D_2} / \left(1 - \frac{1}{\frac{D_3}{D_2}} \right) \right) \times 4 \sin \alpha_2 .$$

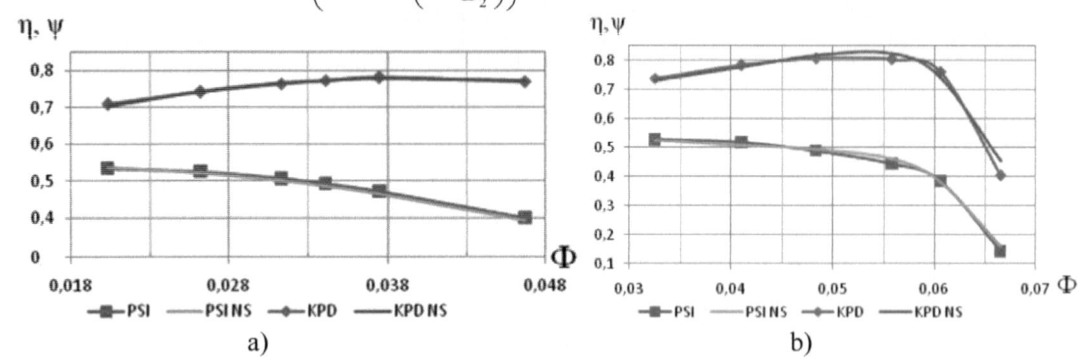

Figure 4. Performances of the stage: a) $\overline{b}_2$ =0,0484, β_{bl2}=37 °. Mu = 0,36.

b) $\overline{b}_2$ =0,039, β_{bl2}=52,1 °. Mu = 0,794

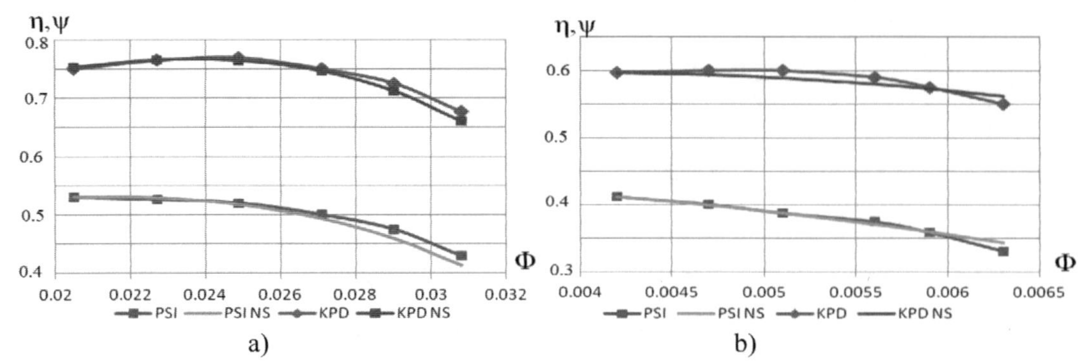

Figure 5. Performances of the stage: a) $\overline{b}_2$ =0,0345, β_{bl2}=48,5 °. Mu = 0,785.

b) $\overline{b}_2$ =0,010, β_{bl2}=35 °. Mu = 0,589.

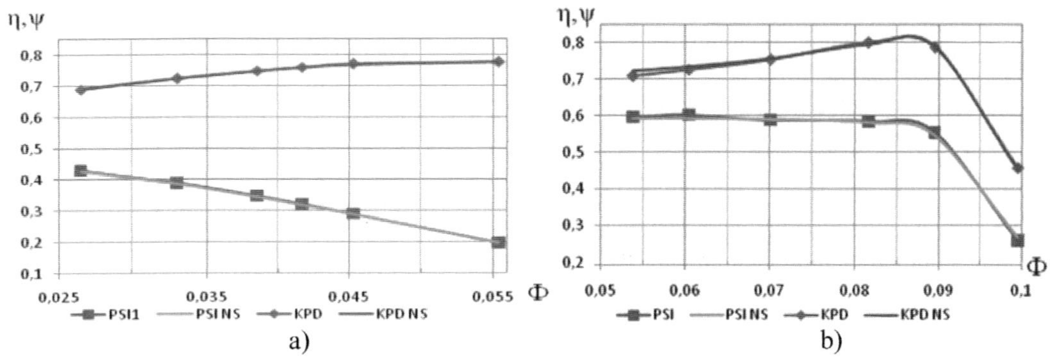

Figure 6. Performances of the stage: a) $\bar{b}_2$=0,063, β_{bl2}=22,5 °. Mu = 0,36.

b) $\bar{b}_2$=0,056, β_{bl2}=85,5 °. Mu = 0,92

For identification of models of friction coefficient λ and loss coefficient ς data on VLD with the relative radius 1,60 were chosen. CFD performance calculations were made at velocity coefficient (compressibility criterion) 0,39; 0,64 and 0,82. Corresponding Reynolds numbers are 6×10^6, $9,2\times10^6$ and $10,4\times10^6$. The flow angle α_2 changed with a step 5°. The maximum and minimum values of the arguments used when forming selection of basic data are specified in table 2.

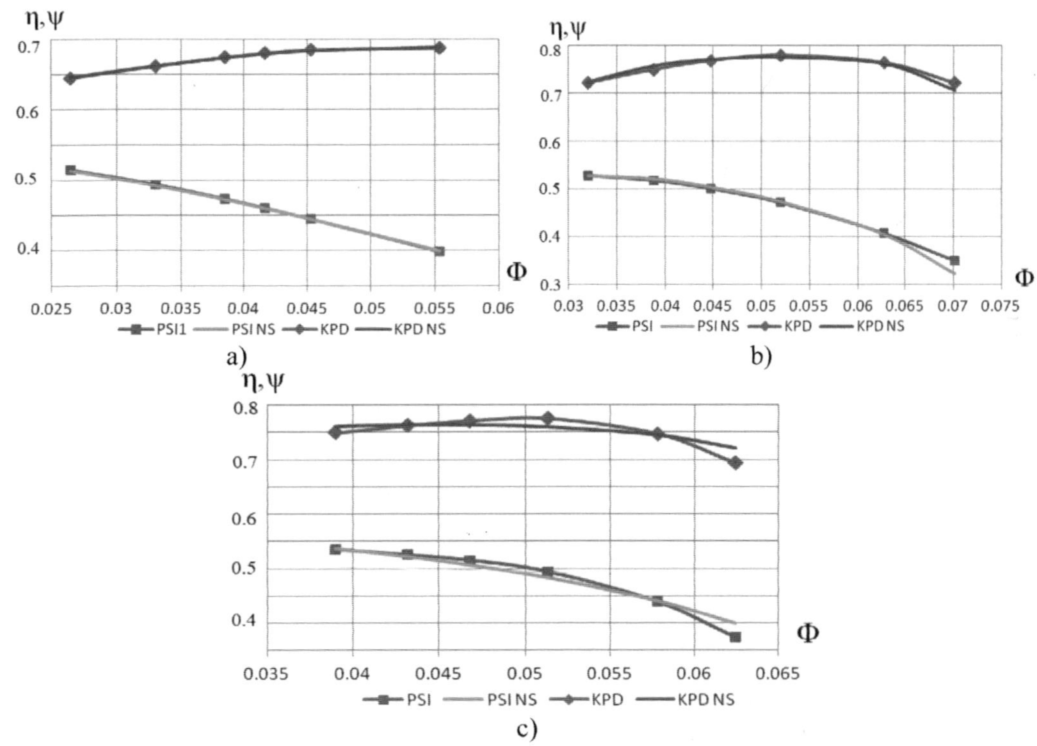

Figure 7. Performance of the stage $\bar{b}_2$=0,063, β_{bl2}=37 ° at different Mach numbers.
a) Mu = 0,36, b) Mu = 0,648, c) Mu = 0,864

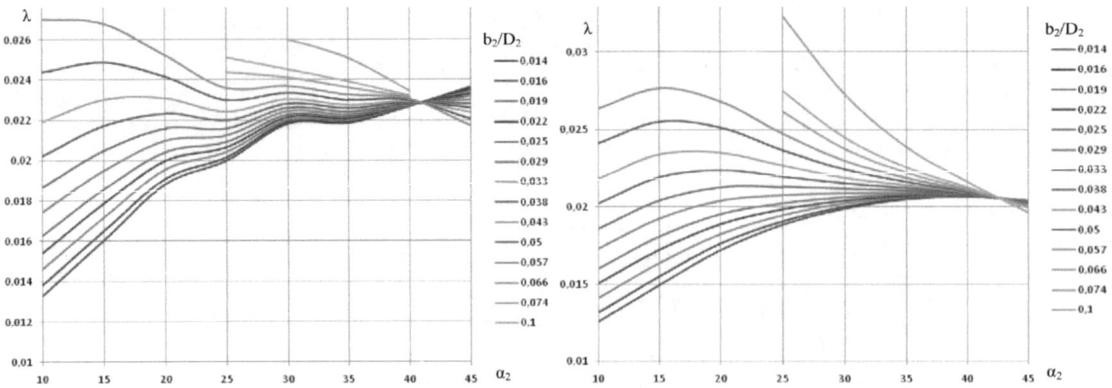

a) b)

Figure 8. a) Friction loss coefficient $\lambda = f(\alpha_2, \frac{b_2}{D_2}, M_{c2}, \mathrm{Re}_{c2})$ $M = 0,39$, $\mathrm{Re} = 6,2\times10^6$,

b) Friction loss coefficient $\lambda = f(\alpha_2, \frac{b_2}{D_2}, M_{c2}, \mathrm{Re}_{c2})$ $M = 0,82$, $\mathrm{Re} = 10,42\times10^6$

Thus, for training of mathematical models, a sample of 308 value vectors was formed. At an early stage some models with various configurations (1-3 layers, 10-30 neurons) were tested. After training networks with the minimum mean square error were chosen. These are two-layer networks, with one neuron in an output layer and 10 neurons in the hidden layer (for $\lambda = f(\alpha_2, \frac{b_2}{D_2}, M_{c2})$ – 25 neurons), wherein function of activation of neurons is a logical sigmoid.

Calculation results of friction coefficient λ are presented in Figure. 8 a, b. The mean squared error is 2.15%. Mathematical models for coefficient of loss coefficient ζ were similarly developed, the mean squared error is 2.45%. Results of modeling are shown in Figure 9 a, b.

Table 2. Range of VLD parameters

Parameter	Minimum	Maximum
$\bar{b}_2$	0,014	0,1
α_2	10	45
λ	0,0125	0,0367
ζ	0,0263	0,4968

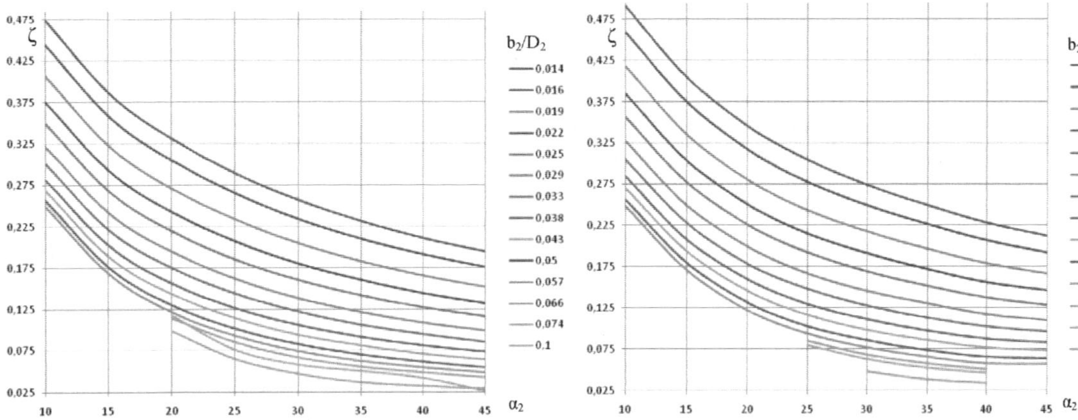

a) b)

Figure 9. a) VLD loss coefficient $\zeta = f(\alpha_2, \frac{b_2}{D_2}, M_{c2}, \mathrm{Re}_{c2})$ $M = 0{,}64$, $\mathrm{Re} = 9{,}2 \times 10^6$;

b) VLD loss coefficient $\zeta = f(\alpha_2, \frac{b_2}{D_2}, M_{c2}, \mathrm{Re}_{c2})$ $M = 0{,}39$, $\mathrm{Re} = 6{,}2 \times 10^6$.

Coefficients of weight and bias show interrelation between neurons and extent influence of parameters on function, characterizing model. Tables of coefficients are not presented due to their bulkiness, but only the data of the analysis of weights of the received models are shown. The weights averages on all neurons for each entrance argument are presented in table 3.

Table 3. Average values of weights for input parameters

Function	Average weight on all neurons for parameter			
	Re_{c2}	M_{c2}	$\bar{b}_2$	α_2
$\lambda = f(\alpha_2, \frac{b_2}{D_2}, M_{c2}, \mathrm{Re}_{c2})$	2,57	2,08	3,39	2,22
$\zeta = f(\alpha_2, \frac{b_2}{D_2}, M_{c2}, \mathrm{Re}_{c2})$	2,27	1,97	4,17	1,44

Average weight coefficients are of the same order. The weight coefficient for $\bar{b}_2$ is the biggest.

Conclusion

Practical application of neural models for the calculations of the compressor characteristics is reduced to the data input in model in the same format and an order which was set at identification. Input of the basic data can be made for calculations manually or by loading of previously prepared tables. The values of friction coefficient calculated based on model and the coefficient of losses can be shown also in the form of tables or matrices. These data can be kept or imported for further work with them in other programs.

IOP Conf. Series: Materials Science and Engineering **232** (2017) 012046 doi:10.1088/1757-899X/232/1/012046

The obtained data on the calculation errors by means of the neural networks allows drawing a conclusion about the practical importance of the received mathematical models, as a tool for generalization and the analysis of the data.

Thus, the adapted and trained, artificial neural networks represent the paralleling systems, capable of training by the analysis the positive and negative impacts. Experiments with neural networks, in relation with the modeling of characteristics of centrifugal compressors, showed the encouraging results that speak about prospects of this direction, especially at the description of processes with a large number of arguments. The authors are ready to provide the interested persons with the above-described neural network models for use in calculations.

References

[1] Seleznev K., GalerkinY..Centrifugal compressors. – Leningrad. - 1982. – P. 271. [Russian].

[2] Galerkin, Y., Soldatova, K. Operational process modeling of industrial centrifugal compressors. Scientific bases, development stages, current state. Monograph.Sankt-Peterburg. - SPbTU. – 2011. [Russian].

[3] Galerkin Y. Turbo compressors. LTD information and publishing center. - Moscow. – 2010 [Russian].Galerkin Y. Turbo compressors. // LTD information and publishing center. - Moscow. – 2010 – P. 650 [Russian].

[4] Drozdov Y., Lunev. A. Application of mathematical model of a centrifugal stage at design and processing of experimental data. Compressors and pneumatics. – 2010. – No. 1.[Russian].

[5] Lunev A., Vyachkilev O., Drozdov Y. Optimization of parameters of a number of centrifugal compressor stagess for gas-compressor unit superchargers on the basis of the identified mathematical model. Theses of the X international scientific and technical conference on compressor equipment. – Kazan. – 1995. [Russian].

[6] Japikse, D. Design system development for turbomachinery (turbopump) designs - 1998 and a decade beyond. JANNAF Conference. // Cleveland. – Ohio. – July 15 -17. – 1998.

[7] http://www.pcaeng.co.uk/software/ PCA Engineers Limited.

[8] Kruglov V., Borisov V. Artificial Neural Networks: Theory and Practice. - GoryachayaLiniya-Telekom. – Moscow. - 2002. – P. 382. [Russian].

[9] Osovskij S.Neural Networks for Information Processing. – M.:Financy I Statistica. - 2002 . – P. 344. [Poland].

[8] Galerkin, Y.B., Soldatova, K.V., Drozdov, A.A. Improvement of method to calculate industrial centrifugal compressors performance. Compressors & Pneumatics.№ 8. - 2013. – P. 24-32. [Russian].

[9] Soldatova, K.V., Rekstin A.F. Engineering methods of design of turbocompressors. Compressors & Pneumatics.- № 5. – 2013. - P. 18-22. [Russian].

[10] Soldatova, A. Drozdov, O. Solovieva. Approximation of numerical experiment of centrifugal compressor stage vaneless diffuzer performance. // Compressors and pneumatics. – 2015. - № 1. – P. 32-36. [Russian].

IOP Conf. Series: Materials Science and Engineering **232** (2017) 012047 doi:10.1088/1757-899X/232/1/012047

Analysis of Three-dimension Viscous Flow in the Model Axial Compressor Stage K1002L

K Tribunskaia and Y V Kozhukhov

Department of Compressor, Vacuum and Refrigeration Engineering Saint-Petersburg Polytechnical University, Russia, Saint-Petersburg

Abstract. The main investigation subject considered in this paper is axial compressor model stage K1002L. Three simulation models were designed: Scheme 1 – inlet stage model consisting of IGV (Inlet Guide Vane), rotor and diffuser; Scheme 2 – two-stage model: IGV, first-stage rotor, first-stage diffuser, second-stage rotor, EGV (Exit Guide Vane); Scheme 3 – full-round model: IGV, rotor, diffuser.
Numerical investigation of the model stage was held for four circumferential velocities at the outer diameter (U_{out}=125,160,180,210 m/s) within the range of flow coefficient: $\varphi = 0.4 - 0.6$. The computational domain was created with ANSYS CFX Workbench.
According to simulation results, there were constructed aerodynamic characteristic curves of adiabatic efficiency and the adiabatic head coefficient calculated for total parameters were compared with data from the full-scale test received at the Central Boiler and Turbine Institution (CBTI), thus, verification of the calculated data was carried out. Moreover, there were conducted the following studies: comparison of aerodynamic characteristics of the schemes 1, 2; comparison of the sector and full-round models.
The analysis and conclusions are supplemented by gas-dynamic method calculation for axial compressor stages.

Nomenclature

R- rotor;
S – stator;
IGV – inlet guide vane;
EGV – exit guide vane.

c_z=local axial velocity; ms^{-1}
F= flow path area; m^3
H = head; kgs^{-1}
$\bar{m}$ = flow rate; $kgm^{-1}s^{-1}$;
n = compressor speed (rotation velocity); s^{-1}; (rev min^{-1};)
P = static pressure; Pa
r = radius; m
$\bar{r} = \frac{r}{r_c}$ = relative radius;
U = local blade velocity; m/s (In paper: U_{out} blade velocity at the outer diameter);
η = efficiency; efficiency coefficient;
Ω - degree of reaction;

IOP Conf. Series: Materials Science and Engineering **232** (2017) 012047 doi:10.1088/1757-899X/232/1/012047

φ = stage flow coefficient (In paper: φ stage flow coefficient calculated for total parameters);
ψ = head coefficient (In paper: ψ^*_{ad} adiabatic head coefficient calculated for total parameters)
ε = degree of turbulence

Subscripts
ad = adiabatic;
av = average
c = calculated parameter;
out = outer(in stator diameter);
T = theoretical;
1,2 = first stage and second stage in case of rotor and stator.
Superscripts
*- total parameters

1. Introduction

1.1. Full-scale experimental equipment
The full-scale investigation conducted at the Central Boiler and Turbine Institution (CBTI) was carried out for axial compressor stage`s with flow path designed constant outer diameter and variable inner diameter (type I) and degree of reaction 1.0.

The test stand was completed with compressor rotor (cantilevered disk) rotated by an electric motor via a step-up gear. The disk was made compound, it allows attaching a variety of blades and changing the angle of installation. The compressor stator has an annular air intake from surrounding space and two removable guide vanes to model various types of stages: intermediate, first and last one. The vertical flange of the compressor connects with the horizontal pipe. The throttle valve was installed at the end of the pipe allowing to measure the hydraulic resistance of the air intake system (pipe + throttle). The throttle aperture (flow nozzle) was installed in the same pipe, which allowed measuring the flow rate through the compressor. The changing of the impeller rotation speed was achieved by changing the of electric motor speed. Rotor speed was defined by a mechanical or electrical tachometer. The gas temperature was measured with a thermometer at the inlet, and a thermometer or thermocouple at the outlet. Atmospheric pressure was measured by barometer. The static gas pressure at the internal surfaces of the cylinder was determined using a U-shaped manometer through holes specially drilled. The pressure and the gas velocity value and direction in the gaps measured with special aerodynamic devices introduced into the flow path through the hole in the cylinder wall between the blades before and after the stage are. For the characteristics of the intermediate stage used measurements of the velocity field in front of the impeller and after guide vane [4].

The experimental data was presented by the adiabatic head coefficient calculated for total parameters:

$$\psi^*_{ad}(\varphi) = \frac{H^*}{\frac{u^2_{out}}{2}} \tag{1}$$

and the adiabatic efficiency:

$$\eta^*_{ad}(\varphi) = \frac{H^*}{H_T{}^*} \tag{2}$$

calculated for stage flow coefficient [1,3]:

$$\varphi = \frac{\bar{m}}{\rho^* F u_{out}} = \frac{c_{Z\,av}}{u_{out}}. \tag{3}$$

1.2. Computing model
Based on the research description made by of CBTI [1, 3] four 3D blade models were designed, namely IGV, rotor (R), diffuser (S), EGV. The models geometry were developed in Creo (PTC). This paper presents the investigation of the three calculation schemes: **scheme 1**, consisting of IGV, R1,

IOP Conf. Series: Materials Science and Engineering **232** (2017) 012047 doi:10.1088/1757-899X/232/1/012047

S1, R2, EGV (Fig. 1); **scheme 2**, consisting of IGV, R, S (Fig. 2); **scheme 3** full-round model, also consisting of IGV, R, S (Fig. 3). The stage geometry parameters are given in the Table 1.

Table 1. Model stage geometric parameters

Flow path parts	Dimension	IGV	R	S	EGV
Outer diameter	mm	250	250	250	250
Inter diameter	mm	125.0	126.1	130.3	130.3
Length of the blade at the entrance	mm	62.5	61.6	59.85	59.85
Length of the blade by the section	mm	12.50	12.32	11,97	11.97
Blade chord	mm	30	30	30	30
Number of blades in the crown	pcs	26	21	28	26
Radial clearance	mm	-	0,57	-	-
Axial clearance	mm	24	24	24	0

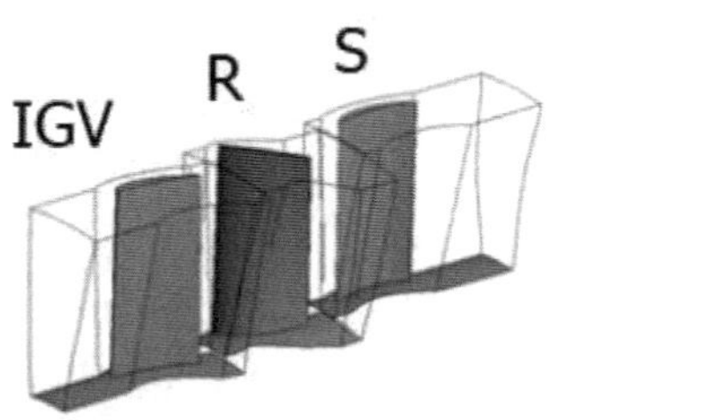

Figure 1. Scheme 1: Flow path consisted of IGV, R, S

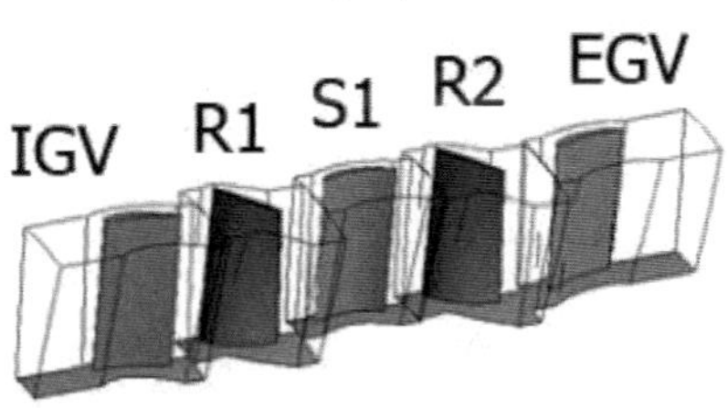

Figure 2. Scheme 2: Flow path consisted of IGV, R1, S1, R2, EGV

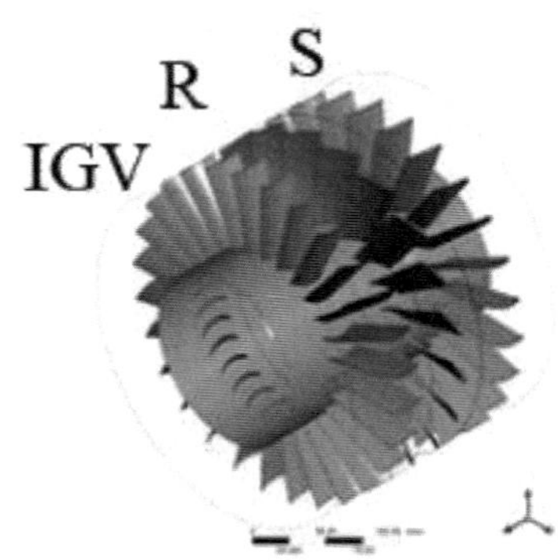

Figure 3. Scheme 3: Full-round model; flow path consisted of IGV, R, S

The calculations were obtained with ANSYS Workbench 14.0. For simulation of the computing domain it was used a high sample rate mesh with the number of nodes up to 500 000 per each sector of blade row. The degree of turbulence at the inlet of the computational domain was taken $\varepsilon = 5\%$. Total pressure and total temperature at the inlet, the mass flow rate at the outlet were taken as the boundary conditions.

Model of turbulence used in the study is SST (Shear-Stress-Transport)

2. Numerical method and the obtained results

2.1. Comparison of CFX and full-scale experiments

The First part of the investigation includes the comparison between the results received from sector CFX calculation (Scheme 1) and the full-scaled CBTI experiment.

The main calculated parameters were measured before the first rotor and after the following diffuser (in the section where the flow has not turn in the axial direction).

The stable work zone was accepted as a part of characteristic near the max efficiency mode, where the estimated parameters are different from the maximum no more than 5%.

The relative error, expressed in a percentage was determined as:

$$\delta\psi^*_{ad} = \left|\frac{\psi^*_{ad\ full-scaled} - \psi^*_{ad\ CFX}}{\psi^*_{ad\ full-scaled}}\right| \cdot 100\% \quad (4)$$

$$\delta\eta^*_{ad} = \left|\frac{\eta^*_{ad\ full-scaled} - \eta^*_{ad\ CFX}}{\eta^*_{ad\ full-scaled}}\right| \cdot 100\% \quad (5)$$

Fig. 4-5 illustrates the set of aerodynamic adiabatic efficiency characteristic and adiabatic total head coefficient ($\eta^*_{ad};\psi^*_{ad}$) (1), (2) depending on the stage flow coefficient (3).

Based on the obtained results there were concluded that with the increase of blade local velocity at the outer diameter (from 125 to 210 m/s) the investigated parameters ($\eta^*_{ad};\psi^*_{ad}$) are offset from the stable work zone to the right by 10% on average, the point of maximum efficiency also moves to the right.

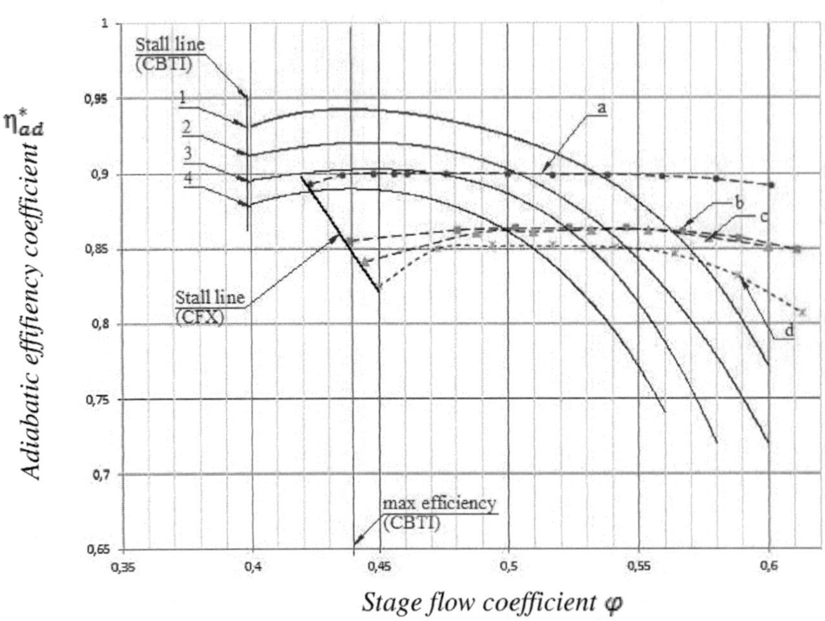

Key for figures 4-5:

Full-scaled (CBTI):
1.U_{out}=125 m/s;
2.U_{out}=160 m/s;
3.U_{out}=180 m/s;
4.U_{out}=210 m/s

CFX (Scheme1):
a.U_{out}=125 m/s;
b.U_{out}=160 m/s;
c.U_{out}=180 m/s;
d.U_{out}=210 m/s

Figure 4. Adiabatic efficiency characteristics for the full-scale and the CFX experiments

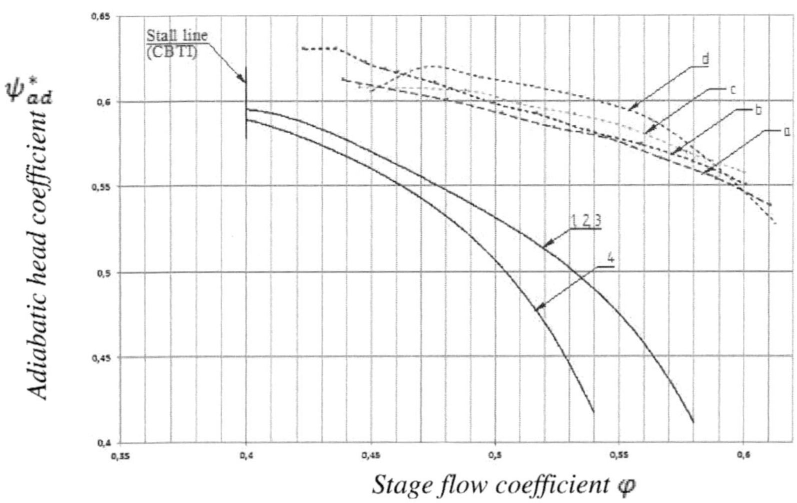

Figure 5. Adiabatic head coefficient calculated for total parameters for the full-scale and the CFX experiments

Max numerical values of aerodynamic characteristics are less dependent on blade velocity at the outer radius (U_{out},) i.e., compressor speed, this is indicated by "flattening" of the characteristics on the chart. Thus, the obtained values of the efficiency and head coefficient for the U_{out} = 160, 180, 210 m/s are quite close. There is overestimation of the CFD-calculation of head coefficient: the difference from the full-scale experiment is 15% on average in the stable work zone.

2.2. The impact of multi-stage on the characteristics of the first stage

In multi-stage compressor the crowns following the first stage impact on it operating [5]. The main task of this investigation was to determine the impact of additional crowns on the aerodynamic characteristics of the first stage in the CFX experiment and compare findings of one-stage (Scheme 1) and two-stage (Scheme 2) models. The reference sections for Schemes 1, 2 were coincide.

Figures 6-9 depict the aerodynamic characteristics: (η^*_{ad};ψ^*_{ad},) marked by the surge line; the vertical lines intersect the characteristics at the maximum efficiency point (direct mode).

Key for figures 6-9
1 – Full-scaled, CBTI
2 – Scheme 1, CFX
3 – Scheme 2, CFX

| - Stall line | - line of efficiency max for CFX ¦ - line of efficiency max for full-scale

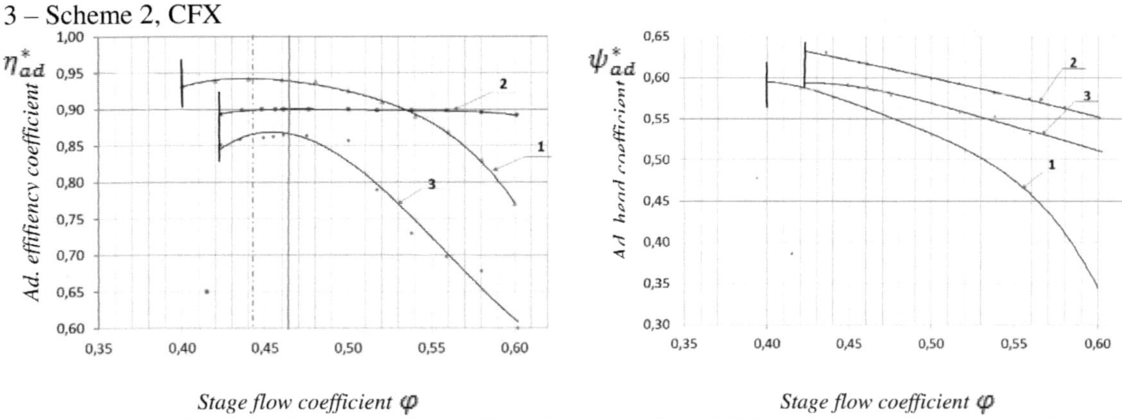

Figure 6. Aerodynamic characteristics of CFX (Scheme1,2) and full-scaled experiments calculated for

$U_{out} = 125\ m/s$

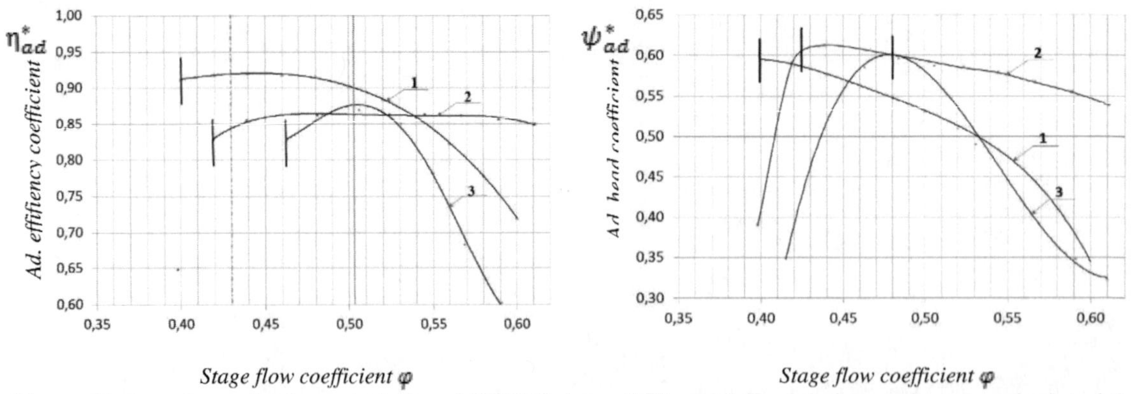

Figure 7. Aerodynamic characteristics of CFX (Scheme1,2) and full-scaled experiments calculated for $U_{out} = 160\ m/s$

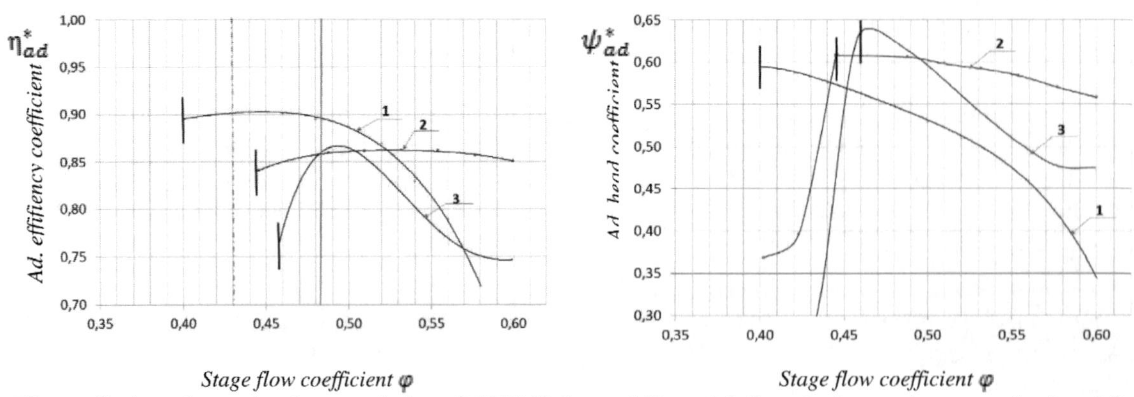

Figure 8. Aerodynamic characteristics of CFX (Scheme1,2) and full-scaled experiments calculated for $U_{out} = 180\ m/s$

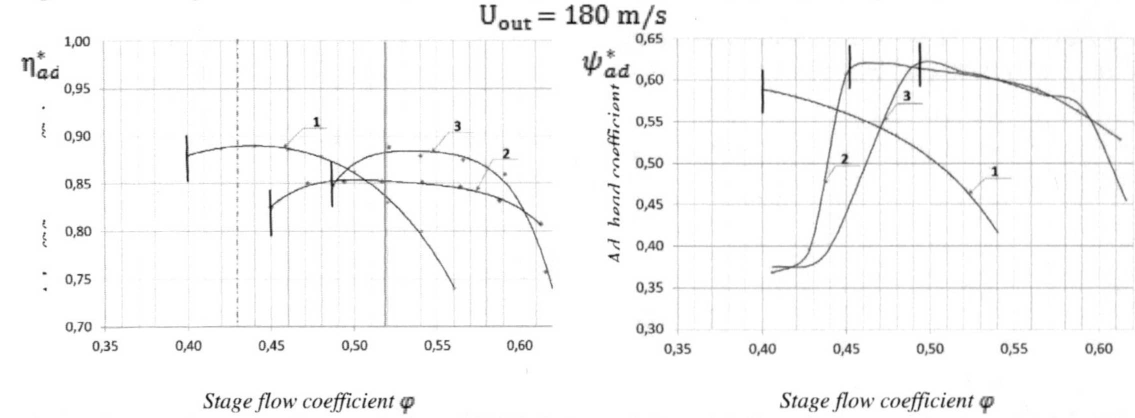

Figure 9. Aerodynamic characteristics of CFX (Scheme1,2) and full-scaled experiments calculated for $U_{out} = 210\ m/s$

Table 2 shows the max efficiency coefficient on each direct mode for CFX sector schemes and full-scaled investigations.

Table 2. Comparison of CFX (Schemes 1, 2) and full-scale investigations

Blade local velocity at the outer diameter, U_{out}, m/s	φ	Full-scale		CFX, Scheme 2		CFX, Scheme 1	
		η^*_{ad}	ψ^*_{ad}	η^*_{ad}	ψ^*_{ad}	η^*_{ad}	ψ^*_{ad}
125	0.461	0.940	0.576	0.865	0.589	0.900	0.616
160	0.506	0.899	0.520	0.87	0.587	0.863	0.592
180	0.488	0.896	0.550	0.865	0.614	0.860	0.606
210	0.521	0.830	0.470	0.888	0.609	0.852	0.608

One-stage model (Scheme 1) has larger stable work zone in contrast to the two-stage model (Scheme 2). The efficiency characteristics appearance (Scheme 1) more similar with the characteristic obtained in the full-scale experiment CBTI. However, the overestimation of the adiabatic head coefficient occurs. For the Scheme 1 max efficiency values are closer correlated with the full-scale experiment data. The difference between them on U_{out} = 125,160,180 m/s modes is less than 7%. Similarly, for U_{out} = 210 m/s mode it is less than 2%. Difference of ψ^*_{ad} values varies from 6 to 30%. With the outer blade velocity increasing the surge line offsets to the right for all sector schemes, reach large displacement for Scheme 2 (Table 3).

For the Scheme 2 max efficiency values at the U_{out}= 160,180 m/s modes are less than full-scale experiment at the same modes nearly by 3%. At the same time for U_{out}= 125, 210 m/s modes this difference is up to 7,5%. From the over hand the distinction between adiabatic head coefficient max value runs up with outer blade velocity increasing from 6% for U_{out}= 125m/s to 30% for U_{out}= 180 m/s.

Table 3. Comparison of flow coefficient on the surge edge for full-scale and CFX (Scheme1,2)

Blade local velocity at the outer diameter, U_{out}, m/s	Flow coefficient φ on the stall line		
	Full-scale	CFX	
		Scheme 1	Scheme 2
125	0.4	0.425	0.425
160	0.4	0.419	0.465
180	0.4	0.445	0.485
210	0.4	0.450	0.486

2.3. Comparison of aerodynamic characteristics of the sector and the full-round models

According to the axial symmetry of axis turbomachines, it is usual practice to solve the problem of flow simulation by examining only one sector. However, since a certain approach to the exact solution of this problem is not found, researchers are increasingly turning to the full-round models. [2] Figures 10-13 show the characteristics of efficiency and adiabatic head coefficient for the sector and the full-round models.

The difference between full-round and sector models max efficiency coefficient (η^*_{ad}) values is 3% at the each direct mode. In the same for adiabatic head coefficient this value is up to 5% (Table 4). This indicates that full-round and sector calculation schemes are quite close and sector models can substitute full-round for initial examination of workflow.

Key for figures 10-13

1 – Full-scaled, CBTI
2 – Scheme 1, CFX
3 – Scheme 2, CFX

| - Stall line | - line of efficiency max for CFX ¦ - line of efficiency max for full-scale

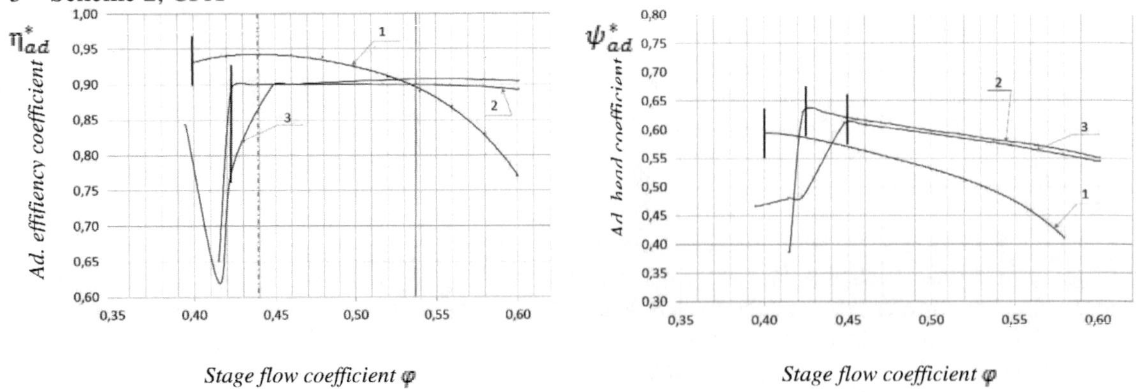

Figure 10. Aerodynamic characteristics of CFX (Scheme1,3) and full-scaled experiments calculated for $U_{out} = 125$ m/s

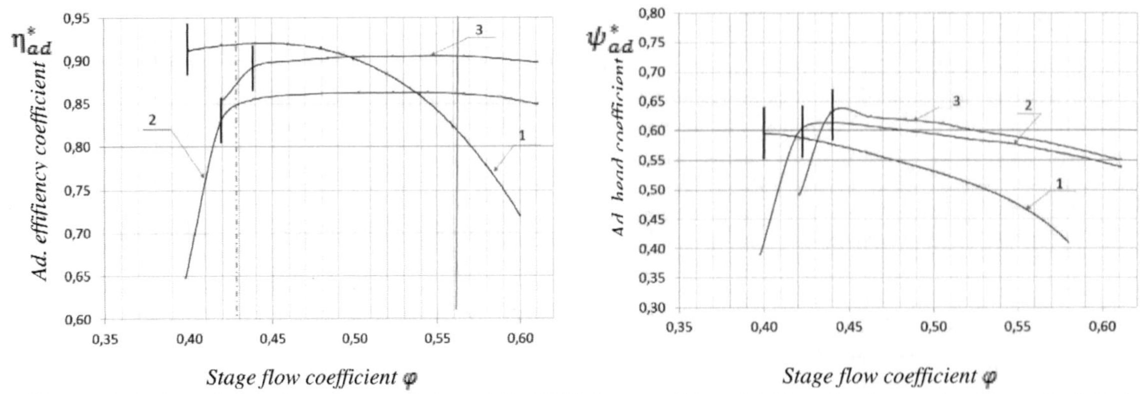

Figure 11. Aerodynamic characteristics of CFX (Scheme1,3) and full-scaled experiments calculated for $U_{out} = 160$ m/s

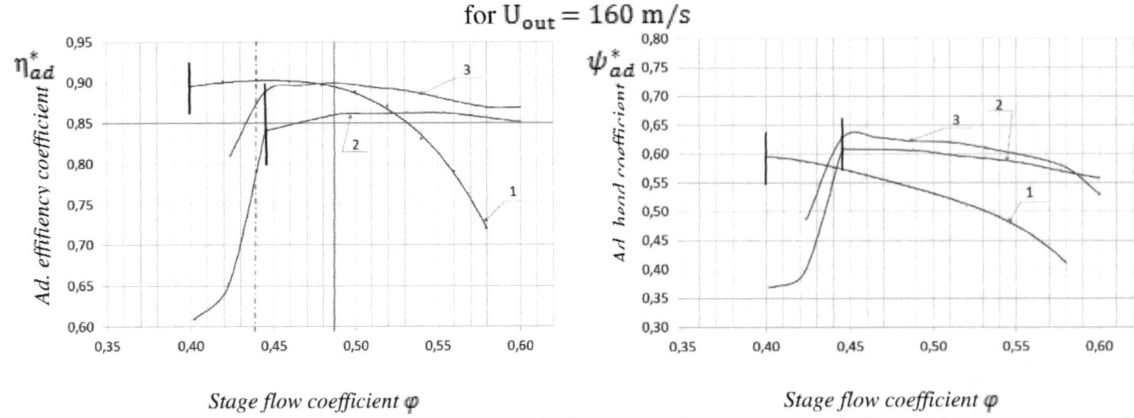

Figure 12. Aerodynamic characteristics of CFX (Scheme1,3) and full-scaled experiments calculated for $U_{out} = 180$ m/s

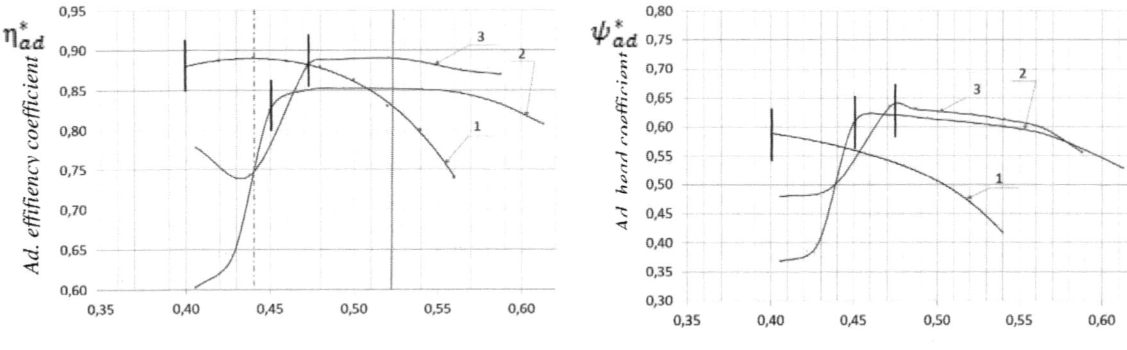

Figure 13. Aerodynamic characteristics of CFX (Scheme1,3) and full-scaled experiments calculated for $U_{out} = 210\ m/s$

Full-round scheme findings are very close to the sector ones, however the values obtained per fill-round model are closer to the full-scale CBTI experiment [1,3]. At the indirect modes full-round values difference reach 5% for efficiency coefficient and 10% for adiabatic head coefficient.

From the other hand the solution of the one design point takes two timed longer than sector solution and 10 times more computer memory space.

Table 4. Comparison of the full-scale and the CFX experiments.

Blade local speed at the outer diameter, U_{out}, m/s	φ	Full-scale		Scheme 3		Scheme 1	
		η^*_{ad}	ψ^*_{ad}	η^*_{ad}	ψ^*_{ad}	η^*_{ad}	ψ^*_{ad}
125	0.538	0.900	0.499	0.907	0.577	0.899	0.582
160	0.567	0.820	0.460	0.905	0.580	0.861	0.566
180	0.488	0.892	0.54	0.899	0.622	0.860	0.606
210	0.521	0.830	0.470	0.890	0.621	0.851	0.600

3. Results and Discussion

Thus, on the basis of the obtained results can be drawn the following conclusions:

The aerodynamic characteristic set of efficiency coefficient (η^*_{ad}) and adiabatic head coefficient (ψ^*_{ad}) for the sector CFX schemes are closely match at the design modes (efficiency max value) and in the stable work zones. The differences between them 7% on average. However, at the critical modes and at the characteristic ends they are quite distinct due to next stage impact and raising of the backlog angles with flow rate increase.

There is the offset of stable work zone to the right; stage parameters are less dependent on increase of the outer blade velocity U_{out}, whereby the characteristics "flattering" to each other; maximum efficiency points are shifted to the right as well.

The miscalculation between the full-scale and CFX experiments is slightly dependent on the outer blade velocity. The calculated efficiency value on average 5% less than full-scaled, however, the head coefficient up to 13% higher.

IOP Conf. Series: Materials Science and Engineering **232** (2017) 012047 doi:10.1088/1757-899X/232/1/012047

There is overestimate of the backlog angles by the ANSYS CFX software for the two-stage model that allows to conclude: that the smaller the blade rows is calculated together, the more accurate calculation is obtained.

Overall, the CFX experiment showed the miscalculation in the stable work zone for up to 6% for efficiency and up to 18% for the adiabatic head coefficient. This suggests the possibility of taking such numerical experiments for stage flow analysis, but with the insertion of the variance for axial compressor stages in the design conditions (within the stable operation), however, the application of this method in critical modes will not give accurate results.

Turbulence model used in the research is SST, but the calculation of one point cannot give a fair view of the impact of the turbulence model on the obtained characteristics.

References

[1] Buinovskia L N Comparison of experimental and calculated aerodynamic characteristics of axial compressors Kotloturbostroenie. Proceedings CBTI 51. Gas turbines. Ed. Acad. B. E. Stechina, et al. - Leningrad, 1964. 10 P.
[2] Galerkin Y B, Soldatova K V, Titensky V N Theory, calculation and design of compressor dynamic machines. Turbochargers. - Textbook. - St. Petersburg: SSTU, 2007. - 142 p.
[3] The methodology for calculating the aerodynamic flow of the axial compressor for fixed installations. RTM 24.020.17-73. - Moscow, 1973 - 205 p.
[4] Guidelines for the calculation of aerodynamic flow of the axial compressors. Atlas initial stages. - Leningrad, 1957 - 8 p
[5] Seleznev K P, Podobuev Y S Theory and calculation of axial and centrifugal compressors. - Leningrad: Engineering literature, 1957 - 395 p.

Analysis of casing treatment`s impact on the axial compressor model stage characteristics

K Tribunskaia and Y V Kozhukhov

Department of Compressor, Vacuum and Refrigeration Engineering Saint-Petersburg Polytechnical University, Russia, Saint-Petersburg

Abstract. There are special requirements for the compressors of aircraft engines. They must ensure maximum efficiency in a maximally large stable work zone Due to a high pressure ratio these stages are more susceptible to the losses from radial clearance. One of the approaches to reduce such losses is the application of above-rotor devices. In the following study there is considered the impact of such treatments on the compressor stage performance. Despite the fact that there is a sufficient amount of research about this issue, their results are contradictory. The use of these devices can affect the characteristics of compressor stage performance both positively and negatively. This study was conducted using the methods of computational fluid dynamics and was based on the NASA Rotor-37 geometry model stage. Results were obtained through the comparison of the characteristics of stages with and without above-rotor devices.

Nomenclature

TC – Tip Clearance
K_y *– Stall Margin (SM)*
$\bar{m}$ *– Mass flow rate*
π^* *– Total pressure ratio*
k – Specific heat ratio
P – Total pressure
T – Total temperature

Subscripts

pk - Peak Efficiency Point
st - Near Stall Point
SC - Smooth casing
CT - Casing treatment

1. Introduction

The purpose of this work is to explore the influence of the casing treatment on the transonic axial compressor stage performance. In reference to the above-rotor devices it should be understood that casing treatments are used in the stage.

During nearly half a century, a large number of above the rotor casing treatments were designed and tested [1, 3, 4, 6, 7]. The main goal of such improvements was to provide the largest stable work zone for transonic axial compressor stages with the minimizing of efficiency losses. According to the

IOP Conf. Series: Materials Science and Engineering **232** (2017) 012048 doi:10.1088/1757-899X/232/1/012048

researcher`s surveys, the more efficient a casing treatment`s stability margin, the less effective the compressor stage work [1].

High pressure ratio in transonic and supersonic stages provokes leakages through the TC from the suction side. This gap between the casing and the rotor blade is one of the most critical factors, because the efficiency losses from the leak can be from 0.3 to 1.1% of stage efficiency [2].

It should also be noticed that the physical phenomena occurring in the flow path of the transonic compressor such as blade tip stall and tip blockage stall are very dangerous and will negatively affect the compressor performance. [3]

Aerodynamic overloading on the blade of the compressor is increased and the risk of rotation stall or surge. Nowadays there are distinguished several methods that use enhancements to the stator ways to solve the problem of increasing the stability of the compressor. They are divided into active and passive methods. Passive methods include creating a different casing treatments configurations: grooves, honeycomb seals, etc.

M. Hembera (etc.) [3] distinguish two categories of the casing treatments according to their shape:

- Circumferential treatment such as circumferential grooves, which are typically axis-symmetric
- Axial treatments such as axial slots, which are non-axis - symmetric

According to the investigation [1] tip injection is another way to increase the stall margin (SM).

Thus, any above-rotor devices must be designed to exclude any low-energy zones (stagnation and blockage zones).

The following activities can minimize the above-mentioned negative effects:

- reducing radial clearance
- the movement of low-energy zone leakages, when you use casing treatments according to the principle of honeycomb seals (axial slots)
- the deflection of the flow direction parallel to the main flow, using, for example, the peripheral annular groove.

2. Numerical method

In terms of given task, the two performance parameters was selected: stall margin and adiabatic efficiency. The stall margin is calculated as following for each configuration:

$$K_y = \left[\frac{\overline{m}_{pk} \cdot \pi^*{}_{st}}{\overline{m}_{st} \cdot \pi^*{}_{pk}} - 1\right] x100 \tag{1}$$

The adiabatic efficiency is calculated as following for each configuration:

$$\eta_{ad} = \frac{\left(P_{t,out}/P_{t,in}\right)^{k-1/k} - 1}{\left(T_{t,out}/T_{t,in}\right) - 1} \tag{2}$$

3. Investigation of the smooth casing

The study presents the investigation of the transonic compressor stage based on the NASA rotor 37 profile geometry (Fig.1). The main design parameters are shown in Table 1. The computational domain (Fig. 2) is divided into two passages – rotor and stator respectively. Design parameters and experimental data were taken from the NASA report [5].

The stator is represented by a smooth surface. The gap between the upper edges of the rotor blades is 0.356 mm.

The impeller peripheral speed is 17188 rpm, pressure ratio (measured at total parameters) is 2.106 and the value of the adiabatic efficiency is as 88.9% at the design point (at the estimated mass flow of 20.19 kg/s).

Table 1. Stage design parameters

Total temperature, K	483.35
Total pressure, kPa	195. 218
Rotation speed, rpm	17 188
Mass flow rate ($\overline{m}$), kg/s	20.19
Tip Mach Number	0.9
Number of rotor blades	36
Number of stator blades	46

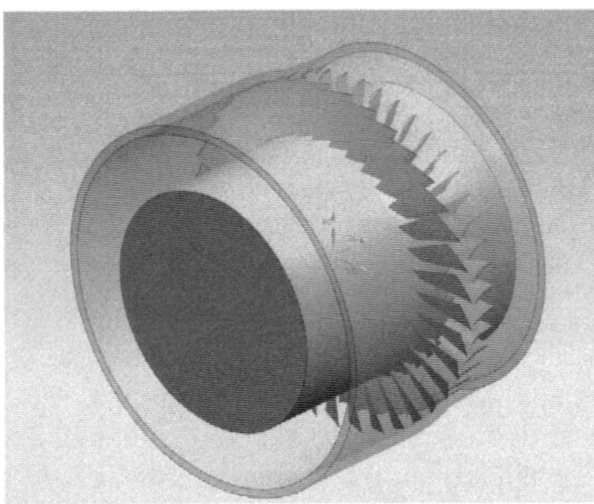

Figure 1. Solid model NASA rotor 37 stage

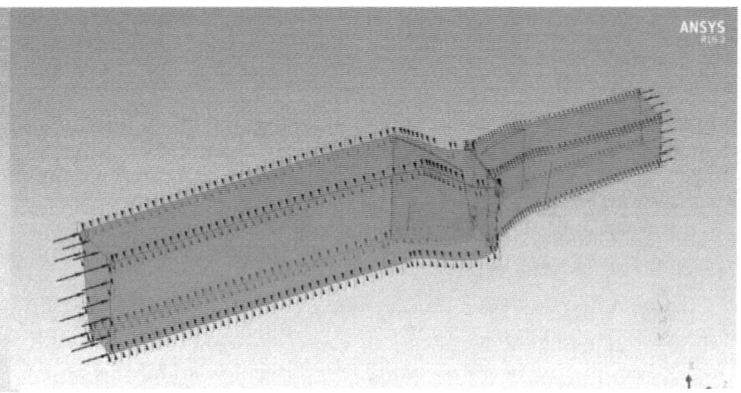

Figure 2. Computational domain

3.1. Mesh description

A structured grid is constructed with the ANSYS Turbogrid. A grid-dependency test is performed for the reference design with the various number of nodes. The result of test shows that the "curse" grid contained 530 000 nodes gives less accurate parameter`s values than the "detail" grid with total number of nodes 970 000 for each domain (stationary and rotational).

3.2. Boundary conditions

The working fluid was considered as ideal gas.
The solid boundaries in the computation domain are considered as adiabatic with the hydraulically smooth walls. The periodic boundaries are set at the blade passage interface, and TC was modeled along the passage. Inlet conditional: total pressure, total temperature (Tab.1). The SST turbulence model was used.

3.3. Verification of the received data

10 different modes were investigated for the validation of the obtained results. The stall line was adopted at the normalized mass flow value $\overline{m}_{st}/\overline{m}$=0.91 kg/s. The SM parameter (1) achieved the value 13.56%.

The numerical results are in good agreement with experimental data. Average difference between parameter is up to 3% (Fig.3).

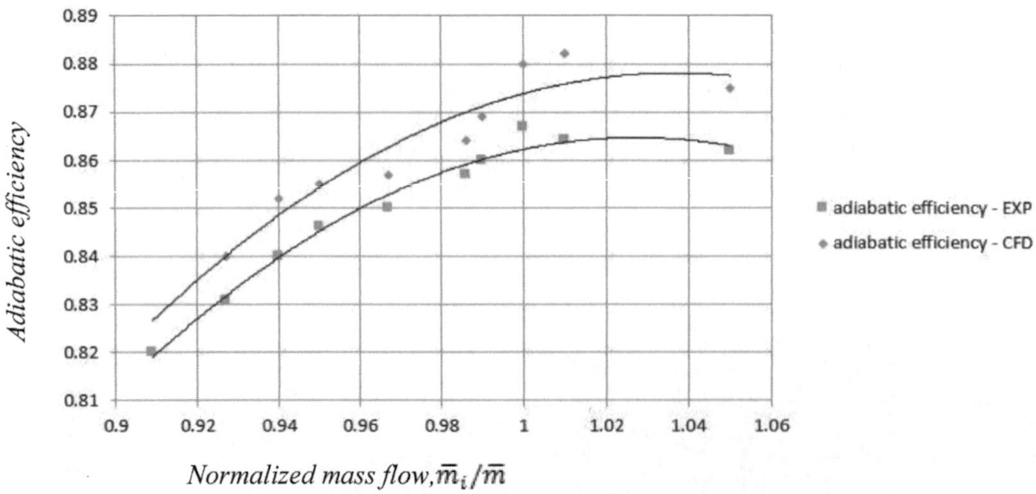

Figure 3. Verification of the CFD results

4. Investigation of the casing treatment

4.1. Selection of casing treatment configuration

In the article of 1984 Fujuta and Takata [6] considered multiple configurations of casing treatment: axial-skewed and axial-radial slots (Fig.4-5). The result of this study indicates that skewed axial slots casing treatment is the best way to provide the margin of stability. However, the circumferential grooves provides less efficiency losses. Thus, to ensure the stable operation with the least loss of efficiency, it is necessary to combine these two types. According to the Cevik`s [7] survey the triangle shape is the optimal to achieve acceptable SM improvement and prevent the large efficiency losses.

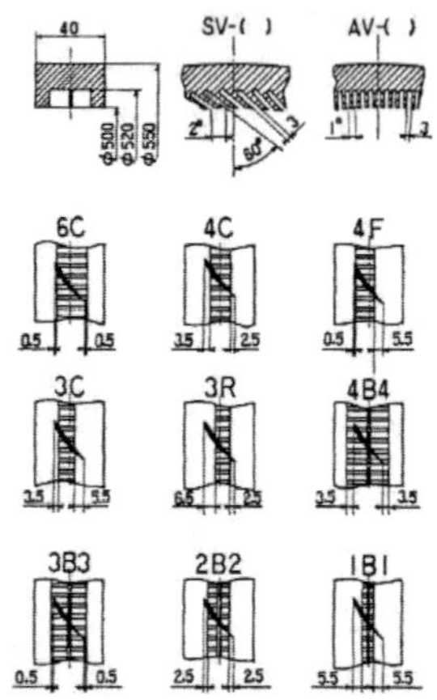

Figure 4. Axial-skewed and axial-radial slot configuration by Fujuta and Takata (1984) [6]

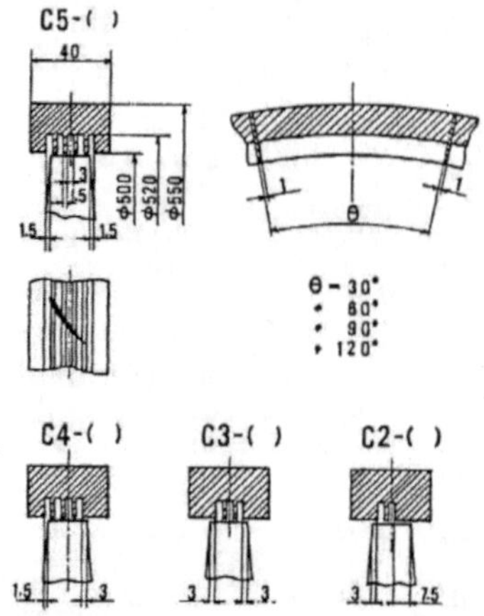

Figure 5. Circumferential groove configurations by Fujuta and Takata (1984) [6]

Selection of casing treatment geometry for the study was by virtue the above-mention and other findings. Another important chose factor was simplicity – the design should be suitable and applicable for the existing geometry with a smooth rotor. For this work was selected annular grooves, triangular shape (Fig.6).

The appearance of investigated casing treatment is shown on Fig.7 shows. Table 2 reflects the main geometric parameters.

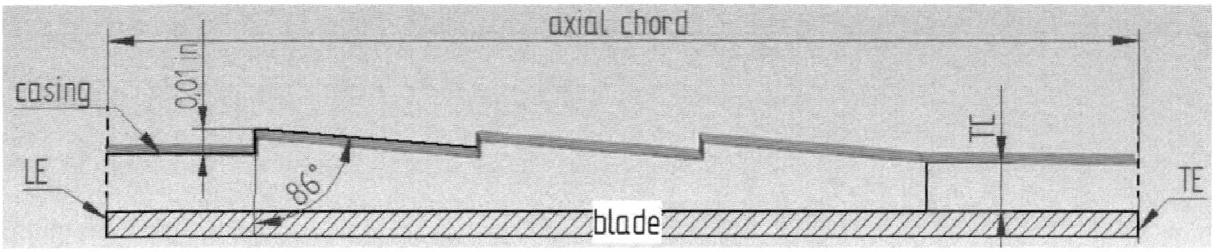

Figure 6. Casing treatment. Front view

Figure 7. Position of the casing treatment above the rotor blade row

Table 2. Design parameters for the stage with casing treatment

Shape	*Triangle*
Depth	*1x10^{-3} inch*
Width	*0,6 Axial Chord*
Groove	*3*

4.2. *Computational model*

The structured mesh was generated in ICEM CFD and it was implemented in previous computation domain. The general grid interface (GGI) was used to connect the grooves to the main domain.

5. Results and discussion

Table 3. Obtained performance parameters for the smooth casing and casing with grooves

Parameters	Smooth casing	Casing with grooves
SM, %	13.56	15.63
Adiabatic efficiency peak,	0.866	0.822

Figure 8, 9 indicated the Mach number counters on the pre-stall mode ($\overline{m}_{st}/\overline{m}$=0.91 kg/s). Noticeably that there are a low speed region (Fig.8). In case of casing treatment this low-energy zone is not so strong (Fig.9) and virtually are distinguishable.

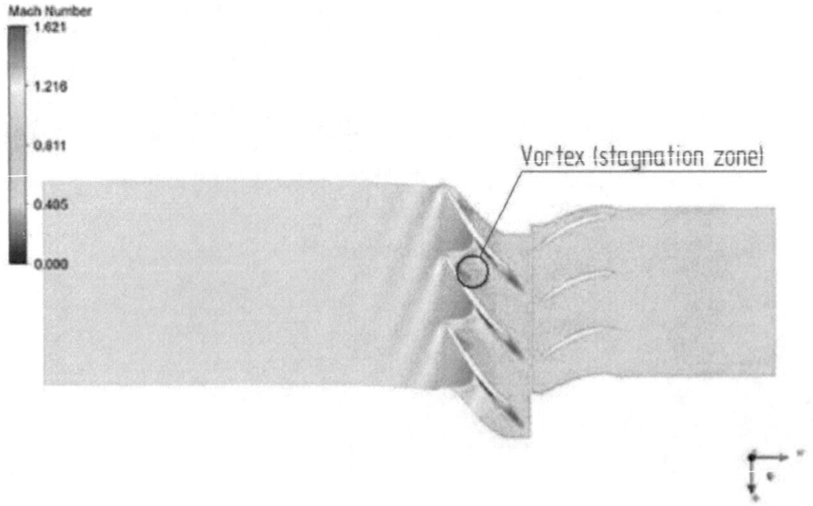

Figure 8.
Smooth rotor
Mach number distribution in blade-to-blade plane at the 98% span
$\bar{m} = 0,91 \cdot \bar{m}_{max}$

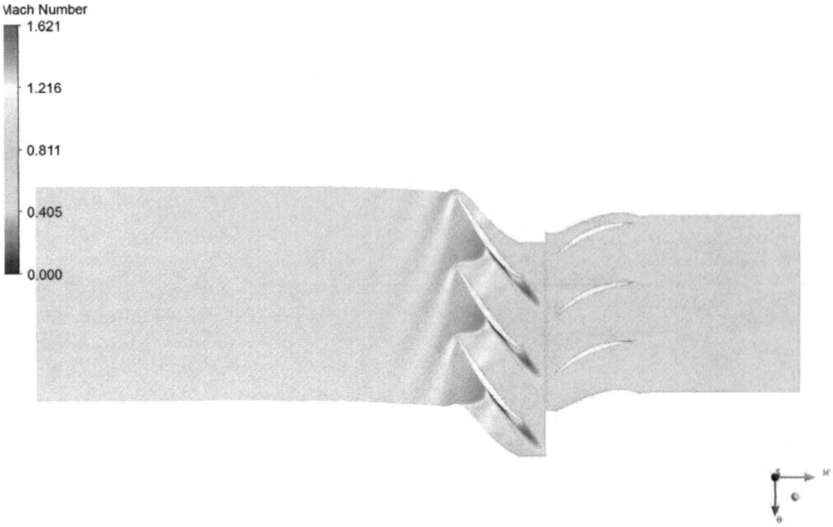

Figure 9.
With casing treatment (casing treatment hidden)
Mach number distribution in blade-to-blade plane at the 98% span
$\bar{m} = 0,91 \cdot \bar{m}_{max}$

6. Conclusions and future works

Needless to say, casing treatments are a good way to improve transonic stage performance. Circumferential grooves are the better choice as their application on an existing rotor design entails a much lower performance penalty than axial slots. [4]

According to obtained results the SM improvement can achieve up to 13 %. The efficiency of such treatment was increased by 0.5%, however, the system remained at the same level. This is also a positive indicator, because there are no efficiency losses by using such treatments.

Also, it must be mentioned that this treatment reduces the mass of the casing, which is a benefit for aircraft design.

Optimization of groove shape is also an interesting and important field of investigation. Moreover, tip injection is the most promising way to have an influence on stationary zones and likewise requires special attention.

IOP Conf. Series: Materials Science and Engineering **232** (2017) 012048 doi:10.1088/1757-899X/232/1/012048

References

[1] Kim Jin-Hyuk, Dae-Woong Kim, and Kwang-Yong Kim 2013 *Aerodynamic optimization of a transonic axial compressor with a casing groove combined with tip injection.* Institution of Mechanical Engineers, Part A: Journal of Power and Energy vol. 227 **8** : 869-884.

[2] Seleznev K P, Podobuev Y S 1957 *Theory and calculation of axial and centrifugal compressors.* - Leningrad: Engineering literature

[3] Hembera M., Kau, H. P., & Johann, E. 2008. *Simulation of casing treatments of a transonic compressor stage*. International Journal of Rotating Machinery

[4] Beheshti B. H., et al. 2005 *Performance enhancement in transonic axial compressors using blade tip injection coupled with casing treatment.* Proceedings of the Institution of Mechanical Engineers, Part A: Journal of Power and Energy 219.5: 321-331.

[5] Reid L., Moore R. D. 1978 *Design and overall performance of four highly loaded, high speed inlet stages for an advanced high-pressure-ratio core compressor.*- NASA-TP-1337, E-9302,122

[6] Fujita H., Takata H. 1984 *A study on configurations of casing treatment for axial flow compressors* Bulletin of JSME..№. 230. – C. 1675-1681.

[7] Cevik M. 2013 *Axial Compressor Gas Path Design For Desensitization of Aerodynamic Performance and Stability to Tip Clearance.* – ÉcolePolytechnique de Montréal

Performance Measurements and Mapping of a R-407C Vapor Injection Scroll Compressor

Domenique Lumpkin[1,3], Niklas Spielbauer[2], Eckhard Groll[3]
[1]Purdue University, School of Civil Engineering, West Lafayette, Indiana, USA
[2]Karlsruhe Institute of Technology Department of Mechanical Engineering Karlsruhe, Germany
[3]Purdue University, School of Mechanical Engineering, Ray W. Herrick Laboratories, West Lafayette, Indiana, USA

Email: Dlumpkin@purdue.edu

Abstract. Environmental conditions significantly define the performance of HVAC&R systems. Vapor compression systems in hot climates tend to operate at higher pressure ratios, leading to increased discharge temperatures. Higher discharge temperatures can lead to higher irreversibilities in the compression process, lower specific enthalpies differences across the evaporator, and possibly a reduction in the compressor life due to the breakdown of the oil used for lubrication. To counter these effects, the use of economized, vapor injection compressors is proposed for vapor compression systems in high temperature climates. Such compressors are commercially available for refrigeration applications, in particular, supermarket refrigeration systems. However, compressor maps for vapor injection compressors are limited and none exist for R-407C. Through calorimeter testing, a compressor map for a single-port vapor injection compressor using R-407C was developed. A standard correlation for mapping single-port vapor injection compressors is proposed and validated using the compressor test results. The system and compressor performance with and without vapor injection was considered. As expected, with vapor injection there was a reduction in compressor discharge temperatures and an increase in the system coefficient of performance. The proposed dimensionless correlation is more accurate than the AHRI polynomial for mapping the injection ratio, discharge temperature, and compressor heat loss. The predicted volumetric efficiency values from the dimensionless correlation is within 1% of the measured valued. Similarly, the predicted isentropic efficiency values are within 2% of the measured values.

1. Introduction

It is well documented that heat sink and source temperatures strongly influence vapor compression system performance. HVAC&R systems operating in extreme temperature climates are required to operate at higher pressure ratios. The system refrigeration capacity and efficiency decline with higher condensing temperatures and lower evaporating temperatures. Lower system performance is attributed to an increase in irreversibilities during compression, and higher compressor discharge temperatures. Economized, refrigerant vapor-injection (VI), vapor compression cycles have been considered promising solutions for reducing the compressor discharge temperature for systems with large compression ratios (Winandy et al., 2002; Wang et al., 2009). This is especially important because higher discharge temperatures can result in additional wear on the compressor, reducing the compressor reliability and life. Vapor-injection during the compression process has also been proven to improve the refrigeration capacity and coefficient of performance (COP) (Wang et al., 2009).

Typically, the AHRI polynomial equations, Equation (1), capture the compressor behavior as a function of test pressure ratios specified by the manufacturer. These equations are then used to predict the compressor power input and suction mass flow rate at the operator's desired operating conditions within the working envelope. The AHRI standard polynomial works best for single-stage compressor with no refrigerant injection. The AHRI polynomial does not capture the performance of a vapor-injection scroll compressor. This polynomial neglects the impact of the injection fluid on the compressor performance. As vapor-injection scroll compressors are commercially available and implemented in current systems, the proposed dimensionless correlation, Equation (2), is a method for mapping vapor-injection scroll compressor performance.

This work presents a new compressor map available for a R-407C vapor injection scroll compressor for use in an air-conditioning system. R-407C is an HFC refrigerant that has been used as a replacement for R-22 in residential and commercial air-conditioners and heat pumps and some existing medium-temperature applications since 2010. The vapor-injection compressor performance was compared with the baseline performance for the same compressor with the injection port closed. The coefficient of performance was defined using Equation (8).

Thus far, economized, vapor injection research has considered the economization method, the injected refrigerant phase, and the port location. The characteristics of a two-stage heat pump system in hot climates was investigated by Wang et al., 2009 using both flash tank (FTC) and internal heat exchanger (IHX) configurations for vapor-injected compression. Compared to a conventional system using a single-stage compressor with the same suction volume, there was a cooling capacity gain of 15% and a COP improvement of 2% at an ambient temperature of 46.1°C. It was concluded that the maximum cooling COP improvement ranged from 2-4% depending on the ambient conditions. Also, the vapor injection almost equally affected the capacity and power consumption. At an ambient temperature of -17.8°C the heating capacity improvement was 30% with a maximum COP improvement of 23% which was achieved by the FTC. Both the IHX and FTC configurations showed an improvement from the baseline. The maximum COP improvement from baseline was achieved by the FTC at the -17.8°C ambient. The IHX configuration enabled a wider operating range of the injection pressure than that of the FTC due to its freedom of setting the injected refrigerant superheat at the injection port. A TXV was suggested for simple and effective control of the injected refrigerant and the FTC required a larger TXV at the evaporator inlet.

An experimental performance analysis of different hermetic scroll compressors with vapor injection, liquid injection and no injection was conducted by Winandy et al, 2002. This work generated a vapor-injection and a liquid-injection compressor maps for the refrigerant R-22. A semi-empirical model that was able to compute variables of first importance, such as the mass flow rate, the electric power consumption and the discharge temperature, as well as secondary variables, such as suction heating-up, discharge cooling-down, and ambient losses, was validated using 45 experimental results.

Navarro et al., 2013 characterized a vapor-injection scroll compressor in a heat pump at low, intermediate, and high pressure and temperature conditions. A correlation for the injection mass flow rate as a function of external conditions and the intermediate pressure was developed. The VI compressor behavior was compared with the same compressor with the injection port closed and to a single-stage compressor of the same displacement. Compared to the single-stage compressor, the heating capacity and COP increased up to 20% and 10%, respectively. The primary performance differences were shown at higher pressure ratios. The VI compressor isentropic and volumetric efficiency were lower for low and medium pressure ratios and had negligible differences at higher pressure ratios. The discharge temperature decreased, especially at extreme conditions, with more than a 10 K difference. This enables a wider operating map and improved compressor life and reliability. Regarding the injection state, the injection superheat did not influence the system COP and the intermediate pressure improved the cycle performance until wet injection occurred.

Dimensionless PI-type mappings for vapor-injected compression for system simulation were introduced by Bach et al. 2015 and are the basis of the correlation presented here. In Bach et al.'s work, the differences of in-system compressor performance testing using a cold-climate heat pump with a dual-port vapor injected compressor were compared to compressor performance testing using a hot-gas bypass test stand and a calorimeter. The main differences between in-system testing and test stand testing can be found in suction and injection superheat values, discharge temperatures, and range of injection flow rates. More mapping points were created from in-system testing whereas the test stand data was limited to two discharge pressures. The Buckingham-PI theorem was used to create a dimensionless output group and an appropriate dimensionless input group. Mappings of volumetric and overall isentropic efficiencies, compressor heat loss, injection flow rates, and discharge pressure were developed as the output group. It was found that inclusion of injection superheats did not lead to significant improvement of accuracy for the mappings. In-system data alone did not always result in reasonable predictions of the mapped test stand data due to the large difference of the test matrices and operating conditions. The reverse was also true and therefore, combining all data sets for finding mapping coefficients lead to more generally applicable mappings.

2. Experimental Setup

Economized vapor injection systems are vapor compression systems modified to include a flash tank (open economizer) or intermediate heat exchanger (closed economizer) for system performance improvement. The closed economizer increases the subcooling of the refrigerant exiting the condenser outlet resulting in a specific enthalpy decrease at the evaporator inlet leading to increased cooling capacity. The refrigeration cycle is represented in Figure 1 and was the basis of the experimental setup in a calorimeter originally constructed by Moesch et al [3].

An existing calorimeter was modified to include an internal heat exchanger as shown in Figure 2 using a plate heat exchanger as the closed economizer and downstream extraction of the injected refrigerant. The modification was integrated after the filter drier in the liquid line of the calorimeter unit using existing oil sample ports. An electronic expansion valve (EXV) was used to control the injection flow as it extended the range of injection superheat control and enabled two-phase injection. A solenoid valve was installed before the EXV to protect the compressor from two-phase refrigerant entering the compressor chamber. Two service valves at the upper connections of the heat exchanger allowed a pressurization and evacuation of the modification section independent from the main system. The compressor vibration was reduced by a vibration absorber before the injection port to insulate the modification. All modification related components and pipes were placed inside the compressor chamber to reduce the overall volume added to the existing test setup, thus reducing the additional refrigerant charge. To reduce the impact of the chamber temperature on the liquid subcooling and the injection superheat, all modification related pipes and components within the compressor chamber were insulated. The instrument measurement accuracies, testing procedures and test results are compliant to the ANSI/ASHRAE Standard 23.1-2010 [1]. A schematic of the instrumentation and piping diagram indicating all significant features is shown in Figure 3.

The operating temperatures are measured with T-Type thermocouples, which were either fully immersed in the refrigerant flow (TC-1 to TC-8B) and water flow (TC-9, TC-10) or closely attached to the measured ambient temperature (TC-11, TC-12). Four temperature sensors (TC-13 to TC-16) were added for measurement of the economized, vapor-injected refrigerant.

The system pressures were measured using high-accuracy absolute capacitance pressure transducers with full scale ranges of 0 to 689 kPa for low-pressure measurements (PE-1 and PE-6), 0 to 3,447 kPa for the secondary calorimeter pressure measurement (PE7) and 0 to 6,895 kPa for high pressure measurements (PE-2 to PE-5). Three pressure sensors (PE-8, PE-9) were added for refrigerant pressure measurements of the economized, vapor-injected refrigerant. The pressure transducers provided an output signal of 4 to 20 mA, which was translated to a 1 to 5 V signal by using a 250 Ohm resistor parallel to each output.

The water flow through the condenser was measured using a turbine flow meter with a measurement range of 0 to 4.542 m^3/h. The output signal is a frequency with a maximum of 1,800 Hz. The frequency signal is converted into a 4 to 20 mA signal by the flow computer. The process water control in the condenser was modified by installing an additional 15.875 mm ball valve to raise and maintain condensing temperatures in the air-conditioning temperature envelope.

The compressor chamber temperature was set to 35 °C, the fan was set to its maximum speed, the value of the suction superheat is adjusted to be 11.1 K, and the voltage for each phase of the compressor was set to 230V at 60 Hz.

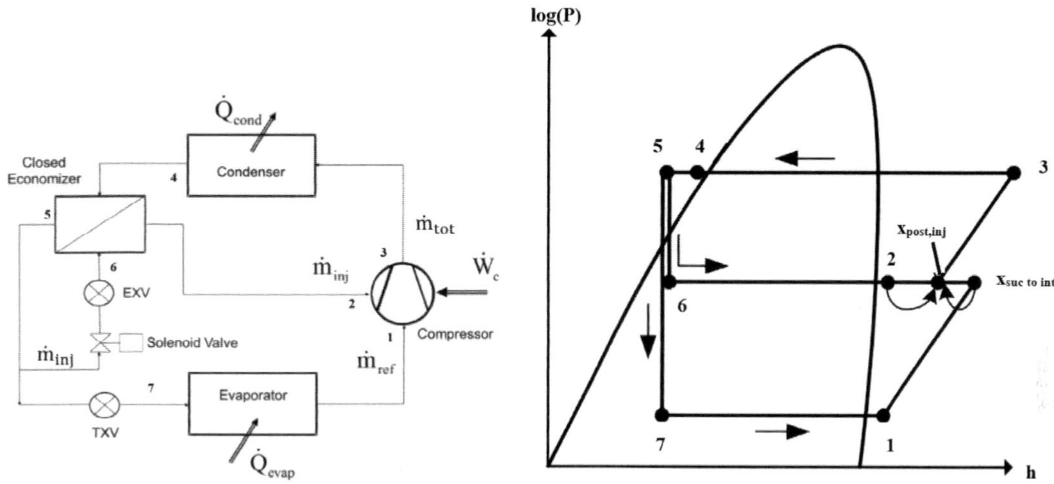

Figure 1. Schematic of an economized vapor injection vapor compression system with downstream extraction and the corresponding log(P)-h diagram

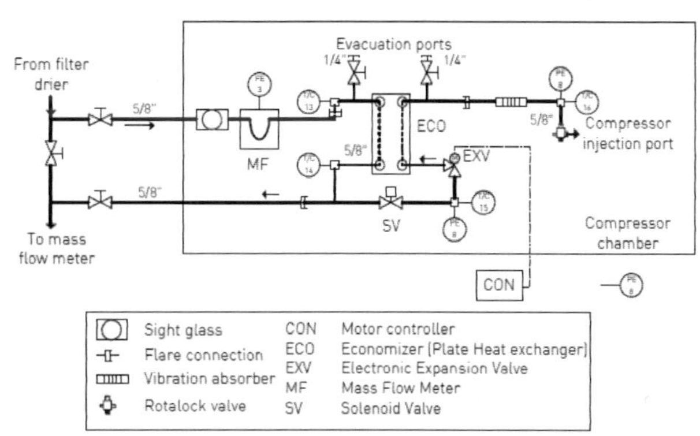

Figure 2. Instrumentation and piping diagram of the vapor injection modification

The primary method for the refrigerant flow measurement was a Coriolis-effect mass flow meter. The flow meter was connected to a transmitter, which translates a mass flow rate of 0 to 544 kg/h to a frequency signal 0 to 2,000 Hz. A transmitter translated the frequency to a 4 to 20 mA signal. The refrigerant mass flow rate was confirmed with the secondary calorimeter method based on ANSI/ASHRAE Standard 41.9-2010.

The air flow in the compressor chamber (AV-1) was measured using an air velocity transmitter, which was a hot wire anemometer with a range of 0 to 5.08 m/s and an output signal of 4 to 20 mA.

The power consumption of the compressor was measured using a watt/watthour transducer with a full-scale range of 0 to 20 kW. The transducer uses two power meters which are set up as an Aron circuit and enable the measurement of the effective power consumption for the 3-phase power supply. The compressor power consumption is displayed in W and its total energy consumption in Wh. The compressor voltage and current were measured only for the first phase using a current transducer and a voltage transducer with full scale ranges of 0 to 20 A and 0 to 600 V. The power consumption of the secondary calorimeter heater is measured using a watt/watthour transducer with a full-scale range of 0 to 18 kW. The output of the power, voltage and current transducers is a 0 to 1 mA signal, which is transformed to 0 to 5 V using a 5,000 Ohm resistor.

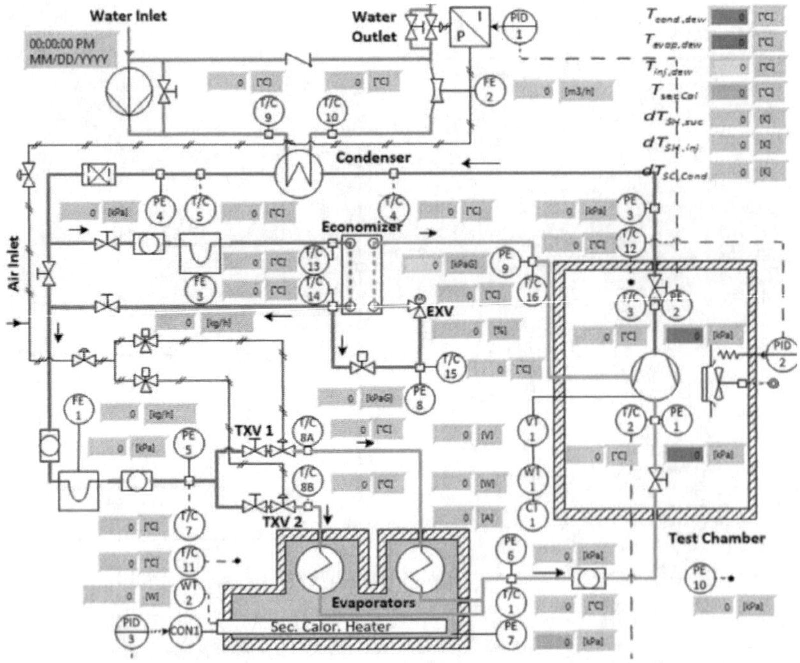

Figure 3. Full piping and instrumentation diagram

3. Compressor Map and Performance Results

The R-407C vapor-injection compressor map is presented in Table 1. The injection ratio ranged from 0.09 to 0.28 (9). The compressor input power ranged from 3.2 to 5.6 kW. The isentropic efficiency, Equation (4), ranged from 64 to 67%. The volumetric efficiency, Equation (3) ranged from 88 to 93%.

Table 1: R407C compressor Map at a fixed injection superheat of 5.56 K

P_{EVAP} (kPa)	P_{COND} (kPa)	$\dot{m}_{ref}$ (kg/s)	$\frac{\dot{m}_{inj}}{\dot{m}_{ref}}$	$\dot{W}$ (W)	T_{DIS} (K)	P_{INJ} (kPa)	T_{INJ} (K)	T_{SUC} (K)
445.4	2221	0.06426	0.2824	4815	366.0	859.1	298.2	283.2
488.1	2221	0.07031	0.2688	4870	363.5	813.6	296.2	286.2
591.6	2222	0.08618	0.2105	5003	360.4	935.6	301	291.8
444.0	1690	0.06577	0.1839	3812	351.2	661.9	289.8	283.3
538.5	1686	0.08013	0.1415	3929	348.5	775.7	294.8	288.9
329.6	1460	0.04838	0.2031	3239	350.8	584.7	286	275.2
536.4	1459	0.08013	0.09434	3501	343.3	830.8	296.9	289
539.2	1934	0.07938	0.1714	4369	355.6	915.6	300.2	289.2
403.3	1938	0.05897	0.2436	4218	360.7	753.6	293.5	280.3
364.7	1938	0.05292	0.2714	4179	363.5	704	291.7	277.9
593.6	2518	0.08543	0.2478	5588	368.0	1095	306.4	292

The baseline compressor performance is compared to the vapor-injection compressor performance in Table 2 with the measured condensing pressures. The reduction in the discharge temperature ranged from 0 to 3.3 K. The highest reduction in discharge temperature occurred at the largest pressure ratio occurring at 322 K condensing and 267 K evaporating temperatures. The highest COP improvement of almost 9.9% occurred at this pressure ratio as well. Conversely, there was virtually no reduction in discharge temperature and no COP improvement at the lowest pressure ratio at 277.7 K evaporating and 311.4 K condensing temperatures. The largest reduction in isentropic efficiency was 2.86%. The volumetric efficiency reduction ranged from 0 to 1.56%. The reduction in isentropic efficiency is due to the additional power required for vapor-injected compression. The difference in volumetric efficiency was minimal but is likely due to rounding and measurement uncertainties.

Table 2: Closed-port (baseline) and open-injection port compressor performance comparison

P_{EVAP} (kPa)	P_{COND} (kPa)	T_{DIS} (ΔK)	COP (Δ%)	η_{isen} (Δ%)	η_{vol}(Δ%)
536.4	1459	0.7	0.0	-2.29	0.10
329.6	1460	-2.6	7.2	-0.67	-0.07
444.0	1690	-1.9	2.1	-0.61	0.05
538.5	1686	-1.0	0.9	-1.86	-0.17
539.2	1934	-0.3	1.9	-2.56	-0.39
403.3	1938	-3.0	8.0	-1.42	-0.12
364.7	1938	-3.3	9.9	-1.17	-0.06
593.6	2518	-2.2	6.9	-2.50	-0.86
445.4	2221	-3.2	9.4	-1.78	-1.01
488.1	2221	-3.0	4.7	-1.10	-1.56
591.6	2222	-1.6	2.8	-2.02	-0.97

4. Mapping of Results

The AHRI polynomial as shown in Equation (1) is typically used to predict the power input, suction mass flow rate, and the refrigerating capacity for positive displacement compressors and compressor units. Its mapping coefficients C_1 through C_{10} are generated based on manufacturer data. The user can then predict their compressor behavior at specified operating conditions.

$$\dot{W}, \dot{Q}, \dot{m}_{ref} = C_1 + C_2 T_{evap} + C_3 T_{cond} + C_4 T_{evap}^2 + C_5 T_{evap} T_{cond} + C_6 T_{cond}^2 + C_7 T_{evap}^3 + C_8 T_{cond} T_{evap}^2 + C_9 T_{evap} T_{cond}^2 + C_{10} T_{cond}^3 \quad (1)$$

The Buckingham-PI theorem as shown in Equation (2) was used to map the compressor performance for the single-speed R-407C scroll compressor. Appropriate dimensionless input groups were considered for each normalized output PI group listed in Equation (3). This correlation considers the compressor power supply frequency $f_{nominal}$ as well as the suction state, injection mass flow rate and discharge pressure. Definitions for the abbreviations are documented in the nomenclature. The injection state was normalized by the suction state. The discharge and suction states were normalized by the appropriate refrigerant critical point values. The compressor chamber temperature was considered by normalizing the suction temperature to this temperature.

The enthalpy difference $\Delta h_{suc,sh}$ at the compressor inlet is characterized by the difference between the measured inlet enthalpy and the saturated liquid enthalpy at the same pressure. The definition of the volumetric efficiency was defined as the ratio of the actual compressor suction volume flow rate to the maximum theoretical compressor suction volume flow rate as shown in Equation (3). The theoretical suction volume flow rate was defined as the product of the maximum volume of the suction process V_{disp} and the compressor rotation rate N_c. The isentropic efficiency was defined as the ratio of the actual compressor power consumption to the power consumption needed for an adiabatic and reversible process from the suction port to the discharge port as shown in Equation (4). The isentropic injection specific enthalpy $h_{inj,isen}$ was defined assuming isentropic compression from the suction state to the intermediate pressure. The specific enthalpy of the refrigerant entering the compressor defines h_{suc}. The specific enthalpy of the refrigerant vapor after mixing between the intermediate pressure flow and the partially compressed flow at state $h_{inj,isen}$ is $h_{post,inj}$ documented in Equation (5). The specific enthalpy of the fluid at the compressor discharge pressure following an isentropic compression of the refrigerant vapor from state $h_{post,inj}$ results in $h_{dis,isen}$. The compressor heat loss is estimated from an energy balance assuming a control volume around the compressor surface as shown in Equation (6). The heat loss ratio f_q, Equation (7), was then defined as the ratio of the compressor heat loss to the measured power input. The mapped coefficients for the discharge temperature at the compressor outlet, the injection mass flow rate, the overall (isentropic) efficiency, volumetric efficiency, and heat loss of the compressor are highlighted in Table 3. The coefficients were generated using the Variable Metric optimization method for minimization in Engineering Equation Solver (Klein, 2012).

$$\pi_{out} = C_0 \frac{f_{power}}{f_{nominal}}^{C_1} \frac{\dot{m}_{inj}}{\dot{m}_{ref}}^{C_2} \frac{P_{inj}}{P_{suc}}^{C_3} \frac{T_{suc}}{T_{amb}}^{C_4} \frac{P_{dis}}{P_{suc}}^{C_5} \frac{\Delta h_{suc,sh}}{\Delta h_{suc,fg}}^{C_6} \frac{P_{dis}}{P_{crit}}^{C_7} \frac{P_{suc}}{P_{crit}}^{C_8} \frac{T_{suc}}{T_{crit}}^{C_9} \quad (2)$$

$$\eta_{vol} = \frac{\dot{m}_{ref} v_{suc}}{V_{disp} N_c} \quad (3)$$

$$\eta_{isen} = \frac{\dot{m}_{ref}(h_{inj,isen} - h_{suc}) + \dot{m}_{tot}(h_{dis,isen} - h_{post,inj})}{\dot{W}} \quad (4)$$

$$h_{post,inj} = \frac{\dot{m}_{ref} h_{inj,isen} + \dot{m}_{inj} h_{inj}}{\dot{m}_{tot}} \quad (5)$$

$$\dot{m}_{ref} h_{suc} + \dot{m}_{inj} h_{inj} - \dot{m}_{tot} h_{dis} + \dot{W} - \dot{Q}_{loss} = 0 \quad (6)$$

$$f_q = \frac{\dot{Q}_{loss}}{\dot{W}} \quad (7)$$

$$\mathrm{COP} = \frac{\dot{Q}}{\dot{W}} \quad (8)$$

$$\text{Injection ratio} = \frac{\dot{m}_{inj}}{\dot{m}_{ref}} \quad (9)$$

The mapped coefficients for the dimensionless correlation for the different aspects of the compressor performance are presented in Table 3 along with the R^2 value for comparison with the AHRI polynomial performance. The proposed dimensionless correlation is more accurate than the AHRI polynomial for mapping the injection ratio, discharge temperature, and compressor heat loss as shown in Figures 4, 5 and 8, respectively. The predicted volumetric efficiency and isentropic efficiency from the AHRI-polynomial is slightly more accurate. However, the predicted volumetric efficiency values from the dimensionless correlation is within 1% of the measured valued as shown in Figure 6. Similarly, the predicted isentropic efficiency values are within 2% of the measured values as shown in Figure 7.

Table 3: Mapping results for the compressor performance

Coefficients	Mapped dimensionless group $\frac{\dot{m}_{inj}}{\dot{m}_{ref}}$	$\frac{T_{dis}}{T_{crit}}$	η_{isen}	η_{vol}	f_q
C_0	0.1087	0.7637	1.237	1.217	0.9727
C_1	1	1	1	1	1
C_2	0	-0.117	0.4838	-0.0004	-0.9649
C_3	-0.7606	0.0109	0.03063	-0.00329	-1.694
C_4	0.2705	-0.1385	0.2827	-0.1713	0.9395
C_5	1.443	0.714	-0.1018	0.2913	1.758
C_6	-0.6861	0.7784	1.033	-0.4828	0.6728
C_7	0.7626	0.103	-0.7837	-0.369	1.47
C_8	0.316	0.3898	0.3174	0.3454	0.6815
C_9	-0.3924	0.1167	-0.2301	0.1807	-1.147
Dimensionless Correlation R^2 value	99.5	99.99	87.86	96.77	98.17
AHRI-polynomial R^2 value	99.29	99.88	91.93	99.75	88.66

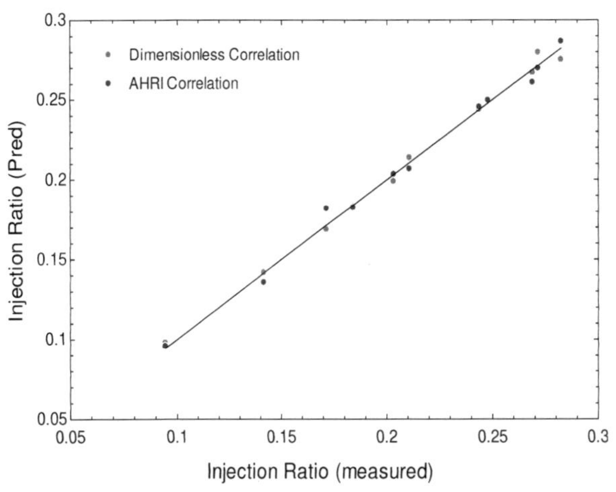

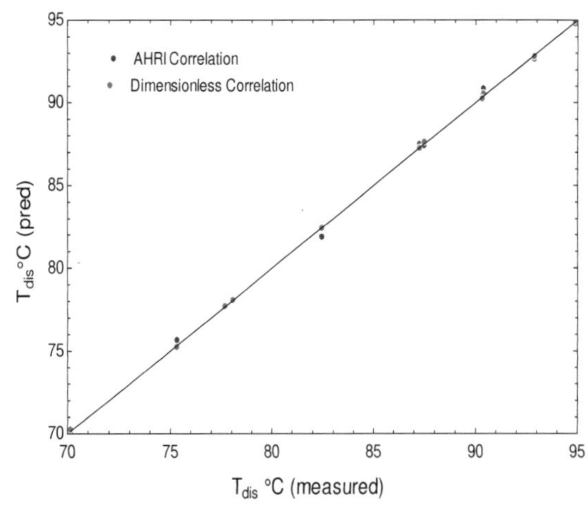

Figures 4 and 5. Injection ratio parity plot (left) and the discharge temperature parity plot (right)

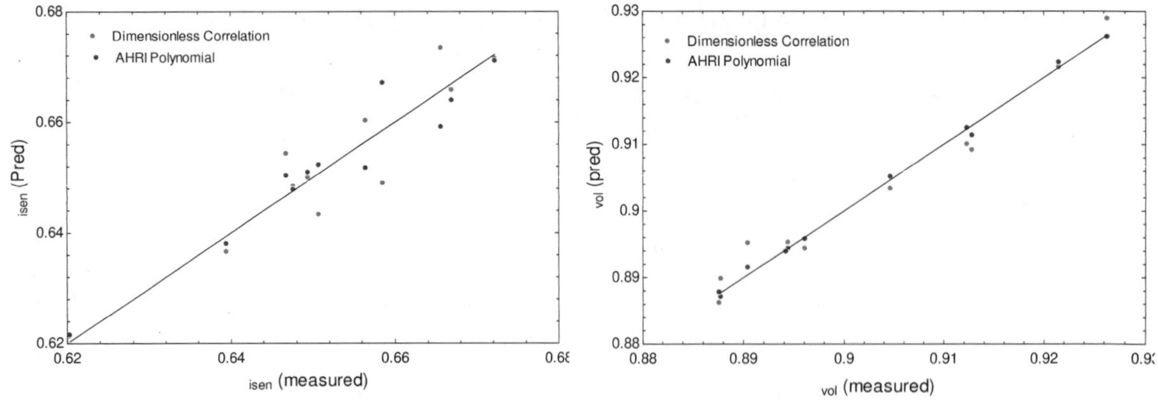

Figures 6 and 7. Isentropic efficiency parity plot (left) and volumetric efficiency parity plot (right)

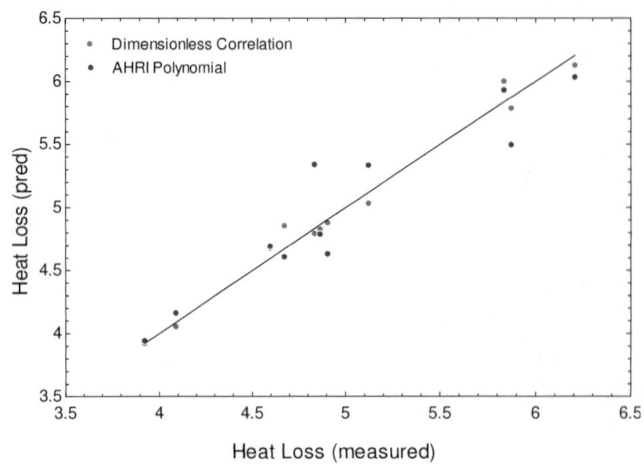

Figure 8. Heat loss parity plot

5. Conclusion

A new correlation for mapping single-port vapor injection compressors is proposed and was validated using compressor test results resulted obtained with a standard calorimeter. The new correlation is expected to more accurately predict compressor performance that the AHRI-polynomial vapor-injection compressors including variable speed compressors than the AHRI-polynomial. An R-407C vapor-injection scroll compressor map was generated and a dimensionless correlation for mapping results was presented and compared to the standard AHRI-polynomial.

6. References

[1] Hall, R. L., Douglas, J. L., Irons, M. T., Leyderman, A. D., Scott, J. P., Slayton, C. R., … Peterson, J. C. (2010). ASHRAE STANDARD Methods of Testing for Rating the Performance of Positive Displacement Refrigerant Compressors and Condensing Units that Operate at Subcritical Temperatures of the Refrigerant. Ashrae Standard.

[2] Klein, S. A. (2012). Engineering equation solver, academic commercial version 9.434-3d. F-Chart Software, LLC. Madison, WI.

[3] Moesch T W, Bahman A M, Groll E A, 2016 Performance Testing of a Vapor Injection Scroll Compressor with R407C, *Proc. Int. Comp. Eng. Conf. (Purdue)*

[4] Navarro, E., Redón, A., Gonzálvez-Macia, J., Martinez-Galvan, I. O., & Corberán, J. M. (2013). Characterization of a vapor injection scroll compressor as a function of low, intermediate

and high pressures and temperature conditions. *International Journal of Refrigeration*, *36*(7), 1821–1829. http://doi.org/10.1016/j.ijrefrig.2013.04.022

[5] Standard, A. R. I. (2004). Standard For Performance Rating Of Positive Displacement Refrigerant Compressors And Compressor Units, *540*.

[6] Wang, B., Shi, W., Han, L., & Li, X. (2009). Optimization of refrigeration system with gas-injected scroll compressor. *International Journal of Refrigeration*, *32*(7), 1544–1554. http://doi.org/10.1016/j.ijrefrig.2009.06.008

[7] Wang, X., Hwang, Y., & Radermacher, R. (2009). Two-stage heat pump system with vapor-injected scroll compressor using R410A as a refrigerant. *International Journal of Refrigeration*, *32*(6), 1442–1451. http://doi.org/10.1016/j.ijrefrig.2009.03.004

[8] Winandy, E. L., & Lebrun, J. (2002). Scroll compressors using gas and liquid injection: Experimental analysis and modelling. *International Journal of Refrigeration*, *25*(8), 1143–1156. http://doi.org/10.1016/S0140-7007(02)00003-8

7. Nomenclature

Symbol	Definition	Unit		Subscripts	
H	enthalpy	(kJ/kg)		comp	compressor
$\dot{m}$	mass flow rate	(kg/s)		evap	evaporator
S	entropy	(kJ/kg-K)		cond	condenser
P	pressure	(kPa)		crit	critical
$\dot{W}$	Power	(W)		dis	discharge
C	mapping coefficients	(-)		disp	displacement
T	temperature	(°C)		loss	loss
Q	cooling capacity	(W)		inj	injection
η	efficiency	(–)		isen	isentropic
V	specific volume	(m^3/kg)		post	after injection
COP	coefficient of performance	(–)		ref	compressor inlet state
V	volume	(m^3)		sh	superheat
IHX	intermediate heat exchanger	(–)		sub	subcooled
				suc	suction
				tot	total, discharge mass flow rate
				vol	volumetric

IOP Conf. Series: Materials Science and Engineering **232** (2017) 012050 doi:10.1088/1757-899X/232/1/012050

Numerical analysis of the transient flow in a scroll refrigeration compressor

SUN Shuaihui[1,2], WU Kai[1], GUO Pengcheng[1,*], LUO Xingqi[1]
1. State Key Laboratory Base of Eco-hydraulic Engineering in Arid Area, Xi'an University of Technology, Xi'an, Shaanxi 710048, China
2. State Key Laboratory of Compressor Technology, Hefei, Anhui 230031, China

*E-mail address: guoyicheng@126.com

Abstract. In the present paper, the CFD technology is adopted to simulate the working process of a scroll refrigeration compressor with R22 as working fluid. The structural grids in the scroll compressor were updated continually during the solving process to cope with the movement boundaries of the fluid domain. The radial meshing clearance was 0.008 mm which was the same with that in the real prototype. The pressure, velocity and temperature distribution in chambers of compressor were computed. Also, the transient mass flux diagrams were calculated out. The results indicated that the pressure was asymmetrical in the two symmetrical suction chambers, because the suction port and passage were not absolutely symmetrical. The gradient of temperature was great in each working chamber due to leakage flow. Velocity vector distribution was asymmetrical in each pair of working chamber owing to the movement of orbiting scroll; the flow was complicated in the central working chamber. The movement of the orbiting scroll had different influence on the vortexes formation in each pair of compression chamber. The inlet and outlet mass flux fluctuated with the crank angle obviously. Because of the 'cut-off' of the refrigeration fluid in the suction chamber when the crank angle was larger than 220°, the inlet mass flux decreased remarkably. Finally, some useful advices were given to improve the performance of the scroll refrigeration compressor.

1. Introduction

Scroll compressors were widely used in small refrigeration systems and air conditioners, and it has become one of the most common used refrigeration compressors. The investigation of scroll compressors was essential to improve its performance.

In recent years, computational fluid dynamics (CFD) technology has gradually become a reliable means to analyze working process of the scroll refrigeration compressor. Cui [1] simulated the three dimensional transient flow in a scroll refrigeration compressor and found that the mass, velocity, pressure, and temperature were asymmetric in each pair of working chamber. Furthermore, the CFD method has been used to simulate the suction process [2] and the discharge process [3] to study the fundamental mechanism of them, and supposed the optimum proposal. Wang et al. [4] designed a multiphase scroll pump and analyzed the effects of gas volume fraction and rotation speed on the performance of scroll pump using the CFD method. Yue et al. [5] established the transient model of the scroll vacuum pump on the basis of dynamic mesh technology with the fluent code; three cases including under-compression, slight over-compression and over-compression were calculated and compared with the experimental results; the maximum error reached 30%. Hesse and Andres [6] used the Twinmesh and CFX to simulate the scroll vane pump with leakage transiently; the results indicated that the simulation can capture the working mechanism and flow conditions in the dry scroll vacuum pump. Ding et al. [7] had done a CFD simulation of an oil-flooded scroll compressor with VOF model and compared the simulation results with the experimental results, and proposed that the method is robust. Chang et al. [8] employed a CFD method for investigating the thermal-hydraulic behavior of the scroll type expanders; the unstructured grid was used and the unbalance pressure distribution had been found in the expanding process. Picavet and Angel [9] studied the scroll compressor with two intermediate discharge valves using CFD method. The flow field of the scroll chamber and the flow domain of the valve had been displayed and analyzed. Liu et al. [10] carried out a numerical simulation of the whole compression process in scroll compressor. The simulated mass flow rate and volumetric efficiency agreed well with the experimental ones. The forces were also calculated with the flow field data, and the non-uniform distribution of pressure caused the fluctuation of gas forces. Sun et al. [11, 12] simulated the unsteady flow of the scroll pump with dynamic mesh model. The results indicated that there was a high speed jet at the meshing clearance and the cavitation generated at negative pressure region in the downstream area.

Although many researchers have investigated the working process of the scroll compressor with the help of CFD methods and obtained some achievements, few researchers study the inner flow characteristics and mechanism in a scroll compressor thoroughly. In the present paper, the radial clearance is set to be 0.008 mm, which is same with that in the actual prototype. The inner flow characteristics and mechanism in a scroll refrigeration compressor were displayed and analyzed in detail. Eventually, some useful advices are given to instruct the optimization and improve the performance of scroll refrigeration compressors.

2. Methodology

2.1. Physical model and grids generation

Table 1. The main parameters of the scroll refrigeration compressor

Parameters	Value
Thickness of scroll vane/mm	3.6
Height of scroll vane/mm	40
Numbers of circle	2.75
Theoretical suction volume/mL	82.4

Table 1 shows the main parameters of the scroll refrigeration compressor. As shown in Figure 1, flow domain includes inlet, working chambers and outlet. The axial clearance was neglected because of the application of the sealing strip the axial flexible adjusting mechanism in the actual scroll compressor prototype. The radial clearance was set to be 0.008 mm. The tangential leakage caused by the radial clearance had a great influence on the performance of scroll refrigeration compressor. Figure 2 shows the radial clearance and tangential leakage in the clearance.

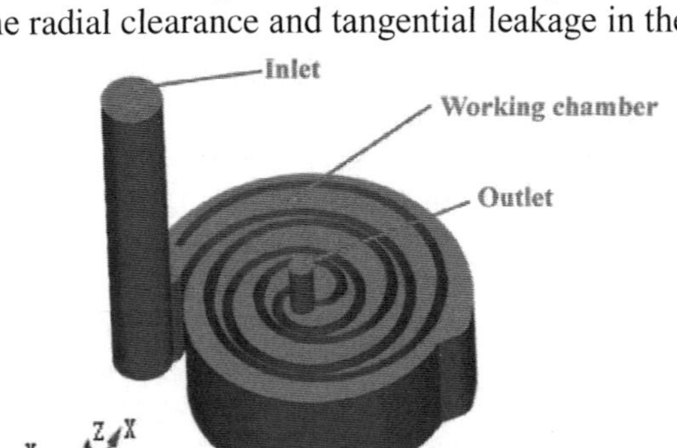

Figure 1. The flow domain of scroll compressor

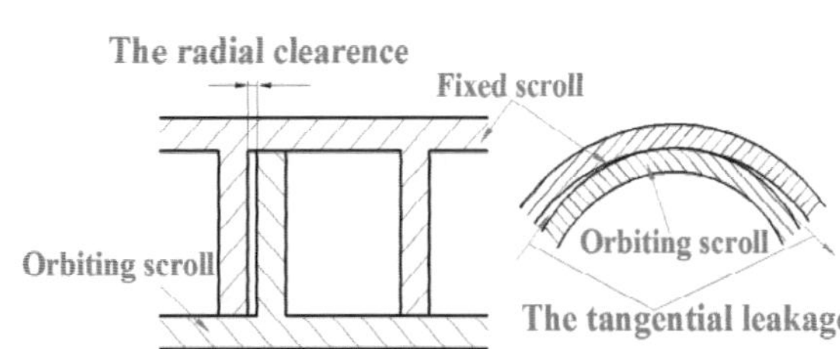

Figure 2. Radial clearence and tangential leakage

The hexahedral structured grids were generated in the compressing domain of scroll refrigeration compressor. The Cartesian grids were used in other domains [13]. Figure 3 and Figure 4 show the grids of the scroll refrigeration compressor. The unstructured grids were commonly used in the numerical analysis of the existing literatures. It was difficult to guarantee the orthogonality of unstructured grids in the radial clearance when the orbiting scroll moved. The convergence and accuracy of model were poor. In this paper, because the hexahedral structural grids were generated and updated as the orbiting scroll moved, there were many layers in the radial clearance and the number of layers remained unchanged during the solved process of the model.

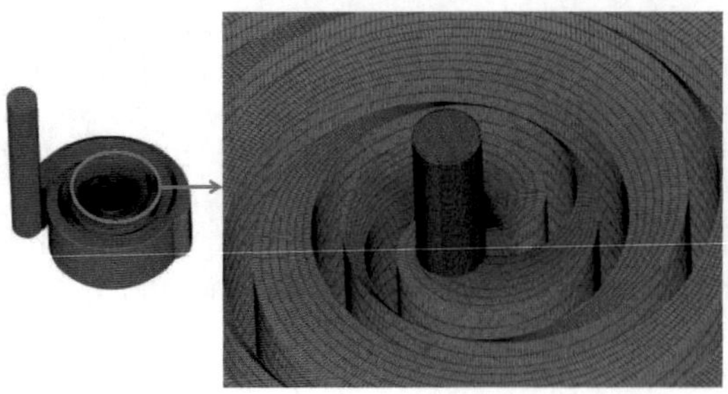

Figure 3. Overall grids

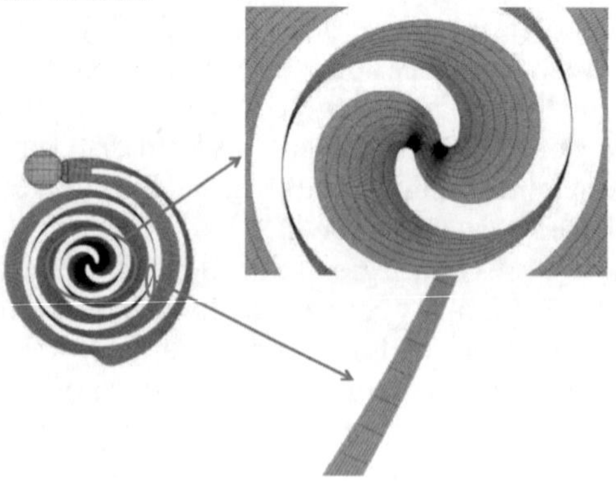

Figure 4. Grids at radial clearance

2.2. Boundary conditions and calculated case

Table 2. Boundary conditions

Parameter	Option
Working fluid	R22
Rotational speed/(r/min)	2880
Conditions of sidewall	Adiabatic/ No-slip
Rotational direction	Clockwise
Suction pressure/MPa	0.627
Suction temperature/K	307.7
Discharge pressure/MPa	2.146

Table 2 is the operational conditions and model setting. The turbulence model was standard k-ε model and the standard wall function method was adopted in the simulation. The time step was 5.787×10^{-5} s when the crank shaft rotated 1°, and the crank angle was defined as θ. Furthermore, Pressure inlet and pressure outlet were adopted for the pressure boundary conditions. Inlet specified temperature and outlet fully developed were adopted for the heat inlet and outlet conditions.

According to table 2, eight groups of grids were adopted as the grid independence check, from 1000,000 to 660,000, respectively. Based on the consideration of calculation, performance of the computer and the computing time, the total number of cells was 300,000 with 11 layers in the radial direction and 20 layers in the direction of height.

3. Simulation and discussion

In this section, on the basics of the simulated results, the inner flow characteristics and mechanism in the scroll refrigeration scroll would be studied in detail. To analyze the result of numerical simulation, the crank angle θ was defined as 0° when suction process began.

3.1. Pressure distribution

IOP Conf. Series: Materials Science and Engineering **232** (2017) 012050 doi:10.1088/1757-899X/232/1/012050

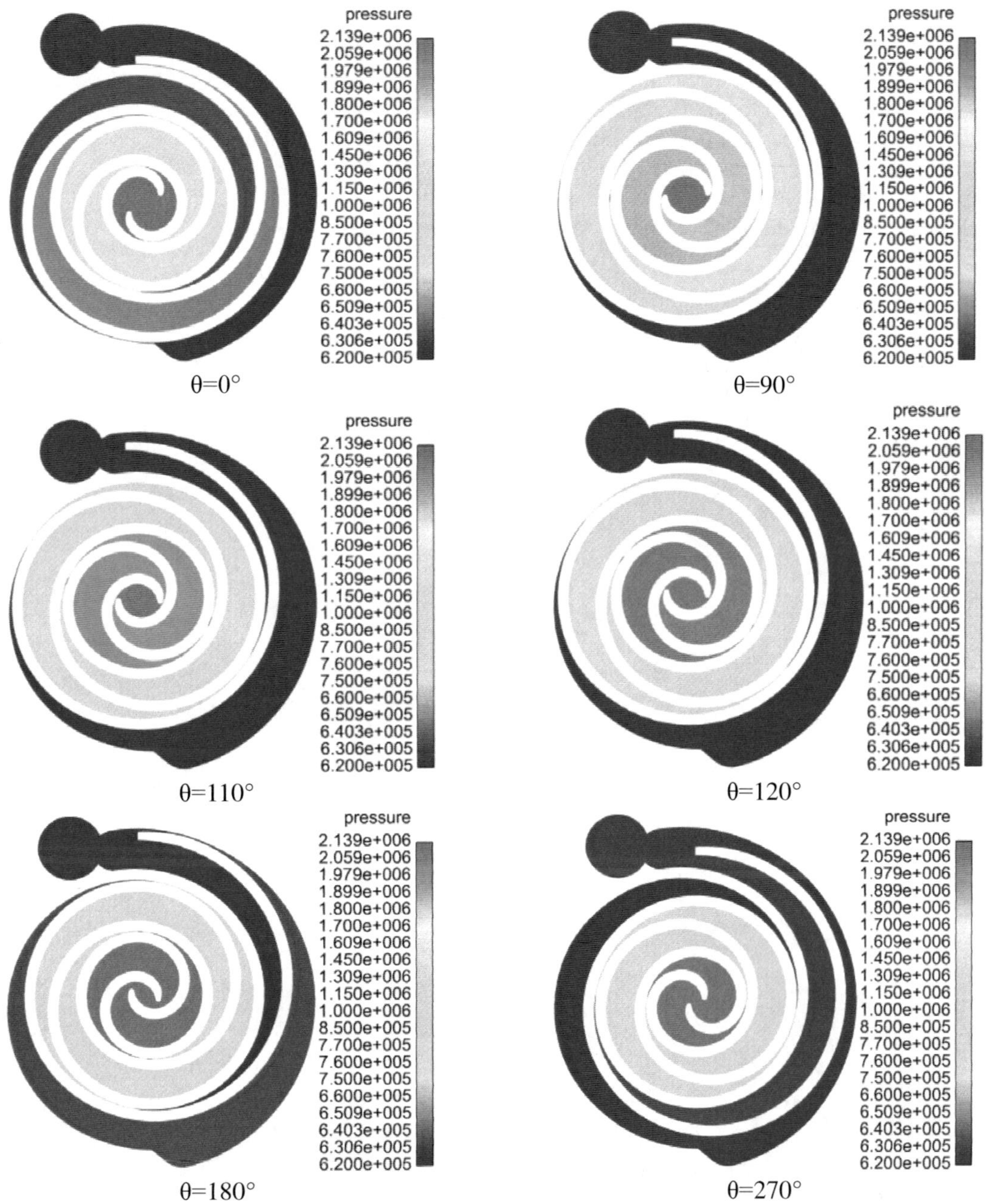

Figure 5. Pressure distribution with crank angle on a cross section area perpendicular to the z axis

Figure 5 shows pressure distribution with crank angle on a cross section area perpendicular to the z axis. Because the suction port and passage were asymmetrical, the pressure distribution would not be symmetrical in the two suction chambers. The asymmetric phenomena were obvious when the crank angle was 90°. The pressure in the working chamber was almost homogeneous except at the meshing clearance. In flow domains of the radial clearance, the gradient of pressure was great, because the refrigerant fluid leaked from high pressure chamber to the low pressure chamber through the radial clearance.

3.2. Temperature distribution

Figure 6 shows the temperature distribution with crank angle on a cross section area perpendicular to the z axis. In every pair working chamber, the temperature distribution was inhomogeneous. The gradient of temperature was great in a working chamber. The temperature of fluid near the radial clearance was higher than that in other flow domain in a working chamber. The reason for that was the leakage flow brought a large amount of heat from the high pressure chamber.

Comparing the Figure 5 with 6, it was found that the pressure in the working chamber was almost homogeneous while the temperature in the working chamber was inhomogeneous. The reason was that the velocity of heat conduction was lower than that of pressure transmission.

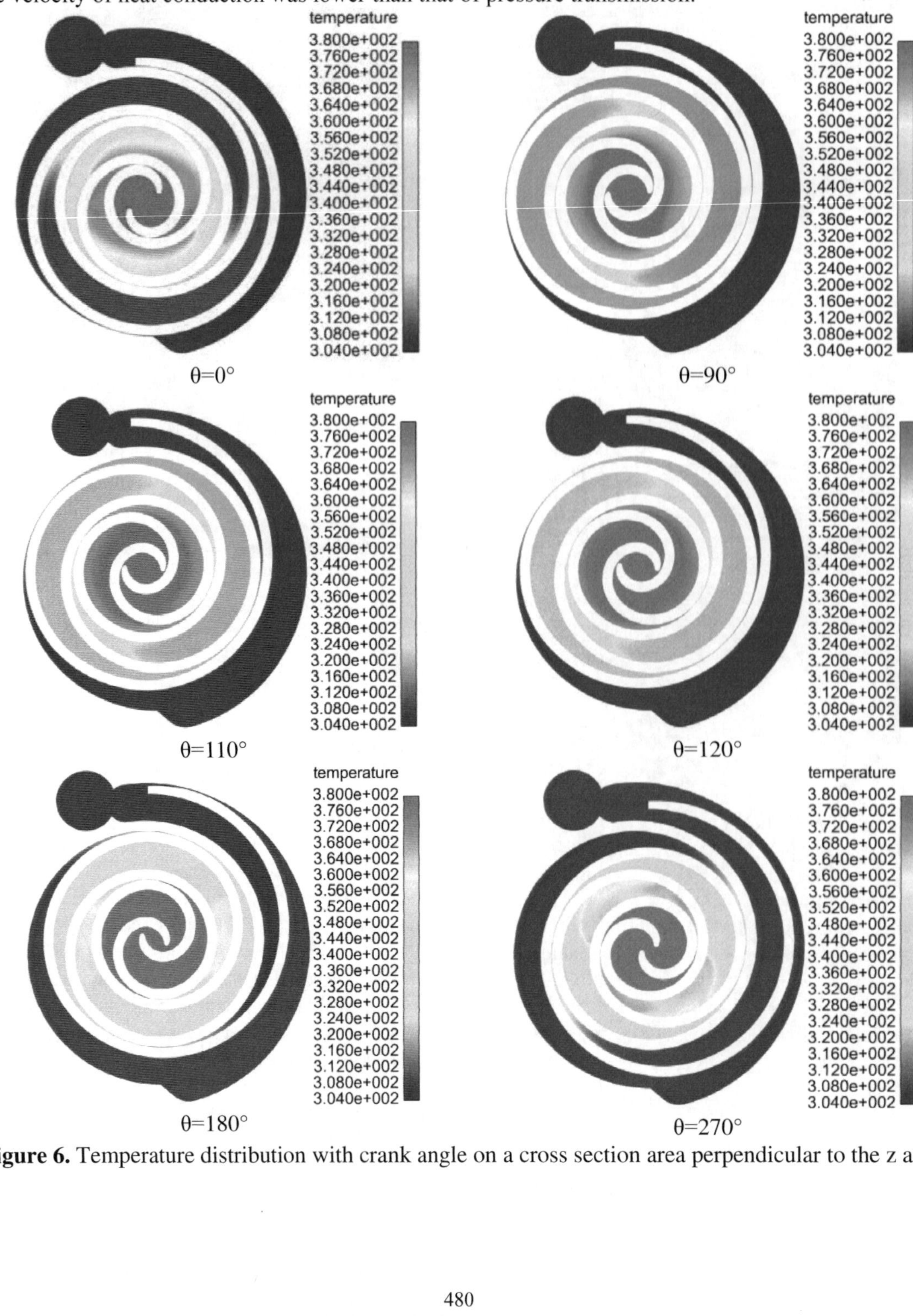

Figure 6. Temperature distribution with crank angle on a cross section area perpendicular to the z axis

IOP Conf. Series: Materials Science and Engineering **232** (2017) 012050 doi:10.1088/1757-899X/232/1/012050

3.3. Velocity vector distribution

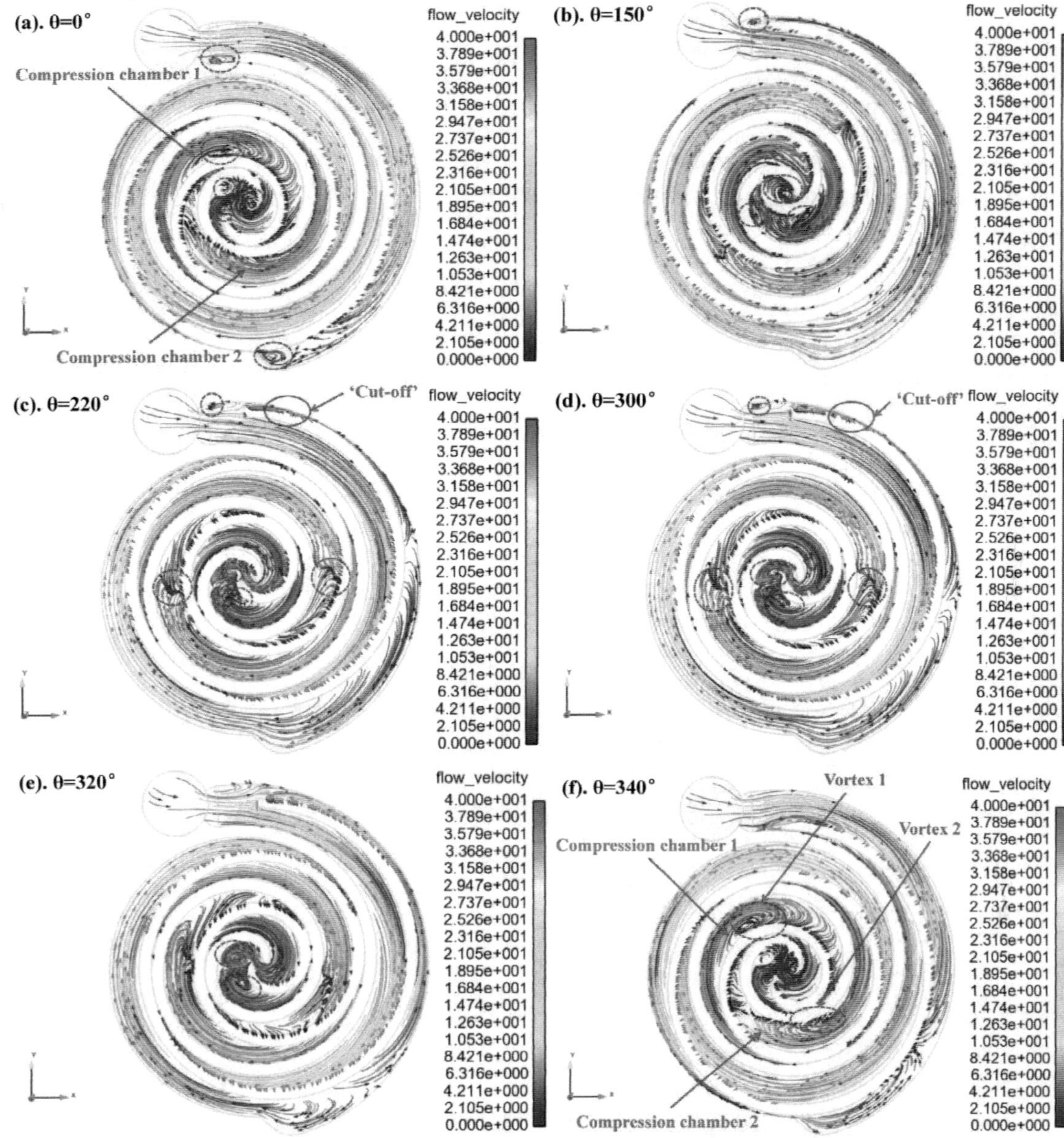

Figure 7. Velocity vector distribution with crank angle on a cross section area perpendicular to the z axis

Figure 7 shows velocity vector distribution with crank angle on a cross section area perpendicular to the z axis. To display the velocity vector distribution in different flow domains preferably, the velocity was set as the variable and the upper limit of velocity was set to be 40 m/s. However, the maximum velocity was close to 200 m/s at the radial clearance in the simulation. In the section, the velocity vector distribution during the suction process, compression process and discharge process were analysed in detail, respectively.

3.3.1. Suction process

As Figure 7(a) shows, a small amount of fluid leaked out from the suction chambers because of the pre-compression of the refrigerant fluid. The leakage flow mixed with the suction flow, so two vortexes formed at the exit of the suction chamber.

As Figure 7(c) and (d) shows, the refrigerant fluid in suction chamber was cut off owing to the movement of orbiting scroll. As a result, a part of refrigerant fluid flowed back, and the other part of refrigerant fluid flowed into the inner suction chamber. Predictably, the inlet mass flux of the scroll compressor would decrease when the crank angle was larger than 220° and the volumetric efficiency would decrease owing to the 'cut-off' of the refrigerant fluid.

3.3.2. Compression process

As Figure 7(f) shows, the compression chamber 1 and the compression chamber 2 were a pair of working chamber. Meanwhile, the vortex 1 formed in the compression chamber 1 was larger than the vortex 2 formed in the compression chamber 2. It was because the movement of the orbiting scroll contributed to the vortex formation in the compression chamber 1 and restricted to the vortex formation in compression chamber 2.

3.3.3. Discharge process

The flow regime in the central working chamber was complicated all the time. Figure 7(a) and (b) show that there were three vortexes in the central working chamber from 0° to 150°. With the movement of orbiting scroll, the refrigerant fluid in the central working chamber was squeezed, so the vortex at the tip of the orbiting scroll moved to the outward and mixed with another vortex. As a result, as Figure 7(c)-(e) show, there were two vortexes in the central working chamber from 220° to 320°.

4. Mass flux curve and discussion

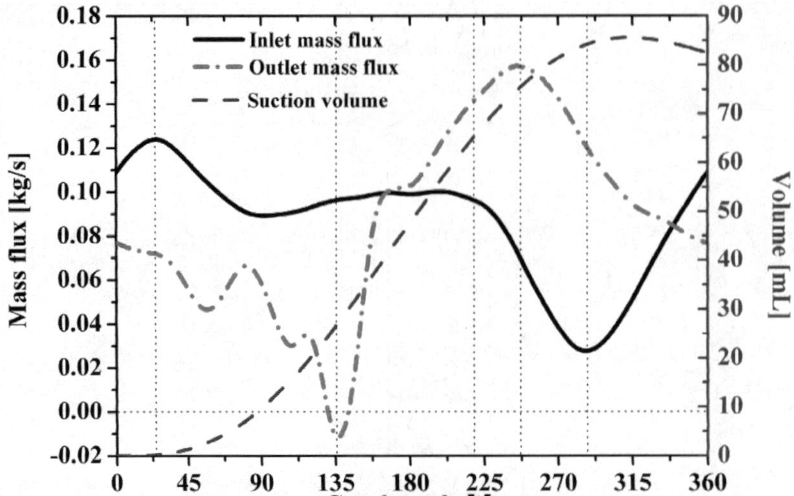

Figure 8. Mass flux with crank angle

Figure 8 shows the variation of mass flux with the crank angle in one cycle. The inlet and outlet mass flux fluctuated with the crank angle remarkably. The inlet and outlet integral average mass flux were 0.0866 kg/s and 0.0876 kg/s, respectively. The relative error was less than 1.2%; hence, it could be regarded as mass conservation during the simulation.

For the inlet mass flux curve, the maximum value happens at 25°, and the minimum value exists at 290°. When the crank angle was 220°, the inlet mass flux began to decrease. As shown in Figure 8, the suction volume in the suction chamber usually increased with the crank angle from 0° to 310° and then decreased, and so the mass flux was expected to increase with the increment of the suction chamber volume and decrease with the reduction of the volume. However, the inlet mass flux curve showed different trends. The influence of the movement of the orbiting scroll and flow passage on the suction process was the main reason for the difference. When the crank angle was 220°, the 'cut-off' caused by the movement of the orbiting scroll prevented the refrigerant fluid flowing into the suction chamber, so the inlet mass flux began to decrease and reached the minimum value at 290°.

As shown in Figure 8, the outlet mass flux decreased in general from 0° to 111.5°. When the crank angle was 111.5°, the central working chamber connected with the near working chamber and the discharge process began. However, at the moment, because the scroll refrigeration compressor was under-compression under the current operation condition, the backflow formed and the minus outlet mass flux reached a minimum at 135°. After 135°, the pressure in the central working chamber increased gradually with decrement of the discharge volume, and reached the discharge pressure when the crank angle was about 150°. Therefore the backflow phenomenon disappeared after this moment. Subsequently, the volume of the central working chamber decreased because of the squeezing of the orbiting scroll, so the pressure in the central working chamber increased. Therefore, the outlet mass flux increased from 150° to 240° and reached the maximum at 240°.

5. Efficiency

Volumetric efficiency and isentropic efficiency are two key performance parameters in a scroll refrigeration compressor. The volumetric efficiency is defined as:

$$\eta_v = \frac{\dot{m}_{ac}}{\dot{m}_{id}}$$

in which: $\dot{m}_{ac}$ is the actual mass flux, $\dot{m}_{id}$ is the ideal mass flux. The isentropic efficiency is defined as:

$$\eta_i = \frac{p_{is}}{p_{ad}}$$

in which: p_{is} is the isentropic theory compression work, p_{ad} is the adiabatic indicated work. η_v and η_i can be calculated by the simulated results, which were 94.35% and 80.57%, respectively. Because the axial clearance was neglected in the simulated model, the volumetric efficiency was slightly higher.

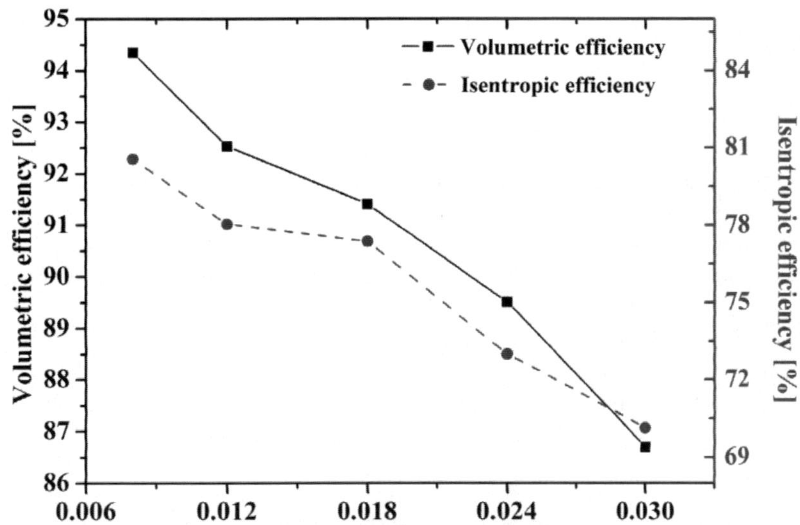

Figure 9. Volumetric efficiency with radial clearance value

To investigate the influence of the radial clearance value on the volumetric efficiency and the isentropic efficiency, other four groups of radial clearance value had also been simulated, which were from 0.012 mm to 0.030 mm. Figure 9 shows the volumetric efficiency and the isentropic efficiency with the radial clearance value. It could be found that the volumetric efficiency and the isentropic efficiency decreased with the increasing of the radial clearance value. When the radial clearance value increased from 0.008 mm to 0.030 mm, the volumetric efficiency decreased from 94.35% to 86.69% and the isentropic efficiency decreased from 80.57% to 70.13%.

6. Conclusion

In present paper, on the basics of the simulated results, the inner flow characteristics and mechanism in a scroll refrigeration compressor were studied in detail. Some conclusions are summarized as follows:

1. Pressure distribution was asymmetrical in the two symmetrical suction chambers. The pressure in the working chamber was almost homogeneous except at the meshing clearance. In a working chamber, the temperature distribution was inhomogeneous because of the leakage.
2. The inlet and outlet mass flux fluctuated with the crank angle remarkably. Because of the 'cut-off' in the suction flow passage, the inlet mass flux decreased remarkably when the crank angle was larger than 220°. Hence, it was essential to optimize the suction port and flow passage to improve the volumetric efficiency. When the discharge process began, the backflow formed and the outlet mass flux reached a minimum value at 135°. A check valve should be installed to prevent the backflow.
3. Velocity vector distribution was asymmetrical in each pair of working chamber owing to the movement of orbiting scroll, the flow conditions was complicated in the central working chamber all the time. In addition, the movement of the orbiting scroll contributed to the vortex formation in the compression chamber1 and restricted the vortex formation in compression chamber 2.
4. When the radial clearance value increased from 0.008 mm to 0.030 mm, the volumetric efficiency decreased from 94.35% to 86.69% and the isentropic efficiency decreased from 80.57% to 70.13%. It indicated that the value of the radial installation clearance had a great influence on the volumetric efficiency and the isentropic efficiency of the scroll refrigeration compressor.

Acknowledgments

This work was supported by the National Natural Science Foundation of China [Grant numbers: 51406163], China postdoctoral science foundation [Grant numbers: 2015M580866] and State Key Laboratory of Compressor Technology [Grant numbers: 2014011059]. The supports are gratefully acknowledged.

References

[1] Cui M 2006 Numerical study of unsteady flows in a scroll compressor *Journal of Fluids Engineering* **128** 947-955

[2] Cui M and Sauls J 2006 Investigation of suction process of scroll compressors *In Proceedings of International Compressor Engineering Conference at Purdue* (Purdue: Purdue Univeristy) C110

[3] Cui M 2007 Comparative study of the impact of the dummy port in a scroll compressor *International Journal of Refrigeration* **30** 912-925

[4] Wang J, Zha H, McDonough J and Zhang D. 2015 Analysis and numerical simulation of a novel gas-liquid multiphase scroll pump *International Journal of Heat and Mass Transfer* **91** 27-36

[5] Yue X, Lu Y, Zhang Y, Ba D, Wang G and Zhang Z 2015 Computational fluid dynamics simulation study of gas flow in dry scroll vacuum pump *Vacuum* **116** 144-152

[6] Hesse J and Andres R 2016 CFD simulation of a dry scroll vacuum pump including leakage flows *In Proceedings of International Compressor Engineering Conference at Purdue* (Purdue: Purdue Univeristy) 1515

[7] Ding H and Jiang Y 2016 CFD simulation of an oil flooded scroll compressor using VOF approach *In Proceedings of International Compressor Engineering Conference at Purdue*

(Purdue: Purdue Univeristy) 1616

[8] Chang J, Chang C, Hung T, Lin J and Huang K 2014 Experimental study and CFD approach for scroll type expander used in low-temperature organic Rankine cycle *Applied Thermal Engineering* **73** 1444-1452

[9] Picavet A and Angel B 2016 Numerical simulation of the flow inside a scroll compressor equipped with intermediate discharge valves *In Proceedings of International Compressor Engineering Conference at Purdue* (Purdue: Purdue Univeristy) 1177

[10] Liu S, Kang X, Shan C and Hu Y 2014 Numerical simulation of three-dimension unsteady flow in the compression chambers of a scroll compressor *In Proceedings of International Compressor Engineering Conference at Purdue* (Purdue: Purdue Univeristy) 1200

[11] Sun S, Wu K, Huang Y, Guo P and Luo X 2017 Numerical simulation on the flow field and cavitation in a scroll hydraulic pump *Journal of Drainage and Irrigation Machinery Engin* **35** 100-105

[12] Sun S, Guo P, Huang Y and Luo X 2016 Numerical analysis on the cavitation and unsteady flow in a scroll hydraulic pump Materials Science and Engineering (IOP) **129** 012028

[13] Gao H, Ding H and Jiang Y 2015 3D Transient CFD Simulation of Scroll Compressors with the Tip Seal *In Proceedings of International Compressor Engineering Conference at Purdue* (Purdue: Purdue Univeristy) **90** 012034

IOP Conf. Series: Materials Science and Engineering **232** (2017) 012051 doi:10.1088/1757-899X/232/1/012051

Oil flow at the scroll compressor discharge: visualization and CFD simulation

Jiu Xu[1]

Pega Hrnjak[1,2,*]

[1] University of Illinois at Urbana-Champaign, Mechanical Science and Engineering, Urbana, IL, USA

[2] CTS, Creative Thermal Solutions, Urbana, IL, USA

pega@illinois.edu

Abstract. Oil is important to the compressor but has other side effect on the refrigeration system performance. Discharge valves located in the compressor plenum are the gateway for the oil when leaving the compressor and circulate in the system. The space in between: the compressor discharge plenum has the potential to separate the oil mist and reduce the oil circulation ratio (OCR) in the system. In order to provide information for building incorporated separation feature for the oil flow near the compressor discharge, video processing method is used to quantify the oil droplets movement and distribution. Also, CFD discrete phase model gives the numerical approach to study the oil flow inside compressor plenum. Oil droplet size distributions are given by visualization and simulation and the results show a good agreement. The mass balance and spatial distribution are also discussed and compared with experimental results. The verification shows that discrete phase model has the potential to simulate the oil droplet flow inside the compressor.

1. Introduction

Oil plays an important role in cooling, lubrication and sealing of the compressors along with noise reduction. Oil is necessary for conventional compressors while oil circulating in vapor compression cycle has adverse effects on heat transfer and pressure drop. Even application of magnetic bearings and new material that reduce the need for lubrication to the level that compressors are oil free resulted in increasing number of non-lubricated machines, their number is still very small. Therefore, engineering attempts in design of vastly dominant lubricated compressors are made to keep the oil in the compressor and reduce the oil circulation ratio (OCR) in the system. Separating the oil from the high pressure refrigerant vapor and draining it back to the compressor is a feasible way to achieve this goal. For oil separator design, it is important to know the key oil flow parameter at the compressor discharge, including oil droplet size distribution, oil droplet velocity and vapor velocity. Furthermore, considering the compactness requirement of the compressor, oil separation structure integrated inside the compressor housing is preferred. These factors justify a closer look at the oil flow inside the compressor plenum.

For oil flow inside the compressor, flow visualization is a useful experimental method to understand the realistic condition. Toyama et al. [1] described methods to quantify oil droplet behavior using high-speed photography in the scroll compressor shell via sight glasses. Besides the size distribution of oil droplets, it is concluded that the mean diameter of oil droplets decreases as the flow speed of the refrigerant gas increases. Wujek and Hrnjak [2] developed visualization techniques to quantify the annular mist flow at compressor discharge pipe. The size and velocity of each oil droplet captured by the high speed camera can be determined by video processing. It shows that the annular-mist flow is developing along the discharge pipe of a swash-plate compressor with R134a and PAG46 oil. Zimmermann and Hrnjak [3-4], showed that the most significant source of oil droplets in a scroll compressor is the break-up process of oil film between valve and valve seat during the opening and closing. The visualization of the reed valve in the scroll compressor shows the valve periodical movement and oil atomization process. Xu and Hrnjak [5-6] measured the OCR in a non-invasive way based on the oil annular mist flow visualization in the compressor discharge tube. Effect of focal depth is considered during the video processing to provide good estimation of number of oil droplets to provide amount of oil in that form in a specific control volume. Although flow visualization provides good image of what happens inside the compressor shell, the method is restricted by narrow vision and inaccuracy of modified space. Videos provide elements for reasonable assessment of droplet flow velocity in two dimensional case, but it is more difficult to say more about third dimension as well as velocities of vapor phase.

Computational Fluid Dynamics (CFD) is a strong tool to numerically simulate even the complicated two-phase flow when number of droplets is small and they are well dispersed Yokoyama et al. [7] analyzed the oil flow in a CO_2 rotary compressor shell by CFD. Numerical simulations shows that flattening the lower balance weight or installing a rotating disk over upper balance weight can reduce the OCR and the results are verified by system experimental results. Gao et al. [8] conducted a simulation to get the dynamic trajectories and separation performance of oil droplets in oil-gas cyclone separator. Noh et al. [8] realized high-frequency visualization of the oil droplets in the compressor shell and simulated the separation efficiency of vane type separators. Xu and Hrnjak [9] validate the CFD simulation model using visualization results for oil droplet flow inside the scroll compressor plenum.

Based on previous research we can say: 1) The source of the oil droplets is the reed valve (if the compressor has it); 2) The flow pattern of refrigerant-oil mixture at the compressor discharge pipe is annular-mist flow; 3) Visualization techniques can be used to quantify the oil flow inside the compressor housing and in the compressor discharge tube; 4) CFD simulation has the potential to simulate the misty oil flow accurately in the compressor or the oil separator.

However, few literatures have combined the visualization result and CFD results together to give validated conclusions. So, our objective is to analyze the oil droplet flow in the scroll compressor by both visualization and simulation. Discrete phase model in CFD is used to predict the oil droplet trajectory in the scroll compressor plenum. The simulation results are validated with visualization results from high speed camera and mass balance. Furthermore, based on the validated simulation, it is meaningful to explore the effect of oil droplet size and potential oil separation structures that can be integrated into scroll compressor plenum.

2. Visualization

2.1. Experimental facilities

The oil flow inside the compressor plenum is visualized by high speed camera. In order to get clear images of the oil droplets inside the plenum, some modifications are made to the upper part of the scroll compressor plenum. As shown in Figure 1(a), initially mounted four side glasses were not sufficient and the transparent cylindrical ring was added between the original compressor housing. The transparent ring can hold around 1 MPa pressure at the compressor discharge. The visibility through the transparent ring is highly dependent on the surface finish. Two flanges with O-rings and bolts are used to seal the chamber. As shown in Figure 1(b), a plastic screen is placed in front of the compressor valve in order to eliminate the misty flow in front of the valves so that a clear vision can be achieved when capturing the video of oil near the reed valve. The particular compressor used in this project has three discharge valves, two of which are closed intentionally for easier visualization. Modification mentioned above makes some change to the real geometry and operating condition of the scroll compressor, but it is necessary cost which makes visualization possible.

Videos are taken by the high speed camera with different lenses. Strong background light source is needed for enough brightness for a clear image. In- front light is needed when capturing the movement

IOP Conf. Series: Materials Science and Engineering **232** (2017) 012051 doi:10.1088/1757-899X/232/1/012051

of the reed valve. Camera frame rate needs to be higher than 1000 fps in order to capture the movement of oil droplets.

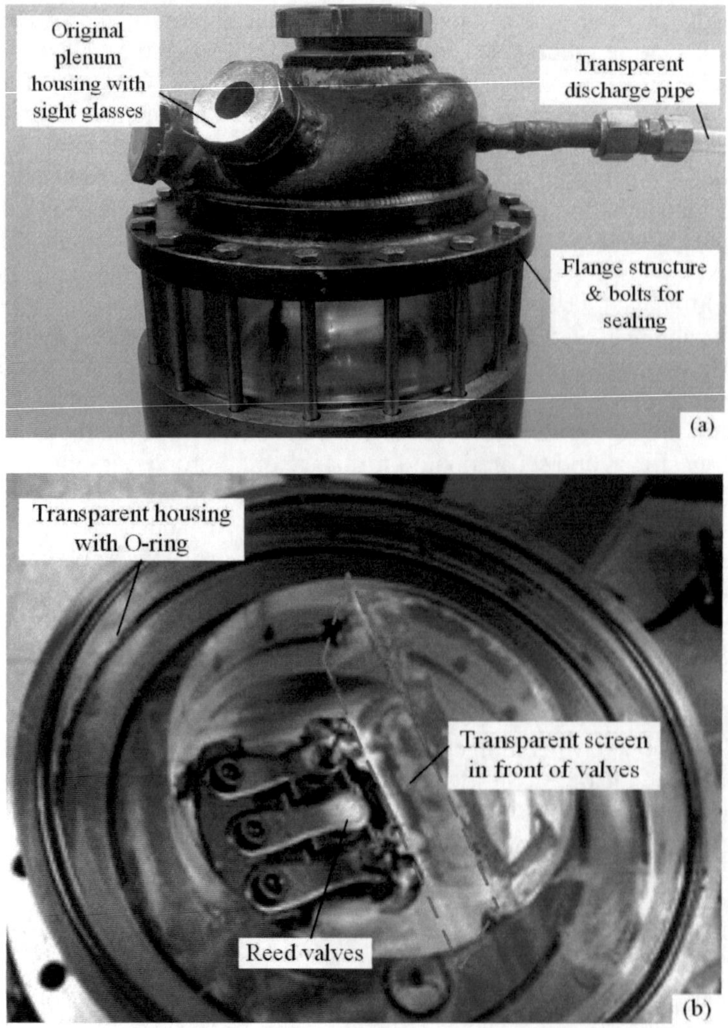

Figure 1. Modification of a scroll compressor plenum for visualization: (a) outer view; (b) inner view

2.2. Visualization of the oil flow

Videos of the oil flow are taken at three locations around the compressor: the center reed valve, the vicinity area in front of the valve and the discharge pipe. The locations are marked with frames in Figure 4 (geometry for CFD simulation). The shape of the plenum after modification is not exact the same as the original shape but the flow condition is expected to be similar under real conditions.

The video of the reed valve is taken by putting the focus directly on the valve tip. As shown in Figure 2(a), the valve opening process can be observed from about 45°.

Zimmermann and Hrnjak [4] have shown that the break-up process of the oil film between valve and valve seat generates oil droplets. A cloud of oil droplets injected by the high pressure of the compressed refrigerant as the valve opens. Video processing method is used to quantify the initial velocity of oil droplets injected to the plenum during the oil film break-up.

Figure 2(b) shows a sample frame from a video taken at the valve vicinity by focusing on the transparent screen in front of the valve. It is processed to estimate the size distribution of the oil

IOP Conf. Series: Materials Science and Engineering **232** (2017) 012051 doi:10.1088/1757-899X/232/1/012051

droplets injected in the valve opening. In the plenum chamber, the oil droplets are moving in all directions at a relatively low velocity. The droplet size can be better estimated than the droplet velocity because it is difficult to give the three-dimensional velocity by a series of planar images. Therefore, oil droplet size distribution is the key parameter that is used to compare visualization results and simulation results.

Figure 2(c), shows the frame taken in transparent discharge pipe located around 20 centimetres from the compressor discharge port. The view in a raw form presents droplets (dark spots) behind the film with waves. During quantification and analysis several filtering procedures are conducted to better capture and measure the size, number and velocity of droplets. . Since the flow is strongly one dimensional droplet velocities determined are closer to the actual values. When estimating the total number of droplets in the tube authors have first determined the depth of view, then number of droplets in the view volume and their diameter and its distribution and then the total mass in the droplet form. Details were presented by Xu and Hrnjak's recent published paper [10].

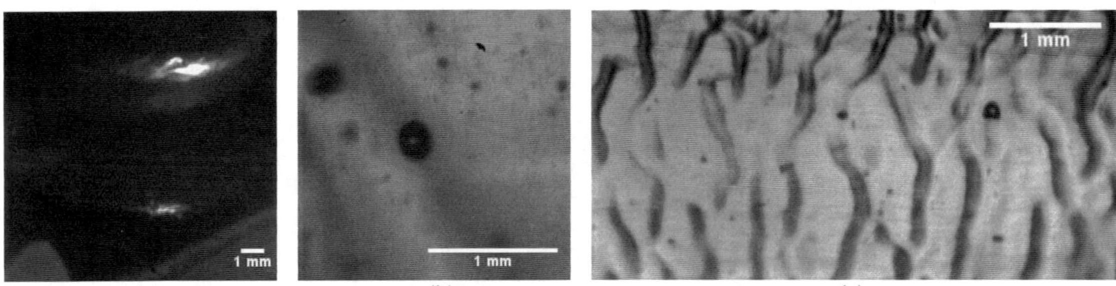

(a) (b) (c)

Figure 2. Sample video frames taken in different locations inside the compressor: (a) center reed valve; (b) valve front; (c) discharge pipe

2.3. Results from video processing

The videos at different locations in the plenum were processed following the procedure developed by Wujek and Hrnjak [3]. The fundamental principles are generally described as following:

- Eliminate the background/foreground wavy pattern (seen in Fig. 2c) by subtracting the average pixel value in period of time;
- Estimate the droplet size based on steepest grayscale gradient method, which detects the boundary between dark droplet image and the light background. The size of the droplets that are out-of-focus is corrected based on estimated depth of the view field;
- Compare two consecutive images in small sections to interrogate droplet displacement;
- Screen out the velocity vector using algorithm as in Particle Image Velocimetry (PIV). Numerous erroneous velocity vectors are generated throughout this process and the two-dimensional cross-correlation is used to screen out true velocity vectors.

The oil size distribution at the valve front and discharge pipe for the same compressor at different motor speed are shown in Figure 3. The curve shows the diameter-specific mass flow as a function of oil droplet diameter. The integration of each curve is equal to the mass flow rate captured by the camera under one specific working condition.

Comparing the curves under different compressor speeds (from 30 Hz to 60 Hz), it is intuitive to conclude that higher compressor speed leads to more oil mass flow rate. Also, the average diameter of the oil droplets at the valve is larger than that of the oil droplets in the discharge pipe. The droplet size distribution results from the visualization will be discussed together with the CFD simulation results in "3.7 Validation of the model".

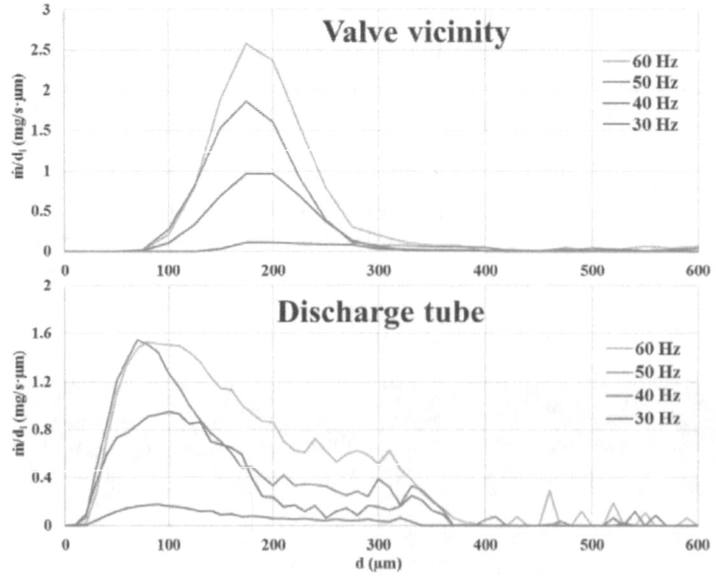

Figure 3. Oil size distribution by video at the vicinity of the valve and at discharge tube 20 cm downstream of the discharge under different compressor speed

3. CFD simulation

3.1. Discrete phase model

Discrete phase model (DPM) in computational fluid dynamics is usually used to simulate one fluid phase dispersed in another continuous fluid phase in a Lagrangian frame of reference. Discrete phase model is commonly used to simulate the liquid/solid particle motion in a continuous fluid. An important preliminary assumption made in the discrete phase model is that the dispersed phase occupies a low volume fraction. In this research, oil flow inside compressor discharge chamber is considered to be misty and oil volume ratio is always less than 5%, which is consistent with the DPM assumptions.

In this research, we use the Euler-Lagrange approach. The fluid phase is treated as a continuum by solving the time-averaged Navier-Stokes equations, while the dispersed phase is solved by tracking a large number of droplets through the calculated flow field. The dispersed phase can exchange momentum with the fluid phase.

3.2. Simulation geometry and mesh

The flow region of the modified compressor plenum is depicted in Figure 4. Due to the visualization requirement, extra space is added to the original plenum, including two sight glasses and a cylindrical space surrounded by the polycarbonate ring.

The inlet of the flow region is set as the surface between the center reed valve and the valve base. Droplet injection is applied to the inlet surface and the details will be discussed in "3.3 Droplet injection". The outlet of the flow region is the cross section of the discharge pipe.

Mesh independence is also carried out in order to achieve a balance between discretion error and calculation time. As shown in Figure 5, six different cases with different mesh numbers are tested and the average velocity in the discharge pipe is used as a reference to check the mesh independence. In the final simulation, the mesh number is around 1 million and the mesh type is tetrahedron with acceptable skewness and mesh quality.

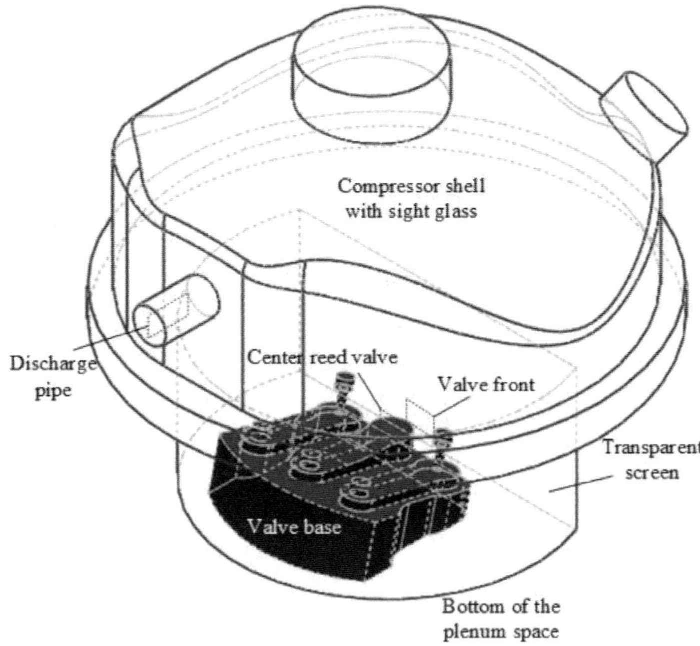

Figure 4. Geometry of flow region used for CFD simulation and locations of video capture

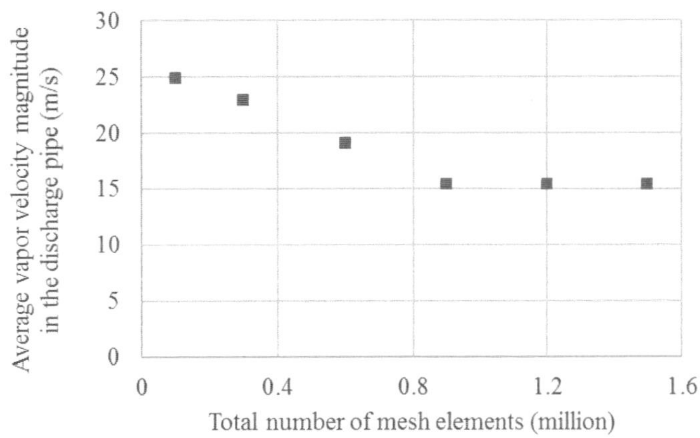

Figure 5. Mesh independence validation using average velocity in the discharge pipe as a reference

3.3. Droplet injection

For discrete phase model, discrete phase is introduced in the simulation by define injection. As shown by visualization, oil droplets are generated by the valve opening process. Droplet pulsating injection instead of dynamic mesh is used to for simulation because the plenum space is the main research object and the simplification of the opening of the valve can save a lot of calculation resources. Droplet size distribution of the injection is obtained from the video processing results at the valve vicinity as shown in Figure 3. Droplet injection direction and vapor phase velocity is got from simplified reed valve model from Zimmermann and Hrnjak [4].

3.4. Boundary conditions

In the discrete phase model, injection function set the source of discrete phase (usually droplets or particles). The injection is released from the inlet surface of discharge valve in the direction normal to the inlet surface. The distribution of injection velocity and size distribution can be decided by the results from video processing.

The boundary condition is simplified to four different cases presented in Figure 6: trap, escape, reflect and wall film. Trap condition means that the droplet will be absorbed by the wall while escape condition allows the droplet pass through when they hit the boundary. A reflect wall will give an opposite velocity vector to the incoming droplet based on momentum balance. Wall-film model is most realistic case which allows a single component liquid drop to impinge upon a boundary surface and form a thin film.

For the CFD simulation carried out in this study, discrete phase boundary condition of all the inner wall of the compressor plenum is set as "wall film". Inlet is set as injection as described in section 3.3 while the outlet of the discharge pipe is set as "escape".

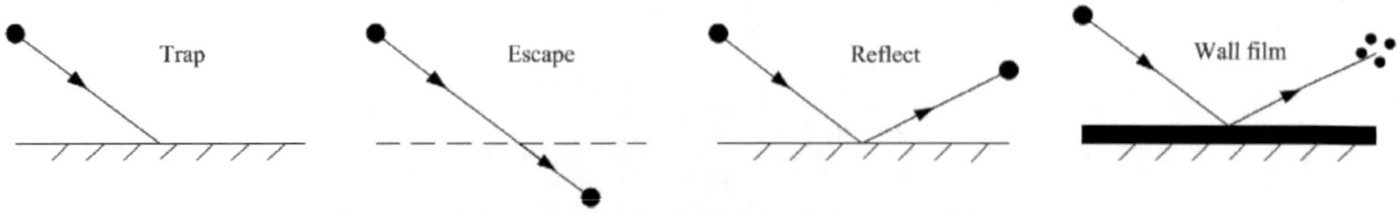

Figure 6. Mechanisms for DPM boundary conditions

3.5. Solver settings

The effect of turbulence is considered and k-ε model is used in the flow calculation. The simulation takes SIMPLE as the scheme and uses second order for pressure discretization and second order upwind for momentum discretization. The time step for transient calculation should be as small as possible but here it is set as 0.001s for reasonable computational time.

3.6. Results

The trajectories of oil droplets can be calculated by discrete phase model using ANSYS Fluent. When the oil droplets hit the boundary, they will either bounce back or splash into more droplets or be absorbed by the wall (presenting absorption in the film). Some of the oil droplets can and will leave the plenum and are recorded by the monitor at the compressor discharge tube. Figure 7 shows the trajectories in the plenum space of a small sample of tracked oil droplets, colored by droplet diameter. It can be seen that larger droplets tend to settle down in the film at the bottom of the compressor while smaller droplets tend to flow along with the vapor stream.

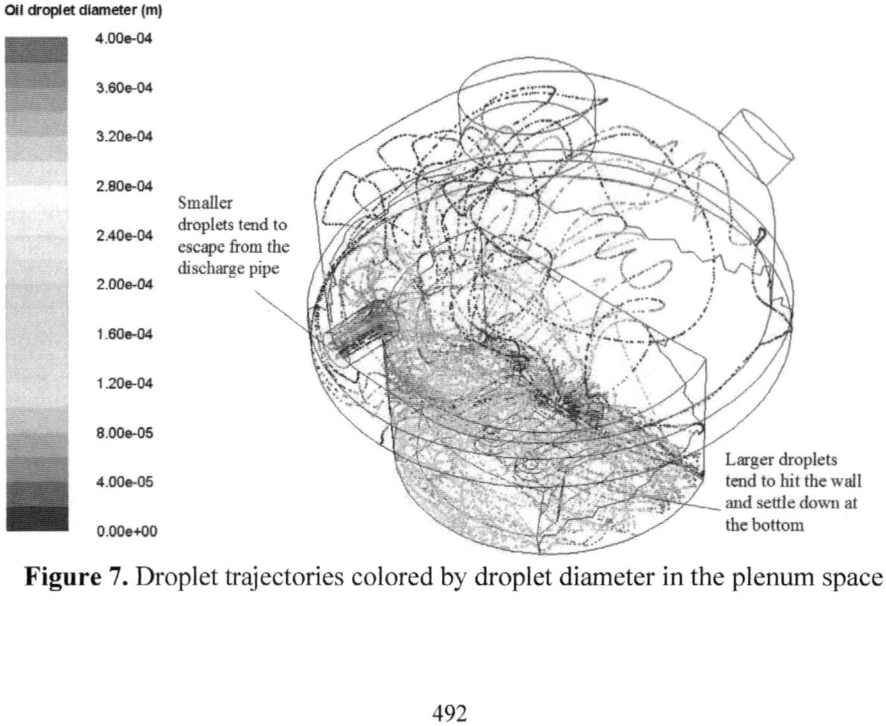

Figure 7. Droplet trajectories colored by droplet diameter in the plenum space

Figure 8 presents the mass balance of the oil in droplet form in the plenum in time. The solid line shows the inlet of oil droplets from the valve opening. The influx curve looks linear but it is actually periodic according the compressor frequency. The dashed line presents the flow out (Escaped through the discharge line). Equal gradients of solid and dash line indicate the steady state in the mass of droplets in the plenum. The dash dot line with its constant value shows the time when flow out became equal to the inlet to the plenum. Constant value of the dot line indicates that as much of the oil is absorbed in the film is removed from it by entrainment. There are three fates that a droplet will meet: flow with the refrigerant vapor, accumulate in the wall film, or escape from the discharge tube. The mass conservation holds when all the oil droplets injected into the domain will either stay in the domain or escape from it. The curve of the droplets of different fates in Figure 8 has shown the mass conservation generally holds.

In principle we would like to see accumulation to grow in time (separation) and to be taken out back to compressor suction.

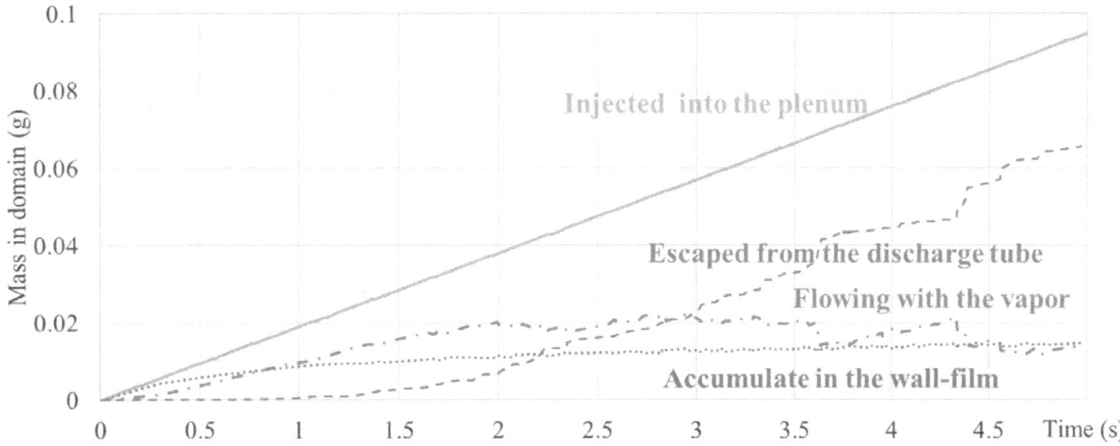

Figure 8. Droplet mass in plenum domain during the start up (0s to 5s)

3.7. Validation of the model

In order to validate the CFD simulation, besides mass balance oil droplet size distribution is chosen as the key parameter to compare visualization results and CFD results. This is mainly because droplet size distribution is the most reliable and statistically representative parameter that we can get from both approaches and it is important for oil separator design. The logic and process of comparison and validation are shown graphically in Figure 9. Video taken near the valve is used to give the as the input parameters for injection, like oil droplet size and velocity. The vapor velocity is given by the pulsating flow model mentioned in 3.3. Based on the input parameters, transient CFD calculation can monitor the oil droplets that escaping out of the compressor plenum and get a droplet size distribution. On the other side, the video taken at the discharge tube is processed to give a droplet size distribution. These two distributions are compared to see if the CFD results can agree with the video results.

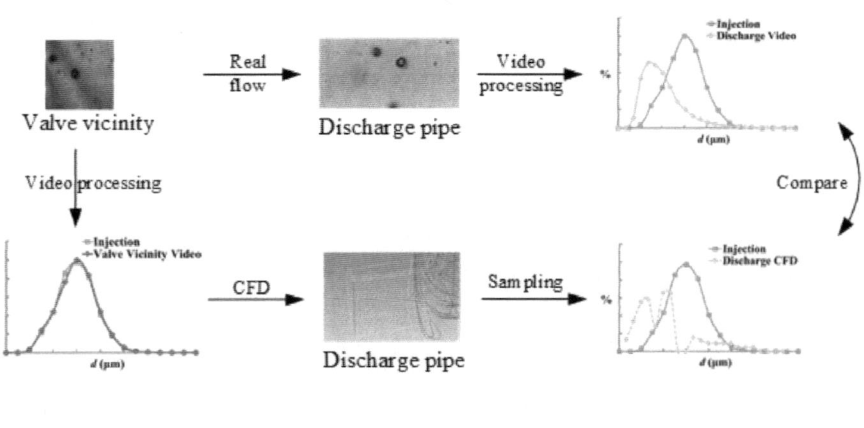

Figure 9. The process of validation of CFD results based on droplet size distribution from visualization

Both experiments and simulation are carried out at different compressor speed in steady state. Figure 10 shows the comparison of droplet size distribution between CFD results and video results. The curve represents the injected oil droplet size distribution, which is set as the input information of the injection in the simulation. What we want to compare is the distribution curve by CFD (dash line) and the distribution curve by video (solid line) in each plots. By comparison, it shows the CFD results can match well with video results at 30, 40 and 50 Hz. The discrepancy of the two results at 60 Hz may be explained by high OCR and high mass flow rate of oil injected by valve frequent opening will lead to a high possibility of collision between droplets. More droplets collision will make smaller droplets coalesce into bigger ones. Since the interaction between droplets are not considered in the simulation, it is possible that the simulation underestimate the amount of large droplets and overestimate the amount of smaller droplets at the compressor discharge.

The results show that CFD simulation is able to give a qualitatively prediction of the droplet size distribution at the compressor outlet. From the comparison between droplet size distribution of the injection and that of the discharge, it is clear that large droplets are more likely to hit the wall and splash on the wall film to generate more small droplets. Smaller droplets are less affected by gravity and inertia so that they can catch up with the vapor flow to escape the compressor and form the annular-mist flow in the discharge tube.

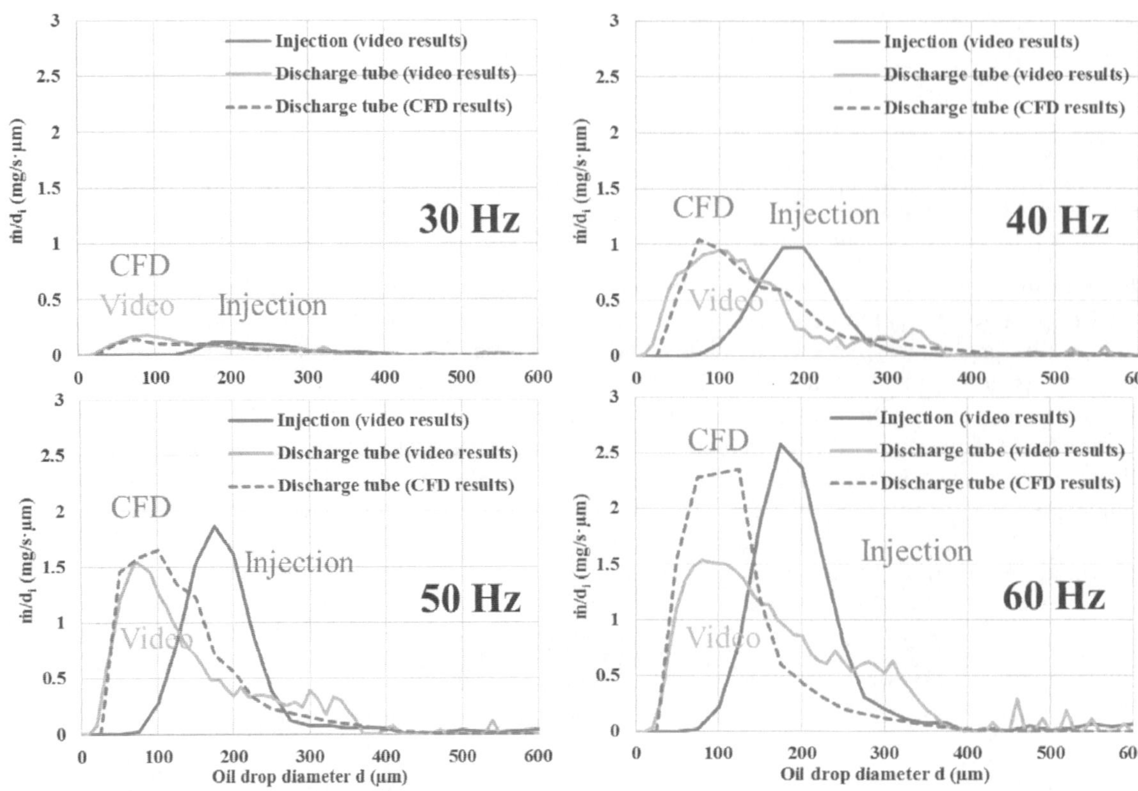

Figure 10. Comparison of oil droplet size distribution by video or CFD at the scroll compressor discharge under different compressor running speed

4. Summary

This paper presented the oil flow inside the discharge plenum of scroll (but could be any) compressor using both visualization technique and CFD simulation. Comparing visualization and simulation results, we can conclude that CFD model provided reasonably good description of the process inside the discharge plenum.

- Visualization of oil flow through discharge reed valve provides droplet size distribution and droplet velocity, which are used as the important input of the CFD simulation.
- Videos of oil droplet flow are taken in front of the valve and in the discharge tube. Video processing techniques can estimate droplet size and velocity based on the high speed video captured.
- Discrete phase model can be utilized to simulate the oil droplet trajectory inside compressor plenum. CFD simulation is validated by comparison of oil drop size distribution at the compressor discharge tube.
- CFD simulation also presents that mass conservation holds, which means oil mass flow rate at the discharge valve is the same as that at the compressor discharge.
- Higher compressor speed generates more oil droplets. Large droplets tend to retain in the plenum while small droplets are more likely follow the refrigerant vapor flow to escape. The discharge plenum space of a vertical scroll compressor has the potential to separate large oil droplets.

Acknowledgement

The authors would like to acknowledge the companies who sponsor of the Air Conditioning and Refrigeration Center for their technical and financial support to this work. The authors are also thankful for Augusto Zimmermann who made the compressor prototype and built a solid foundation of the visualization work.

References

[1] T. Toyama, H. Matsuura, and Y. Yoshida, "Visual Techniques to Quantify Behavior of Oil Droplets in a Scroll Compressor," in *International Compressor Engineering Conference. Paper 1788*, 2006.
[2] S. Wujek and P. Hrnjak, "Mist to annular flow development quantified by novel video analysis. ACRC Report TR285," Urbana, 2011.
[3] A. J. P. Zimmermann and P. S. Hrnjak, "Visualization of the Opening Process of a Discharge Reed Valve in the Presence of Oil," in *International Compressor Engineering Conference. Paper 2369*, 2014.
[4] A. J. P. Zimmermann and P. S. Hrnjak, "Oil flow at discharge valve in a scroll compressor.," *Proc. Inst. Mech. Eng. Part E J. Process Mech. Eng.*, vol. 229, no. 2, pp. 104–113, 2015.
[5] J. Xu and P. Hrnjak, "Refrigerant-Oil Flow at the Compressor Discharge," in *SAE Technical Paper 2016-01-0247*, 2016.
[6] J. Xu and P. Hrnjak, "Oil Flow Measurement at the Compressor Discharge," in *International Compressor Engineering Conference. Paper 2506*, 2016.
[7] T. Yokoyama, K. Shingu, S. Sekiya, H. Maeyama, and N. Nishimura, "CFD Analysis inside a CO2 Rotary Compressor Shell to Improve Oil Separation and Reduce the Shell Size," in *International Compressor Engineering Conference. Paper 2074*, 2012.
[8] X. Gao, J. Chen, J. Feng, and X. Peng, "Numerical and experimental investigations of the effects of the breakup of oil droplets on the performance of oil–gas cyclone separators in oil-injected compressor systems," *Int. J. Refrig.*, vol. 36, no. 7, pp. 1894–1904, 2013.
[9] J. Xu and P. Hrnjak, "Visualization and Simulation of Oil Flow in a Scroll Compressor Plenum Corresponding author," *Int. Compress. Eng. Conf. Pap. 2507*, 2016.
[10] J. Xu and P. Hrnjak, "Quantification of flow and retention of oil in compressor discharge pipe," *Int. J. Refrig.*, 2017.

IOP Conf. Series: Materials Science and Engineering **232** (2017) 012052 doi:10.1088/1757-899X/232/1/012052

CFD simulation of a dry scroll vacuum pump with clearances, solid heating and thermal deformation

A Spille-Kohoff, J Hesse, R Andres and F Hetze
CFX Berlin Software GmbH, Berlin, Germany

E-mail: andreas.spille@cfx-berlin.de

Abstract. Although dry scroll vacuum pumps (DSPV) are essential devices in many different industrial processes, the CFD simulation of such pumps is not widely used and often restricted to simplified cases due to its complexity: The working principle with a fixed and an orbiting scroll leads to working chambers that are changing in time and are connected through moving small radial and axial clearances in the range of 10 to 100 μm. Due to the low densities and low mass flow rates in vacuum pumps, it is important to include heat transfer towards and inside the solid components. Solid heating is very slow compared to the scroll revolution speed and the gas behaviour, thus a special workflow is necessary to reach the working conditions in reasonable simulation times. The resulting solid temperature is then used to compute the thermal deformation, which usually results in gap size changes that influence leakage flows. In this paper, setup steps and results for the simulation of a DSVP are shown and compared to theoretical and experimental results. The time-varying working chambers are meshed with TwinMesh, a hexahedral meshing programme for positive displacement machines. The CFD simulation with ANSYS CFX accounts for gas flow with compressibility and turbulence effects, conjugate heat transfer between gas and solids, and leakage flows through the clearances. Time-resolved results for torques, chamber pressure, mass flow, and heat flow between gas and solids are shown, as well as time- and space-resolved results for pressure, velocity, and temperature for different operating conditions of the DSVP.

1. Introduction

Dry scroll vacuum pumps (DSPV) are essential devices in many different industrial and academic processes. The electronic and semiconductor industry is the biggest application area of vacuum pumps, for example plasma etching and physical vapor deposition in manufacturing displays or computer chips. Vacuum pumps are also used in food industry for packaging technology or freeze-drying, in metallurgy for degassing of melt or inside a coating line. For research vacuum technology is used for electron microscopes or mass spectrometer. There are several different types of compressor mechanisms. The scroll pump is constructively the simplest solution for a vacuum pump. The advantages are high performance down to 10^{-3} mbar, low cost, easy maintenance, low noise and small vibrations compared with other kinds of dry vacuum pumps.

The scroll vacuum pump is quite complex from the fluid dynamics point of view. The driven orbiting scroll wrap changes the position relative to the fixed scroll wrap which results in a time changing working chamber volume. The fluid is compressible and therefore acoustic aspects could be important. Because of the vacuum characteristics the transition from the continuous flow regime to the molecular regime might have to be taken into account.

The Computational Fluid Dynamics (CFD) computation of scroll machines requires accurate grid generation because the fluid flow is transient and depends on the scroll position. Furthermore, the grid needs a high resolution especially in the gaps and their vicinity. Here, we used TwinMesh to generate high quality grids in a short time.

This paper is based on [3] which includes CFD simulation results of a dry scroll vacuum pump. In [4], we simulated the fluid region between the solid scrolls only, as done in [3], but added axial gaps to the simulation domain. Both [3] and [4] neglected the solids but set a thermal boundary condition on the surface of the solid scrolls towards the transported gas. [3] suggested a fixed temperature of 65°C at all solid surfaces but already mentioned that this causes calculation errors and the value should depend on the process parameters, in their case on suction pressure and rotation speed.

In this paper, we added the solid scrolls as given in a sketch in [2] with conjugate heat transfer between compression chamber and scrolls and a convective boundary condition at the outer surface of the scrolls. As a main advantage we get consistent temperatures in the solids depending on process parameters, and we can use the temperature distribution for the calculation of thermal deformation of the scrolls. The main problem is that the timescale of solid heating is very much longer than that of scroll rotation and fluid dynamics. Thousands of rotations would be necessary to capture the heating up to operating conditions, needing a huge amount of simulation time. Therefore the solid heating was accelerated by an increased time step size in the solids.

2. Geometry

The working chamber of the scroll pump is a simple cylindrical volume including one fixed and one orbiting scroll wrap. The inlet and outlet volume is simplified as a pipe, which could have an influence on the DSPV performance in comparison to the real geometry. One important geometrical parameter of DSVP is the volume ratio, which is the ratio of the suction chamber volume to the discharge chamber volume. Here a volume ratio of 2.0 is used corresponding to the involute starting angle Φ_S of 3π like in [3] and [1], leading to a design pressure ratio of 2.64. The pressure inside the working chamber is analyzed on four different locations (orbiting angle of 300°, 570°, 840° and 1110°) next to the fixed scroll wrap, which are the pressure transducer locations of [1]. The used axial clearance size is 30 microns. The specified axial clearance in [1] of 300 microns must be a mistake.

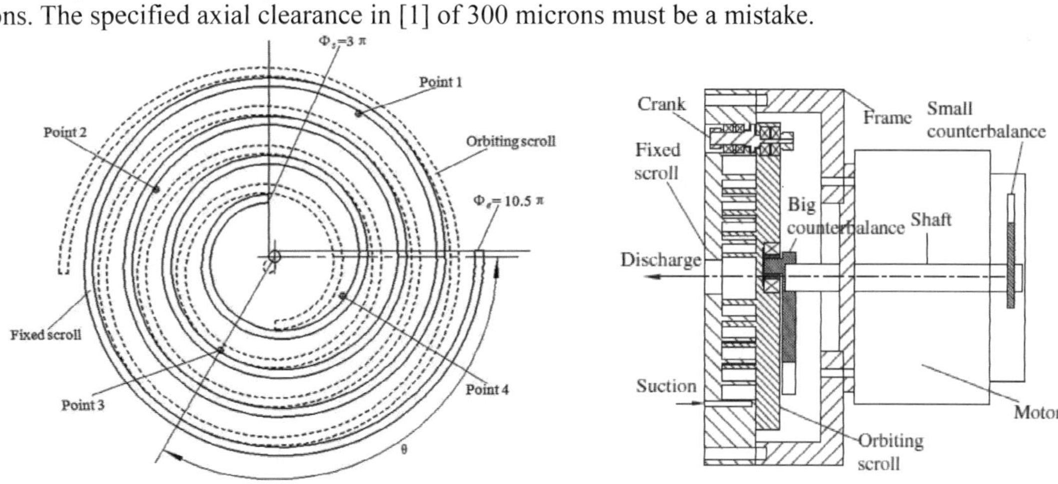

Figure 1: Geometry of the scroll wraps [3] and total assembly [2].

Table 1: Basic parameters of the DSVP.

Radius of the basic circle	3 mm	Height of the scroll wrap	30 mm
Pitch of involute	18.85 mm	Loops of scroll	5.25
Initial angle of involute	40°	Axial clearance	0.03 mm
Thickness of scroll wrap	4.19 mm	Minimum radial clearance	0.036 mm

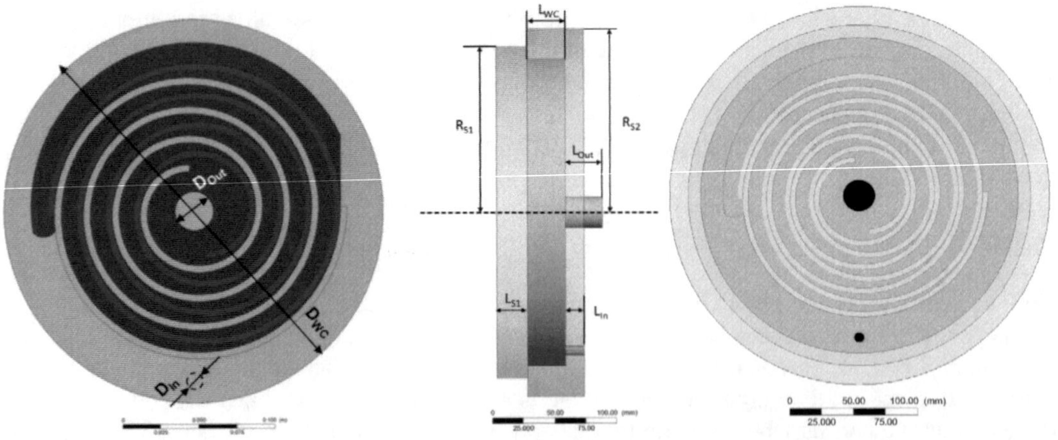

Figure 2: Geometry of the fluid domain (left), side and top view onto fluid and solid domains (middle and right) with dimensions in Table 2.

Table 2: Parameters of the simulation domain.

L_{In}	15 mm	D_{In}	8 mm
L_{WC}	30 mm	D_{WC}	250 mm
L_{Out}	60 mm	D_{Out}	25 mm
L_{S1}	25 mm	R_{S1}	135 mm
		R_{S2}	150 mm

3. Meshing

The discretization of the fluid volume is the main challenge to get reliable simulation results. The time changing volume has to be represented by a numerical mesh including all gaps. There are several methods to model the chamber volume with different restrictions.

3.1. Chamber Modelling for Positive Displacement (PD) Machines

The **overlapping-meshes** method is the simplest approach to investigate the rotor influence on the fluid flow inside a PD machine. The fluid domain of the PD machine is meshed regardless of the rotor geometry, i.e. the whole fluid domain is filled up with grid elements (as if the rotors were removed, called background mesh) whereas a second mesh describes the rotor geometry. For the coupling of both meshes, two main approaches exist. In the **immersed solid method**, the second mesh defines the solid region only; from this, at each time step the CFD code determines the region of overlap of solid and fluid meshes and applies momentum sources to force the fluid there to follow the motion of the solid rotor. In the **overset mesh method** (Chimera), the second mesh describes the fluid region around the solid up to an overset boundary; each time step, the nodes of both meshes are categorized as "dead" (no intersection, no fluid flow), "solve" (solution of conservation equations), and "donor" (gives solution) and "receptor" (gets solution from donor) between. The advantage of these methods is that grid generation for the fluid region is easy since the shape of the rotors and of the gaps need not be taken into account, and that the initial grids can be used for the whole simulation as remeshing or mesh deformation are avoided. However, one of the main disadvantages is that most CFD solvers limit the choice of physical and numerical models, e.g. only incompressible single-phase flows. Furthermore, a very high number of elements is necessary to accurately resolve the gaps; since the gap position varies with time the background grid also has to be refined circumferentially everywhere the gap will be during the rotor motion. For immersed solid method, the solution quality in the near wall region lacks of boundary resolution and turbulent wall functions since there is no wall in the simulation but a rotating fluid. The overset mesh method often lacks of conservation due to interpolation between the meshes, and a severe time step restriction due to problems when nodes change between "dead" and "solve" in one time step.

The **deform-and-remesh method** rotates the rotor each time step and tries to deform the fluid grid to represent the new shape of chambers and gaps. Due to the big shape changes of the fluid regions in PD machines, the deformation often causes a bad quality grid or even an illegal grid. Therefore an algorithm is required that generates a new mesh during run-time depending on the rotor position and local mesh quality. As an alternative, grids may be prepared manually for certain positions before (key-frame grids). Compared to the immersed solid method, the rotor surface is accounted for in a way that the near wall region can be treated more accurately. Also, compressible or multi-phase flows can be calculated. The main disadvantage is that remeshing in small gaps is unavoidable for almost every angle increment, but every remeshing requires an interpolation of the results from the old mesh onto the new mesh, which results in interpolation errors. Furthermore, the remeshing method must use an automatic grid generation program. Such programs often use unstructured mesh elements that result in high element numbers or in poor mesh quality within the gaps. This method is not efficient for a CFD simulation of a DSPV.

The **customized grid generation** can avoid the aforementioned disadvantages. Since each new rotor position results in a change of the fluid domain, the positions for each time step need to be meshed individually. These pre-generated meshes are read in at run-time into the CFD solver. If all these meshes share the same topology, i.e. the number of vertices and their connection to elements stay the same, and if the change in vertex positions from one mesh towards the next is quite small, interpolation of results is not necessary. Instead, the mesh deformation is taken into account via the Leibniz rule in the partial differential equations describing the fluid flow. The demand for a constant mesh topology can be fulfilled most easily by a block-structured mesh of hexahedral elements. The computational effort is minimized and the mesh quality is high leading to reliable simulation results.

In this project, TwinMesh was used to pre-generate the hexahedral meshes for the deforming fluid region between fixed and moving scroll of the DSVP.

3.2. Customized Grid Generation of Scroll compressor with TwinMesh

TwinMesh is a meshing software for PD machines with two axially parallel rotors, with complex rotor geometry, i.e. continuous (e.g. lobe pump), discontinuous (e.g. screw compressor) or with a single rotor (e.g. scroll compressor, eccentric screw pump). CAD data of the rotor and casing curvature can be imported and is used for the structured grid generation. The topology of a scroll is meshed using O-type grid around the rotor curvature. Figure 3 shows the TwinMesh GUI in the case of a scroll pump. TwinMesh generates 2D meshes for each rotation angle of the rotor with the same topology and node numbers including a refined boundary layer resolution layers towards rotor and housing walls. The meshes are smoothed with an explicit and iterative method to reach homogeneous node distribution and orthogonality. The resulting mesh of the chamber volume can be analyzed by quality criteria, such as minimum element angle, aspect ratio and volume change of cells. The 2D meshes are connected to get the 3D volume mesh. The 3D meshes are exported from TwinMesh for each rotor position (i.e. the meshes cover one complete cycle), where an angle increment of 3° is used for this simulation of the DSVP, i.e. 120 meshes for one scroll motion cycle.

3.3. Simulation Domain

The simulation domain is split into six volumes to use the adequate meshing strategy for each volume (Figure 4): The two solids, the deforming fluid region between the scrolls, the stationary fluid region between outer solid and scroll with inlet and outlet pipes, and the two fluid regions in the axial gaps between scroll and opposite disc (one stationary and one orbiting).

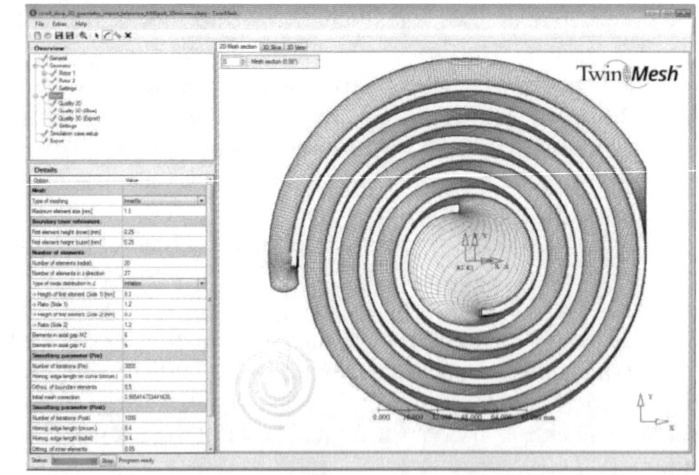

Figure 3: TwinMesh GUI (2D cross section) for a scroll pump.

The meshes of the deforming fluid region are created with TwinMesh whereas all other meshes are generated with ANSYS Meshing as extruded meshes with hexahedrons and wedges. The orbiting solid consists of 72,000 elements, the stationary solid of 47,000 elements. The stationary fluid region consists of 190,000 elements, each axial gap of 75,000 elements, and the deforming fluid region of 2,000,000 hexahedrons (20 in radial direction between the scrolls, 50 in axial direction). In sum, the partial differential equations are solved on 2,788,821 nodes in 2,516,640 hexahedrons and 3508 wedges. We used a reasonable mesh resolution and grid size, but so far haven't done a grid-independence study.

Figure 4: Geometry and meshes of simulation with solids: Orbiting solid scroll (left), deforming fluid region between orbiting and fixed scroll and stationary fluid region with inlet pipe (middle in dark grey), and fixed scroll with borings for inlet and outlet (right).

4. Simulation Setup

The commercial CFD solver ANSYS CFX is used for the simulation. The meshes for the six different parts are connected with GGIs (Generalized Grid Interfaces); the motion of the orbiting solid and of the axial gap between the orbiting scroll and the stationary disc is described with expressions, the deformation of the fluid region between the scrolls by the reading of the pre-generated TwinMesh meshes at run-time via a FORTRAN routine. The GGIs are automatically updated each time step to consider the time dependent connections.

The working fluid of the DSVP is air which is modelled as ideal gas, the discs and scrolls consist of steel. Energy conservation is solved in all regions, additionally Navier-Stokes equations with SST turbulence model in fluid regions. Though the DSVP is a vacuum pump, the lowest pressure of 17 kPa used here gives a Knudsen number of 0.011 so that continuum approach and no-slip boundary conditions at walls are still usable; for lower pressures, the usage of a slip model is possible.

We simulated the DSVP for three process parameters given in [1-3]: Suction pressures of 17 kPa, 42 kPa and 95 kPa at a fixed discharge pressure of 95 kPa at a rotational speed of 1704 rpm. Since design pressure ratio is 2.64, we expect under-compression for 17 kPa, slight over-compression for 42 kPa and high over-compression for 95 kPa.

For the energy equation, 20°C inlet temperature is used at inlet pipe. Due to the compression in the DSVP, temperature rises with pressure in the working chamber and heat flows into the solids. At the outside of the solid discs, we assume a convective boundary condition by setting a heat transfer coefficient of 16.8 W/(m^2 K) towards an outside temperature of 20°C. Some solid components as shaft and frame are neglected since full CAD data was not available. To get comparable solutions to [3] and [4], we made further simulations where we neglected all solids and fixed the wall temperatures in the fluid regions to 65°C, for 17 kPa suction pressure also for 45°C and 85°C.

Time step size of the transient simulations is 0.293 ms due to 120 meshes for one cycle of oscillation time 35.2 ms at 1704 rpm. The advection scheme is high resolution and the transient scheme is second order backward Euler. With this setup, simulation time for one oscillation cycle of the DSVP is 16 hours on 8 cores of Intel(R) Xeon(R) E5-2637 with 3.5 GHz with a demand for 35 GB RAM if solved in double precision. The residuals converged to a root mean square value below 0.001; improved convergence is possible by using a smaller time step size, i.e. more meshes for one rotation.

Simulation was run over a lot of rotations until a periodic state was reached. Since the heating of the solid structures (discs and scrolls) is slow compared to cycle time, i.e. in the range of 10 minutes compared to 35 ms for one cycle, thousands of cycles would be necessary with simulation time of several months. Since only the periodic state in working condition is of interest here, we accelerated the solid heating by an increased time step in the solids (up to a factor 10,000) and decreased this factor in steps to 1 when the temperatures in the solids reached their final level. With this approach, we got stagnant solid temperatures and periodic results in 10 to 20 cycles, i.e. one to two weeks of simulation time for each case.

5. Simulation results

In this paper, we extended the simulations described in [4] by taking into account the solids (discs and scrolls) and their temperature field. Therefore, we focus on the effect on working chamber pressures, on the temperature distribution in the solids and on their thermal deformation. Results for pressure, velocity and temperature distributions in the fluid region were given in [4] and did not change qualitatively. Figures 5 and 6 show working chamber pressures over orbiting angle for the suction pressures 17 kPa, 42 kPa and 95 kPa. In Figure 5, experimental results and ideal process curves from [2] are compared to our simulations without solids, i.e. with wall temperature set to 65°C, and with solids. The ideal process assumes adiabatic compression without gap losses.

With 17 kPa, the scroll pump works in under-compression mode, i.e. the pressure in the working chamber is not increasing to discharge pressure before the chamber opens to discharge pipe at 90° orbiting angle. The simulations capture the pressure increase well; the chambers are heated by the warmer walls or the warm solids and pre-filled from discharge side through the axial and radial gaps, so that real pressure is higher than that from ideal process. When the chamber opens to discharge side, chamber is filled from there until discharge pressure is reached. In the ideal process, filling occurs suddenly, whereas in the real process, filling is slower due to the slow opening of the gap between scrolls. In the experiment, pre-filling is stronger so that working chamber pressure is higher when chamber opens to discharge side. Most probably, the axial and / or radial gaps in experiment are greater than the 30 μm used in the simulations. The simulations with and without solids show almost no difference since the assumed wall temperature of 65°C is approximately the solid temperature.

While the 42 kPa case has small over-compression in ideal case, experiment and simulations show almost 95 kPa in working chamber at opening to discharge pipe. For the over-compression case with suction pressure 95 kPa in Figure 6, no experimental data was given. Here, differences between the simulations without and with solids occur since the solid temperature does not reach 65°C for this case. In all simulations, oscillations can be seen in the pressure of the working chamber just after opening to the discharge pipe. This comes from the usage of standard, i.e. reflective opening boundary conditions at the pipe end; Figure 6 shows simulation results for 17 kPa suction pressure with three different pipe lengths: Oscillations with small period for the 60 mm pipe, oscillations with larger period for a 1 m pipe, and no oscillations for a 100 m pipe.

Deviations between experiments and simulations come from (a) not fully described geometry in [1-2], (b) incorrect gap sizes of 30 μm in the simulations due to measurement errors and thermal deformation, (c) influence of inlet and outlet boundary models, and (d) pressure measurement by pressure transducers with finite size and transducer film instead of monitor point data in the simulation.

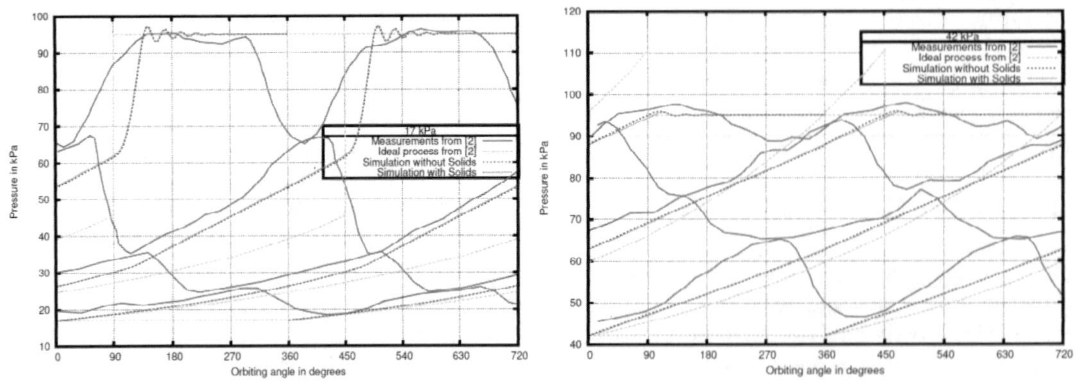

Figure 5: Chamber pressure for suction pressure 17 kPa (left) and 42 kPa (right) at 1704 rpm: Comparison of experimental and ideal process data from [2] with simulations without solids (wall temperature 65°C) and with solids. Differences between both simulations are very small (see text).

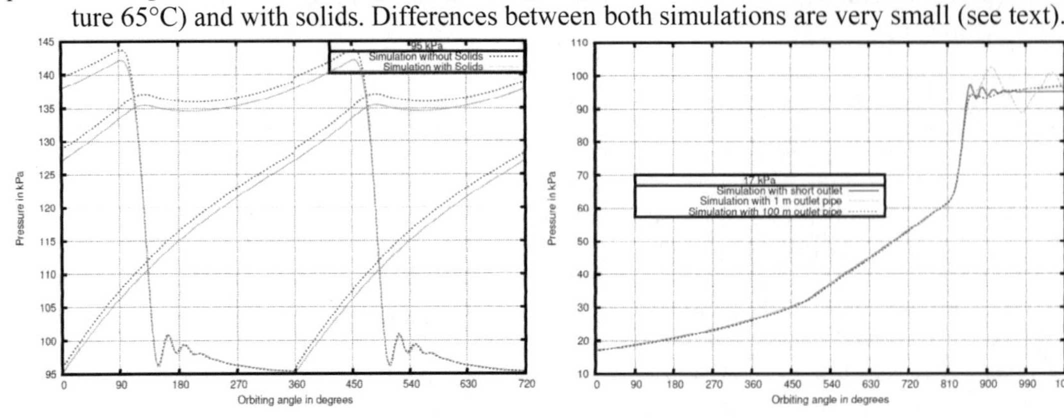

Figure 6: left: Comparison of chamber pressure for suction pressure 95 kPa at 1704 rpm for simulation without solids (wall temperature 65°C) and with solids;
right: Comparison of chamber pressure for suction pressure 17 kPa at 1704 rpm for simulation without solids (wall temperature 65°C) with different outlet pipe lengths to prevent reflection.

Table 3: Comparison of mass flow rate and power consumption for different suction pressures and temperature boundary conditions in the simulations.

Suction pressure in kPa	Wall temperature in °C (no solids)	Heat transfer coefficient in W/(m² K)	Mass flow rate in g/s	Power in W
17	45	-	0.803	113.9
17	65	-	0.770	114.4
17	85	-	0.727	114.1
17	-	8.4	0.707	114.0
17	-	16.8	0.738	114.1
17	-	33.6	0.797	114.4
42	65	-	1.83	93.6
42	-	16.8	1.91	96.5
95	65	-	4.02	50.8
95	-	16.8	4.31	51.2

Table 3 shows mass flow rates and power consumption for all simulated cases: Mass flow rates decrease with wall temperature in cases without solids, and increase with the heat transfer coefficient for the outer solid walls in cases with solids, since higher heat transfer coefficients lead to colder solids. The temperature distribution on the fluid side of both discs and scrolls can be seen in Figures 7 to 9 for a heat transfer coefficient of 16.8 W/(m² K). While the temperatures reach 63 to 94°C for 17 kPa suction pressure, in the 42 kPa case the solids are 43 to 60°C warm, and in the 95 kPa case only 28 to 39°C with the warmest region in the middle of the scrolls and not, as in the other cases, at the innermost ends towards discharge pipe. Since this case shows strong over-compression up to 140 bar, the gas decompresses towards the discharge pipe and therefore becomes colder again.

This temperature distribution leads to a thermal deformation of discs and scrolls; since some parts of the solid geometry were neglected (e.g. shaft and frame), adequate boundary conditions for the thermal deformation simulation were needed. At the red lines in Figure 10, we fixed the position of the centre of gravity for these boundary areas. The resulting deformation for 17 kPa suction pressure and heat transfer coefficient of 16.8 W/(m² K) is shown in Figure 10 also; the effect on the axial gap sides between the scrolls and the adjacent disc is shown in Figure 11: While the gaps near the discharge pipe decrease by 10% from 30 µm to 27 µm, they increase at suction side to 36 µm.

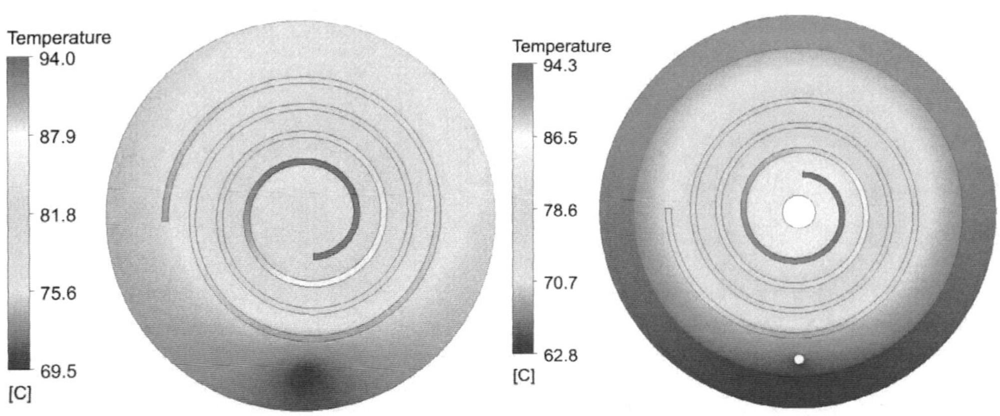

Figure 7: Temperature on surface of orbiting disc and scroll (left) and on stationary disc and scroll (right) for suction pressure 17 kPa for heat transfer coefficient 16.8 W/(m² K).

IOP Conf. Series: Materials Science and Engineering **232** (2017) 012052 doi:10.1088/1757-899X/232/1/012052

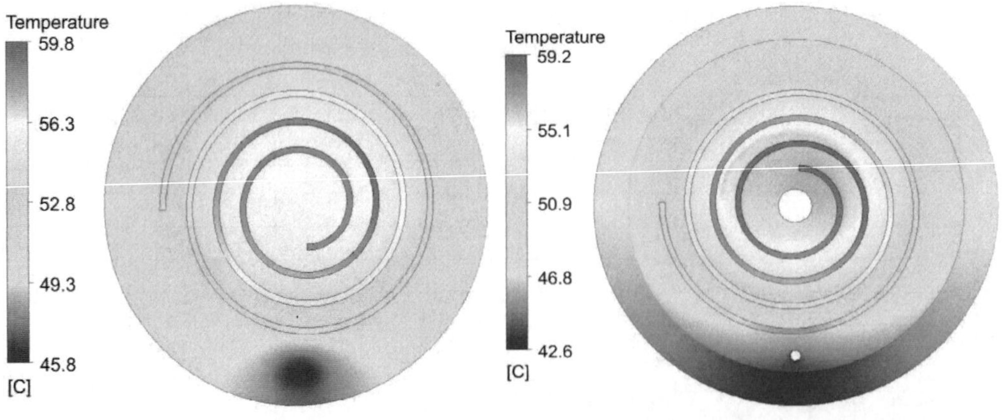

Figure 8: Temperature on surface of orbiting disc and scroll (left) and on stationary disc and scroll (right) for suction pressure 42 kPa for heat transfer coefficient 16.8 W/(m² K).

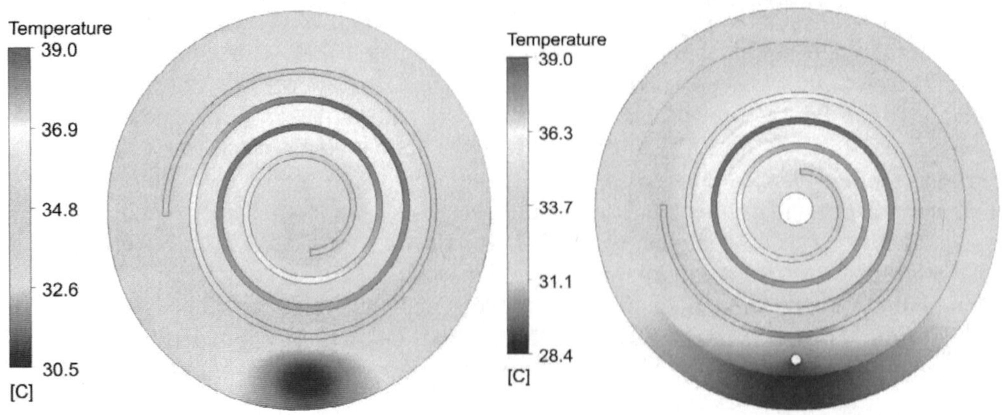

Figure 9: Temperature on surface of orbiting disc and scroll (left) and on stationary disc and scroll (right) for suction pressure 95 kPa for heat transfer coefficient 16.8 W/(m² K).

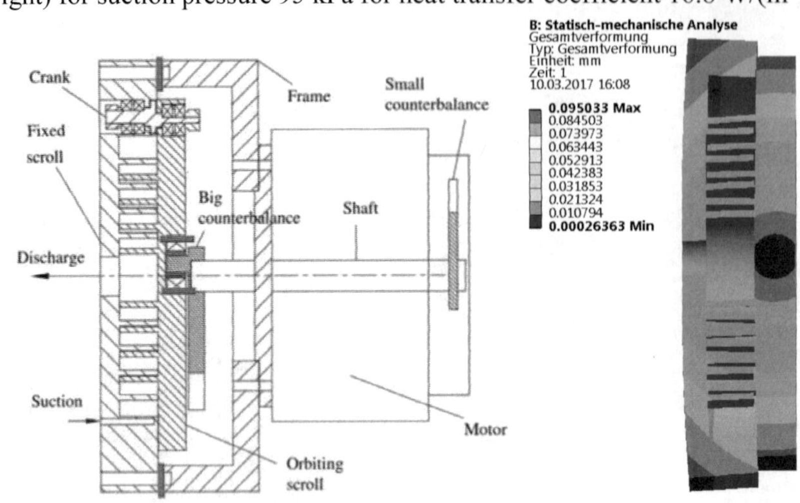

Figure 10: Sketch of location of boundary conditions for the simulation of thermal deformation (left, red lines), and total thermal deformation calculated for the case with suction pressure 17 kPa.

IOP Conf. Series: Materials Science and Engineering **232** (2017) 012052 doi:10.1088/1757-899X/232/1/012052

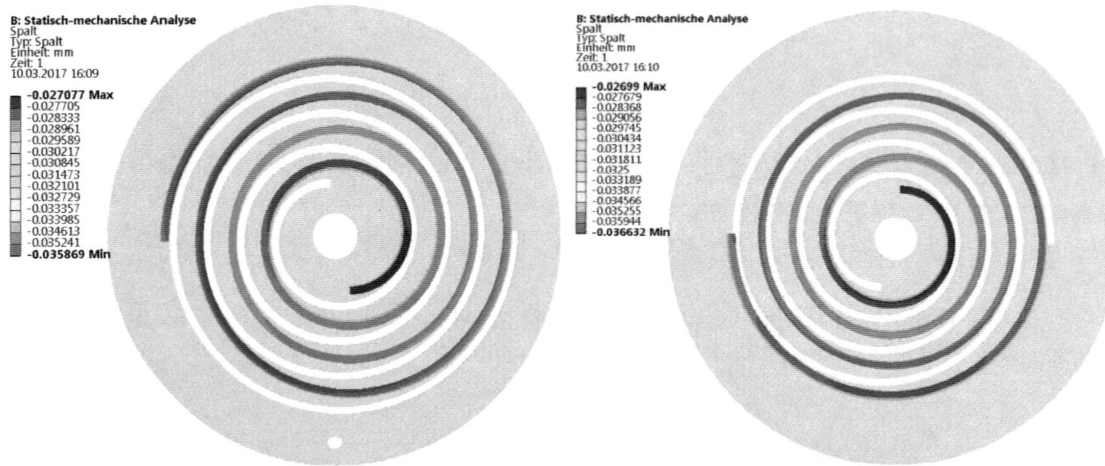

Figure 11: Height of axial gap between the orbiting scroll and the stationary disc (left) and between the stationary scroll and the orbiting disc (right); gap height before thermal deformation was 30 μm.

6. Conclusions and outlook

This paper shows CFD results for a dry scroll vacuum pump including axial gaps and solid components as scrolls and disc. Different suction pressures from 17 kPa to 95 kPa were applied using an inlet temperature of 20°C. The CFD results of the working process are compared with ideal process data and experimental data. In addition, average values of the mass flow and power consumption are shown for different suction pressures. For the CFD simulation ANSYS CFX was used and the meshes for the deforming working chamber were created with the meshing software TwinMesh, all other meshes with ANSYS Meshing.

The comparison of the CFD results with the experimental data indicates that the working mechanism and flow conditions within the dry scroll vacuum pump are well captured by the simulation. The heat transfer from the process gas to the casing and the leakage flow through axial and radial gaps has a main impact on the increase of the pressure over the rotation angle. The differences between experiment and simulation are mainly caused by geometrical uncertainties and model assumptions.

The inclusion of the solids with conjugate heat transfer allows a consistent simulation for different process parameters without the necessity to adjust the assumed wall temperature manually as in the cases without solids. The temperature distribution in the solid components is non-uniform with a higher temperature in the scrolls, especially in the center of the DSVP near discharge. The thermal deformation changes gap sizes by 10 to 20%.

We used a reasonable mesh resolution and grid size in this paper, but plan further mesh refinements to estimate grid dependencies of the solutions. Further investigation will adjust the meshes in the CFD simulations according to the thermal deformation to account for the change of gap sizes. The usage of a slip boundary condition for lower suction pressures towards the ultimate pressure is also planned.

References

[1] Li Z, Li L,Zhao Y, Bu G and Shu P 2010 Theoretical and experimental study of dry scroll vacuum pump *Proceedings of the Vacuum 84*, (415-421), Elsevier

[2] Li Z, Li L, Zhao Y, Bu G, Shu P and Lui J 2010 Test and analysis on the working process of dry scroll vacuum pump *Proceedings of the Vacuum 85*, (95-100), Elsevier

[3] Yue X, Lu Y, Zhang Y, Ba D and Wang G 2015 Computational fluid dynamics simulation study of gas flow in dry scroll vacuum pump *Proceedings of the Vacuum 116*, (144-152), Elsevier

[4] Hesse J and Andres R 2016 CFD simulation of a dry scroll vacuum pump including leakage flows *Proceedings of 23rd International Compressor Engineering Conference*

IOP Conf. Series: Materials Science and Engineering **232** (2017) 012053 doi:10.1088/1757-899X/232/1/012053

Grid generation methodology and CFD simulations in sliding vane compressors and expanders

Giuseppe Bianchi[1,2*], Sham Rane[1], Ahmed Kovacevic[1], Roberto Cipollone[3], Stefano Murgia[4], Giulio Contaldi[4]

[1]City University of London, United Kingdom
[2]Brunel University London, United Kingdom
[3]University of L'Aquila, Italy
[4]Ing. Enea Mattei S.p.A., Italy

*giuseppe.bianchi@brunel.ac.uk

Abstract. The limiting factor for the employment of advanced 3D CFD tools in the analysis and design of rotary vane machines is the unavailability of methods for generation of computational grids suitable for fast and reliable numerical analysis. The paper addresses this challenge presenting the development of an analytical grid generation for vane machines that is based on the user defined nodal displacement. In particular, mesh boundaries are defined as parametric curves generated using trigonometrical modelling of the axial cross section of the machine while the distribution of computational nodes is performed using algebraic algorithms with transfinite interpolation, post orthogonalisation and smoothing. Algebraic control functions are introduced for distribution of nodes on the rotor and casing boundaries in order to achieve good grid quality in terms of cell size and expansion. In this way, the moving and deforming fluid domain of the sliding vane machine is discretized and the conservation of intrinsic quantities in ensured by maintaining the cell connectivity and structure. For validation of generated grids, a mid-size air compressor and a small-scale expander for Organic Rankine Cycle applications have been investigated in this paper. Remarks on implementation of the mesh motion algorithm, stability and robustness experienced with the ANSYS CFX solver as well as the obtained flow results are presented.

1. Introduction

Sliding vane machines are widely employed in industry nowadays. Typical applications where these devices are one of the leading technologies are compressed air systems, automotive thermal management (oil and cooling circuits), oil and gas applications. Nonetheless, unlike other positive displacement technologies that can rely on the design tools at the state of the art to improve and accelerate R&D and product development tasks, sliding vane machine designers cannot entirely benefit from these resources. In fact, the limiting factor for the employment of advanced 3D CFD tools in the analysis and design of rotary vane machines is the unavailability of methods for generation of computational grids suitable for fast and reliable numerical analysis.

In the literature, scientists and engineers have been investigating sliding vane devices both through experimental and numerical approaches. In particular, the modeling approaches that have been developed mostly consider lumped parameter formulations of the conservation equations that undoubtedly benefit of reduced computational cost [1-4]. On the other hand, they do not allow possibilities of identifying and optimizing key issues in positive displacement machines such as the leakage paths, which in these models are schematized as equivalent orifices [5].

In recent years, the lack of a standardized and parameterized approach to discretize the fluid volume that stator, rotor and blades identify in a sliding vane machine has been started to be addressed by an increasing number of research works. Most of the developed procedures focused on expanders for waste heat to power generation based on an Organic Rankine Cycle architecture (ORC). For instance, Kolasiński and Błasiak performed experimental and numerical studies on a micro ORC sliding vane expander with circular stator and radial inlet and outlet ports [6] while Montenegro et al., investigated a sliding vane expander with an elliptic stator [7]. Working fluids were R123 and R245fa while CFD solvers employed in the afore mentioned studies were ANSYS CFX and OpenFOAM respectively. In both the research works, correction terms had to be introduced to deal with inaccuracies in the rotor grids. In addition to these research works, in commercial software such as Simerics PumpLinx or ANSYS there are templates on vane machines [8, 9]. However, they introduce a rather severe approximation when considering a flat blade profile. In fact, together with the stator, the tip of the blade forms a converging-diverging nozzle whose local curvature affect the overall leakage.

Among the available approaches for grid deformation, user defined nodal displacement was found to be the most suitable strategy to create a set of grids representing nodal locations for each time step externally, prior to the numerical flow solution in the calculation domain [10, 11]. User defined nodal displacement methodology was successfully applied to screw machine geometries and led to the development of two different strategies for grid generation. An algebraic approach was pursued by the research group at City, University of London [12] while a hybrid differential approach was considered at the University of Ghent [13]. Both the grid generation methods allowed to perform three-dimensional single and multi-phase simulations on screw compressors and expanders in recent years. In particular, the algebraic approach revealed more suitable for extension of grid generation to variable lead and variable profile rotors, while the hybrid differential approach resulted in better grid smoothing and adaptability to variety of rotor profiles such as hook and claw type positive displacement machines [11].

Regardless of the technology, positive displacement machines involve similar issues in the development of a deforming grid generation i.e. from clearance regions to the main core the aspect ratio of the geometry is 1000th of order of magnitude, the computational cells need to maintain continuity through these highly deforming volumes and the cell structure needs to be maintained in order to avoid solution interpolation from one time step to the other. Hence, in the current study some of the algebraic algorithms developed for screw machinery have been applied to a sliding vane machine geometry and this led to the development of a novel general grid generation methodology. This innovative meshing procedure was tested with reference to two applications that highly differ for dimensions, operating parameters, working fluids and mode of operation. In particular, test cases presented in this paper are an oil injected air compressor and an ORC expander. After recalling the main steps of the grid generation for vane machines, the paper aims at highlighting similarities and differences occurred in the two simulation sets.

2. Grid Generation Procedure

In a sliding vane geometry, the working chamber is formed by the volume entrained between the housing, the rotor and the end plates. As the rotor carrying the set of vanes turns, each vane divides the full volume into successive chambers that undergo suction, compression/expansion and discharge phases. At the tip of each vane there is a thin gap that needs to be treated as the leakage. Even though the change in volume of this gap is very small compared to the core region, because of the eccentricity between rotor and stator, the relative shape of these gaps change with rotor positions. Altogether, the combined core and leakage gap form a O grid topology. Several algorithms for the structured meshing of O type domain topology are available in practice showed that algebraic algorithms with transfinite interpolation, post orthogonalisation and smoothing are the most efficient ones for applications like twin

IOP Conf. Series: Materials Science and Engineering **232** (2017) 012053 doi:10.1088/1757-899X/232/1/012053

screw compressors where a number of mesh files are required to be generated and supplied to the CFD solvers in order to treat mesh deformation [14]. Similar algebraic grid generation principle was used in this work for the vane geometry rotor domain mesh.

Flow chart in Figure 1 lists the main stages of the grid generation procedure. A standalone parametric geometry program built as a Matlab application is used to produce the rotor profile along with the vane at the initial rotor position. For a given set of rotor, stator diameters and number of vane with vane thickness, the program provide coordinate of the points on the chamber boundaries. This profile is an input to the meshing algorithm. Successive rotor positions appear to be rigid body rotations for the rotor while for the fluid volume that is deforming, it is a complex boundary motion where rotor rotates, stator is stationary and each of the vane is undergoing a general body motion. For simplification, blade tip clearance gap changes were not accounted in the presented analysis, but the meshing algorithm allows the possibility to further introduce a gap variation function.

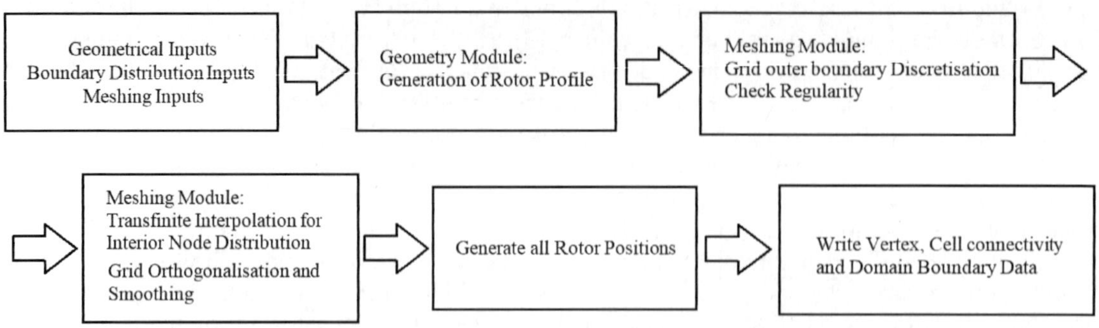

Figure 1. Vane geometry mesh generation procedure

The first stage in the meshing procedure is to discretize the boundaries. At this stage, inputs regarding the number of nodes and parameters for various control functions are taken from the user. The boundary discretization produces the distribution into the topology of an 'O' grid, as shown in Figure 2.b. An 'O' structure for the rotor mesh was purposely selected because it avoids any inaccuracies that can be introduced due to a non-matching mesh connection between core and leakage regions. The sudden transition from the leakage gaps to the core was handled by the introduction of stretching functions that gradually flare the radial mesh lines from leakage gap into the core as seen in Figure 2.c and Figure 2.d.

At the position where the core region topology does not remain rectangular, as it happens when the vane slides inside the rotor, a simplification has been introduced to convert the vane side walls into short segments on the rotor surface. It is anticipated that the leakage flow will not be influenced by this simplification as the gap clearance does not change but it significantly eases the O grid mesh generation.

The second stage in mesh generation is distribution of interior nodes. Transfinite interpolation, orthogonalisation and smoothing as described in [14] were used in these procedures to get a good quality hexahedral cell structure as seen in Figure 2.e. Once the 2D mesh is generated for all rotor positions, the mesh is assembled into 3D volume in the required format of the solver (Figure 2.f). In the current implementation there is possibility to export the mesh in solvers such as ANSYS CFX, ANSYS FLUENT, STAR CCM+ and Simerics PumpLinx. The entire meshing procedure was implemented in the software package SCORG™ developed at City, University of London [16].

The rotor domain is eventually connected to suction and discharge ports via non-conformal interfaces, as usually occurs in CFD analysis on positive displacement machines [12]. Ports meshing is usually carried out using standard tools and with reference to a tetrahedral cell structure to better capture the real geometry.

In the presented analysis, ANSYS CFX solver was used to calculate the oil injected flow fields in the air compressor in the first test case and a real gas fluid definition was used in the calculation of a ORC expander in the second test case. ANSYS CFX provides a model called Junction Box Routine that is a user defined library to specify mesh deformation from custom applications such as SCORG. The solver updates the nodes coordinates from set of pre-generated coordinate files after every crank angle step (or its submultiples). Solver time step size finally results from the selected crank angle step and revolution speed of the rotor.

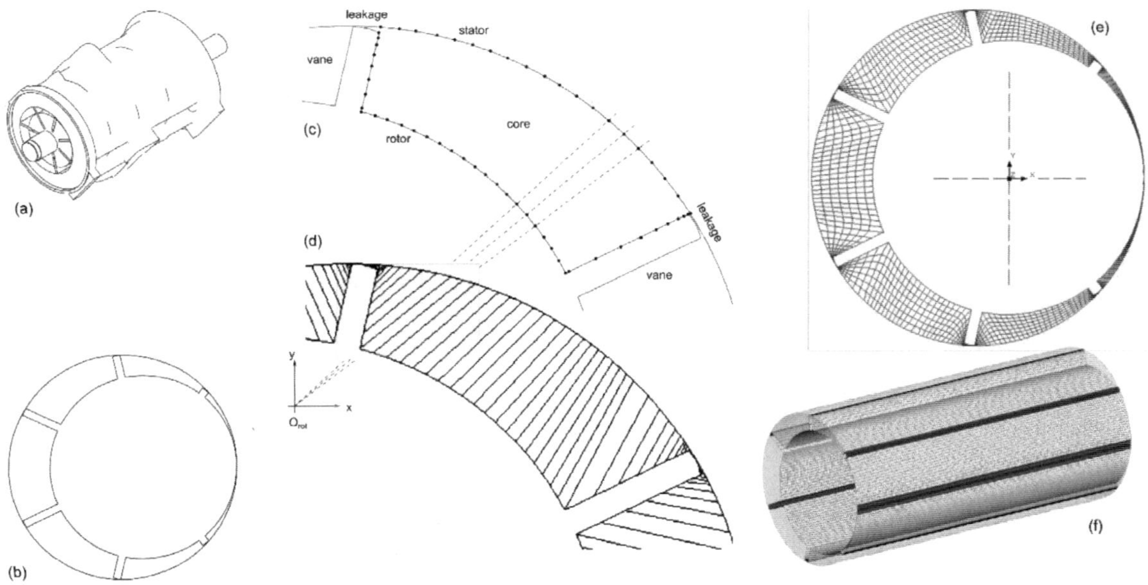

Figure 2. Grid generation procedure: (a) CAD drawing, (b) boundary generation, (c) boundary discretization, (d) stretching function applied to stator, (e) internal node distribution, (f) 3D grid

3. Numerical setups

Two test cases were analyzed using the new grid generation procedure developed in this work as represented in Table 1.

Table 1. Summary of geometrical and mesh specifics

	Oil injected Air Compressor		ORC Expander	
Rotor Diameter	195 mm		65 mm	
Axial Length	540 mm		60 mm	
Tip Clearance	50 μm		10 μm	
	Ports	Core	Ports	Core
Cell type	Tetrahedral	hexahedral	Tetrahedral	hexahedral
Node count (Million)	0.353	0.4735	0.135	0.157
Maximum aspect ratio	221	138	23	228
Minimum orthogonality	8.6	14.9	10.0	11.6

3.1. Case I: Oil injected air compressor.

The operating conditions that were considered as benchmark for the numerical results were 980 RPM as revolution speed and 8.70 bar_a as outlet pressure. Inlet pressure was 1.01 bar_a, inlet temperature 25.4 °C. The oil mass flow rate obtained from measurements was 2.9 kg/s and was used to specify the oil injection pressure so as to achieve this mass injection. Oil injection pressure was set at 6.6 bar_a, oil injection temperature was 65.0 °C. Under these conditions, the measurements showed that air mass flow rate was 0.203 kg/s, compressor indicated power was 68.2 kW and shaft power recorder was 82.8 kW. Table 1 presents a summary of the geometry and CFD model mesh data for the case.

3.2. Case II: Real Gas, ORC application expander.

Table 1 also presents a summary of the geometry and CFD model mesh data for the expander case. From the tests performed in refs. [17, 18], the operating conditions that were considered as benchmark for this case were 1551 RPM as revolution speed and 12.11 bar_a as inlet pressure with 90.5 °C as inlet gas temperature. Outlet pressure was 4.57 bar_a. Aungier Redlich Kwong (ARK) real gas model was used to specify the coefficients of the polynomials to define material properties and equation of state. Coefficients for specific heat and critical state temperature and pressure were supplied to generate a thermodynamic table in the temperature range of 273 and 400 K, while the pressure range was 2.0 and 60.0 bar_a.

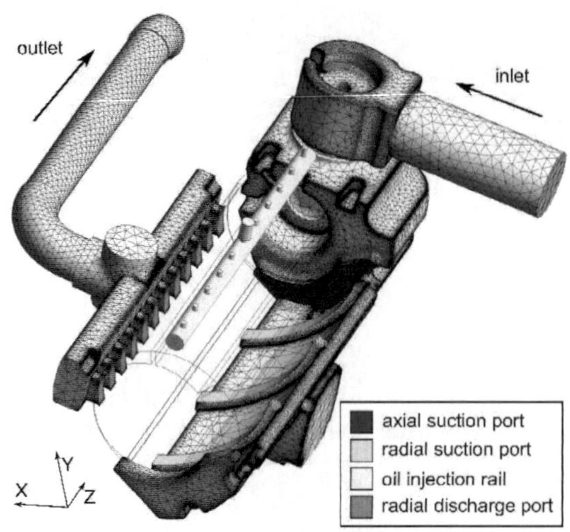

Figure 3. Compressor ports mesh

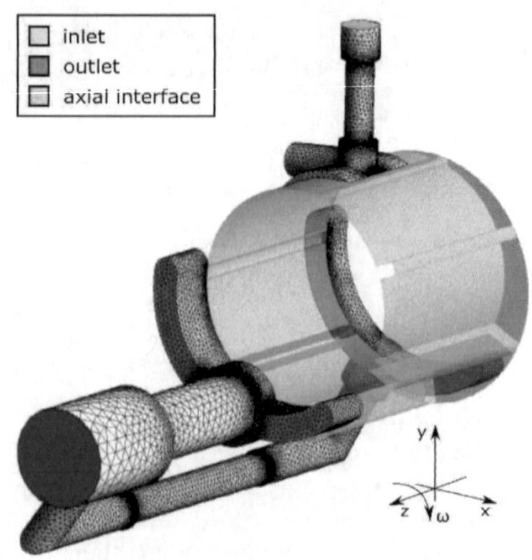

Figure 4. Expander ports mesh

Figure 3 shows the tetrahedral mesh used for the static ports of the oil injected compressor. Suction port, discharge port interface with the deforming rotor domain and the oil injection rail have been highlighted as in the colour scheme. Low pressure inlet and high pressure outlet boundaries are also seen. Similarly, Figure 4 represents the tetrahedral mesh used for the static ports of the real gas ORC expander. High pressure inlet, Low pressure outlet and the axial interface of low pressure port with the deforming rotor domain has been highlighted by the colour scheme. Radial interface of the high pressure port with the rotor is not seen. All these interfaces have been defined in the ANSYS CFX solver as non-conformal conservative flux type of fluid-fluid interfaces.

Table 2 presents the details of the CFD solver numerical setup. Most settings remained common between the two cases, except for the working fluid and solver relaxation parameters. Additionally since compressor case was solved as an oil injected flow, an eulerian-eulerian two phase solution was calculated. Air was the primary phase and constant density oil as liquid was set as the secondary continuous phase. Further details of oil injected compressor modelling can be found in Rane et al (2016). In terms of difficulty of solution, the compressor flow poses a challenge. In the compressor, the leakage flow through the tip gaps is in opposite sense to the main positive flow. Because of this and also due to the presence of oil phase, a gradual rise in discharge pressure was required to be specified. At the same time the solver relaxation factors were of lower order. On the other hand even though ARK real gas equation of state was defined with fixed table limits, a relatively easy convergence of the solver was observed. This could be due to the lower time step size and also the leakage flow sense in the same direction as the main gas flow. A relatively higher solver relaxation factor could be specified in this case.

Table 2. Details on the numerical setups in the ANSYS CFX solver

	Air Compressor	ORC Expander
Mesh deformation	User defined nodal displacement via junction box routines in FORTRAN	
Mesh in ports	Tetrahedral with boundary layer refinements (Generated by ANSYS pre-processor)	
Turbulence model	SST – k Omega (Standard Wall Functions)	
Inlet boundary condition	Opening (Specified total pressure and temperature)	
Outlet boundary condition	Opening (Static pressure In case of backflow used as total pressure and temperature)	
Control volume gradients	Gauss divergence theorem	
Advection scheme	Upwind	
Pressure-Velocity coupling	Co-located layout (Rhie and Chow 4th order)	
Transient scheme	Second order (Fully implicit Backward Euler)	
Transient inner loop coefficients	Up to 20 iterations per time step	
Convergence criteria	r.m.s residual level 1e-03	
Working fluid	Air Ideal Gas + Liquid oil	R236fa (NIST coefficients ARK table)
Relaxation parameters	Solver relaxation fluids (0.1)	Solver relaxation fluids (0.4)

4. Results and Discussion

4.1. Case I: Oil injected air compressor.

Figure 5 reports the pressure contours in a cross section of the machine at 75 mm with respect to a full axial length of 540 mm. The sealing of the high pressure port from the suction port is apparent from the pressure distribution.

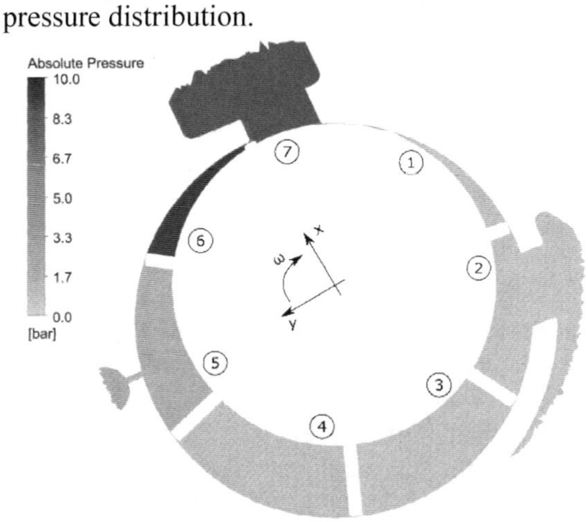

Figure 5. Compressor pressure - axial cross section at 14% of rotor length

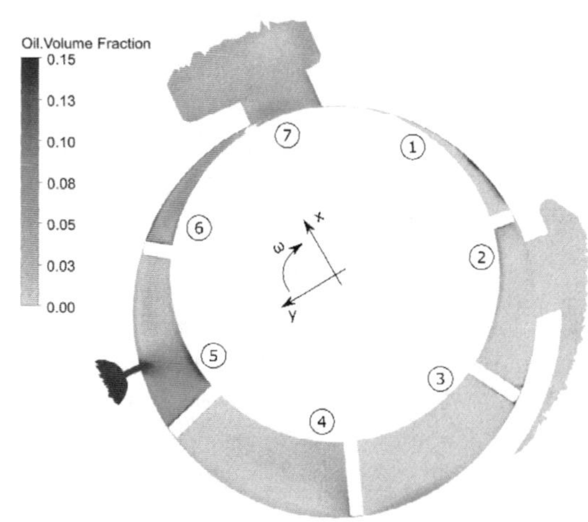

Figure 6. Compressor oil volume fraction - axial cross section at 14% of rotor length

The last 6th chamber is open to the discharge pressure of 8.7 bar_a but shows a slightly higher internal pressure of about 10.0 bar_a due to continued compression of the gas being discharged. Figure 6 reports the distribution of oil volume fraction in the same cross section at the same rotor position. From the leakage paths, oil is getting passed into the 1st suction chamber on the low pressure side. High oil concentration is visible in the injection rail and a radial jet of oil flow is visible in the 5th compression chamber. The pressure in this chamber is about 4.0 bar_a. About 6 – 8% oil volume fraction is observed in the discharge port.

Figure 7.a presents the distribution of gas temperature in this cross section while Figure 7.b is a 3D distribution of gas temperature with an oil iso-surface with 10% volume fraction superimposed in greyscale colors. Despite the sealing action of the lubricant, the clearance at the vane tip induces a backflow driven by adverse pressure gradient in opposite direction to the counter clockwise revolution speed sense in Figure 7.b. In turn, the overall flow field in the compressor vane is affected by a recirculation region whose magnitude is gradually lowered as the leakage flow interacts with the main one (which is aligned with the rotation sense). This phenomenon contributes to enhance the heat transfer between the compressing air and the oil that, after the radial injection shown in Figure 7.a, spreads along the leading side of the vane and mixes with air. The oil accumulation on the walls of the rotor, i.e. on the metallic surfaces which define the vane in reality, can be seen in the bottom left of Figure 7.b. These oil layers allow to lower friction losses between surfaces, namely stator with blade tip and blade side walls with rotor slots.

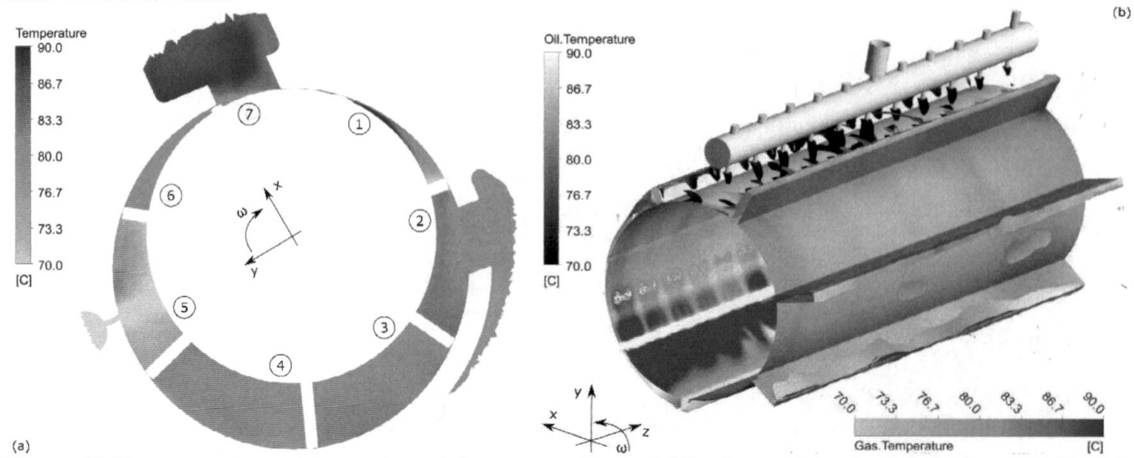

Figure 7. Compressor gas temperature: (a) cross section at 14% of rotor length, (b) 3D inner wall with injection rail and iso-surface having an oil volume fraction of 10%

4.2. Case II: Real Gas, ORC application expander.

Figure 8.a reports the pressure distribution of gas in case of the ORC expander simulation. Pressure from the high pressure port is at 12.11 bar_a and 90.5 °C. With expansion, a smooth decrease in gas pressure is visible across the chambers. Finally, pressure is at lower level of 3 bar_a from the 4th chamber onwards, when low pressure port is exposed. Figure 8.b represent the pressure distribution on the 3D rotor surface along with the vanes. A uniformity of pressure distribution is seen is individual chambers. Across the vane tip leakages, pressure gradients are observable. Similarly, Figure 9.a reports the gas temperature in the cross section and Figure 9.b presents gas temperature on the 3D rotor surface. In the cross section gradual reduction in gas temperature is visible. On the 3D rotor surface in chambers 1 and 2, a small non-uniformity of gas temperature distribution is seen due to the length of the intake port not covering the full rotor length and presence of axial low pressure ports on the two axial ends of the rotor.

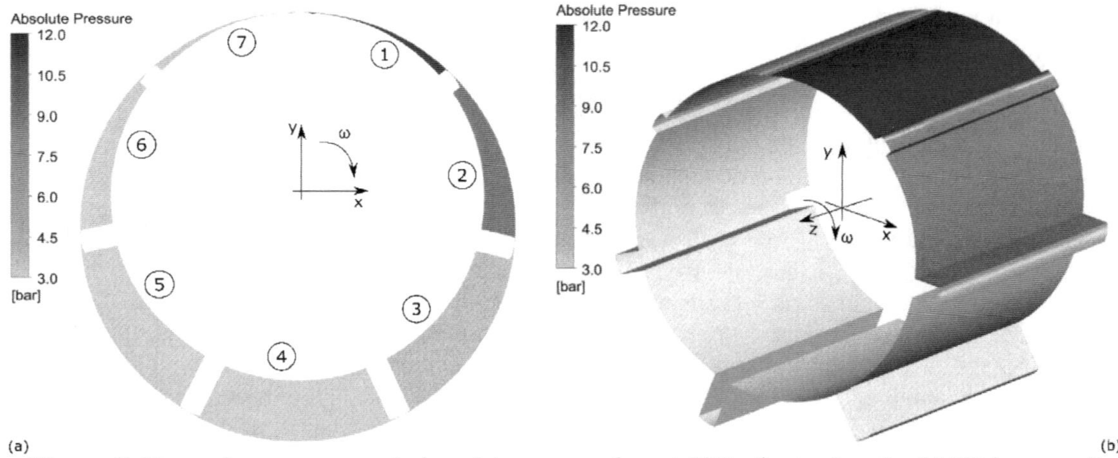

Figure 8. Expander pressure variation: (a) cross section at 50% of rotor length, (b) 3D inner wall

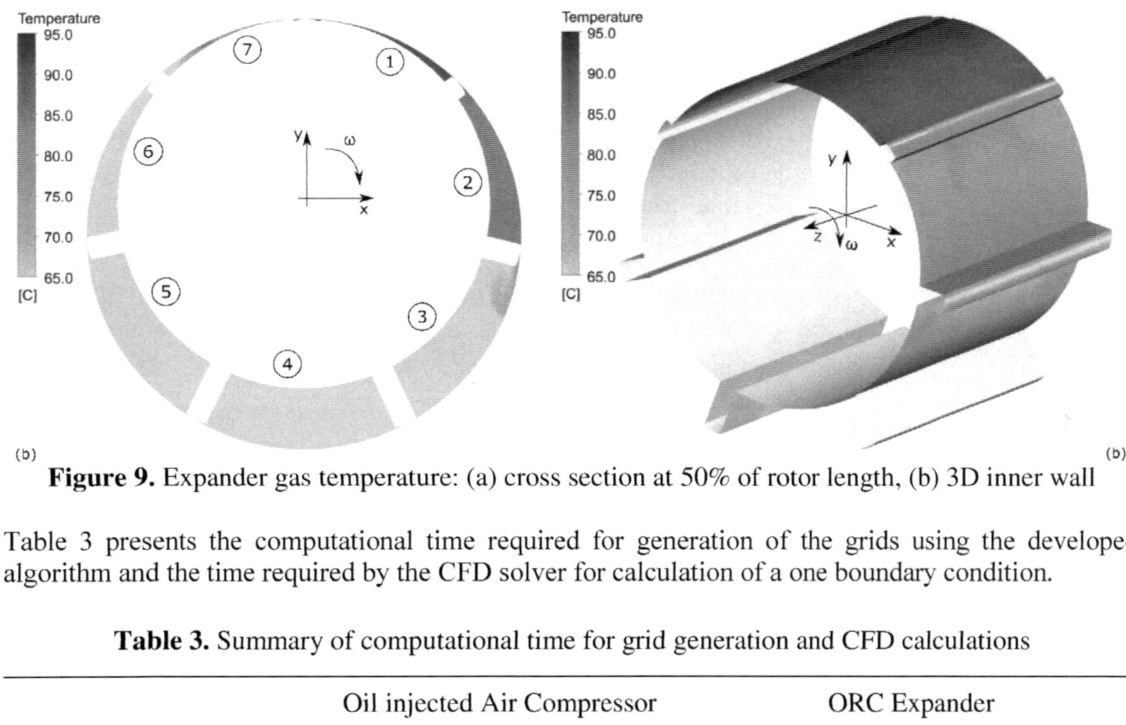

Figure 9. Expander gas temperature: (a) cross section at 50% of rotor length, (b) 3D inner wall

Table 3 presents the computational time required for generation of the grids using the developed algorithm and the time required by the CFD solver for calculation of a one boundary condition.

Table 3. Summary of computational time for grid generation and CFD calculations

	Oil injected Air Compressor		ORC Expander	
Generation for 720 positions				
2D cross section grid	120 seconds		120 seconds	
3D grid	900 seconds		720 seconds	
CFD solver computational time per case				
	Count	Time	Count	Time
Per time step	5 convergence iterations per step	700 seconds	3-5 convergence iterations per step	50 -55 seconds
Cyclically stabilised flow	1440 time steps	160 hours	1440 time steps	4 hours

5. Conclusions

The research work presented in this paper is the development of a robust and fast, deforming grid generation procedure for CFD analysis of sliding vane machines. Available literature suggested that only limited approaches to solving a full 3D numerical simulation in vane machines is currently available and the major limitation was found to be in the generation of deforming grids for the rotor domain. The developed grid and numerical methodology was eventually tested using the ANSYS CFX solver. Two vital test cases have been reported in the paper. First case is an industrial sliding vane oil injected air compressor and the second case is an ORC application, R236fa operated expander.

Multiphase simulations on the oil injected air compressor were successful in revealing the non-uniformity of gas temperature and oil distribution during the operation of the machine. Such a full scale model allowed to track a recirculation region induced in the vane flow field by the leakage flow through the clearance at the tip. Furthermore, the convective heat transfer with air showed the cooling capabilities of the lubricant even when supplied into the vanes as liquid jets. On the other hand, the oil temperature rise decreased its dynamic viscosity and, in turn, its hydrodynamic potential to prevent dry contact between blades and stator.

Real gas ARK model was successfully applied in the ORC expander analysis. The simulations provided information on the expander filling losses, internal expansion phenomenon and performance parameters could be evaluated at various operating conditions. Although due to the scope of this paper, flow results only from one of the selected operating conditions have been presented here.

These new procedures for grid generation of sliding vane machines will allow further design improvements and performance prediction of the machines with design modifications can be performed better. The effects of a variable tip clearance due to oil temperature and blade dynamics will be taken into account in future upgrades of the numerical methodology.

Acknowledgement
The work documented in this report has been funded by the 2015 Scholarship of the Knowledge Center on Organic Rankine Cycle technology (www.kcorc.org), the organization formed by the members of the ORC Power Systems committee of the ASME International Gas Turbine Institute (IGTI).

References

[1] O. Badr, S.D. Probert, P. O'Callaghan, Multi-vane expanders: Vane dynamics and friction losses, Applied Energy, Volume 20, Issue 4, 1985, Pages 253-285, ISSN 0306-2619, DOI:10.1016/0306-2619(85)90018-2.

[2] Tramschek, A. B., and M. H. Mkumbwa. Mathematical modelling of radial and non-radial rotary sliding vane compressors, International Compressor Engineering Conference at Purdue University, 1996, West-Lafayette (IN), United States, URL: docs.lib.purdue.edu/icec/1151/

[3] G. Bianchi and R. Cipollone, Theoretical modeling and experimental investigations for the improvement of the mechanical efficiency in sliding vane rotary compressors, Applied Energy, Volume 142, 15 March 2015, Pages 95-107, ISSN 0306-2619, DOI: 10.1016/j.apenergy.2014.12.055.

[4] G. Bianchi and R. Cipollone, Friction power modeling and measurements in sliding vane rotary compressors, Applied Thermal Engineering, Volume 84, 5 June 2015, Pages 276-285, ISSN 1359-4311, DOI: 10.1016/j.applthermaleng.2015.01.080

[5] Osama Al-Hawaj, Theoretical modeling of sliding vane compressor with leakage, International Journal of Refrigeration, Volume 32, Issue 7, November 2009, Pages 1555-1562, ISSN 0140-7007, DOI:10.1016/j.ijrefrig.2009.07.005.

[6] P. Kolasiński and P. Błasiak, Experimental and Numerical Analyses on the Rotary Vane Expander Operating Conditions in a Micro Organic Rankine Cycle System, Energies, Volume 9, 2016, ISSN 1996-1073, DOI: 10.3390/en9080606.

[7] G. Montenegro and A. Torre et al, Evaluating the Performance of a Rotary Vane Expander for Small Scale Organic Rankine Cycles Using CFD tools, Energy Procedia, Volume 45, 2014, Pages 1136

-1145, ISSN 1876-6102, DOI:10.1016/j.egypro.2014.01.119.
[8] Simerics PumpLinx, URL: www.simerics.com/pumplinx
[9] ANSYS CFX, URL: www.ansys.com/products/Fluids/ANSYS-CFX
[10] S. Rane, A. Kovacevic, N. Stosic, & M. Kethidi, (2013), Grid deformation strategies for CFD analysis of screw compressors, International Journal of Refrigeration, Volume 36, Issue 7, Pages 1883-1893, ISSN 0140-7007, DOI: 1016/j.ijrefrig.2013.04.008.
[11] S. Rane, Grid Generation and CFD analysis of Variable Geometry Screw Machines, 2015, PhD Thesis, City University London, URL: openaccess.city.ac.uk/id/eprint/15129.
[12] A. Kovacevic and N. Stosic and I. Smith, Screw Compressors - Three Dimensional Computational Fluid Dynamics and Solid Fluid Interaction, Springer, 2006, ISBN: 978-3-540-36302-6
[13] Vande Voorde, John, Jan Vierendeels, and Erik Dick. 2004. “A Grid Generator for Flow Calculations in Rotary Volumetric Compressors.” In Proceedings of the Fourth European Congress on Computational Methods in Applied Sciences and Engineering.
[14] Kovačević A., 2005. Boundary Adaptation in Grid Generation for CFD Analysis of Screw Compressors, Int. J. Numer. Methods Eng., Vol. 64: 401-426.
[15] SCORG, URL: http://pdmanalysis.co.uk/
[16] S. Rane, A. Kovacevic, and N. Stosic, (2016), CFD Analysis of Oil Flooded Twin Screw Compressors, International Compressor Engineering Conference – Purdue. Paper 2392.
[17] R. Cipollone and G. Bianchi and D. Di Battista, et al., Mechanical Energy Recovery from Low Grade Thermal Energy Sources, Proceedings of the 68th Conference of the Italian Thermal Machines Engineering Association, Energy Procedia, Volume 45, 2014, Pages 121-130, ISSN 1876-6102, DOI: 10.1016/j.egypro.2014.01.014
[18] G. Bianchi, Exhaust Waste Heat Recovery in Internal Combustion Engines - Development of an ORC-based power unit using Sliding Vane Rotary Machines, 2015, ISBN 978-88-87182-69-9, DOI: 10.13140/RG.2.1.4095.7282 (PhD thesis)

IOP Conf. Series: Materials Science and Engineering **232** (2017) 012054 doi:10.1088/1757-899X/232/1/012054

An intracooling system for a novel two-stage sliding-vane air compressor

Stefano Murgia[1,*], Gianluca Valenti[2], Ida Costanzo[1], Daniele Colletta[1], Giulio Contaldi[1]

[1] Ing. Enea Mattei S.p.a., Italy

[2] Politecnico di Milano, Dipartimento di Energia, Italy

*stefano_murgia@mattei.it

Abstract. Lube-oil injection is used in positive-displacement compressors and, among them, in sliding-vane machines to guarantee the correct lubrication of the moving parts and as sealing to prevent air leakage. Furthermore, lube-oil injection allows to exploit lubricant also as thermal ballast with a great thermal capacity to minimize the temperature increase during the compression. This study presents the design of a two-stage sliding-vane rotary compressor in which the air cooling is operated by high-pressure cold oil injection into a connection duct between the two stages. The heat exchange between the atomized oil jet and the air results in a decrease of the air temperature before the second stage, improving the overall system efficiency. This cooling system is named here *intracooling*, as opposed to intercooling. The oil injection is realized via pressure-swirl nozzles, both within the compressors and inside the intracooling duct. The design of the two-stage sliding-vane compressor is accomplished by way of a lumped parameter model. The model predicts an input power reduction as large as 10% for intercooled and intracooled two-stage compressors, the latter being slightly better, with respect to a conventional single-stage compressor for compressed air applications. An experimental campaign is conducted on a first prototype that comprises the low-pressure compressor and the intracooling duct, indicating that a significant temperature reduction is achieved in the duct.

Nomenclature

Symbols

β	pressure ratio	(-)
γ	cone spray angle	(rad)
μ	viscosity	(Pa s)
ρ	density	(kg m^{-3})
σ	surface tension	(N m^{-1})
P	power	(kW)
SMD	Sauter Mean Diameter	(m)
T	temperature	(K)
V	volume	(m^3)
d	diameter	(m)
p	pressure	(Pa)
q	size distribution parameter	(-)

Subscript

el	electrical
g	gas
id	ideal
in	inlet
l	liquid
or	orifice
out	outlet
tot	total

1. Introduction

Intercooling is an established practice in *compression technology* for reducing discharge temperature and power requirement [1,2]. As known, intercooling is the cooling of the compressed gas between two compression stages by way of a *heat exchanger* employing typically water or air as coolant. Overall, thermal power is transferred from the gas to the *environment*.

The present work proposes a similar yet not identical concept: *intracooling*, which is the cooling of the compressed gas between two compression stages by way of spraying a liquid coolant in the gas flow *without* separating that liquid prior to entering the next compression stage. The liquid shall be separated at the end of the last compression stage. Overall, thermal power is transferred from the gas to the liquid, but not to the environment. When applied to an *oil-flooded* compressor, such as a sliding-vane air compressor considered here, the lubricant itself can be used as sprayed liquid. In this case, intracooling with respect to intercooling allows for fewer pieces of equipment, as shown by Figure 1, and it allows for oil to be available at the highest pressure, a condition that may be exploited for a better injection. Figure 1 depicts the intracooled layout and compares it against an intercooled one: the low-pressure oil separator as well as the intercooling heat exchanger are substituted by the *intracooling duct*. For sake of clarity, the authors will refer to the low-pressure stage as LP and to the high-pressure stage as HP.

The first objective of this work is to predict the thermodynamic performance of an intracooled compressor, taking a single-stage compressor and an intercooled compressor as references. This analysis is executed with an in-house software developed for sliding-vane machines [3,4], but the outcomes are valid for the generic oil-flooded compressor. Moreover, the second objective is to prove the concept on a preliminary prototype that comprises only the LP compressor and the intracooling duct (while the HP compressor will be coupled in the next phase of the project).

To the knowledge of the authors, the concept of intracooling has never been formulated before and, hence, this work proposes a novel terminology as well as novel numerical and experimental analyses.

The following sections describe: the rationale of intracooling, its numerical modelling, the test case adopted for the investigation, the experimental setup for the first prototype, the numerical as well as experimental results, and lastly the conclusions.

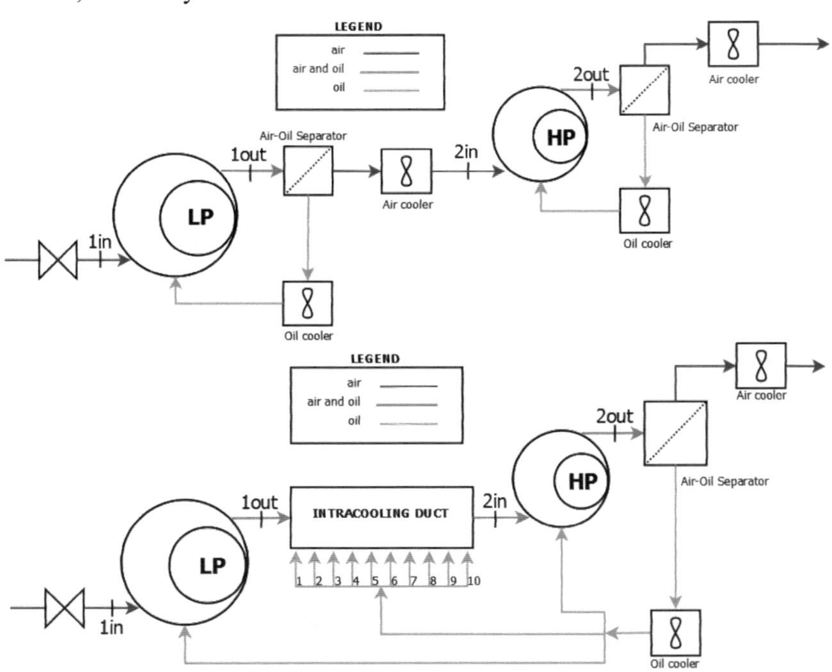

Figure 1. Application of intercooling (top) and intracooling (bottom) to oil-flooded (positive-displacement) compressors, such as sliding-vane air compressors as considered in this work (LP indicates low pressure, HP high pressure).

2. Rationale of intracooling

The well-known fundamental notion of both inter- and intracooling states that the lower the initial temperature of a gas compression, the lower the compression work; hence, an overall compression shall be divided into two (or even more) stages between which the gas is cooled by some mean in order to reduce the overall work requirement. In intercooling, the gas cooling is achieved by a heat exchanger; in intracooling, instead, the gas cooling is achieved by spraying cooled lubricant oil as small droplets into the gas flow within the intracooling duct. The sprayed oil is not separated from the gas, but it is let in the flow for lubricating the next compressor. Oil is separated from the gas only at the discharge of the second (last) compression stage.

The oil spraying in the compressors and in the intracooling duct can be accomplished by *pressure-swirl nozzles*. Hence, the oil sprays in intracooling may not cool the gas with the same effectiveness of the heat exchanger in intercooling. However, the availability of oil at the highest pressure leads to a fine and beneficial atomization in the LP compressor and in the intracooling duct.

3. Numerical modelling

Considering a two-stage compression, the *pressure ratio of the first stage*, β_{stage}^{id}, minimizing the overall compression work is:

$$\beta_{stage}^{id} = \sqrt{\beta_{tot}\left(\frac{T_{2in}}{T_{1in}}\right)^{\frac{\gamma}{\gamma-1}}} \tag{1}$$

where β_{tot} is the total compression ratio, T_{1in} [K] and T_{2in} [K] are the inlet temperatures of the LP and HP stages respectively and γ is the gas heat capacity ratio. If the air temperatures at the inlet of each stage are the same, the ideal compression ratio of a single stage is equal to the square root of the total compression ratio.

Once established the compression ratio of each stage, it is possible to proceed to the compressors sizing. The heat transfer between gas and liquid within the process affects directly the compressor performance and dimensioning. About this topic, a detailed simulation code has been developed by the authors, as reported in the previous works [3,4]. It is based on the following assumptions:

- the gas behaves as an ideal gas and the liquid as an incompressible fluid;
- the gas and the liquid are characterized by a constant specific heat;
- the gas is non-soluble into the liquid and the liquid is non-volatile into the gas;
- the compression is adiabatic toward the surroundings;
- the droplets are non-deformable spheres and the coalescence of droplets is neglected;
- the temperature variation inside the droplets is neglected.

The process is discretized so that gas and liquid temperatures and the gas pressure within the compression chamber can be calculated in every time step of the process. The chamber volume is calculated by geometrical means as a function of the angular position.

The injection system for the lubricating oil is based on both *plain orifices* and *pressure-swirl nozzles* to exploit the thermal effect of sprayed oil inside the compression chamber. The injection system is characterized in terms of number and position of the injectors as well as their jet characteristics.

The oil sprayed through pressure-swirl nozzles splits into a number of droplets of different dimensions affecting strongly the heat transfer from the air to the oil. The probability of a droplet size within a liquid jet, $f(d)$ [-], is described by the Rosin-Rammler relation as detailed by Mugele and Evans (1951):

$$f(d) = q\frac{d^{q-1}}{SMD^q}e^{-\left(\frac{d}{SMD}\right)^q} \tag{2}$$

where d [m] is the droplet diameter, SMD [m] is the Sauter Mean Diameter, defined as the diameter of a sphere that has the same volume-to-surface area ratio of the entire droplet sample, and q [-] is the shape parameter providing a measure of the spread of droplet sizes. In its turn, q is taken equal to 3.5 (Lefebvre, 1989) and SMD computed via the relation by Wang and Lefebvre (1987):

$$SMD = 4.52\left(\frac{\sigma_l\, \mu_l}{\rho_g\, \Delta p}\right)^{0.25}\left[2.7\left(\frac{d_{or}\, \dot{m}_l\, \mu_l}{\rho_l\, \Delta p}\right) cos\frac{\gamma}{2}\right]^{0.25} + 0.39\left(\frac{\rho_l\, \mu_l}{\rho_g\, \Delta p}\right)^{0.25}\left[2.7\left(\frac{d_{or}\dot{m}_l\, \mu_l}{\rho_l\, \Delta p}\right) cos\frac{\gamma}{2}\right]^{0.75} \quad (3)$$

where σ_l is the liquid surface tension [N m^{-1}], Δp the differential pressure [Pa], ρ the density [kg m^{-3}], d_{or} the nozzle/orifice diameter [m], γ the cone spray angle [rad], $\dot{m}_l$ and μ_l are the liquid mas flow rate [kg s^{-1}] and viscosity [Pa s]. The droplet diameter depends strictly on the nozzle differential pressure, which is calculated as the difference between the injection pressure and the pressure inside the compression chamber at the moment of injection.

The same model, properly adapted, is used to determine the heat transfer inside the intracooling duct. The main differences between the two systems are: the compression chamber is a closed system to mass transfer where the geometrical volume decrease causes an increase of the gas pressure; the intracooling duct is a system open to mass transfer where pressure is, as a first approach, assumed constant and where the gas volume decrease thanks to the cooling operated by the sprayed oil.

In case of inter-cooling, the pressure at the end of the first compression stage has to be opportunely increased because of the pressure drop on the cooling fan. In case of intra-cooling no pressure drop occurs into the connection duct, leading to a perfect matching between the two stages.

The second stage sizing is based on the compatibility between the volume of air-oil mix coming from the duct and the volume of the first chamber available in the second stage. Once the compressors geometries are fixed, the simulation allows to calculate the indicated power, as the area enclosed by the pressure-volume curve and the mechanical power of each compression stage. Finally, the overall input power, $P_{el,tot}$ [kW] is calculated as:

$$P_{el,tot} = P_{el,LP} + P_{el,HP} + P_{el,FAN} \quad (4)$$

where $P_{el,LP}$ [kW] and $P_{el,HP}$ [kW] are the low and high pressure input power respectively and $P_{el,FAN}$ [kW] is the electric power adsorbed by the fans, calculated specifically for each fan showed in Figure 1.

4. Case study

The case study is a *large-size sliding-vane air compressor* operating as LP stage. It is characterized by a hybrid injection system comprising plain orifices along the compressor on the stator and three couples of pressure-swirl nozzles on the end plates. The intracooling duct is directly connected to the compressor discharge and is equipped with ten pressure-swirl nozzles, equally spaced on the tube length (Figure 2). Downstream the intracooling duct a cyclonic air-oil separator is installed. An auxiliary pump is used to simulate the effect of the HP compressor for lubricant injection. Before being injected, the lubricant is water-cooled in a plate heat exchanger.

Figure 2. The first prototype comprising the intracooling duct (black) and the LP compressor (red). This prototype is taken as case study for the analysis.

IOP Conf. Series: Materials Science and Engineering **232** (2017) 012054 doi:10.1088/1757-899X/232/1/012054

5. Experimental testing

An experimental campaign is conducted to verify the intracooling effectiveness. The *prototype* is equipped with the instruments necessary to measure and control the process (Table 1): temperatures and pressures of air and oil in all key points of the circuit, the injected oil flow rate, and the electrical power.

The experimental data are acquired via the National Instruments cDAQ-9178 and processed via the NI Signal Express 2016. The experimental procedure comprises the simultaneous startup of the compressor and the auxiliary pump. The air delivery pressure is set to the chosen value acting on a valve. Once the system is in steady state conditions, the measurements are recorded.

The *reference configuration* for the experimental campaign is that with the intracooling deactivated. Consequently, the pressure swirl nozzles are activated progressively. Thirteen different configurations are tested, characterized by different number and positions of active nozzles, but at the same pressure and temperature injection conditions. The parameter used to evaluate the intracooling effectiveness is the temperature difference between the intracooling duct inlet and outlet.

Lastly, the thermal distribution on the intracooling duct is visualized by an infrared camera.

Table 1. Test rig instrumentation.

Instrument	Measured quantity	Absolute uncertainty	Unit
T-type thermocouple	Temperature	0.5	°C
Pressure transducer	Relative pressure	4000	Pa (gauge)
Oil flow meter	Volume flow rate	5	$l\ min^{-1}$
Power analyzer	Active power	0.2% reading	kW

6. Results and discussions

Three compression systems are investigated numerically: a conventional single-stage, an intercooled and an intracooled compressor. The overall pressure ratio as well as the air and liquid inlet temperatures are the same for all cases. Simulations of intercooled and intracooled compressors are executed, using the detailed simulation code based on the numerical model presented in Section 3. , where the compressors geometries are established in order to guarantee the same air mass flow rate and total amounts of injected oil in both the systems. This quantity is higher in two-stage compressors with respect to the single-stage because all compressors must be correctly lubricated. The main inputs and outputs are showed in Table 2.

As shown by the numerical results, the potential energy saving of the two-stage compression systems, compared to the single-stage, is up to 10%. The performance of the intercooled and the intracooled compressors are comparable, with a better result for the intracooled compressor (-2.3% of total input power). The compression power of the LP stage is slightly higher in the intercooled system. Despite the amount of oil injected into the compressor is the same, the lower injection pressure causes a coarser atomization that generates larger droplets dimension, as explained in Section 3. In this condition, the heat exchange between air and oil is penalized leading to an increase of air temperature and pressure as well as a greater consumed power.

The two systems differ also in terms of intermediate pressure between the two stages. On one hand, β_{stage}^{id} of the intercooled LP stage is lower than the intracooled one, because of the air temperature differences at the inlet of both the HP stages. On the other hand, the intercooled system is affected by the air pressure drop across the intercooler that influences the intermediate pressure between the two stages. This affects directly the intercooled HP stage that has a lower inlet pressure compared with the intracooled one, and consequently, a higher adsorbed power. Furthermore, despite the higher initial compression temperature in the HP intracooled compressor, the amount of liquid deriving from the intracooling duct allows to stabilize the thermal process since the intake port closing. In this system, the role of heat carrier fluid imputable to the lubricant is amplified and optimized to cool the air during the overall compression process. Lastly, the fan powers are similar between the two compression systems.

These numerical results indicate that oil atomization via pressure-swirl nozzle is an interesting technology from the input power perspective not only for the process occurring within a compressor, but also between two compressors.

Table 2. Numerical simulation results (NP stands for "not present").

	Unit	Conventional single-stage	Intercooled two-stage	Intracooled two-stage
	Inputs			
Air mass flow rate $\dot{m}_g$	kg s^{-1}	0.3	0.3	0.3
Motor speed	rpm	1000	1000	1000
LP Inlet gas temperature, T_{1in}	°C	40	40	40
HP Inlet gas temperature, T_{2in}	°C	NP	40	45
Liquid injection temperature	°C	40	40	40
Total compression ratio β_{tot}	-	8	8	8
	Outputs			
Total oil volume flow rate $\dot{m}_l$	l min^{-1}	112.8	188.6	187.2
LP stage compression power $P_{el,LP}$	kW	96.8	46.5	46.2
LP stage fan power $P_{el,LP\ FAN}$	kW	1.3	0.5	NP
HP stage compression power $P_{el,HP}$	kW	NP	40.9	39.0
HP stage fan power $P_{el,HP\ FAN}$	kW	NP	0.6	1.2
Total compression power $P_{el,LP+HP}$	kW	96.8	87.4	85.2
Total fan power $P_{el,LP+HP\ FAN}$	kW	1.3	1.1	1.2
Total input power $P_{el,tot}$	**kW**	**98.1**	**88.5**	**86.4**

Table 3 and Figure 3 report the results of the experimental campaign conducted on the first prototype of the intracooled compressor by varying the number and the position of the activated nozzles.

Table 3. Configurations of the intracooled duct differing by activated nozzles (RC stands for reference configuration, C1 to C12 to configuration from 1 to 12).

Cases	Active Nozzles 1	2	3	4	5	6	7	8	9	10	Number of total active nozzles
RC											0
C1	■										1
C2										■	1
C3					■						1
C4	■	■	■								3
C5								■	■	■	3
C6				■	■	■					3
C7	■				■				■		3
C8	■	■	■	■	■						5
C9						■	■	■	■	■	5
C10	■		■		■		■		■		5
C11	■		■	■	■	■	■		■		7
C12	■	■	■	■	■	■	■	■	■	■	10

Figure 3. Measured temperature difference between inlet and outlet of the intracooled duct depending upon the investigated configuration.

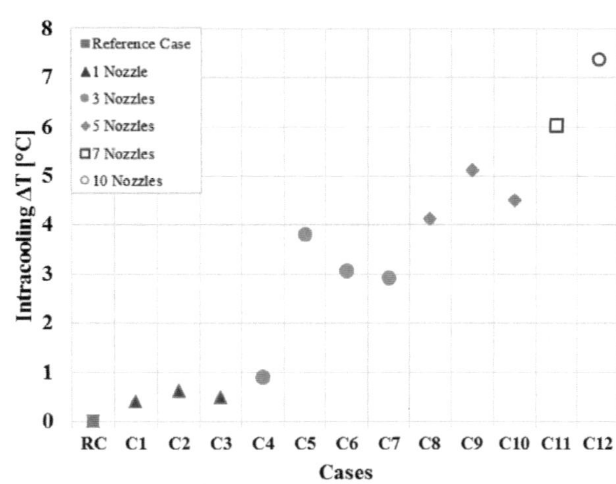

The temperature difference between the intracooling duct inlet and outlet is strongly influenced by the injected oil flow rate: the higher is the number of active nozzles, and thus the flow rate, the higher is the temperature decrease along the duct. In particular:

- Cases C1 thru C3 comprise only one activated nozzle: the intracooling effectiveness is low and the position of the injector has a minor influence;
- Cases C4 thru C7 comprise three activated nozzles: the major amount of liquid injected into the duct leads to a higher cooling effect;
- Cases C8 thru C12 comprise five, seven and ten activated nozzles: the intracooling effect is higher and higher as the number of nozzles increase.

The cases with three and five nozzles show a better effectiveness when the last nozzles are activated (C5 and C9). The magnitude of the phenomenon is greater for C5 and will be investigated in the future.

In brief, the best temperature reduction is 7.4°C for case C12. The infrared analysis confirms it: the temperature distribution for the reference case, RC, and the best case, C12, are shown in Figure 4.

Figure 4. Infrared images of the intracooling duct for the reference case, RC, on the left and the best case, C12, on the right.

7. Conclusions

This work presents the novel concept of intracooling for reducing the overall compression requirement in a similar, yet not identical, fashion to intercooling. The concept of intracooling consists in exploiting part of the amount of oil needed to the high-pressure compressor, not only to guarantee the lubrication effect and the internal sealing, but also for enhancing the air cooling between the two stages.

This work draws the following conclusions.

- The intracooled two-stage compressor is a novel system promising an energy saving up to 10% with respect to a conventional single-stage compressor for typical air compression applications.
- The performance of the intercooled and the intracooled compressors are numerically comparable, with a better result for the intracooled system (-2.3% of total input power).
- The low-pressure stage of the intracooled compressor shows a better compression process compared to the intercooled one, due to the higher injection pressure that results in a finer injected oil atomization; in terms of absorbed power, the first stage of both inter- and intracooled compressors are slightly similar.
- The high-pressure stage has a lower adsorbed power (39.0 kW) compared with the intercooled compressor (40.9 kW); it presents a higher inlet temperature and pressure compared with the intercooled one; the liquid coming from the intracooling duct allows to stabilize the thermal process since the intake port closing.
- The measured intracooling effectiveness depends strongly on the number of active nozzles and, consequently, on the oil flow rate as well as on their position.
- The highest intracooling effect, resulting in a measured temperature reduction of 7.4 °C, can be achieved by activating all ten nozzles.

The future work will focus on the design of the high pressure stage and a control system that governs the nozzles activation in order to maintain constant the temperature at the inlet of the second stage.

IOP Conf. Series: Materials Science and Engineering **232** (2017) 012054 doi:10.1088/1757-899X/232/1/012054

References

[1] Brown R N 1997 *Compressors – Selection and sizing* (Houston: Gulf Professional Publishing)

[2] Bloch H P 2006 *A practical guide to compressor technology - 2nd Edition*. (Hoboken: John Wiley & Sons).

[3] Valenti G, Colombo L, Murgia S, Lucchini A, Sampietro A, Capoferri A and Araneo L 2013 Thermal effect of lubricating oil in positive-displacement air compressors *Applied Thermal Engineering* **51**(1-2). 1055-1066

[4] Valenti G, Murgia S, Costanzo I, Contaldi G and Valenti A 2016 Modeling And Testing The Thermal Effect Of Lubricating Oil Sprayed In Sliding-Vane Air Compressors Using Pressure-Swirl Nozzles, 23rd International Compressor Engineering Conference at Purdue University, 2016, West-Lafayette (IN), United States

[5] Cipollone R, Bianchi G, Di Battista D, Contaldi G, Murgia S. Model based design of an intercooled dual stage sliding vane rotary compressor system, 9th International Conference on Compressors and their Systems at University City of London, 2015, London, United Kingdom

IOP Conf. Series: Materials Science and Engineering **232** (2017) 012055 doi:10.1088/1757-899X/232/1/012055

Thermal modelling of a dry revolving vane compressor

K T Ooi [1,3] and **K T Aw**[2,4]

[1,2] School of Mechanical and Aerospace Engineering, Nanyang Technological University, 50 Nanyang Avenue, Singapore 639798

[3] mktooi@ntu.edu.sg
[4] ktaw1@e.ntu.edu.sg

Abstract. The lubricant used in compressors serves to lubricate, to seal the gaps to reduce internal leakage and to a certain extent, to cool. However, a lubricant free compressor is attractive if lubricants become a source of contaminant, or in areas where the compressor needs be placed under any orientation, such as those in military or portable computing.

In this paper, a thermal model for a dry revolving vane compressor is presented. This thermal model sets out to predict the steady-state operating temperatures of the compressor components. The lumped thermal conductance method was employed. The results of the components temperature will be presented and discussed. A high potential for overheating is observed at the shaft bearings.

1. Introduction

In a lubricant free compressor, dry sliding friction between the moving components would generate excessive heat in the compressor which may compromise the reliability of the compressor.

The lumped thermal conductance method was employed for thermal modelling of various compressors [1–3] including the rolling piston compressor [4]. As the RV compressor shares many similarities with that of the rolling piston, this same method would be used for modelling the heat transfer between the components in the compressor prototype.

2. Assumptions and sub-division of components

The six major components in the compressor are subdivided into 14 elements with simple geometric shapes, as shown in Figure 1. The rotor elements r2, r3 and r4 are concentric.

The following assumptions are made for the thermal model:

- Steady-state condition with isothermal elements
- Since the eccentricity is small (7.5 mm) as compared to the housing diameter (160.0 mm) all cylindrical components/holes are assumed to be concentric.
- Thermal contact resistance between steel components are negligible
- Radiation heat transfer among compressor components is negligible

Existing convection correlations at the solid-fluid interfaces [5–10] are used. Thermal contact and dry sliding friction models from literature [11, 12] will be employed for modelling heat generation and transfer at rubbing interfaces.

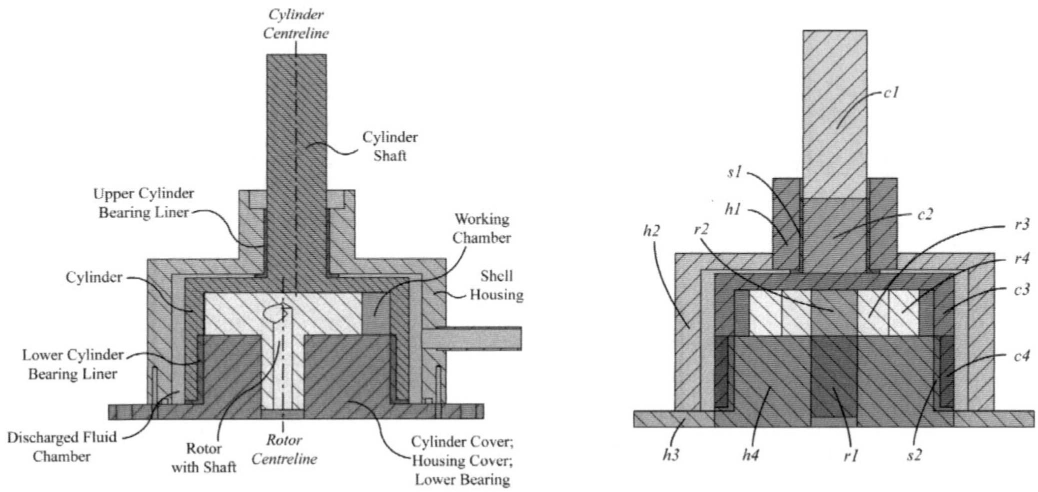

(a): Prototype Cross-section **(b): Cross-section with Simplified Geometry**

Figure 1: Subdivision of RV Components

3. Heat transfer model

3.1 Heat Transfer Equations

Applying the steady state First Law of Thermodynamics [13] to the solid element and ignoring thermal expansion, yields heat transfer equation (1).

$$\sum_i h_i A_i \Delta T_i + \dot{Q}_{gen} = \sum_o h_o A_o \Delta T_o \tag{1}$$

The one-dimensional heat transfer equations [5] are written as Equation (2) for conduction, convection and radiation respectively.

$$\dot{Q} = hA\Delta T \tag{2}$$

where, for each mode of heat transfer, h is shown in equation (3) to (5).

$$h_{cond} = \frac{k}{\Delta l} \text{ or } \frac{k}{r_1 \ln\left({}^{r_2}/_{r_1}\right)} \tag{3}$$

$$h_{conv} = \frac{k}{l}\mathrm{Nu} \tag{4}$$

$$h_{radn} = \varepsilon\sigma F_{i\to j}\left(T_i + T_j\right)\left(T_i^2 + T_j^2\right) \tag{5}$$

3.2 Thermal contact model for dry sliding interfaces

The total dry frictional heat generated between two surfaces, 1 and 2 can be obtained by equations (6) to (8) [11, 14]:

$$\dot{Q}_f = F_f v_{sl} = F_f \omega_{shaft} r_{shaft} \tag{6}$$

$$\xi = \frac{R_{sl,2}}{R_{sl,1} + R_{sl,2}} \tag{7}$$

$$\left.\begin{aligned} \dot{Q}_{f,1} &= \xi \dot{Q}_f \\ \dot{Q}_{f,2} &= (1-\xi)\dot{Q}_f \\ \dot{Q}_f &= \dot{Q}_{f,1} + \dot{Q}_{f,2} \end{aligned}\right\} \tag{8}$$

The correlation for the dimensionless sliding thermal contact resistance for two rough surfaces in dry sliding contact [14], based on the material characteristics, sliding velocity, and profile of the surface

IOP Conf. Series: Materials Science and Engineering **232** (2017) 012055 doi:10.1088/1757-899X/232/1/012055

asperities, is shown in Equation (9), while Equation (10) is the corresponding thermal contact resistance value.

The asperity profile of a typical surface from the turning process [14] shown in Figure 2, and can be modelled by assuming uniform asperities with each asperity having the same height and width, separated at a constant distance as illustrated in Figure 3.

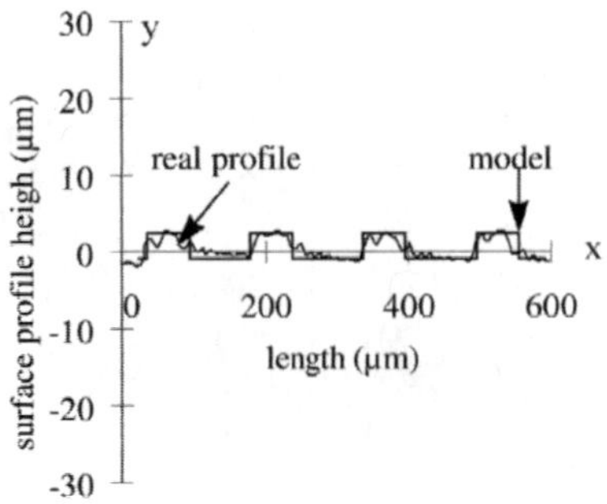

Figure 2: Aspertiy Profile of a Turned Surface [11]

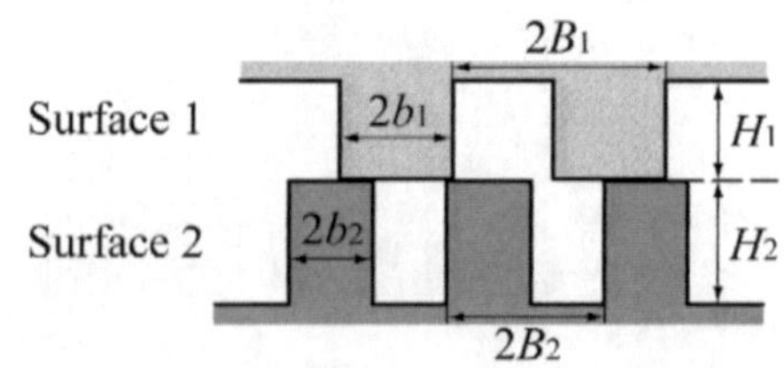

Figure 3: Assumed Asperity Profile at Interface

$$R^*_{sl,i} = 0.93\left(\frac{1}{b^*_i} - 1\right) \cdot H^*_i + 0.4\left(\frac{1}{v^*_i}\right)^{0.24} \cdot \left(\frac{1}{b^*_i} - 1\right)^{1.45}, \quad i = 1,2 \tag{9}$$

$$R_{sl,i} = \frac{2B_i R^*_{sl,i}}{k_i} \tag{10}$$

where

$$v^*_i = \frac{2v_{sl}B_i\rho_i c_{p_i}}{k_i} \tag{11}$$

$$H^*_i = \frac{H_i}{2B_i} \tag{12}$$

$$b^*_i = \frac{b_i}{B_i} \tag{13}$$

Polyetheretherketone (PEEK) in its pure form and bearing grade PEEK are used. A comparison between the sliding thermal contact resistances of the PEEK materials and AISI 4140 steel is shown in Figure 4.

The surface properties are assumed as shown in Table 1 and the material properties are listed in Table 2.

Table 1: Asperity Properties

Property	**Value, µm**
Height, *H*	0.6
Width, 2*b*	0.5
Periodic Separation, 2*B*	1.0

Table 2: Material Properties

	Pure PEEK	**Bearing Grade PEEK**	**AISI 4140 Steel [15]**
Thermal Conductivity, W m^{-1} K^{-1}	0.27	0.82	42.7
Density, kg m^{-3}	1310	1440	7700
Specific Heat Capacity, J kg^{-1} K^{-1}	1100	1100	473

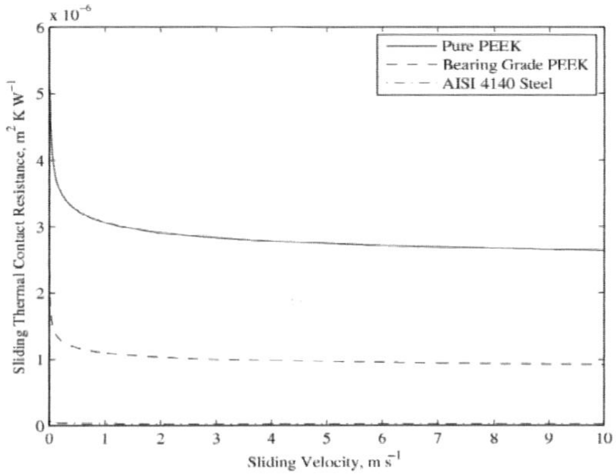

Figure 4: Comparison of Thermal Contact Resistance Between PEEK Materials & Steel

It is noted that realistic deviations from the assumed asperity dimensions would not cause the sliding thermal contact resistance of steel to have drastic changes.

From Figure 5 the heat generation coefficient for steel can be approximated to 0.99 and 0.975 when rubbing with pure PEEK and bearing grade PEEK, respectively. Hence PEEK appears to absorb negligible heat as compared to steel. Therefore, it is assumed that all friction heat is absorbed by steel. In addition, friction heat at the bearings is assumed to be split evenly between the steel shaft and bearing surfaces due to free rotation of the PEEK bearing liner.

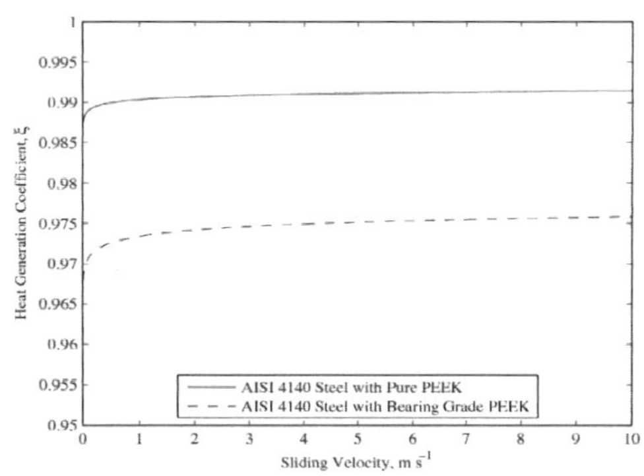

Figure 5: Heat Generation Coefficient for AISI 4140 Steel.

The steady-state temperatures of the PEEK bearing liners would be the average of the steel shaft and bearing temperatures.

3.3 Other assumptions and heat transfer correlations

The relevant heat transfer correlations for horizontal surfaces, vertical surfaces and cylindrical surfaces of the elements are presented:

i. Free convection heat transfer correlations for housing shell and base cover elements – the Nusselt number correlation for free convection [16] on the upper surface of the housing shell and lower surface of the base cover is given by Equations (14) and (15), respectively. The outer vertical surface can be treated as that of a vertical plate in which its Nusselt number [17] is given by Equation (16).

$$\mathrm{Nu} = \begin{cases} 0.54\mathrm{Ra}_{l_{ch}}^{1/4}, & 10^4 \le \mathrm{Ra}_{l_{ch}} < 10^7 \\ 0.15\mathrm{Ra}_{l_{ch}}^{1/3}, & 10^7 \le \mathrm{Ra}_{l_{ch}} < 10^{11} \end{cases} \tag{14}$$

$$\mathrm{Nu} = 0.27\mathrm{Ra}_{l_{ch}}^{1/4}, \quad 10^5 \le \mathrm{Ra}_{l_{ch}} \le 10^{10} \tag{15}$$

$$\mathrm{Nu} = \begin{cases} 0.68 + \dfrac{0.67\mathrm{Ra}_{l_{ch}}^{1/4}}{\left[1+\left(\dfrac{0.492}{\mathrm{Pr}}\right)^{9/16}\right]^{4/9}}, & 0 < \mathrm{Ra}_{l_{ch}} \leq 10^9 \\ \left[0.825 + \dfrac{0.387\mathrm{Ra}_{l_{ch}}^{1/6}}{\left(1+\left[\dfrac{0.492}{\mathrm{Pr}}\right]^{9/16}\right)^{8/27}}\right]^2, & \mathrm{Ra}_{l_{ch}} > 10^9 \end{cases} \tag{16}$$

ii. The in-chamber convective heat transfer shown in Equation (17) by Liu and Zhou [7], was tested by Tan and Ooi [6] to be the most accurate for the RV compressor and is thus chosen.

$$\mathrm{Nu} = 0.75\overline{\mathrm{Re}}^{0.8}\overline{\mathrm{Pr}}^{0.6} \tag{17}$$

iii. Heat transfer correlations for Taylor-Couette flow are adopted for modelling the convective heat transfer in the discharge chamber between the cylinder and housing walls as shown in Equation (18) [8]. The flow in the annular gap at the endfaces is turbulent and therefore, the turbulent convective heat transfer of a flow over a flat plate can be expressed as shown in Equation (19) [5]. The speed of the flow is assumed to be the mean of the cylinder and housing wall speeds and the length would be circumferential length of the gap.

$$\mathrm{Nu} = 0.409\left(\frac{\mathrm{Ta}}{C_{geo}}\right)^{0.241}, \quad 1700 \leq \left(\frac{\mathrm{Ta}}{C_{geo}}\right) \leq 10^7 \tag{18}$$

$$\mathrm{Nu}_{gap} = 0.037\mathrm{Re}_{gap}^{4/5}\mathrm{Pr}^{1/3} \tag{19}$$

where

$$\mathrm{Ta} = \frac{\omega_{crit}^2\rho^2 r_{avg} r_{gap}^3}{\mu^2} \tag{20}$$

$$\omega_{crit} = \frac{\pi^4\mu^2 r_{avg}}{s r_{gap}^3 r_i^2} \tag{21}$$

$$C_{geo} = \frac{1}{s}\left(\frac{\pi^4}{1697}\right)\left(1-\frac{r_{gap}}{2r_{avg}}\right)^{-2} \tag{22}$$

$$s = 0.0572\left(1-\frac{0.652r_{gap}}{r_i}\right) + 0.00056\left(1-\frac{0.652r_{gap}}{r_o}\right)^{-1} \tag{23}$$

$$r_{gap} = r_o - r_i \tag{24}$$

$$r_{avg} = \frac{r_o + r_i}{2} \tag{25}$$

$$\mathrm{Re}_{gap} = \frac{\rho\omega_{avg} r_{avg} l_{gap}}{\mu} \tag{26}$$

$$l_{gap} = \pi(r_o + r_i) \tag{27}$$

iv. Convective heat transfer on rotating cylinder shaft – for a high-speed rotating cylinder in quiescent air, it can be assumed that the convective heat transfer coefficient depends on only the Reynolds number [6, 7]. The correlation by Becker [9] can be used and the average Nusselt number is expressed as shown in Equation (28):

$$\mathrm{Nu} = 0.119\mathrm{Re}_r^{2/3}, \quad 800 \leq \mathrm{Re}_r = \frac{2\rho\omega_{shaft} r_{shaft}^2}{\mu} \leq 10^5 \tag{28}$$

v. Radiation heat transfer to surroundings – the emissivity for polished metal surfaces ranges between 0.10 and 0.40 depending on the grade of finishing. For the polished steel surface of the housing shell, its emissivity shall be estimated to be the average of the given range (ε = 0.25). Assuming a small object radiating in an enclosure would give a view factor at unity [5], therefore, the radiation heat transfer can be expressed as shown in Equation (29) with ε = 0.25.

$$h_{radn} = 0.25\sigma(T_w + T_\infty)(T_w^2 + T_\infty^2) \tag{29}$$

3.4 *Heat Transfer in the Rotor*

The rotor component is made from PEEK material which has very high sliding thermal contact resistance as compared to steel. Hence, the friction heat that is absorbed by the rotor has to be dissipated in the suction line or working chambers.

3.4.1 *Convective heat transfer at rotor shaft tip*

One end of the rotor shaft is exposed to ambient and it can be considered as a rotating disc:

$$\mathrm{Re}_{disc} = \frac{\rho \omega_{disc} r_{disc}^2}{\mu} \tag{30}$$

As the diameter of the shaft is small (Ø23.0 mm), the rotational Reynolds number at 3000 rev min^{-1} is only 2600. Hence, the flow regime at the shaft tip is laminar. As shown by Cobb and Saunders [18] the heat transfer is best modelled by the correlation given by Wagner [19]:

$$\mathrm{Nu} = 0.335\mathrm{Re}_{disc}^{0.5}, \qquad \mathrm{Re}_{disc} < 2.4 \times 10^5 \tag{31}$$

3.4.2 *Mass flow rate in rotor*

The Reynolds number for flow of the fluid inside the rotor can thus be calculated by Equation (32) and varies up to a value of 16,000 for a rotation speed of 3000 rev min^{-1} with volumetric efficiency at unity.

$$\mathrm{Re}_{suc} = \frac{2\rho v_{suc} r_{pt,suc}}{\mu} = \frac{2\dot{m}_d}{\pi r_{pt,suc}\mu} \tag{32}$$

3.4.3 *Convective heat transfer in shaft*

The intake of fluid through the rotor first goes through the rotating shaft and is classified as an axial flow through a rotating duct. The rotation of the shaft affects the flow of the fluid; the rotating walls help reduce turbulence in the flow at the vicinity of the wall and thus heat transfer decreases with higher rotation speeds [20–22]. Seghir-Ouali et al. [22] provide the heat transfer correlations for axial flow inside a rotating cylinder duct with a constant heat flux in Equations (33) and (34).

$$\begin{gathered}\mathrm{Nu} = 0.01963\mathrm{Re}_{suc}^{0.9285} + 8.5101 \times 10^{-6}\mathrm{Re}_{\omega}^{1.4513}, \\ \text{for } 0 < \mathrm{Re}_{suc} < 3 \times 10^4, 1600 < \mathrm{Re}_{\omega} \le 2.77 \times 10^5\end{gathered} \tag{33}$$

$$\mathrm{Nu} = 2.85 \times 10^{-4}\mathrm{Re}_{\omega}^{1.19}, \qquad \mathrm{Re}_{\omega} > 2.77 \times 10^5 \tag{34}$$

where Re_ω is the rotational Reynolds number given as follows:

$$\mathrm{Re}_{\omega} = \frac{2r_{pt}^2 \omega_{shaft} \rho}{\mu} \tag{35}$$

3.4.4 *Convective heat transfer in rotor*

It is an internal laminar flow in the suction line inside the rotor with uniform wall temperature. The heat transfer coefficient is thus given in Equation (36).

$$\mathrm{Nu} = 3.66, \ \mathrm{Re}_{suc} < 10^4 \tag{36}$$

3.5 Linear Algebra Equations for Component Elements

With 12 elements, there will be 12 simultaneous equation, the equations can be arranged into a matrix and solved as a linear algebra equation to obtain the temperatures.

As an example, for the cylinder element $c1$ shown in Figure 1(b), it is affected by free convection and heat transfer from element $c2$. Hence, the heat transfer to and from the element can be written as shown in Equation (37) which is rearranged into Equation (38).

$$H_{c1,c2}(T_{c1} - T_{c2}) + H_{c1,\infty}(T_{c1} - T_{\infty}) = 0 \tag{37}$$

$$\left(H_{c1,c2} + H_{c1\infty}\right)T_{c1} + H_{c1,c2}T_{c2} = H_{c1,\infty}T_{\infty} \tag{38}$$

where,

$$H_{c1,c2} = H_{c2,c1} = \frac{k_{steel}A_{c1,c2}}{l_{c1,c2}} = \frac{2k\pi r_{shaft}^2}{l_{c1} + l_{c2}} \tag{39}$$

$$\begin{aligned} H_{c1,\infty} &= \frac{k_{air}A_{c1,\infty}}{l_{c1,\infty}}\left(0.119\mathrm{Re}_r^{2/3}\right) + 0.25A_{c1,\infty}\sigma(T_{c1} + T_{\infty})(T_{c1}^2 + T_{\infty}^2) \\ &= \frac{2k_{air}\pi r_{shaft}l_{c1}}{l_{c1}}\left[0.119\left(\frac{2\rho\omega_{shaft}r_{shaft}^2}{\mu}\right)^{2/3}\right] \\ &\quad + 0.25\sigma(T_{c1} + T_{\infty})(T_{c1}^2 + T_{\infty}^2)\left(2\pi r_{shaft}l_{c1}\right) \end{aligned} \tag{40}$$

Due to the insulating PEEK bearing liners separating the cylinder from the housing shell, there is no coupling of temperatures between these two components. Furthermore, the rotor is also considered to be a separate system due to the insulating property of PEEK for heat conduction. Hence, each of the components will have their own separate matrices as shown in Equations (41) – (43). $\boldsymbol{T}$ represents the component element temperatures, $\boldsymbol{X}$ represents the heat transfer relations and $\boldsymbol{Y}$ represents the loss to surroundings and heat source terms.

$$\boldsymbol{X}_c\boldsymbol{T}_c = \boldsymbol{Y}_c \tag{41}$$

$$\boldsymbol{X}_r\boldsymbol{T}_r = \boldsymbol{Y}_r \tag{42}$$

$$\boldsymbol{X}_h\boldsymbol{T}_h = \boldsymbol{Y}_h \tag{43}$$

Equations (41) to (43) are then solved for their respective $\boldsymbol{T}$ matrices which contains all the elements' temperature values for each component using Gaussian Elimination. The solution would give the steady-state operating temperatures of the components for the RV compressor prototype.

3.6 Preliminary Analysis

The operating conditions are shown in Table 3. It is assumed that there is no internal leakage occurring between the working chambers.

Table 3: Operating Conditions and Ambient Temperature

Operating Speed, rev min^{-1}	1,000 / 2,000
Working Fluid	Air
Suction Pressure, bar (abs)	1.0
Suction Temperature, °C	27.0
Discharge Pressure, bar (abs)	2.0 / 5.0
Coefficient of Friction	0.4
Ambient Temperature, °C	27.0

The temperatures of the components at a discharge pressure of 2 bar (abs) are presented in Figure 6. The temperature of the upper PEEK bearing liner sleeve $s1$ can be taken as the mean temperature of elements $c2$ and $h1$ which is 58.5°C. Similarly, the temperature of the lower PEEK bearing liner sleeve $s2$ is taken to be the mean temperature of elements *c4* and *h4* which is 72.4°C. For 5 bar (abs) discharge pressure in Figure 7, the respective temperatures of the upper and lower bearing sleeves are 120.5°C and 149.5°C, respectively.

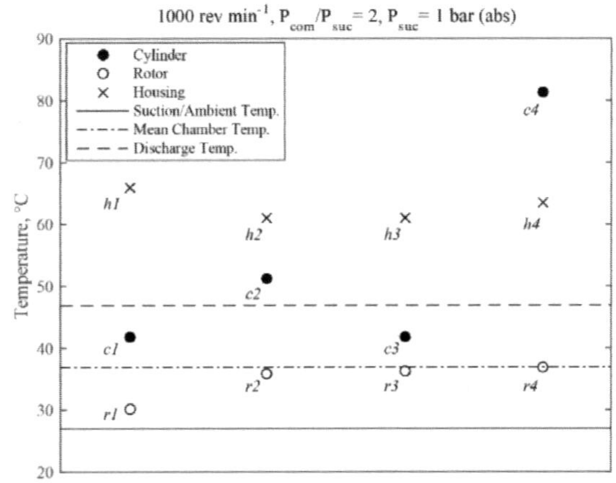

Figure 6: Predicted Steady-state Component Temperatures for 1 bar Pressure Difference

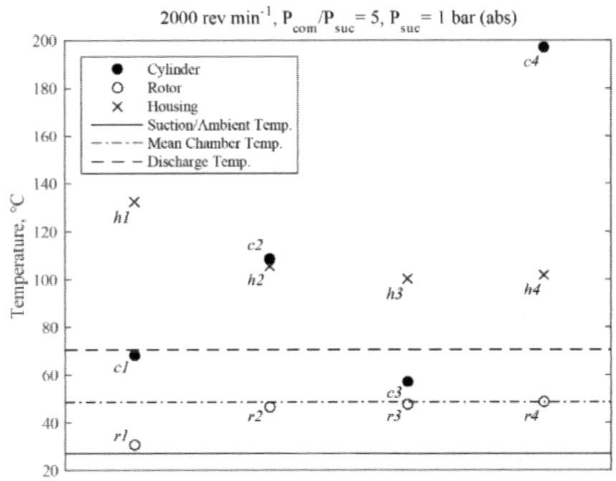

Figure 7: Predicted Steady-state Component Temperatures for 4 bar Pressure Difference

From the analysis, it is noted that the steady-state operating temperature of the rotor would be approximately the same as the average chamber temperature as the rotor has poor thermal conductivity. The rotor shaft (*r1*) would be at a lower temperature due to enhanced heat dissipation when the shaft rotates.

For the cylinder, there is a wide variation between the different elements. Friction heating at the shaft elements (*c2*, *c4*) result in higher temperatures while the rotation of the cylinder enhances heat dissipation in elements *c1* and *c3*. The very high temperature at shaft element *c4* is due to its large diameter in which there is more friction heat.

The housing shell is cooled by free convection on the external surfaces and by the discharge fluid on the inner surfaces. Thus, its temperature variation is not as significant.

The current analysis was conducted under ideal conditions with no internal leakage. Under practical operating conditions, the circulation of leakage fluid between the working chambers would no doubt cause the average chamber temperature and discharge temperature to rise, thus increasing the temperatures of the compressor. The trend would be akin to the increase in the element temperatures between Figure 6 and 7 in which Figure 7 shows a higher chamber temperature.

4. Conclusion

The thermal modelling of an oil-free compressor is presented using the lumped thermal conductance method and a preliminary analysis of the operating temperature was conducted, albeit assuming ideal conditions with no internal leakage. A summary of the analysis is as follows:

- The steady-state operating temperature of the rotor is close to that of the mean chamber temperature. On the other hand, the walls of the suction line in the rotor shaft are rotating, resulting in enhanced heat dissipation and thus, lower temperature values for the rotor shaft.
- The lower cylinder bearing has the highest temperature due to high friction loss caused by a large bearing diameter. On the other end, the continuous flow of working fluid into and out of the cylinder chamber helps to cool the overall component.
- Internal leakage under practical operating conditions would cause the chamber temperature to increase which consequentially increases the temperature of the components.
- Excessive friction heating at the shafts can be mitigated by redesigning the shafts to have smaller diameters and the use of ball/roller bearings instead of plain journal bearings

Nomenclature

A	Area (m^2)
F	Force (N)
$F_{i \to j}$	View factor (–)
h	Heat transfer coefficient (W m^{-2} K^{-1})
k	Thermal conductivity (W m^{-1} K^{-1})
l	Length (m)
Nu	Nusselt number (–)
Pr	Prandtl number (–)
Q	Heat (J)
R	Thermal resistance (m^2 K W^{-1})
Ra	Rayleigh number (–)
Re	Reynolds number (–)
r	Radius; radial distance (m)
T	Temperature (K)
v	Velocity (m s^{-1})
ε	Emissivity (–)
σ	Stefan-boltzmann constant (W m^{-2} K^{-4})
ω	Angular velocity (rad s^{-1})

References

1. Todescat, M. L., Fagotti, F., Prata, A. T., & Ferreira, R. T. S. (1992). Thermal Energy Analysis in Reciprocating Hermetic Compressors. In International Compressor Engineering Conference.

2. Weber, G. C., Lajus, F., Ribas, F. A., & Fonseca, J. N. (2010). A Simplified Thermal Model for a CO2 Compressor. In International Compressor Engineering Conference.

3. Deschamps, M. J. O. and M. C. D. and C. J. (2015). Thermal modelling and analysis of an oil-free linear compressor. IOP Conference Series: Materials Science and Engineering, 90(1), 12016.

4. Ooi, K. T. (2003). Heat transfer study of a hermetic refrigeration compressor. *Applied Thermal Engineering*, *23*(15), 1931–1945.

5. Incropera, F. P., Bergman, T. L., DeWitt, D. P., Lavine, A. S., (2011). *Fundamentals of Heat and Mass Transfer* (6th ed.). Wiley.

6. Tan, K. M., & Ooi, K. T. (2011). Heat transfer in compression chamber of a revolving vane (RV) compressor. *Applied Thermal Engineering*, *31*(8–9), 1519–1526.

7. Liu, R., & Zhou, Z. (1984). Heat Transfer Between Gas and Cylinder Wall of Refrigerating Reciprocating Compressor. In International Compressor Engineering Conference.

8. Gardiner, S. R. ., & Sabersky, R. . (1978). Heat transfer in an annular gap. *International Journal of Heat and Mass Transfer*, *21*(12), 1459–1466.

9. Becker, K. M. (1963). Measurements of convective heat transfer from a horizontal cylinder rotating in a tank of water. *International Journal of Heat and Mass Transfer*, *6*(12), 1053–1062.

10. Özerdem, B. (2000). Measurement of convective heat transfer coefficient for a horizontal cylinder rotating in quiescent air. *International Communications in Heat and Mass Transfer*, *27*(3), 389–395.

11. Bardon, J. P. (1994). Bases physiques des conditions de contact thermique imparfait entre milieux en glissement relatif. *Revue générale de thermique*, *33*(386).

12. Chantrenne, P., & Raynaud, M. (1997). A microscopic thermal model for dry sliding contact. *Int. J. Heat Mass Tran.*, *40*(5), 1083–1094.

13. Cengel, Y. A., & Boles, M. A. (2006). *Thermodynamics: An Engineering Approach* (5th ed.). McGraw Hill.

14. Chantrenne, P., & Raynaud, M. (2001). Study of a macroscopic sliding contact thermal model from microscopic models. *International Journal of Thermal Sciences*, *40*(7), 603–621.

15. EFunda. (n.d.). eFunda: Properties of Alloy Steel AISI 4140. Retrieved July 23, 2014, from http://www.efunda.com/Materials/alloys/alloy_steels/show_alloy.cfm?ID=AISI_4140&prop=all&Page_Title=AISI 4140

16. McAdams, W. H. (1954). *Heat Transmission* (3rd Ed.). New York: McGraw-Hill.

17. Churchill, S. W., & Chu, H. H. S. (1975). Correlating equations for laminar and turbulent free convection from a vertical plate. *International Journal of Heat and Mass Transfer*, *18*(11), 1323–1329.

18. Cobb, E. C., & Saunders, O. A. (1956). Heat Transfer from a Rotating Disk. *Proceedings of the Royal Society of London. Series A, Mathematical and Physical Sciences*, *236*(1206), 343–351.

19. Wagner, C. (1948). Heat Transfer from a Rotating Disk to Ambient Air. *Journal of Applied Physics*, *19*(9).

20. Reich, G., Weigand, B., & Beer, H. (1989). Fluid flow and heat transfer in an axially rotating pipe -I. Effect of rotation on turbulent pipe flow. *International journal for Heat and Mass transfer*, *32*(9), 551–562.

21. Reich, G., Weigand, B., & Beer, H. (1989). Fluid flow and heat transfer in an axially rotating pipe -II. Effect of rotation on laminar pipe flow. *International journal for Heat and Mass transfer*, *32*(9), 563–574.

22. Seghir-Ouali, S., Saury, D., Harmand, S., Phillipart, O., & Laloy, D. (2006). Convective heat transfer inside a rotating cylinder with an axial air flow. *International Journal of Thermal Sciences*, *45*(12), 1166–1178.

IOP Conf. Series: Materials Science and Engineering **232** (2017) 012056 doi:10.1088/1757-899X/232/1/012056

Improved design method of a rotating spool compressor using a comprehensive model and comparison to experimental results

Craig R Bradshaw[1], Greg Kemp[2], Joe Orosz[2] and Eckhard A Groll[3]

[1] Building and Environmental Thermal Systems Research Group, Mechanical and Aerospace Engineering at Oklahoma State University, 218 Engineering North, Stillwater, Oklahoma 74074, USA

[2] Torad Engineering LLC, Cumming, GA 30040, USA

[3] Ray W. Herrick Research Laboratories, Mechanical Engineering at Purdue University, 177 S. Russell St. West Lafayette, IN 47907, USA

E-mail: craig.bradshaw@okstate.edu

Abstract.
An improvement to the design process of the rotating spool compressor is presented. This improvement utilizes a comprehensive model to explore two working fluids (R410A and R134a), various displaced volumes, at a variety of geometric parameters. The geometric parameters explored consists of eccentricity ratio and length-to-diameter ratio. The eccentricity ratio is varied between 0.81 and 0.92 and the length-to-diameter ratio is varied between 0.4 and 3. The key tradeoffs are evaluated and the results show that there is an optimum eccentricity and length-to-diameter ratio, which will maximize the model predicted performance, that is unique to a particular fluid and displaced volume. For R410A, the modeling tool predicts that the overall isentropic efficiency will optimize at a length-to-diameter ratio that is lower than for R134a. Additionally, the tool predicts that as the displaced volume increases the overall isentropic efficiency will increase and the ideal length-to-diameter ratio will shift. The result from this study are utilized to develop a basic design for a 141 kW (40 tonsR) capacity prototype spool compressor for light-commercial air-conditioning applications. Results from a prototype compressor constructed based on these efforts is presented. The volumetric efficiency predictions are found to be very accurate with the overall isentropic efficiency predictions shown to be slightly over-predicted.

1 Introduction

The rotating spool compressor is a novel rotary compressor mechanism most similar to the sliding vane compressor. Primary differences are described by Kemp *et al.* [5] and include three key differences from a sliding vane compressor, as shown in Figure 1.

- The vane is constrained by means of an eccentric cam allowing its distal end to be held in very close proximity to the housing bore while never contacting the bore.
- The rotor has affixed endplates that rotate with the central hub and vane forming a rotating spool.
- The use of dynamic sealing elements to minimize leakage between the suction and compression pockets as well as between the process pockets and the compressor containment.

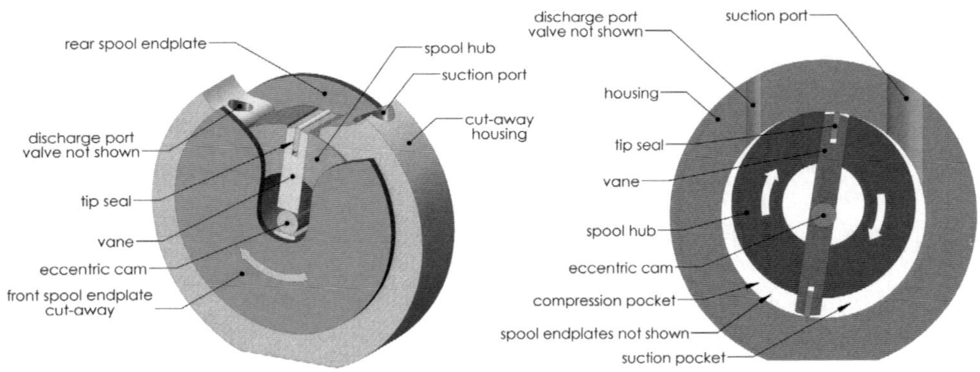

Figure 1. Cutaway view of rotating spool compressor mechanism with key components highlighted.

The movement of the rotor is purely rotary with only the vane and tip seals performing any oscillating movement. The eccentric cam will force the movement of the vane to oscillate by twice the eccentricity during a single rotation. The tip seals will oscillate relative to the vane two times per rotation by an amount proportional to the ratio of diameters of the rotor to the housing bore. The tip seal movement amount is roughly an order of magnitude smaller than the eccentricity and follows a sinusoidal path.

1.1 Previous work on the spool compressor

Since its introduction in 2008 the spool compressor has been improved and better understood. A comprehensive model was developed by Bradshaw and Groll [2] to deepen the understanding of the device, holistically. This holistic approach has allowed for better prediction of non-intuitive behavior of the spool compressor especially applied to applications and working fluids without experimental experience. The model includes 10 leak paths, 8 heat flow paths, port geometries, as well as oil solubility and frictional component sub-models.

Additionally, sub-components of the spool compressor have been studied in detail. Bradshaw [1] presented a study on the analytical model developed to represent the tip seal in the spool compressor. The model was developed based on the tip seal dynamic model coupled with hydrodynamic lubrication theory. Kemp *et al.* [4] presented iterations on the spool seal design and described the critical nature of the design of this component.

Bradshaw *et al.* [3] recently presented the development of a loss analysis to describe the 5th generation spool compressor. This analysis utilized high-speed pressure measurements to determine the indicated, or flow, losses within the device. Coupled with analysis of the frictional losses an complete loss Pareto was generated. This analysis has resulted in a deep understanding of the location of potential improvements as well as an improved comprehensive modeling tool which accounts, more accurately, for the various frictional losses within the device. The resulting improvements using this analysis were also presented which showed a significant improvement in performance between the 5th and 6th generation compressor. These results echo the improved performance presented by Orosz *et al.* [9].

This work will utilize the comprehensive model from Bradshaw and Groll [2], and its improvements, to study the spool compressor operating on R134a with a variety of displacements to determine an appropriate size and shape for a R134a air-conditioning compressor. Additionally, a comparison of designs for R410A and R134a is presented which highlights the utility of the comprehensive design tool for various applications. This is accomplished by examining how the efficiencies of the compressor change with changes in the fundamental geometric parameters which define the shape of the spool compressor; the eccentricity ratio and the length-to-diameter ratio.

2 Parametric Analysis

The geometry of the spool compressor is described analytically in Bradshaw and Groll [2]. This work shows that the basic planar geometry of the spool compressor can be defined with the stator radius, R_s, the rotor radius, R_r, and the eccentricity, e. These inputs are used to calculate the planar area's of each of the three working chambers within the mechanism at any rotor angle. To calculate the total volume of a spool compressor pocket the planar area's are multiplied by the stator height, h_{stator}. This leaves

$$V_{disp} = fcn(R_s, R_r, e, h_{stator}) \tag{1}$$

where V_{disp} is the maximum chamber volume over one complete rotation. This suggests that for a desired displacement, there are four degrees of freedom which can influence change to the displacement. However, by introducing two dimensionless variables, the eccentricity ratio (ϵ), and the length to diameter ratio ($^L/_D$), the number of degrees of freedom of this system is reduced.

The eccentricity ratio is defined as,

$$\epsilon = \frac{R_r}{R_s} \tag{2}$$

while the length to diameter ratio is defined as,

$$\frac{L}{D} = \frac{h_{stator}}{2R_s}. \tag{3}$$

By introducing these variables the chamber volume function can now be formulated as,

$$V_{disp} = fcn(R_s, \epsilon, ^L/_D). \tag{4}$$

Therefore, three of the four parameters above are needed to define the geometry of the spool compressor. Since displacement is generally dictated by the cooling requirement it will be used along with the eccentricity ratio and length-to-diameter ratio to define the basic compressor shape. Figure 2 shows the basic dimensions of the spool compressor and two examples of the basic spool compressor geometry which have the same displacement.

2.1 Compressor displacements

The displacement is a function of the working fluid and cooling capacity required at a particular operating condition. This study will compare the model predicted efficiencies of a fixed capacity on two working fluids, R410A and R134a, as well as 5 displaced volumes on R134a to study the influence of the displaced volume on the compressor efficiency.

For all studies the operating conditions will be fixed at an evaporating temperature of 40°F (4.4^{o}C) and a condensing temperature of 100°F (38°C). Table 1 provides the calculated pressure ratios and calculated displaced volumes for the proposed compressors assuming a 90% volumetric efficiency.

The calculated displacement represents the volume required to process the required volumetric flow rate of working fluid to generate the desired cooling capacity. This assumes an ideal refrigeration cycle with 10R (5.6K) of subcooling and 15R (8.3K) of compressor superheat. The volumetric efficiency and speed are fixed and used as an estimate though it is acknowledged that these values will vary based on operating condition and compressor geometry.

IOP Conf. Series: Materials Science and Engineering **232** (2017) 012056 doi:10.1088/1757-899X/232/1/012056

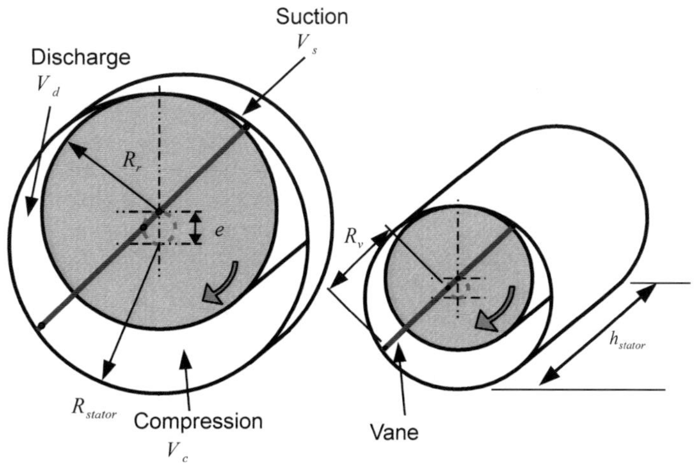

Figure 2. Spool compressor basic dimensions shown on two designs with the same displaced volume and eccentricity ratio with different length to diameter ratio.

Table 1. Table of calculated compressor displacements required to achieve various cooling loads used for study.

Fluid	**PR**	$\dot{Q}_{cool}$	η_{vol}	**Speed**	$V_{disp,tot}$	V_{disp}
-	-	**tonsR (kW)**	-	**rpm**	cm^3/rev	$cm^3/pocket$
R410A	2.49	5 (17.6)	0.9	1750	112	56
R134a	2.79	5 (17.6)	0.9	1750	254	127
R134a	2.79	20 (70.3)	0.9	1750	1014	507
R134a	2.79	40 (141)	0.9	1750	2029	1014
R134a	2.79	80 (281)	0.9	1750	4057	2029

2.2 Manufacturing considerations
One of the key leakage paths in the spool compressor is the Top Dead Center (TDC) gap. This is the gap formed by the small gap between the rotor and cylinder block. To improve leakage, and reduce manufacturing variability, the rotor is inset into the cylinder. This inset requires that a portion of the cylinder is manufactured with a discontinuous profile. The portion of the profile adjacent to the rotor is cut with a radius which is equal to the rotor. This is highlighted in Figure 3 which shows how the rotor is inset into the cylinder to increase the leakage path length of the TDC region. The remaining portion of the cylinder is cut at a different radius. The section of the cylinder with the same radius as the rotor creates a region where the two components come in close contact. However, it elongates the leak path so it can be assumed the leak path is closer to two long plates in close contact opposed to a tangential line contact without this feature.

While this inset improves the leakage performance for a given compressor it is subject to the ability to manufacture the parts. This manufacturing ability changes with the eccentricity and length-to-diameter ratio selected. The length-to-diameter ratio has the most significant influence on this gap. More specifically the stator height, h_{stator}, as the stator radius has minimal influence. As the stator height increases the ability to maintain the squareness of the rotor, relative to its centerline, and the total indicated runout become more challenging. Similarly, when boring the cylinder block the ability to maintain the squareness of the TDC inset relative to its centerline becomes more challenging as the stator height increases. The

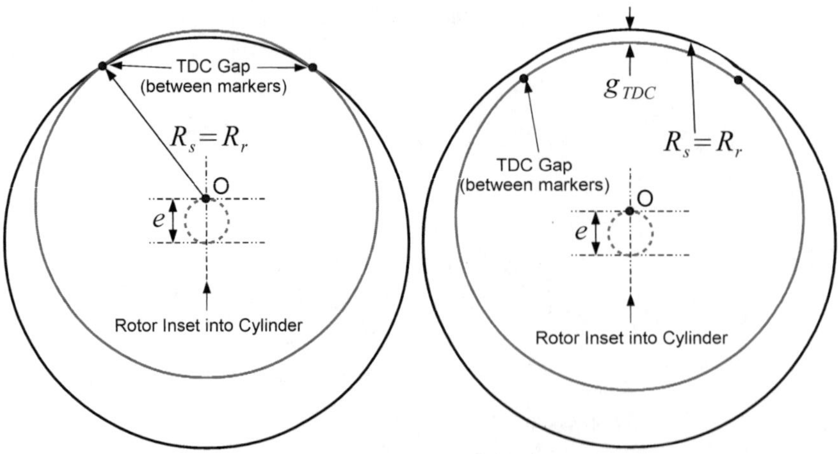

Figure 3. A rotor and cylinder schematic highlighting the theoretical rotor/stator overlap (left) and an actual TDC region and the inset feature of the rotor into the cylinder (right).

result is, for a given displacement, the TDC leakage gap will increase as the length-to-diameter increases. The amount of the deviation varies with manufacturing process used and thus is also a function of the volume of parts produced. Assuming the rotor is turned and OD ground and the cylinder is bored on a high-accuracy horizontal mill, it is estimated that the TDC leakage gap will vary based on the following expression:

$$g_{TDC} = a + bh_{stator} \tag{5}$$

where a is $0.0005in$ ($12.7\mu m$) and b is $0.0001in/in$ ($1\mu m/cm$) and h_{stator} is in inches (cm). This estimation is based on average production machine tolerance at production volumes appropriate for light-commercial compressors (personal communication, Montgomery, 2015 [7]).

Additionally, the eccentricity ratio is limited by design constraints as well. The theoretical minimum eccentricity is 0.5 if the vane were an infinitely thin line. However, the actual geometry dictates that the vane width must be equal to at least twice the eccentricity (e) to ensure the vane can be successfully actuated by the eccentric cam. As the eccentricity is reduced this requires that the slot for the vane also increase in size, reducing strength of the rotor. Additionally, the eccentric bearing increases in diameter with eccentricity which further reduces material in the rear of the rotor. Below an eccentricity ratio of 0.81 the rotor strength was considered suspect for all displacements explored in this study. Therefore, while the model will predict a result for eccentricity ratios below 0.81, the study limits itself to this value at a minimum.

3 Comparing R410A to R134a

Using the displaced volumes presented above in Table 1 for the 5 ton (17.3 kW) compressors, a comparison between an optimum design on R410A and R134a is presented. For each displacement the model is exercised for ϵ values from 0.81 to 0.92 and $^{L}/_{D}$ values from 0.4 to 3 with fixed operating conditions described in Section 2.1. Both analysis are conducted using properties from the same lubricant (68 ISO POE oil) but mixture properties which correspond to the paired refrigerant. Figure 4 shows the overall isentropic and volumetric efficiencies predicted by the model for the 5 ton (17.3 kW) displacements on both R410A and R134a for all eccentricity ratios and length-to-diameter ratios. Where the overall isentropic efficiency is defined as the isentropic compressor work, at a measured mass flow of refrigerant, relative to the total compressor input power (shaft or electrical). The volumetric efficiency of a compressor is defined as the ratio of measured compressor mass flow rate relative to the maximum possible mass flow rate.

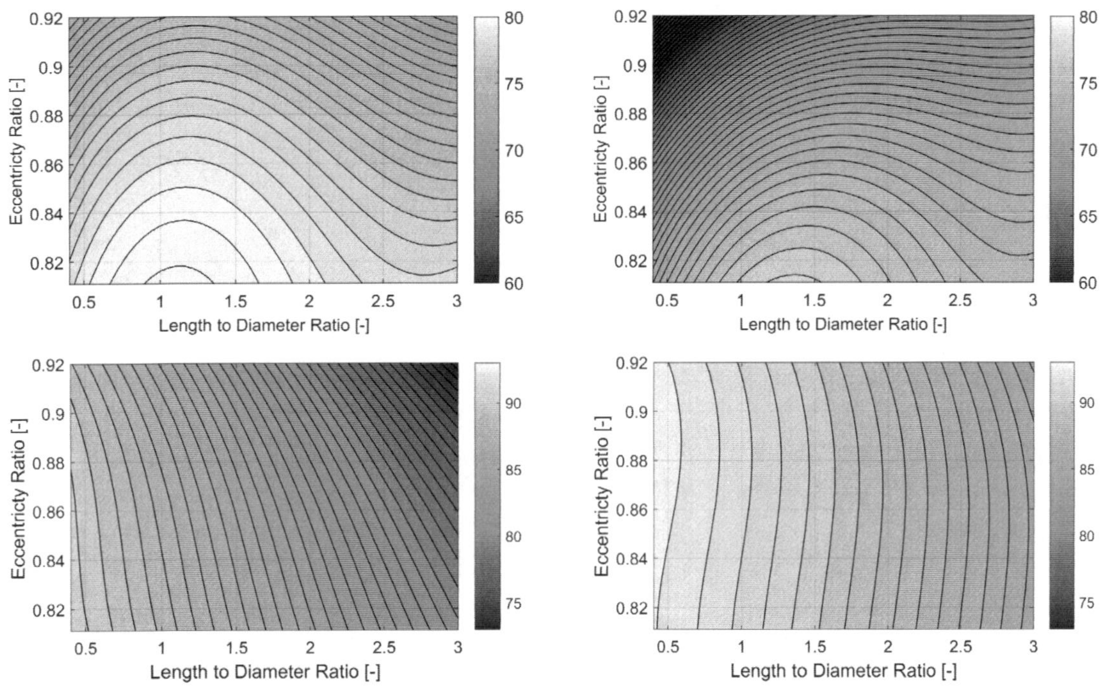

Figure 4. Overall Isentropic Efficiency of 5 ton (17.3 kW) compressor on R410A (top-left) and R134a (top-right) and Volumetric Efficiency on R410A (bottom-left) and R134a (bottom-right) for various eccentricity and length-to-diameter ratios.

The overall isentropic efficiency is significantly higher for the R410A simulation at its peak (~5% points). However, it should be noted that the peak efficiency locations are at different length-to-diameter locations. For R410A the peak efficiency is predicted at length-to-diameter ratio of about 1.05 whereas the R134a peak is predicted to be at roughly 1.4 length-to-diameter ratio. Both fluids are predicted to have the highest efficiency at the lowest eccentricity possible (0.81).

Looking at the figures from the lowest eccentricity ratios to the highest there is generally a decrease in efficiency. The eccentricity ratio has minimal influence on the volumetric efficiency (i.e. leakage) so the decrease in efficiency is a result in changes of frictional losses. At the lowest eccentricity ratio the cylinder diameter (and thus spool seal) diameter is small but the movement of the vane is high. At the highest eccentricity ratio the cylinder diameter is large but the vane movement is low. These results suggest that the spool compressor is more sensitive to the spool seal friction compared with the vane friction. Therefore, based on the results here it suggests for this displacement it would be advantageous to design a compressor with the smallest eccentricity ratio possible. The R134a has a much larger range of efficiencies across a change in eccentricity ratios. In contrast, the accompanying volumetric efficiency of the R134a results only varies from about 82% to 92%, compared with roughly 73% to 91% for R410A. This suggests the R134a design is more sensitive to changes in frictional loads compared with a R410A design.

Examining each figure from lowest length-to-diameter ratio to the highest the tradeoff between friction and leakage can be seen. At the lowest length-to-diameter ratios the cylinder diameter (and thus spool seal diameters) are forced to increase non-linearly. This creates a lot

IOP Conf. Series: Materials Science and Engineering **232** (2017) 012056 doi:10.1088/1757-899X/232/1/012056

of friction per unit displacement and thus to minimize seal friction the suggestion would be to increase the length-to-diameter ratio as high as possible. However, as the length-to-diameter ratio is increased the influence of the TDC manufacturing constraints become apparent and the increase in leakage begins to overwhelm the decrease in friction. This causes a peak value within the ranges explored which is different for each fluid. For R410A the peak being lower suggests that this fluid creates an increase in sensitivity to leakage. Conversely, R134a is less sensitive to the length-to-diameter leakage which pushes the peak to a higher length-to-diameter peak value.

4 Scaling analysis

The extension of the previous analysis for various displacements for R134a is presented next. The displacements were previously provided in Table 1 and operating conditions defined in Section 2.1. The objective is to further the understanding of the tradeoffs in design when scaling to various sizes of compressor. Figure 5 shows the results of the study of the four displaced volumes.

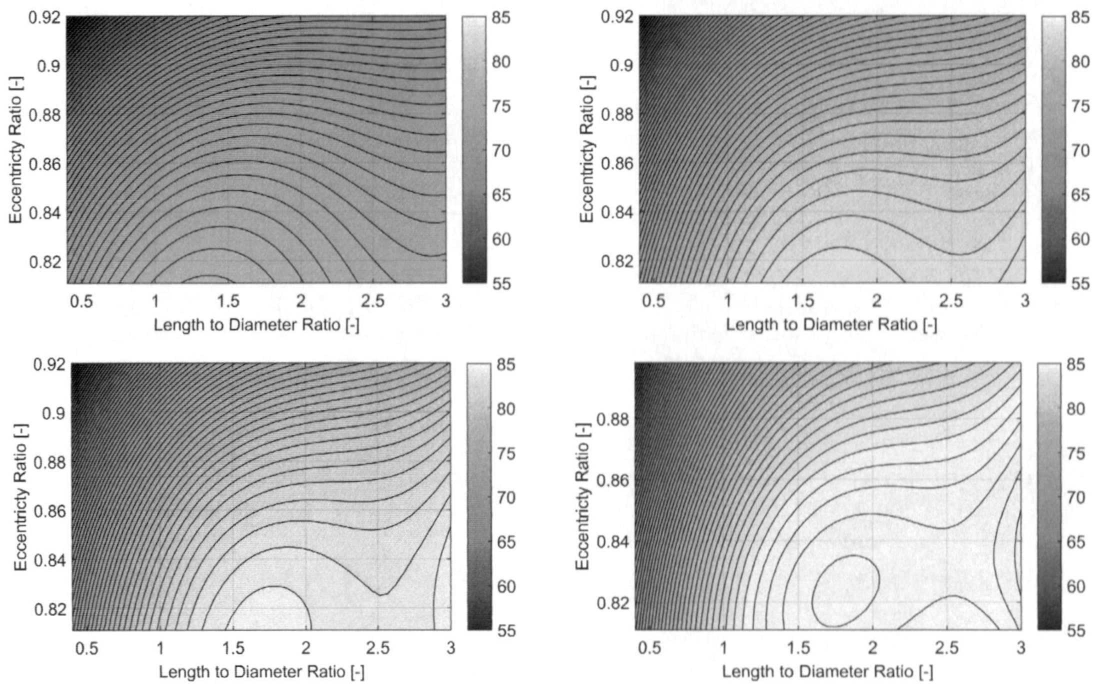

Figure 5. Overall Isentropic Efficiency of R134a compressors at 5 ton (17.3 kW), 20 ton (70.3 kW), 40 ton (141 kW), and 80 ton (281 kW) at various eccentricity and length-to-diameter ratios.

The trends are similar to the trends presented in the previous section for any given displacement. However, as the displaced volume increases the overall isentropic efficiency increases and the optimum location changes. The overall isentropic efficiency increases from a peak of roughly 77% at 5 tons (17.3 kW) to a peak of 86% at 80 tons (281 kW). This is a result of processing dis-proportionally more volume per unit frictional load as the compressor increases in displacement. While this is typical of most compressor technologies, the amount of improvement varies. The optimum location shifts from a length-to-diameter ratio of roughly 1.4 to 1.85 as the displaced volume increases. This is a result of a reduction in the sensitivity to frictional components at this displacement compared with the smaller displacements. However, as the displacement increases the peak eccentricity ratio increases, suggesting the vane friction becomes more important as the displacement increases. The peak overall isentropic efficiency and accompanying length-to-diameter ratio and eccentricity ratio are tabulated in Table 2.

Table 2. Final estimated peak efficiencies and key geometric parameters.

$\dot{Q}_{cool}$	$^{L}/_{D}$	ϵ	$\eta_{o,is}$
tonsR (kW)	-	-	%
5 (17.6)	1.4	0.81	77%
20 (70.3)	1.75	0.81	83%
40 (141)	1.82	0.815	85%
80 (281)	1.85	0.825	86%

5 Design of a R134a prototype compressor for light-commercial air-conditioning

Using the analysis from the previous section a design for a light-commercial air-conditioning compressor is presented. The 40 ton (141 kW) size was selected as its overall isentropic efficiency is predicted to be reasonably high and the displaced volume represents a displaced volume that is difficult to find efficient products in the market. The typical alternatives, screw compressors are generally not as efficient at this displaced volume and scroll compressors are not either [6].

From the data above the most efficient design for a 40 ton (141 kW) compressor is provided in Table 3.

Table 3. Final preliminary design parameters for light-commercial compressor prototype.

Parameter	Units	Value
ϵ	-	0.811
$^{L}/_{D}$	-	1.82
R_r	in (cm)	2.21 (5.60)
R_s	in (cm)	2.72 (6.9)
h_{stator}	in (cm)	9.890 (25.1)
$\eta_{o,is}$	%	84.6
η_{vol}	%	92.6
Discharge Temp	F (C)	128 (53.3)
Massflow	lbm/min (kg/min)	118 (53.5)
Shaft Power	kW	23.5
% Carryover Vol.	%	1.76

6 Experimental results from 40 ton prototype spool compressor

Orosz *et al.* [8] recently presented details surrounding the results of performance testing of a 40 ton prototype spool compressor that was constructed based on the results of this modeling effort. Additional results are included Figure 6 which provides results of overall isentropic and volumetric efficiency of the prototype at various condensing and evaporating temperatures.

The design conditions reflected in the study from Section 4 corresponds to a pressure ratio (PR) of roughly 2.8 at a condensing temperature of 100 °F (37.8 °C). At these conditions the prototype compressor has measured efficiencies of 94.5% and 79% volumetric and overall isentropic efficiencies, respectively. Compared against the model predicted values of 92.6% and 84.6% and can be noted that the volumetric efficiency prediction is slightly lower than actual whereas the overall isentropic efficiency is over-predicted.

The volumetric efficiency is considered sufficiently accurate for this situation, whereas the overall isentropic efficiency prediction is a bit far off. The hypothesis for this the lack of inclusion of a dynamic valve model in the study presented in Section 4. The magnitude of valve losses in the spool compressor previously measured by Bradshaw *et al.* [3] suggest that the valve losses in this prototype could easily account for the roughly 5% discrepancy in overall isentropic efficiency estimation.

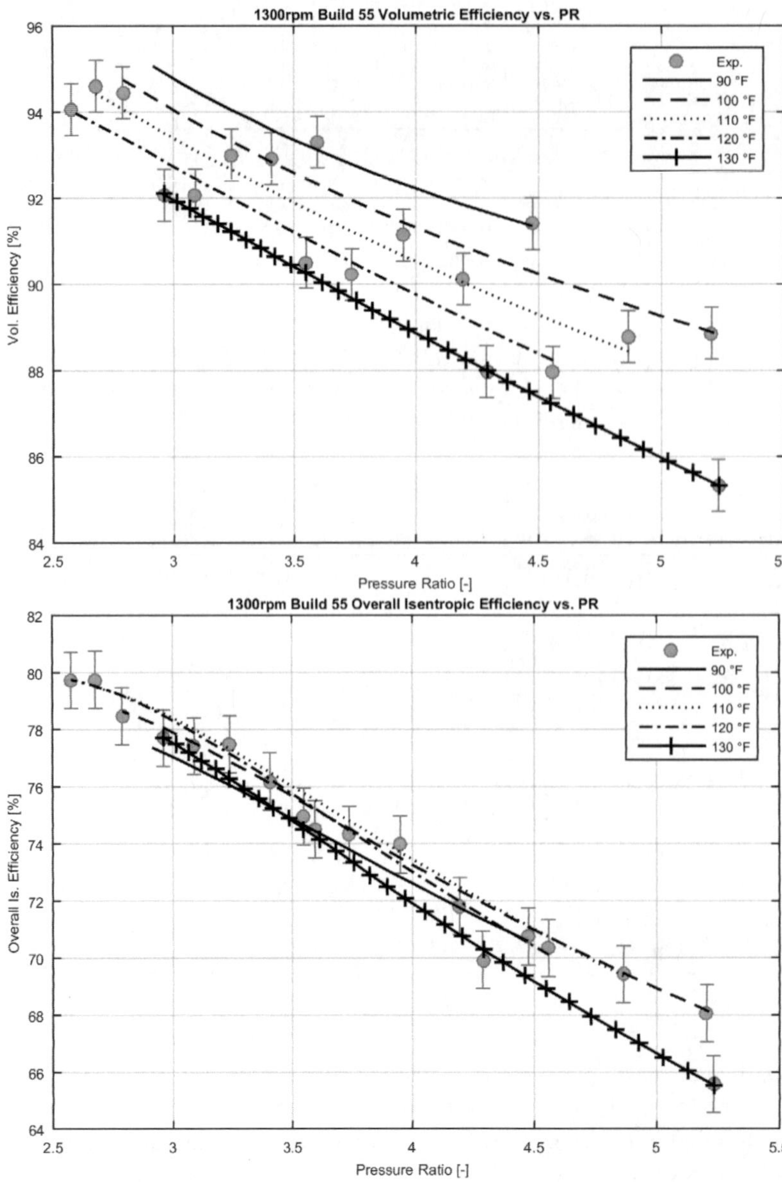

Figure 6. Volumetric (top) and Overall Isentropic (bottom) Efficiency of 40 ton (141 kW) prototype compressor tested at 1300 rpm shaft speed at various condensing and evaporating temperatures.

7 Conclusions

A design methodology utilizing a comprehensive compressor model is presented which culminates in a preliminary design of a 40 ton (141 kW) air-conditioning compressor. The study examined the utility of the approach when applied to different refrigerants as well as examining various displaced volumes. It was found that the tool provided great insight into the appropriate basic dimensions of a compressor that would have otherwise been difficult to obtain through trial and error.

Experimental results from a prototype constructed based on this study is also presented and discussed. It was found that the volumetric efficiency predictions of the modeling study matched well at the design conditions. The overall isentropic efficiency was found to be slightly over-predicted. It is hypothesized that this is a result of the lack of a dynamic discharge valve model in the scaling study.

Nomenclature

g_{TDC}	Gap between rotor and cylinder at TDC region	(m)
h_{stator}	Cylinder height	(m)
$^L/_D$	Length-to-diameter ratio	(-)
PR	Pressure Ratio	(-)
$\dot{Q}_{cool}$	Cooling capacity	(kW)
R_r	Rotor radius	(m)
R_s	Cylinder radius	(m)
$V_{disp,tot}$	Total compressor displaced volume for single shaft revolution	(cm^3/rev)
V_{disp}	Compressor displaced volume for single pocket	$(cm^3/pocket)$

Greek

ϵ	Eccentricity ratio	(-)
η_{vol}	Volumetric efficiency	(-)
$\eta_{o,is}$	Overall isentropic efficiency	(-)

References

[1] Bradshaw, C. Spool compressor tip seal design considerations. In *8th International Conference on Compressors and their Systems* (2013), p. 341.

[2] Bradshaw, C. R., and Groll, E. A. A comprehensive model of a novel rotating spool compressor. *International Journal of Refrigeration 36*, 7 (2013), 1974 – 1981. New Developments in Compressor Technology.

[3] Bradshaw, C. R., Kemp, G., Orosz, J., and Groll, E. A. Development of a loss pareto for a rotating spool compressor using high-speed pressure measurements and friction analysis. *Applied Thermal Engineering 99* (2016), 392–401.

[4] Kemp, G., Orosz, J., Bradshaw, C., and Groll, E. A. Spool seal design and testing for the spool compressor. In *International Compressor Engineering Conference at Purdue University* (2012), no. 1259.

[5] Kemp, G. T., Garrett, N., and Groll, E. A. Novel rotary spool compressor design and preliminary prototype performance. In *International Compressor Engineering Conference at Purdue University* (2008), no. 1328.

[6] Kreith, F. *The CRC handbook of thermal engineering.* Springer Science & Business Media, 2000.

[7] Montgomery, J. Dir. of mfg. torad engineering, machine tolerances for spool compressor in light-commercial application. 2015.

[8] Orosz, J., Bradshaw, C., Kemp, G., and Groll, E. Updated performance and operating characteristics of a novel rotating spool compressor. In *International Compressor Engineering Conference* (2016), no. 1377.

[9] Orosz, J., Kemp, G., Bradshaw, C. R., and Groll, E. A. An update on the performance and operating characteristics of a novel rotating spool compressor. In *International Compressor Engineering Conference at Purdue University* (2014), no. 1378.

Comparison of energy-efficiency and size of portable oil-free screw and scroll compressors

Olly Dmitriev, Prof. Ian MacDonald Arbon

Vert Rotors UK Ltd, Edinburgh, United Kingdom

olly@vertrotors.com

Abstract. This paper presents test data and evaluates if conical screw compressors can become a preferred alternative to scroll compressors in small oil-free duties from 0.04 to 15kW. The conical screw compressor is a new modification of the conventional screw compressor. A 2kW water-injected conical compressor demonstrated a 34% better energy efficiency than a scroll compressor of similar capacity. At 8 bar(g) load, the conical machine used 13% more energy and produced 42% more flow. Size of the conical screw package is 46% of the scroll package. The miniature conical screw package, at 40W rated power, achieved a pressure of 8 bar(g) in water-injected operation. It can be concluded that the conical screw compressor would be an attractive alternative in micro and small oil-free applications.

1. Introduction

Reducing power consumption for oil-free compressors is crucially important as electricity consumption typically represents up to 80% of the total lifecycle cost of the compressor. Furthermore, transmission losses may be reduced by replacing large compression plant with smaller compressors closer to end-users. However, these small compressors need to demonstrate high energy efficiency.

High energy efficiency in oil-free air is achieved with water-injection. It is known from other compressor manufacturers that compression with water-injection provides around 20% energy saving in comparison with dry-running, or adiabatic compression.

Water injection has become a standard option in compressors for the medical and food industries, with larger screw machines from 15 to 110kW. Examples include single screw compressors from CompAir (DH series, from 15kW to 110kW), and twin-screw machines from Atlas Copco (AQ 15-55 VSD, from 15kW to 55kW). However, in smaller sizes both twin-screw and single screw machines have not been able to achieve the required efficiency, due to high leakage through clearance in the "blow hole", which is a design feature of such compressors [1]. Typically, this leakage is reduced by injecting oil into the compression chambers. Since the kinematic viscosity of water is two orders of magnitude lower than that of typical compressor oil, water does not provide the level of sealing required by the twin-screw and single screw compressors[1].

[1] At 40 °C, water has a kinematic viscosity of 0.658 mm^2/s, typical screw compressor oils have kinematic viscosity in the range of 32-64 mm^2/s [2,3]

Because of this limitation, the market for portable oil-free air compressors <15kW is not served by screw compressors. Instead, this market is typically served by dry-running (adiabatic) scroll or reciprocating piston compressors, which are known to be less efficient.

As discussed by Dmitriev and Arbon in detail in [4], the conical screw compressor does not have the "blow hole" and therefore can reach effective sealing between compression chambers even with water. This paper presents results of the most recent work: a 2kW model of the conical compressor (codenamed VERT.80) with water-injection demonstrated a 34% better energy efficiency than a scroll compressor of a similar capacity. The conical screw package drew 13% more power from the mains supply and produced 42% more flow. The size of the conical screw package is 46% of that of the scroll package. This new machine can be a promising solution, especially when serving as a local source of oil-free air.

2. About the conical screw compressor

The conical screw compressor is a modification of the screw compressor, which is different from the well-known twin-screw type. In the twin-screw compressor the helical rotors run on parallel axes enclosed in the casing; N Stošic writes that two such rotors effectively form a helical gear [5]. In the conical screw compressor, the male rotor is enclosed in the female screw rotor, they revolve on eccentric axes and form a cycloidal gear [4], see Fig.1.

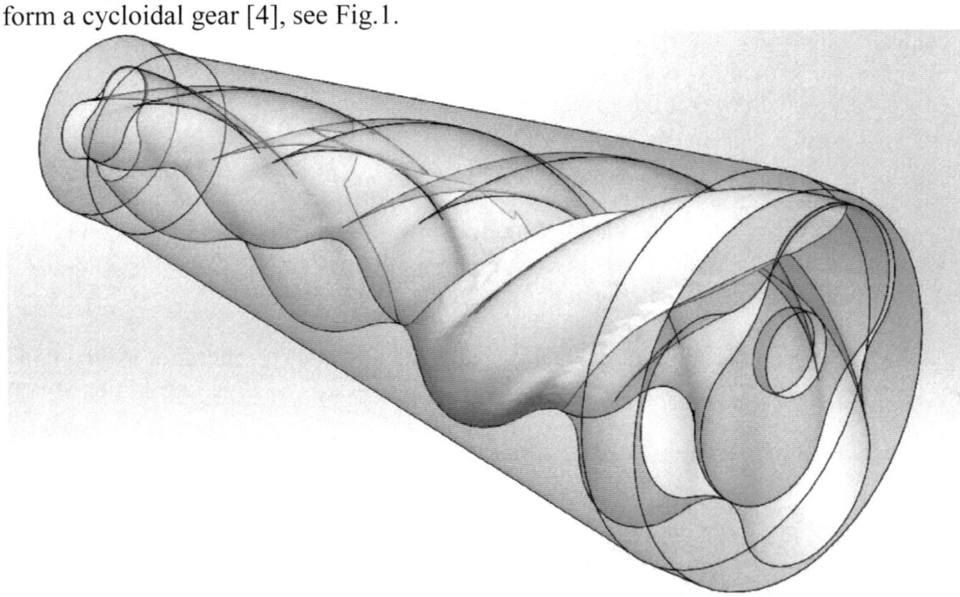

Figure 1. The male, epicycloidal (yellow) and the female, hypocycloidal (light grey) conical screws. Compression chamber is highlighted in light blue.

The concept of the conical screw compressor was first described by Chang of Robbins & Myers, Inc., USA in 1956 [6]. Manufacturing of this compressor required mathematical modelling, computer-aided design (CAD) and manufacturing methods which were beyond the reach at the time. Such methods were developed by Dmitriev and Tabota at Vert Rotors UK Ltd, UK, and in 2014 the first working model of a conical screw compressor was presented at 12th European Fluid Machinery Congress in 2014 [7]. The conical compressor was developed further, and a miniature modification was presented by O.Dmitriev, Tabota and Arbon at 9th Conference on Compressors and their Systems in 2015 [8].The model presented in 2014 operated with oil injection. In 2016, at the 13th European Fluid Machinery Congress, Dmitriev and Arbon presented a 2kW conical screw compressor with water-injection [4].

Figure 2.Bare-shaft conical screw compressor VERT.80

The conical compressor consists of two conical screw rotors. The male epicycloidal rotor is positioned inside the female hypocycloidal rotor. At each planar cross-section they create a cycloidal gear. The rotors revolve on fixed offset axes.

At suction, a volume of air enters the compression chamber at the large end of the cone. When the rotors revolve, the chamber is isolated and the air is trapped inside the closed volume. Following the rotation of the rotors, this closed volume is propelled towards the small end of the cone. The diameter of the compression chamber reduces progressively, the volume decreases and the pressure increases. High-pressure gas is discharged from the small end of the cone.

The authors have been working on evaluating efficiency of oil-free operation of the conical screw compressor with water injection, in comparison with other oil-free compressors.

3. Test of the 2kW conical screw compressor

3.1. Test setup

The conical screw compressor package is built with the bare-shaft compressor element, as shown in Fig.3 below.

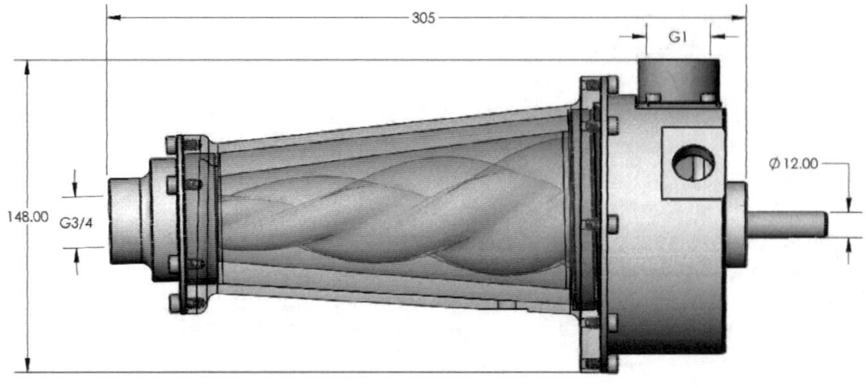

Figure 3.Dimensional drawing of the bare-shaft conical screw compressor VERT.80

Two oil-free, floor-mounted compressor packages of similar capacity were compared: a 2kW water-injected conical screw compressor, VERT.80, and a 1.5kW scroll compressor, Atlas Copco SF 1, see Fig.4.

Figure 4.Scroll compressor Atlas Copco SF 1 (left) and conical screw compressor VERT.80 (right)

The discharge pressure of both compressors was set with a pressure relief valve, Norgren V74G-4GK-NMN, connected to the discharge port of the package. When the pressure relief valve connected to the scroll compressor was loaded to 8 bar(g), the compressor switched off automatically. Because of this, the scroll compressor was characterised at 7.5 bar(g) load.

The test measured the actual electric power drawn by the compressor assembly from the mains supply, instead of nominal shaft power. The power was measured with Digital Power Meter ABB C13 LCD.

Flow rate was measured with air flow meter SMC PFMB7501-F04- E (rated flow rate 1 to 500 l/min).

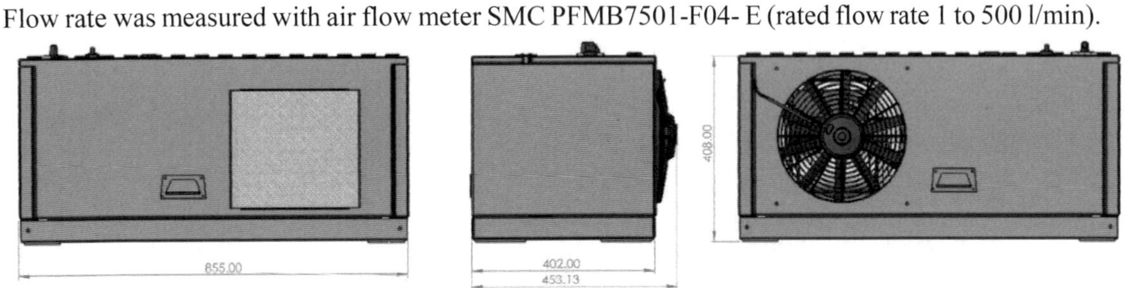

Figure 5. Dimensions of the conical screw compressor VERT.80

Table 1. Evaluated compressor packages

Compressor Package Designation	Working element	Dimensions	Weight	Type of operation	Motor speed r/min	Rated motor power
SF 1	Scroll	590x600x850 mm	120kg	Dry-running	2885	1.5kW
VERT.80	Conical screw	855x402x408 mm	90kg	Water injected	2920	2kW

IOP Conf. Series: Materials Science and Engineering **232** (2017) 012057 doi:10.1088/1757-899X/232/1/012057

3.2. Test results

The conical screw compressor was loaded to 8 bar(g), and performed at specific power of 13kW/m^3/min. The scroll compressor was loaded to 7.5 bar(g) (because at 8 bar(g) it switches off automatically), and operated at specific power of 19.7kW/m^3/min. In this comparison the conical compressor demonstrated 34% better energy efficiency.

The size of the conical screw compressor package (as shown on Figure 5) is 855 x 402 x 408 mm (0.14m^3 volume), and weight is 90kg. The size of the scroll compressor package was 590 x 600 x 850 mm (0.3m^3 volume) and weight was 120kg[2]. This comparison shows that the volume occupied by the conical screw compressor package is 46% of the volume of the scroll package, and weight of the conical screw package is 25% less than the scroll package.

As a third comparison, the flow rate and power of both compressors were analysed. According to the brochure for the scroll compressor, the flow rate was 2.9 l/s, or 174 l/min, installed motor power was 1.5kW, and maximum pressure was either 8 bar(g) or 10 bar(g). The scroll compressor was loaded to 7.5 bar(g) by setting a pressure relief valve, and measurements showed the following:

(A) the flow rate at 7.5 bar(g) was 127 l/min. The observed flow rate was lower than specified in the brochure, and it was understood that the brochure provided information for an unloaded compressor.

(B) the power drawn from the mains supply was 2.5kW. It was understood that the power of 1.5kW described in the brochure was for an unloaded compressor, possibly excluding the electric consumption of other components of the package.

The conical screw compressor was loaded to 8 bar(g) by setting the pressure relief valve, and the following observations were made:

(A) the air flow rate was 220 l/min;

(B) the actual power drawn from the mains supply was 2.86kW.

It should be noted that the water flow rate (in the closed loop system) was 13-15 l/min.

It can be concluded from the third test that the conical screw compressor used 13% more power than the scroll compressor. The power consumption of both compressor packages is in reasonably close proximity to each other. The flow rate of the conical screw machine was 42% greater than one of the scroll machine.

Summary of the test data is shown in the Table 2 below.

Table 2. Test results

Compressor package	Pressure load	Flow rate	Power consumption	Specific power
SF 1	7.5 bar(g)	127 l/min	2.5kW	19.7 kW/m3/min
VERT.80	8 bar(g)	220 l/min	2.86kW	13 kW/m3/min

3.3. Conclusions of the test

It can be concluded that tests of the water-injected compressor VERT.80 follow the theoretical assumption that water-injected compressors are more energy efficient than adiabatic compressors.

[2] Dimensions were measured, weight was taken from atlascopco.com

The test demonstrated that conical screw compressors are capable of meeting the market requirements for portable oil-free compressors sub-15kW.

The dimensions and volume of the conical screw compressor package were dramatically smaller than ones of the scroll package, which may be beneficial for certain applications. Examples of such applications include breathable air systems in oil & gas offshore platforms where size is highly important.

3.4. Discussion and further work

TheVERT.80 tested is currently development package, and it includes several non-optimised parts. Because of this, the authors are working to change non-optimal components, before the assembly is ready for serial production. For example, many components of the bare-shaft compressor assembly were produced from 316 stainless steel. The casing and primary separator were produced from extruded aluminium, and some other parts from brass. As such, this combination of materials, in contact with water, induces galvanic corrosion, with oxides clogging filters very quickly. Further development work will include anodising aluminium parts, and replacing brass with stainless steel.

Further, integrating the primary separator with water filter and water mist separator can help to replace tubes and fittings, and reduce the size of the package even further.

4. 8 bar(g) pressure achieved with the miniature 40W screw compressor

4.1. Results of production and testing

Further work was undertaken to miniaturise the conical compressor and see a practical limit of the screw compressor technology. A new compressor assembly with 40mm-long male and female conical screws was designed and manufactured. Swept volume of this compressor was 0.4 cc. A drawing of this miniature screw compressor codenamed VERT.04 is shown in Fig.6 below.

Figure 6. Miniature conical screw compressor VERT.04

The machine recorded a power consumption of 36-40W (24V outrunner brushless DC motor, current 1.5A).

A water injected package was developed, which included three modular blocks: (a) a module with the bare-shaft conical screw compressor and motor, (b) a module with electrical components and heat exchanger, (c) a primary water separator also working as a pressure vessel, see Fig.7

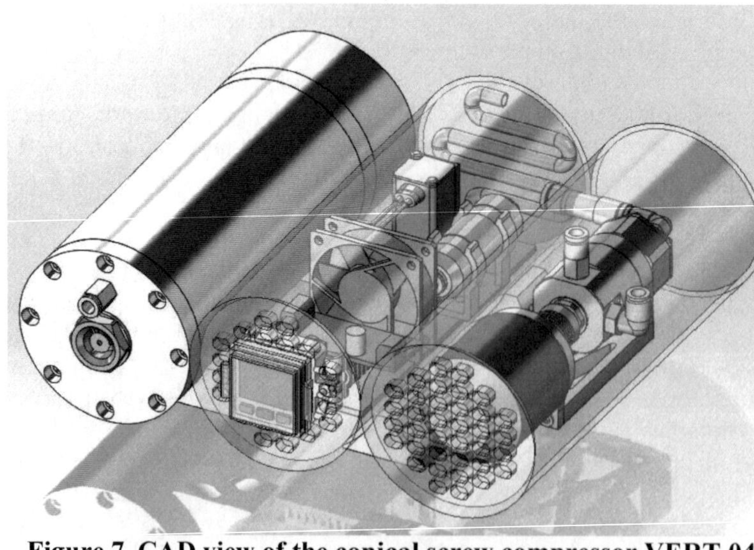

Figure 7. CAD view of the conical screw compressor VERT.04: Water separator / pressure vessel (left), module with heat exchanger and electronics (centre), module with compressor (right)

Working with water=injection, the miniature compressor VERT.04 produced a maximum discharge pressure of 8 bar(g) (or 115 psi(g)).

Figure 8. Test results showing 115.2 psi(g) / 8 bar(g)

4.2. Conclusions and discussion

The miniature VERT.04 is a working prototype and will be developed further before it is ready for the market.

The tests demonstrated that it the conical screw compressor is the smallest known rotary compressor that can achieve pressures such as 8 bar(g), with very low vibration and noise, which is very beneficial for medical or laboratory applications with sensitive equipment.

The pressure of 8 bar(g) demonstrated by the VERT.04 is considerably higher than pressures produced by other known miniature rotary compressors (1-1.5 bar(g) for rotary vane machines). While dry-running reciprocating piston compressors are capable of achieving pressures of 8 bar(g), they generate significant vibrations. Also, the thermodynamics of adiabatic compression dictate that the temperature of such machines escalates rapidly. The benefit of achieving pressures such as 8 bar(g) with water-injection is low temperature of the package, which supports much longer duty cycles

References (in order of appearance):

[1] Buckney D, Kovacevic A, Stosic N (2011), Consideration of clearances in the design of screw compressor rotors, 7th International Conference on Compressors and their Systems, City University London
[2] Kestin J, Sokolov M, Wakeham W A (1978), Viscosity of Liquid Water in the Range of -8oC to 150oC, J. Phys. Chem. Ref. Data, Vol. 7, No. 3, 944
[3] Product Information, Klüber Summit SH 32, 46, 68, 100
[4] O.Dmitriev, I.Arbon (2016), A compact water-injected screw compressor with high volumetric efficiency, 13th European Fluid Machinery Congress, The Hague, The Netherlands
[5] N.Stošic (1998), On gearing of helical screw compressor rotors, Proc Instn Mech Engrs Vol 212 Part C
[6] S.S.L.Chang (1956), Cone Type Compressor, US Patent 2,733,854
[7] O.Dmitriev, E.Tabota (2014), A working conical screw compressor, 12th European Fluid Machinery Congress, Edinburgh, UK
[8] Dmitriev O, Tabota E, Arbon I M, A miniature Rotary Compressor with a 1:10 compression ratio, 2015 IOP Conf. Ser.: Mater. Sci. Eng. 90 012055

IOP Conf. Series: Materials Science and Engineering **232** (2017) 012058 doi:10.1088/1757-899X/232/1/012058

Internally geared screw machines with ported end plates

M G Read, I K Smith, N Stosic

Centre for Compressor Technology, City, University of London, Northampton Square, EC1V 0HB

E-mail: m.read@city.ac.uk

Abstract. It is possible to design cylindrical helical gearing profiles such that an externally lobed inner gear rotates inside an internally lobed outer gear while maintaining continuous lines of contact between the gears. The continuous contact between the inner and outer rotors (analogous to the main and gate rotors in a conventional screw machine) creates a series of separate working chambers. In this type of machine the rotors have parallel axes of rotation, and if both rotors are free to rotate about their own axes, these axes can be fixed in space.

The use of ported end plates is proposed to control the period during which fluid is allowed to enter or leave the working chambers of the internally geared screw machine. As with conventional screw machines, these internally geared rotors can then be used to achieve compression or expansion of a trapped mass of fluid, and the machine geometry can be designed in order to optimise performance for particular applications.

This paper describes the geometrical analysis of some simple rotor profiles and explores the effect on rotor torques for particular applications of this novel screw configuration.

1. Introduction

Commonly used pumps such as that shown in Figure 1 consist of an internally geared outer rotor and an externally geared inner rotor. The rotor profiles must be carefully shaped in order to maintain continuous contact points between the rotors during rotation [1–5]. The meshing between these gears creates a series of separate working chambers, and the volume of each working chamber varies with the angular position of the rotors. Such internally geared positive displacement pumps, also known as gerotor or internal lobe pumps, have straight cut rotors. The aim of this paper, however, is to investigate a novel configuration of internally geared screw machine in which the inner and outer rotors have helical profiles with continuous lines of contact upon relative rotation of the rotors. Open-ended versions of such screw machines can be used to compress or expand a working fluid by employing screw threads of axially varying pitch or profile on the rotors. This creates an enclosed working chamber between the rotors whose volume varies with rotation. However, in order to achieve compression or expansion in a machine using helical rotors of constant profile and pitch, it is proposed that a stationary end plate is located at the high-pressure end of the rotors. For a compressor application, this 'high-pressure end plate' has a port that is shaped to only allow fluid to leave the working chamber of the machine once the required volume ratio of compression has been achieved. If required, a low-pressure end plate can also be located at the low-pressure end of the rotors to ensure that the working chamber is closed once the maximum volume is reached. Hence, each end plate has a port section for admitting

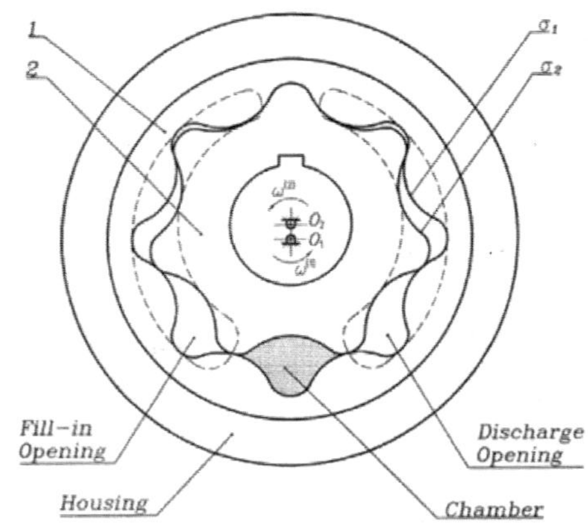

Figure 1: Illustration of a conventional Gerotor type internally geared pump [4]

the working fluid to, or discharging the working fluid from, the screw machine. The end plate(s) and port(s) are arranged to achieve compression or expansion of the working fluid upon relative rotation of the rotors, and an example of this configuration is shown in Figures 2 and 3.

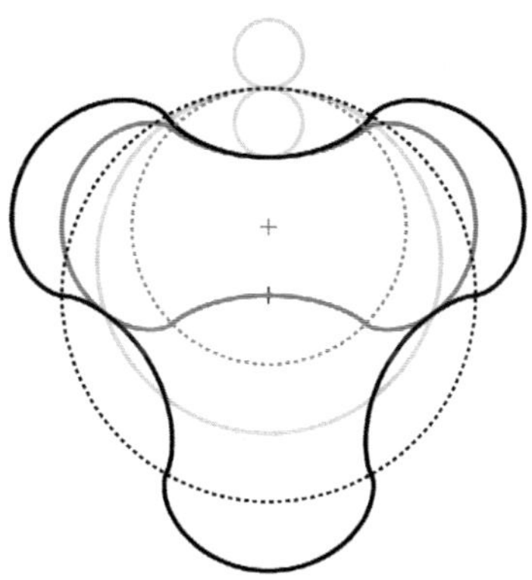

(a) Rotor profiles and locii of contact points

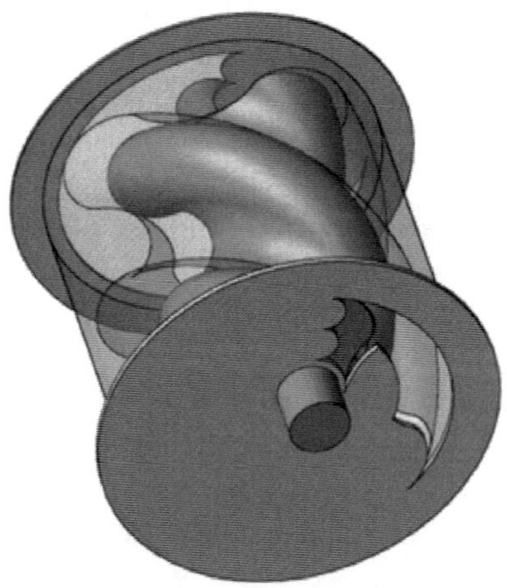

(b) Internal screw machine model

Figure 2: Rotor profiles and CAD model of internally geared screw machine with $N_i = 2$, $\rho_e/E = 0.5$, $L/2\rho_{o,max} = 1.55$ and $\phi_{o,w} = 300°$. The internal rotor is shown in orange, the external rotor is transparent, and the stationary end plates are illustrated

This helical internally geared rotor configuration has a number of potential advantages compared to a conventional screw machine:

- Elimination of rotor-to-casing tip clearances present in conventional screw machines, as the outer rotor encloses the working chamber. The two rotors of the internally geared machine effectively share the same tip leakage areas.
- Elimination of the blow hole area characteristic of conventional machines, and a resulting reduction of the interlobe sealing line length.
- Reduced geometric constraints on the size of bearings that can be used on both rotors.

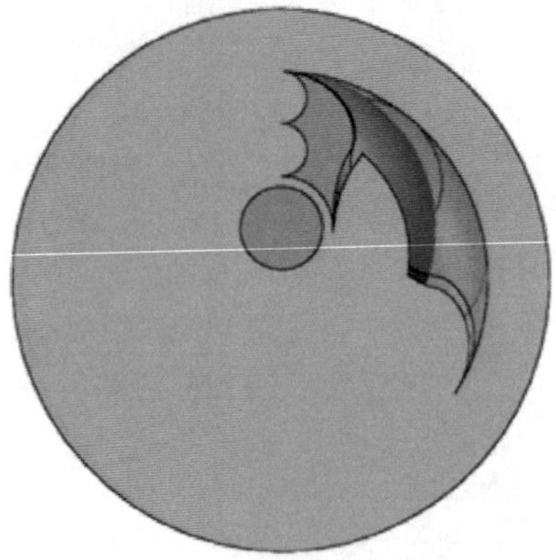

(a) High pressure port geometry

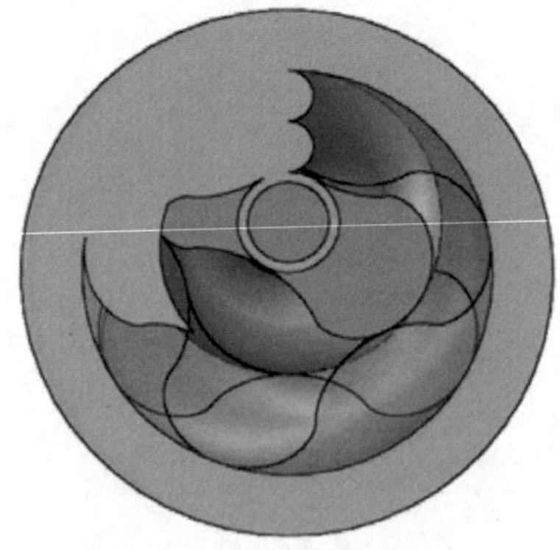

(b) Low pressure port geometry

Figure 3: Details of the port shapes for the high and low pressure end plates designed to achieve $\epsilon_v = 4.5$ for the rotors shown in Figure 2

- Reduced viscous drag due to the much lower relative speed at the contact lines between rotors, as compared to the rotor-casing leakage lines in a conventional machine.
- Reduced distortion due to thermal expansion, as the rotation of both rotors will lead to a uniform circumferential temperature in each during operation, with temperature variation limited to the axial direction.

2. Geometry of simple rotor profiles for internally geared machines

Internally geared positive displacement machines require continuous contact points between inner and outer rotors during rotation about parallel axes. This allows the formation of enclosed working chambers whose volume varies with the angular position of the rotors. A wide range of rotor profiles can be generated that meet this requirement. In this paper, simple geometries consisting of epicycloids and hypocycloids are used to illustrate this requirement of continuous contact between rotors and to investigate machine performance. This requires that the inner rotor has one fewer lobe than the outer rotor. The equations defining coordinates of epicycloid and hypocycloid profiles are shown in Equations 1 and 2. While cycloids are used in this paper to illustrate the analysis of internally geared machines, it should be noted that there is considerable scope for optimisation using other methods of rotor profile generation.

$$\begin{bmatrix} x_e(\theta) \\ y_e(\theta) \end{bmatrix} = \begin{bmatrix} (\rho_b + \rho_e)cos(\theta) - \rho_e cos(\theta(\rho_b/\rho_e + 1)) \\ (\rho_b + \rho_e)sin(\theta) - \rho_e sin(\theta(\rho_b/\rho_e + 1)) \end{bmatrix} \tag{1}$$

$$\begin{bmatrix} x_h(\theta) \\ y_h(\theta) \end{bmatrix} = \begin{bmatrix} (\rho_b - \rho_h)cos(\theta) + \rho_h cos(\theta(\rho_b/\rho_h - 1)) \\ (\rho_b - \rho_h)sin(\theta) - \rho_h sin(\theta(\rho_b/\rho_h - 1)) \end{bmatrix} \tag{2}$$

Continuous contact between rotors can be achieved using composite profiles which can be created by combining sections of epicycloid and hypocycloid profiles. The profiles are defined by the radius of the base circle, ρ_b, the number of lobes required on the rotor, N, and the radius of the epicycloid generating circle, ρ_e, where $0 \leq \rho_e \leq \rho_b/N$. By considering the circumferences of the epicycloid and hypocycloid generating circles, it can be seen that the radius of the hypocycloid generating circle is:

$$\rho_h = \rho_b/N - \rho_e \tag{3}$$

The maximum and minimum radii of the rotor are as follows:

$$\rho_{max} = \rho_b + 2\rho_e, \quad \rho_{min} = \rho_b - 2\rho_h = \rho_b(1 - 2/N) - 2\rho_e \tag{4}$$

It is clear that when $\rho_e = 0$ or $\rho_h = 0$, the rotor is a pure hypocycloid or epicycloid respectively.
In order to achieve the required meshing condition between the inner and outer rotors the following conditions are required:

$$\rho_{b,o}/N_o = \rho_{b,i}/N_i, \quad \rho_{e,o} = \rho_{e,i}, \quad N_o = N_i + 1 \tag{5}$$

The offset distance between the axes, E, can then be defined as follows:

$$E = \rho_{max,i} - \rho_{min,o} = \rho_{b,i} + 2\rho_{e,i} - \rho_{b,o}(1 - 2/N_o) - 2\rho_{e,o} \tag{6}$$

Rearranging Equation 6 and substituting for the relationships in Equation 5 results in the following expression:

$$\rho_{e,o}/E = \rho_{e,o}/(\rho_{e,o} + \rho_{h,o}) \tag{7}$$

It is clear from Equation 7 that the possible values of the ratio $\rho_{e,o}/E$ range between zero and one, corresponding to the cases when the rotors are pure hypocycloids and epicycloids respectively. In all the examples presented in this paper, the geometric parameters for the outer rotor have been chosen in order to achieve a maximum profile radius of 0.5 m, and the profile is therefore bounded by a circle of diameter 1 m. Hence the results can be scaled to find the geometric characteristics of any specified rotor size. As seen in Figure 4, the inner rotor must be scaled such that $\rho_{o,min} = (\rho_{i,min} + E)$ in order to allow the rotors to mesh and achieve the required continuous contact. The meshing condition requires that the normal to both rotor profiles must pass through the pitch point of the rotors, and is used to identify the location of the contact points for a given angular position of the rotors. The locii of these contact points during rotation can therefore be found. These are shown in Figure 4, along with the pitch circles of the rotors which are shown to be the same as the base circles used to generate the epicycloid and hypocycloid curves. Note that all profiles are shown with the angular position of the outer rotor, $\phi_o = 0$, which has been chosen as the position where rotor contact occurs at the minimum radius of both rotors.

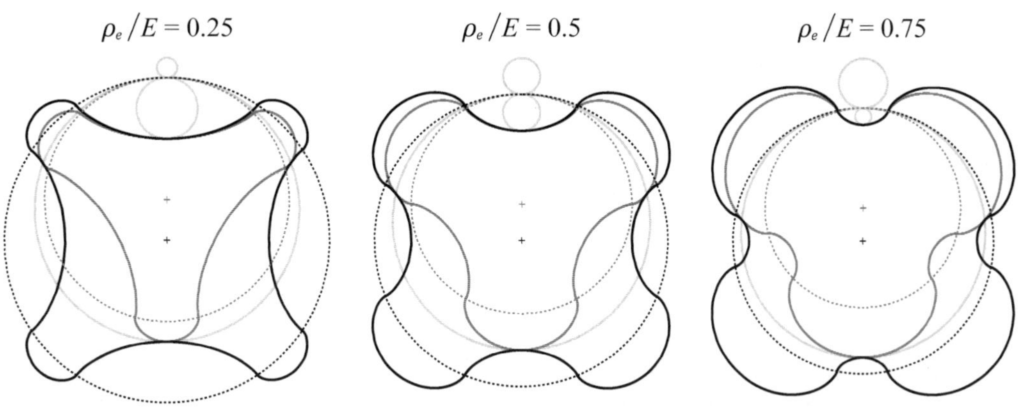

Figure 4: Inner (red) and outer (black) rotor profiles for values of $\rho_{e,o}/E = 0.25, 0.5$ and 0.75, showing locii of contact points during rotation (green) and rotor pitch circles (dotted lines)

3. Geometric properties

When considering the 2-dimensional rotor profiles generated in Section 2, it is possible to identify the angular position of the rotors at which two contact points coincide. Rotation relative to this point creates an area between the two contact points, bounded by sections of the inner and outer rotor profiles. This is illustrated for various rotor positions in Figure 5. For a given rotor position, the area of working chamber is shown in grey, while the leading and trailing contact points are shown as circular blue markers. The line connecting these represents the projected area per unit rotor length over which the pressure acts on the rotors. The perpendicular distance between the line of action of the net force and the axes of rotation are shown on the x-axis relative to each rotor centre. The calculated working chamber area is shown in Figure 6 as a function of the outer rotor position for the three rotor profiles shown in Figure 4.

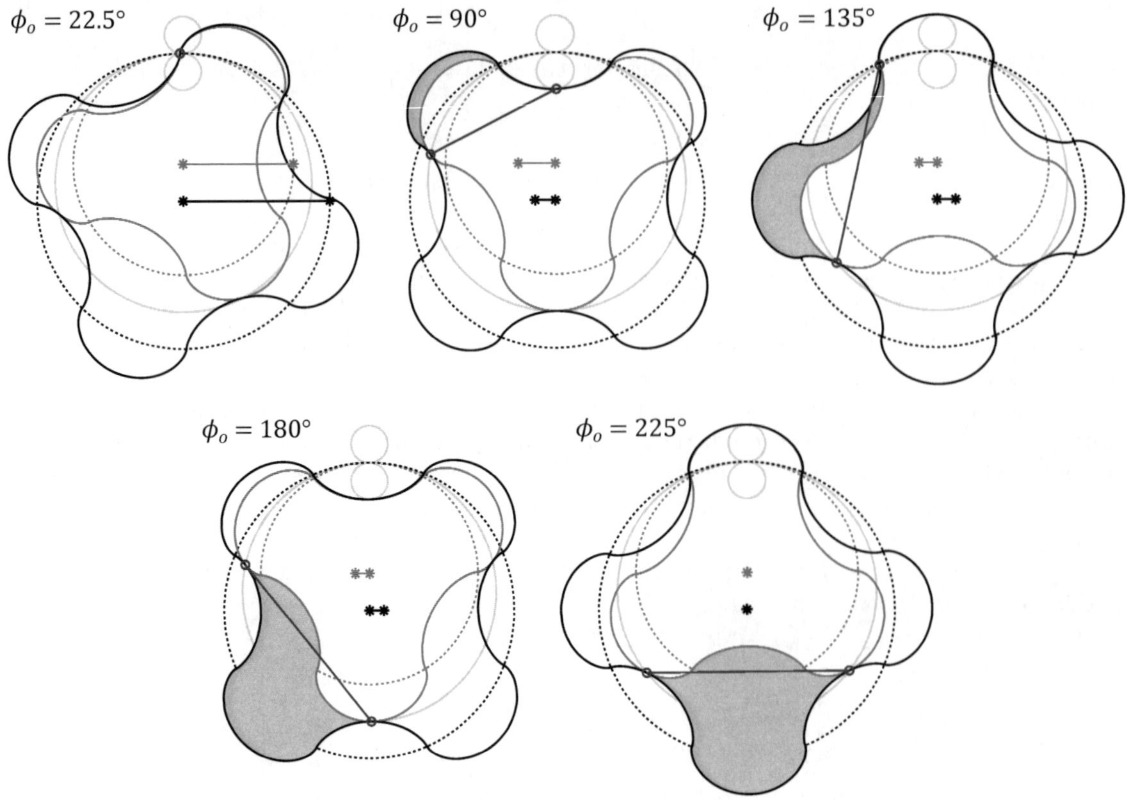

Figure 5: Geometry of rotors with $N_i = 3$ and $\rho_e/E = 0.5$, showing working chamber area, contact points, projected area for pressure force and perpendicular distance between the line of action and the rotor centres

The sliding velocity is defined as the velocity of the outer rotor relative to the inner rotor at the contact point, and is also a function of the rotor geometry. For the particular geometry shown in Figure 5, the sliding velocity of the leading and trailing contact points of the highlighted working chamber are given in Figure 7a. The maximum sliding velocity is also shown in Figure 7b for profiles with $N_i = 3$ and $0 \leq \rho_e/E \leq 1$.

For a straight rotor machines, the working chamber volume at a given angular position is then simply equal to the cross-sectional area of the working chamber multiplied by the rotor length. For the internally geared screw configuration, the working-chamber area must be integrated along the length of the helical rotors in order to find the total working chamber volume for a

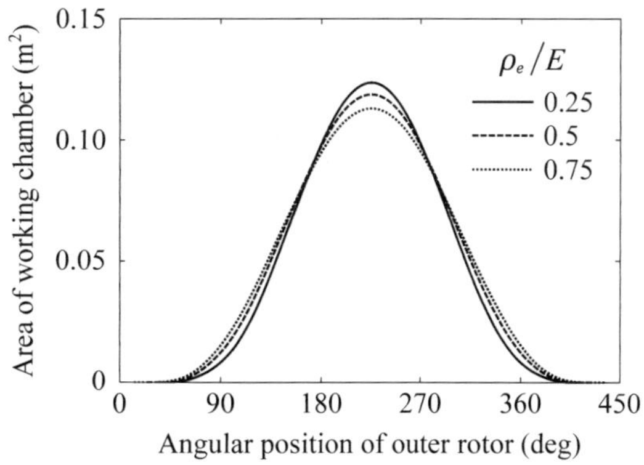

Figure 6: Cross sectional area of working chamber as a function of the angular position of the outer rotor, ϕ_o, for the profiles shown in Figure 4

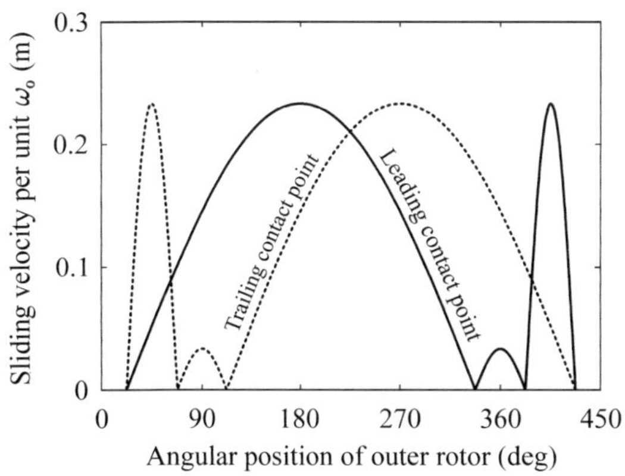

(a) Normalised sliding velocity as a function of the angular position of the outer rotor for the profile shown in Figure 5

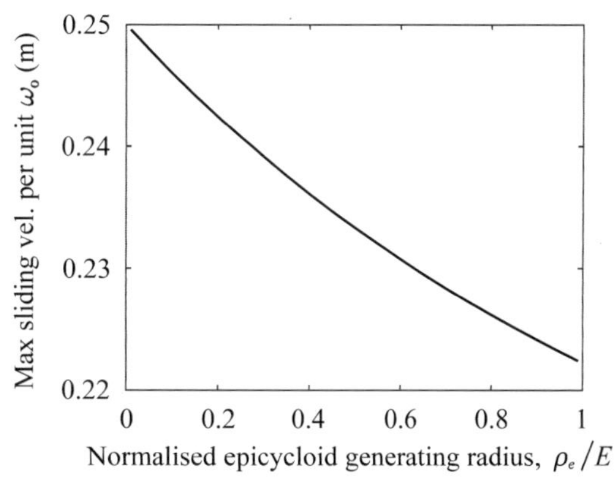

(b) Maximum normalised sliding velocity as a function of the normalised epicycloid generating radius, ρ_e/E, when $N_i = 3$

Figure 7: Sliding velocity of rotor contact points normalised by angular velocity of the outer rotor, ω_o

given rotor position. As for conventional screw machines, this requires the wrap angle of the rotors to be defined along with the length [6]. Using the values specified in Table 1 and the profiles shown in Figure 4, the volume of the working chambers has been calculated as shown in Figure 8a. This allows the swept volume per revolution of the machine to be calculated for a range of profiles and rotor lobe numbers, as shown in Figure 8b.

Table 1: Characteristics of internal screw geometry.

Parameter	Value
Outer rotor length-diameter ratio, $L/2\rho_{o,max}$	1.55
Outer rotor wrap angle, $\phi_{o,w}$	300°

An important characteristic of a positive displacement machine is the built-in volume ratio,

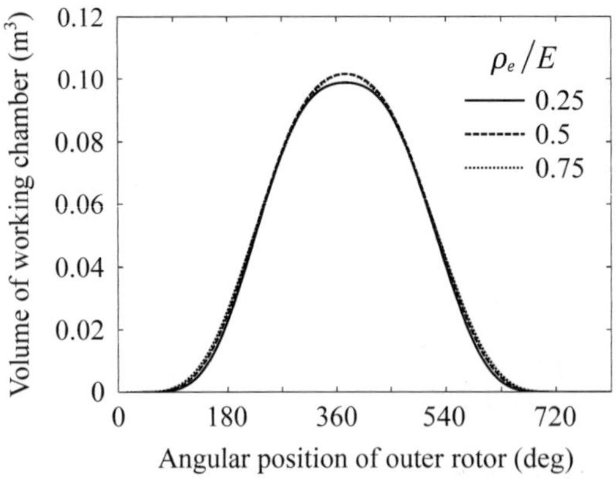

(a) Volume of a single working chamber as a function of the angular position of the outer rotor for the profiles shown in Figure 4

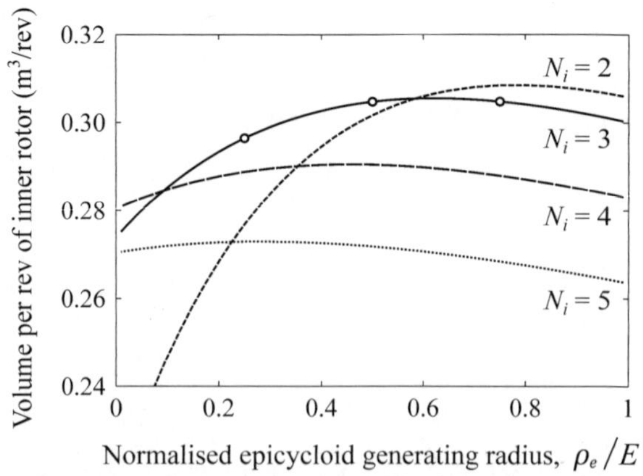

(b) Swept volume per revolution of inner rotor as a function of ρ_e/E for rotors with $N_i = 2, 3, 4$ and 5. Circular markers indicate the three cases illustrated in Figures 4

Figure 8: Volume of working chamber and machine

ϵ_v, defined as the ratio of the maximum-to-minimum volumes of the working chamber that occur while it is closed off from both ports. The angular position of the rotors at which the ports open and close can therefore be defined once the working chamber volume is characterised. A 3D model of an internally geared screw machine with stationary end plates to allow compression or expansion of the working fluid is shown in Figure 2. The rotor position shown is such that a working chamber is slightly open to the high pressure port; in a compressor this corresponds to the early stages of discharge of high pressure gas, or the final stages of suction for an expander. The shape of the ports in the high and low pressure end plates is shown in more detail in Figure 3. The shape of the ports is defined by the shape of the rotor profiles and the locii of the contact points. Defining the ports as shown in Figure 3 eliminates any open paths between an enclosed working chamber and the high and low pressure ports until required for suction or discharge. In reality the necessary clearance gap between the stationary end plates and the rotors will results in end face leakage, as in a conventional screw machine.

4. Operational characteristics

An important factor to consider in selecting rotor profiles is the torque generated on both rotors by the pressure variation in the working chamber during a compression or expansion process. In order to minimise the contact forces that occur between the rotors, the net torque exerted on the non-driven rotor by the working fluid should be minimised. This ensures that power is transferred directly from the driven rotor to the working fluid, with minimal power transferred between the rotors. From the 2D analysis described in Section 3 we can consider a thin slice of the rotor and find the projected area and direction of the pressure force created by the working fluid on the two rotors. The direction of the force allows the torque exerted on the two rotors to be calculated. By integrating these forces and torques along the axial length of the working chamber, the instantaneous torque, axial force and radial force can be calculated for a given rotor position [6]. The net torques and forces generated during a complete cycle can then be found, and the net effect of all working chambers can be considered.

4.1. Analysis of rotor torque for compressor application

Assumptions for working chamber pressure and rotor torque calculations:

- Working fluid is an ideal gas ($\gamma = 1.4$)
- Isentropic compression/expansion process
- Discharge pressure equal to working chamber pressure after compression/expansion
- No pressure drop occurs during filling and discharge of working chamber
- No leakage between inner and outer rotors or between end plates and rotor end faces

An example of the results from this initial analysis of the internally geared screw machine are presented below. The pressure variation in a single working chamber and the resulting rotor torque variation with rotor position are shown in Figure 9, while the net effect of all working chambers on the rotor torque is shown in Figure 10.

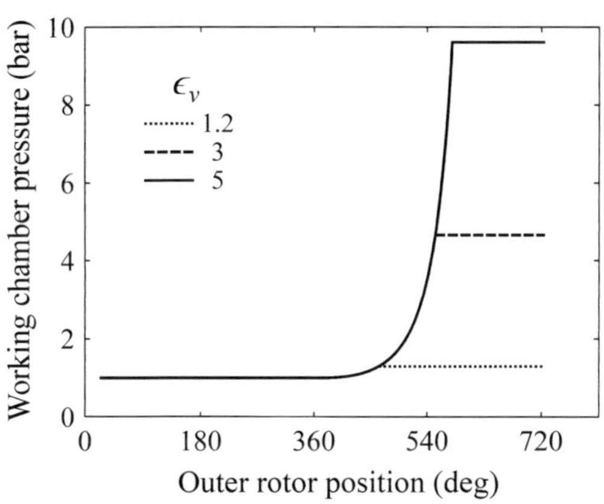

(a) Working chamber pressure

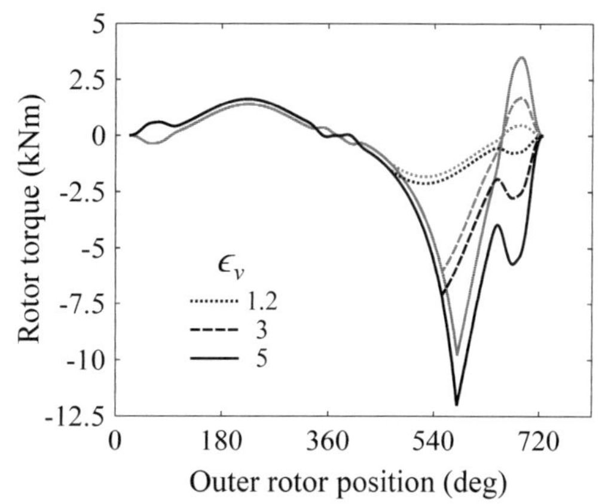

(b) Rotor torques for single working chamber

Figure 9: Working chamber pressure and corresponding rotor torques as a function of rotor position for internally geared screw compressor with $\rho_e/E = 0.5$ and $p_{suc} = 1$ bar, for the three cases when $\epsilon_v = 1.2$, 3 and 5 (note that in (b) red and black lines correspond to inner and outer rotor respectively)

The maximum and minimum net torque exerted on each rotor during operation can therefore be found as a function of ϵ_v and the rotor profiles. This has been calculated for a range of conditions with $1 \leq \epsilon_v \leq 5$ and $0 \leq \rho_e/E \leq 1$, and the resulting maximum and minimum net torque for each rotor are shown as contour plots in Figure 11. These contour plots allow a rotor profile to be selected in order to minimise the net torque on the undriven rotor, and thereby limit the force occurring at the contact points during operation. This is illustrated in Figure 12, which shows the rotor torques that occur when using the limiting values of ρ_e/E that achieve an instantaneous net torque of zero on the undriven rotor at some point during operation.

5. Conclusions

The main conclusions from the current study are as follows:

- Constant profile and pitch internally geared screw machines can achieve compression or expansion via the use of stationary ported end plates.
- Helical rotors with simple composite epicycloid-hypocycloid profiles can achieve the required continuous meshing condition.
- For a fixed machine size, the rotor profile influences the swept volume and torque exerted on the rotors during operation.

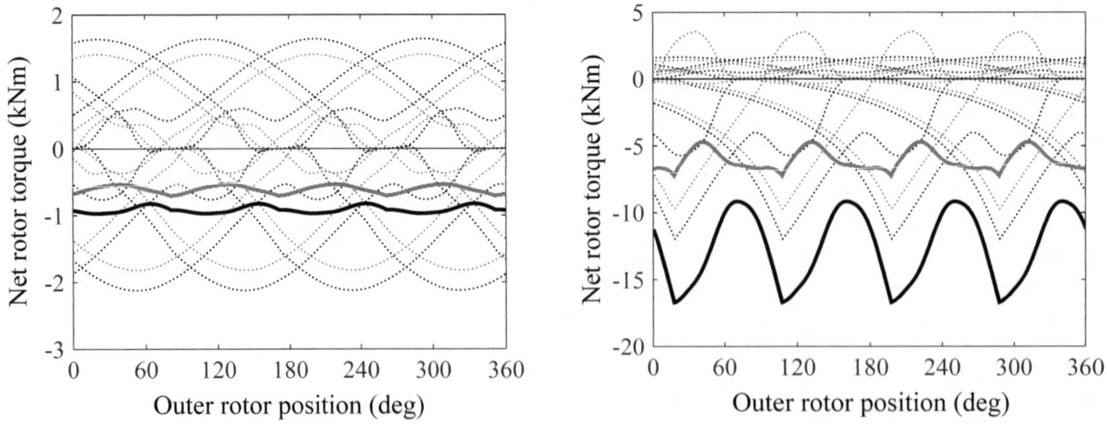

(a) Net torque when $\epsilon_v = 1.2$

(b) Net torque when $\epsilon_v = 5$

Figure 10: Torque from all working chambers (dashed lines) and resulting net torque on inner and outer rotors (solid lines) as a function of outer rotor position for internally geared screw compressor with $\rho_e/E = 0.5$, $p_{suc} = 1$ bar and the specified values of ϵ_v

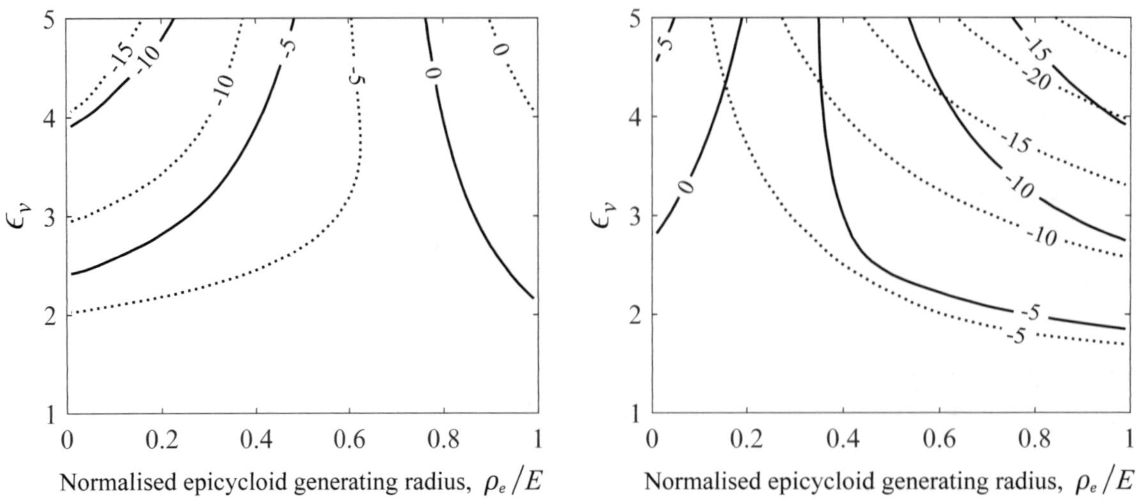

(a) Inner rotor

(b) Outer rotor

Figure 11: Maximum (solid lines) and minimum (dashed lines) net torque (kNm) on inner and outer rotors as a function of ρ_e/E and ϵ_v for the compressor (note: $N_i = 3$, $p_{suc} = 1$ bar and $p_{dis} = p_{wc,max}$ as shown in Figure 9)

- The rotor profile and choice of driven rotor can be selected to minimise torque on the undriven rotor.
- For the simple composite cycloid profiles with $N_i = 3$ and for the operating conditions considered, the ratio of undriven-to-driven rotor torques is minimised when ρ_e/E is in the range $0.76 - 0.9$, in which case the outer rotor should be driven.
- If the inner rotor is driven the maximum undriven-to-driven torque ratio is significantly higher, leading to higher contact forces between the rotors.

Future work will focus on comparing the geometric and performance characteristics of internally

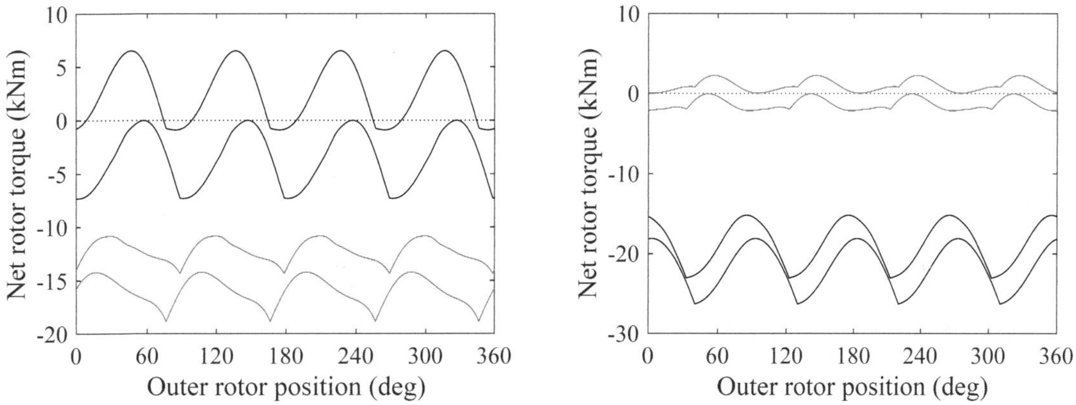

(a) $\rho_e/E = 0 - 0.19$ (b) $\rho_e/E = 0.76 - 0.90$

Figure 12: Range of values of ρ_e/E that achieve a net torque of zero on a) the outer rotor and b) the inner rotor during a compression cycle when $N_i = 3$, $p_{suc} = 1$ bar and $\epsilon_v = 5$

geared configuration with conventional screw machines. This will be achieved via:

- Calculation of axial and radial bearing loads.
- Characterisation of sealing line lengths and leakage flows.
- Characterisation of port areas as a function of working chamber volume.
- Investigation of other internally geared rotor profiles that achieve the required condition of continuous contact.

Nomenclature

E	Distance between rotor axes (m)	ρ	Radius (m)
L	Rotor length (m)	θ	Cycloid generating angle (rad)
N	Number of lobes on rotor	ϕ_w	Wrap angle of rotor (deg)
p	Pressure (bar)	ϵ_v	Built-in volume ratio
ϕ	Rotor angular position (deg)		

References

[1] Colbourne, J.R., 1974. The geometry of trochoid envelopes and their application in rotary pumps. Mechanism and Machine Theory, 9(3-4), pp.421-435.
[2] Beard, J.E., 1985. Kinematic analysis of gerotor type pumps, engines, and compressors, PhD thesis.
[3] Beard, J.E., Yannitell, D.W. and Pennock, G.R., 1992. The effects of the generating pin size and placement on the curvature and displacement of epitrochoidal gerotors. Mechanism and Machine Theory, 27(4), pp.373-389.
[4] Vecchiato, D., Demenego, A., Argyris, J. and Litvin, F.L., 2001. Geometry of a cycloidal pump. Computer methods in applied mechanics and engineering, 190(18), pp.2309-2330.
[5] Litvin, F.L. and Fuentes, A., 2004. Gear geometry and applied theory. Cambridge University Press.
[6] Stosic, N., Smith, I. and Kovacevic, A., 2005. Screw compressors: mathematical modelling and performance calculation. Springer Science & Business Media.

Working principle of new multi-chamber rotary compressor

M Lin[1,4], Y D Lim[2,5] and K T Ooi[3,6]

[1,2,3] School of Mechanical and Aerospace Engineering, Nanyang Technological University, 50 Nanyang Avenue, Singapore 639798

[4] ylin021@e.ntu.edu.sg, [5] ylim046@e.ntu.edu.sg, [6] mktooi@ntu.edu.sg

Abstract – A new multi-chamber rotary compressor (MCRC) is introduced in this paper. This new compressor is designed for compactness and energy efficient, with potential applications in refrigeration systems. In this paper, the design of the new compressor will be presented. Various components and their functions together with the overall advantages will be shown and discussed. In its basic form, it is a rotary compressor with two chambers working simultaneously to increase its output by maximizing the available working space and hence reduces material usage. The basic working principles will also be discussed and its advantages will be presented in comparison with other existing rotary compressors.

Keywords - compressor; efficiency; refrigeration; thermodynamics

1. Introduction
In the past three decades, depletion of ozone layer and global warming are the environmental issues that world leaders tried to overcome. To overcome the environmental issues, Montreal Protocol [1] aimed at phasing out the substances causing ozone depletion was signed and it was followed by Kyoto Protocol [2] and Paris Agreement [3], to tackle the climate change issues.

The applications of cooling and heating systems worldwide result in significant usage of energy. It is estimated that about 32% [4] of the world energy demanded has been used in cooling and heating in the building sector alone. The use of cooling and heating systems is expected to rise. Thus, there is a need to increase the energy efficiency of the cooling and heating systems. Since the compressor is used in almost all these systems, and it is also the device which consumes the most energy in the systems, it is therefore the objective of this paper to introduce a new compressor which is more compact, and hence saving materials usage and at the same time, it is energy efficient.

2. Literature review
In this section, articles relevant to the design and development of the new multi-chamber rotary compressor (MCRC) are reviewed.

2.1. Rolling piston compressor

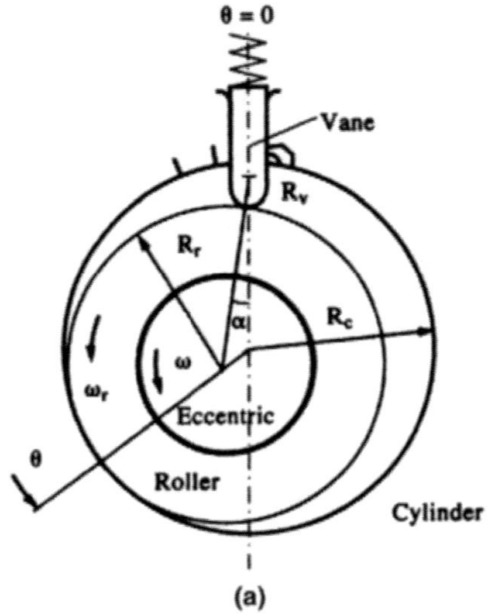

Figure 1. Schematic model of rolling piston compressor [8]

Figure *1* shows a schematic view of a rolling piston compressor [8]. It consists of a rolling piston which is the roller in Figure *1*, a cylinder which holds the roller and a vane-spring mechanism. When the shaft operates, the rolling piston that is mounted on the shaft eccentrically rotates in the cylinder, rolling against the inner wall of the cylinder [9]. During the rotation of the rolling piston, there will be space formed between the vane, the cylinder and the rolling piston. The changes in the size of the space results in suction, compression and discharge of the working fluid.

Yanagisawa and Shimizu [9] examined various frictional losses with varying loading force and sliding velocity in the rolling piston type compressor and found that the vane tip frictional loss is greatly influenced by the changes in sliding velocity on the vane tip. Moreover, it was observed that the tangential force at contact point has a great influence on the vane side frictional loss. Liu and Kosco investigated [10] rolling piston compressor with different tilt angle of vane. It was observed that the vane tip normal force has a minimum value at a certain positive angle, with the vane friction loss remains unchanged. This indicates that a tilted vane only enhances durability and reliability of the vane and piston, whereas the mechanical efficiency is not improved.

Bae et al. [11] introduced the preventing techniques of vane jumping and visualized the vane jumping using image processing technique. Light vane and long spring were suggested to prevent vane jumping. Vane jumping causes loud noise and most importantly, leakage between the suction chamber and compression chamber. The possible leakage paths are found at the contact point between piston and the cylinder, gap due to vane jumping, clearances between upper and lower faces of the rolling piston compressor and clearance between the vane and the vane slot [12-14]. Out of these, leakage through the clearance between piston and bearing and between the vane and the vane slot have the most significant effect on input power [15].

Although rolling piston compressor is currently the most widely used compressor, it can be further improved by optimizing the design dimensions or modify from the current design such as utilizing the volume occupied by the rotor in a conventional rolling piston compressor.

2.2. Swing compressor

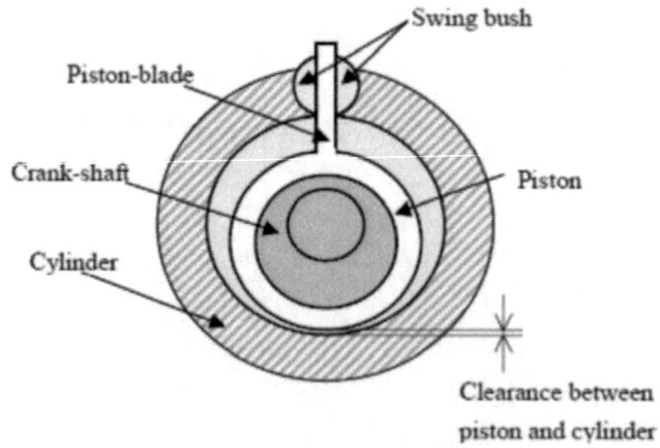

Figure 2. Schematic view of a swing compressor [16]

A schematic view of swing compressor is shown in Figure 2. The design of a swing compressor is basically similar to a rolling piston type, but the rotary structure is changed to swing structure to improve the durability and the lubricating ability [17]. Masuda et al. [17] performed theoretical simulation and experimental investigation to compare rotary compressor and swing compressor. The total efficiency of latter was found 5~9% higher than the rotary type in middle to low speed and 2% higher in high-speed range. Furusho et al. [18] conducted numerical and experimental study on swing compressor characteristic and concluded that the total efficiency is higher than rotary type in capacity range of 15.5kW. The study also stated that different refrigerants should be used in different capacity range to maximize the efficiency of swing compressor. In addition, Hu et al. [19] introduces swing vane compressor (SVC) and reported that the mechanical efficiency in SVC is higher than sliding vane compressor under same rotation speed while the total frictional loss in SVC is 65% less than sliding vane compressor.

In conclusion, from the literatures review, the design of the compressor tends toward common features as following:

- Fewer components and simpler geometries for easier fabrication and lesser production cost.
- More compact in size to allow it for more applications other than air-conditioning, such as in electric vehicles.
- Fewer moving parts to reduce the frictional loss.
- Smoother and quieter operation.

Rotary type compressors are widely used in the current refrigeration systems. With the assistance of the computational power, a simpler geometry, smaller size, lower frictional loss and lower noise level compressor can be designed and developed.

3. Conceptual model of Multi-Chamber Rotary Compressor (MCRC)

The conceptual model of MCRC, which includes the main components, assembly and working principle are introduced.

3.1 Components of MCRPC

3.1.1. Outer cylinder

The major components of MCRC comprise an outer cylinder with the split bush attached to it, a hollow cylinder that is connected to a vane and an inner cylinder. The outermost part of MCRC is the outer cylinder, where the other components of MCRC are contained within it. The inner wall of the outer cylinder is in contact with the hollow cylinder. Split bush is placed in the expanded slot on the outer cylinder to allow the sliding motion of the vane between the slot of the split bush as shown in Figure 11 and Figure 4. The slot is expanded to avoid a direct contact of the vane with the outer cylinder, to reduce the frictional loss.

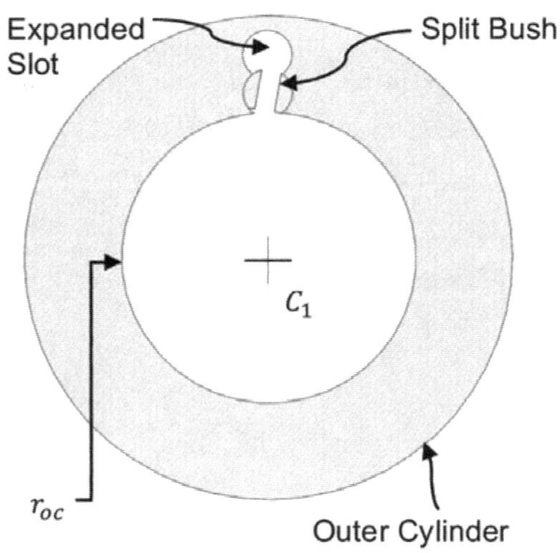

Figure 3. Top sectional view of split bush and outer cylinder of MCRC

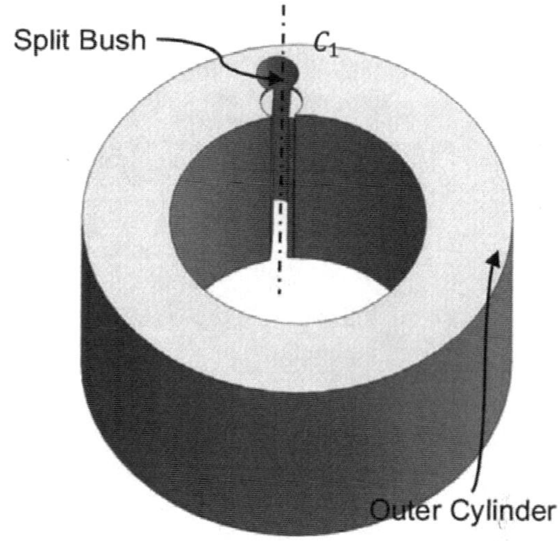

Figure 4. Isometric view of split bush and outer cylinder of MCRC

3.1.2. Hollow cylinder

The outer surface of the hollow cylinder as shown in Figure 5 and Figure 6 is radially attached to a vane named vane 1 such that it is a one-piece component. The inner space of the hollow cylinder can be separated into inner compressor chamber and slot for cam-shaft. The geometry constraint of the hollow cylinder is its outer surface must get in contact with the inner surface of the outer cylinder, while maintaining a contact between the inner surface of the hollow cylinder and the surface of the inner cylinder. Due to this geometry constraint, the geometrical dimensions such as the thickness of the of the hollow cylinder (t_h) must be carefully defined to ensure smooth operations. In addition, the width of vane 1 is determined by the space between the split bush at the outer cylinder. The gap between the vane and split bush must be carefully defined to refrain from having too much direct contact which will result in high frictional loss or low contact level which will cause the jumping of the vane.

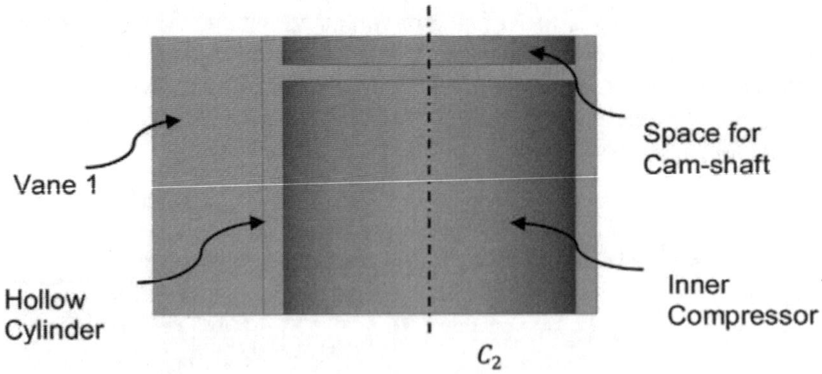

Figure 5. Cross-sectional side view of hollow cylinder and vane 1

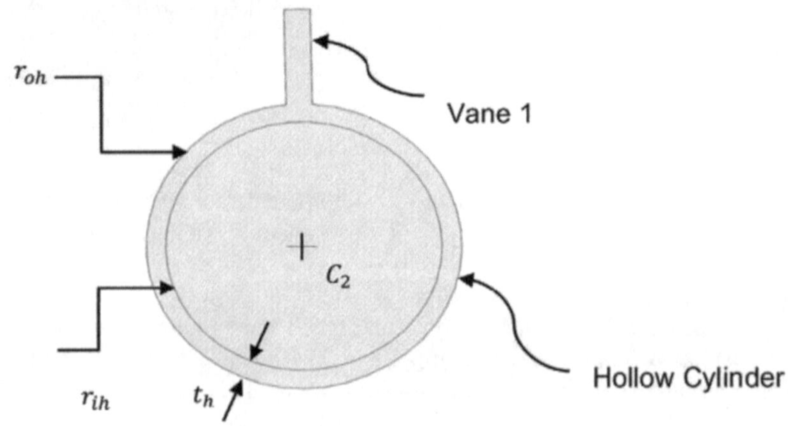

Figure 6. Top sectional view of hollow cylinder and vane 1

3.1.3. Inner cylinder

The innermost part of MCRC is the inner cylinder, where the centre axis of inner cylinder (C_1) coincides with the centre axis of the outer cylinder as shown in Figure 7. Another vane namely vane 2 is attached to the inner cylinder through a spring. The other end of vane 2 is in contact with the inner surface of the hollow cylinder. A slot is created from the outer surface of the inner cylinder towards the centre of the inner cylinder to contain the spring and vane 2.

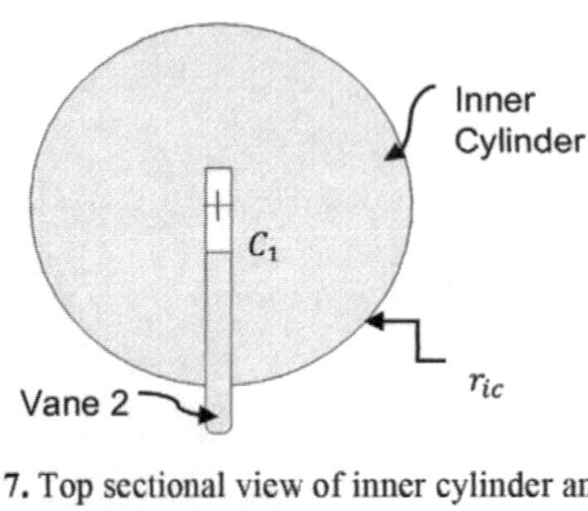

Figure 7. Top sectional view of inner cylinder and vane 2

During operation, vane 2 slides in and out of the slot while the spring exerts a force to push the vane to ensure that it is always in contact with the inner surface of the hollow cylinder. The tip of vane 2 that is in contact with the inner surface of the hollow cylinder is curved to allow for a smoother sliding and to reduce frictional loss at the contacting point.

3.2 Assembly of MCRC

The working chambers are formed by assembling the outer cylinder, the hollow cylinder and the inner cylinder on a base plate as shown in Figure 8. For the outer compressor, vane 1 and the contact line between the inner surface of outer cylinder and the outer surface of hollow cylinder have divided the working chambers into suction chambers and compression chambers. Similarly, the working chamber of the inner compressor is divided by the contact lines between inner surface of hollow cylinder, the inner cylinder surface and vane 2. The suction ports and discharge ports are located at the base plate as shown in the same figure.

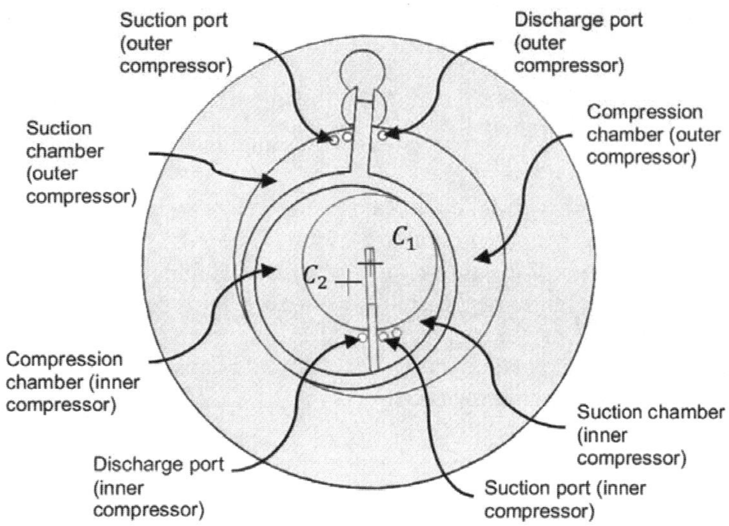

Figure 8. Schematic diagram of the assembly of MCRC

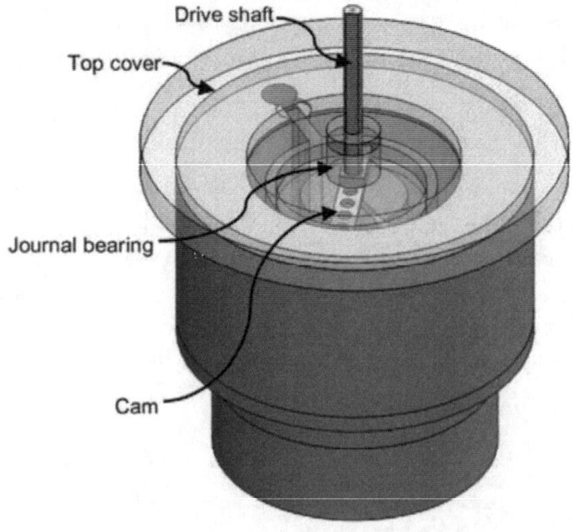

Figure 9. Isometric view of MCRC

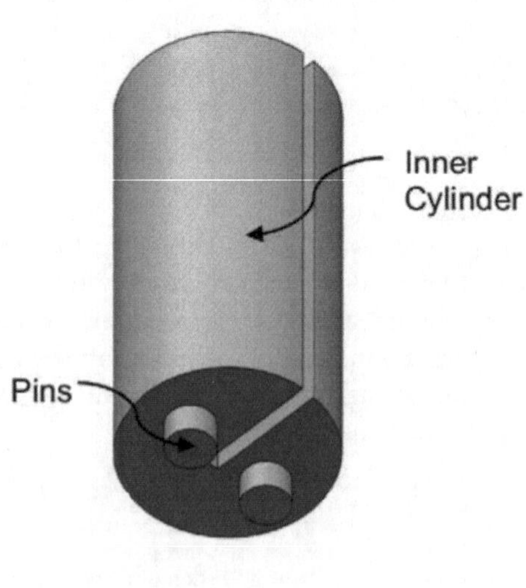

Figure 10. Features at the bottom of inner cylinder

Figure 9 shows a cam-shaft which will be connected to the prime mover drives the compressor. The shaft is held by a journal bearing and is coupled to a motor. When the cam-shaft rotates, it rotates the hollow cylinder and results in the rolling motion of the hollow cylinder. Since the inner cylinder and outer cylinder are stationary, the centre of the hollow cylinder (C_2) is rotating about the point C_1. This rotation causes vane 1 to slide along the slot of the split bush and rotates about the centre axis of the split bush. The inner cylinder is held stationary by the two pins as shown in Figure 10. Noted that both outer and inner compressors operate simultaneously.

From the assembly, C_1 and C_2 is separated by a distance e, which is defined as Equation (1)

$$e = r_{ih} - r_{ic} = r_{oc} - r_{oh} \tag{1}$$

where,
r_{ih} is the inner radius of the hollow cylinder (m)
r_{ic} is the radius of the inner cylinder (m)
r_{oc} is the radius of the outer cylinder (m)
r_{oh} is the outer radius of the hollow cylinder (m)

The outer radius of the hollow cylinder r_{oh} must be specified based on the following equation

$$r_{oh} = \frac{1}{2}(r_{oc} + r_{ic} + t_h) = e + r_{ic} + r_{ih} + 2t_h \tag{2}$$

where,
t_h is the thickness of the hollow cylinder (m)

The inner radius of the hollow cylinder can then be defined as

$$r_{ih} = r_{oh} - t_h \tag{3}$$

Hence, the independent dimensions r_{oc}, r_{Ic} and t_h must be pre-determined to define the dependent dimensions e, r_{oh} and r_{ih}.

3.3 Working principle of MCRC

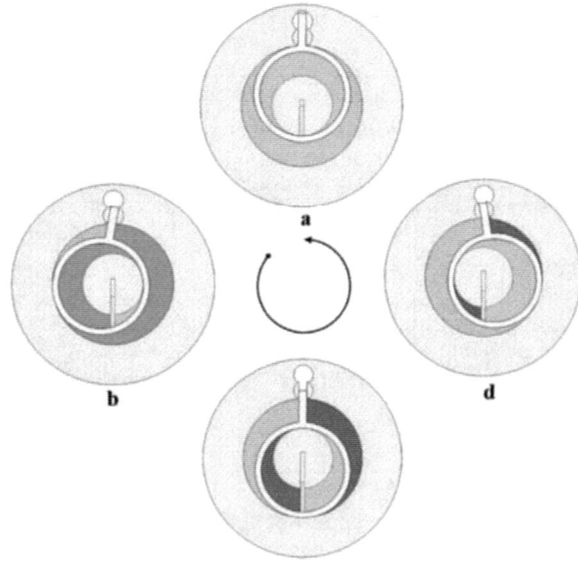

Figure 11. Operating process of MCRC

Figure 11 shows the suction, compression and discharge processes of the MCRC. Figure 11a indicates the end of suction process and the start of compression process. As the shaft rotates and the hollow cylinder moves, the volumes of the compression chambers of both outer and inner compressor decrease and hence resulting in the compression of the working fluid. Simultaneously, low pressure working fluid is being drawn into the suction chambers of both compressors, which are shown in Figure 11b and 11c. When the pressure of the working fluid in the compression chamber reaches the desired discharge pressure, the reed valve will open the discharge process begins, which is somewhere between Figure 11d and Figure 11a. While the low pressure working fluid is being drawn to the compressor, the compressed working fluid is being discharged uninterruptedly until the hollow cylinder back to the position as presented in Figure 11a.

The refrigerant will enter the compressor via two streams, inner compressor and outer compressor. The refrigerant of each stream will first go through the suction chamber and then the discharge chamber. At the end of the discharge process, working fluid from both inner and outer discharge chambers will be discharged to a reservoir which will then lead to condenser through a pipe. The ratio of flow rate in the case will be determined by the rate of change of the volumes of inner and outer chamber.

4. Discussion

A qualitative comparison on the basic information between various types of rotary compressors and MCRC is shown in table 1.

Table 1. Comparison between various types of rotary compressors and MCRC [20].

Feature	Rolling Piston Compressor	Rotary Sliding Vane Compressor	Swing Compressor	MCRC
Number of major components	4 components	At least 3 components	3 components	5 components

Simplicity of components geometry	Cylindrical profile for major components	Cylindrical or oval for major components	Cylindrical profile for major components	Cylindrical profile for major components 1
Size	Larger for a given working fluid capacity	Larger for a given working fluid capacity	Larger for a given working fluid capacity	Smaller for a given working fluid capacity
Working fluid capacity	Lower fluid intake for a given size and mass	Lower fluid intake for a given size and mass	Lower fluid intake for a given size and mass	Higher fluid intake due to more working chambers
Frictional Losses	High due to vane tip friction and vane jumping	High due to vane tip frictional loss	Lower as compare with RP and RSV compressor	Lower as compare with RP and RSV compressor

The current stage of study is unable to provide experimental result and comprehensive theoretical model. In future work, comparison will be made more accurate, the mechanical efficiencies and compressor performance will be evaluated after complete theoretical and experimental work are carried out.

5. Conclusion

A new compressor namely multi-chamber rotary compressor (MCRC) is introduced. The basic components of MCRC comprise of an outer cylinder, a hollow cylinder and an inner cylinder. The shapes of these components are mostly in cylindrical for the ease of manufacturing. The assembly of these components forms inner and outer "compressors" working simultaneous. As compared to the rolling piston compressor, the device utilizes the space that was originally occupied by the roller, hence this new design is expected to be more compact.

However, at the current stage of study, the model is assumed to be going through adiabatic process, perfect sealing and no internal leakage, and frictional losses are not accounted in the mathematical model yet. In real life situation, the model is expected to experience internal leakage, heat transfer losses and frictional losses. The objective of this paper is to introduce the working principle and design, which takes the limitations of previous rolling piston compressors and swing compressors into account. Most of the parts are cylindrical which are reasonably easy to manufacture and it is forecasted to be more energy efficient.

Acknowledgment

We wish to acknowledge the funding for this project from Nanyang Technological University under the Undergraduate Research Experience on CAmpus (URECA) programme.

References

[1] "The Montreal Protocol on Substances that Deplete the Ozone Layer," ed. Kenya: United Nations Environment Programme Ozone Secretariat, 1987.

[2] "Kyoto Protocol to the United Nations Framework Convention on Climate Change," ed. Kyoto: United Nations, 1997.

[3] TheGuardian, "Paris climate change agreement enters into force," Available: https://www.theguardian.com/environment/2016/nov/04/paris-climate-change-agreement-enters-into-force

[4] D. Ürge-Vorsatz, L. F. Cabeza, S. Serrano, C. Barreneche, K. Petrichenko, "Heating and cooling energy trends and drivers in buildings," *Renewable and Sustainable Energy Reviews*, vol. 41, pp. 85-98, 2015

[5] CASCOUSA. (2016, 08-July). *Types of Compressors*. Available:http://cascousa.com/compressed-air-101/types-of-compressors/dynamic-displacement-compressors/

[6] L. Li, Y. Zhao, B. Guo, P. Shu, J. Shen, and S. He, "Wrap of cylinder and its effect on main features of rotary vane compressor for automobile air conditioning system," *International journal of refrigeration,* vol. 26, pp. 566-574, 2003.

[7] G. Bianchi and R. Cipollone, "Theoretical modeling and experimental investigations for the improvement of the mechanical efficiency in sliding vane rotary compressors," *Applied Energy,* vol. 142, pp. 95-107, 2015.

[8] K. T. Ooi, "Design optimization of a rolling piston compressor for refrigerators," *Applied Thermal Engineering,* vol. 25, pp. 813-829, Apr 2005.

[9] T. Yanagisawa and T. Shimizu, "Friction losses in rolling piston type rotary compressors. III," *International journal of refrigeration,* vol. 8, pp. 159-165, 1985.

[10] Y. Liu and J. Kosco, "Vane Dynamics Analysis of a Tilted Vane Rotary Compressor," 1998.

[11] J. Bae, J. Kim, B. Lee, T. Noh, and S. Byun, "Vane jumping in rotary compressor," 1998.

[12] P. Pandeya and W. Soedel, "Rolling piston type rotary compressors with special attention to friction and leakage," 1978.

[13] C. Costa, R. Ferreira, and A. Prata, "Considerations about the leakage through the minimal clearance in a rolling piston compressor," 1990.

[14] T. Yanagisawa and T. Shimizu, "Leakage losses with a rolling piston type rotary compressor. II. Leakage losses through clearances on rolling piston faces," *International journal of refrigeration,* vol. 8, pp. 152-158, 1985.

[15] J. Lee, "Performance analysis of rolling piston type rotary compressor," 1988.

[16] T. Ohkawa, E. Kumakura, H. Higashi, K. Sakitani, M. Higuchi, H. Taniwa, *et al.*, "Development Of Hermetic Swing Compressors For CO2 Refrigerants," 2002.

[17] M. Masuda, K. Sakitani, Y. Yamamoto, T. Uematsu, and A. Mutoh, "Development of swing compressor for alternative refrigerants," 1996.

[18] K. Furusho, T. Okawa, K. Saitoh, and K. Sakitani, "Numerical and Experimental Investigation of Swing Compressor Characteristics," 1998.

[19] X. Hu, Z. Qu, X. Yang, and J. Sun, "Theoretical study on frictional losses of a novel automotive swing vane compressor," *International Journal of Refrigeration,* vol. 36, pp. 758-767, 2013.

[20] K. S. Yap, K. T. Ooi, and A. Chakraborty, "Introduction of the Novel Cross Vane Expander-Compressor Unit for Vapour Compression Cycle," 2014.

Performance evaluation of a novel dual vane rotary compressor

A Shouman[1], A El Dein Hussin[2], A Hamed[2], M Serag El Din[3], N Mahmoud[4] and A El Baz[5]

[1] Emeritus Prof. of Mechanical Eng., New Mexico State University, Las Cruces, New Mexico.
[2] Lecturer, Faculty of Engineering, Ain Shams University, Cairo, Egypt
[3] Research Student, Faculty of Engineering, Ain Shams University, Cairo, Egypt
[4] Professor, Faculty of Engineering, Ain Shams University, Cairo, Egypt
[5] Professor, Faculty of Engineering, The British University in Egypt.

Email: ahmed.elbaz@bue.edu.eg

Abstract: Sliding-vane rotary compressors have considerable advantages in comparison with other positive displacement compressors in terms of volumetric efficiency, and continuity of flow without pulsations. However, sliding-vane rotary compressors have relatively low pressure ratio. This paper presents a novel design of a two-stage sliding-vane rotary compressor in attempt to elevate the pressure ratio of that type of compressors. The basic design features of the novel design are illustrated along with the adopted design geometry after studying the effects of several geometrical factors and their practicality. After several trials, the first prototype of the two-stage sliding-vane rotary compressor has been manufactured and tested experimentally to evaluate its performance. The paper reports the preliminary results of the flow rate, pressure rise and power consumption at different rotating speeds. Comparison of the performance of each stage is also included and the benefits of using the second stage are highlighted. The paper also summarizes the design challenges encountered during its manufacturing in addition to the challenges to reduce inter-stage leakage and the applied solutions.

1. Introduction

Air compressors are widely used in industrial processes such as petrochemicals, water treatment or food industry or as a control fluid for instance in pneumatic systems. Compressed air systems are also considered a main consumer of electricity in several applications. Therefore, air compressors are receiving continuous developments in order to improve their efficiencies and reduce overall plant electricity bill. In many applications, rotary compressors are usually privileged because they can match both pressure level and flow rate requirements within a wide range of electrical power from a few to hundreds of kilowatts. They also have advantages over other positive displacement compressors related to continuity of flow without pulsations. Furthermore, they present a favorable weight to power ratio. Moreover, rotary compressors are characterized by low specific energy consumption per unit mass of fluid inducted. The sliding-vane compressor (SVC) belongs to that family. It comprises a cylindrical rotor with longitudinal slots, within which individual sliding vanes are located. The cylindrical rotor is offset within a cylindrical stator, in which it rotates. As the rotor is running, the

IOP Conf. Series: Materials Science and Engineering **232** (2017) 012060 doi:10.1088/1757-899X/232/1/012060

sliding vanes are pushed out towards the cylindrical wall of the stator due to the centrifugal force and forming closed volume cells.

As a result of the eccentricity of the rotor axis the volume of each cell gradually decreases and thus air is compressed, see Fig. 1. The SVC is thus a single stage machine which can be used to compress gases to about 12 bar. To obtain higher pressures, two or more compressors are connected in series. Inter-cooling may be used to minimize compression work and approach isothermal compression work. Cipollone et *al.*, [1] presented a mathematical model for two sliding-vane compressors connected in series with intercooler. They reported that a two-stage compressor with inter-cooling might potentially lead to overall electrical savings up to 9.5%.

To the knowledge of the authors, multi-stage SVC such that the compression is performed inside the same compressor is not commercially available yet. Such design concept may reduce the specific energy cost especially when intercooler is utilized between the stages. Furthermore, that concept may lower the weight to power ratio. Valente and Villante [2] proposed a design for a single rotor with two-stage sliding-vane compressor, Fig. 2. Their method aimed to solve crucial difficulties related to performance matching between the two stages such as maintaining same pressure level at the discharge port opening. However, no physical prototype was produced for testing their proposed design.

The first author of the present paper filed a patent in 1999 [3]. The patent is a two-stage dual-vane compressor such that there is an elliptical inner stator and an elliptical outer stator. Eight sliding vanes are mounted in slots within a cylindrical rotor, such that the vanes can rotate by the rotor between the two stators. Meanwhile, each vane has the ability to slide radially through its slot in the rotor, see Fig. 3. The proposed first stage is located in the outer part, while the second stage is associated with the inner space of the device. Both stages are proposed as dual compressor with path A and B. For each path of A and B, an external pipe connects the outlet of the first stage (from the side) to the inlet of the second stage (perpendicular to the page). The current paper explores the design of a two-stage sliding-vane compressor. First, the basic geometry is introduced with geometric parameters involved. This is followed by details of the manufacturing process of the first prototype of this device. The last part presents the test apparatus and measurements obtained to evaluate the performance of the prototype at different operating conditions.

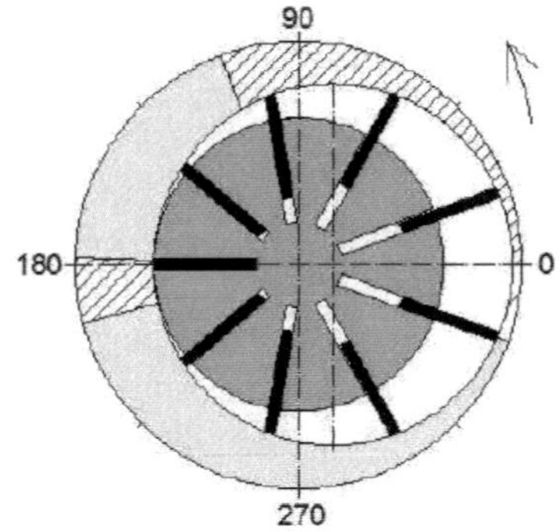

Fig.1: Typical single-stage SVC, [1].

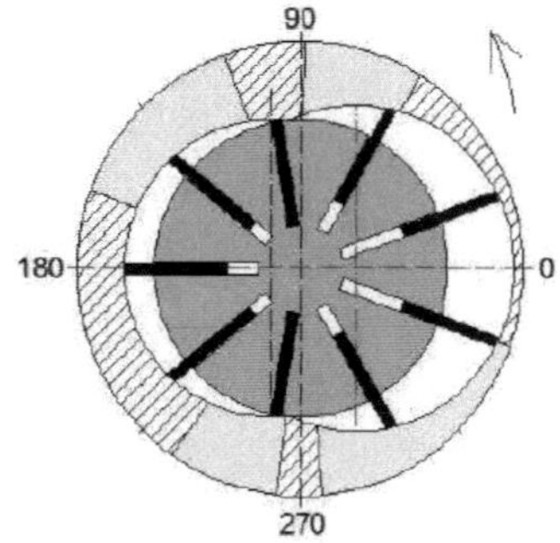

Fig.2: Conceptual design of two-stage SVC, [2].

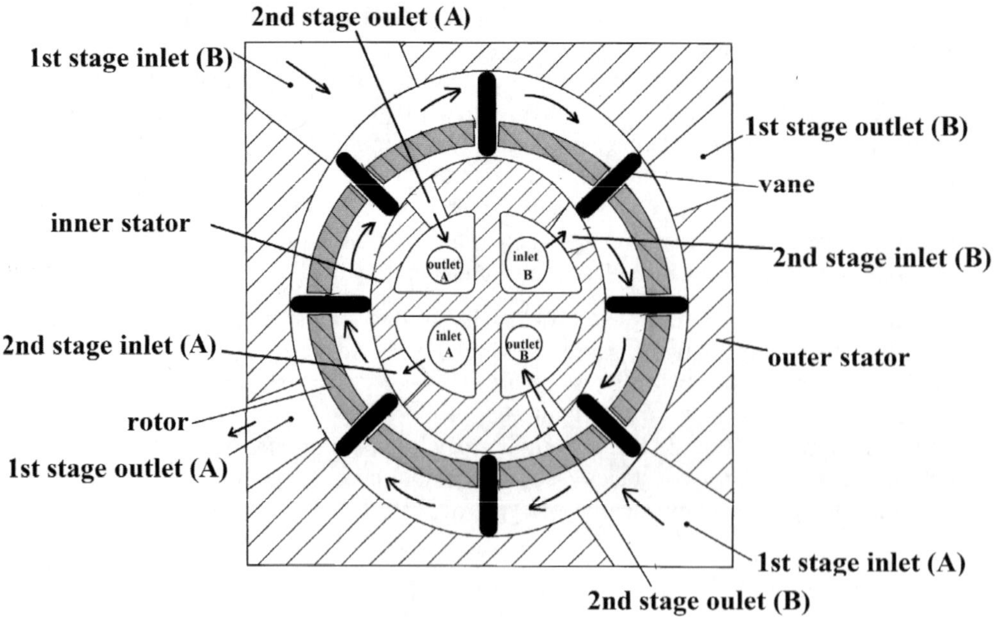

Fig. 3: Patent of a two-stage dual vane compressor, [3].

2. Two-Stage dual vane compressor

The proposed design of the patent was altered to avoid internal leakage, which may occur between the second stage and the first stage via the gap between the vane and its slot, see Fig. 3. This was one of the challenges encountered during the compressor manufacturing. The new design of the two-stage dual vane compressor consists of two concentric circular stators with a circular rotor mounted eccentrically inside, see Fig (4). The circular stators and rotor are easier to manufacture and must provide better contact angle with the rotating vanes. This is an essential requirement to reduce leakage between compressor chambers formed between successive vanes. Separate vanes are employed for each stage. Rotor slots, which house the vanes, are CNC machined along the outer and inner diameter of the rotor. A working cell exists between an adjacent pair of vanes, the rotor, stator surfaces and end plates. The volume of the cell changes as the rotor rotates. The value of the pressure at the end of the compression process is predetermined by suction pressure and the compression ratio. Fig. 5 shows the first prototype of the compressor with its cover removed. Each vane is pushed towards its stator surface to keep them contact using helical springs inserted inside the vane slots. Suction and discharge ports of the first stage are grooved in the outer stator while those for the second stage are grooved in the inner stator. The ports locations are selected such that the maximum cell volume position corresponds to end of suction port closing angle while the minimum volume position corresponds to the discharge port closing. The discharge port opening angle should correspond to the position where the cell pressure reaches a value just above the nominal discharge pressure. Fig. 6 and Fig. 7 show the port positions for the two stages. Discharge from the first stage is transferred to the second stage through an external pipe connected to the inner stator side. No inter-cooling is considered in the current work which only aims at proofing the new design concept. In future stages of the current work, air/water intercooler shall be employed between the two stages. The main dimensions of the prototype developed in the current work are listed in table.1

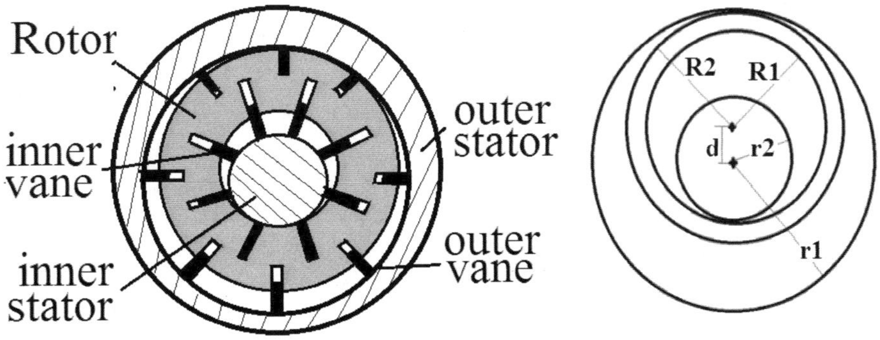

Fig. 4: Basic dimensions of the new design of the two-stage SVC.

Fig. 5: The first prototype of the dual vane, two-stage SVC.

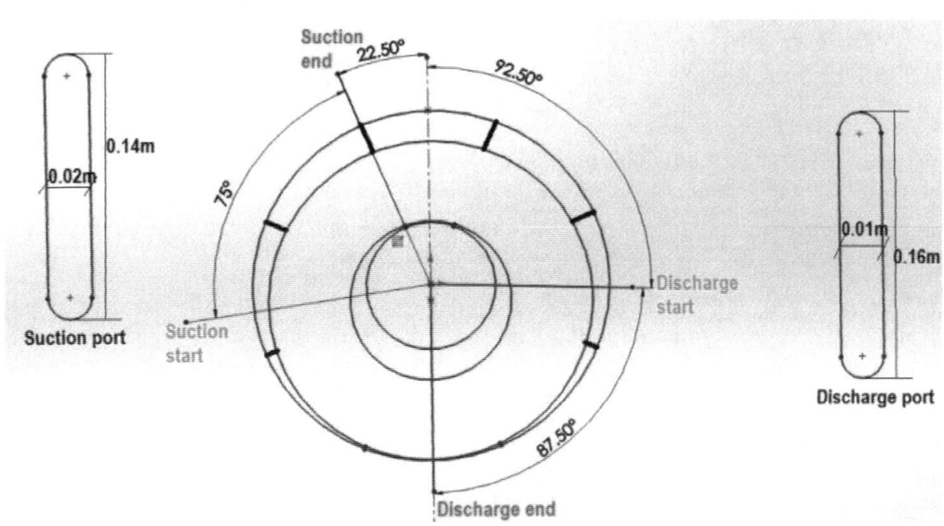

Fig. 6: Locations of suction and discharge ports of the first stage.

IOP Conf. Series: Materials Science and Engineering **232** (2017) 012060 doi:10.1088/1757-899X/232/1/012060

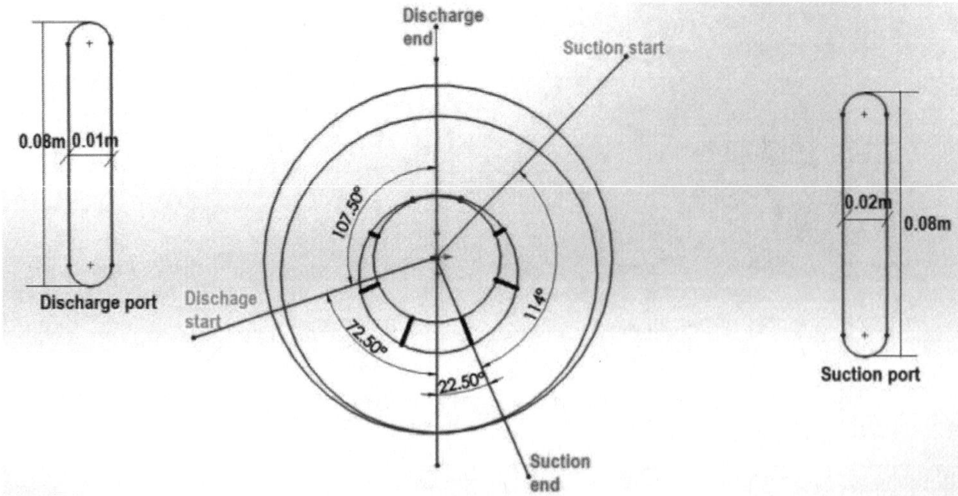

Fig. 7: Locations of suction and discharge ports of the second stage.

Table 1: Dimensions of prototype

	r_1	R_2	L	No of vanes	Eccentricity, d	Vane thickness
First stage	110 mm	100 mm	60 mm	8	10 mm	8 mm
Second stage	50 mm	60 mm	60 mm	8	10 mm	8 mm

The outer stator has an inner radius, r_1, of 110 mm while the rotor outer radius, R_2, is 100 mm. The eccentricity between outer stator and rotor, d, is 10 mm. The same eccentricity is maintained between the inner stator and rotor. The thickness of each of the compressor rotor and stator, L, is 60 mm. The values of these parameters were selected after parametric studying of the effects of each factor on the compressor performance and the manufacturing constrains such as the maximum and minimum eccentricity and the length and thickness of the vane to bear stresses and keep contact with its stator.

The theoretical flow rate of the compressor can be calculated using the maximum cell volume matching with the end of suction port. The cell total angle is 45° for 8 vanes. The theoretical discharge is obtained as:

$$Q_{th} = V|_{\theta suc} . z . \frac{S}{60} \quad (1)$$

Where, $V_{\theta suc}$, is the maximum cell volume, z is the number of cells, and S is the rotational speed in rpm. The volumetric efficiency, η_{vol}, of the compressor is the ratio between the measured flow rate, $\dot{Q}_{act}$, at suction to the theoretical flow rate, $\dot{Q}_{th}$, at the corresponding speed of rotation:

$$\eta_{vol} = \frac{\dot{Q}_{act}}{\dot{Q}_{th}} \quad (2)$$

Based on the prototype dimensions and position of ports, the theoretical compression ratio for first stage is 2 while for second stage is 1.5, assuming isentropic compression.

3. Cell Thermodynamics

The compressor cell acts as a variable volume that interacted with adjacent cells and with the rest of the compressor parts through the suction and discharge ports. A one-dimensional model is used to evaluate the compressor geometry, also to calculate volumetric flow rate which will be delivered by

the compressor, cell volume as per rotational angle to predict the discharge angle for each stage and indicated power so a compatible electric motor can be selected correctly.
The mass flow rate, $\dot{m}_{in}$, can be calculated as:

$$\dot{m}_{in} = \rho_{suc} . \dot{Q}_{th} \tag{3}$$

where, ρ_{suc} , is the air density at suction, and is calculated using the equation of state.
By applying the isentropic compression law, the pressure of discharge cell, p_{dis} , is calculated:

$$p_{dis} = p_{suc} \left(\frac{V|_{\theta_{in}}}{V|_{\theta_{out}}}\right)^k \tag{4}$$

where, k, is the air specific heat ratio and, p_{suc} , is the suction pressure. The discharge pressure can also be calculated as function of rotation angle, [4-6]:

$$\left(\frac{p_{dis}}{p_{suc}}\right)^{1/k} = \frac{2}{1 + \cos\theta - dr \times \sin\theta^2} \tag{5}$$

where, dr, is the relative eccentricity d/r_1. The cell volume, V_θ , can be calculated as function of rotation angle, [6]:

$$V_\theta = R \times d \times L\left\{\beta + 2 \times \sin\frac{\beta}{2} \times \cos\theta + \frac{1}{2} d \times r \times \sin\beta \times \cos 2\theta - \frac{1}{2} d \times r \times \beta\right\} \tag{6}$$

where, $\beta = \frac{2\pi}{z}$
Hence, the indicated power required, P_{ind}, can be obtained as, [5]:

$$P_{ind} = z\,\omega \int p\,dV \tag{7}$$

where z is the number of compressor vanes and ω is the rotational speed in rad/s.

3. Design challenges

3.1 Leakage:

Leakage potential in this design is located at two main locations; first, the contact point between end plates and rotors where high compressed air in second stage tends to return back to first stage. The front plate was designed to be in continues contact with rotor to prevent leakage where friction is also taken into account. Second, the contact point between second stage vane blades and inner stator. The vane blade tip of the second stage was machined with an inclination allowing the vane to be in continuous contact with the outer surface of second stage inner stator and helical springs were installed inside rotor slots between vanes and rotor; by applying these solutions leakage was eliminated between adjacent cells.

3.2 Friction and vane dynamics

Friction occurs at several locations in the two-stage SVC; between shaft and stators, side wall end plates, rotor and blades of the two stages. The friction power consumed in the two-stage compressor is expected to be higher than that in conventional single stage SVC. Many forces act on the vane blade during rotation especially during compression phase. The blade in first stage has sliding motion combined with rotation motion. The rotational motion generates a centrifugal force which forces the vane outside the rotor beside spring force. During compression phase, a pressure difference acts on the blade between adjacent cells generating a pressure force. In the second stage, the same motions and forces mentioned in first stage act on the blades of the second stage except springs force which act in opposite direction where they force blades to slide inward to overcome centrifugal forces in order to keep blades in continuous contact with second stage stator to prevent leakage between adjacent cells.
Oil lubrication was used to minimize friction forces between moving parts and fixed parts. Oil was injected to inlet air by gravity and recovered from second stage outlet through an oil separator.

4. Experimental test setup

The prototype of the two-stage dual vane compressor was tested at the fluid machines laboratory of the Faculty of Engineering, Ain Shams University. The compressor was driven by a 5 kW 3-phase induction motor using a V-belt, Fig. 8. The electric motor was driven by a frequency inverter through which rotational speed was changed. The rotational speed was kept below 1000 rpm in the initial testing. Compressor suction is equipped with oil injection system. An oil separator at compressor discharge was used to separate oil and return it to the oil reservoir again, Fig. 9. Air discharge was connected to a storage tank from which air can be consumed at the required pressure and flow rate via a needle valve, through which the compressor was tested at two discharge pressure; 4 bar and 5 bar absolute. At the inlet and outlet of each stage, pressure transducers and temperature sensors, RTD type were used to measure pressures and temperatures, respectively. Air flow rate was measured using sharp edged orifice plate, in compliance with ISO 5167. The orifice plate was installed in the suction line to the compressor. A torque meter was used to measure torque and rotational speed of the compressor. Two sets of experiments were carried out on the compressor. In one set, the single stage was active only while the second stage was not activated by discharging the outlet of the first stage to the tank directly. In the second set of tests both stages were activated.

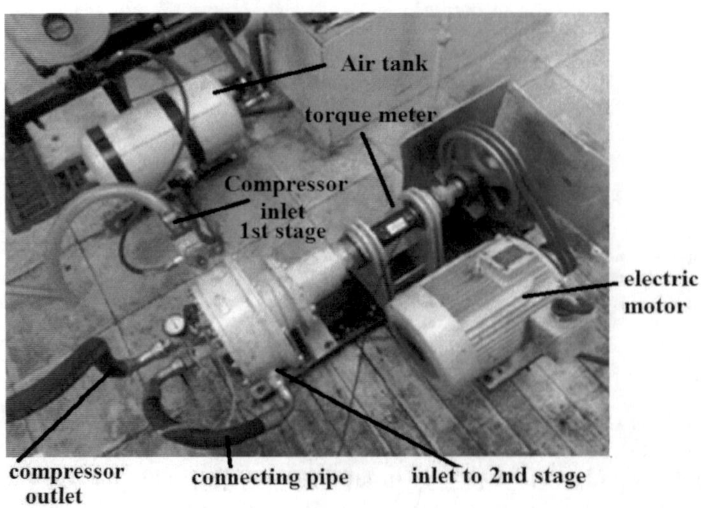

Fig. 8: Photographic illustration of the experiment test rig.

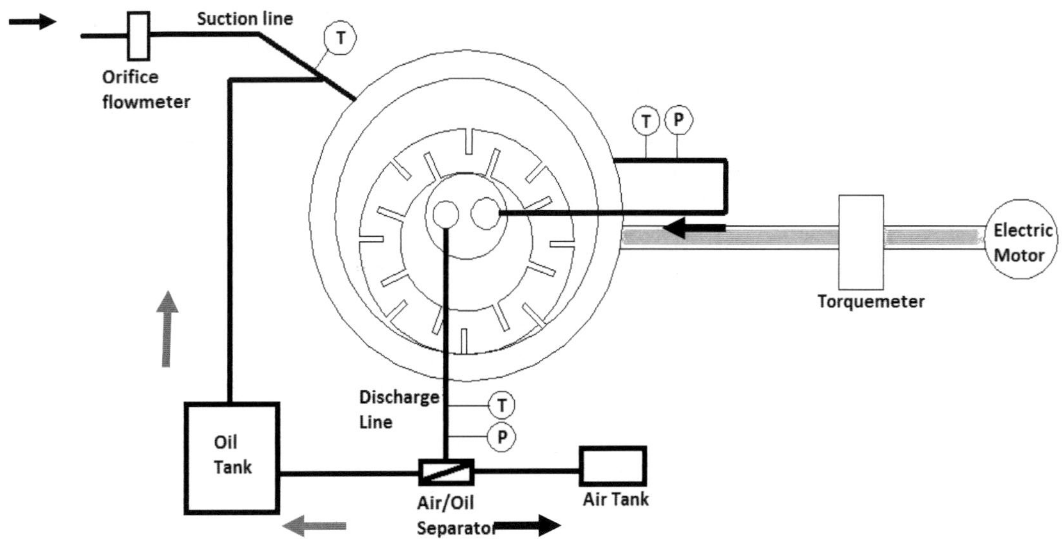

Fig. 9: Schematic of the experiment test rig showing locations of pressure sensor P and temperature sensors T.

5. Results and Discussion

Figure 10, shows the measured variation of inlet air volume flow rate versus rotational speed, at a discharge pressure of 4 bara, in the two cases; single stage active and both stages active. The flow rate increases linearly with rotational speed which is typical for SVC. The actual flow rate is less than the theoretical flow rate at each operating speed, which indicates relatively low volumetric efficiency. The calculated average volumetric efficiency for single stage operation was 76% while being 66% for two stage operation. When the two stages were activated, the flow rate was less than that of the single stage, at the same rotational speed and discharge pressure. The same behavior is shown in Fig. 11, when the discharge pressure was 5 bara. This may be related to a mismatch between the chamber volume at second stage during suction and the flow rate delivered from first stage. Another explanation for the reduced flow rate for two stage operation could be the internal leakage from second stage to first stage.

Figure 12, shows the measured variation of mechanical power input to the compressor with rotational speed for single stage operation and two stage operation and a discharge pressure of 5 bara. The power input increases nearly linearly with rotational speed. The measured power consumption for single stage operation is approximately 8% higher than two stage operation for the same discharge pressure and different rotational speeds. The power consumption at 750 rpm is 5.2 kW for two stage operation while being 5.5 kW for single stage operation. This difference is not significant since similar friction losses are present in both cases. However, a slight drop of power consumption is expected for two stage operation. For single stage operation, higher resisting forces are expected to act on compressor vanes due to the relatively higher size of vanes. The two stage operation, part of the pressure rise occurs in the second stage which has smaller vanes and resisting forces on smaller vanes may contribute to reduced power consumption.

The estimated efficiency, calculated as the ratio between output fluid power and input mechanical power, is approximately 20%. This low efficiency is attributed to high friction forces resulting from friction between rotating parts and end plates and between rotor vanes and stators. The end plates are forced against the rotating parts by springs to minimize leakage in this initial design which causes continuous rubbing between two rows of vanes with these end plates. We have doubts that these contact surfaces have any lubricating oil to reduce frictional forces.

Another reason for the increased compressor power consumption could be attributed to high oil level inside the compressor. The high oil level imposes high torque on compressor vanes as well. Therefore, the observed low efficiency can be improved if detailed design tips of the compressor are improved to minimize these losses. However, the present results can be considered as a good proof of concept of the dual vane, two-stage SVC. In the future upgrade other methods shall be applied to reduce oil level. Also, a sight glass shall be added to monitor oil level.

6. Conclusions

The current research evaluates the performance of a prototype of a two-stage, dual vane SVC with conventional oil injection at suction line and without intercooler between the two stages. Volumetric flow rate showed similar characteristics of two stage, dual vane design and single vane design. The volumetric efficiency is low due to internal leakage which can be reduced if better design of end plate is adopted. Overall compressor friction losses in the two-stage design is contributing to the high mechanical power requirements observed in the experiment and hence low compressor efficiency. Despite the several weak points observed in the first prototype performance, the present results can be considered a good proof of concept of dual vane, two stage siding vane compressor design.

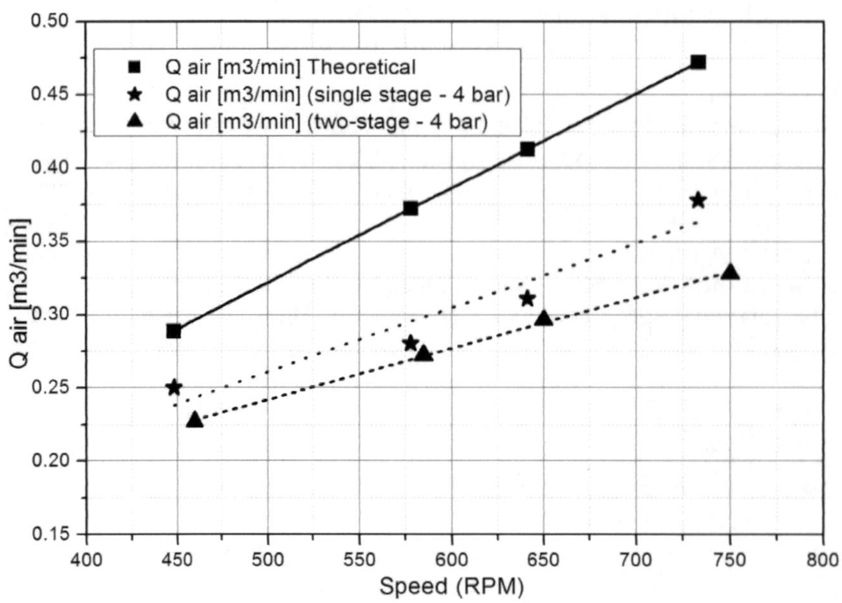

Fig. 10: Measured variation of inlet air flow rate with rpm for a discharge pressure of 4 bara.

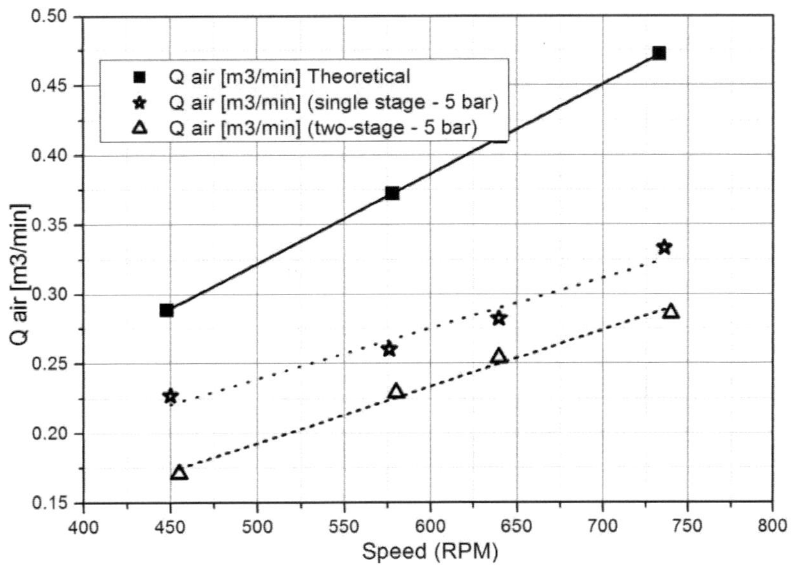

Fig. 11: Measured variation of inlet air flow rate with rpm for a discharge pressure of 5 bara.

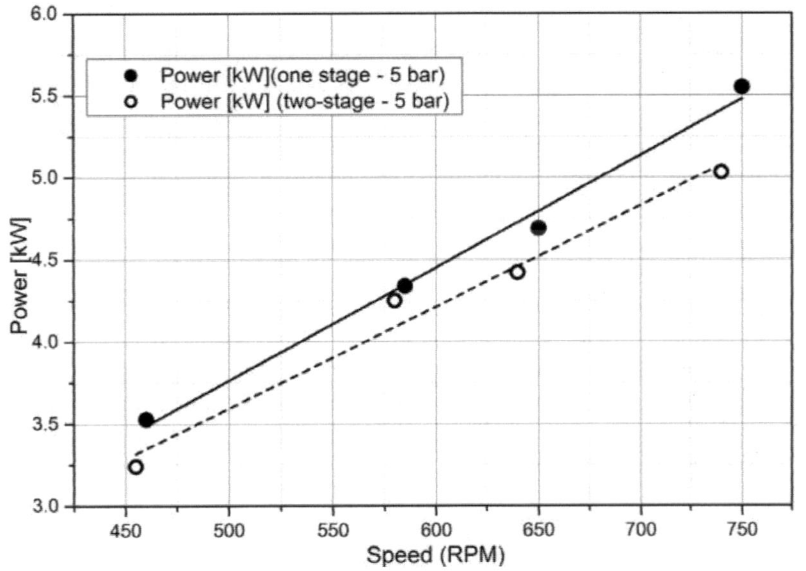

Fig. 12: Comparison between measured variation of mechanical power with rpm for single stage and two stage operation. Discharge pressure = 5 bara.

Acknowledgment
The present work has been funded by the Science and Technology Development Fund (STDF) project number 11871-ERAfrica 27.

IOP Conf. Series: Materials Science and Engineering **232** (2017) 012060 doi:10.1088/1757-899X/232/1/012060

Nomenclature

d	Eccentricity (mm)	β	$2\pi/z$
dr	Relative eccentricity=d/r_1	η_{vol}	Volumetric efficiency (%)
L	Compressor radial length (mm)	ρ_{suc}	Air density at suction port (kg/m^3)
m_{in}	Mass flow rate at suction (kg/s)	ω	Rotational speed (rad/s)
p	Pressure (bar)		
P_{ind}	Indicated power (kW)		
Q	Volumetric flow rate (m^3/s)		
r_1	First stage stator inner radius (mm)		
r_2	Second stage outer radius (mm)		
R_1	Rotor inner radius (mm)		
R_2	Rotor outer radius (mm)		
S	Compressor shaft speed (RPM)		
V	Volume (m^3)		
V_θ	Cell volume		
z	Number of vanes		

References

[1] R Cipollone, G Bianchi1, D Di Battista1, G Contaldi, S Murgia, “Model Based Design Of An Intercooled Dual Stage Sliding-Vane Rotary Compressor System”, 9th International Conference on Compressors and their Systems, IOP Publishing IOP Conf. Series: Materials Science and Engineering 90 (2015) 012035.

[2] Valente, R. and Villante, C., "On the Optimal Design of One-Rotor Two-Stages Rotary-Vane Compressors" (2008). International Compressor Engineering Conference, Purdue, July 14-17. Paper 1895.

[3] A Shouman (1999) “Multiple Purpose Two Stage Rotating Vane Device”, Patent No. 5,855,474, Date of Patent (Jan. 5, 1999).

[4] G Bianchi and R Cipollone (2015) “Theoretical Modeling And Experimental Investigations For The Improvement Of The Mechanical Efficiency In Sliding Vane Rotary Compressors”, Applied Energy 142 (2015).

[5] Tramschek, A. B. and Ooi, K. T. (1992), “Effects of Port Geometry, Dimensions and Position on the Performance of a Rotary Compressor”, International Compressor Engineering Conference.Paper 912.http://docs.lib.purdue.edu/icec/912

[6] Aradau, D. and Costiuc, L., “Friction Power in Sliding Vane Type Rotary Compressors”, (1996). International Compressor Engineering Conference Paper 1357, http://docs.lib.purdue.edu/icec/1357.

IOP Conf. Series: Materials Science and Engineering **232** (2017) 012061 doi:10.1088/1757-899X/232/1/012061

Performance improvement of air source heat pump by using gas-injected rotary compressor

B L Wang*, X R Liu, Y C Ding and W X Shi

Department of Building Science, Tsinghua University, Beijing 100084, China

*Wangbl@tsinghua.edu.cn

Abstract. Rotary compressor is most widely used in small capacity refrigeration and heat pump systems. For the air source heat pump, the heating capacity and the COP will be obviously degraded when it is utilized in low temperature ambient. Gas injection is an effective method to enhance its performance under those situations, which has been well applied in air source heat pump with scroll compressor. However, the development of the gas injection technology in rotary compressor is relatively slow due to limited performance improvement. In this research, the essential reason constraining the improvement of the gas injection on the rotary compressor and its heat pump system is identified. Two new injection structures for rotary compressors has been put forward to overcome the drawback of traditional injection structures. Based on a verified numerical model, the thermodynamic performance of air source heat pumps with the new gas-injected rotary compressor are investigated. The results indicate that, compared to the air source heat pump with the traditional gas injected rotary compressor, the new injection structures both can enhance heating capacity and COP of the air source heat pump obviously.

1. Introduction

Rotary compressors have been widely applied in small capacity air conditioners and heat pumps due to its advantages such as high efficiency, low cost and high reliability. In 2014, more than 100 million rotary compressors have been produced in China. However, the rotary compressor will face the substantial performance degradation when it is adopted in air source heat pump (ASHP) working in low temperature ambient, just like the scroll compressor encountered in light commercial applications. The heating capacity and COP will quickly decrease with the descending of the outdoor temperature, which has become a main technical barrier restraining the further application for ASHP in cold region. Aiming at the problems, many solutions have been proposed for the scroll compressor and the screw compressor, such as cascade cycle, two-stage compression, vapor injection, and so on. Previous many investigations confirmed that the vapor injection technology can improve the heating capacity and COP in low ambient temperature by enhancing the refrigerant mass flowrate and decreasing the under compression loss [1-3]. Consequently, the vapor injection technology has been widely adopted in scroll compressor currently due to both its technical and economic effectiveness.

For rotary compressors, the two-stage compression has been used for years [4, 5]. But different from the traditional two-stage compression system using two separated compressors, the two compression cylinders of rotary compressor always are installed in one shell and through the same crankshaft, which is called twin-cylinder rotary compressor. The twin-cylinder structure solves a lot of critical operation problems two-stage compression system encountered when two separated single-cylinder rotary compressors are adopted, such as easier oil balance, lower noise and vibration due to more balanced structure. However, it has been proofed that the performance of the twin-cylinder rotary compressor is lower than the single-cylinder one under the temperate working conditions [6], such as low compression ratio air conditioning working condition, because of twice discharge loss and fixed volume ratio of two cylinders. As a result, the vapor injection in single cylinder, instead of the two-stage compression, was considered in rotary compressors to realize higher thermodynamic performance both in wild and mild ambient. Besides, compared to the twin-cylinder rotary compressor, the single-cylinder rotary compressor has lower cost, better reliability, and smaller size.

In this paper, the drawbacks of traditional injection structures of rotary compressor are analyzed, two kinds of new injection structures for rotary compressors are developed, and finally their thermodynamic performance are investigated.

2. Drawbacks of traditional injection structures of rotary compressor

According to the different position of the injection port, the single rotary compressor has two types of traditional injection structures, the cylinder injection, in which the injection port was opened on the cylinder, and the end-plate injection, in which the injection port was opened on the end-plate. Apparently, for the cylinder injection, when the rolling piston rotates after the injection port, the injection port will connect to the suction pocket, which results in the injected refrigerant gas was injected into the suction pocket and it will flow back into suction tube as Figure 1(a). The back-flowing refrigerant will largely decrease the mass flowrate of suction, and finally weakens the performance enhancement by gas injection. Previous research [7] showed that 23.3%~29.3% injected refrigerant flowed back into the suction tube and the heating capacity of single-cylinder rotary compressor with traditional cylinder injection is only 5.6%~14.4% higher than the common single-cylinder compressor, which is obviously lower than the 20%~25% improvement of twin-cylinder rotary compressor over the common single-cylinder one. That could be the main reason why the twin-cylinder rotary is more widely used than the gas-injected single-cylinder in low temperature ASHP.

For the end-plate injection, the location of the injection port are diverse. Figure 1(b) shows one of the typical location. It's easy to find that the back-flowing of the injected refrigerant still existing in that end-plate injection structure, and the negative effect is even more serious due to a longer back-flowing time span compared to the cylinder injection.

As a conclusion, developing gas injection system without back-flowing for rotary compressor is quite important for pursuing high heating performance.

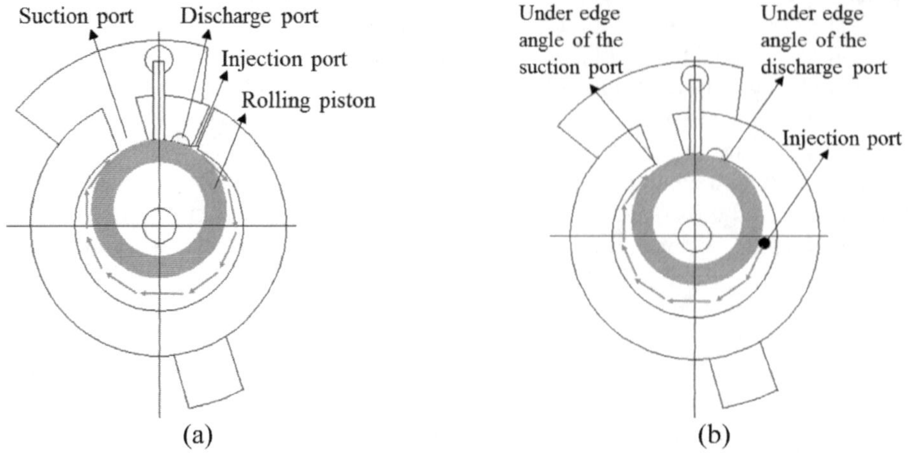

Figure 1. Traditional injection structures for single rotary compressors: (a) cylinder injection structure; (b) end-plate injection structure.

3. Blade injection structure and its performance

A novel injection structure is proposed in Figure 2, in which the injection passage is screwed in the middle of the blade and the injection port is opened oriented to the discharge port. The new injection structure also need to mill a platform on the blade where the spring valve plate and the lift limiter are installed.

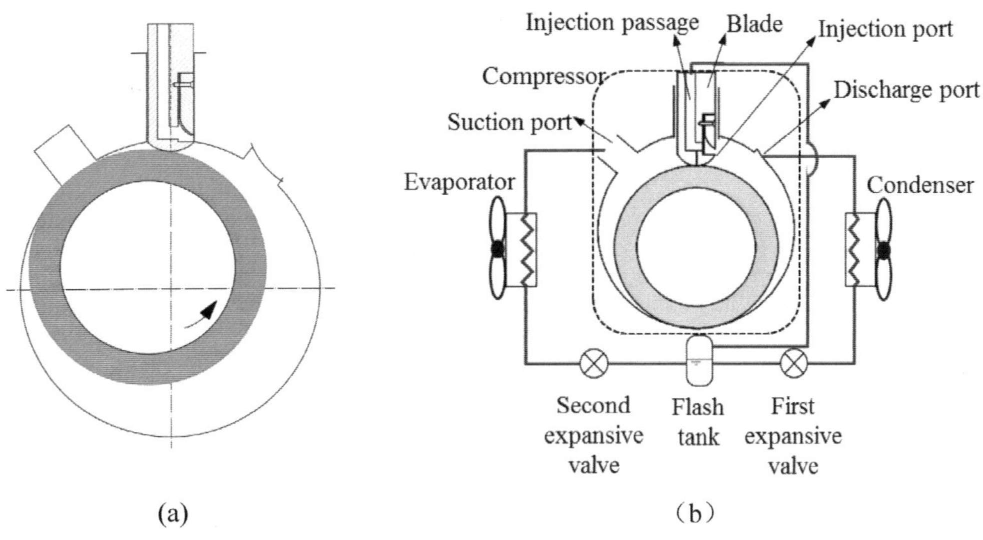

Figure 2. Rotary compressor and ASHP with blade gas injection: (a) compressor; (b) ASHP.

The detailed working process of the rotary compressor with the new injection structure can be explained as following. Before the suction pocket is isolated from the suction tube, the blade don't protrude enough and the injection port is still sealed by the wall of the blade house. Under the press of the spring valve, the injection port is closed, and the injection process does not start. When the suction pocket is isolated from the suction tube, the suction process finishes. At the same time, the blade moves down where the lower edge of the injection port exactly reveals, the injection port connects to the compression pocket and then the spring valve plate opens as the injection pressure is greater than the compression pocket pressure, starting injection. With increase of the rotation angle and increase of the compression pocket pressure, as the compression pocket pressure equals the injection pressure, the spring valve plate closes and the injection process ends. Hereafter, the rolling piston rotates to the bottom center, the initial discharge and the end of discharge, respectively. The injection spring valve plate keeps close.

Therefore, the rotary compressor with the blade injection structure can totally avoid the back-flowing of the injection refrigerant into the suction tube. At the same time, it can achieve injection at the first time when the suction process finishes, and so it can achieve the maximum of injection mass flow. Moreover, the volume between the injection port and the spring valve plate at the injection structure has no influence upon the volumetric efficiency. Because the volume does not connect to the suction tube, it has no influence on the compressor suction mass flow-rate from the suction tube.

Based on a validated model [8], the performance of the ASHP with blade gas injected is compared with the traditional cylinder injection structure.

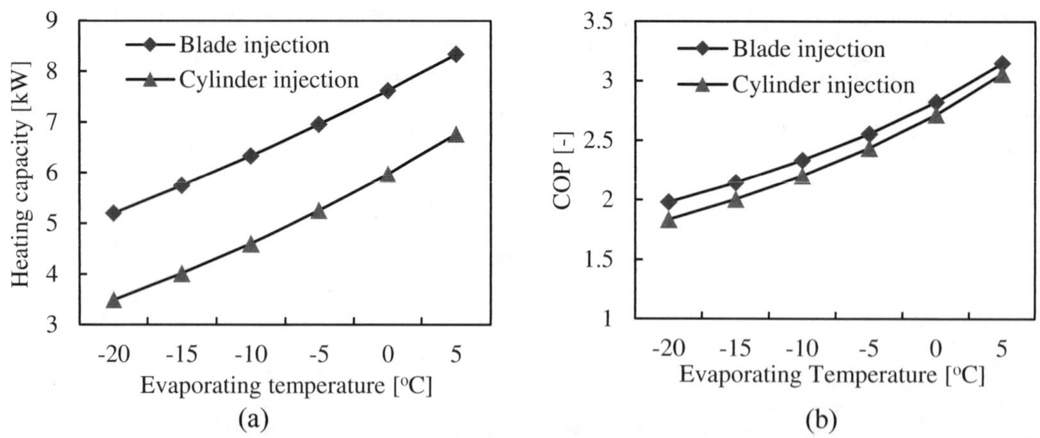

Figure 3. Performance improvement of ASHP by blade gas injection compared to the traditional cylinder injection: (a) heating capacity; (b) COP.

Figure 3 shows the heating capacity and COP of a rotary compressor different working conditions. Comparing to the rotary compressor with traditional cylinder injection structure, the heating capacity and COP of the rotary compressor with the blade injection structure are enhanced by 23.1~48.9% and 3.2~8.0%, respectively. That means the rotary compressor with the new injection structure can solve the degradation of the heating capacity and COP in low ambient temperature to a large extent.

4. No-back-flowing End-plate injection structure and its performance

Apparently, the back-flowing of injected refrigerant is inevitable for the cylinder injection structure because the gas injection port must lower than the discharge port. But for the end-plate injection, the conclusion is not so sure. Firstly, the research attempts to analyze the possible location without back-flowing and the size of the injection port for the end-plate injection.

According to the rotation mechanism of the rolling piston, the necessary and sufficient conditions for the end-plate injection structure avoiding the refrigerant gas back-flowing the suction tube completely are: (1) the injection port cannot connect with the inner circle of the rolling piston at any time to avoid large influence on the compression work and lubricant performance, which mathematically means the injection port should NOT be set in the union set of inner circle of the rolling piston at all rotating angle; (2) the injection port cannot connect with the suction pocket at any time, which mathematically means the injection port should NOT be set in the union set of the suction pocket at all rotating angle; (3) the injection port connects with compression pocket at some time, which mathematically means the injection port must be set in the union set of the compression pocket at all rotating angle.

Obviously, to avoid the back-flowing of the injected refrigerant, the eligible injection area equals to the intersection of the area which is confirmed by the above three conditions.

As a conclusion, the no-back-flowing injection port area does exist. It is close related to the parameters of the compressor and can be expressed by mathematic equations. The area is a half crescent close to the discharge port as Figure 4.

Based on the validated model of previous research [8], the performance of the ASHP with No-back-flowing end-plate injection is also compared with the traditional injection structure.

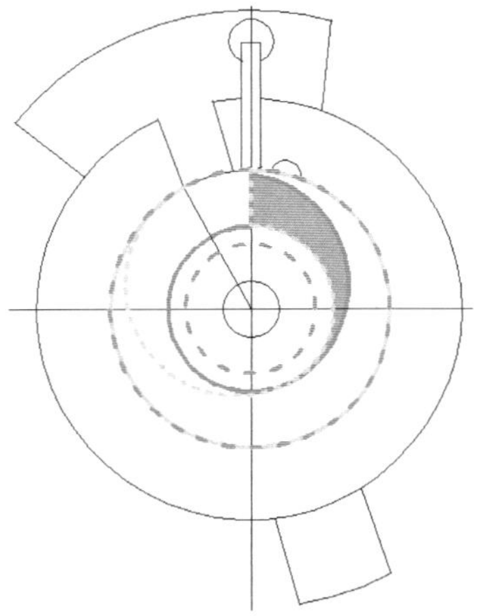

Figure 4. No-back-flowing end-plate injection port area

Figure 5 show the heating capacity and COP of the new no-back-flowing end-plate injection rotary compressor compared to the traditional end-plate injection. The heating capacity for the novel injection system has a greater improvement over 7.8%~11.6% compared to the traditional end-plate injection system as Figure 5(a). The COPs of two end-plate injection system is shown as Figure 5(b). The air conditioner employing the no-back-flowing end-plate injection rotary compressor has a higher COP compared to the system with the traditional end-plate injection rotary compressor over 0.6%~2.9%.

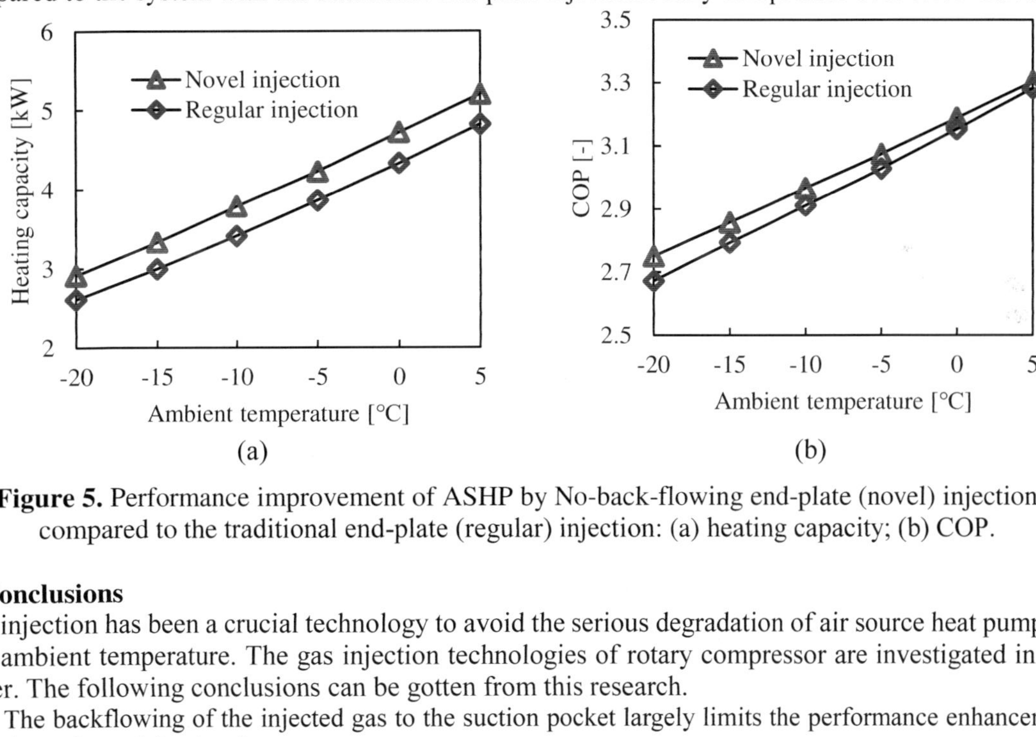

Figure 5. Performance improvement of ASHP by No-back-flowing end-plate (novel) injection compared to the traditional end-plate (regular) injection: (a) heating capacity; (b) COP.

5. Conclusions

Gas injection has been a crucial technology to avoid the serious degradation of air source heat pumps in low ambient temperature. The gas injection technologies of rotary compressor are investigated in this paper. The following conclusions can be gotten from this research.

1) The backflowing of the injected gas to the suction pocket largely limits the performance enhancement through gas injection for rotary compressor;
2) A blade injection structure and a new end-plate injection structure, which both can avoid backflowing of injected gas, are proposed and researched;
3) Compared to the traditional cylinder injection structure, the heating capacity and COP of rotary compressor with the blade injection structure is enhanced by 23.1~48.9% and 3.2~8.0%, respectively;
4) Compared to the air conditioner with the regular end-plate rotary compressor, the air conditioner with the no-back-flowing rotary compressor can enhance the heating capacity and COP by 7.8%~11.6% and

IOP Conf. Series: Materials Science and Engineering **232** (2017) 012061 doi:10.1088/1757-899X/232/1/012061

0.6%~2.9%, respectively.

6. References

[1] E. L. Winandy, J. Lebrun, Scroll compressors using gas and liquid refrigerant: experimental analysis and modeling. International Journal of Refrigeration 25 (2002) 1143–1156.
[2] B. Wang, W. Shi, L. Han, Optimization of refrigerant system with gas-injection scroll compressor. International Journal of Refrigeration 32 (2009) 1544–1554.
[3] H. G. Wu, X. Y. Peng, Z.W. Xing. Experimental study on p–V indicator diagrams of twin-screw refrigeration compressor with economizer. Applied Thermal Engineering, 24 (2004) 1491–1500.
[4] J.Heo, M. W. Jeong,C. Baek, Comparison of the heating performance of air-source heat pumps using various types of refrigerant injection. International Journal of Refrigeration 34 (2011) 444–453.
[5] S. X. Xu, G. Y. Ma, Experiment study on two-stage compression refrigerant/heat pump system with dual-cylinder rolling piston compressor. Applied Thermal Engineering, 62 (2014) 803–808.
[6] Q. L. Jia, L. W. Feng, G. Yan, Experimental Research on Heating Performance of Rotary Compression System with Vapor Injection. Journal of Refrigeration 36 (2015) 65–70. (In Chinese)
[7] X. R. Liu, B. L. Wang, W. X. Shi, P. L. Zhang, A novel vapor injection structure on the blade of a rotary compressor, Applied Thermal Engineering 100(2016) 1219-1228.
[8] B.L. Wang, X.R. Liu, W.X. Shi. Comparative research on air conditioner with gas-injected rotary compressor through injection port on blade. Applied Thermal Engineering 106 (2016) 67–75.

Acknowledgments

This research is funded by the Innovative Research Groups of the National Natural Science Foundation of China (Grant No. 51521005) and the China National Key R&D Program "Solutions to heating and cooling of buildings in the Yangtze river region" (Grant No. 2016YFC0700304).

Modeling and analysis of an open-drive Z-compressor

Davide Ziviani, Eckhard A. Groll

Purdue University, School of Mechanical Engineering, Ray W. Herrick Laboratories, West Lafayette, IN, USA

E-mail: dziviani@purdue.edu, groll@purdue.edu

Abstract. A rotary Z-compressor prototype for compressed air applications has been developed and tested. The Z-compressor working process resembles the one of a two-stage rolling piston compressor where the stages are phased by half rotation. In contrary to the traditional rolling piston design, the vane in a Z-compressor is positioned parallel to the main shaft and not perpendicular. In order to understand the impact of leakage and frictional losses and improve the design of such machine, a mechanistic model has been developed to include governing equations within the working chambers (i.e., two suction chambers and two compression chambers), leakage flow models, detailed mechanical analysis, one-degree of freedom valve model, in-chamber heat transfer and an overall energy balance of the compressor shell. The model has been validated with preliminary experimental data and then exercised to identify the potential performance improvements over a range of clearances and working conditions.

1. Introduction

This paper analyzes an air-compressor prototype that belongs to the family of rotary compressors. A cut-away view of the compressor is shown in Figure 1. To be noted is that the design concept resembles a two-stage rolling compressor and presents two sets of compression and suction chambers with a half revolution shift between the upper and lower level. The volume of the chambers is the space between inner and outer cylinder walls and the height of the chambers changes periodically as function of the crank angle. The element that separates upper and lower chambers is named blade. Additionally, the chambers on the same side of the blade are separated by means of a point of zero height with the housing and by a sliding vane. Due to the shape of blade and inner cylinder from a front view (see right side of Figure 1), the compressor is denoted as Z-compressor.

The Z-compressor was first analyzed by Jovane [1] in his dissertation. Experimental work done on a prototype showed that the Z-compressor had low vibration and noise effects, but its performance was heavily affected by leakages and friction losses. A model of the dynamics of the Z-compressor was developed in order identify the important sources of losses, such as leakage, friction, and heat transfer and to optimize the geometry. The model was validated with internal and external experimental data generated using R410A as the working fluid. From the numerical results, it was concluded that volumetric efficiency and overall isentropic efficiency were lower compared to the efficiency of a rolling piston rotary compressor [2] with the exception of the indicated efficiency which was comparable.

This paper focuses on an improved Z-compressor design in which the blade is part of the outer cylinder. A mechanistic model has been developed to break down leakage, fric-

IOP Conf. Series: Materials Science and Engineering **232** (2017) 012062 doi:10.1088/1757-899X/232/1/012062

tional and heat transfer losses and further employed to investigate potential improvements.

Nomenclature

A	Area (m^2)
h	Specific enthalpy (kJ/kg)
L	Length (m)
m	Mass (kg)
$\dot{m}$	Mass flow rate (kg/s)
N	Rotational speed (-)
p	Pressure (kPa)
$\dot{Q}$	Heat rate (kW)
Re	Reynolds number (-)
T	Temperature (K)
u	Specific internal energy (J/kg) (-)
v	Specific volume (m^3/kg)
V	Volume (m^3)
$\dot{V}$	Volumetric flow rate (m^3/s)
$\dot{W}$	Power (W)
η	Efficiency (-)

Subscript

amb	ambient
dis	discharge
eff	effective
fr	friction
in	inlet
is	isentropic
JB	journal bearing
m	mechanical
oa	over-all
out	outlet
sh	shaft
suc	sucion
TB	thrust bearing
th	theoretical
v	volume

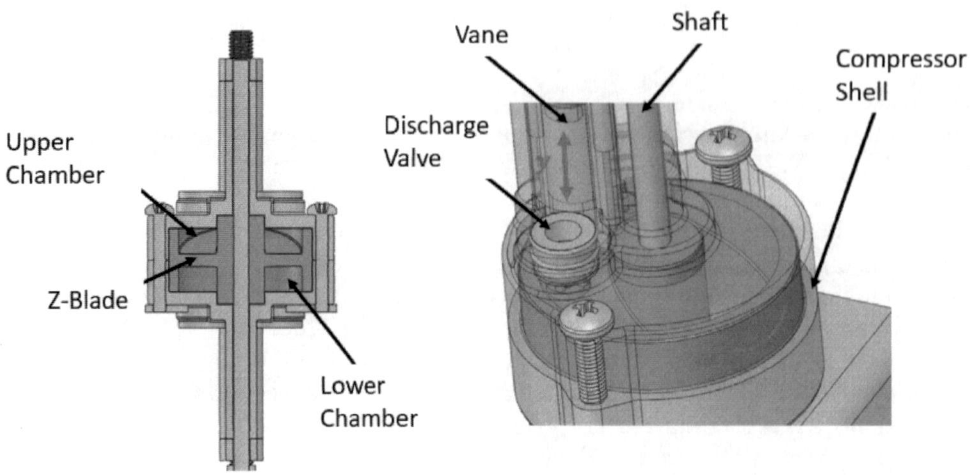

Figure 1. Cutaway view (left) and transparent 3D view (right) of the Z-compressor.

2. Geometric Characteristics

The Z-compressor design considered in this work is an open-drive one. On each side of the compressor there are two ports allowing for suction and discharge processes. Reed valves are installed on the discharge ports. The main design difference with respect to the original Z-compressor analyzed by Jovane [1] is the blade integrated with the outer cylinder which avoids direct leakage paths from upper to lower chambers and vice-versa as shown in Figure 1.

The working process is similar to a twin rolling piston rotary compressor in which two compression processes occur in one rotation. Four working chambers can be identified: upper suction and compression chambers and lower suction and compression chambers as seen in the cutaway view shown in Figure 1. Differently from a rolling piston rotary compressor, the Z-blade allows for volume variations. Two sliding vanes separate suction and compression chambers located on the same side of the Z-blade. The volume of each chamber is the space between inner and outer cylinders and the height of the chambers is dictated by the height of the portion of the Z-blade in contact with the vane, y_b. The length of each portion of the vanes into the working chambers is given as:

$$y_{\text{up}}, y_{\text{low}} = \frac{h_b}{2}\left[1 \mp cos(\theta)\right] \tag{1}$$

where h_b is the maximum height of the Z-blade, the minus sign is used for the upper chambers (up) and the plus sign for the lower chambers (low) and θ is the rotation angle measured from the point of contact between the Z-blade and the vane to the point at with the Z-blade has zero height with respect to the housing.

The volume of the suction and compression chambers in both sides of the z-blade are obtained by defining a maximum chamber volume as

$$V_{max} = \frac{\pi}{2} h_b \left(r_o^2 - r_i^2\right) \tag{2}$$

where r_o and r_i are the inner and outer radius of the Z-blade, and the vane volume within the chamber which is a function of the height of the chambers and the shape of the vane, i.e., $V_v = f(y(\theta),\text{geo})$. Therefore, the upper suction and compression chambers are given by,

$$V_{s,up} = \frac{V_{max}}{2\pi}\left(\theta - sin\theta\right) - V_{v,up}(\theta) \tag{3}$$

$$V_{c,up} = \frac{V_{max}}{2\pi}\left(2\pi - \theta + sin\theta\right) - V_{v,up}(\theta) \tag{4}$$

The lower suction and compression chambers are obtained from Equation 3 and Equation 4 by including the half revolution shift:

$$V_{s,up} = \frac{V_{max}}{2\pi}\left(\pm\pi + \theta - sin(\theta \pm \pi)\right) - V_{v,up}(\theta \pm \pi) \tag{5}$$

$$V_{c,up} = \frac{V_{max}}{2\pi}\left(2\pi \mp \pi - \theta + sin(\theta \pm \pi)\right) - V_{v,up}(\theta \pm \pi) \tag{6}$$

where the upper sign on Equation 5 and Equation 6 is used for $\theta \leq \pi$ and the lower sign for $\theta > \pi$.

The main dimensions of the Z-compressor are reported in Table 1. The evolution of the working chamber volumes and their derivatives are shown in Figure 2 and they can be integrated into the compression process governing equations.

Table 1. Main baseline compressor dimensions.

Component	Dimension
Shaft radius (r_s)	1.6 mm
Blade inner diameter radius (r_i)	5 mm
Blade outer diameter radius (r_o)	13 mm
Maximum chamber height (h_b)	10 mm
Vane thickness (t_v)	1.5 mm

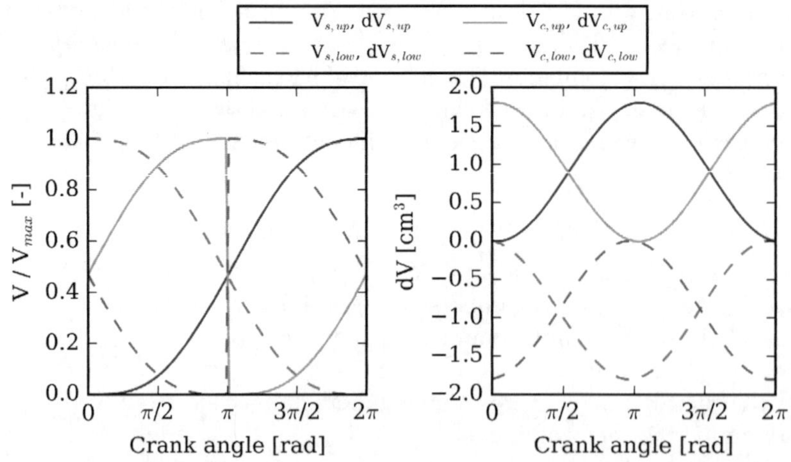

Figure 2. Chamber volume curves and their derivatives as function of the crank angle.

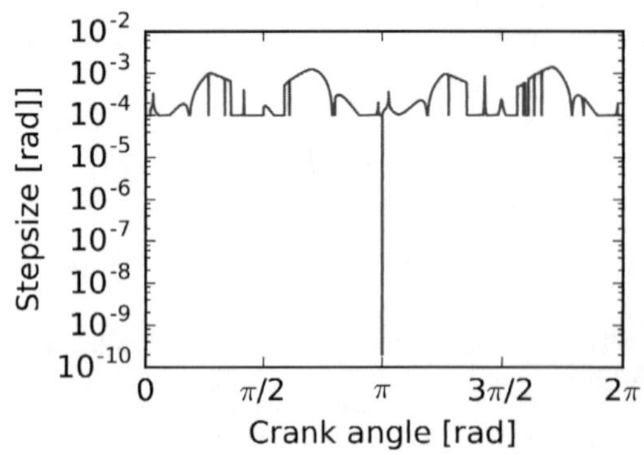

Figure 3. Evolution of the integration step size for Z-compressor simulation (T_{in}= 25 °C, p_{in}= 100 kPa, T_{out}= 115 kPa).

3. Governing Equations

The Z-compressor is divided into five control volumes (CVs) of which four are associated with upper and lower suction and compression chambers and their volumes change according to the rotation angle (see Figure 2). A fifth static control volume is introduced to account for the space between the rotating cylinder and the external housing as well as the space where the vanes are installed. A system of conservation equations is applied simoultaneously to each control volume to obtain the change in thermodynamic states. The properties within a control volume are assumed to be uniform, i.e. gradients within compressor chambers are neglected. As CoolProp [3] is used to retrieve the thermophysical properties of air, two independent thermodynamic properties need to be selected to derive the conservation equations of mass and energy. Temperature and density have been chosen. For each control volume, the general conservation of mass equation can be written as:

$$\frac{d(\rho V)_{CV}}{d\theta} = \frac{1}{\omega} \sum_i \dot{m}_i \tag{7}$$

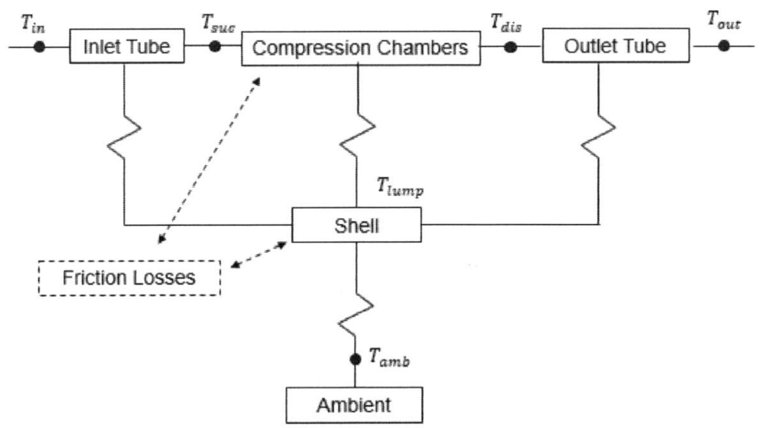

Figure 4. Schematic of the compressor overall energy balance.

The general conservation of energy in each chamber of the compressor can be expressed as:

$$\omega \frac{d}{d\theta}(\rho u V)_{CV} = \dot{Q} + \dot{W}_{CV} + \sum_i (\dot{m}h)_i \tag{8}$$

where the specific internal energy is obtained as $u = h - pv$, the boundary work rate of the control volume can be expressed as $\dot{W}_{CV} = -\omega p(dV/d\theta)$ and $\dot{Q}$ is the heat transfer rate interactions between the components of the compressor chambers and the working fluid. By expanding the left hand side of Equation 8 according to [4], the derivative of temperature with respect to the crank angle can be obtained:

$$m_{CV} c_v \frac{dT}{d\theta} = -T\left(\frac{\partial p}{\partial T}\right)_v \left[\frac{dV}{d\theta} - v\frac{dm_{CV}}{d\theta}\right] - h\frac{dm_{CV}}{d\theta} + \frac{\dot{Q}}{\omega} + \frac{1}{\omega}\sum_i (\dot{m}h)_i \tag{9}$$

where the mass in the control volume is given by $m_{CV} = (\rho V)_{CV}$.

The mass flow rate through the discharge valves are estimated by applying a one-degree-of-freedom dynamic model to the reed valves [5] having two operating modes, i.e., pressure dominant when the net force on the valve is dominated by the pressure difference across the valve, and mass flux dominant when the gas velocity is the dominant force.

As shown in Figure 2, the lower chambers are phased by half rotation. As a consequence, there are two discontinuities in the volume curves when the rotation is at π. During the integration of the differential governing equations, such discontinuities would not allow the convergence of the code. This numerical issue has been handled by coupling an adaptive step-size solver with a step-callback function that helps the solver to step through the discontinuity by taking a sufficiently small step before and after the discontinuity. This numerical technique can be applied to other compressor types for example spool compressors [6], where a similar situation occurs. An example of step size variation with the crank angle can be seen in Figure 3. The Z-compressor involves continuous interactions between upper chambers and lower chambers via leakage paths. The stiffness of the governing equation is not constant and the adaptive solver decreases the step size to maintain the error within a certain threshold. At π, an even smaller step is enforced to step through the discontinuities of the control volumes and continue the integration.

IOP Conf. Series: Materials Science and Engineering **232** (2017) 012062 doi:10.1088/1757-899X/232/1/012062

Table 2. Leakage path description.

Path	Description	Area
L1	Leakage through the clearance of the vane and the bearing slot from the shell to the compression chamber	Constant
L2	Leakage through the clearance of the vane and the bearing slot from the shell to the suction chamber	Constant
L3	Leakage through the flat area of the blade	Constant
L4a,b	Leakages through the clearance of the vane and cylinder	Variable
L5	Leakage through the clearance of the vane and the blade	Variable
Lup	Leakages from the upper compression chamber to the lower suction chamber and from the upper compression chamber to the lower compression chamber	Constant
Ldown	Leakages from the lower compression chamber to the upper suction chamber and from the lower compression chamber to the upper compression chamber	Constant

The compressor model is closed by enforcing a steady-state overall energy balance with the shell of the compressor as shown in Figure 4:

$$\dot{Q}_{amb} + \sum_i \dot{W}_{i,fr} + \sum_i \dot{Q}_i = \epsilon \qquad (10)$$

where ϵ is the residual to be minimized, $\dot{Q}_{amb}$ is the heat transfer between the shell and the ambient, $\sum_i \dot{W}_{i,fr}$ is the sum of the mechanical and friction losses and the general $\dot{Q}_i$ term represents the heat transfer interaction between the gas and the shell during suction and compression processes.

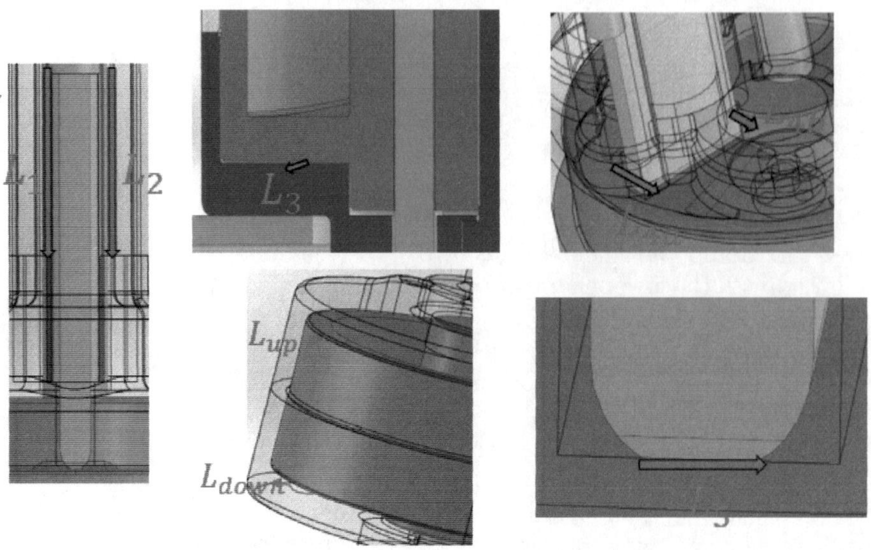

Figure 5. Leakage paths of the Z-compressor.

3.1. Leakage Flow Model

The Z-compressor presents several leakage paths. For the current design, the main leakage paths are identified in Figure 5. Furthermore, a more detailed description of each leakage path can be found in Table 2. By analyzing the leakage paths, it can be found that most of them present lengths that are relatively long compared to the cross-sectional area. In addition, the leakage path flow area is not always constant. Assuming isentropic flow of perfect gas will overestimate the leakage flows. To improve the leakage flow predictions, a detailed 1D-flow with friction analysis has been employed to obtain the mass flow rate estimations across each path. Coupling the three conservation equations, mass, momentum and energy, a general expression for the system of differential equations along the leakage path can be expressed as in [7],

$$\begin{bmatrix} 0 & 0 & 1/V & 0 & 1/\rho \\ 0 & 1 & \rho V & 0 & 0 \\ 0 & 0 & V & 1 & 0 \\ -\left(\frac{\partial h}{\partial T}\right)_\rho & 0 & 0 & 1 & -\left(\frac{\partial h}{\partial \rho}\right)_T \\ -\left(\frac{\partial p}{\partial T}\right)_\rho & 1 & 0 & 0 & -\left(\frac{\partial p}{\partial \rho}\right)_T \end{bmatrix} \begin{bmatrix} dT/dx \\ dp/dx \\ dV/dx \\ dh/dx \\ d\rho/dx \end{bmatrix} = \begin{bmatrix} -\frac{1}{A}\frac{dA}{dx} \\ -\frac{\rho V^2}{2}\frac{4 f_F}{D_h} \\ 0 \\ 0 \\ 0 \end{bmatrix} \tag{11}$$

where f_F is the Fanning friction factor for flow between infinite plates [7] and V is the mean fluid velocity. Solving the system of equations shown in (11) within the compressor model is computationally very expensive and also makes the convergence of the model more difficult to reach. Therefore, to predict the leakage flows in positive displacement machines, an isentropic flow model is generally adopted as it is simpler. However, it is known that such model can significantly overpredict the flow rate [7]. Therefore, the detailed 1D-flow model is used to determine a correction factor between the mass flow rate of leakage using the detailed flow with friction model and the mass flow rate predicted using a simpler isentropic flow model. For a given configuration, the ratio of the isentropic nozzle mass flow rate prediction to that of the detailed model is defined by

$$M = \frac{\dot{m}_{\mathrm{nozzle}}}{\dot{m}_{\mathrm{fr}}} = f\left(Re, L^*, \delta^*\right) \tag{12}$$

where the dimensionless characteristic length and dimensionless gap width are given by $L^* = L/L_0$ and $\delta^* = \delta/\delta_0$ with L_0 and δ_0 being the reference length and gap width values. For brevity, only the case in which the detailed model is applied to the variable leakage path L_4 is shown. In particular, Figure 6(a) shows the generated points of the mass flow rate correction factor M for different boundary conditions (pressure ratio across the path) and gap widths as function of the Reynolds number. The calibration results of Equation 12 is reported in Figure 6(b).

The correction factors are then used to compute the mass flow rate in and out each control volume during the integration of of Equation 7 and Equation 9.

3.2. Friction Losses

The average total shaft power is given by the sum of the compression power, $\overline{\dot{W}}_{\mathrm{pV}}$, and the total friction power losses, $\overline{\dot{W}}_{\mathrm{fr,tot}}$, due to contacts among the different parts,

$$\overline{\dot{W}}_{\mathrm{sh}} = \overline{\dot{W}}_{\mathrm{pV}} + \overline{\dot{W}}_{\mathrm{fr,tot}} \tag{13}$$

A detailed force and moments analyses have been included into estimate the friction losses. In particular, once the pressure distribution within the working chambers is known radial and axial loads on blade, shaft and bearings can be obtained as well as the associated frictional torques. Each frictional torque multiplied by the rotational speed gives the friction loss power.

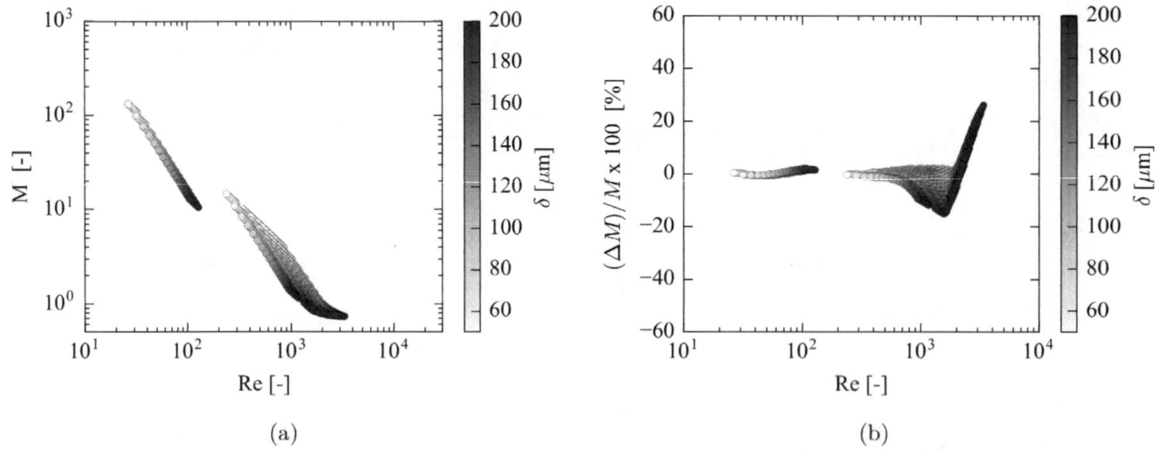

Figure 6. (*a*) Ratio of isentropic nozzle model to detailed model for radial leakage as a function of Reynolds number and gap width (color) ;(*b*) Error in leakage flow correction term as a function of Reynolds number and gap width (color).

A comprehensive description of such approach can be found in [1]. The average total friction loss power is given as:

$$\overline{\dot{W}}_{\rm fr,tot} = \sum_i \dot{W}_{\rm i,fr} = \overline{\dot{W}}_{\rm fr,pt} + \overline{\dot{W}}_{\rm fr,v} + \overline{\dot{W}}_{\rm fr,flat} + \overline{\dot{W}}_{\rm fr,cyl-cyl} + \overline{\dot{W}}_{\rm fr,cyl-flat} + \overline{\dot{W}}_{\rm fr,TB} + \overline{\dot{W}}_{\rm fr,JB-low} + \overline{\dot{W}}_{\rm fr,JB-up} \tag{14}$$

where $\overline{\dot{W}}_{\rm fr,pt}$ is the friction power associated with the tangential component of the pressure forces acting on the Z-blade, $\overline{\dot{W}}_{\rm fr,v}$ is the total friction power due to the contact between vane and blade on both sides of the blade, $\overline{\dot{W}}_{\rm fr,flat}$ is the frictional loss due to the flat portions of the Z-blade, $\overline{\dot{W}}_{\rm fr,cyl-cyl}$ is the friction loss between the outer cylinder wall and the compressor inner shell, $\overline{\dot{W}}_{\rm fr,cyl-flat}$ is the friction losses induced by the upper and lower flat circular crowns of the outer cylinder and $\overline{\dot{W}}_{\rm fr,TB}$, $\overline{\dot{W}}_{\rm fr,JB-low}$, $\overline{\dot{W}}_{\rm fr,JB-up}$ are the losses related to thrust bearing, lower and upper journal bearings, respectively.

4. Results and Discussion

The detailed compressor model developed is used as a tool to investigate the internal behavior of the compressor and to quantify the sources of losses. Figure 7(a) shows an example of pressure traces with respect to the rotation angle. The plot of the internal pressure of the compressor during the working process helped adapting the valve parameters to achieve the required operating points.

Starting from the baseline design, i.e. #1, the model has been exercised to improve such design. A total of five possible prototypes have been identified and they are described in Table 3. As a first step, the Z-compressor shaft structure has been rethought to reduce friction losses, i.e. design #2. A more compact design led to significantly reduce the friction losses between the vane and the blade as well as eliminating the need for two journal bearings, as shown in Figure 7(b). Next, by keeping the same design configuration as design #2, three additional designs have been defined by considering different clearance gap sizes. A nominal operating point has been chosen to carry out the calculations: suction temperature 23 °C, suction temperature 100

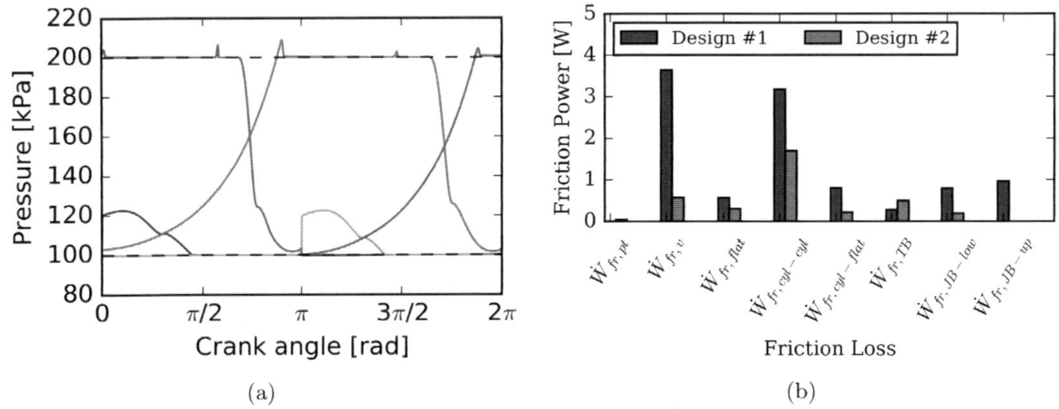

Figure 7. (*a*) Evolution of the pressure within the working chambers throughout one rotation. The colors of the solid lines are identical to the ones in Figure 2; (*b*) Comparison of the friction loss contributions between design # 1 and the improved design # 2. The other designs are equivalent to design # 2 in terms of shaft configuration..

kPa and discharge pressure 115 kPa. The results of the average mass flow rates through each path for all the five designs are grouped in Figure 8. The different gap sizes are associated with different manufacturing processes and costs.

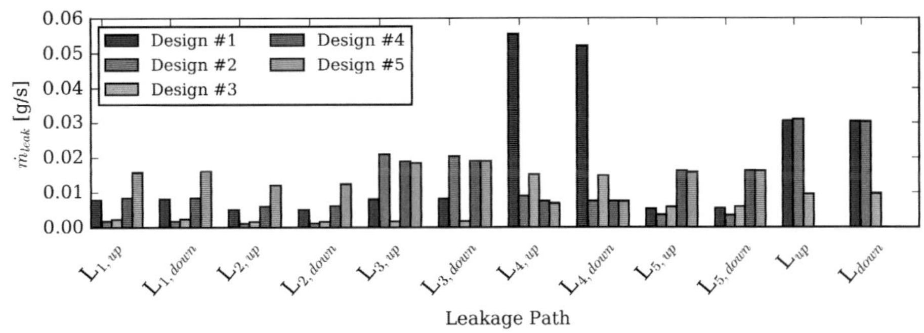

Figure 8. Leakage mass flow rate distribution among all the paths for different Z-compressor designs.

The improvements or degradations of the compressor performance are determined by calculating the volumetric efficiency, as shown in Figure 9. In the baseline design, all the gaps are kept the same leading to a volumetric efficiency close to 40%. A new vane design and reduced gaps of L_1 and L_2 of design #2 allowed to improve the volumetric efficiency by almost 50%. The best performance has been obtained with design #3 in which the gaps have been reduced to minimum and with a thicker vane. In the last two designs, a piston-ring has been added to eliminate the leakage flows between upper and lower chambers. However the gap sizes have been relaxed. As a result, design #4 reached a 51% volumetric efficiency while design #5 showed a 15% improvement over the baseline design. The parametric analyses have been further used to define a design #6 with tip seals for prototyping .

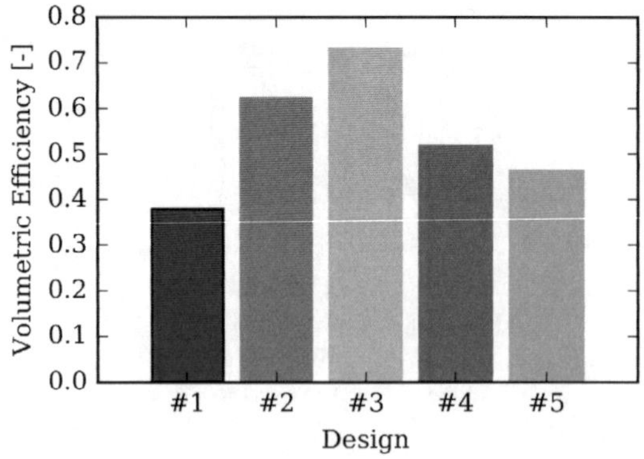

Figure 9. Comparison of volumetric efficiency obtained with the different Z-compressor designs.

Table 3. Description of the different designs analyzed.

Design	Description
# 1	Baseline (as Figure 1)
# 2	Shorter vane geometry and modified shaft configuration (one journal bearing and one thrust bearing); L_1 and L_2 gaps reduced by a factor of 5
# 3	Same as # 3 with thicker vane and minimum gap size everywhere
# 4	Same a #2 ; gaps as baseline and added piston-ring to reduce L_{up} and L_{down}
# 5	Same as # 4 with L_1 and L_2 gaps increased by a factor of 2 compared to baseline

5. Conclusions

A comprehensive mechanistic model of a Z-compressor has been developed to identify the losses of a baseline prototype. From the experimental work, friction losses and leakage flows were detrimental to the performance of such compressor.

The model has been used to evaluate five potential designs to improve the baseline performance. A new shaft configuration and vane design led to decrease the friction losses by approximately 70%. By reducing the clearance gaps, the volumetric performance of the compressor improved by up to 90%. However, due to manufacturing limitations, a optimum solution has been found by compromising between design and gap sizes, which led to developing design #6.

References

[1] Jovane M 2007 *Modeling and analysis of a novel rotary compressor* Ph.D. thesis Purdue University

[2] Mathison M M, Braun J E and Groll E A 2008 *HVAC&R* **14** 719–748

[3] Bell I H, Wronski J, Quoilin S and Lemort V 2014 *Industrial & Engineering Chemistry Research* **53** 2498–2508 (*Preprint* `http://pubs.acs.org/doi/pdf/10.1021/ie4033999`)

[4] Bell I 2011 *Theoretical and Experimental Analysis of Liquid Flooded Compression in Scroll Compressors* Ph.D. thesis Purdue University URL `http://docs.lib.purdue.edu/herrick/2/`

[5] Bradshaw C, A G E and Garimella S 2011 *International Journal of Refrigeration* **34** 63–73

[6] Bradshaw C R and Groll E A 2013 *International Journal of Refrigeration* **36** 1974–1981

[7] Bell I H, Groll E A, Braun J E and Horton W T 2013 *International Journal of Refrigeration*

IOP Conf. Series: Materials Science and Engineering **232** (2017) 012063 doi:10.1088/1757-899X/232/1/012063

Maintenance free gas bearing helium blower for nuclear plant

Dr A Molyneaux[1], Prof M Harris[2], Prof S Sharkh[3], S Hill[2] and T de Graaff[4]

[1] OFTTech Ltd, 49 Station Road, Polegate,, East Sussex, BN26 6EA, UK
[2] Independent consultant, UK
[3] Professor of Power Electronics Machines and Drives, University of Southampton, University Road, Southampton, SO17 1BJ, UK
[4] de Graaff Engineering, Boldistrasse 14, 5415 Rieden, Switzerland

Abstract. This paper describes the design, testing and operation of novel helium blowers used to recirculate the helium blanketing gas in the nuclear reactor used as a neutron source at the Institut Laue Langevan, Grenoble, France. The laser sintered shrouded centrifugal wheel operates at speeds up to 45000 rpm supported on helium lubricated hydrodynamic spiral groove bearings, and is driven by a sensorless permanent magnet motor. The entire machine is designed to keep the helium gas (polluted by a small amount of D2O) out of contact with any iron or copper materials which would contribute to the corrosion of parts of the circuit. It is designed to have zero maintenance during a lifetime of 40,000 hours of continuous operation. This paper will describe the spiral groove journal and thrust bearings. Design and manufacture of the 1 kW motor and centrifugal wheel will be explained including their CFD and FEA analyses. Measurements of rotor displacement will be presented showing the behaviour under factory testing as well as details of the measured centrifugal wheel and motor performances. Two machines are incorporated into the circuit to provide redundancy and the first blower has been in continuous operation since Jan 2015. The blower was designed, manufactured, assembled and tested in the UK using predominantly UK suppliers.

1. Background

The nuclear plant in Institut Laue Langevan, Grenoble is a small heavy water cooled high flux reactor of only 58 MW thermal power used to produce neutrons for research purposes [1]. It operates 220 days a year producing neutrons for research focused primarily on fundamental science in a variety of fields: condensed matter physics, chemistry, biology, nuclear physics and materials science.

The plant has been in operation since 1971 and is due to continue operating for at least another 10 years which may be extended depending on demand. The heavy water is covered (blanketed) by a layer of Helium gas that has to be continuously cooled and treated to remove any heavy water ions. This helium gas must not contain any copper or iron ions as their presence would contribute to corrosion of parts of the reactor vessel, nor is it acceptable to have oil contaminating the circuit which, amongst other issues, would foul the catalytic converters and heat exchangers.

The original helium circulators were manufactured in the early 1970s but were overhauled 15 years ago, with the replacement of the inverter drives. A recent review of safety in French nuclear industry resulted in a requirement to provide modern helium blowers to replace the original ones – even though these latter were still functioning perfectly well on the original gas bearings. The original gas bearings, supporting a regenerative blower, were of the spring preloaded tilting pad type that were common in many nuclear circulating pumps and blowers in the 1960s and 70s. Decker [2] reviews 22 different installations in USA, Germany and England.

OFTTech Ltd are specialists in designing and using spiral groove gas bearings which have many advantages compared to tilting pad ones.

The basic requirement of the blower is as follows:

Flow rate: 20 m^3/hr
Pressure rise 30 mbar
Electrical power input 1 kW

The flow and pressure requirement only require tens of Watts but there is a client requirement that the motor be a 1 kW one for safety considerations. The final design is shown in figure 1. The helium inlet and outlet ports are at the top and an air cooling jacket provides cooling of the stator winding.

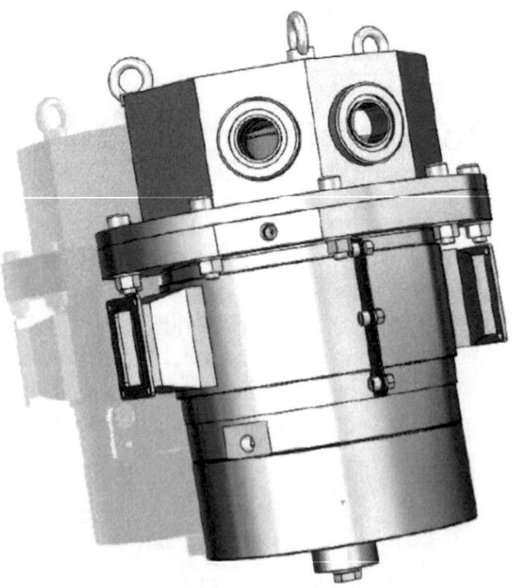

Figure 1: The Helium Blower

2. Blower wheel

The maximum rotational speed of the blower was fixed at 40000rpm to enable a design that could be easily assembled and replaced in service. This leads to a low specific speed (0.21 Balje's definition) and hence a large diameter radial impeller with small blade heights – the impeller diameter is 110mm with blade height varying from 3mm down to 1.2mm. At this low specific speed the fluid friction and disc windage losses are high, leading to relatively low efficiency – however, the shaft power required at design point was only of the order of 50W and, given the 1kW motor capability, high efficiency was not a prime requirement.

The aerodynamic design of the impeller was carried out iteratively using Ansys-CFX with the turbomachinery enhancements BladeGen and TurboGrid. The design philosophy was to employ highly backswept blades (outlet angle -65° from the radial direction) in order to achieve a very stable operating characteristic. The general view of the impeller and design point velocity triangles are given in figure 2.

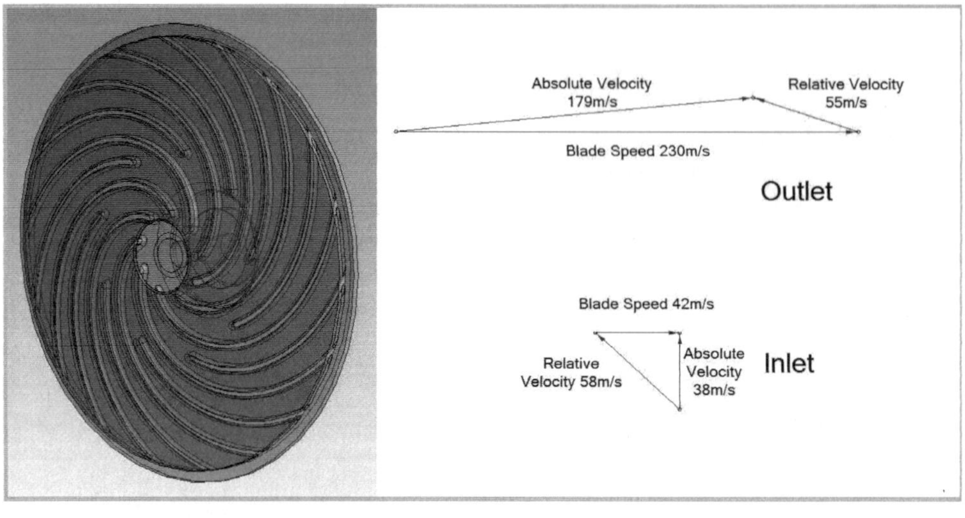

Figure 2. General View and Velocity Triangles

A shrouded design was employed primarily to minimise the axial thrust of the impeller, and an axial labyrinth seal employed at the inlet to control the leakage flow. Figures 3 and 4 show predicted pressure contours and blade loading at design point – the loading diagram shows the smooth velocity distribution obtained, with minimal over-acceleration and modest diffusion levels. For simplicity, the splitter vanes were specified exactly halfway between the main blades and with the same blade angle distribution – this resulted in lower loading on the splitter, as indicated by the two innermost lines in figure 4. The final design had nine full and nine splitter blades.

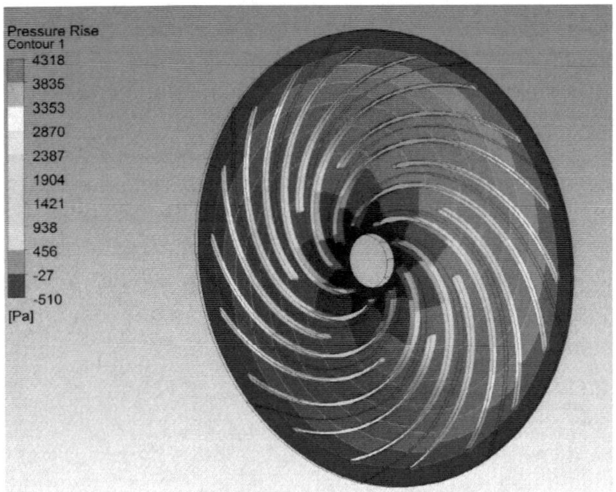

Figure 3. CFD Predicted Pressure Contours

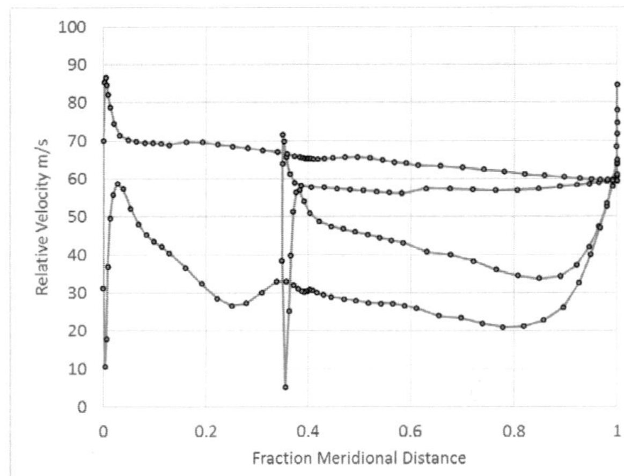

Figure 4. Blade Loading

Initial studies using simple one dimensional methods suggested that the addition of a volute at the impeller exit would be of marginal benefit as the friction losses in the small passage would cancel out any static pressure recovery due to flow deceleration. Therefore the design was specified with no diffusion system – this also enabled the use of the existing inlet and outlet port positions so that the machine could comply with the need to fit to existing pipe connections.

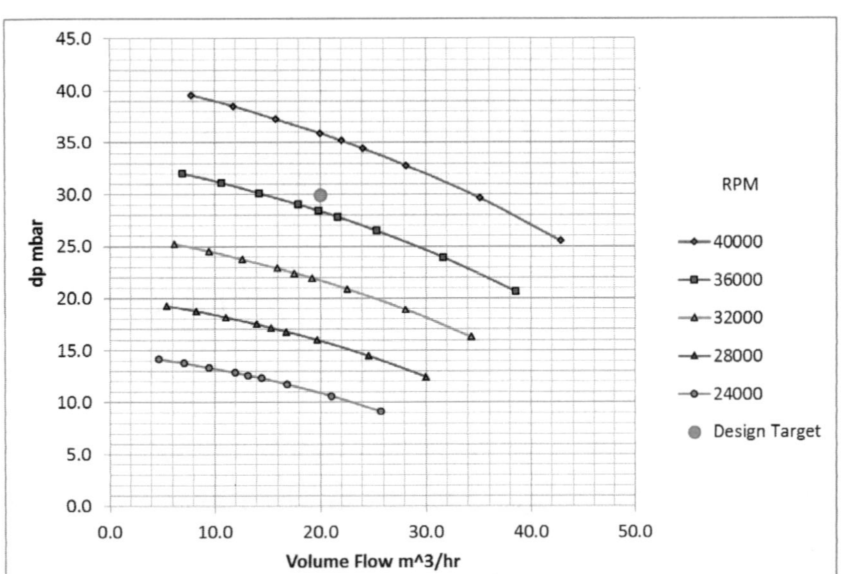

Figure 5. Predicted Characteristic

A full performance characteristic was predicted using CFX, and the results corrected for disc windage and labyrinth seal leakage (figure 5). It can be seen that the design was predicted to achieve higher than target pressure rise at 40000rpm – this was deliberate given the nature of the development program and some of the uncertainties in achievable clearances and roughness effects. It can also be seen that converged CFD solutions were obtained at very low flow rates, which suggested that the required stable operating characteristic should be achieved. It can be seen that the predicted characteristic indicated that the target flow and pressure rise should be achieved at 37000rpm, which gave a useful speed margin.

The mechanical design of the impeller was carried out alongside the aerodynamic design. Stainless steel 316L was specified by the customer. The disc taper and shroud geometry was carefully optimised to minimise deflections and stress levels, and to ensure that disc natural frequencies were well above the running speeds. As the design featured a shroud, it was decided to adopt Direct Metal Laser Sintering (DMLS) to 'print' the impeller and hence achieve a shrouded design without the need for joining. One downside of using this process is the difficulty of achieving smooth internal surfaces. However, the passage Reynolds Number (based on passage hydraulic diameter and relative velocity) is between 3000 (inlet) and 1200 (exit), so the flow is arguably laminar, and roughness effects should be small. It was therefore decided to not finish the internal passages (by abrasive flow machining) unless testing indicated a performance shortfall.

The final peak stress levels in the impeller were at the junction of the blades and the inner diameter of the shroud (figure 6). Due to the lack of previous experience of using the DMLS process the design was conservatively engineered to have a peak stress at 40000rpm of just 65% of the manufacturer's quoted yield stress.

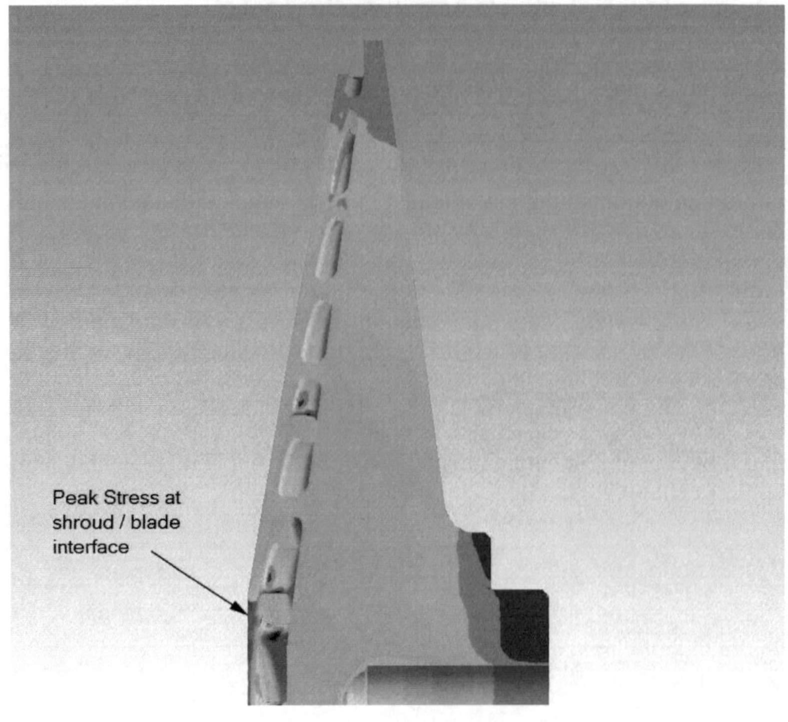

Figure 6. Peak Stress Levels at Blade / Shroud Intersection

3. Permanent Magnet Motor

A permanent magnet synchronous motor topology was chosen. To ensure that the helium is not contaminated by iron or copper from the motor active components, a 0.2 mm thick stainless steel (316L) can was used to enclose the stator. The canned stator was potted in epoxy inside a finned stainless steel housing. The magnets were glued inside the tungsten carbide shaft tube; an ASI4010 (EN19) bolt passes through the centre of the magnets.

A KEB (KEB Automation KG) sensorless control PWM drive is used to drive the motor. A sine filter is used to reduce current ripple and associated losses in the machine. It also helps with improving EMC (ElectroMagnetic Compatibility)

Placing the magnets inside a tungsten carbine tube, and using a conducting stator can made the design of the machine challenging. The former meant the effective airgap of the machine was relatively large: the magnet OD is 20 mm and the effective airgap is 6.5 mm (tungsten carbide has the same permeability as air and hence the effective airgap includes both the clearance gap and the tungsten carbide shaft tube). As a result, there is a significant amount of leakage flux as shown in Figure 8, which shows flux distribution obtained using a finite element analysis model of half the machine.

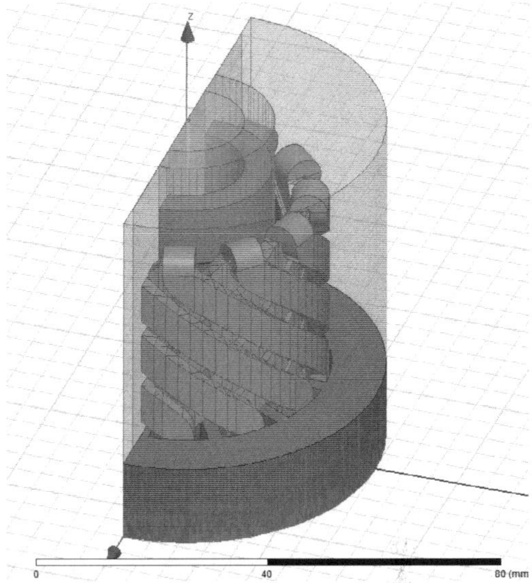

Figure 7. Finite Element 3D Model

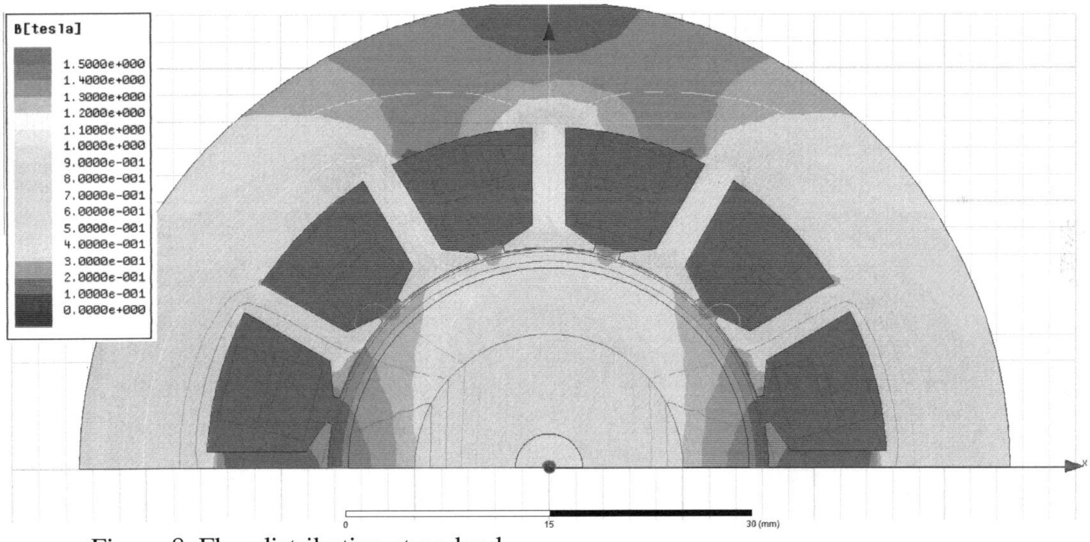

Figure 8. Flux distribution at no-load.

Transient three-dimensional FEA including the PWM drive circuit (see Figure 8) and rotation was used to analyse the machine, estimate losses and quantify its performance. To reduce core losses, the flux density in the laminations was kept well below saturation, at about 1T. This resulted in core losses of about 30W. The copper losses are about 3.5 W. However, these are dwarfed by losses in the can which needs to be as thin as possible as the losses are proportional to its thickness. The eddy current losses in the 0.2 mm can used in the final design are about 130 W (Figure S3). Manufacturing and fitting such a thin can is challenging. Care needed to be taken that the can does not short circuit the laminations and that the gaps between the can and the teeth are filled with thermally conducting epoxy to ensure heat removal.

The Synthet-PM propriety software [3] was used to conduct the initial motor sizing studies as well as carrying out thermal analysis of the final machine design (using FEA loss results). The thermal analysis estimated that the maximum winding temperature is about 140 degrees C, which means that using class H insulation (180 degrees rating) will have a theoretical life of about 320,000 hours, which is well above the target.

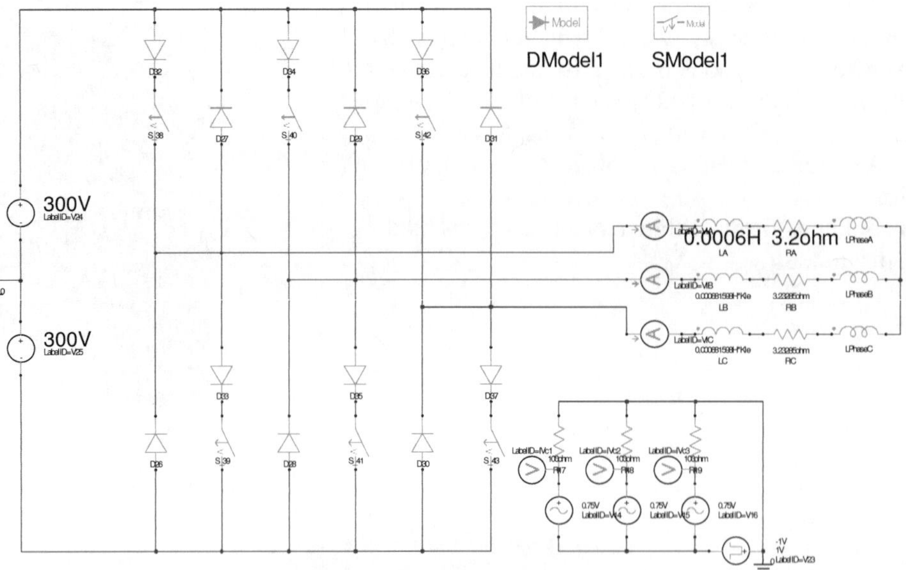

Figure 9. External circuit of the PWM drive used in the 3D FEA model

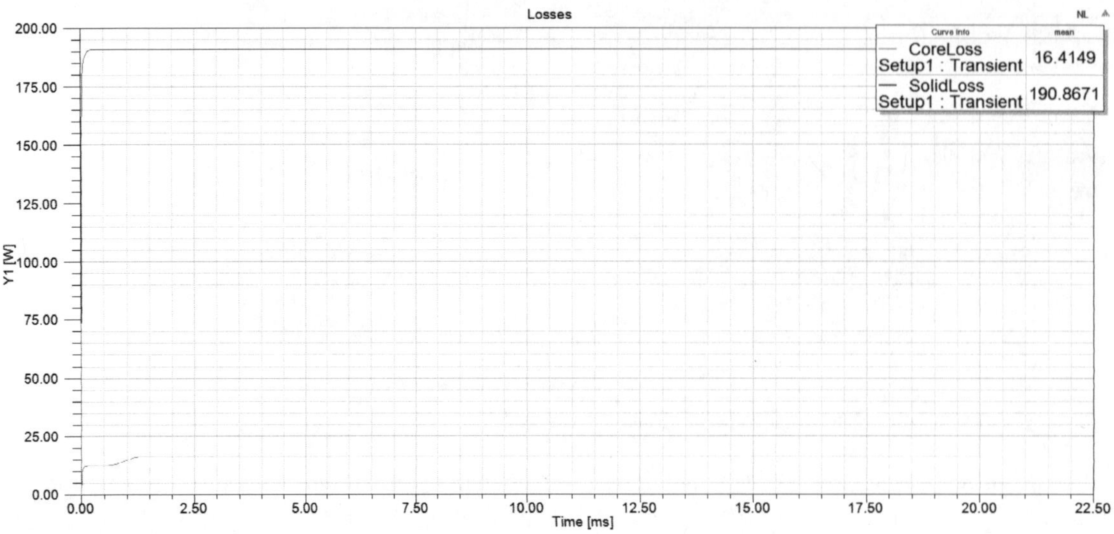

Figure 10. Core and can losses calculated using FEA.

4. Gas Bearings

Spiral groove gas bearings are one of the types of aerodynamic (hydrodynamic, self acting) bearings that has proven successful in the market place. The others being foil bearings (which have lower radial stiffness and wear/fatigue issues) and spring loaded tilting pad bearings (require individual tuning and often have higher starting torque and poor low speed performance). OFTTech has concentrated on spiral groove bearings since the early 1980s. [4].

The key feature of spiral groove bearings are the shallow grooves (of the order of 2 to 4 times greater than the clearance) that create the pressures inside the bearing due to rotation of one of the surfaces (either surface can rotate, it's the relative motion that is needed). The spiral groove thrust bearing is thought to have been invented in the UK during the 1940s but was classified by the AERE [5] for many years.

Figure 11. Journal and thrust bearing bearing grooves.

The two patterns may or may not meet in the middle depending on the optimal performance in the application. These grooves are sometimes called herringbone groove bearings. The right-hand image shows thrust bearing grooves for an inward pumping thrust bearing. Similarly, as for the journal bearings a second set of grooves can be added on the inside that acts to limit or prevent through flow – the choice comes down to the application requirements.

These bearings are usually made of hard materials to ensure low wear and good structural performance in order to keep the surfaces parallel under the small clearances usually necessary for gas bearings. Tungsten carbide and silicon nitride are two of the most common materials now used, boron carbide was the primary choice for the successful navigation gyroscopes used for many years [6]. Clearances are typically of the order of 5 to 15 μm radially, with grooves 10 to 40 μm. The use of cylinders and flat disks makes the manufacture easier than if the complex conical or hemispherical forms are used. Modern grinding techniques allows direct grinding to size although for prototypes the classic finish lapping method can be used.

Surface coatings are often used but only two have proven successful: MoS2 (sputtered) and DLC (amorphous carbon); both with thicknesses of 2-4 μm applied after final sizing providing low friction during start up and overload.

The grooves can be made by many methods: photo etching, grinding, ECM, EDM, ION beam, pressing, grit blasting or by laser. The latter is the only solution we now use as the groove geometry is excellent and relatively low cost.

Recent examples of the use of OFTTech spiral groove gas bearings are in laser blowers [7] and a Brayton cycle cryogenic space cooler [8], which has been in operation on spiral groove gas bearings since 2002: and a new application will involve a turbo expander cryocooler on Neon lubricated spiral groove bearings [9].

The design of aerodynamic gas journal bearings is much more difficult than bearings lubricated with liquids because the compressibility of the gas makes the bearing behaviour frequency dependant. This means the designer has to examine all ranges of operating speeds and frequencies of excitation to ensure the bearings are stable under any excitation. The usual form of instability seen with gas bearings is half speed whirl seen at frequencies between 20 and 45% of the synchronous running speed, and is usually fatal with the bearing crashing within seconds. The parameter usually used to describe the stability of gas journal bearings is the critical mass [10], defined simply as that mass the bearing will support before experiencing half speed whirl. In addition to the gas bearing stability we have to examine the rotordynamic behaviour to check system stability. The analysis of gas bearing behaviour was performed using a numerical method following the narrow groove theory [11, 12]; this design method has been under continual improvement and checking at OFTTech since its origins at the University of Southampton [13].

The journal bearing performance was optimised using a simple hill climbing procedure (although a genetic algorithm is available for simply designs) resulting a journal bearings with a 30 mm diameter, 30 mm long and grooves 17 μm deep.

The basic performance of the journal bearings is shown in figure 12 showing the stability (critical mass, kg), power (Watts), stiffness (steady state, MN/m) and damping (synchronous speed, Ns/m). with variation of radial clearance. Here one sees an optimal clearance around 9.5 μm radially. Increasing the clearance to 11 μm reduces the stability by a factor over 100: but even with this larger clearance the stability margin is still ~5 (critical mass/actual rotor mass per bearing).

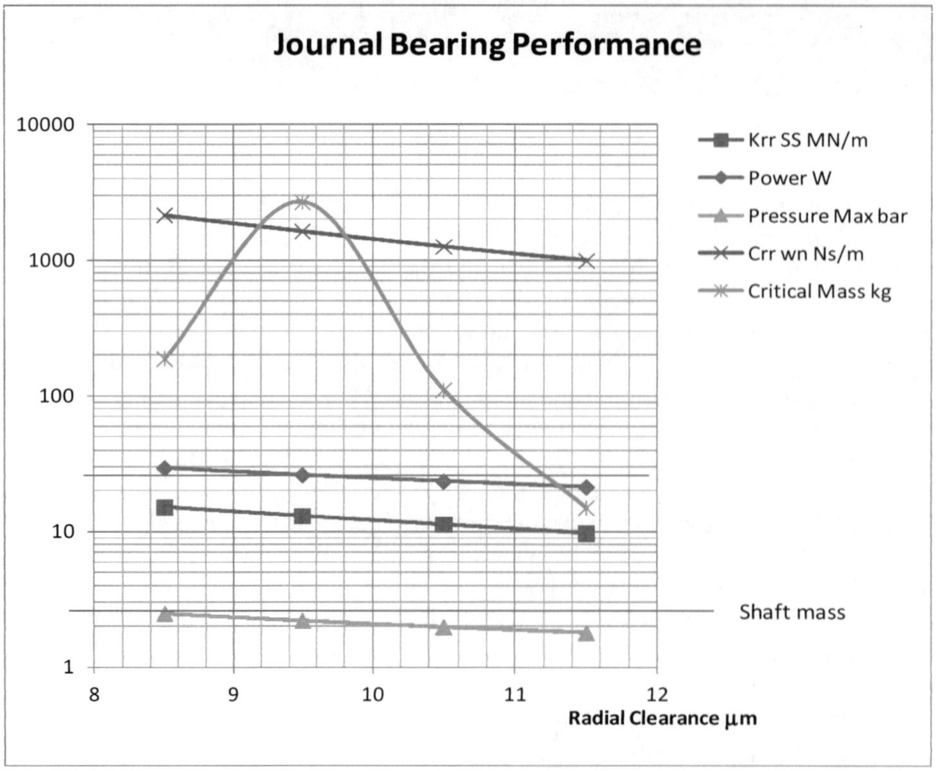

Figure 12: Journal bearing theoretical performance (40,000 rpm, helium lubricant at 1 bar, groove depth 17 μm).
Krr = resulting radial stiffness at steady state conditions.
Crr = resulting damping at the bearing natural frequency.

The thrust bearing has also been optimised with the result shown in figures 13 and 14. Figure 13 shows the variation of thrust bearing performance with speed and figure 14 shows the force generated with variable clearance. The result is that the thrust bearing will operate at an axial clearance of ~24 μm. when supporting the shaft mass.

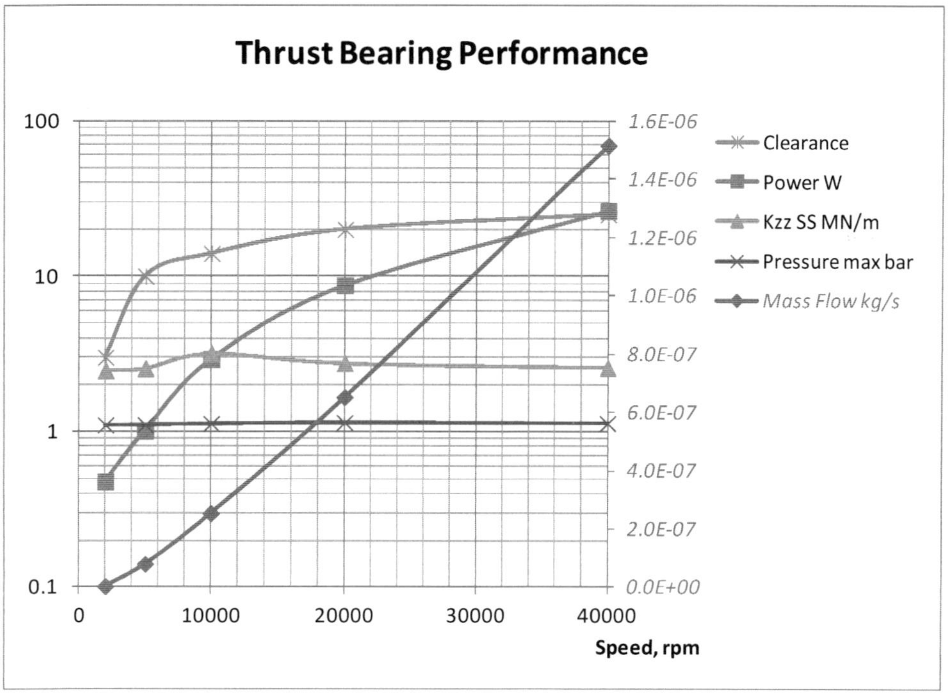

Figure 13 Thrust bearing theoretical performance (40,000 rpm, helium lubricant at 1 bar, groove depth 34 μm).
Kzz = axial stiffness at steady state conditions.
Clearance = axial lift under shaft mass, μm

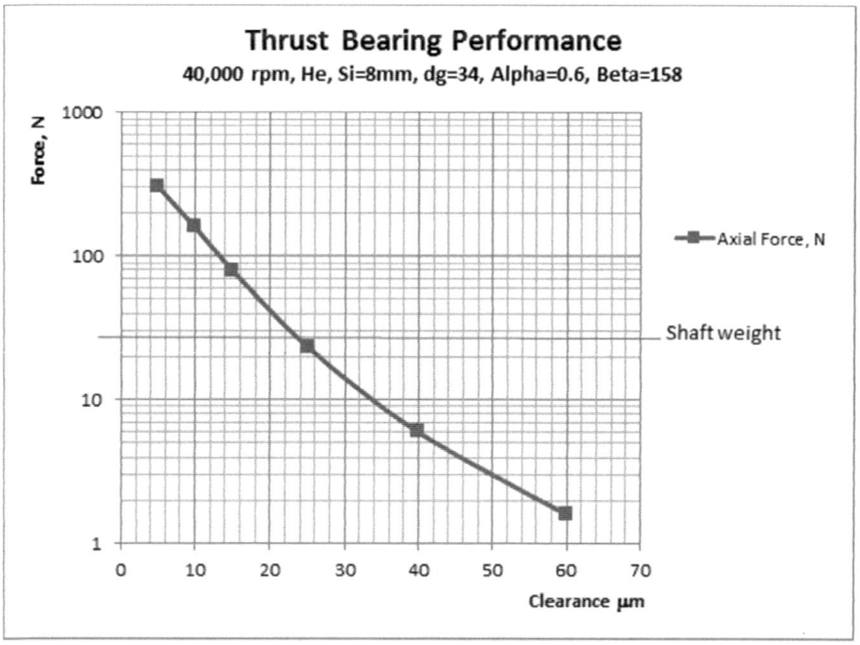

Figure 14: thrust bearing theoretical axial force capability (40,000 rpm, helium lubricant at 1 bar, groove depth 34 μm)

Figure 15 shows the internal view of the assembly showing the two journal bearings and single sided thrust bearing at the top under the wheel. The blower is mounted vertically such that a single sided thrust bearing can be used.

a) Blower wheel
b) Thrust bearing
c) Motor stator and rotor
d) Cooling fins
e) Lower journal bearing

The design is such that all moving parts can easily be replaced from the top. The motor is air cooled with fins being machined into the housing. There are no welds on this design avoiding any need for Non Destructive Testing and has been designed according to the European pressure vessel design standard EN13445-3:2009 including a pressure test to 5 bar which is a factor of safety in excess of 100 compared to the system nominal pressure

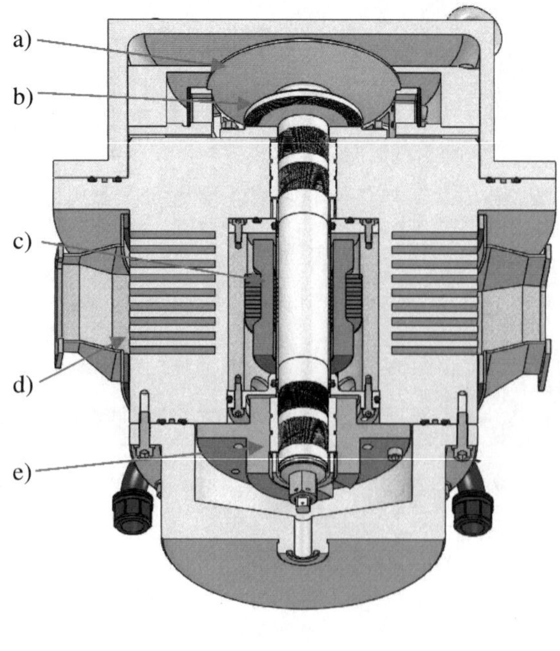

Figure 15. The internal layout of the helium blower

There are no parts requiring maintenance or service and it is expected that the units will run for the 40,000 hrs design life unless contaminated (for example water or other fluids).

5. Test Results

The rotordynamic analyses undertaken shows that the system is stable for all operating speeds. Figure 16 shows the first critical speed at 7762 rpm with the wheel motion being the primary action.

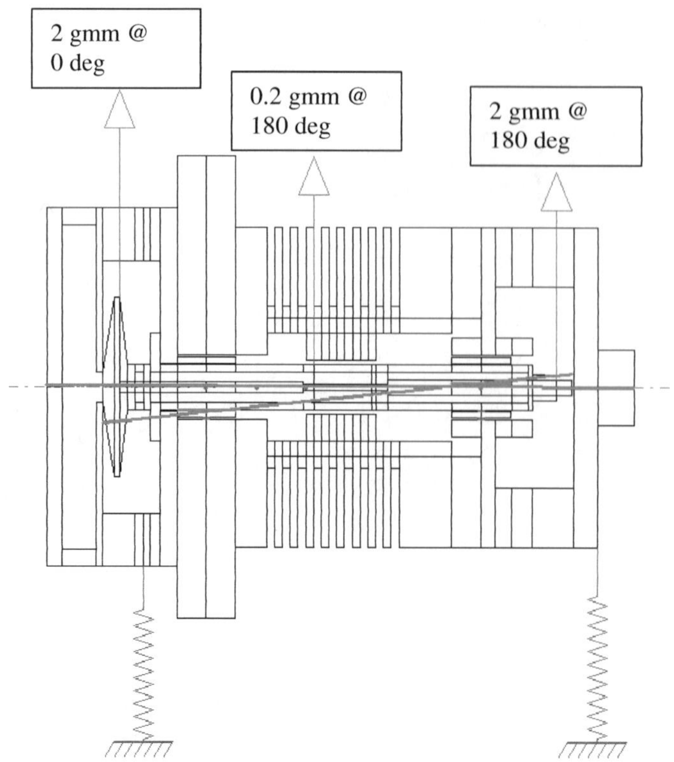

Figure 16. Theoretical calculation of the critical speed – the conical rigid mode at 7762 rpm.

The most important behaviour to consider to enable correct balancing is the response to out of balance. Figure 17 shows the response to 2 gmm on the wheel and 2 gmm on the other end of the shaft. As can be see the peak displacements are only 3 μm – much less than the bearing radial clearance.

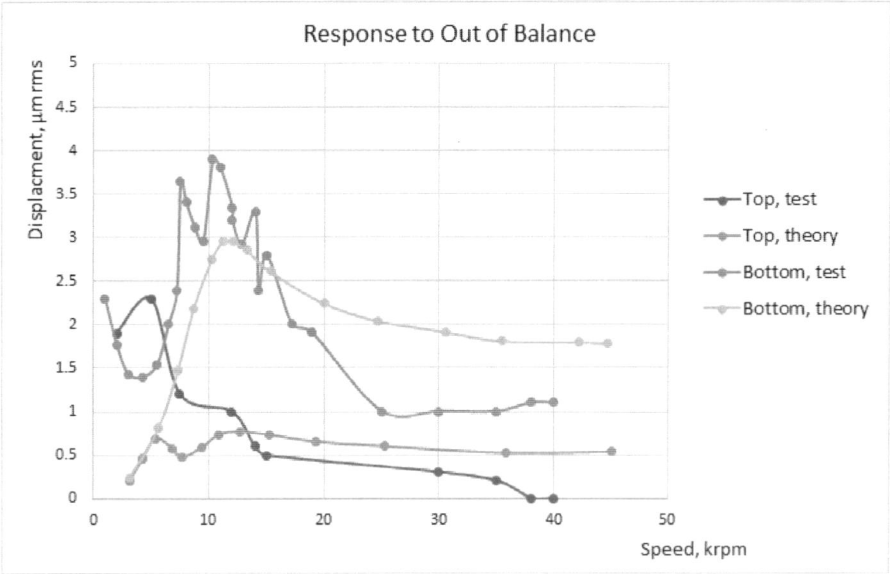

Figure 17. Response to Out Of Balance (OOB). The theoretical predictions use an OOB of 2 gmm at each end of the shaft. The rotor was balanced to <2 gmm at each end. The displacements, both theoretical and experimental, are RMS values.

The grooves are manufactured using laser machining and results in excellent quality grooves with uniform depth, accurate angle and width ratio. The profile is shown in figure 18. The sloping sides of the grooves are intentional to ensure correct laser machining.

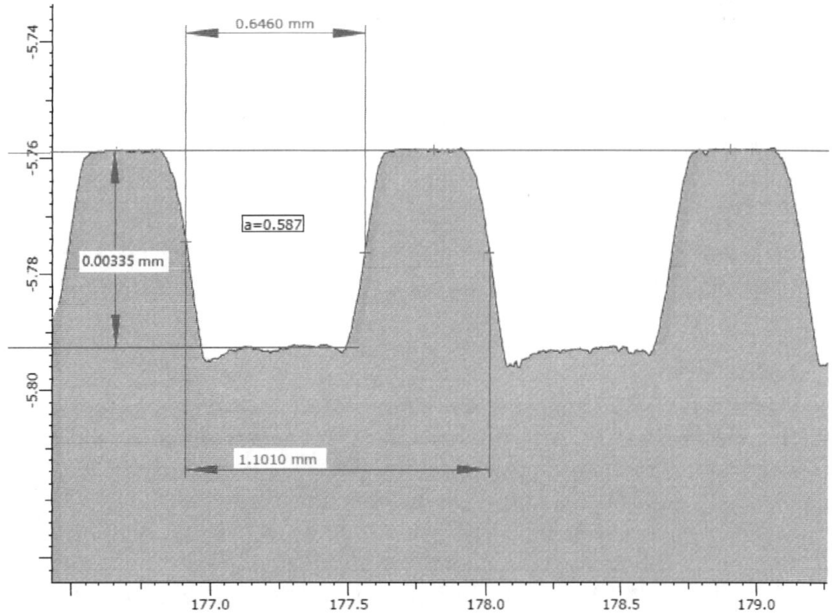

Figure 18. Profile across the thrust bearing grooves machined by laser.
Axes are mm. The mean groove depth is 33.5 μm with tiny variations.

The final performance of the blower was tested by constructing a closed-loop helium-filled circuit with appropriate valves, orifice plate and coolers. Full characteristics were obtained right down to zero flow, with no evidence of increased vibration levels due to stall / surge. Figure 19 shows the test results at 37000rpm for the two delivered units, overlaid with the CFD-derived prediction. It can be seen that the agreement with prediction is good at design flow, but the tests indicate a slightly steeper slope to the characteristic.

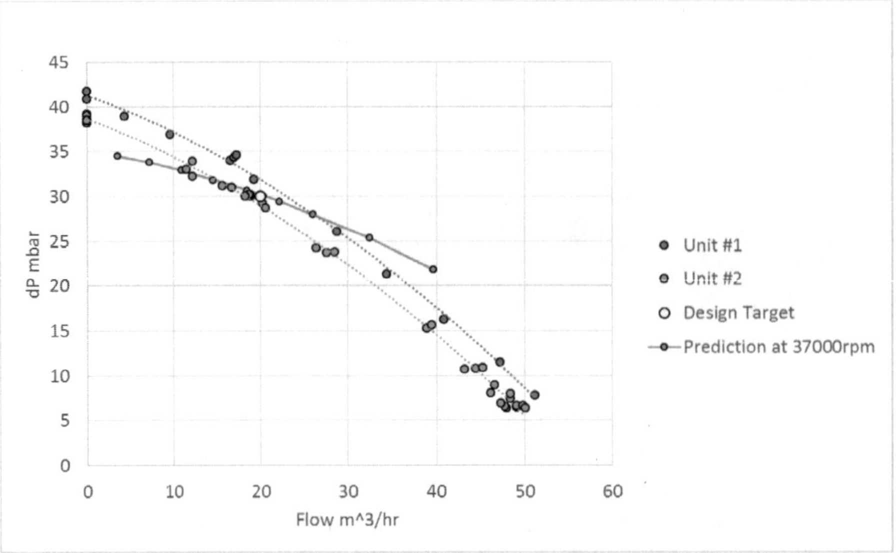

Figure 19. Test Results showing the predicted and experimental performance of the blower wheel (test at 37,000 rpm, 1 bar pressure, helium lubricant).

6. Concluding Remarks

A helium circulator running at 40,000 rpm on spiral groove helium lubricated gas bearings has been in continual operation for over a year with no downtime and no detectable increase of vibration or power required to drive it.

The spiral groove gas bearings have proven to be robust and reliable and the tested performance matches well with the computer predictions.

A permanent magnet motor mounted inside the tungsten carbide gas bearing shaft with a stainless steel can to isolate the motor winding from the helium has shown performance close to the prediction and temperatures measured enar the can correspond well with FEA prediction.

The laser sintered shrouded blower wheel has demonstrated performance within a few% of the CFD predictions.

References

[1] Institut Laue Langevan , Grenoble, France https]//www.ill.eu/about0/what-is-the-ill/
[2] Decker O 1976. Advanced high pressure high temperature gas bearing circulators. Proc 7th International Gas Bearing Symposium. Paper C2. Cambridge, UK.
[3] www.ofttech.com (http]//www.ofttech.com/services)
[4] Molyneaux, A K 1989 The Use of Spiral Groove Gas Bearings in a 350 000 rpm Cryogenic Expander. Tribology Transactions. Vol. 32 , Iss. 2, 1989
[5] Whipple R T P 1951 Theory of the spiral groove thrust bearing with liquid or gas lubricant. Report AERE 622. Available from the OSTI at https]//www.osti.gov/scitech/biblio/4359630.pdf

[6] Beardmore G 1973 Gyroscopes including gas lubricated bearings. Patent US3747417 Smiths Industries Ltd.
[7] Molyneaux A K, Lehr A, Bon P 2010 Gas laser having radial and axial gas bearings. Patent WO 2011095400 A3
[8] Jiménez J A 2002 MELFI (The Minus Eighty degree celsius Laboratory Freezer for ISS). ESA bulletin 109. Available at www.esa.int/esapub/bulletin/bullet109/chapter5_bul109.pdf
[9] Tanchon J, Lacapere J, Molyneaux A K, Harris M, Hill S, Tirolien T. A 2016 Turbo-Brayton cryo-cooler for future European observation satellite generation. Proc. International Cryocooler Conference, Inc., Boulder, CO
[10] Pan C 1964 Spectral Analysis of gas bearing systems for stability studies. Report MTI 64TR58. Available from the DTIC at www.dtic.mil/dtic/tr/fulltext/u2/610872.pdf
[11] Vohr J H, Pan C 1968 Design data—gas lubricated spin axis bearings for gyroscopes Report No. MTI-68TR29 (Mech. Tech. Inc.).
[12] Bonneau D, Absi J 1994 Analysis of Aerodynamic Journal Bearings With Small Number of Herringbone Grooves by Finite Element Method.. J. Tribol 116(4), 698-704
[13] Molyneaux A K 1983 The design of lubricant recirculating hydrodynamic spiral groove bearings using grease lubricants. PhD Thesis. University of Southampton

IOP Conf. Series: Materials Science and Engineering **232** (2017) 012064 doi:10.1088/1757-899X/232/1/012064

Simulation study of multi-chamber rotary compressor

Y D Lim[1], M Lin[2] and K T Ooi[3]

[1,2,3] School of Mechanical and Aerospace Engineering, Nanyang Technological University, 50 Nanyang Avenue, Singapore 639798

[1] ylim046@e.ntu.edu.sg, [2] ylin021@e.ntu.edu.sg, [3] mktooi@ntu.edu.sg

Abstract – A new multi-chamber rotary compressor (MCRC) has been designed and the simulation result is presented in this paper. MCRC utilizes the space within the cylinder and it has the advantage of being more compact. Mathematical models which include geometrical, thermodynamics, mass flow and discharge valve have been formulated to evaluate the performance of MCRC. Parametric studies have also been carried out to determine the effects of design parameters such as suction and discharge ports size, valve length, valve thickness and valve width on compressor performance. In this paper, mathematical models will be presented and the predictions of the model will be shown and discussed.

Keywords – multi-chamber; simulation; mathematical model; rotary; MCRC

1. Introduction

In order to reduce the energy consumption by AC&R [1-6], a new multi-chamber rotary compressor (MCRC) has been designed. This new machine improves the compactness of conventional rolling piston compressor by introducing more working space at a given overall volumetric size of the compressor. In its basic form, MCRC comprises an outer cylinder, a hollow cylinder, an inner cylinder and a vane as shown in Figure 1 and 2.

The working principle of MCRC is illustrated in Figure 3. As the shaft rotates, the volumes of the working chambers vary and result in suction, compression and discharge process. During the operation, the working fluid is being drawn into both outer and inner suction chambers. Meanwhile the working fluid in both outer and inner compression chambers are being compressed and discharged when the pressure reaches the desired discharged pressure.

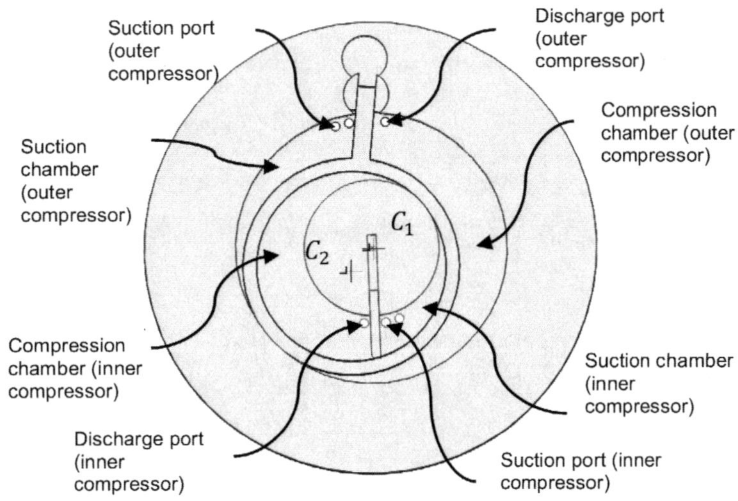

Figure 1. Schematic diagram of the assembly of MCRC

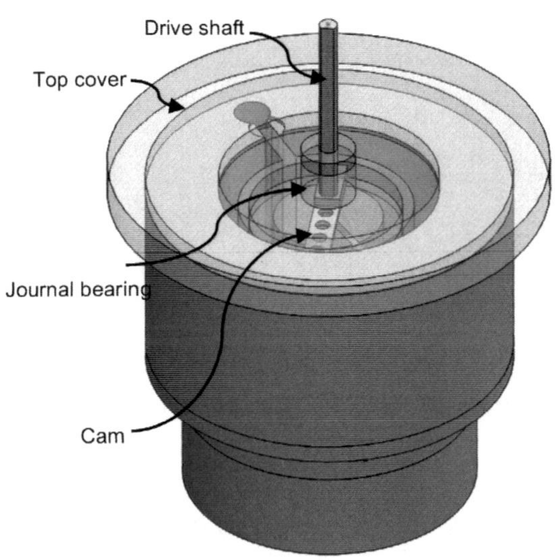

Figure 2. Isometric view of MCRC

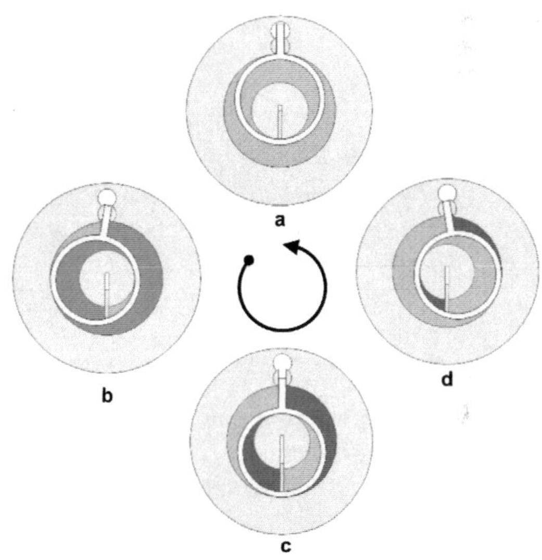

Figure 3. Operating process of MCRC

In this paper, mathematical models which include geometrical, thermodynamics, mass flow and discharge valve are formulated to evaluate the performance of MCRC. The effect of various design parameters on the compressor performance will also be presented and discussed.

2. Mathematical model

2.1. Geometrical Model

In a compressor, the thermodynamics properties of the working fluid are driven by the variation of working chamber volumes. The mathematical expression of outer compressor suction chamber volume, V_{suc1}, compression chamber volume, V_{com1}, inner compressor suction chamber volume, V_{suc2}, and compression chamber volume, V_{com2}, are derived as following.

$$V_{suc1}=\frac{1}{2}\left[r_{oc}{}^{2}\theta_{1}-r_{oh}{}^{2}\theta_{3}-e(l_{v1}+r_{oh})sin\theta_{3}-l_{v1}w_{v1}+\frac{1}{4}w_{v1}{}^{2}tan\gamma_{1}\right]l_{com1} \quad (1)$$

$$V_{com1}=\pi(r_{oc}{}^{2}-r_{oh}{}^{2})l_{com1}-\frac{1}{2}\left[r_{oc}{}^{2}\theta_{1}-r_{oh}{}^{2}\theta_{3}-e(l_{v1}+r_{oh})sin\theta_{3}+l_{v1}w_{v1}-\frac{1}{4}w_{v1}{}^{2}tan\gamma_{1}\right]l_{com1} \quad (2)$$

$$V_{suc2}=\frac{1}{2}\left[r_{ih}{}^{2}\theta_{2}-r_{ic}{}^{2}\theta_{1}-e\left(l_{v2}+r_{ic}-\frac{1}{2}w_{v2}\right)sin\theta_{1}-w_{v2}\left(l_{v2}-\frac{1}{2}w_{v2}\right)-\frac{1}{4}w_{v2}{}^{2}\left(\frac{\pi}{2}-\gamma_{2}\right)\right]l_{com2} \quad (3)$$

$$V_{com2}=\pi(r_{ih}{}^{2}-r_{ic}{}^{2})l_{com2}-\frac{1}{2}\left[r_{ih}{}^{2}\theta_{2}-r_{ic}{}^{2}\theta_{1}-e\left(l_{v2}+r_{ic}-\frac{1}{2}w_{v2}\right)sin\theta_{1}+w_{v2}\left(l_{v2}-\frac{1}{2}w_{v2}\right)+\frac{1}{4}w_{v2}{}^{2}\left(\frac{\pi}{2}+\gamma_{2}\right)\right]l_{com2} \quad (4)$$

where,

$$l_{v1}=-r_{ih}-t_{h}+(e^{2}+r_{oc}{}^{2}-2er_{oc}cos\theta_{1})^{\frac{1}{2}} \quad (5)$$

$$\theta_{3}=cos^{-1}\left(\frac{r_{oc}{}^{2}-(r_{oh}+l_{v1})^{2}-e^{2}}{2e(r_{oh}+l_{v1})}\right) \quad (6)$$

$$l_{v2}=\frac{1}{2}w_{v2}-r_{ic}-ecos\theta_{1}+\left[(r_{ih}-\frac{1}{2}w_{v2})^{2}-e^{2}sin^{2}\theta_{1}\right]^{\frac{1}{2}} \quad (7)$$

$$\theta_{2}=cos^{-1}\left(\frac{e^{2}+\left(r_{ih}-\frac{1}{2}w_{v2}\right)^{2}-\left(r_{ih}+l_{v2}-\frac{1}{2}w_{v2}\right)^{2}}{2e\left(r_{ih}-\frac{1}{2}w_{v2}\right)}\right) \quad (8)$$

$$\gamma_{1}=\begin{cases} cos^{-1}\left(\frac{r_{oc}{}^{2}+(r_{oh}+l_{v1})^{2}-e^{2}}{2r_{oc}(r_{oh}+l_{v1})}\right); & 0\leq\theta_{1}\leq\pi \\ -cos^{-1}\left(\frac{r_{oc}{}^{2}+(r_{oh}+l_{v1})^{2}-e^{2}}{2r_{oc}(r_{oh}+l_{v1})}\right); & \pi\leq\theta_{1}\leq2\pi \end{cases} \quad (9)$$

$$\gamma_{2}=\begin{cases} cos^{-1}\left(\frac{(r_{ih}-\frac{1}{2}w_{v2})^{2}+\left(r_{ih}+l_{v2}-\frac{1}{2}w_{v2}\right)^{2}-e^{2}}{2(r_{ih}-\frac{1}{2}w_{v2})\left(r_{ih}+l_{v2}-\frac{1}{2}w_{v2}\right)}\right); & 0\leq\theta_{1}\leq\pi \\ -cos^{-1}\left(\frac{(r_{ih}-\frac{1}{2}w_{v2})^{2}+\left(r_{ih}+l_{v2}-\frac{1}{2}w_{v2}\right)^{2}-e^{2}}{2(r_{ih}-\frac{1}{2}w_{v2})\left(r_{ih}+l_{v2}-\frac{1}{2}w_{v2}\right)}\right); & \pi\leq\theta_{1}\leq2\pi \end{cases} \quad (10)$$

2.2. Thermodynamic Model

During operation, the change of the volume of each working chamber results in the variation of properties of the working fluid. From the First Law of Thermodynamics [7], an energy rate balance of a control volume with the assumption that the fluid properties are uniform and have no spatial variation within the working chamber, gives

$$\frac{dE_{cv}}{dt}=\frac{dQ_{cv}}{dt}-\frac{dW_{cv}}{dt}+\sum_{i}\frac{dm_{i}}{dt}\left(h_{i}+\frac{v_{i}{}^{2}}{2}+gz_{i}\right)-\sum_{o}\frac{dm_{o}}{dt}\left(h_{o}+\frac{v_{o}{}^{2}}{2}+gz_{o}\right) \quad (11)$$

The rate of work done due to change in chamber volume is given as

$$\frac{dW_{cv}}{dt}=p_{cv}\frac{dV_{cv}}{dt} \quad (12)$$

The total energy of the control volume $\boldsymbol{E_{cv}}$ consists of internal energy, kinetic energy and gravitational potential energy which can be presented by

$$\boldsymbol{E_{cv}=m_{cv}\left(u_{cv}+\frac{v_{cv}{}^{2}}{2}+gz_{cv}\right)} \quad (13)$$

Substituting equation (12) and equation (13) into equation (11), assuming adiabatic and neglect all the potential energy and kinetic energy terms the following equation is obtained

$$m_{cv}\frac{du_{cv}}{dt}+u_{cv}\frac{dm_{cv}}{dt}=-p_{cv}\frac{dV_{cv}}{dt}+\sum_{i}\frac{dm_{i}}{dt}h_{i}-\sum_{o}\frac{dm_{o}}{dt}h_{o} \quad (14)$$

solving this equation for internal energy together with the derivative of density in equation (15), respective thermodynamic properties can be obtained

$$\frac{d\rho_{cv}}{dt} = \frac{1}{V_{cv}}\frac{dm_{cv}}{dt} - \frac{m_{cv}}{V_{cv}^2}\frac{dV_{cv}}{dt} \tag{15}$$

The rate of change of mass of the control volume can be expressed as [8]

$$\frac{dm_{cv}}{dt} = \sum_i \frac{dm_i}{dt} - \sum_o \frac{dm_o}{dt} \tag{15}$$

2.3. Mass Flow Model

The mass flow rate through the suction port and discharge port of both outer and inner compressor is formulated as following

$$\frac{dm}{dt} = C_d \rho_{2s} A_{flow} \sqrt{2(h_1 - h_{2s})} \tag{16}$$

where C_d is given a value of 0.6 [9] and A_{flow} is defined as the port area.

2.4. Discharge Valve Response Model

The valve used in the discharge is an automatic reed valve. In MCRC, there are two discharge valves for both outer and inner compression chambers.

Assuming that the reed behaves like a beam, the free body diagram of an infinitesimal small element of reed is shown in Figure 4.

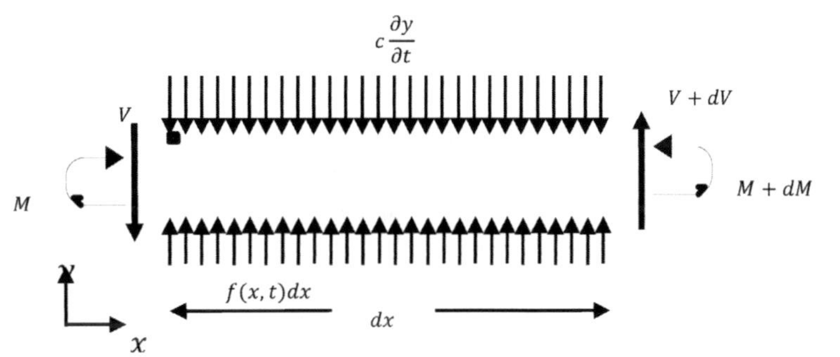

Figure 4. Free body diagram of an infinitesimal small element of valve

From elementary flexural theory [10], the following equations can be obtained

$$E_v I_v \frac{\partial^4 y}{\partial x^4} + \rho A_c \frac{\partial^2 y}{\partial t^2} + c\frac{\partial y}{\partial t} = f(x,t) \tag{17}$$

Equation (17) is a linear, non-homogeneous equation with second order in time and fourth order in space. It is influenced by pressure excitation and the body force as shown at the right-hand side of the equation. Since it is a distributed system, the valve motion is analysed with free vibration and forced vibration approaches.

2.4.1. Free Vibration Analysis

For free vibration, equation (17) is simplified as following

$$E_v I_v \frac{\partial^4 y}{\partial x^4} + \rho A_c \frac{\partial^2 y}{\partial t^2} + c\frac{\partial y}{\partial t} = 0 \tag{18}$$

At certain time, the valve motion displays a unique profile and this profile does not change with time but the amplitude does. Hence, a separable solution is introduced, which is expressed as

$$y(x,t) = f(x) \cdot g(t) \tag{19}$$

Substituting equation (19) into equation (18), applying method of separation of variables and dividing the whole equation by $\rho A_c f g$, the following equation is obtained

$$\frac{E_v I_v}{\rho A_c}\frac{1}{f}\frac{d^4 f}{dx^4} = -\frac{1}{g}\frac{d^2 g}{dt^2} - \frac{c}{\rho A_c}\frac{1}{g}\frac{dg}{dt} = p^2 \quad (20)$$

where p^2 is the constant of separation and equation (20) is actually two ordinary differential equations which can be solved with the boundary conditions. By introducing $a^2 = \frac{E_v I_v}{\rho A_c}$ and $\beta^4 = \frac{p^2}{a^2}$, at k[th] mode of vibration, the solution is as follows

$$y_k(x,t) = A_k e^{(\zeta p_k t)}\left[(sin\beta_k x - sinh\beta_k x) - \frac{sin\beta_k L_v + sinh\beta_k L_v}{cos\beta_k L_v + cosh\beta_k L_v}(cos\beta_k x - cosh\beta_k x)\right]$$

$$\left[\frac{\zeta p_k g_k(0) + \dot{g_k}(0)}{q_{d,k}} sin q_{d,k} t + g_k(0) cos q_{d,k} t\right] \quad (21)$$

The damped frequency $q_{d,k}$ is defined as $q_{d,k} = p_k\sqrt{1-\zeta^2}$, where p_k is the natural frequency at k[th] mode, which is represented by

$$p_k = \beta_k{}^2\sqrt{\frac{E_v I_v}{\rho A_c}} \quad (22)$$

where β_k can be solve from the characteristic equation as below

$$cos\beta_k L cosh\beta_k L = -1 \quad (23)$$

2.4.2. Forced Vibration Analysis

For forced vibration analysis, under the assumption of variable separable form and taking the principle of superposition into account, the solution can be expressed as

$$y(x,t) = \sum_{k=1}^{\infty} f_k(x) g_k(t) \quad (24)$$

The orthogonal natural modes $f_k(x)$ are normalized to satisfy orthonormality condition

$$\int_0^{L_v} f_s(x) f_k(x) dx = \delta_{ks} \quad \text{and} \quad \int_0^{L_v} f_s(x)\frac{d^4 f_k(x)}{dx^4} dx = \left(\frac{p_k}{a_k}\right)^2 \delta_{ks}$$

where δ_{ks} is the kronecker delta. Substituting equation (24) into equation (17), divide the whole equation by ρA_c and multiply the equation by $f_s(x)$, integrate and apply the orthonormality condition, the following equation is obtained

$$\frac{d^2 g_k}{dt^2} + \frac{c}{\rho A_c}\frac{dg_k}{dt} + p_k^2 g_k = \int_0^{L_v} f_k(x)\left(\frac{f(x,t)}{\rho A_c}\right) dx \quad (25)$$

The right-hand side of the equation is known as modal forces. In the case of MCRC, the force term $f(x,t)$ refers to the force per unit length, which is valid for the part where the valve is exposed to working fluid pressure only. As the natural modes $f_k(x)$ is normalized, a relationship is found for the amplitude A_k

$$A_k{}^2\int_0^{L_v}\left[(sin\beta_k x - sinh\beta_k x) - \frac{sin\beta_k L_v + sinh\beta_k L_v}{cos\beta_k L_v + cosh\beta_k L_v}(cos\beta_k x - cosh\beta_k x)\right]^2 dx = 1 \quad (26)$$

Considering that the discharge port diameter is small as compared to the valve length, and let $f(x,t) = \frac{[P_c(t) - P_{disc}]A_f(y)}{\Delta x}$, equation (25) becomes

$$\frac{d^2 g_k}{dt^2} + \frac{c}{\rho A_c}\frac{dg_k}{dt} + p_k^2 g_k = f_k\left(x + \frac{\Delta x}{2}\right)\frac{\Delta P(t) \cdot A_f(y)}{\rho A_c} \quad (27)$$

A general second order ODE solution cannot be applied in equation (27) as the chamber pressure is coupled with the valve response. By rewriting equation (27) into two first order ODEs as shown in equation (28) and equation (29), combining with thermodynamics model and solve both equations simultaneously, the modal coordinate as well as the displacement of the valve can be obtained.

$$\frac{d\phi_2}{dt} = \phi_1 \quad (28)$$

$$\frac{d\phi_1}{dt} = N_k - 2\zeta p_k \phi_1 - p_k{}^2 \phi_2 \quad (29)$$

where ζ is given a value of 0.004 [11], $\phi_1 = \frac{dg_k}{dt}$, $\phi_2 = g_k$ and $N_k = f_k\left(x + \frac{\Delta x}{2}\right)\frac{\Delta P(t) \cdot A_f(y)}{\rho A_c}$.

3. Operating specifications

The geometrical, kinematics, thermodynamics, the mass flow and the discharge valve response models are solved numerically using the 4th order of Runge-Kutta method [12]. These models are integrated concurrently with FORTRAN and the instant thermodynamics properties of the working fluid are obtained from REFPROP database provided by NIST [13]. The following assumptions are made in this numerical simulation

- Adiabatic condition throughout the process.
- Frictional losses are not considered.
- Perfect sealing and no internal leakage.
- Constant shaft angular velocity.

The operating specifications is summarised in Table 1.

Table 1. Operating specifications

Operating Conditions	
Working fluid	R-32 (Difluoromethane) [14]
Operating speed (ω_1)	3000 rpm
Saturated evaporating temperature	7.2 °C [15]
Saturated condensing temperature	54.4 °C [15]

4. Results and Discussion

The normalized variation of volumes, pressure-volume (PV) diagram and variation of mass in MCRC are shown in Figure 5, Figure 6 and Figure 7 respectively. The pressures in the working chambers remain reasonably constant during the suction process and the mass increases gradually. Compression process begins at the end of suction process, where the working fluid pressure increases and the mass remains constant since there is no more working fluid flowing into the compression chamber under the assumption of perfect sealing. After the pressure of the working fluid reaches the nominal discharge pressure, it does not discharge immediately. Additional pressure builds up is required to overcome the valve stiffness and discharge only when the valve is uncovered. It is observed that both outer and inner compressors behave similarly therefore only the results of outer compressor will be shown and discussed.

4.1. Effect of Suction Port Size

The suction ports of MCRC are without a suction valve and the working fluid in suction chamber is always in contact with the upstream working fluid through the suction port. Figure 8 and Figure 9 show the effects of suction port size on the compressor performance. When the suction port diameter is 6.0 mm, the suction pressure is far below the nominal suction pressure and the mass intake is also lower as compared with larger suction port size. This implies a higher suction loss when the suction port is smaller. Hence, the size of the suction port must be appropriately selected.

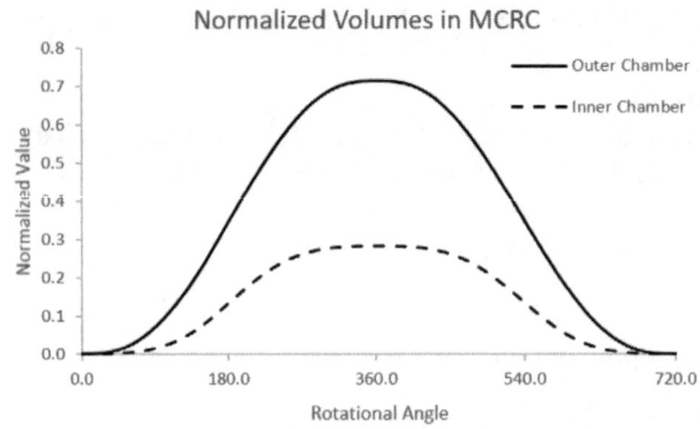

Figure 5. Normalized variation of volumes in MCRC

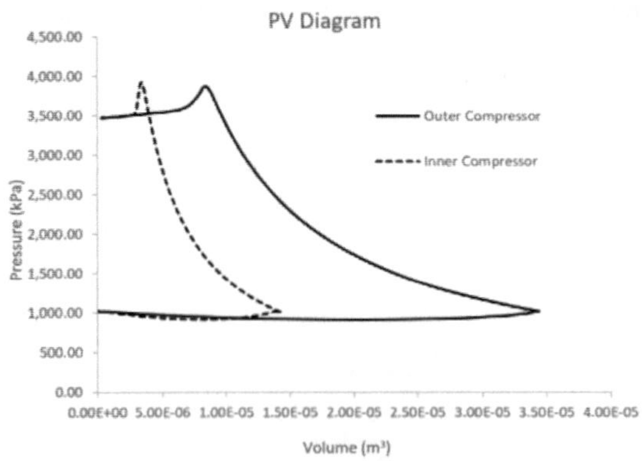

Figure 6. Pressure-volume diagram for MCRC

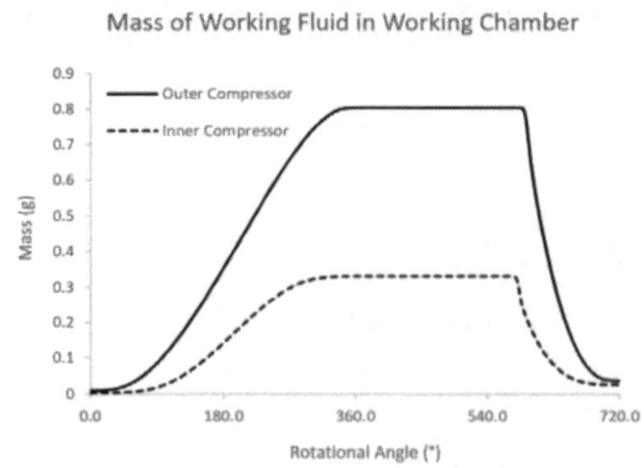

Figure 7. Variation of mass of working fluid in working chambers

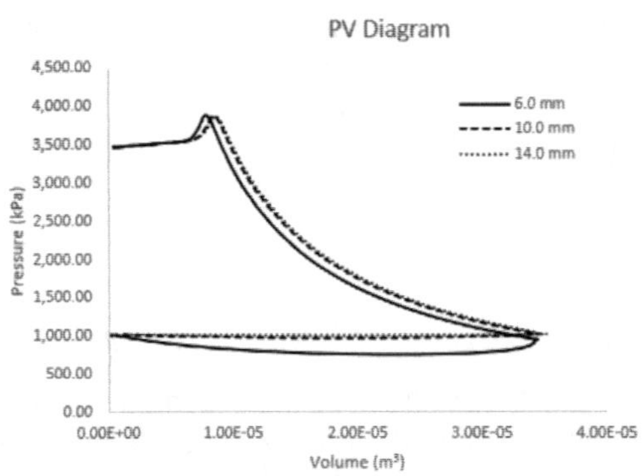

Figure 8. Effect of suction port size on outer compressor performance (pressure-volume)

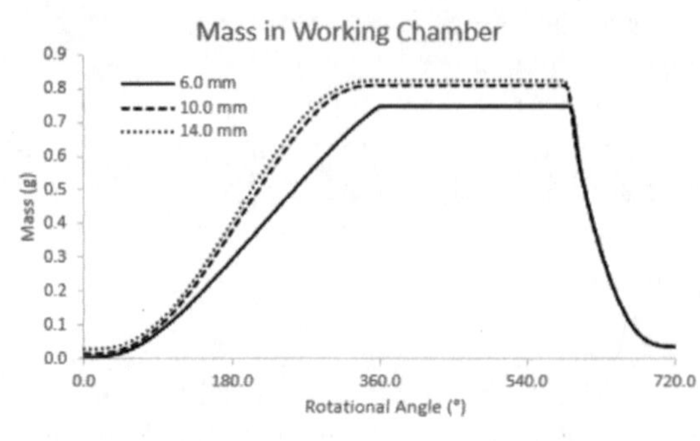

Figure 9. Effect of suction port size on outer compressor performance (mass)

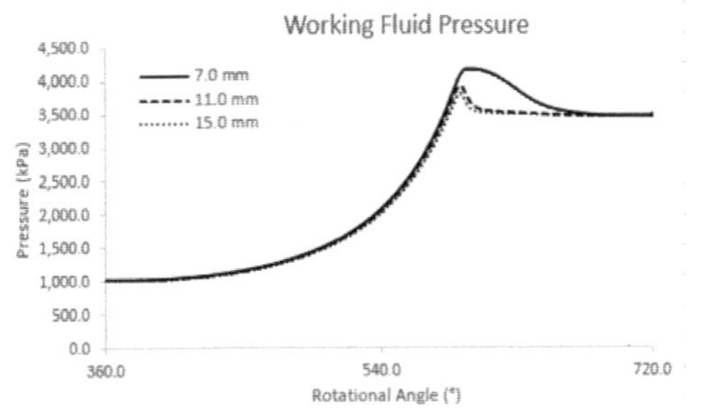

Figure 10. Effect of discharge port size on outer compressor performance (pressure)

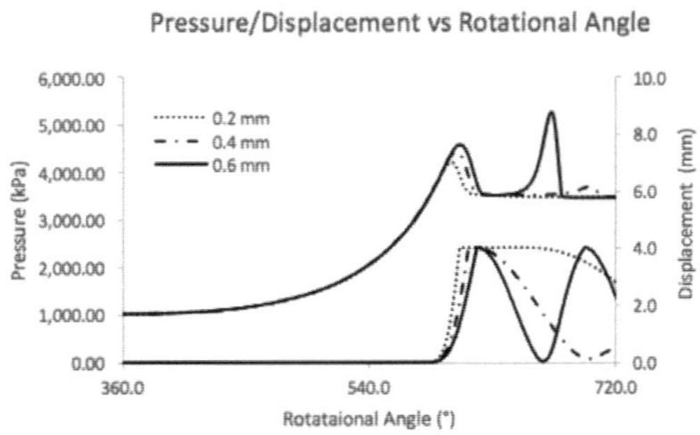

Figure 11. Effect of valve thickness on outer compressor performance

4.2. Effect of Discharge Port Size

A discharge port is always required to be sufficiently large to allow for ease discharge of the working fluid. Also, a larger port size avoids unnecessary over-pressure which may cause discharge valve failure and higher discharge loss. Figure 10 shows the variation of pressure under different discharge port size. The pressure "hump" occurs before the valve opening gets lower when the port size increases, and hence results in a lower discharge loss.

4.3. Effect of Valve Thickness

It is found that as the valve thickness increases, valve stiffness increases. Figure 11 shows the effect of valve thickness on compressor performance. Thicker valve reed also caused a higher discharge loss.

5. Conclusion

The initial simulation study of MCRC has been carried out which includes geometrical model, kinematics model, thermodynamics model, mass flow model and valve response model. Parametric study shows that larger suction and discharge port sizes are preferred in order to reduce the suction and discharge losses. Lastly, a thicker valve reed results in stiffer valve, which has incurred larger discharge loss and vice-versa.

In this paper, the embodiment design of MCRC has made use of the volume which is occupied by the roller in a rolling piston compressor, which is expected to be smaller in size as compared with the conventional rotary compressors. As MCRC is still under early stage development, the compressor performance as well as the corresponding efficiencies are not considered. In fact, more internal leakage flows and frictional losses will certainly play a role by having multi working chambers. For example, the frictional loss between the drive shaft and the hollow cylinder due to the surface in contact. More mathematical models such as frictional losses model, internal leakage model, heat transfer model and oil lubrication network model will be implemented in the future in order to make a more comprehensive theoretical model. With such a complete theoretical model, the comparison between MCRC and existing compressors will be more reasonable. The accuracy of the model will then be checked with the experimental result in the future.

Acknowledgment

The authors would like to acknowledge the assistance from students, technical staff and the financial support from Nanyang Technological University.

IOP Conf. Series: Materials Science and Engineering **232** (2017) 012064 doi:10.1088/1757-899X/232/1/012064

References

[1] F. Dianshu, B. K. Sovacool, and K. M. Vu, "The barriers to energy efficiency in China: Assessing household electricity savings and consumer behavior in Liaoning Province," *Energy Policy,* vol. 38, pp. 1202-1209, 2010.

[2] Y. Kondou, A. Murata, K. Yamada, K. SAKAMOTO, and W. ZHOU, "Study on electricity consumption for domestic refrigerators, air conditioners, and water dispenser in Guangzhou (China) urban area based on questionnaire data," *日本エネルギー学会誌* vol. 90, pp. 554-561, 2011.

[3] E. M. Authority. (2015, 16-Jun-2016). *Singapore Energy Statistics 2015*. Available: https://www.ema.gov.sg/statistics.aspx

[4] E. E. Singapore. (2012, 16-June-2016). *Energy Consumption by Appliance Type*. Available: http://www.e2singapore.gov.sg/

[5] N. C. C. Secretariat. (2016, 16-June-2016). *Climate Change and Singapore*. Available: https://www.nccs.gov.sg/climate-change-and-singapore/domestic-actions/reducing-emissions/households

[6] U. S. E. I. Administration. (2009, 16-Jun-2016). *Household Energy Use in New York*. Available: http://www.eia.gov/

[7] M. J. Moran and H. N. Shapiro, *Fundamentals of engineering thermodynamics*, 6th ed.: Hoboken, N.J. : Wiley, c2008., 2008.

[8] T. L. Bergman, F. P. Incropera, D. P. DeWitt, and A. S. Lavine, *Fundamentals of heat and mass transfer*: John Wiley & Sons, 2011.

[9] C. L. Hollingshead, M. C. Johnson, S. L. Barfuss, and R. Spall, "Discharge coefficient performance of Venturi, standard concentric orifice plate, V-cone and wedge flow meters at low Reynolds numbers," *Journal of Petroleum Science and Engineering,* vol. 78, pp. 559-566, 2011.

[10] L. G. Jaeger, *Elementary Theory of Elastic Plates: The Commonwealth and International Library: Structures and Solid Body Mechanics Division* vol. 2: Elsevier, 2013.

[11] G. Prater and W. P. Hnat, "Optical measurement of discharge valve modal parameters for a rolling piston refrigeration compressor," *Measurement,* vol. 33, pp. 75-84, 2003.

[12] S. C. Chapra and R. P. Canale, *Numerical methods for engineers* 6th ed.: Boston : McGraw-Hill Higher Education, c2010., 2010.

[13] E. W. Lemmon, M. L. Huber, and M. O. McLinden, "NIST Standard Reference Database 23: Reference Fluid Thermodynamic and Transport Properties-REFPROP," 9.0 ed. Gaithersburg: National Institute of Standards and Technology, 2010.

[14] L. DAIKIN INDUSTRIES. (2017, 11-Feb). *R-32 Next-Generation Refrigerant*. Available: http://www.daikin.com/about/why_daikin/benefits/r-32/

[15] A. Standard, "Performance Rating Of Positive Displacement Refrigerant Compressors and Compressor Units," 2004.

IOP Conf. Series: Materials Science and Engineering **232** (2017) 012064 doi:10.1088/1757-899X/232/1/012064

Nomenclature

A	Area [m^2]
A_c	Cross sectional area of valve [m^2]
A_f	Effective force area on discharge valve [m^2]
A_{flow}	Fluid flow area [m^2]
A_k	Amplitude of kth mode of vibration [-]
c	Damping coefficient [Ns/m^2]
C_d	Coefficient of discharge [-]
e	Distance between outer/inner cylinder centers and hollow cylinder centre [m]
E	Summation of kinetic energy, potential energy and internal energy [J]
E_v	Modulus of elastic of valve material [Pa]
$f_k(x)$	Valve response profile at kth mode [m]
$f(x,t)$	Valve response profile at kth mode [m]
$g_k(t)$	Variation of valve response profile amplitude at kth mode [-]
h	Specific enthalpy of working fluid [J/kg]
h_{2s}	Specific enthalpy of working fluid at downstream under isentropic condition [J/kg]
I_v	Moment of inertia of valve [m^4]
K	Flow coefficient [-]
l	Length [m]
L_v	Length of valve [m]
m	Mass [kg]
M	Bending moment [Nm]
N_k	Modal force [N]
p_{cv}	Pressure in control volume [Pa]
p_k	Natural frequency of the valve at kth mode [rad/s]
P	Pressure [Pa]
$q_{d,k}$	Damped response frequency of valve at kth mode
Q	Heat transfer [J]
r	Radius [m]
s	Entropy [J/kg·K]
t	Time [s]
T	Temperature [K]
u	Internal energy [J/kg]
v	Velocity [m/s]
V	Volume [m^3] / Shear force [N]
w	Width [m]
W	Work done [J]
x	Location along the valve longitudinal axis [m]
y	Valve displacement [m]
z	Elevation [m]
α	Angular acceleration [rad/s^2]
δ_{ks}	Kronecker delta [-]
η_s	Isentropic efficiency [-]
ζ	Damping coefficient [-]
ρ	Density [kg/m^3]
ω	Angular velocity [rad/s]
γ	Subtended angle [rad]
ΔP	Pressure difference [Pa]

IOP Conf. Series: Materials Science and Engineering **232** (2017) 012065 doi:10.1088/1757-899X/232/1/012065

A new Wankel-type compressor and vacuum pump

D W Garside

Engineer and Director, Epitrochoidal Compressors Ltd, (ECL) 74, Longlands Rd, Slaithwaite, Huddersfield, HD7 5DR, UK

Email: davidwgarside@btinternet.com

Abstract: When the Wankel principles were first published in the early 1950s most of the initial work was aimed at developing a ***compressor***. At that time many of the characteristics appeared to promise a superior machine than hitherto known. However, all the early designs resulted in a high value for the minimum clearance volume (CV) and this problem was never overcome.

Knowledge now gained from the development and manufacture of the Wankel ***engine*** has enabled the evolution of a new compressor concept where the rotor flank, radially very close-fitting over its central area, provides gas sealing with the housing bore. The rotor has an increased radial clearance towards the apices which makes the machine practical to manufacture. The 'nesting' of the rotor flank with the housing bore at the end of the exhaust stroke results in an extremely small CV.

This machine promises to possess an exceptional combination of all the attributes which are important in achieving ***high energy efficiency*** in positive-displacement compressors and vacuum pumps:

- near-zero CV
- low mechanical friction losses
- low internal gas leakage (assisted via oil flooding)
- high volumetric efficiency.

In addition it is compact, lightweight, vibration-free, consists of few components, and can be built in any chamber size.

The Paper discusses the features and characteristics of the design.

1. Introduction

1.1 Origin of the innovation

In January 2014 the **Daily Telegraph [1]** published a half-page article describing a new type of compressor. This article triggered the question why the Wankel principle, which possesses so many relevant advantages as a positive displacement machine, has so far only been exploited as an engine and has never progressed towards being a successful compressor.

A search was made through all the Technical Papers and all the published Patent Applications relating to Wankel type compressors dating from about 1953 to 2013. The high value of the CV issue dominated the various designs which had been considered and worked upon. The early attempts by Borsig, NSU, Neumag et al all appeared unsatisfactory. In attempting to increase the rotor R/e ratio (R = rotor radius, e = shaft eccentricity) and thereby improve the fundamental geometric compression ratio, thus minimising the CV, these designs all resulted in small diameter eccentric shafts and stationary gears. These components proved to be insufficiently robust for a practical industrial machine **[2]**.

Furthermore, the resulting CV of these designs was still too high for good efficiency, and the high R/e value employed also resulted in the machines not being particularly compact.

The later-developed **Ogura Wankel [3]** of the 1980s, which was manufactured for vehicle air conditioning applications, used a typical engine-design value for the R/e ratio. This resulted in a smaller and more robust machine but it possessed the extremely large and penalising CV value of 16 %. Nevertheless, probably because of its light weight, freedom from radial vibration and good mechanical efficiency **[4],** it secured a market share for a period.

1.2 The new concept

Study has now led to the GW design (designation as in Garside-Wankel), which exploits the principle whereby a rotor with a uniquely shaped flank profile will provide:

- gas sealing between a moving point in the central region of the rotor flank and the epitrochoidal
- bore
- a close-to-zero value for the resulting final CV
- and in combination with greater radial clearance towards the rotor apices will be practical to manufacture.

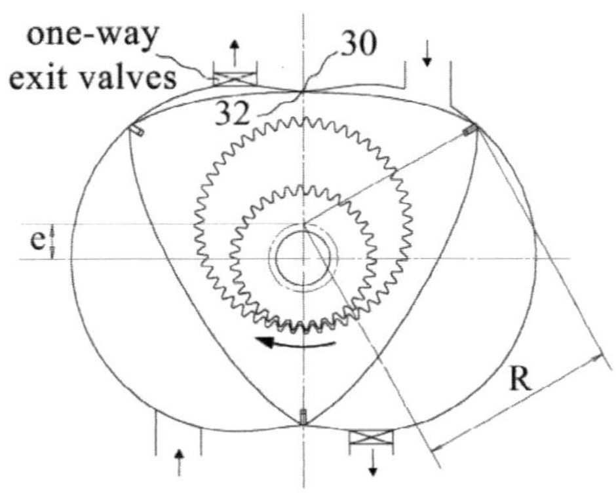

Figure 1 A cross sectional view through the machine with rotor at 'engine' TDC (top dead centre) position.
It will be noted that gear backlash does not affect the radial clearance between 30 and 32 in this TDC position because the point 32 will move merely laterally. At other angular positions the gear backlash plus gear angular location tolerances require that the rotor flank shape is reduced in radial size slightly, to avoid risking impact of the apices with the housing bore.

1.3 A Note on the fundamental sealing quality of the Wankel

Of all the known positive-displacement types of compressor, only the reciprocating and the Wankel possess an adequate sealing system which enables them to operate ***as an IC engine.***

Gas temps of up to 2000 C and gas pressures of around 50 bar exist in such machines.

IOP Conf. Series: Materials Science and Engineering **232** (2017) 012065 doi:10.1088/1757-899X/232/1/012065

To create a 4-stroke IC engine, the sliding vane (SV), screw, scroll and Lontra type compressors for example, would generally require two machines built in some type of back-to-back arrangement.
The gas sealing quality of all these designs would be quite incapable of operating efficiently with such temperatures and pressures.
This comment referring to the IC engine is made to illustrate that the fundamental sealing capabilities of the Wankel (whilst not as good as a reciprocating engine using piston rings in a circular bore) are of such quality that it should have a *relatively* easy task when operating as a ***compressor***. Certainly this should be the situation when operating with oil flooded air.

2. Discussion of the major design characteristics

2.1 Minimum Clearance Volume
A study of the Animated Model [5] illustrates how the rotor movement provides close to zero CV because the trailing part of the rotor flank 'nests' with the housing bore 60° ATDC (shaft angle) at the end of the exhaust stroke.

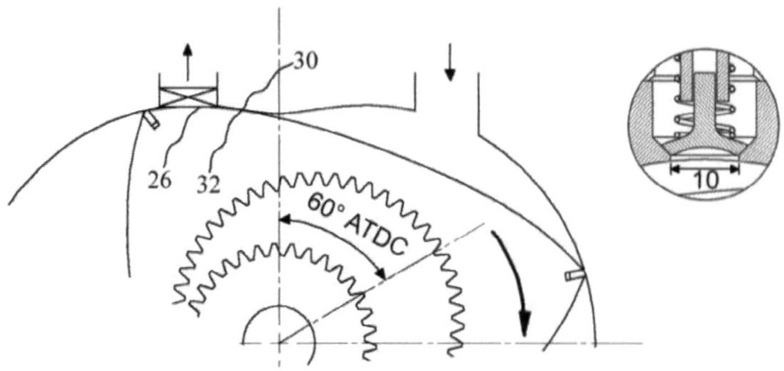

Figure 2. Nesting of the rotor flank with the housing bore at the end of the exhaust stroke; with an inserted view of an exit valve (three are fitted transversely)

The CV (or 'dead volume') in this rotor position consists of the small working clearance between the rotor and housing and small recesses under the outward-opening poppet valves.
The total volume in chamber 26 is less than 0.5% of the intake chamber volume.
In operation, oil is likely to largely fill this cavity

2.2 Gas sealing quality
There are three potential routes for leakage to take place:

2.2.1 Leakage via the radial clearance between the rotor flank and housing bore. The animated model shows how a moving point (32) on the rotor flank continually maintains a close radial clearance with a specifically related moving point (30) of the housing bore during the phase from about 60° BTDC to 60° ATDC of the eccentric shaft movement.
The exhaust stroke is essentially completed by 60° ATDC.
In Wankel ***engines*** the radial clearance between the rotor flank and the bore is not an important consideration, being typically about 0.5 mm . In the central part of the flank there is a combustion recess of course. ***In a compressor*** the clearance needs to be ***much*** smaller, certainly less than 0.1 mm to avoid excessive leakage over the rotor flank during the compression / exhaust stroke.
Calculations indicate that even with 0.05 mm gap the sealing with oil-free air would not be satisfactory but with oil present, as is commonly used in many types of air compressors, leakage per cycle would be less than about 1 % of the free-air intake volume.

With current standard numerically controlled (NC) machines it would not be too onerous to provide a gap of 0.05 ± 0.04 mm. The small leakage over the rotor flank is likely to consist mostly of oil which will have been deposited or centrifuged on to the trochoid bore. Leakage will result largely from the rotor surface movement 'dragging' the oil through rather than viscous flow due to the pressure differential.
From 60° ATDC onwards the rotor flank increased radial clearance to the housing bore no longer affects the gas leakage, this being then controlled by the apex seals.

2.2.2 Leakage via the axial gap between sides of rotor and side plates. A novel design feature in the GW compressor is that the conventional gas side seals, as used in a Wankel engine rotor, are eliminated. The available pressurised oil supplied from the receiver vessel is utilised to completely fill and pressurise the internal cavities of the rotor.
Alternatively, a separate pump may be used to supply the oil.
Centrifugal force on the oil inside the rotor will ensure that the oil pressure at the outer perimeter of the rotor internal cavity will always be higher than the maximum pressure of the air in the working chambers. Some small quantity of this oil, due to the pressure differential, will then continually and intentionally leak ***outwards*** into the working chambers via the axial clearances.
This feature will eliminate ***all*** gas leakage out from the working chambers via the sides of the rotor.
The design has other advantages:

- The manufacturing cost of 30 gas sealing components and their associated machined slots is avoided.

- The mechanical sliding friction of those side seals is eliminated.

- The radial space required by the side seals is not needed. Hence an exceptionally low R/e value for the geometry of the rotor can be selected. This results in a more compact and lighter weight rotor and resulting machine, with further friction savings due to reduced seal travel distances and velocities.

Note: In an Wankel epitrochoidal mechanism the value of R/e can be readily assessed from an axial assy drawing by viewing the closeness of the rotor ring gear PCD to the rotor flank. These coincide when R/e = 5.0 .

2.2.3 Leakage past the apex seals. Similar design apex seals operate satisfactorily in an IC engine. Hence, in an "oil flooded" compressor duty, any gas leakage via the apex seals should be negligible. Wear rates can be expected to be low, as is the case with SV (sliding vane) type compressors, because all metal-to-metal moving contacts will have a hydrodynamic oil film present.

2.3 Mechanical Friction Losses

Comparison with an SV compressor of similar capacity shows that the total surface area swept by the apex seals of the GW in unit time is only about 8% of the area swept by the vanes in the SV (assuming that both are operating at the same rpm). This low figure occurs because the rotor of a GW is only revolving at 1/3 the speed of the eccentric shaft, there are only 3 apex seals (typically 7 or 8 vanes in the SV) and the rotor is axially narrower and employs shorter axial length seals than in the SV type. Furthermore, the lightweight apex seals of the GW are not subjected to significant forces and nominally are stationary in their slots, whereas the heavy vanes of the SV are subject to a high outward centrifugal force. Those vanes are also oscillating radially in their slots with a simultaneous bending moment being imposed.
Each factor of swept area, sliding velocity, radial load, and vane-to-slot movement is more energy consuming in the SV than in the GW.
The GW does have a disadvantageous friction feature in that the rotor gears are operating in 'solid' oil

IOP Conf. Series: Materials Science and Engineering **232** (2017) 012065 doi:10.1088/1757-899X/232/1/012065

which is not ideal. But the gears are open at both axial sides for the 'trapped oil' to escape and the meshing velocities are not high. Special tooth forms can be used.
The likely reduction of gear meshing noise may be valuable.

2.4 Volumetric Efficiency (VE)
Wankel ***engines***, which possess an unrestricted inlet port and a sinusoidal intake volume change, have opportunities for dynamic inlet tuning and typically achieve a VE of over 110% despite the induction air undergoing some heating and expansion during the induction stroke.
As a compressor the VE is likely to be close to 100%, significantly better than all other compressor types which suffer from re-expansion of the CV gas, some leak-back of previously compressed and heated air and generally possess untunable inlet systems.
VE per se does not affect energy efficiency although it does result in a smaller and lighter machine, which possesses a greater throughput for given leakage and friction losses, etc. Consequently those specific loss factors are improved.

3. Minor features of the design

3.1 The primary innovation of the rotor providing gas sealing at a moving point on the rotor flanks does result in unbalanced gas pressure loads on the compressing flank. This imposes a significant cyclic torque load on the gears as illustrated below:

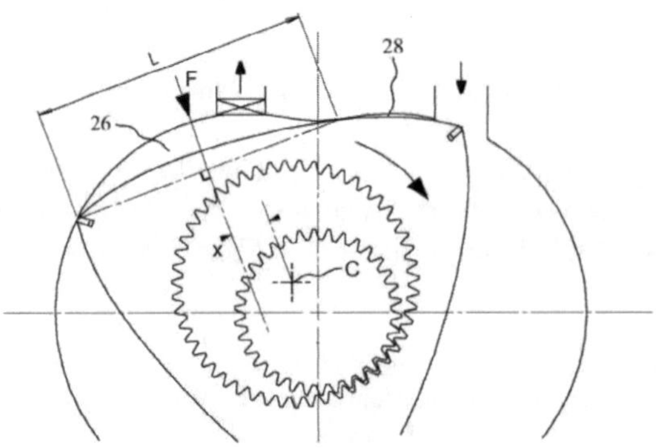

Figure 3. Intermittent torque loading on the gears results from the uneven gas pressure on the rotor flank. Pressure in chamber 26 is higher than in 28 resulting in Force F producing a torque Fx about the centre C of the rotor

The issues are further discussed in the patent document **[6].**
A twin gear arrangement with gears on either side of the rotor is proposed for higher pressure applications.
Life calculations indicate that:

- a single gear pair of unhardened steel would be satisfactory for up to about 6 bar inlet-to-outlet differential
- use of hard steel would allow up to 12 bar
- and double these values with a twin gear arrangement

3.2 Manufacture of the gears
The first prototype required that twin ring gears were shaped separately from each side.
A later design enables either single or twin gears to be wire cut or broached through in a single pass.

3.3 Exit 1-way valves are required in the GW unlike the SV, screw, and scroll types.
There is sufficient space to install multiple small and light-weight poppet valves capable of passing the required volume flow with fast response and low pressure drop characteristics.
The GW has an advantageous characteristic in that the exhaust stroke occupies a long duration of 330° of eccentric shaft rotation, of which 140° are utilised to eject the final 10% of the chamber volume.
As an example, with 7 bar compression, the exhaust phase in a reciprocating compressor occupies only the final 50° before minimum volume is reached. The GW uses 150° for this exhaust phase, greater by a factor of 3.
The exit valves can therefore be smaller, or involve lower pressure loss, or the operating speed can be higher. For this reason it is considered that the unusual combination of 'oil flooding' in a machine fitted with exit valves is unlikely to present major problems.
The exiting oil will be beneficial in relation to damping the movement of the valves, lubricating and sealing the seats and cooling. Three 10 mm dia poppet valves made from 'Torlon' engineering plastic, which possesses low density as well as high strength plus wear and temperature resistant characteristics, are fitted to the proof-of-concept machine in an axial row. The weight of the valves is only 0.38 gm each.

3.4 Manufacturing cost aspects
The scroll type compressor is currently manufactured economically, despite the requirement for very small radial clearances being essential at six 'sealing points' simultaneously **[7].** Computer-controlled machine tools create precise surface geometry maintaining tolerances measured in microns. The GW only requires a single such 'sealing point'. So, if found to be advantageous, selective assembly perhaps based only on housing bore, rotor OD and shaft throw dimensions could be used to achieve an extremely small gap at this single point.
The scroll and GW machines are basically similar in that they both involve a rotor mounted on an eccentric shaft rotating inside a housing. The manufacturing costs of the scroll compressor are regarded as being low. Hence the costs of the GW should also be acceptable.

3.5 A 'free' addition to the swept volume
The feature of the GW design which gives a near zero CV results from the rotor continuing to rotate past the normal 'engine TDC' position and then nest with the housing. A favourable outcome is that the TDC clearance volume (illustrated in Fig 1) is now ***added to*** the conventional rotary engine 'swept volume' calculation to give an increased volume to the intake chamber.
To clarify, there is an addition of 7.5% intake volume bonus for a given size of machine. Again, this offers size, weight and small energy efficiency advantages.
The 7.5% figure is the fundamental TDC clearance of the 2:3 type geometry when the value of $R/e = 5.3$ **[8]** .

3.6 Part-load operation.
Throttling the intake of the SV type compressor, in particular, involves a significant reduction in energy efficiency. The GW has space available on the intake port sides to incorporate two (or more) alternative-position later-closing inlet ports, a more efficient way of reducing the output to match varying flow requirements. This may allow use of a lower cost and more efficient fixed-speed electric drive motor for some applications.

4. Application of the GW as a vacuum pump
The attributes of the GW are particularly advantageous for some vacuum pump duties:

- the near-zero CV combined with good gas sealing will allow a high vacuum to be achieved in a single stage machine
- the low mechanical friction losses become more significant in contributing to energy

- efficiency when the air work energy may be very small.
- the machine can be built in unlimited size without incurring vibration problems.

An early type of rotary piston vacuum pump is shown below:

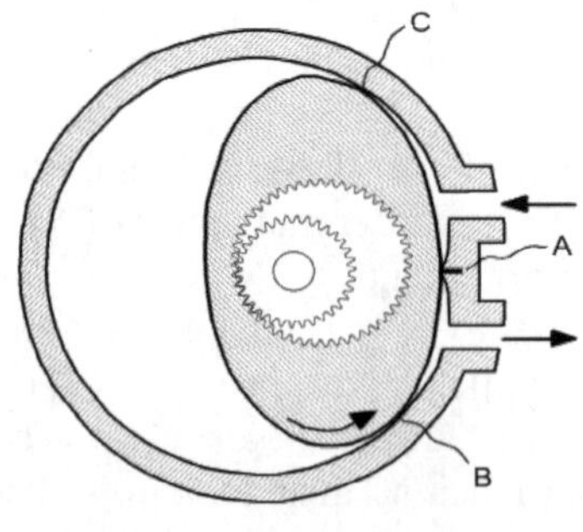

Figure 4 Axial view of a ***hypotrochoidal*** type rotary piston vacuum pump

There are no gas seal pieces in the rotor, simply one 'stationary' seal (A) in the housing at the 3 o'clock position above. Hence it is essential that the rotor has very close 'sealing' clearances with the housing bore ***simultaneously at two points,*** B & C, a difficult manufacturing requirement (particularly in earlier days) bearing in mind that both the gear backlash and the angular positioning accuracy of the gears would affect these clearances. It was perhaps inevitable that this machine was found to be uneconomic to manufacture and it is no longer produced.
However, the descriptions of this machine **[9]** did state that it had advantages over alternative vacuum pumps in that:

- the volume of the pumping chamber with respect to the volume of the entire machine (the so-called 'specific volume') was twice that of the competing 'plunger piston type' vacuum pumps
- it could run at twice the speed because of the low vibration and thus was about four times smaller
- the low vibration eliminated the need for heavy foundations.

5. Some comparisons of the GW with existing compressors

5.1 The CV (or 'dead volume')
This is smaller than other types of competing compressors.

5.2 Gas sealing
The sealing quality potential of the GW offers superior sealing to existing rotary type compressors and may match the best reciprocating type. It should be superior to the screw type, which possesses very long sealing lines, as well as having the 'blow holes'. It should be superior to the scroll compressor which has many more leakage paths. It certainly should be superior to the SV, which has leakage past the axial ends of the rotor and of the vanes, and under the vanes.

5.3 Mechanical friction losses

a) ***Reciprocating compressors:*** An advantage of the GW is the absence of piston side thrust, as well as no requirement for inlet valves with the associated pumping loss. As described in 2.2.2 the GW has zero rotor side mechanical friction, only a small viscous shearing loss. The size and weight of the GW will be lower and consequently will use smaller bearings. Overall friction will be lower, as assessed in more detail in **[6].**

b) The ***scroll compressor*** rotor has a high axial load combined with continual axial face sliding movement which will result in significant energy loss.
c) The ***screw type*** has very long sealing line lengths incurring high speed viscous oil shearing. If no oil is present at any point along these sealing lines, there will be leakage losses.
d) The ***SV*** is discussed in 2.3

The reduction of mechanical friction in the GW and therefore energy / heat input will lower slightly the compression index and hence provide a small efficiency gain.

6. Testing

In 2014 the design and detailing of a first proof-of-concept 7.5 kW, 7 to 10 bar, 2 x 178cc machine was completed by ECL (Epitrochoidal Compressors Ltd).
At 2900 rpm this machine has an intake swept volume of 1.11 m^3 per minute.
Agreement was reached with a UK compressor manufacturer that they would:

a) Procure the parts to build one machine
b) Test as a 6 to 10 bar air compressor on their existing test rig.

This testing is still in hand.
Procurement of 3 further sets of parts by ECL is also now in hand in order to allow broadening of the testing effort. One of these machines will be configured as a vacuum pump.

7. Conclusions

- ***As an IC engine,*** the Wankel has major advantages in comparison with the reciprocating engine in relation to size, weight, vibration, mechanical and volumetric efficiency; but a ***major*** deficiency in gas sealing quality, plus an inferior combustion chamber shape.
  As a compressor, the GW design retains all the above advantages but evades the two deficient areas. In fact it is the inferior *'combustion chamber shape'* which the GW design has advantageously exploited to create the important near-zero CV feature.
- Published Papers **[10]** compare the energy efficiency of SV versus screw, reciprocating versus scroll; etc., etc. . The conclusions appear to be that no one type is much superior in this aspect to any other. Analysis of the CV, gas sealing, friction and volumetric efficiency attributes indicates that the GW may possess a 15 to 25 % higher energy efficiency than the SV; and also therefore similarly higher than other types of existing air compressors.
  A 7.5kW, 8 bar machine may therefore expect to achieve 75 % or so isentropic efficiency.
  To achieve this performance will require development; and probably the high machining precision as is routinely used for the manufacture of scroll compressors.
- A viewing of the area of the GW intake chamber's maximum-size axial profile (noting that there are two intakes per shaft revolution) compared to the complete machine (the ratio which indicates the so-called 'specific volume'), emphasises how much smaller the GW machine will be relative to other positive displacement machines.
- The GW machine is lighter-weight and consists of fewer components than most other compressor types. Hence, in the longer term and when it is produced in volume, it may well be lower cost to manufacture.

Acknowledgements
A GB patent application was filed in July 2014 (granted as B2528309); and the corresponding PCT version (GB2015 / 052040) in July 2015.
Note that the patent describes both the 2:3 (two lobed housing with 3 cornered rotor); and 1:2 type machines.
This Paper concentrates on the former only, this being the version with the larger number of potential applications.
ECL is a UK private limited company
75% of the shares are owned by D W Garside,
25% of the shares are owned by J C Thurner,

IOP Conf. Series: Materials Science and Engineering **232** (2017) 012065 doi:10.1088/1757-899X/232/1/012065

Reference

[1] Daily Telegraph, UK. Page B8, January 20, 2014

[2] Wrede F, Kruse H, 1986 "Recent status of trochoidal compressors for heat pumps in Germany", *International Compressor Engineering Conference at Purdue*

[3] Ichiro Ogura 1982 "The Ogura-Wankel compressor – application of Wankel rotary concept as automotive air conditioning compressor", *Society of Automotive Engineers (SAE 820159)*

[4] Kaiser H, Kruse H, 1984 "An investigation into reciprocating and rotary refrigeration compressors", *International Compressor engineering conference at Purdue*

[5] ECL: 2014 Animated model of 2:3 and 1:2 compressors, (see website *under construction*)

[6] PCT Patent application GB 2015 / 052040

[7] Carrier Corporation, Oct 2004 "Scroll Compressors"

[8] Ansdale R.F, 1968, page 47, "The Wankel RC engine", Iliffe books Ltd, London

[9] Oerlkon Leybold "Fundamentals of vacuum technology", Kat No 199 90

[10] Grace I, Datta D, Tassou S.A, 2002 "Comparison of hermetic scroll and reciprocating compressors operating under varying refrigerant charge and load", *International Compressor Engineering Conference at Purdue*

IOP Conf. Series: Materials Science and Engineering **232** (2017) 012066 doi:10.1088/1757-899X/232/1/012066

Dynamic modelling and PID loop control of an oil-injected screw compressor package

G W Poli[1, 3], W J Milligan[2] and P McKenna[3]

[1.] Howden Compressors, Old Govan Road, Renfrew, UK
[2.] Glasgow Caledonian University, Glasgow, UK
[3.] Author to whom any correspondence should be addressed
gary.poli@howden.com

Abstract. A significant amount of time is spent tuning the PID (Proportional, Integral and Derivative) control loops of a screw compressor package due to the unique characteristics of the system. Common mistakes incurred during the tuning of a PID control loop include improper PID algorithm selection and unsuitable tuning parameters of the system resulting in erratic and inefficient operation.
This paper details the design and development of software that aims to dynamically model the operation of a single stage oil injected screw compressor package deployed in upstream oil and gas applications. The developed software will be used to assess and accurately tune PID control loops present on the screw compressor package employed in controlling the oil pressures, temperatures and gas pressures, in a bid to improve control of the operation of the screw compressor package. Other applications of the modelling software will include its use as an evaluation tool that can estimate compressor package performance during start up, shutdown and emergency shutdown processes.
The paper first details the study into the fundamental operational characteristics of each of the components present on the API 619 screw compressor package and then discusses the creation of a dynamic screw compressor model within the MATLAB/Simulink software suite. The paper concludes by verifying and assessing the accuracy of the created compressor model using data collected from physical screw compressor packages.

1. Overview

Modern screw compressor packages contain a multitude of instruments, control valves and components which influence and change the operational characteristics of the compressor package. The complex design of a screw compressor package poses huge challenges in relation to achieving stable process output, as the measured operational variables (e.g. operating temperatures, pressures and flows) must be closely monitored and controlled to ensure that the compressor package is operating within its designed limits. By monitoring these variables it is possible to tune Proportional, Integral and Derivative (PID) loop controllers that govern the operation of the compressor through the actuation of valves and motors. The PID controller aims to match the measured process variables on the screw compressor package to set points that are predefined by the system operator. However, if a PID loop is improperly tuned the controller will struggle to accurately match the measured process variable to the chosen set point, which can result in unstable and inefficient operation of the screw compressor package [1].

The aim of the project is to create a dynamic mathematical model of an oil injected screw compressor package manufactured by Howden Compressors. The initial mathematical model of the screw compressor package will look to estimate critical values of the screw compressor such as oil and gas system pressures, temperatures and flows. Once the mathematical model of the screw compressor system has been created the next step is to incorporate control valve actuation, including closed loop

PID control of the valves. By including the control valves and their PID control loops it will be possible to assess how the compressor package will react to changes in the process within the screw compressor. This will then allow for the operation of the modelled screw compressor package to be tuned in the same fashion that a physical package would be tuned.

Initially the specification of the mathematical model will be created and a compressor model will be developed within MATLAB and Simulink based on this specification. The model will then be verified against test data acquired from an operational Howden screw compressor package. This will ensure that the developed modelling software is able to accurately estimate each chosen parameter of the screw compressor and that the initial requirements for the model have been fulfilled.

2. Dynamic Compressor Package Model

The motivation behind this project is to create a physically accurate screw compressor model that will allow for PID loop tuning of a modelled screw compressor package to happen before the physical screw compressor package has been commissioned. By doing this time spent commissioning a specific screw compressor package at site will be decreased as it will be possible to apply PID controller settings to the control valves that were estimated to be suitable by the compressor modelling software. It is estimated that over 75% of PID loops within the process industry are poorly tuned, resulting in increased process variability which is detrimental to the operation of the plant [2]. The developed software model will aim to decrease the likelihood of a poorly tuned PID control loop being present on a selected compressor package by pre-emptively estimating the PID loop settings. The dynamic compressor model will also provide insight into pre-start, start, shut down and emergency shutdown conditions for the compressor package, without physically operating a compressor. This will allow for a risk free environment where PID control settings can be optimised without the possibility of a poorly tuned PID loop damaging physical components on a compressor package.

2.1. Dynamic Model Definition

Initially, the physical components on the dynamic screw compressor package will be split into individual modules from which the full compressor package model will be constructed. The model will focus on accurately representing the fundamental components on the compressor package using differential equations and look-up tables. Figure 1 outlines the key screw compressor components that are modelled, including the model inputs, design variables and outputs. Each of the compressor components within figure 1 will be modelled within MATLAB/Simulink, which will include representing each component within a function block. By modelling each component within its own MATLAB/Simulink function block, it is possible to create a customisable reference component library that will be used as the starting point for each individual dynamic compressor model. The user will define and customise each component block so it shares the same parameters as the physical component that it models.

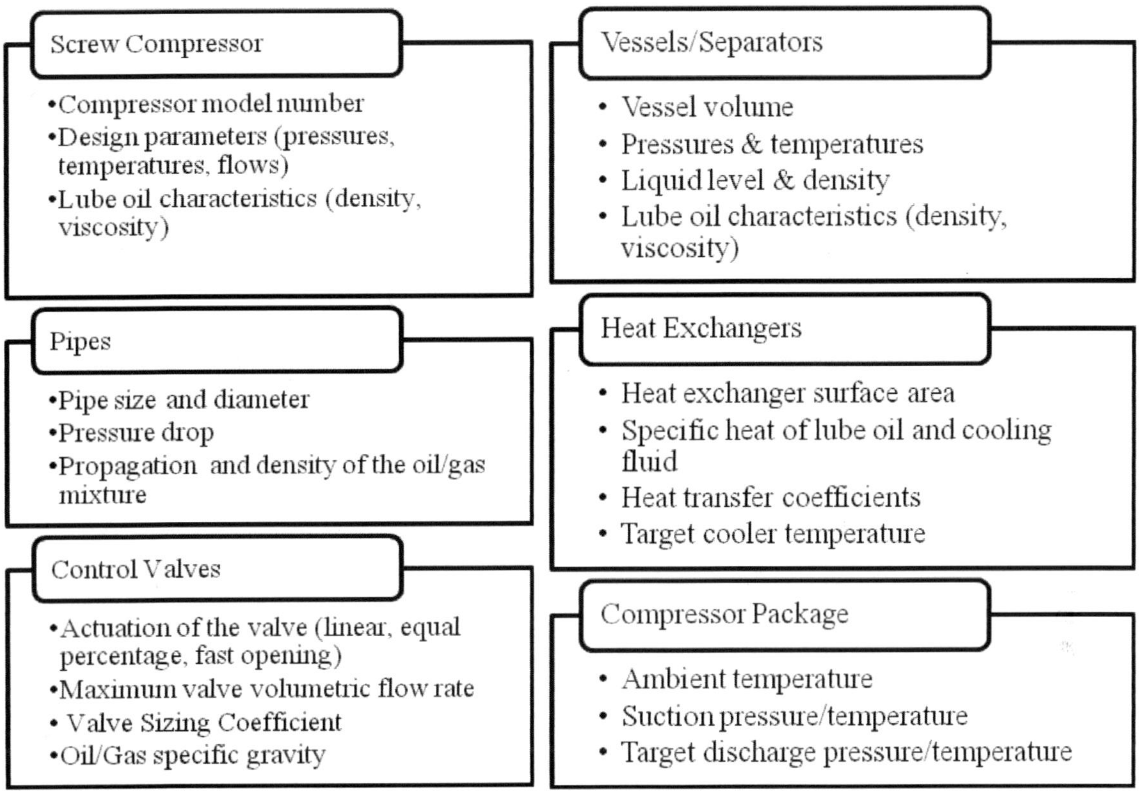

Figure 1. Key Screw Compressor Components

The key components listed within figure 1 are critical and must be accurately modelled within MATLAB/Simulink if the resulting dynamic model is to be reliable and accurate. With the critical components and their variables defined it is paramount that each component is modelled within its own function block. The function blocks for each of the compressor components that will be considered within the dynamic model are shown in Figure 2. These function blocks must be arranged in conjunction with one another so that they accurately represent the physical dynamics of the compressor system that they are being modelled after. The easiest way to achieve this is to use the compressor piping and instrumentation diagrams as a reference to construct the dynamic model.

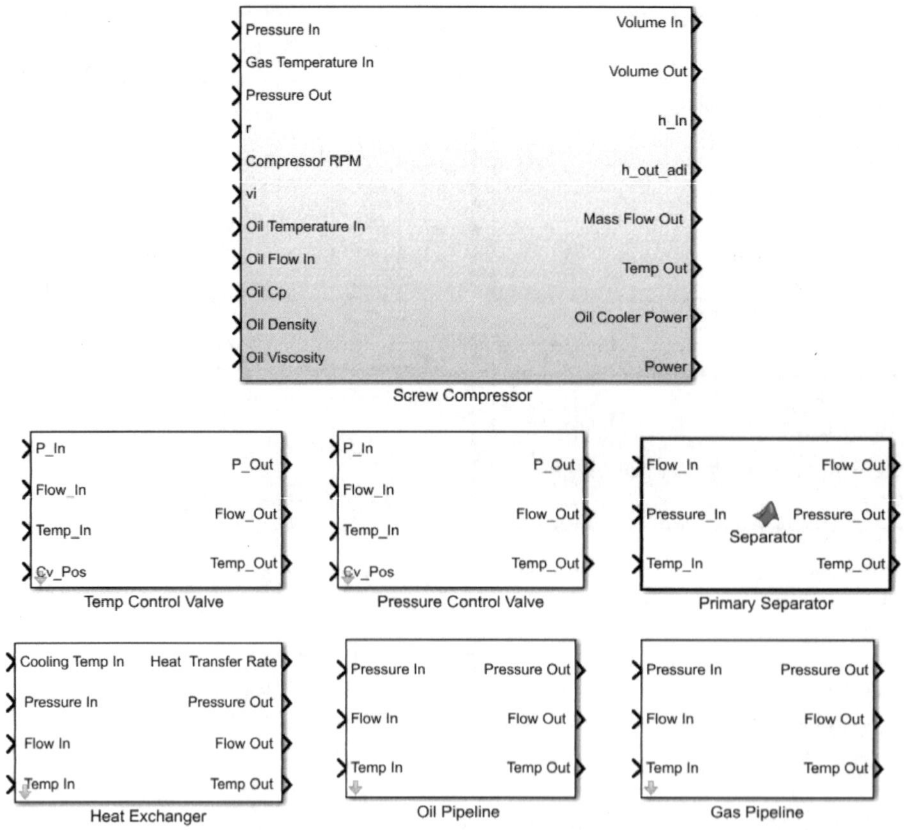

Figure 2. Simulink Component Library

2.2. Dynamic Model Construction

Construction of the screw compressor dynamic model aims to recreate the physical parameters of the selected screw compressor package. To initialise a new screw compressor package model the parameters of the screw compressor need to be defined, which are outlined in figure 1.

The standardised dynamic model for each of the screw compressors manufactured by Howden Compressors will be used as a starting point for each new screw compressor package model. The standardised model will be modifiable which will allow each module to represent a number of different compressor models and setups, which all have a variation of different design and operational characteristics. Defining the compressor model is the starting point for each dynamic model, with each modular component added onto the compressor model. To assist in the creation of the model the piping and instrumentation diagram for the specific compressor that is being modelled can be used as a reference. Figure 3 illustrates the piping and instrumentation diagram for the lubrication oil system of the screw compressor package with the screw compressor, main drive motor, lubrication oil air blast cooler, primary separator, control valves and lubrication oil pumps all represented within the system. Each of the components is modelled using the information included within the manufacturer supplied datasheets, which will ensure that each model accurately reflects the physical parameters of each component.

IOP Conf. Series: Materials Science and Engineering **232** (2017) 012066 doi:10.1088/1757-899X/232/1/012066

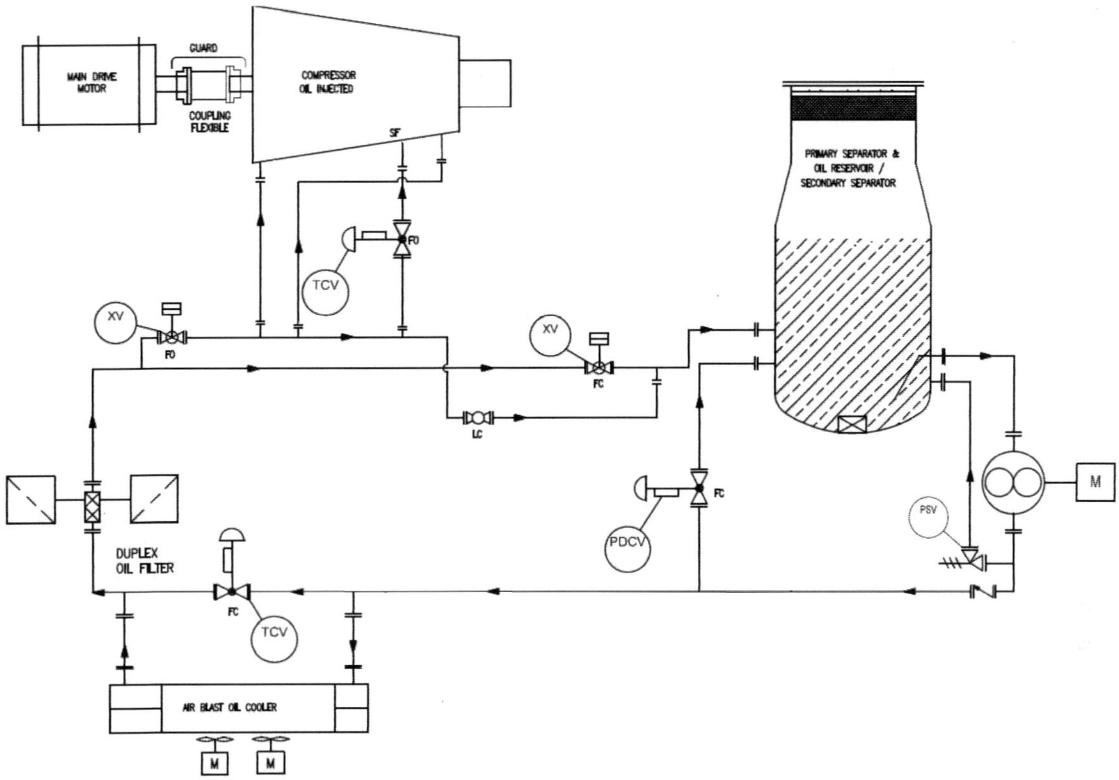

Figure 3. Screw Compressor Lubrication Oil System

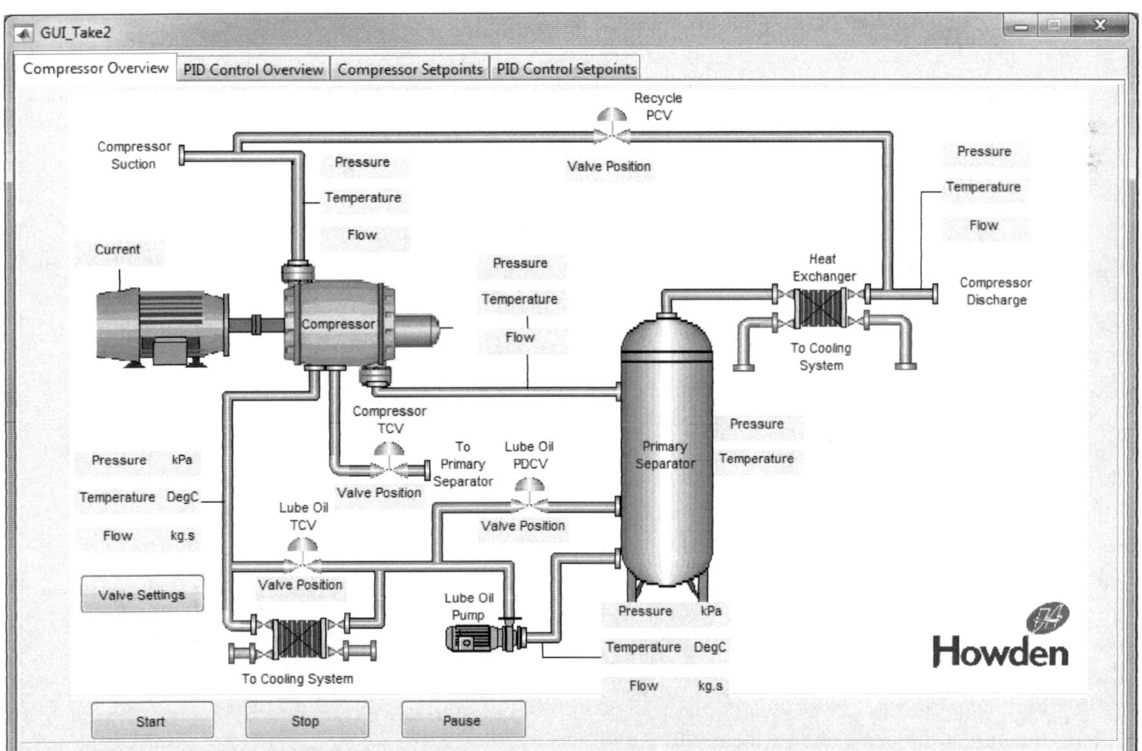

Figure 4. Compressor Package Lubrication Oil System MATLAB/Simulink Representation

Figure 4 is the lubrication oil system shown in Figure 3, recreated within the MATLAB/Simulink software package, where each main component within the system is represented by its own unique function block. This includes separate module blocks for the heat exchanger, compressor, oil pipelines and temperature control valves. To improve the usability of the software, the dynamic model has been integrated into a graphical user interface (GUI) that allows the end-user to easily input variables into the dynamic compressor model without altering any of the underlying code contained within the function blocks. The results from the simulation are then displayed on the GUI and also saved into a spreadsheet for analysis purposes. The next step in this process is to ensure that the model in Figure 4 is reliable and can be used with confidence to estimate the operation of the lubrication oil system. To achieve this it is essential that each modular model is independently verified.

3. Model Verification & Validation

Verification of the screw compressor model is a key step in ensuring the accuracy and reliability of the created model. Model verification ensures that it meets the requirements expected of the model in relation to its functionality and more importantly confirms model reliability and accuracy. It is a key step in the process of mathematically creating a dynamic model as it verifies not only the functionality of the model, but also the structure of the model [3].

The model will be validated against real world operational data to assess its ability to accurately model the operation of the screw compressor package and to ensure that the model is reliable [4]. Validation data will be collected from screw compressor package tests including the QGC package string test and tests of compressor systems at the Howden Compressors test facility. Figure 5 illustrates the process undertaken to collect validation data from the screw compressor package. Field equipment measurement data is collected and processed via the PLC into Logix5000 where it is then stored as an Excel .xlsx file. This information can then be exported into the MATLAB workspace where it will be used to validate the output from the dynamic model.

Figure 5. Data Collection Process for Model Validation

To gather data that is suitable for the purpose of system validation it is essential that the required test conditions are met by the screw compressor during the period of data collection from the screw compressor package. Changes in compressor operation should be instigated by the operator of the screw compressor system by altering temperature and pressure set points within the screw compressor package (via the control panel HMI or PLC) in what is known as a 'bump test'. To instigate this test a significant change in the control system set points needs to take place, which will result in the actuation of temperature and pressure control valves that look to match the new set points set by the system operator. By executing the bump test the dead time within the compressor system will become apparent. The dead time within the system refers to the time taken for a process variable to change its value in relation to the change in an output from a control valve or motor [5]. The dead time of a system can greatly increase the difficulty of tuning a control loop, as the added lag into the system can be challenging for the PID loop controller to manage. It is essential that pressure, temperature and flow changes are recorded at multiple points on the screw compressor during this test so that the reaction of the screw compressor system can be recorded and identified and then used to validate the created mathematical model. If possible, it is advisable that the system inputs, outputs and control valve actuation are recorded as this will ensure that any changes within the system can be easily identified and monitored during system analysis.

3.1. Model Validation Method

For model validation purposes the assumed 'bump test' on the compressor system is defined as change in operation of the valves between start up and continuous operation compressor conditions. This allowed for the screw compressor model to be validated against a dataset that would exhibit the full dynamic range of the physical screw compressor package. The validation data that was used was collected over a five hour period from a single stage oil injected screw compressor package that consists of a suction knock-out vessel, fan cooled lube oil/gas heat exchanger and a combined primary and secondary oil separator and installed on the skid.

To give an example of the model validation process, the dynamic heat exchanger model was assessed by using the compressor output lubrication oil temperature test data as the input into the heat exchanger model. As before, the heat exchanger model has been programmed to model the physical parameters of the heat exchanger installed on the compressor that the test data was collected from. If the model is accurate, the output lubrication oil temperature from the heat exchanger model will mimic the cooled lube oil measurements recorded during the package string test. To enhance the control capabilities of the heat exchanger model a temperature control valve model with identical parameters as the physical temperature control valve that governed the output of the reference heat exchanger has been added to the system. The temperature control valve model will have its actuation controlled by the data collected from the physical control valve positioner data. This will ensure that the actuation of the control valve within the model is precisely mimicking the movement of the physical control valve on the package during the string test. The positioner is controlled by a fully tuneable PID controller during normal operation of the simulation. With the inclusion of the temperature control valve, full control over the temperature of the lubrication oil within the system is possible. The MATLAB/Simulink model used to validate the heat exchanger model against the validation data is shown in Figure 6. To achieve this, the validation data is used as the lubrication oil temperature input into the heat exchanger. The control valve that regulates the flow of oil into the heat exchanger model will then alter the position of the valve with the aim of keeping the temperature of the oil at the chosen set point of 47°C. Inputs pressure and flow into the control valve are constant and their values are matched to what would be expected during normal operation. During the execution of the Simulink model the output temperature and control valve position feedback data are recorded, which will be used to assess the performance of the compressor model.

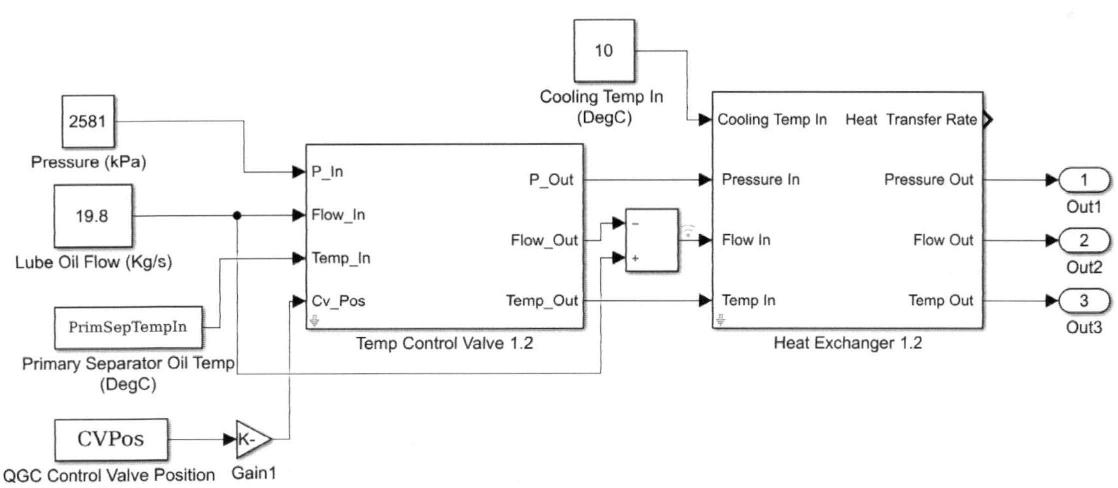

Figure 6. Simulink Heat Exchanger Model

IOP Conf. Series: Materials Science and Engineering **232** (2017) 012066 doi:10.1088/1757-899X/232/1/012066

3.2. Dynamic Heat Exchanger Model Validation

To validate the model, the output from the heat exchanger function block will be compared and analysed against the measured output from the physical heat exchanger. This is a critical step as it will allow for the performance of the control valve and heat exchanger system to be quantified. Figures 7 and 8 illustrate the control valve actuation, lubrication oil temperature input, physical heat exchanger output and model heat exchanger output data.

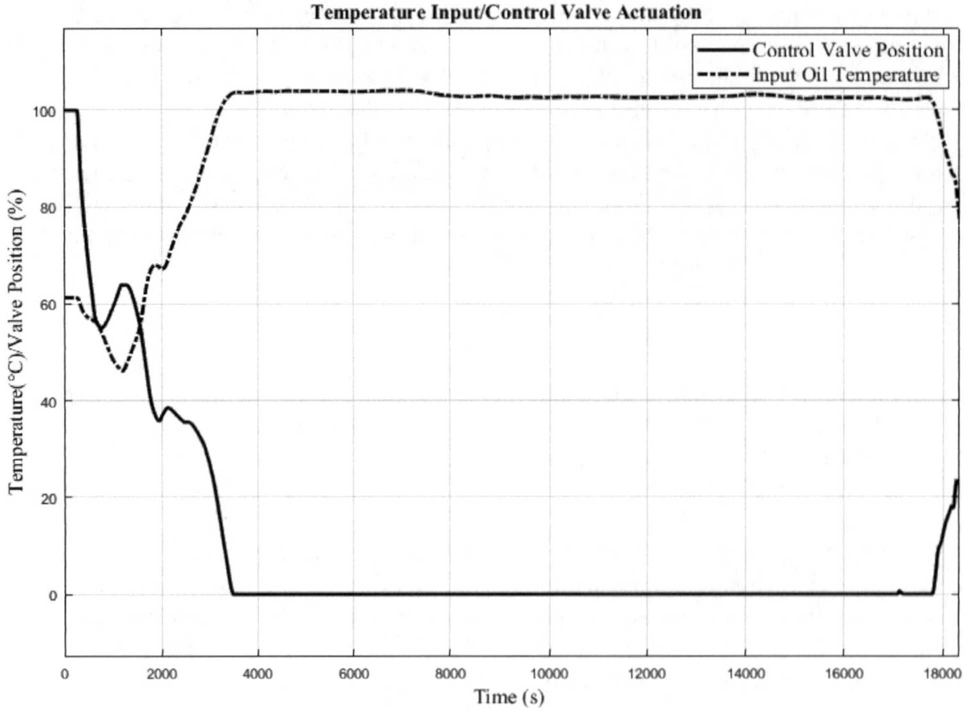

Figure 7. Lubrication Oil Temperature Input and Control Valve Position

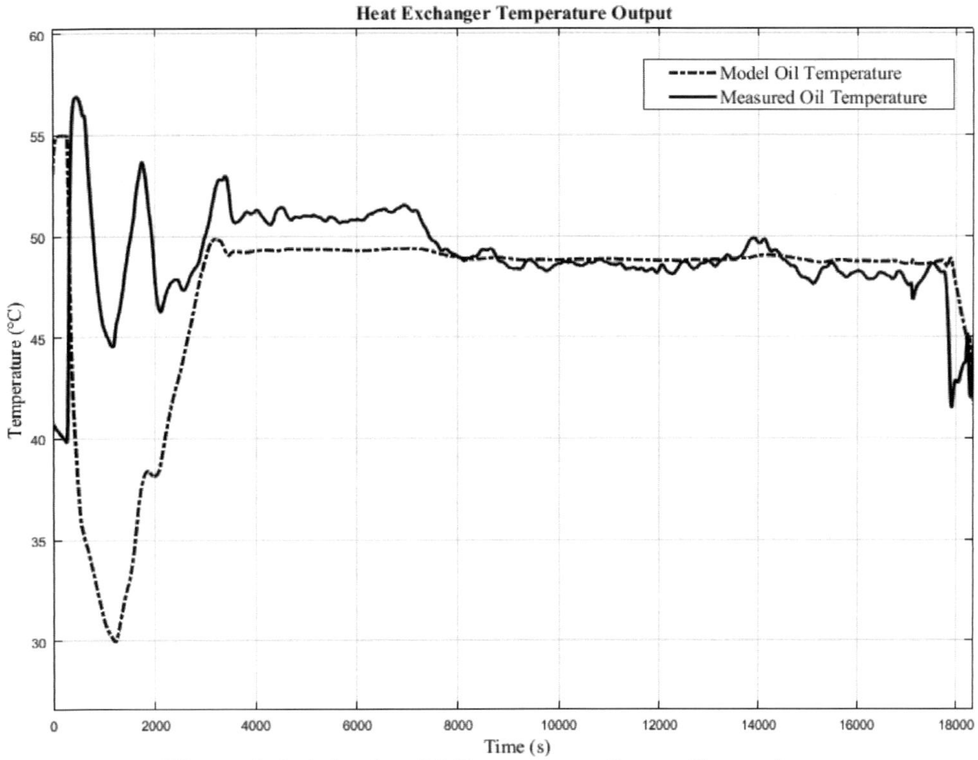

Figure 8. Lubrication Oil Temperature Output Comparison

The two stages of the compressor test can be seen in figure 7 and 8 with the start up phase occurring between 0 – 4000 seconds where the primary separator lubrication oil and oil cooler output temperature will continue to fluctuate until it reaches a steady state value. During this time the temperature control valve is actuating in a bid to maintain an oil temperature set point of 47°C. The oil temperature reaches a steady state after 4000 seconds. At first glance, Figures 7 and 8 show the control valve model and heat exchanger model are capable of producing outputs that are comparable to those recorded from the physical heat exchanger and control valve components installed on the compressor package. To fully quantify and assess the accuracy of the heat exchanger and control valve models the output data from the model will need to be compared to collected validation data. The root mean squared error (RMSE) between the validation data and model output is found, which will return the measured total error as a square root of the sum of the model variance and the bias [6]. This will provide a quantified measurement of how strongly related the output of the model is to the collected test data.

RMSE is defined in equation (1)

$$RMSE = \sqrt{\frac{\sum_{i=1}^{n}(y_i - x_i)^2}{n}} \quad (1)$$

Where: n = number of samples, y_i = measured value and x_i = model estimated value.

Equation (1) returns an average RMSE for the heat exchanger and temperature control models of 5.4062 for the duration of the string test. This RMSE improves to 1.1903 if the initial start up phase of the compressor between 0-4000 seconds is ignored. This is vital information as it shows

IOP Conf. Series: Materials Science and Engineering **232** (2017) 012066 doi:10.1088/1757-899X/232/1/012066

that the model is capable of accurately predicting temperature values during steady state operation however, it is weaker at predicting temperature output during start up operation which will need to be improved in later iterations of the model.

The model validation process will be repeated to assess the accuracy of each of the compressor components in a bid to ensure that the output from each of the models is reliable and can be used to model the full screw compressor package. Accuracy within each model is paramount as it will allow for the most reliable model to be created.

4. Conclusions

The dynamic screw compressor package model is a useful tool that will give engineers the ability to predetermine PID control loop settings to be used that will control the actuation of the pressure and temperature control valves on the screw compressor. This will decrease the time spent by engineers trying to finely tune the PID control parameters and hence speed up the time spent commissioning a screw compressor package whilst also ensuring stable operation of the screw compressor package. The dynamic compressor model is also useful in the design stage of the compressor package, as it can be used to predict the performance of a component or the performance of the compressor package as a whole. Further development of the screw compressor model will need to take place in a bid to improve accuracy, with additional analysis into the validation of the compressor model using multiple datasets from compressor tests.

References

[1] A. O'Dwyer, Handbook of PI and PID controller tuning rules, Singapore: B & JO Enterprise, 2006.

[2] G. Buckbee, "The 6 most common PID configuration errors: how to find and fix them," PlantTriage, Expertune, 2008.

[3] Barlas, Yaman, "Model validation in system dynamics," 1994. [Online]. Available: http://www.systemdynamics.org/conferences/1994/proceed/papers_vol_1/barla002.pdf. [Accessed 26 January 2017].

[4] J. Minihan, "System validation and verification," [Online]. Available: https://www.resourcegroup.co.uk/media/1897/system-validation-and-verification.pdf.

[5] J. Smuts, "Dead time vs. time constant," Opticontrols , 21 June 2011. [Online]. Available: http://blog.opticontrols.com/archives/499. [Accessed 1 December 2016].

[6] National Institute of Standards and Technology, "Root mean squared error," 10 August 2016. [Online]. Available: http://www.itl.nist.gov/div898/software/dataplot/refman2/auxillar/rms.htm. [Accessed 17 February 2017].

IOP Conf. Series: Materials Science and Engineering **232** (2017) 012067 doi:10.1088/1757-899X/232/1/012067

System solution to improve energy efficiency of HVAC systems

L Chretien[1], R Becerra[2], N P Salts[3] and E A Groll[4]

[1] Senior Research and Development Design Engineer, Regal, 1946 W. Cook Rd., Fort Wayne, IN, USA
[2] Research and Development Manager, Regal, 1946 W. Cook Rd., Fort Wayne, IN, USA
[3] Graduate Research Assistant, Purdue University, Ray W. Herrick Laboratories, West Lafayette, IN, USA
[4] Reilly Professor of Mechanical Engineering, Purdue University, Ray W. Herrick Laboratories, West Lafayette, IN, USA

E-mail: ludovic.chretien@regalbeloit.com

Abstract. According to recent surveys, heating and air conditioning systems account for over 45% of the total energy usage in US households. Three main types of HVAC systems are available to homeowners: (1) fixed-speed systems, where the compressor cycles on and off to match the cooling load; (2) multi-speed (typically, two-speed) systems, where the compressor can operate at multiple cooling capacities, leading to reduced cycling; and (3) variable-speed systems, where the compressor speed is adjusted to match the cooling load of the household, thereby providing higher efficiency and comfort levels through better temperature and humidity control. While energy consumption could reduce significantly by adopting variable-speed compressor systems, the market penetration has been limited to less than 10% of the total HVAC units and a vast majority of systems installed in new construction remains single speed. A few reasons may explain this phenomenon such as the complexity of the electronic circuitry required to vary compressor speed as well as the associated system cost. This paper outlines a system solution to boost the Seasonal Energy Efficiency Rating (SEER) of a traditional single-speed unit through using a low power electronic converter that allows the compressor to operate at multiple low capacity settings and is disabled at high compressor speeds.

1. HVAC system overview and component characteristics

Vapor compression air conditioning and heat pumping systems used in US residential habitations typically comprise the following elements: an evaporator coil with a fan, a condensing coil with a fan, a compressor, and an expansion device. These systems are used in conjunction with a thermostat that regulates operation in order to maintain the requested temperature.

The following sections will outline the key characteristics of single-speed systems, that represent the vast majority of residential installations, and variable-speed systems.

IOP Conf. Series: Materials Science and Engineering **232** (2017) 012067 doi:10.1088/1757-899X/232/1/012067

2. Single-speed system overview and characteristics

Single-speed systems use a fixed-speed compressor. When the HVAC unit is designed to operate on single-phase power, the compressor motor is generally a Permanent Split Capacitor (PSC). When the HVAC system is designed to operate on three-phase power, the compressor motor is typically a three-phase induction motor.

In US residential applications, systems traditionally range from smaller units providing two tons (6.7 kW) of cooling capacity to larger units providing up to five tons (17 kW) of cooling, depending on the size of the habitation. These systems are run with single-phase power.

2.1 Performance of PSC motors

A PSC motor is a two-phase induction motor that can run on single-phase power. The motor comprises two sets of windings, a main winding and a start winding. In order for a PSC to operate on single-phase power, the start winding of the motor is connected in series with a run capacitor. This series connection creates an out of phase voltage with respect to the main winding voltage that allows the motor to accelerate from standstill to full load without the use of power electronics circuitry.

Figure 1 below depicts the torque characteristics and machine current for a PSC from standstill to no load.

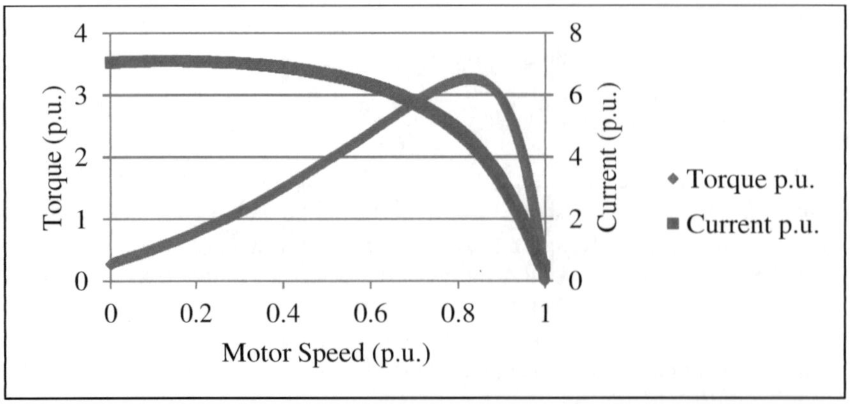

Figure 1. PSC torque and current versus speed.

As can be seen on Figure 2, a PSC motor has very limited starting torque and high inrush current when coming up to speed from standstill. As a result, single-stage systems need protection mechanisms against locked rotor currents that can reach up to ten times normal operating currents.

Figure 2 describes a typical efficiency curve for a PSC motor. As can be seen, the motor peak efficiency is achieved at a particular load point, typically the full-load operating point of the system, and both slight increases and decreases in load negatively affect motor performance. Additionally, it can be seen that operation at lower speeds, also called high motor slip operation, dramatically reduces efficiency.

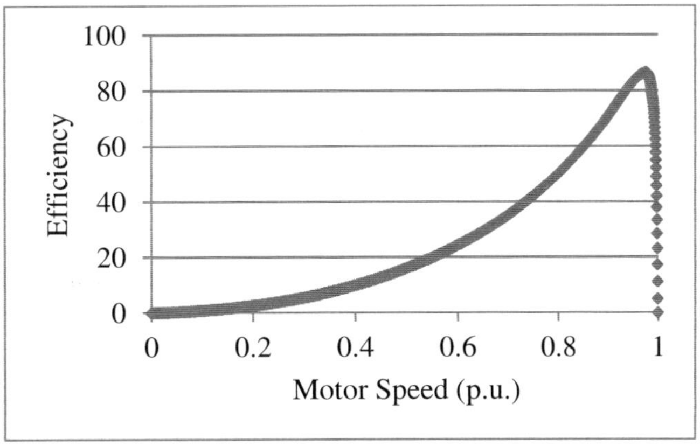

Figure 2. PSC efficiency versus speed.

2.2 Single-speed system operation

For single-speed systems, the thermostat controls the power to the compressor motor. The compressor motor is cycled according to the temperature setting of the thermostat.

Cycling is necessary because the compressor operates at single-speed and the system always provides high cooling capacity. While high cooling capacity is necessary and useful on warm and humid days, for example in the summer, it is excessive on milder days, and the system needs to stay powered only for short bursts of time to avoid excessive cooling of the habitation.

However, there exist constraints on the cycling ability of single-stage systems. Indeed, because of the limited starting torque of the PSC motor, the duty cycle for cycling operation has to be limited. When the HVAC unit is operating, the system refrigerant is compressed and a pressure differential is created between the suction and discharge ports of the compressor. This pressure differential subsists and slowly decays when the compressor turns off. The pressure differential immediately upon compressor shut off exceeds the maximum starting torque of the PSC. Under such conditions, the PSC compressor would be unable to restart, and would enter a locked rotor condition. As a consequence, single-speed systems have built in features, also called interlock mechanisms, that prevent successive and immediate restarts of the compressor.

Cyclical operation of the HVAC system prevents tight regulation of the temperature inside habitations and also negatively affects indoor air dehumidification.

2.3 Efficiency rating for single-speed systems

In the US, the energy efficiency of an HVAC system in cooling mode is measured through its Energy Efficiency Ratio (EER) and Seasonal Energy Efficiency Ratio (SEER). The EER is calculated by taking the ratio between the cooling capacity the system is able to deliver in Btu/hr with the total electrical power consumed in W. The total electrical power consumed comprises the indoor blower input power, the outdoor condenser fan input power, and the compressor motor input power.

The SEER for a single-speed system is calculated according to the ANSI/AHRI Standard 210/240 for the Performance Rating of Unitary Air-Conditioning and Air-Source Heat Pump Equipment [1]. The standard defines required and optional tests at various indoor and outdoor air temperatures. Some of the tests are performed during continuous operation of the unit while others are performed under cyclical operation. The cyclic tests allow to determine the cyclic degradation coefficient C_d. Table 1 below outlines the tests for a single-speed unit.

IOP Conf. Series: Materials Science and Engineering **232** (2017) 012067 doi:10.1088/1757-899X/232/1/012067

Table 1. Single-speed system test points. 80.0ºF / 67.0ºF indoor air.

Test condition	Outdoor air temperature ºF	Compressor Speed	Cooling air volume rate
A (steady)	95.0/75.0	Max.	Max.
B (steady)	82.0/65.0	Max.	Max.
C (steady)	82.0/-	Max.	Max.
D (cyclic)	82.0/-	Max.	Max

Once the tests in Table 1 are completed, the SEER rating is calculated according to Equation (1) below.

$$SEER_{single\ speed} = EER_B \times (1 - 0.5 \times C_d) \quad (1)$$

3. Variable speed system overview and characteristics

Variable speed HVAC systems achieve higher SEER rating than single-speed systems through using high efficiency variable-speed motors for the compressor, the indoor blower and the outdoor fan, that can adjust their speed of operation to match the load. Additionally, these systems may use an Electronic Expansion Valve (EEV) to achieve greater controllability of the refrigerant cycle.

3.1 Performance of motors used in variable-speed systems

To achieve best in class efficiency, variable-speed systems typically use brushless permanent magnet motors mated with electronic controllers. Brushless permanent magnet motors, also known as Brushless Direct Current motors (BLDC) can achieve higher power densities and efficiencies compared to PSC motors as there are no rotor losses during operation. Figure 3 below contrasts efficiency and operating area between BLDC and PSC.

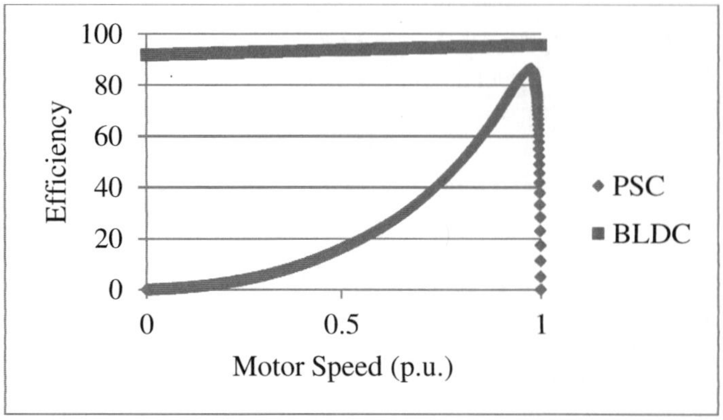

Figure 3. PSC and BLDC efficiency versus speed.

3.2 Electronic circuitry required to run variable speed motors

Variable speed operation of a BLDC motor is accomplished by using an electronic drive. The role of the electronic drive is to convert the fixed voltage and frequency of the utility power source into a source of variable voltage and variable frequency. Figure 4 details the traditional elements of a variable speed drive.

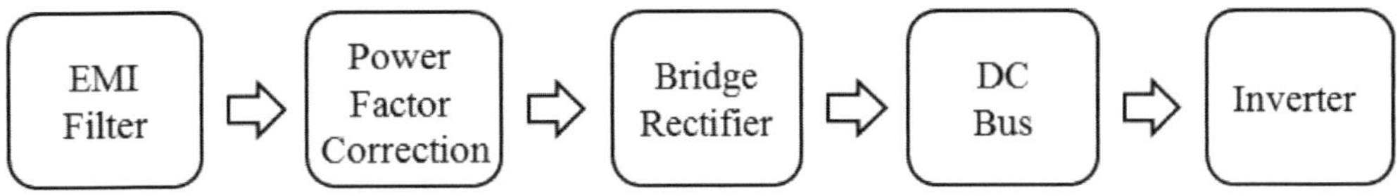

Figure 4. Block diagram of a variable speed drive.

The electronic drive transforms the AC voltage from the utility power source into a DC voltage. The energy is stored by large electrolytic capacitors. An inverter stage composed of power electronic switches transforms the DC voltage into AC voltages with amplitudes and frequencies adjusted using Pulse Width Modulation (PWM) techniques.

The electronic drive also contains a power factor correction (PFC) unit to condition the current drawn from the utility source into a sinusoid. Input power factor represents the ability to extract optimal power from the source. The power factor of a PSC motor under load typically reaches above 90%. An electronic drive without a power factor correction unit on the other hand is generally limited to 60%. Under these conditions, a PSC compressor consuming 4 kW would draw 18 Amps from the source when a BLDC compressor consuming 4 kW would draw 28 Amps. Due to limitations in available currents in residential habitations, it is therefore necessary for variable-speed systems to include PFC units to run the compressors at full power.

The power electronic switches in the PFC stage and in the inverter stage introduce losses from conduction and switching of electric current, as well as Electro Magnetic Interference (EMI). Therefore, to compare system efficiency between PSC and BLDC system, it is necessary to take into account the efficiency of the electronic drive. Typically, for HVAC units ranging from two tons (6.7 kW) to five tons (17 kW) of cooling, the efficiency of electronic drives reaches 92 to 93%. Finally, the losses in power electronic components generate heat that needs to be dissipated outside the drive, and present cooling challenges, especially when the HVAC system operates at high power and high ambient. Under high ambient conditions, the thermal management of the electronic drive sometimes requires derating and reduction of compressor speed, thereby reducing the total available cooling capacity when it is most needed.

3.3 Variable-speed system operation

In a typical variable speed system, a thermostat sends digital information regarding the ambient temperature and the set point selected by the user. Based on the temperature difference between commanded and measured temperatures, as well as other specific modes'of operation, the thermostat sends a cooling request to the system controller of the HVAC unit. The system controller then selects the optimum operating speed for the compressor, the indoor fan and the outdoor fan based on the thermostat request and the information measured from multiple sensors such as refrigerant pressure or temperature sensors.

Figure 5 shows a typical operating profile for the system cooling stage following a thermostat cooling call. Figure 6 illustrates the effectiveness of temperature regulation between single-speed and variable-speed systems.

IOP Conf. Series: Materials Science and Engineering **232** (2017) 012067 doi:10.1088/1757-899X/232/1/012067

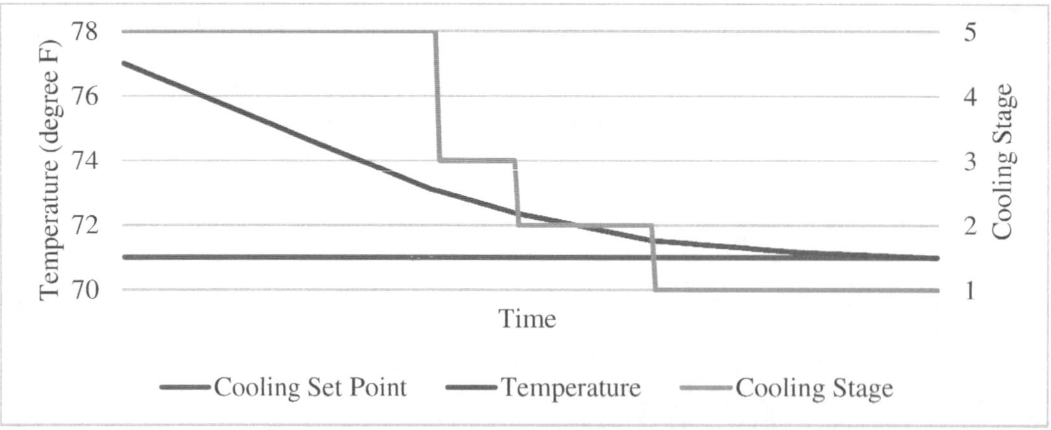

Figure 5. Typical cooling operating profile.

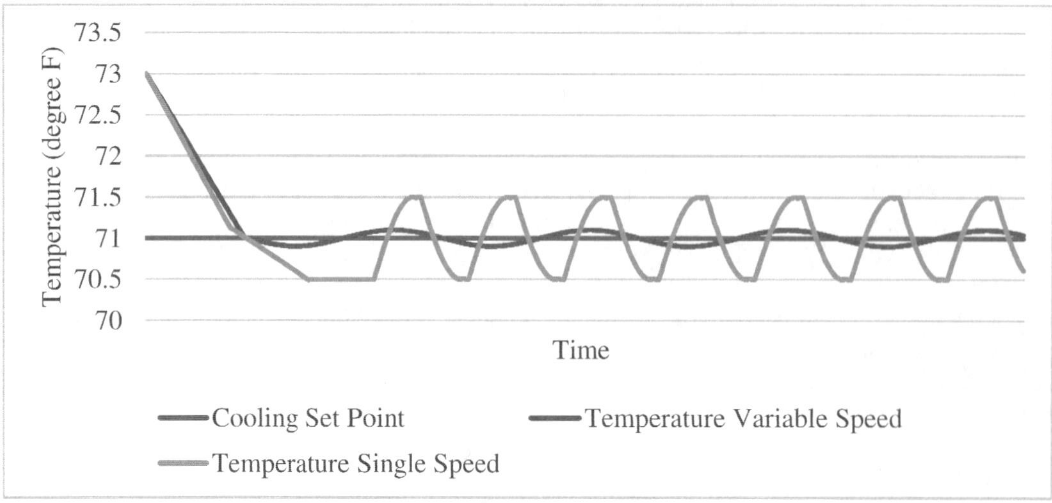

Figure 6. Temperature control of single stage and variable speed units.

As can be seen in Figures 5 and 6, the compressor runs at high speed and the HVAC system delivers high cooling under high temperature differential. As the measured temperature approaches the user set point, the system controller reduces compressor speed to match the cooling load of the building, thereby reducing cycling requirements and optimizing temperature regulation and ultimately user comfort. Typically, variable speed compressors operate in a 900 to 1200rpm range for cooling stage 1, up to a 3600 to 4200rpm range for cooling stage 5. The operating speeds between stages 1 and 5 are generally evenly spaced, yet other speed distributions are possible depending on the programming of the HVAC system controller.

3.4 Efficiency rating for variable speed system

As for single-speed systems, the SEER for variable-speed systems is calculated according to the ANSI/AHRI Standard 210/240. Table 2 below summarizes the tests to calculate the SEER for variable-speed systems.

Table 2. Variable-speed system test points. 80.0°F / 67.0°F indoor air.

Test condition	Outdoor air temperature °F	Compressor Speed	Cooling air volume rate
A_2 (steady)	95.0/75.0	Max.	Max.
B_2 (steady)	82.0/65.0	Max.	Max.
E_v (steady)	87.0/69.0	Intermediate	Intermediate
B_1 (steady)	82.0/65.0	Min.	Min
F_1 (steady)	67.0/53.5	Min.	Min
G_1 (steady)	67.0/-	Min.	Min
I_1 (cyclic)	67.0/-	Min.	-

For variable-speed systems, it is necessary to perform tests at three distinct compressor speeds. The relationship between compressor speeds follows Equation (2).

$$\text{Speed}_{\text{intermediate}} = \text{Speed}_{\text{min}} + \frac{\text{Speed}_{\text{max}} - \text{Speed}_{\text{min}}}{3} \tag{2}$$

In addition to varying the compressor speed, it is possible to vary the indoor and outdoor fan speeds to further adjust the cooling capacity of the system.

4. Proposed system solution: combination of single speed and variable speed components

One of the key elements to improve the SEER of HVAC units is to adjust the compressor speed to match the cooling capacity to the cooling load. However, the use of an electronic drive may be detrimental to efficiency and lead to thermal management challenges at full load operation. To address some of these drawbacks, a system solution using a standard single-speed compressor and an electronic drive is proposed.

4.1 System overview and benefits

The proposed system solution resides in the combination of a single-speed PSC compressor with an electronic drive. The system will operate on the inverter drive when the cooling demand and cooling load are reduced, and run directly from utility power at high load. A block diagram for this solution can be seen in Figure 7 below.

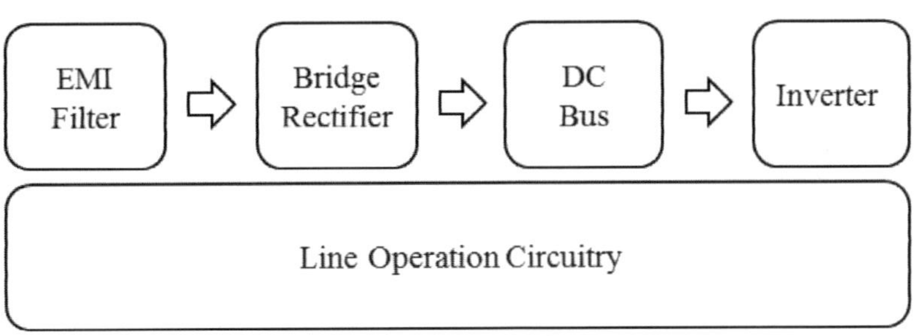

Figure 7. System solution to convert fixed-speed system into variable speed.

The proposed combination provides substantial benefits over traditional single-speed systems and variable-speed systems altogether. First, the ability for the compressor to operate directly from the utility source at high power solves several challenges encountered with variable-speed systems. Indeed, as the electronic drive is disabled at high loads, there is no concern for power factor or EMI compatibility, and there are no losses associated with power electronics, so the full load electrical efficiency of the proposed solution is comparable to the one of variable speed compressors burdened by their electronic losses. Additionally, as the drive is only used at reduced output power, the size and thermal management of electronics is simplified. However, a few limitations exist compared to fully variable speed systems. First, as the standard PSC compressor operates on utility power at full load, the maximum achievable compressor speed remains unchanged. Second, at lower operating speeds, the combination of a PSC on an electronic drive remains less efficient when compared to the efficiency of a BLDC on an electronic drive.

Adding an electronic drive to a PSC compressor also yields significant benefit over single-speed systems. First, the cooling capacity can be modulated to match the cooling load, resulting in higher comfort for the user through improved temperature control and dehumidification. Moreover, the PSC motor used in conjunction with the electronic drive can achieve substantially higher starting torque, removing the need for time delays between successive restarts of the compressor. Finally, the electronic drive allows precise acceleration of the compressor resulting in lower inrush current and audible noise when the air conditioner starts.

4.2 Psychrometric chamber results for a five tons HVAC unit

The proposed system solution combining a production standard PSC compressor from a single-speed system with an electronic drive has been tested using a five tons (17 kW) split-system air conditioning unit in two side-by-side psychrometric chambers. Test procedures have followed the standard testing outlined in ANSI/AHRI 210/240.

Tables 3 and 4 summarize the findings for a system using a 5-tons (17 kW) scroll compressor.

Table 3. Single-speed system test points. PSC Scroll Compressor.

Test condition	EER	SEER
A (steady)	12.08	14 @ $C_d = 0.11$
B (steady)	14.81	

Table 4. Variable-speed test points. PSC Scroll compressor.

Test condition	EER	Compressor speed (RPM)	SEER
A_2 (steady)	12.08	3525	
B_2 (steady)	14.81	3525	
E_V (steady)	14.20	2445	15.37 @ $C_d = 0.11$
B_1 (steady)	15.73	1905	
F_1 (steady)	20.55	1905	

Tables 5 and 6 summarize the findings using a 4.5 tons (15.3 kW) PSC rotary compressor.

Table 5. Single-speed system test points. PSC Rotary compressor.

Test condition	EER	SEER
A (steady)	11.50	13.08 @ $C_d = 0.11$
B (steady)	13.84	

Table 6. Variable speed test points. PSC Rotary Compressor.

Test condition	EER	Compressor speed (RPM)	SEER
A_2 (steady)	11.50	3525	
B_2 (steady)	13.84	3525	
E_V (steady)	14.82	2085	16.44 @ $C_d = 0.11$
B_1 (steady)	18.52	1365	
F_1 (steady)	26.50	1365	

As can be seen in Tables 5 and 6, it is possible to achieve between 1.5 and 3.5 SEER improvement by using the proposed variable-speed drive on the PSC compressor of a single-speed system. A base air conditioner rated at SEER 14 could improve up to SEER 16.5. Using this technology, the full-load efficiency remains unaffected by the drive performance, and the use of an electronic drive allows for low speed operation where the overall system runs much more efficiently.

5. Conclusion

The addition of a low power electronic drive to enable variable-speed operation of a fixed-speed compressor used in single-speed systems results in SEER improvements ranging from 10% to 25% depending on the compressor technology, without affecting the EER value at full load. Future work will include a detailed investigation of SEER improvements using this technology in conjunction with a variable airflow rate indoor system, yet preliminary results suggest that a 30% SEER improvement over the baseline is achievable.

It is expected that this technology could also positively impact the Heating Seasonal Performance Factor (HSPF), which is the pendant metric to SEER and is used to measure the efficiency of heat pumps.

References

[1] ANSI/AHRI Standard 210/240 with Addenda 1and 2 (Formerly ARI Standard 210/240) Performance Rating of Unitary Air-Conditioning & Air-Source Heat Pump Equipment.

IOP Conf. Series: Materials Science and Engineering **232** (2017) 012068 doi:10.1088/1757-899X/232/1/012068

The process of gas-dynamic design of pneumatic braking system using the baseline compressor

Y Novikova[1], G Popov[1], E Goriachkin[1], O Baturin[2], V Zubanov[1]

[1] Junior Researcher, Samara National Research University, Samara, Russia
[2] Associate Professor, Samara National Research University, Samara, Russia

E-mail: y.d.novikova@gmail.com

Abstract. The article presents the results of work on the design of the air brake for testing of industrial gas turbine engines with free turbine. Designing of the air brake was performed on the basis of existing units using the program CFD - simulation Numeca FineTurbo. During the design the air brake arrangement was determined, which allows to utilize the required power to the shaft of the free turbine, increases stall margin of the air brake by waisting of the meridional flow channel. It was also made designing of the outlet guide vane to remove the residual twist. Unified nozzle also was designed to provide the air brake work at necessary points on the characteristic.

1. Introduction

Testing of gas-turbine engines (GTE) is an essential part of their life cycle. Tests are carried for the newly created engines, and for the serially produced engines before their delivery to the customer.

During testing the industrial GTE, designed to drive electric generators, compressors pumps and etc. there is a problem of utilization and measurement of generated power.

Air brake device is an air compressor [1]. Air brake utilizes the power generated by GTE by compressing the air. Creation and design of compressors is a difficult and expensive task [2]. This explains the high cost of GTE air brake devices. However, if existing compressor is use as the air brake, it can significantly reduce the cost of developing of the air brake, and also reduce the time of its creation. This was the way it was decided to go in design of the air brake for testing of industrial gas turbine engines in for the JSC "Kuznetsov [3].

As part of the work it was required to design the air braking system for a family of industrial GTE. The air brake intended to dispose of power on the free turbine shaft. According to the technical task the air brake had to provide power utilization of 25 MW at a rotor speed of 5000 rev / min and the power of 32 MW at the rotor speed of 5500 rev / min. To reduce the cost of development it was supposed to complete the project on the basis of the existing compressor (the base compressor) (Figure 1). Base compressor is a low-pressure compress (LPC) or of a bypass turbojet engine, developed and produced by JSC Kuznetsov. One of the key requirements in the development of the air brake was a minimal redesign of the base compressor, as well as the minimum production of new parts.

The development of the air brake was carried out on basis of the base compressor. For this reason it was decided to use CFD-modeling method, which was implemented in the software package NUMECA FineTurbo [4].

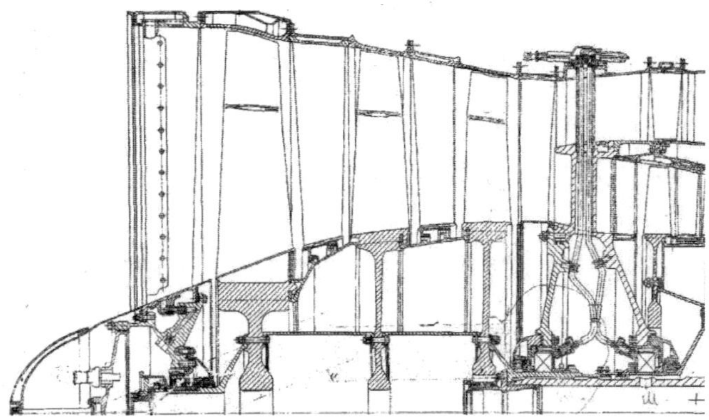

Figure 1. Geometry of the base compressor.

2. Numerical model of the base compressor and its validation

Before using CFD-programs for research, as a rule, verification of calculation models is performed [5]. For this reason, at the first step of the work the validation of numerical models was performed. Validation of numerical models is performed due to comparison between the calculated characteristics of the base compressor and experimental characteristics.

The numerical model of the base compressor was created in the software NUMECA FineTurbo [6] and it consisted of rotor blade domains (RB1, RB2, RB3), guide vanes domains (GV1, GV2, GV3, GV3a), inlet guide vanes domain (IGV), and bearing strut domains in the outer and inner duct (Figure 2). The created model took into account the radial clearance, as well as anti-vibration flange at the rotor blades.

The block-structured grid was made in the program Numeca Autogrid 5. Element size, which is closest to the walls of the computational domain (blade surfaces, struts, meridional shrouds, etc.), is provided the value of the parameter y+ equal to1. The elements number of calculated mesh of the rotor blade domains was about 840 thousand elements, mesh of the guide vanes domains - about 300 thousand elements. The total size of the grid of the entire model was 5.5 million items (Figure 3).

The turbulence model Spalart-Allmaras was used during calculation. The calculation of rotor blade domains was carried out in a rotating coordinate system. The rotational speed was 5000 rev / min. To transfer data between domains the interface Full Non Matching Mixing Plane was used, which averaged flow parameters in the circumferential direction.

At the inlet to the calculated model the parameters of air were set: the total pressure of 101325 Pa and the total temperature 288,15 K. At the outlets of the first and second ducts the massflow rates were set which correspond to the required characteristics point (Figure 2).

Using the created computational model the characteristics of the base compressor were calculated (Figure 4, 5). According to the results of the characteristics calculation, these dependencies were plotted:

1. The compressor total pressure ratio in the first duct of the base compressor vs total massflow rate $\pi_I^* = f(G_{A\Sigma g})$ (curve I contour - figure 4);
2. The compressor total pressure ratio in the second duct of the base compressor vs total massflow rate $\pi_{II}^* = f(G_{A\Sigma g})$ (curve II contour - figure 4);
3. The efficiency of the first duct vs the total massflow rate $\eta_I^* = f(G_{A\Sigma g})$ (curve I contour - figure 4);
4. The efficiency of the second duct vs the total massflow rate $\eta_{II}^* = f(G_{A\Sigma g})$ (curve II contour - figure 4);
5. The total capacity of the air brake vs the relative massflow rate $N = f(G_{A\Sigma g})$ (figure 5).

Plotted calculated dependencies were compared with experimental data. By comparing the results the following conclusions were made:

1. The characteristics obtained by calculation, repeat the qualitative nature of the experimental characteristics;
2. Quantitative error in determining the capacity is 10%;
3. Quantitative error in determining the flow rate is 4.6%;
4. Created computational model can be used in research on the condition of correction of the calculated characteristics on the value of the identified errors.
5. Base compressor does not provide the required power utilization at the operation mode 5000 rev / min.

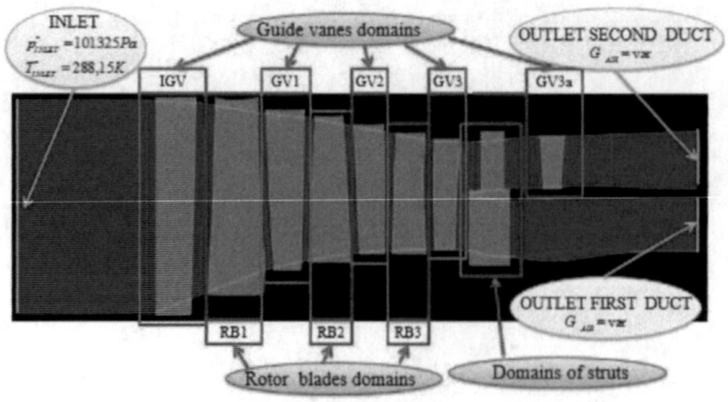

Figure 2. Computational domain of the base LPC.

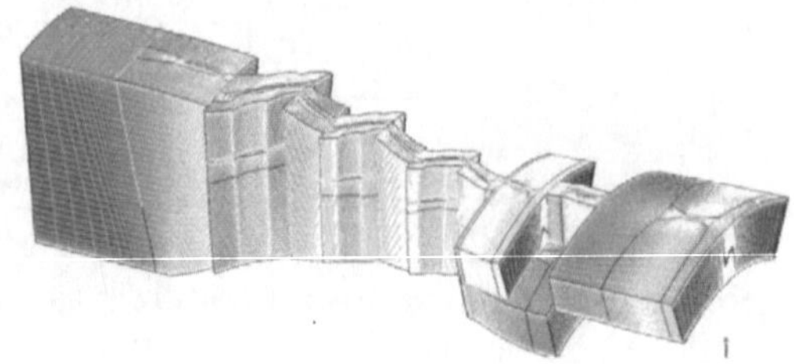

Figure 3. Finite element mesh for the computational model of the base LPC.

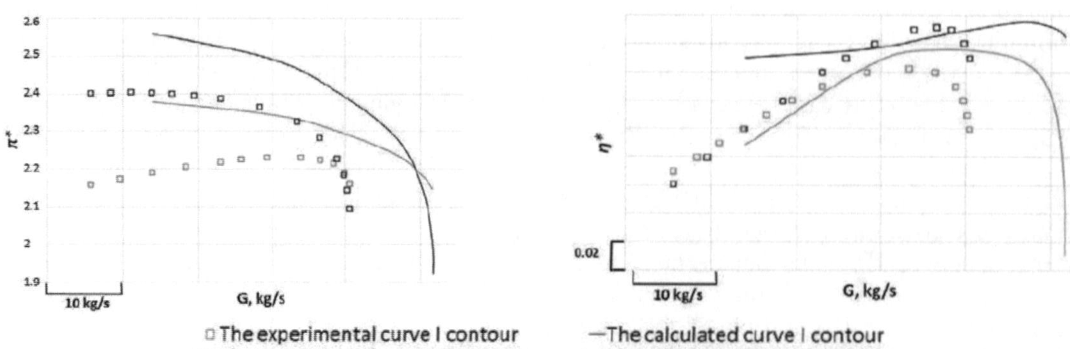

Figure 4. Head characteristics and efficiency for the base LPC.

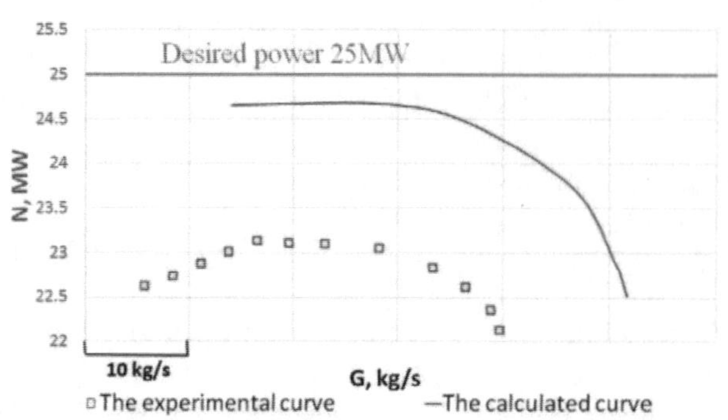

Figure 5. The dependence of the air brake total power vs the relative massflow rate for the base LPC.

3. Four - stage air

Since the base compressor is not able to provide the required power utilization at operation modes of 5000 rev / min and of 5500 rev / min (Figure 5) it was decided to make its modernization, by adding of the additional booster stages to it. For this reason, it was decided to make four stage air brake by duplication of the third stage to reduce the cost of construction (AB_RB4). However, the design of the fourth guide vanes was the same with the design of the third guide vanes. The blades of a fourth rotor blade were obtained by cropping of the third blade wheel so that the shroud of a flow channel of the fourth stage was cylindrical (AB_RB4_CD). This design allows producing the new stage from already available components with minimal modifications.

Designed variant of air brake is shown on Figure 6.

The characteristics were calculated for this variant (Figure 7, 8).

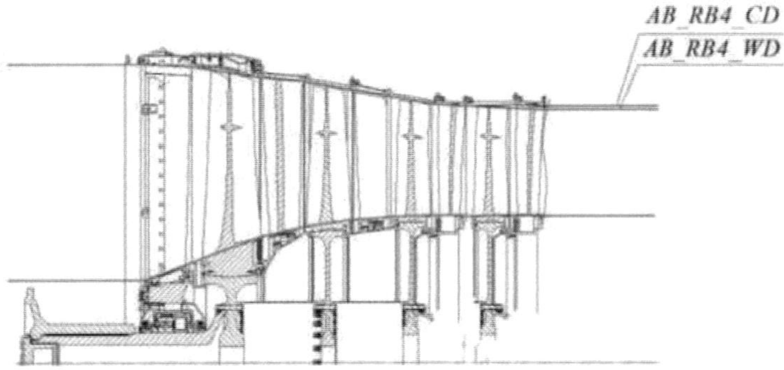

Figure 6. Four stage air brake.

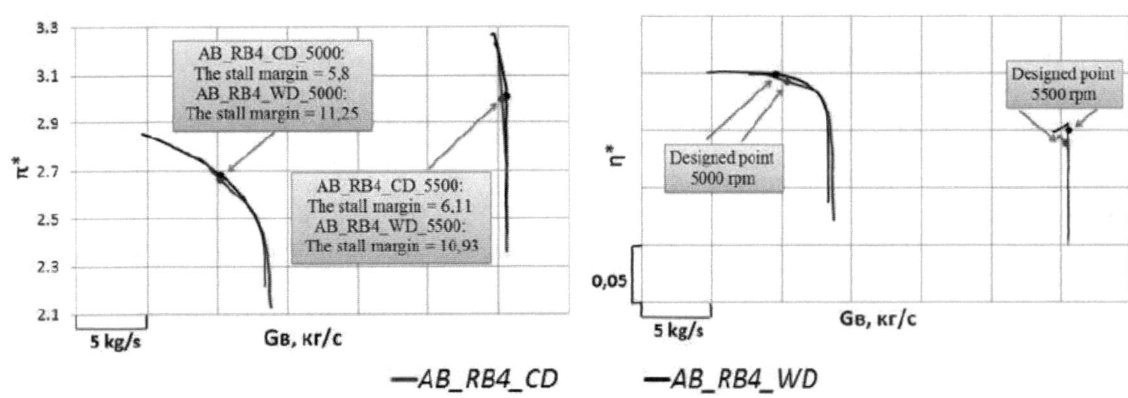

Figure 7. Head characteristics and efficiency of the air.

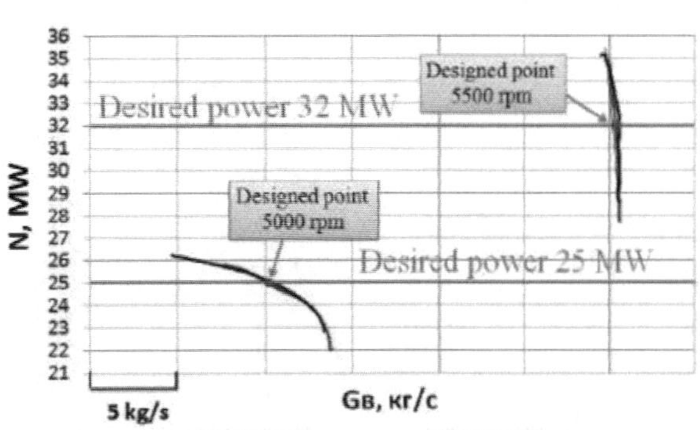

Figure 8. Characteristics of the four stage air brake.

Figure 7 shows that the air brake variant designed by duplication of the last stage, allows to utilize the required power, but it has a low stall margin. At the operation mode of 5000 rev/min the factor of the stall margin is 5,8 %, at the operation mode of 5500 rev/min factor of the stall margin is 6,11 %.

Analysis of flow structure shows that there is flow separation the hub of fourth rotor blade (Figure 9). This flow separation also expands also to the fourth GV, apparently, it is the reason of the stall margin limitation.

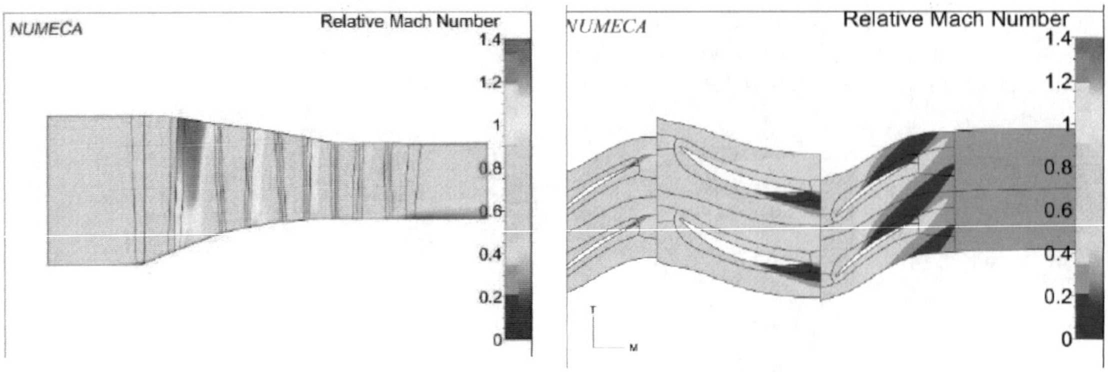

Figure 9. The flow picture of the flow in the hub section of the air brake fourth stage (at operation mode of 5500 rev / min at the operating point).

It was decided to bring waisting of the shroud of the flow channel of the fourth stage. Waisting was carried out by changing shroud of a flow channel of the fourth stage. Hub contours was remained unchanged. Designed air brake arrangement with waisting of the flow channel is shown in Figure 6 (AB_RB4_WD).

Figure 7, 8 shows plots of the characteristics comparing of the air brake variant with waisting of the flow channel and without it.

As it can be seen from the comparison of the characteristics which is illustrated in Figure 7, the stall margin was increased to an acceptable level of 11,25 % at the operation mode of 5000 rev / min and up to 10,93% at the operation mode of 5500 rev / min by the waisting implementation. Improving of the air brake performance by the waisting implementation is also confirmed by the analysis of fields of calculated Mach numbers on the hub section (Figure 10). The flow separation was eliminated.

Air brake with waisting of the flow channel was selected for further investigation. This variant has one disadvantage. The existing third GV should be modified for addition of fourth GV, but it allows providing required stall margin.

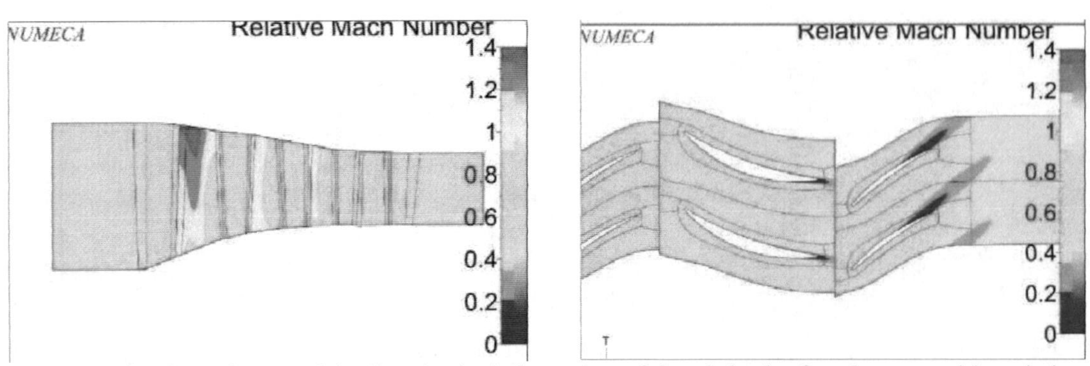

Figure 10. The flow picture of the flow in the hub section of the air brake fourth stage with waisting of the flow channel (at operation mode of 5500 rev / min at the operating point).

IOP Conf. Series: Materials Science and Engineering **232** (2017) 012068 doi:10.1088/1757-899X/232/1/012068

4. Designing of the outlet guide vane

The four-stage air brake with waisting of the flow channel allows to utilize the required power at the operation modes. However, in this configuration, there is a residual twist angle of the flow beyond the last guide vanes. The angle of the flow core in the section at fourth GV outlet at both operation modes is about 31 degrees.

Residual twist beyond the fourth GV is undesirable because it can lead to the instability work of the output devices of the air brake and test rig as a whole. It was decided to design the Outlet Guide Vane (OGV) for elimination of residual flow swirling.

To reduce OGV production costs it was decided to make it with constant cross-section height. Upon request, GV was made of 4 mm thick plate by stamping and subsequent processing of edges. In order to prevent the flow breakdown from the GV back not only in the flow core, but also in the hub and the shroud areas, it was decided to make OGV with an inlet vane angle of 36 degrees. Output vane angle was determined by the condition that the residual flow twist beyond OGV should not exceed 3 degrees. Considering the flow lagging it was decided to make the OGV output guide vanes angle equal to -2 degrees. Based on above requirements, the design of OGV was perfomed. Flow straightener cross section shown on Figure 11. The number of blades was chosen by the condition of optimum blade profiles density of mesh [7] and it was equal to 56.

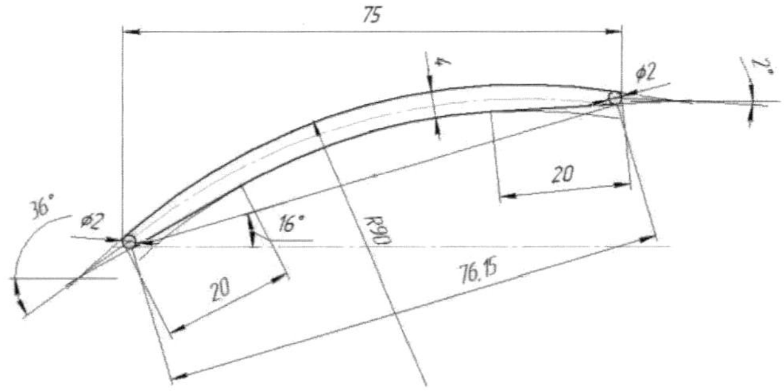

Figure 11. Geometry of the designed OGV.

The meridional profiles of the OGV were determined by design considerations. Thus, the inclination angle of the hub shroud was equal to 7 degrees to provide placing underneath it the required number of unloading disks and cavities. Unloading disks and cavities requireB to place beyond the 4th blade wheel of the air brake in order to reduce the axial force acting on the ball bearing. The angle of the shroud was chosen to be 10 degrees for the narrowing of flow channel. Created air brake configuration is shown in Figure 12.

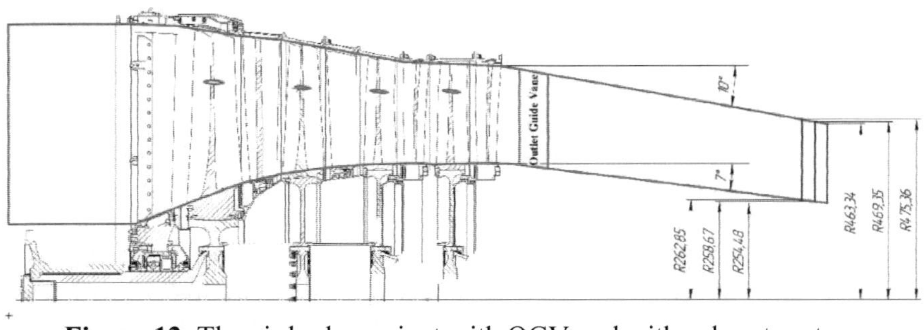

Figure 12. The air brake variant with OGV and with exhaust system.

In order to evaluate the gas-dynamic parameters of the air brake with OGV the calculation of its performance was made (Figure 13, 14). Analysis of the performance showed that this air brake configuration has stall margin 10.66 at 5000 rev / min and 9.01 at 5500 rev / min. However, the angle of the flow beyond OGV was 3 degrees.

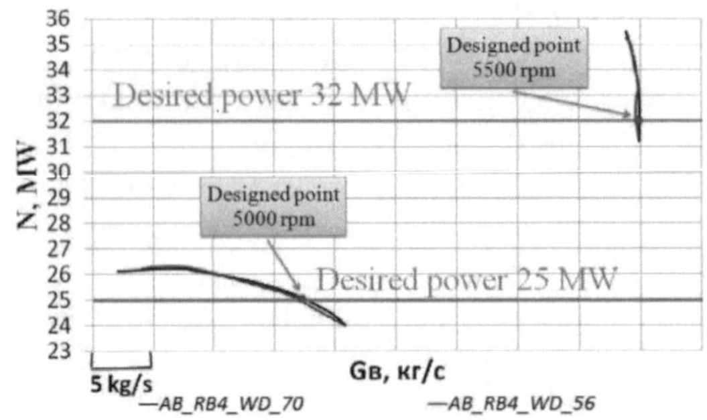

Figure 13. Characteristics of the four stage air with different numbers of OGV blades.

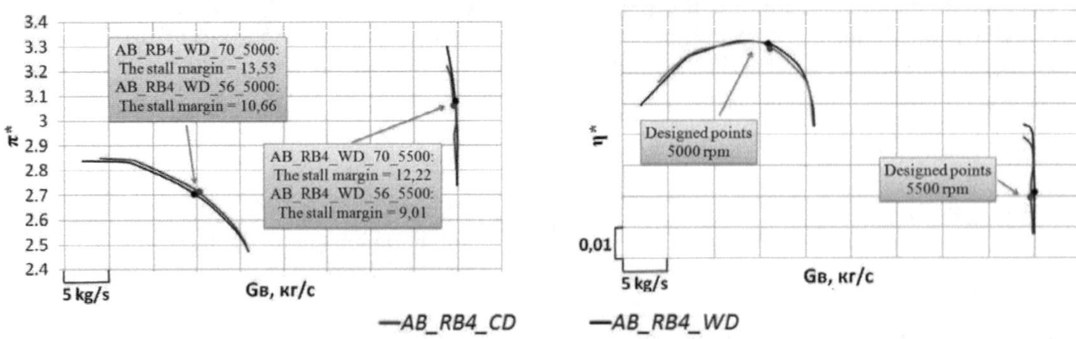

Figure 14. Head characteristics and efficiency of the air brake with different numbers of OGV blades.

Analysis of the flow structure in the air brake showed that the air brake arrangement configuration AB_RB4_WD_56 has flow separation on the hub of the OGV (Figure 12). It has been suggested that this flow separation reduces the stall margin of the air brake. It was decided to increase blade number of OGV to 70 (the air brake configuration AB_RB4_WD_70.) for reducing flow separation. Actually, this decision led to the increase in losses in the straightener (generally, profile losses), but in terms of the air brake it is not critical, because its mission is energy utilization.

Analysis of the air brake AB_RB4_WD_70 characteristics showed that the increase in number of OGV allows to increase stall margin up to 13.53% and 12.22% on operation modes 5000 rev/vin and 5500 rev/min. It also and reduce the flow separation breakdown in the OGV. Moreover, increasing the number of OGV blades has reduced cross-sectional flow angle at the output from the air brake. The output flow angle of the air brake was equal to 0 degrees.

5. Designing of the air brake exhaust system

The variable static pressure at outlet of the air brake calculation model was set during calculating of the characteristics of the air brake in parts 3 and 5 of this paper. In real operating conditions the ambient pressure will not change and will be equal to atmospheric pressure [9]. For this reason, for the air brake work it is necessary to design the special exhaust system. Mission of the exhaust system is to provide the required level of backpressure at the outlet of the air brake. The backpressure is required for the required power utilization by the air brake (the air brake operation in the required points on the characteristics).

In the case of considered configuration of the air brake the simplest exhaust device is a conical converging nozzle with a central body (Figure 12). When the nozzle works at the supercritical drop (the pressure ratio is more than 1.86) at the nozzle exit (narrowest cross section is Ni) the flow speed becomes equal to the speed of sound, and the air flow rate through the nozzle reaches a maximum possible value and the increase of the drop at the nozzle the flow rate does not increase. Thus, the flow rate is set by the constant flow rate through the entire system as a whole: and through the nozzle and through the air brake.

For the design of the nozzle it was necessary to determine the its area of the narrowest cross section Ni by condition of the supercritical flow of air from nozzles at each of the two operation modes (N5000 of rotational speed 5000 rev / min, N5500 of rotational speed 5000 rev / min) (1). So, the total pressure, total temperature, mass flow rate was measured at the outlet from the air brake (in cross section S) at the operating points at rotational speeds of 5000 rev / min and 5500 rev / min. Measurement of flow parameters was performed using computational models described in the part 4 of this paper. The values of the measured parameters are shown in Table 1.

Table 1. The results of the parametric IPC model calculation.

n, rev/min	T_{S1}*, K	p_{S1}*, Pa	G_a, kg/sec	N, MW
5000	395.3	268597	246.85	25
5500	411.1	291132	274.97	32

The area of the narrow section at each operation mode was determined by the formula:

$$F_C = G_{S1} \cdot T_C^{*\frac{1}{2}} \cdot \left(m_g \cdot p_C^* \cdot q(\lambda_C) \cdot \mu_C\right)^{-1}. \quad (1)$$

Outer and inner diameters of the nozzle with a central body considering the meridional profiles designed in the previous step were determined after determining the areas. The calculated areas of the nozzles for each operation mode and outer and inner diameters are given in Table 2. The figure 12 shows the arrangement of the calculated nozzles together with the air brake.

Table 2. Throat areas and diameters of the air brake nozzle

n, rev/min	F_C, m^2	D_h, m	D_w, m
5000	0.471	0.509	0.927
5500	0.493	0.526	0.951

Since the cross-sectional areas for operation modes of 5000 rev / min and 5500 rev/ min are not significantly different (less than 5%), it was decided to design a one, unified nozzle (Figure 12). Use of the one unified nozzle though move working points of the air brake further from originally design points on performance map, but it can reduce the manufacturing cost (it is necessary to produce not two nozzles, but one), and it simplifies the process of the engine tests with the various powers (not need to change the exhaust nozzle during the engine replacement or the switching the operation mode at the testing engine) [10].

The combined CFD - model of the air brake with a nozzle (Figure 15) was created to check the correctness of the decision about the using of the one, unified exhaust nozzle. Designed model included the air brake, the nozzle and attached volume for correct flow modeling in the nozzle. At the inlet of the calculation model (the inlet of the air brake) the total pressure of 101325 Pa and total temperature 288.15 K were set. At the outlet of the model static pressure is equal to atmospheric pressure was set. (Figure 15). Calculation was made for two operation modes of 5000 rev / min and 5500 rev / min.

As a result of the calculations the integral parameters of the air brake at the two operational modes, considering combined its work with the nozzle were determined. Obtained operating points of the air brake with the nozzle at operation modes of 5000 rev / min and 5500 rev / min were plotted to the corresponding characteristics (Figure 17). From the analysis of the figure it can be concluded that:

1) the air brake power at operation mode 5000 rev/min is slightly less than design power and is equal to 24.8 MW;

IOP Conf. Series: Materials Science and Engineering **232** (2017) 012068 doi:10.1088/1757-899X/232/1/012068

2) the air brake power at operation mode 5500 rev/min is slightly more than design power and is equal to 32.3 MW.

The resulting deviation of the working capacity from the designed is feasible and it can be taken into account during testing of engines.

The figure 16 shows the averaged the field of Mach numbers in the circumferential direction in the coupled air brake system with nozzle at the operation mode of 5500 rev / min. From the analysis of the fields of Mach numbers it can be seen that at the nozzle outlet the flow speed is equal to the speed of sound M = 1.

The unified nozzle was designed. This nozzle allows to provide air brake operation on required modes.

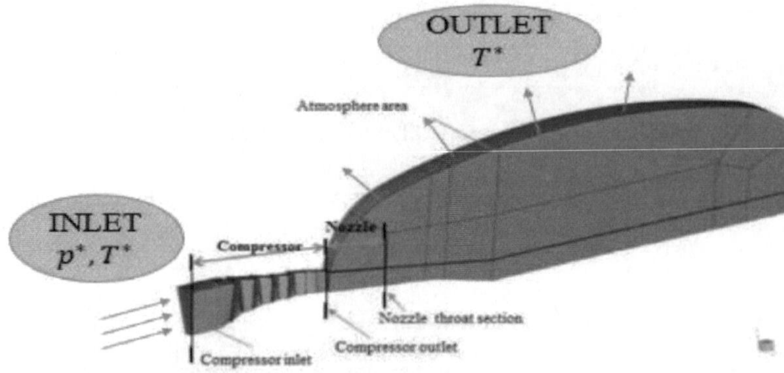

Figure 15. Calculated coupled model with the air brake exhaust device.

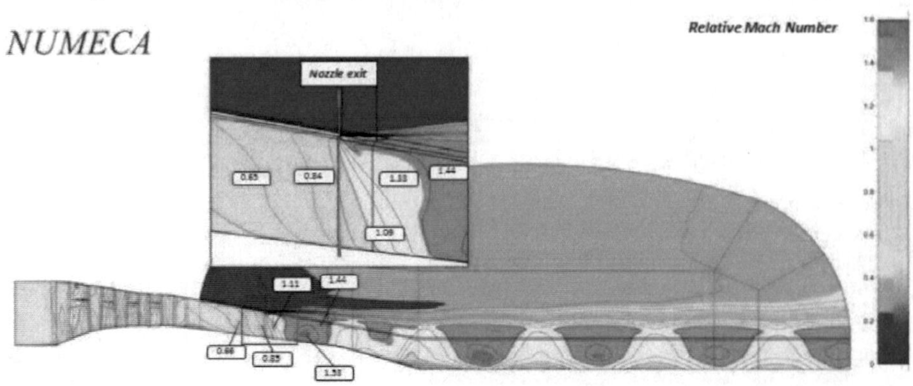

Figure 16. The field of the calculated relative Mach number of the coupled model of the air brake and exhaust nozzle.

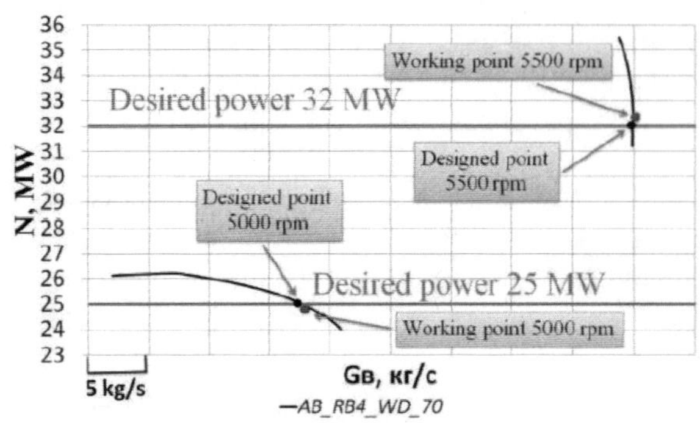

Figure 17. The air brake power characteristics with the account of the exhaust arrangement.

IOP Conf. Series: Materials Science and Engineering **232** (2017) 012068 doi:10.1088/1757-899X/232/1/012068

Conclusion

The design of the air brake was created for industrial GTE tests with the power of 25MW and 32 MW. The development of the air brake had a number of features:

1) the development was carried out using CFD - program Numeca FineTurbo. However, there was not any preliminary calculations of the turbomachinery using 1D / 2D – methods;
2) the development was carried out, mainly, using the existing parts and assemblies.

These features helped to reduce the development time, and reduce the cost of development and production of the air brake.

During the designing of the air brake the following problems have been successfully solved:

1) the workflow model of the three-stage LPC was created and its validation was made;
2) stall margin of the air brake were increased by waisting of the meridional flow channel;
3) OGV was profiled which eliminates the twist beyond the air brake;
4) unified nozzle was designed which provides the air brake work at necessary points on the characteristics.

References

[1] Korotov, M. V., "Method for testing gas-turbine engine and device for its realization", Patent of Russia Federation, RU 2318195, Date of publication 20.06.2003.

[2] Kuzmenko M.L., Egorov I.N., Shmotin Yu. N., Kretinin G.V., Fedechkin K.S., 2010, "Optimization Studies of the Aircraft High-Bypass ratio Turbo Engine Fan,", Conference ASME Turbo Expo 2010, Glasgow, Scotland, UK, June 14-18.

[3] Kuzmichev, V S., Rybakov, V N., Tkachenko, A. Y, Krupenich, I. N., 2014, "Optimization of working process parameters of gas turbine engines line on the basis of unified engine core", ARPN Journal of Engineering and Applied Sciences, 9(10), pp. 1873-1878.

[4] NUMECA, "User Manual AutoGrid5" Release 8.4, NUMECA.inc., Belgium, January 2008.

[5] Zubanov VM., Shabliy L.S. and Krivcov A.V, 2014, "CFD-Modeling of Powerful Screw Centrifugal Kerosene Pump". ASME Paper No. GTINDIA2014-8145

[6] Kuzmenko M.L., Egorov I.N., Shmotin Yu. N., Chupin P.V., Fedechkin K.S., 2006, "Multistage axial flow compressor optimization using 3D CFD code,", 6th ASMO UK/ISSMO conference on Engineering Design Optimization, Oxford, UK, 3-4 July, 2006.

[7] Belousov, A. N., 2003, "Teorija i raschet aviacionnyh lopatochnyh mashin", Samara, Samarskii Dom Pechati p. 336c.

[8] Lewis, R. I., 1996, "Turbomachinery performance analysis" Publisher: Elsevier Science & Technology Books Pub, p. 329

[9] Kuz'michev, VS., Tkachenko, A.Y, Krupenich, I.N. and Rybakov, VN., 2014, "Composing a virtual model of gas turbine engine working process using the CAE system "ASTRA"," Research Journal of Applied Sciences, 9(10), pp. 635-643.

[10] Zubanov V. M., Shabliy L. S., Krivcov A. V, 2014, "Centrifugal kerosene pump CFD-modeling". Research Journal of Applied Sciences, 9(10), pp. 629-634.

IOP Conf. Series: Materials Science and Engineering **232** (2017) 012069 doi:10.1088/1757-899X/232/1/012069

The methodology for the existing complex pneumatic systems efficiency increase with the use of mathematical modeling

A.M. Danilishin[1], S.V. Kartashov[1], Y.V. Kozhukhov[1] and E.G. Kozin[2]

[1] Peter the Great St.Petersburg Polytechnic University
[2] SUE «Peterburgskiy metropoliten»

E-mail: kozhukhov_yv@mail.ru

Absrtact. The method for the existing complex pneumatic systems efficiency increase has been developed, including the survey steps, mathematical technological process modeling, optimizing the pneumatic system configuration, its operation modes, selection of optimal compressor units and additional equipment. Practical application of the methodology is considered by the example of the existing pneumatic systems underground depot reconstruction. The first stage of the methodology is the survey of acting pneumatic system. The second stage of technique is multivariable mathematical modeling of the pneumatic system operation. The developed methodology is applicable to complex pneumatic systems.

Introduction

The increase efficiency technique of complex pneumatic systems is developed. It includes inspection stages, mathematical modeling of engineering procedure, pneumatic system optimization, operational mode, optimum compressor installations selection and service equipment. In terms of energy efficiency and reliability of compressed air supply the offered technique makes it possible to develop the optimum scheme for the existing pneumatic systems at their reconstruction and again designed pneumatic systems of industrial facilities.

The first technique stage is technological audit of the operating pneumatic system [1]. Inspection consists of collecting statistical data, visual and tool inspection, definition of the pneumatic system actual configuration. The second stage is multiple mathematical modeling of pneumatic system work. Calculations are carried out on the mathematical model, which underwent validation on the operating pneumatic system. The system is formed from the certain pipeline sections of compressed air supply with their optimum diameters calculation. The optimum configuration is defined with topology of the area and operation requirements. Numerical work modeling of pneumatic system separate elements is carried out using computational gas dynamics methods. In addition, their optimum configuration in terms of technological and economic problems is selected. At the final stage the scheme of a pneumatic system with the compressor and service equipment selection is formed. It takes into account compressor reservation and recommendations about the operation modes.

Subject of research

Practical technique application is considered on the example of the St. Petersburg Metropolitan depot pneumatic systems reconstruction. The St. Petersburg Metropolitan incorporates 6 depots, each one is equipped with a low pressure pneumatic system (up to 10 bars). Depot pneumatic systems represent an object with the following characteristics and features of operation. Pneumatic system powered by compressed air from the main compressor station. The Compressor Station (CS) with hangar compressors placing is located in single heated machine hall and equipped with 4 or 5 volume action compressors. By now all previously set many years ago piston compressors replaced with modern oil-injected screw compressors. Each depot has various compressed air consumers with different requirements to its pressure, flow and quality. The main groups of compressed air consumers in each depot and their main characteristics are shown in table 1.

Table 1. The main groups of compressed air consumers in each depot and their main characteristics.

Consumer	**Type of use**	**Usage time**	**Pressure (not less) bar; cleaning class; dew-point temperature, °C**
Train depot	Filling wagons with compressed air, airflow technical, pneumatic tools, pneumatic testing equipment stands	year-round	7; 1; -40
Track fan	Snow blowdown of railroad switches in the automatic and manual modes	snow period	6; 1; -40
Technical services and divisions	Technical blowing, pneumatic tools, machines and etc	year-round	6; 1; -40

The necessary condition is any time compressed air supply warranty of the required consumer parameters on important sectors regardless operation modes at other consumption points. The most responsible in terms of the passenger transport safety is the train depot where the refueling is carried out with compressed air of wagon brake systems. Because the bringing into work of the braking system, opening and closing doors, and others provided with compressed air.

Technological audit

A pneumatic system technological audit called also by pneumoaudit is the integral stage at its reconstruction, at normal working capacity restoration of the existing or a new pneumatic system construction. Pneumoaudit essence is to conduct comprehensive, detailed, instrumental and statistical state survey and pneumatic system in general and objects particularly of their collaboration in a single system. Based on such an audit at the following stages are offered measures for improving the existing pneumatic system operation or developing a new basic scheme of air supply and air preparation, which will improve the reliability and energy efficiency of compressed air considering consumers pneumatic operation features. At the stage of a pneumatic parameters statistical survey documentary data about compressor settings consumption of energy, air consumption, technical air parameters, operation (seasonality) modes, which need to be considered when calculating pneumatic system efficiency collected and analyzed. Then, the experimental researches of workflows and acting pneumatic system modes held. Visual and instrumental observations are conducted according to the specially developed for the object program and the method of examination. The basis of the program and the survey methodology are the provisions contained in the sources [2, 3]. The inspection technique consists in carrying out visual assessment with photo video fixing of the actual pneumatic system wear, measurement and registration of pneumatic system characteristics by stationary and portable devices on the set modes of behavior. A tool measurements series of pneumatic system work parameters carried out on various modes. The generation balance and compressed air consumption is analyzed on condition guaranteed provision needs of resource consumers. In conducting research pressure values in pneumatic control points are measured, the values costs are determined, the pressure change dynamics at operating

IOP Conf. Series: Materials Science and Engineering **232** (2017) 012069 doi:10.1088/1757-899X/232/1/012069

change modes for different consumption branch of system is fixed and air leaks on defective places of a pneumatic system are measured.

At the time of carrying out inspection in all depots pneumatic system had the general collector to which the compressor station gives compressed air. From the general collector there is an air bleeding to each of consumers who have different requirements to pressure and quality of compressed air. In this regard, air consumption by any consumers exerts impact on parameters of the given air to other consumers. It can negatively affect the functioning certain consumers up to consequences in the form of failures in the rolling stock work on the line. Let us consider pneumoaudit results on the example of one-surveyed depots – «Nevskoye» depot as inspection results are in a varying degree characteristic also of other surveyed depots. It should be noted that now for «Nevskoye» depot all specified defects of a pneumatic system condition are eliminated by results of the described works. As a visual inspection result of pneumatic system, the depot "Neva" found to be satisfactory pipeline condition from the compressor station before entering into depot building, on washing where the drainage of railway carriage by compressed air after washing is made, and to a railway upshot, characterized by lack of escapes. The unsatisfactory condition of the line pipes found in the train depot building with constant significant leaks about 8 m3 / min (see. below Fig. 4) due to leakage of the connecting device to the pneumatic electric trains brake system. In addition, considered unsatisfactory electro-valves state in the automatic blowing shooter system with a constant leak value, determine the value of which is difficult due to the absence of valves.

Inspection by a tool mode of generation system and compressed air distribution for needs of «Nevskoye» depot was made on these actual conditions:

1) The mode No. 1 – the consumer: only the depot building, the size of inlet pressure to the building of depot was measured and an exit from compressor station, the expense from escapes was also measured in the depot building. The schedule of pressure drop is represented in fig. 1;
2) The mode No. 2 - the consumer: escapes in the depot building, the inlet pressure size was measured in the depot building and an exit from compressor station, the volume of escapes was determined (by an expense) in the depot building, air supply in depot is stopped. Within 10 minutes pressure falls from value of 7,0 bars to 6,0 bars;
3) The mode No. 3 - the consumer: escapes in the washing building, the size of inlet pressure on washing was measured, the expense was measured, air supply on washing is stopped. Within 5 minutes pressure falls from value of 6,3 bars to 5,6 bars;
4) The mode No. 4 - the consumer: depot building and a railway upshot without arrows blowing, the size of inlet pressure to the depot building, an exit from compressor station, on a traveling fan was measured, the expense was measured. The schedule of pressure drop is represented in fig. 2;
5) The mode No. 5 - the consumer: the depot building, a railway upshot at the engaged automatic arrows blowing, the size of inlet pressure to the depot building, an exit from compressor station, on a traveling fan was measured, the expense was measured. The schedule of pressure drop is represented in fig. 2;
6) The mode No. 6 – the consumer: the depot building, a railway upshot at the engaged automatic and manual arrows blowing, the size of inlet pressure to the depot building, an exit from compressor station, on a railway upshot was measured, and the expense was measured. The schedule of pressure drop is represented in fig. 2;
7) 7 The mode No. 7 – the consumer: the depot building, a railway upshot at the engaged automatic arrows blowing was measured, the new receiver of 3,2 m^3 is connected, the size of inlet pressure to the depot building, an exit from compressor station, on a railway upshot and the expense was measured. The schedule of pressure drop is represented in fig. 3;
8) The mode No. 8 - the consumer: depot and washing building, the size of inlet pressure on washing was measured, the expense was measured. Within 7 minutes pressure falls from value of 7,0 bars to 6,0 bars.

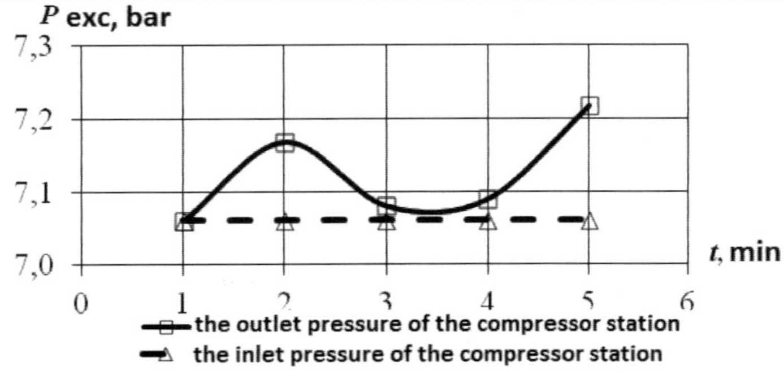

Fig.1. Pressure drop on the mode №1: the consumer is only the depot building

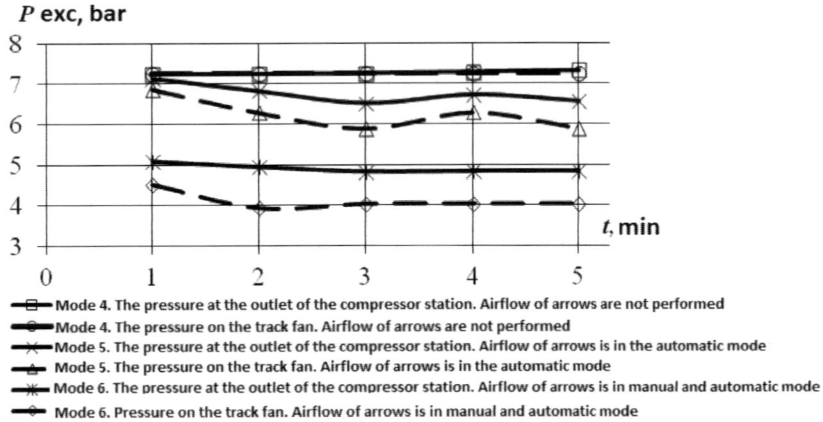

Fig.2. The pressure drop in the modes: №4 - leakage for the track fan; №5 - track fan with automatic blasting; №6 - track fan with manual and automatic blasting

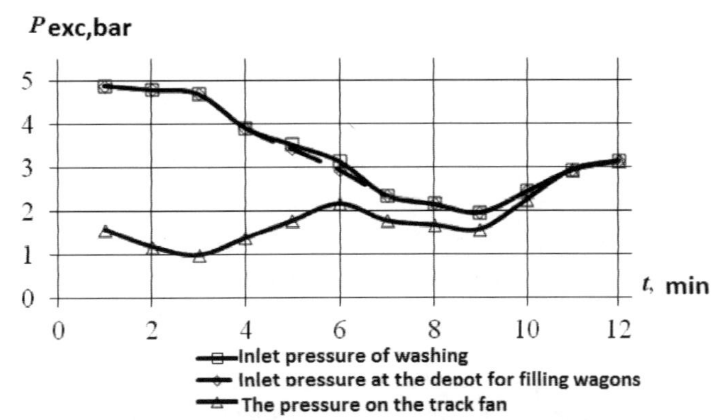

Figure 3. Pressure drop on the mode №7: track fan with automatic switches and blasting work by a compressor connected to the air tanks with a capacity of 3.2 cubic meters

From the analysis of graphs, the following conclusions can be made:

1. According to the general recommendations the value of the pressure drop in the section of the pipeline from the compressor station to the consumer must not exceed 0.1 bar. In the studied areas the exceeding of this value has been recorded at startup of the airflow fan of the arrows from 0.5 to 1.0 bar (Fig. 2). For threads depot the pressure drop is within 0.1 bar (Fig. 1). For washing thread manometer has not been installed directly at the consumer due to the absence of the mounting fitting in the pipeline. However, by the assess of the manometer, which is set for the washing thread immediately following branching to the sink the pressure drop at a given part of the pneumatic pipeline between the manometers is less than 0.1 bar.

2. At the maximum air consumption on the directional fan the pressure decreases across the network below the normal value for a significant - about 2 - 5 bar depending on the combination of the included consumers(figure 3.). This makes it impossible to fully fill the braking system of the wagon, which can lead to failure of work of the rolling-stock on the line when the air pressure is insufficient for uncompressing brake blocks after a stop at the station.

3. For all pneumatic pipelines consumers except for washing, there are recorded air leakages.
From this analysis it follows that the system of pneumatic pipelines airflow fan generates resistance above the recommended values and the pneumatic system generally is organized without taking into account the features of using the technology resources by the consumer and requirements to ensure the required reliability parameters of compressed air.

In Figure 4 shows the dependence of generation of compressed air by the compressor station at the time when the washer does not work, the fan is turned off, and the depot is not loaded to full capacity (not 100%). Data has been recorded at a minute interval

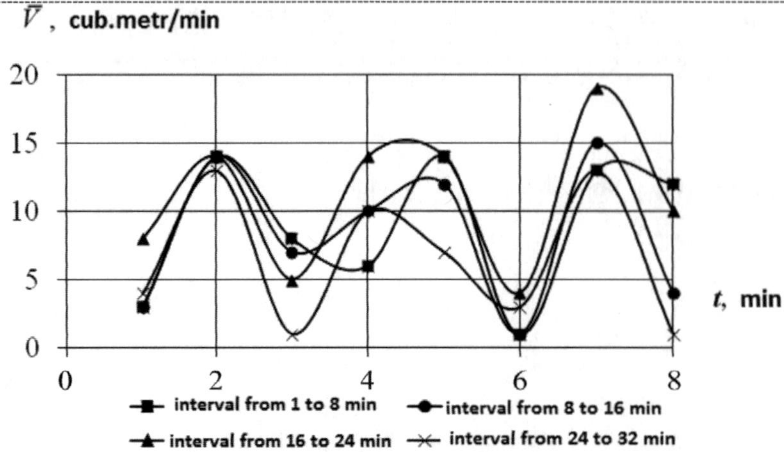

Figure 4. The cycle diagram of generation when working at the depot, the wash does not work, the fan is turned off.

The analysis of the graphs in Fig. 4 shows that at this mode consumption figures are having pulsating character with traceable cyclic pattern. This curve form can be explained by the fact that there is considerable leakage of the network, which is an average value of about 8 m3 / min. It is known that in the general purpose air networks in the presence of only one pipe hole about 5 mm in diameter leads to leakage to 2 m3 / min and 10 kW power lost [2].
Leakage value given to the conditions of the suction can be determined by the formula:

IOP Conf. Series: Materials Science and Engineering **232** (2017) 012069 doi:10.1088/1757-899X/232/1/012069

$$V_{leak} = \frac{V_p(p_{max} - p_{min})}{t}$$

Where V_p - the total volume of the network, determines the amount of pipelines, m^3; p_{max}, p_{min} - maximum and minimum pressure in the system, a bar; t - duration of the deflation process, min.

Thus, from Fig. 4 it follows that pneumatic system works in the following mode. The compressor supplies compressed air to the network to set pressure, in the absence of compressed air consumers (fully or partially), the compressor switches to the control mode with of compressed air discharge into the atmosphere.

After the pressure in the network decreases till the minimum established value at the compressor, the latter switches to the nominal mode of generation and again supplies air to the network. The graph (Fig. 4) the frequency of such work is approximately 2-3 minutes. Thus, for a long time, compressed air is only consumed to fill leaks.

Figure 5 shows graphs of the four-year period the electricity consumption of the compressor station from 2010 to 2013.

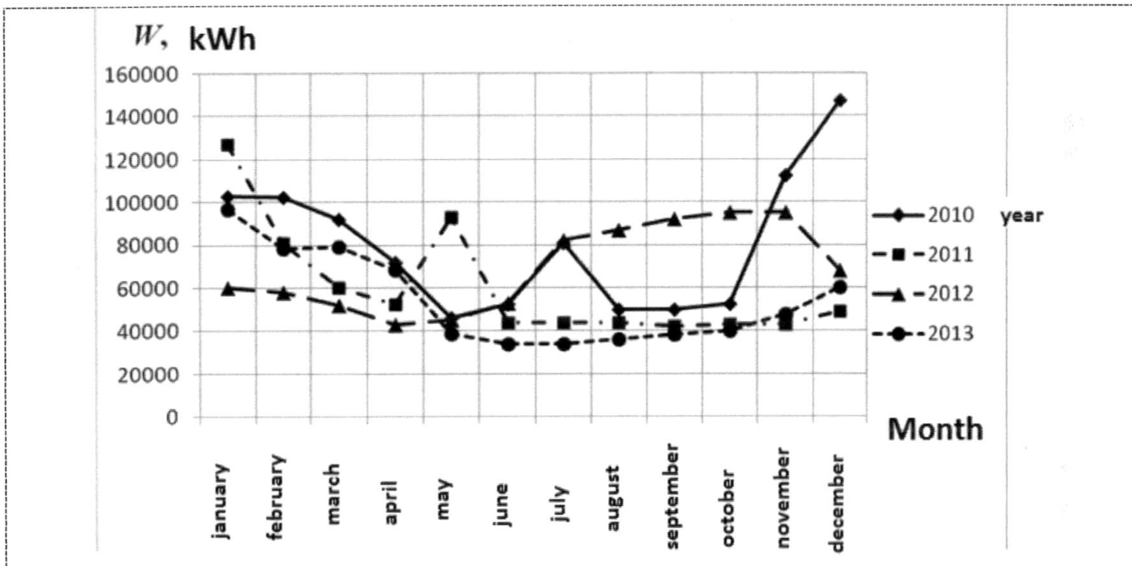

Figure 5. The monthly energy consumption by years

From the graphs in Fig. 5 it is clear that, depending on the season, the consumption of electric energy may vary by several times. The number of simultaneously working compressors varies by month and can be calculated on the basis of the data of Fig. 5 according to the formula:

H = E / Ek,

where H - the number of compressors; E - actual energy consumption (kW × h × number of days in the month × 24 hours a day); Eck - Nominal power consumption with one compressor (Ek = 110 kW × h × number of days in the month × 24 hours a day.).

The calculation shows that, in general, one compressor is working, and sometimes in winter the second one was activated. This is due to the increased air intake fan on a snowy day.

An important indicator in assessing the pneumatic mode is its maximum peak load when switched on simultaneously all consumers of compressed air. The warranty of supplying all consumers with simultaneous operation of the air with necessary parameters ensures the reliability of the entire pneumatic system. Based on these data, there were obtained the highest possible average hourly expenditure of compressed air in three consumers (see. Table 2) and suggested compressor station equipment with redundancy for depot "Neva" (see. Table 3). It is recommended to use the central control system (CCS), possess the function coordination of the compressors work together to regulate the work of the compressor station.

Table 2. Maximum hourly flow rate

Consumer	The nature of using	The maximum flow rate, m^3 / hour (m^3 / min)
Train depot	Filling electric brake systems, technological needs. Not allowed excessive pressure drops below 6.5 bar, high demands for clean air	480 (8)
Track fan	A rinsing of turnout switchers	1800 (30)
Motodepot	Drying wagon, technological needs	1020 (17)
Total consumption		3300 (55)
Total from the perspectives of the depot and the provision development		4320 (72)

Table 3. Offered equipment of compressor stations by compressor units

Consumer	Compression type	Max Prod. m3 / min	Drive power kW	System of regulation
Train depot	screw oil-filled	12	75	Frequen. conversion
Track fan	screw oil-filled	29	160	Frequen. Conversion., Periodic starts and stops
Motodepot	screw oil-filled	19	110	Frequen. conversion
Reserve	screw oil-filled	19	110	Periodic starts and stops

The described technological audit carried out for all the metro depot considering the characteristics of each of them on the composition of the equipment and engineering process. The next step in offered the methodology is to develop a concept of air supply to the definition of the optimal pipeline diameter, topology map, taking into account the operating conditions at the facility based on the energy efficiency requirements and the reliability of the air supply.

Mathematical modeling

At this stage, there are holding the multivariable configuration and settlement and analysis of operation compressed air networks based on the data obtained by using statistical and instrumental examination, developing a mathematical model of pneumatic systems. In connection with the various parameters of air consumed by consumers and the inadmissibility of pressure drop on the branch of the refueling wagon (Fig. 3), it is proposed the division of pipelines to consumers. In this way, the supply of compressed air to each of the consumers is done by isolated from other consumers pipeline and manufactured from a single compressor. Air supply remains centralized. In this case pneumatic system is configured so that if necessary either compressor can be replaced by any other without necessarily reserve by appropriate combinations of open and closed valves. The implementation of the described concept provides localization of all compressors in the same service room - already equipped with a specialized site. Therefore there is no need to conduct a detailed analysis of the local localization of compressors (modular accommodation) for consumers, as it is in connection with a separate service facilities, will make further organizational expenses in the process of controlling the operation of pneumatic system, not to mention the preparation of deployment sites compressor units directly from consumers. Following the regulatory documents [4 - 6], the implementation of the solution described in the technical re-equipment provides localization contours pipelines of all compressors in the same service rooms, while providing for each consumer the desired class of air cleanliness and control flexibility of pneumatic system as a whole.

Example for the proposed reconstruction of the concept of air supply is shown in Figure 6.

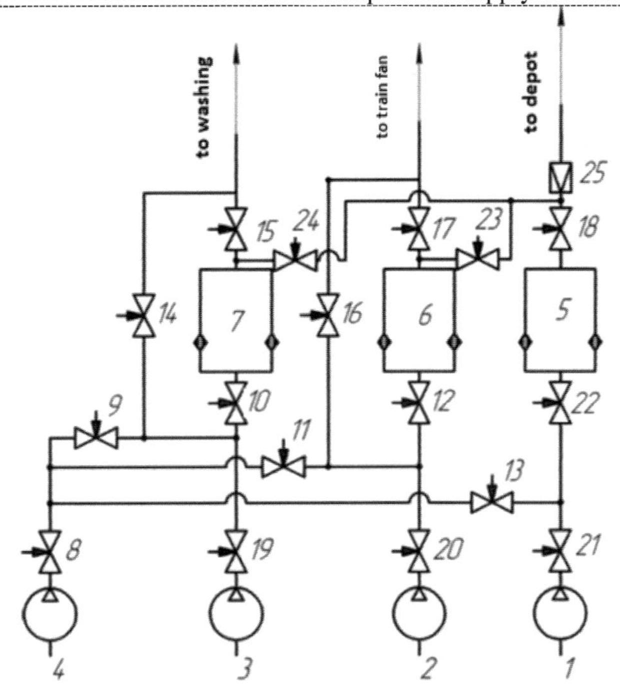

Figure 6. The basic diagram of the air supply for reconstruction
1, 2, 3 - the main compressor units with frequency converters, and built-in coolers and moistures; 4 - Reserve compressor unit; 5, 6, 7 - drying units with a fine filter; 8 - 24 - stop valves; 25 - Check Valve

It creates a mathematical model developed to optimize the air supply circuit, the main provisions of which are listed below.

In order to determine the loss ΔP in the area, the following formula (1), combining the amount of friction loss and the amount of loss $\sum_{i=1}^{i} \Delta P_{Mi}$ due to local resistance (valves, faucets, outlets, a sudden expansion, sudden narrowing, etc.):

$$\Delta P = \Delta P_{fr} + \sum_{i=1}^{i} \Delta P_{Mi} \text{, [Pa] (1)}$$

In order to determine friction the loss following formula is used Darcy-Weisbach [2]:

$$\Delta P_{fr} = \frac{\lambda \cdot k \cdot l \cdot c^2 \cdot \rho_\tau}{2 \cdot d_{ps}} \text{, [Pa] (2)}$$

In the formula (2) λ is the coefficient of resistance, l - length of the section of the pipeline, [m]; c - velocity of the medium in the pipeline, [m/s], ϱ_τ-line pressure, [Pa], D_{ps} - the diameter of the pipeline section, [m].

To calculate the coefficient of resistance is used empirical formula Blasius (3) and Nikuradze (4) designed to calculate a hydraulically smooth pipes:

$$\lambda = \frac{0,3164}{\mathrm{Re}^{0,25}} \quad (3)$$

$$\lambda = 0,0032 + 0,221\,\mathrm{Re}^{-0,237} \quad (4)$$

$$\mathrm{Re} = \frac{c \cdot d_{ps}}{\nu} \quad (5)$$

ν- the kinematic viscosity of the medium, [m^2/s].

Equation (3) is used in the case of contact with the Reynolds number in the range from 3000 to 7000, the formula (4) when getting the Reynolds number in the range from 10^5 do10^7.

The flow velocity is determined from the mass flow equation:

$$c = \frac{\overline{m}}{\rho_\tau F} = \frac{4 \cdot \overline{m}}{\rho_\tau \pi d_{ps}}, \text{ [m/s] } (6)$$

In order to determine the loss of local resistance using a modified formula of Darcy-Weisbach:

$$\Delta P_{м} = \zeta \cdot k \frac{c^2 \cdot \rho_\tau}{2}, \text{ [Pa] } (7)$$

In the formula (7) ζ - the local resistance coefficient is determined by empirical dependencies of the manual on water connections I.E Idelchik [7].

Because the at the site there are long term operation pipes, and it is impossible to reliably determine the roughness of the inner surface of the tube, changing in course of time due to corrosion and wear and tear of the material, then for the above formulas is introduced the correction factor for the roughness of the k, defined as a result of instrumental examination optimized pneumatic systems (in this case, its value is k = 1,22). As the pipes wear out in course of time, this factor will allow an increase in pressure in advance to prevent losses in the pipeline sections.

Based on the above key provisions there is developed the program code of the mathematical model written in C ++. In the program code there are implemented data entry modules, graphical results of the analysis modules (Fig. 7). Progress of work with the program is conducted as follows:

1. The program loads pneumatic configuration with an indication compressed gas of interest to consumers.
2. In the window "calculation" is performed on the data input pressure at the outlet of the compressor station and flow rates of interest to the consumer.
3. In the "Results" is produced the issuance of calculation indicating the differential pressure between the compressor station and the consumer.
4. In the pneumatic system configuration changes are made (configuration, length and diameter of the pipeline) and the calculation is made again to achieve an acceptable level of losses.

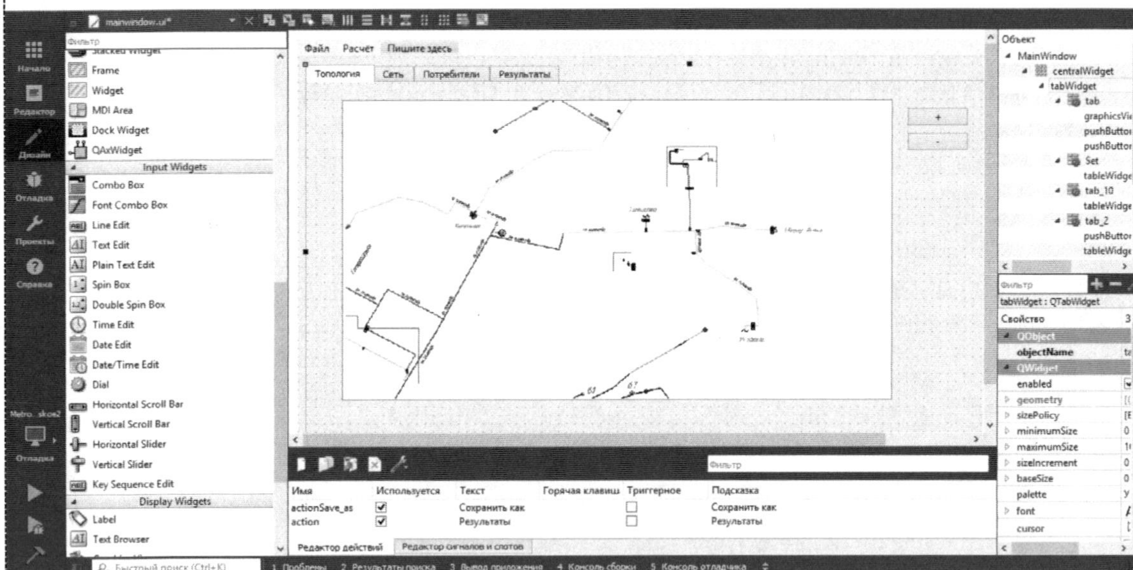

Figure 7. The windows of the program for pneumatic system data entry

Further, if necessary, is produced entirely mathematical modeling of the pneumatic or its components separately using computational techniques of gas dynamics software package Ansys CFD by calculation with a viscous gas. Modeling elements are subject fittings, hose blowouts switches paths nozzle apparatus blower hoses, etc. By results of modeling there is produced optimization of these elements.

Conclusions

The proposed method of increasing the efficiency of existing and newly designed complex pneumatic systems with the use of mathematical modeling is used at the stage of pre-design engineering of industrial facilities. The technique allows to develop a pneumatic system that ensures energy efficiency reliability and flexibility of its exploitation and the subsequent modernization in connection with the expansion of production.

References

[1] Kozin E. G., Kozhukhov Y. V. *"Technological audit pneumatic system as a step in the process of its reconstruction"* // Scientific and technical statements STU. Vol. 2 (243) / 2016. Ed. STU. Pp 34 - 44.

[2] Weyrauch A. N., Izmailov R. A., Fotin B. S. "*Compressor stations*": tutorial // Leningrad Polytechnic Institute - L., 1990 .- 84 c.

[3] Sadowski N. I., Kozhukhov Y.V. *"Theory, calculation and design of the dynamic compressor. Test compressors in parallel and series operation. Collaboration compressors"*: Proc. allowance. - SPb.: Publishing house of the Polytechnic. University Press, 2014. - 40 p.

[4] GOST 28567-90. Interstate standards. Compressors. Terms and Definitions. Compressors. Terms and definitions. - M .: Standartinform, 2005. - 30 liters.

[5] PB 03-581-03. Rules devices and safe operation of stationary compressor units, air ducts and pipelines. - M .: STC "Industrial safety". 2008 - 23.

[6] PB 03-576-03. Rules for operation and safety of the vessels working under pressure. - Gosgortechnadzor Russia. M .: PIO MBT. 2003. - 89 liters.

[7] Idelchik I. E.*"Manual hydraulic resistance"* / Ed. M.O. Shteynberg.- 3rd ed., Rev. and dop.- M.; Mechanical Engineering, 1992.

IOP Conf. Series: Materials Science and Engineering **232** (2017) 012070 doi:10.1088/1757-899X/232/1/012070

CFD analysis of linear compressors considering load conditions

Sanghyun Bae[1,2] and Wonsik Oh[1,3]

[1]Department of Refrigerator Compressor, L&A Research Centre, LG Electronics, 51, Gasan digital 1-ro, Geumcheon-gu, Seoul, Republic of Korea

[2]s.bae@lge.com, [3]wonsik.oh@lge.com

Abstract. This paper is a study on computational fluid dynamics (CFD) analysis of linear compressor considering load conditions. In the conventional CFD analysis of the linear compressor, the load condition was not considered in the behaviour of the piston. In some papers, behaviour of piston is assumed as sinusoidal motion provided by user defined function (UDF). In the reciprocating type compressor, the stroke of the piston is restrained by the rod, while the stroke of the linear compressor is not restrained, and the stroke changes depending on the load condition. The greater the pressure difference between the discharge refrigerant and the suction refrigerant, the more the centre point of the stroke is pushed backward. And the behaviour of the piston is not a complete sine wave. For this reason, when the load condition changes in the CFD analysis of the linear compressor, it may happen that the ANSYS code is changed or unfortunately the modelling is changed. In addition, a separate analysis or calculation is required to find a stroke that meets the load condition, which may contain errors. In this study, the coupled mechanical equations and electrical equations are solved using the UDF, and the behaviour of the piston is solved considering the pressure difference across the piston. Using the above method, the stroke of the piston with respect to the motor specification of the analytical model can be calculated according to the input voltage, and the piston behaviour can be realized considering the thrust amount due to the pressure difference.

1. Introduction

Recently, global depletion of resources and global warming have accelerated, and the importance of energy saving is emphasized in many industries. It is a relevant step to reduce household appliances energy consumption to meet the regulation in decreasing the global energy consumption. Household refrigerator must operate 24 hours a day and this requires the largest proportion of energy consumption in the house. Furthermore, the compressor consumes the most of the energy required by the household refrigerator. Thus, various kinds of the research and development are continuously carried out to increase the efficiency of the compressor in the household refrigerator manufacturers and related universities.

An analytical approach using CFD is one of the efficient tools in research and development of refrigerator compressor. CFD analysis helps predict the phenomena occurring in compressor sensitive to heat or fluid flow, and identify the variations by easily changing parameters. In addition, by using the fluid structure interaction (FSI) technique, the behaviour of the valve is simulated to develop an

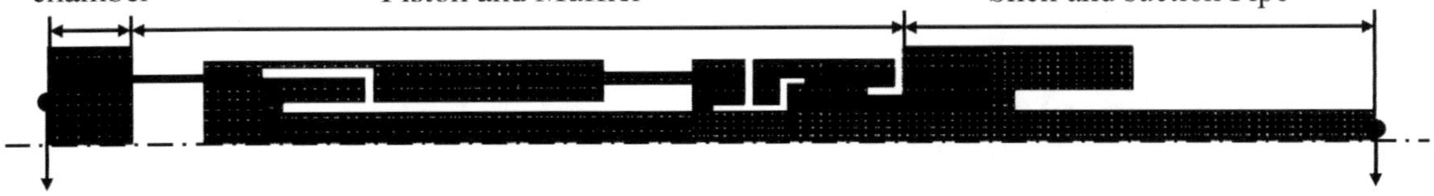

Figure 1. Fluid domain in CFD analysis of the linear compressor.

analysis technique that is closer to the actual phenomenon. Oh et al. (2016) studied the effects of CFD on the muffler shape of a linear compressor. The influence of the temperature and pressure of the refrigerant on the compressor efficiency according to the change of the muffler shape was analysed and the design direction was suggested. Suh et al. (2006) analysed the suction valve movement and the pressure of the refrigerant before compression process according to the muffler shape of the linear compressor using FSI technique. The effect of muffler frequency and acoustic supercharging effect on compressor efficiency according to muffler shape was shown. Kim et al. (2011) studied the capacity variable characteristics of linear compressors. It was analysed through the numerical method by equation and conducted experiments with the prototype compressor. They explained linear compressors require a new design approach because the stroke mechanism changes as load conditions change by the free piston system.

In this study, the UDF is used for the piston behaviour in the flow analysis of the linear compressor. In the UDF, the stroke is calculated by two coupled equations, the mechanical equation of the piston and the electrical equation of the motor. When the sine wave voltage is applied, the two coupled equations are solved using the Runge-Kutta method and the velocity of the piston is calculated. In the above process, stroke shift, which is characteristic of linear compressors, can be reflected.

2. Numerical Methodology

In this study, the mechanical equations of the piston and the electric equations of the motor are used to realize the behaviour of the piston over time in the transient analysis, and the two equations are coupled.

$$m\ddot{x} + c\dot{x} + kx = f + \alpha\dot{q}$$

$$L\ddot{q} + R\dot{q} + \frac{1}{C}q + \alpha\dot{x} = V$$

Where, forcing term of the mechanical equation is reflected in the force applied to the piston by the compressed gas, which plays an important role in simulating the actual behaviour of the linear compressor piston. There are two advantages to solve the two coupled equations to realize the behaviour of the piston. Firstly, it is possible to realize the behaviour of the piston. Unlike a reciprocating compressor (using crankshafts and rods), the stroke of a piston is not constrained, and a linear compressor is subjected to a force in the opposite direction of compression due to the difference between the pressure in the piston-cylinder compression chamber and the pressure inside of the piston. Stroke shift occurs when the piston is pushed backward. This explains the stroke applied gas force and the stroke neglected gas force are not equal at the same input voltage, and in the former case, the stroke is shifted backward. In this method, similar to the actual piston can be reflected in the analysis, and furthermore, the maximum stroke and the maximum cooling capacity in a limited geometry can be calculated. Secondly, the electrical work can be calculated. If the stroke of a piston is implemented as a sinusoidal wave, the efficiency is calculated by using the compression work as input of the compressor. On the other hand, when the electrical equations are solved together, the power consumed by the motor can be calculated, and the efficiency of the compressor can also be calculated using the

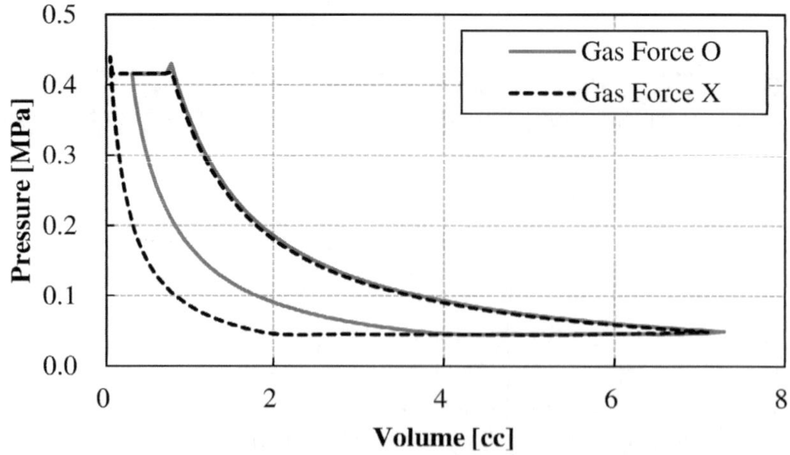

Figure 2. P-V Diagram with and without gas force.

electric power. It shows that the efficiency of the compressor and the motor is changed according to the specification of the motor. This provides the optimum point of the motor.

ANSYS Fluent V.15 was adopted for CFD simulations. Two-dimensional axisymmetric domain was used to reduce the number of cells and to simplify the structure. As shown in Figure 1, the domain is considered to be the back of the shell where the refrigerant entering through the suction pipe diffusion, the interior of the piston, and the compression chamber. The interpretation domain was assumed to be unsteady, viscous flow, compressible real gas model and κ-ε turbulence model. We set the simulation fluid as a real-gas-isobutane based on physical property in Refprop. A square cell was used and dynamic mesh including layering was used to implement motion of the piston. Numerical values used in the boundary conditions were obtained by experiments. The boundary condition corresponding to the suction pipe was 298 K temperature, 48,000 Pa pressure, and mass flow inlet condition was given. For the fully open discharge valve, the temperature 348 K, the pressure 415,000 Pa and the pressure-outlet condition was given. The temperature of the surface of the parts such as piston, muffler, shell, etc. was given as the boundary condition obtained from the experiment. The SIMPLE scheme is used by the pressure-velocity coupling algorithm. We apply a second upwind line to the interpolation between the node and the node. The total numbers of nodes and elements in the fluid model were about 65,000 and 128,000, respectively. Each time step needed about 5e-5 seconds to run calculation.

The operating process of the compressor applied to the CFD analysis consists of compression, discharge, expansion and suction. Discharge occurs when the discharge valve opens when the pressure in the compression chamber reaches P_d. Likewise, suction is achieved by opening the suction valve when the pressure in the compression chamber reaches P_s. Where P_d and P_s are the pressures corresponding to the temperatures of the condenser and the evaporator for a particular operating condition of the refrigerator. Compression and expansion processes are polytropic processes. The valve is assumed to be a fully-open, fully-closed ideal valve. The actual suction port is eccentric in the central axis of the piston, so that the area is the same as the donut shape in order to reflect this in the axis-symmetry.

Table 1. Result comparison with and without gas force

	Cooling Capacity (W)	Compression Work (W)	C.O.P.
With gas force	116.6	36.3	3.21
Without gas force	199.5	62.7	3.18

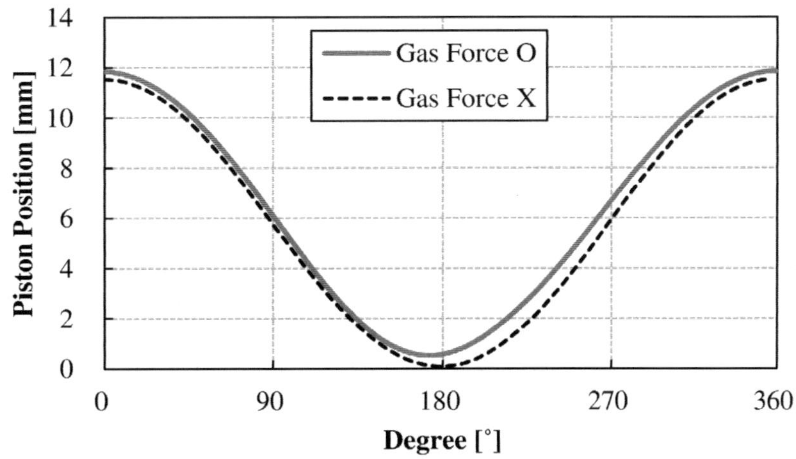

Figure 3. Stroke curve with and without gas force.

3. Result and Discussion

3.1. Effect of gas force on P-V diagram

In this section, we compare the case where the gas force acts and the case where the gas force does not act. In both cases, the stroke is calculated by the coupled two equations, the mechanical equations of the piston and the electrical equations of the motor, and depends on the presence or absence of forcing term f by the gas in the mechanical equation of the piston. And the same voltage was applied. Figure 2 is a pressure-volume (P-V) diagram for the compression chamber of a piston-cylinder. The P-V curve with a gas force is shifted to the right for a P-V diagram without a gas force as presented in Figure 2. We can see that the bottom dead centre (BDC) that starts compression and the top dead centre (TDC) that ends discharge move to the right. It can be seen that the characteristics of the linear compressor, in which the piston is subjected to the force in the opposite direction of compression by the gas force and the stroke is shifted, is well reflected by the analysis. As the P-V diagram changes, the cooling capacity and compression work also change, as shown in Table 1. When the gas force is applied, both the cooling capacity and the compression work are seen to decrease because the stroke is reduced by the gas force. Theoretically, even though the stroke changes, C.O.P., which is calculated using compression work, does not change, but it can be a necessary result to predict the maximum cooling capacity of a linear compressor through the above discussion. Due to the free piston, which is a characteristic of a linear compressor, pushing back occurs when a gas force is applied and piston can move a maximum stroke greater than twice the compression chamber distance corresponding to the initial position of the piston.

3.2. Effect of gas force on piston behaviour at same input voltage

Figure 3 shows the behaviour of the piston with and without gas force. The behaviour of the piston by the gas force is not perfect sinusoidal form. This is similar to the case where the angular velocity is not the same in the compression and expansion process of the reciprocating compressor. The compression process was slightly shorter and the expansion process was slightly longer as shown in Figure 3. This means that the gas force is reflected, the phase of TDC becomes faster. Also, the difference in TDC position is larger than BDC position. The centre positions of the strokes are 6.2 mm and 5.8 mm, respectively, when gas force is applied and when it is neglected. It shows the centre position is pushed BDC direction by 0.4 mm as the gas force acts. If the diameter of the pistons becomes larger or the spring stiffness becomes weaker, the centre position will be pushed more. Also, the larger the pressure differences between the condenser and the evaporator, the greater the pushed back amount. This can be an important factor in configuring the geometry of a linear compressor and can provide the information required to match the compressor and the refrigerator. When designing the initial position

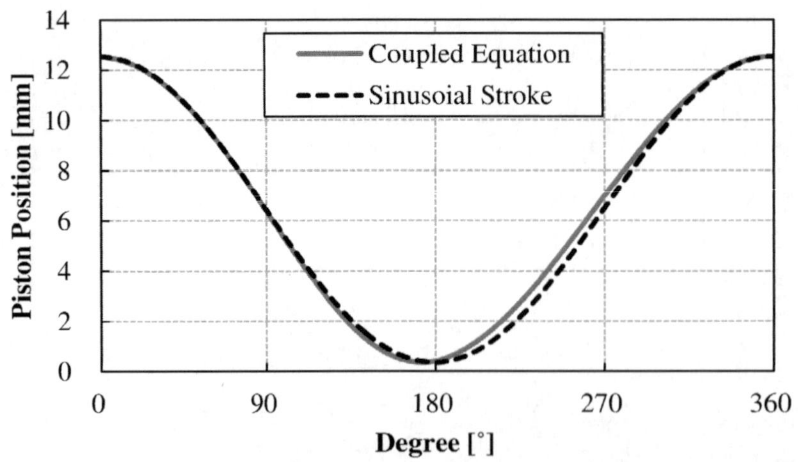

Figure 4. Piston positions about two cases. (calculated by coupled equations, and sinusoidal stroke)

of the piston and the stroke amplitude required for the maximum cooling capacity, it is necessary to reflect the amount of push back by the gas force, so that mechanical collision and reliability deterioration can be prevented.

3.3. Effect of gas force on piston behaviour at same amplitude stroke

In the previous section, we compared the results with and without the gas force on the same voltage in calculating stroke using coupled electrical and mechanical equations. In this section, the results are compared by giving a sinusoidal stroke waveform with the same amplitude as the stroke calculated by the two coupled equations, the mechanical equation and the electrical equation. In both cases, compression work, and C.O.P. were the same. In the compression chamber, if the piston positions are the same, pressure is the same due to the polytropic process, regardless of the stroke waveform. Figure 4 shows the stroke curves for two cases calculated by coupled equations, and sinusoidal stroke. While the TDC phase of the sinusoidal stroke is 180 degrees, the TDC phase of the coupled equation stroke is slightly faster. In addition, the waveform of the stroke obtained from coupled equation is distorted rather than a perfect sine form. In this paper, although the suction valve and the discharge valve are considered to be ideally opened and closed, the phase of the TDC is important considering the system that can reflect the actual valve behaviour. If the TDC phase becomes faster, the discharge valve will open quickly. And, as the phase of TDC reaches faster, both of the time required for backward movement of piston and required for suction becomes longer. This may affect the time at which the suction valve opens and closes. Designing the suction valve, it is necessary to consider the shift of the TDC phase, in particular to match the time of opening and closing the valve. In the future, the system reflected the actual valve behaviour will be further analyzed.

3.4. Motor efficiency calculation through electric work

The compression work can be obtained by the P-V diagram, when it is assumed that the motion of the piston is represented as a sinusoidal wave via CFD analysis of the linear compressor. The efficiency of the compressor with the compression work as input includes the following two types of loss. The first is the pressure drop caused by the structural reason and the flow loss due to the vortex in the refrigerant passage. The second is the heat loss due to the temperature of the boundary condition. This does not include motor loss. On the other hand, the electrical work can be estimated by the input voltage and the calculated current value, when it is assumed that the stroke is implemented by solving the mechanical and the electrical equations. The motor efficiency is 91.7% in the specification of the linear compressor applied to the analysis in this paper. If the stroke analysis method using the coupling equation is used, it is helpful to study the motor efficiency and specification for the optimum operating point.

4. Conclusion

In this paper, we introduce the technique of solving the mechanical equation of the piston and the electrical equation of the motor which are coupled through the Runge-Kutta method and setting the behaviour of the piston. This leads us to the following conclusions.

- The actual behaviour of the piston differs from the perfect sine curve, which should be reflected in the analysis.
- To give the behaviour of the piston including the force, should solve the mechanical equations of the piston and the electrical equation of the motor.
- The interpretation of the stroke shift allows for more accurate information on designing and optimizing the linear compressor.
- Computing and designing the maximum cooling capacity of the compressor without considering the gas force can cause mechanical collision and cause reliability problems.
- The electric work and the motor efficiency can be found by solving the electrical equation together.

Nomenclature

m	mass of system
c	total damping coefficient of system
k	stiffness of system
α	thrust constant
f	gas force
q	quantity of electric charge
L	inductance of motor
R	resistance of motor
C	capacitance of motor
V	input voltage

References

[1] Suh K, Heo D and Kim H 2006 *Int. Compressor Engineering Conf. at Purdue*
[2] Oh H, Gong S, Oh W and Park K 2016 *Int. Compressor Engineering Conf. at Purdue*
[3] Kim J K, Roh C G, Kim H and Jeong J H 2011 *Int. J. of Refrigeration 34* 1415-1423

IOP Conf. Series: Materials Science and Engineering **232** (2017) 012071 doi:10.1088/1757-899X/232/1/012071

Characteristics between the meshing pairs with different envelope profile in single screw compressors

R Huang[1]*, F Liu[2], T Li[3] and Q Feng[2]

[1] Institute of water Resources and hydro-electing Engineering, Xi'an University of Technology, Xi'an, Shaanxi, China
[2] School of Energy and Power Engineering, Xi'an Jiaotong University, Xi'an, Shaanxi, China
[3] College of Water Resources and Architectural Engineering, Northwest A&F University, Yangling, Shaanxi, China

E-mail*: huangruirxf@163.com

Abstract. Single screw compressors have been used in various industrial fields. However, because the star-wheel teeth are easy to wear, the market for the development of single screw compressors is limited. In order to extend the service life of the star-wheel, researchers have developed different kinds of star-wheel tooth profile, such as single line envelope profile, single column envelope profile, and multi-column envelope profile. These profiles greatly affect the lubrication characteristics between the star-wheel teeth and the screw grooves. In this article, the lubrication characteristics between the meshing pairs with different envelope profiles are analyzed. Results show that the pressure peak of the single line envelope profile, single column envelope profile, and multi-column envelope profile are 3.23×10^5Pa, 3.38×10^5Pa, and 4.31×10^5Pa, respectively. This means that the multi-column enveloped meshing pair can resist the biggest external impact load. The deviation angle (γ) of the single line envelope profile, single column envelope profile, and multi-column envelope profile are 0.0139°~0.0286°, 0.0225°~0.0306° and 0.0122°~0.0262°, respectively. Thus, the self-balancing ability of the multi-column enveloped meshing pair is the strongest, and the oil film thickness on both sides of the multi-column enveloped star-wheel tooth is the most reasonable, which indicates a good lubrication state during operation, that is, longer operation life of the star-wheel teeth.

1. Introduction

The single screw compressor (SSC) is a positive displacement rotary compressor used in various industrial fields, such as refrigeration, air/gas compression, and chemical engineering systems. A typical structure of a SSC is shown in figure 1, and the main components include two star-wheels, a screw rotor and a casing. As it can be seen from the figure 1, the two star-wheel rotors are symmetrical about the center axis of the screw. Because of the symmetrical structure, SSC has the advantages of rotor force balance, no clearance volume, high speed with small loading, simple structure. One star-wheel and the screw constitute a couple of meshing pair. The key technologies of a SSC are the meshing pair's profile design and manufacturing. If the design of the profile is not reasonable, the star-wheel tooth will be under high relative sliding speed between the tooth flank and the screw groove flank. At present, various of the meshing pair profiles have been published [1], but because of the restrictions of machining method, the widely used profiles are still single line envelope profile, single column envelope profile, and multi-column envelope profile [1] which was proposed in 2011 and developed with its manufacturing method in recent years. The three types of profiles are shown in figure 2.

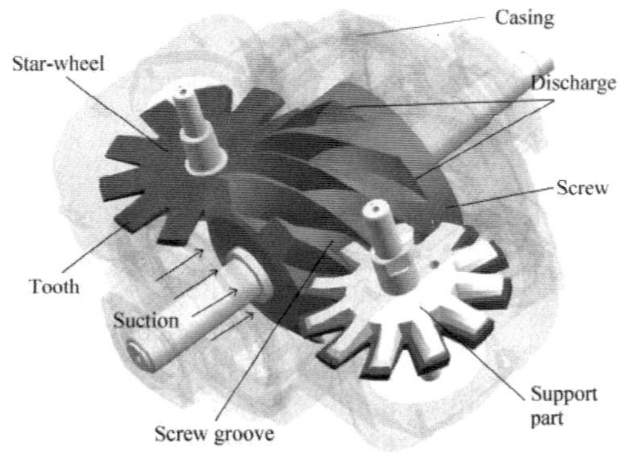

Figure 1. Typical structure of a single screw compressor.

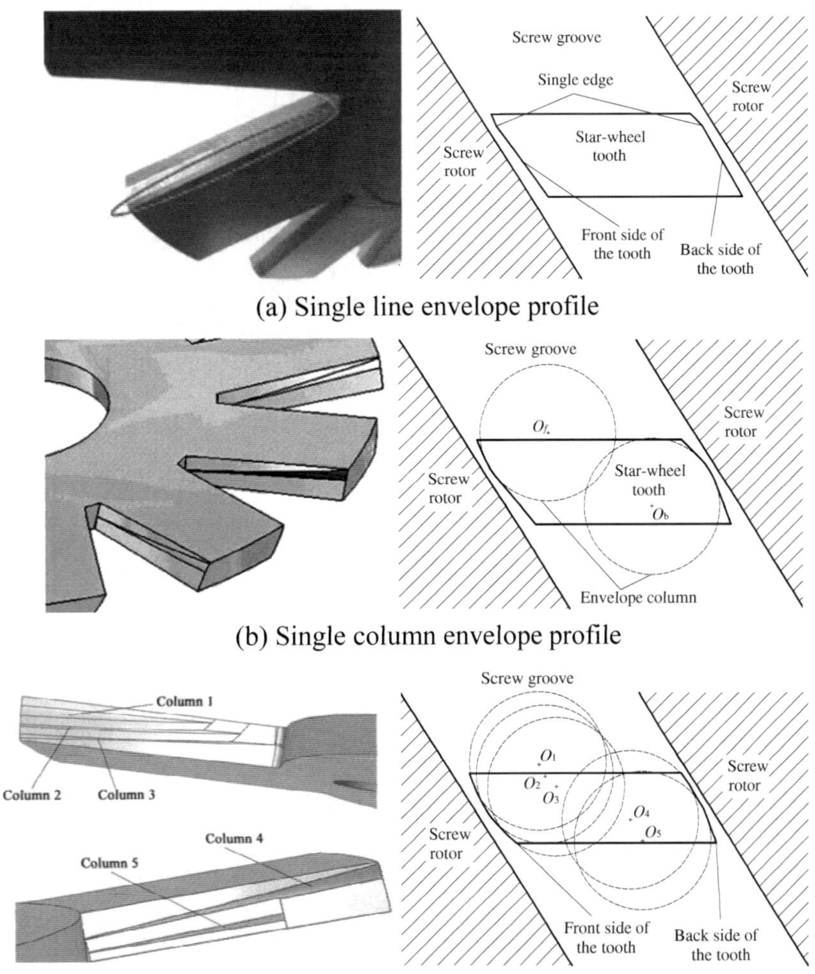

(a) Single line envelope profile

(b) Single column envelope profile

(c) Multi-column envelope profile

Figure 2. Three types of the meshing pair profiles.

IOP Conf. Series: Materials Science and Engineering **232** (2017) 012071 doi:10.1088/1757-899X/232/1/012071

According to the hydrodynamic lubrication theory, the geometry between the star-wheel tooth flank and the screw groove flank is directly related to the lubrication state between the meshing pair, and the lubrication state can greatly affect the wear resistance of tooth flank [2-4]. Therefore, it is very important and urgent to carry out the research on the lubrication characteristics between the meshing pairs with different envelope profile in SSCs.

In 1982, Jin Guangxi [5] recognized the importance of the hydrodynamic lubrication and used the Matin equation to calculate the minimum oil film thickness between the meshing pair. In 1986, Post and Ziwaan [6] studied the lubrication characteristics in single screw compressor, and they proposed that the fluid inertia must be considered in the oil film force calculation because the meshing pair has a high relative velocity. Post and Ziwaan used and recommended B.E Lauder sliding bearing lubrication calculation model (including the inertia term) [7] in their paper. Liu Yu [8] calculated the lubrication angle in a meshing period. He proposed that the lubrication angle varies from 80° to 90°, and the minimum value appears when the tooth of the star-wheel just engaged with the groove, the maximum value appears at the high pressure side of the compression chamber. The calculated results show that the hydrodynamic lubrication state of the high pressure side is always superior to the low-pressure side in SSCs. In recent years, some researchers mentioned that study on the hydrodynamic lubrication state can improve the meshing pair profile [9], but the work is not carried out in depth.

In this paper, B.E Lauder bearing lubrication calculation model is adopted, and the lubrication characteristic between the star-wheel tooth flank and the screw groove flank with different envelope profiles are compared and analyzed.

2. Oil film force analysis

The geometry between the meshing pair, shown in figure 3, is a long narrow channel. The narrow channel shows geometrical convergence at the oil inlet section, and shows geometric divergence at the oil outlet section. It is generally believed that the oil film pressure only appears in the convergence area of the channel, while in the divergent region the oil film collapsed [9]. Therefore, oil film pressure on the star-wheel tooth side is produced only in the convergence area, and the divergent region only bears the suction pressure of the SSC, because the divergent area of the tooth flank is exposed in the suction chamber. Besides the oil film pressure, the oil friction force, which caused by the shearing action of the lubricating oil, also acts on the tooth side. However, according to the literature [3], the frictional force is much smaller than the oil film force, so the friction force is ignored in the calculation process in this paper.

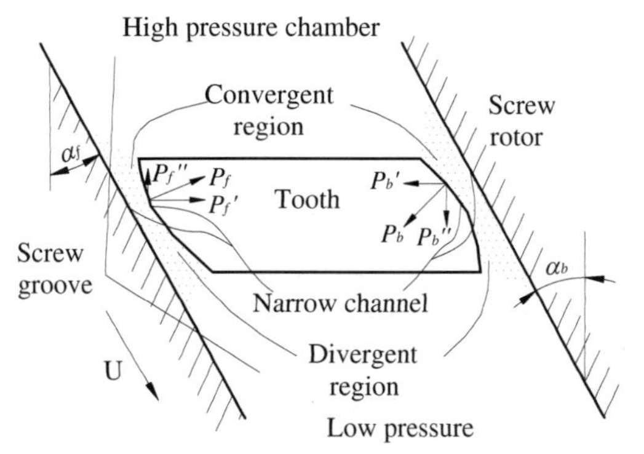

Figure 3. Cross-sectional view of the clearance between a tooth and a groove.

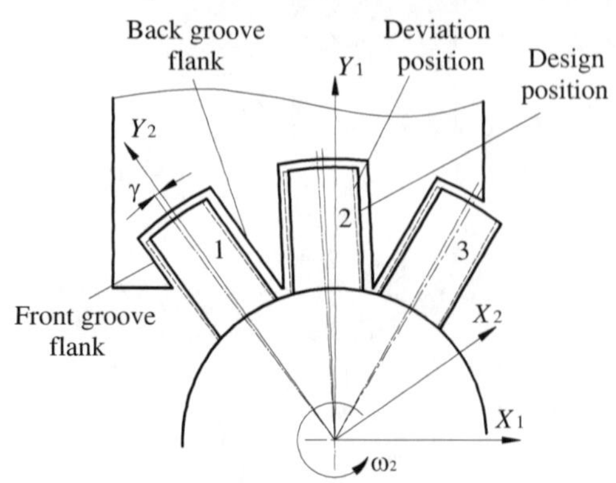

Figure 4. Deviation angle.

During meshing, both the front side flank and the back side flank of the tooth bear the oil film forces. Hence, theoretically, the total oil film torque acting on the tooth is the sum of the torques produced by the oil film pressure in the two sides narrow channels. Under the action of the oil film torque, the star-wheel tooth will be pushed to the direction which the oil film force is smaller, and then the tooth deviation occurs. The deviation of the tooth changes the geometry between the meshing pair, and then changes the oil film forces. When the sum of the oil film torques is equal to zero, the star-wheel tooth will stay in a balanced position. The balanced position is called deviation position in figure 4. Therefore, the deviation angle (γ), which between the tooth centerline when the tooth is in the middle of the groove (design position) and the centerline when the tooth is in the deviation position, is used to investigated the unbalanced degree of the oil film torques, shown in figure 4.

3. Lubrication model

B.E. Lauder sliding bearing lubrication model [7] is used in this paper, and the coordinate system is established in figure 5. Due to the fact that the tooth flank thickness is much smaller than the screw groove length, the cross-sectional curve of the screw groove can be set as the tangent line of the envelope column at the meshing point. As shown in the figure 5, the upper surface represents the star-wheel tooth flank, the lower surface represents the screw groove flank.

Because of the complex geometry shape between the meshing pair, the continuous variation of the sliding velocity and the inclusion of the inertial item, it is very difficult to solve the two-dimensional B.E. Lauder lubrication model. Model should be reasonably simplified. Compared the characteristic width (*B*) of the lubricating oil flow path, the tooth length is much longer. So, the tooth flank is divided into differential elements along the tooth length direction, and in any differential element, dP/dz is much smaller than dP/dx. It follows that the oil film pressure gradients in z direction are neglected. The distribution of oil film pressure in each differential elements is solved, then the pressure distribution over the entire tooth flank can be obtained.

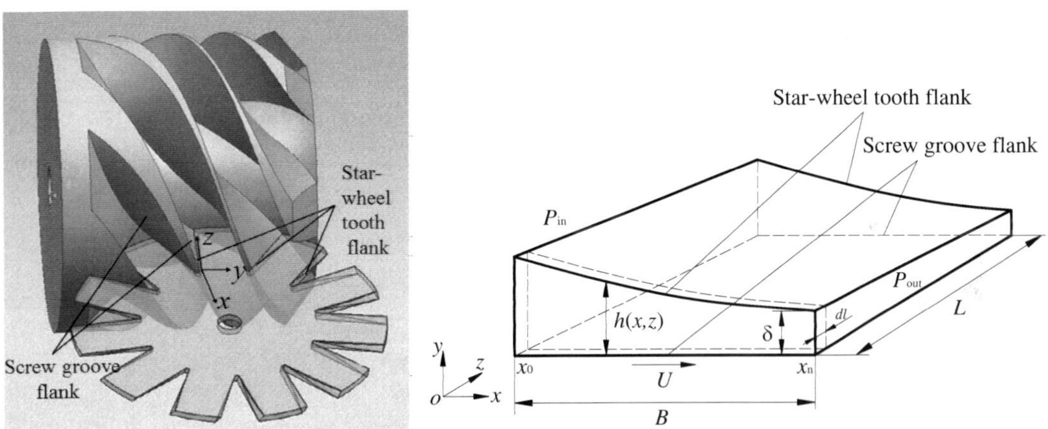

(a) Coordinate system in 3D geometry (b) Geometric parameters in the coordinate system

Figure 5. Geometrical configuration and coordinate system.

The one-dimensional B.E. Lauder lubrication model is expressed in the following equation:

$$\frac{dP(x)}{dx}=\frac{12\mu}{h(x)^2}\left(\frac{U}{2}-u_m\right)+\frac{\rho}{h(x)}\left(\alpha u_m^{\ 2}-\beta U^2\right)\frac{dh(x)}{dx} \quad (1)$$

where, α=1.2; γ=0.133; $u_m=\frac{1}{\rho h}\int_0^h u dx$; $h(x)$ is a function of the minimum film thickness δ and the profile design parameters. δ can be derived from the deviation angle γ (refer to the equation (11) and equation (12)).

The continuity equation is expressed as follows:

$$\frac{d}{dx}\left[\rho u_m h(x)\right]=0 \tag{2}$$

where the term of $\rho u_m h(x)$ is not varying with x, so it was set to be a constant value M.

Substituting equation (2) into equation (1) and simplifying:

$$\frac{dP(x)}{dx}=\frac{6\mu U}{h(x)^2}-\frac{12\mu M}{\rho h(x)^3}+\alpha\frac{M^2}{\rho h(x)^3}\frac{dh(x)}{dx}-\frac{\rho}{h(x)}\beta U^2\frac{dh(x)}{dx} \tag{3}$$

The boundary condition is expressed as:

$$\begin{cases} x=x_0 & P(x)=P_{in} \\ x=x_n & P(x)=P_{out} \end{cases} \tag{4}$$

where P_{in} is equal to the instantaneous gas pressure $P(\phi_2)$ in the compressing chamber. P_{out} is the suction pressure in the suction chamber, because the divergent area of the tooth flank is exposed in the suction chamber.

The pressure distribution along the groove surface is derived by integrating equation (3):

$$P(x)=P_{out}+\int_{x_0}^{x}\frac{6\mu U}{h(x)^2}dx-\frac{12\mu M}{\rho}\int_{x_0}^{x}\frac{1}{h(x)^3}dx-\alpha\frac{M^2}{2\rho}\left(\frac{1}{h(x)^2}-\frac{1}{h(x_0)^2}\right) \\ -\rho\beta U^2\left[\ln h(x)-\ln h(x_0)\right] \tag{5}$$

where,

$$M=\frac{\frac{12\mu}{\rho}\int_{x_0}^{x_n}\frac{1}{h(x)^3}dt-\sqrt{\left(\frac{12\mu}{\rho}\int_{x_0}^{x_n}\frac{1}{h(x)^3}dt\right)^2+\frac{2\alpha}{\rho}\left(\frac{1}{h(x_n)^2}-\frac{1}{h(x_0)^2}\right)N}}{\frac{\alpha}{\rho}\left(\frac{1}{h(x_0)^2}-\frac{1}{h(x_n)^2}\right)}$$

$$N=\int_{x_0}^{x_n}\frac{6\mu U}{h(x)^2}dx-\rho\beta U^2\left[\ln h(x_n)-\ln h(x_0)\right]+P_{out}-P(\phi_2)$$

The oil film pressure $P(x)$ calculated by equation (5) is perpendicular to the star wheel tooth flank. However, the calculation of oil film moment only need to consider the pressure component (refer to figure 3, P_f' or P_b') which parallel to the upper tooth surface. In this paper, $P(x)$ is decomposed, and the component pressure is integrated along the tooth thickness, then the oil film forces can be obtained:

$$F_{oij}'=\cos\alpha\cdot\int_{x_0}^{x_n}P(x)dx \tag{6}$$

where F_{oij}' is the oil film forces acting on a differential element of the tooth flank.

It can be seen from figure 6, in most cases, only part of the tooth flank involved in the engagement during meshing, while the remaining tooth flank is exposed in the suction chamber. Therefore, the forces acting on the tooth flank can be divided into three parts: the oil film pressure caused by the lubricant film in the convergent region, the suction pressure on the engaged tooth flank in the divergent region, and the suction pressure on the unengaged segment flank. Thus, the torque on a tooth is the sum of the torques produced by the pressures mentioned above.

The torques acting on the front flank and the back flank of a tooth can be expressed respectively:

$$T_{fi}(\phi_2)=\int_{l_{fi}(\phi_2)}^{L} F_{foij}'\cdot r_{2j}(l)dl+\int_{l_{fi}(\phi_2)}^{L} F_{fgi}\cdot r_{2j}(l)dl+\int_{0}^{l_{fi}(\phi_2)} F_{fgi}'\cdot r_{2j}(l)dl \quad (7)$$

$$T_{bi}(\phi_2)=\int_{l_{bi}(\phi_2)}^{L} F_{oijb}'\cdot r_{2j}(l)dl+\int_{l_{bi}(\phi_2)}^{L} F_{bgi}\cdot r_{2j}(l)dl+\int_{0}^{l_{bi}(\phi_2)} F_{bgi}'\cdot r_{2j}(l)dl \quad (8)$$

where $l_{fi}(\phi_2)$ and $l_{bi}(\phi_2)$ is the unengaged segment length.

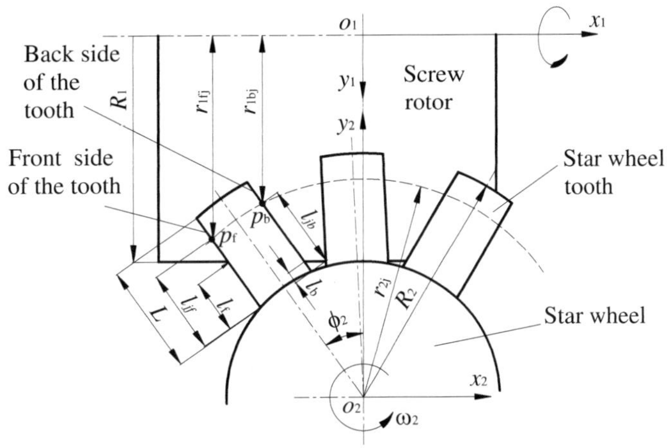

Figure 6. Geometric relations between star-wheel teeth and screw rotor.

The total torque on a tooth is indicated as follows:

$$T_i(\phi_2)=T_{fi}(\phi_2)+T_{fi}(\phi_2) \quad (9)$$

In this article, the torque acting on the front tooth flank is set to be negative, and the torque acting on the back tooth flank is set to be positive.

During the operation, three or four teeth of the star-wheel simultaneously participate in the engaging. Therefore, the sum of T_i (composite torque) should be obtained as follows:

$$T(\phi_2)=\sum_{1}^{n} T_i(\phi_2) \quad (10)$$

where, $n = 3\sim4$.

When the star-wheel tooth is operated at the design position, the calculation flow chart of $P(x)$ and $T(\phi_2)$ is shown in figure 7.

If the composite torque $T(\phi_2)$ is not equal to zero, the star-wheel tooth will deviate from the design position until the composite torque is balanced. The calculation flow chart of the deviation angle γ is shown in figure 8.

When the deviation angle γ is calculated, the oil film minimum thickness of each elements along the tooth flank can be obtained as follows:

$$\delta_f = \cos\alpha_f \left(\sigma - \frac{\gamma \cdot \pi}{180} \cdot r_{2j}\right) \tag{11}$$

$$\delta_b = \cos\alpha_b \left(\sigma + \frac{\gamma \cdot \pi}{180} \cdot r_{2j}\right) \tag{12}$$

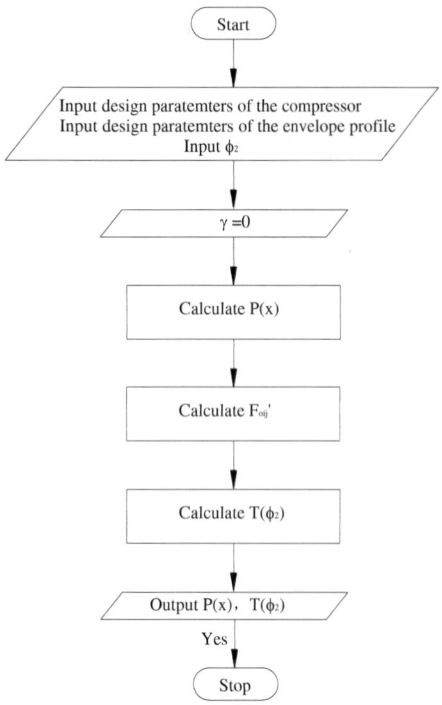

Figure 7. Calculation flow chart of $P(x)$ and $T(\phi_2)$.

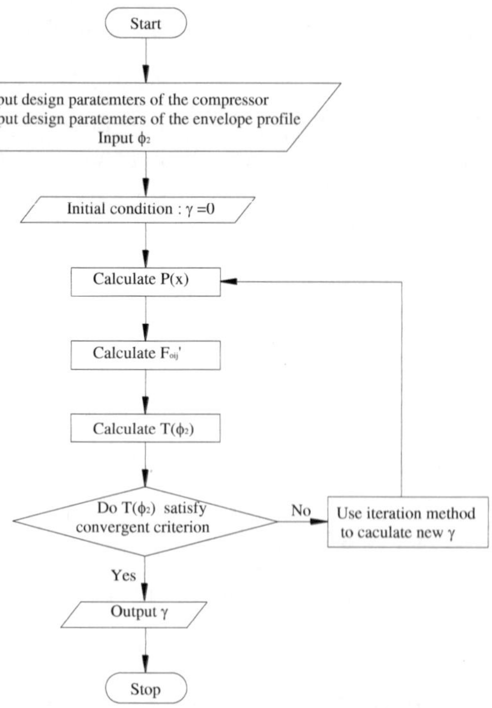

Figure 8. Calculation flow chart of deviation angle γ.

4. Calculation results

In this paper, the meshing pairs with different profiles in an oil-flooded SSC with 2.6 m^3/min displacement was selected as the study object. The major design parameters of the SCC are:

Table 1. Dimensions of the sing screw compressor.

Parameter	Value	Parameter	Value
Diameter of screw rotor d_1/mm	136	Tooth length L/mm	35
Diameter of star wheel d_2/mm	148	Central distance of the compressor A/mm	108
Tooth width b/mm	20	Discharge pressure P/ 10^5 Pa	8
Design clearance width σ/mm	0.04	Motor rotation speed /r·min^{-1}	2970

In this compressor, when the star-wheel tooth rotation angle ϕ_2 is 12.7°[1], the compression chamber instantaneous pressure is 2.88×10^5Pa, and star-wheel tooth is in the middle of the screw groove, the oil film pressure distribution along the front tooth flank (t denotes the tooth thickness direction, and l denotes the tooth length direction) with different envelope profiles are shown in figure 7, where l=0 mm represents the tooth root, l=35 mm represents the tooth tip, t=0 mm represents the lower surface of the tooth, t=8 mm represents the upper surface of the tooth.

When the star-wheel tooth rotation angle ϕ_2 is 12.7°, the unengaged segment of the tooth is 0~4.4 mm, and the engaged segment is 4.4~35 mm. The convergent region of the engaged part of the tooth flank bears oil film pressure, the remaining part of the tooth flank (both engaged part and unengaged part) bears suction pressure. It can be seen from figure 9, the pressure peak of the single line envelope profile, single column envelope profile, and multi-column envelope profile are 3.23×10^5Pa, 3.38×10^5Pa, and 4.31×10^5Pa, respectively. It shows that the multi-column envelope profile can establish larger hydrodynamic oil film pressure peak than the other two profiles.

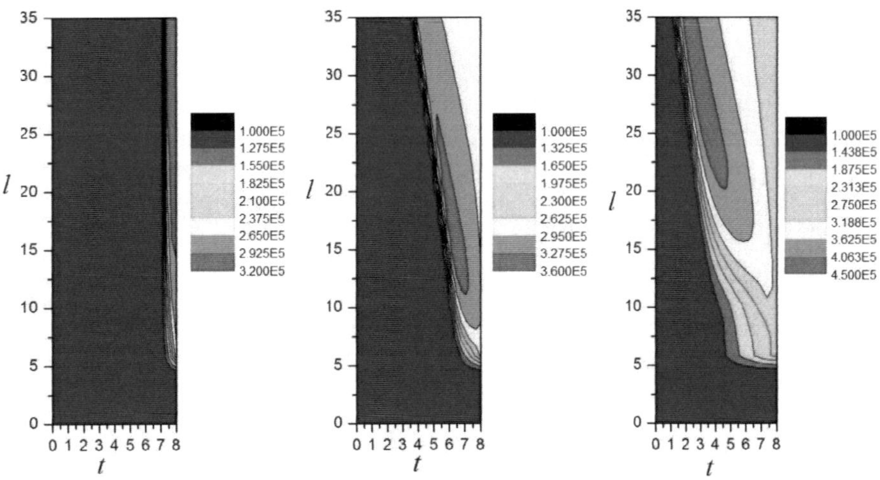

(a) (b) (c)

Figure 9. Oil film pressure distribution along the tooth flanks with different profiles. (a) single line envelope profile; (b) single column envelope profile; (c) multi-column envelope profile.

In addition, the single line envelope profile have the smallest hydrodynamic oil film action area, while the multi-column envelope profile have the largest oil film action area. Therefore, the oil film at the multi-column envelope profile flank can generate the largest bearing capacity.

[1] In this compressor, when the star-wheel tooth rotation angle is about 12.7°, the meshing points of the front tooth flank with multi-column envelope profile is on the surface of column 3, then the difference of the oil film pressure distribution of the three kinds of profiles are most obviously.

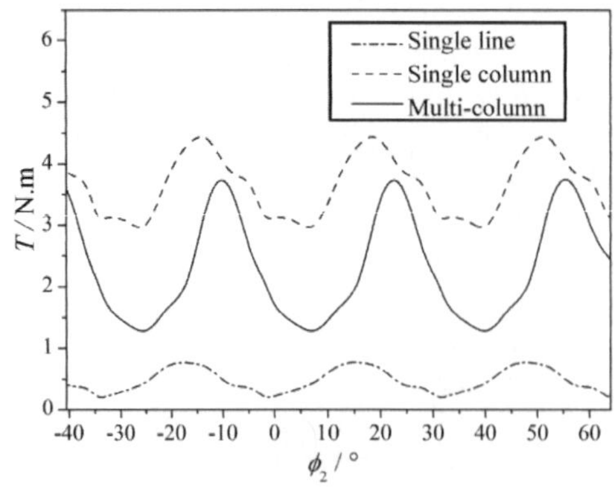

Figure 10. Composite oil film torques of the star-wheel teeth with different profiles.

It can be seen from figure 10, the composite oil film torque of the single line envelope profile, single column envelope profile, and multi-column envelope profile are 0.22 N·m~0.77 N·m, 2.92 N·m~4.5 N·m, 1.3 N·m~3.7 N·m, respectively. The composite oil film torques of the single line envelope profile is the most balanced, and that of the single column envelope profile is the most unbalanced. At the beginning of the multi-column envelope profile's design, measures to enhance the oil film torques on the front tooth side were taken into account. Two columns surfaces were added to the front side flank of the tooth to increase the oil film action area, which effectively reduce the composite oil film torque's imbalance.

Figure 11 shows the variations of γ when the composite torques are balanced in a meshing period. The γ of the single line envelope profile, single column envelope profile, and multi-column envelope profile are 0.0139°~0.0286°, 0.0225°~0.0306°, 0.0122°~0.0262°, respectively. It shows that the self-balancing ability of the multi-column envelope profile is the strongest, and it can easily balance the composite torques through adjusting the tooth deviation angle γ. The self-balancing ability of the single line envelope profile and single column envelope profile is relatively poor, and it requires a large adjustment of the tooth deviation angle to make the torques balanced.

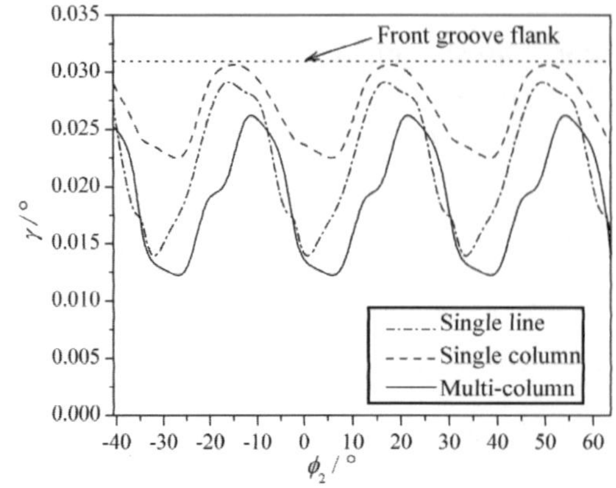

Figure 11. Variations of γ when the composite torques are balanced.

5. Conclusions

In this paper, B.E. Lauder bearing lubrication model is used to analyze the lubrication characteristics of meshing pairs with single line envelope profile, single column envelope profile and multi-column envelope profile, and the following conclusions can be obtained.

1) Compared with the single line envelope profile and the single column envelope profile, the oil film pressure action area and the peak pressure distribution of the multi-column envelope profile

IOP Conf. Series: Materials Science and Engineering **232** (2017) 012071 doi:10.1088/1757-899X/232/1/012071

is the largest, which indicates that the oil film torque is the largest too. In other words, the oil film bearing capacity of the multi-column enveloped tooth is the biggest, which can resist more stronger impact load from the screw rotor to the star wheel teeth.

2) The calculation results of the composite oil film torques and the deviation angle γ show that the single line enveloped tooth has the smallest composite oil film torques and weak self-balancing ability, the single column enveloped tooth has the largest composite oil film torques and the weakest self-balancing ability, while the multi-column enveloped tooth has middle composite oil film torques and the strongest self-balancing ability. It is illustrated that the meshing pair with multi-column enveloped profile can make both side of the star-wheel teeth operated under a good hydrodynamic lubrication state, thus reducing the wear of the star-wheel teeth and prolonging the teeth operation life.

Appendix

Nomenclature

P	Pressure (Pa)
P_{in}	Lubricating oil inlet pressure (Pa)
P_{out}	Lubricating oil outlet pressure (Pa)
$P(\phi_2)$	Instant gas pressure in compressing chamber (Pa)
U	Relative sliding velocity (m/s)
T	Torque (N·m)
u_{m}	Axial mass-mean velocity (m/s)
Fo	Oil film force (N)
Fg	Gas force (N)
L	Length of the star-wheel tooth
B	Distance from the lubricating oil inlet to the meshing point (mm)
R_1	Radius of the screw rotor (mm)
R_2	Radius of the star-wheel (mm)
p	Instant meshing point on arbitrary cross section along the length of the tooth
$r_{2\mathrm{j}}$	Distance from a meshing point p_f (p_b) to the star-wheel rotation centre
h	Oil film thickness (mm)
r	Radius (mm)
b	Tooth width (mm)
t	Tooth thickness (mm)
Subscripts	
i	Different tooth
j	Meshing point in every element of the tooth flank
b	Back side of the tooth
f	Front side of the tooth
o	Lubricant
g	gas
Greek symbols	
γ	Deviation angle (°)
μ	Dynamic viscosity (Pa·s)
ρ	Density of the lubricant (kg/m^3)
φ	Rotation angle (°)
α	Groove inclined angle (rad)
δ	Minimum oil film thickness (mm)

IOP Conf. Series: Materials Science and Engineering **232** (2017) 012071 doi:10.1088/1757-899X/232/1/012071

References

[1] Cun Z, Weifeng W, Quanke F *et al* 2011 *Forming method of the non-equal multi-column tooth profile in single screw compressors* 2011. 201110314977.7
[2] Zhenglin L 2009 *Tribological theory* (Beijing: Higher Education Press)
[3] Dacheng S 2009 *Lectures on lubrication mechanics* (Beijing: China Friendship Publishing Company)
[4] Kanters AFC, 1990 *On the caculation of leakage and friction of reciprocating elastometric seals* (AZ Eindhoven, Netherlands: Technische Universiteit Eindhoven)
[5] Jin G X 1982 Study of the profile and hydrodynamic lubrication of a mono-screw compressor *Journal of Xi'an Jiaotong University* **16**(06) pp 75–83 (in Chinese)
[6] Post W and Swaans M 1986 Computer simulation of the hydrodynamic lubrication in a single screw compressor *Proceedings of Purdue compressor technology conference* pp 334–48
[7] Lauder B E and Leschziner M 1978 Flow in finite-width, thrust bearings including inertial effects I-laminar flow *Transactions of the ASME Journal of Lubrication Technology* **100**(7) pp 330–38
[8] Liu Y 2007 Research on the meshing pairs lubricating characteristic of a single screw compressor, based on I-DEAS *Compressor Technology* **202**(2) pp 23–25
[9] Elsharkawy Abdallah A., 2001 On the hydrodynamic liquid lubrication analysis of slider/disk interface *International Journal of Mechanical Sciences* **43** pp177–92

Investigation on the electromagnetic centring technique in compressor with labyrinth seal structure

W Zhang, C Feng, J Cheng, Q Feng and W Wu

Xi'an Jiaotong University, Xi'an 710049 China

E-mail: weifengwu@xjtu.edu.cn

Abstract. At present, the piston of compressors with labyrinth seal structure generally runs eccentrically, which causes uneven radial clearance, serious leakages and lower volumetric efficiency. This has become an urgent problem in the development of labyrinth compressors. In this study, electromagnetic levitation technology was introduced to achieve concentric centering between the piston and cylinder, and the conventional cantilever structure for the piston centering was replaced by a simple support structure using the through-piston rod. Furthermore, the simulation model of the electromagnetic centering system was established and the experimental prototype was built. The mathematical simulation model was verified by comparing simulated and tested results. Then, the centering effect of the system was assessed and the variation of the leakage in the compressor was studied by models using dynamic mesh technology. The results showed that the radial clearance between piston and cylinder can be maintained in the range of -0.3 mm to 0.3 mm through the electromagnetic centering control. In addition, the inner leakage of the compressor was quite appreciable without the electromagnetic control. However, it was reduced by 1.8 times with the introduction of the electromagnetic control. Thus, it can be concluded that the precise centering between the piston and the cylinder can be achieved by the introduction of the electromagnetic centering technique.

1. Introduction

In a labyrinth piston compressor, the labyrinth seal is adopted between piston and cylinder mirror, piston rod and cylinder-seat [1,2]. This sealing structure not only can ensure the compressed gas oil free, but also avoid the production of dust and debris in the cylinder. More importantly, if the gas is flammable and explosive, or toxic, the compressor has high safety and reliability [3]. Currently, with the development of the petroleum and chemical industry, the labyrinth piston compressor has been widely used in oil refining, pharmaceutical, metallurgy, military and other industries.

Because of using labyrinth seal in the process of design and manufacture of reciprocating compressor, the serious inner leakage and low volumetric efficiency are the major obstacles for wide application of the labyrinth piston compressor. The design of the labyrinth structure, the eccentric running of the piston, the design of radial clearance between piston and cylinder wall and other factors can cause the labyrinth piston compressor serious inner leakage [4,5]. Among of them, maintain coaxial centering between piston and cylinder is the most critical problem. In the current labyrinth piston compressor, the piston is always guided by the crosshead and the guide bearing which causes piston's eccentrically running and friction between the piston and cylinder in the actual working process [6]. Therefore in order to avoid friction between piston and cylinder wall, the larger value of radial clearance between piston and cylinder wall should be taken. But the larger value of radial clearance between piston and cylinder caused the compressed gas serious inner leakage which caused the volumetric efficiency low and made a great waste of energy [7].

In order to solve these problems, many scholars have done a lot of research related to the labyrinth's structure, the piston shape, and installation position of the guide bearing, etc. But the problems of piston eccentric operation and lower volumetric efficiency of the compressor still has not been effectively resolved [8~12].

To solve these problems, through an in-depth and detailed analysis, this thesis try to adopt the reciprocating magnetic levitation technology to realize the concentric centering of the labyrinth piston. So that the clearance between the piston and the cylinder mirror can be remarkable reduced without scraping, and the inner leakage will be greatly decreased and the volume efficiency of the compressor will be greatly improved. By looking up a large number of the relevant research information, the magnetic levitation technology has been successfully applied in turbine machinery, but there are still no reports on the application of magnetic levitation technology to the reciprocating machinery [13~15]. The research topic of this thesis not only relates to the radial movement of the piston, but also related to the steady flow in the labyrinth clearance, and it is more related to the study on nonlinear electromagnetic dynamics of reciprocating magnetic levitation technology.

The main research works and main results obtained in this thesis are as follows: the electromagnetic levitation technology was first used to realize concentric centering between the piston and cylinder and the transfer function model of the centering systems was derived and the simulation model of the electromagnetic centering system was established by matlab/simulink. And the traditional cantilever structure for the piston centering was transform into the simple support structure by using the through-piston rod; Considering the clearance between crosshead and slide-way, piston rod and guide bearing, the radial deflection equation of the piston with crank angle was deduced through the movement relationship of the piston and the piston rod system; In order to verify the model correctness and centering effect of the electromagnetic centering system, an experimental platform of the electromagnetic centering system based on the core of DSP was designed and built; An unsteady flow-field and leakage analysis model of multi-labyrinth was established by the commercial FEM software, and the modified turbulence model (RNG)k-ε was applied to investigate the changing regularity of the pressure distribution and flow pattern.

2. Theories and models of the investigation

2.1. Electromagnetic guiding theory

At present, the cantilever structure of labyrinth piston compressor is main used to suppressing the piston's eccentric running. The radial clearance exists between the crosshead and the slide-way, also between the guide bearing and the piston rod. As mentioned above, it is very difficult for the proposed cantilever structures to realize the piston's precision centering. In order to solve this problem, an electromagnetic guiding system is introduced in this paper.

The new structure adopts the through-piston rod as shown in figure 1. This kind of structure can ensure that the piston and piston rod with the same radial eccentric motion law. On the top of the cylinder set up the radial magnetic orientation structure around the guide bar to restrain and control the guide bar radial eccentric movement. The improved structure can not only realize coaxial centering of the piston and piston rod but also simplifies the traditional cantilever structure.

Electromagnetic centering control mode is heavily dependent on the attraction of the electromagnetic coil on the guide bar, and according to the actual location of the guide bar, energized by adjusting the control current of coil to control the size of the force on the guide bar.

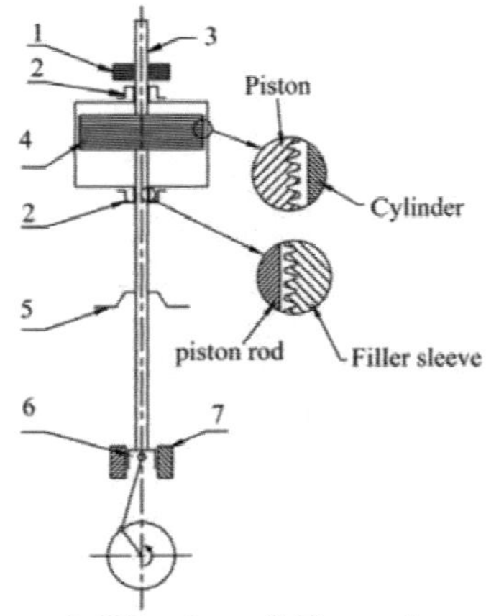

1. Guiding sleeve 2. Filler sleeve 3. Upper piston rod 4. Piston
5. Oil retainer 6. Crosshead 7. Slide-way

Figure 1. The diagram of piston centering by electromagnetic guiding system.

The active electromagnetic control system adopt in this paper is mainly composed of displacement sensor, power amplifier, PID, the electromagnetic coil, controlled object, etc. They form a closed loop system. Displacement sensor is mainly used for real-time detection of controlled object, feedback control signal to control system. PID is mainly to control the operation and output feedback signal. According to the different functions, the design of electromagnetic centering control system includes the design of the electromagnetic actuator and control system design. Electromagnetic actuator mainly includes power amplifier and the radial electromagnetic coil two parts as shown in figure 2. The reasonable parameters design of radial electromagnetic coil structure is one of the prerequisites to ensure that bearing capacity and realize stable suspension of the object. According to the guide bar's motion characteristics, the axial reciprocating motion do not control and there are two degrees of freedom need to control in the radial direction of the guide bar. In order to realize the active control in the direction of the degrees of freedom, every radial degree of freedom needs a set of actuators, which makes the guide bar by electromagnetic attraction on positive and negative direction in each radial degree of freedom. In this article, the electromagnetic coil using the structure of circular groove and the round wire, as shown in figure 3. There are eight poles in electromagnetic coil and the adjacent two make a group. Therefore, the electromagnetic coil is actually made up of four pole pairs and there are two pole pairs in the direction of each degree of freedom.

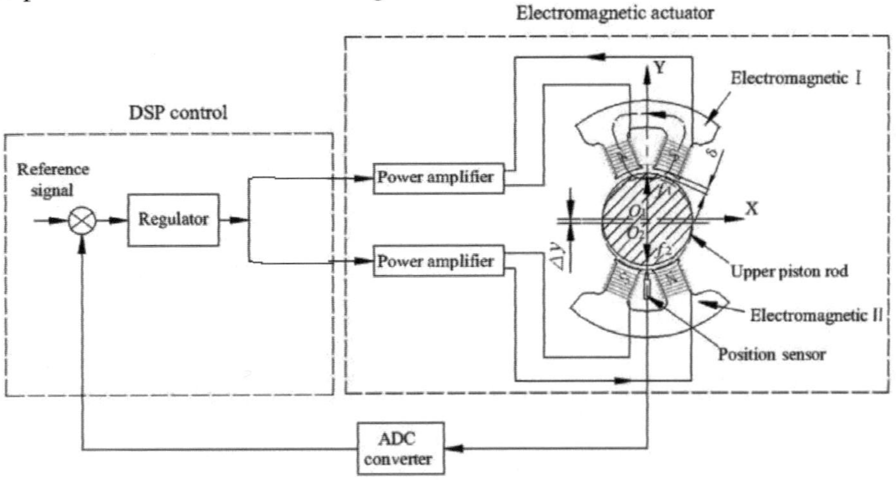

Figure 2. The working principle of electromagnetic control system.

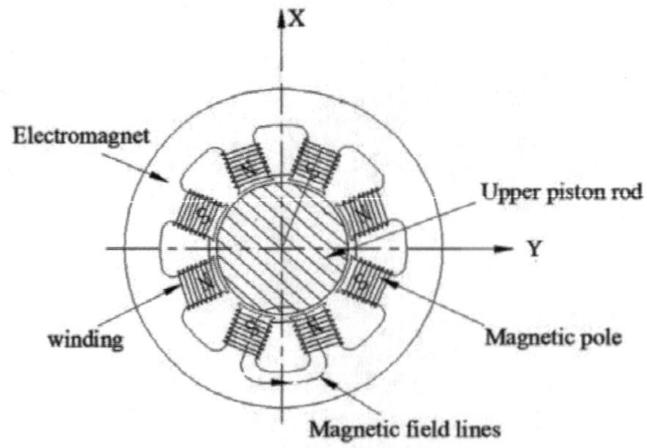

Figure 3. The radial electromagnetic coil section.

Assume the air gap of magnetic coil is uniform, according to the field theory and virtual displacement principle, the electromagnetic force is given in equation 1. Where N is the coil number of turns, I is the current, δ is air gap. According to the equation 1.

$$f = \frac{\partial W}{\partial \delta} = \frac{B^2 A}{\mu_0} = \frac{1}{4}\mu_0 N^2 A \cdot \frac{i^2}{\delta^2} \tag{1}$$

2.2. The leakage analysis model

In labyrinth compressor, the labyrinth contains two parts, one part is a labyrinth seal between the piston and cylinder, the other is a labyrinth packing between the piston rod and seal. The two parts of the sealing principle is the same. When leaking gas flow within the labyrinth channel, its axial flow speed is greater than the circumferential speed along ring groove. As a result, the circumferential flow's influence on the leakage gas can be neglected.

According to the labyrinth symmetric structure, this paper established a two-dimensional labyrinth leakage flow field analysis model by using FLUENT software as shown in figure 4. Due to the numerous model of labyrinth chamber, each chamber will produce eddy, in order to guarantee the accuracy of the calculation results, this paper USES the (RNG) k - ε turbulence model as given in equation 2. This model consider turbulent eddies and medium viscous force in the standard model of energy dissipation equation k - ε, improve the calculation accuracy.

In the process of calculation, the finite volume method is adopted to discrete calculation equation, and mass conservation, momentum conservation equation and energy conservation equation using second order format.

IOP Conf. Series: Materials Science and Engineering **232** (2017) 012072 doi:10.1088/1757-899X/232/1/012072

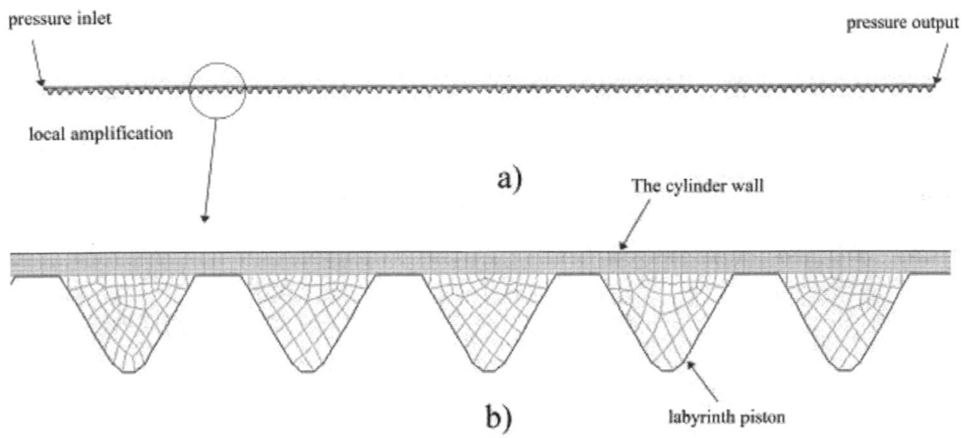

Figure 4. The finite element model.

$$\begin{cases} \frac{\partial}{\partial t}(\rho k)+\frac{\partial}{\partial x_i}(\rho k_i \mu_i)=\frac{\partial}{\partial x_i}\left[\partial_k \mu_{eff}\frac{\partial k}{\partial x_j}\right]+G_k+G_b-\rho\varepsilon-Y_M+S_k \\ \frac{\partial}{\partial t}(\rho\varepsilon)+\frac{\partial}{\partial x_i}(\rho\varepsilon\mu_i)=\frac{\partial}{\partial x_i}\left[\partial_k \mu_{eff}\frac{\partial k}{\partial x_j}\right]+C_{1\varepsilon}\frac{\varepsilon}{k}(G_k+C_{3\varepsilon}G_b)-C_{2\varepsilon}\frac{\varepsilon^2}{k}-R_\varepsilon+S_\varepsilon \end{cases} \quad (2)$$

3. Results and discussion

At present, the mainly method to analyze the electromagnetic field is finite element method, which basic idea is discrete the model calculation area fist, then joint variation principle and piece-wise interpolation method to calculate maxwell's equations of each discrete area boundary conditions. In this article, we use the software ANSOFT to analyze the electromagnetic field of radial electromagnetic coil. To simplify the calculation, this paper ignores the magnetic flux leakage and built two-dimensional finite element model of electromagnetic coil according to the established model symmetry, which is shown in figure 5. Figure 6 shows the electromagnetic coil magnetic field lines distribution.

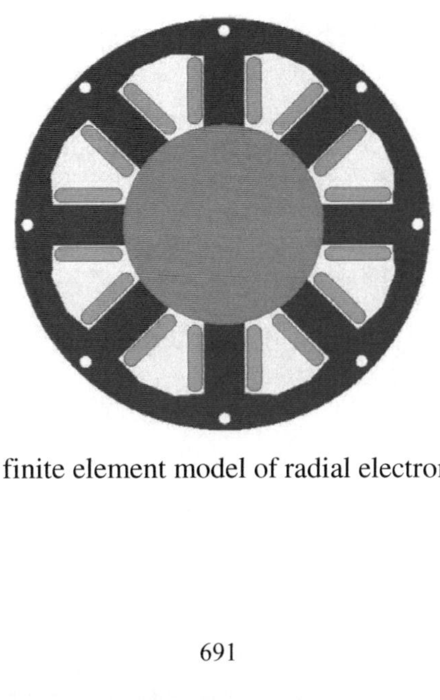

Figure 5. The finite element model of radial electromagnetic coil.

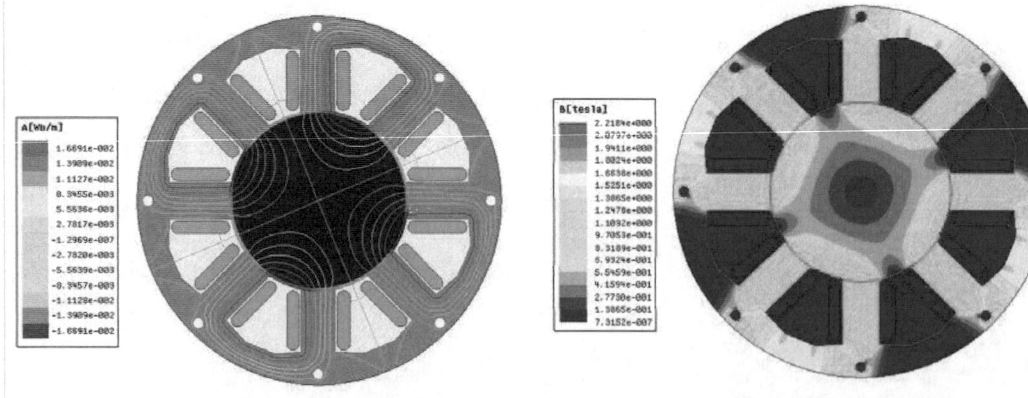

The distribution of magnetic filed lines   The distribution of Magnetic induction intensity

Figure 6. The distribution of magnetic characteristics.

In order to improve the dynamic characteristics of electromagnetic centering control system, this article introduce the variable speed into control system and take variable speed as the intermediate variable design the multivariable nonlinear decoupling feedback control strategy. Form the current, speed, displacement of three closed loop control system as shown in figure 7. In order to further test the electromagnetic centering control system reliability, we set up a test bench. We tested the current loop, speed loop and displacement loop of control system to make sure the correctness of the mathematical model. First of all, 1A was given to current loop and recorded the output current waveform, then compare with the actual result. The result shows in figure 8. Experiment and simulation results of overshoot amount are around 16% and adjust time is about 8 ms. There steady-state value both reached the expected ones and fit better. As shown in figure 9, after applying external disturbances on the controlled object, the given speed change and decay over time until to reach steady state due to the changes of radial displacement. In the whole process, the actual speed was controlled by displacement's difference, and always can be in accordance with the given speed. The figure 10 shows that the experiment and simulation results of the displacement of controlled object fit well under the action of external force.

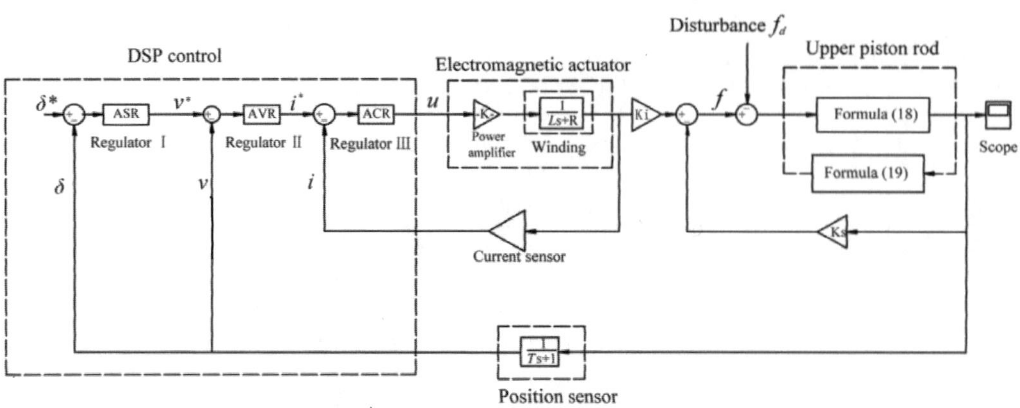

Figure 7. The simulation model of electromagnetic guiding system.

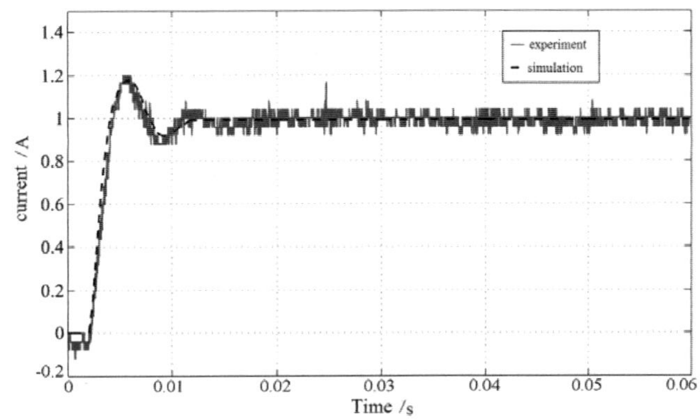

Figure 8. The step response characteristic of current closed loop

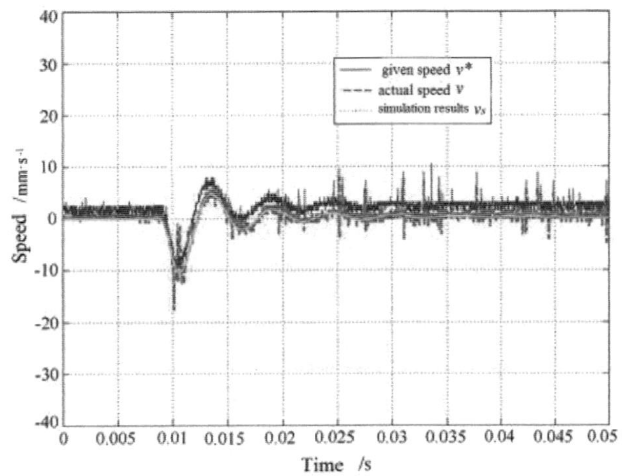

Figure 9. The curve of response of speed

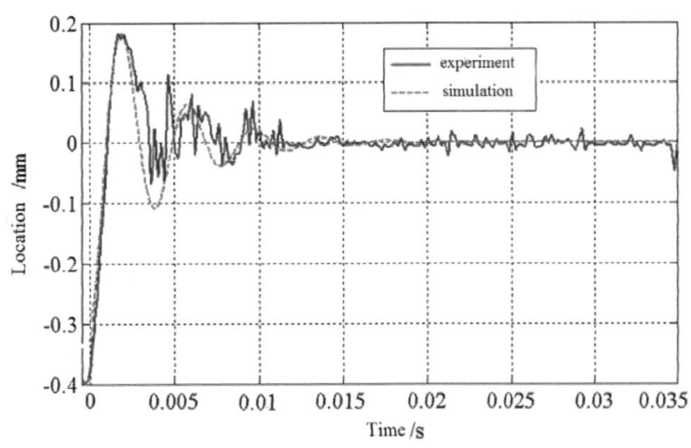

Figure 10. The curve of response of displacement

Based on the dynamic mesh technique and regarded the eccentric trajectory as the boundary conditions of piston motion, this paper analyzed the leak amount change of the fourth column piston before and after the centering. The curve of the leakage before and after the electromagnetic center under the different pressure ratio is shown in figure 11. The leakage reduced 1.8 times after applying electromagnetic centering. On the premise of applying electromagnetic centering, the leakage ratio of 0.1 mm clearance is relatively reduced more than 80% compared with 0.3 mm clearance.

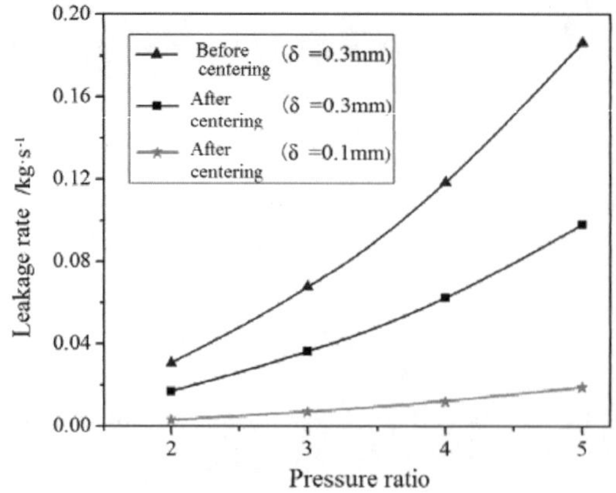

Figure 11. The leakage before and after centering

4. Conclusion

In a labyrinth piston compressor, the non-contact labyrinth seal is adopted between piston and cylinder mirror, which cause large clearance and serious leakage. The electromagnetic levitation technology was used to realize concentric centering between the piston and cylinder. And the traditional cantilever structure for the piston centering was transform into the simple support structure by using the through-piston rod. Established the three closed loop centering control system of current、speed and displacement and mathematical simulation model of electromagnetic centering control system. The experimental and simulated results showed that the electromagnetic centering control system can realize coaxial centering of the piston and cylinder. Fluent software analysis was used to study the electromagnetic control affects on the quantity of leaking. The results showed that by using electromagnetic centering technology and decreasing clearance between in cylinder wall and piston seal, the internal leakage of labyrinth reciprocating compressor can be reduced significantly and volumetric efficiency improved greatly.

References

[1] Vetter H 1972 The Sulzer oil-free labyrinth piston compressor, *Proceeding of the 1984 International Compressor Engineering Conference* (Purdue university: W Lafayette, IN, USA) pp 221–28
[2] Krik R and Guo Z 2009 Influence of leak path friction on labyrinth seal inlet swirl, *Tribology Transactions* **52**(2) pp 139–145
[3] Luo Y S 1973 The summarize of the oil-free lubrication technology of reciprocating piston compressor abroad, *Fluid Machinery* **6** pp 29 –43 (in Chinese)
[4] Graunke K and Ronnert J 1984 Dynamic behavior of labyrinth seals in oil-free labyrinth-piston compressor, *Proceeding of the 1984 International Compressor Engineering Conference* (Purdue University: W Lafayette, IN, USA) pp 7–15
[5] Pei Z J 2012 *Compressor engineering handbook* (Beijing: China Petrochemical Press) (in Chinese)
[6] Guoqing W and Hongzhao L 2009 Dynamic wear research of clearance joint in crank-rod mechanism, *Journal of Mechnical Science and Technology* **23**(11) pp 2830–38
[7] Soto E and Childs D 1998 Experimental rotordynamic coefficient results for a labyrinth seal with and without shunt injection and a honeycomb seal, *Journal of Engineering for gas Turbines and Power-transactions of the ASME* **121**(1) pp 153–59
[8] Feng Q K and Zhang Y Y 1997 The labyrinth leakage flow calculation and experimental study of

IOP Conf. Series: Materials Science and Engineering **232** (2017) 012072 doi:10.1088/1757-899X/232/1/012072

the labyrinth piston compressor, *Xi'an Jiaotong University* **31**(5) pp 97–102 (in Chinese)

[9] Tang H N, Wang S J and Dong X Y 2015 Stability analysis and modification of guide support system of piston rod in labyrinth reciprocating compressor, *Proceedings of the 2015 international conference on electrical automation and mechanical engineering (EAME2015) (Phuket, Thailand)* pp 196–200

[10] Yang H 2002 *Research on guiding and centering of labyrinth piston* (Tianjin, Institute of Chemical Machinery and Automation Co. Ltd) (in China)

[11] Hu D, Jia L and Yang L 2014 Dimensional analysis on resistance characteristics of labyrinth seals, *Therm Sci* **23**(6) pp 516–22

[12] Kim T S and Cha K S 2009 Comparative analysis of the influence of labyrinth seal configuration on leakage behavior, *Mech Sci Technology* **23**(11) pp 2830–38

[13] Bjork R 2016 The magnetic properties of the hollow cylindrical ideal remanence magnet, *Journal of Magnetism and Magnetic Materials* **416** pp 321–24

[14] Balandin D V and Kogan M M 2011 Motion control for a vertical rigid rotor rotating in electromagnetic bearings, *Journal of Computer and Systems Sciences International* **50**(5) pp 683–97

[15] Kaya Y, Cansiz A, Yildizer I *et al* 2013 Driving mechanism of a superconducting magnetic bearing system, *Journal of Superconductivity and Novel Magnetism* **26**(4) pp 1233–39

IOP Conf. Series: Materials Science and Engineering **232** (2017) 012073 doi:10.1088/1757-899X/232/1/012073

Reduction of gas flow nonuniformity in gas turbine engines by means of gas-dynamic methods

V Matveev[1], O Baturin[2], D Kolmakova[3], G Popov[4]

[1] Professor, Samara National Research University, Samara, Russia
[2] Associate Professor, Samara National Research University, Samara, Russia
[3] Junior Researcher, Samara National Research University, Samara, Russia
[4] Junior Researcher, Samara National Research University, Samara, Russia

E-mail: kolmakova.daria@gmail.com

Abstract. Gas flow nonuniformity is one of the main sources of rotor blade vibrations in the gas turbine engines. Usually, the flow circumferential nonuniformity occurs near the annular frames, located in the flow channel of the engine. This leads to the increased dynamic stresses in blades and as a consequence to the blade damage. The goal of the research was to find an acceptable method of reducing the level of gas flow nonuniformity as the source of dynamic stresses in the rotor blades. Two different methods were investigated during this research. Thus, this study gives the ideas about methods of improving the flow structure in gas turbine engine. On the basis of existing conditions (under development or existing engine) it allows the selection of the most suitable method for reducing gas flow nonuniformity.

1. Introduction

Circumferential nonuniformity of gas flow is one of the main sources of turbomachinery rotor blades destruction [1-5]. There are several reasons for this phenomena: violation of axial symmetry at turbomachinery inlet; presence of guide and nozzle vanes in the flow passage of turbomachinery; impact of load-bearing elements (annular frame struts and bearings); operation of blow-off system and air bleeding system in compressor; operation of bringing fuel nozzles of combustion chambers; distortion of load-carrying casing; warping of flame tubes of combustion chambers [1].

Due to the circumferential nonuniformity of gas flow, the RW blades are under higher dynamic stresses. All this leads to the oscillations of the blades and its damage.

Several approaches are implemented in practice to reduce the level of gas flow nonuniformity. The main idea of these methods is rather common. It is to eliminate the negative effect of nonuniformity on the problem domain of the turbomachinery, which can be achieved in two ways. Firstly, it is possible to modify the design of turbomachinery itself by changing the axial clearance between the RW and GV or by changing the stagger angles and pitch of GV blades to non-uniform around the circumference [6-8]. Secondly, the modifications can be implemented for the annular frame, for example by changing the number of struts or using the inclined struts.

The main drawback of theses both ways is that it is difficult to use them for existing engines. These approaches require a large number of significant changes in the engine construction.

This article is dedicated to reducing the circumferential nonuniformity of gas flow form the annular frame of axial compressor. In turn, the goal of this research was to find the ways of the flow nonuniformity reduction, which would lead to a decrease in the level of dynamic stresses in RW blades located in front of the annular frame, but would not require a fundamental engine design changing.

Both ways of reduction of flow nonuniformity including variations in design of compressor and frame were considered in the paper. The main advantage of this research is in detailed consideration of the reasons of negative effects of flow nonuniformity in present case, as well as possible ways to reduce the nonuniformity and results that can be obtained through the use of these approaches.

2. Motivation of the research. Description of the compressor

The motivation for this study was the real-life problem of the rotor blades' destruction of the fifth rotor wheel of intermediate pressure compressor (RW5 in the Figure 1). This compressor is part of a gas turbine engine NK-36ST (Figure 1), which has been designed and manufactured at Samara enterprise of aviation and space propulsion engineering [9].

The cause for the blade destruction is circumferential flow nonuniformity caused by struts of GTE middle annular frame. The struts of annular frame have different thickness and are distributed with different angular displacement relative to each other (Figure 2). The struts cause the local areas of high pressure, passing through which the impeller blades experience the dynamic impact. This impact leads to blades' forced vibrations and high dynamic stresses (Figure 3).

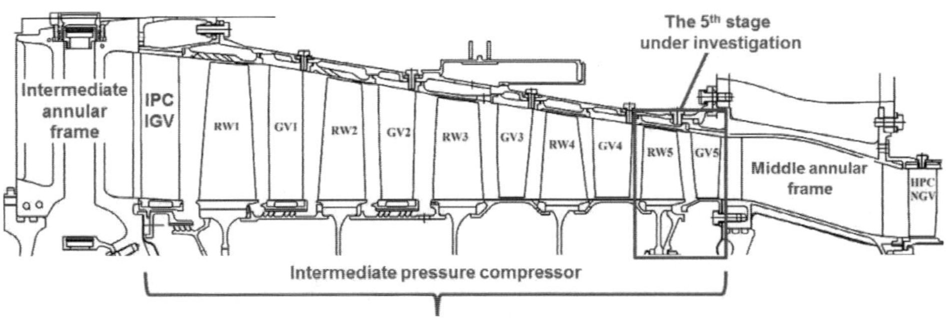

Figure 1. Scheme of the flow section of the test compressor

As a consequence of this problem, the blades of RW5 had mid-span shroud. However, this solution has several drawbacks [10]: the shroud increases the weight of a compressor; shrouded blades have low efficiency and life cycle, and greater complexity in manufacturing.

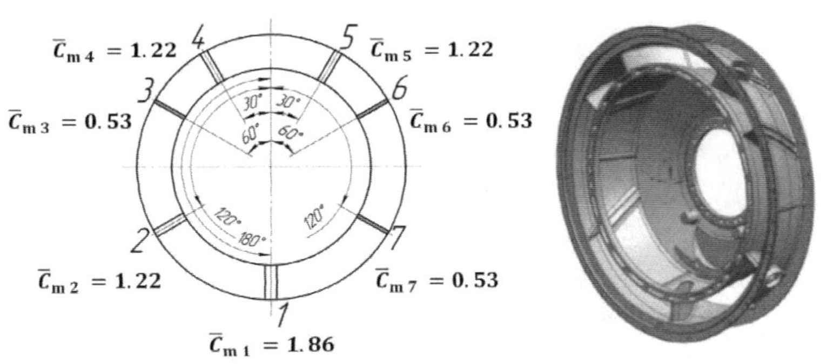

Figure 2. The appearance of the middle engine annular frame

3. Method of calculation of the level of gas flow nonuniformity

Calculation method of the gas flow nonuniformity level after the rotor blades of the fifth IPC stage was developed at the initial stage. This method consists of two steps (Figure 4).

Step 1. CFD simulation of the sector IPC model was performed at the main GTE modes.

Sector IPC model consists of the following domains: intermediate annular frame, all blade rows of IPC, middle annular frame, and inlet guide vanes of high-pressure compressor. All blade domains were simulated as one blade passage with periodical boundary conditions; domains of annular frames were simulated as a sector of one strut.

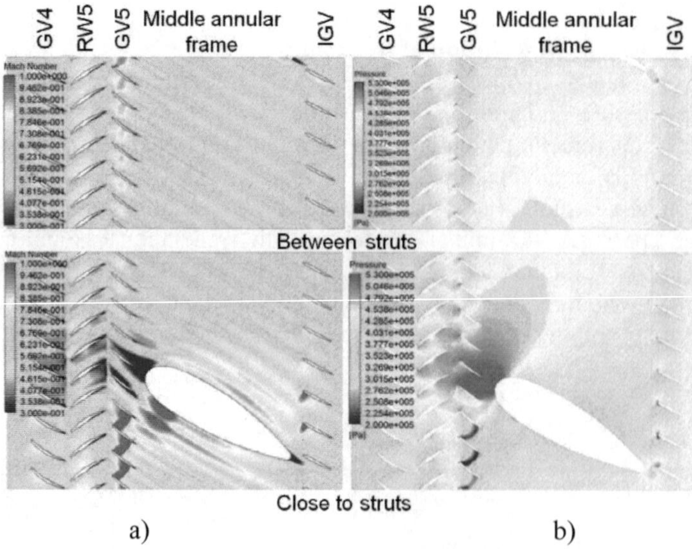

Figure 3. Mach number (a) and pressure (b) fields between struts and close to struts

To transfer the flow parameters between the domains, the *Mixing Plane* interface averaging the flow parameters in the circumferential direction was used [11].

The aim of this calculation was to determine the distribution of the flow parameters along the height of flow section (total pressure, total temperature and flow angle) in the section behind the RW4. These parameters will be used as boundary conditions for subsequent calculations.

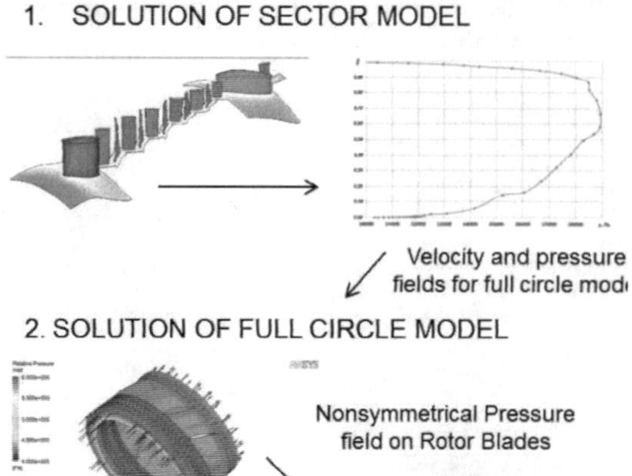

Figure 4. Scheme of method for calculation compressor rotor blades forced oscillations

Step 2. Calculation of the "full circle" model was performed.

The reason for using the full circle model for the further calculations is the desire to detect the flow circumferential unevenness, which arises from the presence of annular frame struts, unevenly located along its circumference in the compressor passage.

Full circle model consists of blade row domains of the fourth GV, the fifth RW and GV, the middle annular frame and HPC IGV. Gas dynamic load acting on the all blades as well as static pressure distribution along the circumference of the impeller are determined in this calculation.

Finite volume mesh of the full circle model was created in the module *AutoGrid 5* of software package *Numeca FineTurbo* [11]. Parameters of finite volume mesh and configuration of computational models have been chosen according to the recommendations given in [12-14]. Computational models have been verified in studies [6] in the evaluating of the vibration level of the rotor blades. This shows the reliability of calculations carried out in the described study.

Total pressure and total temperature, and flow direction at the inlet obtained from sector model calculations [14] and outlet static pressure were set as boundary conditions.

To transfer parameters between rotating and stationary domains of the full circle model, the standard interface *Frozen Rotor* was used. This interface transmits the flow parameters between domains without their averaging in the circumferential direction in contrast to the *Stage* interface [15]. Thus, it can be used to calculate the circumferential flow nonuniformity.

4. Gas-dynamic methods to reduce gas flow nonuniformity

4.1 Modifications of Compressor Design

One of the methods of reducing the circumferential nonuniformity is to use guide vanes with a different circle pitch and stagger angle in front of annular frame struts [7-8]. It possible to redistribute the flow between the different blade passages and to adjust the position of high pressure zones [16].

To implement the proposed method, a parametric model of the GU5 was created. All the guide vanes were divided into 7 groups according to the location of seven struts of the annular frame. Changes in the stagger angles and pitch were performed within each group. With the introduction of different stagger angles, the minimum number of variable vanes was a key factor, since the production of a large number of blades with different geometry greatly increases the manufacturing charges.

The stagger angle of GV of the fifth stage has not been changed for vanes that are located in the symmetry plane of the strut, the first and the last vane in the group. Changing in the stagger angles within the groups was carried out according to the linear law. The vanes located on the opposite sides of the symmetry plane of the strut were rotated in opposite directions relative to the initial position (Figure 5). When the vanes are located closer to the strut, they were rotated by a larger angle. If the vane was rotated for closing (an increase in the stagger angle), there is sign "+" before the angle if the vane was rotated for the opening (a decrease in the stagger angle), there is sign "-" before the angle. It should be noted, that stagger angels were measured from the leading edge plane.

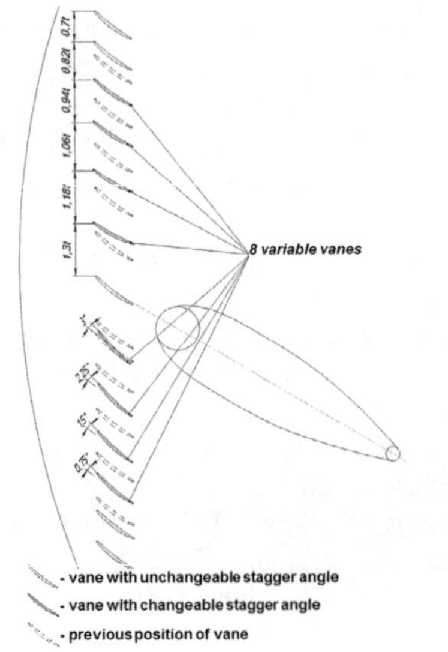

Figure 5. Scheme of GU vanes position change

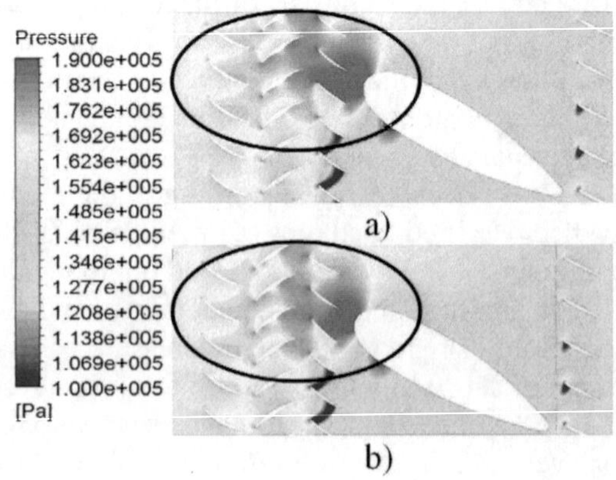

Figure 6. Static pressure distribution of basic variant (a) and GU variant No. 9 (b)

In case of pitch change, the number of variable vanes was not restricted. Different pitch was set within -0.35...+0.35 of the initial pitch. The sign "-" indicates that the pitch between vanes is decreased, while the sign "+" indicates the pitch increase. The number indicates the maximum increase (decrease) in the pitch between the vanes in the group in relative values from the initial pitch with evenly spaced vanes. The position of extreme vanes in the groups has not been changed with the introduction of different pitch. The law of the pitch changing was also linear.

There were investigated 11 different configurations of the fifth GU with circumferentially different stagger angles and pitch (Table 1).

Table 1. The results of the parametric IPC model calculation.

No. of variant	Parameter of different stagger angles, maximum stagger angles (No. of blades) for groups:			Parameter of alternating blade pitch for the groups:					**Number of variable blades**	**Dynamic stresses MPa**
	2, 5, 6	1, 3, 7	4 (3)	1, 7	2, 6	3	4	5		
1	0	0	0	0	0	0	0	0	0	86.7
2	0	0	0	0.3	0.3	0.3	0.3	0.3	0	115.98
3	3 (6)	3 (6)	3 (8)	0	0	0	0	0	42	43.791
4	**3 (6)**	**3 (6)**	**3 (8)**	**0.3**	**0.3**	**0.3**	**0.3**	**0.3**	**42**	**31.877**
5	3 (6)	6 (6)	9 (8)	0	0	0	0	0	42	37.536
6	0	0	0	0.35	0.35	0.35	0.35	0.35	0	86.95
7	0	0	0	-0.3	-0.3	-0.3	-0.3	-0.3	0	105.94
8	0	0	0	-0.15	-0.15	-0.15	-0.15	-0.15	0	137.741
9	**6 (2)**	**6 (2)**	**6 (2)**	**0**	**0**	**0**	**0**	**0**	**14**	**46.737**
10	0	6 (2)	6 (2)	0	0	0	0	0	8	57.214
11	6 (2)	6 (2)	6 (4)	0	0	0	0	0	16	44.426

Analysis of the results showed that the use of circumferentially different stagger angles and pitch makes it possible to achieve a significant reduction in the level of gas flow nonuniformity, and. The positive impact of circumferentially different stagger angles and pitch on the value of dynamic stresses can be explained from gas-dynamic point of view. Flow structure becomes more uniform, with less high pressure zones due to the redistribution of the flow (Figure 6). In addition, the values of pressure peaks behind the RW5 have been reduced (Figure 7). In turn, these improvements lead to the decrease in dynamic stresses in rotor blades (Table 1). That was confirmed in [17] with the help of method of calculation of dangerous harmonic amplitudes proposed by researches A.I. Ermakov and A.O. Shklovets.

Thus, the introduction of circumferentially different stagger angles and pitch of GU vanes located in front of the annular frame reduces the circumferential flow nonuniformity. However, the implementation of this method leads to a complication of manufacturing technology, since it will be necessary to produce the guide vanes using not versatile technology in this case.

In addition, due to the change in the vane pitch, a new technology for grooving will be required.

4.2 Modification of Annular Frame Design

There are experimental studies that prove that the airfoil cascade, which is located in front of the cylinder, streamlined by the flow, enhances the upstream transmission of the high pressure zone [16].

The proposed method for reducing the circumferential nonuniformity of the gas flow is based on the results of the study [16]. It is in distancing the leading edge of annular frame struts (in this case it is the cylinder) from the trailing edge of GV blades located upstream (in this case it is the cascade) in order to neutralize the negative effects. At the same time, an important requirement is the need to maintain the internal cavities of the annular frame struts for placement of the engine system.

Struts can be divided into three groups according to the maximum thickness of profile Cm: thick (strut No. 1), medium (struts No. 2,4,5) and thin (struts No. 3,6,7) (Figure 2).

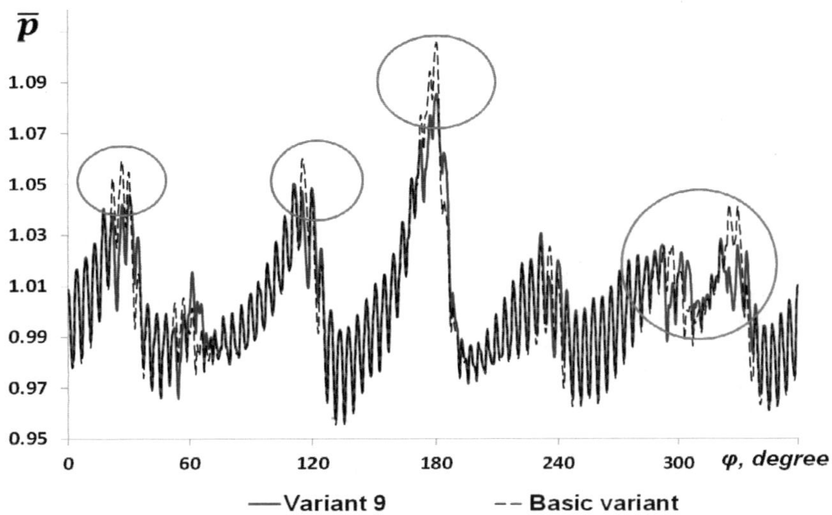

Figure 7. Graphs of the relative static pressure behind the fifth impeller of the basic variant (a) and GU variant No. 9 (b)

To implement the chosen method of the nonuniformity reduction, the parametric models were developed for each type of strut. The models make it possible to adjust the position of the leading edge, pressing the profile of the strut, but maintaining the thickness and the rest of the profile by specifying the desired value of shifting (Δb) (Figure 8). Parametric models take into account the limitations related to the location of the engine systems in the struts.

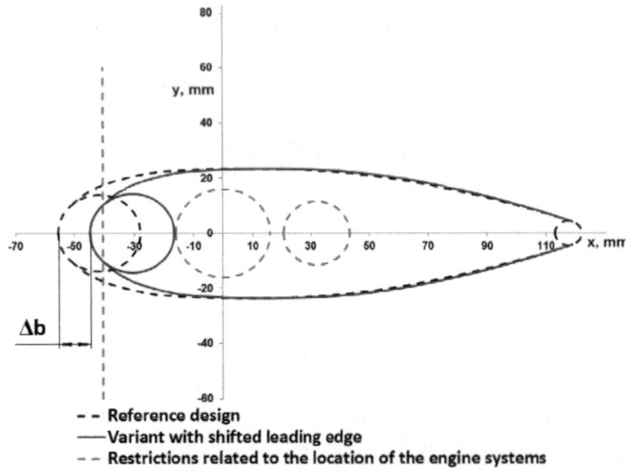

Figure 8. The algorithm of shifting the leading edges of struts in case of a thick strut

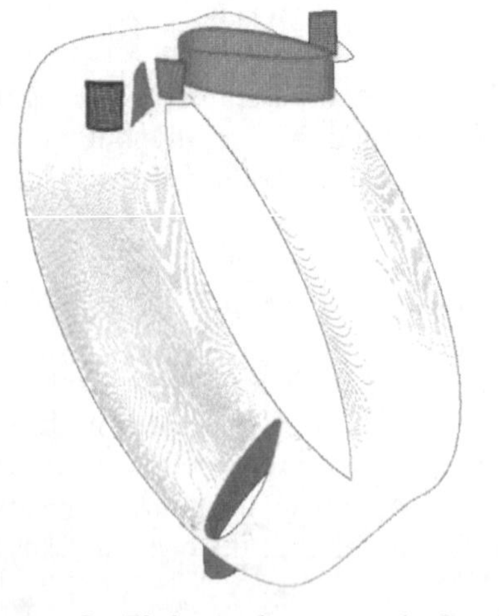

Figure 9. Finite volume mesh for the elementary model with thick struts

To analyze the effect of leading edge shifting on the circumferential nonuniformity of the gas flow, the set of computational models of the compressor was created using the parametric models of struts. All of them were full circle models and were consisted of five domains: GV of the fourth stage, RW and GV of the fifth stage, middle annular frame, and IGV of HPC. At the same time, the annular frame contained only two struts disposed facing each other with the same thickness in each model (Figure 9). Such models were called elementary, and were created to assess the contribution of each type of strut in the total circumferential flow non-uniformity. In other words, they were used to calculate the pressure peaks occurring opposite each type of strut.

Moreover, leading edge shifting of struts with $\overline{\Delta b} = 0$, $\overline{\Delta b} = 0.2$, $\overline{\Delta b} = 0.4$, $\overline{\Delta b} = 0.6$, were set for each type of struts in different models. Thus, the total number of computational models was 12.

The calculation and analysis of the results were also carried out in Ansys CFX using the supercomputer of SSAU "Sergei Korolev".

Computational models were verified in [6,17] while assessing the level of vibrations in rotor blades. This indicates the reliability of calculations described in this paper.

ehe circumferential distributions of relative static pressure in the section behind the RW for each type of strut with all values of leading edge shifting were obtained. It should be noted that obtaining of the pressure distribution only for one model took a long time, both in terms of creation model, and the calculation process. The total time for testing of just one variant is approximately 12 hours.

These distributions were averaged to obtain the so-called elementary peaks (Figure 10). In addition, the level of mean pressure after the RW (red line) was correlated with zero of y-axis for a visual representation of the elementary peaks.

Figure 11 shows a field of static pressure distribution for case of thick strut with of $\overline{\Delta b} = 0$.

It is obvious from the graphs form Figure 10 that annular frame struts lead to the appearance of complex forms of pressure peaks after the RW. The peaks have not only the maximum jump h (positive upper part of the graph), but also go to the negative part of the graph into the decline h_1. A similar pattern is seen in Figure 11 by following in the positive direction of the angle φ.

The complex peak shape can be caused by the rotation of the RW blades in the area of the circumferential nonuniformity of gas flow.

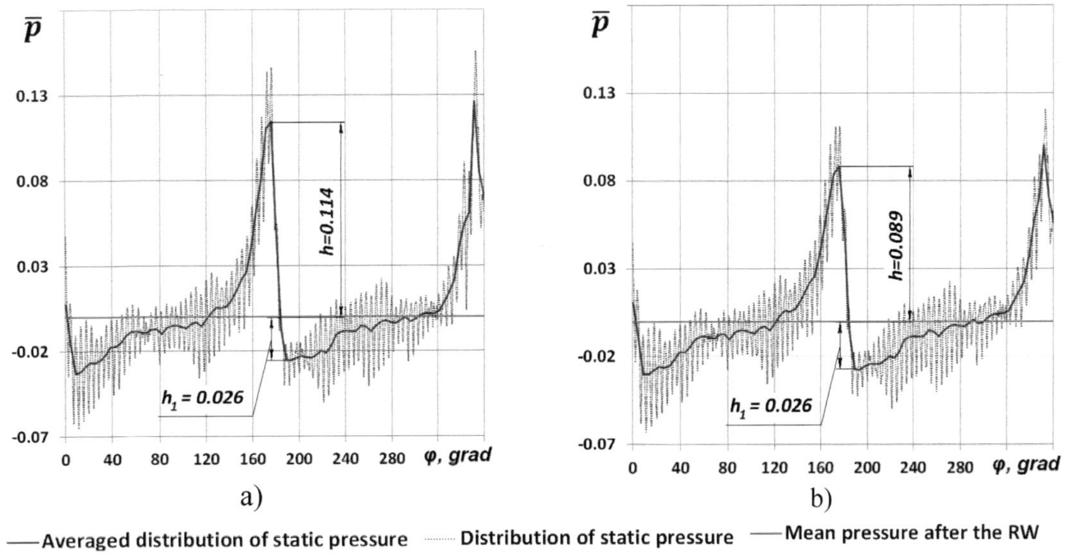

Figure 10. The distribution of the relative static pressure in case of thick strut with the leading edge shift of $\overline{\Delta b} = 0$ (a), $\overline{\Delta b} = 0.6$ (*b*)

Maximal jump of pressure peaks h decreases (upper part of graph) while the value of the leading edge shifting is increasing. It is worth noting that the maximum failure of pressure peaks h_1 (the bottom part of the graph) does not depend on the shift of the leading edge and it is constant in all charts. Similar results were obtained for other types of struts.

To verify the adequacy of the obtained dependences, the graph from the individual elementary peaks of each type of strut was assembled and compared with the graph for the reference design of seven-strut annular frame (Figure 12a).

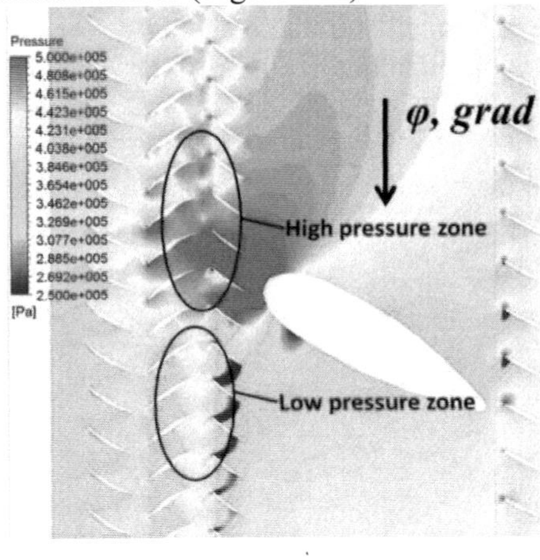

Figure 11. The field of static pressure for elementary model with a thick strut with shifted leading edge of $\overline{\Delta b} = 0$

Also, the distribution of relative static pressure after the RW of the similar compressor, but with experimental 13-strut annular frame was obtained in previous studies [17]. Thick and medium struts in this annular frame correspond to the same struts in seven-strut annular frame, and thin struts have the twice smaller thickness than the thin strut in seven-strut annular frame.

A similar investigation aimed on the identification of elementary peaks was conducted for that annular frame. Comparison of chart from the elementary peaks with the chart received in the calculation of the reference design of 13-strut annular frame is shown in Figure 12b.

In general, graphics derived from the elementary peak, describe the pressure distribution after the RW well. The results can be used for further research stages.

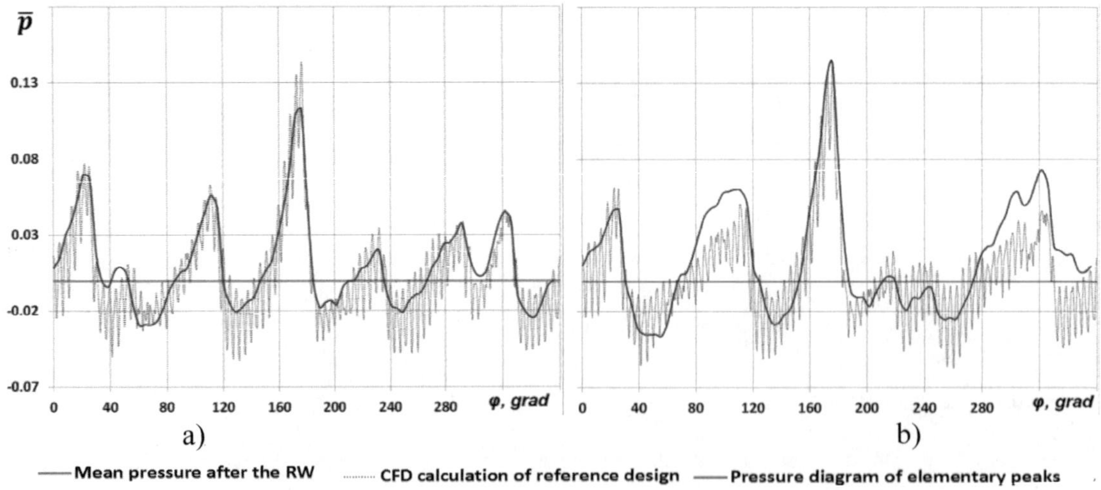

Figure 12. Comparison of the graph of relative static pressure after RW for reference design of seven-strut (a), 13-strut annular frame (b) and graphic obtained by elementary models

4.2.1 Searching for the equations of the peak height dependence on the leading edge shifting. In order to use the obtained results, it is necessary to describe the dependence of pressure peaks on the shift of the strut leading edge via equations. That was the next step of the study.

Quadratic equations were derived for each type of strut using the regression analysis techniques, describes the effect of the shift of the leading edge (Δb) on the pressure peak height (h). Equation 1 corresponds to the thick strut, Equation 2 to the medium strut and Equation 3 to the thin one.

$$h = 0.001\Delta b^2 - 0.135\ \Delta b + 7.5 \quad (1)$$

$$h = 0.001\Delta b^2 - 0.135\ \Delta b + 7.5 \quad (2)$$

$$h = 2.64 \cdot 10^{-18}\Delta b^2 - 0.02\ \Delta b + 2 \quad (3)$$

Thus, the so-called surrogate model was created, by means of which the value of pressure peak behind the RW can be predicted at the preliminary stage of investigation under certain provisions of the leading edges of the annular frame struts.

Application of leading edge shift of annular frame struts allows for not only quantitative changes in the values of pressure peak behind the impeller (Figure 13). The proposed method also provides better quality structure of the flow in the compressor: field becomes more uniform, high-pressure zone reduces its impact on the blade rows, located upstream.

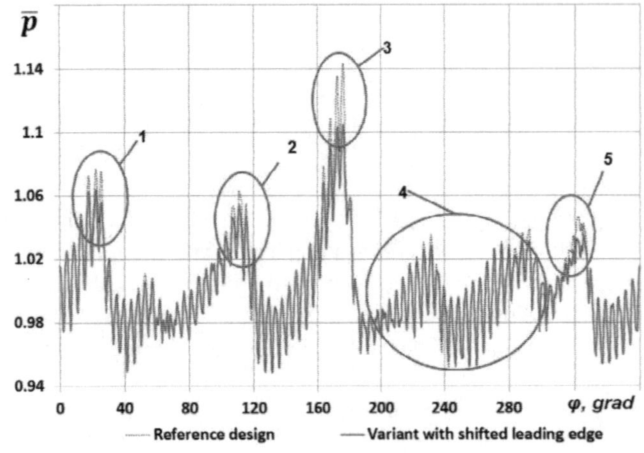

Figure 13. Circumferential distribution of the relative static pressure behind the RW for the reference design of the 7-strut annular frames and annular frame with shifted leading edges

The received regression equations were used for optimization of the location and shift of annular frame struts to reduce the dynamic stresses on the rotor wheel of the fifth stage of the IPC [18].

Conclusion

Several approaches to decrease the gas flow nonuniformity in the axial compressor arise because of the presence of annular frame have been developed as a result of conducted research. Each of the proposed approaches made it possible to achieve the goal that was to reduce the level of pressure peaks after the rotor blades of the last compressor stage.

Firstly, it was suggested to eliminate the negative influence of flow nonuniformity by modifying the guide vanes of the last compressor stage. It was found that the different blades stagger angles and alternating blade pitch implementation allows "flattening" the circumferential variation in the section following RW5. In addition, more periodic and uniform flow field is achieved for all specific areas.

Secondly, the circumferential unevenness of gas flow in the compressor can be reduced by a minimum interference in design. It is enough to alienate its source (the support rack) from the elements, experiencing the negative impact of circumferential unevenness of gas flow (rotor blades located upstream). This is accomplished by shifting the leading edges of the support racks.

The advantage of the second method is that it makes possible to predict the pressure peak level behind the RW to the particular design of the support racks without the long process of preparation of the computation model, calculation and post-processing of the results by using surrogate models. This can accelerate the design process of new compressors and design refinement of existing units.

The obtained results were used to optimize the design of the support to reduce the dynamic stresses in the RW5. The surrogate model was integrated into the strength module. Pressure peaks were decomposed to Fourier series, for further calculation of dangerous harmonics.

Acknowledgments

The work was financially supported by the Ministry of education and science of Russia in the framework of the implementation of the Program of increasing the competitiveness of SSAU among the world's leading scientific and educational centers for 2013-2020 years.

References

[1] Ivanov V P 1983 *Kolebaniya Rabochego Kolesa Turbomashin (Vibrations of Turbomachinery Impeller) (*Mashinostroenie, Moscow) p 244
[2] Vorob'ev Yu S 1988 *Kolebaniya Lopatochnogo Apparata Turbomashin (Vibrations Turbomachinery Blading)* (Kiev, Nauk. Dumka) p 244
[3] Cohen H, Rogers G F C and Saravanamuttoo H I H 1996 *Gas Turbine Theory. 4-th edition* (Longman Group Limited)
[4] Logan E Jr 2003 *Handbook of Turbomachinery* (CRC Press; 2 edition)
[5] Carter T J 2005 Common failures in gas turbine blades *Engineering Failure Analysis* **12(2)** 237–247
[6] Kolmakova D, Popov G, Shklovets A and Ermakov A 2014 Techniques and Methods to Improve the Dynamic Strength of Gas Turbine Engines Compressor Rotor Wheels *Proc. of the ASME 2014 Gas Turbine India Conference* GTINDIA2014-8203
[7] Sladojević I, Sayma A I and Imregun M 2007 Influence of stagger angle variation on aerodynamic damping and frequency shifts *Proc. of the ASME Turbo Expo* **5** 683–700
[8] Yang Y, Yang A, Dong R, Chen E and Dai R 2012 Influence of stagger angle on aerodynamic sound performance of compressor cascade *Hangkong Xuebao/Acta Aeronautica et Astronautica Sinica* **33(4)** 588-596
[9] JSC "Kuznetsov", Accessed July 20, (2016). http://www.kuznetsov-motors.ru/en
[10] Cumpsty N A 2004 *Compressor Aerodynamics* (Krieger Publishing Compan, Edition 2)
[11] NUMECA, User Manual AutoGrid5 Release 8.4, NUMECA.inc., Belgium, January 2008.
[12] Matveev V N, Popov G M, Goryachkin E S and Smirnova Y D 2014 Effect of Accounting of Air

Bleed from the Flow Passage of the Multi-Stage Axial Low Pressure Compressor on its Design Performances *Research J. of Applied Sciences* **9(11)** 784-788

[13] Popov G, Goryachkin E, Baturin O and Kolmakova D 2014 Development of Recommendations on Building of the Lightweight Calculation Mathematical Models of the Axial Turbines of Gas Turbine Engines *Int. J. of Engineering and Technology* **6(5)** 2236-2243

[14] Ermakov A, Shklovets A, Popov G and Kolmakova D 2014 Investigation of the effect of the gas turbine compressor supports on gas flow circumferential nonuniformity *Research J. of Applied Sciences* **9** 684-690

[15] ANSYS® Release 12.0, Help System, Ansys CFX-Solver Modeling Guide, ANSYS, Inc.

[16] Saren V 1984 *Flow around irregular lattice of plates placed in front of the cylinder Technical Report* (Moscow: Central Institute of Aviation Motors) p 36

[17] Popov G, Kolmakova D, Shklovets A and Ermakov A 2015 Optimization of the axial compressor flow passage to reduce the circumferential distortion *9th International Conference on Compressors and Their Systems*

[18] Kolmakova D, Popov G, Shklovets A and Ermakov A 2016 Rational Design of Gas Turbine Engine Compressor to Provide the Required Dynamic Strength Level of Rotor Blades *Proc. of the ASME Turbo Expo GT2016* GT2016-56991

IOP Conf. Series: Materials Science and Engineering **232** (2017) 012074 doi:10.1088/1757-899X/232/1/012074

Numerical Simulation on Flow and Heat Transfer Performance of Air-cooler for a Natural Gas Storage Compressor Unit

Liu Biyuan[1], Zhang Feng[2], Ma Zenghui[2], Zheng Zilong[2], Feng Jianmei[1]

[1] Xi'an Jiaotong University, People's Republic of China

[2] No1. Gas Production Plant of Xin Jiang Oilfield Company

Corresponding author: jmfeng@mail.xjtu.edu.cn

Abstract: Heat transfer efficiency has been a key issue for large size air coolers with the noise reducers used in natural gas storage compressor unit, especially operated in summer with cooling air at a high temperature. The 3-D numerical simulation model of the whole air cooler was established to study the flow field characteristic with different inlet and outlet structures by CFD software. The system pressure loss distributions were calculated. The relationship was obtained among heat exchange efficiency, resistance loss, and the structure of air cooler, the results presented some methods to improve cooling air flow rate and heat exchange efficiency. Based on the results, some effective measures were proposed to improve heat exchanger efficiency and were implemented in the actual operation unit.

Key words: Natural gas compressor, exhaust temperature, air cooler, heat transfer efficiency

1. Introduction

This Natural Gas storage station consists of 11 sets of compressor units and their corresponding 11 air cooler units. After the actual production, the noise value of the gas storage station exceeds the national standard [1]. Therefore, actions were taken to reduce the noise of compressor plant and air cooler unit. The exhaust temperature of natural gas compressors aims to be not higher than 65 ℃, however, the exhaust temperature of compressor working is up to 72 ℃, has exceeded the original design requirements. And during the period of gas injection, the units are going to experience the condition of high temperature in summer. With the increase of discharge pressure, the exhaust temperature of compressor will further increase and affect the stable operation of compressor. At the same time, the working condition range of the compressor of the gas storage is very wide, the designing exhaust pressure of the compressor is 16.0~32.0MPa, and the wide load range of the working condition also affects the stable operation of the compressor and the air cooler unit. In the prior technology, the important measure for reducing the exhaust temperature of the natural gas compressor is to adopt an air cooler. The air cooler not only needs low maintenance cost, but also does not have the scaling problem compared with water cooling system, and also has a longer service life. Therefore, there is great significance to study the main factors that affect the heat transfer efficiency of the air cooler unit and to improve the structure of the air cooler for improving heat transfer efficiency.

Air cooling technology has been widely used in many compressor units. The research published by scholars Bergles [2] is the most representative. The software ANSYS CFD was used to simulate the heat transfer of air cooler and the flow field of compressor plant. Duvenhage [3] simulated 3-D flow field of air cooler, considering the horizontal wind along the long axis direction and the longitudinal direction. Staden [4] established mathematical model to study the effects of ambient crosswind on the performance of direct air cooling system. The domestic researchers of air cooler in distribution and heat transfer enhancement carried out a large number of numerical simulations and experimental researches. Liu Zhanbin [5] studied heat transfer process of fin tube numerically and experimentally. Structures of fin tubes were optimized to obtain the best heat transfer performance. Cheng Yuanda [6] studied fin tubes for direct air-cooler numerically, using porous model to replace the concept of the actual fin. Different structure parameters affected heat transfer efficiency and flow field of air cooling unit were obtained.

Now CFD technology is the main means to study performance of air cooler. Based on the establishment of a 3-D model of air cooler unit working in gas storage system, the flow field and temperature field under different conditions of air cooler were used to analyze the key factors affecting the air cooling heat exchanger efficiency [7]. Measures were proposed for improving performance of air cooler and bring down gas injection temperature lower than 65 ℃.

2. **Physical model and simulation method**

2.1. Physical Model

The physical model of an air cooling unit of the gas storage compressor was established as shown in **Figure 1**. The model was established with a ratio of 1:1 based on the air cooler units are running. In order to ensure the precision of the numerical calculation, the outlet part of air cooler mathematical model was lengthened. The air cooler unit include the first exhaust gas cooler and the second exhaust gas cooler. The finned-tube of the first stage cooler is arranged staggered with three rows and of the second stage cooler is with four rows.

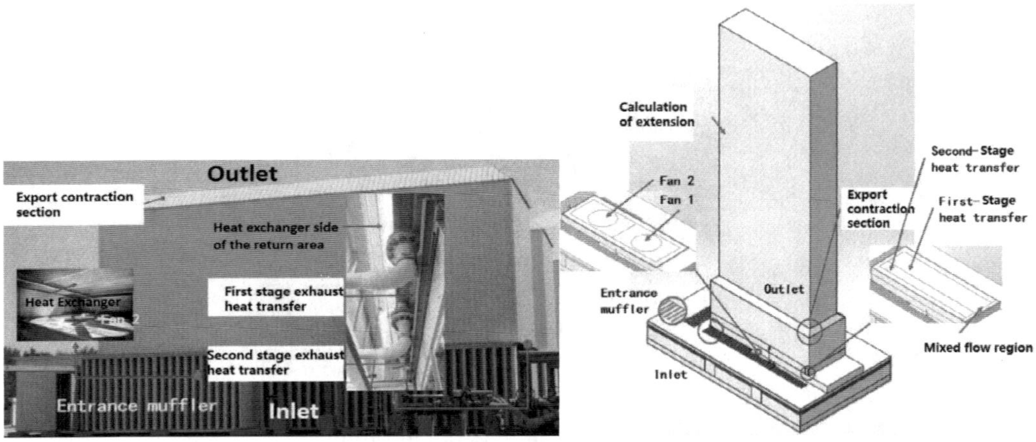

Figure 1: Air cooler unit for natural gas compressor of gas storage

In this paper, several models were established to compare with the basic model. Because of the limitation of this article, the specific model structures were not presented one by one, and the structural differences between the models were listed in **Table 1**:

Table 1: Structure of air cooler and the corresponding model number

Type	Entrance muffler	Mixed flow region	Export contraction section
Model 1（Basic model）	√	√	×
Model 2（Entrance muffler）	×	√	×
Model 3（Mixed flow region）	√	×	×
Model 4（Export contraction section）	√	√	√

2.2. *Air Cooler Model Mesh Generation and Grid Independence Verification*

There are big differences among each part of the air cooling unit structure size, and some regions were in irregular shape. So, tetrahedral mesh and hexahedral mesh were used to mesh the model. Hexahedral mesh was adopted in relatively regular shape part and tetrahedral mesh was adopted in area with complex shape. In the entire process the calculate condition and the accuracy are taken into account to choose an appropriate grid size, at the same time to meet a good continuity when data transferring.

In order to verify the mesh independence, the air cooler models were divided into 2070000, 3300000, and 4640000 grids, respectively. The mass flow rate of the two fans, the mass flow of the cooling air inlet and the air pressure of the cooling air upstream of the finned tube bundle (y=3755mm) were monitored. The results were shown in **Table 2**. The 3300000 meshes size model was used for simulation, as shown in **Figure 2**.

Table 2: Grid independence verification results

Grids number and the cooling air flow rate（kg/s）	Fan 1	Fan 2	Inlet
2070000	116.36	117.15	230.5
3300000	115.82	116.17	229.88
4640000	115.71	116.41	229.62

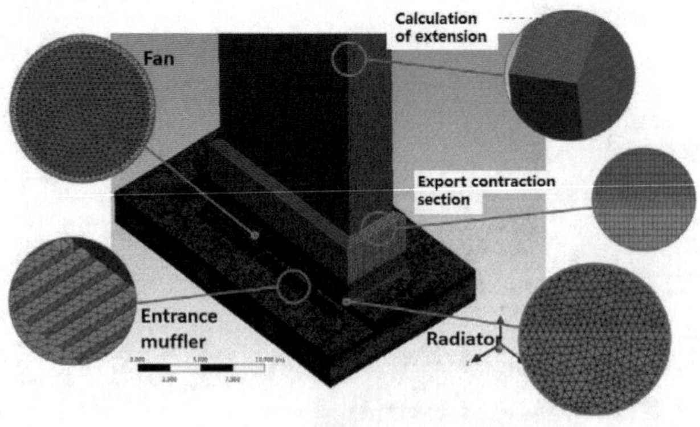

Figure 2: Air cooler model meshing

2.3. *Simulation methods*

The turbulence model has a significant influence on numerical results. Therefore, it is of great importance to select an appropriate turbulence model for computation. And there is a complex three-dimensional turbulent flow in the air cooler unit. As a turbulence model, the standard k-ε model has its own superiority in predictions of flow and heat transfer.

The air fluid has temperature-dependent physical properties. In order to obtain the heat exchange performance of the air cooler unit, energy equation was used.

2.4. *Boundary conditions*

In the calculation of the flow field and heat transfer in the air cooling unit, the boundary conditions should be set for the calculation area.

(1) Velocity entrance boundary conditions and pressure outlet boundary condition were set to simulate the resistance loss of each part of the air cooler unit.

(2) Positions of the fan and heat exchangers are treated as internal planes to analyze the resistance loss of air cooler internal flow field.

(3) The Fan model and the Radiator model are used to analyze heat transfer condition of air cooler. Meanwhile, pressure inlet boundary and pressure outlet boundary were set for the air cooler model.

The whole simulation of air cooling units is concerned with the overall heat transfer performance rather than the flow tube internal heat exchange. Therefore, the model is simplified reasonably. Finned tubes were treated as a radiator plane, in which the relationship of practical pressure loss and the heat exchanger coefficient with the velocity were arranged in the plane. The flow resistance and loss coefficient are related to the dynamic pressure of fluid. In this paper, the flow resistance loss curve of the first stage air cooler [6] is $P = 0.1624 + 12.505v + 0.9972v^2$, the flow resistance loss curve of the second stage air cooler [6] is $P = 0.17864 + 13.7555v + 1.09692v^2$. The total convective heat transfer coefficient of the heat exchanger [7] is $k = 276.23 + 39.028v - 8.5439v^2 + 0.7057v^3$.

Fan was treated as a performance curve using lumped parameter [8]. It is assumed as an infinite thin plane. The air pressure is going to increase when going through the fan. The discontinuous pressure rise is defined with the velocity of air. The fan performance curve [6] is $P = 360.875 + 11.013v - 1.84v^2 + 0.01993v^3$ in this paper.

Considering the effect of tangential velocity on the flow field, the rotating speed of the fan model was set up. The radial velocity was considered as zero, and there is a linear relationship [6] with tangential velocity and the radius $v_{\theta} = fr$, in which $f = 10.09$.

3. Results and analysis

3.1. *Effects of local Structure on Flow and Heat Transfer*

The influence of the local structure of air cooler units on the flow resistance loss was investigated, by comparing the cooling air flow rate of different models, The larger the cooling air flow, the smaller the resistance of the air cooler. The influence of the cooling air flow rate affects the heat transfer capacity when air goes through the fined tube with different velocity.

As results shown in Table 3, the cooling air flow rate increased 1.04% when there was no entrance mufflers. The flow rate of air decreasd 1.26% when there exists export contraction section. Therefore, the entrance muffler and the export contraction section of the air cooler would increase the resistance loss of the air cooler and reduce the flow rate of the cooling air.

Table 3: Cooling air flow rate of different models

Type	Mass flow rate kg/s			Increase
	Inlet	Fan1	Fan2	(%)
Model 1 (Basic model)	229.88	115.82	117.47	-
Model 2 (Entrance muffler)	232.26	116.95	116.17	1.04
Model 3 (Mixed flow region)	232.29	115.96	116.33	1.05
Model 4 (Export contraction section)	226.98	114.76	116.22	-1.26

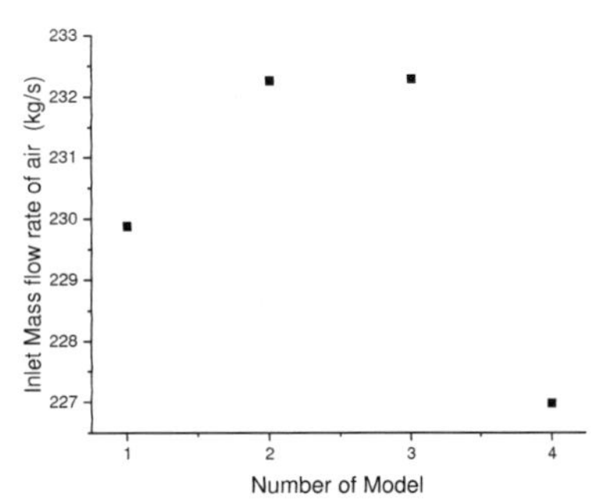

Figure 3: Inlet mass flow rate of cooling air of different models

The mixed flow region was found in the air cooling unit of the natural gas storage(as shown in Figure 1). To present the influence of mix flow region on the flow rate and temperature distribution, the temperature and velocity distribution of the surface under mixed flow region 50mm and the fan import surface were analyzed, by comparing with Model 1 (basic model) and Model 3 (mixed flow region).It can be seen in **Figure** 34. When there existed the mixed flow region in air cooling unit, the average temperature of the surface under mixed flow region was 320.1K, which was 18.27℃ higher than the model when the mixed flow region was blocked. Fan 1 is closer to the mixed flow region, so higher temperature air flow will go through the fan and enter the air cooling system again. The inlet air temperature of fan 1 is 301.22K, which is 1.05K higher than that of no mixed flow region. At the same time, higher temperature air back into the fan 1 would decline the fan performance. The air flow rate of two fans were lower by 3.5% than that has no mixed flow region.

Figure 4 showed the temperature distribution on the plane z = 0mm of Model 1 (basic model) and the velocity distribution of the two fans plane.

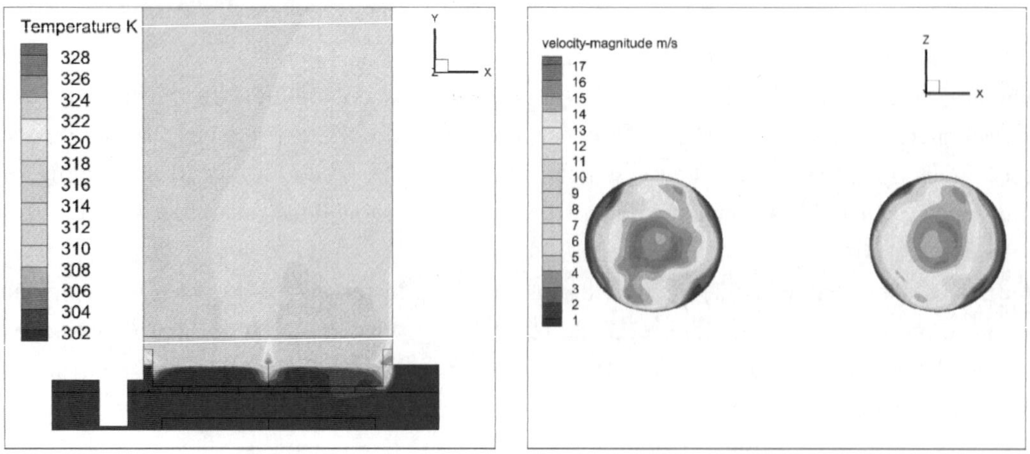

（a）Temperature distribution of plane Z = 0mm （b）Velocity distribution of two fans

Figure 4: Model 1（Basic Model）

Figure 5 showed the temperature distribution on the plane of the mixed flow region of Model 1 (Basic model, with mixed flow region) and Model 3 (No mixed flow region) .

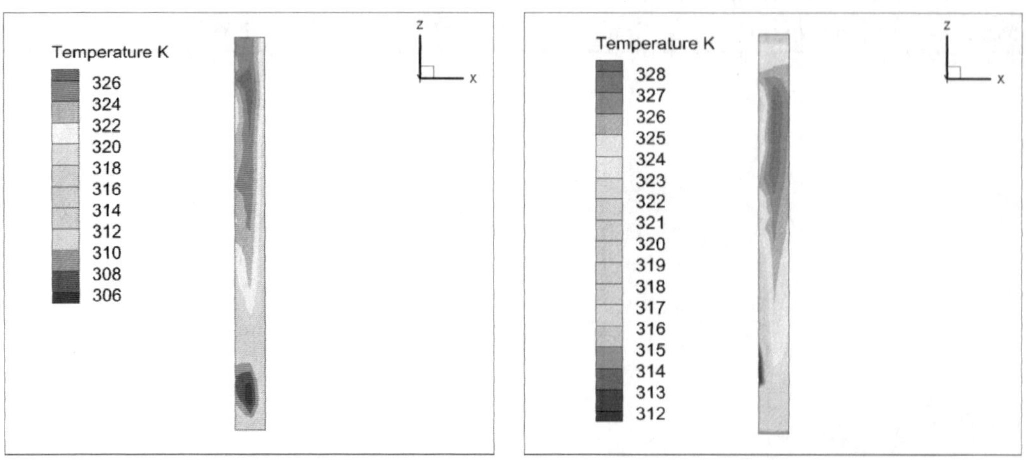

（a）Model 1（basic model） （b）Model 3（no mixed flow region）

Figure 5: Temperature distribution on the plane of the mixed flow region

3.2. Effects of finned tube fouling on flow and heat transfer performance

At site, it was found that there is heavy fouling on the outer surface of finned tubes in the working air cooler unit. Therefore, finned tubes with 0.1mm to 0.5mm thickness of fouling were simulated to study the effects of fouling on the flow field and heat transfer performance of air cooler unit, using the Model 1 as the studied model . The efficiency of heat exchange of Model 1 was simulated for the fouling thickness of 0.1mm, 0.2mm, 0.3mm, 0.4mm, 0.5mm,respectively.

According to the fan performance curve and the first-stage heat exchanger loss resistance curve, the performance curve was shown in Figure 6. When the fouling thickness of heat exchanger increased from 0.1mm to 0.5mm, the resistance loss increased by 4 times[9]. The results showed that the flow rate of cooling air reduced and the performance of fan decreased.

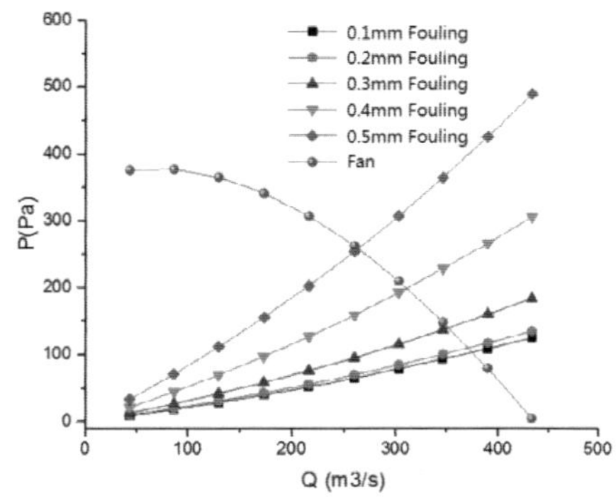

Figure 6 The performance of the fan and the heat exchanger under various fouling thickness

The fouling on the finned tube narrowed down the channel secton, which increased the cooling air velocity and made the resistance loss of air cooler unit increased. Meanwhile, the increased velocity would enhance the heat transfer coefficient. So, it's necessary to consider many factors in the finned tube outside to study how the heat transfer coefficient changed. The heat transfer coefficient with the finned tube fouling thickness increase showed a slightly increasing, but the fan performance reduced greatly. The decrease of flow rate of the cooling air weakened the heat transfer coefficient increase. The resultant effect behaved as the heat exchange capacity of finned tubes decreased.

Figure 7 showed the temperature distribution of plane z=0mm and the surface of finned tubes.

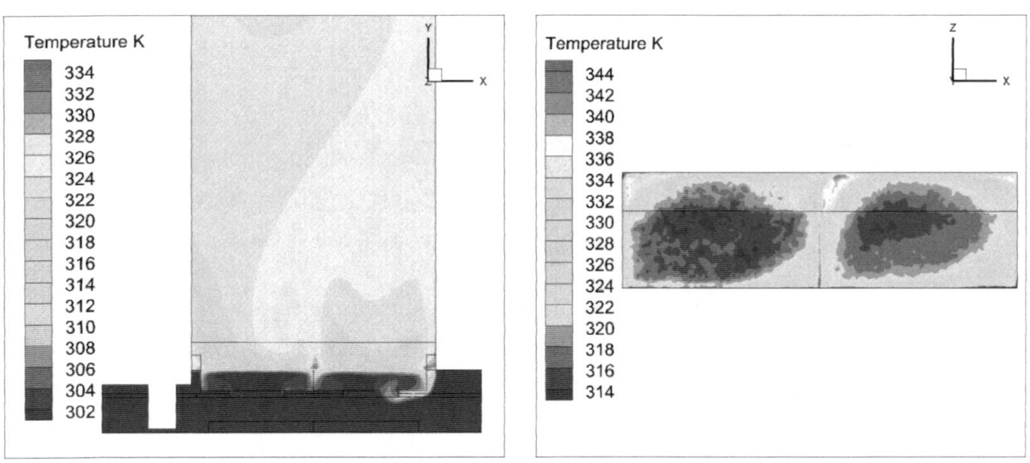

（a）The plane Z=0mm （b）The heat exchanger surface

Figure 7: The temperature distribution using the model 1-0.5mm fouling

3.3. Effects of Ambient Temperature on Flow and Heat Transfer

The physical properties of gas affected the resistance loss of system and fan performance. According to environment temperature range of the actual operating condition, this paper simulated the flow field and heat transfer performance of air cooling unit operating under different environment temperature from 243 K to 323 K every 10 K based on Model 1.

With the temperature of inlet cooling air increasing, the cooling air density decreased. According to the density effects of fan[10], when the density of the gas flowing through the system decreased, the fan power and fan pressure would be reduced. When the inlet temperature increased from 243 K to 323 K, the cooling air flow rate reduced by 13.47%.

Figure 8 showed the temperature distribution in the z=0mm plane and the velocity distribution in the surface of two fans. When the ambient temperature in winter is 243K, compared with air cooler working in summer at 323 K, the cooling air flow into the air cooler unit increased by 21.87%. Because of the high temperature in summer and the low temperature in winter, there is a big difference in the physical properties of the air. When the air cooler operates in high temperature, the cooling efficiency of the air cooler will be reduced. Therefore, optimizing the structure of the air cooler and improving the performance of the fan is more important to improve the heat transfer efficiency of the air cooler in the summer with high ambient temperature.

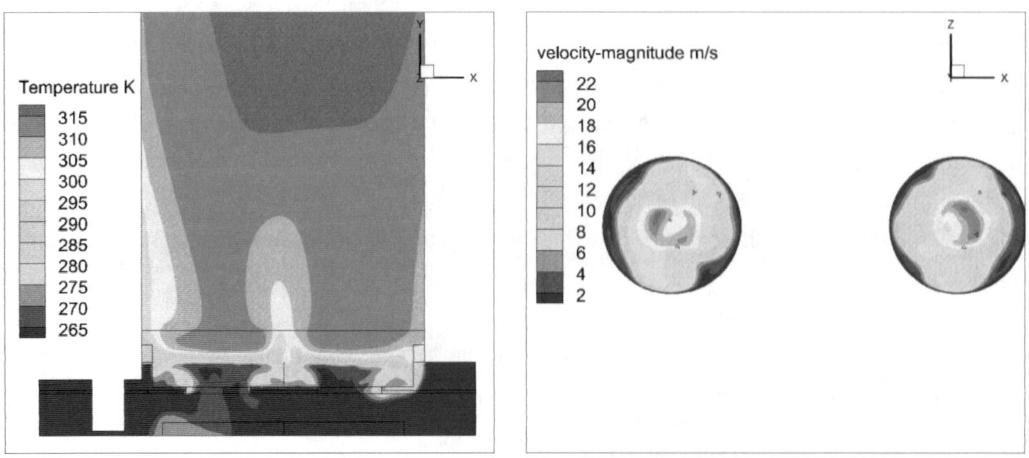

(a) Temperature distribution of plane Z = 0mm (b) Velocity distribution of fans

Figure 8: Model 1- the inlet temperature of air is 263K

3.4. The effects of fan blades number on flow and heat transfer performance

When the number of fan blade increases, the pressure-rise of fan usually improved and the running noise reduced. This paper simulated the fan with 8 blades and 10 blades to study the flow field and heat transfer performance of the air cooling unit.

Refer to different conditions in Section 3.2, simulation of the 10-blade fan was done under different fouling thickness to study the flow and heat transfer performance of air cooling unit. The same fouling thickness was considered with numbers in Section 3.2. The simulation results are shown in **Figure 9**. When the fouling was 0.5mm, the air flow rate of 10-blade fan was 8.72% higher than that of 8 blade fan. **Figure 10** showed the velocity distribution of the fan with 8 blades and the 10 blades in the fan plane.

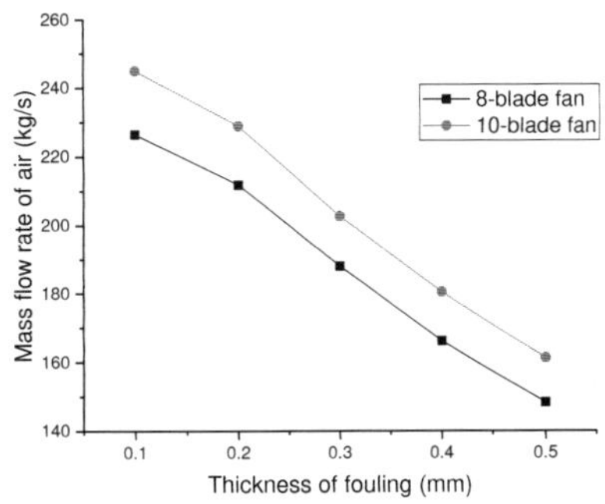

Figure 9: Flow rate of air cooler with 8 blades and 10 blades fan

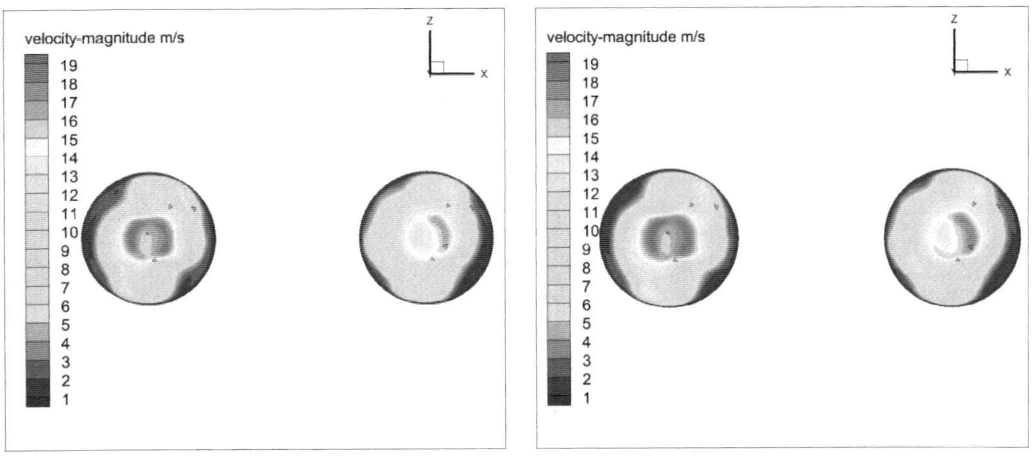

（a）8-blade fan （b）10-blade fan

Figure 10: Velocity distribution on fan plane

The fan of 2# air cooler unit at site is the fan with 8 blades, and the 1# air cooler unit is the fan with 10 blades. The outlet temperature of 1# and 2# air cooler unit were measured at the same time. As shown in **Table 4**, the outlet temperature of the air cooler unit of the 2#unit is 5-8℃ lower than that of the 1# unit.

Table 4: Temperature of out exhaust of model with 8 blades and 10 blades (unit ℃)

Set No.	20:00	23:00	2:00	5:00	8:00	11:00	14:00	17:00
1#	70.51	69.62	68.60	66.54	64.90	67.60	67.58	67.92
2#	65.49	63.42	62.79	60.20	59.08	62.46	62.51	62.72

According to the results, when the number of blades of the cooling fan changed from 8 to 10, the flow rate of cooling air increased by more than 8% under the same condition. The main reason is that when the number of blades increased, the performance curve of the fan improved and pressure rise easily overcame the resistance loss of the air cooler.

4. Conclusions

The numerical simulation model of the whole air cooling unit was established, and the influence of local structure and environment temperature on the performance of fan and heat exchanger was simulated and compared. The main conclusions were as follows.

1) Air cooler entrance muffler has little effect on the flow properties of air cooler. The cooling air flow was only reduced by 1.02% when there are mufflers at the entrance. The air flow rate can be increased by 1.28% when the outlet contraction section was removed. The air flow rate increased by 0.73% when the mixed flow region beside heat exchange was blocked.

2) When the fouling thickness of finned tubes were increased from 0.1mm to 0.5mm, the resistance loss of finned tubes was 3.94 times than that of no fouling, which led the cooling air flow rate reduce by 27.55%, and resulted in decreasing of efficiency of heat exchange. Therefore, cleaning fouling can effectively improve the heat transfer efficiency of air cooling unit.

3) When the ambient temperature increased from 243 K to 323 K, cooling air flow rate reduced by 17.94% and the flow performance of two fans reduced by about 17%, which resulted in lower efficiency of air cooling unit. Cooling air flow rate increased by about 8% when the blade number increased from 8 to 10, which could effectively improve the heat exchange capacity of air cooler.

References:

[1] Environmental Monitoring In China. Emission standard for industrial enterprises noise at boundary[S]. 2008.

[2] E B A. Heat transfer enhancement second generation heat transfer technology[C]. Proctor. 10th UIT National Heat Transfer Conference. 1992: 3-21.

[3] Duvenhage K K D G. The influence of wind on the performance of forced draught air-cooled heat exchangers[J]. Journal of Wind Engineering & Industrial Aerodynamics,1996,62(2-3):259-277.

[4] Staden M P V P L. Numerical modelling of the effects of ambient conditions on large power station air cooled steam condensers[J]. American Society of Mechanical Engineers Fluids Engineering Division Fed,1995,39:387.

[5] Liu Zhanbin. Numerical simulation and experimental research of heat transfer process of fin tube [D]. Xi'an University of Technology,2008.

[6] Cheng Yuanda. Numerical investigation on the heat transfer and fluid flow characteristic of finned-tube heat echanger and direct air cooled condenser unit [D].Beijing Jiaotong University,2008.

[7] Bu Yongdong. Flow and heat transfer characteristics and air-side flow field optimization of air-cooledcondenser cell [D]. North China Electric Power University,2013.

[8] Ding Liyuan. Numerical simulation and analysis on Internal Flow Field of Air-cooled Unit with Deflectors [D]. North China Electric Power University,2014.

[9] Chen Wenfei. Analysis on influence characteristic of dust accumulation for air-cooled condenser[D]. North China Electric Power University,2013.

[10] Xu Kuichang. The handbook of fan[M]. China Machine Press,2011.

IOP Conf. Series: Materials Science and Engineering **232** (2017) 012075 doi:10.1088/1757-899X/232/1/012075

Effect of end-wall riblets on radial turbine performance

M A Khader[1], A I Sayma[2]
City, University of London, UK

Email: [1]Mahmoud.khader.1@city.ac.uk, [2]a.sayma@city.ac.uk

Abstract. This paper presents a detailed study of the impact of manufacturing residual riblets at the rotor hub surface of a radial inflow turbine on the flow within the rotor passages and their contribution to drag reduction. Numerical analysis has been used to study the effects of those features at design point conditions. Riblets with different height and spacing have been examined to determine the riblet geometry where the maximum drag reduction is achieved. The relative height of the riblets to rotor inlet blade height was introduced to generalise the results. At the end of this study the results were compared with the available data in literature. It was found that the introduction of riblets could reduce the wall shear stress at the hub surface, while they contribute to increasing the streamwise vorticity within the rotor passage. For the geometries tested, the minimum drag was achieved using riblets with relative height $h_{rel} = 2.5\%$ equivalent to 19.3 wall units. The results revealed that the spacing between riblets have a minor effect on their performance, this is due to the size of the streamwise vortex above the hub surface which will be discussed in this work.

1. Introduction

Flow within the radial turbine rotors is highly three-dimensional and is combined with strong secondary flows, which become more complicated near the passages walls. The interaction between the turbulent boundary layer and the solid surfaces affects the boundary layer development and causes energy losses due to skin friction drag. Surface roughness is one of the important factors that affect the flow within the boundary layer. Therefore it is necessary to study the effects of the manufacturing residual riblets on the losses within the turbine rotor. The improved understanding could lead to the introduction of intentional riblets that are increasingly becoming feasible with the fast development of three dimensional printing technology of components. Different studies have proved that smooth surfaces do not necessarily have less drag than rough surfaces [1, 2, 3, 4]. Riblets are surface structures arranged in the streamwise direction to help in reducing wall shear stresses. They can have different shapes and arrangements as shown in figure 1. Walsh [3] studied the effect of different riblet shapes on drag reduction over flat plate surface, where 8% drag reduction was achieved using V-shaped riblets. By adding V-shaped riblets to the blades of axial compressor cascade, Oehlert et al. [5] achieved 3.6% reduction in total pressure loss. Feng et al. [6] conducted an experiment to test the effectiveness of riblets in drag reduction where they achieved 10% drag reduction by adding the riblets to the pressure surface of an axial compressor cascade. They also noticed that this reduction in drag is insensitive to the flow incidence angle. On the contrast, Bushan et al. [7] found that adding riblets to a straight rectangular duct has a negative impact on drag. In their experiment, blade shape riblets with different heights and spacing were attached to the duct wall, and the pressure drop was measured across the duct for range of flow velocities. For all riblets arrangements and along the speed range the pressure drop was higher compared to the surface without riblets.

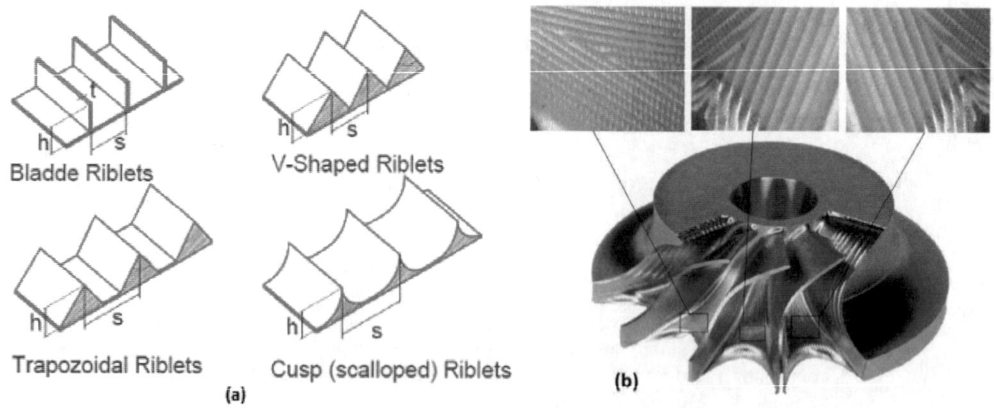

Figure 1: (a) Different riblet configurations (b) Turbine rotor with three different hub surface finish

Zhang et al. [8] explored the effects of adding riblets to an axial turbine hub cascade experimentally. The grooves of the Riblets were machined in the streamwise direction. The study revealed that riblets can reduce pressure side of the horse-shoe vortex, and alleviate the growth of the passage vortex. Kim et al. [9] introduced a single rectangular fence in a 90° turning to mimic the effect of riblets on turbine passage. The mass-weighted average total pressure loss in the duct with fence was reduced by 8.6% relative to the duct without the fence. Miao et al. [10] found that introducing riblets with different shapes and density to a 90° duct alleviates the stream wise vorticity compared to smooth walls duct. Riblets concept has been also applied to radial turbomachines. Where Lei et al. [1] tested a centrifugal compressor with different machining residual riblets heights on the impeller hub. They found that using riblets decreases wall shear stress which reaches its minimum value for riblets of height 0.58 relative to impeller exit blade width.

Walsh [3] investigated the effectiveness of several riblet geometries on drag reduction where he found that the highest drag reduction could be achieved using riblets with cusp shape. In a different study, Oehlert et al. [5] found that riblets with blade shape has greater effect on drag reduction over the cusp shaped riblets. Nonetheless, blade riblets can neither be machined nor they would withstand high stresses. Moreover, the residual surfaces resulted from the actual machining process of the turbine rotor are very similar to cusp riblets. Figure 1b shows rotor sample machined using ball end cutter milling with different surface finish, where cusp riblets are clear in the magnified surfaces. So, for practical reasons, the present work focused on studying the effect of cusp shape riblets.

To the knowledge of the authors, the effects of the residual surface structures on the radial turbines performance were not addressed before, therefore it was important to include riblets with different geometry in the study. It is understood from the literature that the main geometrical parameters that affect the riblets performance are the height of the riblets and the spacing between them. Therefore, riblets with different heights (h) and spacing (s) were introduced to the hub surface of a turbine rotor designed for 6 kWe micro gas turbine. The turbine stage was then analysed numerically at design point. Table 1 lists the turbine operational conditions and figure 2 shows the main geometrical parameters for the rotor.

Table 1: Turbine specifications

Parameter	Value
Total Pressure Ratio (bar)	3
Turbine inlet total temperature (K)	1073
Turbine mass flow rate (kg/s)	0.09
Power generated by turbine (kW)	21
Rotational speed (rpm)	130,000

To generalise the results of the study, the dimensionless parameter h_{rel} was introduced to characterise the size of the riblets. h_{rel} is defined as the percentage ratio of the cusp height h to the rotor inlet blade height b (figure 3). The space between riblets was defined using the polar angle (Ø) between the adjacent riblets as the area of the passage is changing along the streamwise direction. Thus, the distance s can be calculated at certain radius of the hub as:

$$s = r_{hub}\emptyset \quad (1)$$

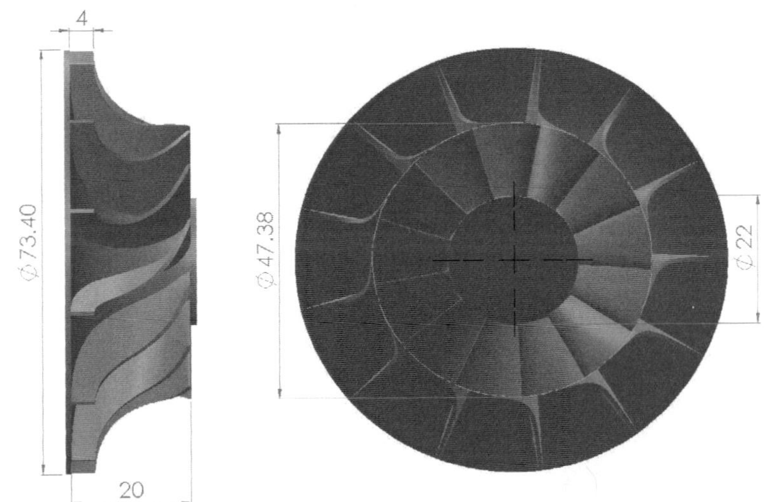

Figure 2: rotor dimensions [mm]

The detailed dimensions for the riblets used in this study are summarised in Table 2.
The computational domain included a single rotor passage and single Nozzle Guide Vanes (NGVs) passage. The commercial tool ANSYS TurboGrid was used to generate hexahedral structured mesh for the stator vane, where ANSYS Workbench was used to generate unstructured mesh for the rotor. CFD computations were performed using ANSYS CFX 15.0. CFX is Reynolds Averaged Navier Stokes (RANS) based solver which solve the flow equations using finite volume formulation. The flow was assumed to be fully turbulent due to high Reynolds number in the rotor passage. The Reynolds number was calculated assuming a rectangular duct, therefore using the passage hydraulic diameter, the Reynolds number at the design point was 27000.

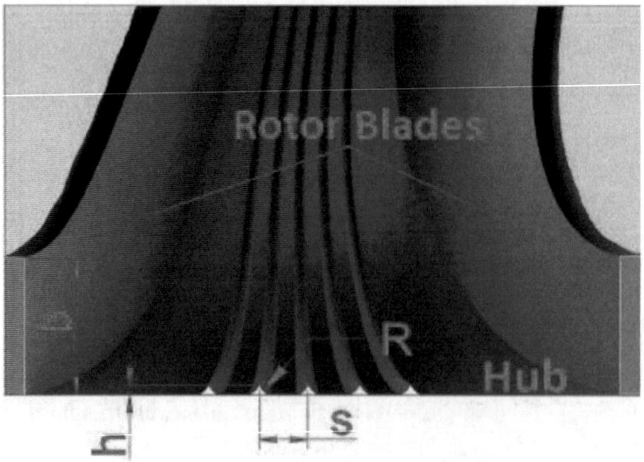

Figure 3: Riblets terminology

Dai et al. [11], found that turbulence intensity along the rotor passage does not exceed 7%, so 5% has been used in this work. The Shear Stress Transport (SST) turbulence model was used as it showed a good agreement with the experimental data for radial turbines in previous studies [12, 13].

Non-slip boundary condition was set for the passage surfaces. To solve the boundary layer accurately, the near wall mesh was refined to achieve first node non-dimensional distance $y^+ = 1$. Stage connection (mixing plane) was used between the rotor and stator, where all fluxes are averaged through the interfaces surfaces. Uniform total pressure and temperature were set at the domain inlet and uniform static pressure at the domain outlet. As one vane and one rotor blade passage were used, rotational periodicity was imposed on the periodic boundaries. Using this setup, CFX was run for a steady state solution.

Table 2: riblets dimensions

h [mm]	**h_{rel} [mm]**	**R [mm]**	**s [3 riblets] $\emptyset = 4.45°$**	**s [5 riblets] $\emptyset = 2.2°$**	**s [10 riblets] $\emptyset = 1.1°$**
0.06	1.5%	0.8	$0.078r_{hub}$	$0.039r_{hub}$	$0.019r_{hub}$
0.1	2.5%	0.6	$0.078r_{hub}$	$0.039r_{hub}$	$0.019r_{hub}$
0.2	5%	0.6	$0.078r_{hub}$	$0.039r_{hub}$	$0.019r_{hub}$

Grid independence check was performed for both numerical domains (smooth and ribbed rotors). For the rotor with riblets, the grid independence study was performed on the rotor with the highest cusp and smallest space as it requires the highest quantity of grid nodes to model those features. Using five grids of increasing refinement, the absolute vorticity, mass flow rate, flow velocities and angels at rotor inlet and outlet were calculated and compared. The For the domain with riblets, the girds density varied between 400,000 and 2 million nodes, and for the rotor without riblets, the this was between 0.4-1.7 million nodes. Figure 4 shows the error percentage in calculating the vorticity referring to the value generated using the finest grid. Vorticity showed the highest sensitivity to grid refinement, this is due to the small turbulent structures that needed fine mesh to be captured. For the domain with riblets the error is reduced to less than 1% with the mesh of 1.2 million nodes, and for the domain without riblets the error is reduced to less than 0.9% using the grid with 900k nodes. The final mesh for both rotors is shown in figure 5.

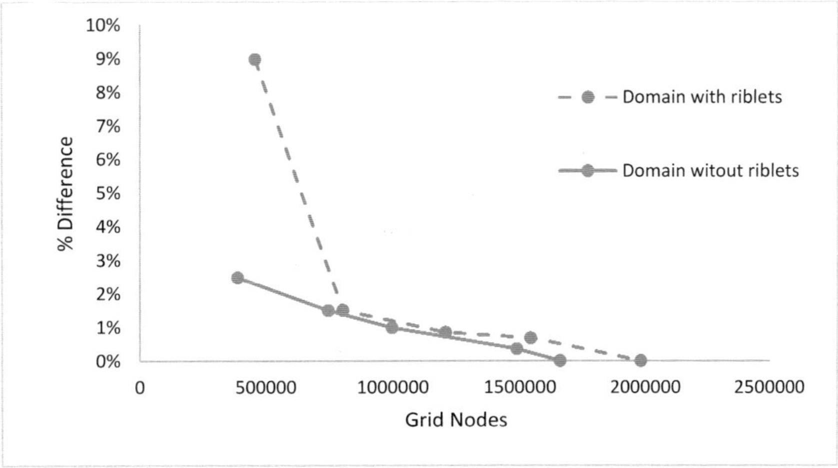

Figure 4: Grid independence study for computational domain with riblets

2. Validation of the CFD methodology

To verify the accuracy of the numerical method used, flow solutions were compared with experimental data. Experimental studies describing the flow within the turbine rotor are scarce. One of the main reasons for that, is difficulty to insert measuring instruments and the lack of the optical access to the rotor passage. Also, the high operating temperatures is another obstacle in front of using different flow visualisation techniques. Moreover, most of the published experiments does not have an adequate information to regenerate the geometry. Jones [14] provided the experimental data and the geometry description for the turbine used in Sundstrand Power Systems T-100 Multipurpose Power Unit. The turbine generates 50 hp at pressure ratio 5.5 and a turbine inlet temperature 1056 k while running at 105,600 rpm which is very close to the turbine used in this study.

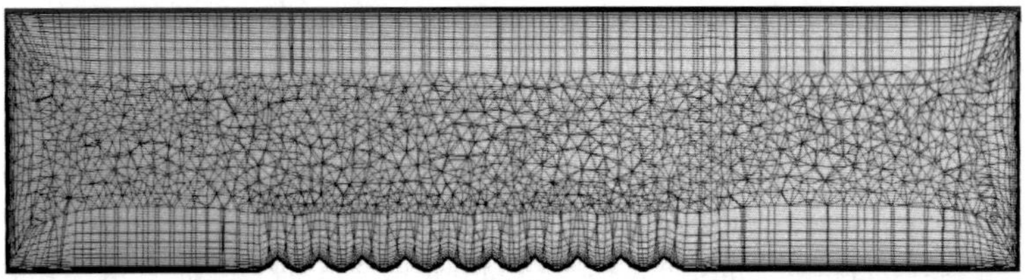

(a)

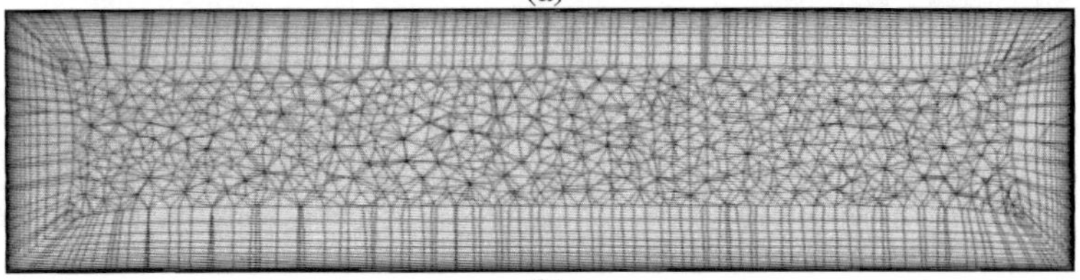

(b)

Figure 5: Computational grid (a) rotor passage with riblets (b) rotor passage without riblets

The absolute flow angles at the rotor exit are compared in figure 6a, where CFD results show a good agreement with the experimental results. However, CFD predictions deviate from the experimental near the tip region. This difference could be due to the rotor back-face clearance flow which was not included in the numerical domain. Sun et al. [15] noticed similar deviation in the flow angle predicted using CFD when they studied the effects of back-face clearance gap on the flow in the turbine passage. Figure 6b shows a comparison between CFD and test rig results for the span-wise distribution of meridional velocity at the rotor exit. It can be noticed that the difference between the values is almost constant, which means that this deviation is related to the over estimation of the mass flow in the CFD analysis.

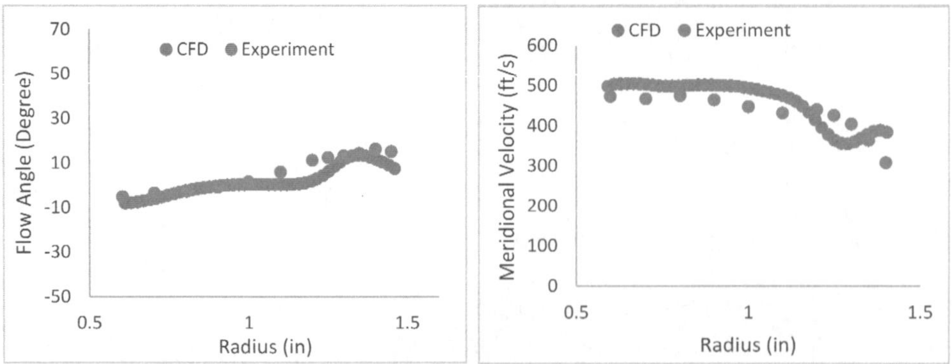

(a) (b)

Figure 6: Spanwise distribution of (a) flow angle (b) meridional velocity at rotor discharge

This could be because of the assumption of smooth profile of the turbine components, which will reduce the boundary layer blockage allowing more flow to pass. Moreover, during the test, the flow is expected to leak through the back-face seal of the turbine which could reach up to 1% of mass flow. Generally, the simulation results agree with the experimental data. Therefore, the present CFD methodology was considered to be suitable to study the flow behaviour through the turbine rotor, and is capable to demonstrate the effect of adding riblets on the turbine performance.

3. Results and Discussion

This section presents a description for the secondary flow motion in radial turbine passage, operating mechanism of riblets and comparison of different riblets geometries on drag reduction.

3.1 Secondary Flow in the Radial Turbine Rotor

In the literature; authors divided secondary flow characteristics in a turbine rotor into three categories based on their location along the passage. At the inlet section where the flow is purely radial, Coriolis acceleration has a very large effect on the low momentum fluid at the hub and shroud; therefore it acts to move flow from the Pressure Surface (PS) to the Suction Surface (SS) where the reduced static pressure ($P_r = P_s - \frac{1}{2}\rho\omega^2 r^2$) is minimum. Along the rotor bend, where the flow starts to change its direction from radial to axial, secondary flows are generated because of both passage rotation and curvature. Coriolis acceleration is responsible for the movement of the low energy fluid from PS to the SS at the hub and shroud surfaces. Also, the passage curvature acts to move the low energy fluid at the pressure and suction surfaces from hub to shroud. For a fluid particle that is moving along the bend with a particular velocity there is a pressure gradient acting normal to its direction of motion to balance the centrifugal force on it. This pressure gradient is constant along the spanwise direction while the near wall fluid particles have a lower velocity than particles in the free stream. This difference in velocity causes the near wall particles to move toward the inner curve to reduce the radius of curvature [16]. Towards the end of this section, the secondary vortices at the hub start to move towards the top middle of the passage.

At the exducer section, the rotor passage is curved in the tangential direction. This curvature causes the low momentum fluid at the hub and shroud to swipe towards the SS where the curvature radius is minimum. Also, the radial component of the Coriolis acceleration causes a movement of the low energy fluids at the PS and SS from hub to shroud. In this section the secondary vortices accumulate in middle and top parts of the passage close to the suction surface. Figure 7 illustrates the motion of the low momentum flows at the different sections of the rotor passage [17].

IOP Conf. Series: Materials Science and Engineering **232** (2017) 012075 doi:10.1088/1757-899X/232/1/012075

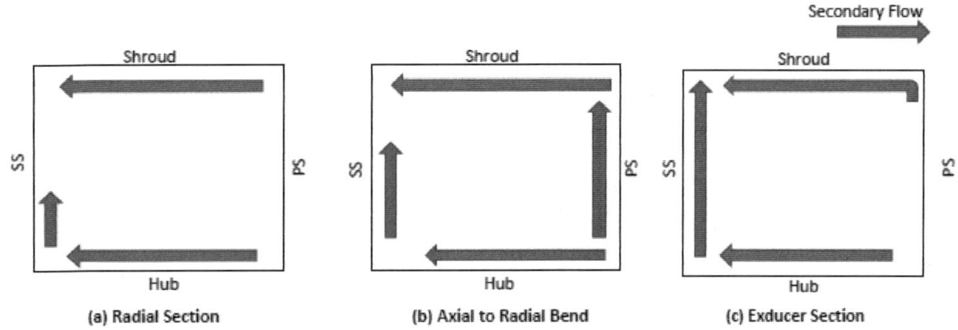

Figure 7: Secondary flow motion along turbine rotor passage

3.2 Riblets Operating Mechanism

To understand the operating mechanism of riblets, the flow structure is compared between ribbed and smooth hub rotor passage. In this section the smooth hub surface was compared with surface with 5 riblets of h_{rel} 2.5%. Figure 8 shows the cross-flow motion at the hub surface at 40% cross plane. It can be noticed that the cross-stream flow interacts with the riblets at the hub surface impeding it movement and generating secondary vortices that prevent the streamwise vortex from moving inside riblets valley. This will retain the slow fluids inside riblets valley thus the momentum exchange in the boundary layer is reduced. The interaction between the stream wise vortex and the low momentum fluid close to the hub causes high local shear at the rotor hub surface. By introducing riblets to the wall, this vortex is shifted from the wall to interact with the riblets tips.

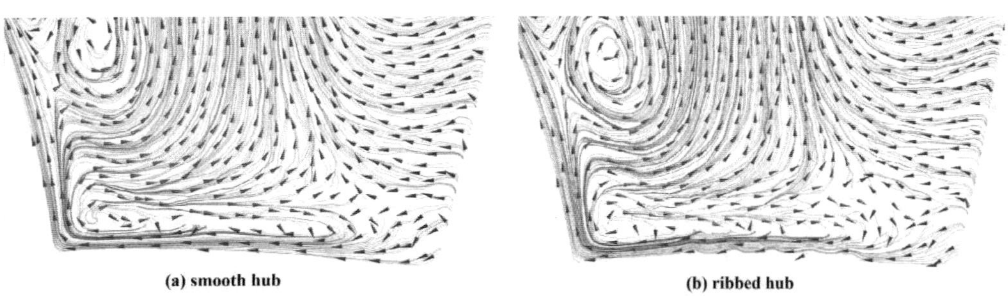

Figure 8: effect of riblets on cross stream motion of the low energy fluids near the hub surface

This interaction generates secondary vortices which weaken the streamwise vortex by interfering with the cross-stream motion of the low momentum fluids at the hub surface and keeps the streamwise vortex away from the wall. This reduction of the cross-flow fluctuation and shifting the streamwise vortex away from the wall reduces the turbulent momentum transfer which results in reducing skin friction. The limiting streamlines (velocity streamlines at a distance normal to the wall approaching zero) at the hub surface in figure 9a shows the strong relative motion of the low energy fluids at the hub surface. Introducing riblets to the hub surface (figure 9b) form an obstacle which retards the cross-stream motion and tries to align them in the streamwise direction.

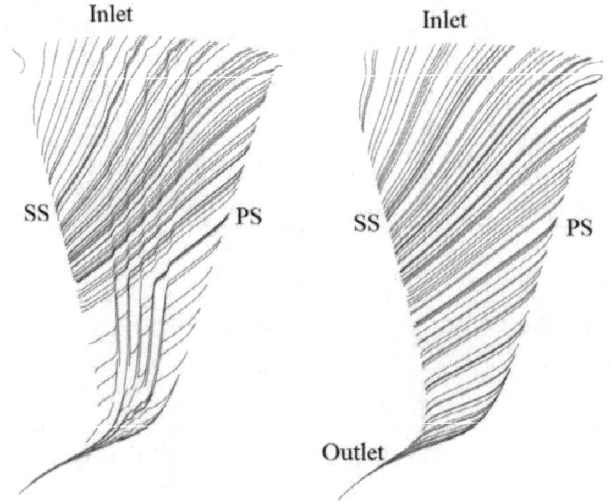

(a) Hub with Riblets (b) Smooth Hub

Figure 9: Limiting streamlines at the rotor hub

In figure 10, the Q-criterion method was used to visualise the streamwise vortex over smooth and ribbed hub surfaces at 40% cross plane. As it can be seen; the interaction between the cross flows and the riblet tips generates secondary vortices that weaken the streamwise vortex and form a barrier that lifts it from the hub surface.

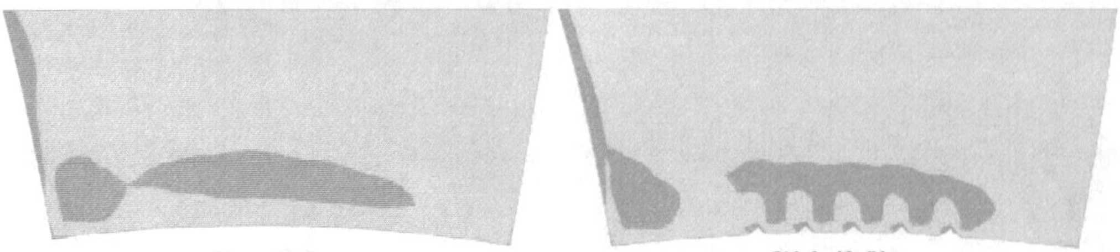

(a) smooth hub **(b) hub with riblets**

Figure 10: Stream-wise vortex structure over hub surface

The mass averaged vorticity normalised by multiplying by (Inlet blade height (b)/ Inlet flow velocity) is presented in figure 11. The vorticity value shows a noticeable reduction inside the riblets valley compared to the smooth surface which confirms that riblets weaken the streamwise vortex close to hub surface. However, the vorticity at the outer region of the riblets valley is slightly higher because of the secondary vortices generated at the riblets tip. At the tip region, the vorticity reaches its maximum value in the region where the secondary vortices are located. Nonetheless the secondary vortex close to riblet tip does not have a significant effect on increasing skin friction as the vortex size is relatively small and it interacts with very small area at the peak of the riblet.

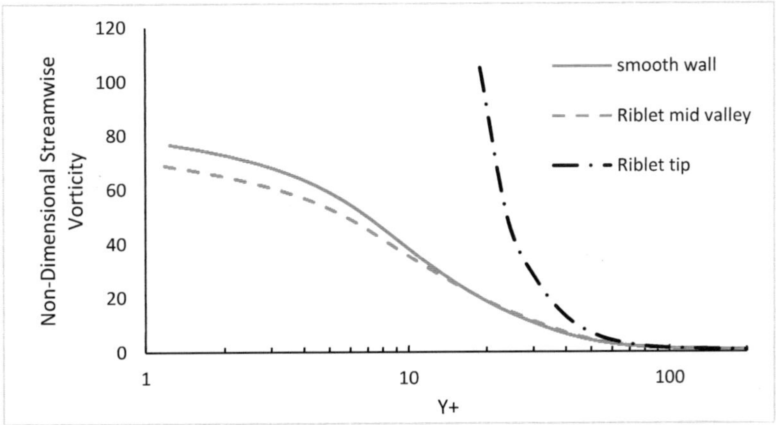

Figure 11: mass averaged streamwise vorticity at 40% stream-wise location

To evaluate the effect of increasing the vorticity at riblets tips on skin friction, the wall shear was calculated along the hub surface at the 40% plane. Figure 12 reveals that increasing the vorticity at riblets tips increases the wall shear at small areas around the tips, while the wall shear inside the riblets valley is reduced keeping the average shear below that for the smooth surface.

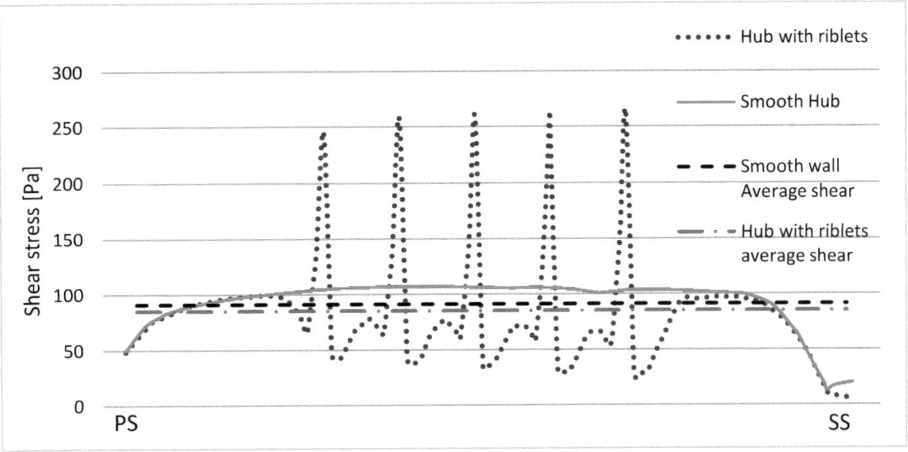

Figure 12: wall shear at the hub surface calculated at 40% stream-wise location

The turbulent kinetic energy profiles over smooth hub, riblets valley and riblet tip are plotted in figure 13. It is obvious that introducing the riblets to the hub surface damps the velocity components fluctuations leading to reduction in turbulent kinetic energy. Chun et al. [4], Lee et al. [18] and Duan [19] have spotted similar drop in turbulent kinetic energy when riblets were introduced to flat plate surfaces.

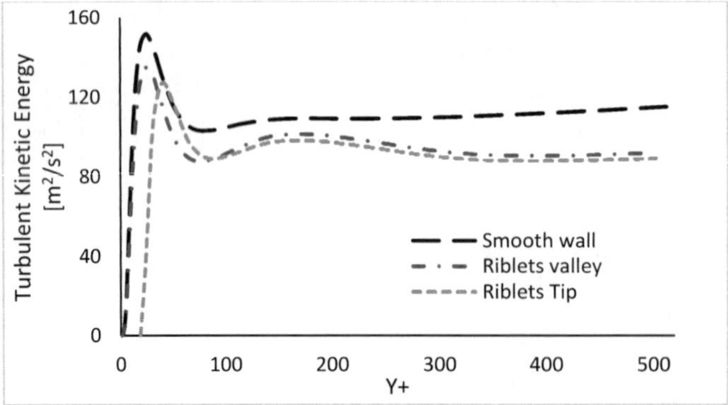

Figure 13: comparison of turbulent kinetic energy for both smooth rotor and rotor with riblets

3.3 Riblets Geometry Effect

In this section the effect of different riblets geometries (refer to table 2) on drag reduction is examined and compared with the smooth wall results. Figure 14 compares the area averaged shear at the hub surface for the riblets with different height and spacing. The results reveal that the spacing between riblets has a minor effect on the wall shear stress reduction and the key parameter that affects the riblets performance is their height. The highest drag reduction was achieved using riblet with relative height 2.5%, whilst increasing riblet height to 5% has a negative impact, where drag increases at the hub surface.

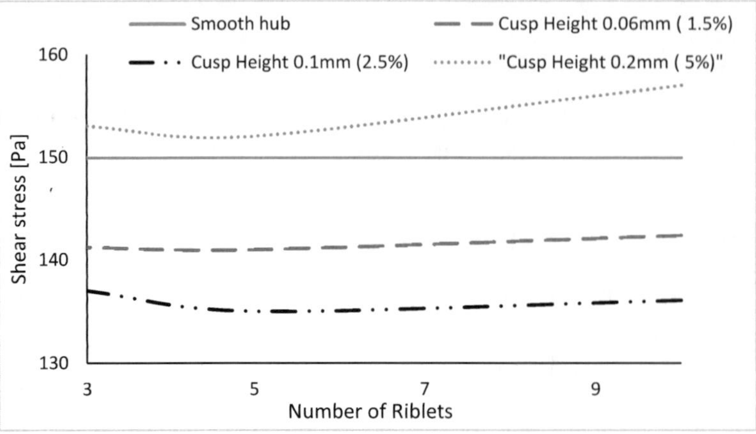

Figure 14: Wall shear stress comparison between different riblets geometries

Most of the research studies performed on the riblets set geometrical conditions on riblets performance. According to those studies; riblets reduces drag for normalised spacing $s^+ < 30$ and normalised height $h^+ < 25$ wall units. Figure 15 presents the change in drag ($\Delta\tau$ defined in eq. 2) for riblets with different spacing s^+ and constant height ($h_{rel} = 2.5$, $h^+ = 19.3$).

$$\Delta\tau = (\tau_r - \tau_s)/\tau_s \qquad (2)$$

Where, τ_s and τ_r are the wall shear stresses at the smooth and ribbed hub surfaces respectively. The graph shows that the spacing s+ has a negligible effect on riblets performance, it shows also that riblets can reduce drag for high values up to 300 wall units. This conflicts with the values in literature and the value presented here is mainly due to the type of the flow. Most of the available experimental data that limits riblet spacing to 30 wall unit are done for external flows. This is directly related to the streamwise vorticity at body surface, where size of the streamwise vortices for external flows is about 30 wall units in diameter. Therefore, increasing the space between riblets to more than that will provide the freedom for the streamwise vortex to move freely in the area between them. This will cause increment in area exposed to high speed cross flow which in turn increases skin friction [18]. On the other hand, the streamwise vortex at the turbine hub occupies almost the whole passage width, therefore increasing the spacing s^+ didn't change the performance of the riblets (see figure 11).

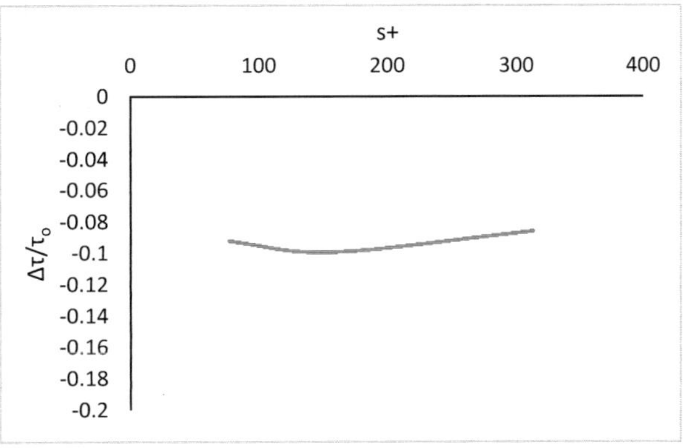

Figure 15: drag reduction for different riblets spacing

Figure 16 presents the change in drag for different riblets height h^+ at constant rib spacing ($s = 0.039r_{hub}$, $s^+ = 158$). Wall shear using riblets is reduced until reaching the rib height ($h^+ = 19.3$) where the wall shear is minimum, while increasing riblets height further leads to increase the wall shear. A similar trend for riblets performance within internal flows was spotted by Lei et al. [1], where the drag reduction was found to be related to riblet height. Increasing riblets height increases the generation of secondary vortices which increases the down washing motion of the high-speed fluids around riblet tip causing more drag.

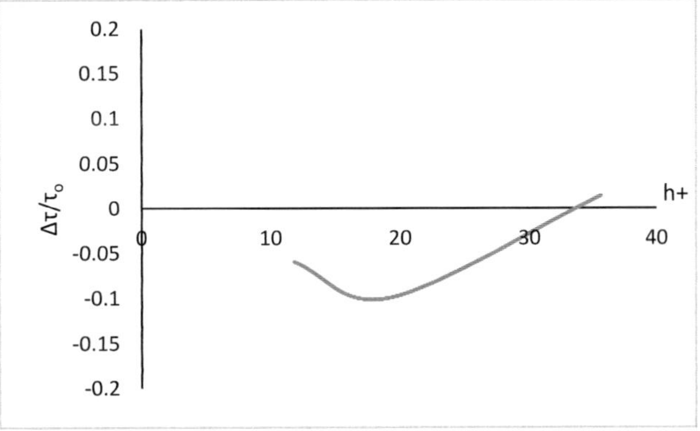

Figure 16: drag reduction for different riblets spacing

Figure 17 shows the mass averaged streamwise vorticity along the turbine passage. The graph shows that increasing the height of the riblets increases the streamwise vorticity along the passage. This explains the performance results in figure 18 where the turbine with riblets of height $h_{rel} = 1.5\%$ has a higher efficiency than $h_{rel} = 2.5\%$ even the latest shows better performance in reducing drag.

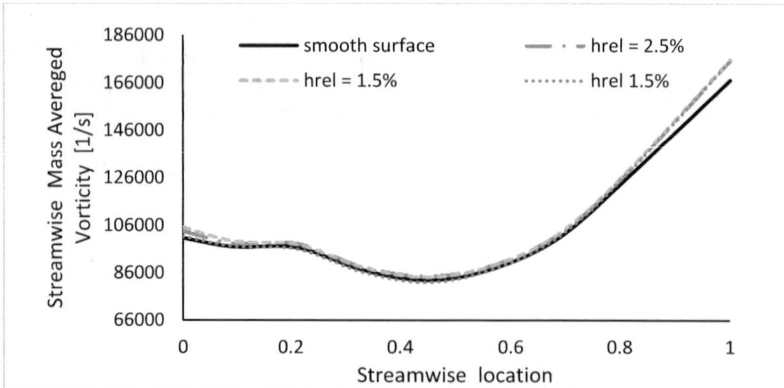

Figure 17: Mass averaged streamwise vorticity along rotor passage

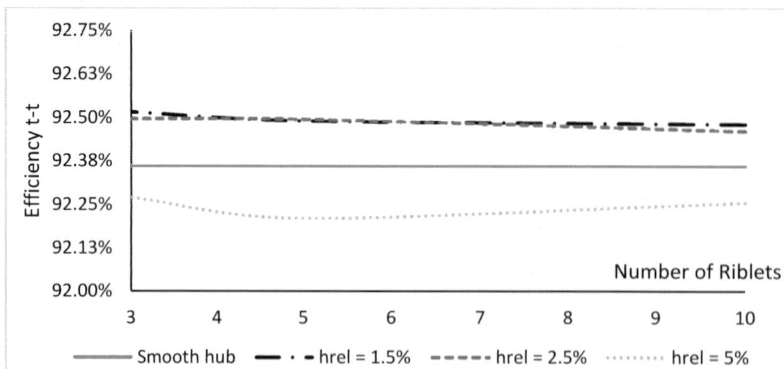

Figure 18: Turbine performance for different riblets geometry

Conclusion

In this work the effects of residual riblets on radial turbine performance were analysed numerically. It was found that residual riblets reduce the cross-stream motion of the low momentum fluids leading it to move from pressure to suction side of the rotor passage and separate the streamwise vortex from interaction with hub surface. This entrains low speed fluid in the riblets valley which in turn reduces the momentum transfer in the boundary layer causing reduction in wall shear stress.

The spacing between riblets was found to be of secondary effect on their performance in drag reduction as the streamwise vortex occupies most of the passage length. Riblets height was found to be the main parameter affecting drag reduction, where the maximum drag reduction found to occur with riblets with h_{rel} = 2.5%. Introducing riblets to hub surface is beneficial for drag reduction while it increases the streamwise vorticity in the rotor passage and the maximum performance was achieved for the turbine with the shortest riblets (h_{rel} = 1.5%). Future work will focus on studying the effect of riblets on the turbine performance at off-design conditions.

Nomenclature

Y^+	Non-dimensional wall distance: $Y^+ = \frac{Yu_\tau\rho}{\mu}$
s+	Non-dimensional spacing between riblets: $s^+ = \frac{su_\tau\rho}{\mu}$
h^+	Non-dimensional riblets height: $h^+ = \frac{hu_\tau\rho}{\mu}$
u_τ	Friction velocity: $\sqrt{\tau_o/\rho}$
τ_o	Wall Shear

ρ	Density
ω	Rotational speed
r	radius
P_s	static pressure
P_r	reduced static pressure

References

[1] D. Lei and L. Cao, "Effects of Residual Ribles of Impeller's Hub Surface on Aerodynamic Performance of Centrifugal Compressors," *Engineering Applications of Computational Fluid Mechanics,* vol. 9, pp. 99-113, 2005.

[2] M. Walsh, "Turbulent Boundary Layer Drag Reduction Using Riblets," in *20th Aerospace Sciences Meeting*, Orlando, 1980.

[3] M. Walsh, "Riblets as a Viscous Drag Reduction Technique," *AAIA Journal,* vol. 21, pp. 485-486, 1983.

[4] H. Chun, O. El-Samni and H. Yoon, "Drag Reduction of Turbulent Flow over Thin Rectangular Riblets," *International Journal of Engineering Science,* vol. 45, pp. 436-454, 2007.

[5] K. Oehlert and J. Seume, "Exploratory Experiments on Machined Riblets on Compressor Blades," in *ASME Fluids Engineering Division*, Miami, 2006.

[6] C. Fang , C. Zhang and T. Ping, "An Experimental Investigation of Loss Reduction with Riblets on Cascade Blade Surface and Isolated Airfoils," in *ASME Gas Turbine Division*, New York, 1990.

[7] B. Bhushan and B. Dean, "The Effect of Riblets in Rectangular Duct Flow," *Applied Surface Science,* vol. 258, no. 8, pp. 3936-3947, 2012.

[8] Z. Q, X. Miao, L. Wang, J. H and H. Qi, "Application of Riblets on Turbine Blade End-wall Secondary Flow Control," *Journal of Propulsion and Power,* vol. 31, pp. 1578-1585, 2015.

[9] K. Kim, J. Cho, J. Kim and E. Jeong, "Controlling the Secondary Flows Near Endwall Boundary Layer Fences in a 90° Turning Duct Using Approximate Optimization Method," *Journal of Mechanical Science and Technology,* vol. 25, pp. 2025-2034, 2011.

[10] X. Miao, Z. Sun, Q. Zhang and C. Atkins, "End-wall Seconary Flow Control Using Engineering Residual Surface Structure," in *ASME Turbo Expo*, Seoul, 2016.

[11] J. Dai, N. Ijichi, H. Tange, H. Shibata, H. Tamaki and S. Yamaguchi, "Comparison of internal flow field between experiment and computation in a radial turbine impeller," *JSME international Journal,* vol. 47, pp. 48-56, 2004.

[12] J. E. Bardina, P. G. Huang and T. J. Coakley, "Turbulence Modeling Validation, Testing, and Development," Ames Research Center, Moffett Field, California, 1997.

[13] Q. Deng, J. Niu and Z. Feng, "Study on the leakage flow characteristics of radial inflow turbines at rotor tip clearance," *Science in China Series E: Technological Sciences,* vol. 51, 2008.

[14] A. Jones, "Design and Test of a Small, High Pressure Ratio Radial Turbine," in *ASME int. gas turbine congress*, Netherlands, 1994.

[15] Z. Sun, C. Guo, H. Chen and C. Tan, "Aerothermal Investigation of Backface Clearance Flow in Deeply Scalloped Radial Turbines," *Journal of Turbomachinery,* vol. 135, 2013.

[16] E. M. Greitzer, C. Tan and M. B. Graf, Inetrnal Flow: Concepts and Applications, Cambrige University Press, 2007.

[17] M. Zangeneh-Kazemi, W. Dawes and W. R. Hawthorne, "Three Dimensional Flow in Radial-Inflow Turbines," in *Gas Turbine and Aeroengines Congress*, Amsterdam, 1988.

[18] S. J. Lee and S. H. Lee, "Flow Field Analysis of Turbulent Boundary Layer Over A Riblet Surface," *Experiments in Fluids Springer,* vol. 30, pp. 153-166, 2001.

[19] L. Duan, "Effects of Ribelts on Skin Friction in High-Speed Turbulent Boundary Layers," in *AIAA 50th* , Nashville, 2012.

IOP Conf. Series: Materials Science and Engineering **232** (2017) 012076 doi:10.1088/1757-899X/232/1/012076

Non-symmetric approach to single-screw expander and compressor modeling

Davide Ziviani[1,2], Eckhard A. Groll[2], James E. Braun[2], W. Travis Horton[2], M. De Paepe[1], M. van den Broek[1]

[1]Department of Flow, Heat, and Combustion Mechanics, Ghent University - UGent Sint-Pietersnieuwstraat 41, 9000 Gent, Belgium

[2]Purdue University, School of Mechanical Engineering, Ray W. Herrick Laboratories, West Lafayette, IN, USA

E-mail: davide.ziviani@ugent.be, groll@purdue.edu, jbraun@purdue.edu, wthorton@purdue.edu, martijn.vandenbroek@ugent.be, michel.depaepe@ugent.be

Abstract. Single-screw type volumetric machines are employed both as compressors in refrigeration systems and, more recently, as expanders in organic Rankine cycle (ORC) applications. The single-screw machine is characterized by having a central grooved rotor and two mating toothed starwheels that isolate the working chambers. One of the main features of such machine is related to the simultaneous occurrence of the compression or expansion processes on both sides of the main rotor which results in a more balanced loading on the main shaft bearings with respect to twin-screw machines. However, the meshing between starwheels and main rotor is a critical aspect as it heavily affects the volumetric performance of the machine. To allow flow interactions between the two sides of the rotor, a non-symmetric modelling approach has been established to obtain a more comprehensive model of the single-screw machine. The resulting mechanistic model includes in-chamber governing equations, leakage flow models, heat transfer mechanisms, viscous and mechanical losses. Forces and moments balances are used to estimate the loads on the main shaft bearings as well as on the starwheel bearings. An 11 kWe single-screw expander (SSE) adapted from an air compressor operating with R245fa as working fluid is used to validate the model. A total of 60 steady-steady points at four different rotational speeds have been collected to characterize the performance of the machine. The maximum electrical power output and overall isentropic efficiency measured were 7.31 kW and 51.91%, respectively.

1. Introduction

Single-screw machines are widely employed as compressors and more recently as expanders in organic Rankine cycle applications [1, 2].

One of the main features of such machines is that the meshing between the main rotor and the starwheels (or gaterotors) generates identical working chambers on both sides of the main rotor with benefits on the loads acting on the main shaft bearings, as shown in Figure 1. The common approach to model the single-screw machine is to consider only one side of the compressor or expander with multiple control volumes according to the number of grooves [3]. The authors proposed a comprehensive mechanistic model of a single-screw expander accounting for the wrapping of the grooves at the end of expansion process [4].

Nomenclature

c_v	Specific Heat at Constant Volume (J kg^{-1} K^{-1})
D	Diameter (m)
h	Specific Enthalpy (kJ/kg)
	Heat Transfer Coefficient (kW/m^2K)
i	Transmission Ratio (-)
m	Mass (kg)
$\dot{m}$	Mass Flow Rate (kg/s)
p	Pressure (kPa)
$\dot{Q}$	Heat Rate (kW)
T	Temperature (K)
u	Specific Internal Energy (kJ/kg)
v	Specific Volume (m^3/kg)
V	Volume (m^3)
$\dot{W}$	Power (W)
$\mathbf{v}$	Velocity Vector (m/s)
w	Tooth width (m)
z	Number of Grooves or Teeth (-)
η	Efficiency (-)
ρ	Density (kg/m^3)
θ	Rotation Angle (rad)
ω	Angular Speed (rad/s)

Subscript

amb	ambient
CV	Control Volume
ex	discharge
exp	expander
el	electric
eff	effective
fr	friction
g	groove
loss	losses
mech	mechanical
nom	nominal
su	suction
sh	shaft

In the case of an expander, the discharge process is important to evaluate under or over expansion losses.

The majority of the research works have considered the expansion process until discharge opening without including the complete evolution of each groove. The authors extended the original mechanistic model to include the full evolution of each groove during one working cycle in [5, 6]. Preliminary CFD results of a single-screw expander under transient conditions showed that the pressure distributions at the suction inlet are not identical due to the internal ducting pressure drops as well as because the two sides of the rotor are not fully isolated [7]. For these reasons, a non-symmetric approach to model single-screw machines is proposed within this work with the aim of estimating the indicated power (p-V work rate), the friction losses and the mechanical efficiency. A SSE with an 11 kWe generator, shown in Figure 2, with a 122 mm rotor is used as reference case. The expander performance was characterized my using R245fa as working fluid.

2. Mechanistic Model

The single-screw expander is divided into ten control volumes (CVs) of which six are associated with working chambers (i.e., grooves) and their volumes change according to the rotation angle. The remaining static control volumes are associated with the common inlet, the two ducts leading to the suction ports and the discharge shell volume. The different control volumes are visible in Figure 1. A polygon approach is applied to both starwheels to obtain the tooth areas mating with the corresponding grooves, as shown Figure 3. The polygon approach is also used to obtain all the geometric information of the expander such as volume curves, porting areas, leakage path lengths and groove surfaces. Such procedure is outlined in details in [5, 8]. The complete volume curve evolution as well as the areas of suction and discharge ports are reported in Figure 4.

To handle the complexity of the geometry model and the multi-control volumes, a multi-layer solution scheme has been developed. The conceptual schematic of the algorithm is shown in Figure 5. Once the simulation is initialized with the boundary conditions and input data, the geometric properties calculated are stored. For each control volume cycle integration is carried out by using an adaptive solver that adjusts the step taken according to the stiffness of the

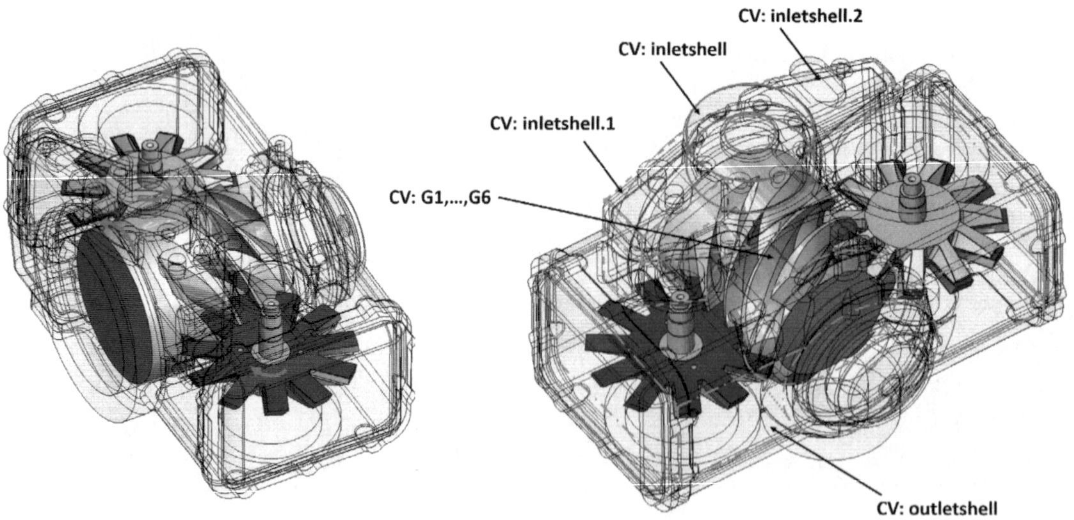

Figure 1. Transparent 3D view of the single-screw machine: with rotating directions (left) and identifications of the control volumes.

Figure 2. 11 kWe single-screw expander considered in the present study.

Table 1. Main geometric parameters of 11 kWe single-screw expander.

Engaging ratio	[-]	11/6
D_{sr}	[mm]	122
D_{sw}	[mm]	132
$V_{g,max}$	[cm^3]	57.39
$r_{v,built-in}$	[-]	5.3
L_{rotor}	[m]	121

problem. The cycle includes suction and discharge flow models, inlet tube models accounting for pressure drops and heat transfer with the housing walls, in-chamber heat transfer and the state variable derivatives. Once the cycle has converged, forces, moments, boundary work rate and friction losses are computed and an overall energy balance is enforced.

2.1. Governing Equations

The conservation equations of mass and energy are applied simoultaneously to each control volume to obtain the change in thermodynamic states. The properties within a control volume are assumed to be uniform, i.e. gradients within compressor chambers are neglected. Temperature and density have been chosen as independent thermodynamic properties to take advantage of the equation of state implementation in the CoolProp library [9]. For each control volume, the general conservation of mass equation can be written as:

$$\frac{d(\rho V)_{CV}}{d\theta} = \frac{1}{\omega}\sum_i \dot{m}_i \tag{1}$$

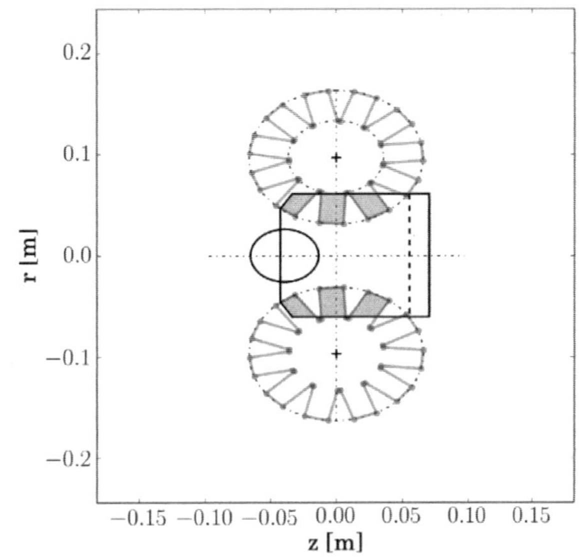

Figure 3. Planar representation of single-screw geometry with polygon approach (adapted from [6]).

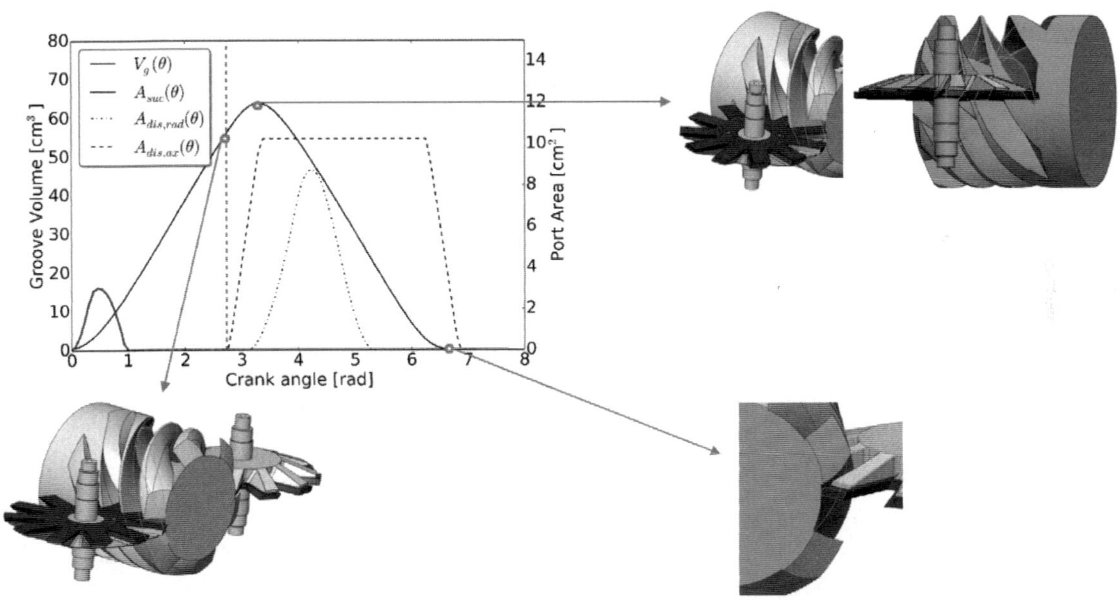

Figure 4. Groove volume and port area evolution with respect to the main rotor crank angle. 3D CAD views help visualize discharge opening, maximum groove volume and discharge end situations.

The general conservation of energy in each chamber of the compressor can be rearranged to obtain the derivative of temperature with respect to the crank angle [10]:

$$m_{CV} c_v \frac{dT}{d\theta} = -T \left(\frac{\partial p}{\partial T}\right)_v \left[\frac{dV}{d\theta} - v\frac{dm_{CV}}{d\theta}\right] - h\frac{dm_{CV}}{d\theta} + \frac{\dot{Q}}{\omega} + \frac{1}{\omega}\sum_i (\dot{m}h)_i \tag{2}$$

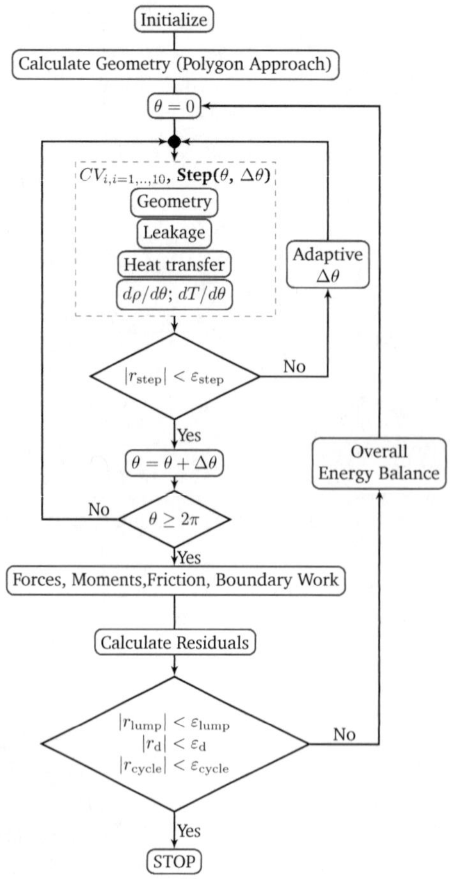

Figure 5. Solution algorithm of the single-screw mechanistic model.

where the mass in the control volume is given by $m_{CV} = (\rho V)_{CV}$.

In a single-screw machine, there are nine main leakage paths as described by Ziviani et al. [4]. The leakage paths can be divided into three types:

- constant clearance gap: leading and trailing edges of the grooves, gap between starwheel disk and housing and end of groove band;
- variable clearance gap: sealing lines of the teeth.
- orifice type: blow holes between tooth and groove due to wear of the profile or non-perfect milling process.

An isentropic flow through nozzle with proper flow coefficient is used to model gaps associated with wear, as they are quite difficult to accurately predict. For the remaining leakage paths, the flow model is still used, but a detailed 1D compressible flow model [11] has been employed to estimate the correction factor, M, between isentropic flow and flow with friction:

$$M = \frac{\dot{m}_{\text{nozzle}}}{\dot{m}_{\text{1D,fr}}} = \frac{a_0\left(L/L_0\right)^{a_1}}{a_2\left(\delta/\delta_0\right)+a_3}\left[\xi\left(a_4 Re^{a_5}+a_6\right)+\left(1-\xi\right)\left(a_7 Re^{a_8}+a_{10}\right)\right]+a_{10} \tag{3}$$

where the dimensionless characteristic length and dimensionless gap width are given by $L^* = L/L_0$ and $\delta^* = \delta/\delta_0$ with L_0 and δ_0 being the reference length and gap width values.

The average boundary work rate generated by the expander is calculated as

$$\overline{\dot{W}}_{pV} = \frac{\omega}{2\pi}\int_0^{2\pi} p(\theta)\frac{\mathrm{dV}}{d\theta}d\theta \tag{4}$$

The average mechanical power output of the expander is obtained by subtracting the mechanical losses from the boundary work rate:

$$\dot{W}_{\mathrm{mech}} = \overline{\dot{W}}_{pV} - \dot{W}_{\mathrm{mech,loss}} \tag{5}$$

where the mechanical losses are obtained by solving the forces and moments balance equations for starwheel and main rotor [6], shown in Figure 6. The mechanical torque of the main shaft is determined by decomposing the gas force distribution along the groove on the Y-Z and X-Z planes [12].

Thus, the mechanical efficiency of the expander is calculated as the ratio of the mechanical shaft power to the boundary work rate (or indicated power):

$$\eta_{\mathrm{mech}} = \frac{\dot{W}_{\mathrm{mech}}}{\overline{\dot{W}}_{pV}} \tag{6}$$

The overall energy balance of the expander housing is written as,

$$\dot{W}_{\mathrm{mech,loss}} + \sum_i \dot{Q}_i(T_{\mathrm{lump}}) = 0 \tag{7}$$

where T_{lump} is the mean temperature of the lumped mass of the expander shell and $\dot{Q}_i$ represents all the heat transfer interaction terms between the different parts of the expander and the shell. In the present model, a single-lumped temperature is used and Equation (6) reduces to:

$$\dot{W}_{\mathrm{mech,loss}} + \overline{\dot{Q}}_{CV,wall} + \dot{Q}_{inletshell,wall} + \dot{Q}_{outletshell,wall} + UA_{\mathrm{amb}}\left(T_{\mathrm{amb}} - T\right) = 0 \tag{8}$$

where the average value of the overall heat transfer coefficient UA_{amb} has been determined by calibrating a semi-empirical model of the expander on the experimental data [13].

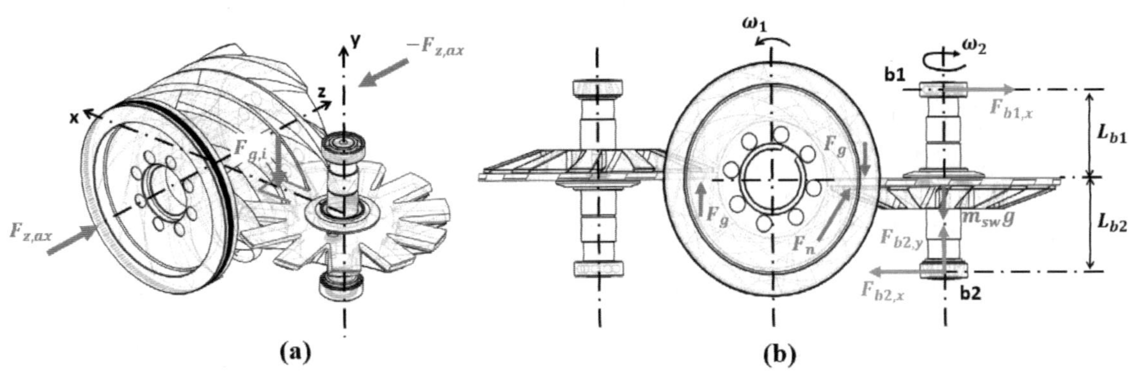

Figure 6. Schematic of the forces on main rotor and starwheels (adapted from [6]).

Table 2. Minimum and maximum values of the measured and derived variables for the single-screw expander with rotational speeds of 2000 rpm and 3000 rpm running with R245fa.

	$p_{su,exp}$ (kPa)	$T_{su,exp}$ (°C)	$p_{ex,exp}$ (kPa)	$T_{ex,exp}$ (°C)	$\dot{m}_r$ (kg/s)	r_p (-)	r_v (-)	$\dot{W}_{exp,el}$ (W)	$\varepsilon_{is,oa,exp}$ (-)	φ_{FF} (-)
Min	566	106.7	120	75.5	0.12	3.71	3.76	1283	0.2058	0.938
Max	1230	124.9	232	104.6	0.37	7.26	7.74	7364	0.5191	1.231

3. Experimental Data

The experimental characterization of the considered single-screw expander has been discussed in [13]. For the sake of compactness, the minimum and maximum values of the measured and calculated variables for two rotational speeds are reported in Table 2. The expander was directly connected to a generator and inverter. The electrical power output as a function of the expansion specific volume ratio is shown in Figure 7. The maximum electric power output measured is 7.364 kW at an overall isentropic efficiency of 51.91%. To be noted is that there are two groups of data at 3000 rpm due to variations of the condensing temperatures.

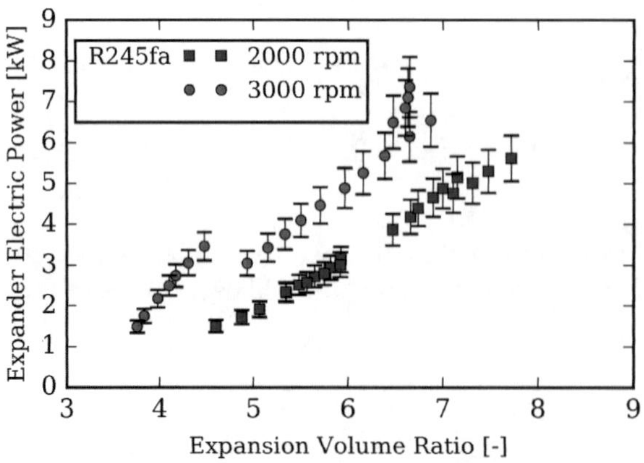

Figure 7. Expander electric power out as function of the expansion volume ratio for R245fa at different rotational speeds. The steady-state points at 2000 rpm and 3000 rpm are displayed. Uncertainty bars of the power output are overlaid.

4. Results and Discussion

The usefulness of the mechanistic model is related to its capability of estimating the mechanical losses associated with both friction as well as leakage flows.

As only the electrical power out has been measured, the combined electro-mechanical and heat losses can be estimated. An overall energy balance on the expander and electric equipment is performed and the sum of heat losses and electro-mechanical losses, $\dot{W}_{el-m,loss}$, can be calculated as:

$$\dot{Q}_{amb,exp,loss} + \dot{W}_{el-m,loss} = \dot{m}_r\left(h_{su,exp} - h_{ex,exp}\right) - \dot{W}_{el,grid,exp} \tag{9}$$

where the enthalpy difference across the expander is obtained by the measured temperature and pressure values upstream and downstream the expander. The resulting losses are shown in Figure 8(a).To be noted is that the energy balance is closed with significant losses compared to the power output. Moreover, the total losses decreases with the increase of the pressure ratio.

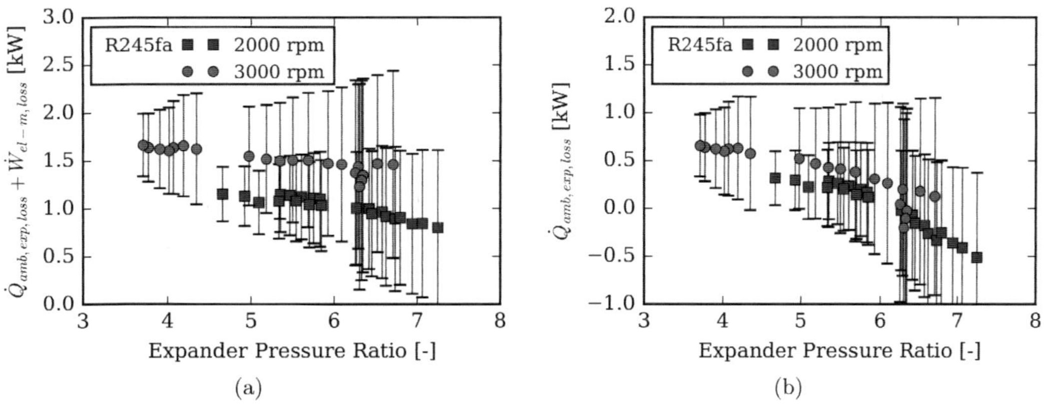

Figure 8. (*a*) Overall energy balance of the expander; (*b*) estimation of the expander heat losses through the housing.

In fact, as the power output increases, the efficiency of the electric equipment increases as well. In addition, the contribution of the decrease in electrical-mechanical losses is more significant compared to the heat losses through the housing since the inlet temperature of the working fluid was kept close to 125 °C for the majority of the experimental points.

As different source of losses are included in such energy balance, the performance map of generator and inverter have been determined in order to isolate the heat losses through the expander housing. This is accomplished by performing an energy balance only on the expander:

$$\dot{Q}_{amb,exp,loss} = \dot{m}_r \left(h_{su,exp} - h_{ex,exp}\right) - \dot{W}_{mech} \tag{10}$$

The heat rate loss values and the associated uncertainties are reported in Figure 8(b). For some of the experimental points, the heat losses are slightly negative. This fact is certainly associated with the accuracy of the performance map used to back up the expander shaft power.

The mechanistic model is employed here to estimate the mechanical shaft power and to obtain the actual mechanical efficiency by including the friction power losses associate with the bearings and friction contacts between teeth and grooves. Therefore the mechanistic model is first validated with a set of experimental data. The resulting parity plots between calculated and experimental points are shown in Figure 9.

The validated model has been exercised to calculate the boundary work rate and the friction losses for each of experimental conditions. The results are reported in Figure 10(a). In particular, the filled markers represents the boundary work rate for two rotational speeds. By applying Equation 4, suction pressure losses are accounted for. The indicated efficiency contributions can be estimated by including the leakage losses. The total friction losses are overlaid in the same figure with empty markers. To be noted is that the friction losses represent a significant contribution to decrease the expander power output. In fact, the mechanical efficiency is estimated to be around 70% for expander pressure ratios between 5 and 7, near optimum operating conditions for the given internal volume ratio and internal pressure drops [2]. Such high friction losses have been associated with the non-sufficient lubrication between tooth and starwheel as well as the bearings. After a detailed inspection, the injection orifices were found partially occluded. Additionally, the tooth profiles were worn out, suggesting high wear.

To further validate the mechanistic model, dynamic pressure transducers have been installed in different points of the housing to capture the entire expansion process. The experimental work is currently ongoing on a new set of rotor and meshing pair of starwheels.

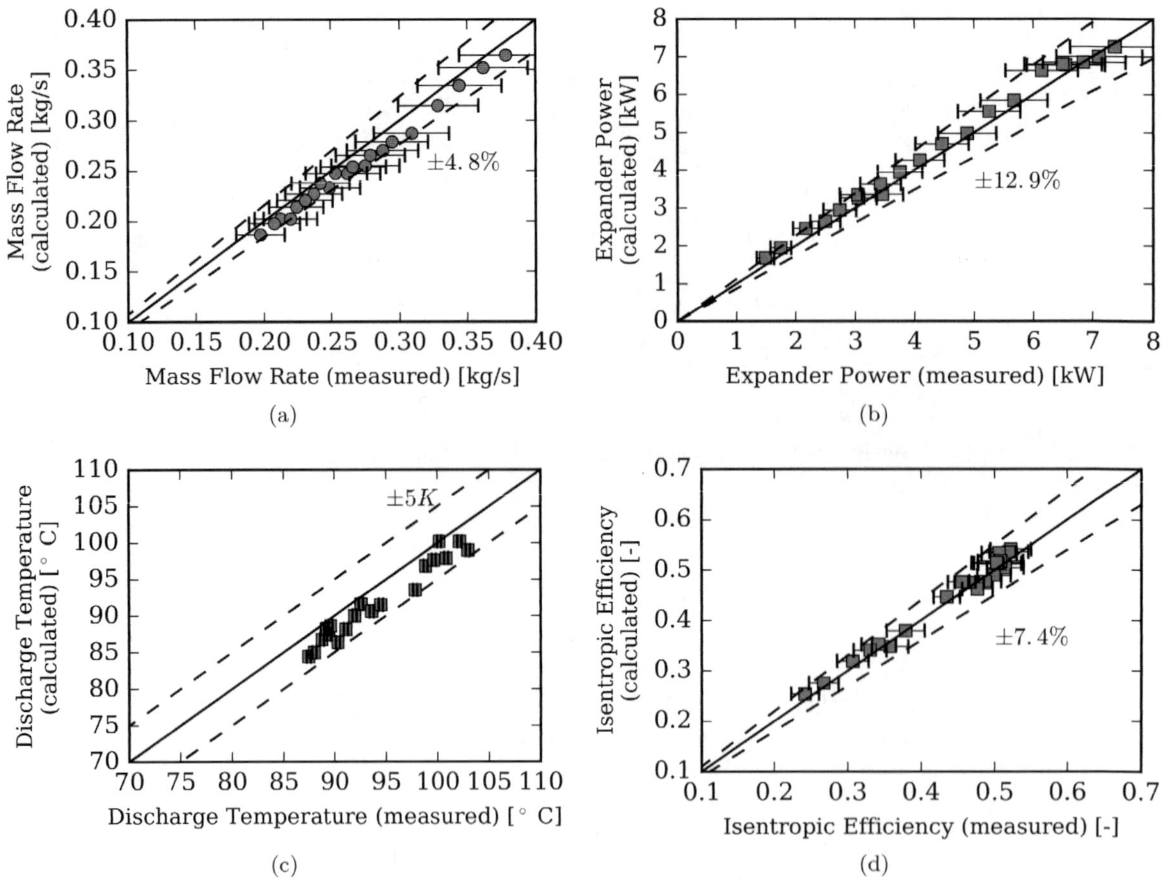

Figure 9. Non-symmetric single-screw expander model parity plots: (a) mass flow rate; (b) electric power output; (c) expander discharge temperature; (d) overall isentropic efficiency. A total of 21 steaty-state points are shown at 3000 rpm. Uncertainty bars are over laid. The maximum relative error is reported in each plot.

5. Conclusions

In this paper, a non-symmetric approach to modeling single-screw machines has been presented. Such approach accounts for the working process occurring on both sides of the rotor and possible interactions between the working chambers. A single-screw expander in an ORC application has been used to validate the model and to estimate the total friction losses and expander mechanical efficiency. the model predicted a maximum mechanical efficiency of approximately 70%. The trade-off between complexity of the model and accuracy will be evaluated by employing dynamic pressure sensors to record the indicated diagram of the expander.

Acknowledgments

The authors would like to acknowledge Sergei Gusev for the extensive experimental campaign carried out as well as thank the Center of High Performance Building (CHPB) for sponsoring the project entitled "Development of general purpose simulation tools for positive displacement compressors".

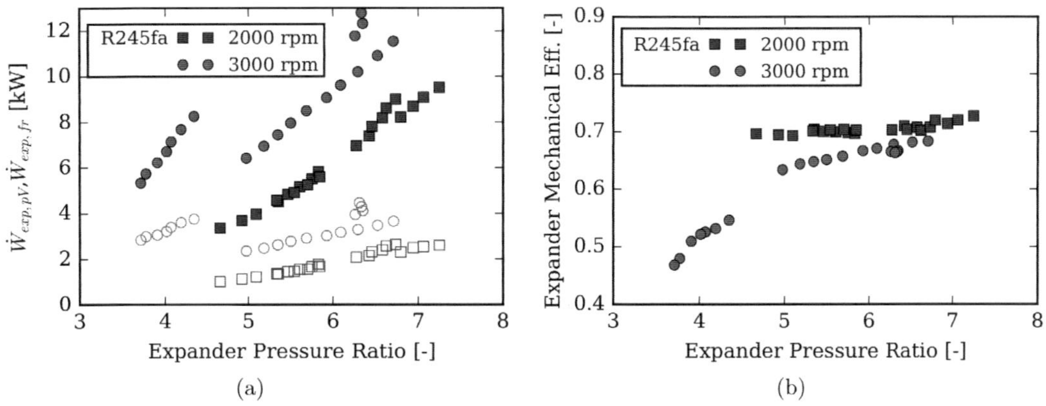

Figure 10. (*a*) Calculated boundary work rate (filled markers) and friction power losses (empty markers) as a function of the applied pressure ratio; (*b*) calculated mechanical efficiency of the expander.

References

[1] Wei Wang B L, Wu Y T, Ma C F, J-F W, Zhang L, Li C, Zhao Y K and Zhi R P 2016 *Energy* **116** 43–52

[2] Ziviani D, Sergei S, Lecompte S, Groll E A, Braun J E, Horton W T, van den Broek M and De Paepe M 2017 *Applied Energy* **189** 416–432

[3] Bein T W and Hamilton J F 1982 Computer modeling of an oil flooded single screw air compressor *International Compressor Engineering Conference* paper 383

[4] Ziviani D, Bell I H, De Paepe M and van den Broek M 2014 Comprehensive model of a single screw expander for orc-systems applications *22th Int.Compressor Engineering Conf. at Purdue* 1506

[5] Ziviani D, Bell I, M D P and van den Broek M 2015 Update on single-screw expander geometry model integrated into an open-source simulation tool *9th Int. Conf. on Compressors and their Systems, City University of London, London* 39

[6] Ziviani D, Bell I H, De Paepe M and van den Broek M 2016 Mechanistic model of an oil-flooded single-screw expander *23th Int.Compressor Engineering Conference at Purdue* 1486

[7] Ziviani D, Suman A G J, Pinelli M, De Paepe M and van den Broek M 2016 Cfd approaches applied to a single-screw expander *23th Int.Compressor Engineering Conference at Purdue* 1488

[8] Ignatiev K M 2012 Approach to the numeric geometry analysis of positive displacement compressors, its application to a single screw compressor simulation and verification by experiment *International Compressor Engineering Conference* 2059

[9] Bell I H, Wronski J, Quoilin S and Lemort V 2014 *Ind. Eng. Chem. Res.* **53** 2498–2508

[10] Bell I H 2011 *Theoretical and Experimental Analysis of Liquid Flooded Compression in Scroll Compressors* Ph.D. thesis Purdue Univeristy

[11] Bell I H, Groll E A, Braun J E and Horton W T 2013 *Int. J. Refrigeration* **36** 1965–1973

[12] Wu Y R and Tran V T 2016 *International Journal of Refrigeration* **65** 111–128

[13] Ziviani D, Sergei S, Lecompte S, Groll E A, Braun J E, Horton W T, van den Broek M and De Paepe M 2016 *Applied Energy* **181** 155–170

Theoretical investigation of flash vaporisation in a screw expander

Hanushan Vasuthevan, Andreas Brümmer

TU Dortmund University, Chair of Fluidics
Leonhard-Euler-Str. 5, 44227 Dortmund, Germany

hanushan.vasuthevan@tu-dortmund.de, andreas.bruemmer@tu-dortmund.de

Abstract. In the present study flash vaporisation of liquid injection in a twin screw expander for a Trilateral Flash Cycle (TFC) is examined theoretically. The TFC process comprises a pressure increase in the working fluid, followed by heating the liquid close to boiling point. The hot liquid is injected into the working chamber of a screw expander. During this process the pressure of the liquid drops below the saturation pressure, while the temperature of the liquid remains virtually constant. Hence the liquid is superheated and in a metastable state. The liquid jet seeks to achieve a stable state in thermodynamic equilibrium and is therefore partially vaporised. This effect is referred to as flash vaporisation. Accordingly, a two-phase mixture, consisting of vapour and liquid, exists in the working chamber. Thermodynamic simulations were carried out using water as the working fluid for representative screw expander geometry. The simulations presented are performed from two different aspects during the filling process of a screw expander. The first case is the vaporisation of the injected liquid in a state of thermodynamic equilibrium, whereby the two-phase mixture is treated entirely as a compressible and homogeneous gas. The second case considers flashing efficiency. It describes the quantity of flashed vapour and consists of a liquid and vapour domain. Both models are compared and analysed with respect to the operational behaviour of a screw expander.

1. Introduction

The Trilateral Flash Cycle (TFC) is a suitable opportunity for power recovery from low-temperature sources such as geothermal sources or industrial waste heat. Compared to the Clausius Rankine Cycle or the Organic Rankine Cycle the Trilateral Flash Cycle does not need vaporisation or superheating of the working fluid. The TFC contains basically only liquid heating close to boiling point and vaporisation during expansion into the wet vapour region, which is referred to as flash vaporisation. Flash vaporisation occurs if a liquid undergoes a sudden reduction in pressure, while the temperature is virtually constant. In this state the temperature is higher than the saturation temperature at its current reduced pressure. Hence the liquid is superheated and in a metastable state. In order to regain a more stable state in thermodynamic equilibrium, part of the liquid is vaporised, which can occur explosively and violently depending on boundary conditions. Consequently a two-phase mixture with high liquid content exists during the expansion process. For this expansion process a screw expander is suitable. Unlike turbomachines, screw expanders are able to expand multiphase fluids with high liquid content. In the past, screw expanders in a TFC were primarily examined experimentally. A literature survey is given in [1]. Detailed methods or models for the design of a two-phase screw expander for a trilateral flash cycle are not available, although this technology seems to have a high efficiency potential. The comprehension of the inner process in a liquid injected two-phase screw expander with flash

vaporisation is an important requirement if the efficiency is to be maximised. One essential part of a screw expander working cycle is the filling process, which involves hot liquid injection into the working chamber. During this injection via a nozzle a superheated liquid jet is formed, which is partially vaporised in the working chamber due to flash vaporisation.
The main objective of this paper is to investigate the flash vaporisation of a superheated liquid jet in a twin screw expander theoretically. For this purpose fundamentals of the flash vaporisation process will be described first. For the modelling of the flash vaporisation of a liquid jet a simple empirical model will be used. The simulations are carried out with the working fluid water for a single representative screw expander geometry. They are performed ideally without consideration of gap flows, to investigate the effect of the liquid jet on the working cycle. Two cases will be considered in the simulation. The first case considers the flash vaporisation in thermodynamic equilibrium. In this case the two-phase mixture is treated as a homogeneous and compressible gas. The second case includes flashing efficiency, which describes the vaporised mass flow rate, whereby the phases are considered separately with liquid and homogeneous wet-vapour domains. Finally, the results of the simulations will be described and discussed.

2. Flash vaporisation of superheated liquid jets

Flash vaporisation of superheated liquids has a variety of practical applications. One application among others is for example the desalination of salt water, whereby steam is generated and condensed to remove minerals from salt water. Further applications are e.g. fuel injection in engines, drying processes such as paper drying, geothermal power plants generating steam for turbines, etc. [2]. Compared to the targeted use of flash vaporisation these less controlled processes can be undesirable and hazardous, causing the accidental relaxation of liquids which can lead to explosions in piping systems, valves or vessels. One important example is the loss of coolant accident (LOCA) in nuclear power plants, whereby the pressure in the cooling system drops suddenly and the superheated coolant is vaporised explosively. The various desirable and undesirable practical applications of flash vaporisation have led to many publications. An extensive literature survey of flashing liquid jet studies is given in [3]. In the following the focus is on the essential fundamentals and modelling methods for the flash vaporisation of superheated liquid jets.

2.1. Fundamentals

A superheated liquid jet can be formed when the liquid is injected through a nozzle into a gaseous chamber with a pressure lower than the inlet pressure. During injection and at the nozzle exit the pressure drops while the temperature of the liquid is nearly constant. Consequently the temperature of the liquid is higher than its saturation temperature corresponding to the chamber pressure. Vice versa the pressure of the chamber is lower than the saturation pressure corresponding to the liquid temperature. This situation is shown in figure 1 in a pressure-temperature diagram. As a consequence the liquid is brought into a metastable state. The infiltration of the liquid into the metastable state can be described as superheating ΔT_{sh}. In equation (1) the general definition for the superheating is given as the difference between the liquid temperature and the saturation temperature at the ambient pressure. For the liquid entering the chamber, the superheating is only a function of the chamber pressure which varies with time, when the inlet temperature of the liquid is constant. The maximum superheating for a given chamber pressure can be achieved if the inlet temperature is equal to the saturation temperature, which corresponds to the inlet pressure.

$$\Delta T_{sh} = T_{liq} - T_{sat}(p_{ch}) \tag{1}$$

Under specific conditions without physical disturbances the metastable state can be maintained to a certain extent, depending on boundary conditions. In most practical situations the liquid goes through a transient vaporisation process to achieve thermodynamic equilibrium. During this process the temperature of the liquid decreases due to the energy consumption required for vaporisation. While the superheating of the injected liquid is only a function of the time-dependent chamber pressure, the superheating of the liquid jet in the chamber depends on the local liquid temperature of the jet and the chamber pressure. Both these parameters vary with time.

The flash vaporisation process of the superheated liquid jet can be divided into four steps [4]. The first step is the nucleation process, which is a concentration process of molecules of the new or vapour phase to stable nuclides. The second step is the bubble growth of the stable nuclides. After that, the bubble growth within the jet leads to its breakup or atomisation, which leads further to ligaments and droplets. The last step is the vaporisation of the droplets.

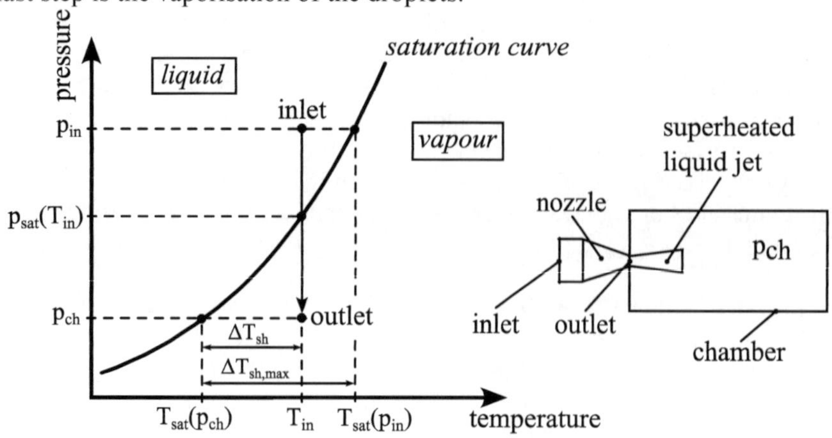

Figure 1. Change of state in a pressure-temperature diagram for a nozzle flow with liquid

2.2. Modelling methods

There are different modelling methods for flash vaporisation with variable modelling depth, which can be adapted for different numerical calculation methods. In general, flash vaporisation can be divided into two modelling sections: atomisation and vaporisation. Modelling flash atomisation delivers information about the droplets such as distribution, diameter, number, temperature and velocity. The droplets are then adapted into the modelling of droplet vaporisation. To model atomisation, empirical correlations derived from experiments can be used. Another approach with higher modelling depth is to simulate the nucleation process by means of the classical nucleation theory and subsequent consideration of bubble growth and break-up to acquire information about droplets. Existing vaporisation models of droplets are based essentially on heat fluxes between droplets and their surroundings. These models also contain empirical correlations in order to determine, for example, heat transfer coefficients. A valuable review for modelling of atomisation and vaporisation is given in [4] and [5]. Another approach with a lower modelling depth is to determine the vaporised mass due to the flash vaporisation directly in terms of flashing efficiency, which is derived from experimental results. Miyatake presented such a model with an empirical equation for the working fluid water which depends entirely on superheating [6]. This model is used and adapted for the simulation of vaporisation in a screw expander (Section 3).

3. Modelling of the flash vaporisation in a twin screw expander

In this section modelling of flash vaporisation in a screw expander by means of an empirical model is described. Therefore first of all the fundamentals of the screw expander are described. Next, the modelling procedure for the screw expander is explained.

3.1. Fundamentals

Twin screw expanders are rotary displacement machines with spirally shaped intermeshing rotors enclosed in a housing. Clearances between the rotors within the enclosed housing form the working chambers, whose volume changes periodically due to the rotation of the rotors. The progression of the chamber volume as well as idealised inlet and outlet areas for the present simulations, as functions of the male rotor angle, are shown in figure 3 (Section 4.1). The working cycle is divided into three steps. The first step is the filling process, where the working fluid from the high pressure side is transferred to the working chamber, while the volume of the working chamber increases. After the filling process ends there is no connection between the high pressure side and the working chamber. Thus, the working chamber - except for clearances - is enclosed from the environment. At this point the expansion process begins and proceeds as the rotors continue to turn. This process ends when the working chamber reaches its maximum volume. The working chamber is then connected to the low pressure side via the outlet area. At this point the discharge of the working fluid begins, while the chamber volume decreases during the third phase of the working cycle. The working cycle ends with the disappearance of the chamber volume.

3.2. Modelling procedure

For the simulation and its thermodynamic modelling a one-chamber model is used, where the rotors of a twin screw expander (male rotor and female rotor) are considered together. The calculations are mainly based on the conservation of mass and energy which are derived separately for a homogeneous gas phase and a liquid phase. This approach offers the option of examining thermodynamic non-equilibrium states within working chambers. The conservation equations for a stationary system are given in [1]. The basic simulation procedure is as follows: For a given time step or male rotor angle the geometry of the screw expander will be considered as a frozen model. It therefore follows that chamber volumes and flow areas between working chambers and ports are defined. Within this time step numerical integrations of the conservation equations are performed. In the present paper the simulations are carried out ideally without clearances. Furthermore heat flows between fluid and solid components are ignored. In the simulations the inlet flow, outlet flow and the flash vaporisation of the injected liquid within a working chamber are considered. In the following, modelling of the essential processes is described.

3.3. Liquid injection

The filling process involves the injection of a hot liquid from the high pressure side into a chamber volume via a nozzle. The calculation of the mass flow rate is carried out by means of the Bernoulli equation for steady and incompressible flows from point 1 (reservoir upstream of the nozzle) to point 2 (nozzle outlet exit), which is given in equation (2), ignoring geodetic height differences and friction losses.

$$\frac{p_1}{\rho} + \frac{c_1^2}{2} = \frac{p_2}{\rho} + \frac{c_{2,is}^2}{2} \tag{2}$$

Kinetic energy in point 1 can be ignored. At point 2 pressure p2 is assumed to be equal to the pressure in the working chamber which is to be filled. The pressure in the working chamber changes with time or male rotor angle, which leads to a step response and also a change in the nozzle exit velocity in the simulation. With equation (2) the theoretical maximum injection mass flow rate can be determined with density ρ and nozzle exit area A. Taking losses from point 1 to point 2 into account will reduce the theoretical maximum mass flow rate. Therefore the losses are included with a flow coefficient α. The actual mass flow rate for the injection of liquid into the working chamber can be calculated with equation (3).

$$\dot{m}_{inj} = \alpha \dot{m}_{th} = \alpha \rho A \sqrt{\frac{2}{\rho}(p_1 - p_2)} \tag{3}$$

3.4. Flash vaporisation of superheated liquid

The liquid in the working chamber is vaporised partially due to flash vaporisation. The modelling of flash vaporisation is performed with an empirical equation for the working fluid water which was developed by Miyatake [6]. Miyatake examined the transient spray flash evaporation characteristics of a desalination system experimentally. Although the time-scale used by Miyatake is far greater than the time-scale of a typical screw expander working cycle, this model will be used to examine the effects in a screw expander due to its simplicity and general application. The model contains a flashing efficiency, which is defined as the ratio of the vaporised mass flow rate to the maximum mass flow rate. In equation (4) the definition and the empirical equation for the flashing efficiency are given, which depends entirely on the superheating. Miyatake figured out that this equation corresponds quite well with his experiments for various nozzle velocities, nozzle lengths and nozzle diameters. With increasing superheating the flashing efficiency tends towards one. Furthermore equation (4) is only valid for superheating equal to or greater than one.

$$\eta_{flashing} = \frac{\dot{m}_{vap}}{\dot{m}_{vap,max}} = 1 - \left[1 + 2{,}5\left(\frac{\Delta T_{sh}}{K} - 1\right)\right]^{-1} \quad \forall\, \Delta T_{sh} \geq 1K \tag{4}$$

The theoretical maximum mass flow for the vapour generation can be calculated with equation (5), which is derived from conservation of energy. Here it is assumed, that the vaporised mass flow is in a saturated state. The maximum mass flow of vapour is generated when after vaporisation the liquid reaches the saturation temperature in thermodynamic equilibrium, corresponding to its chamber pressure.

$$\dot{m}_{vap,max} = \dot{m}_{liq} \frac{h_{liq}(T_{liq}) - h_{liq}(T_{sat}(p_{ch}))}{h_{vap,ch}(T_{sat}(p_{ch})) - h_{liq}(T_{sat}(p_{ch}))} \tag{5}$$

In figure 2 flash vaporisation in a chamber is illustrated. The chamber contains two general domains. The first domain is the vapour domain, which is in a single state and can be a superheated gas or a homogeneous wet-vapour mixture. This domain is treated entirely as a compressible gas and is calculated in thermodynamic equilibrium. The second domain is the liquid domain, which is incompressible and occupies the chamber volume in part. This domain contains a certain number of liquid elements n, which results from the liquid injection. For every time step in the simulation during the filling process a liquid element is created. Each liquid element has a different individual state. The detailed conservation equations of mass and energy for both domains which are used for the simulation, are given in [1]. In the following the calculation procedure for flash vaporisation within the chamber, with mass and energy transfer between the two domains for a time increment Δt, is described without further analysis.

Within the time increment every liquid element of the liquid domain is considered. First the superheating for a liquid element is determined, which depends on the pressure in the chamber and the temperature of the liquid element. For liquid elements with superheating greater than one, the flashing efficiency is calculated, which is then multiplied with the maximum mass flow rate in equation (5) to determine the actual mass flow rate of vaporisation from a liquid element. The vaporised mass of a liquid element is calculated with numerical integration of the mass flow rate of vaporisation and a time increment as shown in equation (6). With equation (7) the heat due to vaporisation can be calculated, assuming that the vaporised mass is in a saturated state. In the case of superheating lower than one, the appropriate liquid element will not be considered.

$$\Delta m_{vap,n}(t_j) = \dot{m}_{vap,n}\Delta t = \eta_{flashing,n}(t_j) m_{liq,n}(t_j) \frac{h_{liq,n}(T_{liq,n}(t_j)) - h_{liq}\left(T_{sat}(p_{ch}(t_j))\right)}{h_{vap}\left(T_{sat}(p_{ch}(t_j))\right) - h_{liq}\left(T_{sat}(p_{ch}(t_j))\right)} \quad \Delta T_{sh,n} \geq 1K \tag{6}$$

$$\Delta Q_{vap,n}(t_j) = \Delta m_{vap,n}(t_j) h_{vap}\left(T_{sat}\left(p_{ch}(t_j)\right)\right) \quad \forall\, \Delta T_{sh,n} \geq 1K \tag{7}$$

Due to the vaporisation of liquid elements and the resulting energy flow from liquid elements to the vapour domain, the temperature of liquid elements drops. If the superheating is high and the flashing efficiency is near one, the temperature of the liquid element may drop below the vapour domain temperature. In this case a heat flow would occur from the vapour domain to the liquid element. A heat flow model for this case is not considered explicitly, but instead it is assumed that the affected liquid element is atomised very finely and has sufficient heat transfer to be homogeneous and in thermodynamic equilibrium within the vapour domain. Thus, the liquid elements involved are transferred completely to the vapour domain. In this case the transferred mass and heat to the vapour domain is given in equation (8) and equation (9).

$$\Delta m_{vap,n}(t_j) = m_{liq,n}(t_j) \ \forall \ T_{liq,n} < T_{vap} \tag{8}$$

$$\Delta Q_{vap,n}(t_j) = U_{liq,n}(t_j) \ \forall \ T_{liq,n} < T_{vap} \tag{9}$$

The mass transfer within a chamber and between the two domains is performed with the calculated vapour mass, which is subtracted from the corresponding liquid element as given in equation (10). The sum of the vapour mass from all liquid elements under consideration is added to the vapour domain as shown in equation (11).

$$m_{liq,n}(t_j) = m_{liq,n}(t_{j-1}) - \Delta m_{vap,n}(t_j) \tag{10}$$

$$m_{vap}(t_j) = m_{vap}(t_{j-1}) + \sum_n \Delta m_{vap,n}(t_j) \tag{11}$$

Furthermore the transferred heat due to the vaporisation between the two domains is dealt with in equations (12) and (13).

$$U_{liq,n}(t_j) = U_{liq,n}(t_{j-1}) - \Delta Q_{vap,n}(t_j) \tag{12}$$

$$U_{vap}(t_j) = U_{vap}(t_{j-1}) + \sum_n \Delta Q_{vap,n}(t_j) \tag{13}$$

It should be noted that the above equations only describe the mass and heat transfer due to the flash vaporisation between the liquid and vapour domains. For the simulation of the screw expander the complete mass and energy conservation equations are used, as given in [1].

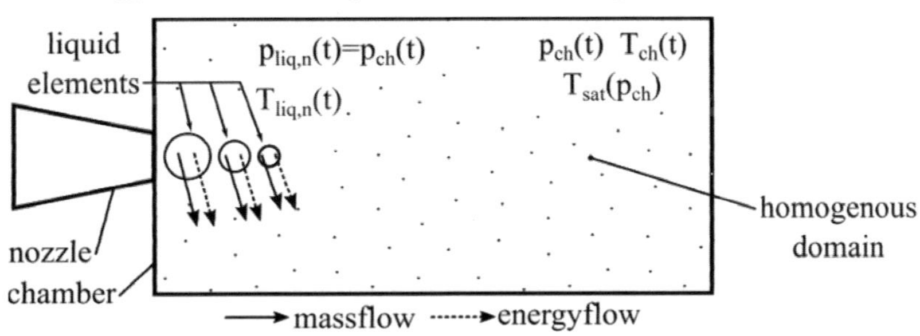

Figure 2. Liquid injection and flash vaporisation in a chamber

3.5. Discharge Process

In this section the calculation of the discharge flow through the outlet area from the chamber (ch) to the low pressure (index lp) side is described. It is assumed that in the outlet area the state is equal to the state on the low pressure side. The isentropic discharge velocity from the vapour domain at the outlet area is calculated by means of the conservation of energy with equation (14). For the discharge process the liquid elements can be joined together to one element due to equal specific energy of each liquid element. Furthermore it is assumed, that the velocity from the liquid domain to the outlet area is equal to the velocity from the vapour domain. For the calculation of the velocity in case of a choked flow reference to [1] is made.

$$c_{vap,out,is}(t_j) = \sqrt{2 \cdot \left(h_{vap,ch}(t_j) - h_{lp}\right)} = c_{liq,out,is}(t_j) \tag{14}$$

For the calculation of the mass flow, the flow area for the two domains is different. Here, it is assumed that the flow area is divided, corresponding to the volume of the domains in the chamber. With the volume fraction of the vapour domain in the chamber, which is shown in equation (15), the mass flows for both domains are calculated. In equation (16) and equation (17) the actual mass flow is shown, resulting from the multiplication of theoretical maximum (isentropic) mass flow and flow coefficient, taking losses into account.

$$\varepsilon(t_j) = \frac{V_{vap}(t_j)}{V_{ch}(t_j)} \tag{15}$$

$$\dot{m}_{vap,out}(t_j) = \alpha\dot{m}_{vap,out,is}(t_j) = \alpha\varepsilon(t_j)A(t_j)\rho_{vap}(t_j)c_{vap,out,is}(t_j) \tag{16}$$

$$\dot{m}_{liq,out}(t_j) = \alpha\dot{m}_{liq,out,is}(t_j) = \alpha(1-\varepsilon(t_j))A(t_j)c_{liq,out,is}(t_j) \tag{17}$$

4. Simulation

First of all the boundary conditions and the assumptions for the simulations are presented. Subsequently the results are shown and explained.

4.1. Boundary conditions and assumptions

In figure 3 the geometry of the simulated machine is given. The inlet and outlet areas are idealised as rectangular functions. The flow coefficients for each flow are set at one and treated as ideal. Compared to typical screw expanders the inlet area is designed as a circular area due to the innovative consideration that the liquid is injected via a circular nozzle by means of an injection system. The diameter of the nozzle exit is d_{nozzle} = 5 mm. Injection is started at a male rotor angle of 35° to avoid direct injection onto the front surface of the rotors. Injection is concluded at 97°, which is equal to an internal volume ratio of v_i = 8. The simulations are performed ideally without clearances. Furthermore heat flows due to a temperature difference between phases and solid parts are not considered. Two calculation models are analysed. The first model consists of one homogeneous vapour domain which is treated as a compressible gas. The state of this domain can be superheated or in the wet-vapour region and is calculated in thermodynamic equilibrium. This model will be abbreviated as HTEM (Homogeneous Thermodynamic Equilibrium Model) in the following. The second model consists of a liquid domain with a certain number of liquid elements and a homogeneous vapour domain, which is described in section 3.4. This model, which deals mainly with mass and heat transfer between the two domains due to flash vaporisation, will be abbreviated as FVM (Flash Vaporisation Model) in the following. Furthermore, any interactions between the injected liquid and rotor or casing walls are not taken into account. Simulations were carried out for the following conditions:

- inlet: $p_{in} = 10\cdot10^5$ Pa T_{in} = 453 K
- outlet: $p_{out} = 1\cdot10^5$ Pa T_{out} = 373 K
- tip speed: $u_{male} = 30$ m s^{-1}

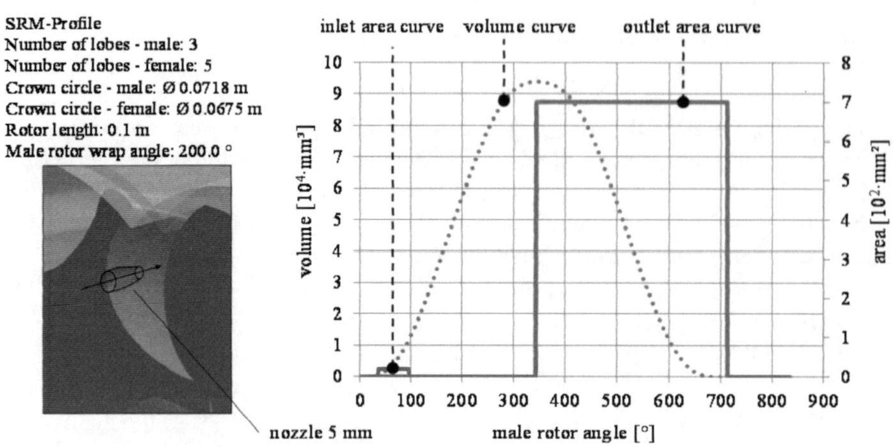

Figure 3. Screw expander geometry with volume and area curves

4.2. Results

In figure 4 the pressure curve depending on the male rotor angle for HTEM and FVM is shown. The dashed line represents the pressure curve from HTEM and the continuous line shows the pressure curve from FVM. Integral values are also shown in the diagram. It can be seen that internal power, internal work and internal isentropic efficiency are slightly reduced for the FVM compared to the HTEM. The difference in values between the two models is less than 3%.

Nevertheless the essential differences between the two models are still worth describing. The progression of mass in the chamber and the steam content of the vapour domain during the working cycle for both models can be seen in figure 5. The FVM includes the liquid and vapour domain, compared to the HTEM, which only includes one vapour domain. During the injection and expansion process it can be seen that the mass of the liquid domain decreases, while the mass of the vapour domain increases, due to flash vaporisation and mass transfer between the domains. The sum of the liquid domain mass and the vapour domain mass remains constant during the expansion process. For the HTEM the mass in the chamber increases during injection and remains constant during the expansion process up to the discharge process. The mass of the vapour domain for the FVM is smaller than for the HTEM because the liquid elements are in a separated domain. The volume of the vapour domain remains nearly constant for both models due to the high density of liquid water. Therefore the density of the vapour domain for the FVM is smaller, thus influencing the pressure in the chamber. Due to the lower density of the vapour domain the pressure progression in the chamber is slightly smaller during the injection and expansion process for FVM compared to HTEM. This results in the reduced internal work for FVM.

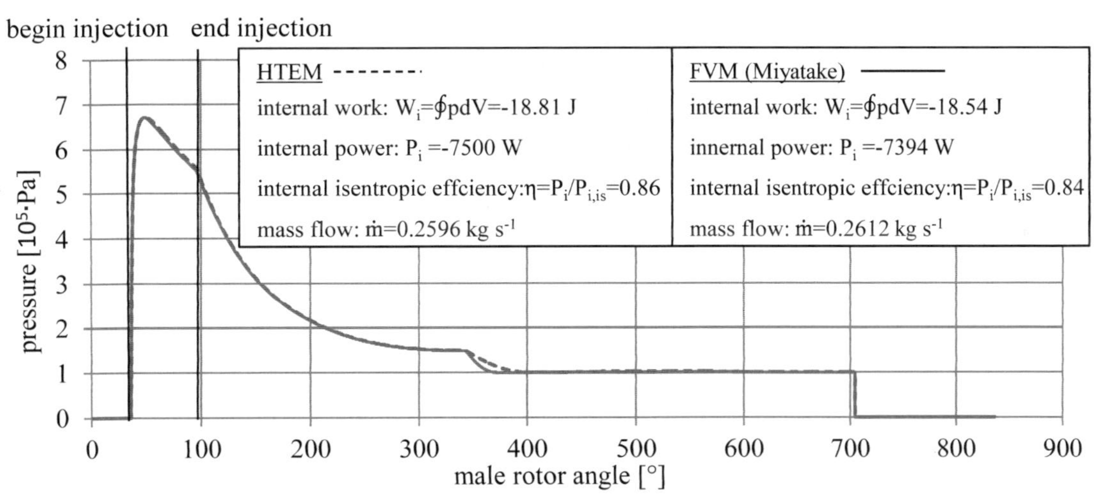

Figure 4. Pressure curve depending on male rotor angle for the homogeneous thermodynamic equilibrium model and flash vaporisation model (Miyatake)

In figure 5 it can be seen by means of the steam content, that the fluid of the vapour domain for both models is a wet vapour mixture, except for the beginning of injection for the FVM, where the state of the fluid is saturated vapour. Furthermore it can be seen, that the steam content of the vapour domain is higher for the FVM compared to HTEM. The reason for this is that beside the vapour domain a liquid domain exists, where the liquid elements remain. Due to the higher steam content of the vapour domain for the FVM, the specific enthalpy in the chamber is higher compared to the HTEM. Therefore a higher velocity is calculated for the discharge flow by means of conservation energy. The consequence of this can be seen at the beginning of the discharge process from a male rotor angle of 345° in figure 4. In case of the FVM, the pressure decreases faster to the low pressure side than in the case of HTEM. This increases the internal work for the FVM compared to HTEM.

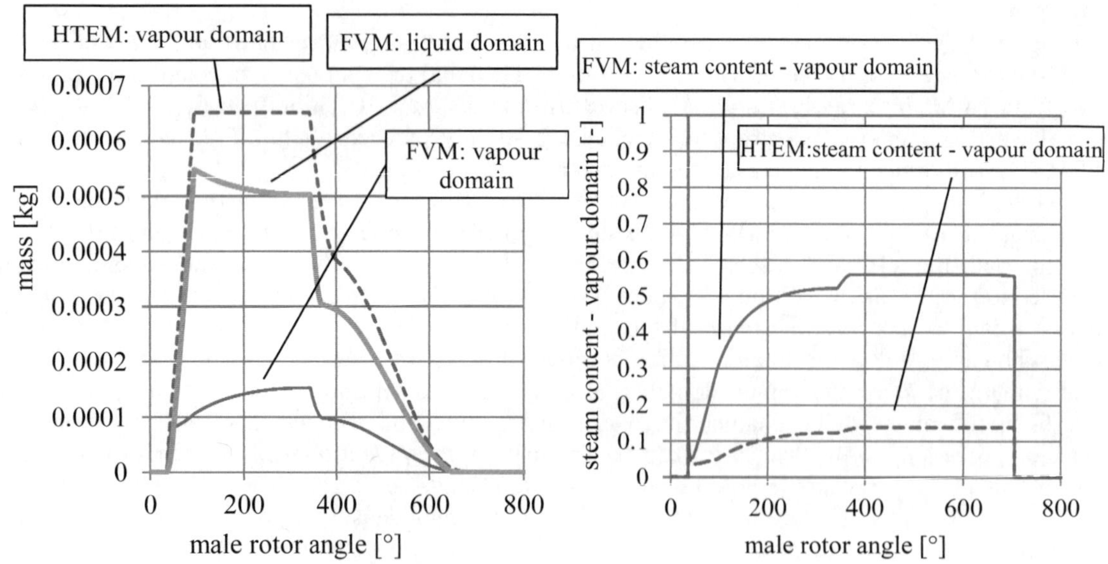

Figure 5. Mass progression for different domains and steam content depending on male rotor angle

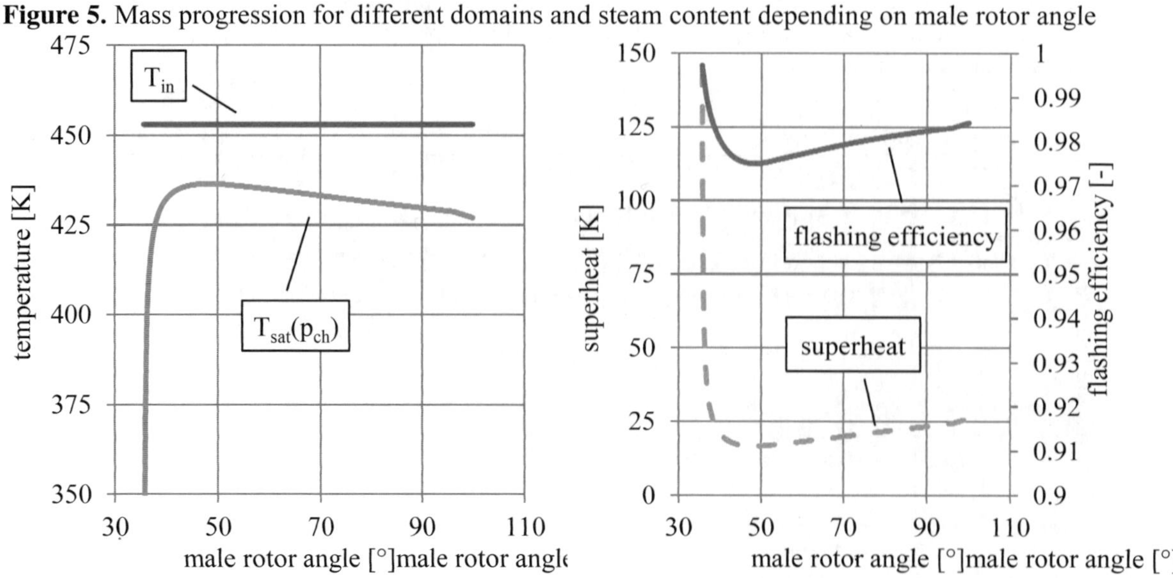

Figure 6. Left: Injected liquid temperature, saturation temperature, depending on male rotor angle
Right: Flashing efficiency and superheating of injected liquid depending on male rotor angle

In the following the reason for the small difference of the pressure angle curve in figure 4 between both models will be outlined by means of the results from FVM. In figure 6 temperature curves and flashing efficiency for the liquid entering the chamber are shown. As injection begins, the pressure inside the chamber is low, so that the corresponding saturation temperature T_{sat} is also low, while the temperature of the injected liquid T_{in} remains constant at the moment of entering the chamber. So there is high superheating for the liquid entering the chamber, which leads to high flashing efficiency. The pressure increases rapidly at the start of injection due to the small chamber volume, leading to smaller flashing efficiency for the liquid entering the chamber. From a male rotor angle of about 50° the pressure decreases due to the volume increase during injection. Hence the saturation temperature for the corresponding chamber pressure decreases as well. This leads to increased superheating and flashing efficiency for liquid entering the chamber at the end of the injection process. The flashing efficiency during injection is very high, which leads to vaporisation close to thermodynamic equilibrium.

5. Conclusions

The main objective of the present paper is to examine the flash vaporisation of a superheated liquid jet in a screw expander. For this purpose, a representative screw expander geometry was simulated. Two different cases were considered. In the first case, as a reference, the screw expander was simulated with one chamber domain, where the working fluid was treated entirely as a homogeneous compressible gas in thermodynamic equilibrium. The second case consists of two domains and considers flashing efficiency, which was derived from experiments [6] and describes the amount of vaporised substance depending on superheating. Although the model has not yet been confirmed for the time scale of a screw expander, it has been used to analyse the basic effects. The simulations were carried out ideally with a one chamber model without clearances in order to examine in isolation the effect of the superheated liquid jet in a screw expander. Furthermore, heat flow and momentum transfer between working fluid and solid parts were not considered. For the point of operation under examination it has been shown that internal work and internal isentropic efficiency are higher for the first case. In the case of flashing efficiency less mass is vaporised compared to the first case, which influences the pressure progression and leads to reduced internal work. Furthermore, at the beginning of the discharge phase, low pressure is achieved faster in the second case, which results in an increase in internal work. To sum up, the difference between the integral values in the two cases is less than 3%, which results from high flashing efficiency during injection.

List of symbols and abbreviations

Symbols

A	cross section	(m^2)
c	velocity	(ms^{-1})
h	specific enthalpy	(Jkg^{-1})
$\dot{m}$	mass flow	(kgs^{-1})
p	pressure	(Pa)
T	temperature	(K)
t	time	(s)
V	volume	(m^3)
u	circumferential speed	(ms^{-1})
v_i	internal volume ratio	(-)
α	flow coefficient	(-)
ε	volume fraction	(-)
η	efficiency	(kgm^{-3})
ρ	density	

Abbreviations

ch	chamber
i	internal
in / out	inlet / outlet
inj	injection
is	isentropic
j	simulation step
liq	liquid
lp	low pressure side
max	maximum
n	number of liquid elements
sat	saturation
sh	superheat
th	theoretical
vap	vapour

References

[1] Vasuthevan H and Brümmer A *23rd International Compressor Engineering Conference at Purdue* paper 1432
[2] El-Fiqi A K, Ali N H, El-dessouky H T, Fath H S and El-Hefni M A 2007 Desalination **206** 311-321
[3] Polanco G, Hold A E and Munday G 2010 *Journal of Hazardous Materials* **173** 2-18
[4] Kamoun H 2014 *Flash-atomization and vaporisation at near vacuum conditions-Dissertation University of Stuttgart*
[5] Aggarwal S K and Peng F 1995 *Journal of Engineering for Gas Turbines and Power* **117** 453-461
[6] Miyatake O, Koito Y, Tagawa K and Maruto Y 2001 *Desalination* **137** 157-166

IOP Conf. Series: Materials Science and Engineering **232** (2017) 012078 doi:10.1088/1757-899X/232/1/012078

Effects of spontaneous condensation in steam-driven screw expanders

Manuel Grieb and Andreas Brümmer
TU Dortmund University, Chair of Fluidics
Leonhard-Euler-Str. 5, 44227 Dortmund, Germany
E-mail: manuel.grieb@tu-dortmund.de, andreas.bruemmer@tu-dortmund.de

Abstract. Condensation and its effects on turbo machinery operation are well understood and have been widely investigated. However, only little scientific work on condensation in positive displacement machines has been published. Although, depending on machine type, high expansion rates and, as a consequence, significant supersaturation can be achieved for working fluids with a negative saturation vapour curve. In this paper the effects of spontaneous steam condensation in screw expanders are discussed. Classical nucleation theory is used for the thermodynamic simulation of operational behaviour. The study shows at which point during the expansion phase spontaneous condensation can be expected, and typical nucleation rates are determined. Since the calculation of condensation requires a suitable fluid model providing consistent data for the metastable region, the advantages of Pollak's equation of state over the IAPWS property formulation are demonstrated. The impact of released latent heat during expansion on chamber states is depicted. Furthermore, a comparison of purely metastable expansion with equilibrium expansion is provided in order to show the full range discrepancies.

1. Introduction

Condensation phenomena occur not only in everyday reality such as in the formation of clouds, but in a wide range of technical applications of all kinds. In particular, the development of droplets in gaseous fluids has been investigated in supersonic nozzles with regard to wet steam operation in the low pressure stages of steam turbines. However, very little scientific work on condensation in positive displacement machines has been published. Since screw expanders are very robust, which permits operation with a considerable amount of liquid enclosed in working chambers, they are highly suitable for use with only slightly superheated vapours or even two-phase fluids. Thus, utilising screw expanders for power generation in Rankine cycles with low temperature heat sources is a promising concept. However, for performance prediction and correct design it is necessary to take condensation into account during simulation. Here, a first approach is presented for the application of classical nucleation theory to the thermodynamic simulation of screw expander operation.

2. Condensation phenomena

Condensation generally describes the phase change of a fluid from gas to liquid, whereby a significant release of latent heat occurs. This process is further differentiated in filmwise or dropwise condensation on cooled solid surfaces and spontaneous condensation, which is characterised by the formation of droplets in a gaseous phase. While surface condensation involves heat transfer through the solid surface, which is eventually covered by a condensate film, latent heat from spontaneous condensation is directly transferred to the surrounding gas phase.

The simulation of spontaneous condensation, which is the main subject of discussion in this paper, requires three models: first a formulation is required for the properties of condensing fluid that allows for the calculation of supersaturated vapour states; next, a model is needed to calculate emerging condensate nuclei per volume; and finally a model is required to describe the growth of droplets, thus quantifying mass- and heat exchange between the droplets and the vapour phase.

What is today known as classical nucleation theory (CNT) is based on the work of Volmer and Weber, Farkas, Becker and Döring, Zeldovich as well as Frenkel [1, 2, 3, 4, 5]. A detailed overview of the development and improvement of classical nucleation theory is given by Bakhtar et al. [6]. It is the most commonly used model for condensation calculation. Experimental validation of CNT was carried out in cloud chambers [7, 8, 9] and also in divergent supersonic nozzles, which allow for faster expansion and thus correspond to grid flows in turbo machinery [10, 11, 12, 13]. Fundamental work was carried out by Gyarmathy who presented an analytical approach for the calculation of condensation phenomena in nozzles with constant relaxation speed which allowed for efficiency predictions of wet steam turbines [14]. In addition, he developed a calculation model for the growth of small droplets which is also used in this paper [15].

2.1. Spontaneous condensation

Spontaneous condensation is defined as the formation of a disperse condensed phase in a supersaturated (subcooled) gas phase [16]. This metastable state is maintained only for a limited period of time and is attained when the binodal curve of both liquid and gaseous phase is crossed e.g. by expanding saturated or slightly superheated steam in fluid machinery. Supersaturated states of vapour are described by saturation S with p_V being the pressure of vaporous components and $p_{V,sat}$ the saturation pressure of condensable components corresponding to vapour temperature T_V [17].

$$S = \frac{p_V}{p_{V,sat}(T_V)} \tag{1}$$

Supersaturation ($S > 1$) is the first of three consecutive steps of spontaneous condensation. The second process is nucleation where clusters, which are small assemblies of bound monomers, emerge. With increasing supersaturation formation probability of stable clusters grows, and a significant number of clusters form spontaneously, acting as droplet nuclei. The final step in spontaneous condensation is droplet growth, whereby further monomers attach to existing droplets and clusters while released latent heat successively reduces the subcooling of the gas phase.

Besides this homogeneous mechanism taking place in pure vapour, droplet formation also occurs on wettable impurities, typically at lower supersaturation. This heterogeneous condensation affects the spontaneous condensation process, especially at a high particle number density of impurities [17]. The point of noticeable condensation onset is also referred to as the Wilson point in the literature.

In order to determine the course of spontaneous condensation, the first requirement is a calculation of the nucleation rate, which describes formation of stable clusters in terms of time and volume. Classical nucleation theory is based on an equilibrium model which allocates one cubic nuclei to each metastable fluid state [16, 1, 2, 3]. The first formulation of nucleation rate J was later improved by adding factors for non-isothermality Θ and non-equilibrium Z [17, 18]. The product of the condensation coefficient α_c and the first bracketed expression (collision probability) gives the number of colliding monomers with the critical cluster, given the particle density n_V, mean thermal velocity $\bar{c}_V$ and cluster surface area O^*. The exponential expression denotes the formation probability of critical clusters with enthalpy of formation ΔG^* and Boltzmann constant k.

$$J = \Theta\, Z\, \alpha_c \left(\frac{n_V\, \bar{c}_V\, O^*}{4} \right) n_V \exp\left(-\frac{\Delta G^*}{k\, T_V} \right) \tag{2}$$

Enthalpy of formation and ideal radius of critical clusters r^* are highly dependent on supersaturation and surface tension of liquid phase σ. Here, an approach is used to calculate σ without considering curvature radius of the droplets [19]. In the formulation v_L is the specific volume of liquid and R is the specific gas constant of condensing vapour.

$$r^* = \frac{2\, \sigma\, v_L}{R T_V\, \ln S} \tag{3}$$

As long as the vapour phase continues to exist in a supersaturated state, critical clusters grow swiftly by absorbing monomers, while releasing latent heat to the vapour phase. This exchange of mass and energy was described by Gyarmathy [15] for pure vapour and vapour gas mixtures where assumption were made in order to simplify the complex progress (e.g. homogeneous droplet state, no droplet interaction, no slip, phase equilibrium at droplet interface). The change in droplet radius over time t for a pure vapour is hence described in the following. In this formulation Λ_V is the mean free path, λ_V is the thermal conductivity of the vapour phase, and Δh_v is the specific latent heat. Temperature and supersaturation are determined for vapour outside the droplet near field. Gyarmathy's formulation for droplet growth is based on complete heat exchange between droplet and vapour phase and considers both Knudsen and continuum range.

$$\frac{\mathrm{d}r}{\mathrm{d}t} = v_L\, \frac{1 - \frac{r^*}{r}}{r + 1.59\, \Lambda_V}\, \frac{\lambda_V\, R T_V^2\, \ln S}{(\Delta h_v)^2} \tag{4}$$

2.2. Implementation of classical nucleation theory for screw expander simulation

CNT is used to model droplet formation during steam expansion in screw expanders. An idealised expander cycle is analysed in order to show the full impact on operation, and to isolate condensation from typical loss effects such as leakage and throttling.

An ideal isobaric chamber filling is assumed, which reduces occurrence of condensation strictly to the expansion phase, which is assumed to be isentropic. Furthermore, isobaric discharging at a constant back pressure is postulated. The condensation model is implemented in self-developed simulation software. Since the calculation of nucleation rate and droplet growth depend highly on supersaturation, a suitable fluid model, which will be discussed later, is vital. Here, Pollak's fundamental equation for water and steam is implemented [20]. It generates fluid state data with temperature and density input. Hence, root finding features are used for iterative calculation with differing input pairs. This provides highly accurate fluid data but requires high computation time.

An overview of the simulation process is given in **figure 1**. Based on the expander's chamber volume history, the expansion phase is divided in calculation steps with a selected time step. During each time step pressure-volume work is calculated initially with current isentropic expansion coefficient. Updates of chamber pressure and temperature are calculated, based on changes in internal energy due to the pressure-volume work being performed. The updated chamber state is checked for supersaturation, and eventually the calculation routine for classical nucleation theory is deployed. Provided that the nucleation rate generates a sufficient number of nuclei, a new droplet set is initiated and needs to be considered during the remaining part of the simulation process. For a given expansion start near the saturation curve, the number of droplet sets is in the range of five percent of the calculation steps. After generation

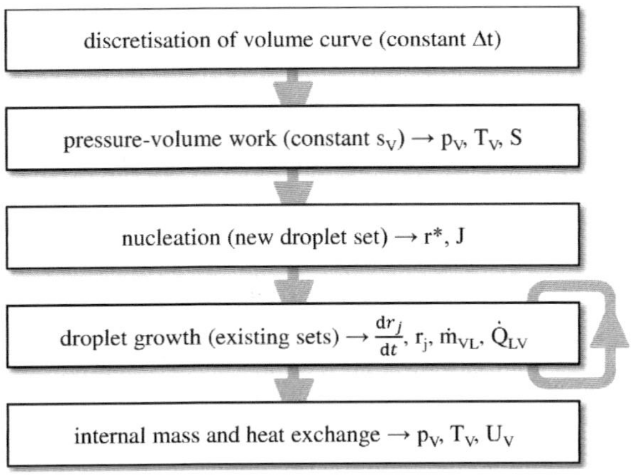

Figure 1: Simulation process with classical nucleation theory.

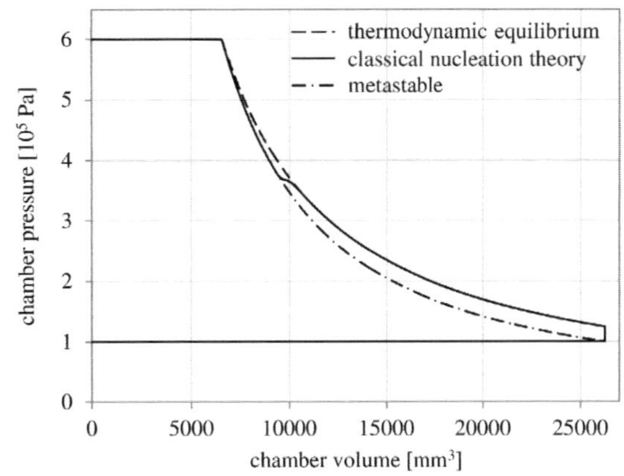

Figure 2: Calculation model dependent chamber pressure history.

of new nuclei, droplet growth of all existing droplet sets is calculated according to the current chamber state. Mass and heat exchange between the vapour and liquid phase are determined, based on the change in droplet radius over time. Chamber pressure, temperature and specific internal energy for the time step in question are finally calculated at a constant chamber volume. Indicated work $W_i = \oint p\,\mathrm{d}V$ and indicated power $P_i = n\, z_{male}\, W_i$ are later determined by the integration of pressure history with respect to the chamber volume. This approach allows for implementation in more complex multi-chamber simulation, where initial changes of states are not only affected by performed pressure-volume work, but also by internal leakages, which result in higher simulation quality.

2.3. Simulation parameters

The calculation model presented in this study requires several numerical parameters as well as machine parameters. The condensation coefficient in equation 2 for nucleation rate is $\alpha_c = 0.7$ for all presented data. In the formulation for droplet growth (equation 4) the selected droplet radius needs to be slightly larger than the critical value for initial growth rate of emerged critical clusters. Here r^* is increased by $0.1\,\%$, which has no significant effect on simulation results. With respect to time-dependent droplet growth, the expansion phase is discretised with a defined time step Δt and the droplet growth equation is linearised. A sensitivity analysis with regard to maximal supersaturation during expansion and indicated power P_i implies $\Delta t = 1.0 \cdot 10^{-7}\,\mathrm{s}$ being sufficiently small for the parameter range in question.

Relevant information on the machine geometry, which determines the volume curve and links tip speed with rotational speed, is shown in **table 1**.

3. Expander cycle with spontaneous condensation

The influence of spontaneous condensation on an ideal expander cycle is shown in **figure 2**. The following observations refer to expander tip speed $u_{male} = 80\,\mathrm{m\,s^{-1}}$, inlet pressure $p_{in} = 6.0 \cdot 10^5\,\mathrm{Pa}$, with no superheating and outlet pressure $p_{out} = 1.0 \cdot 10^5\,\mathrm{Pa}$. Ideally, chamber pressure equals inlet pressure during the filling phase. Isentropic expansion starts when the chamber volume reaches the start value defined by the built-in expansion volume ratio. At this point the high pressure control edge seals the chamber under examination. When maximum volume is reached the chamber passes the low pressure control edge, and ideally isochoric pressure compensation takes place. The outlet phase is assumed to be isobaric at outlet pressure level.

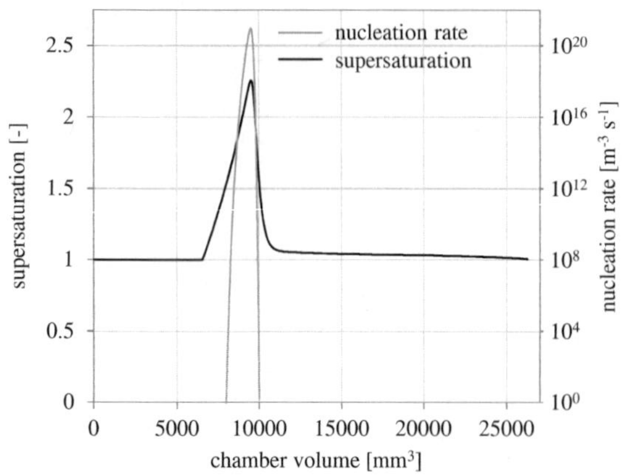

Figure 3: Supersaturation and nucleation rate during expansion.

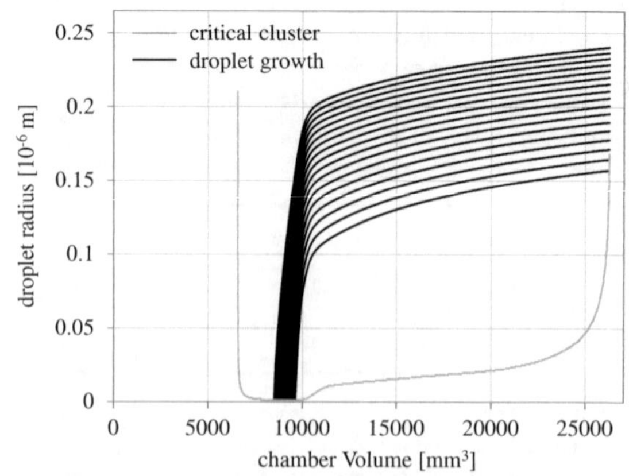

Figure 4: Critical cluster radius and growth during expansion for exemplary droplet sets.

Since the expander is fed with saturated steam, the wet steam region is reached instantly when pressure drops at the start of expansion. If fluid states are calculated in thermodynamic equilibrium (which is the borderline case of infinitely slow expansion), optimal chamber pressure is reached and thus maximum pressure-volume work is obtained. If no phase transition is taken into account and the working fluid is dry supercooled steam, pressure develops at a lower level during expansion. The depicted metastable fluid states correspond to the borderline case of infinitely fast expansion, where all condensation phenomena are suppressed. It should however be noted that at the latest when the spinodal curve is reached, and metastability is no longer in force, condensation must occur.

The continuous line in figure 2 describes pressure history calculated with classical nucleation theory. At the start of expansion the steam is in metastable (dry) state, and due to low supersaturation, the nucleation rate is low, see **figure 3**. The nucleation rate raises rapidly with increasing supersaturation, until large numbers of emerging droplets release latent heat to the vapour phase, thus reducing supersaturation. The continuing expansion phase is characterised by low supersaturation, since heat and mass exchange between growing droplets and vapour take place without delay. The course of droplet growth during expansion is depicted in **figure 4**. The continuous grey line marks the critical radius of emerging nuclei, which is characterised by vertical asymptotes for $S = 0$. In general, the critical radius is small compared to droplet size at the end of expansion. As supersaturation decreases rapidly after reaching a global maximum,

Table 1: Expander data

characteristic	value
rotor profile	SRM A
number of lobes	4 (male) / 6 (female)
maximum chamber volume	$26254\,\text{mm}^3$
expansion volume ratio	4.0
rotor tip diameter	58.9 mm
axis to axis distance	46.2 mm

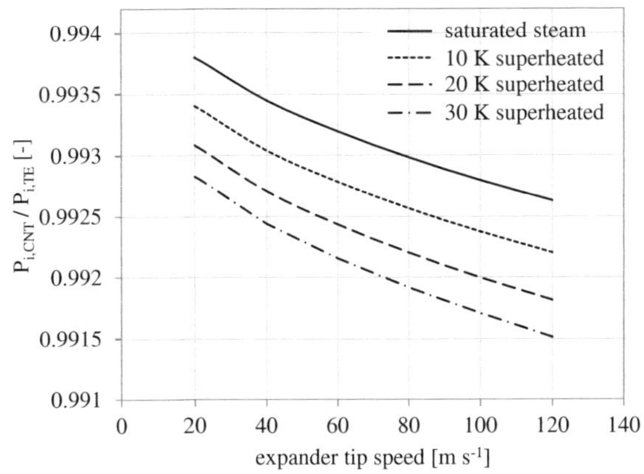

Figure 5: Effect of spontaneous condensation on ideal indicated power, $p_{in} = 6 \cdot 10^5\,\text{Pa}$, $p_{out} = 1 \cdot 10^5\,\text{Pa}$.

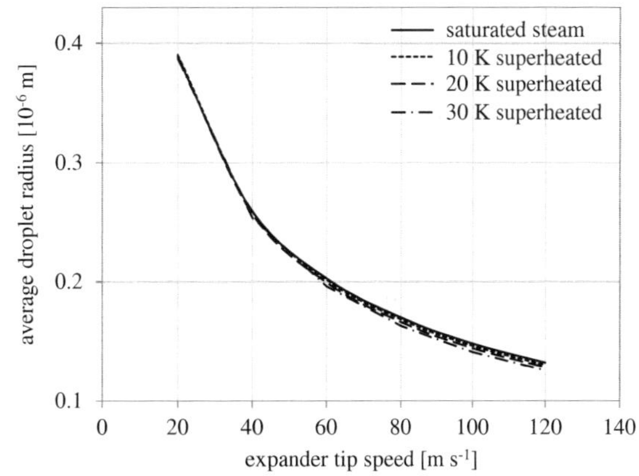

Figure 6: Average droplet size at end of expansion phase, $p_{in} = 6 \cdot 10^5\,\text{Pa}$, $p_{out} = 1 \cdot 10^5\,\text{Pa}$.

the critical radius increases with ongoing expansion. This causes droplets that did not grow larger than the current critical radius to reduce in size (see formulation for droplet growth) and eventually reevaporate. Hence, the formation of stable droplets is limited to a relatively small window during the expansion phase. Continuous black lines show the course of droplet growth for representative sets. Weighting with the number of droplets per set z_j makes it possible to calculate an average droplet size $\bar{r}$.

$$\bar{r} = \frac{\sum (r_j\, z_j)}{\sum z_j} \tag{5}$$

For the given example, the average droplet size is $\bar{r} = 1.7 \cdot 10^{-7}\,\text{m}$, which indicates a shift of emphasis towards droplets emerging in range of maximum supersaturation.

4. Influence of expander operation on spontaneous condensation

Figure 5 shows a comparison of indicated power for a process with spontaneous condensation and a process in thermodynamic equilibrium for inlet pressure $p_{in} = 6 \cdot 10^5\,\text{Pa}$ and back pressure $p_{out} = 1 \cdot 10^5\,\text{Pa}$. Expander tip speed is varied from $u = 20$ to $120\,\text{m}\,\text{s}^{-1}$. Superheating is displayed from 0 K to 30 K. With increased superheating spontaneous condensation takes place later during expansion phase since the saturation curve is crossed at a later point in time and at lower pressure. However, a greater influence on pressure history by means of spontaneous condensation is determined for superheated steam. The calculated nucleation rates decrease with increasing superheating, and thus higher supersaturation is reached. Higher supersaturation and consequently lower pressure-volume work is also observed at higher expander speed. This results from smaller time frames for heat and mass exchange between droplets and the vapour phase. In general it needs to be noted that spontaneous condensation only has minor effects on ideal pressure history and thus expander power.

CNT provides valuable information on droplet size and distribution, see **figure 6**. It can be shown that the effect of superheating on droplet size at the end of the expansion phase is negligible. Droplet size is mainly affected by expander speed, and average droplet radius also decreases with faster expansion. Within this study, droplet size is in the range of 0.1 to $0.5 \cdot 10^{-6}\,\text{m}$, which, according to Gyarmathy's studies, indicates relatively uniform droplet distribution and deceleration times in the range of $10^{-6}\,\text{s}$, which leads to practically no slip between droplets and vapour [14]. This furthermore allows for the correct choice of two-phase

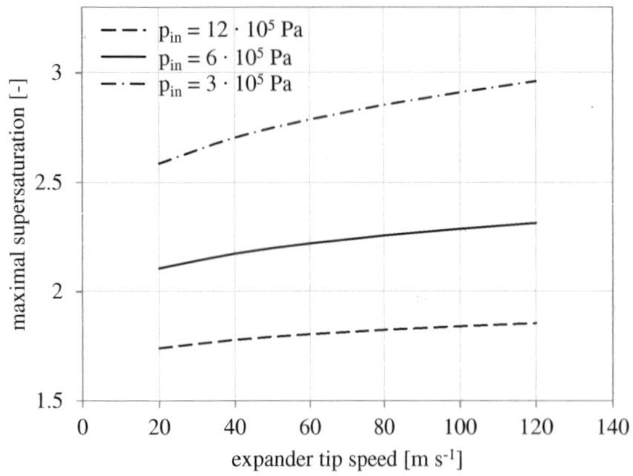

Figure 7: Maximal supersaturation during saturated steam expansion, $\Pi = \frac{p_{in}}{p_{out}} = 3.0$.

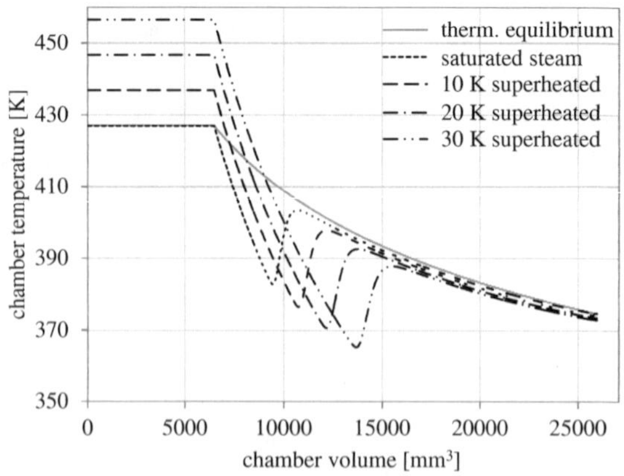

Figure 8: Temperature history during expansion, $p_{in} = 6 \cdot 10^5\,\mathrm{Pa}$, $p_{out} = 1 \cdot 10^5\,\mathrm{Pa}$ $u_{male} = 80\,\mathrm{m\,s^{-1}}$.

flow models for gap and port flows for future detailed simulative investigations of steam-driven screw expanders.

Despite having only a minor impact on chamber pressure, detailed modelling of condensation process illustrates the correct development of temperature during expansion. Maximum supersaturation for variations of expander speed and inlet pressure is depicted in **figure 7** for saturated steam. A major impact factor on supersaturation during expander operation is the negative slope of the saturation curve, which decreases for water at low pressures. Expanders operating with low pressure steam are more likely to be affected by condensation effects than expanders operating at high pressure levels. Moreover, only consideration of intermittent supersaturated vapour states allows for correct modelling and analysis of heat transfers during expander operation. **Figure 8** shows the temperature history during expansion for variations in inlet superheating. The temperature calculated in thermodynamic equilibrium is depicted for saturated steam only for reasons of clarity. It is noted that assuming change of state in thermodynamic equilibrium for complete expander process might lead to defective temperature course.

5. Property formulations for metastable fluid states

Calculation of time-dependent spontaneous condensation process requires a suitable fluid model to provide accurate data for the metastable region. Non-ideal fluid data is normally calculated with fundamental equations, which are adapted to extensive experimental data. However, it seems there are no experimental data on subcooled steam, as common measurement principles affect metastable fluid states (e.g. surface condensation on temperature sensors). As a consequence, property formulations for superheated steam need to be extrapolated in the metastable region. **Figure 9** shows isobars calculated with various fluid models [20, 21]. Metastable isobars show continuous progression, whereas isobars in the thermodynamic equilibrium feature the familiar discontinuity when passing the saturation curve. It is noted that extrapolation of the IAPWS equation for the pressure in question is limited to a maximum supersaturation of $S_{max} = 2.8$. The depicted low gradients of metastable isobars calculated with IAPWS equation are characteristic for all pressure ranges. Since the measurement of higher supersaturation than maximum values calculated with IAPWS formulation is documented in the literature (e.g. in [11]), extensive extrapolation is not recommended. However, excellent

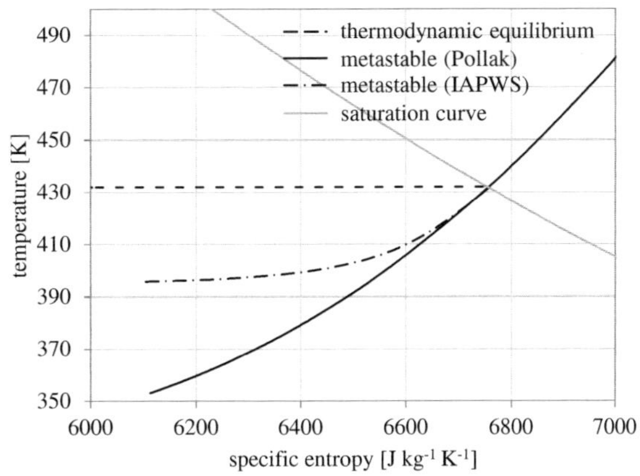

Figure 9: Comparison of isobars in T-s-diagram for different fluid models, $p = 6 \cdot 10^5$ Pa.

accuracy is achieved for steam data in thermodynamic equilibrium.

A slightly less complex approach for a canonical equation of state was presented by Pollak and is used for this investigation [20]. As depicted in figure 9 it shows reasonable behaviour when applied to supersaturated states and coincides with alternative formulations like Young's simplified equation while offering higher accuracy and wider range [22]. Advantages over real gas equations (e.g. Van-der-Waals, Redlich-Kwong) are the provision of accurate data for both liquid and vapour phase, and the ability to determine all required data by derivates of just one equation for specific free energy.

6. Outlook

Classical nucleation theory provides an initial model for the prediction of spontaneous condensation in screw expander chambers. The calculation of subcooled steam states permits future modelling of heat transfer to expander machine parts at non-equilibrium temperature levels. Future work will focus on this external heat exchange and eventually combine mist formation and surface condensation on machine parts. It is noted that latent heat transferred to machine parts is lost and cannot raise the temperature of expanding vapour. Furthermore, a calculation of droplet size and formation time in expander chambers with CNT lays the foundation for detailed droplet simulation in order to map their trajectory, and where appropriate identify collisions with expander surfaces.

In addition, CNT is applicable to compressible vapour flows in screw expanders. For example, considerable supersaturation might occur during inlet phase, especially at the beginning and end of filling process, due to small port cross sections and relatively high pressure ratios at the opening. The same effect might take place in gap flows when working chambers with high pressure differences are connected. It is a debatable point whether or not the supersaturation period in these accelerated flow is long enough to induce spontaneous condensation, since the steam is decelerated again in the chamber.

The performance of screw expanders is significantly influenced by internal leakages. High flow velocities in the gaps lead to high heat fluxes. If the surface temperature is below the vapour temperature film- or dropwise condensation might occur. Future work should focus on leakage flow in order to analyse vapour states and identify potential condensation phenomena on expander surfaces.

IOP Conf. Series: Materials Science and Engineering **232** (2017) 012078 doi:10.1088/1757-899X/232/1/012078

7. Conclusion

Spontaneous condensation during the expansion phase of screw expander was modelled with classical nucleation theory providing a first simulation tool for droplet formation in screw expanders. Despite showing only a minor effect on indicated power, the study provides valuable information on temperature development during expansion and subcooling during the period after the saturation curve is crossed. Furthermore, droplet size is determined, which allows for the detailed simulation of interaction between the vapour and liquid phases. Typical droplet size for screw expander in question is between 0.1 and $0.4 \cdot 10^{-6}$ m. One major influence on droplet size is expander speed. Maximum supersaturation is in the range of 1.8 to 3.0. The level of supersaturation during expansion is mainly affected by pressure boundary conditions.

References

[1] Volmer M and Weber A 1926 Keimbildung in übersättigten Gebilden *Zeitschrift für physikalische Chemie* **119**

[2] Farkas L 1927 Keimbildungsgeschwindigkeit in übersättigten Dämpfen *Zeitschrift für physikalische Chemie* **125**

[3] Becker R and Döring W 1935 Kinetische Behandlung der Keimbildung in übersättigten Dämpfen *Annalen der Physik* **24**

[4] Zeldovich Y B 1942 Theory of formation of new phase *Cavitation, JETP* **525**

[5] Frenkel J 1946 *Kinetic Theory of Liquids* (Oxford: Clarendon Press)

[6] Bakhtar F, Young J B, White A J and Simpson D A 2005 Classical nucleation theory and its application to condensing steam flow calculations *Proc. Instn. Mech. Engrs.* **219** 1315–1333

[7] Schmitt J L, Adams G W and Zalabsky R A 1982 Homogeneous nucleation of ethanol *J. Chem. Phys.* **77** 2089–2097

[8] Adams G W, Schmitt J L and Zalabsky R A 1984 The homogeneous nucleation of noane *J. Chem. Phys.* **81** 5074–5078

[9] Ehrler F, Repple K H, Schüßler J, Treffinger P and Wright W 1996 Special cloud chambers for investigation into the time-behaviour of homogeneously nucleated spontaneous condensation *Experiments in Fluids* **21** 363–373

[10] Barschdorff D, Hausmann G and Ludwig A 1976 Flow and drop size investigation of wet steam at sub- and supersonic velocities with the theory of homogeneous condensation *Pr. Inst. Maszyn Przeplwowych* **241** 70–72

[11] Moses C A and Stein G D 1978 On the growth of steam droplets formed in a laval nozzle using both static pressure and light scattering measurements *Journal of Fluids Engineering* **100** 311–322

[12] Bakhtar F and Tochai M T 1989 Nucleation phenomena in flowing high-pressure steam: experimental results *Proc. Instn. Mech. Engrs.* **203** 195–200

[13] Bakhtar F and Zidi K 1989 Nucleation phenomena in flowing high-pressure steam, part 2: theoretical analysis *Proc. Instn. Mech. Engrs.* **204** 233–242

[14] Gyarmathy G 1962 *Grundlagen einer Theorie der Naßdampfturbine* (Zürich: Juris-Verlag)

[15] Gyarmathy G 1963 Zur Wachstumsgeschwindigkeit kleiner Flüssigkeitstropfen in einer übersättigten Atmosphäre *Z. Angew. Math. Phys.* **14** 280–292

[16] Volmer M 1939 *Kinetik der Phasenbildung* (Dresden, Leipzig: Theodor Steinkopff) Die Chemische Reaktion

[17] Ehrler F and Schaber K 2013 *Spontane Kondensation und Aerosolbildung* (Berlin, Heidelberg: Springer-Verlag) pp 1053–1080 VDI-Wärmeatlas 11th ed

[18] Feder J, Russel K C, Lothe J and Pound G M 1966 Homogeneous nucleation and growth of droplets in vapours *Adv. Phys.* **15** 111–178

[19] IAPWS 2014 Revised release on surface tension of ordinary water substance *http://www.iapws.org*

[20] Pollak R 1974 *Die thermodynamischen Eigenschaften von Wasser - dargestellt durch eine kanonische Zustandsgleichung für die fluiden homogenen und heterogenen Zustände bis 1200 Kelvin und 3000 bar* (Bochum: Ruhr-Universität Bochum)

[21] Wagner W and Pruß A 2002 The IAPWS formulation 1995 for the thermodynamic properties of ordinary water substance for general and scientific use *J. Phys. Chem. Ref. Data* **31** 387–535

[22] Young J B 1988 An equation of state for wet steam for turbomachinery and other flow calculations *Journal of Engineering for Gas Turbines and Power* **110** 1–7

List of symbols

Symbol	Unit	Property
$\bar{c}$	$\mathrm{m\,s^{-1}}$	mean thermal velocity
J	$\mathrm{m^{-3}\,s^{-1}}$	nucleation rate
k	$\mathrm{J\,K^{-1}}$	Boltzmann constant
$\dot{m}_{VL}$	$\mathrm{kg\,s^{-1}}$	condensation mass flow
n	$\mathrm{m^{-3}}$	particle density
n	$\mathrm{s^{-1}}$	expander speed
O	$\mathrm{m^2}$	surface
p	Pa	pressure
P	W	power
$\dot{Q}_{LV}$	W	latent heat flow
r	m	radius
R	$\mathrm{J\,kg^{-1}\,K^{-1}}$	specific gas constant
s	$\mathrm{J\,kg^{-1}\,K^{-1}}$	specific entropy
S	–	supersaturation
t	s	time
T	K	temperature
u	$\mathrm{m\,s^{-1}}$	tip speed
U	J	internal energy
v	$\mathrm{m^3\,kg^{-1}}$	specific volume
V	$\mathrm{m^3}$	volume
W	J	work
z	–	number of
Z	–	non-equilibrium factor
α_c	–	condensation coefficient
ΔG	J	enthalpy of formation
Δh_v	$\mathrm{J\,kg^{-1}}$	specific latent heat
Δt	s	time step
λ	$\mathrm{W\,m^{-1}\,K^{-1}}$	thermal conductivity
Λ	m	mean free path
Π	–	pressure ratio
σ	$\mathrm{N\,m^{-1}}$	surface tension
Θ	–	non-isothermal factor

Subscript	Property
CNT	classical nucleation theory
i	indicated
in	inlet
j	control variable
L	liquid
max	maximum
out	outlet
sat	saturation
TE	thermodynamic equilibrium
V	vapour
$*$	critical cluster

IOP Conf. Series: Materials Science and Engineering **232** (2017) 012079 doi:10.1088/1757-899X/232/1/012079

Three-dimensional numerical simulation and experimental validation of flows in working chambers of roots blowers with backflow design

S K Sun, Q Zhou, J Wen, X Y Peng*

School of Energy and Power Engineering, Xi'an Jiaotong University, No.28 West Xianning Road, Xi'an, Shaanxi, P.R.China, 710049.

*Email: xypeng@mail.xjtu.edu.cn

Abstract. the backflow design is one way to reduce the noise in the Roots blower. In this paper, the three-dimensional computational fluid dynamics (CFD) models were established to simulate the flows in the working chambers of the Roots blower with and without backflow design, using the dynamic mesh method. And then, the variations of pressure in the cylinder of the Roots blower with backflow design were tested in experiment to validate these CFD models. The result shows the CFD models were appropriate. The backflow could reduce the amplitude of the pressure pulsation in the outlet of the Roots blower more than 50%, but the volume efficiency also be reduced by 3.5%.

1. Introduction

The Roots blower is a positive displacement machine that is widely used in many fields, including mining, oil refining, the chemical industry or in chemical and environmental protection. The Roots blower has double rotors, with two or more lobes rotating in opposite directions in the cylinder. These double rotors are non-contacted and driven by a couple of synchronized gears. The rotors and the cylinder wall consist a working chamber which turn on and off periodically to transport fluid from the inlet to outlet. The pressure of the fluid does not rise in the transport process. The high pressure fluid in the outlet pocket flow back to compress the fluid in the working chamber at the moment the working chamber connected to the outlet pocket.

Many researchers have focused their attention on the profile design and meshing analysis of Roots blowers. Mimmi [1] optimized the parameters of the screw rotor based on a mathematical model of the lobe profiles. Valdès [2] used the semi-empirical Knudsen–Dong law to predict the conductance of each clearance within a few percentage points for a wide range of pressures covering the transient flow regime. Tong and Yang [3,4,5] explored a new method of profile design by deriving a function to help choose the most efficient lobe parameters for high-sealing Roots blowers, according to the given flow rate functions. Burmistrov [6] proposed an angular coefficients method to calculate channel conductance in Roots blowers, based on the similarities between the laws of radiation and those of the diffusive reflection of molecules. Wang [7] designed a tooth profile consisting of five arcs, which obtained a better sealing performance due to its longer and narrower clearance. Yao [8] designed a novel three-lobe helical rotor for the Roots blower, with the cross section designed as a combination of concave arcs, convex arcs and cycloidal curves. The simulated cutter trajectory of the new rotor was then analyzed. Yao also devoted much effort to investigating the characteristics of the flows and leakages in the rotor's working cycle. Hwang [9,10] proposed a new rotor profile with a variable trochoid ratio, and investigated how to achieve higher volumetric efficiency and higher sealing performance. Chiu [11] developed a new profile by using an elliptical roulette, and then compared the performances of the new and the traditional profiles.

With the development of computational fluid dynamics (CFD), growing numbers of researchers have focused on the numerical analysis of the Roots blower. Joshi [12] assumed that clearance between the impellers was almost inevitable during operation, and therefore built a static model to simulate the leakage through the clearance. Huang [13] used the k-ε model to simulate a three-lobe Roots blower, and compared the results of this simulation with those from semi-empirical formulas in terms of their non-uniformity of outlet flows. Li [14] investigated the effects of the rotor's pressure angle on the flow characteristics by using a two-dimensional turbulence model. Hsieh [15,16] compared the differences between cylindrical and screw-type Roots blowers in terms of average outlet flow rates and instant outlet flow rates, based on a CFD case, namely BlowerLinx. Hsieh also analyzed the effects of the rotor phase angle in multi-stage Roots blowers with serial or parallel connections in terms of their flow rates and pulsation.

The Roots blower often working with big noise. The pressure pulsation is the major reasons of the aerodynamic noise, and backflow design is a good way to reduce the airflow pulsation. In this study, we established three-dimensional computational fluid dynamics (CFD) models of the Roots blower with and without backflow to analyze the flow character in the working circle. An experimental test rig was built with a novel Roots blower which the size identical to the simulated models. The pressures distribution in the working chamber and the inlet/outlet pockets were measured with five pressure sensors, and both the pressure and the flow rate results were used to verify the simulation result. Based on the CFD analytical and experimental investigations, the differences between the Roots with and without backflow design were discussed.

2. Numerical simulation

2.1 The structure of the Roots blower.

There are two parallel shafts rotating synchronously in opposite directions in a Roots blower, each of which is supported by two bearings, and one rotor and one gear are assembled on it. In a traditional Roots blower, the rotors were set in the middle of the shafts and the bearings were assembled on the two

IOP Conf. Series: Materials Science and Engineering **232** (2017) 012079 doi:10.1088/1757-899X/232/1/012079

side of the rotors. But in the Roots blower we used, the bearings were set at the two side of gears, as shown in Fig. 1. All of the bearings were set on one side of the rotors, so that only a front plate was needed on the free side of the rotors to cover the cylinder. The clearance in the cylinder could be easily measured by open the front plate. Furthermore, the pressure sensors could be conveniently installed on the front plate to measure the pressure in the cylinder. There are eight backflow holes at the cylinder wall, through which the higher pressure gas can improve the pressure in the working chamber at the position of 120° of the rotor's angle, four on the upper working chamber and the other four on the downer working chamber. In this paper, the four horizontal backflow holes were closed by valves, so only four vertical backflow holes worked.

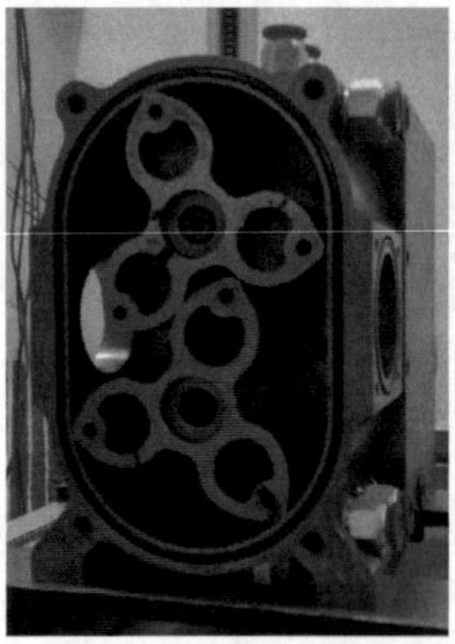

Fig. 1 The structure of the Roots blower.

The Roots blower used in this study was a three-lobe blower. The diameter of the cylinder was 52.5 mm, and the length was 70 mm. The center distance of the two rotors was 70 mm, the profile of which consisted of six parts, as follows: two lines, three arcs and one involute [17]. The clearance between the two rotors was 0.14 mm, and that between the rotor and cylinder was 0.07 mm. The clearance between the rotor and the front plate was 0.11 mm, and that between the rotor and the back plate was 0.03 mm.

2.2 The geometry and grid of the Roots blower with backflow structure

In this study, the numerical simulation is the fluid dynamics analysis. So the fluid space was chosen as the computational domain. The clearances between the rotors and cylinder consist of two parts: circumferential and radial directions. The clearance of the radial directions was ignored because it is hard to achieve by dynamic mesh method. But the clearance between the rotors and cylinder were increased to 0.15mm to keep the total leakage no change. Onto the above-described geometry, inlet/outlet piping was added. The geometrical model of the Roots blower was established. The flow field consisted of the cylinder, inlet piping, inlet buffer area, outlet piping, outlet buffer area and backflow piping.

The grids of the three-dimensional Roots blower with backflow design was shown in Fig. 2. The inlet piping, inlet buffer area, outlet piping and outlet buffer area used hexahedral elements, and the cylinder used triangular prism elements, and the pre-flow piping used tetrahedral mesh. The total number of elements was 813221. The grids of the Roots blower without backflow was the same to the model with backflow except the interfaces between the backflow piping and cylinder were set to wall.

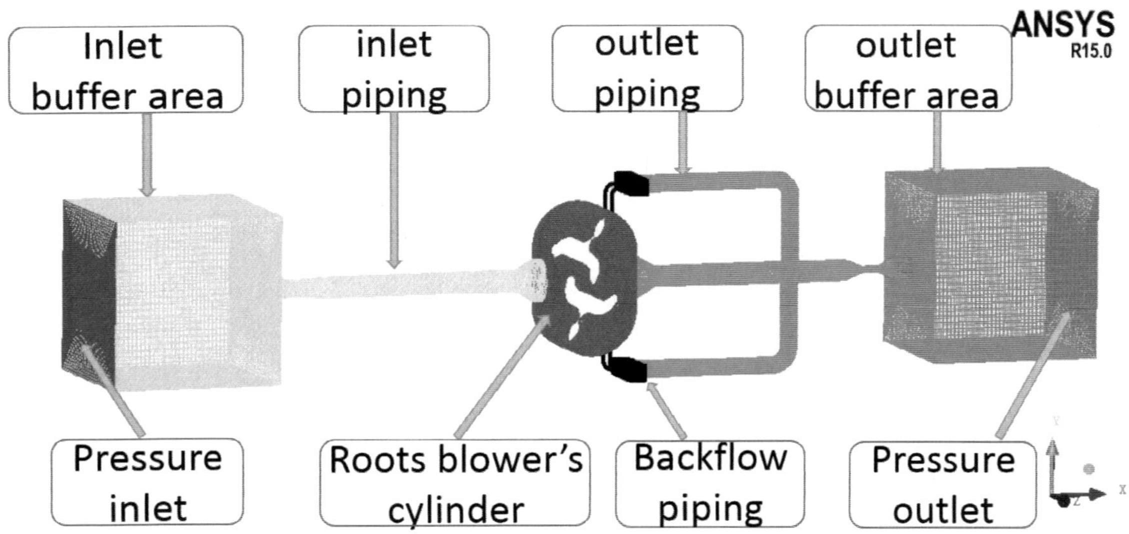

Fig. 2 The geometry and grid of the Roots blower.

2.3 Governing equations and boundary conditions

In the numerical simulation based on the CFD software Fluent, the governing equations included continuity, momentum and energy conservation equations as follows [18]:

$$\frac{\partial p}{\partial t} + \nabla \cdot (\rho \vec{v}) = 0 \tag{1}$$

$$\frac{\partial}{\partial t}(\rho \vec{v}) + \nabla \cdot (\rho \vec{v} \vec{v}) = -\nabla p + \nabla \cdot \bar{\bar{\tau}} + \rho \vec{g} + \vec{F} \tag{2}$$

$$\frac{\partial}{\partial t}(\rho E) + \nabla \cdot \big(\vec{v}(\rho E + p)\big) = -\nabla \cdot \left(\sum_j h_j J_j \right) + S_h \tag{3}$$

Where p was the static pressure, $\bar{\bar{\tau}}$ was the stress tensor, and $\rho \vec{g}$ and $\vec{F}$ were the gravitational body force and the external body force, respectively.

The realizable k-ε model was chosen for the turbulence equation because it had been validated for rotating shear flows and separated flows [18]. The near-wall treatment used the scalable wall function. The pressure-inlet and pressure-outlet boundary conditions were adopted. The gauge pressure of both inlet and outlet were set to 0 kPa. Although the outlet pressure was set to 0kPa, the average pressure in the outlet pocket of the Roots blower was 50 kPa. It is because there was a narrow pipe which throttle the fluid to increase the pressure before the fluid reached the outlet buffer area. The fluid was air with properties as idea-gas. The connections between adjacent parts used interface method to transfer data.

The working process of a Roots blower is a transient problem, so it is necessary to bring in the dynamic mesh. The local-remeshing and the spring-based smoothing dynamic mesh methods were adapted in the cylinder areas. A UDF (user defining function) was programed to drive the rotors rotating as rigid bodies.

The space conservation equations of the dynamic meshes is an integral form of the conservation equation for a general scalar ϕ on an arbitrary control volume V, whose boundary is moving can be written as [18]

$$\frac{d}{dt}\int_{V}\rho\phi dV+\int_{\partial V}\rho\phi(\vec{u}-\vec{u}_g)\cdot d\vec{A}=\int_{\partial V}\Gamma\nabla\phi\cdot d\vec{A}+\int_{V}S_{\phi}dV \quad (4)$$

where ρ is the fluid density, $\vec{u}$ is the flow velocity vector, $\vec{u}_g$ is the mesh velocity of the moving mesh, Γ is the diffusion coefficient, S_ϕ is the source term of ϕ, ∂V is used to represent the boundary of the control volume V.

The time step was determined by $\Delta t=\frac{CFL\bullet\Delta x}{\lambda_{max}}$, where CFL was the Courant number, Δx was the grid distance, and λmax was the maximum velocity scale. Therefore, the time step was set up as 1 x 10^{-5} s.

3. The experiment on the Roots blower

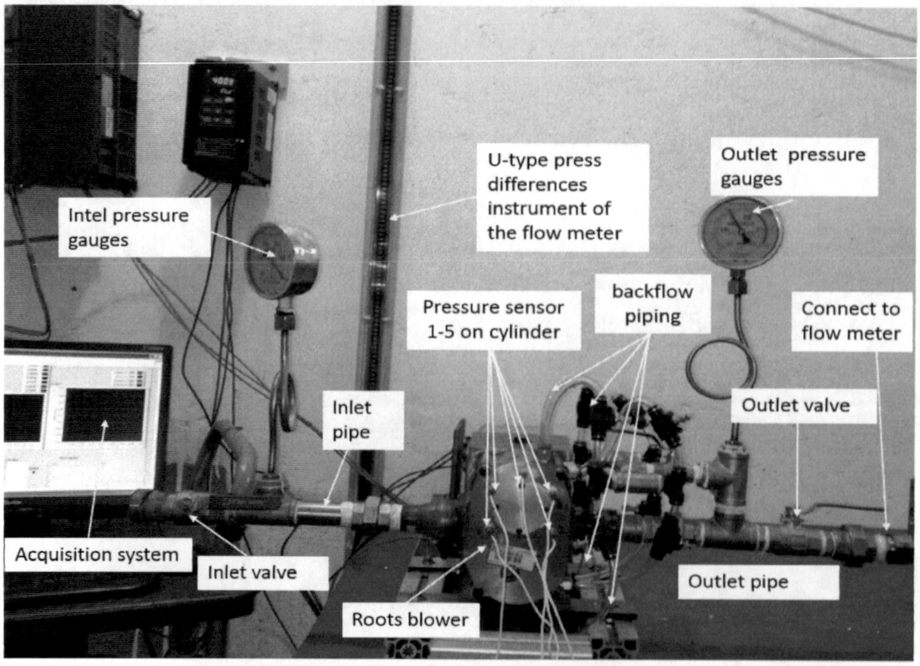

Fig. 3 The experimental equipment.

The equipment for the experimental investigation of the Roots blower was shown in Fig. 3. This equipment is similar to the simulation model. It consisted of a motor, a Roots blower, pressure gauges, inlet/outlet piping, valves, a flow meter, pressure sensors, backflow piping and an acquisition system. The pressure gauges and valves could measure and control the inlet and outlet pressure of the Roots blower. The pipe behind the outlet valves was connected to a flow meter. There were eight backflow pipes connect to the outlet piping, each of them contain a valve to control whether this pipe is on work. In order to accord to the simulation models with backflow, four of the valves were closed. When the data of the Roots without backflow, the eight valves were all closed.

In this research, the purpose of the experiment was to verify whether the CFD simulation could provide an enough accuracy result for the Roots blower with backflow. So the pressure distribution in working chamber and inlet/outlet pocket was need to compare with the simulation one. Therefore five sensor were installed on the front plate of the Roots blower. Fig. 4(a) shows the exactly position of these sensors. Sensor 1 was to measure the pressure in the inlet pocket, sensor 2, 3 and 4 were to measure the pressure in the working chamber, sensor 5 was to measure the pressure in the outlet pocket. All the sensor were installed around the upper rotor, because the double rotors were fitted nearly symmetrically except a difference of 60° in the phase angle.

The rotor rotated 300° during a working circle as shown in Fig. 4(b-g). The working chamber of Roots blower closed at 120° and opened at 180°. The available test angle of sensors were shown in Table. 1. So the pressure of the working chamber could be measured at any time.

Table.1 The available test angle of the pressure sensors.

Sensor number	angle
Sensor 1	0° ~ 60°
Sensor 2	60° ~ 120°
Sensor 3	120° ~ 180°
Sensor 4	180° ~ 240°
Sensor 5	240° ~ 300°

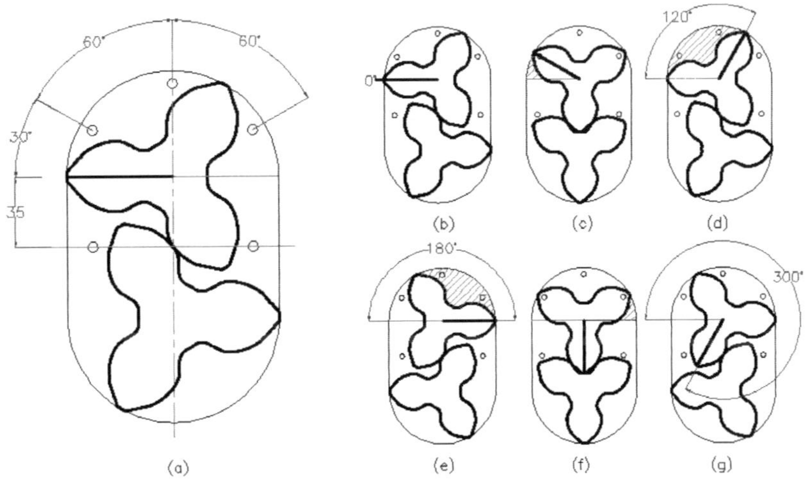

Fig. 4 The positions and the available test angles of the pressure sensors.

With the experimental equipment introduced above, the dynamic pressure in the working chamber, inlet and outlet pocket of the Roots blower could be acquired. The experimental data were obtained under the following conditions: The motor was rotating at 2400 rpm. The valve in the inlet piping was fully open, and the valve in the outlet piping was adjusted to ensure that the outlet pressure was 50 kPa. The measuring range of the pressure sensors was 0~3.5 bar (absolute pressure), and the sample rate of the sensors was 10 kHz.

4. The results and discussions

4.1 Comparison of experimental and simulation results for the flowrate

The numerical simulation performed in this study involved a transient problem with an initial value. Before the simulation achieved a periodic steady state, it was necessary to calculate a large number of time steps. One of the choices for determining if the simulation was steady was to decide whether the difference in the average mass flow between the inlet and outlet during a working cycle could be ignored. Generally, the simulation could be considered steady if the difference was lower than 5%. Table. 2 shows the average mass flow rate for the two cases in the third working cycle, from which we could determine that the simulations had been periodically steady while the calculations were run to the third period.

Both the mass flow rates of the simulation and the experiments were shown in Table. 2. It shows the simulation results was 6.5% lesser than the experiments one for Roots without backflow and 4.0% lessor for Roots with backflow. Both the simulations results were lesser than the experiments results, it is because it hard to achieve accurate measurements of the size of clearance. Consider the differences were small, the results of the simulation were accredited. The Table. 2 also shows the mass flow rates of the Roots blower with backflow was lesser than the one without backflow. It is because the pressure of the working chamber of the Roots blower with backflow was higher than the blower without backflow. The higher pressure between the working chamber and the inlet lead to more leakage, which reduce the mass flow rate on the outlet. The mass flow rate of the Roots blower with backflow was about 3.5% lesser than the one without backflow in this study.

Table.2 The average mass flow rates of the simulations and the experiment.

	Experiment(kg/s)	CFD simulation(kg/s)	
		inlet	outlet
Roots without backflow	0.01475	0.01554	0.01579
Roots with backflow	0.01423	0.01456	0.01504

4.2 Comparison of experimental and simulation results for the variations in pressure

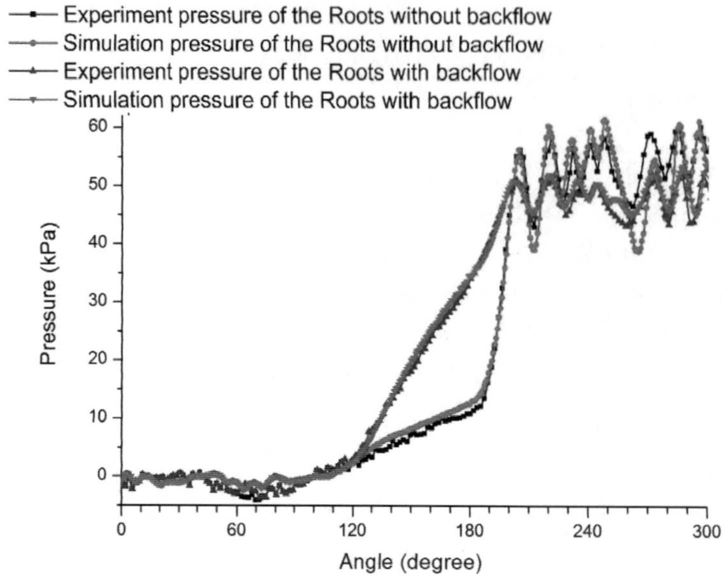

Fig. 5 The variations in pressure of the rotor's angle in a working cycle.

According to the available test angles of the pressure sensors as shown in Table.2, the variations of pressure

IOP Conf. Series: Materials Science and Engineering **232** (2017) 012079 doi:10.1088/1757-899X/232/1/012079

in the working chamber, with the rotor rotating angles, are shown in Fig. 5. The results between the experiments and the simulation for both the Roots blower with and without backflow agreed with each other well. It is said that the CFD simulation results were believable.

Fig. 5 shown the variations in pressure for the Roots with backflow were almost the same to the Roots without backflow from 0° to 120°, because the working chamber was connected to the inlet pocket and disconnect to the backflow piping in this time, so the impact of the backflow could hardly show up. But since 120°, the pressure variation is difference. For Roots blower without backflow, from 120° to 180°, the working chamber was closed, the higher pressure air leak from outlet pocket through the clearance between the rotor and the cylinder to the working chamber, which made the pressure in the working chamber increased slowly. Since 180°, the working chamber connected to the outlet pocket, the air in the outlet pocket flow back to the working chamber because the pressure in the working chamber was lower than the outlet. The direction of the flow back was opposite to the rotor rotating, which caused impact and pulsation of the air. The amplitude of pressure pulsation was about 25kPa. For Roots blower with backflow, from 120° to 180°, the working chamber was disconnect to the inlet pocket and connect to the backflow piping, the higher pressure air flow into the working chamber and increased the pressure in the chamber. When the working chamber connect to the outlet pocket, the pressure in the working chamber was close to the outlet pocket. So the pressure pulsation in the outlet pocket was smaller than Roots blower without backflow, it is only 10kPa. Both the simulation and the experiment shown the backflow can reduce the amplitude of pressure pulsation of the Roots blower.

4.3 The pressure field

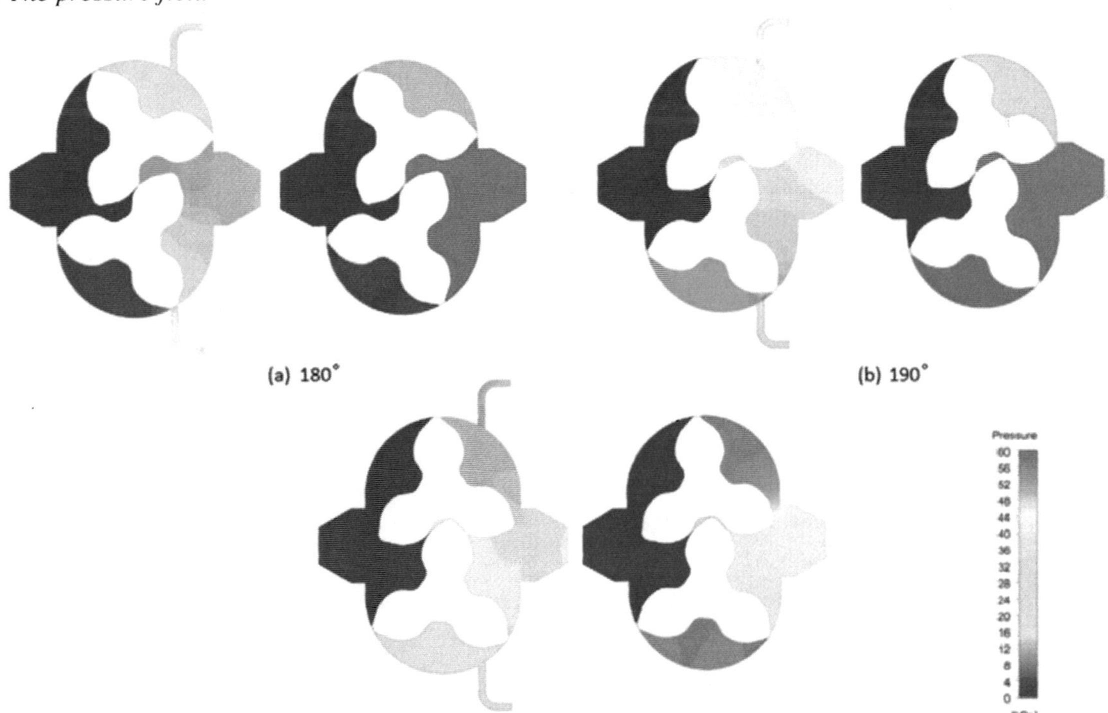

Fig. 6. The comparison of the Roots blower with and without backflow on pressure field.

The symmetry plane were choosen as the observe plane to study the difference between the Roots blower with and without backflow on the pressure field. Fig. 6 shows the pressure field of these two case.

In the Roots blower without backflow, the pressure difference between the working chamber and outlet pocket was about 40kPa at 180° (Fig. 6(a)). When the working chamber connected to the outlet pocket, the large pressure difference run the air in the out pocket flow back to the working chamber which reverse the air flow in the working chamber and lead to air impact at 190° (Fig. 6(b)). The air in the working chamber was compressed by the flow back air impact, and the pressure in the outlet pocket reduced because the air flow back to the working chamber at 200° (Fig. 6(c)), so a new pressure difference was generated. The direction of the air flow was changed again. But in the Roots blower with backflow, the pressure difference was much less than Roots without backflow. It is benefit from the backflow piping help the pressure in the working chamber close to the outlet pocket, which reduce the pressure impact and the amplitude of the pressure pulsation.

5. Conclusions

Based on the CFD analytical and experimental investigations presented in this study, the following conclusions may be drawn:

(1) The backflow structure can help reduce the pressure pulsation in the outlet of the Roots blower. The amplitude can be reduced more than 50%.

(2) The backflow structure could reduce the mass flow rate of the Roots blower, because the higher pressure difference between the working chamber and the inlet pocket generated more leakage. The volume efficiency could be reduced about 3.5%.

(3) The main reason of the pressure pulsation in Roots blow was the air impact caused the flow back from outlet pocket to the working chamber.

This study's procedure for the analysis of the Roots blower with backflow provides a good basis for further consideration of leakage and noise reduction.

Funding

This study was supported by the National Natural Science Foundation of China (Research Project 51306136).

References

[1] Mimmi G, Pennacchi P, Mimmi G. 1999, Analytical model of a particular type of positive displacement blower. ARCHIVE Proceedings of the Institution of Mechanical Engineers Part C Journal of Mechanical Engineering Science, 213(5):pp.517-526.

[2] Valdès. L.C, Barthod. B, Perron. Y.L, 1999, Accurate prediction of internal leaks in stationary dry Roots vacuum blowers. Vacuum, 52 (4), pp.451–459.

[3] Tong S H, Yang D C H. 2000, On the generation of new lobe blowers for higher blowering flowrate. Mechanism & Machine Theory, 35(35): pp. 997-1012.

[4] Yang D C H, Tong S H. 2002, The specific flowrate of deviation function based lobe blowers - derivation and analysis. Mechanism & Machine Theory, 37(10): pp. 1025-1042.

[5] Tong S H, Yang D C H. 2005, Rotor profiles synthesis for lobe blowers with given flow rate

functions. Journal of Mechanical Design, 127(2): pp. 287-294.

[6] Burmistrov. A, Belyaev. L, Ossipov. P, Fomina. M, Khannanov. R, 2001,Combined experimental and calculation study of conductance of Roots blower channels. Vacuum, 62 (4) pp.331–335

[7] Wang P Y, Fong Z H, Fang H S. 2002, Design constraints of five-arc Roots vacuum blowers. Proceedings of the Institution of Mechanical Engineers Part C, 216(2): pp. 225-234.

[8] Yao L, Ye Z, Dai J S, 2005, et al. Geometric analysis and tooth profiling of a three-lobe helical rotor of the Roots blower. Journal of Materials Processing Technology, 170(1–2): pp. 259-267.

[9] Hwang Y W, Hsieh C F. 2006, Study on high volumetric efficiency of the Roots rotor profile with variable trochoid ratio. Proceedings of the Institution of Mechanical Engineers Part C Journal of Mechanical Engineering Science, 220(9): pp. 1375-1384.

[10] Hsieh C F, Hwang Y W. 2007, Study on the high-sealing of Roots rotor with variable trochoid ratio. Journal of Mechanical Design, 129(12): pp. 1278-1284.

[11] Hsieh C F. 2015, A new curve for application to the rotor profile of rotary lobe blowers. Mechanism & Machine Theory, 87: pp. 70-81.

[12] Joshi A M, Blekhman D I, Felske J D, et al. 2006, Clearance analysis and leakage flow CFD model of a two-Lobe multi-recompression heater. International Journal of Rotating Machinery, (1023-621X): pp. 10.

[13] Huang Z F, Liu Z X. 2009, Numerical study of a positive displacement blower. ARCHIVE Proceedings of the Institution of Mechanical Engineers Part C Journal of Mechanical Engineering Science, 223(10): pp. 2309-2316.

[14] Li Y B, Jia K, Meng Q W, et al. 2013, Flow simulation of the effects of pressure angle to lobe blower rotor meshing characteristics. 6th International Conference on Blowers and Fans with Compressors and Wind Turbines, Materials Science and Engineering, 52(3):032022.

[15] Hsieh C F, Deng Y C. 2015, A design method for improving the flow characteristics of a multistage Roots blowers. Vacuum, 121: pp. 217-222.

[16] Hsieh C F, Zhou Q J. 2015, Fluid analysis of cylindrical and screw type Roots vacuum blowers. Vacuum, 121: pp. 274-282.

[17] XY Peng ea al. 2000, Design improvement on involute profile of roots blower Rotors. Compressor blower & fan technology 2000(3): 3–5. (In Chinese)

[18] FLUENT 15.0, 2013 User help.

Numerical Analysis of an Oil-free Twin Screw Compressor Using 3D CFD and 1D Multi-chamber Thermodynamic Model

Stuart Kennedy[1], Maria Wilson[1], Sham Rane[2]

[1] Howden Compressors, Old Govan Road, Renfrew, PA4 8XJ, UK

[2] City, University of London, EC1V 0HB, UK

Email: stuart.kennedy@howden.com

Abstract. The application of three-dimensional computational fluid dynamics in twin-screw compressors provides an outstanding opportunity for developers to gain an understanding of the complex internal flow phenomena occurring within the machine. Equipped with this knowledge, design parameters, such as clearances and port geometry can be optimised, to enhance performance. However, as with all modelling, be it numerical or analytical, a high degree of certainty in the accuracy of the results is necessary.

This paper presents the results of a study of oil-free twin screw compressor in which the results of two modelling techniques are compared. The modelling techniques used are a one dimensional thermodynamic chamber model and a three dimensional computational fluid dynamic model. The paper presents an overview of an oil-free twin screw compressor machine, before describing important operating characteristics and the modelling techniques used. To validate, both models are compared against historical test data, this validation indicated the chamber model is more accurate. Following this, the focus will be on the comparison of key performance indicators, including, volume flow rate, volumetric efficiency, indicated power, and discharge temperature at varying duty points. The paper concludes that the difference between both models decreases as the compressor operating speed increases, although the level of variance is dependent on pressure ratio.

1. Introduction

The operating principle of the twin-screw compressor remains practically identical to that initially developed in the late 19th century. Development of the technology over the following 140 years has been steady, primarily coinciding with advancements in manufacturing techniques and computer processing capabilities. The latter has brought about an evolution of highly sophisticated mathematical modelling techniques, such as Finite Element Analysis (FEA), Computation Fluid Dynamics (CFD), and more specific to twin-screw compressors, the development rotor profile generation, grid generation and thermodynamic modelling software. With this, the modern-day screw compressor design engineer has a plethora of design tools at their finger-tips.

This paper presents the results of a study in which a selection of these tools are used to calculate performance characteristic of an oil-free screw compressor at specified speeds and pressure ratios.

The tools used are; 3D CFD modelling and thermodynamic chamber modelling. A particular focus of the study will be on both the methodology undertaken and results obtained through CFD modelling, as this will be utilised extensively in future development work.

1.1 Oil-free Twin Screw Compressor - Overview

As the name suggests, in an oil-free twin screw compressor the working chamber is free from oil. This is opposed to the oil injected range of machines, where the oil is used to lubricate, control temperature and reduce internal leakages in the working chamber. Oil-free machines are typically used in industries where the working fluid must be free of any containments. Eliminating oil from the rotor chamber adds in own complexities and challenges, such as sealing arrangements, timing gears and increased operating temperatures. Additionally, as oil is absent from the working chamber, the induced efficiency losses that occur due to oil shear are avoided, meaning the oil-free screw compressor can operate at significantly higher tip speeds. With this, aspects such as gas velocities, vibration, and critical speeds must then be considered.

2. Modelling Methodologies

The performance characteristics of the machine can be calculated numerically, through CFD modelling or specially designed thermodynamic chamber modelling tools - such as that used in SCORG™ [1]. There are tremendous differences in the modelling methodologies. The greatest is the processing time and power required to obtain the results given the specified boundary conditions. Where the chamber model can output predictions in seconds, the CFD model would take days, ultimately due to quantity of calculations required. Additionally, to perform these calculations, a large amount of high specification processing power is required, whereas the chamber model can operate on a desktop machine. Further to this, the CFD model will provide an extensive range of results, including flow phenomena that can be numerically and visually interrogated; as opposed to the chamber model which outputs single calculated values relating to performance. This is not to say a particular model is superior, simply their use should be considered on a case to case basis.

2.1. Thermodynamic Chamber Model (SCORG™)

One-dimensional mathematical modelling of screw machines has been used to great effect for many years. An accurate prediction of many aspects of machine performance can be obtained from the models which simultaneously solve thermodynamic processes, based on ideal or real gas laws, with the geometric parameters of the machine, including chamber volume, port areas and sealing line lengths. From the many modelling methodologies, that presented by Stosic et al [2] forms the foundation of the model used in this study. Described as multi-chamber thermodynamic model, the screw machine is divided into sub-domains representing each stage of the compression process. The challenge of computing time dependant flow variables within the control volumes is overcome by using a combination of differential equations to solve the rate of change of mass and internal energy, and algebraic equation to solve other internal flow behaviour such as velocity and leakages.

From a user perspective, the complexities involved in the computation of the aforementioned flow variables are hidden. As much as it is recommended to have a theoretical understanding of the model, developers have excelled in creating a tool that is as much user friendly as it is reliable. This study will utilise the thermodynamic model inherent to SCORG™.

2.2. CFD Model

CFD modelling is rapidly being adopted in all areas of modern engineering practice to solve complex flow and heat transfers problems. An obvious application of this methodology is in twin screw compressors. However, the challenges involved in modelling and computing the complex physics and transient nature inherent to the machine have proven difficult to overcome. Nonetheless, through extensive and committed research, the development of design tools and users obtaining a greater understanding of the problems, it is now possible to undertake these simulations with a certain degree of confidence that a valid solution can be obtained. To obtain this, a host of 3D modelling and analytical tools are employed for pre-processing, solving and post-processing, these involve; 3D modelling, mesh creation, rotor grid generation and CFD software.

2.2.1. Rotor Grid Generation. The helically formed rotors - and the continuously varying volume that is formed between the male, female rotor rotors, and casing bore as they rotate - has offered the greatest challenge to developers for many years. Publications on techniques and methodologies are in abundance and presenting these is outside the remit of this study. However, techniques summarised in Kovacevic et al [3] provide an excellent overview of developments as well as describing the authors of the publication own methodology. This analytical grid generation procedure is presented in detail by Kovacevic [4] and expanded upon by work presented by Rane [1].This methodology is at the core of grid generation software SCORG^{TM} - the tool used in this study.

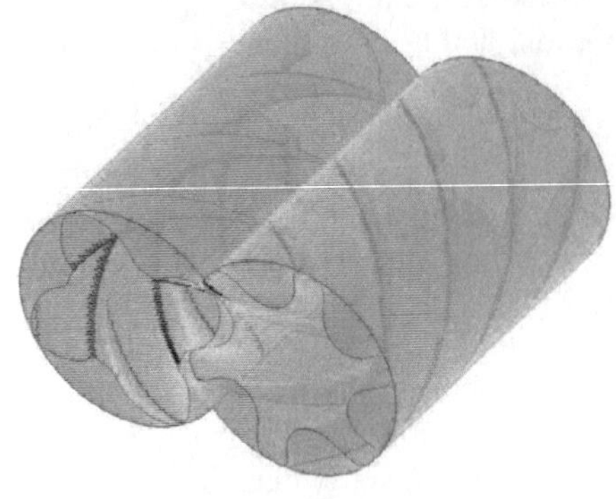

Figure 1: 3D Rotor Mesh

Understanding the flow characteristics of the working fluid, or fluids, in the chamber is of great importance. Therefore, the grid used in the calculations must be of excellent quality to assure accuracy. The design tool, SCORG^{TM}, provides the necessary means to achieve this. Dividing the male and female rotor into two sub-domains and then into a series of 2D domains, a numerical mesh for each rotor position is generated. This is then translated into a 3D mesh by positioning these in series, axially along the rotor length, an example of which, and those used in this study are shown in Figure 1. Furthermore, using a conformal boundary map at the sub-domain interfaces allows the male and female rotors to be merged as one single domain.

These are 'N' profile rotors, which are generated using data points imported from rotor profile generation software SCORPATH^{TM,} as described by the authors [2]. This compatibility between software offers many benefits, of which the most important from a analysis perspective is the profiles used in the model are identical to those in the physical machine.

3. Case Study

The case study presents a comparison of calculated performance characteristics obtained from the described tools. The approach undertaken is to initially determine the variation between actual test data and each method, thus gaining an overall understanding of the level accuracy. Following this, the set-up of both models will be described with particular attention taken to the CFD model set-up. Finally, each model will be used to predict performance at a range of operating conditions allowing the variance to be evaluated.

3.1. Machine Configuration

This case study involves a common machine configuration of 4 male and 6 female rotor lobes. Both rotors are of equal outer diameter at 255mm with an L/D ratio of 1.65, equating to a rotor length of 420.75mm. Typically, machines are defined by their built-in volume ratio (Vi), in this case this is 1.9. The specific duty of the physical compressor has not been defined. Therefore this study will focus on a range of speeds and pressure ratios of what are considered normal operating conditions. This will provide an extensive overview of the machine performance at various conditions.

3.2. Model Validation Study

An extensive study undertaken by Rane [1,6] presents a high level of accuracy between CFD modelling and test data is achievable when using a single domain rotor grid. To obtain additional confidence an attempt was made to validate this study's CFD set-up. In the absence of an identical prototype test machine, historical test results were examined to extract comparable performance data. The measured variables available are limited, therefore the volume flow rate is identified as the most suitable to validate the CFD model. Additionally, this data relates to machines operating with symmetric rotor profiles, therefore the rotor geometry/mesh in the CFD model is modified accordingly to reflect the compressor configuration.

In addition to validating the CFD model, the opportunity is taken to compare the volume flow rates predicted by the chamber model, thus gaining an overall evaluation of the accuracy of each method. The machine was tested at a constant speed of 7385 rpm at 0.988barA suction pressure with pressure ratios of 1.52 and 1.93. The measured and calculated volume flow rates are shown in Figure 2a, with the associated error in Figure 2b.

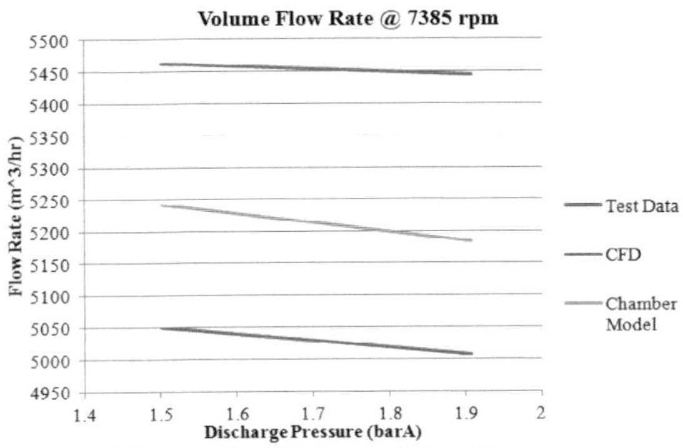

Figure 2a: Volume Flow Rate

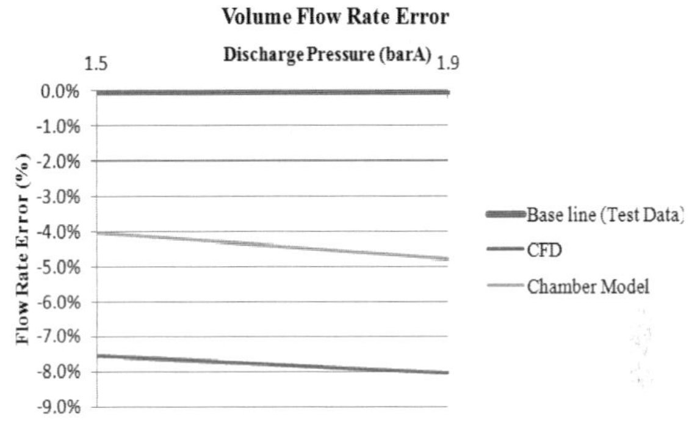

Figure 2b: Calculated Error

Given the results, it can be established that in both models there is an under prediction of volume flow rate. The chamber model calculates approximately 4%-5% less flow. The CFD results indicate an under-prediction of the volume flow by approximately 8%. It can be identified that in both cases, the results have the correct trend, with decreasing flow rates at increasing discharge pressures, this reflects what is theoretically expected.

From the several factors that may be accountable for the degree of error identified in the CFD model, the most likely reasons are: rotor mesh quality, the machines physical geometry and its alteration as it responds to heat transfer. Firstly, to ensure numerical accuracy, as well as capturing flow characteristics more precisely, the rotor mesh must be generated to an acceptable level of quality and quantity. Therefore, a full sensitivity study examining the change in calculated results with consecutive cases of grid refinement will be undertaken in future work.

Secondly, the rotor thermal expansion that will occur during operation and the consequential modification to internal clearances introduces additional factors of uncertainty (also applicable to chamber model). The complex interaction between the working fluid and the rotor is currently not captured in the model, which adds a degree of inconsistency in the validation study. Options available to resolve this are to include fluid-solid interaction in the model, or one could calculate theoretical thermal expansion of the rotors and model the revised clearances, the latter will be the focus of a future study.

3.3. Aspects of the CFD Model

By taking advantage of the functionality of a range of tools as shown in Table 1, the CFD model is created. As listed, the calculations are completed by the ANSYS CFX solver [7], chosen due to its compatibility with SCORG^{TM}, and it's known reliability and robustness and in computing highly deforming mesh type models. As with all computational analysis, it is critical the mesh quality is of a high standard and undertaking a full mesh dependency study is prudent. However, due to the time restrictions, this has been omitted from the current study. The authors are reliant on effectiveness of the tools used and previous experience to create a high quality mesh.

Stage	Part	Software
Pre-processing	3D Geometry	Autodesk Inventor
		ANSYS Design Modeller
		ANSYS SpaceClaim
	Meshing	ANSYS Meshing
		ANSYS ICEM
		SCORG
	Set-up	ANSYS CFX
Solver		ANSYS CFX
Post-processing		ANSYS CFX
		Microsoft EXCEL

Table 1: Design Tools

3.3.1 Rotor Mesh As discussed in Section 2.2.1, SCORG™ is used to generate the 3D rotor fluid domain. The tool allows the user to define certain mesh parameters to calculate a suitable mesh. Key parameters used to generate the mesh are defined in Table 2 as represented in Figure 3.

Parameter	Description	Input
A	Circumferential Divisions	40
B	Radial Divisions	7
C	Angular Divisions	40
D	Interlobe Divisions	60
E	Type of Distribution	Casing to Rotor - Conformal

Table 2: 2D Grid Set-up

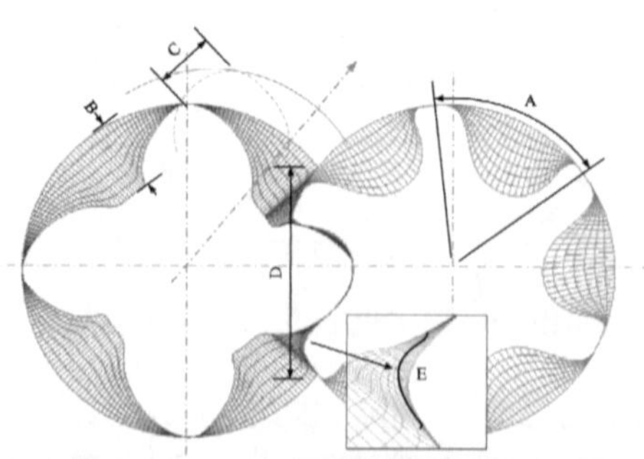

Figure 3: 2D Grid Parameters

A review of the mesh quality determines the adequacy of the above mesh, the statistics of the 2D and 3D rotors summarised at Table 2. These indicate a good overall quality rotor mesh, with <1% of the cells in the 3D mesh not meeting the recommended orthogonality and expansion factor quality criteria. Additionally this illustrates the 2D mesh quality depreciates during transformation to a 3D mesh - ultimately it is this 3D mesh quality calculated by the CFD solver that is important.

	Orthogonality			**Expansion Factor**			**Aspect Ratio**		
	Good	Acceptable	Poor	Good	Acceptable	Poor	Good	Acceptable	Poor
2D	99%		<1%	96%		4%	98%		2%
3D	72%	27%	1%	99%	1%	<1%	100%		

Table 3: Rotor Mesh Quality

3.3.2 Stationary Domains Mesh Figure 4a illustrates the 3D model used in the simulation. Due to the complex geometry, it was decided to deconstruct the stationary domains into simpler geometry, or sub-domains, to allow, when possible, an effective hexahedral cell structure to be created, as shown in Figure 4b. This does however mean additional interfaces must be added at each sub-domain - meaning improved processing speed may come at a cost of numerical accuracy. However, the overall global mesh quality calculated indicates the risk of numerical error is minimised.

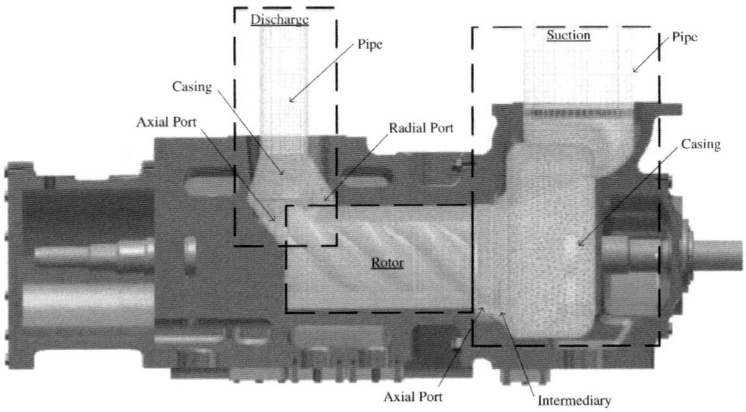

Figure 4a: Flow Domain (Green) and Screw Compressor

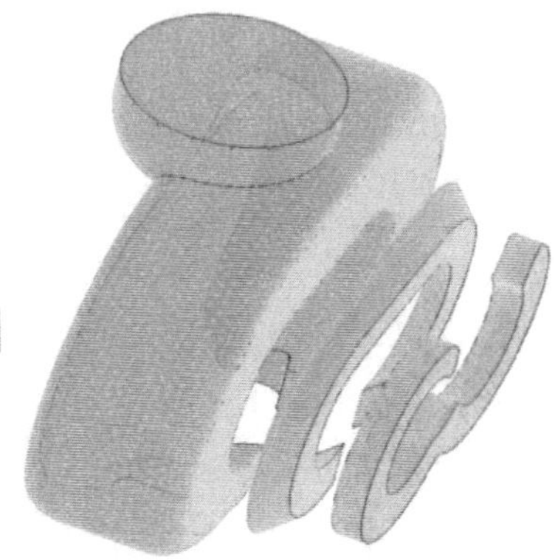

Figure 4b: Suction Sub-domains

3.3.3 CFD Boundary Conditions To provide a extensive list of results, performance is interrogated at a total of 9 duty points. This is achieved by applying discharge pressures of 2.0, 2.5 and 3.0 barA, to represent under, optimum and over compression, respectively, and varying speed of 4000, 5500 and 7000rpm at each of these pressure points - as summarise in Table 4

Speed (RPM)	Working Fluid	Suction Pressure (bar A)	Discharge Pressures (bar A)			Suction Temperature (K)
4000	Air	1.0	2.0	2.5	3.0	293.0
5500						
7000						

Table 4: Case Set-up

Run	$P_{Discharge}$	Speed
1	2.0	4000
2	2.0	5500
3	2.0	7000
4	2.5	7000
5	2.5	5500
6	2.5	4000
7	3.0	4000
8	3.0	5500
9	3.0	7000

Table 5: Simulation Workflow

Due to the varying speed and discharge pressures, the order in which the calculations are undertaken must be carefully considered to minimise numerical instability. It is predicted that the change in discharge pressure will effect solver stability more than a change of speed, therefore it is decided to calculate a specified discharge pressure and increase/decrease the speed, then change the discharge pressure and repeat. It is key to obtain a balance between numerical accuracy, stability and time taken to solve. Based on experience gained from previous studies, it is decided that each set-up will be calculated for 6 male rotor rotations (960 time steps) to gain stability, then 1 further rotation in which the results are recorded - the run sequence is shown in Table 5.

3.3.4 Solver Parameters The calculations are performed by an Intel Xeon X5650 processor, with 48 GB of available RAM. The solver, running parallel with 8 available 2.66GHz cores, used a High Resolution Advection Scheme and Second Order Backward Euler Transient scheme with a SST - k Omega turbulence model. For convergence, the maximum number of co-efficient loops is set to 10 iterations per time step, with convergence criteria of $1e^{-04}$. For stability, it is necessary to apply under-relaxation factors to solve momentum and continuity equations.

3.4 Thermodynamic Model

As discussed in Section 2.1, the thermodynamic chamber model used is that inbuilt in SCORG™. Additionally, as the same software is used to create the rotor mesh, the process of obtaining thermodynamic results is extremely efficient, as the rotor geometry and internal clearances used in the calculations are already defined. As this is a study into oil-free machines, the thermodynamic model set-up is somewhat simplified. Furthermore, as the CFD does not calculate power losses due to bearings and seals, another element of the set-up is removed. This means only a few variables are defined for each calculation, those being rotor speed, operating pressures and the working fluid properties. The boundary conditions listed in Table 4 are used to calculate the results for 9 of the cases; typical results presented by the thermodynamic model are; Volume Flow, Mass Flow, Indicated Power, Volumetric Efficiencies, Specific Power and Gas Discharge Temperatures.

4.0 Results

The results present the performance calculations, applying the boundary conditions listed above, from both models. From here a clear comparison can be drawn, the variance calculated and conclusions drawn.

4.1 Mass Flow Imbalance - CFD model

To ascertain a degree of certainty of numerical accuracy in the CFD model prior to comparing with the chamber model, the mass flow imbalance between the suction and discharge boundaries is calculated. The error percentage between these is listed below in Table 6. As shown the error is <2% in all cases, showing good balance between what is entering and leaving the domain, thus it can be concluded that the quality of the mesh has resulted in high numerical accuracy.

	Pressure Ratio		
Speed	**2:1**	**2.5:1**	**3:1**
4000	0.1%	0.3%	0.01%
5500	0.3%	-1.2%	0.7%
7000	1.2%	-0.6%	2.0%

Table 6: Mass Flow Imbalance

4.2 Volume Flow Rate and Volumetric Efficiency

The achievable volume flow rate of a machine is a key component in the selection process. Therefore, it is extremely advantageous to have the capabilities to accurately predict this, given a specified rotor profile, built-in volume ratio, and at a range of pressure ratios and speeds. Firstly, focussing on the calculated volume flow rates, Figure 5a shows the change in volume flow rate with speed for each pressure ratio. As shown, the overall trend of the volume flow rates in each model are a well matched. However, when comparing the variance between CFD results and the chamber model, as shown in Figure 5b, there is a recognisable difference in prediction of flow rate, which is particularly evident at lower speeds. For example, at 4000 rpm, it can be observed that the variance increases from 8.6% at P.R. 2:1 to 15.8% at P.R. 3:1. Interestingly, in all cases as the speed increases, this variance between models decreases. Also worth highlighting is from 5000 rpm to 7000 rpm, the variance at each pressure ratio is relatively stable. This shows clearly CFD model will predict flow rate more accurately at higher speeds.

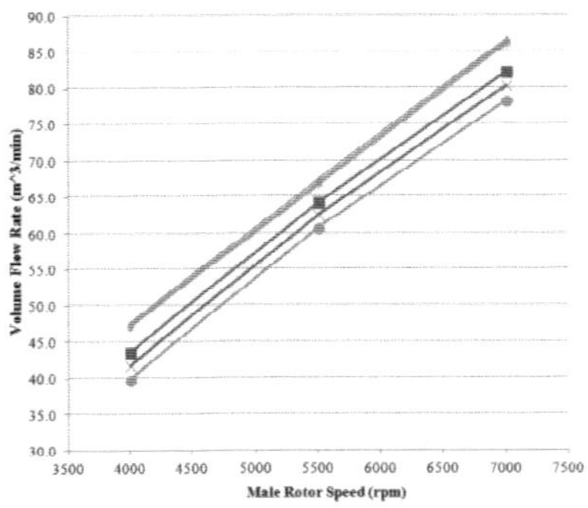

Figure 5a: Volume Flow Rate

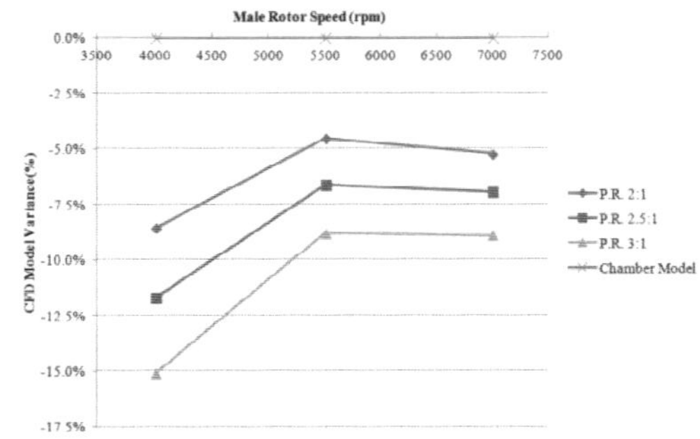

Figure 5b: Volume Flow Rate Model Variance

The volumetric efficiency is a key performance indicator of the screw machine, defined as the ratio actual volume flow rate to the theoretical capacity. With reference to Figure 6, there is a clear trend in the behaviour of each model. Both the chamber and CFD models predict the volumetric efficiency to increase as the speed increases - this behaviour reflects what is theoretically expected. At all pressure ratios, the variance between models decreases as the speed increases which reflects the overall behaviour of the volume flow rate results. Thus, it is established that at low pressure ratios and high rotational speeds, both models are closely matched.

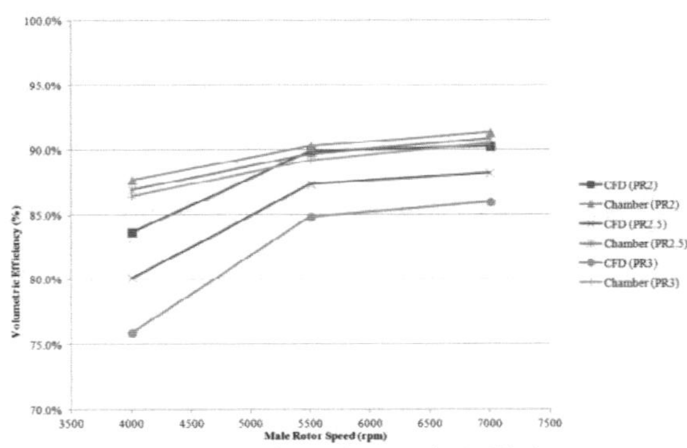

Figure 6: Volumetric Efficiency

4.3 Indicated Power and Adiabatic Efficiency

Indicated power is defined as the rate of work transferred from the screws to compress the gas, commonly calculated directly from the pressure-volume curve, or using angular velocity and torque. In this study, power consumed by bearings and seals is neglected, as these losses cannot be calculated by means of CFD.

With reference to Figure 7a, as expected, the indicated power increases with speed and pressure ratio. At all pressure ratios, it can be identified that the variation between the predicted indicated power increases as speed increases. The CFD model predicts greater indicated power than the chamber model in all cases; ranging from 4-6kW at 4000 rpm to approximately 19kW at 7000 rpm.

An additional performance characteristic of interest is specific power. The specific power describes the power consumed to deliver a cubic metre per minute of working fluid (air). Therefore, a low specific power is a useful indicator of efficient machine performance. Figure 7b illustrates the trends of both models. It can be identified that the difference in models is less than 0.5 kW/m^3/min in all cases.

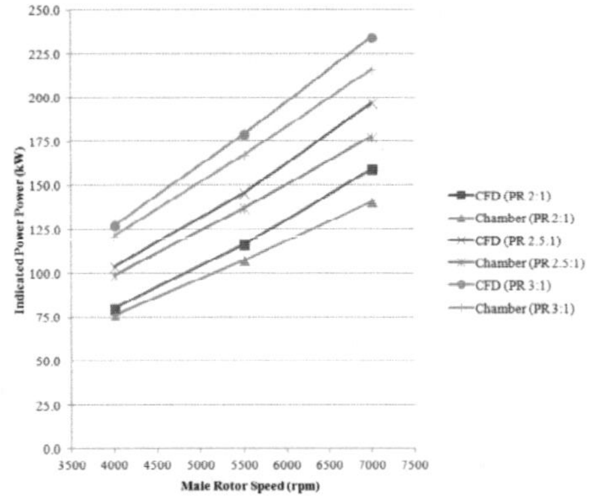

Figure 7a: Indicated Power

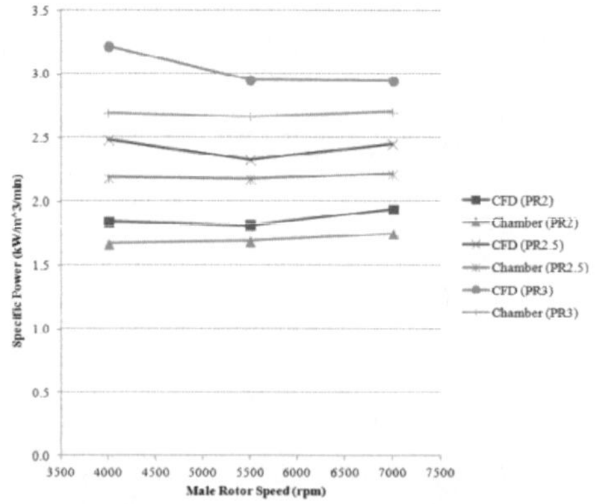

Figure 7b: Specific Power

4.4 Gas Discharge Temperature

As discussed in Section 1.1, gas temperature is a critical operating parameter, therefore having the ability to predict this is extremely valuable. The omission of oil in the rotor chamber will naturally result in high gas temperatures, that will over time transfer heat to the compressor components. In this study the fluid domain walls, such as those of the rotor bore and the discharge casing, are adiabatic. Therefore, conjugate heat transfer between the fluid and structure is not computed. Although this fluid-structure interaction would be most desirable, it does not devalue knowing only the gas temperature. As this can be used in theoretical calculations for heat transfer, and subsequent structural analysis for deformation.

Additionally, the isentropic relationship between pressure and temperature ultimately means the greater the pressure ratio, the greater the gas discharge temperature. Historically, this has imposed operating limitations to oil-free twin screw compressors.

The results in Figure 8a present the average gas temperatures (over speed range) for each pressure ratio. As shown in both models, the trend reflects that expected, however there is a noticeable variation. The CFD model predicts a higher gas temperature compared to the thermodynamic chamber model - this deviation ranges from 15-30 Deg. C or 15-17% as shown in Figure 8b

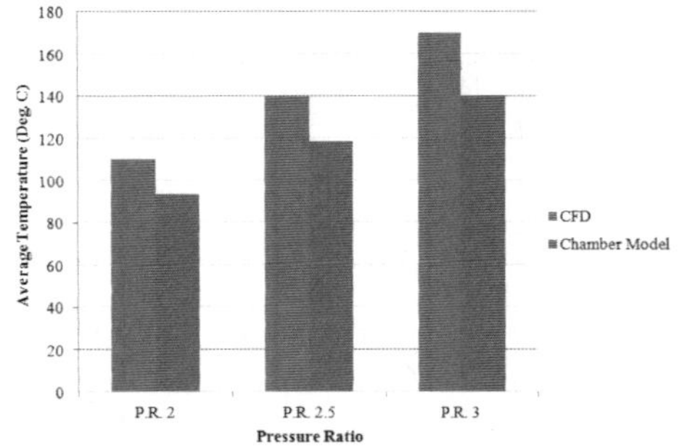

Figure 8a: Gas Discharge Temperature

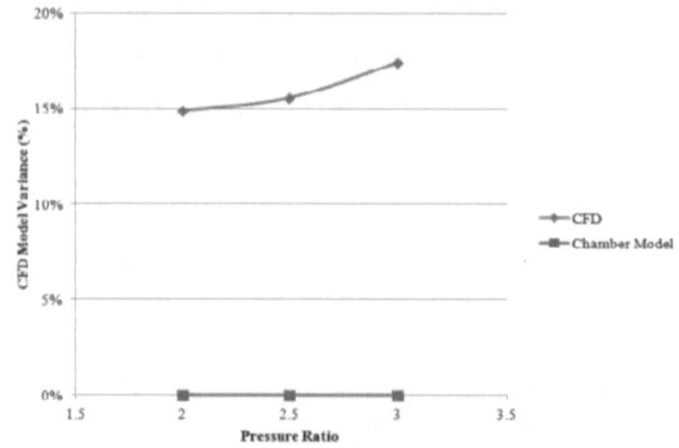

Figure 8b: Temperature Variance

5.0 Conclusion

This paper presents the comparison of performance characteristics predicted by numerical CFD and thermodynamic modelling techniques typically used to evaluate twin-screw compressors. Through the validation study undertaken, it is recognized the thermodynamic chamber model prediction is closer to actual machine performance than the calculations of the CFD model. It is with this it is assumed the chamber model is the bench mark in which the CFD is compared to at various duty points. Given the results, the following can be concluded:

- Volume Flow Rate - the CFD model under predicts volume flow rate at lower speeds. As operating speed increases, there is a noticeable improvement in closeness between both models.

Furthermore, it is identified the variance increases as operating pressure ratio increases. Thus, it can be concluded in this case, CFD provides more accurate results at higher operating speed and lower pressure ratios.

- Power - As the operating speed increases, the difference in calculated indicated power increases. With the CFD model predicting higher indicated power in all cases. As the specific power is a function of volume flow rate, this is impacted by the increased accuracy at higher speeds as described above.
- Discharge Temperature - the CFD model over predicts gas temperature at discharge, in all cases. The variance in results also increases as the pressure ratio increases.

It can be established from the results presented that the thermodynamic chamber model calculates consistent results, with identifiable trends in flow rate, power and temperature. In contrast, the CFD model is sensitive to operating conditions, both the operating speed and pressure. As a consequence, the variation between models is inconsistent. Both the magnitude and inconsistency of the variance between the models highlights the importance of undertaking a grid sensitivity study, with the aim of reducing the variance.

Overall, the study undertaken is successful in calculating performance characteristics using numerical and thermodynamic modelling techniques. It is also worth highlighting the comprehensive results obtained from the CFD simulation provides the opportunity to visually and numerically examine internal flow phenomena. Therefore, making it an extremely advantageous tool in the optimisation and development of twin screw machines.

Having now created a functional CFD model provides the opportunity to undertake further work which will involve a grid sensitivity study, additional validation and investigating the impact modifying internal clearances due to thermal expansion has on performance.

References

[1] Rane, S. and Kovacevic A., (2017). *Algebraic generation of single domain computational grid for twin screw machines Part I – Implementation*, Advances in Engineering Software 107 (2017) 38–50

[2] Stošic N., Smith I.K. and Kovacevic A., (2005). *'Screw Compressors: Mathematical Modelling and Performance Calculation'*, Monograph, Springer Verlag, Berlin, June 2005, ISBN: 3-540-24275-9.

[3] Kovacevic A., Stošic N. and Smith I. K., (2007). *Screw compressors - Three dimensional computational fluid dynamics and solid fluid interaction*, ISBN 3-540-36302-5, Springer-Verlag Berlin Heidelberg New York.

[4] Kovacevic A., 2002. *'Three-Dimensional Numerical Analysis for Flow Prediction in Positive Displacement Screw Machines'*, Ph.D. Thesis, School of Engineering and Mathematical Sciences, City University London.

[5] Rane, S., (2015). *'Grid Generation and CFD analysis of Variable Geometry Screw Machines'*, Thesis, City University London.

[6] Rane, S. and Kovacevic A., (2017). *Algebraic generation of single domain computational grid for twin screw machines Part II – Validation*, Advances in Engineering Software 109 (2017) 31–43

[7] ANSYS CFX v17.2. (2017). *ANSYS Inc.*

Numerical investigation of refrigeration machine compressor operation considering single-phase electric motor dynamic characteristics

Y Baidak[1] and V Smyk[2]

[1]Professor of Thermodynamics and Renewable Energetics Department, Odessa National Academy of Food Technologies, Odessa, Ukraine
[2]post-graduate student, Odessa National Maritime Academy, Odessa, Ukraine

E-mail:kozak_admin@ukr.net

Abstract. Using as the base the differential equations system which was presented in relative units for generalized electric motor of hermetic refrigeration compressor, mathematical model of the software for dynamic performance calculation of refrigeration machine compressors drive low-power asynchronous motors was developed. Performed on its ground calculations of the basic model of two-phase electric motor drive of hermetic compressor and the proposed newly developed model of the motor with single-phase stator winding, which is an alternative to the industrial motor winding, have confirmed the benefits of the motor with innovative stator winding over the base engine. Given calculations of the dynamic characteristics of compressor drive motor have permitted to determine the value of electromagnetic torque swinging for coordinating compressor and motor mechanical characteristics, and for taking them into consideration in choosing compressor elements construction materials. Developed and used in the process of investigation of refrigeration compressor drive asynchronous single-phase motor mathematical and software can be considered as an element of computer-aided design system for design of the aggregate of refrigeration compression unit refrigerating machine.

1. Introduction

Simulation of dynamic processes that affect the operation of key facilities involved in the automated production, namely electromagnetic devices, electric motors, electromagnetic switchgear, transformers that form the modern automated electric drive of compression refrigeration equipment and determine the conditions of their operation while changing modes of operation – is an important basis for creation of more advanced production process automated control systems. At the stage of dynamics simulation the importance should be given to solve issues which lay the field of electromagnetism, thermal physics, and mechanics. The mechanics of the object, despite the fact that determines the final characteristics of automated production, is the most influential (*Dote, 1990*). Structural materials from which the elements of automated devices are made have to be pushed nearer to their limits. Simulation of dynamic processes that define the operation of production facilities is a multi complex, multifactorial problem and therefore the issue of solving tasks of dimensions number reducing by identifying the most influential output / input criteria and factors as well as disclosure of connections and relationships between them is a priority one *(Guinee, R.A. & Lyden, C.,1998; 2001)*.

Mathematical model used as a basis for the dynamic characteristics determination of the electromechanical energy conversion process in time in the electrical motor of the refrigeration machine compressor drive, despite of its complexity, allows deeper address matters related to processes, operational factors and modes of any electromechanical energy converter. The solution of differential equations granted for instantaneous values of variables can give answers to questions that can not be obtained through the use of equivalent circuits, vector variables and their graphs *(Persson, E.K. & Buric, M., 1976)*.

Recently in the practice of engineering calculations a wide variety of multipurpose software for calculation of dynamic processes in automated electric drive is applied, e.g. MatLab - Simulink addition of Power System Blockset manufacturer Math Work *(Beucher, O., Weeks M. (2008), Ong, Chee-Mun. (1998); Chaturvedi Devenndra, K. (2010), Giurgiutiu, V., Lyshevski, S. E., (2009), Karris, Steven T. (2008)*. However, despite their extensive capabilities, the licensed product usage is always a difficulty. Mastering software products of leading world manufacturers requires much time, knowledge and skills.

Electric asynchronous motor with short circuit rotor is the most widespread actuator in many automated production systems and a lot of researchers were involved into the development of its mathematical or "generalized" theory, with its further implementation to the study of dynamic performance characteristics *(Ansari A. et.al. (2010), Galan, N., Mammadov A. I., (2006), Leonhard, W. (2001)*. The issue of "generalized" theory development arose as an attempt to combine in a single unit previously known and innovative methods of electric machines theoretical analysis: the method of symmetrical components, two reactions, rotating magnetic fields. "Generalized" theory clearly demonstrated that a set of circuits that are mutually and relatively moving may be replaced by equivalent mutually fixed electric circuits. Since mathematical tools at this level are special, it is useful to consider the relationships between the general theory of electrical machines and classical approach to the analysis of dynamic characteristics of single-phase induction motor with short circuit rotor on the example of refrigeration machine compressor motor drive.

2. Simulation of refrigeration compressor drive electrical motor operational characteristics in dynamics

2.1. Differential equations of the dynamics of short-asynchronous motor

The actual two-phase asynchronous electromotor with short circuit rotor rotating relatively fixed coils of the stator winding w_a, w_d, made of unequal number of turns and arranged in the stator bore under the solid angle of their magnetic axes shift α, β at 90 electrical degrees, Fig.1, while simulation of operating dynamic characteristics can be provided as a joint system of electrical balance equations, which take into consideration the rotor rotation. Bringing rotor winding resistances to the stator winding should also be considered, as well as currents and voltages of one stator coil winding to another through transformation ratio of turns of the coils $k = w_a / w_d$.

The equations system of electrical balance of asynchronous motor with short circuit rotor, which is provided relatively to voltages and currents vectors in natural or converted coordinates - with respect to the fixed stator coordinate system $(\alpha, j\beta)$, and with respect to the movable coordinate system of the rotor (d, jq), in which the voltage of short circuit rotor $\overline{U}_{p_{d,q}} = 0$, is given by

$$\left.\begin{aligned} \overline{U}_{c_{\alpha,\beta}} &= \bar{i}_{c_{\alpha,\beta}} r_c + \frac{d\overline{\Psi}_{c_{\alpha,\beta}}}{dt}; \\ 0 &= \bar{i}_{p_{d,q}} r_p + \frac{d\overline{\Psi}_{p_{d,q}}}{dt}. \end{aligned}\right\} \quad (1)$$

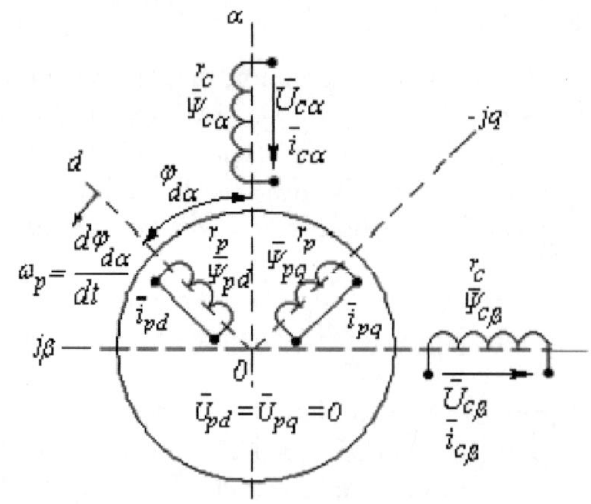

Figure 1. Two-phase asynchronous electromotor with stator and rotor coils in rectangular system of coordinates

In equation (1) $\overline{U}_{c_{\alpha,\beta}}$ – vector of coils voltage supply $w_{a\alpha}, w_{d\beta}$ windings w_c of the stator; $\bar{i}_{c_{\alpha,\beta}}, \bar{i}_{p_{d,q}}$ – vectors of currents in coils of the stator winding and rotor winding rods; r_c, r_p – active resistances of stator and rotor. The joint solution of equations system (1), from which those relating to the stator are given in a fixed coordinate system, and for rotor - in the moving one, is impossible and therefore it is appropriate to rearrange the equation of rotating rotor electrical circuit voltages for stator fixed coordinate system. In this case, it is considered conventionally that the electric motor rotor is rotating, for example, counterclockwise, and solid angle of shift of rotor and stator winding axis $\varphi_{d,\alpha}$, Fig.1, at any time is considered by multiplying equation components of rotor voltages on the phase operator of its axis rotation relative to stator axis, i.e. $1 \cdot e^{-j\varphi_{d,\alpha}}$. Then the equation of voltages of short circuit rotor, reduced to fixed stator axes is given by

$$0 = \bar{i}_{p_{d,q}} \cdot e^{-j\varphi_{d,\alpha}} r_p + e^{-j\varphi_{d,\alpha}} \frac{d\overline{\Psi}_{p_{d,q}}}{dt}. \tag{2}$$

In turn, the rotor winding flux linkage $\overline{\Psi}_{p_{d,q}}$, which is provided relatively to coordinate system (d, jq), in the coordinates $(\alpha, j\beta)$ is $e^{j\varphi_{d,\alpha}} \overline{\Psi}_{p_{\alpha,\beta}}$

and, accordingly, the rotor voltage equation (2) is converted to the form

$$0 = \bar{i}_{p_{d,q}} e^{-j\varphi_{d,\alpha}} r_p + e^{-j\varphi_{d,\alpha}} \frac{d(e^{j\varphi_{d,\alpha}} \overline{\Psi}_{p_{\alpha,\beta}})}{dt} = \bar{i}_{p_{d,q}} e^{-j\varphi_{d,\alpha}} r_p + \frac{d\overline{\Psi}_{p_{\alpha,\beta}}}{dt} + j\overline{\Psi}_{p_{\alpha,\beta}} \frac{d\varphi_{d,\alpha}}{dt},$$

or

$$0 = \bar{i}_{p_{\alpha,\beta}} r_p + \frac{d\overline{\Psi}_{p_{\alpha,\beta}}}{dt} + j\overline{\Psi}_{p_{\alpha,\beta}} \frac{d\varphi_{d,\alpha}}{dt}.$$

Thus the system of equations of electrical balance of asynchronous motor with short circuit rotor, written for vectors of voltages and currents in the transformed coordinates to stator fixed coordinate system $(\alpha, j\beta)$, is given by

$$\left.\begin{aligned} \overline{U}_c &= \bar{i}_c r_c + \frac{d\overline{\Psi}_c}{dt}; \\ 0 &= \bar{i}_p r_p + \frac{d\overline{\Psi}_p}{dt} + j\overline{\Psi}_p \omega_p; \end{aligned}\right\} \tag{3}$$

where $\omega_p = \dfrac{d\varphi_{d.\alpha}}{dt}$ – the angular speed of the rotor rotation; $j\overline{\Psi}_p\omega_p$ – electromotive force of rotor winding, which allows its rotation in the stator fixed coordinate system.

To simplify the expressions, the voltages concerning the stator and rotor are given in the joint coordinate system of the stator (3), and reducing the number of unknown variables we replace the flux linkage vectors included in their composition, by corresponding dependencies from creating currents, namely by expressions:

$$\left.\begin{aligned} \overline{\Psi}_c &= \bar{i}_c L_c + \bar{i}_p e^{-j\varphi_{d.\alpha}} L_m; \\ \overline{\Psi}_p &= \bar{i}_p L_p + \bar{i}_c e^{j\varphi_{d.\alpha}} L_m, \end{aligned}\right\} \tag{4}$$

where $L_c = L_{c\sigma} + L_m, L_p = L_{p\sigma} + L_m$ – full inductances of stator and rotor windings with taking into consideration the self- and mutual inductance L_m at the coincidence of their axes $(\alpha, j\beta)$ i (d, jq). In the same way as at bringing equations of rotor electrical circuit voltages to stator fixed coordinate system, we perform reduction of rotor winding flux linkage (4), included in the moving coordinate system (d, jq) to the fixed stator coordinate system $(\alpha, j\beta)$ by multiplying by the phase rotation operator $1 \cdot e^{-j\varphi_{d.\alpha}}$ coordinates of windings of rotor and stator. We obtain

$$\overline{\Psi}_p e^{-j\varphi_{d.\alpha}} = \bar{i}_p e^{-j\varphi_{d.\alpha}} L_p + \bar{i}_c e^{j\varphi_{d.\alpha}} e^{-j\varphi_{d.\alpha}} L_m,$$

or in general form and in the common coordinate system $(\alpha, j\beta)$

$$\left.\begin{aligned} \overline{\Psi}_c &= \bar{i}_c L_c + \bar{i}_p L_m; \\ \overline{\Psi}_p &= \bar{i}_p L_p + \bar{i}_c L_m. \end{aligned}\right\} \tag{5}$$

Making the replacement of values vectors of stator windings voltage supply, currents and flux linkages by their projections on the axes $(\alpha, j\beta)$, as in $\bar{i}_c = i_{c_\alpha} + ji_{c_\beta}$, we obtain the ultimate system of equations of two-phase short-asynchronous motor (3) with five indeterminate - four current projections of the stator, the rotor on the axes and the angular velocity of the rotor, and known - supply voltage and windings resistance:

$$\left.\begin{aligned} U_{c_\alpha} &= i_{c_\alpha} r_c + \frac{d\Psi_{c_\alpha}}{dt}; \\ U_{c_\beta} &= i_{c_\beta} r_c + \frac{d\Psi_{c_\beta}}{dt}; \\ 0 &= i_{p_\alpha} r_p + \frac{d\Psi_{p_\alpha}}{dt} + j\Psi_{p_\alpha}\omega_p = i_{p_\alpha} r_p + \frac{d\Psi_{p_\alpha}}{dt} + \Psi_{p_\beta}\omega_p; \\ 0 &= i_{p_\beta} r_p + \frac{d\Psi_{p_\beta}}{dt} + j\Psi_{p_\beta}\omega_p = i_{p_\beta} r_p + \frac{d\Psi_{p_\beta}}{dt} - \Psi_{p_\alpha}\omega_p, \end{aligned}\right\} \qquad (6)$$

in which the projections of flux linkage vector of stator and rotor windings on the axis at a fixed stator coordinate system are

$$\left.\begin{aligned} \Psi_{c_\alpha} &= i_{c_\alpha} L_c + i_{p_\alpha} L_m; \\ \Psi_{c_\beta} &= i_{c_\beta} L_c + i_{p_\beta} L_m; \\ \Psi_{p_\alpha} &= i_{p_\alpha} L_p + i_{c_\alpha} L_m; \\ \Psi_{p_\beta} &= i_{p_\beta} L_p + i_{c_\beta} L_m. \end{aligned}\right\} \qquad (7)$$

The system (6) consisting of four equations can be solved provided the addition of a fifth equation to determine the angular speed of the rotor rotation ω_p .

2.2. Mathematical model of direct start of asynchronous motor compressor drive

Two-phase asynchronous electromotor is an electromechanical transducer of electrical energy, the rotor rotation of which is characterized by the motion equation of the type

$$J\frac{d\omega_p}{dt} + M_{on} = M_{eM},$$

where $J = 0{,}0862\, GD^2 n^2$ - moment of inertia of the rotor and associated with it, the reduced moment of mechanism inertia; GD^2 – non-jumping moment, reduced to the motor shaft; n – rotor speed; M_{on} - torque of resistance on the rotor shaft, which includes loading torque and mechanical losses; M_{eM} - electromagnetic torque.

Electromagnetic torque is determined by vector addition of solid vector of stator winding flux linkage and solid vector of stator current, i.e. in the form $M_{eM} = \mathrm{Re}\left[j\Psi_c \times \hat{i}_c\right]$ where $\hat{i}_c$ – complex conjugate vector of stator current. If vector addition is provided as a projection of vectors on the axes of the stator winding *(α,jβ)*, will be given by

$$M_{eM} = \left|\left(\Psi_{c_\alpha} + j\Psi_{c_\beta}\right)\cdot\left(i_{c_\alpha} + ji_{c_\beta}\right)\right| = \Psi_{c_\alpha} i_{c_\beta}[1, j] + \Psi_{c_\beta} i_{c_\alpha}[j, 1] = \Psi_{c_\alpha} i_{c_\beta} + \Psi_{c_\beta} i_{c_\alpha}. \qquad (8)$$

Transforming regarding the current components the equations system of flux linkage rotor and stator winding vector projections (7), shown on the axes of the stator fixed coordinate system, we obtain

$$\left.\begin{aligned}i_{c_\alpha} &= (\Psi_{c_\alpha} L_p - \Psi_{p_\alpha} L_m)/(L_c L_p - L_m^2);\\ i_{c_\beta} &= (\Psi_{c_\beta} L_p - \Psi_{p_\beta} L_m)/(L_c L_p - L_m^2);\\ i_{p_\alpha} &= (\Psi_{p_\alpha} L_c - \Psi_{c_\alpha} L_m)/(L_c L_p - L_m^2);\\ i_{p_\beta} &= (\Psi_{p_\beta} L_c - \Psi_{c_\beta} L_m)/(L_c L_p - L_m^2).\end{aligned}\right\} \quad (9)$$

Through expressions substitution of currents (9) to the electrical balance equations system (6) recorded according to the second Kirchhoff's rule, and to the electromagnetic torque expression (8), we obtain a system of five inhomogeneous differential equations of the first order, which already have their own solution, i.e.

$$\left.\begin{aligned}\frac{d\Psi_{c_\alpha}}{dt} &= U_{c_\alpha} - r_c(\Psi_{c_\alpha} L_p - \Psi_{p_\alpha} L_m)/(L_c L_p - L_m^2);\\ \frac{d\Psi_{c_\beta}}{dt} &= U_{c_\beta} - r_c(\Psi_{c_\beta} L_p - \Psi_{p_\beta} L_m)/(L_c L_p - L_m^2);\\ \frac{d\Psi_{p_\alpha}}{dt} &= -\omega_p \Psi_{p_\beta} - r_p(\Psi_{p_\alpha} L_c - \Psi_{c_\alpha} L_m)/(L_c L_p - L_m^2);\\ \frac{d\Psi_{p_\beta}}{dt} &= \omega_p \Psi_{p_\alpha} - r_p(\Psi_{p_\beta} L_c - \Psi_{c_\beta} L_m)/(L_c L_p - L_m^2);\\ \frac{d\omega_p}{dt} &= \frac{1}{J}\left[\Psi_{p_\beta}\left(\Psi_{p_\alpha} L_c - \Psi_{c_\alpha} L_m\right)/(L_c L_p - L_m^2) - \Psi_{p_\alpha}\left(\Psi_{p_\beta} L_c - \Psi_{c_\beta} L_m\right)/(L_c L_p - L_m^2) - M_{eM}\right]\end{aligned}\right\} \quad (10)$$

Given in the form of expression of the derivative $y_i' = f(y_i)$ with five unknown values $i = 5$, i.e.: $\Psi_\alpha, \Psi_{c_\beta}, \Psi_{p_\alpha}, \Psi_{p_\beta}, \omega_p$ and expected decision in the form of time dependence $y = f(t)$, where $t -$ time, the system can be solved by any method, including a numerical method of Runge-Kutta according to the scheme

$$y_{i(n+1)} = y_{in} + (K_{i1} + 2K_{i2} + 2K_{i3} + K_{i4})/6, \quad (11)$$

where $n -$ integration step; $i -$ target value, so component ratios are related by:

$$K_{i1} = h \cdot f(y_{in}),$$

$$K_{i2_i} = h \cdot f(y_{in} + K_{i1}/2),$$

$$K_{i3} = h \cdot f(y_{in} + K_{i2}/2),$$

$$K_{i4} = h \cdot f(y_{in} + K_{i3}).$$

where $h -$ period of time or integration step. At the beginning of equation (11) solution, for the moment of time $t = 0, n_2 = 0$, the initial values of unknown values y_{i1} should be chosen which correspond to starting of the engine – i.e. zero ones, because the currents, flux linkage are absent, but are those according to which later instantaneous values of their speeds $y_{i(n+1)}$ will be calculated.

IOP Conf. Series: Materials Science and Engineering **232** (2017) 012081 doi:10.1088/1757-899X/232/1/012081

3. Results

3.1. The numerical calculations of dynamic characteristics and analysis of transition processes in the motor of refrigeration machine compressor drive

Determination of dynamic characteristics of an asynchronous motor with short circuit rotor based on the mathematical model in the form of equations system (10), using the Runge-Kutta method (11) is performed on the base of developed program in the Visual Fortran Pro, v.6.1. software (*Baidak Yu., 2011*). Time dependences were built for stator and rotor currents, rotating torque depending on the angular speed of the stator windings magnetic field rotation and acceleration characteristics of the base motor with two-phase double-layer winding and the new one, with improved power characteristics taking into consideration the calculations results of asynchronous capacitor motor magnetic equilibrium inhomogeneous differential equations system were determined. On the Figure 2 and Figure 3 summarized in time calculation results for currents, electromagnetic torque and rotor speed for new and base engines are shown.

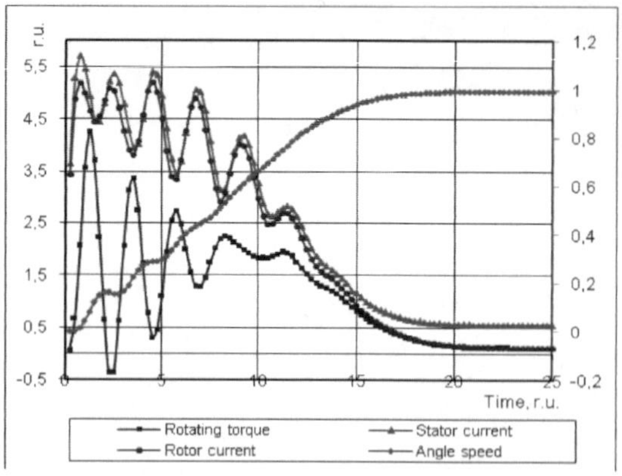

Figure 2 Summarized in time results of calculation of currents, electromagnetic torque and rotor speed of the base motor

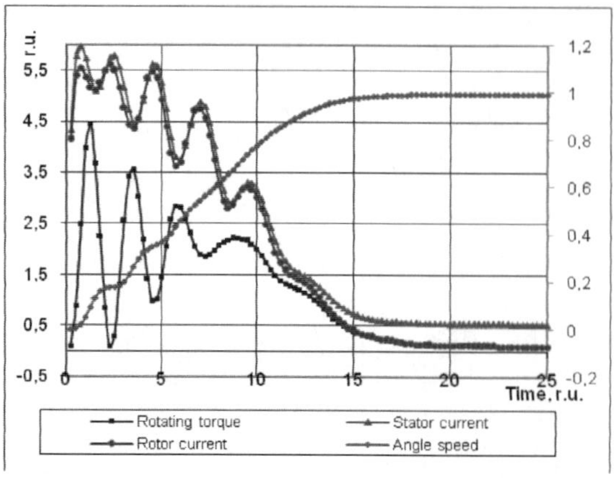

Figure 3 Summarized in time results of calculation of currents, electromagnetic torque and rotor speed of the new motor

General view of the characteristics allows to determine that time dependencies of stator and rotor currents as well as of electromagnetic torque reach the default values at significant influence of their too complex free components.

At the beginning of simulation of dependencies when $t = 0,\ \omega = 0,$ the program is performing calculations on direct start of the motor without taking into consideration winding inductance, i.e. when $(\Psi_{\alpha}, \Psi_{c_{\beta}}, \Psi_{p_{\alpha}}, \Psi_{p_{\beta}}, \omega_{p}) = 0$. Later, an abrupt increase of stator and rotor currents is observed (9), dependent on them flux linkage (10) and dependent on both of them electromagnetic torque (8).

Their dynamic dependencies have at the beginning of the movement a large amplitude of oscillations that affect the angular speed of the rotor, braking it. Speed characteristics of both motors have several sites with the loss of momentum dialed by rotor, that is connected not only with a sharp decrease of electromagnetic torque, but even with a change in its direction to opposite braking and this is while simultaneous increasing of current in the stator and rotor. The reason for this phenomenon should be considered as a sharp decrease of flux linkages of stator and rotor windings, equation (10).

Moreover, the dynamic electromagnetic torque at some moments of time significantly exceeds the maximum moment of static mode. But unlike the static moment, it takes into consideration not only the equivalent circuit parameters of the motor, but also the time of load and inertia of the rotor. The length of electromagnetic transient and the dispersal of the motor largely depend from them. In contrast, the length of the electromechanical process mainly depends on the winding inductance and remains unchanged. Thus, we should define that an essential impact on the duration of the transition process during rotor rotational speed increase has instantly undamped magnetic field generated by currents of stator and rotor and is determined by their flux linkage.

Further analysis of the dynamic characteristics of electromagnetic torque and rotor angular frequency shows that the number of torque oscillations in the new motor with a single-phase winding is less than in the basic one, despite the fact that their amplitude is slightly higher. This makes high-speed angular frequency characteristics of the new engine more rigid, and the transition process shorter. The amplitude of the current oscillations in the new motor is less, this fact positively affects on the operation of the power supply. Reaching 75% of the rotor speed in the new motor is performed for 0.03 seconds, while at the base one - with 0,036 sec.

If we assume that the quadratic current pulse during the transition process - starting of electric motor results in adiabatic temperature increase of its windings, and the additional coil of base motor, which is made of winding wire with diameter of 0.41 mm with cross-section of 0.132 mm^2 and a valid current 6 A/mm^2, for the period of time 0.036 seconds , at the start of the motor will be hotter on 19°C. Provided that the stipulated duration of motor start is 0.3 seconds - with a nominal static moment of load, excess of temperature of additional windings will approach to its critical value in the class of wire insulation thermal resistance, namely 154°C.

At the same time, additional coil of new motor, made of wire of cross section 0.26 mm^2 according to the scheme given in (*Baidak Yu., 2011*), will increase its temperature only on 3.85°C or in long mode start-up - on 38.5°C that almost does not affect its insulation properties and, as a consequence, the reliability of the engine. Speed characteristics also show that achieving the 99% nominal revolutions of base motor rotor is carried out within 0.065 seconds and 0.055 seconds for the new motor that should be considered as an advantage. In general, except above-mentioned, that is necessary to consider during the implementation of energy saving measures, both motors on dynamics satisfy the conditions of their use as a drive of refrigeration machine compressor.

Conclusion

Based on a system of differential equations, suggested in relative units for generalized electric machine, mathematical model of the programs for calculation of the dynamic performance of refrigeration machine compressors low-power asynchronous motors drive was developed and proposed. This model is invariant for any asynchronous motors, a numerical model of which is adequate to the test sample. Performed on its basis comparative analyses of the basic model of two-phase electric motor of hermetic compressor drive and a new model of motor with single-phase stator winding, which is an alternative to industrial motor winding, confirmed the benefits of the new motor with single-phase stator winding over the base engine.

The calculations of the dynamic characteristics of the motor of compressor drive permit even at the theoretical level determine the value of electromagnetic torque jumps to coordinate mechanical characteristics of the compressor and the motor, and for taking them into account in choosing construction materials for the compressor elements. Developed and used at research of single-phase asynchronous motor of compressor drive math- and software should be considered as a constituent part of computer-aided design of refrigeration equipment.

References

[1] Dote Y 1990. Servo Motor and Motion Control using Digital Signal Processors, PHI.

[2] Guinee R and Lyden C 1998 Accurate Modelling and Simulation of a High Performance Brushless DC Motor Drive System for Industrial Applications, *Proc. of the IASTED International Conference Applied Modelling and Simulation,* Honolulu, HawaiiMore references

[3] Guinee R and Lyden C 2001 Motor Parameter Identification using Response Surface Simulation and Analysis, Proc. of American Control Conference, ACC-2001, VA, USA.

[4] Persson E and Buric M 1976 Mathematical Modelling and Simulation of High Performance Brushless DC Motors, *IMCSD* 4th annual symp.

[5] Beucher O and Weeks M 2008 *Introduction to MATLAB & Simulink: a project approach*. Hingham, Massachusetts New Delhi, Inf. Sci. Press LLC Publ., 390 p.

[6] Ong Chee-Mun 1998 *Dynamic simulation of electric machinery using MATLAB/Simulink*. Upper Saddle River, New Jersey, Prentice Hall PTR Publ., 627 p.

[7] Chaturvedi Devenndra K 2010 .*Modeling and simulation of systems using MATLAB and Simulink*. New York, CRC Press Taylor & Francis Group Publ., 711 p.

[8] Giurgiutiu V and Lyshevski S 2009 *Micromechatronics modeling, analysis, and design with MATLAB*. New York, CRC Press Taylor & Francis Group Publ., 920 p.

[9] Karris S 2008 *Introduction to Simulink with engineering applications*. Orchard Publications,. 716 p

[10] Ansari A and Deshpande D 2010 Mathematical Model of Asynchronous Machine in MATLAB Simulink" *I J E S T* Vol. 2(5)

[11] Galan N Mammadov A 2006 The building system and the mathematical model of axial air-gap three-phased asynchronous motor, Taormina – Italy, SPEEDAM .

[12] Leonhard W 2001. Control of Electrical Drives, 3rd ed., SpringerVerlag, New York

[13] Baydak Yu et.al 2013 Energy Efficient Single-Phase Electric Motor of Domestic Refrigeration Devices Motor-Compressor Aggregate. *Proceedings of the 8th International Conference on Compressors and Coolants, Papernicka* – Smolenice, Slovak Republic.

[14] Baydak Yu 2011 Asynchronous condenser engine Pat. 63322 Ukraine, МПК^{7} HO2K 17/08. №u201101642; appl. 14.02.11; published. 10.10.11, Bul. №19.

IOP Conf. Series: Materials Science and Engineering **232** (2017) 012082 doi:10.1088/1757-899X/232/1/012082

Numerical Simulation and Experimental Validation of Failure Caused by Vibration of a Fan

Qiang Zhou[1], Wu Han[1], Jianmei Feng[1,2], Xiaohan Jia[1,2], Xueyuan Peng[1,2]

[1] Xi'an Jiaotong University, Xi'an, Shaanxi, China, People's Republic of

[2] National Engineering Research Center of Fluid Machinery and Compressors, Xi'an, Shaanxi, China, People's Republic of

E-mail: xypeng@mail.xjtu.edu.cn

Abstract. This paper presents the root cause analysis of an unexpected fracture occurred on the blades of a motor fan used in a natural gas reciprocating compressor unit. A finite element model was established to investigate the natural frequencies and modal shapes of the fan, and a modal test was performed to verify the numerical results. It was indicated that the numerical results agreed well with experimental data. The third order natural frequency was close to the six times excitation frequency, and the corresponding modal shape was the combination of bending and torsional vibration, which consequently contributed to low-order resonance and fracture failure of the fan. The torsional moment obtained by a torsional vibration analysis of the compressor shaft system was exerted on the numerical model of the fan to evaluate the dynamic stress response of the fan. The results showed that the stress concentration regions on the numerical model were consistent with the location of fractures on the fan. Based on the numerical simulation and experimental validation, some recommendations were given to improve the reliability of the motor fan.

Key words: motor fan; modal analysis; finite element analysis; modal test

1. Introduction

The reciprocating compressor units driven by the motor are widely used in petrochemical fields. The oscillatory torques on the compressor crankshaft will induce torsional vibration of the compressor shaft system which mainly consists of the crankshaft, coupling, motor shaft and the components mounted to the shafts. With the development of large-scale and high-speed natural gas reciprocating compressor, the torsional vibration problems in compressor shaft system become more and more serious. As an auxiliary equipment on the motor shaft, the motor fan with a number of blades has high probability of fracture failure due to oscillatory torques and centrifugal forces[1].

As a method to study the dynamic characteristics of the structure, modal analysis plays an important role in the analysis and solution of engineering vibration and fault diagnosis[2]. It is very important to study the vibration characteristics of the fan to ensure that it is not abnormal vibration in the working speed range. Yang Wen studied the response characteristics of fan by using numerical simulation method[3]. Srinivasan summed up the impact on the vibration problems of the gas turbine blades[4].

IOP Conf. Series: Materials Science and Engineering **232** (2017) 012082 doi:10.1088/1757-899X/232/1/012082

Poursaeidi analyzed the effect of natural frequency on the failure of the compressor blades[5]. Cheng studied the vibration and stress caused by centrifugal and aerodynamic loads of axial fan blades by using the finite element analysis and experimental measurements[6].
In this paper, the root cause of fracture failure of a motor fan in a reciprocating compressor unit were investigated. Vibration modes of the fan were studied by numerical simulation and experimental validation. The Campbell diagram was used to study the possibility of the low-order resonance of the fan. On the base of modal analysis, a torsional vibration analysis of the shaft system of the reciprocating compressor unit was conducted to obtain the working torques of the fan, and the harmonic response analysis of the fan was performed to predict the dynamic stress of the fan under the acting of working torques. Based on the investigation of vibration modes and stress response, the root cause of fracture failure were identified, and some recommendations were proposed to solve the failure problem of the motor fan and improve the reliability of the reciprocating compressor unit.

2. Numerical simulation

2.1. Geometrical model

The motor fan discussed in this paper was used in a natural gas reciprocating compressor unit. The compressor was a four-throw and balanced-opposed reciprocating compressor, and was driven by a motor, as shown in figure 1. The fan is located at the end of motor shaft. The rated speed of the compressor is 995 r·min^{-1} and the first order excitation frequency is 16.6 Hz. The material of the fan is aluminum alloy, and the outer diameter of the fan is 910 mm. A total of 12 blades are evenly distributed, with the thickness of 3 mm. Geometrical model of the fan is shown in figure 2. In the regular inspection, it was found that fracture failure occurred on the blades and base plates of the fan, as shown in figure 3. The cracks initiated from the root of blades and propagated to the middle of blades. After the replacement of new fans in the natural gas reciprocating compressor unit, the fan still have failure fracture problems during the short-term operation, which is in line with the characteristics of torsional resonance damage. In order to identify the root cause of the fan failure problem, the vibration modal analysis of the fan was carried out by finite element simulation and experimental validation.

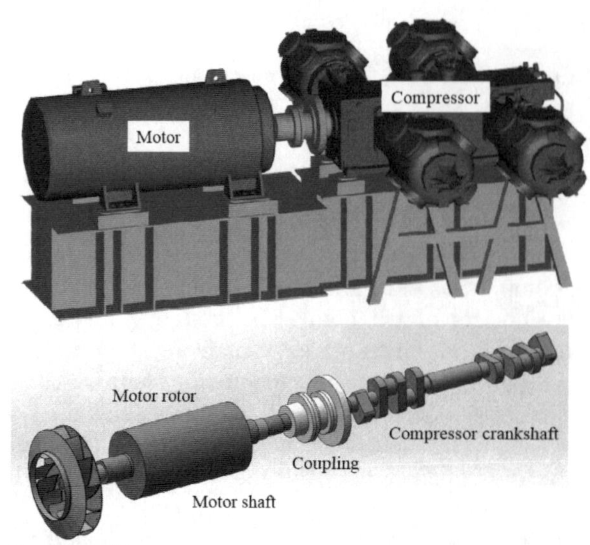

Figure 1. Shaft system of the reciprocating compressor unit.

Figure 2. Geometrical model of fan.

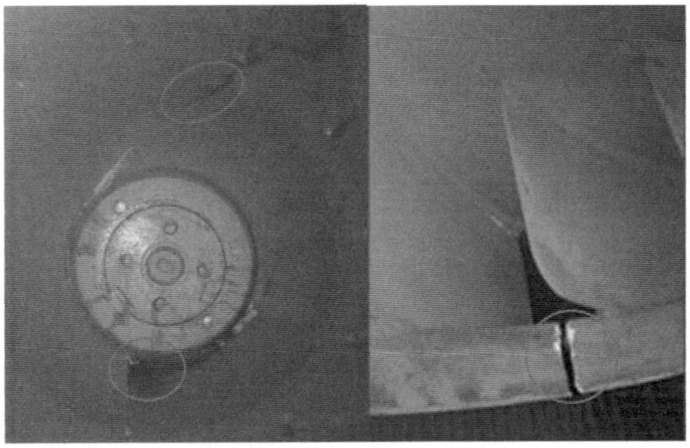

Figure 3. Fracture of fan.

2.2. Numerical modelling

2.2.1 Meshing. After simplifying the geometrical model, it is necessary to mesh the model. For some non-key research areas, the simulation results are not sensitive to the density of the mesh. So it can be relatively sparse grid to reduce the amount of calculation and do not affect the final result. And for the focus of the study areas or narrow irregular areas, the mesh is need to be fine, because the number and quality of the grid will directly affect the simulation results. Finally, the tetrahedral element is used for meshing. For the shaft plate of the fan, the grid size of 10 mm was set. For the fan and support plate parts, grid size is set to 2.5 mm due to the thin structure. Through the grid-independent verification, the effect of the calculation frequencies is less than 1% after further increasing the number of grid. A total of 456003 nodes and 187984 elements were generated in the numerical model of the fan.

2.2.2 Boundary condition. The various components of the fan are connected by welding. In the pre-treatment of finite element analysis, the contact area of each component is treated with spot weld, so that the load transfer between the various parts of the model is closer to the actual situation. Considering the constraints of the actual operation of the fan, when the fan is subjected to finite element modal analysis, the shaft of the fan and the keyway side wall are fixed. The damping of the fan is small, and its influence on the natural frequencies of the fan is negligible, so it is not to be considered. The finite element model of the fan is shown in figure 4.

IOP Conf. Series: Materials Science and Engineering **232** (2017) 012082 doi:10.1088/1757-899X/232/1/012082

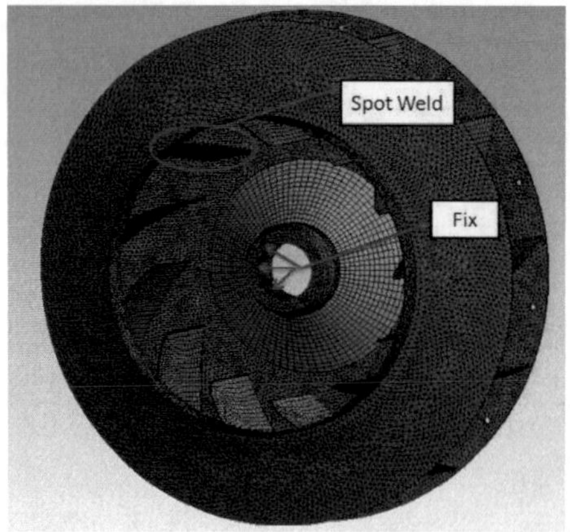

Figure 4. Finite element model of the fan.

3. Experimental modal analysis

3.1. Modal test

In order to verify the reliability of the finite element analysis results, it is necessary to carry out the modal test of the fan. The experimental modal analysis is based on the input and output of the system, that is, the excitation and response data collected, to make modal parameter identification[7]. The experimental modal analysis method is divided into multiple input single output method, single input multiple output method and multiple input multiple output method. In this paper, single point excitation and multiple point response method (SIMO) is selected. The test system of the fan is shown in figure 5. The entire test system consists of four parts, which are excitation equipment, sensors, acquisition equipment and analysis software. When excitation equipment stimulate the vibration, the sensors pick up the vibration signal into the computer analysis software by acquisition equipment. Finally, the natural frequency, vibration mode, damping and other modal parameters of the fan can be obtained.

In the experimental modal analysis, there are two kinds of structural support types, which are the free suspension and the original boundary conditions of the specimen. This experiment focuses on the resonance frequency and vibration reduction technique of the analysis structure. And the boundary condition of the actual constraint was simulated. According to the selected modal test method, the 086D50 type hammer of PCB company, the 356B18 three-way acceleration sensor, the Dasp-V10 intelligent data acquisition and signal analysis system provided by The Beijing Oriental Vibration and Noise Technology Research Institute were used as test equipments. The experimental modal test of the fan is shown in figure 6.

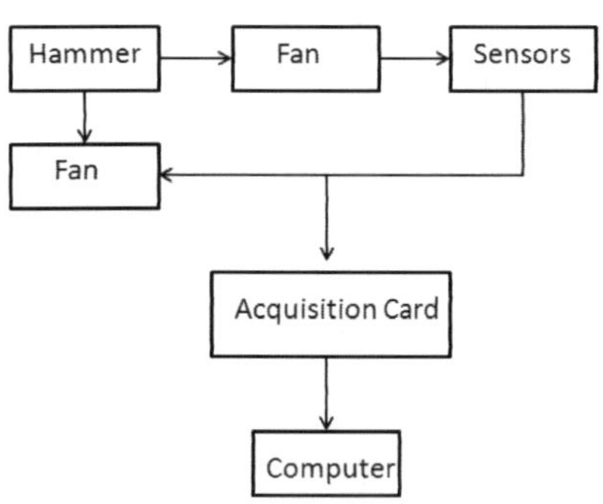

Figure 5. Modal test system.

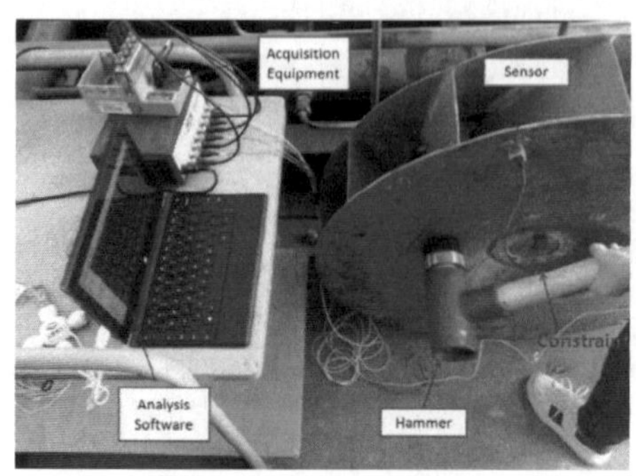

Figure 6. Modal test.

3.2. Excitation point and response point selection

The selection of excitation and response point is very important in the modal test. The fan is a continuous vibration system. The inertia, elasticity, damping and motion of the system depend on the spatial coordinates. So it is impossible to obtain the continuous distributed response measurement and the infinite number of characteristic solutions. Therefore, in the vibration analysis and parameter identification, infinite degree of freedom of the continuous vibration system is limited to a limited degree of freedom of the vibration system. And a total of 56 vibration points of the fan on both sides of the upper and lower sides of the distribution were selected. Vibration point distribution is shown in figure 7. The principle of selecting excitation point is to make the exciting force easy to transfer to the various parts of the structure and to avoid the vibration nodes and the weak structure[8], the test excitation point is to be determined by the reference value of the simulation results. And through trying to knock, the different excitation of the time domain waveform and spectrum are compared. After several trials, the 17th point was selected as the excitation point, and the direction was the vertical direction.

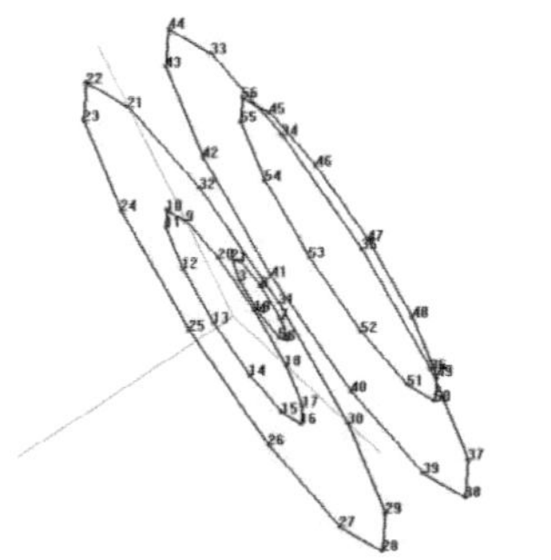

Figure 7. Modal test model.

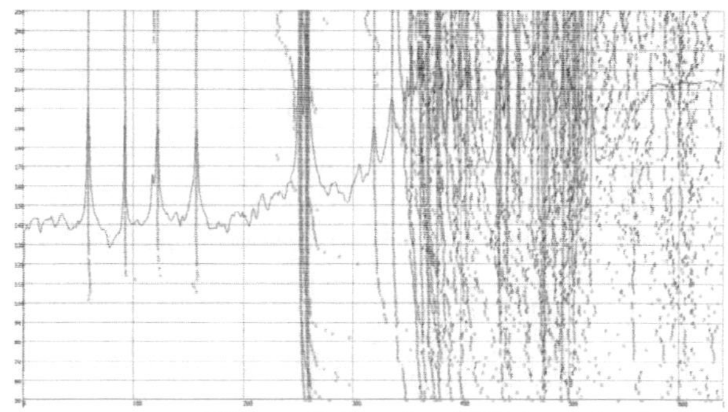

Figure 8. Result of ERA method.

3.3. Modal parameter identification

The main task of modal parameter identification is to determine the modal parameters of the vibration system from the measured data, including the modal natural frequencies, modal damping ratio, modal mass, modal stiffness and mode[9]. The current parameter identification is mainly divided into frequency domain method and time domain method. Aiming at the this modal test, the morphological parameters of the frequency domain method can be overcame by using the feature system algorithm (ERA method), which is useful for the small damping and modal coupling system, because that can identify dense modes and heavy roots. The result of ERA method is shown in figure 8.

4. Results

4.1. Natural frequencies and vibration mode

The first eight orders of vibration modes of the fan were obtained by numerical and experimental modal analyses. The natural frequencies of the fan are shown in table 1. Limited to the length of the article, the finite element analysis and test results of the most important first three order modes are listed, the contrasts of the first three modes are shown in Figure 9, 10 and 11 respectively.

It can be seen from the table 1 that the calculated results are corresponded to the experimental data, and the natural frequencies calculated by the finite element model agree well with the measured frequencies. The absolute error of the corresponding frequency is below 10%. From the analysis, the modes reflect the structural distortion very well. The first order and the second order are the overall bending mode, while the third order are the composite mode of the bending and torsional, indicating that the established finite element model reflects the actual structure well. The reliability of the finite element modal analysis is verified.

Table 1. Natural frequencies of calculated and measured.

Order	Calculated(Hz)	Measured(Hz)	Difference(%)
1	55.70	59.07	5.71
2	55.72	60.41	7.76
3	103.79	94.60	-9.71
4	243.67	250.00	2.53
5	244.09	250.08	2.39
6	245.05	252.15	2.81
7	255.66	258.44	1.07
8	256.10	258.58	0.95

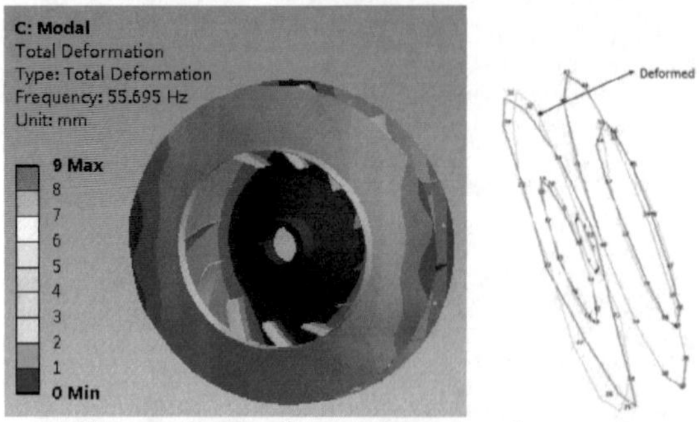

Figure 9. The first-order vibration mode.

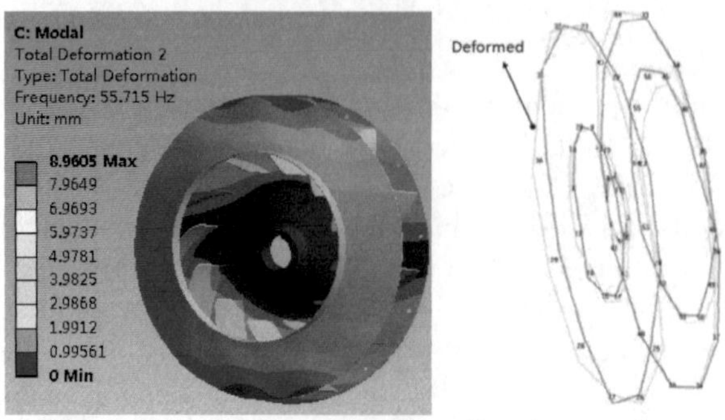

Figure 10. The second-order vibration mode.

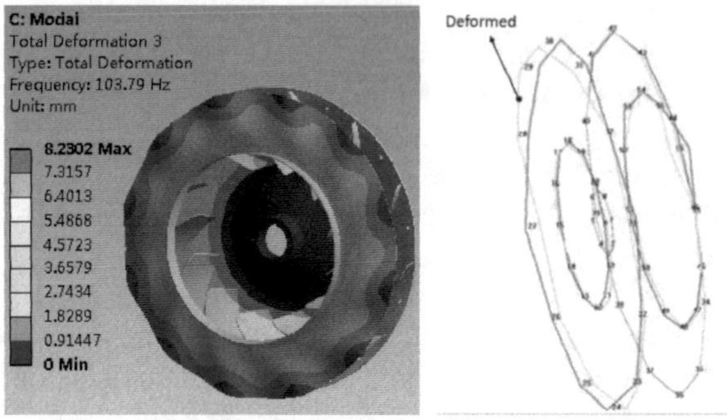

Figure 11. The third-order vibration mode.

4.2. Campbell interference diagram
From the finite element simulation and the experimental modal analysis results, combined with the interference pattern of the fan vibration mode shown in figure 12, the first order natural frequency of the fan is close to the speed of 3 times under the constraint condition, therefore the first order bending mode of the fan may be excited in the actual case, causing the failure of the fan. Despite that the first order mode is the bending mode, there are bending vibrations in the compressor shaft system and the fan will have a certain bending load. Furthermore, the vortex formed by the air flow in the fan will also cause the force to the fan. At the same time, the third order natural frequency of the fan is close to the speed of 6 times, and the bending and torsional modes are shown in figure 11.

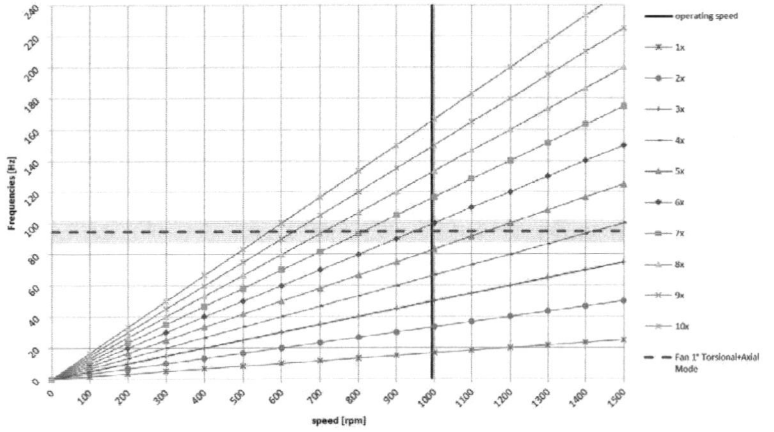

Figure 12. Campbell interference diagram.

5. Dynamic stress analysis

5.1. Torsional vibration analysis of the compressor shaft system
To obtain the working torques of the fan, the torsional vibration analysis of shaft system of reciprocating compressor units were conducted. The rotating components of the compressor shaft system mainly include the compressor crank, flexible coupling, motor shaft and motor external fan. The reciprocating components including the piston, piston rod and crosshead also have effects on the torsional characteristics of the compressor shaft system. The mass-elastic model of the compressor shaft system was established, as shown in figure 13. The rotating components with large inertia were simplified as lumped mass elements, and rotating components with small inertia were simplified as elastic elements. The equivalent inertia of reciprocating components were simplified by the principle of equal kinetic energy in a cycle. And the shaft system of the reciprocating compressor unit was simplified as 29 lumped masses elements connected with each other by elastic elements. The inertia and stiffness of each section of the compressor shaft system are provided by the manufacturers of the compressor, coupling and motor.
Without considering damping, The torsional vibration analysis was conducted to obtain torsional natural frequencies (TNFs) and mode shapes of the compressor shaft system. The results showed that the first two order TNFs of the compressor shaft system were 92.19 Hz and 191.00 Hz. The first order TNF is appropriately positioned between the fifth and sixth order of run speed, and the torsional resonance of the shaft system will not occur.
The forced response analysis calculates the dynamic torsional response of the compressor shaft system. A modal damping factor of 2% was utilized. The input torques were applied to the mass-elastic model, and the deformation, stress and torques at all sections of the mass-elastic model were obtained. The working torques of the fan were extracted, and the Fourier decomposition was performed to obtain the torque harmonics at each order of run speed, as shown in figure 14.

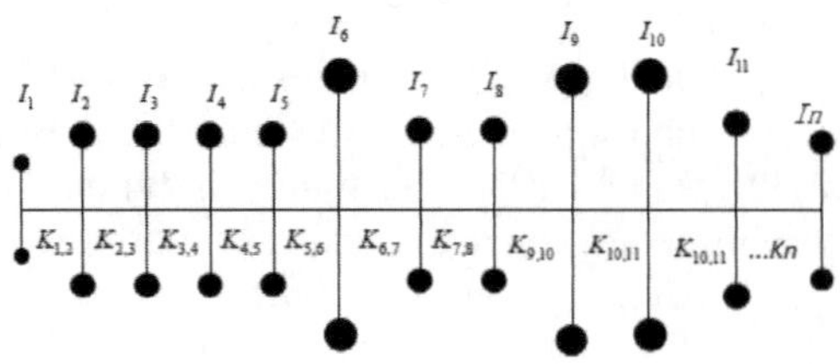

Figure 13. Mass-elastic model of the compressor shaft system.

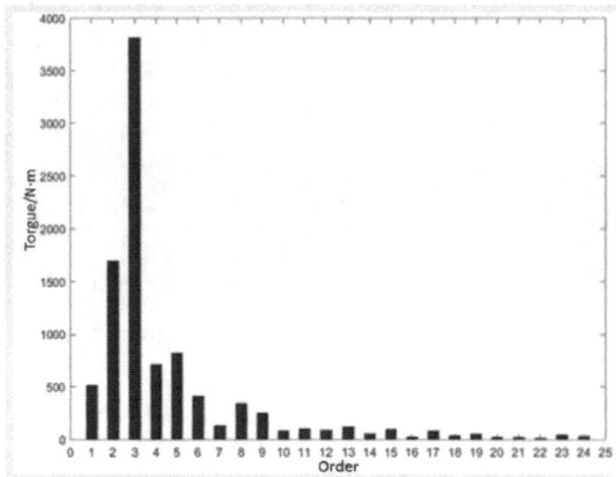

Figure 14. Working torques of the fan.

5.2. *Harmonic response analysis of the fan*

Since the first natural frequency of the fan was in the range of +10% of the sixth order of excitation frequency 99.5 Hz, namely 89.55-109.45 Hz, the low-order resonance was likely to occur. The harmonic response analysis of the fan was carried out to study dynamic stress response of the fan. The sixth order torque harmonic was applied to outer surface of the hub, with the use of the mode superposition method. The damping ratio obtained from the modal test was used for the harmonic response analysis.

After harmonic response analysis of the fan, the stress distribution at the third order frequency was obtained, as shown in figure 15. It can be seen from the figure that the stress concentration areas coincide with the fan's crack locations, which further confirms the possibility of torsional resonance. Based on the above analysis, it was indicated that the excessive dynamic stress at the root of blades caused by torsional resonance attributed to the fatigue fracture of the fan blades.

Figure 15. Stress distribution of the third order natural frequency.

6. Summary and recommendations

This paper presents the root cause failure of the fan blades of a natural gas reciprocating compressor unit. The results show that the first order natural frequency of fan is close to the speed of 3 times, the natural frequency of third order is close to 6 times, which is easy to cause low-order resonance, and the third order mode of bending and torsional is the main reason. The torque data obtained by a torsional vibration analysis of the compressor shaft system was exerted on the numerical model of the fan to evaluate the dynamic stress response of the fan. The results of the dynamic response analysis of the fan shows that the fan stress concentration regions coincide with the fan crack locations.

It is recommended to adjust natural frequencies of the fan by increasing bending and torsional stiffness or decreasing the weight of the fan to avoid the low-order resonance. At design stage of the compressor, it is necessary for compressor manufacturers to conduct modal analysis, torsional vibration analysis and dynamic response analysis of auxiliary components mounted to the shafts like the motor fan. The torsional natural frequencies of the auxiliary components should be away from compressor excitation frequency to avoid torsional resonance induced by working torques and insure the reliability of the reciprocating compressor unit.

References

[1] Zhang Q and Zhang Y 2011 Modal analysis of high bypass turboengine fan blade *J. Shenyang Aerospace University.* **28** 14

[2] Sitarz P and Powałka B 2018 Dual ant colony operational modal analysis parameter estimation method *J. Mechanical Systems and Signal Processing.* **98** 231

[3] Yang W 2007 Study on dynamic response characteristics of wide chord hollow fan blade *J. Aeronautical Dynamics.* **22** 444

[4] Arinivasin A V 1997 Flutter resonant vibration characteristics of engine blades *J. Engineering for Gas Turbines and Power.* **119** 742

[5] Poursaeidi E and Babaei A 2012 Effects of natural frequencies on the failure of R1 compressor blades *J. Engineering Failure Analysis.* **25** 304

[6] Xu C, Amano R S, Lee E K 2004 Investigation of an axial fan blade stress and vibration due to aero dynamic pressure field and centrifugal effects *J. International Journal Series*: B **47** 75

[7] Xu X and Yan H 2011 Model test and static test for wind turbine blade *J. Aerospace Materials & Technology.* **25** 44

[8] Shen S, Ying H, Lei S 2000 Modal test for yellow river bridge exciting with hammer *J. Vibratio n Engineering.* **13** 492

[9] Mottershead J E and Friswell M I 1993 Model updating in structural dynamics: a survey *J. Sound and Vibration.* **167** 347

Study on the CO_2 electric driven fixed swash plate type compressor for eco-friendly vehicles

Donglim Nam[a], Kitae Kim[a], Jehie Lee[b], Yunki Kwon[c], Geonho Lee[d,*]

[a] R&D Center of Doowon Heavy Industrial Co., Ltd., Anseong, Republic of Korea
[b] Department of Information and Communication, Doowon Technical University, Anseong, Republic of Korea
[c] Department of Computer Aided Design, Doowon Technical University, Anseong, Republic of Korea
[d] Department of Building Engineering Services, Doowon Technical University, Anseong, Republic of Korea

ghlee@doowonhi.com

Abstract. The purpose of this study is to experiment and to performance analysis about the electric-driven fixed swash plate compressor using alternate refrigerant(R744). Comprehensive simulation model for an electric driven compressor using CO_2 for eco-friendly vehicle is presented. This model consists of compression model and dynamic model. The compression model included valve dynamics, leakage, and heat transfer models. And the dynamic model included frictional loss between piston ring and cylinder wall, frictional loss between shoe and swash plate, frictional loss of bearings, and electric efficiency. Especially, because the efficiency of an electric parts(motor and inverter) in the compressor affects the loss of the compressor, the dynamo test was performed. We made the designed compressor, and tested the performance of the compressor about the variety pressure conditions. Also we compared the performance analysis result and performance test result.

1. Introduction

Since the end of the 19th century, carbon dioxide was used a refrigerant in ship' refrigerators and other stationary systems. After CFCs and HCFCs refrigerants, which became major refrigerants, was introduced, and carbon dioxide were declined from the market. However, environmental issues as the ozone depletion and global warming was caused by using CFCs and HCFCs refrigerants. Lorentzen et al.[1] rediscovered the carbon dioxide as a suitable refrigerant. Carbon dioxide is a good refrigerant which is environmentally friendly and has excellent thermal properties.[2] Since he reasoned with several possible trans-critical cycles with carbon dioxide for vehicle air condition system, many studies have been conducted on the compressors and systems for carbon dioxide refrigerant. [3-6]

Many studies have been carried out on the steady-state simulation models for reciprocating compressors. Many analysis models for electric driven compressor did not consider or calculated the efficiency of the motor using simple relationship. However, because the electric driven compressor for vehicles is equipped with a motor and an inverter therein, it has to consider the accurate efficiency of electric parts in the compressor. And the efficiency of electric parts is changed by the rotational speed and the compression ratio.

This study considered the efficiency of electric parts in the electric driven fixed swash plate compressor including a motor and an inverter for a more accurate performance analysis. The efficiency of electric parts is carried out by the dynamo test and the test results are used in the performance analysis. The efficiency of the electric parts is measured with the rotational speed by the load through the dynamo test. The power of the compressor is calculated by a polynomial relationship using the dynamo test results. And the performance analysis results are compared with the performance evaluation results of the prototype compressors using CO_2.

2. Numerical simulation model

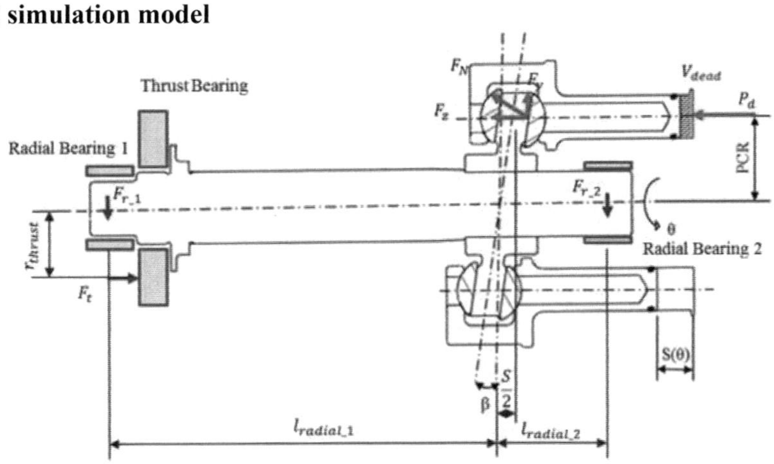

Fig. 1 Schematic of the geometry and kinematics of the fixed swash plate

A schematic of the geometry and kinematics of the fixed swash plate type compressor is shown in fig.1. The Swash plate is fixed on the crank-shaft and the piston with the shoes installed on the swash plate. The piston reciprocates in accordance with the rotation of the shaft, wherein the piston stroke, S is expressed as a function of the shaft angel, θ as follow:

$$S(\theta) = \mathrm{PCR}\tan\beta\,(1+\cos\theta)$$

Where, PCR is the pitch circle radius of the pistons and β is the angle of swash plate.

The volume inside the cylinder is calculated by:

$$V(\theta) = S(\theta)\cdot A_p + V_{dead}$$

Where, V_{dead} is the dead volume inside the cylinder and A_p is the area of the piston.

Taking the cylinder enclosure as the boundary of the control volume. Based in the law of mass conservation, the mass change of refrigerant in the cylinder is determined:

$$\frac{dm_c}{dt} = \frac{dm_{suc}}{dt} + \frac{dm_{li}}{dt} - \frac{dm_{dis}}{dt} - \frac{dm_{lo}}{dt}$$

The conservation equation of energy for the cylinder contents can be written by the first law of thermodynamics.

$$\delta Q = \delta U + \delta W - [dE_i - dE_o]$$

Therefore, the temperature in the cylinder as a function of time:

$$\frac{dT_c}{dt} = \frac{\frac{dQ_c}{dt} - \frac{dW_f}{dt} + \left[\frac{dm_{suc}}{dt} \cdot (h_{suc} - h_c) + \frac{dm_{li}}{dt} \cdot (h_{li} - h_c)\right] + m_c \cdot \frac{dv_c}{dt}\left[v_c \cdot \left(\frac{\partial P_c}{\partial v_c}\right)_T - \left(\frac{\partial h_c}{\partial v_c}\right)_T\right]}{m_c \cdot \left[\left(\frac{\partial h_c}{\partial T_c}\right)_v - \left(\frac{\partial P_c}{\partial T_c}\right)_T \cdot v_c\right]}$$

The heat transfer from the refrigerant between the gas and the cylinder wall is calculated using the heat transfer coefficient, proposed by Fernanda et al.[12]

$$\frac{dQ}{dt} = hA_{heat}\left(T_{cyl}(t) - T_s\right)$$

$$h = \frac{k}{D_p} aRe^b Pr^c$$

Where T_s is surface temperature of the compression chamber walls, A_{heat} is the area of the gas chamber, including the piston and end cap, T_{cyl} is the average gas temperature in the cylinder, Nu is the Nusselt number, h is the convective heat transfer coefficient, k is the thermal conductivity of the cylinder, $Pr\left(= {C_p\mu}/{k}\right)$ is the Prandtl number, Re is the Reynolds number, D_p is the characteristic length (Cylinder diameter), and a, b, and c are constants. The characteristic velocity for the Reynolds number is the piston mean velocity.

$$\bar{V}_p = 2Sf$$

$$V_c = \frac{|\dot{m}(t)|}{\rho(t)A_p}$$

Where, f is the rotational speed of the compressor, $\dot{m}$ is the mass flow rate through the valves, and $\rho(t)$ is the gas density in the cylinder.

Due to the distinct phenomena taking place in the compression cycle, different definitions for the Reynolds number ($Re = \frac{\rho VD}{\mu}$) and different values for a, b, and c were specified for the compression, discharge, expansion and suction processes, as indicated in Table 1.

Table 1. Reynolds number and constants for each process of the compression cycle

Process	Reynolds number	Constants values
Suction	$Re = \frac{\rho(t)D(\bar{V}_p + 2\bar{V}_p^{-0.4}V_c(t)^{1.4})}{\mu(t)}$	$a = 0.08,\ b = 0.9,\ c = 0.6$
Compression	$Re = \frac{\rho(t)D\bar{V}_p}{\mu(t)}$	$a = 0.08,\ b = 0.8,\ c = 0.6$
Discharge	$Re = \frac{\rho(t)D(\bar{V}_p + \bar{V}_p^{0.8}V_c(t)^{0.2})}{\mu(t)}$	$a = 0.08,\ b = 0.8,\ c = 0.6$
Expansion	$Re = \frac{\rho(t)D\bar{V}_p}{\mu(t)}$	$a = 0.12,\ b = 0.8,\ c = 0.6$

In convective heat transfer, the Churchill–Bernstein equation is used to estimate the surface averaged Nusselt number for a cylinder in cross flow at various velocities. This equation is named after Stuart W. Churchill and M. Bernstein, who introduced it in 1977. This equation is also called the Churchill–Bernstein correlation.[11]

$$\overline{Nu_D} = 0.3 + \frac{0.62 Re_D^{1/2} Pr^{1/3}}{[1 + (0.4/Pr)^{2/3}]^{1/4}} \left[1 + \left(\frac{Re_D}{282000}\right)^{5/8}\right]^{4/5} \qquad Pr Re_D \geq 0.2$$

The Churchill–Bernstein equation can be used for any object of cylindrical geometry in which boundary layers develop freely, without constraints imposed by other surfaces.

The swash plate compressor is used for the inlet valve and discharge valve, the opening and closing action of the valve has a significant effect on the performance of the compressor. This study were modeled as vibration problems of one degree of freedom model to analyze the behavior of the valve such as respectively for the inlet and outlet valve.

$$\ddot{y} = \frac{\Delta P A_f}{m} - 2\xi \omega_n \dot{y} - {\omega_n}^2 y$$

Where, y is the displacement of the valve, m is the mass of the valve, c($= 2m\xi\omega_n$) is the damping coefficient, and k($= m{\omega_n}^2$) is the stiffness. ΔP is the pressure difference across the valve(the difference between the upstream pressure and the downstream pressure). A_f is the effective force area of the valve.

The mass flow rate through the discharge valve and the suction valve is usually expressed that would occur for an isentropic flow condition of an ideal gas.

$$\frac{dm_{valve}}{dt} = C_{valve} A_e P_u \sqrt{\frac{2k}{(k-1)RT_u}\left\{\left(\frac{P_d}{P_u}\right)^{\frac{2}{k}} - \left(\frac{P_d}{P_u}\right)^{\frac{k+1}{k}}\right\}}\,, \quad \left(\frac{P_d}{P_u}\right) \geq \left(\frac{k+1}{2}\right)^{\frac{k}{1-k}}$$

$$\frac{dm_{valve}}{dt} = C_{valve} A_e P_u \sqrt{\frac{2k}{(k-1)RT_u}\left\{\left(\frac{k+1}{2}\right)^{\frac{2}{1-k}} - \left(\frac{k+1}{2}\right)^{\frac{k+1}{1-k}}\right\}}\,, \quad \left(\frac{P_d}{P_u}\right) < \left(\frac{k+1}{2}\right)^{\frac{k}{1-k}}$$

Where, the coefficient C_{valve} is 0.58 for the suction valve and 0.6 for the discharge valve.[10] A_e is the effective flow area of the valve.

Gas leakage through the piston ring is usually expressed that would occur for an isentropic flow condition of an ideal gas.

$$\frac{dm_{gap}}{dt} = C_{gap} A_{gap} P_u \sqrt{\frac{2k}{(k-1)RT_u} \cdot \left\{\left(\frac{P_d}{P_u}\right)^{\frac{2}{k}} - \left(\frac{P_d}{P_u}\right)^{\frac{k+1}{k}}\right\}}\,, \quad \left(\frac{P_d}{P_u}\right) > 0.54$$

$$\frac{dm_{gap}}{dt} = C_{gap} A_{gap} P_u \sqrt{\frac{k}{RT_u}\left(\frac{2}{k+1}\right)^{\frac{k+1}{k-1}}}\,, \quad \left(\frac{P_d}{P_u}\right) \leq 0.54$$

Where the discharge coefficient C_{gap} is assumed to be 0.86.[9] The subscripts u and d is defined as upper and down, respectively.

The torque due to the frictional force between the swash plate and the shoes relative to the normal force by the internal pressure in the cylinder.

$$T_{S\_S} = PCR\mu F_N$$

Also it is added to the inertia force caused by the mass of the piston and the shoes.

$$T_{f\_t} = \mu \text{PCR}\left\{A_p \sum_{n=1}^{N}\left(\frac{P_{cyl_N} - NP_s}{cos\beta}\right) + (m_p + 2m_s)\sum_{n=1}^{N} S''(\theta)_N\right\}$$

Thus, a compressor power by the internal pressure and the inertia force as follows:

$$P_f = 2\pi \frac{N}{60} T_{f\_t}$$

The compressor power by the frictional resistance between the piston ring and the cylinder can be expressed by the speed of the piston and the friction force of the piston ring due to in-cylinder pressure.[7] The friction force between the piston ring and the cylinder is represented by the area and pressure on the pressure.

$$F_{p\_f} = \mu_{p\_f} P_{cly} A_{p\_r}$$
$$P_{ring} = \frac{3F_{p\_f}S'}{2\pi}$$

The compressor power by the bearings is obtained by calculating the force on each bearing to the shaft as the force and moment balance equations.

$$\sum F_y = 0\,, \qquad F_y + F_{r\_1} + F_{r\_2} = 0$$
$$\sum F_z = 0\,, \qquad F_z + F_{thrust} = 0$$
$$\sum M = 0\,, \qquad F_z PCR + F_y \frac{S}{2} + F_{r\_1} l_{r\_1} + F_{r\_2} l_{r\_2} + F_{thrust} r_{thrust} = 0$$

The force on each of the radial bearings of both end of the shaft is as follows:

$$F_{r\_2} = \frac{(PCR - r_{thrust})F_z + \left(\frac{S}{2} - l_{r\_1}\right) F_y}{l_{r\_1} - l_{r\_2}}$$
$$F_{r\_1} = -(F_y + F_{r_2})$$

The compressor power by the frictional resistance of the radial bearings is indicated by the rotational angular velocity and the friction torque.

$$T_{r\_1} = \mu_{r\_1} r_{r\_1} F_{r\_1}$$
$$T_{r\_2} = \mu_{r\_2} r_{r\_2} F_{r\_2}$$
$$T_{thrust} = \mu_{thrust} r_{thrust} F_{thrust}$$
$$T_{total} = T_{r\_1} + T_{r\_2} + T_{thrust}$$

Thus, the compressor power by the bearings is as follows:

$$P_{bearing} = 2\pi \frac{N}{60} T_{total}$$

Therefore, the shaft power of the compressor can be determined by:

$$P_{shaft} = P_f + P_{bearing} + P_{ring} + P_{comp}$$

The compressor input power can be determined by the shaft power and the efficiency of the electric parts(inverter and motor). Then, the compressor input power is derived as follows:

$$P_{input} = \frac{P_{shaft}}{\eta_{elec}}$$

Bin Yang [7] is to calculate the efficiency of the motor by Hubacher's study[8]. It was proposed to the iterative method by the input power. However, this equation can't be applied to any motors. In this study for a more accurate prediction, the power loss by the efficiency of the electric parts can be determined by the dynamo test results.

3. Dynamo Test

Fig. 2 is a device configuration for the dynamo test of the motor and the inverter. The equipment has the dynamometer(Magtrol, 1WB65) with accuracies of $\pm 0.3\%$ respectively and the controller for measuring a torque and giving a load on the test motor. It is installed the test inverter to drive the test motor. And the can board and the computer to translate the can communicate instructions and responses was installed. The power meter was installed to measure the power before and after the inverter with accuracies of $\pm 0.1\%$ respectively. It can calculate the efficiency of the inverter, and it is possible to calculate the efficiency of the motor from the dynamo test result. The dynamometer is measured the torque, the phase current and the phase voltage of the test motor according to the load at constant speed.

Dynamometer test method is as follows. The motor and inverter to provide for the test is drive at a constant rotational speed. The dynamometer increases the torque by the break. This acts as a load on the motor. It measures the input power of the motor according to increase of the load at a constant rotational speed. The torque to be applied in the dynamometer is the shaft power of the compression section. Therefore, we can calculate the efficiency of the motor and inverter according to the various shaft load at each rotational speed.

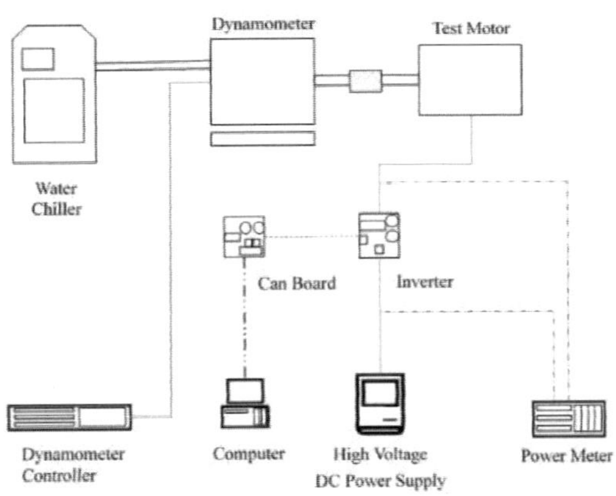

Fig. 2 Schematic of dynamo test rig

The dynamo test results can be represented by the efficiency of the electric parts according to the shaft power at the rotation speed. In case of low speed, the efficiency of the motor and inverter was lower. The efficiency is measured up to 78.9% at 2000rpm. And it is measured up to 84.4% at 3000rpm. As the increase of the rotational speed, its efficiency is increased. In more than 6000rpm measured more than 90% efficiency.

The efficiency of the electric parts was approximated by the following 4th order polynomial equation. Table 2 is constants of the 4th order polynomial equations which approximates the shaft power and the efficiency of the electric parts according to each rotational speed.

$$\eta_e = a + bP_{shaft} + cP_{shaft}^{\ 2} + dP_{shaft}^{\ 3} + eP_{shaft}^{\ 4}$$

Therefore, the efficiency of the electric parts can be calculated by the compression power (shaft power) in the calculated performance analysis of the mechanical parts. The total power of the electric driven compressor can be calculated in consideration of the electric parts efficiency. Thus, the compressor performance contributed more accuracy by using the electric parts efficiency.

Table 2. Constants of the efficiency equation for the electric parts

Speed [rpm]		2000	3000	4000	5000	6000	7000	8000	8600
Constants	a	6.6933E-01	7.3146E-01	6.1707E-01	7.0600E-01	5.4996E-01	6.6170E-01	6.1125E-01	5.4863E-01
	b	4.7397E-04	2.3404E-04	4.4717E-04	2.3303E-04	4.7095E-04	2.5128E-04	3.3320E-04	3.9498E-04
	c	-6.1928E-07	-1.7103E-07	-2.8080E-07	-1.0872E-07	-2.2525E-07	-9.7034E-08	-1.4086E-07	-1.6658E-07
	d	3.1113E-10	5.1302E-11	7.4179E-11	2.2118E-11	4.5657E-11	1.6752E-11	2.6103E-11	3.0924E-11
	e	-5.7696E-14	-5.9301E-15	-7.1889E-15	-1.7049E-15	-3.3388E-15	-1.1062E-15	-1.7830E-15	-2.1119E-15

4. Experimental equipment and electric driven compressor using R744

To evaluate the performance of the proto type compressor using the R744 was constructed hot-gas cycle. A schematic of the test rig for R744 compressor is shown in Fig. 3. The electric driven compressor is used which includes an AC/DC converter for operating the compressor. The experimental setup utilized thermocouples, pressure transducers, mass flow meter for refrigerant and oil, another mass flow meter for separated oil and power meter with accuracy of $\pm 0.1\text{K}$, $\pm 0.33\%$, $\pm 0.25\%$, $\pm 0.25\%$, $\pm 0.3\%$ respectively.

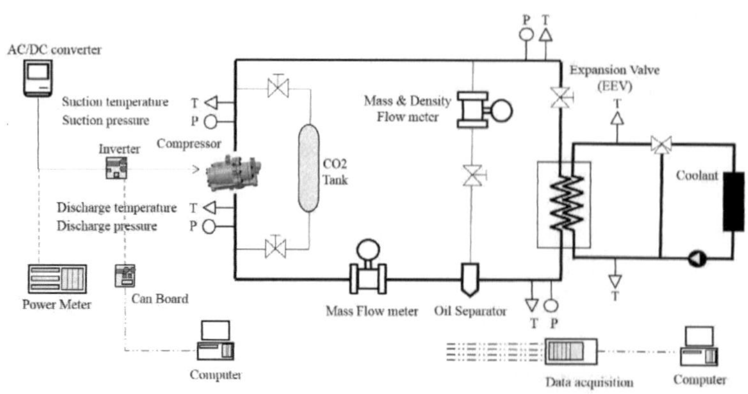

Fig. 3 Schematic of the hot-gas cycle for the test of compressor

Test conditions is fixed at suction pressure 3.92MPa, suction temperature 30℃. Discharge pressure is changed at 7.85, 9.81, 11.77MPa. And discharge temperature, mass flow rate, and input power of compressor are measured by the each sensor.

The compressor using the test is new developed proto compressor of the electric driven compressor for CO_2. The specifications of the compressor used in the experiment were shown in the table 3. The development of a fixed swash plate type compressor is a conventional compression method for using the vehicles. The compressor was equipped with a speed-adjustable BLDC motor. The inverter for motor driving was installed outside to the compressor. It is made up of three pistons and stroke volume of 4.7cc/rev. It's top speed is 8600rpm. In order to prevent leakage for the use of high-pressure refrigerant, it applied the piston ring. The clearance controller was installed in the compressor to consider the effects of dead volume in the small capacity. This can minimize the influence of dead volume because of the structure to adjust the top clearance. It was placed in a thrust bearing with motor rearward to support the axial load of the shaft by the high-pressure refrigerant in the cylinder. Lubricating oil for the compressor is applied the PAG oil. An inverter for driving a motor was used as the high-efficiency inverter applying the maximum torque control. The motor and inverter efficiency through a dynamo test was measured previously.

Table 3 Specifications of electric driven compressor for CO_2

Model	DEC47
Type	Electric driven compressor (3 Piston)
refrigerant	R744 (CO_2)
Capacity	4.7 cm^3/rev
Motor	BLDC motor (max. 7N.m)
Input Voltage	330VDC
Max. Speed	9,000 rpm

5. Comparison of numerical and experimental results

An electric driven CO_2 compressor with fixed swash plate was tested using the hot gas system according to the pressure ratio and the speed of compressor. The experimental results are compared to the theoretical analysis results.

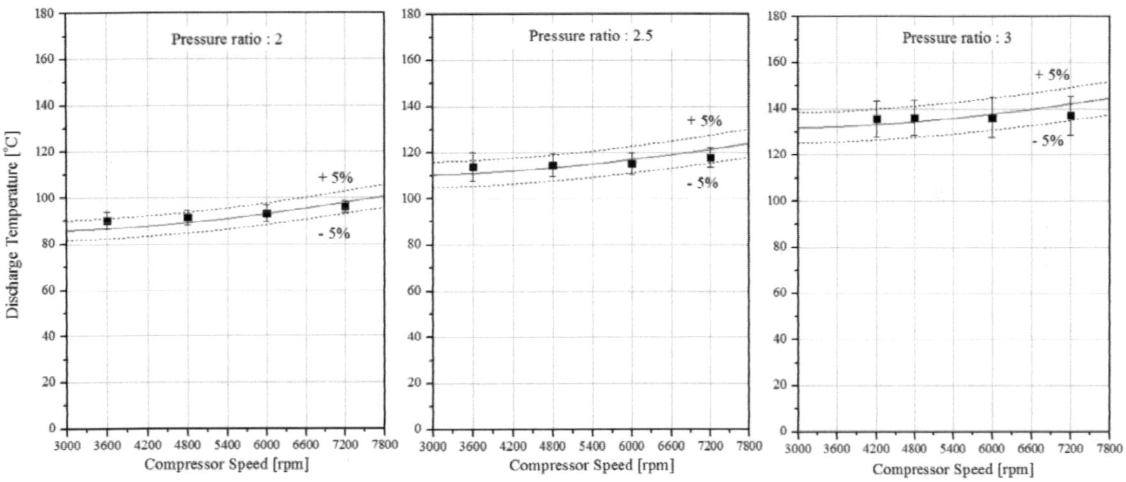

Fig. 4 Comparison of experiment and predicted discharge temperature

Fig. 4 show the deviations between measured values and predicted values for the discharge temperature of the compressor. The red solid line is predicted value, the blue dotted line represents an around of the analysis result of ±5%. And black squares with error bar are experimental results. The suction pressure and temperature is 40kgf/cm^2 and 30°C. The discharge pressure is 80, 100, 120kgf/cm^2. The discharge temperature increase with the pressure ratio. The analysis model predicts the discharge temperature with the error percentage of ±5%.

Fig. 5 show the deviations between measured values and predicted values for the mass flow rate of the compressor. The mass flow rate increase with the speed of compressor and the pressure ratio. The analysis model predicts the discharge temperature with the error percentage of ±5% too.

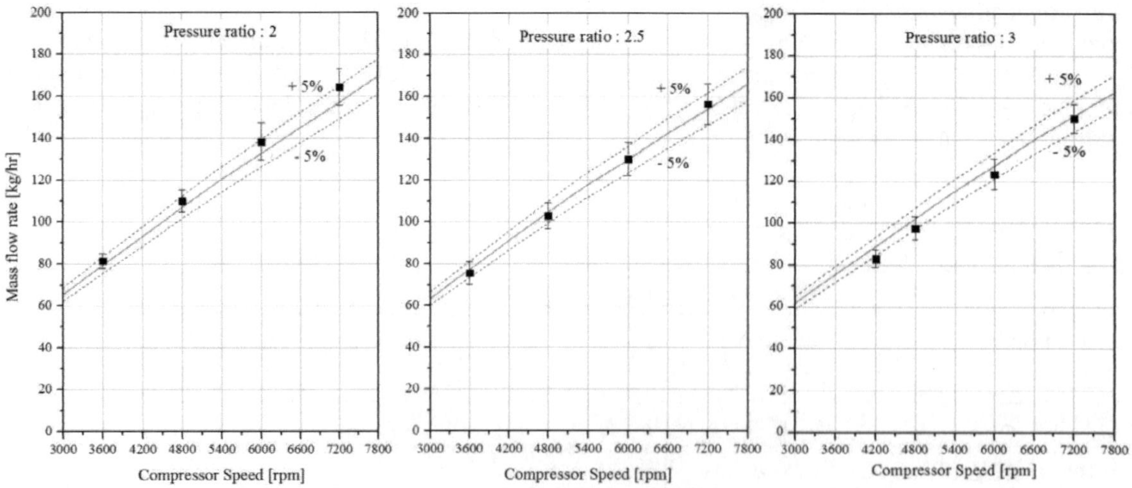

Fig. 5 Comparison of experiment and predicted mass flow rate

Fig.6 show the deviations between measured values and predicted values for the compressor input power. And it show the result to calculate the efficiency of the motor by Hubacher's study[8]. The input power of compressor increase with the speed of compressor and the pressure ratio. The shaft power by the electric parts transforms a portion of the compressor input power. As the comparison results, the exact efficiency of the electric parts for electric driven compressor can provide more accurate prediction. The predicted input power of the compressor agrees quite well with the experimental value, as shown in Fig. 6.

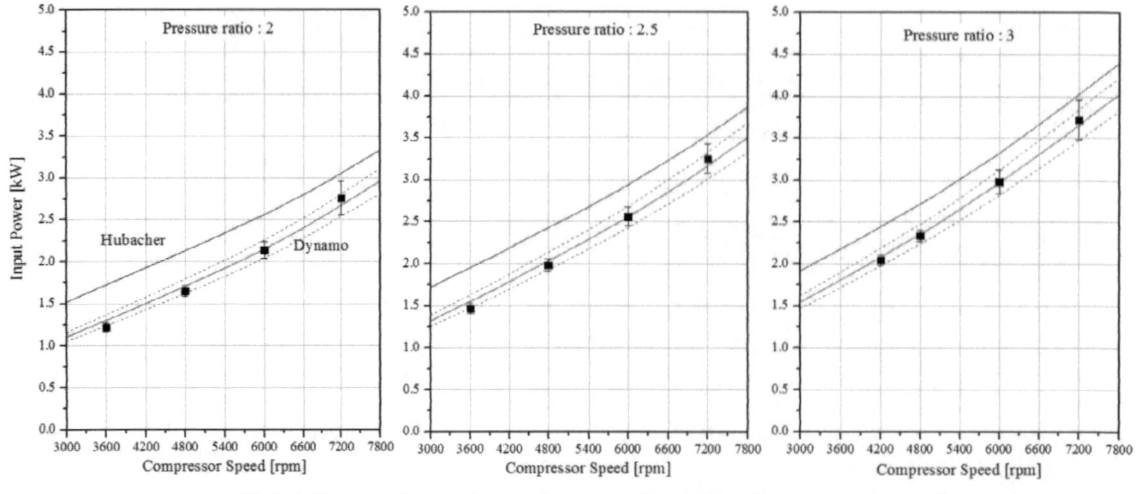

Fig. 6 Comparison of experiment and predicted input power

The accurate prediction result is because the correct efficiency of the electrical components in the compressor by measurement with the load of the compression unit and the rotational speed of compressor through the dynamo test. In general, the industry using a BLDC motor and inverter applied the efficiency of 90% and 95%. In past studies often apply a simple one formula. But the BLDC motor and inverter causes the larger error when interpreting this one formula because it is designed for maximum efficiency out at the design speed and load. In the case of the compressor for the vehicles, the motor is designed to have high efficiency come at a rotational speed of 6000 ~ 8000rpm, shaft power of 2000 ~ 4000W. This is designed in accordance with the driving conditions of the electric driven compressor within the vehicles.

6. Conclusions

A comprehensive simulation model for an electric driven compressor using CO_2 for eco-friendly vehicle is presented. This model consists of a compression model and dynamic model. Compression model included valve, leakage, and heat transfer models. And Dynamic model included frictional loss between piston ring and cylinder wall, frictional loss between shoe and swash plate, and frictional loss of bearings. Especially, the efficiency of an electric parts(motor and inverter) in the compressor affects the loss of the compressor.

In this study, for more accurate predictions, its efficiency was formulation through dynamo test. The efficiency of the electric parts was approximated by the following 4th order polynomial equation. It is constants of the 4th order polynomial equations which approximates the shaft power and the efficiency of the electric parts according to each rotational speed. The efficiency of the electric parts can be calculated by the compression power (shaft power) in the calculated performance analysis of the mechanical parts. The total power of the electric driven compressor can be calculated in consideration of the electric parts efficiency. Thus, the compressor performance contributed more accuracy by using the electric parts efficiency.

References

[1] Lorentzen, G., Pettersen, J., 1992. New possibilities for non-CFC refrigeration. In: Proceedings from International Symposium on Refrigeration, Energy and Enviroment, Trondheim, 147-163.

[2] Lorentzen, G., 1994. The use of natural refrigerants: a complete solution to the CFC/HCFC predicament. International Journal of Refrigeration, 18, 190-197

[3] Brown, J. S., Yana-Motta, S. F., Domanski, P. A., 2002. Comparitive analysis of an automotive air conditioning systems operating with CO_2 and R134a. International Journal of Refrigeration, 25, 19-32.

[4] Groll, E. A., Kim, J. H., 2007. Review of Recent Advances toward Transcritical CO_2 Cycle Technology. HNAC & R Research, Vol 13. 3, 499-520.

[5] Lee, M. Y., Lee, H. S., Won, H. P., 2012. Characteristic Evaluation on the Cooling Performance of an Electrical Air Conditioning System Using R744 for a Fuel Cell Electric Vehicle. Energies, 1371-1383.

[6] Lee, H. S., Lee, M. Y., 2013. Cooling Performance Characteristics on Mobile Air-Conditioning System for Hybrid Electric Vehicles. Advances in Mechanical Engineering. 1-9.

[7] Yang, B., Bradshaw, C. R., Groll, E. A., 2013. Modeling of a semi-hermetic CO_2 reciprocating compressor including lubrication submodels for piston rings and bearings. International Journal of Refrigeration, 36, 1925-1937.

[8] Hubacher, B., 2003. Experimental and Theoretical Performance of Carbon Dioxide Compressors. M.S. thesis, Ray W. Herrick Laboratories, Purdue University, Indiana, USA.

[9] S. Furuhama, T. Tada, 1961, On the flow of gas through the piston-rings(1st Report, The discharge coefficient and temperature of leakage gas), The Japan Society of Mechanical Engineers, Vol. 4, No. 16, 684-690.

[10] Lin, M., Sun, S. Y., 2006, Theory of the Piston Compressor, Xi'an Jiaotong University Press, China.

[11] F.P. Incropera, D. P. Dewitt, 1996, Fundamentals of Heat and Mass Transfer, Fourth Edition, WILEY.

[12] F. P. Disconzi, C.J. Deschamps, E. L. L. Pereira, 2012, Development of an In-Cylinder Heat Transfer Correlation for Reciprocating Compressors, 2012 International Compressor Engineering Conference at Purdue, 1342.

Technology's present situation and the development prospects of energy efficiency monitoring as well as performance testing & analysis for process flow compressors

L Li[1], Y Zhao[1,2], L Wang[1,2], Q Yang[1,2], G Liu [1,2], B Tang [1] and J Xiao[1,2]

[1]State Key Laboratory for Compressor Technology, Hefei General Machinery Research Institute, Hefei 230031, Anhui Province, China

[2]Hefei General Environmental Control Technology Co., Ltd., Hefei 230031, Anhui Province, China

Abstract. In this paper, the background of performance testing of in-service process flow compressors set in user field are introduced, the main technique barriers faced in the field test are summarized, and the factors that result in real efficiencies of most process flow compressors being lower than the guaranteed by manufacturer are analysed. The authors investigated the present operational situation of process flow compressors in China and found that low efficiency operation of flow compressors is because the compressed gas is generally forced to flow back into the inlet pipe for adapting to the process parameters variety. For example, the anti-surge valve is always opened for centrifugal compressor. To improve the operation efficiency of process compressors the energy efficiency monitoring technology was overviewed and some suggestions are proposed in the paper, which is the basis of research on energy efficiency evaluation and/or labelling of process compressors.

Key words： Process flow compressor; performance test; energy efficiency monitoring; situation and perspectives

1. Introduction

Process flow compressors are key equipment of processing industries such as petrochemical engineering, coal chemical engineering, etc., and have huge shaft power and flux, complex pipe net and variable working conditions and gas components, which result in much difficulty in the performance testing before ex-factory and energy efficiency evaluating at user field. Processing compressors in petrochemical, coal chemical and metallurgical industries, for examples, the hydrogen compressor in oil refining, the cracking gas, propylene and ethylene compressors in ethylene production, the blower in blast-furnace and the large scale of axial compressor in air-separation system, consume huge energy and have lower operation efficiency [1].

Four types of compressor, reciprocating, screw, axial and centrifugal, are mainly used in the processing industry, and their routine tests in the manufacture's factory are very difficult because of the larger rate of flow and shaft power, complex pipe line, multiple gas constituents and their feature of inflammable, easy explosion and the toxicity. Normally the delivery test only include mechanical running, impeller accelerating and seal trial [2]. So, the compressor performance curve from manufacturer do not reflect their real quality.

However, users need real performances of processing compressors to meet the demand of flow parameter variety. Reliability, higher efficiency and long term running are basic demands of user to processing compressor, but in real operation more attentions are paid to its safety. For an example, in the surge control system of centrifugal compressor the anti-surge valve is generally opened in advance and the compressed gas is forced to flow back to inlet pipe so that the operation efficiency of compressor set is decreased. On the premise of safe and reliability how to keep the compressors operating at higher efficiency at wider range of working conditions is main developing direction of process compressor technologies, to rate real efficiency of process compressor, to grade compressors and to constantly monitor their performances for energy saving operation are also the key point of development.

In this paper it will be introduced that the present situation and development trend of performance worksite test and energy efficiency monitor technique of processing compressors. Take centrifugal compressor as example, the reasons of performances difference between user field and manufacturer factory, surge control and energy consume increase are analyzed. Relative technologies of process compressor will be discussed in the paper too.

2. Performance curves and background of user field test of process compressor as well the results analysis

Energy efficiency evaluation of compressor is basis of energy saving and carbon emission reduction, and also an effect method of impartial assessment of real operation efficiency of compressor set. For general power and refrigeration compressor, the manufacturer and/or the third part can test their real performances at varies operation conditions with different working fluids by setting up test rig because of smaller compressor power, easier operation condition adjustment and simpler gas component. So, similar to the energy efficiency standards of power and refrigeration compressor of Japan, EU and US [3], China also drawn herself energy efficiency standards and began energy efficiency identification of product [4].

For process compressor, so far, there is no complete energy efficiency identification standard in the world because exist many indeterminacy factors in the performance test. The match of compressor performances with parameter change of the process system is a big problem, which results in lower operation efficiency of compressor set. So, it is significant to research performance test and energy efficiency monitoring technique, and to draw energy efficiency standards of process compressor.

2.1. Theoretical performance curves of centrifugal compressor

Delivery quantity, discharge pressure (pressure ratio), power and so on, are characteristic performance parameters of compressor, and are also the design basis of product. Performance curve of centrifugal compressor depend on four relative parameters, the inlet flow rate *Q*s、 the inlet pressure *p*s, the inlet temperature *T*s and the rotation speed of compressor *n*. At constant inlet states (*p*s, *T*s) and *n*, the stage pressure ratio ε and polytrophic efficiency η_{pol} of centrifugal compressor vary with *Q*s, $\varepsilon = f_\varepsilon(Q_s)$ and $\eta_{pol} = f_n(Q_s)$,which can be gotten by numerical analysis and/or theory calculation. The polytrophic efficiency η_{pol} is calculated using equation (1) below [5].

$$\eta_{pol} = \frac{\varepsilon^{(k-1/k)} - 1}{\varepsilon^{(m-1/m)} - 1} \quad (1)$$

Where *k* and *m* are isentropic index and polytrophic exponent respectively, and ε is pressure ratio.

Change trend of theoretical performance curves of centrifugal compressor is shown in figure 1(a). Vary of pressure ratio and efficiency at certain rotation speed with inlet flow rate can be gotten. At certain rotation speed to decrease inlet flow rate the pressure ratio gets large, and as the inlet flow rate is reduced to some value the surge condition of compressor set occurs. Conversely, at certain rotation speed to decrease pressure ratio the inlet flow rate gets large, and as the pressure ratio is reduced to some value the blocking condition of compressor set occurs. The conditions between surge and blocking make sure safety operation of centrifugal compressor.

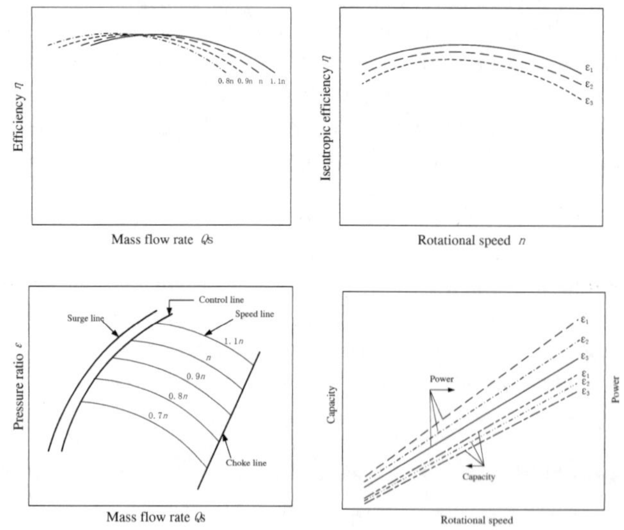

(a) (b)

Figure 1. Performance curve of compressor(a) centrifugal (b) piston and screw

For positive displacement compressor, for example, the piston and screw compressors, whose theoretical efficiency is expressed as isentropic efficiency. The efficiency, power and delivery quantity (inlet flow rate) of piston and screw compressors also vary with rotation speed of compressor, see figure 1(b). At certain pressure ratio the maximum efficiencies of piston and screw compressors correspond to some rotation speed, but nothing to do with delivery quantity, so, the surge and blocking conditions do not occur for positive displacement compressor.

2.2. User field test and results analysis of process compressor

User field test of process compressor not only may get their real operation performance datum to resolve conflict between users and manufacturers or vendors, but provides references for manufacturer to improve product design to adapt real high efficiency and reliability operation demand of compressor.

M Sib Akhtar introduced a Joint Industry Project entitled "Pointing to the optimum performance of centrifugal compressors for oil & gas"[6]. In the project, design and operating data of more than 50 compressors and drivers were obtained, performance and history of the machine was analyzed over an 18-month period, and real performances varying with working conditions was gotten. For simplicity, the pointes expressing efficiency at different operation conditions in Ref. 6 were deleted, only fitted curves were retained, see figure 2.

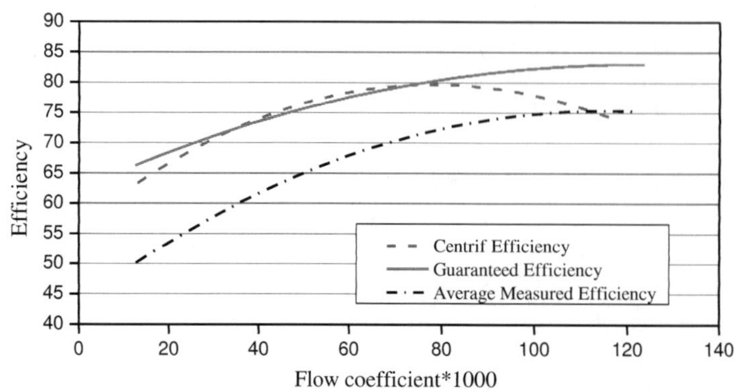

Figure 2. Comparison of compressor performance between operating and guaranteed

From figure 2 we can see that the real efficiencies of compressors are always lower than the values provided or guaranteed by manufacturer, the deviation becomes greater at small flow coefficient. The user field test results prove the significance and difficulty of design and manufacture of centrifugal compressors at small flow coefficient.

Change trends of theoretical efficiency curve of centrifugal compressor in figure 1(a) and figure 2 are alike, which have a little better coincidence with the efficiency change guaranteed by manufacturers especially at large flow coefficient. In other words, it is impossible for manufacturers to provide real operation efficiency of compressor sets because their test in factories are difficult.

Authors of this paper investigated several of petrochemical enterprises in China such as oil refining factory, where there are also performance deviation between compressor real operation and manufacturer's guarantee. To let compressor run safely real working conditions of compressor are usually far away the surge line. Running chart of compressor of an 800 thousand t/a ethylene plant in Shandong is shown in Figure 3, in which red line and blue line are separately the surge line and surge control line of pyrolysis gas centrifugal compressor. Because real operation point of compressor set, the black point in the figure, always keeps away from the surge control line by opening anti-surge valve the operation reliability of compressor is improved but much power be consumed owing to the compressed gas flowing back into inlet pipe.

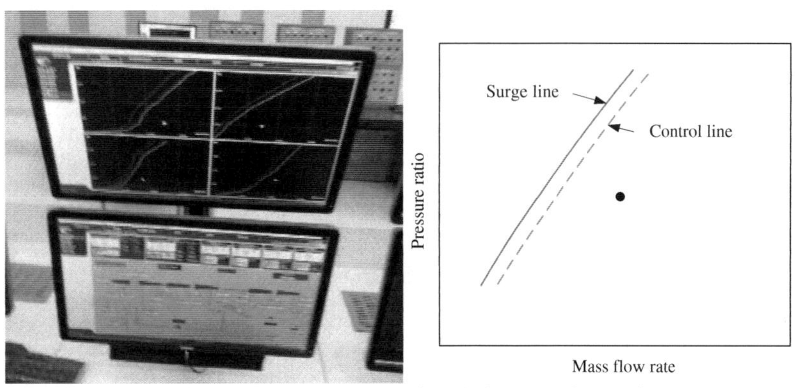

Figure 3. Running chart of compressor of 800 thousand t/a ethylene plant in China

Real operation performance curves of centrifugal compressor not only influence operation reliability of compressor set, for an example, the surge working condition will occur in advance as real flow rate is less than its demand at design conditions, but have a big effect on the energy consumption of compressor set, for instance, much more power is needed because real operation efficiency is lower than the designed value. Data from China petroleum and petrochemical energy saving monitoring centre is shown in table 1.

Table 1. Design efficiency and real operation efficiency of process compressor.

Flow Compressor name & model	Equipment	Design efficiency (%)	Operation efficiency (%)	Main reason
Turbo compressor CT5102	hydrofining	70	32.22	Lower compressor operation load, for instance, the flow rate of C1101 is 61%, the C5102 is lower.
Compressor C5102		75	55.98	
Turbo compressor CT1101	hydrocracking	70	64.71	
Compressor C1101		75	60.09	

2.3. Difficult points of user field test of process compressor and discuss of energy efficiency labeling

The direct object of user field test of process compressor is to verify if the compressor performances guaranteed by manufacturers could be reached at design working condition and real gas operation, moreover, to provide users the actual performances curve of centrifugal compressor including curve trend, working condition points corresponding to smallest and largest flow rate for operation control and adjustment.

User field test of process compressor is different from lab or factory workshop test, field test is difficult and test result is uncertainty. Firstly, design of test system placing is subject to field environment, system parameters selection like that in laboratory is impossible. Pipe placing, operation condition change and gas compositions and so on will influence the accuracy of instrument installment and test, so, the special test plan for particular user field is essential. Secondly, it is difficult to choose the test instrument suited to the variable operation conditions. In addition, severe field environment and maintenance & calibration measure let measuring apparatus being damaged more easily, which influence test precision accuracy.

Field performances test is basis of efficiency evaluation and energy rating label of process compressor. The third part is responsible for the test and needs wide consultation on test plan from manufacturers, users and designing institute, on which typical compressor sets from petrochemical and metallurgy flow equipment can be selected for test, for examples, centrifugal and reciprocating compressor in oil refining factory, the axial compressor in the blast furnace, etc. The performances of the elected compressor sets can be tested according to ASME PTC 10 Performance Test Code on Compressors and Exhausters[7], the main parameters include inlet and outlet pressure and temperature, flow rate, gas compositions, speed, power and so on.

Main thermal parameters of process compressors can be gotten by field test and the performance parameters such as pressure ratio and efficiency of centrifugal compressor can be calculated, on which the curves of Qs-ε and η_{pol}-Qs at given speed can be plotted. At same time, about one year operation data of the compressor sets were recorded and sorted out, which is reference value of efficiency evaluation and energy rating label, see figure 4.

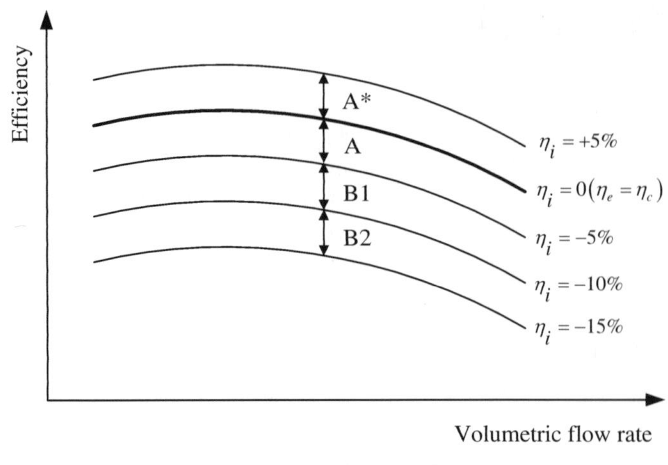

Figure4. Grading of energy efficiency of process gas compressor

For the grading of energy efficiency of process gas compressor shown in figure 4, the fundamental performances curve should be confirmed and approved by vast majority of users and manufacturers first, then energy efficiency grad is formulated according to the deviation between real performance and fundamental performance, which is expressed in equation (2).

$$\eta_i = \left(\frac{\eta_e - \eta_c}{\eta_c}\right) \times 100\% \tag{2}$$

Where, η_i is the efficiency corresponding to the energy efficiency grad of compressor set, η_c is fundamental efficiency, and η_e is real operation efficiency. The percentage of η_e deviated from η_c can be calculated by equation (2), on which the energy efficiency grading could be done, for examples, -5% $\leqslant \eta_i < 0$, labeling as A; -10% $\leqslant \eta_i <$ -5%, labeling as B1; -15% $\leqslant \eta_i <$ -10%, labeling as B2, etc., If 0 % $\leqslant \eta_i <$ 10%, labeling as A*.

In the long run, as the development of networking and intellectualization technique it is possible to do efficiency evaluation and energy rating label on the basis of on line monitoring of real operation performances of process compressor set.

3. Energy efficiency monitoring technique of process compressor

3.1. Present situation of energy efficiency monitoring

Compressor Controls Corporation developed a kind of instantaneous monitoring software CPA [8], the monitoring results on trend of performances change and deterioration of compressors can be displayed in the remote monitoring station window connected to CCC server. CPA may monitor gas pressure and temperature in inlet and outlet of compressors, flow rate and velocity, gas compositions and so on, create models of pressure ratio ε 、energy head H_{pol}、efficiency η_{pol} and energy J with flow rate Q_s 、speed n and gas compositions, for instance, $\varepsilon = f(Q_s, n)$ 、 $H_{pol} = f(Q_s, n)$ 、 $\eta_{pol} = f(Q_s, n)$ 、 $J = f(Q_s, n)$, calculate, monitor and record production capacity and efficiency that reflect performance parameters of compressor operation, and track the deviation situation and trend between real performances and the date from health state model and if it reaches the set threshold alarm will sound.

US OnSet Corporation summarized the defects of now available control system for compressor set [9], the efficiency control problem of application together of positive displacement and turbo compressors, the effect integration of different bands of compressor products, the advanced algorithm for best compressors combination, the complete performance monitoring, and the accurate surge condition forecast. For process compressor the indirect monitoring of power can be realized by direct testing of electric current, but more precise power calculation is necessary.

Ted Gresh introduced an on-line performance monitoring system [10], by which the health condition of compressor can be evaluated by comparing instant performance data with the parameters from factory. The performance calculation software they developed can be used to calculate the input power by using the original parameters such as pressure, temperature, speed, and flow rate of compressor. Operation state change of compressor could be monitored by the records for some time, which is used for specified maintenance and trouble founding, see figure 5.

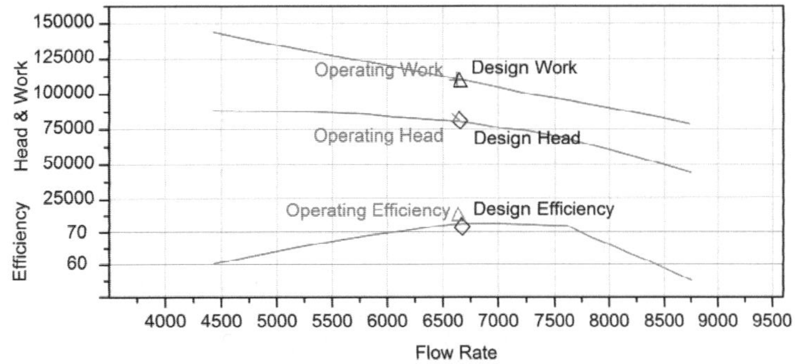

Figure5. Comparison of compressor performance between design and actual in Flexware company

A. Suomilammi and J. Leppäkoski [11] introduced another on-line performance monitor system with fuzzy classification and merging technology. System can show the desired value by using fuzzy clustering prediction model as well acquisition of history data of outlet pressure, temperature and speed of compressor, and as the performance of compressor set seriously deviates from the desired value the maintenance order will be touched off.

ITCC, Integrated Turbine Compressor Control, is a new technique based on advanced technologies such as electronic, communication, control algorithm, and so on. Nowadays, the main functions of integrated control systems for large scale of process compressors include unit interlock ESD (Emergency Shutdown Device), SOE(Sequence of Events) recording, unit control PID (for examples, anti-surge control and speed adjustment control), and routine instruction record, fault diagnosis, etc. Widely used control system in the world include Micro Net TMR of Woodward Corp., Tricon of Triconex Corp., TMR of CCC, FSC of Honeywell, GMR of GE, Trusted of ICS Triplex, and H41/H51 of Hima. Speed adjustment control, process control and compressor control in conventional control system are separate and provided by different venders, while three function control modules and their interrelationship are utilized thoroughly in the ITCC, the positive actions are enhanced and negative actions are eliminated, on which the performance integrated control of increasing operation efficiency and reliability of compressor sets, saving energy consume and fund and minimizing operation and maintenance cost can be realized to the greatest extent. ITCC can decouple anti-surge control returns, and an application case of ITCC in propylene compressor is shown in figure 6 [12]. From it we can see that the compression process of compressor comprises of six sections with each anti-surge control return and share a shaft, and scope of stable operation working conditions can be widened by move left of the new surge curve.

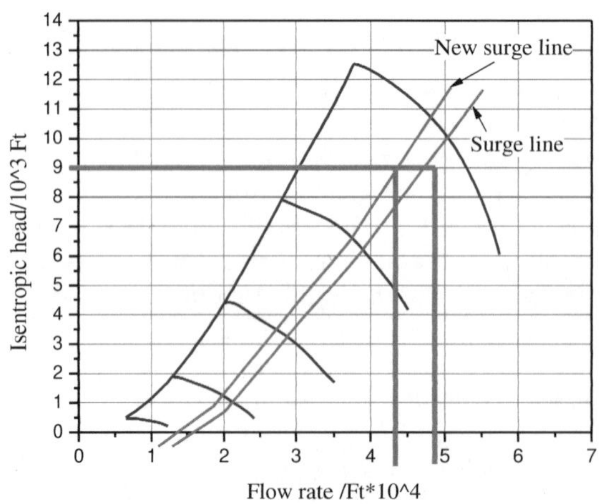

Figure 6. The surge line control of propylene compressor

Irwin Stambler [13] introduced a complex array intelligent tools by which the general health situation of gas turbo system with single cycle and combination cycle can be inspected from points of mechanics and thermodynamics. For compressor performances the operation condition can be assessed by comparison of real data with desired one.

Under the assistance of fund from Chinese Academic of Engineering and National Nature Science Foundation of China authors participate in the present operation situation investigation of compressor sets in petrochemical and metallurgy industries, and mainly focus on sets heath operation and energy efficiency monitoring [14]. Most of large scale of centrifugal compressor and piston compressor sets in the process device in China have been installed on-line operation state monitoring system but no energy efficiency monitoring system.

3.2. Development prospects of Energy Efficiency Monitoring

Energy efficiency monitoring is an important measure to ensure high efficiency operation of compressor sets. The technique development trends of energy efficiency monitor system include: (1) to synchronously test much more parameters, to set reference data base, to create real performance curves of process compressors on the basis of history operation data, and to realize the function parameters such as the power visible and adjustable; (2) to centrally regulate and control a group of compressor sets, to distribute load utilizing intelligent algorithm to ensure all compressor sets in optimum operating condition. To improve operation efficiency of compressor sets and to achieve energy efficiency maximization by depth integration optimization of compressor sets and process system; (3) to increase system efficiency of compressor set by regulating and controlling its auxiliary system, for instance, to adapt the effect of process parameters change on the operation condition by adjusting and controlling intercooler, shaft seal, and so on; (4) to keep compressor sets always run nearby the operation condition with highest efficiency by accurate predict, for an example, to push the surge control line outward the small flow interval by taking different monitoring methods so that stability margin can be widened and energy save operation can be realized for the process compressor sets.

Besides the key points mentioned above, the technique development of energy efficiency monitor of process compressors should consider these factors: (1) the compressed gas leakage result in direct power loss and leakage of process gas probably can cause emergent fatal accident such as explosion, fire, etc., so, to monitor leakage and prevent leak are not only for energy saving but for safe; (2) for noisy working environment it is exigent for developing intelligence detection apparatus to detect leakage point. To accurately fast find the leaking place it is necessary to rationally plan the exploration path and to clearly record the detection place.

4. Conclusions

(1) Full performances test of process compressor in manufacturer workshop is difficult because of variety of the real operation conditions and gas compositions, which results in performance deviation between real operation and the guaranteed data by manufacturers. For centrifugal compressor the performance deviation at small flow rate is large, so manufactures should pay more attention to the design and improvement of its prediction at small flow rate.
(2) Real performances curves of process compressor with operation condition and gas composition can be gotten by the user field test. On basis of test data energy efficiency labeling system can be planned, which is also used by manufacturers to improve compressor set performances and to better meet the need of flow parameters and users.
(3) Energy efficiency monitoring is an effective supplement of safe monitoring of process device. The regulation and control level of process compressor sets operating at full working conditions can be increased by energy efficiency and operation state monitoring so that the compressor operation nearby surge line is also reliable and the power loss owing to anti-surge valve open can be greatly decreased.

5. Appendices

H_{pol}	Energy head	J	Energy, W
k	Isentropic index	m	Polytropic exponent
n	Speed, r/min	ps	Inlet pressure, kPa
Q_s	Flow rate, m^3/min	Ts	Inlet temperature, °C
ε	Pressure ratio	η_{pol}	Polytropicefficiency, %

6. References

[1] Gao Jinji 2013*Research on the current situation and energy saving countermeasures of China's high energy consumption machinery and equipment* (Beijing: Science Press)pp174-188.
[2] API Standard 617 2002 *Axial and centrifugal compressors and expander-compressors for petroleum, chemical and gas industry services* (Washington: American Petroleum Institute)
[3] Standard, ANSI/AHRI 510-2006 *Performance rating of positive displacement ammonia compressor and compressor unit.* (Arlington: Air-Conditioning, Heating and Refrigeration Institute)
[4] GB19153-20092009 *Minimum allowable values of energy efficiency and energy efficiency grades for displacement air compressors* (Beijing: China Standards Press)
[5] Xu Zhong 1998*Principle of centrifugal compressor* (Xi'an: Xi'an Jiaotong University)

[6] Akhtar M S 2006 Determining the real performance of centrifugal compressors operating in oil & gas production facilities *Ninth European Fluid Machinery Congress* Institution of Mechanical Engineers (Hague:Chandos)
[7] ASME PTC 10 1996 *Performance Test Code on Compressors and Exhausters* American Society of Mechanical Engineers (New York: American Society of Mechanical Engineer)
[8] www.cccglobal.com/sites/default/files/documents/CPA_Product_Brochure.pdf
[9] www.energycontroltechnologies.com/docs/ECT-RA-PSUG-Plant-Air-Presentation.pdf
[10] Ted Gresh 2005 Online performance monitoring *Turbomachinery*. **12**28
[11] SuomilammiA andLeppäkoski J 2006 Gas compressor unit performance monitoring using fuzzy clustering *23rd World Gas Conference* (Amsterdam: International Gas Union)
[12] www.ge-energy.com/occn
[13] Irwin Stambler 2006 Diagnostic health monitoring will reduce new technology O&M risk. *Gas turbine world* **22**28
[14] Hefei General Machinery Research Institute 2015 *Countermeasures Research report of energy efficiency monitoring and intelligence of high-end compressor in petrochemical and metallurgy industry* (Hefei:Hefei General Machinery Research Institute)

Acknowledgments
This work was financially supported in part by National Natural Science Foundation of China (51406047, 51405130)and Chinese Academic of Engineering Consulting Project.

Carbon and Energy Saving Financial Opportunities in the Industrial Compressed Air Sector

Diego Vittorini, Roberto Cipollone

University of L'Aquila, Department of Industrial and Information Engineering and Economics, L'Aquila, Italy

diego.vittorini@graduate.univaq.it

Abstract. The transition towards a more sustainable energy scenario calls for both medium-to-long and short term interventions, with CO_2 reduction and fossil fuel saving as main goals for all the Countries in the World. Among all others, one way to support these efforts is the setting-up of immaterial markets able to regulate, in the form of purchase and sales quotas, CO_2 emissions avoided and fossil fuels not consumed. As a consequence, the upgrade of those sectors, characterized by high energy impact, is currently more than an option due to the related achievable financial advantage on the afore mentioned markets. Being responsible for about 10% electricity consumption in Industry, the compressed air sector is currently addressed as extremely appealing, when CO_2 emissions and burned fossil fuels saving are in question. In the paper, once a standard is defined for compressors performances, based on data from the Compressed Air and Gas Institute and PNEUROP, the achievable energy saving is evaluated along with the effect in terms of CO_2 emissions: with reference to those contexts in which mature intangible markets are established, an estimation of the financial benefit from savings sale on correspondent markets is possible, in terms of both avoided CO_2 and fossil fuels not burned. The approach adopted allows to extend the analysis results to every context of interest, by applying the appropriate emission factor to the datum on compressor specific consumption.

1. Introduction

The need for sustainable socio-economic models along with the awareness of the environmental risks related to present development paradigm are currently expressed in terms of fossil fuel consumption and CO_2 emissions reduction by policy makers of all World Countries. According to most recent data, up to 90% final energy uses are covered with fossil fuels, with equal shares for oil, coal and natural gas (i.e. per day, 86.7 million barrels, 10.6 million tons equivalent and 9.2 million barrels). Analysis prove that fossil fuels shortage is a short-to-mean term problem and fix a 53, 55 and 113 years-time horizon for oil, gas and coal exhaustibility, respectively. Based on current understanding, if atmospheric CO_2 concentration (currently stuck at 406 ppm) stabilizes at 450 ppm, a 2.1°C global warming has to be expected by 2100. In order not to overcome this threshold, an 80% reduction of current GHG emissions by 2100 has to be implemented and the massive renewable penetration in present market is left as the only viable option to be implemented in the medium-long term [1, 2, 3, 4].

These are the main reasons why, in almost all World Countries GHG - namely, CO_2 - emissions reduction, energy saving and energy generation from renewables by 2020 have been committed and environmental markets have been launched, on the idea that a financial value can be assigned to immaterial goods, such as avoided CO_2 emissions and fossil fuel saved: once these quantities are certified, quotas can be put on the market and a supply-demand dynamic applies, between "virtuous subjects" -either private or public- and those subject committed to reach emissions and/or energy saving targets. While far from ideal, as the price is not only fixed by a demand-to-supply balance, these markets allow to establish a technological standard and assign a financial value to energy virtuosity. On the same page, Carbon taxes and financial incentives to reward the adoption of high efficiency components have been established.

The compressed air sector (CAS) entirely relies on electrical energy: most recent data [5, 6] fix its impact in terms of electricity consumption at 10% industrial electricity consumption. By including the compressed air energy consumption in other sectors (i.e. residential and transports), this figure doubles, reaching 20% Industry electricity consumption [7, 8, 9]. 30-40% present energy consumption is commonly accepted as an estimation of the achievable electricity saving in the CAS [10, 11, 12, 13]. Once the datum on CO_2 emissions per kWh electricity is known, each context can be assigned with its corresponding quotas in terms of avoided CO_2 and fossil fuel saving.

The paper deals with the CAS Carbon dimension: after a brief overview on the compressed air electricity consumption for different Countries in the World, Carbon emissions associated to present market compressor technology are evaluated and presented as a function of delivered pressure and machine size (i.e. mass flow rate), based on data from the Compressed Air and Gas Institute (CAGI) and PNEUROP for the US and the EU market [14, 15]. The comparison among different machine types (e.g. screw and vanes) allows a quantitative estimation of the potential saving in existing compressors [15, 16]. Such a datum can be reversed in terms of CO_2 and fossil fuel credits, to be traded on environmental markets. Given the intrinsically diversified nature of intangible markets by Country and the different degrees of acceptance and applicability for fossil fuel and CO_2 saving interventions, dependently on their distance to target, the paper accounts for different scenarios, i.e. for different pricing mechanisms for both CO_2 quotas and energy saving. The reference is to the more mature markets toward which all others are orienting in terms of accounting and trading rules.

2. Overview of compressed air electricity consumption

World electricity consumption ranked at 18912 TWh in 2012, 13000 TWh (i.e. up to 70%) of which came from fossil fuels, i.e. oil, gas and coal [3]. Main use for electricity is in Industry and Buildings (either residential or commercial):

- among developed Countries, the US show the lowest share of electricity consumption in the industrial sector (23%), especially if compared to Europe (36%) and Australia (38%);
- in the US, the widest room for saving is in the building sector (72.3% overall consumption);
- absolute values prove that 48.0% Industry electricity consumption (i.e. 3843 TWh) comes from BRICS Countries; China alone represents a 72.7% of this share;
- the economic growth in BRICS Countries hinges on Industry, whose electricity consumption ranks at 68% (China), 60% (South Africa), 44% (Brazil), 46% (Russia) and 44% (India).

Projections on future energy consumption assume 2035 as the reference year: by then, they fix 450 ppmCO_2 as the limit atmospheric concentration and assume up to 40% global electricity consumption to come from BRICS Countries, with China share close to the sum of EU and the US. Hence, energy saving and CO_2 reduction technologies in Industry represent important options, to comply with environmental goals [3, 4].

Compressed air production accounts for a mean 10% overall Industry electricity consumption in developed Countries; when the compression needs from other sectors (e.g. portable tools, pneumatic heating, air conditioning, ventilation) sum up, the consumption doubles and ranks at 20% industrial electricity needs [5, 6, 7]. The transport sector is excluded, since there a direct conversion is done by mechanical energy. Nonetheless, fuel saving and CO_2 emission reduction should account for the contribution from this sector as well. When evaluating the room for improvement related to electricity consumption in the CAS, it is worth observing that a decrease in the compressor-package absorption (i.e. compressor and electric motor) proportionally reflects on the overall energy consumption. The widest room for improvement is in leakages prevention (i.e. air waste reduction from present 20-40% level, due to both the compressor and the piping) and pressure level optimization (e.g. through filter replacement and proper piping design) [7, 8, 9]. Apart from the compressor technology, an interesting feature is the decrease in operating pressure levels, to combine with an upgrade of delivery and end-use systems: the compressor consumption roughly depends on the fourth square of the pressure ratio and the benefit of a 1 bar reduction from a given delivery pressure is expressed by Equation (1), with the adiabatic work as the reference:

$$S_{\beta,\%} = \left[1 - \frac{(\beta - 1)^{\frac{k-1}{k}} - 1}{\beta^{\frac{k-1}{k}} - 1}\right] \cdot 100 \tag{1}$$

A 1 bar reduction, from an 8 pressure ratio leads to a 10% saving. The higher the delivered pressure, the lower the benefit: at 15 bar delivery pressure, the 1 bar reduction leads to a 4% saving. A further beneficial effect associated to pressure reduction is in pipe losses reduction. When 13 bar is assumed as the starting pressure level (i.e. 6.34 kW/m^3/min) the effect of a progressive pressure line decrease on the specific consumption can be seen as a percentage of the specific consumption at 13 bar: moving to 9 bar, a 20% saving is possible. Real absolute values for the specific energy consumption can be obtained by dividing the data by an overall global compressor efficiency [15]. As a rule of thumb, if both compressor and piping leaks are removed, the potential energy saving via compressor replacement can be doubled [5].

A study by the Authors [16] investigates the reduction in energy cost, via the improvement of present compressors technology (i.e. sliding vane rotary machines) and correlates it to the investment increase having the compressor size and type, the operating conditions and a 3 years payback time as parameters. At each efficiency increase the maximum acceptable investment increase grows from light to medium to heavy duty, due to the higher impact the energy cost has on the system financial management. The study accounts for air and water cooled machines as well and the analysis suggests that in big size machines a given performance improvement requires a higher investment in water than in air cooled compressors. The study also makes gives evidence that a further major variable affecting the investment feasibility is the discount factor purchaser can count on, particularly in presence of high financial investments, i.e. big size compressors and heavy duty applications.

3. Compressed air sector Carbon dimension

As a commodity, compressed air is an electricity user, hence values for emissions per kWh electricity allow a direct evaluation of its contribution to CO_2 emissions by Country. Dependently on the adopted energy mix and the reference value for power plants efficiency by context, a high spread among Countries should be expected among both EU States (e.g. Norway 0.01 kgCO_2/kWh vs Greece 0.90 kgCO_2/kWh) and different World Countries.

The average datum for Europe is 0.324 kgCO_2/kWh; apart from Brazil (0.2 times the EU value) BRICS Countries have CO_2 emissions close to (see Russia, 0.9 times the EU value) or above the European datum: 3 times in India and South Africa, 2.3 times in China; specific emissions for Australia and US are 2.6 and 1.8 times those in the EU, respectively [17, 18, 19]. An assessment of compressors performances from many different manufacturers is presented in [15], based on CAGI data, gathered during a certification campaign in the US, and PNEUROP data for the EU. The results are in Figure 1: the volumetric flow rate is evaluated at 1 bar and 293.15 K and values for compressors power refer to unified delivery pressures, to enable a direct comparison among different machines.

Data, referring to air cooled compressors, are presented with no mention to the manufacturers' names and as a function of delivery pressure at rated flow rate and oil cooling medium (i.e. air or water) [14, 15]. Despite the fact that compressors processing high flow rates exhibit a lower scatter in performances among manufacturers if compared to lower-size machines, the great variability in compressors performances suggest that the technology is not aligned to the best standards. For 40-50 m^3/min flow rates and below, 30-40% savings can be achieved by replacing the compressor unit.

Figure 2 reports the specific CO_2 emissions, with reference to the European context (324 gCO_2/kWh), in terms of Carbon per cubic meter of compressed air (ccm). The coefficient of determination R^2 in the regression curves is a statistical measure of how well the regression model approximates ccm of existing machines (real data points). It can be observed that:

- ccm decreases with the flow rate. Specific emissions in air cooling machines, at 5, 20 and 50 m^3/min flow rate are 0.041, 0.038 and 0.036 $kgCO_2/m^3$, whereas in water cooled machines they rank at 0.038, 0.036 and 0.035 $kgCO_2/m^3$, respectively;
- the higher power absorption by the pump for water circulation (water cooled machines) with respect to the fan (air cooled machines) leads to higher specific emissions;
- at 8 bar delivery pressure, air cooled machines have a high technological scatter among manufacturers, (e.g.: at 10 m^3/min, a 0.065 $kgCO_2/m^3$ scatter applies to a 0.036 $kgCO_2/m^3$ standard), hence a wise compressor choice allows huge energy and emissions saving in this flow rate domain. Same goes with water cooled compressors, even if the scatter here is one order of magnitude lower than in air cooled; at higher flow rates, the ccm trend is asymptotic to lower values for both specific emissions and scatter (e.g.: 50 m^3/min, 0.033 $kgCO_2/m^3$ and 0.027 $kgCO_2/m^3$, respectively);
- the higher the delivery pressure, the higher the ccm difference among air and water cooled families, with a strong preference for water cooled machines; for a 9 bar pressure, the scatter further reduces, leading to a compressors population that is tightly distributed around the mean value: for a 30 m^3/min flow rate, a 0.035 $kgCO_2/m^3$, with a 0.002 $kgCO_2/m^3$ scatter is detected. For water cooled compressors, the scatter is negligible yet at 20 m^3/min flow rate.

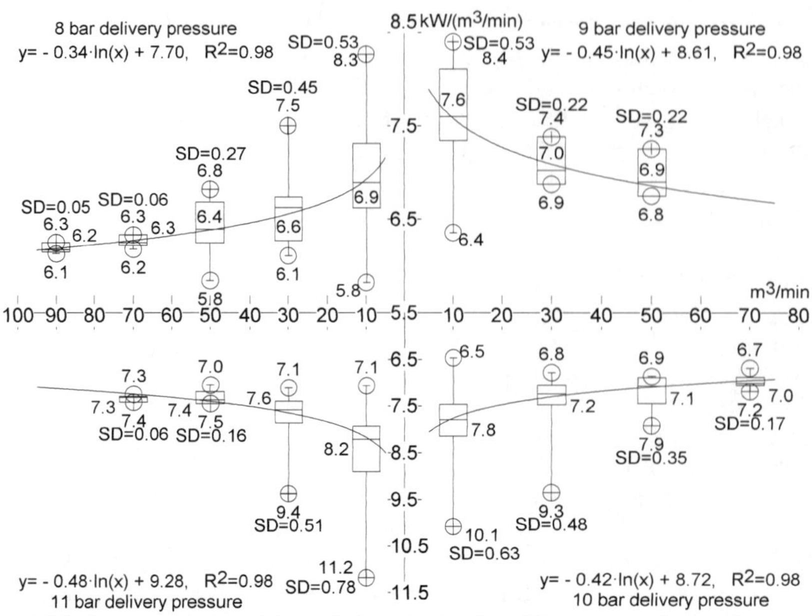

Figure 1. Market compressors (air cooled version) – Specific consumption vs flow rate [15]

It is worth observing that the chart can be referred to other contexts of interest, by simply scaling the numerical data in Figure 2 through the appropriate factor, as previously discussed. Thus the chart provides the metric to read the trend in CO_2 emissions by machine size and type in the compressed air sector in any context of interest.

4. Environmental markets – Carbon and energy saving trading

Presently, the European Emission Trading Scheme (EU-ETS) covers 45% of EU greenhouse gas emissions (roughly a 10-12% the whole emission scenario). The unit exchanged on this market is the ton of CO_2 (tCO_2) avoided and it remained as unit for the others Carbon pricing instruments: covered installations are committed to surrender an allowance for each ton CO_2 they emit during the year. The allowances availability, combined with their tradability, creates a market price for allowances, that motivates committed emitters to choose between abatement options (investments) and purchasing on the market: 6-7 €/tCO_2 is the reference price at the moment, due to the abundance of credits, induced by the economic crisis. Nonetheless actors on the Carbon market have to face price variability for traded CO_2, in both consolidated and newly established markets, dependently on whether fixed (e.g. Carbon taxes) or variable and market-sensitive (national and subnational ETS) fees apply [17, 20, 21, 22], which makes hard to predict Carbon price in both the long and the mean-to-short term [23, 24, 25]. An average 15 €/tCO_2 price can be considered, whilst in the eventuality of more stringent CO_2 commitments, 30-50 €/tCO_2 appears more realistic.

The same principle behind the Carbon market inspires the energy saving market, where fuel saving is exchanged and saving demand, coming from subjects that the law obliges to reach a target, is balanced by supply, provided by subjects who reached certified savings. The units exchanged are the Energy Saving Certificates (ESC), corresponding to a ton of oil equivalent (TOE) saved. When the analysis deals with the saving of electrical energy, the efficiency of the electric generation must be considered and kWh converted into TOE. A known datum is that inside the European context average electric generation efficiency changes, ranking from 30% (France and Sweden) to 46% (Luxembourg and Netherlands); an even higher spread on electric generation efficiency characterizes the worldwide scenario, adding further complexity to the analysis. The UK was the first Country to set up an Energy Savings Certificates scheme, with a cumulative actualized (i.e. accounting for saving devices obsolescence) 130 TWh mandatory saving for energy suppliers [28]. ESC trading in Italy, Great Britain and France, started in 2005, produced a trading unit -White Certificate (WhC)- soon expected to be implemented by other member States. 21 US States have energy efficiency targets, mandatory or voluntary [29]. New South Wales (Australia) ETS program is ready to be launched. Recently, India released a Climate Action Plan that encouraged the creation of a national market for ESCs and China also is debating on the scheme suitability [30]. A differentiation comes out between mature markets, benefitting from a consolidated structure and price dynamics – as in Italy – and new markets – as in UK, France, USA, China, Australia and India - whose full development and exploitation still requires time, but seems very promising, especially considering the growth in energy consumption in key sectors, as Industry and Residential, these Countries are expected to experience in the near future.

National commitments specificity, the need to match the period in which certificates are generated and spent along with the differences in Industry and energy standards among different Countries represent major limitations to the definition of a unified frame for national energy saving markets. The different sensibility to the themes of fossil fuel sources availability and CO_2 emissions increase further adds to this. The former is commonly perceived as a 'questionable matter', mostly due to the lack of a ready-to-use number on sources exhaustibility, whereas with CO_2, the 450 ppm datum provides a clear quantitative indication on the room available for technology development for years to come and gives a better sense of what extent efforts should be pushed to, in order to prevent the overstepping of an assigned threshold. Thus a unitary development of environmental markets is hard to achieve.

In new markets (e.g. United Kingdom, France and Denmark), all savings are calculated and contributed to the first year and the number of certificates (NESC) is given by:

$$N_{ESC} = S \cdot L_{SPAN} \cdot \delta \tag{2}$$

in which S is the energy saving, calculated with respect to a baseline consumption, L_{SPAN} is the technical life for the application considered and δ is the discount coefficient, accounting for technical aging, i.e. for the fact that the saving as declared at the beginning tends to decrease, as the time goes by [31]. These certificates are recognized at the same time the investment takes place.

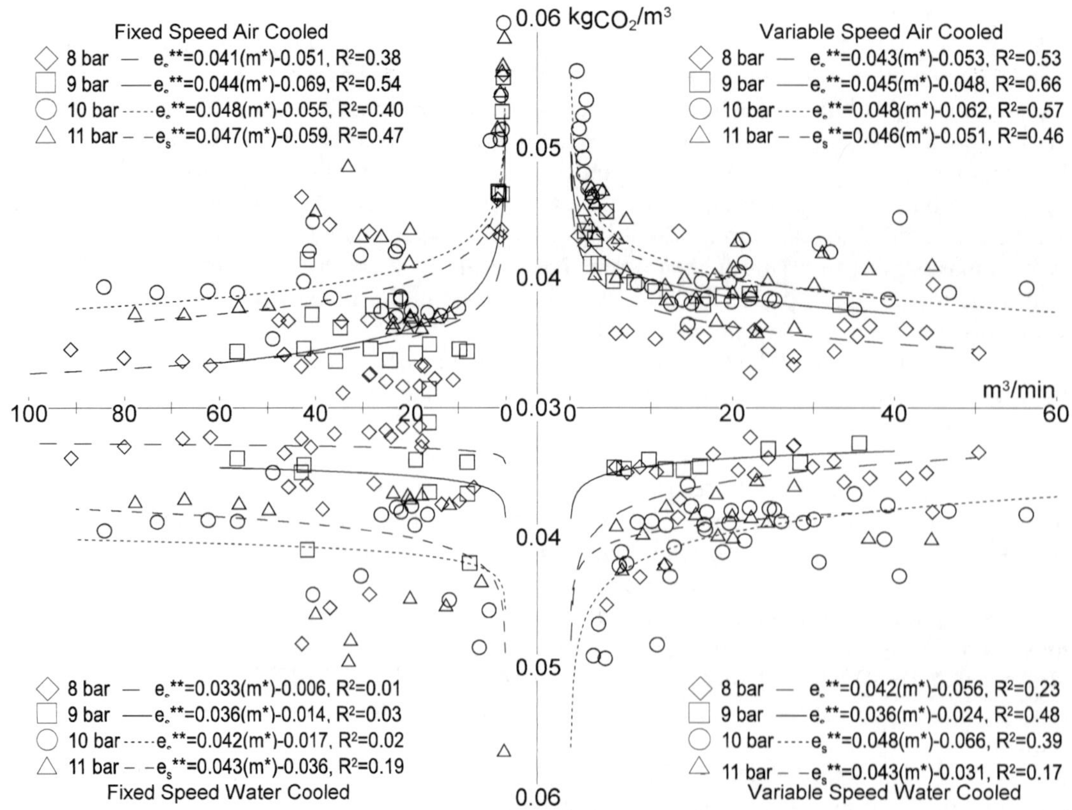

Figure 2. Specific CO_2 emissions per unit mass compressed air in EU

In Italy, certificates are eligible only when energy is saved, i.e. the reward is provided at the same pace the energy savings happen [29]. Hence the uncertainty about the certificates value during years, due to market price fluctuations is the key for ESCs' market success [32] as in various comprehensive analysis [33, 34, 35, 36]. Equation (2) becomes:

$$N_{ESC} = k \cdot S \cdot \tau \tag{3}$$

in which, S is still the saving, k is a factor that converts thermal or electrical saving into TOE. Since certificates lifespan $N_{l,s}$ (i.e. the 5 to 8 years period, during which the certificates are recognized) and the technology lifespan (i.e. the 15 to 20 years period, the saving lasts and after which technology becomes obsolete) are different, a discount factor, τ, to redistribute the overall saving over the lifespan (i.e. to expresses the saving in the overall technological lifespan as a multiple of the first year saving), needs to be introduced [30, 31, 32]:

$$\tau = 1 + \frac{\sum_{j=1+N_{l,s}}^{N_{t,l}} (1-\delta)^j}{N_{l,s}} \tag{4}$$

5. Energy and CO_2 emissions saving – Financial value assessment

A safe estimation of CAS related energy consumption on a European basis by PNEUROP, is 56 TWh, to which both screw and vanes compressors contribute, with 36.58 TWh and 19.17 TWh, for fixed and variable speed drive versions respectively, as in Figure 3 [6].

The compressor replacement, whose potential depends on the efficiency target for compressors and on the specific energy below which a compressor cannot be operated, represents an interesting option in the EU context. An in-depth investigation of market compressors specific consumption is in [15], where ISO 1217 testing conditions (i.e. 0°C and 1 bar) are considered: a 130 Wh/Nm^3 specific energy consumption is associated to 7 bar delivery pressure, with premium machines set at 85 Wh/Nm^3. It has to be observed that these figures are nothing but a general datum, since compressors specific consumption depends on operating conditions, that can significantly vary among industrial applications. A by far more effective approach, when depicting machine replacement-based scenarios, is the one that assumes the compressor adiabatic isentropic efficiency as machine merit criterion. PNEUROP [6] sets a baseline curve for efficiency and foresees that a 0.7-2.1 TWh saving is possible by machine replacement, when fixed speed compressors are considered; the same figure for variable speed compressors is 0.3-1.2 TWh. An overall 1.0-3.3 TWh saving on compressors is thus achievable. When the whole compressed air chain underwent an upgrade process (e.g. compressor and piping leakage fix and pressure losses optimization), a further 2-7 TWh energy saving could be put on the market, for a cumulative 3-10.3 TWh saving. With reference to Italy, a 0.19 TOE/MWh electrical efficiency can be considered; with a 15 years lifespan, a 5 years financial lifespan, and a 0.02 δ, τ in Equation (4) is 2.65 and Equation (3) gives a 1.5-5.3 MTOE saving in the CAS. When an average 100 €/TOE price is assumed, this leads to a 150-530 M€, to split on 5 years financial lifespan.

By accounting for the financial benefit associated with CO_2 trading on the European ETS market, 15-35 M€ (i.e. 15 €/tCO_2 and 1-2.5 MtCO_2 considered as average Carbon price and Carbon saving, respectively) add to the previous figure, leading to 165-565 M€ financial value. In a strict-commitment 50 €/tCO_2 scenario, the financial value of CAS-related Carbon credits and ESCs ranks at 200-655 M€.

A general datum can be provided, to allow a comparison among main immaterial markets, established at the present day, for the given 3-10.3 TWh saving scenario. A direct comparison with a different European context is possible by considering UK: a 18 €/MWh price applies to electricity saving, whereas the Carbon Price Floor fixes the Carbon quotas sale price at 15 €/tCO_2 [37]. Consequently, a 540-to-1854 M€ and a 204-to-700 M€ financial impact comes from energy saving certificates and CO_2 quotas selling, respectively: these two figures add up to a 750-to-2560 M€.

The pace of reform in China's electricity market is gathering speed, thus increasing competition and market awareness of large power users [38]. When a 38.4 €/MWh is considered as the reference, the financial revenue from electricity selling ranges between 1.2 and 4.0 billion Euro. As far as the trading of CO_2 quotas is concerned, the Bejiing Pilot ETS can be safely assumed as the reference frame, in which a 9 €/tCO_2 price applies [17]; this leads to a minimum 189 M€ and a maximum 650 M€ revenue from CO_2 trading. The cumulative financial benefit of compressed air-related Carbon credits and ESCs ranks at 1.4 to 4.7 billion Euro.

Same considerations apply to Australia: apart from regionalism in price assignment [39], 17 €/MWh and 21 €/tCO_2 can be considered as reference prices for electricity saving and avoided CO_2 (according to the Carbon Pricing Mechanism, as in [17]), respectively. The correspondent market benefits are estimated in 500-1700 M€ and 532-1825 M€. The cumulative revenue is in the 1.0-3.5 billion Euro range.

In order to refer previous calculations to India a 39 €/MWh price for saved electricity [40] and a 100 €/tCO_2 sale price for CO_2 quotas on the market [17] are considered. Financial value for electricity saving ranges between 1.2 and 4.0 billion Euro, whereas a 2.9-9.8 billion Euro revenue is achievable by selling CO_2 quotas. The two of them add up to 4.1 to 14 billion Euro cumulative financial benefit on the market.

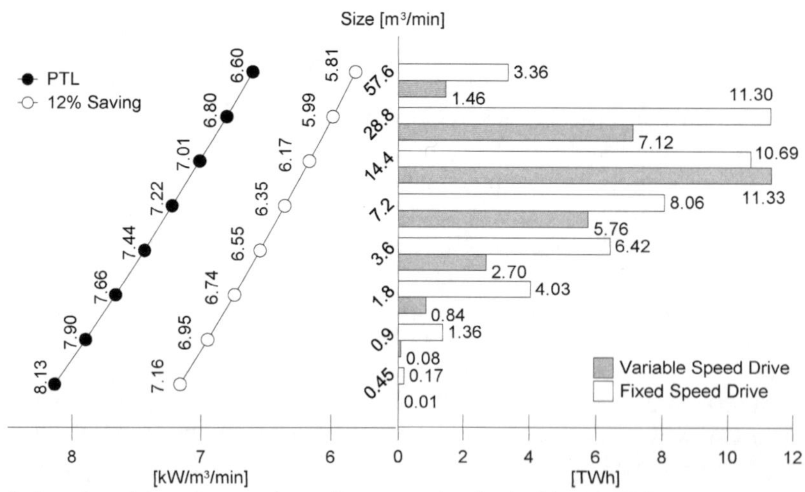

Figure 3. Cumulative electricity consumption – Present technological level (PTL) vs saving scenario [15]

In the US, each member State adopts its own paradigm in terms of immaterial market for both electricity saving and avoided Carbon emissions. Nonetheless, a general datum can be provided on the Carbon market, based on the Cap and Trade initiative, with a 8 €/tCO_2 Carbon price [17, 41]. Once an average 29 €/MWh electricity price is assumed [42], the financial impact on the market is estimated in 880-3015 M€ for electricity and 124-426 M€ for Carbon quotas sale, i.e. a cumulative 1.0-3.5 billion Euro. Such results call for a lobby action in the CAS and the achievable financial benefit encourages the adoption of a synergic activity to promote a new approach to compressed air production and delivery. One last datum can be given, that sums up all elements of the previous analysis: when the overall mean financial saving (430 M€) is referred to the overall electricity saving (4.5 TWh), the specific saving in the EU CAS turns out to be 10 c€/kWh.

The same approach applies to other contexts, even though due to the differences on both the NESC definition and numerical values for prices and discounting coefficient differences shall be expected on the value of specific saving in the CAS. In detail, 40 c€/kWh applies to UK, 70 c€/kWh to China; the value for Australia and US is 50 c€/kWh while it raises at 200 c€/kWh for India.

6. Conclusions

The compressed air industry entirely relies on the use of electrical energy, with shares ranging from few percentage points to values close to 20% overall industrial electricity needs. Given the present efficiency standards, a 25-50% saving is possible by fixing leakages in compressor and piping, optimizing pressure levels and replacing machines. Both CO_2 emissions and fossil fuels use are committed in almost all World Countries - on both a local (national and subnational) and global basis - and traded on immaterial markets assigning saved fossil fuel and avoided emissions with a price. The paper describes the Carbon dimension of present compressor technology and fixes CO_2 emissions per unit of compressed air delivered, based on CAGI and PNEUROP data for the US and EU context, respectively. The approach adopted allows to extend the results of the analysis to every context of interest, by applying the appropriate emission factor to the datum on compressor specific consumption. According to the PNEUROP analysis, 1-3.5 TWh can be saved through machine replacement in Europe, the overall saving almost doubles if leakages fixing and pressure level control take place. In terms of Carbon credits and fossil fuel saving, tradable quantities are 1-2.5 MtCO_2 and 1.5-5.3 MTOE, that refer to a 5 years financial timespan. This datum refers to the European context and based on projections for future trends in environmental markets, fixing 50 €/tCO_2 and 100 €/TOE as good approximations of applicable prices, it corresponds to a 430 M€ cumulative saving (mean value between 200 M€ and 655 M€). As a synthesis datum, the saving in the EU CAS can be estimated in 10 c€/kWh. Such a datum is close to the net electricity cost in many Countries. Such an approach applies to all main Countries in the World and by providing a clear indication on both the mean value for cumulative saving and equivalent electricity cost in the CAS, it represents an effective tool to orient the decision making process. In the UK, a 1.7 billion Euro saving volume can be put on the market, corresponding to an average 40 c€/kWh. Same indicators for China are 3.0 billion Euro and 70 c€/kWh, whereas for Australia and USA they rank at 2.3 billion Euro and a 50 c€/kWh electricity cost. India has the greatest potential for saving, as 9.0 billion Euro saving can be achieved, corresponding to 200 c€/kWh electricity cost.

References

[1] BP Energy Outlook 2035 January 2014 available at www.bp.com/energyoutlook#BPstats
[2] Bergen@Energi AS Energy Outlook 2014 available at https://www.bergen-energi.com/
[3] Energy Outlook 2014 (WEO-2014) available at www.worldenergyoutlook.org/
[4] IPCC Fifth Assessment Report, Climate Change 2013: The Physical Science Basis" (AR5) available at https://www.ipcc.ch/publications_and_data/
[5] Radgen P and Blaustein E. (Eds) 2001 Compressed Air Systems in the European Union. Energy, Emissions, Saving, Potential and Policy Actions LOG_X, ISBN 3-932298-16-0
[6] PNEUROP European Association of Manufacturers of Compressors, Vacuum Pumps, Pneumatic available at http://www.eco-compressors.eu/documents.htm
[7] Saidur R et al. 2010 A Review on Compressed Air Energy Use and Energy Saving J. *of Ren. and Sust. En. Rev.* **14** pp 1135-1153
[8] Compressed Air Energy Input and Useful Energy (Adapted from North West Energy Alliance) ©Copyright 2007 CEA Tech. Inc.
[9] Compressed Air: Opportunities for the Business 2012 © The Carbon Trust 2012
[10] GPG 385 Energy Efficient Compressed Air Systems – Carbon Trust by British Compressed Air Society Ltd 2005
[11] U.S. Department of Energy's Office of Energy Efficiency and Renewable Energy (EERE) Improving Compressed Air System Performance: a Sourcebook for Industry - U.S. Department of Energy's Office of Energy Efficiency and Renewable Energy (EERE) Best Practices and the Compressed Air Challenge® 2003 DOE/GO-102003-1822
[12] Da Cunha F 2007 Compressed Air: Energy Efficient Reference Guide *CEA Technology Inc. (CEATI) Customer Energy Solution Interest Group (CESIG)*
[13] Energy Saving in Compr. Air Sys. — How Norgren is Helping to Improve Energy Efficiency
[14] CAGI Compressed Air Gas Institute Datasheets available at www.cagi.org
[15] Cipollone R and Vittorini D 2014 Energy Saving Potential in Existing Compressors *Proc. Int. Compressor Engineering Conf. (Purdue)* **2379** available at http://docs.lib.purdue.edu/icec
[16] Vittorini D and Cipollone R 2016 Financial Analysis of Energy Saving via Compressor Replacement in Industry *Energy* **113** pp 809-820, http://dx.doi.org/10.1016/j.energy.2016.07.073
[17] World Bank Ecofys and Vivid Economics 2016 State and Trends of Carbon Prizing (Washington DC)
[18] McKynsey&Company 2010 Pathways to a Low Carbon Economy available at http://www.mckinsey.com/globalGHGcostcurve
[19] International Energy Agency 2016 CO2 Emissions from Fuel Combustion available at www.iea.org/statistics/onlinedataservice/
[20] Carraro C. and Favero A. 2009 The Economic and Financial Determinants of Carbon Prices *Czech Journal of Economics and Finance* **59** 5 pp 396-409
[21] De Bruyn S, Nelissen D and Koopman M 2013 Carbon Leakage and the Future of the EU ETS Market - Impact of Recent Developments in the EU ETS on the List of Sectors Deemed to be Exposed to Carbon Leakage *CE Delft* available at www.cedelft.eu
[22] A.Öko-Institut 2012 Strengthening the European Union Emissions Trading Scheme and Raising

IOP Conf. Series: Materials Science and Engineering **232** (2017) 012085 doi:10.1088/1757-899X/232/1/012085

Climate Ambition (Berlin)

[23] Chestney N 2016 UN Carbon Price Forecasts to 2020 Cut Further Point Carbon *Website Reuters Poll* available at http://www.reuters.com/article/us-carbon-poll-idUSBRE89109V20121002

[24] Roches F 2012 Scenarios for the ETS Carbon Prize *Euroelectric Conference "The Future role of EU-ETS"* (Brussels)

[25] von Schemde A and Tennbakk B 2014 TCG-Insight 5-2014: EU ETS Reform is Set to Increase the Cost of Emitting CO2 *Thema Consulting Group* available at http://www.t-cg.no/

[26] Thomson Reuters Commodities Research and Forecasts 2016 Carbon Market Monitor - Review of Global Markets in 2015 and Outlook for 2016-2018

[27] Luckow P, Stanton E A and Fields S 2016 National Carbon Dioxide Price Forecast *Synapse Energy Economics Inc.* available at www.synapse-energy.com

[28] ADEME Energy Saving Certificates 2011-2013 Knowledge for Action available at www.ademe.fr

[29] Liska R and Canavan D 2008 Bottom Line on Energy Savings Certificates available at http://www.wri.org/publication/bottom-line-energy-savings-certificates

[30] Xiaodong W 2013 White Certificates Trading, Green Certificates Trading, Emission Trading: Which One to Choose? *EASCS* available at https://www.thepmr.org/system/files/documents/WBBeijing20Office20coordination20of20trading20schemes20September202013

[31] Togeby M, Mikkelsen K D and James-Smith E 2007 Design of White Certificates Comparing UK, Italy, France and Denmark *EA Energy Analyses* available at http://www.ea-energianalyse.dk/reports/710_White_certificates_ report_19_Nov_07.pdf

[32] ENEA WG on White Certificates 2012 I Titoli di Efficienza Energetica. Cosa Sono e Come si Ottengono i "Certificati Bianchi" Alla Luce della Delibera EEN 9/11 GUIDA OPERATIVA/2 available at http://www.enea.it/it/ produzione- scientifica/ pdf-volumi/v2012-guida-cb2.pdf

[33] De Sanctis L and Ranieri D 2013 Il Meccanismo dei Certificati Bianchi available at http://www.enea.it/it/comunicazione/events/energy-med-2/Napoli12aprile2013bis.pdf

[34] Rezessy S and Bertoldi P 2010 Energy Supplier Obligations and White Certificate Schemes: Comparative Analysis of Results in the European Union Energy Policy **38** 3

[35] Giraudet L G and Finon D 2010 On the Road to a Unified Market for Energy Efficiency: The Contribution of White Certificates Schemes Sustainable Development Series, *nota di Lavoro 132.2010 Fondazione ENI E. Mattei* available at http://www.feem.it/userfiles/attach/201010191056554 NDL2010-132.pdf

[36] Cowart R The Regulatory Assistance Project-Energy Savings Obligations and White Certificates Energy Regulators *Regional Association* available at http://www.raponline.org

[37] Department for Communities and Local Government 2013 Energy Performance of Buildings Directive – Cost Optimal Calculations: UK Report to European Commission

[38] Ng E 2011 China presses ahead with Power Pricing Reform available at http://www.scmp.com/business/companies/article/2018410

[39] Betz R, Jones M C, MacGill I and Passey R 2014 Trading in Energy Efficiency in Australia: What are the Lessons Learnt so Far? *ECEEE Summer Study Proceedings – Rethink, Renew, Restart* **2** 188

[40] Mediratta R K 2016 Role of Power Exchanges in India *Indian Energy Exchange – Integrated Research and Action for Development*

[41] Department of Energy – Office of Energy Efficiency and Renewable Energy 2015 Energy Conservation Program – Energy Conservation Standards for Air Compressors, available at www.regulations.gov/docket?D=EERE-2013-BT-STD-0040

[42] Department of Energy – Office of Energy Efficiency and Renewable Energy 2015 Energy Incentive Program, California available at https://energy.gov/eere/femp/energy-incentive-programs-california

IOP Conf. Series: Materials Science and Engineering **232** (2017) 012086 doi:10.1088/1757-899X/232/1/012086

Turbomachinery group for cold energy concept application

S Alavi[1], A Arpi[2], M Ascani[2], G Cerri[1], L Chennaoui[1] and A Manni[2]

[1] Roma Tre University, Department of Engineering, Roma, Italy
[2] Turboalgor S.r.l., Massa Martana, Perugia, Italy

cerri@uniroma3.it

Abstract. The paper discusses a special attempt to improve COP and compressor capacity. Theoretical and experimental results are presented, and they show a very positive trend. Compressor and Expander Group impellers are selected among the Car Engine Turbocharger Units. Modifications are proposed to achieve the best performance, taking the peculiar characteristics of the refrigerant into consideration. Tests have confirmed the positive saving trend, which can reach up to 22%-24% for an optimized bleed vapour generator and turbomachinery-based booster unit.

1. Introduction

Energy saving, climate changes and environmental pollution call for a decrease in worldwide electricity consumption without reducing the quality level of life and industrial production.
One of the products that is widely used is the cooling power produced by Vapour Compression Refrigeration Plants (VCRPs) powered by electricity.

Reducing electricity consumption as well as the related primary energy and pollutant emissions are important problems, because the electricity consumed worldwide by VCRPs is on the order of 15% - 17%, which contributes some 4% – 5% of global pollutant emissions. In industrialized countries, this consumption reaches 23% – 25%, and in warm countries, it rises to about 40% [1-3]. It can be easily understood that numerous attempts have been investigated and applied toward reducing the electric power consumed per unit of cooling power. These attempts are mainly based on thermal regeneration, which in vapour compression cycles leads to increasing the condensate subcooling through cycle internal processes. Attempts have also been made to reduce the main compressor pressure ratio by power regeneration by means of ejectors or positive displacement machinery [4,5]. Due to the low efficiency of these types of machinery, special attention has not been given to the field.

The Power Regenerated Cycle presented in [2] was investigated in the COLD ENERGY project. This cycle uses turbocompressor and turboexpander groups to increase the VCRP main compressor inlet pressure. The interesting cold energy results have shown improvements of up to 15% in COP for constant cooling power, which has stimulated the industrial production of such units for installation in VCRPs.

This paper deals with the analysis of the cycles and their optimisation to fulfil the minimum electricity consumption per unit of cooling power, which means the best use of the main compressor in terms of increased main compressor cooling power, together with the Coefficient of Performance (COP). The development of the components in pursuing the goal is discussed along with the main results. An industrial solution is presented.

2. Scientific and technical background

The scientific and technical aspects of this work regard the following:

- Improvements that can be achieved, in terms of cooling power and COP, by modifying a vapour compression refrigeration simple cycle (SC) as a reference, adding some devices;
- Optimum sizing of the innovative devices, using economically viable technologies;

2.1. Simple cycle as a reference

A simple cycle has been taken as a reference. Sketches of the plant and of the p-H Chart thermodynamic cycle are shown in Figure 1.

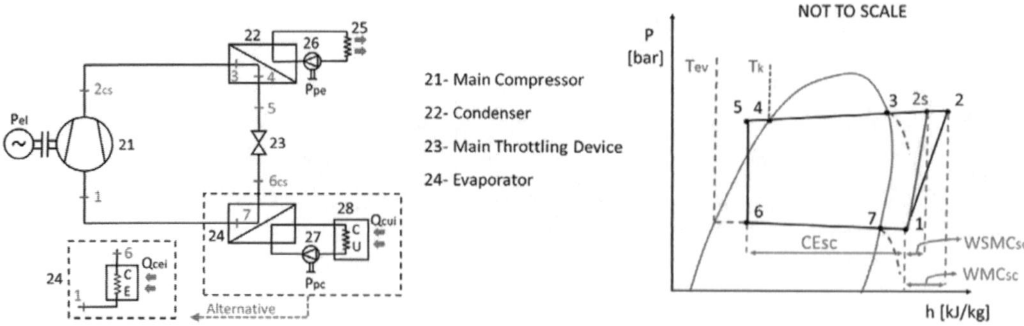

Figure 1. Simple cycle arrangement

Four main components characterize this type of plant: the Main Compressor (MC); the Condenser (K); the Main Expansion Valve (MEV) and the Main Evaporator (ME).

2.1.1. Component models. The SC model is established by a modular approach. Each module corresponds to one component. Details of the models can be found in [3, 6-12].

- The working fluid (WF) is one of the components whose properties are defined by a set of nonlinear equations (WFNLES):

$$\boldsymbol{F}_{WF}\ (T, p, H, s, v, c_p, c_v, c_s, other) = 0 \qquad (1)$$

There is one independent variable for boiling point and dew point lines and two independent variables for liquid and vapour-gas zones.

- Main compressors (MC) are usually utilized in semi-hermetic arrangement as shown in Figure 2a. Internal processes are rather complex: sketches are given in Figures 2b and 2c. The MC model is represented by a MCNLES that establishes correlations between variables:

$$\boldsymbol{F}_{MC}\ (m_{MC}, p_1, T_1, p_2, T_2, \eta_{MC}, \eta_v, P_{MC}, Q_d, V_d, n, P_N, n_N, n_o, P_{MX}, \eta_{em}) = 0 \qquad (2)$$

η_{MC}, η_v, Q_d and η_{em} are expressed by empirical functions. V_d, n_o, n_N, P_N, P_{MX} are data associated with the compressor and electric motor. V_d can be replaced by pressure ratio and efficiency versus mass flow rate map curves for constant speed. T_1, p_1, p_2 are MC boundary conditions that are exchanged in the matching between components.

- The condenser is a three-zone heat exchanger as shown in Figure 3. The model establishes correlations among the variables by the KNLES:

$$\boldsymbol{F}_k(m_k, p_2, T_2, p_3, p_4, p_5, \Delta p_{sh}, \Delta p_k, \Delta p_{sc}, T_3, T_4, T_5) = 0 \qquad (3)$$

$m_k, T_2, p_2, \Delta p_j$ as well as the condensing temperature T_k are given. One of the constraints is:

$$T_3 - T_k = 0 \qquad 4)$$

- The main expansion valve ensures an adiabatic fully dissipative process. The model [6,8] is described by the MEVNLES correlations:

$$\boldsymbol{F}_{MEV}\ (p_5, T_5, H_5, p_6, T_6, H_6) = 0 \qquad (5)$$

- The evaporator is a two-zone heat exchanger as shown in Figure 4. The evaporator steady state model establishes EVNLES correlations between the variables involved in:

$$F_{ev}(m_e, T_{ev}, T_6, H_6, T_7, p_7, H_7, T_1, p_1, H_1, \Delta p_{67}, \Delta p_{71}) = 0 \tag{6}$$

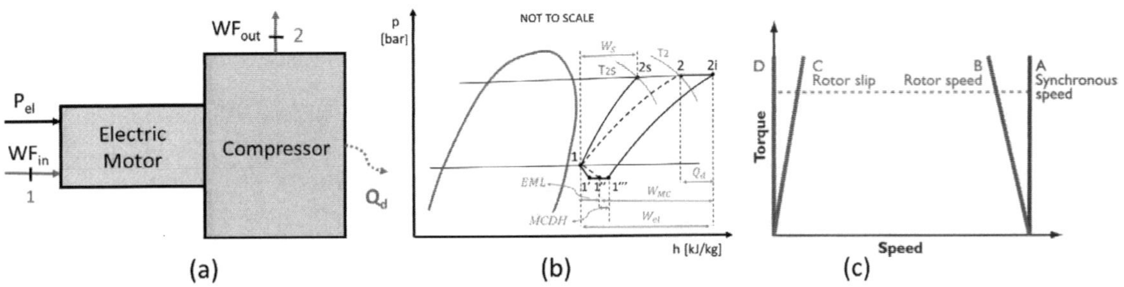

Figure 2. Main compressor model, p-H chart processes, torque vs speed of an induction motor

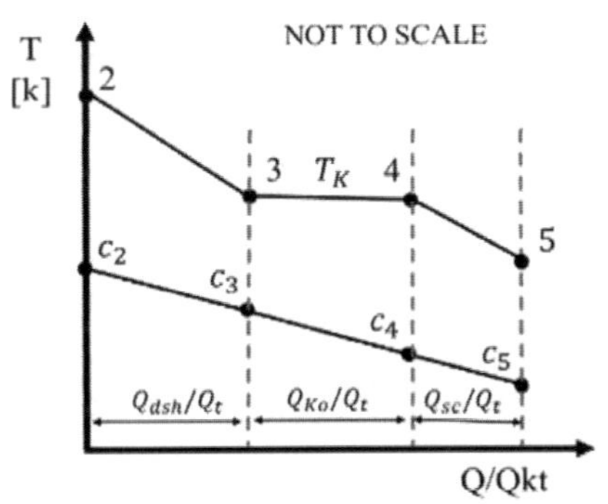

Figure 3. Condenser TQ chart

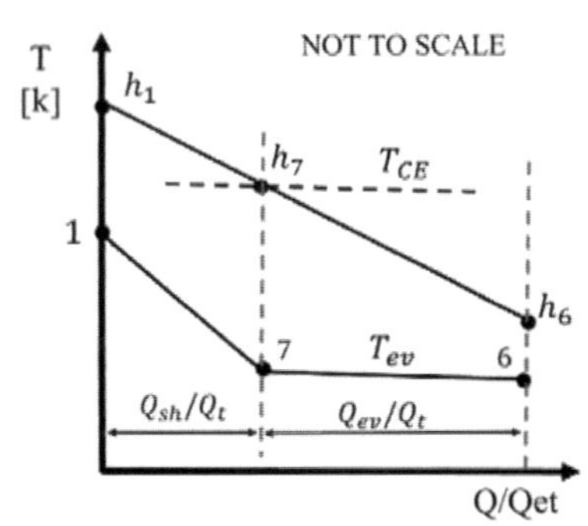

Figure 4. Evaporator TQ chart

2.1.2. The overall SC model. The SC is described by a set of independent nonlinear equations (SCNLES) that is the sum of the (1) - (6) NLES. The number of equations is equal to the number of unknown variables. $T_k, T_{ev}, \Delta T_{sc}, \Delta T_{sh}$, V_d, n_o, n_N, P_N, P_{MX} are the data. There is no DoF. The SCNLES also contains all the empirically defined relationships.

2.2. Power Regenerated VCRP

The most interesting Power Regenerated Cycles (PRCs) that have been investigated in the COLD ENERGY project are shown in Figure 5. Power regeneration is based on internal power cycles (IPCs) that drive compressors to increase the pressure at the main compressor suction. Each IPC uses as working fluid a fraction of the condensate mass flow that is bled and reduced in pressure by the bleed valve (BV). The Bleed Mass Flow (BMF) vaporises, receiving the heat power from the condensate stream, which increases the subcooling in the bleed vapour generator (BVG). Figure 6 shows a diagram of the BVG processes. As shown in Figure 5, the superheated bleed stream (SBS) is expanded in a turbine that drives a compressor where the stream from the main evaporator is compressed before moving towards the main compressor inlet together with the bleed expander exhaust stream.

The model of each Power Regeneration Internal Cycle (PRIC) (index j) is made from j[th] IPCNLES that can be added to the SCNLES and acquiring s_{BVG}, p_{bj} and μ_{bj} as degrees of freedom. The domain of definition of variables and the degree of freedom are bounded by the Approach and Pinch Point Temperature Differences (DTA and DTPP) in the jth BVG and by the data for each compressor expander group, diagrammed in Figure 7. The BVG model establishes a BVGNLES:

$$F_{BVGj}(\mu_{hj},\mu_{cj},T_{5j},T_{6j},T_{bj},H_{bj},p_{bj},T_{bshj},H_{bshj},p_{bshj},DTA_j,DTPP_j,u_{hj},u_{cj},\sigma_{ev},\sigma_{sh},S_{BVG},Architecture,Geometry)=0 \quad (7)$$

μ_{cj} and μ_{hj} being the fractions of the overall condensate mass flow that gives the heat power being subcooled and the fraction of the bleed cold stream receiving the heat power. σ_{ev} , σ_{sh} are the fractions of the overall BGV heat transfer surface area (S_{BVG}) .

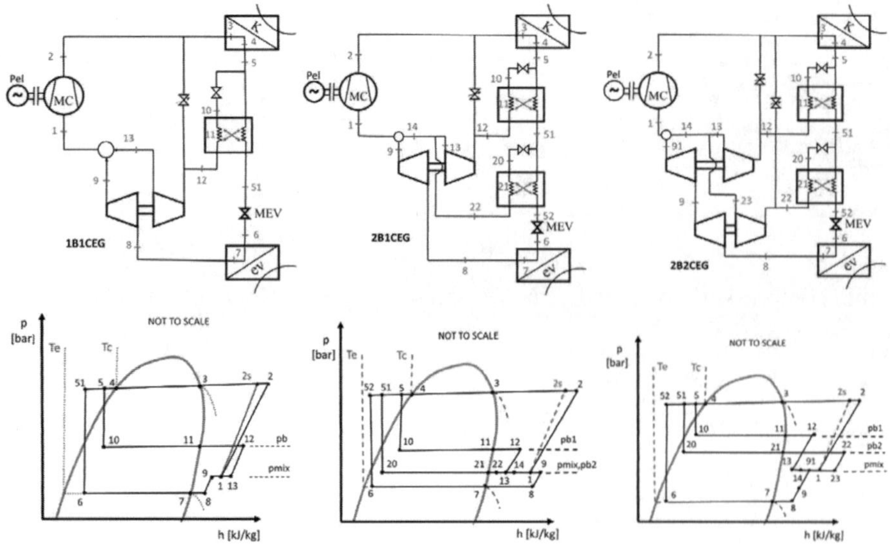

Figure 5. Layouts and p-H chart cycles for three options

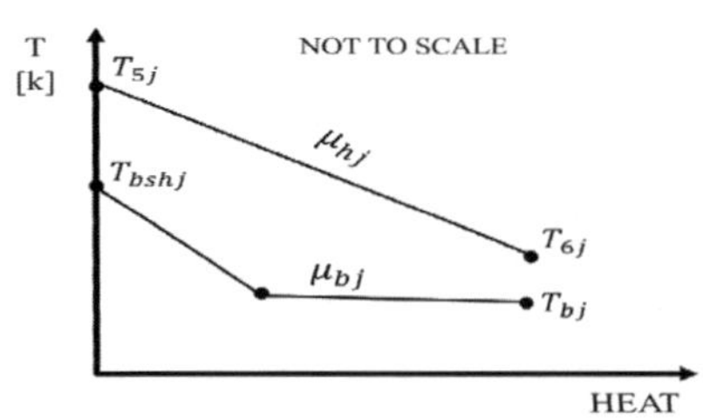

Figure 6. Bleed vapour generator TQ chart

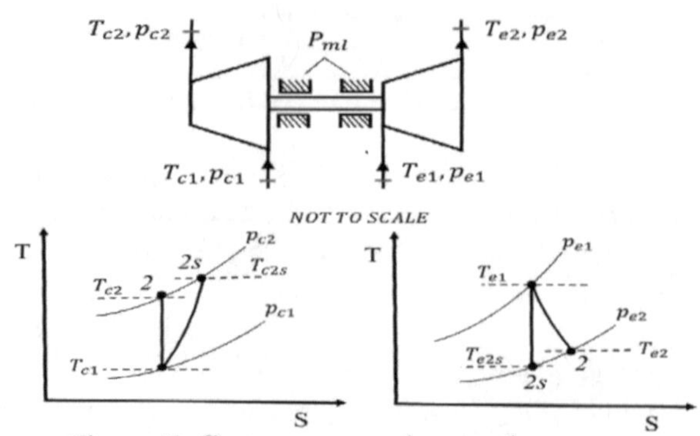

Figure 7. Compressor and expander processes

An additional PRIC submodel is for the Compressor Expander Group (CEG). A sketch of the CEG and TS chart processes is given in Figure 7. CEGNLES establishes the following correlations

$$\mathbf{F}_{CEG}(m_c,m_{e1},T_{c1},p_{c1},H_{c1},p_{c2},T_{c2s},H_{c2s},T_{c2},H_{c2},\eta_{cs},p_{e1},T_{e1},H_{e1},p_{e2},T_{e2s},H_{e2s},T_{e2},H_{e2},\eta_{cs},\eta_{es},\eta_m,P_{cs},P_{es},P_c,P_e,P_{ml})=0 \quad (8)$$

η_{cs},η_{es} and η_m being specified as data or functions.

2.2.1. The overall PRVCRC. The overall Power Regenerated Vapour Compression Refrigerated Cycle (PRVCRC) is represented by an Overall NLES (ONLES) which consists of the sum

$$\mathbf{ONLES{=}SCNLES \cup BVGNLES \cup CEGNLES} \quad (9)$$

BVGNLES and **CEGNLES** being the sum of all the **BVGNLES**(j) and the **CEGNLES**(j). The degrees of freedom for each jth PRIC are S_{BVGj}, p_{bj} and m_{bj} $\forall\ j=1,N_b$. If these $3\cdot N_b$, DoFs are inside the feasibility domain, a solution can be calculated. Of course, an optimization process can be established:

Search S_{BVGj}, $\overline{p}_{bj}$ *and* $\overline{m}_{bj}$ $\forall\ j=1,N_b$

maximize $COP = CP / P_{MC}$ (10)

subject to the ONLES.

Some inequality constraints are restrictions on temperatures and pressures.

Once the compressor and expander impellers have been selected, it is necessary to provide the inlet ducts, the compressor diffuser, the expander nozzle and the expander exhaust diffuser. These subcomponents must be sized and the impeller profiles must be modified to achieve the best performance.

2.2.2. Turbomachinery Selection. The Turbomachinery Selection (TS) model is based on similitude concepts using the Car Engine Turbocharger Technology (CETT) Data Base. Details are given in [3]. This model leads to a TSNLES.

$$\boldsymbol{F}_{TS}\,(n,m_c,T_{c1},p_{c1},H_{c1},H_{c2s},\eta_{cs},\omega_{cs},D_{c2},m_e,T_{eo},p_{eo},p_{e4},D_{e4},\eta_{es},\omega_{es},u/C_{spo})=0 \quad (11)$$

This additional model introduces the compressor specific angular speed ω_{cs} as one more degree of freedom eliminating η_{es} and η_{cs} as inputs. Compressor inducer tip speed is bounded by the choking. The modified ONLES problem (equation (10)) that also contains $\boldsymbol{F}_{TS}$ leads to the optimizing of the PRVCRC performance with the more suitable turbomachinery impellers being selected.

2.2.3. Turbomachinery Sizing. Once the nominal speed, the compressor exducer and expander inducer dimensions have been selected together with the impeller types, the hub and shroud profiles are known. The problem is that of modifying the shroud profiles of the compressor and expander to achieve the best performance, minimizing the losses.

2.2.3.1. Turbomachinery 1D MLM. The shroud profiling of the compressor and expander CETT selected impellers is performed using a 1D Mean Line Modelling approach. The compressor and expander through flow sections are shown in Figure 8. The 1D ML is based on the use of channel sections that are quasi-orthogonal to the channel centreline, see Figure 9. Using the mass and rothalpy conservation, it is possible to establish a set of NLES for any impeller that, for each radius r, correlates:

$$\boldsymbol{F}_i(m,r,h,p,W,b.\Delta h_l)=0 \quad (12)$$

If r and $b(r)$ are given, the h, p and W can be found. Due to complex flow in the channels (an example is given in Figure 10), entropy production is associated with the following losses: distributed friction along the channels; incidence; loading; shock; clearance; mixing; recirculation; windage; leakages; diffuser and collector. Disc windage loss can play an important role when the impeller input power is not high. Loss $\sum \Delta h_l$ can be calculated as an objective function, thus the following problem can be stated:

Search $\overline{b}(r_j)$ $\forall$ j=1, N_r

minimize $fob = \sum \Delta h_l$ (13)

subject to TSNLES.

The solution of this problem provides the best $W(r)$, $p(r)$ and $h(r)$ distributions along the channel centreline and to establish the shroud profile:

$$SP(r_s, z_s) = 0 \quad (14)$$

then the analysis problem can be solved.

IOP Conf. Series: Materials Science and Engineering **232** (2017) 012086 doi:10.1088/1757-899X/232/1/012086

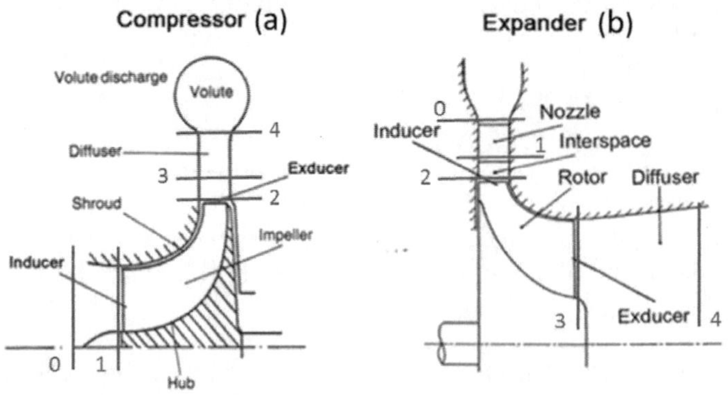

Figure 8. Compressor and expander through flow sections

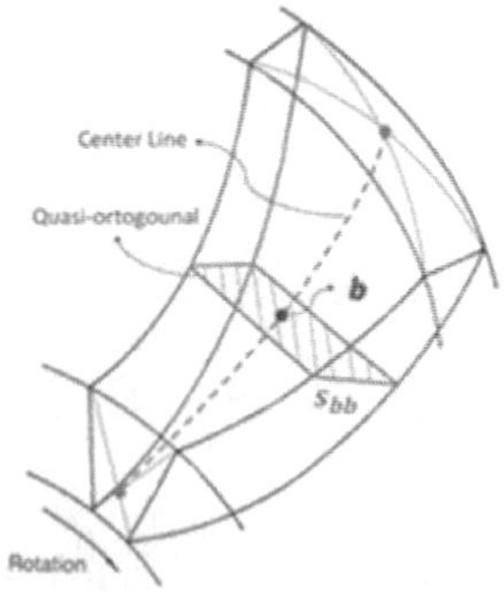

Figure 9. Channel quasi-orthogonal section

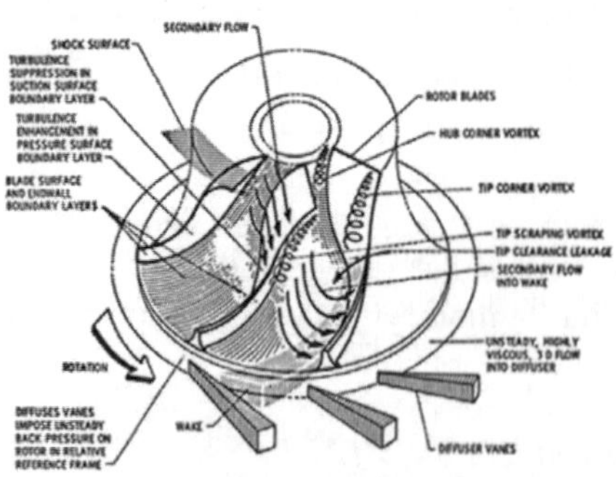

Figure10. Schematic view of the complex flow structure in a radial compressor impeller

2.2.3.2. Compressor Model Characteristics. Modelling the centrifugal compressor impeller flow is a complex task; Figure 10 shows a typical picture. As concerns the exit triangle of velocity, the deviation connected with the shift factor has to be taken into account. The flow blockage has to take the recirculation (wake) zone into consideration.

2.2.3.3. Expander Characteristics. The incidence is the difference between the angle of the flow approaching the rotor channel inlet and the optimal flow angle associated with the impeller and flow field features. The deviation of the flow in respect to the direction of the channel axis must be calculated taking the channel characteristics and the flow features into account. High Mach number and post expansion conditions produce a substantial flow turning. The nozzle-rotor interspace size also plays an important role in the flow direction and losses.

2.2.3.4. Turbomachinery sizing outcomes. Application of the 1D MLM provides the profiles of the compressor and expander impellers as shown in Figure 11. A typical compressor off-design analysis result is given in Figure12.

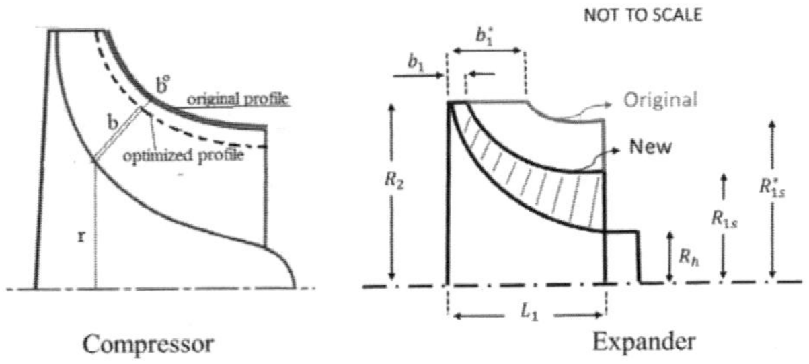

Figure 11. Modified impeller shroud profiles

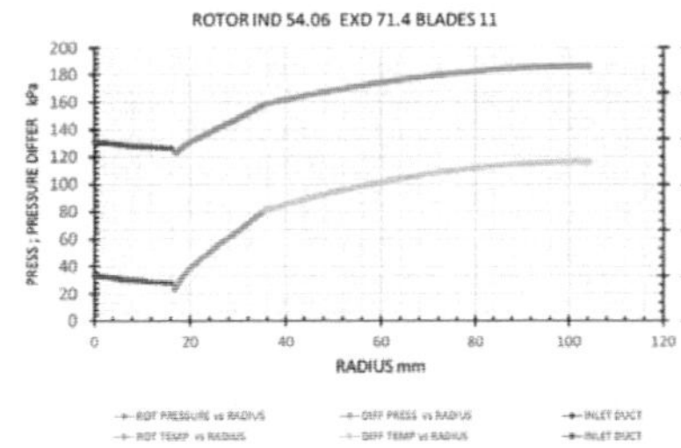

Figure 12. Simulated pressure and temperature distributions from the compressor inlet to the exit

3. Theoretical and experimental results

The PRVCRC has been optimized taking into consideration the use of a V_d = 3 dm^3/turn screw compressor running at a synchronous speed of 3000 rpm. For 2B1G layout with R404a WF having $T_c = +40°C$ and $T_{ev} = -40°C$, $\Delta T_{sc} = 5°C$ and $\Delta T_{sh} = 5°C$. Optimum performance versus p_{b1} is shown in Figure 13, which for constant Cooling Power (CP) shows how the main compressor power decreases, achieving a minimum of about 650–700 kPa bleed pressure. For $V_d \cdot n_o = const$ (i.e. for the specified main compressor) it can be remarked that by increasing the 1st bleed pressure, COP and CP rise until they reach the maximum value, close to 650–700 kPa. The optimum compressor pressure increase is about DPC = 80 kPa, thus the MC suction pressure becomes about 210 kPa, while for the SC it is about 133 kPa. MC pressure ratios are 8.7 and 14 respectively for the optimum PRVCRC and for the SC. For the optimum PRVCRC, the selected CEG speed is about 4 krad/s and impeller tip velocities are 143 m/s and 137 m/s respectively for compressor and expander. The selected shroud impellers have been profiled for the R404a optimum CEG compression and expander processes. Figure 14 shows the machined compressor wheel and Figure 15 shows the original and the machined expander impellers.

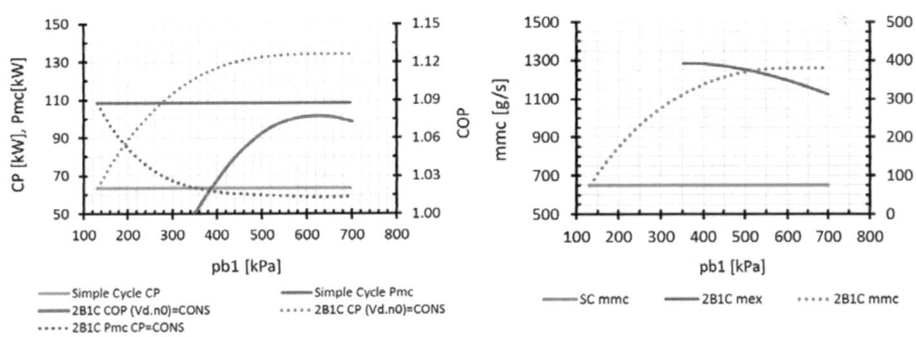

Figure 13. Optimum CP, COP, m_{MC} and m_{ex} versus the 1st bleed pressure

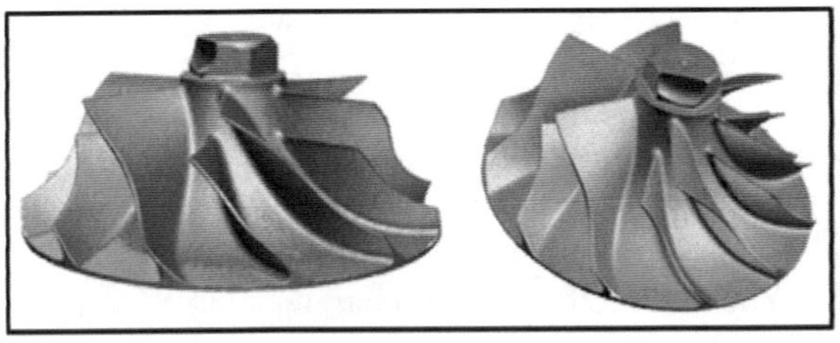

Figure 14. Two views of the machined compressor impeller

Figure 15. Original and machined expander impellers

Figure 16 (a) gives the volute and nozzle arrangements. Figures 16 (b) and (c) show the variable geometry mechanism. The compressor diffuser has been selected as a vaneless one, owing to the de-rated working conditions that could occur. The outer diffuser diameter is 2.8 times larger than the impeller exducer diameter.

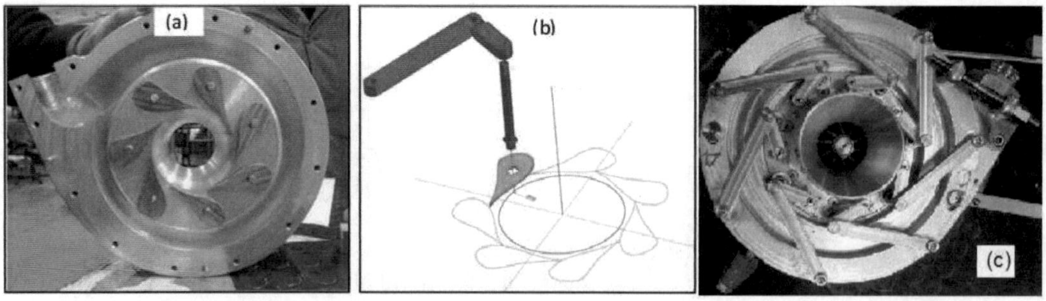

Figure 16. Volute, stator vanes and variable nozzle geometry mechanism

To verify the Power Regeneration Concept by means of IPCs, a Test Bench (TB) for the 2B1G scheme shown in Figure 5 was designed and built. Figure 17 shows the TB built and the CEG inside the hermetic container. The TB consists of four essential elements:

1. Condenser unit including the controlled the condensing water loop;
2. Calorimeters section, in which cooling power produced by the plant is measured, also called the cooling load system;
3. Compressor-expander group (CEG);
4. Cold liquid R404a piping for hybrid ball bearing lubrication purposes.

The TB is provided with special software that ensures the automatic control, measured data acquisition and storage of all parameters related to the overall operation of the refrigerating system, and more specifically of compressor-expander group.

The CEG was installed in this test bench, and some results are summarized in following. Tests have been carried out for different expander nozzle openings using the bypass valve (BPV) to adapt the expander flow rate to what is needed. Figure 18 shows the CEG rotational speed and Compressor Pressure Difference (CPD) versus time (normalised) elapsed during the test. Up to about 4000 times the tests were carried out with no condenser bypass flow rate; the deficiency of the bleed vapour mass flow rate does not allow running over 25 krpm, and consequently the pressure increase through the compressor was not higher than 20 kPa. After the normalised 4000th time the condenser bypass valve was opened, increasing the mass flow rate through the expander and reducing it through the compressor. Thus, the CEG reached an equilibrium at a higher rotational speed. The lack of the bleed vapour is connected to the insufficient heat transfer in the BVG; work is now being done to fix this problem.

Figure 19 gives the expander pressure difference, the compressor pressure difference and the CEG speed versus expander mass flow rate; the measured points and tendency lines are shown. According to theory, Expander Pressure Difference (EPD) and krpm vs Expander Mass Flow (EMF) seem to have a linear behaviour, while for the CPD vs EMF, the figure refers to 40% of the nozzle opening, which corresponds to the design condition for this type of main compressor. The corrected mass flow shows a linear behaviour.

As shown in Figure 20, the EPD vs EMF slope increases for low nozzle openings, and EPD greater than 500kPa is obtained for large mass flow rates that can be obtained with BVG having a large Number of heat Transfer Units (NTU), i.e. high effectiveness. The corresponding rpm are in the range of 38-40 krpm with about 90 kPa of compressor pressure increase, with the pressure ratio being around 1.7.

Figure17. Cold energy test bench photo

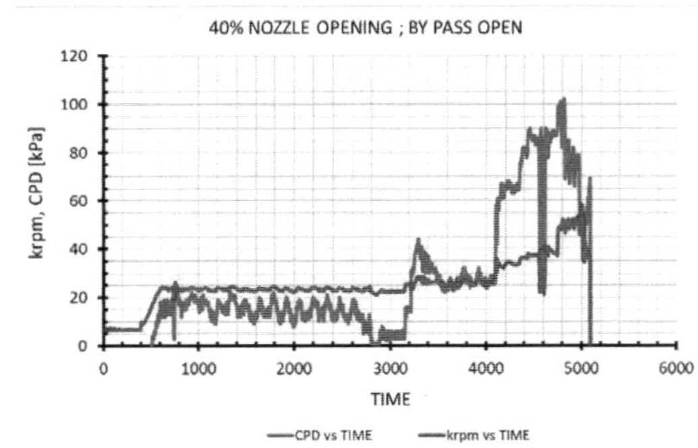

Figure 18. CPD and krpm versus normalised time for 40% nozzle opening

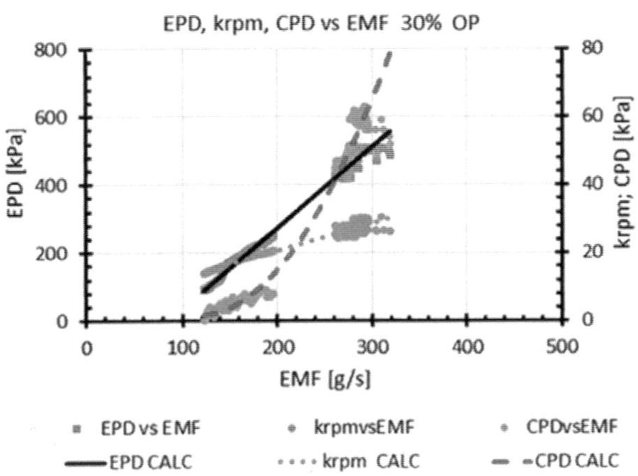

Figure 19. Expander and compressor pressure differences and CEG rotational speed versus expander mass flow rate

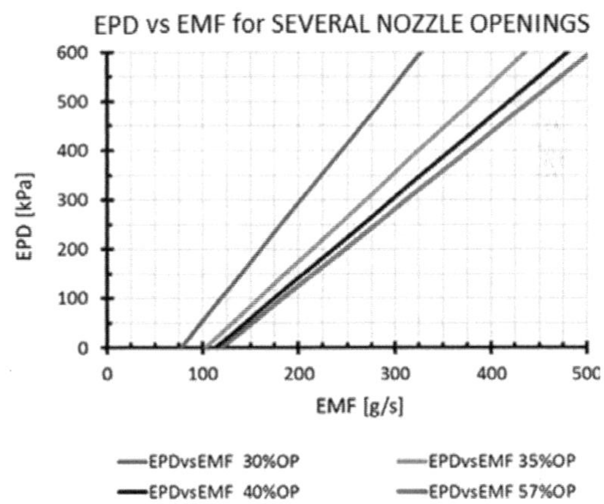

Figure 20. EPD versus Expander mass flow rate for constant nozzle openings

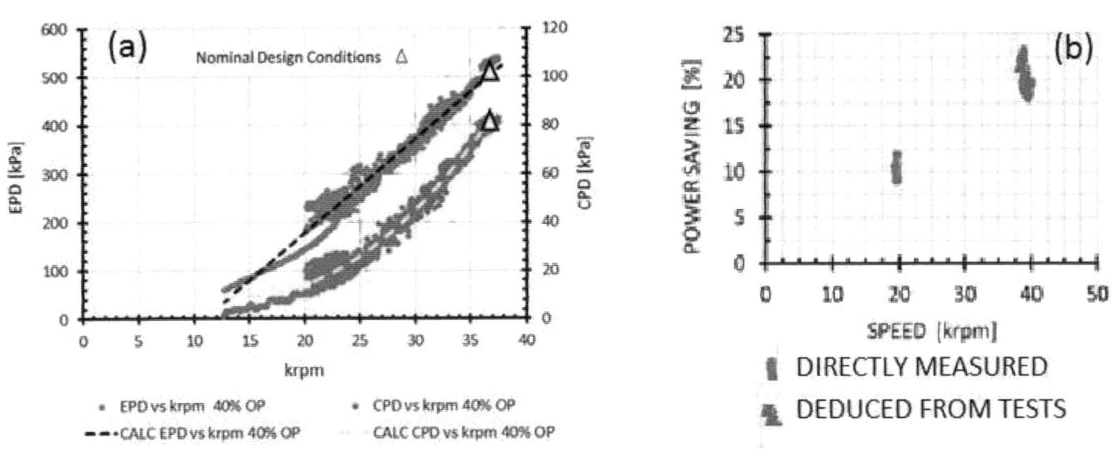

Figure 21 (a) & (b). Expander pressure difference (a) and power saving (b) versus CEG speed

Figure 21 (a) shows the expander and compressor pressure differences versus rpm. The CEG without expander flow is pushed by the main compressor suction action. Increasing the expander mass flow, the velocity increases up to 13 krpm, with no compressor positive pressure difference. For higher rpm, both compressor and expander pressure differences increase up to the design conditions, indicated by yellow triangles.

The analysis of power saving, using 100 kW cooling power as a reference, is given in Figure 21 (b). The measured power saving in respect to the simple cycle is about 12%-14% with the plant running outside of the optimized conditions and with no bypass flow rate. The predicted power saving, using the measured data to deduce the results under optimized conditions corresponding to about 80 kPa of compressor pressure increase, is about 22%-24% and higher than expected by the theoretical analysis.

4. Conclusions

The utilization of an internal power cycle for power regeneration purposes brings about a lowering of the power consumption per unit of cooling power produced by a VCRP. It has been demonstrated that the cycle and components can be optimized up to achieve about 22%-24 % of power saving, with the reference being a simple cycle.

Turbomachinery for the compressor expander group can be selected from the units available in the Car Engine Turbocharger Technology. Modifications to the impeller profiles must be taken into consideration. Tests have confirmed the theoretical approach.

The results obtained by the COLD ENERGY project provided the motivation to continue both in performance improvement and product industrialization. An industrial application is now being investigated and some initial forecasts have been studied and will be demonstrated in dedicated industrial VCRPs.

Acknowledgements

The authors wish to acknowledge the Italian Ministry for the Environment, Land and Sea; Angelantoni Industries; Se.Te.L s.r.l. and Roma Tre University for the COLD ENERGY Project Support. Acknowledgements also to the EC for the Cold Energy ID 737829 Grant.

References

[1] Borlein, 2011 Energy Savings in Commercial Refrigeration Equipment: Low Pressure Control, *White paper, Schneider Electric.*

[2] Ascani M, Cerri G and De Francesco E, 2015 Power Reduction in Vapour Compression Cooling Cycles by Power Regeneration, *Elsevier Ltd, Energy Procedia 81,* 1184 – 1197.

[3] Alavi S.B., Cerri G., Chennaoui L., Giovannelli A. and Mazzoni S., 2015 Optimum Turbomachine Preliminary Selection for Power Regeneration in Vapor Compression Cool Production Plants, *Inter. Journal of Mechanical, Aerospace, Industrial and Mechatronics Engineering,* Vol: 9 (4).

[4] Kairouani L., Elakhdar M., Nehdi E. and Bouaziz N., 2009 Use of Ejectors in a Multi-Evaporator Refrigeration System for Performance Enhancement, *International Journal of Refrigeration,* Vol.32, pp.1173-1185.

[5] Bergander M. J., 2006 Refrigeration Cycle with Two-Phase Condensing Ejector, *International Refrigeration and Air Conditioning Conference,* Paper 748.

[6] Cerri G. and Battisti L., 1994 Valve Control for Optimum Performance in Compression Refrigeration Cycles, *Journal of Heat Recovery Systems & CHP,* Vol. 14 (1), pp.61 – 66.

[7] Cerri G. and Khatri D. S., 1998 A Neural Network Approach in Thermodynamic Process Evaluation, *International Conference on Engineering Application of Neural Network (EANN-98).* Gibraltar, Great Britain, Paper No. 98172.

[8] Cerri G., Battisti L. and Salvini C., 1998 Parametric Analysis of Vapour Compression Refrigeration Plant Steady State Behaviour, *Emerging Trend in Refrigeration & Air Conditioning-IIR Conference,* New Delhi, India.

[9] Boccaletti C. and Cerri G., 1999 A Dynamic Model for Reciprocating Compressor Parametric Analysis, *International Conference on Enhancement & Promotion of Computational Methods in Engineering & Science (EPMESC VII),* Macao.
[10] Navarro H.A. and Cabezas-Gómez L.C., 2007 Effectiveness-NTU Computation with a Mathematical Model for Cross-Flow Heat Exchangers, *Brazilian Journal of Chemical Engineering*.
[11] Cerri G., 1988 Organic Fluid Turbines for Various Engine Power Level Turbochargers, *The 33rd ASME International Gas Turbine and Aeroengine Congress*, Amsterdam, Netherlands, ASME pap. 88–GT–1.
[12] Cerri G., 1987 Organic Fluid Turbocharger for a 2000 kW Diesel Engine, *International Gas Turbine Congress*, Tokyo, Japan.

Nomenclature

Symbols

b	Channel height
cs	Sound Speed
c_p , c_v	Specific Heat Capacity
C_{spo}	Spouting Velocity
D	Diameter
F	Function
H, h	Enthalpy
m	Mass Flow Rate
n	Rotational Speed (rpm)
n_o	Synchronous Speed
N_r	Number of Radial Nodes
P	Power
p	Pressure
Q	Heat
r	Radius
s	Entropy
S	Surface
T	Temperature
u	Heat Transfer Coefficient
v	Specific Volume
V_d	Delivery Volume
W	Work
z	Cartesian Coordinate
Δ	Difference
η	Efficiency
μ	Mass Flow Fraction
σ	Heat Transfer Surface Fraction
ω	Angular Speed

Subscripts

b	blade, bleed
c	compressor, cold
d	dispersed, lost
e	expander
em	electric motor
ev	evaporator
h	hot
i, j, k	order number
k	condenser
m	mechanical
MC, mc	main compressor
ml	mechanical loss
MX	maximum
N	nominal
s	specific, isentropic, shroud
sc	simple cycle
sh	superheated
v	volumetric

Climate Adaptivity and Field Test of the Space Heating Used Air-Source Transcritical CO_2 Heat Pump

Yulong Song, Zuliang Ye, Feng Cao

School of Energy and Power Engineering, Xi'an Jiaotong University,
Xi'an 710049, China

fcao@mail.xjtu.edu.cn

Abstract: In this study, an innovation of air-sourced transcritical CO_2 heat pump which was employed in the space heating application was presented and discussed in order to solve the problem that the heating performances of the transcritical CO_2 heat pump water heater deteriorated sharply with the augment in water feed temperature. An R134a cycle was adopted as a subcooling device in the proposed system. The prototype of the presented system was installed and supplied hot water for three places in northern China in winter. The field test results showed that the acceptable return water temperature can be increased up to 55°C, while the supply water temperature was raised rapidly by the presented prototype to up to 70°C directly, which was obviously appropriate to the various conditions of heating radiator in space heating application. Additionally, though the heating capacity and power dissipation decreased with the decline in ambient temperature or the augment in water temperature, the presented heat pump system performed efficiently whatever the climate and water feed temperature were. The real time COP of the presented system was generally more than 1.8 in the whole heating season, while the seasonal performance coefficient (SPC) was also appreciable, which signified that the economic efficiency of the presented system was more excellent than other space heating approaches such as fuel, gas, coal or electric boiler. As a result, the novel system will be a promising project to solve the energy issues in future space heating application.

1. Introduction

Recently, lots of cities in northern China were invaded by haze once the heating season came. A main cause of the strict dilemma in haze is that plentiful of distributed coal-fired boiler in northern China exhausted a mass of dust and smog to atmosphere. Nowadays, policies were presented in China to promote extensively the alternation of the thermal resource from coal to electricity in order to decrease the environment pollution. Comparing with the electric boiler, the very high system performances of air-source heat pump (ASHP) which can absorb the thermal energy from atmosphere and transport to water makes the ASHP become the best choice to provide thermal energy to users in winter.

However, it is the heat radiator instead of the underfloor heating or fan coil which is still the broadest heating facility in China. The heating conditions of circulating water in heat radiator which are 60~80°C in supply temperature and 40~50°C in return temperature are far beyond the operating conditions of heat pumps using convenient refrigerants which can only provide the water supply temperature of 40~50°C [1]. Fortunately, the transcritical air-source CO_2 heat pump which can provide hot water with temperature of higher than 80°C directly is proper for the working conditions of the heat radiator.

To solve the problem of the space heating used transcritical CO_2 heat pump (SHTCHP) that the system performances deteriorated sharply with the increase in the return water temperature which is 40~50°C in heat radiator, abundant researches were carried out. Wang [2] illustrated that the transcritical CO_2 system could be applied to wonderful thermal resource due to its huge temperature glide in the gas-cooler. However, the COP of the transcritical CO_2 system is quite lower than systems with convenient refrigerants due to the remarkable irreversibility which caused by the superheated vapor horn and the high throttling losses. The conclusion can be obtained by comparing the results from [2] and [1]. Thus, plenty of improvements were worked out to enhance the efficiency of transcritical CO_2 systems by employing various modifications.

Lots of scholars experimentally discussed the performances of the transcritical CO_2 system with and without the IHX in single-stage transcritical CO_2 systems [3-5] as well as similar investigations were carried out to the two-stage systems [6-7]. The performances of the transcritical CO_2 system with an IHX was assessed by [3] in which they found that the system COP was enhanced by up to 10.5% by adding the IHX since the fluid temperature at gas-cooler inlet decreased from 40 to 25 °C. The performances of an IHX based CO_2 transcritical refrigeration prototype at three various evaporating temperatures (-5 °C, -10 °C and -15 °C) and two gas-cooler outlet temperatures (31 °C and 34 °C) were analyzed by [5] experimentally and they found an augment in system cooling capacity of 12% and the system efficiency up to 12%. As for the two-stage compression cycle from [6-7], the simulation research indicated that IHX rose the specific cooling capacity and declined the optimal discharge pressure, and the adoption of the IHX enhanced the system COP up to almost 10%. The effects of the vapor extraction from the vessel and subsequent injection into the refrigeration cycle in a double-stage system were studied by [8] experimentally and they found that the system COP increased by approximately 16.5%.

Another excellent way to enhance the performances of the transcritical CO_2 system is the subcooling technology. The performances of a transcritical CO_2 system with a thermoelectric facility were investigated by [9] theoretically. The results found that the system COP can be increased by up to 25.6%. The energy improvements of the transcritical CO_2 system with a dedicated mechanical subcooling were analyzed theoretically by [10]. The simulation results showed that the enhancements in system COP and system cooling capacity was up to 20% and 28.8%, respectively. Besides, the comparison was carried out by different refrigerants in the subcooling cycle. The combined transcritical CO_2/R134a heat pump system which was dedicated for space heating applications was investigated theoretically and experimentally by [11] and [12], respectively. The results of those studies proposed that the combined transcritical CO_2/R134a system can operate efficiently and provide stably hot water with temperature of higher than 70°C. Furthermore, the combined CO_2/R134a system was compared with the cascade CO_2/R134a system.

As mentioned above, although the transcritical CO_2 system showed a good potential on space heating application and hot water production, there is little literatures can be found to test such a machine in specific conditions of space heating field. In this paper, three prototypes of SHTCHPs which is similar with the plant in literature of [12] are installed and tested in three different cities (Tianjin, Shijiazhuang and Zhangjiakou) with different climate conditions. Further, the heating performances of the prototypes like return water temperature, supply water temperature, heating capacity and COP are recorded and analyzed. Additionally, the effects of defrosting procedure is also discussed.

2. System description

2.1 Field test facility

The proposed transcritical CO_2 heat pump system is a combination form of a standard transcritical CO_2 system and a standard R134a refrigeration system. In this form, the standard R134a refrigeration system play a role as a subcooling machine to provide cooling capacity for the refrigerant CO_2 from the gas-cooler which is not cooled enough. The sketch map of the proposed system which includes a transcritical CO_2 cycle, an R134a cycle and a circulating water system is shown in Figure 1a. The ambient temperature is varied from -20 to 10 °C as well as the feed water (from user) varied from 40 to 50 °C. The feed water flows through the three-way proportional valve and then is split into two parts of streams. The first part of feed water enters the condenser of the R134a subsystem in which it is heated up by the high pressure and high temperature R134a refrigerant before it is channeled into the mixing tank. Another part of feed water is channeled into the evaporator of the R134a subsystem in which it is cooled down by the low pressure and low temperature R134a refrigerant. The cold water in the exit of R134a evaporator is then flowed into the gas-cooler of the transcritical CO_2 cycle in which the cold water sub-cools the CO_2 refrigerant and meanwhile is heated up by the high temperature CO_2. This part of hot water is channeled finally into the mixing tank in which it is mingled to the first part of hot water from the R134a condenser. In this system, circulating water system is used to be an intermediate fluid to link the gas-cooler of the transcritical CO_2 subsystem and the evaporator of the R134a subsystem.

The experimental prototype which includes an R134a subsystem and a transcritical CO_2 subsystem is shown in Figure 1b. Three same prototypes were installed in three different places: a hall of a ski resort where the ambient temperature is low to -20°C in Zhangjiakou City, a government building where the average ambient temperature is almost -10°Cin Shijiazhuang City, and a train station where the average ambient temperature is almost 0°C in the whole heating season in Tianjin City.

The test data acquisition system is comprised by various measurement facilities that can measure the system parameters and maintain the system status. For instance, the ambient temperature and relative humidity are measured by a temperature and humidity recorder. The measurement range of the temperature sensor is -40°C~150°C with the accuracy of ±0.5°C and that of the humidity sensor is 0~100%. T-type thermocouples with the accuracy of ±0.5°C are used to measure the circulated water temperatures (both inlet and outlet). The water flow rates are measured by electromagnetic flowmeters with the measurement range of 0~6m3h-1 and the uncertainty is ±1%. The total electric power dissipation from compressors, fans and water pumps is recorded by an electric power meter. More details of the measurement facilities are showed in Table 1.

Table 1. Details of the measurement facilities

Measuring object	Facilities	Measuring range	Uncertainty
Ambient temperature (°C)	Temperature and humidity record	-40~150	±0.5°C
Relative humidity (%)	Temperature and humidity record	0~100	±5%
Water temperature (°C)	T-type thermocouple	-40~150	±0.5°C
Water volume flow rate (m3h-1)	Electromagnetic flowmeter	0~6	±1% of reading
Power dissipation (kW)	Electric power meter	0~100	±0.5% of full scale
R134a pressure (MPa)	Pressure transmitter	0~6	±1% of reading
CO_2 pressure (MPa)	Pressure transmitter	0~16	±1% of reading

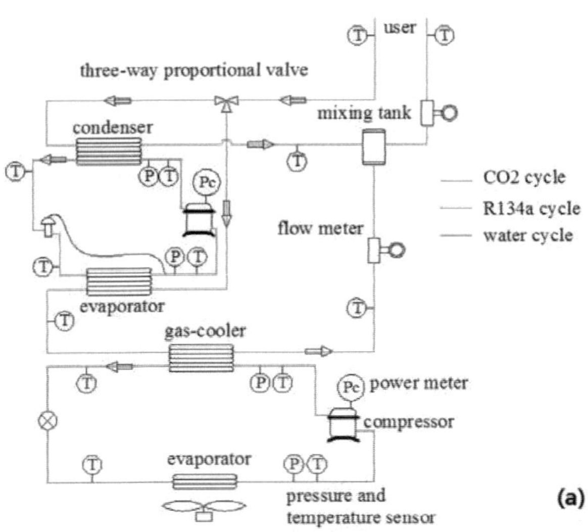

Figure 1a. The sketch map of the space heating used transcritical CO_2 system

Figure 1b. The picture of the experimental prototype

2.2 System analysis and defrost control

In this study, the heating capacity and the power dissipation of the prototype can be measured and then the system COP can be calculated. Since the water volume flow rate and both the inlet and outlet temperatures were measured and recorded, the heating capacity of the prototypes can be calculated from water side accurately, which is written by:

$$Q_i = c_{p,water} \cdot G_i \cdot (t_{out,i} - t_{in,i}) \quad (1)$$

Additionally, in order to assess the system performances of the prototype during all the heating season, the total heating capacity during the total working period τ can be calculated by:

$$Q_{season} = \int_0^{\tau} c_{p,water} \cdot G_i \cdot (t_{out,i} - t_{in,i}) d\tau \quad (2)$$

Since the ambient temperature varies continuously with time under the field test conditions, the instantaneous COP has less significance than the seasonal performance coefficient (SPC) which can be calculated by (all the test data of the prototype will be recorded per 1.5 minutes):

$$SPC = \frac{Q_{season}}{W_{season}} = \frac{\sum_{k=1}^{n}\sum_{h=1}^{j}\sum_{m=1}^{p} Q_{1.5\,min}}{\sum_{k=1}^{n}\sum_{h=1}^{j}\sum_{m=1}^{p} W_{1.5\,min}} \quad (3)$$

where the total power dissipation can be recorded by the electric power meter.

As for the operating logic, the transcritical CO_2 subsystem is the main cycle while the R134a subsystem is the vice-cycle. If the return water temperature is lower than 30°C in which case the CO_2 gas-cooler outlet temperature can be cooled adequately by only the return water, the transcritical CO_2 subsystem is activated to heat the circulating water and the R134a subsystem is unoperated. When the return water temperature reaches 30°C or higher, the R134a subsystem is activated to be the subcooler.

The frosting phenomenon is a serious constraint for all air-source heat pump because the system performances of the heat pump system will be deteriorated sharply with the increase in the thickness of the frost layer. Thus, the defrost method of the proposed system is also significant for the performance assessment in field test conditions. The convenient defrost approach which employ a four way valve to interchange the condenser and evaporator will cause low water supply temperature thereafter decline of indoor temperature during the defrost procedure. Moreover, there is no suitable four way valve which can be used in the transcritical CO_2 system because of the very high operated pressure. Therefore, the hot gas by-pass approach is adopted. The water pump stops immediately and the high pressure and high temperature CO_2 which is discharged from the compressor release little thermal energy to the remained water in the gas-cooler before it flows into the evaporator once the defrost mode is activated. Thus the mass of remained thermal energy in the refrigerant CO_2 which is channeled to the evaporator can be used to melt the frost layer in fins. In this way, the defrost procedure can be controlled stably without a four way valve as well as the supply water temperature and indoor temperature can be kept in constant.

As for the defrosting control logic, the defrosting mode is activated by reaching three judgments simultaneously as follow:

- The ambient temperature is lower than 0°C;
- The operated period is longer than 45 mins;
- The temperature difference between the ambient and the finned tube is higher than 13°C.

On the contrary, the defrosting mode was cancelled by reaching any of three judgments as follow:

- The defrosting period reaches 15 mins;
- The low pressure of the transcritical CO_2 system reaches 3.8MPa;
- The finned tube temperature reaches 7°C.

3. Experimental results and discussion

3.1 Heating performances of the SHTCHP

As mentioned above, three prototypes of SHTCHPs are installed and tested in three different cities (Tianjin, Shijiazhuang and Zhangjiakou) with different climate conditions. The ambient temperature, water temperature, heating capacity and the system COP are recorded and analyzed in this study.

The main parameters and performances of the prototype in Tianjin are shown in Figure 2. The data showed that the ambient temperature kept approximately 0°C during the heating period (almost from 10:00am to 17:00pm) expect for a sudden drop of almost 5°C for almost 1 hour. Meanwhile, the prototype of the SHTCHP was switched to running from almost 10:00am to 17:00pm except for an interruption in 14:30. The return water temperature was remained 40~50°C during most time of the running period, which meant that the water temperature of the whole circulating water system can be heated quickly by the heating activity of the prototype. On the basis of the high return water temperature, the supply water temperature can be heated to 60~70°C once the prototype was running. The return water temperature thereafter the supply water temperature declined slightly due to the increased requirement of the heating capacity which might cause by the sudden drop of the ambient temperature. However, the sudden drop of the ambient temperature effected very little on the system heating capacity as well as the system COP, as shown in Figure 2. On the one hand, the system heating performances should be deteriorated by the drop of the ambient temperature, on the other hand, the heating performances was quite enhanced by the drop of the return/supply water temperature because the decline of the return water temperature decreased the CO_2 gas-cooler outlet temperature, which could improve the system performance significantly. Because of the combination of the two effects, the heating capacity and COP remained almost constant during all the heating period in this day. Additionally, the temperature difference between the supply water temperature and return water temperature was also kept almost constant due to the stable heating capacity and water flow rate.

The main parameters and performances of the prototype in Shijiazhuang are shown in Figure 3. The ambient temperature which was lower than -12°C in morning and evening and higher than -8°C in noon was more regular here than that in Tianjin. During all the running period (9:30am to 17:30pm), the data showed that the return water temperature increased slightly (from almost 35°C to nearly 50°C) from morning to noon due to the continuous heating activity of the prototype. Next, the return water temperature declined sharply first and then kept on constant until the machine was turned off in evening. The sudden drop of return water temperature was caused by the additional heating requirement indoor (a standby fan coil was activated) and the heat releasing capacity of the circulating water was increased. As mentioned above, the decline in return water temperature enhanced the heating performance of the prototype, which can be observed by the trend of COP in Figure 3. Because of the fluctuation of the return/supply water temperature when the water temperature drop occurred, the heating capacity of the prototype which was calculated by the temperature difference of the return/supply water also fluctuated subsequently. If the system performances (COP and heating capacity) can be split into two portions by the fluctuation of the water temperature drop, it can be observed that the system performances increased in the first half with the increase in the ambient temperature and decreased slightly in the second half with the slight decline in the ambient temperature, which is rational with the theory of the air-source heat pump system.

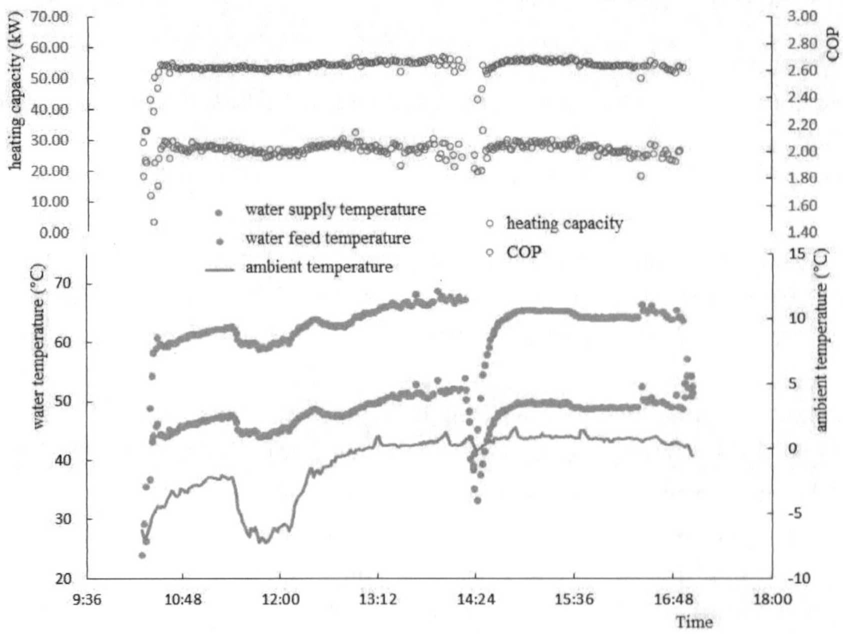

Figure 2. The main parameters and performances of the prototype in Tianjin, China

Similarly, the main parameters and performances of the SHTCHP in Zhangjiakou are shown in Figure 4. It can be seen that the ambient temperature in Zhangjiakou varied in a wide range in this day, which was lower than -15°C in the morning and increased up to -7°C in the afternoon. Very similar to the situation in Shijiazhuang, the frequent switch between start and stop of the standby fan coil caused the fluctuation in the return water thereafter the supply water temperature. However, the system heating capacity approximately increased first and then decreased with the ambient temperature except for several fluctuation, which is rational with the theory of the ASHP. Additionally, it can be noticed that the heating capacity is low while the system COP is quite high during the morning. Because the return water temperature in that time was lower than 30°C and the R134a subsystem was unoperated thereafter the power dissipation was lower. However, with the increase of the return water temperature and the activation of the R134a subsystem, the heating capacity was enhanced and the system COP can be remained at a rational value even the return water temperature is higher than 45°C which can deteriorate the system COP of the convenient transcritical CO_2 system to almost lower than 1.3.

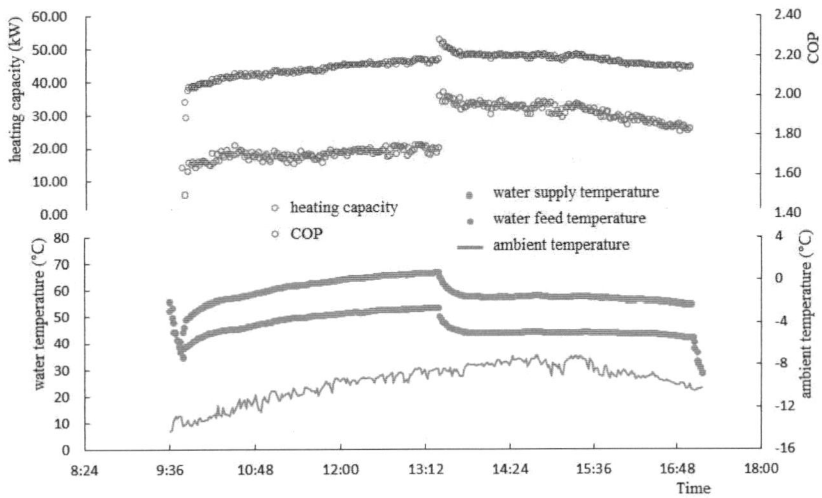

Figure 3. The main parameters and performances of the prototype in Shijiazhuang, China

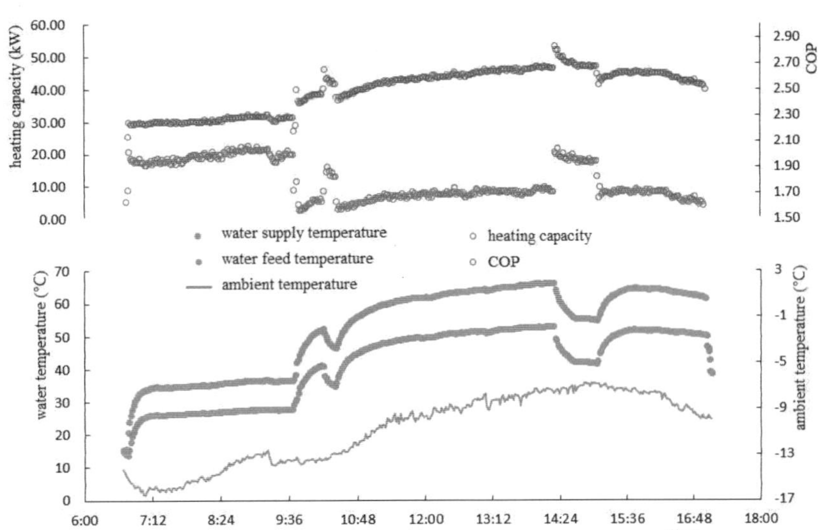

Figure 4. The main parameters and performances of the prototype in Zhangjiakou, China

According to the Figures 3~5, the system COP of the prototypes increased with the increase in ambient temperature and the decline in return water temperature. Besides, the system COP can reach to 2.0 if the operated conditions (ambient and water temperatures) were proper, and the system COP can also remain to more than 1.6 when the ambient temperature is lower than -15°C and the return water temperature is higher than 50°C. The results showed that this system can be used to supply efficiently hot water with more than 70°C in the extreme working conditions. Upon equation (3), the

SPC of the prototypes in three different cities can be obtained as follow: SPC=2.04 for prototype in Tianjin where the average ambient temperature is 1.8°C; SPC=1.81 for prototype in Shijiazhuang where the average ambient temperature is -8.6°C; SPC=1.68 for prototype in Zhangjiakou where the average ambient temperature is -15.2°C during the heating season.

3.2 Effects of the defrosting procedure on system performance

It is well-known that the frost phenomenon and defrosting procedure that is mainly determined by the operated conditions (ambient temperature and relative humidity) could deteriorate the performance of the heat pump system in winter. In this section, the parameters of the prototype during the defrosting procedure were analyzed under different operated conditions.

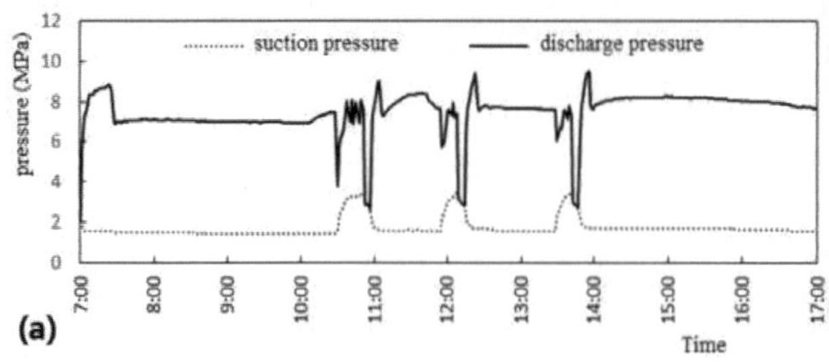

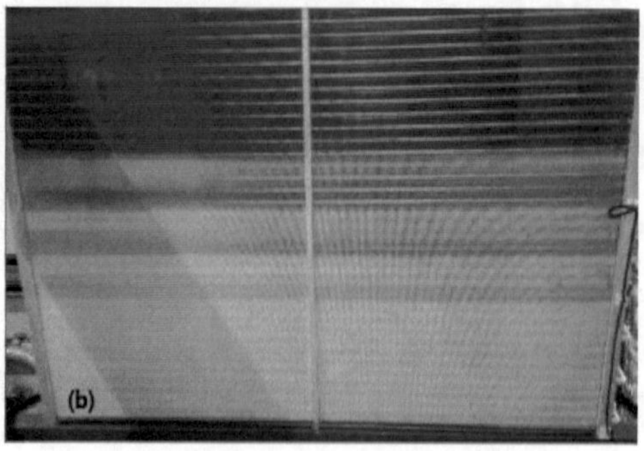

Figure 5a. The CO_2 pressures during several defrosting periods

Figure 5b. The picture of the frosting of system

The discharge and suction pressure of the transcritical CO_2 subsystem during a heating period is showed in Figure 5a. The operated conditions were given as follow: the ambient temperature was low to -20.4°C in the morning and -19.9 in the evening while it was up to -13.9°C in the noon. Besides, the relative humidity was up to 82% in the noon and decreased to lower than 70% in the morning and evening. There were three defrosting procedure occurred during all the heating period in this day, and the times of the three defrosting procedure were gathered from 10:30am to 14:00pm when the ambient temperature and the relative humidity were high. However, although appropriate ambient temperature and high relative humidity are able to cause frequent frosting phenomenon, the proposed system in which the hot gas by-pass defrosting approach was employed could defrost the frosted layer quickly and keep the heat pump system operated stably on the basis of the data in Figure 5a.

4. Conclusion

According to the field test results mentioned above, the space heating used transcritical CO_2 heat pump performed excellent at low ambient temperature condition and provide great potential to replace the coal-fired boiler for space heating application in northern China in winter.

On the basis of the field test data, the system performance increased with the increase in ambient temperature and the decline in return water temperature. The system COP can reach to more than 2.0 if the operated conditions were proper, and it can also remain to more than 1.6 when the ambient temperature is lower than -15°C and the return water temperature is higher than 50°C (water supply temperature can be obtained by higher than 70°C). The seasonal performance coefficient was higher than 2.0 at the average ambient temperature of 0°C.

Besides, the proposed system in which the by-pass defrosting approach was employed could defrost the frosted layer quickly and keep the heat pump operated stably during all the heating period.

References

[1] L. Jin, F. Cao, D. Yang, performance investigation of an R404A air-source heat pump with an internal heat exchanger for residential heating in northern China, International Journal of Refrigeration, 2016, 67, 239-248.

[2] Wang, S.G., Tuo, H.F., Cao, F., Xing, Z., 2013. Experimental investigation on air-source transcritical CO_2 heat pump water heater system at a fixed water inlet temperature. Int. J. of Refrig. 36(3), 701-716.

[3] Aprea, C., Maiorino, A., 2008. An experimental evaluation of the transcritical CO_2 refrigerator performances using an internal heat exchanger, International Journal of Refrigeration, 31, 1006-1011.

[4] Sánchez, D., Torrella, E., Cabello, R., Llopis, R., 2010. Influence of the superheat associated to a semi-hermetic compressor of a transcritical CO_2 refrigeration plant. Appl. Therm. Eng. 30, 302-309.

[5] Torrella, E., Sánchez, D., Llopis, R., Cabello, R., 2011. Energetic evaluation of an internal heat exchanger in a CO_2 transcritical refrigeration plant using experimental data, Int. J. Refrig. 34, 40-49.

[6] Cavallini, A., Cecchinato, L., Corradi, M., Fornasieri, E., Zilio, C., 2005. Two-stage transcritical carbon dioxide cycle optimization: a theoretical and experimental analysis, Int. J. Refrig. 28, 1274-1283.

[7] Cavallini, A., Corradi, M., Fornasieri, E, Zilio, C., 2007. Experimental investigation on the effect of the internal heat exchanger and intercooler effectiveness of the energy performance of a two stage transcritical carbon dioxide cycle, in IIR (Ed.) 22nd International Congress of Refrigeration, Beijing, China, 2007.

[8] Cho, H., Baek, C., Park, C., Kim, Y., 2009. Performance evaluation of a two-stage CO_2 cycle with gas injection in the cooling mode operation, Int. J. Refrig. 32, 40 - 46.

[9] Sarkar, J., 2013. Performance optimization of transcritical CO_2 refrigeration cycle with thermoelectric subcooler, Int. J. Energy Res. 37, 121-128.

[10] Llopis, R., Cabello, R., Sánchez, D., Torrella E., 2015. Energy improvements of CO_2 transcritical refrigeration cycles using dedicated mechanical subcooling, Int. J. Refrig. 55, 129-141.

[11] D. Yang, Y. Song, F. Cao, theoretical and experimental investigation of a combined R134a and transcritical CO_2 heat pump for space heating, 2016, International Journal of Refrigeration, 72, 156-170.

[12] Y. Song, D. Li, D. Yang, performance comparison between the combined R134a/ CO_2 heat pump and cascade R134a/ CO_2 heat pump for space heating, 2017, International Journal of Refrigeration, 74, 592-605.

IOP Conf. Series: Materials Science and Engineering **232** (2017) 012088 doi:10.1088/1757-899X/232/1/012088

The Integration of an API619 Screw Compressor Package into the Industrial Internet of Things

W J Milligan[1,3], G Poli[1], D K Harrison[2]

[1] Howden Compressors, Old Govan Road, Renfrew, UK
[2] Glasgow Caledonian University, Glasgow, UK
[3] Author to whom any correspondence should be addressed

billy.milligan@howden.com

Abstract. The Industrial Internet of Things (IIoT) is the industrial subset of the Internet of Things (IoT). IIoT incorporates big data technology, harnessing the instrumentation data, machine to machine communication and automation technologies that have existed in industrial settings for years. As industry in general trends towards the IIoT and as the screw compressor packages developed by Howden Compressors are designed with a minimum design life of 25 years, it is imperative this technology is embedded immediately.

This paper provides the reader with a description on the Industrial Internet of Things before moving onto describing the scope of the problem for an organisation like Howden Compressors who deploy multiple compressor technologies across multiple locations and focuses on the critical measurements particular to high specification screw compressor packages. A brief analysis of how this differs from high volume package manufacturers deploying similar systems is offered. Then follows a description on how the measured information gets from the tip of the instrument in the process pipework or drive train through the different layers, with a description of each layer, into the final presentation layer. The functions available within the presentation layer are taken in turn and the benefits analysed with specific focus on efficiency and availability. The paper concludes with how packagers adopting the IIoT can not only optimise their package but by utilising the machine learning technology and pattern detection applications can adopt completely new business models.

1. The Industrial Internet of Things (IIoT)

The Internet of Things (IoT) refers to connecting devices over the internet allowing them to communicate with humans, applications, and each other. The term Internet of Things was proposed by Kevin Ashton in 1999[1]. The Industrial Internet is a term coined by GE [2] and refers to the integration of complex physical machinery with networked sensors. Therefore the IIoT is the industrial subset of the IoT that is connecting complex physical machinery and their sensors to the internet to allow them to communicate with humans, various applications and each other. The IIoT sees smart machines that can accurately, consistently capture and communicate data.

The convergence of information technology and operations technology (IT/OT), where the industrial automation components are integrated into the business IT infrastructure is being driven by the decision makers within the business. In a recent study by Microsoft [3] only 4% of IoT initiatives were led by IT personnel. This captured data can enable the decision makers within the companies operating screw compressor packages to pick up on inefficiencies and problems sooner thus saving time and money, make better decisions regarding their packages, improve their process and allow the decision makers within screw compressor manufacturers to develop new business models. However, there are complexities involved in the integration of these two technologies not least the bridging of the old proprietary communications wiring and protocols of the automation world into the more structured and standardized IT world.

The evolution of the compressor control system has seen the change from manual control of motors and valves depending on the operators reading of gauges and sight glasses to rudimentary automatic control systems using relay logic. The advent of programmable logic controllers (PLC) in the 1970's and 1980's allowed the control systems to introduce complex logic into the package control sequence and to communicate with the end users overall factory Supervisory Control And Data Acquisition (SCADA) system or their refinery Distributed Control System (DCS).

Each of the automation houses had developed their own communication protocols and wiring standards such as ControlNet from Rockwell with its co-axial cable or Profibus from Siemens with its twisted pair cable, with only a few becoming adopted industry wide such as Foundation Fieldbus via the FieldComm Group or DeviceNet with the development of the Open DeviceNet Vendors Association. These propriety protocols, often referred to as industrial fieldbuses, were most commonly deployed between individual nodes of a small control system for example between a PLC processor rack and racks of inputs and outputs (IO) that was being controlled by the processor or perhaps between the PLC and a piece of smart plant such as a control valve. Rather than run multiple cables for limit switches, positioners and solenoids a single cable could carry the communication data to and from the valve. Historically the communication between the package PLC and the SCADA or DCS would have been a serial connection communicating at a few thousand bits per second baud rate such as an RS485 Modbus connection. Overtime the RS232, RS422 and RS485 SCADA and DCS connections have been phased out to be replaced with more robust Ethernet connections. More specifically 100Base-Tx Ethernet accounts for about 95% of all Ethernet data transfers [4] and has changed the automation industry with this also now replacing the industrial fieldbuses. The automation houses have realized the advantages of connectivity to the IT infrastructure and have developed Ethernet communications modules that fit into the PLC chassis alongside the IO modules to bridge between the control system proprietary protocols and the IT network. This is the starting point of the OT/IT convergence and has been taking place over the past decade. The IIoT is a further evolution of this convergence with the adopted IT standard working practices being used to present this data to the business decision makers.

2. The API619 Screw Compressor Package

Howden Compressors and complete compressor packages are used in the petroleum, petrochemical, refrigeration, oil and gas, power generation and other markets where performance and reliability are critical. Applications of compressors can include sour gas processing, flare gas recovery, gas reliquification and fuel gas boosting for turbines. Howden designs and manufactures a range of different compressor packages that can be categorised as: screw compressors, reciprocating piston compressors, diaphragm compressors and turbo compressors.

The selection of a reciprocating-, centrifugal, turbo, or screw compressor is based on the flow and pressure requirements that are specified by the client. In overlap areas of the operating envelope the selection of a particular type of compressor is not explicit. In these cases the selection is based on other technical and commercial aspects.

During the detailed design of these packages hazard and operability studies (HAZOP) and Safety Integrity Level (SIL) assessments are held as the severity of an accident if something did go wrong could lead to serious injury or death to one or more people, have a sizable environmental impact and cost hundreds of thousands if not millions of pounds to repair.

3. High Volume Screw Compressor Packages

Remote monitoring has been tried on a single node API 619 screw compressor package deployed in the field by Kobelco EDTI Compressors Inc previously. The application was for five Kobelco KS31 SHE compressors feeding fuel gas into four GE 7FA gas turbines. One of the compressor packages was a spare standby. This was a single site that, if required, an engineer could dial into the system to watch it live or it could be setup to email a supervisor if a particular alarm occurred.

More commonly, remote monitoring has also been rolled out on large scale manufactured machines of the same design. Gardner Denver has developed the ESP 20/20 compressor monitoring system for their air compressors. Ingersoll Rand has developed the IntelliGuard system to do the same and Atlas-Copco has iAccess. It is important to understand these are off the shelf, high volume manufactured, air compressor systems that these companies manufacture. Rolls-Royce have deployed a similar system to remotely monitor engine health albeit on a much larger scale in terms of points measured: air compressor manufacturers measure between five and fifteen whereas Rolls-Royce measure twenty-five. If this is compared with a two stage oil free package currently being manufactured in Renfrew that has one hundred and seventy different measurements being measured at any one time and the next package may be a relatively low spec oil injected machine with thirty measurement points on it. It can be seen that the API619 screw compressor package monitoring system needs to have orders of magnitude amounts of points available from an off the shelf compressor package monitoring system and at the same time also be more dynamic in the ability to chose what points are being measured whereas these companies manufacturing large volumes of compressors, all of the same technology, and all of the same design they only need to design the monitoring system once but roll it out many times and always looking at the same performance data.

The engine health management system from Rolls-Royce should also be considered in terms of criticality: Gardner Denver, Ingersoll Rand and Atlas-Copco manufacture air compressors for industrial processes where failure may well cause loss of production if the compressors cannot be brought back online by the time the reservoir empties whereas the Rolls-Royce system is monitoring aeroplane engines operating at thirty thousand feet with up to hundreds of lives on board. The model developed for critical compressor packages would tend towards the engine systems model rather than the industrial compressor model doe to the number if IO and critical nature of the machines.

4. The Data Journey

The figure below is a graphical representation of the data journey from the point of measurement through the various layers to the business presentation layer. On a typical package the start of this process is at a compressor or motor bearing or a gas pressure or oil temperature. From here the instrument is connected via a junction box located at the edge of the package to the package control system. These instruments are then grouped together and ran back to the PLC on larger multi-core cables to the PLC in a control room. Although the above describes a typical package there can be many different adaptations of the sensors and control system layer with control panels sometimes being mounted on the package rather than in a blast proof control room a few hundred meters away. Or to save copper cabling costs sometimes the multicore home run cables are replaced with fiber optical cable for the non critical signals and the IO is moved to the package junction boxes.

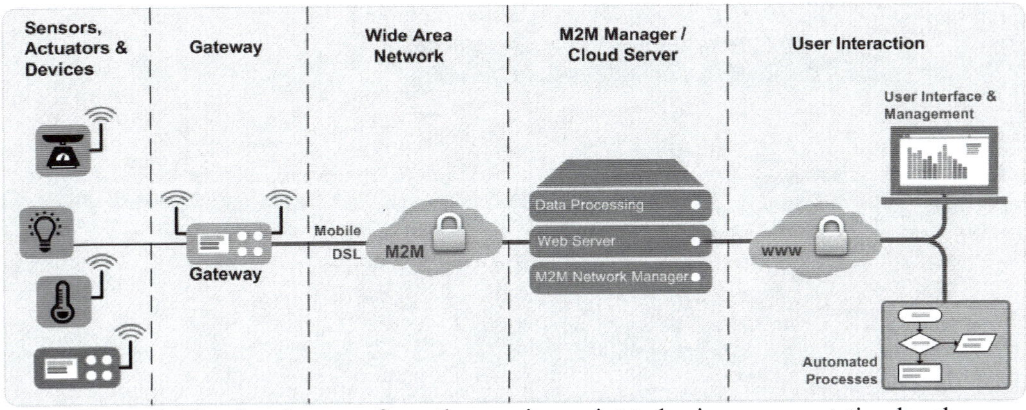

Figure 1, The data journey from the sensing point to business presentation level.

The PLC then meets the Edge. This is the gateway between the OP and the IT worlds or the control system local area network (LAN) and the wide area network (WAN). The data then makes its way into a database that will be hosted either in an on-premises server, in the cloud or a combination of both. The combination or hybrid scenario comes about when sensitive business data cannot be entrusted to third parties but the non-critical non-sensitive data can be stored on external sources. Microsoft Azure and Amazon Web Services are example of cloud based services that IIoT applications are being hosted on. Indeed Rockwell Automation has partnered with Microsoft and are using Azure in their services business and are working together and aligning strategically around technology roadmaps allowing easily development and deployment of IIoT applications using their respective products.

Lastly is the presentation layer where the business decision makers can see and interrogate it. This layer used to be a SCADA system or a DCS where the operators could view the data supplied by the PLC and make decisions based on it. This would all be done in a control room local to the PLC and local to the process. The evolution into the IIoT means this can be the general plant manager at breakfast before coming to work in the morning on his phone or, more importantly for a screw compressor manufacturer, it could be the engineering department interrogating historical data on a particular bearing on a particular size of machine. The different platforms available for the presentation layer all have the ability to drill down into specific areas with, for instance, the ability to trend over a two year period all of the inlet end male rotor bearing temperatures of a particular model of compressor with a few button clicks so see if there is any correlations to be made.

5. Proof of Concept

At the moment an initial single node proof of concept has been developed. The first step in the development of the system was to create a test assembly that was able to simulate the functionality of an operational screw compressor package. This was achieved by using an Allen Bradley ControlLogix L61 PLC which has 16 analogue inputs, 8 analogue outputs, 16 digital outputs and 16 digital inputs that are all used to control and monitor the operation of the simulated screw compressor. The inputs and outputs to the PLC were simulated via an array of variable resistors, I/O switches, LEDs and ammeters which were situated within a separate control panel. To monitor the real time simulated analogue and digital inputs and outputs of the PLC a Human Machine Interface (HMI) was added to the test assembly just as it would have if it were a real package. Screenshots of the HMI program used to monitor the PLC can be seen in the figure below.

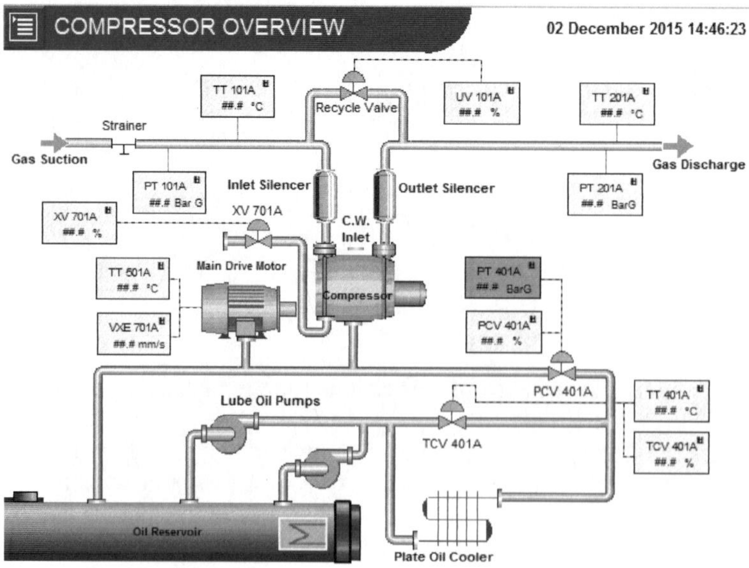

Figure 2, Proof of concept HMI screen.

To add a business presentation layer and data storage functionality to the proof of concept assembly, new software and hardware components were added. After researching available software options it was concluded that FactoryTalk Historian SE [5] would be the most suitable solution to store the historical data from the test assembly as it could easily integrate with the installed Allen Bradley PLC. This software is able to capture process data from the PLC which could be programmed to capture an extensive archive of data that could be used to assess any PLC input or output over a chosen period of time. The FactoryTalk Historian software was installed on a server running Windows Server 2008 R2 which was connected to the PLC via Ethernet TCP/IP that made it possible for the software to collect and store data on any of the inputs and outputs defined in the PLC ladder logic. FactoryTalk Historian can connect to not only Allen-Bradley PLCs but using any 3rd part PLC or data acquisition (DAQ) device by means of an object linking and embedding for process control (OPC) connection. Just another example of the OT/IT convergence but it does allow almost any industrial automation device to put data into the database.

To add analytical functionality to the test assembly FactoryTalk VantagePoint software was added to the server PC. This software expanded the capability of the historian server by allowing anyone that was connected to the server PC to log in to a page on their internet browser via a web address and access the historical data stored by the FactoryTalk Historian software [6]. This software is also capable of producing predetermined reports that can be customised to present data on components that may be deemed the most likely to fail on the compressor. This would give engineers the ability to quickly assess the operational health of any chosen component on the compressor simply by logging into a PC that is connected to the server PC.

To make this test truly represent a screw compressor package installed in a remote location, a mobile GSM modem was integrated into the test assembly via an Ethernet connection. The GSM modem allows for an engineer to connect to the PLC and the condition monitoring system, regardless of their location, allowing them to then carry out data analysis on the industrial machine. It is beneficial to include a remote monitoring system as it allows for the time taken for a system fault to be diagnosed to be minimised, reduces the expenditure on engineers being called out to assess machine faults and reduces the resulting downtime caused by a component failure [7]. The completed machine monitoring test assembly is shown in Figure 3.

Figure 3, the single node proof of concept test assembly.

6. Presentation Layer

The presentation layer of any IIoT platform must be able to perform a variety of roles as this will be used by various levels throughout a business. This layer is responsible for being able to extract the data into well known platforms like Microsoft Excel, SQL Server Reporting and SharePoint to allow the data to be published in business reports and presentation. These reports, trends and dashboards should be able to be pre-configured to save time every time the report is issued. Mobile functionality is also required to allow not just local intranet but web-enabled browsing, with end users being able to create ad-hoc displays to interrogate data and create reports on their device. The presentation layer must have the functionality to provide information in a timely manner, via SMS or email for instance, to allow the business decision makers time to make productive decisions.

The presentation layer must be able to securely link to a variety of databases such as the hybrid on-premises storage databases and cloud storage databases scenario discussed earlier. The platform must also securely allow users only see information only pertaining to them by defining user roles.

Most importantly the presentation layer, and also the data storage solution, must be scalable. This is one of the advantages of the cloud solutions over the on-premises hardware solution. The faceplate template model of the screw compressor package that links to the data must be able to be repeated over and over every time a new package is deployed to the field. Once the initial faceplate template is developed, the Historian server is pointed at the PLC in the field and the standard template is then copied and re-named and is pointed to the tags in the Historian server. This facilitates quick introduction of any new package into the system but requires the PLC programme to have a standard structure of PLC tags so the suction pressure value is always stored in the same tag in every machine same with the discharge pressure etc etc. Figure 4 below demonstrate the data capturing capabilities that are available by using both FactoryTalk Historian and FactoryTalk VantagePoint together in the proof of concept test.

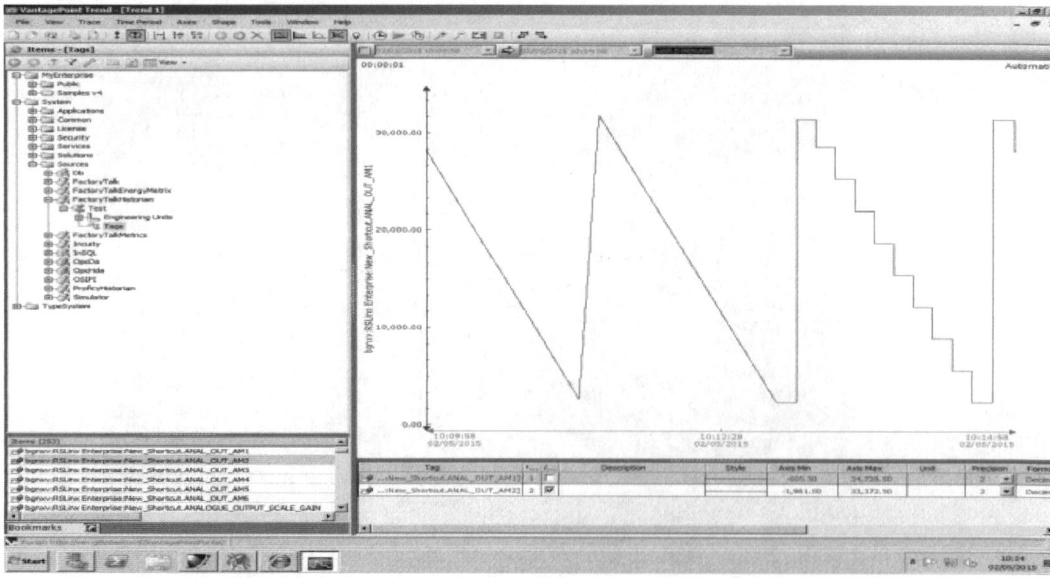

Figure 4, simulated historical data displayed on the presentation software during test.

Many IIoT platforms are introducing augmented reality features that offer exciting possibilities. They propose that a service engineer turns up at a package, opens their phone, points it at the package and a 3-D model of the package appears on the display, they can bring up all the relevant drawings, they can dismantle the 3-D model and see how to get to a particular component and see all of the process values (pressure, temperatures, flows and vibrations etc) superimposed on the model. These are however still in their infancy but small examples of pumps and motorbikes have been produced.

7. Conclusions and Future Work

The proof of concept single node revealed that the technology is available to gather the information right at the front end of a compressor package and make it available at any time to all levels of the business. The hardware installed in automation systems deployed just now already integrate easily with the business IT infrastructure. Even older isolated packages with obscure protocols could be integrated albeit with some design work required. The bespoke and critical nature of these packages means that the model developed by the high volume screw compressor manufacturers would be the wrong model to adopt. The IO count needs to be increased by an order of magnitude and the template for the packages needs to be more dynamic than that model. With this in mind two networked proof of concepts are in the advanced stages of development.

One is a Rockwell Automation platform similar to the single node proof of concept but this time rather than using single server with all of the historian software and presentation and analytic software installed on the one machine the historian software is installed on one server and the presentation and analytic software on a separate server. The flight desk is still used to feed simulated data into the PLC and HMI assembly. The Rockwell networked proof of concept architecture is shown below in Figure 5 with the only difference between the proof of concept and a finalised platform would be that rather than hosting on Howden servers these would be copied and migrated into the Microsoft Azure cloud.

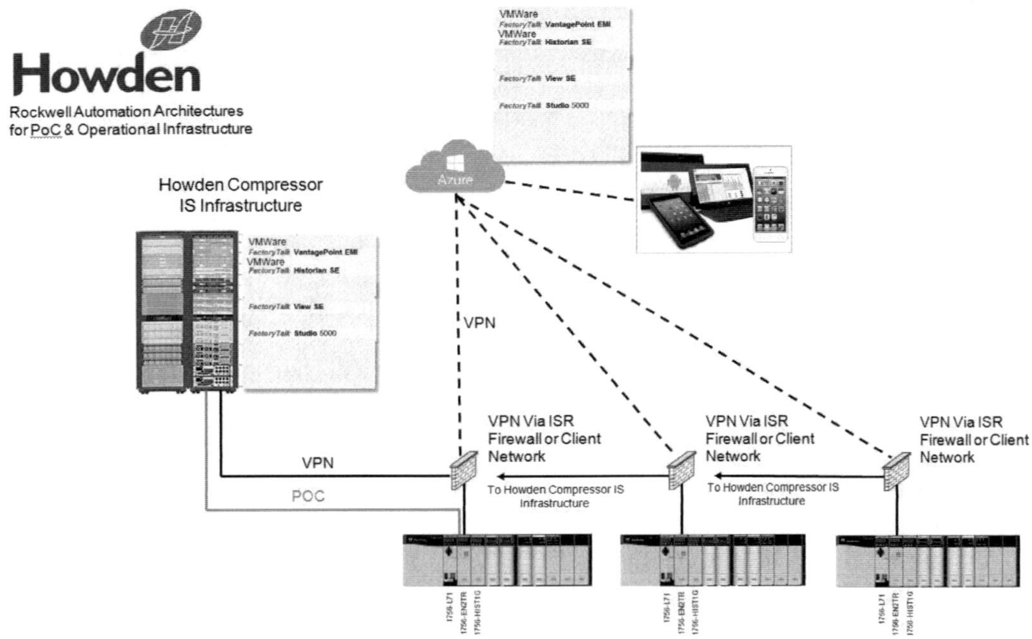

Figure 5, the Rockwell IIoT network proof of concept (courtesy of Rockwell Automation).

The second platform is ThingWorx from PTC deployed by InVMA. It provides IoT specific development tools and capabilities that enable developers to rapidly build and deploy smart, connected solutions for the Internet of Things. ThingWorx is purpose-built for the development of scalable and secure solutions that are simple to create and makes end-user experiences more compelling.

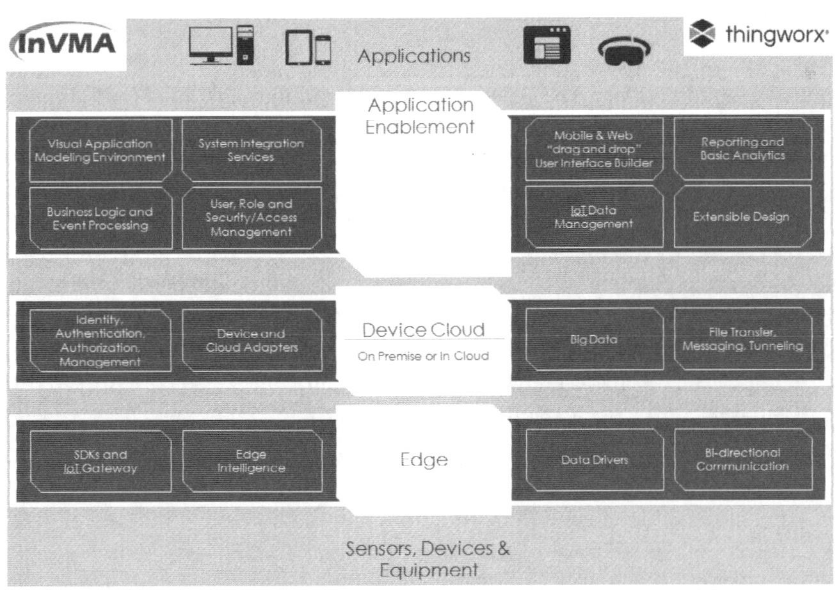

Figure 6, the PTC Thingworx proof of concept in development by InVMA (courtesy of InVMA).

A powerful feature of the ThingWorx capability is its Analytics add-on. Analytics enables developers to quickly and easily add real-time pattern and anomaly detection and predictive analytics to the platforms they build so rather than wait for the platform to tell the end user that the compressor has a high bearing temperature this add-on mines the data and recognizes that the last time this bearing started to trend like this a high temperature alarm was raised in approximately three weeks time and notifys the user prior to the machine going into alarm. Analytics is designed to be used by end users that may not have expertise in complex mathematics, statistics or machine learning with the alternative is manual coding of maintenance of predictive models and recommendations. This requires specific expertise in advanced analytics, statistical modeling, machine learning and system integration.

Both of these platforms offer significant advantages and after the final development and trial period a business case for both will be presented to the Howden Compressors senior management team but it should be noted that the dynamic nature and scalability of these platforms make them ideal not only for the screw compressor business but for capturing the data for all of the air and gas handling equipment manufactured within the group.

The development and implementation of the IIoT platform will transform the compressor packaging industry on three different levels. Operationally the platform will be used to optimise performance of the packages and reduce service costs all for the end users thanks to examples like the production managers and maintenance managers being able to access and interrogate process and machine condition data. Strategically the platform will elevate the manufacturer's relationship with its end users as not only an innovator but also something that will differentiate the manufacturer from their competitors, at least for the time being. Lastly the IIoT platform, if implemented correctly, allows completely different business models to be developed: for instance no longer will an end user pay $2M USD for a compressor package but they will hire it at $8K USD a week so long as the manufacturer guarantees 99% availability with the IIoT platform being used to monitor the health of the package.

References

[1] Ashton, Kevin (22 June 2009). "That 'Internet of Things' Thing, in the real world things matter more than ideas". *RFID Journal* [Accessed 12 February 2017]
[2] Leber, Jessica (2012-11-28). "General Electric's San Ramon Software Center Takes Shape | MIT Technology Review". Technologyreview.com. [Accessed 12 February 2017]
[3] Microsoft (15 July 2015). "Navigating the IT/OT convergence". Available at: https://blogs.microsoft.com/iot/2015/07/15/navigating-the-itot-convergence/#cJHoD8Tsbkc1yHsf.99 [Accessed 12 February 2017]
[4] Park J. and Mackay S. (2003a) Practical Data Acquisition for Instrumentation and Control Systems, 1st ed, Oxford: Elsevier, pp.195-203, ISBN 07506 57960
[5] Rockwell Software, "FactoryTalk Historian SE," Rockwell Software, 2015. [Online]. Available: http://www.rockwellautomation.com/rockwellsoftware/products/factorytalk-historian-se.page. [Accessed 02 December 2016]
[6] Rockwell Software, "FactoryTalk VantagePoint EMI," Rockwell Software, 2015. [Online]. Available at: http://www.rockwellautomation.com/rockwellsoftware/products/factorytalk-vantagepoint.page. [Accessed 02 December 2016]
[7] R. Merritt, "Benefits of Remote Diagnostics and Monitoring," Control Design, 08 August 2005. [Online]. Available: http://www.controldesign.com/articles/2005/287/. [Accessed 07 December 2016]

IOP Conf. Series: Materials Science and Engineering **232** (2017) 012089 doi:10.1088/1757-899X/232/1/012089

Theoretical investigation of vapor mass fraction measurement methods for two-phase injection compression

T W Moesch[1], D Lumpkin[2,3], C Thomas[1], U Hesse[1] and E Groll[3]

[1] Bitzer Chair of Refrigeration, Cryogenics and Compressor Technology, Institute of Power Engineering, Technische Universität Dresden, Dresden, Germany
[2] School of Civil Engineering, Purdue University, West Lafayette, Indiana, USA
[3] School of Mechanical Engineering, Ray W. Herrick Laboratories, Purdue University, West Lafayette, Indiana, USA

E-mail: thomas.moesch@tu-dresden.de

Abstract. The economized vapor injection (EVI) method is used to decrease the discharge temperature of scroll compressors and to improve the cycle performance of vapor compression refrigeration systems by directly injecting refrigerant vapor into the compression chamber. In previous studies, the optimal cycle performance for an air conditioning system using EVI was detected during two-phase injection. However, the exact vapor mass fraction could not be measured directly. For further investigations on two-phase injection systems, accurate vapor mass fraction (VMF) measurements need to be conducted. This paper compares different options for the VMF measurement using pure refrigerants or refrigerant mixtures. It focuses on the achievable accuracy of the VMF measurement, based on the propagation of uncertainty.

1. Introduction

Vapor injection scroll compressors and the economized vapor injection (EVI) method are utilized to enlarge the operation range of single-stage vapor compression heat pump systems. During EVI operation a defined amount of subcooled liquid refrigerant is extracted after the condenser (upstream extraction) or after the economizer (downstream extraction), expanded to the injection pressure and evaporated in the economizer. This leads to a further subcooling of the liquid refrigerant after the condenser. After its evaporation at injection pressure, the refrigerant is directly injected into the compression chamber. Figure 1 shows an EVI cycle with downstream extraction and a heat exchanger economizer that was modelled and tested by Moesch et al. [1]. The authors investigated EVI for air conditioning (A/C) systems using R-407C. Their results indicate that two-phase injection with a high vapor mass fraction (VMF) leads to an optimal performance of the air conditioning system. However, for two-phase operation a reliable VMF measurement method is required. There is already a variety of published VMF methods based on flow measurements and two-phase properties. Wang et al. [2] measured the VMF at the evaporator outlet of an R-134a A/C system using a gas/liquid separator and measuring the saturated gas flow rate. The overall uncertainty of this method was not given by the authors. Mitra et al. [3] used a tunable diode laser to measure the specific volume from which they derived the VMF of a two-phase H_2O flow. Based on their experimental results, the authors claim to achieve 1% accuracy for the derived VMF. However, the uncertainty for different fluids or applications cannot be derived from their data.

To narrow down the achievable uncertainty of different VMF measurement methods and their applicability in regard to the existing EVI setup and future standardized EVI performance tests, the authors conducted a theoretical uncertainty investigation based on the required sensors of each method.

This paper presents four different VMF methods that are based on heat balances, fluid properties and fluid separation. It provides an input on the methods' correlations and their theoretical uncertainties for a given reference point.

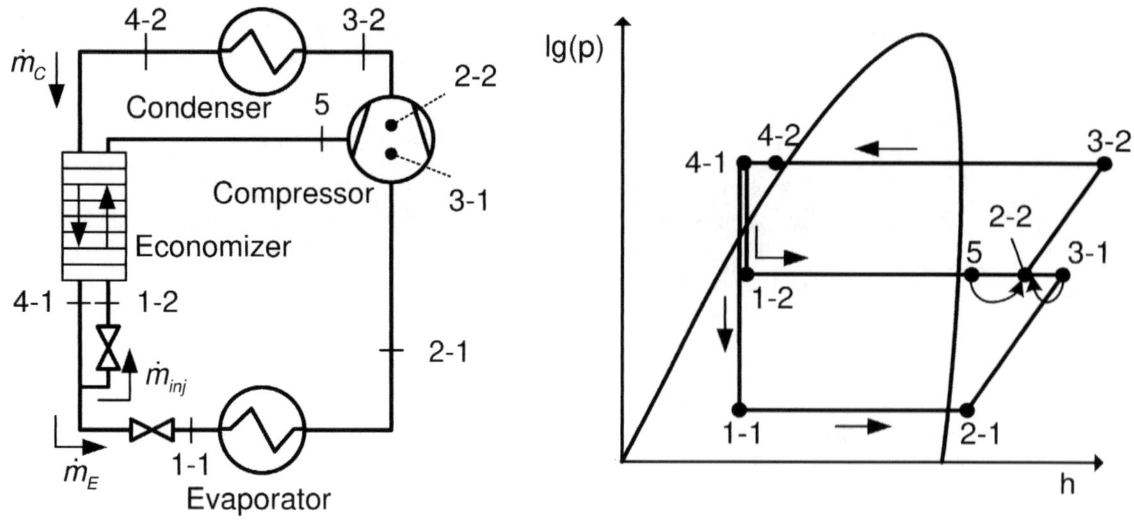

Figure 1. EVI cycle with downstream extraction according to [1] and corresponding lg(p)-h diagram

2. Study on different VMF methods

2.1. VMF measurement methods

The VMF or vapor quality of two-phase flows is generally defined by Equation (1), where $\dot{m}$ represents the mass flow rate of the two-phase refrigerant flow and $\dot{m}''$ is the mass flow rate of the saturated vapor.

$$\dot{x} = \frac{\dot{m}''}{\dot{m}} \tag{1}$$

Another definition of the VMF is based on fluid properties in an equilibrium state and is given by Equation (2), where h is the enthalpy of the two-phase refrigerant flow, h'' is the enthalpy of the saturated liquid refrigerant as h' and the enthalpy of the saturated vapor refrigerant. Strictly speaking this definition is only valid for homogeneous flow approaches, where both fluid phases are assumed to have the same flow velocity, or for stationary applications (e.g. closed vessels or tanks).

$$x = \frac{h - h'}{h'' - h'} \tag{2}$$

The four investigated VMF methods are described in the following subsubsections. The described Calorimeter method and Direct PT method are based on Equation (2), the Direct PD method and the Separation method are based on Equation (1).

2.1.1. Calorimeter Method. A schematic of the Calorimeter method is shown in Figure 2. It uses the available heat exchanger economizer as a secondary fluid calorimeter as described in ASHRAE Standard 41.9 [4]. Instead of a secondary fluid though, which is usually water or brine, the subcooled liquid flow exiting the condenser (state point 4-2) is used to evaporate the extracted refrigerant flow at state point 4-1, which is the injected flow $\dot{m}_{inj}$. The enthalpy of the injected flow is calculated based on an energy balance across the economizer as shown in Equation (3), where h_{inj} represents the enthalpy of the injected two-phase flow, T_{4-1} and T_{4-2} are the measured temperatures, p_{4-1} and p_{4-2} are the measured pressures, $\dot{m}_C$ and $\dot{m}_E$ are the measured mass flow rates, and φ_{amb} is the relative heat loss to ambient.

$$h_{\text{inj}} = \frac{\dot{m}_C \cdot \left[h\left(p_{4-2},T_{4-2}\right) - h\left(p_{4-1},T_{4-1}\right)\right] \cdot \left(1-\varphi_{amb}\right)}{\dot{m}_C - \dot{m}_E} + h\left(p_{4-1},T_{4-1}\right) \tag{3}$$

This method requires two mass flow meters, three pressure sensors and two temperature sensors and is therefore more costly. The relative heat loss to ambient φ_{amb} is not measured and therefore seen as an unknown parameter. This method is not applicable for other economizers, such as flash tanks. However, for EVI applications with closed heat exchanger economizers, it provides a reliable way to determine the injection enthalpy.

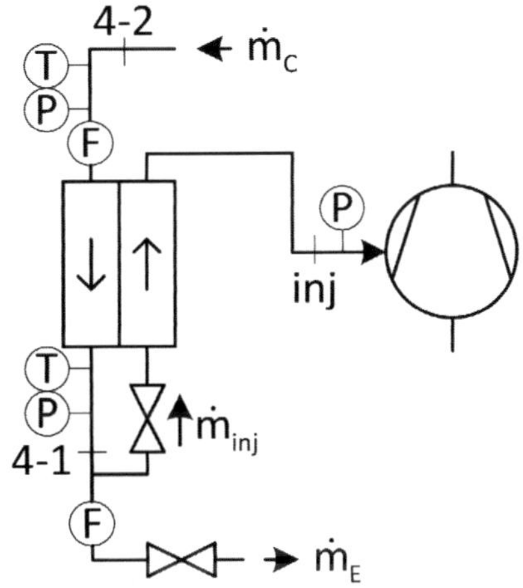

Figure 2. Calorimeter Method

2.1.2. Direct PT Method. The Direct PT method is shown in Figure 3. It uses the temperature glide during the phase change of R-407C. The enthalpy of the injected two-phase flow is directly calculated based on the measured injection pressure p_{inj} and injection temperature T_{inj} (see Equation (4)).

$$h_{\text{inj}} = h\left(p_{\text{inj}},T_{\text{inj}}\right) \tag{4}$$

This method only requires one pressure sensor and one temperature sensor and is therefore the preferable method in terms of cost-effectiveness. It is only applicable to EVI systems with zeotropic refrigerant mixtures with a certain temperature glide and does not interfere with the EVI system.

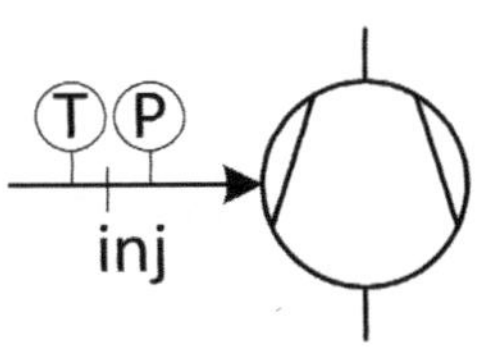

Figure 3. Direct PT Method

2.1.3. Direct PD Method. The Direct PD method uses a similar approach as reported in [3] and measures the density of the two-phase flow as shown in Figure 4. According to [5], the average density of a heterogeneous two-phase flow is defined by Equation (5), where α represents the void fraction, ρ' is the density of saturated liquid refrigerant and ρ'' is the density of saturated vapor refrigerant.

$$\rho_{inj} = \rho''(p_{inj}) \cdot \alpha + \rho'(p_{inj}) \cdot (1-\alpha) \tag{5}$$

The vapor mass fraction or quality for a two-phase flow is defined by Equation (6), where S is called the velocity ratio or slip and represents the ratio of the vapor refrigerant velocity to the liquid refrigerant velocity.

$$\dot{x}_{inj} = \left[1 + \frac{1}{S} \cdot \frac{1-\alpha}{\alpha} \cdot \frac{\rho'(p_{inj})}{\rho''(p_{inj})}\right]^{-1} \tag{6}$$

Merging Equations (6) and (5), the void fraction can be eliminated and the VMF for the Direct PD method can be defined as follows.

$$\dot{x}_{inj} = S \cdot \left[\frac{1}{\rho_{inj}} - \frac{1}{\rho'(p_{inj})}\right] \cdot \left[\frac{S}{\rho''(p_{inj})} + \frac{1-S}{\rho_{inj}} - \frac{1}{\rho'(p_{inj})}\right]^{-1} \tag{7}$$

There are different approaches for the velocity ratio S. Zivi [6] analytically investigated an annular flow based on its minimal total kinetic energy and suggested Equation (8) for the estimation of S.

$$S = \left(\frac{\rho'}{\rho''}\right)^{1/3} \tag{8}$$

Chisholm [7] used an annular flow theory and a homogeneous approach for the fluid density. This resulted in the correlation in Equation (9), where x represents the vapor quality.

$$S = \left[1 - x \cdot \left(1 - \frac{\rho'}{\rho''}\right)\right]^{1/2} \tag{9}$$

This method requires one pressure sensor and one density sensor for two-phase flows. The density measurement is based on a Coriolis flow meter and thus, the costs will be between the Direct PD Method and the Calorimeter Method. For this investigation the velocity ratio S is assumed to be an unknown parameter since the authors have no information on the accuracy of the Equations (8) and (9) This method is applicable for both pure refrigerants and refrigerant mixtures and may be applied to alternative economizers such as a flash tank as well. The pressure loss over the density sensor might interfere with the injection process.

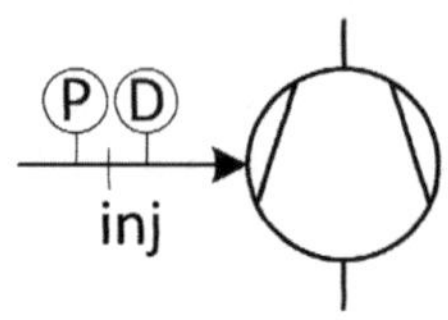

Figure 4. Direct PD Method

*2.1.4. Separation Method.*The Separation Method is similar to the approach presented in [2] and uses an additional vessel to separate the saturated vapor refrigerant flow from the saturated liquid refrigerant flow as shown in Figure 5. This allows a direct measurement of the vapor refrigerant flow.

The VMF of the injection two-phase flow is then defined by Equation (10), where $\dot{V}_{inj,l}$ represents the measured volume flow of the saturated liquid refrigerant, p_{inj} is the measured injection pressure, and $\dot{m}_C$ and $\dot{m}_E$ are the measured mass flow rates.

$$\dot{x}_{\mathrm{inj}} = \frac{\rho''(p_{\mathrm{inj}}) \cdot \dot{V}_{\mathrm{inj,g}}}{\dot{m}_C - \dot{m}_E} \tag{10}$$

This method uses three flow meters and one pressure sensor and its costs are between the Direct PD Method and the Calorimeter Method. It can be applied to any EVI configuration and refrigerant. Due to the separation of the injected flow this method interferes significantly with the injection process. For accurate measurements a stable filling of the separator needs to be guaranteed.

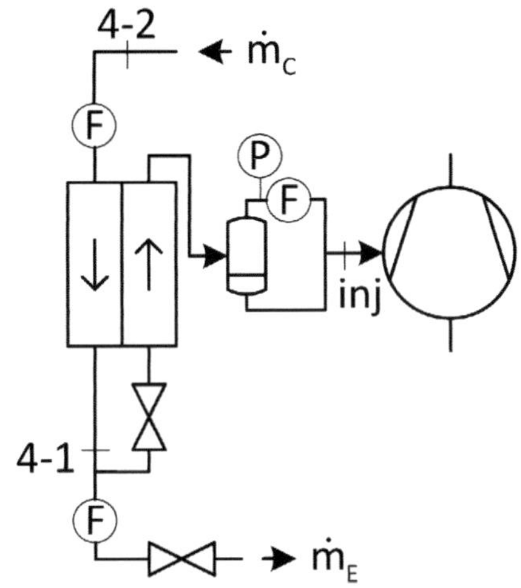

Figure 5. Separation Method

2.2. Approach for uncertainty analysis

The uncertainty of the VMF method results from the uncertainty of each measured parameter and its propagation. The propagation of uncertainty was conducted using the Taylor Series Method (TSM) described by Coleman and Steele [8]. The TSM is defined by Equation (11), where U_x represents the uncertainty of the VMF calculation, $U_{Y,i}$ is the uncertainty of each measured parameter and $\partial x / \partial Y_i$ is the partial derivative of the VMF correlation for each parameter.

$$U_x = \left[\sum_{i=1}^{n} \left(\frac{\partial x}{\partial Y_i} \cdot U_{Y_i} \right)^2 \right]^{0.5} \tag{11}$$

The relative uncertainty contribution of each parameter $\varepsilon_{Y,i}$ is given by Equation (12).

$$\varepsilon_{Y_i} = \left(\frac{\partial x}{\partial Y_i} \cdot U_{Y_i} \right)^2 \cdot \left[\sum_{i=1}^{n} \left(\frac{\partial x}{\partial Y_i} \cdot U_{Y_i} \right)^2 \right]^{-1} \tag{12}$$

The partial derivatives are numerically approximated by the central-difference approach in Equation (13) as defined in [8], where ΔY represents the finite difference, Y_i^0 are the referenced measured parameters, and Z^0 is the referenced unknown parameter.

$$\left. \frac{\partial x}{\partial Y_1} \right|_{Y_2^0,\ldots,Y_n^0 = const.} \approx \frac{x\left(Y_1^0 + \Delta Y_1, Y_2^0, \ldots, Y_n^0, Z^0\right) - x\left(Y_1^0 - \Delta Y_1, Y_2^0, \ldots, Y_n^0, Z^0\right)}{2 \cdot \Delta Y_1} \tag{13}$$

The uncertainty of each parameter $U_{Y,i}$ depends on the measuring device. Table 1 lists the uncertainty range of different sensors based on manufacturer data and European standards ([9], [10]).

Table 1. Uncertainties of various measuring devices

Measured property	Measurement Principle	Measurement range	Uncertainty range
Temperature	Type T thermocouple	-40 to 350 °C [a]	± 0.5 °C to ± 1.0 °C [a]
	Type J thermocouple	-40 to 750 °C [a]	± 1.5 °C to ± 2.5 °C [a]
	Platinum resistance	-196 to 600 °C [a]	± 0.1 °C to ± 0.6 °C [a]
Pressure	Semi-conductor	variable	± 1 % [b]
	Piezo-resistance	variable	± 0.05 % FS to 4.0 % FS [b]
Density	Coriolis	400 to 2500 kg/m³ [b]	± 0.5 kg/m³ to ± 5.0 kg/m³ [b]
Mass flow	Coriolis	variable	± 0.15 % to ± 1.0 % [b]
Volume flow	Float-type	variable	± 1.0 % to ± 4.0 % [b]
	Vortex frequency	variable	± 1.0 % to ± 2.0 % [b]

[a] Based on European standards
[b] Based on manufacturer data

2.3. Reference state

A comparative investigation of the VMF methods was conducted for a mutual reference state for all measured parameters Y_i^0, unknown parameters Z^0, and measurement uncertainties $U_{Y,i}^0$.

The referenced measured parameters are based on the EVI system described in [1] for the refrigerant mixture R-407C, a VMF of x = 0.95, and an injection pressure of p_{inj} = 10 bar(a). Further, the authors assumed a condensing dew point temperature of $t_{C,dew}$ = 50 °C, a subcooling after the condenser of ΔT_{SC} = 2 K, and a minimal temperature difference at the economizer of ΔT_{min} = 2 K. State 4-1 at the economizer outlet can be calculated using Equation (14).

$$h\left(T_{4-1}^0, p_{4-1}^0\right) = h\left(T_{4-1}^0 - \Delta T_{min}, p_{inj}^0\right) \tag{14}$$

The referenced suction mass flow rate $\dot{m}_E^0$ results from Equation (15) and is based on a scroll compressor with a theoretical displacement volume flow of $\dot{V}_{th} = 14{,}1\ m^3/h$, a volumetric efficency of ε_V = 0.9, a evaporator dew point temperature of $T_{E,dew}$ = 5 °C, and a suction superheat of ΔT_{SH} =10 K.

$$\dot{m}_E^0 = \dot{V}_{th} \cdot \varepsilon_V \cdot \rho\left(T_E + \Delta T_{SH}, p_{E,dew}\right) \tag{15}$$

The referenced condenser mass flow rate $\dot{m}_C^0$ is derived from Equation (3) assuming no heat losses and the referenced enthalpy of the injected refrigerant as stated in Equation (16).

$$h_{inj}^0 = h\left(p_{inj}^0, x_{inj}^0\right) \tag{16}$$

The values for the referenced measured parameters Y_i^0 are summarized in Table 2. The refrigerant properties were calculated using RefProp [11].

Table 2. Reference state for measured parameters

p_{4-2}^0, p_{4-1}^0 [bar(a)]	p_{inj}^0 [bar(a)]	T_{4-2}^0 [K]	T_{4-1}^0 [K]	T_{inj}^0 [K]	ρ_{inj}^0 [kg/m³]	$\dot{m}_C^0$ [kg/s]	$\dot{m}_E^0$ [kg/s]	$\dot{V}_{inj,g}^0$ 10^{-4}[m³/s]
19.88	10.00	316.5	293.9	297.2	44.94	0.09675	0.07767	4.226

The referenced unknown parameters Z^0 are representative for ideal conditions. For the Calorimeter Method the reference heat losses are zero ($\varphi_{amb}^0 = 0\ \%$) assuming an ideally insulated heat exchanger economizer and injection line. For the Direct PD Method the reference slip ratio is based on a homogeneous flow ($S^0 = 1$) which suggests that both vapor and liquid travel at the same velocity.

The referenced measurement uncertainties are based on the best sensor uncertainty available in Table 2 and are summarized in Table 3. For this investigation, the pressure range of all pressure sensors is assumed to be 35 bars.

Table 3. Reference state for measurement uncertainties

U_p^0 [bar]	U_T^0 [K]	U_ρ^0 [kg/m³]	$U_{\dot{m}}^0$ [kg/s]	$U_{\dot{V}}^0$ [m³/s]
±0.0175	±0.1	±0.5	$\pm 0.0015 \cdot \dot{m}$	$\pm 0.01 \cdot \dot{V}$

2.4. Approach for comparative investigation
The comparative investigation is divided into the distribution analysis, the sensitivity analysis, and unknown analysis.

The distribution analysis is based on Equations (11) and (12) and calculates the overall uncertainty U_x of each method as well as the relative uncertainty contribution of each sensor type at the reference state 0. The result indicates the importance of each sensor for the VMF measurement as well as the lowest achievable VMF uncertainty for each method.

The sensitivity analysis is based on Equation (11) and compares the impact of a changing measurement uncertainty of a particular sensor type. For the analysis the uncertainties of the investigated sensor is varied within the range of the common uncertainties listed in Table 2 while the other sensor types remain at their referenced uncertainty $U_{Y,i}^0$. This analysis simulates the impact of low accuracy sensors or altering sensor uncertainties on the overall uncertainty of the VMF method.

For the unknown analysis, the unknown parameters are varied within a reasonable range and their effect on the reference state is determined by Equation (17), where Z represents the investigated unknown parameter, are the referenced measured parameters as, and Z^0 is the referenced unknown parameter.

$$\Delta x = x\left(X_1^0,\dots,X_n^0,Z\right) - x\left(X_1^0,\dots,X_n^0,Z^0\right) \tag{17}$$

For the Calorimeter method the relative ambient heat loss is varied within a range of $-5\ \% \leq \varphi_{amb} \leq +5\%$ as both heat gain and heat loss seem reasonable. For the Direct PD method the velocity ratio range is based on the estimated velocity ratios from Equations (8) and (9). Based on the reference injection pressure and the refrigerant R-407C the density ratio ρ'/ρ'' yields to 27.2 which leads to $S_{Zivi} = 3.00$ and $S_{Chisholm} = 5.09$. The velocity ratio S was therefore varied within a range of $1 \leq S \leq 10$.

3. Results of the comparative investigation

3.1. Results of the distribution analysis
The results of the distribution analysis are shown in Figure 6. The figure indicates that only three methods are preferable for the VMF measurement at the EVI system and that each method with mass flow measurements can be improved. At the reference state, the Direct PT Method shows the highest uncertainty with $U_x = \pm 0.021$ kg/kg and is therefore not preferable for the VMF measurement. The uncertainties of the other methods are comparable at the reference state and range from ± 0.0098 kg/kg to ± 0.0134 kg/kg.

The VMF uncertainty of the Calorimeter Method is mainly affected by the uncertainty of the mass flow measurement ($\varepsilon_m = 68.74\%$) and the uncertainty of the temperature measurement ($\varepsilon_T = 31.23\%$). The distribution of the mass flow measurement is caused by injected mass flow which is determined from a mass flow difference measurement. A direct mass flow measurement of the injected mass flow would decrease the overall VMF uncertainty. The distribution of the pressure measurement uncertainty can be neglected for the Calorimeter Method since the calculated enthalpies in Equations (2) and (3) are based on liquid phase refrigerant and thus the enthalpy is mainly temperature dependent. For the Direct PT Method the temperature measurement shows the major distribution to the overall VMF uncertainty with $\varepsilon_T = 73.35\%$ and $\varepsilon_p = 26.65\%$. Compared to the other methods, the pressure has a significant impact on the Direct PT Method due to the temperature glide of R-407C. The uncertainty of the Direct PD method mainly results from the density measurement which distributes $\varepsilon_\rho = 97.4\%$. For the Separation Method both volume flow measurement and mass flow measurement have the main influence on the overall uncertainty with $\varepsilon_V = 50.37\%$ and $\varepsilon_m = 47.97\%$. Again, the mass low measurement distribution can be reduced when measuring the injected mass flow rate directly.

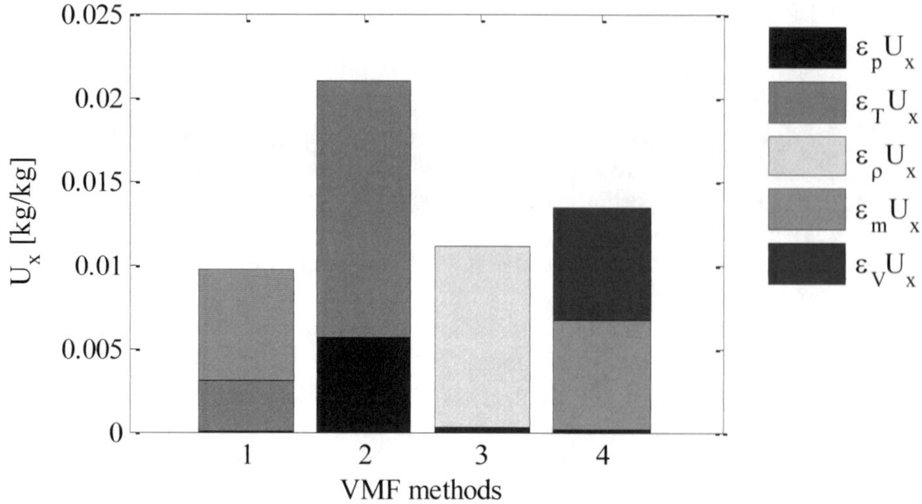

Figure 6. Results of the distribution analysis for the Calorimeter method (1), the Direct PT method (2), the Direct PD method (3) and the Separation method (4)

3.2. Results of the sensitivity analysis

The results of the sensitivity analysis are shown in Figure 7. The figure reveals high accuracy VMF measurements for sensor uncertainties close to the reference uncertainties $U^0_{Y,i}$ and that three methods are very sensitive to increased pressure measurement uncertainties.

The Calorimeter Method is sensitive to changes in temperature and mass flow measurement uncertainties. The VMF uncertainty changes with 0.052/K due to U_T and 0.052/% due to U_m. The sudden rise in uncertainty of the Calorimeter Method is caused by state point 3 and the small subcooling of 2 K, which leads to an apparent two-phase state when using low accuracy pressure sensors ($U_p > \pm 2.5\%$) or temperature sensors ($U_T > \pm 2K$). The Direct PT Method shows the highest sensitivity of all four methods with 0.090/K due to U_T and 0.105/% due to U_p. This high sensitivity is not preferable for accurate VMF measurements and it is caused by the moderate temperature glide of R-407C which is 5.6 K at the referenced injection pressure. However, this sensitivity will be reduced for refrigerants with higher temperature glides as compared to R-407C. The VMF uncertainty of the Direct PD Method varies with 0.022/(kg/m³) for different U_ρ and with 0.035/% for varying U_p. The Separation Method is mainly sensitive to pressure and mass flow measurement uncertainties. Its VMF uncertainty varies with 0.058/% for different U_m, 0.033/% for different U_p and 0.009/% for different U_V. The pressure sensitivity of the Separation Method could be eliminated when using a mass flow meter to measure the saturated refrigerant vapor flow.

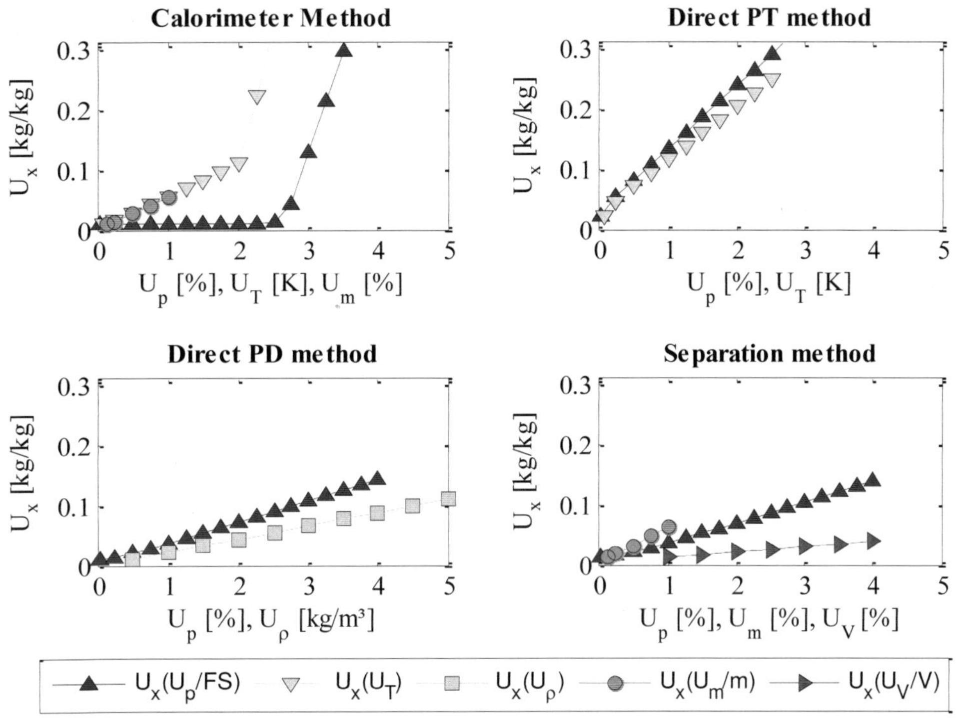

Figure 7. Results of the sensitivity analysis for the investigated methods

3.3. Results of the unknown parameter analysis

The results of the unknown parameter analysis are shown in Figure 8. The unknown heat loss or gain in the Calorimeter method leads to an error of $\Delta x/\Delta\phi_{amb} = \mp 0.009/\%$. Therefore, the heat loss has to be determined during the Calorimeter Method since the error to the VMF measurement is unneglectable. The variation of the velocity ratio S leads to an underestimation of the VMF at the maximum of $\Delta x = 0.05$ kg/kg. This is due to the definition of $\dot{x}_{inj}$ in Equation (6), which yields to $\dot{x}_{inj}=1$ for $S \rightarrow \infty$. For $S = S_{Zivi} = 3.00$, the underestimation is $\Delta x = 0.033$ kg/kg and $S = S_{Chisholm} = 5.09$ leads to an underestimation of $\Delta x = 0.040$ kg/kg. The assumption of a homogeneous flow ($S = 1$) is therefore not sufficient. Without any information on the velocity ratio, the Direct PD Method can only narrow down the range of possible VMF values and is therefore not applicable for accurate VMF measurements.

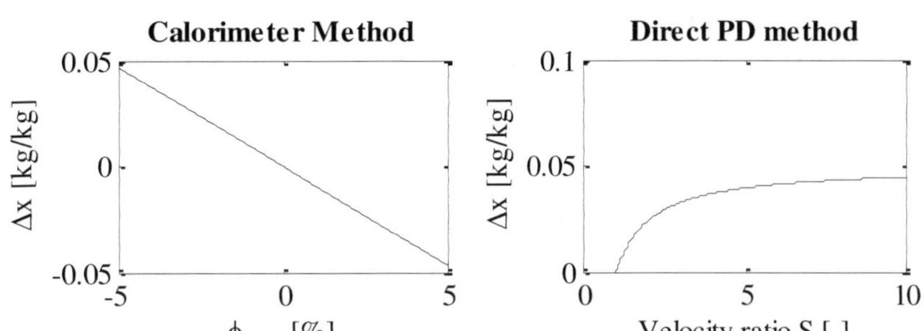

Figure 8. Results of the unknown analysis

IOP Conf. Series: Materials Science and Engineering **232** (2017) 012089 doi:10.1088/1757-899X/232/1/012089

4. Conclusion

The uncertainty investigation in this paper revealed that three of the four investigated VMF methods are generally applicable for VMF measurements. The achievable VMF uncertainty for the investigated reference state and R-407C as the refrigerant ranged from ± 0.0098 kg/kg to ± 0.021 kg/kg. Each method showed different sensitivity to changing sensor uncertainties and generally required high accuracy sensors for low VMF uncertainties. For future two-phase injection operation of EVI systems, the limits of the VMF measurement accuracy need to be taken into account for a reliable injection control.

References

[1] Moesch T W, Bahmann A M, Groll E A, 2016 Performance Testing of a Vapor Injection Scroll Compressor with R407C, *Proc. Int. Comp. Eng. Conf. (Purdue)*

[2] Wang S, Gu J, Dickson T, Dexter J, McGregor I, 2005 Vapor quality and performance of an automotive air conditioning system, *Exp. Therm. Flu. Sci* **30** pp 59-66

[3] Mitra C, Maity S, Banerjee A, Pandey A, Behera A, Jammu V, 2011 Development of Steam Quality Measurement and Monitoring Technique Using Absorption Spectoscopy With Diode Lasers, *IEEE Sensors J.* **11** 5 pp 1214-1219, May 2011.

[4] ASHRAE Standard 41.9-2011, *Standard Methods for Volatile-Refrigerant Mass Flow Measurements Using Calorimeters*, American Society of Heating, Refrigerating and Air-Conditioning Engineers, Inc., 1791 Tullie Circle NE, Atlanta, GA 30329

[5] Schmidt H, 2013,*VDI Wärmeatlas*, Kap. L2.1 Berechnung der Phasenanteile, pp 1285-1292, Springer Verlag, 2013

[6] Zivi S M, 1964 Estimation of Steady-State Steam Void-Fraction by Means of the Principle of Minimum Entropy Generation, *J. Heat Transfer*, Vol. 86, pp. 247-252

[7] Chisholm D, 1972 An Equation for Velocity Ration in Two-Phase Flow, NEL Report 535

[8] Coleman H W, Steele W G, 2009 *Experimentation, Validation, and Uncertainty Analysis for Engineers*, Jon Wiley & Sons, Inc, New Jersey.

[9] DIN EN 60584-1 *Thermoelemente - Thermospannungen und Grenzabweichungen*, DIN Deutsches Institut für Normung e. V., Berlin, Juli 2014

[10] DIN EN 60751 *Industrielle Platin-Widerstandsthermometer und Platin-Temperatursensoren*, DIN Deutsches Institut für Normung e. V., Berlin, Mai 2009

[11] Lemmon, E.W., Huber, M.I., McLinden, M.O., 2013, *NIST Standard Reference Database 23: Reference Fluid Thermodynamic and Transport Properties (RefProp)*, Version 9.1, National Institute of Standards and Technology, Standard Reference Data Program, Gaithersburg.

IOP Conf. Series: Materials Science and Engineering **232** (2017) 012090 doi:10.1088/1757-899X/232/1/012090

Modification of a compressor performance test bench for liquid slugging observation in refrigeration compressors

Max Ola, Christiane Thomas, Ullrich Hesse

Bitzer Chair of Refrigeration, Cryogenics and Compressor Technology
Technische Universität Dresden, 01062 Dresden, Germany

max.ola@tu-dresden.de

Abstract. Compressor performance test procedures are defined by the standard DIN EN 13771, wherein a variety of possible calorimeter and flow rate measurement methods are suggested. One option is the selection of two independent measurement methods. The accuracies of both selected measurement methods are essential. The second option requires only one method. However the measurement accuracy of the used device has to be verified and recalibrated on a regular basis.

The compressor performance test facility at the Technische Universitaet Dresden uses a calibrated flow measurement sensor, a hot gas bypass and a mixed flow heat exchanger. The test bench can easily be modified for tests of various compressor types at different operating ranges and with various refrigerants. In addition, the modified test setup enables the investigation of long term liquid slug and its effects on the compressor. The modification comprises observational components, adjustments of the control system, safety measures and a customized oil recirculation system for compressors which do not contain an integrated oil sump or oil level regulation system. This paper describes the setup of the test bench, its functional principle, the key modifications, first test results and an evaluation of the energy balance.

1. Introduction

To determine the right refrigerant compressor for a given application most of the time one can only rely on data provided by the manufacturers. Since some applications cannot be operated statically over an extended period of time and tend to transcend the limitations of the specified “safe” area of operation, it is reasonable to examine the effects of these “unsafe” operation modes. However, standardized compressor performance test systems are designed to prevent operating modes which could result in a failure of the tested machine.

For the purpose to investigate a compressors possible failure mode due to such overload, the test facility has to be modified especially regarding its underlying system control.

The content of this paper will present all the modifications and measures taken, to provide a test facility which can be operated in different ways beyond the specified operation limits of the compressor. The most important aspect is the provision of liquid refrigerant at the compressor inlet to simulate short and long-term liquid slug. Furthermore the challenge is taken to operate compressors designed without an oil sump in the same test bench, which was designed and used for compressors which rely on an oil separation and return cycle.

IOP Conf. Series: Materials Science and Engineering **232** (2017) 012090 doi:10.1088/1757-899X/232/1/012090

2. Setup of the performance test bench

2.1. Components, pipework and instrumentation

Figure 1 shows the piping and instrumentation diagram of the compressor performance test bench, including the oil recirculation and measurement devices subsequent to their modification.

This particular compressor performance test bench was designed according to DIN EN 13771 [1]. The underlying measurement concept is described in subsection 5.6 of this standard. A refrigerant flow meter is used to determine the gas volume flow rate at the suction or discharge side of the compressor. The use of this method is limited by a few conditions. The first requirement states a maximum of 1.5 % of the oil mass to be contained within the refrigerant gas at the flow rate measuring position. Another requirement is to prevent the refrigerant flow to include any liquid droplets to ensure a measurement in the superheated gas area. This second condition has a special significance for the aspired test of liquid slug, resulting in the situation that no performance measurement can be conducted while providing liquid. Merely the consequences of periodical liquid slug can be investigated with this method.

This setup allows the testing of various types of compressors. First tests were performed on reciprocating piston compressors while current tests are conducted on scroll compressors. The electrical power consumption of these compressors is limited to 30 kW, while most components are designed for a maximal conveyed volume flow of 150 m^3/h. Additionally, tests with highly flammable refrigerants such as butane are possible.

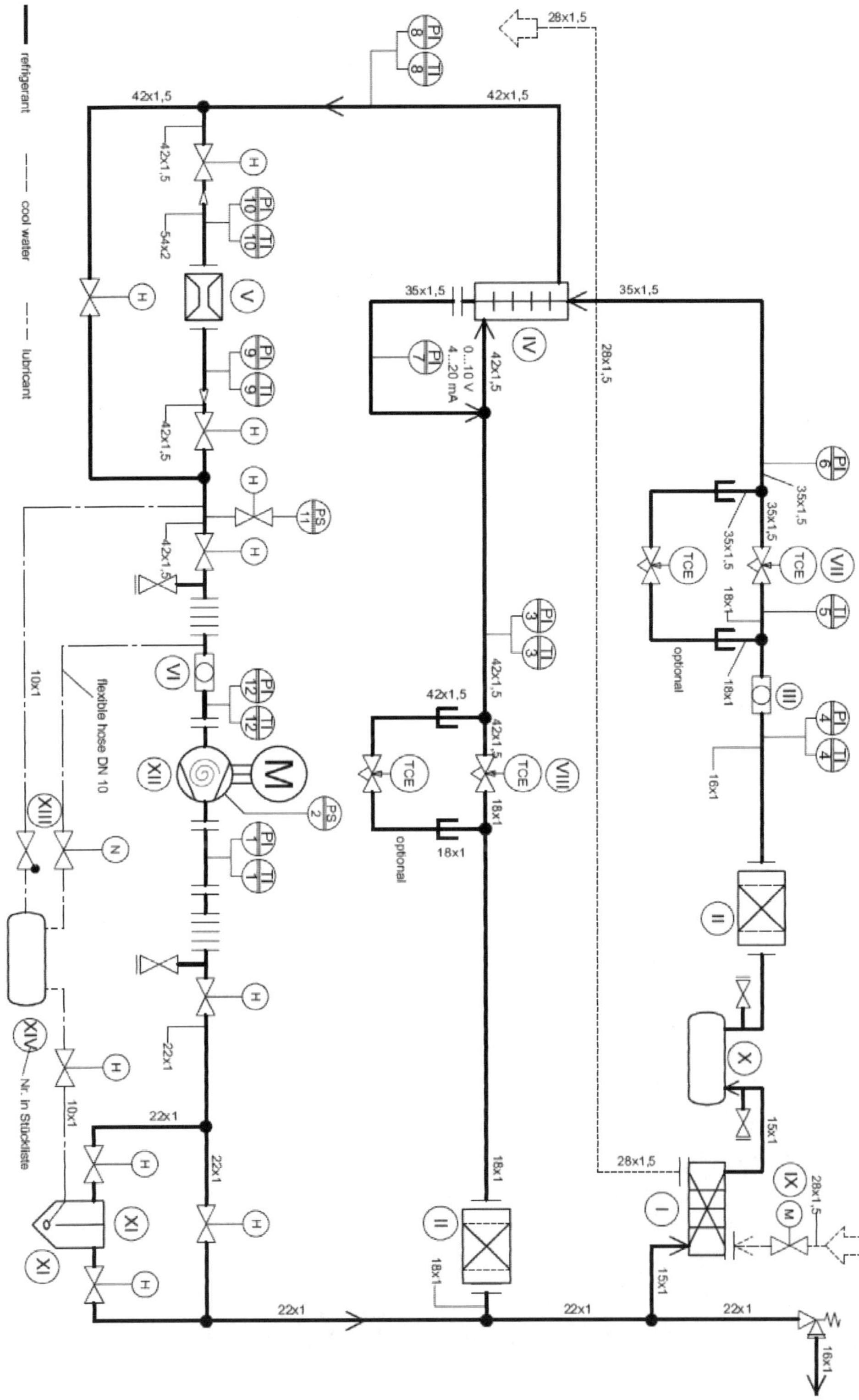

Figure 1. piping and instrumentation diagram, compressor test bench, TU-Dresden

The evaporation pressure and outlet temperature are controlled by two motorized expansion valves located in the condensation segment (VII) and the bypass line for the hot pressurized gas (VIII). The condensation pressure and temperature are set by a valve which controls the flow of cooling water through the condenser (IX). In contrary to the use of a liquid line flow meter, in this set up the condenser does not need to subcool the liquid. A receiver (X) is installed to provide liquid refrigerant to the expansion valve VII.

The evaporation will be achieved by a mixed flow heat exchanger (IV). This module works as an internal heat exchanger where the primary flow from the condenser (6) is evaporated by gaining heat from a mixture of refrigerant consisting of the flow exiting the primary heat exchanger side (7) and the refrigerant flowing through the bypass section (3). This evaporator is used primarily to ensure the superheating at the compressor inlet. To provide liquid refrigerant for liquid slug tests with this kind of configuration a control system is required that is able to forcefully shut off the bypass gas flow.

After being evaporated the combined refrigerant flow from the bypass and condensation segment enters the volume flow meter (V). For an automated operation during the performance tests including temporary liquid slugging, the volume flow meter has to be protected against any liquid refrigerant entering with the gas flow. If liquid refrigerant would get in between the two rotary pistons of the flow meter, an evaporation of the fluid would occur which leads to a deformation and failure of the flow meter. Because of that two shut-off valves have to be operated during the bypass shut off mode as well. These valves will respectively close the inlet line of the flow meter and open a line leading directly to the compressor inlet.

Shock absorbers (flexible lines) are used to reduce the stress from vibrations on the pipework. While most connections between pipes and fittings are soldered to achieve a high overall system leak tightness, the compressor is connected by screwed fittings. The intention is to improve the flexibility of the test bench and enable a fast exchange of test objects.

During the measurement procedure pressure and temperature values are recorded at all times. The system is completely insulated to reduce the influences of any possible heat flow across the system boundaries.

Figure 2. isometric view of the original compressor test bench, TU Dresden

2.2. Key Modifications

The mostly automated testing of compressors requires from the control system to automatically operate the valve needle positions and other variables like the frequency of a variable speed compressor. Since the compressors that are going to be tested on this bench will be operated beyond their specified application limitations the use of a conventional control system is not recommendable and the controller was programmed especially for this performance test bench using the software DASYLab®. The use of a customized control system allows the operator to intervene in the process at any time, improving its flexibility.

In case of the described test bench, the before mentioned manually operated shut-off valves had to be replaced by solenoid valves, which can be automatically controlled according to the operation mode and the refrigerant state at the flow meter inlet. The data acquisition system is a modular designed system to allow an easy extension for future needs.

A frequency inverter was implemented in the system for tests with variable speed compressors. With regard to the controller the speed regulation is based on an analogue output signal. Because the test bench is built for the testing of many different compressors of a wide power range, the frequency inverter is not permanently installed within the power panel but used as an exchangeable component to always provide the best fitting combination with the compressor.

The measurement and control program was adjusted to fit the new requirements of the automated operation. In the basic operation mode the valves, which are addressed by analogue signals, are controlled by PID modules. For the instance of a liquid refrigerant flow at the compressor inlet, the valves need to be forced into specific positions. To enable the liquid refrigerant to flow through the system while generating as little gas as possible at the compressor inlet, the expansion valve VII of the condensation segment has to be fully opened. At the same time the bypass valve VIII has to be closed. This in return will prevent the evaporation of the liquid refrigerant when traversing the mixed flow heat exchanger.

Due to the amount of the liquid stored in the receiver, the described operation of the expansion valves can feed liquid to the suction line for a longer period of time. Meanwhile the before mentioned solenoid valves are activated simultaneously and are preventing the liquid refrigerant from entering the volume flow meter by opening the bypass. Another method to induce liquid to the suction line is to feed the controller with adjusted control parameters. When the saturation temperatures are set and the controller receives signals from the sensors which do not fit the actual state of the refrigerant, the control can be used to create a two phase refrigerant flow. An example is given in Figure 3. To produce a partial liquid flow at the compressor inlet the pressure signal from which the control parameter for the PID module is derived, has to be adjusted to a lower value. At the same time the amount of superheat has to be set to a minimum. The difference between the transferred pressure signal p_{trans} and the real operating pressure p_{real} will then result in a two phase refrigerant flow at the compressor inlet.

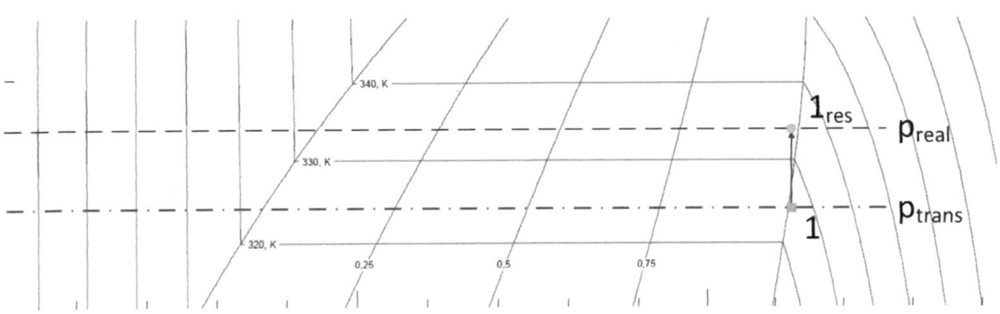

Figure 3. lg(p)-h-diagram, adjusted pressure, 2-phase control method

The sight glass (VI) is used to monitor the state of the refrigerant at the suction side of the compressor and to evaluate the oil return rate qualitatively. The integration of another sight glass within the oil return line is planned to provide an independent verification option for the oil return rate.

However, the most significant modification took place at the oil separation cycle itself. Because of the lack of an oil sump and the corresponding oil level regulator in certain compressor models, the oil

IOP Conf. Series: Materials Science and Engineering **232** (2017) 012090 doi:10.1088/1757-899X/232/1/012090

has to be fed into the suction line directly from the oil receiver (XI). To control the oil return rate a manually operated needle valve (XIII) is used. The valve is supposed to be set to a specific position and not to be changed during testing. The valve itself separates the oil receiver on the high pressure side from the suction line's low pressure level and can influence the controller upon a change of its opening degree. A view of the rebuild suction line is given in Figure 4. Suction line of the test bench including the sight glass, oil return line and suction state sensor.

Figure 4. Suction line of the test bench including the sight glass, oil return line and suction state sensors

3. Experimental results & functionality tests

3.1. First experimental results

Before the modifications were realized the performance test bench was operated to provide a reference measurement. The power consumption of a piston compressor from Bitzer®, the 4NES-20Y-40P was measured at different operation points. Afterwards the measured value was compared to the data sheet of the compressor [2]. To be in compliance with the standard DIN EN 13771, the deviation of the measurement has to be below 4 % from the data sheet. The data was averaged over a runtime of 15 min as required by the standard. The results are shown in Table 1.

Table 1. Performance accuracy of reference measurement, 4NES-20Y-40P, 15 min mean values

Variable	Target value	Measured value
Compressor inlet temperature [°C]	20.00	20.16
Condensation temperature [°C]	70.00	71.05
Evaporation temperature [°C]	15.00	14.17
Power consumption [kW]	13.03	13.73
Rel. error (power consumption) [%]		5.39

The manual control of the expansion valves (VII and VIII) increases the oscillating characteristics of the machine and thereby the settling time of the values. As a consequence the achieved temperature settings have a high deviation from the target values. When a corrected value is used to calculate the power consumption for a condensation temperature of 71 °C the calculated relative error is reduced to 4.7 %. To reach the desired maximum error of 4 % the sensors need to be recalibrated.

In general for a test bench working with a single volume or mass flow meter the standard defines a period of 6 months as the recalibration interval for the flow meter. It is assumed that the prolonged downtime without the recalibration of the bench is the cause for the observed deviations in the measured data. Thus the flow meter has to be recalibrated to reach the desired measurement accuracy.

3.2. Functionality tests

After the performance test of the piston compressor the controller was reprogrammed. The functionality was tested using a relative low capacity scroll compressor of the ELV21 series manufactured by Bitzer. This compressor has no oil sump and a speed control working on either a pulse width modulation or a 0 to 10 V signal.

An important test was the investigation of the PID modules to control the expansion valves. The proportional, integral and derivative terms were specified using iterative experiments. Chart 1 shows a temperature profile for an experiment determining the functionality of the PID controller for the bypass expansion valve. The error band is below 10 % at a target value of 10 °C (s. Chart 2). The mean temperature for the runtime of 15 minutes at the compressor inlet is 10.13 °C. While the accuracy of this value is no big improvement in comparison with the compressor inlet temperature measured during the first experiments with the 4NES-20Y-40P, a very high improvement was realized in regards to the run-in period. It took only about 200 s from standstill until a quasi-stationary process whereas some manually operated tests had taken more than 6 fold of that time. Since the PID control was implemented for both expansion valves as well as for the water flow control valve, the coordination of all the valves had to be adjusted during following experiments. In the end the same accuracy of the controller is desired for the condensation and evaporation temperature without increasing the settling time.

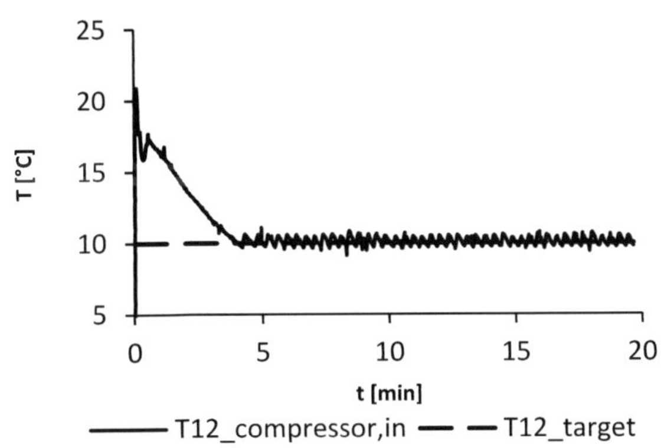

Chart 1. Compressor inlet temperature profile, PID controller bypass expansion valve

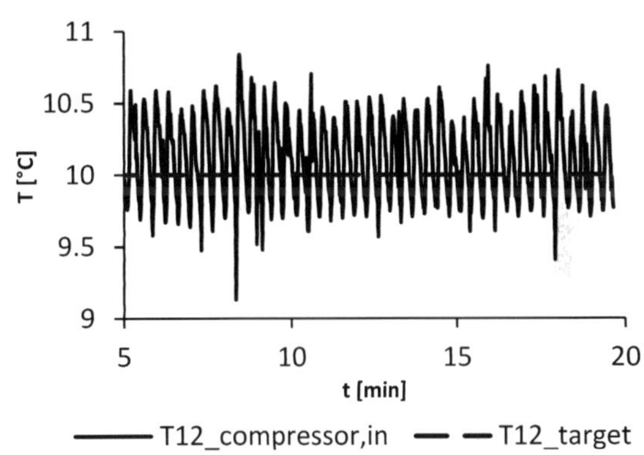

Chart 2. Compressor inlet temperature profile, detailed error band

Other functionality tests included the automatic switch for the relay-controlled solenoid valves as well as for the liquid slug activation process. DASYLab® is a data acquisition and control system based on a visual programming language. Figure 5 shows a section of the program depicting the elements responsible to process the control system.

The control module (Control1) is used to provide a target value for the condensation, evaporation and compressor inlet temperature for the PID controller. However, the frequency signal is predefined by the user and transmitted to the analogue output module as a constant value. The expansion valve signals are forwarded to a formula editor module. A switch (EXV-Switch) is used as a third input variable and primary condition for an if-then-function (EXV If/Then).

The simultaneously reversed change of the magnetic valve positions is achieved by using one switch module (MV-Switch) in combination with a logical link (Bit-Logik00) which operates one of the valves contrarily to the input signal.

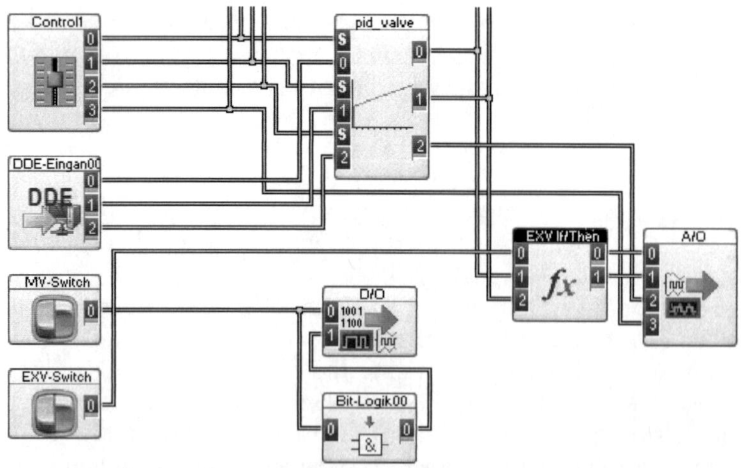

Figure 5. Program section DASYLab®, control system

The DASYLab® program communicates with a Microsoft Excel worksheet. The worksheet is fed with data from the acquisition modules and uses an add-in for the property program Refprop of the National Institute of Standards and Technology (NIST). The measurement data are used in combination with Refprop to determine all other necessary properties (e.g.: specific enthalpy, specific entropy and specific volume).

4. Energy and mass balance

When a compressor performance test facility is operated in compliance with the standard the user also has to calculate the energy and transport values in a specified way. Values defined by the standard [1] are the refrigerant mass flow, cooling capacity, volumetric efficiency/ efficiency of the compressor and the COP of the standardised cycle.

At this test bench with a suction line gas flow meter, the mass flow $\dot{m}_R$ has to be calculated using the measured refrigerant volume flow $\dot{V}_R$ and the mean gas density across the volume flow meter ρ_m:

$$\dot{m}_R = \dot{V}_R \rho_m \tag{1}$$

The cooling capacity $\dot{Q}_0$ is not determined by the actual evaporation capacity of the installed heat exchanger. The same applies to the condensation capacity. To standardise all compressor performance tests, the following calculation uses instead the specific enthalpy of the saturated liquid state at high pressure which is also the evaporator inlet enthalpy $h_{l,h}$. The high pressure is measured at the compressor outlet, neglecting further pressure losses until reaching the expansion valve. The evaporator outlet enthalpy is taken as the measured compressor inlet state.

$$\dot{Q}_0 = \dot{m}_R \left(h_{c,in} - h_{l,h} \right) \tag{2}$$

To calculate the volumetric efficiency λ_c the theoretical volume flow $\dot{V}_{c,th}$ of the tested compressor and the inlet density $\rho_{c,in}$ have to be known.

$$\lambda_c = \frac{\dot{m}_R}{\dot{V}_{c,th}\, \rho_{c,in}} \tag{3}$$

An overall efficiency of the compressor η_c can be calculated with the measured data of the compressor inlet state and the isentropic compressor outlet enthalpy ($h_{c,out,s}$). Furthermore the measured power consumption of the electric motor $P_{c,el}$ is needed to provide the reference value.

$$\eta_c = \frac{\dot{m}_R \left(h_{c,in} - h_{c,out,s}\right)}{P_{c,el}} \tag{4}$$

Finally the coefficient of performance is determined from the cooling capacity and the measured power consumption, according to equation (5).

$$COP = \frac{\dot{Q}_o}{P_{c,el}} \tag{5}$$

5. Conclusion

The modifications of the data acquisition and control system resulted in an overall improvement of controllability of the presented customised performance test bench with newly achieved settling times of about 200 s for a single operation state.

The most important modifications were carried out at the suction line and the oil recirculation system. These modifications were verified in a series of functionality tests and enable the test bench to be used for tests of compressors which are not equipped with an oil sump. An analysis of such a compressor was done after 80 h of operation time. The draining and measuring of the oil load from the compressor did result in the same amount that the compressor was filled with initially by the manufacturer to prevent an unlubricated start.

The operation of compressors without an oil sump results in the demand for special modifications such as monitoring options to assure the lubrication of the compressor at all times. The installation of two sight glasses, one of which is located in the oil recirculation line, can provide the needed verification for the oil return rate. The second sight glass which is located in the compressor suction line allows for the qualitative evaluation of the liquid refrigerant ratio at the suction state when the liquid slug operation mode is engaged. At the same time the newly implemented magnetic valves offer an automatic protection mechanism for the rotary flow meter. These changes to the test bench provide the necessary conditions to conduct liquid slug tests.

With the recalibration of the flow meter the test bench allows for compressor performance tests within the defined tolerance of the standard DIN EN 13771 to be conducted. After the implementation of a frequency converter the test bench can be used for the testing of compressor with variable speed operation modes.

References

[1] Normenausschuss Kältetechnik (FNKä) im DIN Deutsches Institut für Normung e. V. 2003 *Leistungsprüfung und Prüfverfahren Deutsche Fassung EN 13771-1*

[2] Bitzer Kühlmaschinenbau GmbH 2012 *Data Sheet: Semi-Hermetic Reciprocating Compressors – New Bitzer Ecoline R134a/R404A*

IOP Conf. Series: Materials Science and Engineering **232** (2017) 012091 doi:10.1088/1757-899X/232/1/012091

Evaluations of PVE Lubricants for Refrigeration and Air Conditioning system with the Low GWP Refrigerants

Tomoya MATSUMOTO[1], Masato KANEKO[1] and Yasuhiro KAWAGUCHI[2]

[1]Idemitsu Kosan Co.,Ltd., Lubricants Research Laboratory
[2]Idemitsu Kosan Co.,Ltd., Lubricants Department

E-mail: tomoya.matsumoto@idemitsu.com
E-mail: masato.kaneko@idemitsu.com
E-mail: yasuhiro.kawaguchi@idemitsu.com

Abstract. For the prevention of global warming, various low GWP refrigerants (R1234yf, R1234ze, R448A, R449A, R452A, R452B, R454B etc.) are proposed as the alternative of R410A and R404A for refrigeration and air conditioning system. PVE lubricants were developed for refrigeration and air-conditioning system with low GWP refrigerants. In this report, the combinations of the low GWP refrigerants and lubricant were evaluated. The evaluation items are physical properties (miscibility, solubility, viscosity, and electric insulation), thermal stability and lubricity.

1. Introduction
Now, the HFC refrigerants of the ODP (0) are used for the refrigerating equipment. Since the HFC refrigerants had the high GWP, it was set to one of the greenhouse gas by the Kyoto Protocol in 1997, and became a candidate for reduction. At the F gas regulations of Europe, the refrigerants for refrigerator with hermetic compressor will be upper limit of GWP 2500. And, the refrigerants for RAC will be upper limit of GWP 750. In the industry of refrigeration and air conditioning systems, it has been promoting that the study of low GWP refrigerants. [1] The HFO refrigerants (R1234yf and R1234ze) and the HFO blend refrigerants (R448A, R449A, R452A, R452B, R454B etc.) of the low GWP refrigerants are investigated as one of alternative technology of R404A and R410A. Then, it is necessary to evaluate the specifications and performance of these refrigerants and lubricants.
However, it was checked that the present PVE is inferior in the stability of the HFO refrigerants. So, the development of PVE optimized the new stabilizer for improvement in stability.
In this report, it was evaluated that the relationship between various low GWP refrigerants and PVE lubricants. The evaluation items are physical properties (miscibility, solubility, mixture viscosity and volumetric resistivity) and thermal stability and lubricity.

2. Experimental

2.1. Lubricants and Refrigerants
The evaluated lubricants were PVEs. The specifications of PVE are shown table1 in the chemical structure of PVE shown in Figure 1. PVE32A and PVE68A were included the antioxidant, the acid

catcher, and antiwear for R404A and R410A application. The development of PVE32B and PVE68B were added the new stabilizer to PVE32A and PVE68A for HFO application.

Table 2 shows GWP, molecular structure, and refrigerant components of the evaluated refrigerants. The evaluated low GWP refrigerants were seven kinds. Alternative of R404A was considered with R448A, R449A and R452A. Alternative of R410A was considered by R452B and R454B. The evaluation was performed by the method shown in the sub-sections 2.2~2.6 below.

Table 1 Specification of PVEs

Lubricant		PVE32A	PVE32B	PVE68A	PVE68B
Apprication		R404A	HFO	R410A	HFO
Viscosity @40℃ (mm^2/s)		32.40	33.19	66.57	68.41
Viscosity @100℃ (mm^2/s)		5.120	5.260	8.037	8.316
Viscosity Index		78	84	84	88
Density @15℃ (g/cm^3)		0.925	0.936	0.937	0.944
Acid Number (mgKOH/g)		0.01>	0.01>	0.01>	0.01>
additive	antiwear	include	include	include	include
	antioxidant	include	include	include	include
	acid catcher	include	include	include	include
	New stabilizer	-	include	-	include

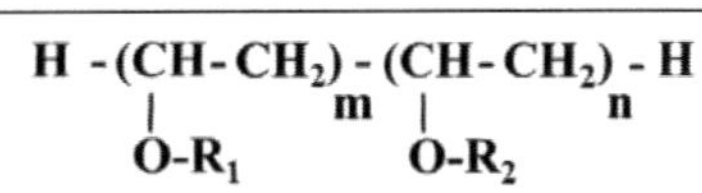

Figure 1 Chemical Structure of PVE

Table 2 Specification of Refrigerants

Refrigerant			R404A	R448A	R449A	R452A	R410A	R452B	R454B	R1234ze	R1234yf
GWP			3920	1273	1397	2140	2090	675	460	6	4
Refrigerant Compoment (wt%)	R32	CH_2F_2		26	24.3	11	50	67	68.9		
	R125	CF_3CHF_2	44	26	24.7	59	50	7			
	R134a	CF_3CH_2F	4	21	25.7						
	R143a	CF_3CH_3	52								
	R1234ze	$CF_3CH{=}CHF$		7						100	
	R1234yf	$CF_3CF{=}CH_2$		20	25.3	30		26	31.1		100

2.2. *Stability*

Stability of lubricants and refrigerants mixture was evaluated with the autoclave. Table 3 shows the testing conditions. The influence of air and moisture is also examined. The evaluated analysis item was the acid value.

Table 3 Stability test conditions

Conditions	Test 1	Test 2
Temperature (℃)	175	175
Test time (h)	336	336
Oil amount (g)	30	30
Refrigerant amount (g)	30	30
Water content (ppm)	50>	500
Air pressure (kPa)	0.7>	13
Catalysts	Fe / Cu / Al	Fe / Cu / Al

2.3. Miscibility
Figure 2 shows miscibility test apparatus and method. [2] The test tube is made of the sapphire and it was filled up with lubricant/refrigerant mixture. A photo-sensor is set up to detect light through the tube. The bath temperature is gradually increased (or decreased). Initially the lubricant and refrigerant mixture is clear. As the temperature increased (or decreased), the mixture becomes cloudy appearance which indicate lubricant/refrigerant separation. The two-phase separation temperature of that mixture is determined by the light transmittance of the photo-sensor. The temperature representing the center of this curve is the critical separation temperature (CST). The mixed quantity of lubricant was measured in the range of 5 – 40 wt%.

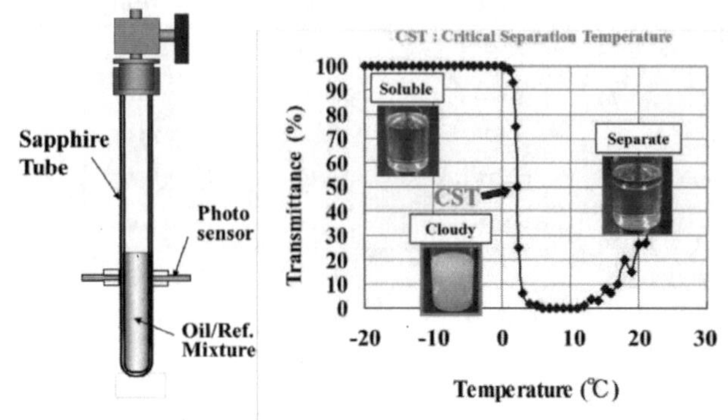

Figure 2 Miscibility test apparatus and method

2.4. Solubility and mixture viscosity
Figure 3 shows the hermetic type viscometer. This apparatus measures both the solubility and mixture viscosity of lubricant/refrigerant mixtures. [3] To measure the viscosity, a capillary-type viscometer in a pressure tight case is used. Solubility was determined by the calculating formula in Figure 3. The Daniel Chart was drawn on measurement data.

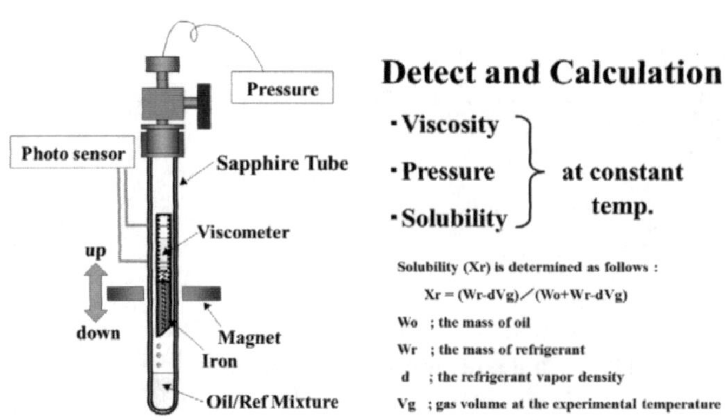

Figure 3 Hermetic type viscometer

2.5. Electric property
Figure 4 shows the hermetic type volumetric resistivity tester. [4] The volumetric resistivity indicates electrical insulation properties of refrigeration lubricant. Volumetric resistivity is a ratio of electrostatic strength to the current density when the direct electric field is impressed to electrode that fills the sample. The volumetric resistivity was measured at varied refrigerant contents (wt%). Test conditions were the voltage (250V) and temperature (20℃).

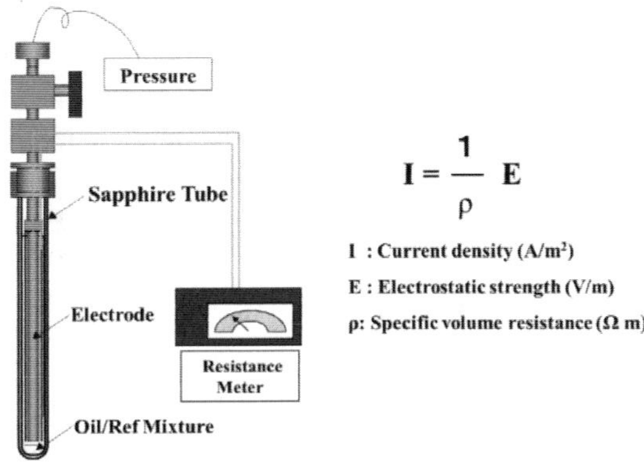

Figure 4 Hermetic type volumetric resistivity tester

2.6. Lubricity

Figure 5 shows the hermetic type Block-on-Ring tester and the lubricating evaluation test conditions. The evaluation was performed by comparison of the wear volume.

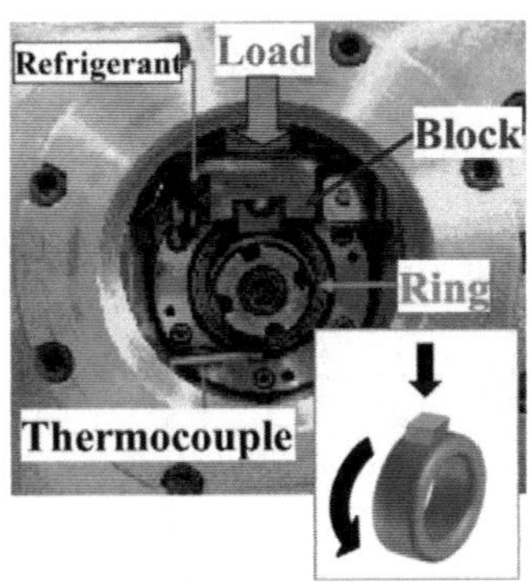

Conditions	Test
Ring material	JIS FC250
Block material	JIS SKH51
Temperature (℃)	50
Test time (min)	60
Oil amount (g)	100
Refrigerant pressure(MPa)	0.3
Load (N)	1000
Speed (m/s)	0.5

Figure 5 Hermetic type block-on-ring tester and wear test condition

3. Results and discussions

3.1. Stability

Figure 6 shows the autoclave tests of PVEs with refrigerants. The acid number did not increase in PVE32A / R404A and PVE68A / R410A on the both test conditions. On the other hand, the acid number increased in PVE32A and PVE68A with HFO refrigerants on the test2 condition. The acid number was controlled PVE32B and PVE68B included new stabilizer. It was found that stability of PVE32B and PVE68B was similar to PVEs / HFC refrigerants. The sludge by new stabilizer wasn't seen after the test.

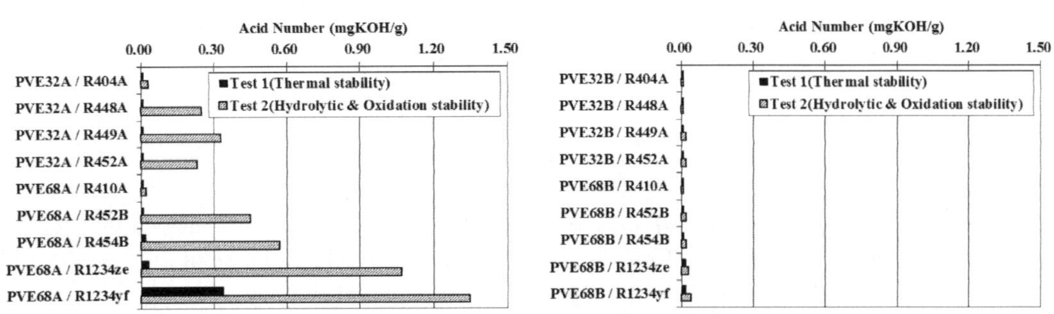

Figure 6 Stability of PVEs with refrigerants

IOP Conf. Series: Materials Science and Engineering **232** (2017) 012091 doi:10.1088/1757-899X/232/1/012091

3.2. Miscibility

Figure 7 shows the miscibility of PVEs with HFC refrigerants (R410A and R404A). At 10wt% of oil content, two-phase separation temperature of PVE32A / R404A and PVE68A / R410A at high temperature were 48℃ and 45℃. Figure 8 shows the miscibility of PVEs with HFO refrigerants (R448A, R449A, R452A, R452B, R454B, R1234ze and R1234yf). HFO refrigerants were more soluble than HFC refrigerants. In particular, the miscibility of PVEs with R1234yf, R1234ze, R448A and R449A were completely soluble within the range of -50℃ to 70℃.

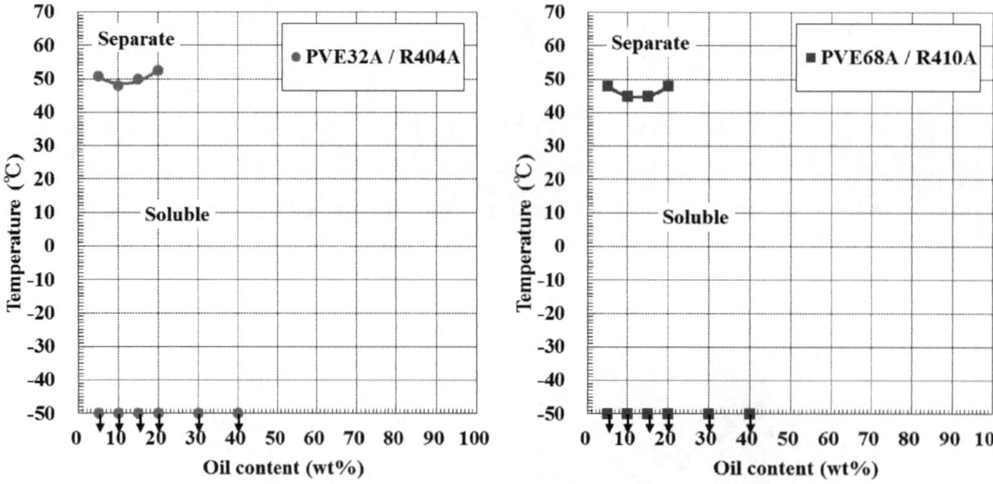

Figure 7 Miscibility of PVEs with HFC refrigerants

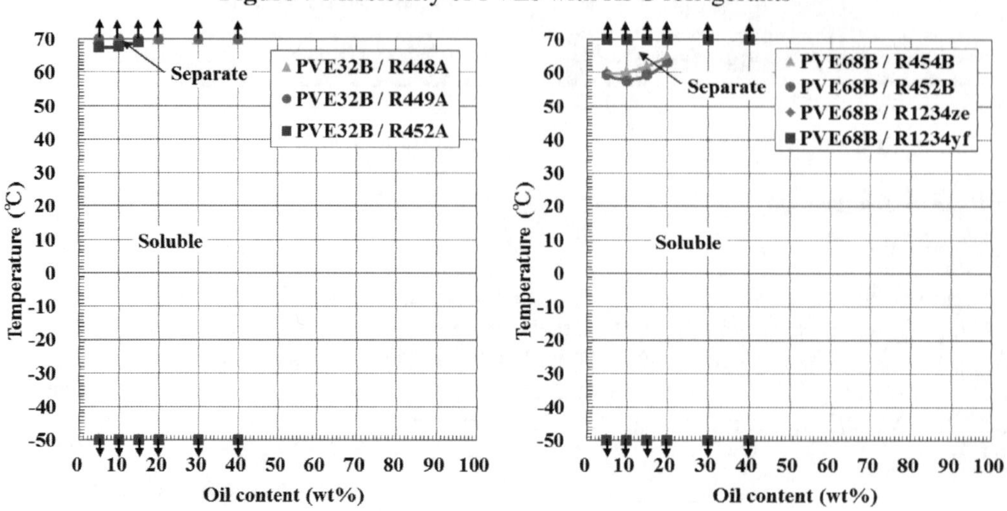

Figure 8 Miscibility of PVEs with HFO refrigerants

3.3. Solubility and mixture viscosity

Figure 9 and 10 show the Daniel Chart of PVEs with refrigerants. As an example, solubility and mixture viscosity of the condition (60℃, 1.5MPa) are shown in table 4.

It was understood that the solubility of PVE32A / R404A was 16%, and that of PVE32B / R1234yf was 50%. So, the solubility of R1234yf was higher than that of R404A. The solubility of PVE32B / R449A and PVE32B / R452A were decreased by PVE32B / R1234yf. On the other hand, the mixture viscosity of R1234yf becomes low for increased solubility.

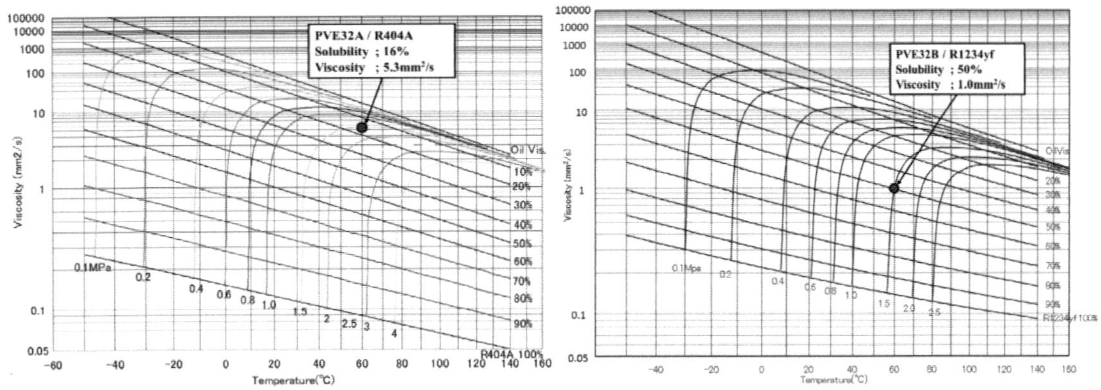

Figure 9 Daniel Chart of PVEs with refrigerants

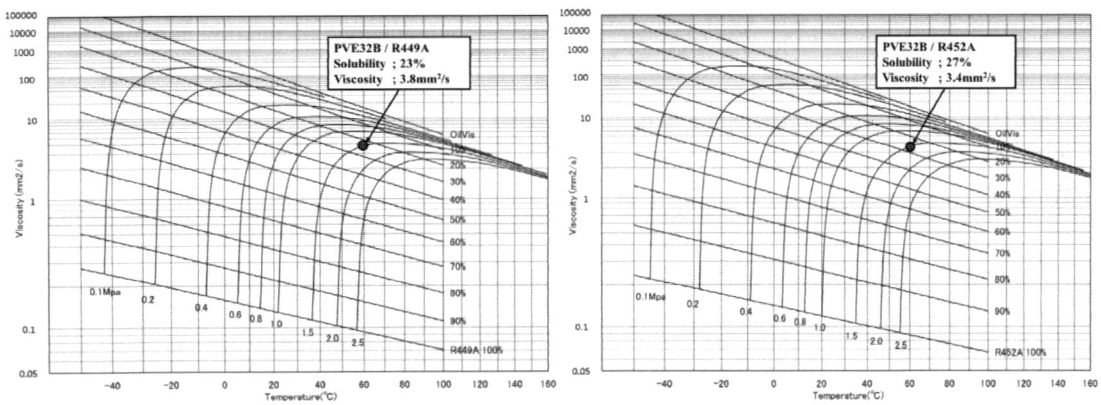

Figure 10 Daniel Chart of PVE32B with refrigerants

Table 4 Solubility and mixture viscosity of PVEs with refrigerants (60℃, 1.5MPa)

Lubricants	Refrigerants	Solubility (wt%)	Viscosity (mm^2/s)
PVE32A	R404A	16	5.3
PVE32B	R449A	23	3.8
PVE32B	R452A	27	3.4
PVE32B	R1234yf	50	1.0

3.4. Electric property

Figure 11 shows the volumetric resistivity of PVEs with refrigerants. The volumetric resistivity of PVE32A and PVE32B were in order of $10^{11}\Omega\cdot m$. It turned out that there is no influence by the new stabilizer. On the other hand, the volumetric resistivity of the refrigerants became high in order of R404A ≈ R449A ≈ R452A < R1234yf. The volumetric resistivity of R449A and R452A didn't increase by R1234yf. The volumetric resistivity of refrigerant mixture decreased with refrigerant content increase. They were slope downward with single logarithmic plot. An order of volumetric resistivity of refrigerants and refrigerant mixture by a constant fraction were the same tendency.

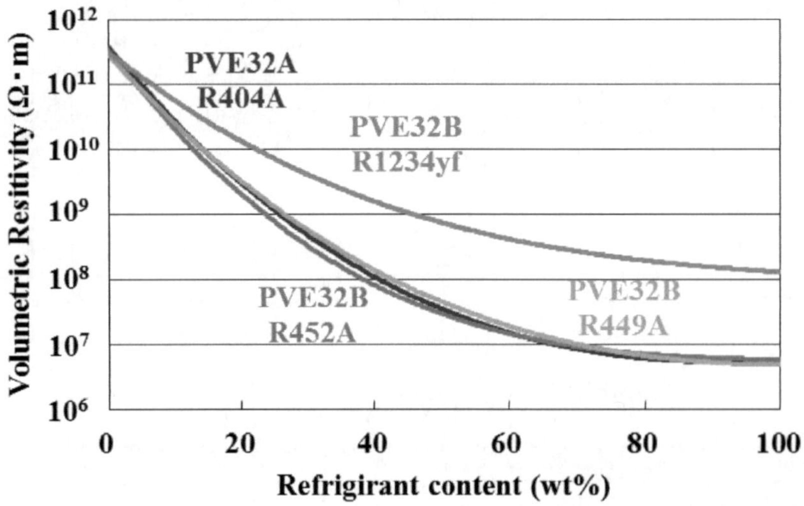

Figure 11 Volumetric resistivity of PVEs with refrigerants

3.5. Lubricity

Table 5 shows lubricity of PVEs with refrigerants. As for PVE32A and PVE32B, PVE68A, PVE68B, there is little wear volume and the difference was not seen. There was no influence by the difference in the refrigerant, and lubricity was good under the examination conditions.

Table 5 Lubricity of PVEs with refrigerants

Lubricants	Refrigerants	Ring Wear (mg)	Lubricants	Refrigerants	Ring Wear (mg)
PVE32A	R404A	1.6	PVE68A	R410A	1.2
PVE32B	R448A	1.5	PVE68B	R452B	1.1
PVE32B	R449A	1.5	PVE68B	R454B	1.1
PVE32B	R452A	1.5	PVE68B	R1234ze	0.9
PVE32B	R1234yf	1.3	PVE68B	R1234yf	0.7

4. Conclusions

It was evaluated that the relationship between the low GWP refrigerants and PVE lubricants. The evaluation items are physical properties (miscibility, solubility, mixture viscosity and volumetric resistivity) and thermal stability and lubricity. The development of PVE32B and PVE68B included new stabilizer for improvement in stability with HFO refrigerants (R1234ze and R1234yf). Furthermore, PVE32B and PVE68B can be used also for lubricant with the HFO blend refrigerants (R448A, R449A, R452A, R452B and R454B).

References

[1] H. Hara and M. Oono, I. Iwata. 2010 *International Symposium on Next-generation Air Conditioning and Refrigeration Technology* NS26.
[2] JP2823123
[3] JP3711303
[4] JP5624782

IOP Conf. Series: Materials Science and Engineering **232** (2017) 012092 doi:10.1088/1757-899X/232/1/012092

Identifying lubricant options for compressor bearing designs

J Karnaz[1], C Seeton[1] and L Dixon[2]

[1]Shrieve Chemical Products, Inc. USA
[2]Shrieve Products International Ltd UK

E-mail: jkarnaz@shrieve.com
E-mail: cseeton@shrieve.com
E-mail: ldixon@shrieve.com

Abstract. Today's refrigeration and air conditioning market is not only driven by the environmental aspects of the refrigerants, but also by the energy efficiency and reliability of system operation. Numerous types of compressor designs are used in refrigeration and air conditioning applications which means that different bearings are used; and in some cases, multiple bearing types within a single compressor. Since only one lubricant is used, it is important to try to optimize the lubricant to meet the various demands and requirements for operation. This optimization entails investigating different types of lubricant chemistries, viscosities, and various formulation options.

What makes evaluating these options more challenging is the refrigerant which changes the properties of the lubricant delivered to the bearing. Once the lubricant and refrigerant interaction are understood, through various test methods, then work can start on collaborating with compressor engineers on identifying the lubricant chemistry and formulation options. These interaction properties are important to the design engineer to make decisions on the adequacy of the lubricant before compressor tests are started.

This paper will discuss the process to evaluate lubricants for various types of compressors and bearing design with focus on what's needed for current refrigerant trends. In addition, the paper will show how the lubricant chemistry choice can be manipulated through understanding of the bearing design and knowledge of interaction with the refrigerant to maximize performance. Emphasis will be placed on evaluation of synthetic lubricants for both natural and synthetic low GWP refrigerants.

1. Introduction

The world sits at a crossroads. Challenges like climate change try to focus on reducing the impact that refrigerants have on the environment while still balancing this with indirect emissions though energy consumption by equipment. As the world grows more accustomed to comfort cooling, domestic refrigeration, food preservation, etc., refrigeration and air-conditioning has become an integral part of how we go about our day. Residential, commercial, and industrial equipment used in refrigeration and air-conditioning products can account for 20-30% of the electrical energy consumption in most regions. In the US three quarters of the homes have some form of air conditioning accounting for 6-8% of energy consumption [1]. Keeping this equipment reliable and energy efficient results in better preservation and comfort, cost savings and environmental benefits. The compressor in these systems is an essential component for performing the required work but it also consumes energy. For example, in restaurants, the electrical energy consumption associated to refrigeration is about 10-16% and for supermarkets 44-60%. In each case, the compressor is accounting for 75-80% of the consumption [2]. These types of values make operation of the compressor critical for maintaining adequate food preservation and distribution along with establishing appropriate levels of energy efficiency.

Today we are facing the additional complexity of evaluating alternate refrigerants that have less impact on environmental concerns like global climate change. As refrigerants are being evaluated, it is important that levels of reliability and performance are maintained or strengthened in systems when compared to refrigerants they are replacing. For the most part, compressors used in applications utilize a lubricant to help protect the bearings and perform other desirable tasks. These lubricants are always interacting with the refrigerant so understanding the impact of this interaction is essential to understanding the impact of performance. Evaluating interaction of the lubricant within compressors is a multistep task starting with the chemistry and property of the lubricant, the type of refrigerant being used, the compressor design, and system operation. Choosing the right lubricant can provide compressor operation that is reliable and efficient in combination with the refrigerant being used [3].

2. Refrigerant Direction

The world is becoming integrated by way of communication, trade, and other activities but there is still a difference when it comes to the use of refrigerants. Countries are not moving at the same pace or in the same direction when it comes to determining what they will chose as a viable refrigerant moving forward. Some developing countries still using R-22 appear to be considering other options instead of R-410A. The Kagali Amendment to the Montreal Protocol will minimize the use of HFC refrigerants in the future but is a phase down not a phase out so numerous refrigerant chemistries could be in play for several years [4]. Natural refrigerant options such as carbon dioxide, ammonia, and propane, though not new, are finding more acceptability and use for their low GWP values and availability. Countries where higher ambient temperature conditions exist are finding it difficult to move away from HCFC refrigerants to HFC refrigerants or to next generation lower GWP refrigerants due to higher operating temperatures and pressures that are generated. Other solutions might be needed for these situations.

Overall refrigerant use on a global outlook still has some uncertainty, but regardless of the refrigerant used the implications this introduces to operation of a compressor and lubricant should be investigated. Refrigerant and lubricant interactions will vary and sometimes significantly based on the chemistry of both parties and therefore must be evaluated on more of a case by case basis. Due to refrigerant properties, it takes advanced knowledge and special equipment/instrumentation to accomplish the tasks.

3. Lubricant and Refrigerant Interaction

Many papers, including those by these authors, describe the various things that should be evaluated for lubricant and refrigerant interaction. Probably the three most critical are: 1) stability; 2) miscibility and 3) working viscosity. Stability is handled by testing per ASHRAE Standard 97 or testing in pressure metal vessels [5]. Miscibility is the ability of the lubricant and refrigerant to maintain a single phase over a range of temperatures and concentrations. Typical miscibility curve is a measurement over a range of lubricant to refrigerant concentrations (1 to 50%) and temperatures that are representative of the system application, for example 50°C to -40°C [6]. When the solution indicates two phase separation, or in some cases a level of cloudiness, this temperature is recorded as the phase separation temperature. When measured over a range of concentrations, the curves that are drawn represent the maximum lower critical solution temperature (LCST) and the minimum upper critical solution temperature (UCST). Determining the working viscosity is more complicated of a procedure and to be done accurately needs special designed equipment to handle the range of operating temperature and pressure while simultaneously measuring the desired parameters. Seeton introduced this approach and equipment which has been implemented by others over the years [7]. Most effective approaches to these measurements require analysis over a range of refrigerant concentrations and temperatures representative of system operation. Once measurements are made, modelling of the results provide coefficients that can be used to input into equations to calculate dilution, density, and viscosity. These coefficients and equations can be incorporated into a solver program that can output values with inputs of temperature and pressure. Figure 1 below is an example of coefficients derived from a test and equations for calculating the required values. This type of information has been an essential tool to assist compressor and bearing design engineers.

coeff #	Pressure	Density	Kin. Visc
a1	6.59421E+00	1.28840E+00	1.61744E+01
a2	-2.08408E+03	-1.17262E-03	-2.54796E+00
a3	1.10193E+05	6.46815E-07	0.00000E+00
a4	4.98235E+00	1.94331E-01	2.41055E+01
a5	-2.42988E+03	7.34878E-04	-4.46803E+00
a6	2.95275E+05	-2.10033E-06	0.00000E+00
a7	5.99507E-01	1.32697E+00	-4.04662E+01
a8	-4.27630E+02	-6.78219E-03	6.79744E+00
a9	5.46739E+04	8.29146E-06	0.00000E+00
rsq	0.999891	0.9996	0.9936

Vapor Pressure:

$$Log_{10}(P) = a_1 + \frac{a_2}{T} + \frac{a_3}{T^2} + Log_{10}(\omega)\left(a_4 + \frac{a_5}{T} + \frac{a_6}{T^2}\right) + Log_{10}^2(\omega)\left(a_7 + \frac{a_8}{T} + \frac{a_9}{T^2}\right)$$

Density:

$$\rho = a_1 + a_2\mathrm{T} + a_3\mathrm{T}^2 + \omega\left(a_4 + a_5\mathrm{T} + a_6\mathrm{T}^2\right) + \omega^2\left(a_7 + a_8\mathrm{T} + a_9\mathrm{T}^2\right)$$

Kinematic Viscosity:

$$\begin{aligned}\mathrm{Log_e}\left(\mathrm{Log_e}\left(\nu + 0.7 + e^{-\nu}K_0(\nu + \varphi)\right)\right) &= a_1 + a_2 Log_e(T) + a_3 Log_e^2(T) \\ &+\omega\left(a_4 + a_5 Log_e(T) + a_6 Log_e^2(T)\right) \\ &+\omega^2\left(a_7 + a_8 Log_e(T) + a_9 Log_e^2(T)\right)\end{aligned}$$

Figure 1: Coefficients and Equations used in Daniel Plot Graphs

4. Compressor Design and System Demand

Understanding bearing lubrication is an important core need for compressor design engineers. Calculations can be made to measure desired film thickness needed for a bearing to sustain an operational lifetime and meet efficiency constraints. Over the years, with the advancement of computer speed and capacity, modelling programs for bearings have become an important step in bearing evaluations with lubricants. Numerous bearing designers offer programs that can determine film thickness measures with inputs of certain parameters. For applications where just the lubricant is present this is usually a straight forward input of bearing sizes, surface finish and particularly the viscosity of the lubricant at conditions like temperature. Most of this information is easily accessible by the bearing design or compressor engineer. But for calculations in refrigeration and air conditioning where the refrigerant now plays a vital role in determining what these operating parameters and subsequent inputted values are, it becomes more burdensome to obtain accurate information. This makes testing as described earlier for lubricant and refrigerant interaction more valued to the designer and today has become as important as asking for information like chemistry and base oil viscosity of a lubricant.

Each compressor type is designed differently and within each design, bearing design can vary. When it comes to evaluating and choosing a refrigeration lubricant, it is essential to make the evaluation in a systematic approach considering the compressor type and internal design. The next few sections will look at various types of refrigeration and air conditioning compressor designs and account for what is important to consider when looking for the appropriate lubricant. We will assume that lubricant and refrigerant interaction has already been evaluated for stability and system compatibility.

4.1 Rotary

Many rotary compressors are used in the air conditioning market in a variety of applications. Particular lubrication challenges within rotary operation is that the lubricant operates under discharge temperature and pressure conditions and the vane-roller bearing interface is predominately in boundary lubrication. The vane-roller interface dynamic is usually not solved by the lubricant viscosity but more so with the incorporation of additives that have extreme pressure (EP) and/or antiwear (AW) functionality [8]. Usually the lubricant viscosity is raised high enough to allow for some small amount of film thickness without degrading efficiency at the other bearings. Operating at higher temperatures typically means

IOP Conf. Series: Materials Science and Engineering **232** (2017) 012092 doi:10.1088/1757-899X/232/1/012092

the dilution factor of refrigerant in the lubricant is moderate but the higher temperature also means the base oil viscosity will be lowered. For rotary compressors, it is beneficial to have a lubricant that can minimize refrigerant dilution (without sacrificing other system requirements), a base oil with a higher viscosity index, a higher pressure-viscosity coefficient (though in the presence of refrigerant, the lubricant α coefficients differences are minimized), and an effective formulated additive package. The other bearings in the compressor are typically journal bearings and thrust surface bearing which can be lubricated adequately with the base lubricant viscosity, but if too high, can result in some frictional drag increase. The bearing material used in most rotary compressors is some form of ferrous based material so additive selection is more straight forward if it remains compatible with the lubricant and refrigerant. The lubricant choice in these compressors is usually a factor of providing enough viscosity to maintain compressor bearing durability but compressor engineers today are trying to push the limits of the lubricant to help achieve better energy efficiency. So a combination of reducing base oil viscosity, lubricant chemistry and additive technology is being challenged to find solutions.

Lubricant Keys:
Rotary compressors over the years with different refrigerants have been consistent on the base oil viscosity requirements which is 68 cSt. These compressor bearing designs are typically challenged by the additive that is used and how this additive is impacted by the bearing material, base oil chemistry, refrigerant, and bearing mechanism. Rotary compressors are more suited for higher evaporator temperature application which limits the effective operating range and range of capacity. These situations usually help to define the need for a limited base oil viscosity range. Rotary compressors are always formulated with an additive package that helps to minimize bearing wear. Over the years the selection of additive chemistry has been critical to maintain levels of effectiveness and compatibility with certain lubricants and refrigerants.

4.2 Reciprocating

Reciprocating compressors are a piston driven compression mechanism with a large variety of sizes, design, and lubrication complexities. Since these compressor types can range from very small capacity hermetic units to large hermetic and semi-hermitic units, the lubricant selection is a broad range of viscosity and chemistry depending on the application and refrigerant type. Bearing design in reciprocating compressors are a variety of types, material, and motions of operation along with the method that lubricant is distributed to the bearing. This complexity makes lubricant recommendation a complicated effort of modelling design and compressor testing. Today with requirements to obtain energy efficiency levels or commercial competitive advantages to providing efficiency most reciprocating compressors are designed to have minimal oil film thickness. This is especially evident in small compressor designs for applications like refrigerators or light commercial applications where small changes to lubrication can influence compressor efficiency and reliability [9]. Figure 2 is a representation of what can happen when a bearing design is not maximized for working viscosity. If viewed as a modified Stribeck Curve, when the working viscosity is too low, energy efficiency is sacrificed due to potential bearing wear. If working viscosity is too high, efficiency suffers due to bearing drag. The concept of lowering the base oil viscosity to achieve better energy efficiency has been utilized for years in the refrigerator compressor design as depicted in Figure 3. As with every bearing design, there are limits to the surface finishes and oil distribution where at some point too low of viscosity results in bearing fatigue and efficiency losses.

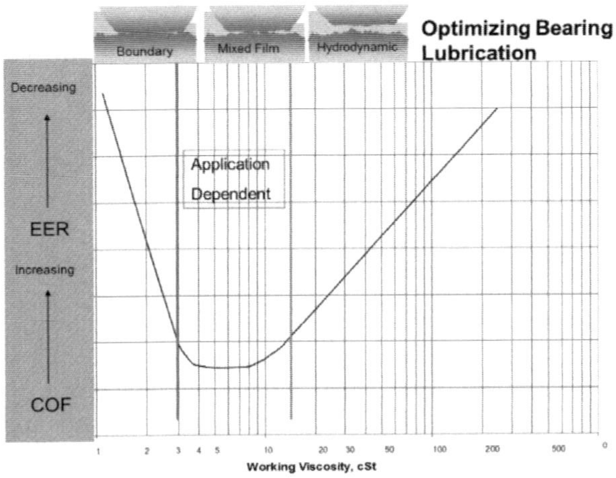

Figure 2: Modified Stribeck Curve depicting Working Viscosity Optimization

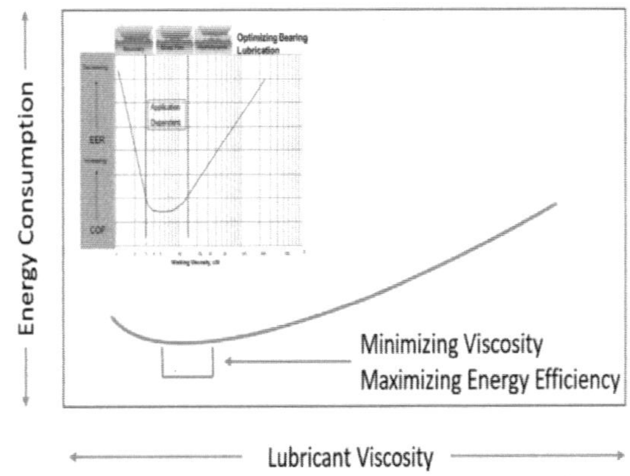

Figure 3: Modulating Viscosity to Enhance Energy Efficiency

Lubricant Keys:
The reciprocating compressor due to its diversity of sizes and applications requires a larger variation of lubricant base viscosities and additive chemistries to cover the range. Lubricant viscosity requirements can range from 3-32 centistokes in small fractional horsepower compressors up to viscosities in the range of 68 to 100 cSt in larger units operating in such applications as supermarkets. Additive chemistry can vary because the bearing material used in compressors can be made of iron, copper or aluminium substrates which will have different interaction properties and additive effectiveness. Since different bearings in reciprocating compressors can be affected by both the refrigerant pressure differential associated with an application and the compression ratio, most of the time the lubricant base oil viscosity is over engineered for larger compressors and stretched to its lower viscosity limit in smaller capacity compressors. Another difficulty with reciprocating compressor lubricant selection is that these compressors operate over a very wide range of low temperature to high temperature evaporator designs and for most cases the same lubricant is desired.

4.3 Scroll
Scroll compressors have found many uses especially in air conditioning applications because of the levels of energy efficiency and sound that can be achieved under certain conditions. Scroll designs have both compressor low side operation or high side operation so this needs to be taken into consideration for lubricant selection. The scroll wraps used in the design as the compression mechanism is a critical component for lubrication and is sometimes equipped with tip seals to enhance lubrication [10]. Other bearing lubrication needed for scroll compressors are journal or sleeve bearings, Oldham rings and sometimes roller bearings. Many scroll designs today operate with R410A refrigerant operating with an ISO 32 lubricant in low side design and up to an ISO 68 lubricant in scroll compressor operating as a high side design. Like with all applications improving energy efficiency is always an ongoing project even with current R410A designs. Today, transitioning away from higher GWP R410A has scroll compressor manufacturers looking at R32 refrigerant which brings about some lubricant selection challenges. These challenges will be discussed later under lubricant options.

Lubricant Keys:
For scroll compressors used in both air conditioning/heat pump applications and in commercial refrigeration, most designs use a lubricant with base oil viscosity of 32 cSt. The operational conditions and design of the compressor usually doesn't call for the need of additives in the oil to enhance or protect lubrication so bearings are typically designed to be optimal with the ISO 32 base oil. The choice of lubricant for scroll compressors is more fixed on the refrigerant and maintaining optimized lubricant-refrigerant interaction properties. A lubricant challenge that sometimes can exist with scroll compressor

IOP Conf. Series: Materials Science and Engineering **232** (2017) 012092 doi:10.1088/1757-899X/232/1/012092

design is that they are capable of handling more liquid refrigerant when compared to reciprocating and rotary compressor designs. This needs to be considered when evaluating working viscosity calculations.

4.4 Screw

Lubrication of screw compressor bearing designs is usually not driven by optimizing the lubricant to provide energy efficiency improvements. The lubricant supply to the screw mechanism and other bearings, usually roller bearings, is maintained at a high enough working viscosity to preserve bearing separation and provide sealing to minimize refrigerant leakage loses. Most screw compressors operate with base oil viscosity above ISO 100 and the working viscosity with some refrigerants have presented some challenges to both providing a high enough base viscosity and using the right lubricant chemistry. Lubrication for screw compressors is managed by an oil separator on the discharge side or an oil sump inside the compressor operating under suction conditions. With either mechanism the working viscosity requirements with the refrigerant is usually a wider viscosity range instead of specific values to achieve minimal bearing film thickness like in other compressor types. Challenges exist with trying to find high enough lubricant viscosity for certain lubricant chemistries. For example, when operating with R22 refrigerant many screw compressors designs had to move from mineral oil type lubricants to synthetic chemistries to obtain a high enough lubricant viscosity. These challenges were easily met but today with refrigerants like R1234zeE another challenge exists because of the high dilution potential that this refrigerant has with numerous types of lubricants. For example, many screw compressors operating with R134a use POE lubricants with viscosities in the range of 100 to 220 cSt, and when evaluated at discharge conditions, give adequate working viscosity. But when the working viscosity of these same lubricants are measured with R1234zeE, the values fall short of what is desired.
Other lubricant chemistry options are needed to solve this issue. Table 1 is a comparison of lubricant chemistries with R134a and R1234zeE at operating conditions showing how alternate lubricant chemistries like polyalkylene glycols (PAG) can match current operation.

Table 1: Screw compressor operating condition comparison

Operating Conditions: 60°C and 8 bar

Lubricant / Refrigerant	Dilution, %	Viscosity, cSt
ISO 170 POE / R134a	15	13
ISO 170 POE / R1234zeE	32	3.5
ISO 100 PAG / R1234zeE	23	12

Lubricant Keys:

Screw compressors fall in-line somewhat with reciprocating compressors when it comes to the range of lubricant viscosities required for various applications. Typically, base oil viscosity range from 68 cSt up to 400 cSt depending on refrigerant, OEM viscosity requirements and application. Since the screw bearings are designed to operate on a large amount of working viscosity, most lubricant recommendations are in the range of 5 to 20 cSt working viscosity for lubrication via the compressor sump and some oil separation on the discharge side. Most screw compressors that are operating with oil supply from the discharge oil separator are aiming for working viscosity in the range of 15 to 50 cSt. Usually these compressors operate with very high viscosity lubricants in the 220 to 400 cSt range or work with oil cooling after the separator to increase working viscosity at refrigerant dilution. Lubricants for screw compressors usually work without the use of additives for wear and friction reduction.

5. Lubricant Options for Low GWP Refrigerants

Based on all the discussions we have had on lubricant-refrigerant interaction and compressor bearing operation, it is valuable to look at what lubricants are best suited to meet these demands. Usually for this exercise it is best to break out by refrigerant type to determine measured interaction properties, then evaluate based on compressor type and specified needs within a system.

5.1 Hydrocarbon Refrigerants

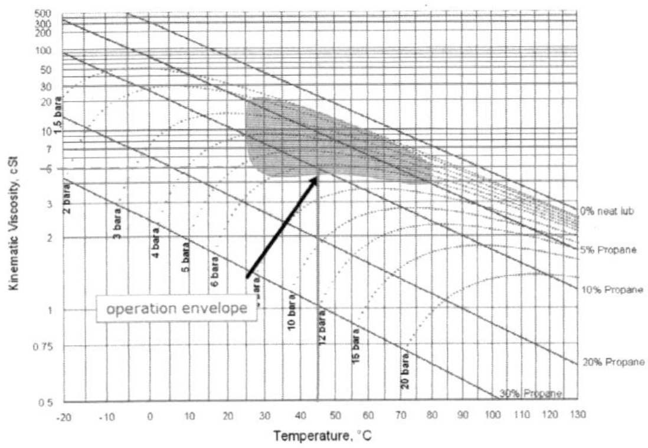

Figure 4: AB/R290

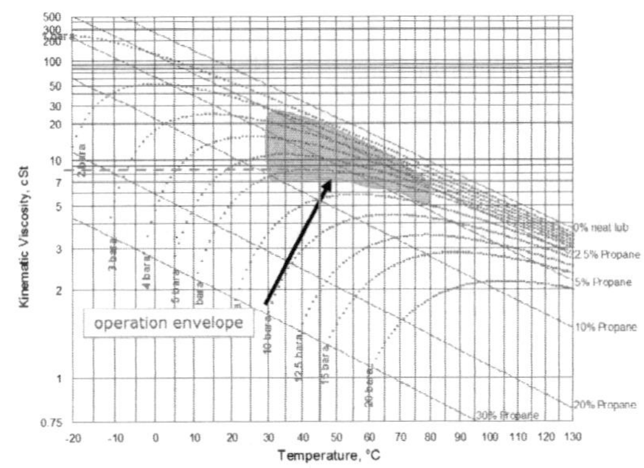

Figure 5: POE/R290

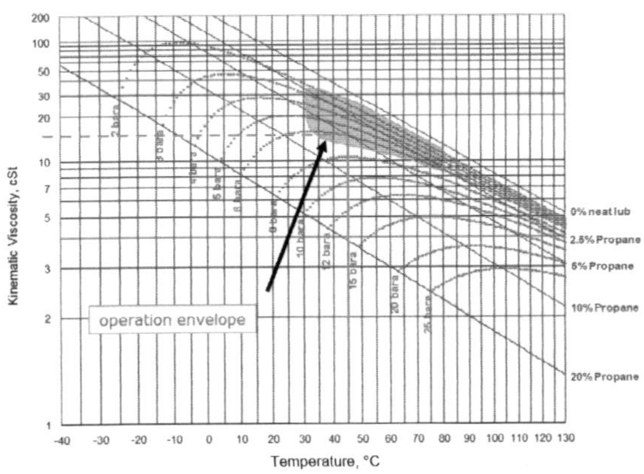

Figure 6: PAG/R290

Figures 4-6 represent Daniel Plots of various labelled lubricant chemistry of the same ISO viscosity grade [11]. The solid and dotted lines represent an oil sump temperature at potential suction pressure and corresponding refrigerant concentration and working viscosity. For this example, the temperature was set at 45°C and the pressure at 6 bara, a condition common in R290 refrigeration units operating in a light commercial unit with a reciprocating compressor.

As the chemistry changes, so does the amount of dilution and the working viscosity. For the AB example, the dilution is around 10% with a viscosity of 5 cSt; for POE dilution at 7.5% and viscosity at 8 cSt and for PAG the dilution is reduced to 5% and viscosity increased to 15 cSt. In some applications, the values associated with the AB or POE might be adequate for the bearing design but there may be instances that this is still too low. The PAG provides the advantage of having a higher viscosity but also lower dilution which can help in applications that are restricted today by charge limitation of flammable refrigerants. It is also easy to adjust the viscosity of the PAG without changing the chemistry since it is a polymer. This can be an advantage when moving to lower starting base oil viscosity which can benefit compressor efficiency.

The same concepts and principles can be used for rotary compressor operation where lower solubility and higher viscosity can be an advantage especially at the vane and roller interface. Typically, these applications will be compared with lubricants with starting viscosities closer to 68 cSt, but once again the advantage the PAG chemistry brings might result in the ability to lower the starting viscosity [12].

5.2 R-32

R-32 refrigerant has emerged as a lower GWP alternative to switching away from R-22 or as a substitute to R-410A. Even though R-32 is 50% of R-410A it has significant differences regarding miscibility and working viscosity compared to R-410A and current lubricants. This has required evaluating different lubricants to meet the desired miscibility and working viscosity for both scroll and rotary compressor

IOP Conf. Series: Materials Science and Engineering **232** (2017) 012092 doi:10.1088/1757-899X/232/1/012092

operation. A novel lubricant chemistry has been developed to try to answer some of the need. A typical lubricant used with R-410 in a scroll compressor will have a miscibility range at around 20% lubricant of 40°C to -25°C. The new lubricant will show similar miscibility with a range of 30°C to -20°C at the 20% lubricant concentration. The working viscosity is another challenge because trying to maintain an equivalent R-410A miscibility with some lubricants has resulted in a working viscosity reduction that is not adequate for effective compressor bearing operation. This new lubricant can answer this problem by matching what is typically measured with R-410A and current lubricant at operational conditions. Table 2 below shows measured PVT data for miscibility, solubility, and working viscosity for:

1) R-410A and a 32 cSt POE lubricant (baseline current market product)
2) R-32 and 32 cSt POE lubricant
3) R-32 and a POE lubricant developed for R-32
4) R-32 and 56 cSt POE lubricant developed for R-32
5) R-32 and new lubricant chemistry
6) R-410A and new lubricant chemistry

Table 2: Lubricant Development for R-32 Refrigerant; Comparison to R-410A Products

Conditions used for R-410A: 39°C and 12.5 bara

Conditions used for R-32: 41°C and 13 bara

Above Combinations	Dilution, %	Viscosity, cSt	Miscibility at 20%, °C
#1	18	7.2	-25
#2	13	6.8	10
#3	14	5.8	-20
#4	15	9.1	-15
#5	15	9.4	-25
#6	20	8.7	-50

The new lubricant chemistry for R-32 shows working viscosity and miscibility that matches or exceeds the standard associated with R-410A. The working viscosity of this product is more equivalent to a higher viscosity POE designated for R-32 and an additional benefit of this product is that it should be effective as an R-410A lubricant.

6. Conclusion

Numerous examples have been presented on compressor type and bearing design associated with the HVAC&R market regarding lubrication. There is much complexity that should be understood when developing and recommending lubricant options for compressor design and system application. This requires experts in the field and specialized equipment to make accurate evaluation and analysis. The refrigerant market is evolving and this requires stability, miscibility, and working viscosity data on each refrigerant and lubricant combination to advance this development and provide the compressor and system engineer the data they require. Some examples were given of current day needs for lubricants for some lower GWP refrigerants that are being evaluated for various markets using different compressor types and the benefits these lubricant options can present. In summary, for certain lower GWP refrigerant options the following lubricant needs are recommended. Lubricants for R-32 refrigerant will require enhancing the miscibility while maintaining the required working viscosity. This will require some lubricant base chemistry changes. Lubricants for HFO-1234zeE will require a lubricant that reduces the high solubility effect produced by this refrigerant and the lubricant should have an increased viscosity index. Lubricants for hydrocarbon refrigerants like R-290 will require reducing the solubility factor of this refrigerant to help maintain working viscosity and this should also help benefit in situations of charge limitation requirements/restrictions.

References

[1] J. Clemente "Global Warming More Air Conditioning and More Energy". Forbes 2015.
[2] R. Lehman "The State of Energy Consumption Legislation for Commercial Reach-in Refrigerators and Freezers and Other Related Products". The International and Refrigeration Air Conditioning Conference. Purdue 2006.
[3] J. Karnaz "Compressor and System Energy Efficiency Improvements Through Lubriant Optimization". 8th International Conference on Compressors and Their Systems. London 2013.
[4] Environmental Investigation Agency. "The Importance of Ambition in the 2016 HFC Phase-Down Agreement" Montreal Protocol Vienna 2016.
[5] ASHRAE Standard 97 "Sealed Glass Tube Method to Test the Chemical Stability of Materials for Use within Refrigeration Systems". 2007.
[6] C. Seeton and J. Karnaz "Thermodynamic and Transport Properties of Lubricant and Refrigerant Mixtures" Oil Management in Compressors and Their Systems Purdue University Short Course. Purdue 2016.
[7] C. Seeton "CO_2-Lubricant Two-Phase Flow Patterns in Small Horizontal Wetted Wall Channels The Effects of Refrigerant/Lubricant Thermophysical Properties". Dissertation Defense ACRC University of Illinois 2009.
[8] J. Karnaz D. Kultgen "Keeping Your Compressor Healthy: Developing the Right Lubricant Formualtion is the Key". 9th International Conference on Compressors and Their Systems. London 2015.
[9] J. Karnaz "Lubricants for Low GWP Refrigerants". The International Symposium on New Refrigerants and Environmental Technology. Kobe 2010
[10] J. Bush J. Elson "Scroll Compressor Design Criteria for Residential Air Conditioning and Heat Pump Applictions, Part I: Mechanics", International Compressor Engineering Conference Purdue 1988.
[11] J. Karnaz "Lubricants and New Refrigerants: System Performance Needs" The International Symposium on New Refrigerants and Environmental Technology. Kobe 2016.
[12] L. Dixon "A Comparitive Review of the Chemistry and Performance of Novel Synthetic Lubricants with CO_2 Refrigerant". Chillventa 2012.

IOP Conf. Series: Materials Science and Engineering **232** (2017) 012093 doi:10.1088/1757-899X/232/1/012093

Refrigeration oils for low GWP refrigerants in various applications

R Saito[1] and S G. Sundaresan[2]

[1]Reserch and Development Center, Japan Sun Oil Company, Ltd.
20 Futamatashinmachi Ichikawa, Chiba 272-0002, Japan
[2]Sundaresan Consulting Services, LLC.
351 West Parkwood Street Sidney, Ohio 45365, USA

[1]E-mail: rei_saito@sunoco.co.jp

[2]E-mail: SonnySundaresan@yahoo.com

Abstract. The practical use of the refrigeration systems is considered as a methods to suppress global warming. The replacement of a refrigerant with a new one that has lower global warming potential (GWP) has been underway for several years. For the application fields of refrigerators, domestic air conditioners, automotive air conditioners and hot water dispensers, the investigation has almost finished. It is still underway for the application fields of commercial air conditioners and chillers, refrigeration facilities for cold storage, etc. And now, the refrigeration system is being applied in various ways to decrease global warming above the generation of electric power with organic Rankine cycle, the binary electric generation with ground source heat pump, and so on. In these situations, various refrigerants are developed and several kinds of suitable refrigeration oils are selected. This paper presents the consideration of suitable refrigeration oil for the various low GWP refrigerants.

1. Introduction

At present, the transition to the low-carbon society is pushed forward in various industries. Main issues considered are the reduction of fossil fuel through the methods of energy-saving and conversion from fossil fuel to natural energy. But the change of refrigerants and the effective utilization of refrigeration cycle are added to those issues in the field of refrigeration.

As the high global worming potentials (GWP) of the refrigerants are thousands of times against carbon dioxide, it seems to be expected the large effect of the carbon reduction by the change of refrigerants form high GWP to low GWP.

Studies had been groping the effective utilization of refrigeration cycle by using the unused thermal energy. One of such is the organic Rankine cycle using thermal deference between waste heat from factory and atmospheric temperature. Another is the high temperature heat pump, which represents a heat pump boiler that makes the high temperature based on the unused thermal energy.

Given such background, we introduce the results of the evaluation of the compatibility of the candidate refrigerants for several applications of various refrigeration cycles and refrigeration oils.

IOP Conf. Series: Materials Science and Engineering **232** (2017) 012093 doi:10.1088/1757-899X/232/1/012093

2. Low GWP refrigerants for air conditioner and suitable refrigeration oil

In comparison with R-32 that is one of the low GWP mild flammable refrigerants adopted for the air conditioner for residential and store, various refrigerants are introduced for the purposes of GWP reduction, retrofit for the devises using R-22, change to nonflammable refrigerants and so on [1,2]. Most of them are the mixtures of HFO refrigerants which have C=C double bond and current HFC refrigerants. And the compatible refrigeration oils are the ester based synthetic refrigeration oil, polyol ester (POE), for current HFC refrigerants.

In Table 1, the solubility of refrigeration oils and low GWP refrigerants prepared from refrigerant suppliers is indicated.

Table 1. Comparison of Solubility of Low GWP Refrigerant with Current Refrigeration Oil

[CST*: Oil/Refrigerant = 1/4 Unit℃]

	Lower CST*	Upper CST*	GWP**	Flammability
POE VG 68 Oil for HFC				
R-410A (Reference)	+8	+39	1924	None
R-32	Insoluble	Insoluble	677	Mild
R-447A	-6	>+70	572	Mild
R-452B	-2	+50	676	Mild
R-454B	-4	+53	467	Mild
R-449C	-45	>+70	1146	None
R1123+R32	-35	+52	370	Mild
POE VG 68 Oil for R-32				
R-410A	-50	+60	1924	None
R-32	-6	+58	677	Mild
Naph VG56 Oil for R-22				
R-22	+12	>+70	1760	None

*: CST is critical solution temperature **: IPCC AR5

According to Table 1, because that R-32 has low solubility with “POE VG68 for HFC” (Current oil for HFC), the new POE refrigeration oil for R-32 has developed.

The refrigerant mixtures of HFO and HFC such as R-452B, R-454B, R-449C and HFO-1123 +R-32 have shown good solubility with same current POE VG68 refrigeration oil.

Table 2 shows the result of the compatibility test of the HFO+HFC refrigerant mixtures prepared from refrigerant supplier with current POE VG68 refrigeration oil for HFC.

Test method of the chemical stability test is almost same as the Ashrae 97. To obtain the acid number of tested oil, using larger glass tube (18 mm OD, 12 mm ID, 200 mm long), changing the amount of oil and refrigerant.

IOP Conf. Series: Materials Science and Engineering **232** (2017) 012093 doi:10.1088/1757-899X/232/1/012093

Table 2. Comparison of Chemical Stability of Refrigeration Oil with HFO mixture Refrigerant

Condition: Sealed Tube Tests, Oil / Ref =5g/1g, 175deg.C, 14days

Refrigeration Oil	POE VG68 for HFC				
Refrigerant	447A	452B	454B	449C	1123+32
Colour, ASTM	L0.5	L0.5	L0.5	L0.5	L0.5
Catalyst					
Fe	Good*	Good*	Good*	Good*	Good*
Cu	Good*	Good*	Good*	Good*	Good*
Al	Good*	Good*	Good*	Good*	Good*
Acid Number, mgKOH/g	0.01	0.01	0.01	0.01	0.01
Deposit	None	None	None	None	None

*: “Good” indicate that the surface of catalyst was almost unchanged.

According to Table 2, HFO refrigerant mixtures evaluated with this test had indicated excellent compatibility with current POE VG68 refrigeration oil.

3. Low GWP refrigerants for refrigerated facilities and suitable refrigeration oil

As the next refrigerant of HCFC refrigerant R-22, HFC refrigerant R-404A has been used in low temperature equipment such as the showcases for cold foods in supermarket, drug store, refrigerating devices for refrigeration vehicle, and so on. However, because this refrigerant has high GWP (3943 from IPCC 5th report), it is studied to be changed to the low GWP refrigerant in some countries. First, R-407F and R-407H developed from the change of formulation of current HFC refrigerant R-407 series are introduced as the refrigerant of GWP reduction to half of R-404A [1]. Next, HFO+HFC refrigerant mixtures are developed and announced. As the mild flammability of HFO refrigerant, these refrigerant mixtures have variety of flammability and GWP.

The result of HFC refrigerants and HFO + HFC mixture refrigerant solubility test with current refrigeration oil POE VG32 for HFC is shown in Table 3.

Table 3. Comparison of Solubility of Low GWPRefrigerant with Current Refrigeration Oil VG32

[CST: Oil/Refrigerant = 1/4 Unit℃]

	Lower CST*	Upper CST*	GWP**	Flammability
POE VG 32 Oil for HFC				
R-404A	<-60	+45	3943	None
R-32	+36	+42	677	Mild
R-448A	-53	>+70	1273	None
R-449A	-53	>+70	1282	None
R-454A	<-60	>+70	237	Mild
R-407F	-38	>+70	1674	None
R-407H	-16	>+70	1378	None
Naph VG32 Oil for R-22				
R-22	-8	>+70	1760	None

*: CST is critical solution temperature **: IPCC AR5

According to Table 3, R-32 can solve only a little with "POE VG32 for HFC" (Current oil for HFC).The refrigerant mixtures of HFO and HFC, such as R-448A, R-449A, R-454A, and HFC refrigerant R-407F and R-407H had shown good solubility with same current POE VG32 refrigeration oil.

Table 4 shows the result of the compatibility test of the HFO + HFC refrigerant mixtures and HFC refrigerant prepared from refrigerant supplier with current POE VG32 refrigeration oil for HFC.

Table 4. Comparison of Chemical Stability of Refrigeration Oil with HFO Refrigerant Mixtures

Condition: Sealed Tube Tests, Oil / Ref =5g/1g, 175deg.C, 14days

Refrigeration Oil	POE VG32 for HFC				
Refrigerant	448A	449A	454A	407F	407H
Colour, ASTM	L0.5	L0.5	L0.5	L0.5	L0.5
Catalyst					
Fe	Good*	Good*	Good*	Good*	Good*
Cu	Good*	Good*	Good*	Good*	Good*
Al	Good*	Good*	Good*	Good*	Good*
Acid Number, mgKOH/g	0.01	0.01	0.01	0.01	0.01
Deposit	None	None	None	None	None

*: "Good" indicate that the surface of catalyst was almost unchanged.

According to Table 4, HFO + HFC refrigerant mixtures evaluated with this test are indicating relatively good compatibility with current POE VG32 refrigeration oil.

On the other hand, Chlorine contained HFO refrigerants HCFO-1233zd (E) and HCFO-1224yd (Z) are proposed for substitute of R-245fa for water cooled chiller [3,4]. These HCFO refrigerants are not destructive to the ozone layer in spite of their chlorine content, because of the destruction of these refrigerants in the atmosphere by the C=C double bond of HFO's feature.Therefore, the Ozone depletion Potential (ODP) of these HCFO refrigerants are nearly zero, and the non flammability is obtained from chlorine.

For the refrigeration oil point of view, with one chlorine in molecular structure of these refrigerants, the mineral oil is also soluble with them. The suitability of naphthenic mineral oil on the solubility is checked and indicated in Figure 1 as an example.

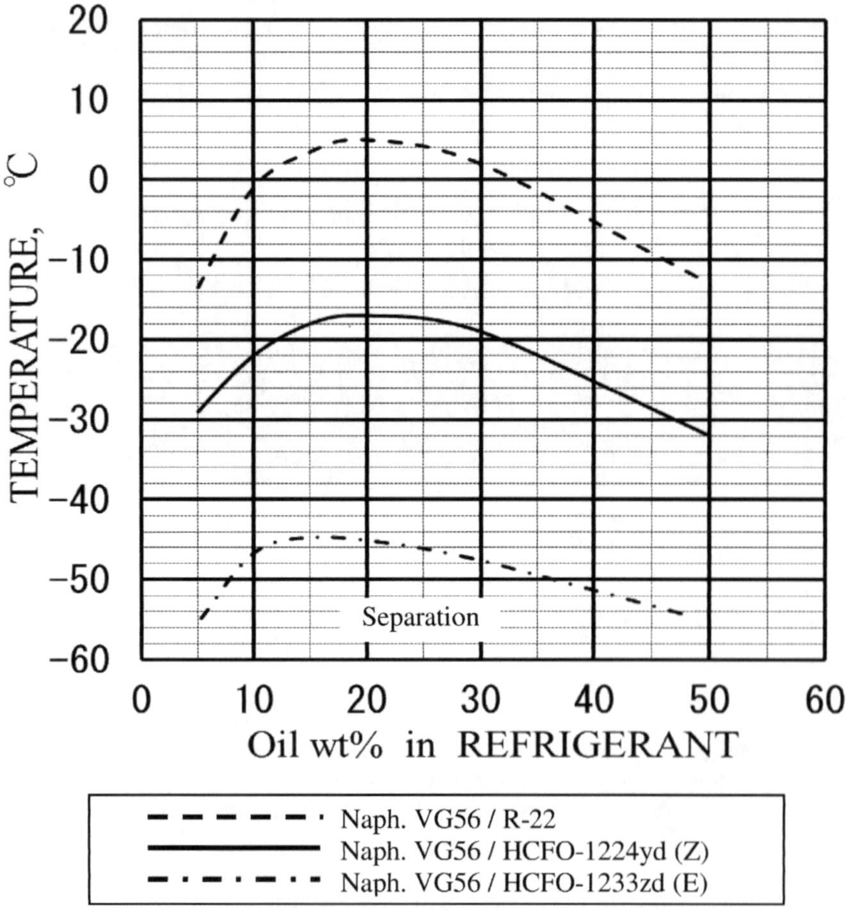

Figure 1. Separation Temperature of Naphthenic Mineral Oil VG56 in HCFO refrigerants

It is evaluated that this naphthenic mineral oil has good stability, with the result indicated in Table 5.

Table 5. Comparison of Chemical Stability of Refrigeration Oil with HCFO Refrigerants

Condition: Sealed Tube Tests, Oil / Ref =10g/2g, 175deg.C, 14days

Refrigeration Oil	NaphVG56for HCFC	NaphVG56for HCFC	POE VG68for HFC
Refrigerant	HCFO-1233zd(E)	HCFO-1224yd(Z)	HCFO-1233zd(E)
Colour, ASTM	L0.5	L0.5	L0.5
Catalyst			
Fe	Good*	Good*	Good*
Cu	Good*	Good*	Good*
Al	Good*	Good*	Good*
Acid Number,mgKOH/g	0.05	0.03	0.11
Deposit	None	None	None

*: "Good" indicate that the surface of catalyst was almost unchanged.

IOP Conf. Series: Materials Science and Engineering **232** (2017) 012093 doi:10.1088/1757-899X/232/1/012093

4. Low GWP refrigerant for high temperature application and suitable refrigeration oil

One of the methods of the progress to the low-carbon society is the effective utilization of refrigeration cycle by using the unused thermal energy of waste heat from factories, etc. The organic Rankine cycle and heat pump boiler have been studied with R-245fa for chiller and R-365mfc for precision cleaning as the refrigerant [5].

Furthermore, it is studied to change these refrigerants to low GWP ones such as the HCFO refrigerants written above as the R-245fa alternative, and HFO-1336mzz (Z) as the R-365mfc alternative.

For the refrigeration oil, it seems that viscosity and chemical stability needs to be improved, high solubility with refrigerant is not necessary. The compatibility of the combination of current naphthenic mineral oil for HCFC / HCFO refrigerant, and current POE oil for HFC / HFO-1336mzz (Z).

In Table 6, the solubility of refrigeration oils and low GWP refrigerants prepared from refrigerant suppliers is indicated.

Table 6. Comparison of Solubility of Low GWP Refrigerant with Current Refrigeration Oil

[CST: Oil/Refrigerant = 1/4 Unit℃]

	Lower CST*	Upper CST*	GWP**	Flammability
Naph.VG100 Oil for HCFC				
HCFO-1233zd(E)	-40	>+80	1	None
HCFO-1224yd(Z)	-7	>+80	<1***	None
POE VG 170 Oil for HFC				
HFO-1336mzz(Z)	<-60	>+80	2	None

*: CST is critical solution temperature **: IPCC AR5 ***: Data from Supplier (Asahi Glass)

The results of the compatibility test of the combination of current naphthenic mineral oil for HCFC / HCFO refrigerant, and current POE oil for HFC / HFO-1336mzz(Z), are shown in Table 7 and Table 8, respectively. Test conditions include the influences of the high temperature, air and water.

Table 7. Comparison of Chemical Stability of Refrigeration Oil for HCFC with HCFO Refrigerants

Condition: Sealed Tube Tests, Oil / Ref =5g/5g, 200deg.C, 14days

RefrigerationOil	NaphVG100 +Additive	NaphVG100 +Additive	NaphVG100	NaphVG100 +Additive
Refrigerant	HCFO-1233zd(E)	HCFO-1233zd(E)	HCFO-1224yd(Z)	HCFO-1224yd(Z)
Air, ppm	<0.1	100	<0.1	100
Water, ppm	<50	100	<50	100
Colour, ASTM	L3.5	L2.5	1.0	L2.5
Catalyst				
Fe	Cu P**	Cu P**	Good*	Good*
Cu	Good*	Good*	Good*	Good*
Al	Good*	Good*	Good*	Good*
Acid Number,mgKOH/g	0.19	0.43	0.05	0.01
Deposit	None	None	None	None

*: "Good" indicate that the surface of catalyst was almost unchanged.

**: "Cu P" indicate the copper plating occurred on the catalyst surface.

Table 8. Comparison of Chemical Stability of Refrigeration Oil for HFC with HFO-1336mzz(Z)

Condition: Sealed Tube Tests, Oil / Ref =5g/5g, 14days

Refrigeration Oil	POE VG220	POE VG220	POE VG220
Refrigerant	HFO-1336mzz(Z)	HFO-1336mzz(Z)	HFO-1336mzz(Z)
Temp. deg.C	200	200	250
Air, ppm	<0.1	100	<0.1
Water, ppm	<100	500	<100
Colour, ASTM	L0.5	0.5	L1.5
Catalyst			
Fe	Good*	Good*	Good*
Cu	Good*	Good*	Good*
Al	Good*	Good*	Good*
Acid Number,mgKOH/g	0.07	0.49	18.43
Deposit	None	None	None

*: "Good" indicate that the surface of catalyst was almost unchanged.

According to Table 7 and 8, bad influences occur under the situations of refrigerant, refrigeration oil, small amount of air and water coexist in the high temperature system. Therefore, the refrigerant have to be selected to match the temperature of the application, and contamination of air and water have to be controlled as little as possible.

5. Conclusion

On the transition to the low carbon society, we introduced the outline of studies of the compatibility of low GWP refrigerant and refrigeration oil from the standpoint of refrigeration oil.

Since the refrigeration cycle has high energy efficiency, it has a possibility to be applied for various purposes in the future. And, it seems to be adopted to control temperature and energy efficiency against the various climate conditions in the world. In addition, if the GWP of refrigerants for each application are reduced, the refrigeration field will be made a great contribution to transition to low carbon society.

References

[1] JASRAE 2016 *Refrigeration* **091** 1062
[2] JRAIA 2014 *Proceedings of the International Symposium on New Refrigerants and Environmental Technology*
[3] JSRAE 2014 *Proceedings of the seminar "The correspondence and the prospects for the latest Issues of refrigerants in 2014"*
[4] JSRAE 2015 *Proceedings of the seminar "The correspondence and the prospects for the latest Issues of refrigerants in 2015"*
[5] JSRAE 2014 *Proceedings of the seminar "The Latest technologies of waste heat recovering instruments"*